T0205736

Lecture Notes in Computer Science   5619

Commenced Publication in 1973
Founding and Former Series Editors:
Gerhard Goos, Juris Hartmanis, and Jan van Leeuwen

Lecture Notes in Computer Science 5619

Commenced Publication in 1973

Founding and Former Series Editors:
Gerhard Goos, Juris Hartmanis, and Jan van Leeuwen

Editorial Board

David Hutchison
Lancaster University, UK
Takeo Kanade
Carnegie Mellon University, Pittsburgh, PA, USA
Josef Kittler
University of Surrey, Guildford, UK
Jon M. Kleinberg
Cornell University, Ithaca, NY, USA
Alfred Kobsa
University of California, Irvine, CA, USA
Friedemann Mattern
ETH Zurich, Switzerland
John C. Mitchell
Stanford University, CA, USA
Moni Naor
Weizmann Institute of Science, Rehovot, Israel
Oscar Nierstrasz
University of Bern, Switzerland
C. Pandu Rangan
Indian Institute of Technology, Madras, India
Bernhard Steffen
University of Dortmund, Germany
Madhu Sudan
Massachusetts Institute of Technology, MA, USA
Demetri Terzopoulos
University of California, Los Angeles, CA, USA
Doug Tygar
University of California, Berkeley, CA, USA
Richard Weikum
Max-Planck Institute of Computer Science, Saarbruecken, Germany

Masaaki Kurosu (Ed.)

Human Centered Design

First International Conference, HCD 2009
Held as Part of HCI International 2009
San Diego, CA, USA, July 19-24, 2009
Proceedings

Volume Editor

Masaaki Kurosu
The Open University of Japan
Center of ICT and Distance Education
2-11 Wakaba, Mihama-ku, Chiba-shi, Chiba 261-8586, Japan
E-mail: masaakikurosu@spa.nifty.com

Library of Congress Control Number: Applied for

CR Subject Classification (1998): A.1

LNCS Sublibrary: SL 3 – Information Systems and Application, incl. Internet/Web
and HCI

ISSN   0302-9743
ISBN-10   3-642-02805-5 Springer Berlin Heidelberg New York
ISBN-13   978-3-642-02805-2 Springer Berlin Heidelberg New York

springer.com

© Springer-Verlag Berlin Heidelberg 2009
Printed in Germany

Typesetting: Camera-ready by author, data conversion by Scientific Publishing Services, Chennai, India
Printed on acid-free paper   SPIN: 12711994   06/3180   5 4 3 2 1 0

Foreword

The 13th International Conference on Human–Computer Interaction, HCI International 2009, was held in San Diego, California, USA, July 19–24, 2009, jointly with the Symposium on Human Interface (Japan) 2009, the 8th International Conference on Engineering Psychology and Cognitive Ergonomics, the 5th International Conference on Universal Access in Human–Computer Interaction, the Third International Conference on Virtual and Mixed Reality, the Third International Conference on Internationalization, Design and Global Development, the Third International Conference on Online Communities and Social Computing, the 5th International Conference on Augmented Cognition, the Second International Conference on Digital Human Modeling, and the First International Conference on Human Centered Design.

A total of 4,348 individuals from academia, research institutes, industry and governmental agencies from 73 countries submitted contributions, and 1,397 papers that were judged to be of high scientific quality were included in the program. These papers address the latest research and development efforts and highlight the human aspects of the design and use of computing systems. The papers accepted for presentation thoroughly cover the entire field of human–computer interaction, addressing major advances in knowledge and effective use of computers in a variety of application areas.

This volume, edited by Masaaki Kurosu, contains papers in the thematic area of Human Centered Design (HCD), addressing the following major topics:

- Usability and User Experience
- Methods and Techniques for HCD
- Understanding Diverse Human Needs and Requirements
- HCD in Industry
- HCD for Web-Based Applications and Services
- User Involvement and Participatory Methods
- HCD at Work

The remaining volumes of the HCI International 2009 proceedings are:

- Volume 1, LNCS 5610, Human–Computer Interaction—New Trends (Part I), edited by Julie A. Jacko
- Volume 2, LNCS 5611, Human–Computer Interaction—Novel Interaction Methods and Techniques (Part II), edited by Julie A. Jacko
- Volume 3, LNCS 5612, Human–Computer Interaction—Ambient, Ubiquitous and Intelligent Interaction (Part III), edited by Julie A. Jacko
- Volume 4, LNCS 5613, Human–Computer Interaction—Interacting in Various Application Domains (Part IV), edited by Julie A. Jacko
- Volume 5, LNCS 5614, Universal Access in Human–Computer Interaction—Addressing Diversity (Part I), edited by Constantine Stephanidis

- Volume 6, LNCS 5615, Universal Access in Human–Computer Interaction—Intelligent and Ubiquitous Interaction Environments (Part II), edited by Constantine Stephanidis
- Volume 7, LNCS 5616, Universal Access in Human–Computer Interaction—Applications and Services (Part III), edited by Constantine Stephanidis
- Volume 8, LNCS 5617, Human Interface and the Management of Information—Designing Information Environments (Part I), edited by Michael J. Smith and Gavriel Salvendy
- Volume 9, LNCS 5618, Human Interface and the Management of Information—Information and Interaction (Part II), edited by Gavriel Salvendy and Michael J. Smith
- Volume 11, LNCS 5620, Digital Human Modeling, edited by Vincent G. Duffy
- Volume 12, LNCS 5621, Online Communities and Social Computing, edited by A. Ant Ozok and Panayiotis Zaphiris
- Volume 13, LNCS 5622, Virtual and Mixed Reality, edited by Randall Shumaker
- Volume 14, LNCS 5623, Internationalization, Design and Global Development, edited by Nuray Aykin
- Volume 15, LNCS 5624, Ergonomics and Health Aspects of Work with Computers, edited by Ben-Tzion Karsh
- Volume 16, LNAI 5638, The Foundations of Augmented Cognition: Neuroergonomics and Operational Neuroscience, edited by Dylan Schmorrow, Ivy Estabrooke and Marc Grootjen
- Volume 17, LNAI 5639, Engineering Psychology and Cognitive Ergonomics, edited by Don Harris

I would like to thank the Program Chairs and the members of the Program Boards of all thematic areas, listed below, for their contribution to the highest scientific quality and the overall success of HCI International 2009.

Ergonomics and Health Aspects of Work with Computers

Program Chair: Ben-Tzion Karsh

Arne Aarås, Norway
Pascale Carayon, USA
Barbara G.F. Cohen, USA
Wolfgang Friesdorf, Germany
John Gosbee, USA
Martin Helander, Singapore
Ed Israelski, USA
Waldemar Karwowski, USA
Peter Kern, Germany
Danuta Koradecka, Poland
Kari Lindström, Finland

Holger Luczak, Germany
Aura C. Matias, Philippines
Kyung (Ken) Park, Korea
Michelle M. Robertson, USA
Michelle L. Rogers, USA
Steven L. Sauter, USA
Dominique L. Scapin, France
Naomi Swanson, USA
Peter Vink, The Netherlands
John Wilson, UK
Teresa Zayas-Cabán, USA

Human Interface and the Management of Information

Program Chair: Michael J. Smith

Gunilla Bradley, Sweden
Hans-Jörg Bullinger, Germany
Alan Chan, Hong Kong
Klaus-Peter Fähnrich, Germany
Michitaka Hirose, Japan
Jhilmil Jain, USA
Yasufumi Kume, Japan
Mark Lehto, USA
Fiona Fui-Hoon Nah, USA
Shogo Nishida, Japan
Robert Proctor, USA
Youngho Rhee, Korea

Anxo Cereijo Roibás, UK
Katsunori Shimohara, Japan
Dieter Spath, Germany
Tsutomu Tabe, Japan
Alvaro D. Taveira, USA
Kim-Phuong L. Vu, USA
Tomio Watanabe, Japan
Sakae Yamamoto, Japan
Hidekazu Yoshikawa, Japan
Li Zheng, P.R. China
Bernhard Zimolong, Germany

Human–Computer Interaction

Program Chair: Julie A. Jacko

Sebastiano Bagnara, Italy
Sherry Y. Chen, UK
Marvin J. Dainoff, USA
Jianming Dong, USA
John Eklund, Australia
Xiaowen Fang, USA
Ayse Gurses, USA
Vicki L. Hanson, UK
Sheue-Ling Hwang, Taiwan
Wonil Hwang, Korea
Yong Gu Ji, Korea
Steven Landry, USA

Gitte Lindgaard, Canada
Chen Ling, USA
Yan Liu, USA
Chang S. Nam, USA
Celestine A. Ntuen, USA
Philippe Palanque, France
P.L. Patrick Rau, P.R. China
Ling Rothrock, USA
Guangfeng Song, USA
Steffen Staab, Germany
Wan Chul Yoon, Korea
Wenli Zhu, P.R. China

Engineering Psychology and Cognitive Ergonomics

Program Chair: Don Harris

Guy A. Boy, USA
John Huddlestone, UK
Kenji Itoh, Japan
Hung-Sying Jing, Taiwan
Ron Laughery, USA
Wen-Chin Li, Taiwan
James T. Luxhøj, USA

Nicolas Marmaras, Greece
Sundaram Narayanan, USA
Mark A. Neerincx, The Netherlands
Jan M. Noyes, UK
Kjell Ohlsson, Sweden
Axel Schulte, Germany
Sarah C. Sharples, UK

Neville A. Stanton, UK

Xianghong Sun, P.R. China

Andrew Thatcher, South Africa

Matthew J.W. Thomas, Australia

Mark Young, UK

Universal Access in Human–Computer Interaction

Program Chair: Constantine Stephanidis

Julio Abascal, Spain

Ray Adams, UK

Elisabeth André, Germany

Margherita Antona, Greece

Chieko Asakawa, Japan

Christian Bühler, Germany

Noelle Carbonell, France

Jerzy Charytonowicz, Poland

Pier Luigi Emiliani, Italy

Michael Fairhurst, UK

Dimitris Grammenos, Greece

Andreas Holzinger, Austria

Arthur I. Karshmer, USA

Simeon Keates, Denmark

Georgios Kouroupetroglou, Greece

Sri Kurniawan, USA

Patrick M. Langdon, UK

Seongil Lee, Korea

Zhengjie Liu, P.R. China

Klaus Miesenberger, Austria

Helen Petrie, UK

Michael Pieper, Germany

Anthony Savidis, Greece

Andrew Sears, USA

Christian Stary, Austria

Hirotada Ueda, Japan

Jean Vanderdonckt, Belgium

Gregg C. Vanderheiden, USA

Gerhard Weber, Germany

Harald Weber, Germany

Toshiki Yamaoka, Japan

Panayiotis Zaphiris, UK

Virtual and Mixed Reality

Program Chair: Randall Shumaker

Pat Banerjee, USA

Mark Billinghurst, New Zealand

Charles E. Hughes, USA

David Kaber, USA

Hirokazu Kato, Japan

Robert S. Kennedy, USA

Young J. Kim, Korea

Ben Lawson, USA

Gordon M. Mair, UK

Miguel A. Otaduy, Switzerland

David Pratt, UK

Albert "Skip" Rizzo, USA

Lawrence Rosenblum, USA

Dieter Schmalstieg, Austria

Dylan Schmorrow, USA

Mark Wiederhold, USA

Internationalization, Design and Global Development

Program Chair: Nuray Aykin

Michael L. Best, USA

Ram Bishu, USA

Alan Chan, Hong Kong

Andy M. Dearden, UK

Susan M. Dray, USA

Vanessa Evers, The Netherlands

Paul Fu, USA

Emilie Gould, USA

Sung H. Han, Korea
Veikko Ikonen, Finland
Esin Kiris, USA
Masaaki Kurosu, Japan
Apala Lahiri Chavan, USA
James R. Lewis, USA
Ann Light, UK
James J.W. Lin, USA
Rungtai Lin, Taiwan
Zhengjie Liu, P.R. China
Aaron Marcus, USA
Allen E. Milewski, USA

Elizabeth D. Mynatt, USA
Oguzhan Ozcan, Turkey
Girish Prabhu, India
Kerstin Röse, Germany
Eunice Ratna Sari, Indonesia
Supriya Singh, Australia
Christian Sturm, Spain
Adi Tedjasaputra, Singapore
Kentaro Toyama, India
Alvin W. Yeo, Malaysia
Chen Zhao, P.R. China
Wei Zhou, P.R. China

Online Communities and Social Computing

Program Chairs: A. Ant Ozok, Panayiotis Zaphiris

Chadia N. Abras, USA
Chee Siang Ang, UK
Amy Bruckman, USA
Peter Day, UK
Fiorella De Cindio, Italy
Michael Gurstein, Canada
Tom Horan, USA
Anita Komlodi, USA
Piet A.M. Kommers, The Netherlands
Jonathan Lazar, USA
Stefanie Lindstaedt, Austria

Gabriele Meiselwitz, USA
Hideyuki Nakanishi, Japan
Anthony F. Norcio, USA
Jennifer Preece, USA
Elaine M. Raybourn, USA
Douglas Schuler, USA
Gilson Schwartz, Brazil
Sergei Stafeev, Russia
Charalambos Vrasidas, Cyprus
Cheng-Yen Wang, Taiwan

Augmented Cognition

Program Chair: Dylan D. Schmorrow

Andy Bellenkes, USA
Andrew Belyavin, UK
Joseph Cohn, USA
Martha E. Crosby, USA
Tjerk de Greef, The Netherlands
Blair Dickson, UK
Traci Downs, USA
Julie Drexler, USA
Ivy Estabrooke, USA
Cali Fidopiastis, USA
Chris Forsythe, USA
Wai Tat Fu, USA
Henry Girolamo, USA

Marc Grootjen, The Netherlands
Taro Kanno, Japan
Wilhelm E. Kincses, Germany
David Kobus, USA
Santosh Mathan, USA
Rob Matthews, Australia
Dennis McBride, USA
Robert McCann, USA
Jeff Morrison, USA
Eric Muth, USA
Mark A. Neerincx, The Netherlands
Denise Nicholson, USA
Glenn Osga, USA

Dennis Proffitt, USA
Leah Reeves, USA
Mike Russo, USA
Kay Stanney, USA
Roy Stripling, USA
Mike Swetnam, USA
Rob Taylor, UK

Maria L.Thomas, USA
Peter-Paul van Maanen, The Netherlands
Karl van Orden, USA
Roman Vilimek, Germany
Glenn Wilson, USA
Thorsten Zander, Germany

Digital Human Modeling

Program Chair: Vincent G. Duffy

Karim Abdel-Malek, USA
Thomas J. Armstrong, USA
Norm Badler, USA
Kathryn Cormican, Ireland
Afzal Godil, USA
Ravindra Goonetilleke, Hong Kong
Anand Gramopadhye, USA
Sung H. Han, Korea
Lars Hanson, Sweden
Pheng Ann Heng, Hong Kong
Tianzi Jiang, P.R. China

Kang Li, USA
Zhizhong Li, P.R. China
Timo J. Määttä, Finland
Woojin Park, USA
Matthew Parkinson, USA
Jim Potvin, Canada
Rajesh Subramanian, USA
Xuguang Wang, France
John F. Wiechel, USA
Jingzhou (James) Yang, USA
Xiu-gan Yuan, P.R. China

Human Centered Design

Program Chair: Masaaki Kurosu

Gerhard Fischer, USA
Tom Gross, Germany
Naotake Hirasawa, Japan
Yasuhiro Horibe, Japan
Minna Isomursu, Finland
Mitsuhiko Karashima, Japan
Tadashi Kobayashi, Japan

Kun-Pyo Lee, Korea
Loïc Martínez-Normand, Spain
Dominique L. Scapin, France
Haruhiko Urokohara, Japan
Gerrit C. van der Veer, The Netherlands
Kazuhiko Yamazaki, Japan

In addition to the members of the Program Boards above, I also wish to thank the following volunteer external reviewers: Gavin Lew from the USA, Daniel Su from the UK, and Ilia Adami, Ioannis Basdekis, Yannis Georgalis, Panagiotis Karampelas, Iosif Klironomos, Alexandros Mourouzis, and Stavroula Ntoa from Greece.

This conference could not have been possible without the continuous support and advice of the Conference Scientific Advisor, Prof. Gavriel Salvendy, as well as the dedicated work and outstanding efforts of the Communications Chair and Editor of HCI International News, Abbas Moallem.

I would also like to thank for their contribution toward the organization of the HCI International 2009 conference the members of the Human–Computer Interaction Laboratory of ICS-FORTH, and in particular Margherita Antona, George Paparoulis, Maria Pitsoulaki, Stavroula Ntoa, and Maria Bouhli.

Constantine Stephanidis

HCI International 2011

The 14th International Conference on Human–Computer Interaction, HCI International 2011, will be held jointly with the affiliated conferences in the summer of 2011. It will cover a broad spectrum of themes related to human–computer interaction, including theoretical issues, methods, tools, processes and case studies in HCI design, as well as novel interaction techniques, interfaces and applications. The proceedings will be published by Springer. More information about the topics, as well as the venue and dates of the conference, will be announced through the HCI International Conference series website: http://www.hci-international.org/

General Chair
Professor Constantine Stephanidis
University of Crete and ICS-FORTH
Heraklion, Crete, Greece
Email: cs@ics.forth.gr

Table of Contents

Part I: Usability and User Experience

Part II: Methods and Techniques for HCD

Part III: Understanding Diverse Human Needs and Requirements

Part IV: HCD in Industry

Part V: HCD for Web-Based Applications and Services

Part VI: User Involvement and Participatory Methods

Part VII: HCD at Work

Part I

Usability and User Experience

Performance-Based Usability Testing: Metrics That Have the Greatest Impact for Improving a System's Usability

Robert W. Bailey[1,3], Cari A. Wolfson[2,3], Janice Nall[3], and Sanjay Koyani[4]

[1] Computer Psychology, Inc - Sandy, Utah
[2] Focus on U! - Tallahassee, Florida
[3] Centers for Disease Control and Prevention - Atlanta, Georgia
[4] Food and Drug Administration - Silver Spring, Maryland
bob@webusability.com, cariwolfson@usabilityfocus.com,
sanjay.koyani@fda.hhs.gov, jnall@cdc.gov

Abstract. Usability testing methods and results have evolved over the last 35 years. With new advancements being introduced every year, it is important to understand the present state of the field and opportunities for further improvement. This paper will detail the research based methods and metrics which are being used to ensure that usability recommendations are data-driven and performance-based. By focusing on the types of usability metrics being captured during usability tests, we will attempt to illustrate how usability researchers can quantifiably measure the performance of a system, use these measurements to make meaningful changes, and subsequently illustrate the improvements in user effectiveness, efficiency and satisfaction.

Keywords: Usability testing, Usability metrics, Effectiveness, Efficiency, Satisfaction, FirstClick, Usability methods.

1 Evolution of Usability Testing Methods

Before looking at the state of usability testing today, it is important to note how usability testing methods have evolved over the course of the last 35 years. At the 1972 Human Factors Society annual conference, Bailey [1] presented a paper that described the process Bell Laboratories had been using to conduct usability testing. At the time, the methods were considered first generation usability testing, wherein participants were tested one at a time with a usability researcher sitting next to them to manually record success and time on task. There were no real-time observers unless they sat quietly in the room behind the participant. Test sessions were videotaped using one camera pointed at the participant's face, hands and keyboard.

In the years following, usability researchers began to conduct testing in test facilities, complete with a one-way mirror for observers. Typical usability testing consisted of one-hour test sessions in which participants would perform a series of tasks, while thinking aloud. These tests focused on users' abilities to successfully complete tasks, with little emphasis on users' efficiency in completing the tasks.

During the test sessions, participants were generally allowed to take as long as they needed to complete a scenario while the usability test facilitator observed. The

M. Kurosu (Ed.): Human Centered Design, HCII 2009, LNCS 5619, pp. 3–12, 2009.
© Springer-Verlag Berlin Heidelberg 2009

facilitator typically recorded comments made by participants, as well as notes about the user's behavior, e.g., frequent use of the 'Back' button.

Much of this usability testing focused simply on determining if participants were able to complete the tasks. The resulting usability reports made suggestions for improvements based on these aspects and focused on many of the qualitative issues discovered during testing.

By today's standards the tests were 'soft', and the test sessions were difficult to replicate, making it almost impossible to conduct valid and meaningful retests. Some tests were so qualitative in nature, that they actually resembled a 'live' heuristic evaluation of the system and focused less on quantitative metrics regarding users' success and efficiency in using a system.

While we do not discount the importance of qualitative observations made by skilled usability practitioners during usability testing, this paper will attempt to illustrate the ways in which these observations can more accurately be quantified and standardized, resulting in higher-quality testing and recommendations, substantiated by meaningful usability metrics. Consistent with Tullis and Albert [2], we attempt to capture the state of quantitative usability metrics that are now (or should be) included in current usability test reports.

2   Focus on Data-Driven Recommendations

Over the past five to ten years, the usability testing process has substantially changed and most likely will continue to change within the next few years.

One of the biggest shifts has been the emphasis on performance-based recommendations in lieu of more qualitative recommendations. In the past, many usability reports focused on recommendations based on the facilitators' observations and qualitative notes; today's usability reports use metrics to substantiate these observations and quantify the performance of a system.

These metrics are due, in part, to advances in technology that now automate much of the data recording and provide new levels of data that were not possible to capture manually. Sophisticated testing tools have been available for the past few years, and have substantially changed the way in which usability tests are conducted. Not only do these tools automatically capture much of the data, they also assist with the analysis of this data, considerably reducing much of the time previously spent calculating success rates, time on task, page views, etc.

The usability testing tool that we used to collect most of the data shown in this report is the Usability Testing Environment (UTE) [3] [4]. The Usability Testing Environment consists of two applications. The first is the UTE *Manager* which helps the usability researcher set up task scenarios (test-items), and pre-test and post-test questions. The UTE *Runner* then automatically administers the test to participants and tracks the actions of participants as they take the test, including clicks, keystrokes, and scrolling. Once the test is completed, the UTE Manager analyzes the results from all participants, and automatically produces a Word-based test report – complete with text, statistics and graphics – as well as an Excel spreadsheet with all of the raw data collected during the testing.

The Usability Testing Environment (UTE), and similar testing tools, have revolutionized usability testing for Web sites and Web applications. UTE has substantially reduced the time required to construct and conduct usability tests, and has improved the usefulness of test results.

3 Using Metrics to Substantiate Usability Recommendations

To develop performance-based usability recommendations, meaningful usability metrics must be consistently captured across all participants. We will share some of the metrics that we have found to have the greatest impact on improving Web sites and Web applications.

4 Important Sets of Data

There are three important, and very useful, sets of data generated by most modern usability tests. First is the performance data, which includes task scenario success, time to complete each scenario, and the number page views required to complete each scenario. Second are the preference data that are generated from questionnaires at the end of each scenario, and/or at the end of the test. Third, are the comments made by participants during the test, and their impressions gathered after they had completed all scenarios, including their overall impressions, and what they liked best and least.

4.1 Success Rates

Success rates are one of the most helpful sets of data associated with each scenario. There are two success rates that are fairly easy to collect, and are very useful. The first is the success rate with the initial or first click [5] [6]. The second is the overall success rate when totally completing each scenario.

The first click success helps us to determine whether or not the participants are starting on the 'right foot'. Whereas, the overall success rate provides an estimate of how successful users were in completing an entire task.

FirstClick© Analyses

One of the most significant and useful advances we have made in our usability testing is to focus on the user's first click in each scenario – particularly when participants are interacting with Web sites. Over the course of our testing, we have noticed that participants' ultimate success with a task was very closely related to what they did on the first page.

We analyzed users' first click success and ultimate success from various tests, across multiple Web sites and found that if the user's first click was correct, the chances of getting the overall scenario correct was .87. On the other hand, if the first click was incorrect, the chance of eventually getting the scenario correct was only .46, which is less than a 50-50 chance of being successful. In general, we found that participants were about twice as likely to succeed if they selected the correct response on the first screen. The correct/incorrect ratio was 1.9, with a range from 1.4 to 2.7.

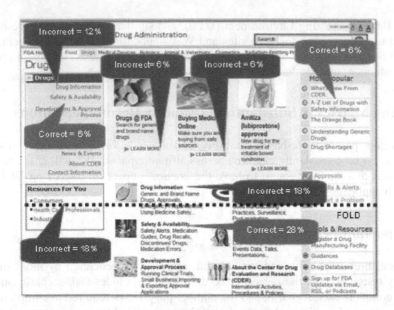

Fig. 1. FirstClick Callouts Shown on a Low-fidelity Wireframe for One Scenario

In presenting this data to design teams, we found that the best approach was to present the data scenario by scenario, and to include callouts to indicate where users clicked and whether the click would be considered a successful first click. To illustrate the data, correct clicks are shown in green, and the incorrect first clicks are in red, with a dotted line to show where the fold appeared during testing.

Ideally, the majority of participants would click on the correct link(s), and there would be few erroneous clicks. However, many times first clicks are made in a variety of different, and unexpected, locations. This data not only helps designers understand where users would look for information, but helps to validate an information architecture at the highest level of a Web site.

One major advantage of doing FirstClick testing is that far more scenarios can be included in a traditional one-hour test. This provides much greater 'task coverage'. Rather than using 10-15 scenarios during the traditional one-hour tests, we have been able to include over 100 FirstClick scenarios during the same period of time. This is significant, as previous studies [7] have found positive correlations between the number of tasks executed by participants and the proportion of usability issues found. In other words, the greater the number of tasks, the larger the number of usability issues identified.

Another advantage to FirstClick testing is the ability to uncover potential usability issues with a minimal number of prototypes and/or lower fidelity prototypes. Typical task scenarios require navigation through multiple pages, whereas, FirstClick tests only require the homepage and/or initial landing pages to be completed for testing.

Overall Task Completion/Success
The second success metric that can be captured is overall success, or simply whether or not a participant was able to successfully complete a task scenario within the given time limit.

In some instances, we are interested in measuring whether or not participants can find information on a site, whereas, in other cases, we want to see if participants can find the information and use that information to answer a multiple choice question correctly. Therefore, we judge success in one of two ways 1) when a user successfully navigates to the correct page or 2) when a user correctly answers a multiple choice question based on the content of a Web site.

For each scenario, success is either correct or not correct (binary). In our experience, success rates that are based on a facilitator's rating, such as correct, partially correct, and failure, can be very subjective and vary across usability practitioners. Therefore, we define success as either successful or not successful.

In interpreting success rates from usability reports, it is important to look at success in context of the scenarios asked. If the overall test-wide success rate is too high (80%-100%), the scenarios *may* have been too easy; if the overall, test-wide success rate is too low (< 50%) then the scenarios *may* have been too difficult.

A typical graph showing the success rate for each scenario is below.

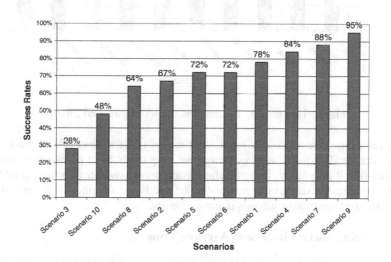

Fig. 2. Success Rates for Scenarios Presented in Order of Worst Performing

Once a test is completed, the scenarios are presented to the design team using a bar graph that shows the least successful scenario first, and the most successful scenario last. This helps to reinforce the idea that when making changes to the site, designers should start by fixing the scenarios that elicit the *worst performance*, and thus have the largest potential to make significant improvements.

4.2 Average Time to Complete Scenarios

The average time taken by participants to complete a scenario can be very informative and can help to measure users' ability to efficiently complete tasks in a reasonable amount of time.

When collecting this data, the average time to perform each scenario is usually measured from when participants see the first page (having already read the scenario), until they complete the scenario. Usually, shorter times indicate a well-designed site that allows users to efficiently complete tasks, whereas longer times *may* indicate that users had trouble with a task.

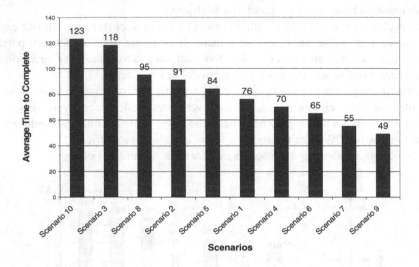

Fig. 3. Time on Task Presented in order of Worst Performing

When analyzing the results, we report the time on task for *successful* scenarios. This is to prevent the data from being eschewed by long times from participants who were not successful. It also allows design teams to determine whether or not the time to successfully complete a scenario is acceptable or needs improvement. A typical graph showing the average time in seconds is shown below.

4.3   Combining Success Rates and Average Time

Frequently, it is useful for designers to see the success rates and average time together. This can be done by providing a graph that combines both the success rate and the average time. Many times, the least successful scenarios take the longest time to perform, and the most successful scenarios take the least amount of time to perform. This provides two good reasons for presenting the results of a usability test in order of the scenarios with the worst performance. This type of graph is shown below.

4.4   Average Number of Extra Page Views

A fourth metric frequently used to evaluate users' efficiency is the number of average page views viewed by participants per scenario. Usually, the fewest number of page views leads to the fastest performance. Over several tests, we calculated a correlation of .82 (p<.0001) between the number of page views and the time taken to successfully complete the scenarios. In other words, average page views and average time to complete a scenario are usually related.

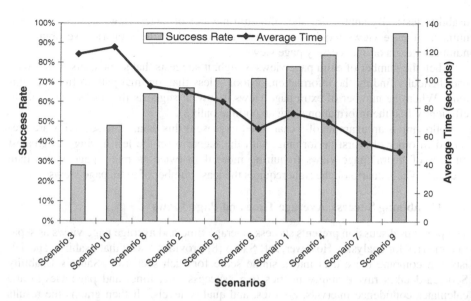

Fig. 4. Scenarios Presented in order of Lowest Success Rate overlaid with Average Time to Complete each Scenario

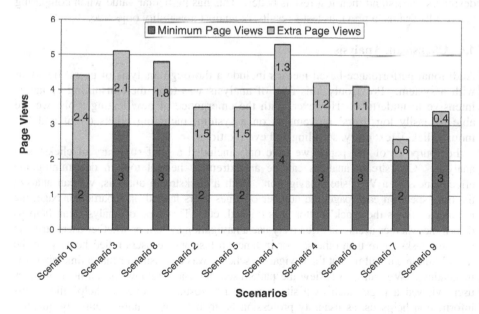

Fig. 5. Scenarios Presented in order of Worst Performing

While measuring the average number of page views per scenario is extremely valuable, we have found that a far more useful metric is a comparison between the average number of page views (for *successful* scenarios) with the minimum number of page views required to successfully complete the scenario (determined before the start of the usability test). By subtracting the minimum number of pages views from the average number of page views required to successfully complete the scenario, we obtain the number of extra or unnecessary page views.

When the number of extra page views is high, it suggests that some users were 'lost', had difficulty finding the information, or took a less than optimal path to find the content. When the number of extra page views is low, it suggests that users were able to efficiently find the information using an optimal path(s).

Following is an example of a chart used to present this data. The scenarios are presented in order of worst performing, with the scenario on the left having the greatest number of 'extra' page views (requiring more than twice as many page views than needed). The scenario on the right requires the least number of extra page views.

4.5 Combining Success, Average Time and Page Views

The previous discussion presents success, average time and average page views as separate metrics for analysis. However, Jeff Sauro [8] provides a way that usability specialists can combine these data into a single score for each scenario. Sauro's Usability Scorecard takes raw usability metrics (e.g., success rate, time, and page views) and calculates confidence intervals, z-scores, and quality levels. It then graphs the results automatically. The primary advantage of using the Usability Scorecard is the built-in confidence intervals. The confidence intervals can be used to help both testers and designers understand their test results better. This has particular value when comparing test results against a previous test's results, or against a usability objective.

4.6 Clicksteam Analysis

Additional performance-based metrics include a thorough analysis of users' behavior with a system. Previsouly, this type of analysis was too time-consuming or labor-intensive to undertake. However, with the emergence of new testing tools, we are able to easily log users' movements on a system, including clicks (right and left mouse clicks), text entry, scrolling and eye fixations.

For purposes of this paper, we have only included a brief summary of clickstream analysis. Clickstream analysis can be an extremely helpful tool in determining the effectiveness of a Web site's navigation. With a clickstream analysis, we can analyze the time spent on each page, the number of times users landed on a particular page, the number of times the 'back' button was clicked, etc. This type of analysis can help us determine whether users can effectively use a navigation system or whether users struggle on some tasks more than others. For instance, a task where users relied heavily on the 'back' button can tell us that the navigation schema was confusing or that a link label was misleading. We can also review the paths users took and review the number of times users viewed a page, such as a site map, search results, or even a 'help' file. This information helps us, as usability professionals, to better and more accurately quantify usability issues.

4.7 Preference Metrics

In addition to some of the above performance metrics, we also supplement our usability analysis with qualitative or preference metrics, including users' comments, concerns, frustrations and suggestions for improvement. In some instances, we evaluate or code user responses into similar categories in order to determine which elements users were commenting on the most.

Where possible, we also try to quantify users' thoughts and opinions with the use of post-scenario questions, as well as post-test satisfaction surveys. For instance, we have very successfully paired each scenario in a FirstClick test with a post-scenario question that asks users "How confident are you that the page you selected would have the information you are looking for?" This type of question allows us to analyze users' success in making a first click, the time it took users to select an item, and users' confidence that they would be able to find the information. By pairing confidence, satisfaction and preference data with performace data, we are able to have a more complete view of a system's usability, and thus, the recommendations for improvement that will have the greatest impact on the usability of that system.

5 Conclusions

By quantifying users' performance with a system, usability professionals can focus on the recommendations that provide the greatest potential for improvement, as well as ensure that we can reliably measure the effectiveness of these improvements in subsequent testing.

To that end, usability practitioners should try to focus first on scenarios with the:

- Lowest success rate
- Slowest average time
- Most extra page views
- Lowest efficiency, and
- Lowest satisfaction rate.

Caution must be exercised when usability reports are based strictly on the opinions of usability researchers or qualitative data gathered from participants' comments.

As practitioners, we must continually strive to quantify and justify the quality of our work and our usability recommendations. It is critical that we avoid making recommendations simply based on our opinions, or a 'live' heuristic review of the site, without clearly identifying those recommendations within our reports. One way that we have used is to relate each recommendation to a specific guideline in the *Research-Based Web Design & Usability Guidelines* book [9]. By working to quantify usability issues, we can ensure that recommendations are data-driven and performance-based, thus increasing the probability that improvements will measurably improve users' effectiveness, efficiency and satisfaction with a system.

References

1. Bailey, R.W.: Testing manual procedures in computer-based business information systems. In: Proceedings of the 16th Annual Meeting of the Human Factors Society, pp. 395–401 (1972)
2. Tullis, T.S., Albert, B.: Measuring the User Experience. Morgan Kaufmann, Boston (2008)

3. Bailey, R.W., Bailey, K.N.: Bailey's 'Usability Testing Environment'. In: Proceedings of the Human-Computer Interaction International Conference, July 22-27 (2005)
4. The Usability Testing Environment, http://www.mindd.com
5. Wolfson, C.A., Bailey, R.W., Nall, J., Koyani, S.: Contextual card sorting (or 'FirstClick' testing): A new methodology for validating information architectures. In: Proceedings of the UPA (2008)
6. Bailey, R.W., Wolfson, C.A., Nall, J.: Revising a Homepage: Applying Usability Methods that Guarantee Success. In: Proceedings of the UPA (2008)
7. Lindgaard, G., Chattratichart, J.: Usability testing: What have we overlooked? In: CHI 2007 Proceedings (2007)
8. Sauro, J.: The Usability Scoreboard, http://www.measuringusability.com/scorecard/login.php
9. Koyani, S.J., Bailey, R.W., Nall, J.R.: Research-Based Web Design & Usability Guidelines, U.S. Government Printing Office (2002, 2004, 2006)

Extending Quality in Use to Provide a Framework for Usability Measurement

Nigel Bevan

Professional Usability Services, 12 King Edwards Gardens, London W3 9RG, UK
mail@nigelbevan.com

Abstract. ISO has recently developed a new more comprehensive definition of quality in use, which has usability, flexibility and safety as subcharacteristics that can be quantified from the perspectives of different stakeholders, including users, managers and maintainers. While this provides a more complete set of requirements for operational use of a product, it also presents new challenges for measurement.

Keywords: Standards, usability, quality in use, requirements, measurement, safety.

1 Usability and Quality in Use

A traditional view of usability that is popular among product developers is that it is the attributes of the user interface that makes a product easy to use. This is consistent with one of the views of usability in HCI, for example in Nielsen's [22] 1993 breakdown where a product can be usable, even if it has no utility:

> system acceptability
> social acceptability
> practical acceptability
> cost
> compatibility
> reliability
> usefulness
> utility
> usability

This view of usability is also consistent with the first ISO definition of usability as part of software quality in ISO/IEC 9126 (1991):

Usability: A set of attributes that bear on the effort needed for use, and on the individual assessment of such use, by a stated or implied set of users.

This definition of user interface usability contrasts with the system perspective of usability defined from an ergonomic point of view in ISO 9241-11 (1998):

Usability: The extent to which a product can be used by specified users to achieve specified goals with effectiveness, efficiency and satisfaction in a specified context of use.

M. Kurosu (Ed.): Human Centered Design, HCII 2009, LNCS 5619, pp. 13–22, 2009.

This wider interpretation of usability was incorporated in the revision of ISO 9126-1 (2001), renamed "quality in use" as it is the user's perspective of the quality when using a product [3]. The software quality characteristics: functionality, reliability, efficiency, usability, maintainability and portability contribute to this quality.

2 The New Definition of Quality in Use

In 2006 the US standard for a Common Industry Format for Usability Test Reports (CIF) [1] was adopted by ISO as part of the revised Software product Quality Requirements and Evaluation (SQuaRE) set of standards [18]. As potential users of the CIF had originally expressed a preference for the term "usability" rather than "quality in use", the ISO 9241-11 definition of usability was retained when the CIF became part of this series.

When the ISO/IEC 9126-1 quality model came to be incorporated in the SQuaRE series (as ISO/IEC 25010), some ISO/IEC National Bodies commented on the discrepancy between the narrow definition of usability inherited from ISO/IEC 9126 and the broader definition in the CIF. But with the higher profile of usability in industry, there was now pressure to align the SQuaRE definition with the CIF, rather than vice versa. This was achieved by renaming the narrower ISO/IEC 9126 concept of usability as operability. This made it possible to define **Usability** as a characteristic of Quality in Use, with sub-characteristics of Effectiveness, Efficiency and Satisfaction. Quality in use in ISO/IEC CD 25010.3 has two further characteristics: **Safety** from 9126-1, and a new characteristic: **Flexibility**. These three characteristics are described in more detail below.

The complete quality in use model is:

Quality in use

Usability	Flexibility
Effectiveness	Context conformity
Efficiency	Context extendibility
Satisfaction	Accessibility
Likability	
Pleasure	Safety
Comfort	Commercial damage
Trust	Operator health and safety
	Public health and safety
	Environmental harm

2.1 Usability as A Characteristic of Quality in Use

In the current draft of ISO/IEC CD 25010.3, usability has the same subcharacteristics as in ISO 9241-11. Two of the subcharacteristics retain the same definitions: **effectiveness** (defined in terms of "accuracy and completeness") and **efficiency** (defined in terms of "resources expended").

But **satisfaction**, which is defined in ISO 9241-11 in terms of "comfort and acceptability of use", has been given a broader interpretation in ISO/IEC CD 25010.3.

As Hassenzahl points out, current approaches to satisfaction typically assess primarily the users' perception of effectiveness and efficiency, so that if users perceive the product as effective and efficient, they are assumed to be satisfied [9]. But there is evidence that fun or enjoyment is an aspect of user experience that also contributes significantly to overall satisfaction with a product [8].

So in order to encompass the overall user experience, satisfaction needs to be concerned with both pragmatic and hedonic user goals.

The pragmatic user goals are:

• Acceptable perceived experience of use (pragmatic aspects including efficiency).
• Acceptable perceived results of use (including effectiveness).
• Acceptable perceived consequences of use (including safety).

In ISO/IEC CD 25010.3 it is suggested that these can be summarised as:

Likability (cognitive satisfaction): the extent to which the user is satisfied with the ease of use and the achievement of pragmatic goals, including acceptable perceived results of use.

Trust (satisfaction with security): the extent to which the user is satisfied that the product will behave as intended and with acceptable perceived consequences of use. Hassenzahl identifies three hedonic goals [10]:

• Stimulation (i.e. personal growth, an increase of knowledge and skills).
• Identification (i.e. self-expression, interaction with relevant others).
• Evocation (i.e. self-maintenance, memories).

To these I would add:

• Pleasurable emotional reactions to the product (Norman's visceral category [23]).

In ISO/IEC CD 25010.3 it is suggested that these can be summarised as:

Pleasure (emotional satisfaction): the extent to which the user is satisfied with their perceived achievement of hedonic goals of stimulation, identification and evocation and associated emotional responses.

In addition ISO/IEC CD 25010.3 includes:

Comfort (physical satisfaction): the extent to which the user is satisfied with physical comfort.

Measuring Satisfaction

Many software developers regard satisfaction as a personal response that cannot be quantified, and in much usability testing only qualitative feedback on satisfaction is obtained. Ad hoc questionnaires are sometime used, but psychometrically designed questionnaires will give more reliable results [12].

Simple questionnaires (such as SUS [5]) just measure the user's assessment of the ease of use. Longer questionnaires can measure more specific aspects, such as affect, efficiency, helpfulness, control and learnability in SUMI [20]. Trust can be measured using the System Trust Scale [19], and pleasure with questionnaires such as Attrak-Diff [11]. There are also a variety of questionnaires for comfort, e.g. [24].

2.2 Flexibility as a Characteristic of Quality in Use

Usability is defined in terms of user performance and satisfaction in a particular context of use. Considering context is important, as a product that is usable in one context of use may not be usable in another context with different users, tasks or environments. However, one consequence is that it may be possible to demonstrate very high usability for a product with carefully selected, but unrepresentative, users, tasks and environments of use. This was one of the motivations for the Common Industry Format: to enable a potential purchaser to judge whether the users, tasks and environments for which usability had been demonstrated, matched their own needs [25]. The Common Industry Specification for Usability – Requirements (CISU-R) went a step further, including a specification of the context of use in which usability is to be achieved as part of the usability requirements [4].

The need to consider the context has been made explicit in the new definition of quality in use in ISO/IEC CD 25010.3. A third characteristic of Flexibility has been included, with subcharacteristics of Context Conformity and Context Extendibility.

Context Conformity is defined as the degree to which usability and safety meet requirements in all the intended contexts of use. This provides the basis for measuring the extent to which usability has been achieved in the intended contexts of use.

Context Extendibility is defined as the degree of usability and safety in contexts beyond those initially intended. Context extendibility can be achieved by adapting a product for additional user groups, tasks and cultures. Context extendibility enables products to take account of circumstances, opportunities and individual preferences that may not have been anticipated in advance. If a product is not designed for context extendibility, it may not be safe to use the product in unintended contexts.

Accessibility is defined as the degree of usability for users with specified disabilities. This definition was used in preference to the ISO 9241-171 definition: "the usability of a product, service, environment or facility by people with the widest range of capabilities", as the ISO 9241-171 definition is stated as an objective that is difficult to quantify.

In an earlier draft, **learnability** was also a subcharacteristic of flexibility. Given the importance of setting objectives for learnability as usability in a learning context (effectiveness, efficiency and satisfaction with achieving learning goals), learnability could be reintroduced, for example as "the extent to which a product can be used by specified users to achieve specified learning goals with effectiveness, efficiency and satisfaction in a specified context of use". This would make learnability analogous to accessibility, with associated product characteristics for "technical learnability".

Measuring Flexibility

If usability requirements have been specified in a format similar to the CISU-R, testable usability objectives for user performance and satisfaction will have been given for specified ranges of context of use (user groups, tasks and environments). These requirements could in principle be evaluated in usability tests. In practice scarce testing resources will have to be prioritised, but identifying the range of intended contexts of use can reveal additional users, tasks and environments that may be a priority for testing.

It is important to specify requirements for flexibility, even if it is not practical to test them. The wider range of user groups, tasks and environments can be translated into additional design requirements for product features needed to support usage for all the identified contexts of use. Context Conformity in contexts that cannot be tested can be assessed by expert judgment.

Context Extendibility is more difficult to specify and measure in advance, as it is concerned with usage in unanticipated contexts of use. Design requirements that facilitate Context Extendibility include designing a product so that it can either be configured for specific needs (e.g. language, culture, task steps), or can be adapted by the user to suit individual capabilities and needs. Products are frequently used for unanticipated purposes (for example, use of Excel for prototyping [2]), so the ability to adapt the product to new needs significantly extends the usability. This can only be properly evaluated by patterns of actual usage, but the potential can be assessed by the extent to which the design is open to configuration and adaptation. As with Context Conformity, identifying requirements for Context Extendibility can have a large impact on design.

Accessibility can be specified by establishing objectives for usability for users with particular types of disabilities, thus giving a design objective for the success rate and productivity to be expected of users with disabilities. This can subsequently be validated by user testing.

2.3 Safety as a Characteristic of Quality in Use

Safety is defined in ISO/IEC CD 25010.3 as the degree of expected impact of harm to people, business, data, software, property or the environment in the intended contexts of use.

While effectiveness and efficiency measure the positive benefits of productivity and goal achievement, the term safety is here interpreted in a broad way to measure the potential negative outcomes that could result from incomplete or incorrect output. For a consumer product or game, negative business consequences may not only be associated with poor performance, but also, for example, with a lack of pleasurable emotional reactions or of achievement of other hedonic goals.

Safety has four subcharacteristics:

- **Commercial damage:** The degree of expected impact of harm to commercial property, operations or reputation in the intended contexts of use. This could include the administrative costs of correcting erroneous output, inability to provide an acceptable service, or loss of current or future sales. Examples include lack of sales due to poor web site design, and the chaos following introduction of a new system to issue passports in the UK [25,21].
- **Operator health and safety:** The degree of expected impact of harm to the operator in the intended contexts of use. The legislation for workstation design in the EU [7] is intended to minimize this risk.
- **Public health and safety:** The degree of expected impact of harm to the public in the intended contexts of use.
- **Environmental harm:** The degree of expected impact of harm to property or the environment in the intended contexts of use.

Specification and measurement of usability should always be considered in conjunction with associated safety risks.

Measuring Safety
It is not easy to measure product safety, but it is usually possible to list the potential adverse consequences of product failure or human error, based on previous experience with similar products or systems. Multiplying the estimated frequency by the estimated impact of the potential failures provides a basis for prioritization of the need to find design solutions to minimize the probability of failures occurring. The solutions can include improving the user interface to reduce the probability of human error.

2.4 Quality in Use for Different Stakeholders

One reason for the popularity of the ISO 9241-11 definition with usability professionals, is that when interpreted from the perspective of the organisation's goals it provides a business rationale for the importance of usability that is more compelling than mere ease of use.

But usability can also be seen from the inside out as meeting the user's goals, rather than the organisation's goals, which takes usability back closer to its original meaning. From this perspective the key element in the ISO definition is satisfaction, and this is one reason for the expanded interpretation of satisfaction in ISO/IEC 25010.3. For an officer worker and their manager cognitive satisfaction may be most important, for a games player or manufacturer, pleasure may be more important.

Measuring Quality in Use from Different Stakeholder Perspectives
Usability and flexibility are measured by effectiveness (task goal completion), efficiency (resources used) and satisfaction. The relative importance of these measures depends on the purpose for which the product is being used (for example in some personal situations, resources may not be important).

Table 1 illustrates how the measures of effectiveness, resources, satisfaction, flexibility and safety can be selected to measure quality in use from the perspective of different stakeholders.

From an organisational perspective, quality in use and usability are about achievement of task goals. But for the end user there are not only pragmatic task-related "do" goals, but also hedonic "be" goals [6]. For the end user, effectiveness and efficiency are the pragmatic goals, and stimulation, identification, evocation and pleasure are the hedonic goals.

From an end user perspective, quality in use can be used to measure the extent to which users achieve their goals within an acceptable time, with hedonic and pragmatic satisfaction and without adverse consequences to the user's health and safety. They may also want to modify the interaction and appearance to suit their individual needs and preferences.

From the perspective of the organization using the product, the measures could be the extent to which the task goals are achieved with acceptable cost, with management satisfaction and with acceptable financial consequences to the business of potential errors. They may also want to be able to customize the system to meet changing needs.

Table 1. Stakeholder perspectives of quality in use

STAKEHOLDER:	End User *Usability*	Usage Organisation *Cost-effectiveness*	Technical support *Maintenance*
GOAL: MEASURES	Pragmatic and hedonic goals	Task goals	Support goals
Effectiveness	User effectiveness	Task effectiveness	Support effectiveness
Resources	Productivity (time)	Cost efficiency (money)	Support cost
Satisfaction	Hedonic and pragmatic satisfaction	Management satisfaction	Support satisfaction
Flexibility	Individualisation	Customisation	Adaptability
Safety	Risk to user (health and safety)	Commercial risk	System failure or corruption

Similarly from a support perspective, the measures could be the extent to which the support goals, including adaptability, are achieved with acceptable cost, with support personnel satisfaction and with acceptable financial consequences to the business of any resulting human or software errors.

3   Measuring Product Attributes or Quality in Use

Table 2 shows how measures of usability and safety are dependent on product attributes that support different aspects of user experience. In Table 2 the columns are the quality characteristics that contribute to the overall user experience, with the associated product attributes needed to achieve these qualities.

The quality characteristics are frequently assessed by direct measurement of product attributes (the ISO/IEC 9126/25010 internal and external measures), for example by using heuristic evaluation, web design guidelines or the Web Content Accessibility Guidelines (WCAG) developed by the WAI [26]. However, these are only a means to an end: acceptable user performance and satisfaction.

The users' goals may be pragmatic (to be effective and efficient), and/or hedonic (stimulation, identification and/or evocation).

The actual experience of usage is difficult to measure directly. The measurable consequences are the user's performance, satisfaction with achieving pragmatic and hedonic goals, comfort and pleasure.

User performance and satisfaction is determined by qualities including attractiveness, functional suitability and ease of use. Other quality characteristics will also be relevant in determining whether the product is learnable, accessible, and safe in use.

Pleasure will be obtained from both achieving goals, and as a direct visceral reaction to attractive appearance [23].

Table 2. Factors contributing to system usability and UX

Quality characteristic	Attractiveness	Functional suitability	Ease of use	Learnability	Technical accessibility	Safety
Product attributes	Aesthetic attributes	Appropriate functions	Good UI design	Learnability attributes	Accessibility attributes	Safe and secure design
Pragmatic do goals		To be effective and efficient				
Hedonic be goals		Stimulation, identification and evocation				
Actual experience	Visceral	Experience of interaction				
Performance measures		Effectiveness and Efficiency: effective task completion and efficient use of time		Learnability: effective and efficient to learn	Accessibility: effective and efficient with disabilities	Safety: occurrence of unintended consequences
Satisfaction measures	Pleasure		Likability and Comfort			Trust

Note that this framework does not include the popular measure of the number of errors made by the user, as the consequences of errors are already incorporated in effectiveness, efficiency and satisfaction. Uncorrected errors will impact on successful task completion, and corrected errors will contribute to task time and reduced satisfaction. However, counts of errors can be a useful indicator of usability problems, particularly when it is not practical to obtain measures of successful task completion.

4 Conclusions

The new definition of quality in use provides a framework for a more comprehensive approach to specifying usability requirements and measuring usability, taking account of the following issues.

1. From which stakeholder perspective(s) (e.g. users, staff and/or managers) does usability need to be specified and measured?
2. What is the scope of the context of use in which it is important to establish usability requirements?
3. What aspects of effectiveness, efficiency and satisfaction are most important, and how can they be measured?
4. What are the acceptable levels of risks of potential adverse consequences to the identified stakeholders resulting from poor usability or inappropriate output?
5. What product attributes are needed to achieve the identified objectives?
6. How can these product attributes be monitored during development?
7. Which are the most important contexts of use in which to validate usability?

References

1. ANSI: Common industry format for usability test reports (ANSI-NCITS 354-2001). ANSI (2001)
2. Berger, N., Arent, M., Arnowitz, J., Sampson, F.: Effective Prototyping with Excel: A practical handbook for developers and designers. Morgan Kaufmann, San Francisco (2009)
3. Bevan, N.: Quality in use: meeting user needs for quality. Journal of Systems and Software 49(1), 89–96 (1999)
4. Bevan, N., Claridge, N., Maguire, M., Athousaki, M.: Specifying and evaluating usability requirements using the Common Industry Format: Four case studies. In: Proceedings of IFIP 17th World Computer Congress, Montreal, Canada, August 25-30, pp. 133–148. Kluwer Academic Publishers, Dordrecht (2002)
5. Brooke, J.: SUS: A "quick and dirty" usability scale. In: Jordan, P., Thomas, B., Weerdmeester, B. (eds.) Usability Evaluation in Industry, pp. 189–194. Taylor and Francis, Abington (1996)
6. Carver, C.S., Scheier, M.F.: On the self-regulation of behavior. Cambridge University Press, Cambridge (1998)
7. CEC: Minimum safety and health requirements for work with display screen equipment Directive (90/270/EEC). Official Journal of the European Communities No L 156, 21/6/90 (1990)
8. Cockton, G.: Putting Value into E-valuation. In: Law, E.L., Hvannberg, E.T., Cockton, G. (eds.) Maturing Usability. Quality in Software, Interaction and Value. Springer, Heidelberg (2008)
9. Hassenzahl, M.: The effect of perceived hedonic quality on product appealingness. International Journal of Human-Computer Interaction 13, 479–497 (2002)
10. Hassenzahl, M.: The thing and I: Understanding the relationship between user and product. In: Blythe, M., Overbeeke, C., Monk, A.F., Wright, P.C. (eds.) Funology: From Usability to Enjoyment, pp. 31–42. Kluwer, Dordrecht (2003)
11. Hassenzahl, M.: AttrakDiff(tm), http://www.attrakdiff.de
12. Hornbæk, K.: Current practice in measuring usability: Challenges to usability studies and research. International Journal of Human-Computer Studies 64, 79–102 (2006)
13. ISO/IEC 9126: Software engineering – Product quality. ISO (1991)
14. ISO/IEC 9126-1: Software engineering – Product quality - Part 1: Quality model. ISO (2001)
15. ISO 9241-11: Ergonomic requirements for office work with visual display terminals (VDTs) Part 11: Guidance on Usability. ISO (1998)
16. ISO 9241-171: Ergonomics of human-system interaction – Part 171: Guidance on software accessibility. ISO (2008)
17. ISO/IEC CD 25010.3: Systems and software engineering – Software product Quality Requirements and Evaluation (SQuaRE) – Software product quality and system quality in use models. ISO (2009)
18. ISO/IEC 25062: Software Engineering - Software product Quality Requirements and Evaluation (SQuaRE)-Common Industry Format (CIF) for Usability Test Reports. ISO (2006)
19. Jian, J.-Y., Bisantz, A.M., Drury, C.G.: Foundations for an empirically determined scale of trust in automated systems. International Journal of cognitive Ergonomics 4(1), 53–71 (2000)

20. Kirakowski, J.: The Software Usability Measurement Inventory: Background and Usage. In: Jordan, P., Thomas, B., Weerdmeester, B. (eds.) Usability Evaluation in Industry. Taylor and Francis, Abington (1996)
21. National Audit Office: United Kingdom Passport Agency: The passport delays of Summer 1999. Publication HC 812 1998-1999 (1999)
22. Nielsen, J.: Usability Engineering. Academic Press, London (1993)
23. Norman, D.: Emotional design: Why we love (or hate) everyday things. Basic Books, New York (2004)
24. Norman, K., Alm, H., Wigaeus Tornqvist, E., Toomingas, A.: Reliability of a questionnaire and an ergonomic checklist for assessing working conditions and health at call centres. Int. J. Occup. Saf. Ergon. 12, 53–68 (2006)
25. Theofanos, M., Stanton, B., Bevan, N.: A practical guide to the CIF: Usability Measurements. Interactions 18(6), 34–37 (2006)
26. Web Accessibility Initiative (WAI): Web Content Accessibility Guidelines 2.0, http://www.w3.org/TR/WCAG20/

Combining Fast-Paced Usability and Scientific Testing to Improve the Lunar Quest Physics Game

Holly Blasko-Drabik, James Bohnsack, and Clint Bowers

University of Central Florida, USA
HollyBlasko@gmail.com, jimisack@gmail.com, bowers@mail.ucf.edu

Abstract. This study focuses on conducting fast-paced in-house testing in combination with user comments and scores on the Questionnaire for User Interface Satisfaction (QUIS) to provide a better educational game. The standard QUIS was shortened and focused to be more sensitive to aspects of our educational game. The game, Lunar Quest, was created at the University of Central Florida as a supplemental instruction tool that would provide students with examples of several different Physics concepts while being an enjoyable and fun game. Overall the modified version of the QUIS was not successful in determining which categories of the game should be targeted, although the open ended questions did help our researchers focus the game redesign and showed improvement throughout the testing period.

Keywords: Usability, QUIS, Serious Games, Learning.

1 Introduction

We have all heard that usability and in-house testing can improve a product, making it more user friendly or effective. The counter argument is that such testing often takes too much time, effort, and money to make it worthwhile. This study focused on conducting fast-paced in-house testing in combination with user comments and scores on the Questionnaire for User Interface Satisfaction (QUIS) to provide a better educational game. The game, Lunar Quest, was created at the University of Central Florida as a supplemental instruction tool that would provide students with examples of several different Physics concepts while being an enjoyable and fun game. Lunar Quest currently consists of 8 mini-games, each pair of games teaching one main concept typically covered in general Physics courses, such as the effects of gravity or kinematics.

The QUIS was specifically designed to determine the user's satisfaction with computer software in 1988 as a nine point Likert scale questionnaire. Users are asked to evaluate each question on this scale based on a pair of adjectives at the anchor points. For example, "Error messages were unhelpful (1) to helpful (9)", if the users felt that the question did not apply to the system then they are asked to circle N/A. Since it was first implemented the measure has been increased to cover many new aspects of technology. The current version of the QUIS is broken down into 12 parts and contains 125 Likert scale question, 12 open ended response questions, and several questions pertaining to the users past experience with the system

M. Kurosu (Ed.): Human Centered Design, HCII 2009, LNCS 5619, pp. 23–26, 2009.
© Springer-Verlag Berlin Heidelberg 2009

or similar systems. Because we wanted to administer the QUIS after each of the mini-games we were looking for a shorter more focused version of the QUIS that could be completed with 5-10 minutes but that would still help us focus our development of the mini-games.

In order to achieve this goal, a modified version of the QUIS (see appendix a), specifically targeting five different categories we believe would impact the learning and usability of the game was created and given to our game testers. These categories were based on the full 12 part QUIS many subsections and questions were removed. It was determined that 1) Overall reaction to the game, 2) Text and Organization, 3) Use of terms and information, 4) Learning, 5) Game features & Capabilities would be targeted. Many questions within these categories were also removed if they seemed confusing to the users or experimenters or did not pertain to features that were included in the games we were testing. The final version of the questionnaire that we used consisted of 28 of the Likert questions. Researchers also asked users to several open ended questions, specifically, what they liked or disliked about the game, how we could make the game more usable, and what would make the game more fun or interesting.

2 Method

Pairs of students were brought into the lab and asked to play the mini-games and evaluate them using the modified QUIS. Ideally, between 10 and 14 students were included in each phase of testing, however, due to the discovery of several major bugs or fixes that took longer than expected to implement, the groups were not equal. Testing was broken up into 5 phases (including the initial pilot testing) and 64 students played the Lunar Quest mini games. In the first period of testing, critical bugs and errors were discovered, recorded, summarized, and provided directly to the game developers. As soon as these changes were made, another small group of subjects were tested and the games were again modified. This allowed the mini-games to be quickly improved while development was ongoing instead of waiting until the games were completed to test them.

3 Results

One-way ANOVAs were conducted looking at the five categories across all eight mini-games. The ANOVAs showed that there was a significant improvement in several of the mini-games, specifically the game features and the way information was organized. Follow up t-tests also showed several significant findings where a decrease in scores between the phases of the game were found, specifically on the learning and capabilities categories.

4 Discussion

Overall, many important issues were discovered early enough in the design process to allow for the games to be improved without a total redesign. By using this method, we were able to identify which specific categories our users found most important, specifically the game features the way information was organized.

Using the modified QUIS allowed us to target suspected features that we believed to be critical in learning the material and concepts explained in the games. However, based on our results we do not believe that the modified QUIS was sensitive enough to accurately determine changes in the usability of the game. The student's comments to the open-ended questions showed a marked improvement in their opinions of the games between many of the phases of testing, however, the data showed that no significant change or an inverse relationship.

Several possible constraints may also have influenced our results. In our efforts to shorten the questionnaire to 5 or 10 minutes, several questions that many have been important to the validity of the overall QUIS may have been removed. Because the game needed to be completed and changed quickly, group sizes varied from 5 to 15, in addition several students data could not be used because they circled the adjectives instead of a number on the 0-9 scale.

Future testing of the Lunar Quest game is currently being conducted looking not only at the usability of the games, but at whether using the game has a positive impact on learning the targeted Physics concepts.

Appendix A

		Overall reaction to the game											
1		terrible	0	1	2	3	4	5	6	7	8	9	wonderful	N/A
2		difficult	0	1	2	3	4	5	6	7	8	9	easy	N/A
3		frustrating	0	1	2	3	4	5	6	7	8	9	satisfying	N/A
4		dull	0	1	2	3	4	5	6	7	8	9	stimulating	N/A
5		rigid	0	1	2	3	4	5	6	7	8	9	flexible	N/A
		Screen											
6	Reading characters on the screen	hard	0	1	2	3	4	5	6	7	8	9	easy	N/A
7	Highlighting simplifies task	not at all	0	1	2	3	4	5	6	7	8	9	very much	N/A
8	Use of bolding	Unhelpful	0	1	2	3	4	5	6	7	8	9	Helpful	N/A
9	Organization of information	confusing	0	1	2	3	4	5	6	7	8	9	very clear	N/A
10	Arrangement of information on screen	Illogical	0	1	2	3	4	5	6	7	8	9	Logical	N/A
11	Sequence of levels	confusing	0	1	2	3	4	5	6	7	8	9	very clear	N/A

		Terminology & Game Information											
12	Use of terms throughout game	inconsistent	0	1	2	3	4	5	6	7	8	9	consistent	N/A
13	Terminology related to task	never	0	1	2	3	4	5	6	7	8	9	always	N/A
14	Position of messages on screen	inconsistent	0	1	2	3	4	5	6	7	8	9	consistent	N/A
15	Prompts for input	confusing	0	1	2	3	4	5	6	7	8	9	clear	N/A
16	Computer informs about its progress	never	0	1	2	3	4	5	6	7	8	9	always	N/A
17	Error messages	unhelpful	0	1	2	3	4	5	6	7	8	9	helpful	N/A
		Learning											
18	Learning to play the game	difficult	0	1	2	3	4	5	6	7	8	9	easy	N/A
19	Exploring new features by trial and error	difficult	0	1	2	3	4	5	6	7	8	9	easy	N/A
20	Remembering directions	difficult	0	1	2	3	4	5	6	7	8	9	easy	N/A
21	Performing tasks is straightforward	never	0	1	2	3	4	5	6	7	8	9	always	N/A
22	Help messages	unhelpful	0	1	2	3	4	5	6	7	8	9	helpful	N/A
23	Instructions	confusing	0	1	2	3	4	5	6	7	8	9	clear	N/A
		Game Capabilities											
24	Game speed	too slow	0	1	2	3	4	5	6	7	8	9	fast enough	N/A
25	System reliability	unreliable	0	1	2	3	4	5	6	7	8	9	reliable	N/A
26	System tends to be	noisy	0	1	2	3	4	5	6	7	8	9	quiet	N/A
27	Correcting your mistakes	difficult	0	1	2	3	4	5	6	7	8	9	easy	N/A
28	Designed for all levels of users	never	0	1	2	3	4	5	6	7	8	9	always	N/A

Considering User Knowledge in the Evaluation of Training System Usability

Clint Bowers[1], Jan Cannon-Bowers[2], and Talib Hussain[3]

[1] Department of Psychology, University of Central Florida, Orlando, FL, USA 32816
[2] Institute for Simulation and Training, University of Central Florida,
Orlando, FL, USA 32816
[3] BBN Technologies, Cambridge, MA, USA
{Clint Bowers,bowers}@mail.ucf.edu, talib@bbn.com

Abstract. A variety of software-based systems are being used as training media. There is not, however, an accepted approach to evaluating the usability of these systems. Traditional usability approaches can be employed with some effectiveness, but they may lack appropriate specificity for use in training. In this paper, we evaluate whether assessing, and remediating, gaps in learner knowledge might be an important addition to training system evaluation. The results suggest that remediating knowledge gaps might lead to more accurate usability conclusions.

Keywords: Learner-centered design, usability, training.

1 Introduction

As the utilization of electronic systems in training becomes more prevalent, it will be important to adopt usability evaluation approaches that yield an accurate portrayal of the system's potential as a training platform. However, according to Squire & Preece, [1], traditional usability evaluation approaches may not be sufficient to address the particular needs of the training community. These authors conclude that there has been a failure to integrate usability evaluation with the science of how people learn. Thus, usability evaluations may yield results that are misleading. Given this problem, some researchers have discussed the notion of "pedagogical usability analysis." This approach extends traditional usability analysis by considering not just the user's ability to interact meaningfully with the system, but the ability of the system to satisfy its intended educational goals [2].

In response to the challenges described above, it has been suggested that the traditional user-centered design approach be altered to "learner-centered design" [3, 4]. A central tenet of this approach is the emphasis on the ability of the system to help the learner get from their current state to their desired state. This design approach and its corollary evaluation approach consider a number of factors (see Ardito et al., 2006 for a review of these issues [5]). However, one feature that all approaches have in common is a need to make some assumptions about the learner's state of knowledge. These assumptions play a critical role not only in the development of training systems, but in the interpretation of usability evaluation data. For example, if a user is unable to

M. Kurosu (Ed.): Human Centered Design, HCII 2009, LNCS 5619, pp. 27–30, 2009.
© Springer-Verlag Berlin Heidelberg 2009

perform a task for which they are presumed to have the requisite knowledge and ability, it would logically indicate a design flaw that might necessitate an expensive redesign. Given the importance of these data, it is important that their validity be established.

In many cases, user knowledge is determined by successful completion of educational experiences. For example, if a student has passed a particular course, it is typically presumed that they have a mastery of materials covered in that course. This approach is used frequently in the usability arena, but there is some reason to suspect that it may not be optimal. Learners frequently receive "passing grades" without acquiring targeted knowledge. Further, even knowledge that was successfully acquired is likely to have been forgotten if the usability testing is conducted well after the educational experience. Consequently, conclusions based on assumed competencies may be incorrect.

A variety of usability evaluation methods have been suggested for educational software [1, 6, 7] . In large part, these approaches all emphasize the manner in which the software presents information, but there is somewhat less emphasis on the existing knowledge of the learner. This is likely because these studies have focused on education of novices rather than the training of participants who are likely to have some, perhaps substantial, prior experience with the course material. In training applications, software designers are likely to make assumptions about the knowledge state of the learner. The accuracy of these assumptions, however, is likely to be an important factor in downstream usability and training effectiveness. Thus, we assert that assessment of learner state, or appropriate remedial activities, may be an important consideration in this type of usability analysis.

To further investigate this possibility, we conducted a study to determine whether faulty assumptions about user knowledge based on experience would influence the interpretation of usability data. The details of this study are detailed below.

2 Participants

A total of 60 recruits at a military recruiting command participated in the study. There were 56 males and 4 females in the sample. The average age of participants was 19.7 years. The users had a wide range of experience with computers, with the majority describing themselves as comfortable with computer applications. All users had completed their initial training and had passed a "capstone exercise" in which targeted skills had been successfully demonstrated within the preceding 6 weeks.

3 Instruments

Users were asked to play a newly developed computer-based simulation that emphasized the control of onboard flooding. The scenario required users to understand ship navigation, proper communication procedures, and repair procedures. All users completed a tutorial that taught the simulation controls. Users then completed a simulated damage control mission similar to their capstone experience. User performance was observed and rated by trained raters. Users also completed a customized version of the Questionnaire for User Interface Satisfaction (QUIS, Chin et al., 1988) and the System Usability Scale (Digital Equipment Corporation, 1986).

4 Method

Users were assigned to one of two groups. All groups received informed consent, the tutorial, the test mission, and the questionnaires. Half of the group was assigned to a "training condition," which quickly reviewed the concepts of ship navigation, a critical ability for the test mission. The training was a brief review (< 5 minutes) which included a text-and-graphics form that demonstrated navigation principles.

5 Results

Statistical analyses of demographic data indicated no significant differences between the training and control groups on any of the QUIS items. Further, independent samples t-tests indicated that there were no differences in the subjective evaluation of system usability. Interestingly, however, the data indicated that the training group made fewer errors on performance tasks involving navigation ($\underline{t} = 2.87$, p < .05). Further, users in the training group were significantly less likely to "fail" on these events (e.g., be unable to accomplish a desired task) ($X^2 = 5.09$, p < .05).

6 Discussion

The results of this study demonstrate that assumptions about user knowledge based on experience may lead to misleading usability conclusions. In the current case, the assumption of competency based on experience (i.e., successful completion of the capstone experience) would have lead us to conclude that the training system was flawed, and would have likely led to a re-design and the delays and costs associated with it. By providing brief, extremely inexpensive training, we were able to reveal that the critical errors were not likely due to a design flaw, but the lack of a critical competency in our users. We were able to correct this deficiency without changing the software at all.

Clearly, there is a need to provide accurate evaluations of training systems. These evaluations will increase in frequency and importance as we grow more reliant upon technology-based education. A critical element in these evaluations will be an understanding of the user's pre-existing knowledge, skills and abilities. These results of the present study suggest that there may be an advantage to evaluating this factor more carefully than is often done. A more thorough evaluation (and remediation when necessary) is likely to result in a more accurate and useful evaluation.

References

1. Squires, D., Preece, J.: Predicting quality in educational software: Evaluating for learning, usability and the synergy between them. Interacting with Computers, 467–483 (1999)
2. Silius, K., Tervakari, A.M., Pohjolainen, S.: A multidisciplinary tool for the evaluation of usability, pedagogicalusability, accessibility and informationalquality of Web-based courses. Proceedings of PEG2003–The 11th International PEGConference: Powerful ICT for Teaching and Learning (2003)

3. Quintana, C., Soloway, E., Norris, C.: Learner-Centered Design: Developing Software That Scaffolds Learning. In: Second IEEE International Conference on Advanced Learning Technologies (ICALT 2001) (2001)
4. Quintana, C., Krajcik, J., Soloway, E.: Exploring a structured definition for leaner-centered design. In: Proceedings of the ICLS 2000 (2000)
5. Ardito, C., Costabile, M.F., De Marsico, M., Lanzilotti, M., Levialdi, S., Plantamura, S., Roselli, T., Rossano, V., Tersigni, M.: Towards Guidelines for Usability of e-Learning Applications. User-Centered Interaction Paradigms for Universal Access in the Information Society (2004)
6. Dringus, L.: An iterative usability evaluation procedure for interactive online courses. J. Int. Instr. Dev. 7(4), 10–14 (1995)
7. Ravden, S.J., Johnson, G.I.: Evaluating usability of human-computer interface: a practical method. Wiley, Chichester (1989)

Engaging Experience:
A New Perspective of User Experience with Physical Products

Chun-Juei Chou and Chris Conley

Institute of Design, Illinois Institute of Technology,
350 N. LaSalle Dr., Chicago IL 60654, USA
cjchou@id.iit.edu, chris@gravitytank.com

Abstract. Engaging experience is a specific kind of experience that a user acquires when and after using a product frequently, intensively, actively, vividly, and completely, etc. For example, if the appearance of a toaster is completely transparent, a user can see how the bread changes from white to brown. The transparent sides stage the toasting process as a visually engaging performance. To expand on engaging experience, this paper presents the definition of engaging experience, example products that engage users, the classification and instinctiveness of engaging experience, and product properties that foster engaging experience.

Keywords: Engaging experience, user experience, product experience, user-product interaction, user-product relationship.

1 Introduction

In this research, engaging experience is defined as a specific kind of experience that a user acquires when and after using a product frequently, intensively, actively, vividly, and completely, etc. For example, if the appearance of a toaster is completely transparent, a user would see how the bread changes from white to dark. The transparent sides *stage* the toasting process as a visually engaging performance. Another example is a tape dispenser with a simple odometer. In addition to acquiring tape, the user would notice how much tape has been dispensed in terms of distance. The odometer makes every pull a vivid experience. These two cases demonstrate how an individual product can engage users.

Engaging experience makes the users' interaction with products somehow more interesting. Engaging experience is a novel research domain in product design. It is different from the study of functionality, usability, aesthetics, interaction and affection. While these current domains are important to satisfying user needs, engaging experience is both different from, but potentially as significant as these domains. The author expects this research leads to a better understanding of how products can be designed to engage users, bringing another dimension of value to product design.

M. Kurosu (Ed.): Human Centered Design, HCII 2009, LNCS 5619, pp. 31–40, 2009.
© Springer-Verlag Berlin Heidelberg 2009

2 Defining Engaging Experience

In <u>The Experience Economy</u>, Pine and Gilmore [1] assert that "when a person buys an experience, he pays to spend time enjoying a series of memorable events that a company stages – as in a theatrical play - to engage him in a personal way". "Staging experiences is not about entertaining customers; it is about engaging them." The Experience Economy concerns how to engage customers in commercial activities. Pine and Gilmore develop the experience realms, shown in figure 1, to illustrate four examples of engaging experience. The two dimensions, active-passive participation and absorption-immersion characterize how people engage. In *active participation*, people become participants, actors or even players rather than bystanders in activities. In *passive participation*, people attend, appear in or are exposed to activities. *Absorption* means that people need to absorb information spread throughout in activities, whereas *immersion* means that people regard the spread information as the atmosphere or background unnecessary to be further processed. According to the experience realms, four examples of engaging experience are provided:

- *Entertainment engagement*: When a person pays to join in an activity, he is entertained by the activity through his senses and participation. The person has little disturbance to the activity, leaving it essentially untouched. Being a volunteer in a magic show, shouting with crowds in a baseball game, and rocking and beating in a Jazz Festival are examples of entertainment engagement.
- *Education engagement*: When a person pays to join in an activity, he creates something as a part of the activity and enjoys the result. For example, in a cooking class, the apprentices join in food preparation, practice cooking and eat delicious cuisine. Each type of participation is education engagement.
- *Escapist engagement*: When a person pays to join in an activity, he escapes temporarily from his normal, daily reality. The person is interested in the activity because it is not always available in his daily life. Examples of environments that enable escapist engagement include casinos, cyberspace, and theme parks, etc.

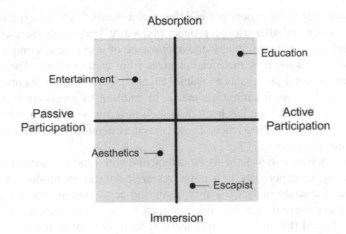

Fig. 1. The experience realms [1]

- *Aesthetic engagement*: When a person pays to join in an activity, he is attracted by an interesting aspect of the activity. The person has little disturbance to the activity, leaving it essentially untouched. Eating in front of a chef who is skillfully processing food, drinking coffee in Starbucks to enjoy the atmosphere, etc. are examples of aesthetic engagement.

In addition, Pine and Gilmore also mention example products that enable engaging experiences. For instance, the Rawlings Sporting Goods Company of St. Louis, Missouri, introduced a ball that makes play-catching more engaging [2]. This "Radar Ball" has a microchip in it that digitally displays how fast the ball has been thrown after each toss. "Radar Ball" makes it affordable to know the throwing velocity. But the real value lies in the new social interaction generated between two children playing catch. From the previous definitions, "Radar Ball" enables entertainment engagement in which two players keep tossing and checking the speed of their fastball.

Another example product involves markers. Sanford scents markers to match 18 different colors [3]. Thus, each marker has its own fragrance. For instance, black has a licorice scent, red is cherry, and green is mint, etc. The scented markers enable aesthetic engagement in which users associate the scents with fruits when they scribble. Studies have shown that scent-cued memories are more emotional than the visual-cued and verbal-cued ones [4][5]. A particular scent can facilitate individuals to directly retrieve specific memories about the original event. Thus, the scented markers may please the users through bringing them back to their past experiences related to fruits.

The two example products suggest that individual products can engage users. They also indicate that engaging experience with products involves at least two activities. One is interacting with a product and the other is a peripheral activity. For example, the scents of markers are associated with fruits. This peripheral activity is instinctive and makes the user-product interaction somehow more engaged. Based on the experience realms and several examples mentioned above, engaging experience in user-product interaction is defined as:

When a user interacts with a product, he also participates in a peripheral activity enabled by the product. The user:

- is entertained by the activity through his senses and participation, or
- creates something as a part of the activity and enjoys the result, or
- escapes temporarily from his normal, daily reality, or
- is attracted by an interesting aspect of the activity.

As a result, the user's experience of using the product is reinforced.

3 Products That Engage Users

Playing digital games provides an exciting and engaging experience. Game stations probably are a good example of products that engage users. However, every digital product including software, programs or even a small LCD can engage users because there are vast possibilities in the virtual world. What is discussed in this section is more about "how a game station engages users without a game disc." That is, the discussion focuses on the physical rather than the digital design of products.

Actually, several concepts related to engaging experience have been applied to product design. Figure 2 presents four products that would engage users. "Baby Plane Spoon" not only makes a baby's meal time more fun but also provides a more certain way to feed a baby [6]. This spoon has wings that realize the imagination of being fed. The baby could be fascinated with the food delivered by "air express" and open his mouth wide for the flying spoon. And interestingly, this product enables *escapist engagement* in which both parent and baby imaginatively escape from routine feeding and eating to playing.

"E-rope" is a conceptual design of an extension cord created by Kang et al. This student design won the bronze prize of 2006 IDEA awards [7]. "E-rope" enables the user to add or reduce sockets as needed. To better accommodate large bulky cords, each socket can be rotated at most 180 degrees so that adjacent sockets are not blocked. Apparently, the modular design is similar to Lego toys that engage the user to build his preferred configuration with creativity and fun. Similar to the "Baby Plane Spoon", this product enables *escapist engagement* in which the user imaginatively escapes from using an extension cord to building toy bricks.

"Salt Glass", a salt shaker produced by droog [8]. This salt shaker is also a double hourglass that displays the flow of salt as the passing of time. This display engages the user as a spectator. It is possible that the user would stare at it for a short while right after shaking salt. Thus, this product can enable *aesthetic engagement* in which the user admires the metaphoric aesthetics of a salt shaker.

"Glass Toaster" is a conceptual design of "Culinary Art Project" developed by Philips [9]. One impressive feature is the transparent glass that makes its internal toasting function visible. Therefore, the user is able to visually participate in the toasting process rather than passively waiting for it to finish. This product enables *aesthetic engagement* in which the user watches how the bread changes from white to brown. This concept is analogous with experience restaurants, in which customers can watch how a chef processes food when tasting delicious cuisines.

The four example products discussed above include both tableware and home appliances, and both existing products and conceptual designs. It suggests that engaging experience has been noteworthy and applied to different kinds of consumer products. More importantly, a product that engages users neither becomes more complicated nor requires additional user involvement. Users can intuitively enjoy engaging without increasing workload or spending time to react. These examples suggest that products can engage users without decreasing their utility.

Fig. 2. Three products that would enable engaging experience: "Baby Plane Spoon" [6], "E-rope" [7], "Salt Glass" [8], and "Glass Toaster" [9]

4 Notions of Engaging Experience

In the previous sections, several example products that engage users were discussed. The discussion implies two notions of engaging experience. In section 4.1, the classification of engaging experience is introduced. In addition, whether a specific engaging experience is instinctive depends on the relevance between the interaction with a product and the peripheral activity. This notion is expanded on in section 4.2.

4.1 The Classification of Engaging Experience

According to several example products discussed previously, engaging experience could be sensory, physical or emotional. *Sensory engagement* takes place when products attract users to perceive. For example, the "Salt Glass" and the "Glass Toaster" visually engage users. Sensory engagement is similar to what tourists do in front of a magnificent view at the top of a skyscraper. It indicates that visitors comprehend the scenery only through senses in the distance, such as seeing and listening. *Physical engagement* takes place when products attract users to act. For example, the "Radar Ball" and the "E-rope" engage users in play. Physical engagement is similar to what a person can do if given a stick. He can become a baseball batter, a conductor directing an orchestra, or even a Jedi Knight. With imagination and enjoyment, physical engagement can be triggered by many artifacts such as products. *Emotional engagement* takes place when products evoke specific feelings that affect users. For example, the scented markers produced by Sanford affect users to associate with fruits. Emotional engagement is similar to how people are affected when inspired from reading a diary, or when given a precious souvenir.

To illustrate a product that enable sensory, physical and emotional engagement, figure 3 shows a filter water pitcher in which there is a fake but cute small fish. This fish looks like it is drifting when in water and lying down when without water. The hypothesis is that a user would like to look over the fish (sensory engagement); deciding whether he needs to refill the water pitcher to keep the fish looking alive. When refilling water, the user feels that he is saving a fish (physical engagement) and being merciful (emotional engagement).

Fig. 3. A water pitcher that enables three types of engaging experience

4.2 Instinctive Engaging Experience

As mentioned in section 2, engaging experience involves two activities. One is interacting with a product and the other is a peripheral activity. The latter must be instinctive so that it makes sense to a user and engages him in sing a product. Significantly, whether a peripheral activity is instinctive depends on the relevance between the two activities. That is, engaging experience must take place instinctively when a user uses a product. Take products in figure 2 as examples. Note that flying a toy plane (peripheral activity) relevant to feeding a baby (using a spoon), playing toy bricks relevant to configuring an modular extension cord, flowing sands in an hourglass relevant to placing a salt shaker on the table, and viewing the toasting function relevant to seeing a toaster. The four examples show how peripheral activities are relevant to using the products and make sense to users.

To illustrate counterexamples, figure 4 shows two products. "Clocky" is a mobile alarm clock [10]. When the clock starts to beep, it runs away in random directions. In this case, looking for a runaway alarm clock is a peripheral activity. But this peripheral activity does not make sense to users not only because users have never turn off an alarm clock in this way but also because it is little relevant to using an alarm clock. That is, the experience of using the clock is hardly reinforced. The other counterexample is "Egg g~g~", salt and pepper shaker [11]. When a person uses it to shake salt, it sounds like a chick. In this case, hearing chirps is a peripheral activity. Although hearing chirps makes sense to users, it is not relevant to using a salt shaker. That is, the experience of using the shakers is hardly reinforced. Although these two products do attract users to respond emotionally, they have less potential to engage users.

Whether a peripheral activity makes sense to users cannot be definitely defined. The reason is the same with why one joke makes sense to one listener who laughs immediately whereas another listener does not respond. Based on the examples and counterexamples discussed, the author suggests three guides for enabling instinctive engaging experience. First, a peripheral activity must be common activities so that it makes sense to users. Second, a peripheral activity must be relevant to using the product. Third, what a user does in a peripheral activity must be relevant to the experience of using the product.

Fig. 4. Two counterexample products that do not engage users: Clocky – run away alarm clock [10] and Egg g~ g~ salt and pepper shaker [11]

5   Product Properties That Foster Engaging Experience

Based on the example products previously discussed, they possess particular properties that play significant roles in fostering engaging experience. Thus, to identify these product properties, the categories of product properties are discussed in section 5.1. Then, the essential elements of engaging experience are discussed in section 5.2. Finally, the mapping between product properties and the elements of engaging experience are established in section 5.3. This mapping indicates how to design products that engage users.

5.1 The Categories of Product Properties

To categorize product properties, this research refers to the following three studies. Jordan [12] asserts that a product can be defined by its properties. Product properties include tangible features and intangible attributes, both of which result in the final product specification. According to Jordan, tangible features are called formal properties, whereas intangible attributes are called experiential properties. *Formal properties* are what can be objectively measured or clearly defined within the design process of a certain product. *Experiential properties*, in contrast, are how users feel when experiencing a product in context. For example, if one functional specification of a toaster is "prevent from over toasting", its formal properties could be "a cancel button and/or multiple-level heat settings", whereas its experiential properties could be "easy to control the heat".

Focusing on the relationships between user and product, Hassenzahl [13] differentiates product attributes into either pragmatic or hedonic one based on users' satisfaction. The level of satisfaction depends on the success of using a product to achieve certain desirable goals. If a user's expectations are confirmed, he will feel satisfied. *Pragmatic attributes* satisfy the fulfillment of a user's behavioral goals, whereas *hedonic attributes* satisfy his psychological well-being. Corresponding to product features, the pragmatic attributes result from, for example, fascinating functionality. The hedonic attributes result from, for example, communicating identification, or provoking memories that a user values.

From a different point of view dealing with product properties, Janlert and Stolterman [14] articulate the character of things. According to their argument, a *role* is a functional specification, whereas a *character* is a meta-functional metaphor. A character is usually created from many *characteristics*. Several related characteristics integrate into a coherent whole of the character. For example, "racing" is a role of a sports car. "Powerful, fast, and colorful" are its characters. "Acceleration, noise, streamline and commercial stickers" are its characteristics. Often, the definition of characteristics is wide so that any kind of specification of an artifact could be a characteristic. The manifestation of a certain characteristic depends on the type of actions, properties, individuals, and situations the artifact is concerned. For example, an individual fully experiences a powerful sports car only when he is driving it at high speed on a racetrack and feeling its roar, vibration and the wind pressure.

For the purpose of this research, six categories of product properties that foster engaging experience are defined. They are (1) interaction, (2) use, (3) function, (4) purpose, (5) appearance, and (6) components. Interaction is referred to what a product property enables the user to do in sense, action or emotion. Use is referred to that a user manipulate or control a product. Specifically, use can be divided into regular and periodical use, both of which imply different occasional actions of use. Function is the task(s) a product intends to perform. Different from function, purpose pertains to practical aspects. For example, the function of a salt shaker is to dispense salt and the purpose is to help the user to move preferable amount of salt into foods. Appearance relates to the aesthetics, styling or exterior appealingness of a product. Components of a product are the exterior elements that the user can reach in sense or action.

5.2   The Elements of Engaging Experience

In terms of theatrics, at least four elements are necessary to compose an activity [1][15]. They are performance, stage, prop and character. To define, a *performance* is the peripheral activity that a product enables to attract or prompt the user. A *stage* is what a product operates or what the user responds to. A *prop* is the product property that enables the user to involve in the performance on the stage. A *character* is who the user plays. For example, in the case of "Glass Toaster", watching bread toasted is a performance. The toaster itself is a stage because it displays how the bread changes from white to brown. The transparent case of the toaster is a prop. The user plays a character who visually engages in the toasting process.

In a different point of view, engaging experiences require peoples' active participation. How people actively participate in an activity can be aligned with three different levels: theme, central activity and supporting activity [1][16]. A *theme* is a subject or a particular idea that runs throughout the user-product interaction. A *central activity* is what the user does in order to experience the theme. A *supporting activity* is what the user does to support the central activity. For example, Disney World is a theme park full of legends and stories for children. Visitors who enjoy in different stories participate in different central activities. Within each central activity, visitors can take pictures with the cartoon character(s), play games or have fun in each cartoon story, each of which is a supporting activity.

Clearly, various activities that engage people can be described by the seven entities mentioned above. They can be used to design/describe a specific engaging experience enabled by a product.

5.3   The Mapping between Product Properties and Engaging Experience

To relate product properties to the elements of engaging experience, the author suggests that the use or interaction with a product corresponds to the performance. The primary function of a product relates to the stage. A particular component or the appearance of a product relates to the prop. And the user, of course, relates to the character. To illustrate, in the case of "Glass Toaster", the transparent glass makes the toasting function visible and therefore the user can visually engage in the toasting process. Correspondingly, the toasting process that attracts users (interaction) is the performance. The visible toasting function is the stage. The transparent glass (appearance) is the prop. And the user is a spectator.

In addition, regarding the theme, central activity, and individual activity, the author suggests what product properties correspond to the theme should be defined by designers. The main purpose of a product corresponds to the central activity. The regular or periodical use corresponds to the supporting activity. To illustrate, in the case of the "E-rope", the modular design allows the user to change the configuration as playing Lego toys. "E-rope" displays a particular configuration that accommodates more bulky cords. Accordingly, playing Lego toys is the theme created by designers. Changing the configuration for bulky cords (main purpose) is the central activity. And whenever the user needs to connect other cord onto "E-rope" (periodical action) that results in changing the configuration is the supporting activity.

The hypothetical mapping between product properties and the elements of engaging experience is established in Table 1. This mapping helps to identify product properties that can foster engaging experience.

Table 1. The Mapping between product properties and the elements of engaging experience

Product Properties	Elements of Engaging Experience
Use or Interaction	Performance
Primary Function	Stage
Component or Appearance	Prop
User	Character
(defined by designers)	Theme
Main Purpose	Central Activity
Regular or Periodical Use	Supporting Activity

6 Conclusion

This paper discusses the fundamental concepts of engaging experience with physical products. Engaging experience is defined as a user, when interacting with a particular product, participates in a peripheral activity enabled by the product. This peripheral activity, as a result, reinforces the experience of using the product. Based on resulting responses, engaging experience can be categorized in three types: sensory, physical and emotional. In addition, engaging experience must be instinctive so that it makes sense to users.

Through the review of example products, the author claims that individual products, as expected, can engage users. Since products are full of tangible features and intangible attributes, designers are able to create a product that engages users in its performance. A product that engages users neither becomes more complicated nor requires additional user involvement. Users can instinctively enjoy engaging without increasing workload or spending time to react. Thus, engaging experience is a new value that a product can bring to users in addition to functionality, usability and aesthetics. To engage users, designers must be concerned with not only product utilities but also compelling user-product relationship. Relating product properties to the elements of engaging experience provide a hypothetical approach for designing products that engage users.

References

1. Pine II, B.J., Gilmore, J.H.: The Experience Economy- Work Is Theatre & Every Business a Stage. Harvard Business School Press, Boston (1999)
2. Radar Ball,
   http://www.bizjournals.com/stlouis/stories/1997/07/21/
   story4.html
3. Scented Markers,
   http://www.sanfordcorp.com/sanford/consumer/jhtml/new-product/
   productdetail.jhtml?attributeId=&nrProductId=SN20002

4. Chu, S., Downes, J.J.: Proust nose best: Odors are better cues of autobiographical memory. Memory and Cognition 30(4), 511–518 (2002)
5. Herz, R.S., Schooler, J.W.: A naturalistic study of autobiographical memories evoked by olfactory and visual cues: Testing the Proustian hypothesis. American Journal of Psychology 115, 21–32 (2002)
6. Baby Plane Spoon, http://www.pylones-usa.com/item.php?item=ST-NAV&product=152
7. E-rope, http://images.businessweek.com/ss/06/06/idea2006/index_01.htm?campaign_id=ds1
8. Salt Glass, http://www.droogdesign.nl/#frameslb=lb.php?f=2&r=35&k=1458,rb=rb.php?f=1&r=107&k=1458
9. Glass Toaster, http://www.design.philips.com/shared/assets/Downloadablefile/Glass_toaster_high-13275.jpg
10. Cloky- run away alarm clock, http://www.nandahome.com/
11. Egg g~ g~, salt and pepper shaker, http://dxioncom.plustonecreative.com/en/collection/egg-en.html
12. Jordan, P.W.: Designing Pleasurable Products- An introduction to the new human factors, pp. 82–124. Taylor and Francis, London (2000)
13. Hassenzahl, M.: The Thing and I: Understanding the Relationship between User and Product. In: Blithe, M.A., Overbeeke, K., Monk, A.F., Wright, P.C. (eds.) Funology: From Usability to Enjoyment, pp. 31–42. Kluwer Academic, Dordrecht (2005)
14. Janlert, L., Stolterman, E.: The Character of Things. Design Studies 18, 297–314 (1997)
15. Falk, J.: Interfacing the Narrative Experience. In: Blithe, M.A., Overbeeke, K., Monk, A.F., Wright, P.C. (eds.) Funology: From Usability to Enjoyment, pp. 249–256. Kluwer Academic, Dordrecht (2005)
16. Gupta, S., Vajic, M.: The Contextual and Dialectical Nature of Experiences. In: Fitzsimmons, J., Fitzsimmons, M. (eds.) New Service Development: Creating Memorable Experiences, pp. 33–51. Sage Publications, Thousand Oaks (1999)

User-Centered Mouse Access Evaluation Design: Windows-Based Simulation Technology

Chi Nung Chu

China University of Technology, Department of Management of Information System,
No. 56, Sec. 3, Shinglung Rd., Wenshan Chiu, Taipei, Taiwan 116, R.O.C
nung@cute.edu.tw

Abstract. This paper introduces a Windows-Based Simulation Technology (WBST) to monitor user's interaction with computer through a mouse. This design could evaluate a client's pointing and selecting proficiency by measuring the cursor movement and motion control. The simulated Windows-based task operations require the client synthesize four basic types of mouse operating skills, including clicking, cursor moving, cursor moving and clicking, and dragging. The WBST can record the positions and responses of a mouse during any specific task. It can also rebuild the recorded results of cursor moving and motion control on the screen. The WBST not only provides the clinical professionals with more detailed information to evaluate the specific difficulties of manipulating mouse for a client, but also allows engineers to design adaptive input device for the people with special needs.

Keywords: Windows-Based Simulation Technology.

1 Introduction

Mouse is the main input device for computer manipulations in the windows-based environment. For those who have difficulties to manipulate the mouse operations, the mouse would be the barrier to prevent them from the opportunities brought by the information technology. A lot of human computer interface studies, which include interacting technique in software and device hardware, explored the behavior of the motion impaired with mouse operations to design and meet what they need [1],[2],[3],[4],[5]. However, the artificial evaluation processes are tedious and hard to collect and analyze exactly what the user struggles with the mouse for some specific tasks. Our design in this study is to simulate what the Microsoft Windows Operating System works. The Windows-Based Simulation Tracing Tool can track and collect information behind the course of mouse operations for any indicated task. With the accurate analysis of quantitative data, the clinicians can efficiently understand the process of any user's cursor movement in different situations such as distances, directions, accuracy, speed, acceleration with any alternative mouse devices. Moreover, electronic engineers can more accurately design the alternative mouse for individuals with different disabilities or limitations to meet the needs in actual Windows environment.

M. Kurosu (Ed.): Human Centered Design, HCII 2009, LNCS 5619, pp. 41–45, 2009.
© Springer-Verlag Berlin Heidelberg 2009

2   Windows-Based Simulation Tracing Tool

The design of Windows-Based Simulation Tracing Tool is a windows-based simulation software design running on the Microsoft Windows platform with resolution
1024*768 (Figure 1). In order to increase the accuracy of evaluation for the mouse
operations, the system simulates four types of functional performance tests which
include "Operation System manipulating", "window operating", "word processing"
and "on-screen keyboard pointing and selecting". Each test consists of several typical
evaluation tasks in Windows platform, some examples are shown in the Table 1. In
the process including any failure during the task, all the locus of movement course
with screen coordinates, operations of mouse and spending time are recorded to the
database system which has been built into the Windows-Based Simulation Tracing
Tool. Therefore the collected data after being analyzed from each subtask could show
the specific difficulties that he/she met during the functional performance. The Windows-Based Simulation Tracing Tool not only reduces the traditional overload of
evaluation process made by human efforts, but also provides the accurate information
for the clinicians and electronic engineers to support people with special needs in the
information and communication technology.

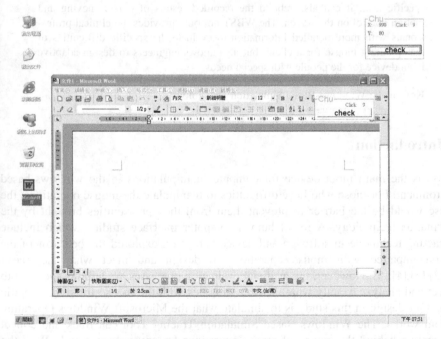

Fig. 1. Windows-Based Simulation Tracing Tool

3   Analysis Module

The Windows-Based Simulation Tracing Tool permits the clinical evaluators setting
up assessment tasks to test the user's mouse performance. The system will record the

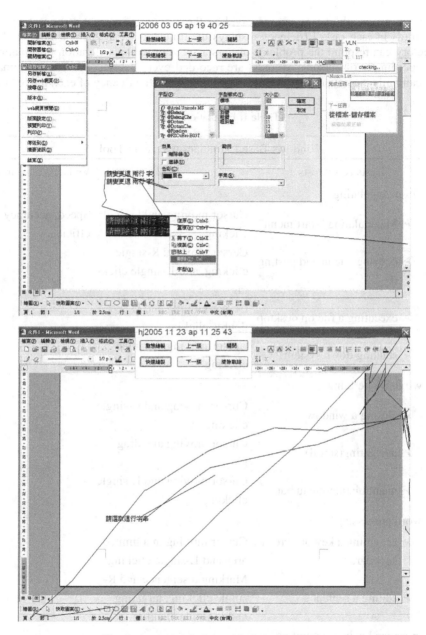

Fig. 2. Analysis Module with dynamic Viewer

detailed responses in database to provide the Analysis Module with dynamic Viewer (Figure 2) to ease the understanding of user's performance and difficulties. Windows-Based Simulation Tracing Tool can measure speed, accuracy and efficiency of each evaluation task. Speed indicates the time spent for accomplishing a single test item.

Accuracy reveals the percentage in dividing the amount of correct responses by the total trails in each task. Efficiency displays the trajectory of the cursor moving. The trajectory can respond the problems of cursor moving with acceleration information. The coordinates of cursor movement are recorded every decisecond. The coordinate data can build the pattern of each movement and probe the fluency of each movement.

Table 1. Evaluation tasks

Windows-Based Simulation Tracing Tool		
Tests & subtests	Tasks	Variables measured
■ OS manipulating		
➤ Manipulating "start menu"	Cursor moving and L-single clicking	Speed, accuracy, efficiency
➤ copying a icon and pasting it	Cursor moving, R-single clicking, and L-single clicking	
➤ executing a file on desktop	Cursor moving and L-double clicking	
➤ shutting down system	Cursor moving, L-single clicking	
■ window operating		
➤ closing a window	Cursor moving and L-single clicking	
➤ navigating(scroll)	Cursor moving and dragging	
➤ manipulating menu bar	Cursor moving and L-single clicking	
■ word processing		
➤ activating a key on screen keyboard	Cursor moving in a limited area and L-single clicking	
➤ editing a sentence	Marking a sentence and R-single clicking, targeting and L-single clicking	
■ on-screen keyboard pointing and selecting		
➤ opening on-screen keyboard	Cursor moving and L-single clicking	
➤ selecting a key	Cursor moving and L-single clicking	

4 Conclusion

The goal of this study design is to provide an efficient and accurate solution for the evaluation of mouse performance that would allow people with disabilities having opportunity to access the computers. A study of Windows-Based Simulation Tracing Tool has been conducted to the clinical service intervention. It is great to start anyone with computer access disability on Windows-Based Simulation Tracing Tool test by giving them an opportunity to find out the adaptive input device. The analyzed quantitative data can efficiently provide the accurate information for the clinicians and electronic engineers to evaluate and design the most suitable alternative mouse to the people with special needs.

References

1. Hwang, F., Keates, S., Langdon, P., Clarkson, J.: Mouse Movements of Motion-Impaired Users: A Submovement Analysis. In: Proceeding of ASSERS 2004, pp. 102–109 (2004)
2. Keates, S., Hwang, F., Langdon, P., Clarkson, P.J.: The Use of Cursor Measures for Motion-Impaired Computer Users. Universal Access in the Information Society 2(1), 18–29 (2002)
3. Mackenzie, I.S., Kauppinen, T., Silfverberg, M.: Accuracy Measures for Evaluating Computer Pointing Devices. In: Proceeding of CHI, pp. 9–15 (2001)
4. Mithal, A.K., Douglas, S.A.: Difference in Mouse Microstructure of the Mouse and the Finger-Controlled Isometric Joystick. In: Proceeding of CHI, pp. 13–18 (1996)
5. Phillips, J.G., Triggs, T.J.: Characteristics of Cursor Trajectories controlled by the Computer Mouse. Ergonomics 44(5), 527–536 (2001)
6. Soukoreff, R.W., MacKenzie, I.S.: Towards a Standard for Pointing Device Evaluation, Perspectives on 27 Years of Fitts Law Research in HCI. Int. J. Human-Computer Studies 61, 751–789 (2004)
7. Chen, M.C., Chu, C.N., Wu, T.F., Yeh, C.C.: Computerized Assessment Approach for Evaluating Computer Interaction Performance. In: Miesenberger, K., Klaus, J., Zagler, W.L., Karshmer, A.I. (eds.) ICCHP 2006. LNCS, vol. 4061, pp. 450–456. Springer, Heidelberg (2006)
8. Zhou, H., Hu, H.: A Survey-Human Movement Tracking and Stroke Rehabilitation. University of Essex, Colchester United Kingdom (2004)
9. Hourcade, J.P., Bederson, B.B., Druin, A., Guimbretiere, F.: Differences in Pointing Task Performance Between Preschool Children and Adults Using Mice. ACM Transaction on Computer-Human Interaction 11(4), 357–386 (2004)

Engaging and Adaptive: Going beyond Ease of Use

Kevin Clark

Content Evolution LLC Worldwide
kevin.clark@mindspring.com

Abstract. Making products and services easier to use is a durable goal, yet will likely be insufficient to meet the expectations of a new generation of customers. This paper suggests "ease-of-use" be augmented with the goal of being "engaging and adaptive" for products, services, and the overall experience people have with organizations that provide them. Being intentional and using design thinking can be used to deliver engaging and adaptive experiences to customers around the world.

1 Introduction

Making products and services easier to use is an accepted goal of human factors and ergonomics professionals. It is also consistent with the practice of universal design. The call for ease-of-use is now being extended to all managed interactions with companies, creating intentional points of contact and designed customer experiences [1]. This progress is welcome and needed to make products and services more accessible to greater portions of humanity.

That said, there is a new customer on the horizon that will have a dramatically different set of expectations. A lowest-common denominator version of interactions, out-of-box impressions, and usage experiences will not be sufficient for this emerging customer.

They will want: <u>engaging</u> and <u>adaptive</u>...
… not just easy to use.

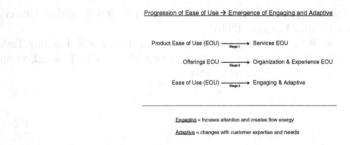

"A feeling of control, a good conceptual model, and knowledge of what is happening are all critical to ease of use," says Donald Norman about products. "When is something difficult? When the controls and actions seem arbitrary, when the system can get itself into peculiar states, peculiar in the sense that the person does not know what it is doing, how it got there, or how to recover." [2]

M. Kurosu (Ed.): Human Centered Design, HCII 2009, LNCS 5619, pp. 46–54, 2009.

During the first stage progression of ease-of-use, learning from product development and customer requirements is now applied to service engagements. Companies make investments in designing service engagements that meet or exceed customer expectations. The notion that the organization should remember the customer becomes known and visible at this stage.

Gerald Zaltman in his book How Customers Think outlines misconceptions that drive managers to "make some predictable errors that can destroy even the most carefully thought-out product launch. These errors fall into three categories: mistaking descriptive information for insight, confusing customer data with understanding, and focusing on the wrong elements of the consumer experience."[3]

During the second stage progression, all offerings (products, services, and customer interactions) are being considered for intentional interaction design. The transformative step is recognition the organization itself needs to be easier to use and more approachable for a variety of constituencies to thrive and help drive the survival and success of all economic actors. Here the total organization experience becomes an imperative.

In the chapter "The Customer is the Product," in the Experience Economy, the authors say "the experiences we have affect who we are, what we can accomplish, and where we are going, and we will increasingly ask companies to stage experiences that change us." Joe Pine and Jim Gilmore continue, "human beings have always sought out new and exciting experiences to learn, grow, develop and improve, mend and reform. But as the world progresses further into the Experience Economy, much that was previously obtained through noneconomic activity will increasingly be found in the domain of commerce."[4]

In stage three, the interaction landscape changes and ease-of-use itself becomes a less relevant idea in achieving customer satisfaction, loyalty, and overall competitive advantage. Ease-of-use eventually becomes a source of dissatisfaction if it means creating a lowest common denominator experience for all customers and constituencies wanting to do business with the organization.

Engaging as an imperative is what we're inheriting from a generation of mobile device users that want to be engaged – resulting from habitual use of instant messaging, twittering, online social networks, and other forms of instant information feedback with emotional content and gratification. It is also being driven by the generation that grew up with video and computer games and new definitions of fun, time, and interaction.

Adaptive is also what we need to reach this new generation of emerging customers, employees, suppliers, and business partners. Adaptive is both a frame of mind – and a set of procedures, practices, and application of technologies to continually change the context of how we engage in ways that are meaningful to customers and constituents.

2 Instructions as Stories

> Instruction manuals are essentially stories.
> How do I use this?
> How do I put this together?
> How do I get there?

For the first-time user, a set of really easy instructions will be appropriate. Yet what are the right instructions set for an experienced user? Will the new generation of customers be satisfied with reading the same story they read last time?

Mihaly Csiksentmihalyi suggests in his book <u>Flow</u> that optimal experience is achieved in a zone that drives focus and attention that exists between the emotional poles of frustration (what Csiksentmihalyi calls anxiety) and boredom.[5]

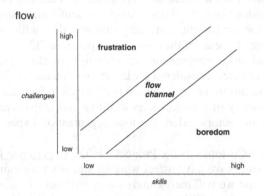

Visual adaptation of <u>Flow: The Psychology of Optimal Experience</u> by Mihaly Csikszentmihalyi, Harper & Row, 1990, pg. 74

Flow is a good guide to the instruction manual of the future.

For products it would have some of the characteristics of games, including some of the emotional characteristics identified by XEODesign to keep customers in an early and engaged flow state (see sidebar). For instance, in the case of electronic devices during the out-of-box, set up, and other first-impression tasks, the touch-points would be intentionally designed to meet the skill level of the customer. It would "inherit" the known skill level of the owner from the previous device interaction history.

For services the organization would have a clear understanding of customer behavior and interaction history. For instance, they would know if a customer has a tendency to be "reactant."[6] When a course of action is suggested, a reactant personality will want to reject that direction; which calls for a different approach to service upgrade suggestions or mid-course contract corrections.

For organizations themselves…

…it is focusing the attention of key constituencies on the ideas and activities that will replicate the memes that sustain survival and health. It is getting more people in the flow channel of an intentionally adaptive and engaging organization experience.

"Attention is like energy in that without it no work can be done…"

"Some people learn to use this priceless resource efficiently, while others waste it. The mark of the person who is in control of consciousness is the ability to focus attention at will, to be oblivious of distractions, to concentrate for as long as it takes to achieve a goal, and not longer" says Csiksentmihalyi.[5, pg. 31]

2.1 Emotion without Story (Sidebar)

XEODesign under the leadership of Nicole Lazzaro conducted research based on player experience from video and computer games in 2004.[7]

XEO created a framework showing how games create engagement, what makes them self-motivating activities without the need for a story. Lazzaro identifies "four keys to emotion without story," using these criteria: "1. What Players Like Most About Playing – 2. Creates Unique Emotion Without Story – 3. Already Present in Ultra Popular Games – 4. Supported by Psychology Theory and Other Larger Studies.[7, pg. 3]

Here is the resulting "4 Fun Keys" model:

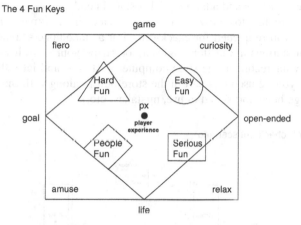

The 4 Fun Keys model visually adapted and reprinted with permission, XEO Design,© Inc.

"The Four Keys unlock emotion with:

1. **"Hard Fun:** Players like the opportunities for challenge, strategy and problem solving. Their comments focus on the game's challenge and strategic thinking and problem solving. This "Hard Fun" frequently generates emotions and experiences of Frustration and Fiero (personal triumph over adversity. The ultimate game emotion. Italian).

2. **"Easy Fun:** Players enjoy intrigue and curiosity. Players become immersed in games when it absorbs their complete attention, or when it takes them on an exciting adventure. These immersive game aspects are "Easy Fun" and generate emotions and experiences of Wonder, Awe, and Mystery."

3. **"Serious Fun** (previously 'Altered States'): Players treasure the enjoyment from their internal experiences in reaction to the visceral, behavior, cognitive, and social properties. These players play for internal sensations such as Excitement, or Relief from their thoughts and feelings.

4. **"People Fun** (previously 'The People Factor'): Players use games as mechanisms for social experiences. These players enjoy the emotions of Amusement, Schadenfreude (gloat over misfortune of a rival. Boasts about player prowess and ranking. German.), and Naches (or Kvell, the pleasure of pride at the accomplishment of a child or mentee. Kvell is how is feels to express this pride in one's child or mentee. Yiddish) coming from the social experiences of competition, teamwork, as well as opportunity for social bonding and personal recognition that comes from playing with others.

Scenario 1: Checking Out
During an ethnographic study of retail customers while at IBM, we observed and interviewed people in a variety of settings. One of them was the self-check-out area of grocery stores. One study participant told us he gets frustrated being behind a novice that's never used the system before. "There should be a line for novices, and a line for experts," he told us.

While it turns out laying out the store this way is impractical, let's consider the request for a moment. What he's asking for is a really good idea. How can we make the store more engaging and adaptive to this stated need?

First, the store has to know if you're a novice or an expert or somewhere in-between when you're queued for check-out. Put a standalone scanner in front of the self-check-out stations in the store and scan or swipe your store loyal card. This action calls up your record in the store computer system – and let's the self-check-out area knows if you've used this part of the store before along with the heuristics of use (previous usage behavior: speed of use, mistakes, etc.).

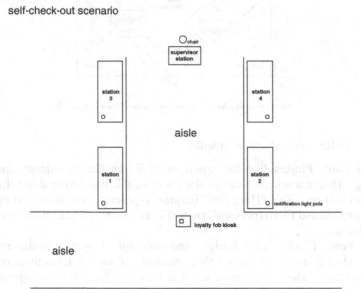

Scenario: A light flashes on pole letting you know that this is the next station ready for you and has loaded your background and preferences. The system greets you by name. If you've never used the system before, it takes the time to walk you through the self-check-out experience. If you're an expert, it bypasses many of the novice queues and instructions (something akin to the speed prompts available on many corporate phone answering systems). This makes the self-check-out area more adaptive.

Using some of "Fun Keys" discovered by Lazzaro, we can also make it more engaging – more like a game. Do you want to get better at the check-out task? For example, tell the system you want time your check out and beat your own previous record. Make checking out on your own less of a chore and more of a game.

Scenario 2: The New Mobile Phone

After months or years of use, a mobile phone should be constantly learning about you and your preferences. It should be building up a digital personality that can be transferred to the next device you own and immediately adapt to your needs – and engage you with new features and functions that take you to the next level of user experience.

For instance, my current phone should know about my voice and text message preferences – know my type-ahead patterns, and pattern mistakes with workarounds. In addition to transferring my address book, the web bookmarks, use information and "digital personality" should transfer forward to the new device.

It should keep you in the flow channel of an optimum out-of-box experience with the new device, bypassing known information that would lead to boredom, and not revealing all new functionality at once that would create frustration.

My phone has an instruction manual that runs 220 pages long. I have surely not discovered all the things it can do, yet it does not engage me in ways to discover new functionality incrementally to continually delight me with the user experience. This is the challenge for mobile device manufacturers and telecommunications carrier partners – to create complete mobility systems that are engaging and adaptive – to meet the needs of a new generation of interactive and game-savvy customers.

2.2   Edible Food vs. Memorable Experiences (Sidebar)

By: Kelly Tierney
Experience Designer & Strategist
IBM Corporation

Having easy to use as the pinnacle of a good customer experience is a bit like a restaurant declaring their goal to be edible food. Memorable experiences, the kind that transform your customers into evangelists, are engaging because they include epochs, rewards, and often a level of challenge – the same elements that are used to create great gaming experiences.

Designer Games, often called Eurogames, are uniquely qualified to help illustrate the elements of great experience design. Their success and longevity is solely based on the quality and richness of the experiences they deliver. They have five unique qualities that are key elements of great experiences. They incorporate very little luck - the players are in control. The players choose their own strategy – each experience is unique. The games are challenging without being discouraging – the system is built on rewards. No player has an advantage – everyone is encouraged to be equally engaged throughout the experience. They take commitment to teach and learn – knowledge share is a part of the pleasure. These elements combined create a journey that ebbs and flows delivering uniquely satisfying experiences for all players.

Other types of games deliver very different experiences. They sometimes focus less on a unique journey but rather rely on luck and the personality of the players to determine the quality of the experience. These games which deliver an unexceptional experience are a lot like the generic airlines: Delta, Continental, and Northwest among a few. When they merge or go bankrupt it goes nearly unnoticed. They deliver a flat experience rather than a journey made to ebb and flow, each customer is a commodity like a box to deliver.

In contrast Designer Games deliver engaging and adaptive experiences that reward the participants, creating a unique experience for each player. The SouthWest airline experience could be compared to this. Each customer is in control of their queue placement and ultimately their seat placement. This creates a slight element of competition and reward. Each flight is unique with the flight crew enabled to deliver an adaptive and special experience each time. To build on this and create more customer evangelists they could reward the 'players' who are behind. If they were to add an advantage, even if small, to those who queued further back they would have a unique reward dynamic creating an incentive and perception of winning even for those in the back of the line.

Even very popular services like Facebook and LinkedIn could be forgotten tomorrow unless they, like SouthWest, put a higher priority on the customer journey. These types of networks have been slow to incorporate meaningful interactions and a sense of challenge and reward. Third party applications are pasted on to fill this void, but ultimately these networks remain at their core directories and email forwarding services. If however they took cues from customer focused products and services, like Designer Games, making the customer journey integral to their mission these companies that would be capable of delivering uniquely satisfying experiences, ensuring their longevity and relevance.

Here is further information on Eurogames/Designer Games:

☐ http://en.wikipedia.org/wiki/Eurogames_(tabletop_games)
☐ http://www.economist.com/world/europe/displaystory.cfm?story_id=12009728

Applying Design Thinking to Your Organization

> "Design is about making intent real.
> "There is plenty of unintentional to go around.
> "When you design, something new is brought into the world with purpose."
>
> [1, pg. 8]

That's how I started an article with co-author Ron Smith, titled "Unleashing the Power of Design Thinking" during the summer of 2008. What we wanted to convey is intentional design methods can be usefully brought to bear all kinds of business decisions.

We say, "In an age of renewed interest in innovation, we suggest the cultivation of a new generation of design patrons who want to collaborate with designers in a new way – business patrons who want to move design strategy and design methods into the mainstream of business thought to accomplish business goals. These patrons would be going to designers not just to acquire the output of well-integrated design, but also use design methods to make business itself more intentional."

We documented ways design thinking is being used to solve real business challenges at IBM – including how new employees are recruited – and the experience clients have when they come visit IBM for business briefings.

"We believe design should move beyond its traditional boundaries to grow (and)…design thinking can help any profession solve problems in innovative ways."

Scenario 3: The Organization Instruction Manual
If we remember what was said at the outset about stage three:

<u>Progression of Ease of Use → Emergence of Engaging and Adaptive</u>

Product Ease of Use (EOU) ――→ Services EOU
                          Stage 1

Offerings EOU ――→ Organization & Experience EOU
             Stage 2

Ease of Use (EOU) ――→ Engaging & Adaptive
                 Stage 3

<u>Engaging</u> = focuses attention and creates flow energy

<u>Adaptive</u> = changes with customer expertise and needs

...you would want to move from merely making your offerings and company easier to use, and make everything you do more engaging and adaptive to meet the needs of an evolving customer marketplace.

To make this real for the organization itself you would need an engaging and adaptive "instruction manual" to navigate the capabilities of the enterprise. This would be a good metaphor to drive future web design for organizations of all types.

Does your "organization instruction manual" on-line tell a compelling and useful story about you?

Does it tell you how to use your company and navigate your richness?

Does it show you how to put together the product or capabilities you need?

Does it show you how to get there?

Does it remember you so it adapts to your needs over time?

Can you apply some of the engaging features of "emotion without story" to punch up some of your web site capabilities so it draws in customers in ways that are more like interactive and adaptive games than simply catalogs of information?

For the first-time someone doing business with you or landing on your web site, a set of really easy instructions will be appropriate.

For people who do business with you regularly or visit you on-line often, especially the new generation of customers we're about to inherit, they won't be satisfied with reading the same story they read last time. Your site should engage and adapt to their needs on a 1:1 basis, not be designed for the lowest common denominator.

Be engaging and adaptive.

Be intentional.

References

1. Clark, K., Smith, R.: Unleashing the Power of Design Thinking (keynote article) Design Management Review (Summer 2008)
2. Norman, D.: The Invisible Computer: Why Good Products Can Fail, the Personal Computer Is So Complex, and Information Appliances Are the Solution, p. 174. The MIT Press, Cambridge (1998)

3. Zaltman, G.: How Customers Think, p. 15. Harvard Business School Press (2003)
4. Pine II, B.J., Gilmore, J.H.: The Experience Economy: Work is Theatre and Every Business a Stage, p. 163. Harvard Business School Press (1999)
5. Csikszentmihalyi, M.: Flow: The Psychology of Optimal Experience. Harper & Row (1990)
6. Chartrand, T.L., Dalton, A.N., Fitzsimons, G.J.: Nonconscious Relationship Reactance, When Significant Others Prime Opposing Goals. Journal of Experimental Social Psychology 43, 719–726 (2007); Department of Psychology and Fuqua School of Business, Duke University
7. Lazzaro, N.: Why We Play Games: Four Keys to More Emotion Without Story. XEODesign, March 8 (2004)

Usability Evaluation of Mp3/CD Players:
A Multi-Criteria Decision Making Approach

Ergün Eraslan

Department of Industrial Engineering, Baskent University,
06590 Ankara, Turkey
eraslan@baskent.edu.tr

Abstract. Globalization and the competition obliged the user-oriented design today. In the last years, usability has become a highly important research subject. Usability, considering user satisfaction along with the user performance, is one of the key factors in determining the success of a product in today's competitive market. Product usability is a prerequisite for high customer satisfaction and future sales of companies. Mp3 players and portable CD players are selected for this study since usability is highly important for them. Designing a usable mp3 or CD player is extremely important for users who have close interaction with them. In this study, 14 different mp3/CD players are selected and their usability is analyzed. The usability criteria used in the mp3/CD players' evaluation are divided into two major categories: performance and emotional expectations. The best alternative is determined with three different multi-criteria decision making methods which are TOPSIS, Analytic Hierarchy Process (AHP), and Fuzzy Axiomatic Design Theory (FADT). Although the same data obtained from semantic differential experiment are used for all multi-criteria decision making methods, different rankings are obtained from each method.

Keywords: Usability, TOPSIS, Analytic Hierarchy Process (AHP), Fuzzy Axiomatic Design Theory (FADT), Semantic Differential Scale.

1 Introduction

User-oriented manufacturing and design became a must because of the competition among the firms. Customer profile has significantly changed. Customers who can buy every product have disappeared. Instead, customers know what they want and choose the products which satisfy their senses and needs in terms of price and design. Consumers want to buy "usable" products which will satisfy their needs and be easy to use.

Usability of a product is a very valuable prerequisite for future sales and high customer satisfaction. Understanding the product features which are significant for the customers and reflecting the feedbacks of the customers to the design and development processes of the product are important. Designers have to evaluate vital points of usability and the satisfaction and delight rose by these vital points carefully.

Usability concept includes a few concepts such as effectiveness, efficiency, and user satisfaction. In its broadest sense, usability is the expression of whether the

M. Kurosu (Ed.): Human Centered Design, HCII 2009, LNCS 5619, pp. 55–64, 2009.
© Springer-Verlag Berlin Heidelberg 2009

product is good enough to meet all needs of the user [1]. User evaluates the product according to the usability measures and decides whether it is usable or not. This decision affects purchasing of the product.

Han et al. (2001) stated that image/ impression and performance dimensions should be considered together to evaluate usability of electronic consumer goods. To be able to model the effects of product interface factors on the product performance and image, performance and image/impression dimensions are detailed [2].

In this study, usability of mp3 (Motion Pictures Experts Group 1 Audio Layer 3) and CD players, used very commonly in last years, are investigated. Mp3/CD players are selected for this study because of the increasing user crowd, high variability of the product and lack of studies about this subject in the literature. 14 mp3/CD players with different dimension and features are determined and usability criteria for these products are determined by a focus group study. These criteria are evaluated by three Multi-Criteria Decision Making (MCDM) Methods to determine the most usable product.

Three MCDM methods are explained in section 2, the products are evaluated and compared by the chosen methods in section 3, and the study is finalized in section 4 with the evaluation of the results.

2 Multi-Criteria Decision Making Methods (MCDMs)

In this study, TOPSIS, Analytic Hierarchy Process (AHP) and Fuzzy Axiomatic Design Theory (FADT) methods are used and they are briefly explained as follows.

2.1 TOPSIS Method

TOPSIS (The Technique for Order Preference by Similarity to Ideal Solution) method is based on that the chosen alternative should have the shortest distance from the positive ideal solution (PIS) and the longest distance from the negative ideal solution (NIS). TOPSIS simultaneously considers distances to not only the positive ideal solution but also the negative ideal solution, and the alternatives are ranked according their relative closeness coefficients. The method begins with the construction of the decision matrix for n alternatives $(a_1, a_2, \ldots, a_n)$ and k criteria $(y_1, y_2, \ldots, y_k)$, and follows 5 steps [3,4]:

Step 1: Normalization of decision matrix y where x_{ij} is the evaluation of alternative i with respect to criterion j.

Step 2: Construct the weighted normalized decision matrix. The weighted values X_{ij} are calculated multiplying with the weights (w_{ij}).

Step 3: Determine the ideal (a^*) and negative-ideal solutions (a^-).

$$a^* = \{x_1^*, x_2^*, \ldots, x_k^*\}. \tag{1}$$

$$a^- = \{x_1^-, x_2^-, \ldots, x_k^-\}. \tag{2}$$

Step 4: Calculate the separation measures from the ideal (S_i^*) and the negative-ideal solutions (S_i^-), respectively, using the n dimensional Euclidean distance. The separation of each alternative from the ideal solutions are given as

$$S_i^* = \sqrt{\sum_{j=1}^{k}(x_{ij} - x_j^*)^2} \,.$$ (3)

$$S_i^- = \sqrt{\sum_{j=1}^{k}(x_{ij} - x_j^-)^2} \,.$$ (4)

Step 5: Calculate the relative closeness index (C_i^*) to the ideal solution. Rank the preference order. Using this index, alternatives are ranked in decreasing order.

$$C_i^* = S_i^- / (S_i^+ + S_i^-) \qquad 0 \le C_i^* \le 1.$$ (5)

2.2 Analytic Hierarchy Process (AHP)

AHP, developed by Saaty, determines the relative importance of a set of factors in a MCDM problem [5]. The process depends on judgments on intangible qualitative criteria alongside tangible quantitative criteria. The AHP method is based on three principles: first, structure of the model; second, comparative judgment of the alternatives and the criteria; third, synthesis of the priorities.

In the first step, the problem is structured in a hierarchy which includes objective, criteria and decision alternatives. The second step is the comparison of the alternatives and the criteria with pairwise comparison matrices. The prioritization procedure starts in order to determine the relative importance of the criteria within each level. The Saaty's 1-9 scale is used for this process. At the last step, the mathematical process commences to normalize and find the relative weights for each matrix.

It should be noted that the quality of the output of the AHP is strictly related to the consistency of the pairwise comparison judgments. The final consistency ratio (*CR*), using which one can conclude whether the evaluations are sufficiently consistent, is calculated as the ratio of the *CI* and the random index (*RI*). The number 0.1 is the accepted upper limit for *CR*. If the final consistency ratio exceeds this value, the evaluation procedure has to be repeated to improve consistency [5].

This method is a common one and is used in several studies from different areas especially in the last two decades.

2.3 Fuzzy Axiomatic Design Theory (FADT)

Axiomatic Design (*AD*) forms a scientific basis to design and improves designing activities by providing the designer with a theoretical foundation based on logical and traditional thought process and tools. *AD* provides a systematic search process through the design space to minimize the random search process and determine best design solution among many alternatives considering all product development decisions.

In the literature, *AD* theory and principles are used to design products, systems, organizations and software [6]. Lo and Helander (2004) proposed axiomatic design as a formal method for usability analysis for consumer products [7]. Karwowski (2005) also emphasized the applicability of axiomatic design for solving complex ergonomics design problems [8].

Design axioms, namely independence axiom and information axiom, are two key stones of *AD*. Independence axiom is related to maintaining the independence of functional requirements (*FRs*), i.e., design solution must be such that each one of *FRs* can be satisfied without affecting the other *FRs*. Therefore, a correct set of design parameters have to be chosen to be able to satisfy the *FRs* and maintain their independences. Information axiom, the second axiom of AD, states that the design with the smallest information content must be chosen among the design solutions which satisfy independence axiom. Information axiom of *AD* is recently used as a Multi Attribute Decision Making (MADM) method [9]. Information content of overall system, I_{sys}, is defined in terms of the probabilities (p_i) of satisfying FR_i below.

$$I_{sys} = \sum_{i=1}^{n} I_i = \sum_{i=1}^{n} \log_2\left(\frac{1}{p_i}\right) = -\sum_{i=1}^{n} \log_2(p_i). \tag{6}$$

Design with the smallest "*I*" is the best design as it requires the least amount of information to achieve design goals. If all probabilities are large enough, near to one, then information content is minimum. Probability of satisfying a *FR* is specified by design range defined by designer and generating ability of the systems range [9, 10].

Many decision making and problem solving tasks are too complicate to be understood quantitatively. Fuzzy set theory resembles human reasoning in its use of approximate information and uncertainty to generate decisions. If system cannot be defined by using traditional quantitative terms, it is more plausible to use fuzzy linguistic terms. Linguistic terms can be transformed into fuzzy numbers. Fuzzy case of information axiom (FIA) is used herein. In fuzzy case, the intersection area of fuzzy numbers which are ranges of system and design is considered.

3 Usability Evaluation of the Mp3/CD Players with MCDM Methods

In this section, the studies to evaluate the usability of mp3/CD players are provided. In this context firstly the focus group study and semantic differential scale (SDS) used in assessment are given and reliabilities of the evaluations are discussed. Then, the usability of the selected 14 products are evaluated and compared with the results of MCDMs.

3.1 Data Collection and System Reliability

Focus group is an application widely used in areas such as human factors, social sciences and market researches. Focus group is composed of people who come together to discuss a specific subject or problem. The information obtained from the group is qualitative and consists of experience, ideas, thoughts and behaviors rather that figures and facts. Synergy among the group members distinguishes the focus group study from other applications [11].

Han et al. (2001) developed a model for the effects of product interface factors of electronic devices on the product performance and image. These dimensions are valid for all electronic consumer devices [2]. When usability study is to be performed for different products, criteria appropriate for the goods should be selected. Focus group study is used to determine the criteria for usability of mp3/CD players. For this purpose, 6 subjects who attended the meetings are selected among people who are either potential user of mp3/CD players. There were 14 mp3/CD players in the meetings and the participants examined them. In addition, all conversations are recorded during the meetings and printed afterwards.

According to the results obtained, the usability criteria specific to mp3/CD players are determined and it is decided to take technical properties into account under performance dimension. It is anticipated that some technical properties such as memory property, quality of the sound, battery life and data processing speed, play an important role in the usability of mp3/CD players. The criteria are given in Figure 1 for the mp3/CD players which are determined in focus group study formed a basis for SDS.

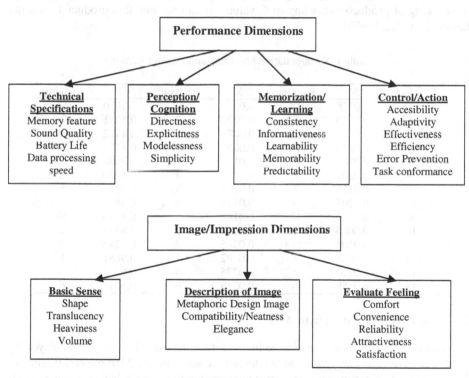

Fig. 1. Criteria Determined for Mp3/CD Players by Focus Group Study

A catalog including the details of each mp3/CD player is prepared. Technical properties which cannot be understood by visual examination such as sound quality, battery life and memory are added to this catalog. There were 22 subjects in SDS and allocated time for each subject is about 50 minutes. A reliability test is performed to confirm appropriateness of SDS results. Internal consistency among the questions is analyzed by computing Cronbach Alpha. Cronbach alpha is 0.94 meaning that the study is reliable according to these results. Confirming that results are appropriate we passed to evaluation stage by multi criteria decision making methods.

3.2 Usability Evaluation with TOPSIS Method

5 steps of TOPSIS method are used to rank 14 mp3/CD players according to 31 usability criteria. Firstly, criteria weights are calculated. All criteria are compared pairwise and the case of a criterion is more important than another is stated with yes or no. Using the number yes values, criteria weights are determined [12].

In the next step, vector normalization is used to compute normalization ratios. The normalized values are multiplied with criteria weights and by this way normalized matrix is calculated.

Positive-Ideal and Negative-Ideal solutions (a^* and a^-) are computed by taking the weighted normal values calculated in the previous step into account. Distance between alternatives is measured to compute the similarity to PIS, the alternative with maximum C_i value is selected as the best product based on all criteria. Table 1 shows the ranking of products according to C_i values. It can be seen that product 12 has the largest C^* value, 0.9681.

Table 1. Differentiation Measures and Product Ranking

Products	S^+	Ranking	S^-	Ranking	C^*	Ranking
1	0.0338	13	0.0081	12	0.1930	12
2	0.0180	7	0.0227	7	0.5571	7
3	0.0330	12	0.0071	13	0.1762	13
4	0.0346	14	0.0060	14	0.1479	14
5	0.0261	11	0.0152	11	0.3688	11
6	0.0201	8	0.0219	8	0.5220	8
7	0.0126	5	0.0274	10	0.6856	5
8	0.0204	9	0.0195	9	0.4887	9
9	0.0227	10	0.0168	5	0.4249	10
10	0.0065	2	0.0351	2	0.8433	2
11	0.0108	3	0.0285	3	0.7245	3
12	0.0013	1	0.0382	1	**0.9681**	1
13	0.0120	4	0.0275	4	0.6854	4
14	0.0158	6	0.0238	6	0.6015	6

3.3 Usability Evaluation with AHP

The criteria which are determined to be important for the usability of mp3/CD players by the focus group study are examined in a hierarchic structure via AHP method (Figure 1). It is assumed that performance and image/impression dimensions have the same effects on electronic devices. Therefore, neither dimension is superior to the other. Besides, there is no relation among their subcriteria.

After AHP hierarchy is composed, it is aimed to determine dominance of each criterion at every level to others. For this purpose, results of the SDS are used. Satisfaction criterion is used as the basis here. Superiorities of the criteria to each other are determined according to the points of the criteria against satisfaction. These superiorities are used compute the weights required for pairwise comparisons.

Consistency ratios of all matrices formed according to these scales and none of them was more than %10. After the control of consistency ratios, relative importance vector is calculated by multiplying the weights with each other in accordance with the hierarchic structure.

The most suitable product for the objective is found when values of the relative importance vector are multiplied with the product points obtained from SDS. Analyzing Table 2, it can be seen that 8.5225 is the greatest value which belongs to product 12.

Table 2. Points of Products

# of Product	Point Values	# of Product	Point Values
1	5.5365	8	7.2623
2	6.8955	9	6.5628
3	5.1271	10	7.9827
4	5.2476	11	7.8769
5	6.2387	12	**8.5225**
6	6.9416	13	7.7513
7	7.5897	14	7.4977

3.4 Usability Evaluation with FADT

To make design analysis using fuzzy axioms first the results of semantic differential study should be converted to fuzzy data. For this purpose, we used the conversion scale which is the most commonly used tool to convert the qualitative data into triangular fuzzy numbers in the literature.

In SDS, every criterion is given a point by each user between 1 and 5 meaning that, bad, average, good, very good and excellent, respectively. Averages of the triangular fuzzy numbers corresponding to these points are computed. These values are determined as the system intervals showing the sufficiency of the current system.

After defining system intervals it is also necessary to determine design intervals to make the calculations. Design interval is determined by the designers, and it can be defined as the features which are demanded in the product or the system. For this purpose, we determined the intervals for every criterion and gave two of them as an example (Table 3). After determining the system and design intervals, information content is computed for every product and for each criterion. Summing the information contents for each product, it can be seen that Product 10 is the best one with the least information content (Table 4).

Table 3. Design Ranges of the Some Criteria

Perception/Cognition		Memorization/Learning	
Directness	Very Good	Consistency	Very Good
Expicitness	Very Good	Informativeness	Very Good
Modelessness	Excellent	Learnability	Very Good
Simplicity	Excellent	Memorability	Very Good
		Predictability	Very Good

Table 4. Total Information Contents

# of Product	Information Content	# of Product	Information Content
1	Infinity	8	Infinity
2	Infinity	9	Infinity
3	Infinity	10	**54.176**
4	Infinity	11	57.562
5	Infinity	12	61.994
6	Infinity	13	60.948
7	60.518	14	Infinity

Table 5. Ranking of the products for MCDMs: Top 5 ranks include the same 5 products(*)

Ranks / Method	TOPSIS Ranking	AHP Ranking	FADT Ranking
1*	12	12	10
2*	11	10	11
3*	13	11	7
4*	10	13	13
5*	7	7	12
6	2	14	-
7	8	8	-
8	14	6	-
9	1	2	-
10	9	9	-
11	3	5	-
12	4	1	-
13	6	4	-
14	5	3	-

3.5 Comparison the Usability for MCDM Methods

The comparison of the results of the MCDMs is given in Table 5. TOPSIS and AHP methods give the ranking for all products while FADT method only gives the top five products. From the ranking it can be seen that top five products are same for all three methods although their orders are different. The orders of the other products are also different for TOPSIS and AHP.

4 Discussion and Conclusion

In this study, a decision approach for selection of usable mp3/CD players is presented. The selection process is based on the comparisons of finite alternatives according to identified usability criteria. For this purpose, 14 mp3/CD players are selected and criteria regarding the usability of mp3/CD players are determined by a focus group study. A semantic differential study is performed using these criteria and results are analyzed by MCDMs. TOPSIS, AHP and FADT methods are used in these analysis.

The data used in the three methods were obtained from the SDS performed before-hand. They were also used with minimum modifications as the methods permitted; however, some differences were observed in the results obtained. In all three methods, there were different rankings for different products. We think the reason for differences in these methods was caused by application formats. The differences are as follows:

In TOPSIS method, two distance ranges are calculated based on the idea of minimum distace to PIS and maximum distance to NIS. However, these two different distance ranges might not generate the same results. Therefore, in TOPSIS method, the result is obtained by examination to determine if there is a more suitable alternative from these two ranges utilizing a closeness index. This method is both easy to apply and is preferred.

In the AHP method, the problem is analyzed in a hierarchical structure and rankings of criteria in relation to each other at each level are determined. Criteria weights are calculated according to the pairwise comparison matrices and the best alternative which belongs to the highest weight is identified. The calculation of the criteria weights depend on the structure of the decision matrix that could alter the rankings in other methods. The ideas which can be objective and subjective can be combined in AHP in the ranking of the criteria which can be evaluated at different levels that in turn affects results when obtaining different rankings.

Using the method of design with fuzzy axioms, analysis is performed over system and design intervals. System intervals are the qualitative data obtained from the evaluations of the products by the users. Design intervals are the qualitative features of the products demanded by the designers or the people evaluating the products. This comparison might be incapable if the evaluation matrix for the alternatives cannot be formed with crisp values. Some criteria could have a qualitative structure or have an uncertain structure which cannot be measured precisely. In such cases, fuzzy numbers can be used to obtain the evaluation matrix. In analyses, the product with the minimum information content is searched for, by taking the intersections of triangular fuzzy regions of the system and design intervals. Different results can be found if design intervals change. For example, according to TOPSIS and AHP, the most usable product is product 12, while according to FADT product 10 is the most usable. It should be noted that although the methods approach the problem differently, the first five product results are the same in all methods.

It is not possible to assert that any one of the methods is better than the others. However, they all have some advantages depending on where they are utilized. When the MCDM methods are evaluated with all aspects we can say that if the given criteria in selection process are qualitative the best application method is AHP. As the AHP method includes the qualitative criteria into the decision processes better. If the criteria are quantitative, the other methods can be used for problem solving but there is no comparative degree to each other.

The fuzzy numbers can be used to obtain the evaluation matrix in all MCDM methods. Besides, the results of other MCDMs frequently using in the literature can be investigated to make a consensus. This will improve the decision making process and will give more accurate results that are the directions in the future research.

References

1. Wixon, D., Ve Wilson, C.: The usability engineering framework for product design and evaluation. In: Helander, M. (ed.) Handbook of Human-Computer Interaction, pp. 665–667. North Holland, Amsterdam (1997)
2. Han, S.H., Yun, M.H., Kwahk, J., Hong, S.W.: Usability of consumer electronic products. International Journal of Industrial Ergonomics 28, 143–151 (2001)
3. Isıklar, G., Buyukozkan, G.: Using a multi-criteria decision making approach to evaluate mobile phone alternatives. Computer Standards and Interfaces 29, 265–274 (2007)
4. Chu, T.C., Lin, Y.C.: A Fuzzy TOPSIS Method for Robot Selection. Advanced Manufacturing Technology, Taiwan (2003)
5. Saaty, T.L.: The Analytical Hierarchy Process. McGraw-Hill International Book Company, USA (1980)
6. Suh, N.P.: Axiomatic design, advances and applications. Oxford University Press, Oxford (2001)
7. Lo, S., Helander, M.G.: Developing a formal usability analysis method for consumer products. In: Proceedings of ICAD, The Third International Conference on Axiomatic Design, vol. 26, pp. 1–8 (2004)
8. Karwowski, W.: Ergonomics and human factors: the paradigms for science, engineering, design, technology and management of human-compatible systems. Ergonomics 48, 436–463 (2005)
9. Kulak, O., Durmusoglu, M.B., Kahraman, C.: Fuzzy multi-attribute equipment selection based on information axiom. Journal of Materials Processing Technology 169, 337–345 (2005)
10. Eraslan, E., Akay, D., Kurt, M.: Usability Ranking of Intercity Bus Passenger Seats Using Fuzzy Axiomatic Design Theory. In: Luo, Y. (ed.) CDVE 2006. LNCS, vol. 4101, pp. 141–148. Springer, Heidelberg (2006)
11. Helander, M.G., Lin, L.: Axiomatic Design In Ergonomics and An Extension of The Information Axiom. J. Eng. Design 13, 321–339 (2002)
12. Hwang, C.L., Yoon, K.: Multiple attribute decision making: methods and applications. Springer, Heidelberg (1981)

From Usability to Playability: Introduction to Player-Centred Video Game Development Process

Jose Luis González Sánchez, Natalia Padilla Zea, and Francisco L. Gutiérrez

Video Games and E-Learning Research Lab. (LIVE) – GEDES Research Group,
Software Engineering Department. University of Granada. E-18071, Spain
{joseluisgs,npadilla,fgutierr}@ugr.es

Abstract. While video games have traditionally been considered simple entertainment devices, nowadays they occupy a privileged position in the leisure and entertainment market, representing the fastest-growing industry globally. We regard the video game as a special type of interactive system whose principal aim is to provide the player with fun and entertainment. In this paper we will analyse how, in Video Games context, Usability alone is not sufficient to achieve the optimum Player Experience. It needs broadening and deepening, to embrace further attributes and properties that identify and describe the Player Experience. We present our proposed means of defining Playability. We also introduce the notion of Facets of Playability. Each facet will allow us to characterize the Playability easily, and associate them with the different elements of a video game. To guarantee the optimal Player Experience, Playability needs to be assessed throughout the entire video game development process, taking a Player-Centred Video Game Design approach.

Keywords: Video Games, Playability, Usability, Interactive Systems, User Experience.

1 Introduction

Throughout history, humans have had the capacity to manage their own leisure time, this being a significant driver in cultural development. Nowadays, video games and entertainment systems collectively make up the biggest industry in terms of turnover, more so than music and cinema. From this we can deduce that video games have become the preferred game of choice, exerting significant social and cultural influence over children, teens and adults [1].

In this paper we present Video Games as interactive systems with special characteristics, with a focus on fun and entertainment that distinguishes them from traditional desktop systems. We analyse why Usability is therefore not sufficient to describe the full User Experience in relation to Video Games. Secondly, we present a definition of Playability to characterise and measure the Player Experience with these kinds of systems. Thirdly, we introduce the notion of Facets of Playability that will allow us to study Playability easily across the different video game elements. Finally we introduce a methodology based on Playability characteristics to develop more effective Video Games, taking a Player-Centred Video Game Design approach.

M. Kurosu (Ed.): Human Centered Design, HCII 2009, LNCS 5619, pp. 65–74, 2009.

2 User Experience in Entertainment Systems

When a *Desktop System* (DS), such as a word processor, is developed, the main objective is that *users can execute a set of tasks*, determined by a clear functional objective, in a predetermined context, for example working in an office. The overall utility of an interactive system has a strong functional component (functional utility) and another component that indicates the means by which users can achieve this functionality. In this context, *Usability* is a measure of product use whereby users achieve concrete objectives in varying degrees of *effectiveness*, *efficiency* and *satisfaction*, within a specific context of use [2]. Developing useable software improves the final quality of the User Experience. *User Experience* can be defined as the combined sensations, feelings, emotional responses, assessments and satisfaction of the user in relation to a system, and their resulting perception of their interaction with it [3, 4].

A *Video Game* can be considered a 'special' interactive system, in that it is used for leisure purposes by users seeking fun and entertainment. A video game is not conceived for the user to deal with daily tasks, like a word processor, but rather it has a very specific objective: *to make the player feel good when playing it*. They are more likely to be diverse and subjective.

We propose that analysing the quality of a video game purely in terms of its *Usability* is not sufficient – we need to consider not only functional values but also non-functional values, given the specific properties of video games. Additional factors to be considered might include, for example: storytelling techniques or character design. In other words, the *Player Experience* could be much more extensive than the *User Experience*. We need to establish a set of attributes and properties to identify the experience of players playing a video game and indicate to us whether a game is 'playable' or not – that is, they will identify the *Playability* of the video game.

Playability is a live topic in the scientific community; it has been studied from different points of view and with different objectives without consensus on its definition or the elements that characterise it. We have identified two specific strands of research: *Playability* as *only Usability* in video games context (understanding and control of the game system), and research based on particular elements of video games. In the former, we note Federoff or Desurvire research [5, 6] focused on applying Nielsen's ideas using Usability heuristic techniques to measure it in video games. There are some interesting studies on how to use heuristic and design guidelines with specific video game elements [7, 8, 9]. In the second line of research, we find references to: *Playability* in the quality of Gameplay [10]; types of challenges [11], the manner of storytelling [12] or the degree of emotion when players play video games [13].

There are few studies focused on defining *Playability* formally, exceptions being Fabricatore's proposition [14] and Järvien's work [15], but without specific reference to *Playability* attributes or properties to characterize it. The former puts forward a Usability quality design model, applied to action video games based on the analysis of some representative game elements. Järvien presents a multifaceted model Theories to identify *Playability* based on famous Csíkszentmihályi's Flow.

The main objective of our work is to propose a precise and complete definition of *Playability*, its attributes and properties and a multifaceted conceptual framework to facilitate the analysis, so as to be able to measure it as part of an in-depth analysis, and factor it in to video game development from the initial design phase onwards to optimise the Player Experience.

3 Playability: How to Characterize the Player Experience

As already stated, the User Experience in a Desktop System has two principal points of view which characterize it: *process* and *product*. They are motivated and enriched by the emotional user reactions, and the perception of non-instrumental qualities [16]. *Playability* is based on Usability, but, in the context of video games, goes much further. Furthermore, *Playability* is not limited to the degree of 'fun' or 'entertainment' experienced when playing a game. Although these are primary objectives, they are concepts so *subjective*. It entails to extend and complete formally the User Experience characteristics with *players' dimensions* using a broad set of attributes and properties in order to measure the *Player Experience*.

We define *Playability as: 'a set of properties that describe the Player Experience using a specific game system whose main objective is to provide enjoyment and entertainment, by being credible and satisfying, when the player plays alone or in company'.*

It is important to emphasise the 'satisfying' and 'credible' dimensions. The former is more difficult to measure in video games than in desktop systems due to the high degree of subjectivity of non-functional objectives. Similarly, the latter depends on the degree to which players assimilate and become absorbed in the game during play – also difficult to measure objectively. Playability is characterised by attributes that exist in Usability but that have different meanings in video game context. For example Effectiveness' in a video game is not related to the speed with which a task can be completed, because typically a player will play for the maximum time possible, this being one of the game's main objectives. *Playability represents the degree to which specified users can achieve specified goals with effectiveness, efficiency and specially satisfaction and fun in a playable context of use*, with an emphasis on the interaction style and plot-quality of the game or the quality of Gameplay. Playability is affected for example, by the quality of the storyline, responsiveness, usability, customizability, control, intensity of interaction, intricacy, and strategy, as well as the degree of realism and the quality of graphics and sound and son on.

Analysing several video games and their different characteristics, we propose a set of seven attributes to characterise Playability, namely: *Satisfaction, Learnability, Effectiveness, Immersion, Motivation, Emotion* and *Socialization*. In Fig 1, we present the Playability Model for Video Games with the different relationships among the Desktop System Usability and Video Game Playability attributes.

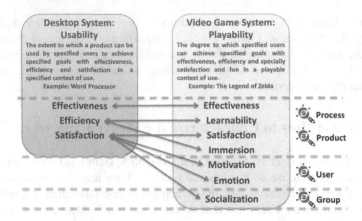

Fig. 1. Correspondence between Playability and Usability attributes

3.1 Playability: Attributes and Properties

Now we will outline in more detail the attributes and some examples of properties of *Playability*, to subsequently measure them:

Satisfaction: We define this as the gratification or pleasure derived from playing a complete video game or from some aspect of it. We characterise Satisfaction using the following properties: *Fun*: the main objective of a video game is to entertain, hence a video game that is no fun to play could never satisfy players. *Disappointment:* we should ensure that players do not feel so disappointment or uneasy when playing a video game that they abandon it altogether. *Attractiveness*: this refers to attributes of the video game that increase the pleasure and satisfaction of the player.

Learnability: We define this as the player's capacity to understand and master the game's system and mechanics (objectives, rules, how to interact with the video game, and so on). We propose the following properties to characterise Learning: *Game Knowledge*: a player's prior knowledge of a video game will influence the degree to which they are affected by the learning curve proposed by the game. *Skill*: it is demonstrated in how the player plays. Once they have understood and assimilated the game's objectives and rules (cognitive skill), how they address the game's challenges to reach the different objectives and rewards is a matter of skill (interactive skill). *Difficulty*: it may be higher or lower depending on how steep the learning curve is, relative to the player's skills and how long they have been playing. A high Difficulty level can provoke a greater effort by the player to learn how to play. *Frustration*: this property is often part of the learning process and is produced by the player's feelings of unease when unable to achieve a particular challenge or objective, or when failing to understand and certain concepts. *Speed*: The speed with which new concepts and contents are introduced into the game directly affects the learning process. *Discovery*: the different game resources support better assimilation of the game's various contents so that the player needs successively less time to improve his abilities to achieve the game's objectives.

Effectiveness: We define this as the time and resources necessary to offer players a fun and entertaining experience whilst they achieve the game's various objectives and reach the final goal. We identify Effectiveness as having the following properties: *Completion*: a video game is more effective if the percentage of Completion is high. In other words, it can be considered effective in that the player found no parts of the game uninteresting. *Structuring*: a video game is well structured elements (that is, where, when and how they appear in the Gameplay) when it achieves a good balance between the various objectives to be achieved and the different challenges to overcome, such that the player remains engaged and enjoys himself throughout the entire game time.

Immersion: We define this as the capacity of the video game contents to be believable, such that the player becomes directly involved in the virtual game world. To characterise Immersion we propose the following properties: *Conscious Awareness*: the degree to which the player is consciously aware of the consequences of his actions in the virtual world, understanding what happens as a result of carrying out a particular action helps the player imagine what to do next and to develop the necessary abilities to overcome challenges. *Absorption*: a player who is completely absorbed is involved in the Gameplay to such a degree that they focus all their all their abilities and attention on overcoming the game's challenges. *Realism*: the more realistic a video game (in the use of controls, presentation of contents, or atmosphere), the greater the Immersion of the player. Realism helps to focus the player on the game's challenges, rules and objectives by making the virtual world as believable as the real world. *Dexterity*: this refers to the player's ability to interact with the game's controls (interactive dexterity) and carrying out different movements and actions in the virtual world in which they are immersed (virtual dexterity). *Socio-Cultural Proximity*: video games have more, or less, immersive efficacy depending on the degree of socio-cultural proximity to the player – appropriate to their age or gender, for instance. The metaphors and atmosphere used in the game, even when realistic, can still reduce Immersion of the player if they do not reflect certain socio-cultural characteristics that the player can identify with.

Motivation: We define this as the set of game characteristics that prompt a player to realise specific actions and continue undertaking them until they are completed. We characterise Motivation as having the following properties: *Encouragement*: the degree of player encouragement is affected by the level of confidence they feel when facing new game challenges and the possibility of reaching new game objectives. *Curiosity*: it can be generated by the inclusion of optional features, objectives and challenges that offer the player the freedom to interact with a greater number of elements and wonder what will come next. *Self-improvement*: it occurs when the player or their character develops their ability and skills – be it to overcome specific challenges, or simply because the player enjoys employing a particular skill. *Diversity*: The number of different elements makes the game more attractive to players and reduces the likelihood of monotony.

Emotion: This refers to the player's involuntary impulse in response to the stimulus of the video game that induces feelings or a chain reaction of automatic behaviours. We characterise Emotion as having the following properties: *Reaction*: the player reacts to a video game because the system is a source of different stimuli. The player's initial reaction may then trigger several types of emotion. *Conduct*: video games are behavioural mechanisms in that they can influence the conduct of the player during Game Time, by leading them through different emotions thanks to the stimuli they provide. *Sensory Appeal*: the game needs to transmit an interest or desire in aesthetic aspects of it to increase the emotion of attraction to the player; it needs to use different sensory channels, e.g. the audiovisual channel, to stimulate the player's senses and enables them to process the game whilst feeling the emotions it induces.

Socialization: We define this as the set of game attributes, elements and resources that promote the social dimension of the game experience in a group scenario. This kind of collective experience makes players appreciate the game in a different way, thanks to the relationships that are established with other players (or with other characters from the game). Socialization is also at work in the connections that players make with the characters of the video game. Examples of this might include: choosing a character to relate to or to share something with; interacting with characters to obtain information, ask for help, or negotiate for some items; and how our influence on other characters may benefit, or not, the achievement of particular objectives. We propose that Socialization has the following properties: *Social Perception*: this is the degree of social activity used and understood by players, who experience a more extensive game in a multiplayer context than they do playing on their own. *Group Awareness*: this refers to the conscious awareness of players of being part of a 'team', and of sharing common objectives, challenges and game elements. Players must understand that they are a part of a group and that the success of the group depends on achieving shared objectives. *Personal Implication*: the player needs to be aware that individual achievement leads to group victory. Hence game resources need to be developed that help raise the player's awareness of their role in the group's success, and their identification with it. *Sharing*: when a player plays within a group, the objectives, de different resources and how they are managed are shared by the group. *Communication*: multiplayer video games should offer communication mechanisms that enable optimal interchange of information among players. *Interaction*: how the rules of the game are perceived by the group or how members will interact to achieve the objectives, that is the way in which characters or players relate to each other allows objectives and challenges to be overcome in different ways according to the interests fostered by interaction among group members. We highlight the following types of interaction: Competitive (when a player plays to achieve personal success, when one player wins the rest of the group generally loses); Collaborative (individual success is replaced by group success. Here the notion of ´team´ applies – the entire group shares and achieves a common goal) and Cooperative (players can have their individual goals whilst forming a group to benefit themselves, thanks to the help of other members, team approach is not essential to achieve the player's objective, rather it arises only circumstantially).

In Fig. 2 we show the Playability Model with the attributes and properties to measure the Player Experience in a video game that we have described previously.

3.2 The Facets of Playability

Playability analysis is a very complex process due to the different perspectives that we can use to analyse the various parts of video game architecture. In this work, we propose a classification of these perspectives based on six *Facets of Playability*. Each facet allows us to identify the different attributes and properties of *Playability* that are affected by the different elements of video game architecture [17]. The first facet is *Intrinsic Playability*: this is the *Playability* inherent in the nature of the video game itself and how it is presented to the player. It is closely related to Gameplay design and Game Mechanic. *Mechanical Playability*: this is the facet related to the quality of the video game as a software system. It is associated to the Game Engine. *Interactive Playability*: this is facet associated with player interaction and video game user interface development. This aspect of *Playability* is strongly connected to the Game Interface. *Artistic Playability*: this facet relates to the quality of the artistic and aesthetic rendering in the game elements and how these elements are executed in the video game. *Intrapersonal Playability* or *Personal Playability*: This refers to the individual outlook, perceptions and feelings that the video game produces in each player when they play, and as such has a high subjective value. *Interpersonal Playability* or *Social Playability*: This refers to the feelings and perceptions of users, and the group awareness that arise when a game is played in company, be it in a competitive, cooperative or collaborative way.

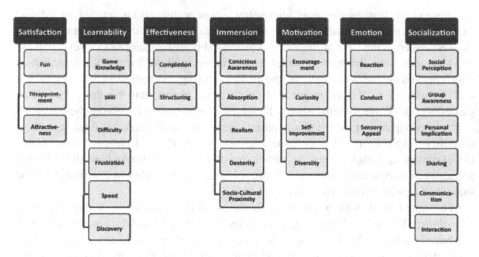

Fig. 2. Playability Model to characterize the Player Experience

The overall Playability of a video game, then, is the sum total of values across all attributes in the different Facets of Playability. It is crucial to optimise Playability across the different facets in order to guarantee the best Player Experience. Fig. 3 represents the different Facets of Playability and their relationship to common elements of a video game.

4 Introduction to Player-Centred Video Game Development

Nowadays, the methodologies used in video game development are similar to those used in software development, but with the addition of certain elements or phases

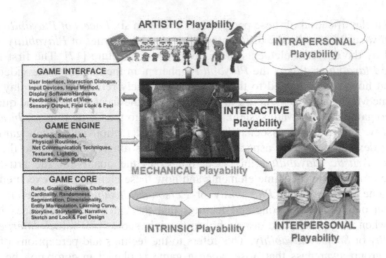

Fig. 3. Correspondence between Facets of Playability and video game elements

reminiscent of film production: screenplay, scenery recreation, virtual world and character design, and so on. Can we guarantee *Playability* in the development phase of every video game? We argue that *Playability* should be factored-in to every phase of video game development in order to guarantee quality. Typically *Playability* is only checked in the test phase of the product, using evaluation techniques to test specific aspects of the video game. However, we assert that the design of video games, as interactive systems, should be focused on users, by involving them directly in the development process, from initial specification through to final test stage Usability Engineering. We propose the use of a Player-Centred Video Game Development approach, using the principles of *Playability* throughout the different phases of development in order to achieve a high level of quality in *Playability,* in the same way as with traditional desktop systems.

We should start with a game specification that includes the *Requirements of Playability* [17] deduced from reference to the *Facets of Playability*, analysing which attribute is affected by which specific video game elements. In the creative video game design phase, we propose the adaptation of *Game Patterns* [18], introducing *Playability* attributes and properties to improve the efficiency and the effectiveness, in *Playability* terms, of these patterns. *Game Style Guides* [19, 20] will be necessary in order to design appropriate and playable elements according to the context of the game or player profiles. Finally, we recommend using *Playability Tests* during the entire development process using *Playability* properties to validate and verify the *Requirements of Playability* and ensure the quality of *Playability* in the final product. By using *Facets of Playability* we can check *Playability* properties in the different phases, for example using facet-by-facet heuristic to test specific video game elements or using more specific tests or metrics that help us to identify which attributes are more relevant in each phase, or to improve the attributes that are most suitable for the nature or genre of the video game or the Player Profile.

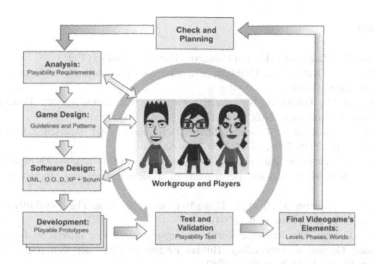

Fig. 4. Correspondence between Facets of Playability and video game elements

5 Conclusions and Future Work

In this paper we have presented video games as special interactive systems developed to entertain the user, concluding that Usability alone is an insufficient measure for determining the full Player Experience.

We have presented the concept of *Playability* as a complement of Usability applied to video games, outlining the attributes that characterise it and their properties, in order to measure and guarantee an optimum Player Experience. To facilitate the analysis of *Playability*, we have proposed the *Facets of Playability* to study every property in each attribute in order to identify the elements necessary to achieve overall *Playability* in different video games. We have shown the importance of Player-Centred Video Game Development wherein *Playability* must be taken into account in every phase of the game development, in order to, amongst other things, anticipate any unexpected or negative results for the developer and guarantee a high quality of playability and improve the Player Experience in the final product.

Currently we are designing a conceptual model of a video game which will enable us to specify and analyse *Playability* characteristics in the design phase, and to incorporate *Playability* techniques into software patterns, style guides and heuristic techniques, thus ensuring optimum *Playability* of the end-product. We are also adapting techniques used in Usability Engineering and User-Centred Design in order to include *Playability* in a quality model to enhance the Player Experience throughout the different phases of video game development.

Acknowledgments. This research is financed by: the Spanish International Commission for Science and Technology (CICYT); the DESACO Project (TIN2008-06596-C02-2); and the F.P.U. Programme of the Ministry of Science and Innovation, Spain.

References

1. Provenzo, E.: Video kids. Harvard University Press, Cambridge (1991)
2. ISO 9241-11: Guidance on Usability, also issued by the International Organization for Standardization (1998)
3. Hassenzahl, M., Tractinsky, N.: User Experience – A Research Agenda. Behaviour and Information Technology 25(2), 91–97 (2006)
4. Effie, L., Virpi, R., et al.: Towards a Shared Definition of User Experience. In: CHI 2008 Procedings, pp. 2395–2398. ACM, New York (2008)
5. Federoff, M.: Heuristic and Usability Guidelines for the Creation and Evaluation of Fun Video Games. Master Thesis, Department of Telecommunications, Indiana University (2002)
6. Desuvire, H., Capñan, M., Toth, J.: Using Heuristic to Evaluate The Playability in Games. In: CHI 2004. ACM, New York (2004)
7. Malone, Thomas, W.: Heuristics for designing enjoyable user interfaces: Lessons from Computer Games. In: Proceedings Human Factors in Computer Systems, Washington, D.C, pp. 63–68. ACM, New York (1982)
8. Shneiderman, B.: Designing for fun: How to make user interfaces more fun. ACM Interactions 11(5), 48–50 (2004)
9. Korhonen, H., Koivisto, E.: Playability Heuristic for Mobile Games. In: MobileHCI 2006. ACM, New York (2006)
10. Rollings, A., Morris, D.: Game Architecture and Design. New Riders Games (2003)
11. Salen, K., Zimmerman, E.: Rules of Play: Game Design Fundamentals. MIT Press, Cambridge (2003)
12. Glassner, A.: Interactive Storytelling: Techniques for 21st Century Fiction. Ak Peters (2004)
13. Lazzaro, N.: Why We Play Games: Four keys to More Emotion without Story. In: Game Developer Conference (2004)
14. Fabricatore, C., Nussbaum, M., Rosas, R.: Playability in Action Video Games: A Qualitative Design Model. Human-Computer Interaction 17, 311–368 (2002)
15. Järvien, A., Heliö, S., Mäyrä, F.: Communication and Community in Digital Entertainment Services. Prestudy Research Report. Hypermeda Lab. University of Tampere (2002)
16. Mahlke, S.: Visual Aesthetics and the User Experience. In: Proceedings: The Study of Visual Aesthetics in Human-Computer Interaction (2008) N: 08292; ISSN: 1862-4405
17. González Sánchez, J.L., Padilla Zea, N., Gutiérrez, F.L., Cabrera, C.: De la Usabilidad a la Jugabilidad: Diseño de Videojuegos Centrado en el Jugador. In: Proceedings of INTERACCION 2008, pp. 99–109 (2008)
18. Björk, S., Lundgren, S., Holopainen, J.: Game Design Patterns. In: Copier, M., Raessens, J. (eds.) Proceedings of Digital Games Research Conference (2003)
19. González Sánchez, J.L., Cabrera, M., Gutiérrez, F.L.: Diseño de Videojuegos aplicados a la Educación Especial. In: Procedings of INTERACCION 2007, pp. 35–45 (2007)
20. Padilla Zea, N., González Sánchez, J.L., Gutiérrez, F.L., Cabrera, M., Paderewski, P.: Design of Educational Multiplayer Video Games. A Vision from Collaborative Learning. In: Noor, A.K., Adey, R.A., Topping, B.H.V. (eds.) Journal: Advances in Engineering Software. Elsevier, Amsterdam (Forthcoming, 2009)

Mapping of Usability Guidelines onto User's Temporal Viewpoint Matrix

Tadashi Kobayashi[1] and Hiromasa Nakatani[2]

[1] Aichi Institute of Technology, Faculty of Computer Science, Toyota,
Aichi, 470-0392, Japan
tadashik@acm.org
[2] Shizuoka University, Department of Computer Science, Hamamatsu, 432-8001, Japan
nakatani@inf.shizuoka.ac.jp

Abstract. There are many sets of usability guidelines that could be used to quantitatively evaluate products or systems. There were, however, no quantitative means so far to evaluate a set of usability guidelines by comparing with another set of usability guidelines of an established reputation. In this paper, a new evaluation method of usability guidelines is introduced and verified as an applicable evaluation method to all kinds of usability guidelines. Our method has characteristics of employing two temporal scales, forming a user's temporal viewpoint matrix with a scale of utilization timeline and a scale of applied principles, as the means of improving the comparison accuracy. By comparing the graph patterns for each scale, we can provide a means of qualitative evaluation of the targeted guidelines; by comparing the computed similarity value of user's temporal viewpoint matrix, we can provide a means of quantitative evaluation of the targeted guidelines.

Keywords: Usability guidelines, quantitative evaluation, temporal viewpoint, usability principles, utilization pattern.

1 Introduction

To what extent usability guidelines as a whole are based on human-centered design is judged qualitatively by the expertise of usability specialists, and there is no research example of a method to do the comparative and quantitative evaluation of a targeted set of usability guidelines and a base set of usability guidelines which have a good reputation. This paper describes a comparative evaluation method for usability guidelines by comparing a targeted set of usability guidelines and a base set of good usability guidelines, based on a methodology that compares both set of guidelines qualitatively by distribution patterns of guideline items and quantitatively by the computation of pattern similarity. This is not to evaluate the value of guideline content, but to find out the undesirable concentrations, distributions, or deficiencies of usability guideline items as human-centered design guidelines.

M. Kurosu (Ed.): Human Centered Design, HCII 2009, LNCS 5619, pp. 75–83, 2009.

2   A Comparative Evaluation Method of Usability Guidelines

This paper describes the comparative evaluation methodology of usability guidelines by combining the two methods from the recent research results about human-centered design methodologies that cover the time flow from the viewpoint of users: One is the methodology to consider the aspect of time flow in using products or systems; the other is the methodology to consider human factors.

2.1   Utilization Time Axis of Products or Systems

Kuramochi [1] analyzed the temporal process of consumer usage of products, and classified the stages between "not using" and "start using" into three stages of 1) "get interested in them," 2) "understand them," and 3) "decide to buy them." Moreover, even if they begin to use products or systems, whether users keep using them or not depends on the user's evaluation of the easy-to-use level. In addition, Ando et al. proposed the idea of a long-term usability [2]. If the temporal stage from "not using" to "start using" and the temporal stage from "start using" to "using for a long time" are organized as the process on a time axis, the distribution pattern of usability guideline items on the time axis is one of the comparison elements. Therefore, a time flow in the use of products or systems is adopted as one of the time axes of the user aspect to compare the usability guidelines in this paper.

2.2   Principle Time Axis of Products or Systems

As usability principles are thought to be representing human factors, another time axis should utilize the aspect of human factors. Kobayashi [3] introduced seven natural usability principles focusing on the human factors: "feel matching," "distance matching," "speed matching," "balance matching," "tangibility matching," "spatial matching," and "reflection matching." Natural usability principles are temporal-based ones, and also a subset of ISO 9241-110 [4]. They are not as usable as ISO 9241-110, because only temporal and partial evaluation is possible in comparison with evaluations done with ISO 9241-110. The covering check of ISO 9241-110 by natural usability guidelines makes the lacking concepts apparent. To cover all the guidelines of ISO 9241-110, natural usability principles should be enhanced to include principles of "time matching," "deduction matching," and "flow matching." [5] Now natural usability principles consist of ten principles.

2.3   User's Temporal Viewpoint Matrix

This section describes a comparative evaluation method of usability guideline sets by introducing User's Temporal Viewpoint Matrix (UTVM) which combines the two time axes adopted in the previous sections. This method maps usability guidelines on the matrix formed by Utilization Time Axis (UTA) and Principle Time Axis (PTA). UTVM can be used to quantitatively perform comparative evaluations.

UTA categories consists of six categories. Usability principles defined in ISO 9241-110 are for office work, and the consideration of principles during the temporal stage between "not using" and "start using" is not enough. This paper divides the

Utilization Time Axis into the six following stages by organizing utilization stages such as "until start using" or "from start using to keep using for a long time": 1) Usability at the first impression, 2) Usability during the trial use, 3) Usability at the decision to introduce, 4) Usability at the start of use, 5) Usability during the continued use, and 6) Usability during the long-term use.

PTA categories should be ordered in a certain pattern. Ten natural usability principles are mapped onto a time axis. No fixed temporal order is set among those principles. Though many distribution patterns can be made, this paper hypothetically adopts the most natural distribution pattern according to the following thinking, to be validated of its appropriateness by the comparison results:

- Feeling, visibility, and distance are the decisive factors when considering the introduction of products or systems
- Speed, time, and space are the problem finding factors during the use of products or systems
- Deduction and balance are the problem solving factors during the use of products or systems
- Flow and reflection are the factors for everyday operation during the long-term use.

Ten principles are set in this order on the PTA axis, and inspections of the comparative evaluation method of usability guideline sets are done.

As both UTA and PTA are defined by time factors, use them to create a User's Temporal Viewpoint Matrix (UTVM) — UTA as the horizontal axis and PTA as the vertical axis. Targeted usability guidelines are mapped onto this matrix, and are compared and analyzed with the pattern of authoritative typical usability guidelines by their position, distribution, and concentration.

2.4   Creating UTVM, PTA, and UTA

When mapping a specific usability guideline onto a UTVM, due to the high abstract nature of ten principles for PTA, the most suitable principle is decided case by case by the person in charge. For example, "Conformity with user expectations," is a very ambiguous and widely usable ISO 9241-110 principle, may be mapped onto any of the ten principles defined for PTA.

More specifically writing, such as "Dialogues should reflect data structures and forms of organization which are perceived by users as being natural," one of guidelines defined for "Conformity with user expectations" principle, is mapped differently onto PTA, depending on the order of mapping check. Therefore, so as not to blur the judgment, keywords that help decide the mapping and the order to consider which principle to apply are provided.

Moreover, as the UTVM is two-dimensional matrix, usability guidelines are easily mapped by classifying them sequentially for each axis; at first temporarily map usability guidelines onto PTA of principle time axis and then classifying them again onto UTA of utilization time axis. Sequential mapping will make it easy to uniquely classify usability guidelines onto UTVM. If the classification is done according to a defined procedure, usability guidelines can be mapped onto UTVM without much effort.

Make a UTVM for ISO 9241-110 as a comparison base with other guideline sets. The results is a large table; each block is filled with many guideline sentences. It is inconvenient, however, to continue the work in the form of characters. From now on, the information in a block, now shown in characters, is replaced by the number of guidelines contained in each block (see the case of ISO 9241-110 shown in the left side of Figure 1 below). Next, another UTVM, each block is filled with numbers, is made for a targeted set of guidelines by following the same procedure with the case of ISO 9241-110. A UTVM has a UTA with six utilization stage items and a PTA with ten principle items. An example of UTVM made for ISO 9241-110 and ISO 9241-10 [6] is shown in Figure 1 below.

ISO 9241-110

	Usability before start using			4. Usability after start using	5. Usability during the continued use	6. Usability during the long-term use	
	1. First	2. Trial	3. Decision				
10 Reflection							0
9 Flow					1	5	6
8 Balance					4		4
7 Deduction					5		5
6 Space				1			1
5 Time				1			1
4 Speed				1			1
3 Distance		4	3	11	3		21
2 Tangibility		1	1	1			3
1 Feeling	2	7	6				15
	2	12	10	15	13	5	57

ISO 9241-10

	Usability before start using			4. Usability after start using	5. Usability during the continued use	6. Usability during the long-term use	
	1. First	2. Trial	3. Decision				
10 Reflection							0
9 Flow						2	2
8 Balance					2		2
7 Deduction				2	9		11
6 Space							0
5 Time				1			1
4 Speed				1			1
3 Distance	1	5	4	13	2		25
2 Tangibility		1	1				2
1 Feeling	1	2	2	1			6
	2	8	7	18	13	2	50

Fig. 1. An example of UTVMs

2.5 How to Compare Pta and Uta

To compare a set of PTAs, a graph is made for each PTA, and two PTA graphs are compared and their visual patterns are analyzed. For example, to make two PTA graphs from Figure 1, a PTA graph for ISO 9241-110 is made from the column of figures in the rightmost column of ISO 9241-110 table in Figure 1, and another PTA graph for ISO 9241-10 is made from the column of figures in the rightmost column of ISO 9241-10 table in Figure 1; then, the two PTA graphs can be compared.

The UTA comparison is made in the same way as PTA; to compare a set of UTAs, a graph is made for each UTA, and two UTA graphs are compared and visual patterns are analyzed. For example, to make two UTA graphs from Figure 1, a UTA graph for ISO 9241-110 is made from the row of figures in the bottom row of ISO 9241-110 table in Figure 1, and another UTA graph for ISO 9241-10 is made from the row of figures in the bottom row of ISO 9241-10 table in Figure 1; then the two UTA graphs can be compared.

2.6 Computation of Similarities between UTVMs, PTAs, and UTAs

Qualitative comparisons can be done by analyzing the distribution patterns of Figure 1 above, by analyzing the difference of two UTA graphs, and by analyzing the difference of two PTA graphs. This paper adopts another quantitative comparison method by computing the similarities between UTVMs. This quantitative method is also applicable to two UTA graphs and two PTA graphs. Similarities can be computed by the following equation. Let $M(i,j)$ be the number of guidelines that corresponds to the UVTM element of ith row and jth column, i.e., ith principle on the PTA axis defined in 2.3 and jth stage on the UTA axis defined in 2.3 respectively. Here, the scope of i and j is: i=1,2,..,10; j=1,2,..,6 respectively. Then, the similarity $C(M_g, M_h)$ between two guidelines, $Mg(i,j)$ and $Mh(i,j)$ is defined as:

$$C(M_g, M_h) = \frac{\sum_i \sum_j (M_g(i,j) - \overline{M_g})(M_h(i,j) - \overline{M_h})}{\sqrt{\sum_i \sum_j (M_g(i,j) - \overline{M_g})^2} \sqrt{\sum_i \sum_j (M_h(i,j) - \overline{M_h})^2}} \tag{1}$$

where $\overline{M}$ is the mean value of $M(i,j)$, i.e., $\overline{M} = \sum_i \sum_j M(i,j)/60$.

Because of $M(i,j) \geqq 0$, $0 \le C(M_g, M_h) \le 1$ is true, and the value of $C(M_g, M_h)$ gets larger in proportion to the similarity between two sets of guidelines. Here, equation (1) is widely employed as the correlation coefficient that represents the similarity between two patterns as shown in [7]. Then, the similarity between two PTAs that represent two sets of guidelines is defined. PTA's pattern $P(i)$ is defined as the projected distribution of $M(i,j)$ distribution onto the vertical axis, and denoted as:

$$P(i) = \sum_j M(i,j) \tag{2}$$

Therefore, the similarity $C(P_g, P_h)$ between PTAs that corresponds to two sets of guidelines of $Mg(i,j)$ and $Mh(i,j)$ is defined as:

$$C(P_g, P_h) = \frac{\sum_i (P_g(i) - \overline{P_g})(P_h(i) - \overline{P_h})}{\sqrt{\sum_i (P_g(i) - \overline{P_g})^2} \sqrt{\sum_i (P_h(i) - \overline{P_h})^2}} \tag{3}$$

where $\overline{P}$ is the mean value of $P(i)$, i.e., $\overline{P} = \sum_i P(i)/10$.

Similarly, the similarity $C(U_g, U_h)$ between UTAs that corresponds to two sets of guidelines of $Mg(i,j)$ and $Mh(i,j)$ is, using UTA's pattern $U(j)$ which is denoted as

$$U(j) = \sum_i M(i,j) \tag{4}$$

defined as:

$$C(U_g, U_h) = \frac{\sum_j (U_g(j) - \overline{U_g})(U_h(j) - \overline{U_h})}{\sqrt{\sum_j (U_g(j) - \overline{U_g})^2} \sqrt{\sum_j (U_h(j) - \overline{U_h})^2}} \tag{5}$$

where $\overline{U}$ is the mean value of $U(j)$ i.e., $\overline{U} = \sum_j U(j)/6$.

Similarities of PTA elements, UTA elements, and UTVM tables are computed by applying equations (1), (3), and (5) to two UTVM tables to compare and analyze two sets of guidelines.

3 Evaluation of Effectiveness of UTVM

This section actually applies the idea of UTVM to the sets of usability guidelines, and verifies the comparative evaluation method of targeted usability guidelines as usable. A combination of usability guidelines is selected to verify the effectiveness of this proposed method: ISO 9241-151 and Research-based Guidelines. Selected reason is as follows: Research-based Guidelines [8] is widely supported and utilized by usability specialists all over the world. ISO 9241-151 [9] is a set of web usability guidelines that was planned to include all the features of Research-based Guidelines. All the details of both sets of guidelines are known and comparable. Therefore, it is possible to correctly interpret and evaluate the differences of visual UTA pattern, visual PTA pattern, and the meaning of its similarity.

In this evaluation, two sets of usability guidelines are used: ISO 9241-151 and Research-based Guidelines. For this combination, a UTVM table is created, and the UTA pattern and the PTA pattern are visually compared. By computing a similarity for this combination, the degree of similarity can be analyzed, and used to prove the appropriateness of the results. If appropriateness is adequately shown, the proposed comparative evaluation method is also proved.

3.1 Visual Pattern Comparison of UTA and PTA

Qualitative comparison of visual pattern for ISO 9241-151 and Research-based Guidelines is made by visually comparing UTA and PTA.

UTA comparison is shown in Figure 2 below. Two things can be pointed out in this graph:

- In Research-based guidelines, the number of guidelines in the "Trial" stage double that of ISO 9241-151.
- In ISO 9241-151, the number of guidelines after the Start stage is relatively large considering the total number of guidelines. No noteworthy differences are found after the Start stage.

These findings match the concept of the Research-based Guidelines that are noted for their familiarity with abundant visuals: Research-based Guidelines provide many high-quality and easily understandable guidelines. On the other hand, ISO 9241-151 does not aim to provide decisive guidelines at the Decision stage; it provides comparatively many guidelines after the Start stage to be usable in practical situations. Therefore, the results show the validity of this comparative evaluation method of two sets of usability guidelines.

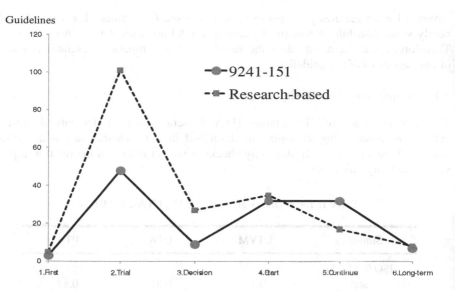

Fig. 2. UTA comparison between ISO 9241-151 and Research-based Guidelines

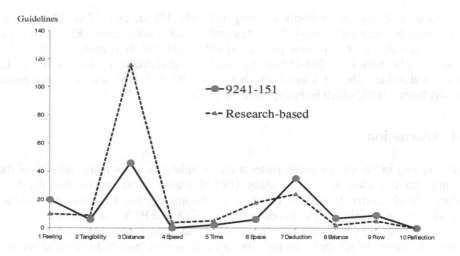

Fig. 3. PTA comparison between ISO 9241-151 and Research-based Guidelines

PTA comparison is shown in Figure 3 below. This is a comparison based on the repertory of usability principles in temporal order. In this graph, it is apparent that Research-based Guidelines are concentrating their efforts in shortening the distance between users and web systems. The rest of principles show nearly the same distribution pattern. This result matches the creation process of ISO 9241-151. In the creation process of ISO 9241-151, the guideline repertory of ISO 9241-151 was always compared with that of Research-based Guidelines so that ISO 9241-151 covers all main features provided in Research-based Guidelines; this resulted in the nearly same distribution pattern of usability guidelines except the "Distance" stage. Therefore, again, the results show the validity of this comparative evaluation method of two sets of usability guidelines.

3.2 Comparison of Similarity

To compare two sets of UTVM tables, UTA elements, and PTA elements, similarities can be computed using the equations described in 2.6. Whether or not the values computed are valid is easily shown by checking the actualities of the relationships of sets of usability guidelines.

Table 1. The similarity among UTVMs, UTAs, and PTAs

Similarity	UTVM	UTA	PTA
ISO 9241-151 and Resarch-based	0.82	0.81	0.83

Table 1 shows the similarities among UTVMs, UTAs, and PTAs. The UTVM similarity between ISO 9241-151 and Research-based Guidelines is 0.82, which is the expected value if the creation process of ISO 9241-151 is considered. The UTA similarity between ISO 9241-151 and Research-based Guidelines is 0.81, which is the expected value. The PTA similarity between ISO 9241-151 and Research-based Guidelines is 0.83, which is the expected value.

4 Discussion

The results in comparing graph patterns and similarities showed the validity of the comparative evaluation method using UVTM matrix. This means that the two-dimensional matrix UTVM is effective in comparing any two sets of usability guidelines. In actuality, UTVM is composed of UTA and PTA, and both of them were proved to be valid by inspecting the set of world-famous usability guidelines. The temporal layout of principles on the principle axis is also thought to be valid because the shape of UTA and PTA has similar characteristics of having two humps.

Applying the comparative evaluation method to use UVTM is not limited to ISO standards. It is applicable to any two sets of usability guidelines as a general comparison method. It should be noted, however, the object of this method is the evaluation of the validity of guideline distribution patterns. It is not intended to evaluate the effectiveness, efficiency, or user satisfaction of usability guidelines. The accuracy or correctness of the targeted guidelines is to be certified beforehand by other usability methodologies.

This methodology can be applied to qualitatively and quantitatively diagnose the appropriateness of following usability guidelines for human-centered design:

- Usability guidelines in general.
- Web usability guidelines.
- Accessibility guidelines.

In addition, following diagnosis is also possible by analyzing graph patterns and similarities:

- What are the key points to improve human-centered design?
- What usability guidelines are, considering the characteristics of the business, to be used or based on?
- What type of considerations is missing or what segment of business is too heavily worked on?

5 Conclusion

This research finds no inconveniences in the layout pattern of natural usability principles on the time axis. Results from employing different patterns remain as a future issue. From the pattern comparison result of PTA, it was found that there were very few usability guidelines that correspond to the "speed," "time," and "space" principles. Research should be done on whether or not these guidelines are necessary. In this paper, human-centered design was considered from the viewpoint of the distance between the system and the products or systems, however, it should be also considered from the viewpoint of the distance among humans, such as users and designers.

References

1. Kuramochi, A.: A Study of Factors that Motivate the Use New Functions of Digital Appliances. Journal of Human Interface Society 7(4), 47–50 (2005) (in Japanese)
2. Ando, A., Kurosu, M.: Concept Framework for the Long Term Usability and Its Measures. In: The Proceedings of the Usability Professionals' Association (2006)
3. Kobayashi, T., Nakatani, H.: A Comparative Evaluation Method of Usability Guidelines by Mapping onto User's Temporal Viewpoint Matrix. Human-centered design 4(2) (2008) (to be published in 2009) (in Japanese)
4. ISO 9241-110, Ergonomics of human system interaction - Part 110: Dialogue principles, Geneva, International Organization for Standardization (2006)
5. Kobayashi: The Effectiveness of Temporal Perspectives for Usability Principles, A Dissertation (2008)
6. ISO 9241-10, Ergonomic requirements for office work with visual display terminals (VDTs). Part 10:Dialogue principles, Geneva, International Organization for Standardization (1996)
7. Gonzalez, R.C., Woods, R.E.: Digital Image Processing, 2nd edn. Prentice Hall, Upper Saddle River (2002)
8. Research-Based Web Design & Usability Guidelines,
   http://usability.gov/guidelines/ (accessed September 17, 2007)
9. ISO 9241-151(2nd CD), Ergonomics of human system interaction - Part 151: Software ergonomics for World Wide Web user interfaces, Geneva, International Organization for Standardization (2006)

A Study on User Centered Game Evaluation Guideline Based on the MIPA Framework

Jinah Lee and Chang-Young Im

373-1 Guseong-dong, Yuseong-gu, Dajeon, Republic of Korea
DMC Lab, Graduate School of Culture and Technology, KAIST
Dong Seoul College, Game Graphic Design
zina777@gmail.com, zinalee@kaist.ac.kr

Abstract. The purpose of this experiment was to identify the relative benefits of the usability checklist and to investigate how the identified usability problems varied by groups. From our experience, there are no structured game frameworks for user interface design. This is why evaluation methods are important in the game development process. The MIPA framework can perform efficient evaluations and correctly identify as many usability defects as possible. Also, accurate evaluations earlier in the design phase can save money and time. Therefore the result is an effective task-oriented usability evaluation checklist that is easy to learn and apply for not only experts but also non experts.

Keywords: MIPA framework, user interface, game design.

1 Introduction

Game designers face the challenge of creating games that can be effectively played, easily learned, and emotionally enjoyed by gamers. With very limited theoretical foundation research on gamers, they have to entirely depend and rely on their intuition and experience. This is why about 80% of games fail on the market each year [1]. While technologies have improved rapidly, game design has evolved slowly.

Given the fact that the game market is so competitive, every aspect of game design and development has been studied carefully to find better ways to design more successful games. It is interesting to notice that other game software industries have invested a lot of time and effort in finding new methods and processes to design and evaluate user interface.

Human Computer Interaction has been a thriving field in recent years where a lot of innovative ideas have been generated. These ideas lead to a variety of processes, methodologies, techniques, and tools being developed and successfully practiced in other game software industries. They have helped developers solve many of their problems which are closely related to the problems of game design. Once they are learned and used by game designers, these processes and methods can greatly improve their work. Game designers need to restructure their design processes, redefine their design strategies, and reorganize their teams to reflect these new ideas.

M. Kurosu (Ed.): Human Centered Design, HCII 2009, LNCS 5619, pp. 84–93, 2009.
© Springer-Verlag Berlin Heidelberg 2009

Therefore the proposed framework is based on game mechanism, game interface, game play, and game aesthetics to understand the user perspective of games. The main purpose of this framework is to 1) bridge the gap between game design, development, and research 2) clarify and strengthen the iterative process 3) make it easier for all parties to decompose, study, and design a broad class of game designs and artifacts. The study analyzes the usability methods used in games and provide insights and guidelines to improve game design in order to sustain and enhance players' motivation. This new approach can be used by researchers to understand design issues seen in other types of specialty software. It can be used in further studies of games and new heuristics can be developed.

2 Background

2.1 Game Design and User Interface

Usability has multiple components and is traditionally involved with these five usability attributes: learnability, efficiency, memorability, errors, and satisfaction. In the case of game usability, effectiveness and efficiency are secondary considerations in relation to satisfaction. Games are about enjoyment rather than efficiency [2]. Player enjoyment is a very important goal for computer games. Csikszentmihalyi [3] found that optimal experience is flow, and classifies them into eight elements. Malone has attempted to develop a set of heuristics to the unique software category of games. The focus of his research was instructional games concerning the development of games with the primary objective of entertaining the user. Since the concept of a game, implies that there is an 'object of the game'[3], or goal, it is not surprising that Myer's study of Game Player Aesthetics [4], found that 'challenge' was the most preferred characteristic of a favorite game. As Karat and Ukelson [5] point out in their discussion of interfaces and motivation, people find satisfaction in mastery of a tool to reach a desired goal and so are willing to invest a great deal of time in doing so. Offering challenge and the opportunity to master a skill seems to provide sufficient motivation for people to engage in games.

Clanton [6] offers a way to encapsulate the different usability issues of games into three areas: game interface, game mechanics, and game play. Game interface is the device through which the player interacts with the game. This includes whatever is used to physically control the game such as a controller, joystick, mouse, or keyboard. Also, it is the visual representation of software controls that players use to set up their games, engage in a tutorial, move through a game, obtain their status in the game, save their games, and exit the game. Game mechanics are the physics of the game, which are developed through a combination of animation and programming. They are used to describe how players interact with rules, game goals, player actions and strategies, and game states. This includes the way the player is allowed to move through the game environment such as walking, running, jumping, driving. Game play is the process by which a player reaches the goal of the game. All three relate to the game being both functional and satisfying and require design and evaluation. This includes the problems and challenges a player must face to try to win the game. Crawford [7] defines game play as pace and cognitive effort, and Shelley [8] agrees by equating fun with interesting decisions having to be made in a required amount of time. Current literature on usability on games presents many heuristics for designing and evaluating games. Although many useful and valid heuristics are presented below in the chart, there is no integrated user centered framework.

2.2 Perspectives on Games

The usability of a game is similar to other software in this manner; the usability of the product cannot be evaluated without taking context into consideration. When working with games, it is helpful to consider both the designer and player perspectives. It helps to observe the small changes in one layer that can cascade into others. As Haddon points out, in the case of computer games there is a thin line between user and designer. The game designer approaches the creation of the game from the Mechanics-end where the designer creates the game. The player experiences the game from the Aesthetics-end where the gamer consumes the game. But as the game progresses, the aesthetics become irrelevant and the player starts to focus on game play, in other words, how the player plays the game. As time passes the player begins to understand the mechanics by analyzing the dynamics to achieve the best understanding of the game. This is the dynamics in MDA model. The MDA framework provides insights into the relations between the formal, algorithmic elements of games and how they are presented to and manipulated by players. Nevertheless, it is a model that does not allow for the description and analysis of a mechanic due to a relative inconsistency in the formulation of the definition.

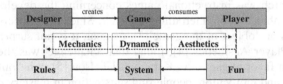

Fig. 1. Designer/Player Perspectives of Game System

Also considering about the user encourages a player centered design. A player centered approach to design can contribute to the success of a project targeting the main player. Game design process needs to consider the profiles of the gamers especially when the game production is not solely intended for entertainment but is also meant to inform, advertise, or educate.

In the end, the player focuses on how to understand and use the mechanics, because they determine what is relevant. However we believe that this is primarily the case when pacing is high, which it often is in FPS games. Games with a lower pacing gives the player time to examine the aesthetics, whereas the high demand for constant response in FPS games, shooters makes the player ignore the aesthetics and focus on the dynamics or mechanics.

2.3 Game Design and User Evaluation

Game designers don't have enough time to research the design methods from scratch. The game designer is not necessarily a graphic designer or a programmer but a person that identifies, develops and refines the game idea, mechanics of gameplay, and

Table 1. Heuristics from Literature

Game Interface	Customizable controls	(Bickford, 1997; Sanchez-Crespo Dalmau,1999)
	Play the game without reading the manual	
	Non-intrusive interface	(Sanchez-Crespo Dalmau,1999)
	Include online help	
	Identify score/status in game	(Malone,1982; Shneiderman,1992)
	Sense of control over the game interfaces	
	Shorten the learning curve	(Sanchez-Crespo Dalmau,1999)
	Support in recovering errors	
	Consistent in control (color, typo, dialog design)	(Sanchez-Crespo Dalmau,1999)
	Players should always know their status and score	
	Minimize menu layers and control options	(Shelley,2001)
	Use meaningful sound feedback	(Norman,1990)
	Do not expect the user to read a manual	(Norman,1990)
Game Mechanics	Immediate feedback to display user control	(Bickford,1997; Malone, 1982; Sanchez-Crespo Dalmau, 1999)
	Sense of control over the game shell(starting, stopping, saving...)	
	Easy to learn and use	
	Get the player involved quickly and easily	(Bickford, 1997)
	Sense of control over the input devices	
	Controls should be intuitive and a natural mapping	
Game Play	Variable difficulty level	(Malone,1982; Norman,1990; Shneiderman,1992)
	Provide new challenges at an appropriate pace	
	Multiple goals on each level	(Malone,1982
	Level of challenge should increase the player	
	Easy to learn and hard to master	(Crawford, 1982; Malone, 1982)
	Overriding goals should be clear and presented early	
	Artificial intelligence should be reasonable yet unpredictable	(Bickford, 1997; Crawford, 1982)
	Should feel viscerally involved in the game	
	Maintain an illusion of winnability	Crawford,1982)
	Provide stimuli that are worth attending to	
	Give hints, but not too many	(Clanton, 1998
	Quickly grab the players' attention and maintain their focus throughout the game	
	Give rewards appropriately	(Bickford, 1997;Clanton, 1998;Shelley, 2001; Shneiderman,1992)
	Pace the game to apply pressure, but not frustrate the player	(Clanton, 1998;Shelley, 2001)
	Allow players to build content/ Make the game replayable Create a great storyline	(Shelley, 2001)

technologies involved in the game. What they can do, however, is to look at related research fields and other software industries and to borrow ideas from them.

User centered design is an established practice in product and digital media design but is not a common practice in game design. HCI is the closest research field to game design. It is the study of how people use computers, and how to design, implement and evaluate computer systems so that can be used easily, effectively, and enjoyably. User testing can be a very important component of good game design, and is often performed when the game ides is already established. Game usability testing has evolved into a more detailed and thorough process. Among these methods, Heuristic Evaluation (HE) and Cognitive Walkthrough (CW) are valuable and can be adapted to help UI game designers.

Heuristic evaluation is a method for structuring the critique of a system using a set of general heuristics [9]. The heuristic evaluation method requires a group of people to act as evaluators and independently critique a system and suggest usability problems. The evaluators use the list of heuristics to generate ideas while critiquing the system. This guideline for heuristic evaluation was considered appropriate for use in the present study as they provide a broad overview of interface design. It is also task-free which allows them to be applied universally to a variety of games.

Cognitive Walkthrough is a method that focuses on evaluating interface design for ease of learning by exploration. Its focus is motivated by the observation that many users prefer to learn software by exploration. This is the case for games, for example when gamers start playing the games they generally know nothing about it. They learn how to play and use the interface by trial and error. Another example is that this can be used to evaluate level designs of action adventure games. Action adventure games consist of a set of levels with various goals.

The player has to explore a level to achieve that goal, then proceed to another level for another goal. This scenario matches the idea behind the cognitive walkthrough very well [9], the intent of which is to evaluate a design for its ease of learning through exploration.

3   Case Study on Game Usability Evaluation

3.1   Procedure

The users were to note any usability errors they found that were out of sequence, confusing, and did not understand or make any sense. The time required to complete the checklist was recorded upon completion (M=60mins, Range 45-75mins), so that the tests could be of reasonable duration. The usability evaluation was conducted by two groups. Group (A) are classified as expert gamers and Group (B) as non-gamers.

Table 2. Classified Participant Group

	Group (A) Expert Evaluation	Group (B) Non-Expert Evaluation
Age	Average 28	Average 28
Sex (male : female)	3 : 3	3 : 3
Game Experience	5~8 yrs.	0~1 yrs.
Game Play	4~7 times/week Every 8 hrs.	0~2 times/week Every 3 hrs.
Game Knowledge	Expert in Online, Console, Mobile	Beginners in Online, Console, Mobile

The materials of the evaluation were World of Warcraft (commonly known as WoW) which is a MMORPG game. The user controls a character avatar within a persistent game world, exploring the landscape, fighting monsters, performing quests, building skills, and interacting with NPCs, as well as other players. The game rewards success with in-game money, items, experience and reputation, all of which in turn allow players to improve their skill and power.

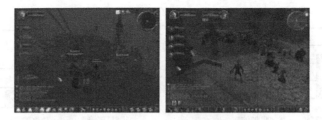

Fig. 2. Gameplay in Usability Evaluation

The user evaluating the checklist looked at the actions for each task and evaluated on the usability problems. The groups played the game and wrote notes on the usability issues they found while playing. The findings were based on the developed framework of game usability heuristics. The groups were told to evaluate the game neither had any specific instructions given on what to focus in the game. But before starting to evaluate the game they were instructed how to play the game and reminded that in games some issues are supposed to be challenging whereas everything else should be as easy as possible.

After the evaluation, the groups presented their findings to the evaluation moderator and discussed the reasons behind the problems, severity classifications and the possible solutions. Then the moderator(evaluation leader) collected the problems. The problems were grouped within predefined categories. After the categorization, similar problems within each category were grouped together. This categorized and grouped list served as the basis for the final framework.

3.2 Results

In this section we describe quantitative results from the case study. First, we examined the total number of problems by severity on a scale of 1(minor) to 7(major) and found the means of the problems per person(mean1) and also the means of the severity of the problem(mean2). The summary of the results are presented in the chart below.

Table 3. Total number of Problems Found

Severity / Evaluation	1 Minor	2	3	4	5	6	7 Major	Mean(1)	Mean(2)
Group (A)	13	2	3	36	48	22	2	21	4.41
Group (B)	9	5	0	23	28	17	1	13.8	4.33

It shows that Group(A) and Group(B) had difference on finding the total number of problems but there was no significant difference in the severity of the problem.

Second, we identified the number of accurate problems detected in tests finding the accurate problem numbers per person(mean1) and the severity of the problem(mean2). Then we discussed the accuracy rate of the two groups on finding the usability problems.

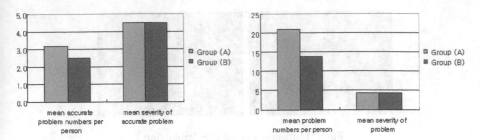

Fig. 3. Means of problem /person and severity of problems & Means of accurate problem /person and severity of accurate problems

Table 4. Accurate Problems Found

Severity Evaluation	1 Minor	2	3	4	5	6	7 Major	Mean(1)	Mean(2)
Group (A)	2	1	0	6	8	4	1	3.14	4.5
Group (B)	1	1	0	5	3	5	3	2.5	4.5

From the above figure and chart, accurate problems found by the expert Group(A) was 3.14 whereas the mean for the non-expert Group(B) was 2.5. There was no significant difference between the severity of the accurate problems of both of the groups which have the means of 4.5.

Table 5. Accuracy Rate

Severity Evaluation	1 Minor	2	3	4	5	6	7 Major	Mean Accuracy
Group (A)	15%	50%	0%	17%	17%	18%	50%	17.62%
Group (B)	11%	20%	0%	13%	18%	18%	0%	18%

When the numbers of the usability problems that did have a difference between the two groups, it was found that there was no significant difference between the accuracy of the expert and non-expert. Also the performed evaluation was based on the concerns of the user's success rates in completing the tasks, and their ability to find the problems. We did not expect to measure user's speed of task performance.

4   MIPA Framework

From the above usability evaluation results of the case study, the MIPA framework has two main roles in the game design process.

First, the framework can serve as a set of game design principles that can be used during the pre-production phase or the formative stages of the game design and development. Second, this can be used to carry out usability evaluations where developers, evaluators, and designers could use to critique the design. This will help developers derive valuable and useful data for the game development. The following detailed elements evaluation elements are shown below.

Function	Description	Elements	Evaluation
G A M E M E C H A N I C S	Physical elements of expressing the combination of animation and programming	Immediate Display	Display immediate visual and auditory feedback of the user control
		Physical	Provide natural weight of the mechanics Provide quantity of motion
		Participation	Rapid, swift, and easy participation
		Learnability	Continuous action mapping and response Short learning time, direct recognition
		Intuitive	Easy control and customizable Natural mapping Expandable in options

Fig. 4. Game Mechanics of MIPA Framework

The mechanics are evaluated on the display, physical attributes, participation, and response in action mapping, easy and customizable controls, and expandable options. Together with the game content (levels, assets and so on) the mechanics support the overall game play dynamics.

Function	Description	Elements	Evaluation
G A M E I N T E R F A C E	Interaction with the game on all devices	Consistency	Consistency in naming, structure, expression. Consistency in icons and look and feel of the style
		Control	Initialization of game control Freedom of game initialization Immediate operation in game control
		Feedback	Easy to remember the number of icon in the screen Communicate the optimal amount of information
		Natural Correspondence	Support a natural response between action and result, control and its effect
		Affordance	Show the actual realistic characteristics
		Mental Model	Easy to remember Appropriate grouping and naming Visualize the leveling structure
		Help / Support	Appropriate help info Appropriate expressions Not too much frequent game hints
		Navigation	Precise leveling structure Appropriate depth and breadth of menus Provide a variety of paths

Fig. 5. Game Interface of MIPA Framework

The interface is classified into the consistency of structures and representation, initialization of game control, freedom of key control, easy memorable information, natural correspondence of the control and its effect and result of the action, affordance of objects, clear and distinct structure, diverse paths of navigation, complete mental model of the internal mechanics of the game.

It is important to recognize that the user interface is closely related to game play. No matter how beautiful the 3D images are, or how involving the story is, without good game play a game definitely cannot succeed. But game play is a rather vague concept and hard to describe. So it is very important to find away to define game play.

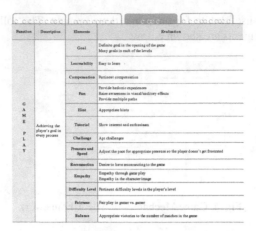

Function	Description	Elements	Evaluation
G A M E P L A Y	Achieving the player's goal in every process	Goal	Definite goal in the opening of the game Many goals in each of the levels
		Learnability	Easy to learn
		Compensation	Pertinent compensation
		Fun	Provide hedonic experiences Raise awareness in visual/auditory effects Provide multiple paths
		Hint	Appropriate hints
		Tutorial	Show interest and enthusiasm
		Challenge	Apt challenges
		Pressure and Speed	Adjust the pace for appropriate pressure so the player doesn't get frustrated
		Reconnection	Desire to have reconnecting to the game
		Empathy	Empathy through game play Empathy in the character image
		Difficulty Level	Pertinent difficulty levels in the player's level
		Fairness	Fair play in gamer vs. gamer
		Balance	Appropriate victories to the number of matches in the game

Fig. 6. Game Play of MIPA Framework

The game play should have a definite goal in the opening part of game and at each level, easy to learn, provide fun and multiple paths, give hints at the right point, offer challenge. Also the player should be at the range of appropriate suspense or tension in order to adjust the pace, feel empathy through game play and characters, have fair game play between the players, and balance of game percentage of victories.

Aesthetics describes the desirable emotional responses evoked in the player, when the player interacts with the game system. The detailed elements are aesthetics of the menu, layout, controls, and also minimize inputs and expression, easy to understand, visual affordance, transparency, and visual metaphor.

Function	Description	Elements	Evaluation
G A M E A E S T H E T I C S	Emotional and visual elements in the game	Aesthetic	Provide aesthetics in menu, layout, control option...
		Concise	Minimum expressions Minimum input
		Visual Affordance	Easy to understand and straightforward
		Transparency	Immersed in the virtual world
		Metaphor	Provide pertinent metaphor

Fig. 7. Game Aesthetic of MIPA Framework

Attractive graphics are important, but it attempts to find predictive metrics of user preferences for esthetic qualities are risky. We know that alignment and grouping is important for rapid performance. Balance and symmetry are classic notions for graphic design, but when do they also increase preference and improve performance Smooth transitions and zooming are enjoyable and helpful, principles of rapid, incremental, and reversible actions with immediate visibility of results, also increases satisfaction and performance

5 Conclusion

The purpose of this experiment was to identify the relative benefits of the usability checklist and to investigate how the identified usability problems varied by groups.

From our experience, there are no structured game frameworks for user interface design. This is why evaluation methods are important in the game development process. The MIPA framework can perform efficient evaluations and correctly identify as many usability defects as possible. Also, accurate evaluations earlier in the design phase can save money and time. Therefore the result is an effective task-oriented usability evaluation checklist that is easy to learn and apply for not only experts but also non experts.

The framework continuously needs to be modified and therefore will be greatly improved. Other case studies in other development game companies could provide valuable data to compare. Ongoing research of this study is to make a evaluation tool based on the MIPA framework.

References

1. Game Software Industry Report. In: Alien Brain Product Catalog, NxN Software (2001)
2. Malone, T.W.: What Makes Things Fun to Learn, A Study of Intrinsically Motivating Computer Games. Ph.D Dissertation, Department of Psychology, Stanford University (1980)
3. Csikszentmihalyi, M.: Flow: The Psychology of Optimal Experience. Harper Perennial, New York (1990)
4. Myers, D.: A Q-Study of Game Player Aesthetics. Simulation & Gaming (1990)
5. Karat, J., Karat, C., Ukelson, J.: Affordances, Motivation, and the Design of User Interfaces. Communications of the ACM 43 (2000)
6. Clanton, C.: An Interpreted Demonstration of Computer Game Design. In: Proceedings of the conference on CHI 1998 (April 1998)
7. Crawford, C.: Washington State University Vancouver. The Art of Computer Game Design (1982)
8. Shelley, B.: Guidelines for Developing Successful Games. Gamasutra (August 2001)
9. Nielsen, J., Molich, R.: Heuristic Evaluation of User Interfaces. In: Proceedings of the ACM SIGCHI Conference on Human Factors in Computing Systems (1990)
10. Desurvire, H., Caplan, M., Toth, J.: Using Heuristics to Improve the Playability of Games. In: CHI Conference (2004)
11. Federoff, M.: Heuristics and usability guidelines for the creation and evaluation of fun in video games. Thesis at University graduate school of Indiana university (2002)
12. Gamasutra, http://www.gamasutra.com/
13. Shneiderman, B.: Human Factors of Interactive Software. In: Designing the User Interface: Strategies for Effective Human-Computer Interaction. Addison-Wesley, Reading (1997)
14. Parush, A., Nadir, R., Shtub, A.: Evaluating the layout of graphical user interface screens: Validation of a numerical computerized model. International Journal of Human-Computer Interaction (1998)
15. Shneiderman, B.: Designing for fun: How to make user interfaces more fun. ACM Interactions 11 (2004)

The Factor Structure of the System Usability Scale

James R. Lewis[1] and Jeff Sauro[2]

[1] IBM Software Group, 8051 Congress Ave, Suite 2227
Boca Raton, FL 33487
[2] Oracle, 1 Technology Way, Denver, CO 80237
jimlewis@us.ibm.com, jeff@measuringusability.com

Abstract. Since its introduction in 1986, the 10-item System Usability Scale (SUS) has been assumed to be unidimensional. Factor analysis of two independent SUS data sets reveals that the SUS actually has two factors – Usable (8 items) and Learnable (2 items – specifically, Items 4 and 10). These new scales have reasonable reliability (coefficient alpha of .91 and .70, respectively). They correlate highly with the overall SUS ($r = .985$ and $.784$, respectively) and correlate significantly with one another ($r = .664$), but at a low enough level to use as separate scales. A sensitivity analysis using data from 19 tests had a significant Test by Scale interaction, providing additional evidence of the differential utility of the new scales. Practitioners can continue to use the current SUS as is, but, at no extra cost, can also take advantage of these new scales to extract additional information from their SUS data. The data support the use of "awkward" rather than "cumbersome" in Item 8.

Keywords: System Usability Scale, SUS, factor analysis, psychometric evaluation, subjective usability measurement, usability, learnability, usable, learnable.

1 Introduction

In 1986, John Brooke, then working at DEC, developed the System Usability Scale (SUS) [1]. The standard SUS consists of the following ten items (odd-numbered items worded positively; even-numbered items worded negatively).

1. I think that I would like to use this system frequently.
2. I found the system unnecessarily complex.
3. I thought the system was easy to use.
4. I think that I would need the support of a technical person to be able to use this system.
5. I found the various functions in this system were well integrated.
6. I thought there was too much inconsistency in this system.
7. I would imagine that most people would learn to use this system very quickly.
8. I found the system very cumbersome to use.
9. I felt very confident using the system.
10. I needed to learn a lot of things before I could get going with this system.

To use the SUS, present the items to participants as 5-point scales numbered from 1 (anchored with "Strongly disagree") to 5 (anchored with "Strongly agree"). If a

M. Kurosu (Ed.): Human Centered Design, HCII 2009, LNCS 5619, pp. 94–103, 2009.
© Springer-Verlag Berlin Heidelberg 2009

participant fails to respond to an item, assign it a 3 (the center of the rating scale). After completion, determine each item's score contribution, which will range from 0 to 4. For positively-worded items (1, 3, 5, 7 and 9), the score contribution is the scale position minus 1. For negatively-worded items (2, 4, 6, 8 and 10), it is 5 minus the scale position. To get the overall SUS score, multiply the sum of the item score contributions by 2.5. Thus, SUS scores range from 0 to 100 in 2.5-point increments.

The ten SUS items were selected from a pool of 50 potential items, based on the responses of 20 people who used the full set of items to rate two software systems, one of which was relatively easy to use, and the other relatively difficult. The items selected for the SUS were those that provided the strongest discrimination between the systems. In the original paper by Brooke [1], he reported strong correlations among the selected items (absolute values of r ranging from .7 to .9), but he did not report any measures of reliability or validity, referring to the SUS as a quick and dirty usability scale. For these reasons, he cautioned against assuming that the SUS was any more than a unidimensional measure of usability (p. 193): "SUS yields a single number representing a composite measure of the overall usability of the system being studied. Note that scores for individual items are not meaningful on their own." Given data from only 20 participants, this caution was appropriate.

1.1 Psychometric Qualification of the SUS

Despite being a self-described "quick and dirty" usability scale, the SUS has become a popular questionnaire for end-of-test subjective assessments of usability [2]. The SUS accounted for 43% of post-test questionnaire usage in a recent study of a collection of unpublished usability studies [3]. Research conducted on the SUS has shown that although it is fairly quick, it is probably not all that dirty. The typical minimum reliability goal for questionnaires used in research and evaluation is .70 [4, 5]. An early assessment of the reliability of the SUS based on 77 cases indicated a value of .85 for coefficient alpha (a measure of internal consistency often used to estimate reliability of multi-item scales) [6, 7]. More recently, Bangor, Kortum, and Miller [8], in a study of 2324 cases, found the coefficient alpha of the SUS to be .91. Bangor et al. also provided some evidence of the validity of the SUS, both in the form of sensitivity (detecting significant differences among types of interfaces and as a function of changes made to a product) and concurrent validity (a significant correlation of .806 between the SUS and a single 7-point adjective rating question for an overall rating of "user friendliness").

Although not directly measuring reliability, Tullis and Stetson [9] provided additional evidence of the reliability of the SUS. They conducted a study with 123 participants in which the participants used one of five standard usability questionnaires to rate the usability of two websites. With the entire sample size, all five questionnaires indicated superior usability for the same website. Because no practical usability test would have such a large number of participants, they conducted a Monte Carlo simulation to see, as the sample size increased from 6 to 14, which of the questionnaires would converge most quickly to the "correct" conclusion regarding the difference between the websites' usability, where "correct" meant a significant t-test consistent with the decision reached using the total sample size. They found that two of the questionnaires, the SUS and the CSUQ [10, 11] met this goal the most quickly, making the correct decision over 90% of the time when $n \geq 12$. This result is implicit evidence of reliability, and also suggests that comparative within-subject summative usability studies using the SUS should have sample sizes of at least 12 participants.

1.2 The Assumption of SUS Unidimensionality

As previously mentioned, there has been a long-standing assumption that the SUS assesses the single construct of usability. In the most ambitious investigation of the psychometric properties of the SUS to date, Bangor et al. [8] conducted a factor analysis of their 2324 SUS questionnaires and concluded, on the basis of examining the eigenvalues and factor loadings for a one-factor solution, that there was only one significant factor, consistent with prevailing practitioner belief and practice.

The problem with this conclusion is that Bangor et al. [8] did not report the possibility of a multifactor solution, especially, the possibility of a two-factor solution. The mechanics of factor analysis virtually guarantee high loadings for all items on the first unrotated factor, so although their finding supports the use of an overall SUS measure, it does not exclude the possibility of additional structure. Examination of the scree plot (see their Figure 5) shows the expected very high value for the first eigenvalue, but also a fairly high value for the second eigenvalue – a value just under 1.0. There is a rule-of-thumb used by some practitioners and computer programs to set the appropriate number of factors to the number of eigenvalues greater than 1, but this rule-of-thumb has been discredited because it is often the case that the appropriate number of factors exceeds the number of eigenvalues greater than 1 [12, 13].

1.3 Goals of the Current Study

The primary purpose of the current study was to conduct factor analyses to explore the factor structure of the SUS, using data published by Bangor et al. [8] and an independent set of data we collected as part of a larger data collection and analysis program [3] that included 324 complete SUS questionnaires. Secondary goals were to use the new data to assess the reliability and, to as great an extent as possible, the validity of the SUS.

2 Factor Analysis of the SUS

At the time of this study, we had collected 324 completed SUS questionnaires from the usability data for 19 usability studies, which was an adequate number for investigating the factor structure of the SUS [5]. Fortunately, Bangor et al. [8] published the correlation matrix of the SUS items from their studies (see their Table 5). It is possible to use an item correlation matrix as the input for a factor analysis, which meant that data were available for two independent sets of solutions – one using the Bangor et al. correlation matrix, and another using the 324 cases from Sauro and Lewis [3].

Having two independent data sources for a factor analysis of the SUS afforded a unique method for assessing the factor structure. It takes at least two items to form a scale, which makes it very unlikely that the 10-item SUS would have a structure with more than four factors. Table 1 shows side-by-side solutions for both sets of data for four, three, and two factors. Our strategy was to start with the four-factor solution (using common factor analysis with varimax rotation), then work our way down until we obtained similar item-to-factor loadings for both data sets. The failure of this approach would be evidence in favor of the unidimensionality of the SUS.

As Table 1 shows, however, the results converged for the two-factor solution. Indeed, given the differences in the distributions and the differences in the four- and three-factor solutions, the extent of convergence at the two-factor solution was striking, with the solutions accounting for 56-58% of the total variance. For both two-factor solutions, Items 1, 2, 3, 5, 6, 7, 8, and 9 aligned with the first factor, and Items 4 and 10 aligned with the second factor. Given 8 items in common between the Overall SUS and the first factor, we named the first new scale Usable. Based on the content of Items 4 and 10 ("I think I would need the support of a technical person to be able to use this system" and "I needed to learn a lot of things before I could get going with this system"), we named the second new scale Learnable. It was surprising that Item 7 ("I would imagine that most people would learn to use this system very quickly") did not also align with this factor, but its non-alignment was consistent for both data sets, possibly due to its focus on considering the skills of others rather than the rater's own skills.

3 Additional Psychometric Analyses

3.1 Item Weighting

Rather than weighting each scale item the same (unit weighting), it can be tempting to use the factor loadings to weight items differentially. Such a practice is, however, rarely worth the effort and increased complexity of measurement. Nunnally [5] pointed out that such weighting schemes usually produce a measurement that is highly correlated with the unweighted measurement, so there is no statistical advantage to the weighting. That was the case with these new Usable and Learnable scales, which had, respectively, weighted-unweighted correlations of .993 and .997 (both $p < .0001$), supporting the use of unit weighting for these scales.

3.2 Scale Correlations

The correlations between the new scales and the Overall SUS were .985 for Usable and .784 for Learnable (both $p < .0001$). Because each of the new scales had items in common with the Overall SUS, this is an expectedly high level of correlation. The correlation between Usable and Learnable was .664 ($p < .0001$). They were not completely independent, but neither were they completely dependent, with shared variance ($R2$) of about 44%. Consistent with the interpretation of the factor analyses, this finding supports both the use of an Overall SUS score and the decomposition of that score into Usable and Learnable components.

3.3 Reliability

For our 324 cases, coefficient alpha for Overall SUS was .92, a finding consistent with the value of .91 reported by Bangor et al. [8]. Coefficient alphas for Usable and Learnable were, respectively, .91 and .70. Even though only two items contributed to Learnable, the scale had sufficient reliability to meet the typical minimum standard of .70 for this type of measurement [4, 5].

Table 1. Four-, three-, and two-factor solutions for the two independent data sets

Bangor et al.

Item	1	2	3	4
Q1	**0.64**	0.19	0.31	0.04
Q2	0.38	0.30	**0.53**	0.25
Q3	**0.66**	0.42	0.31	0.22
Q4	0.22	**0.67**	0.22	0.03
Q5	**0.61**	0.20	0.38	0.00
Q6	0.37	0.32	**0.58**	-0.04
Q7	**0.59**	0.33	0.30	-0.01
Q8	0.41	0.35	**0.52**	0.03
Q9	**0.61**	0.52	0.20	0.10
Q10	0.25	**0.66**	0.25	0.05

Current

Item	1	2	3	4
Q1	**0.65**	0.17	0.19	0.29
Q2	**0.59**	0.43	0.20	0.25
Q3	**0.50**	0.39	0.18	0.47
Q4	0.25	**0.64**	0.07	0.14
Q5	0.32	0.16	0.18	**0.64**
Q6	**0.46**	0.36	0.16	0.35
Q7	0.49	0.28	**0.58**	0.31
Q8	**0.67**	0.34	0.22	0.37
Q9	0.46	0.45	0.14	**0.47**
Q10	0.18	**0.68**	0.45	0.24

Item	1	2	3
Q1	**0.63**	0.19	0.33
Q2	0.41	0.32	**0.49**
Q3	**0.66**	0.42	0.33
Q4	0.22	**0.67**	0.23
Q5	**0.60**	0.19	0.40
Q6	0.35	0.31	**0.59**
Q7	**0.58**	0.33	0.31
Q8	0.40	0.35	**0.54**
Q9	**0.62**	0.52	0.20
Q10	0.25	**0.67**	0.26

Item	1	2	3
Q1	**0.69**	0.20	0.26
Q2	**0.60**	0.46	0.23
Q3	**0.54**	0.43	0.43
Q4	0.27	**0.58**	0.12
Q5	0.33	0.20	**0.71**
Q6	**0.47**	0.38	0.33
Q7	**0.52**	0.45	0.35
Q8	**0.69**	0.38	0.35
Q9	**0.50**	0.46	0.40
Q10	0.24	**0.78**	0.24

Item	1	2
Q1	**0.70**	0.22
Q2	**0.59**	0.38
Q3	**0.71**	0.45
Q4	0.27	**0.69**
Q5	**0.71**	0.23
Q6	**0.58**	0.39
Q7	**0.64**	0.36
Q8	**0.60**	0.41
Q9	**0.60**	0.52
Q10	0.31	**0.69**

Item	1	2
Q1	**0.71**	0.21
Q2	**0.62**	0.46
Q3	**0.69**	0.43
Q4	0.28	**0.58**
Q5	**0.60**	0.26
Q6	**0.58**	0.39
Q7	**0.62**	0.46
Q8	**0.77**	0.38
Q9	**0.64**	0.47
Q10	0.32	**0.79**

Var	3.46	2.12	**Total**
% Var	34.63	21.18	55.81

Var	3.61	2.20	**Total**
% Var	36.07	21.95	58.01

3.4 Sensitivity

To assess scale sensitivity, we conducted an ANOVA with Test as an independent variable with 19 levels (for the 19 tests from which the SUS scores came) and Scale as a dependent variable with 2 levels (Usable and Learnable). To make the Usable and Learnable scores comparable with the Overall SUS score (ranging from 0 to 100), we multiplied their summed score contributions by 3.125 and 12.5, respectively. The resulting scale score for Usable ranged from 0 to 100 in 32 increments of 3.125, and for Learnable ranged from 0 to 100 in eight increments of 12.5. The ANOVA had a significant main effect of Test ($F(18, 305) = 7.73$, $p < .0001$), a significant main effect of Scale ($F(1, 305) = 47.6$, $p < .0001$), and a significant Test by Scale interaction ($F(18, 305) = 3.81$, $p < .0001$). In particular, the significant Test by Scale interaction provided evidence of the sensitivity of the Scale variable. If there had been no interaction, then this would have been evidence that Usable and Learnable were contributing the same information to the analysis. As expected from the factor and correlation analyses, however, the results confirmed the differential information provided by the two scales, as shown in Figure 1 (with the tests ordered by decreasing value of Usable). As expected due to the moderate correlation between Usable and Learnable, when the value of Usable declined, the value of Learnable also tended to decline, but with a different pattern. In most of the studies (except for three cases), the value of Learnable tended to be greater than the value of Usable, but to varying degrees as a function of Test.

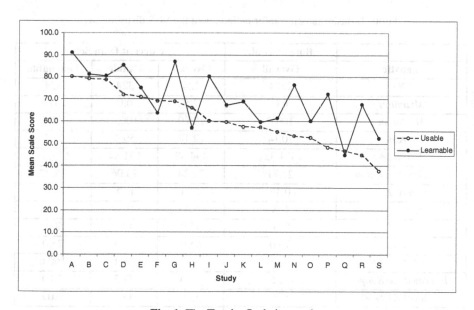

Fig. 1. The Test by Scale interaction

4 The Distribution of SUS Scores

Bangor et al. [8] provided some information about the distribution of SUS scores in their data. Table 2 shows basic statistical information about their distribution and the distribution of our new data. Figure 2 shows a graph of the distribution of the Overall SUS scores from the current data set (for comparison with Figure 2 of Bangor et al.), and the distributions of the Usable and Learnable scores (all set to the same scale).

Of particular interest is that the central tendencies of the Bangor et al. (2008) and our Overall SUS distributions were not identical, with a mean difference of 8.0. The mean of the Bangor et al. distribution was 70.1, with a 99.9% confidence interval ranging from 68.7 to 71.5 [8]. The mean of our Overall SUS data was 62.1, with a 99.9% confidence interval ranging from 58.3 to 65.9. Because the confidence intervals did not overlap, this difference in central tendency as measured by the mean was statistically significant ($p < .001$). There were similar differences (with the Bangor et al. scores higher) for the 1st quartile (10 points), median (10 points), and 3rd quartile (12.5 points). The distributions' measures of dispersion (variance, standard deviation, and interquartile range) were close in value.

As expected, the statistics and distributions of the Overall SUS and Usable scores from the current data set were very similar. In contrast, the distributions of the Usable and Learnable scores were distinct. The distribution of Usable, although somewhat skewed, had lower values at the tails than in the center. By contrast, Learnable was strongly skewed to the right, with 29% of its scores having the maximum value of 100. Consistent with the results of the ANOVA, their 99.9% confidence intervals did not overlap, indicating a statistically significant difference ($p < .001$).

Table 2. Basic statistical information about the SUS distributions

Statistic	Bangor et al. Overall	Current Data Set Overall	Usable	Learnable
N	2324	324	324	324
Minimum	0.0	7.5	0.0	0.0
Maximum	100.0	100.0	100.0	100.0
Mean	70.14	62.10	59.4	72.7
Variance	471.32	494.38	531.54	674.47
Standard Deviation	21.71	22.24	23.06	25.97
Standard Error	0.45	1.24	1.28	1.44
Skewness	NA	-0.43	-0.38	-0.80
1st Quartile	55.0	45.0	40.6	50.0
Median	75.0	65.0	62.5	75.0
3rd Quartile	87.5	75.0	78.1	100.0
Interquartile Range	32.5	30.0	37.5	50.0
Critical Z (99.9)	3.09	3.09	3.09	3.09
Critical d (99.9)	1.39	3.82	3.96	4.46
99.9% CI Upper Limit	71.53	65.92	63.40	77.18
99.9% CI Lower Limit	68.75	58.28	55.48	68.27

Table note: Add and subtract Critical d (computed by multiplying the Critical Z and the standard error) from the mean to get the upper and lower bounds of the 99.9% confidence interval.

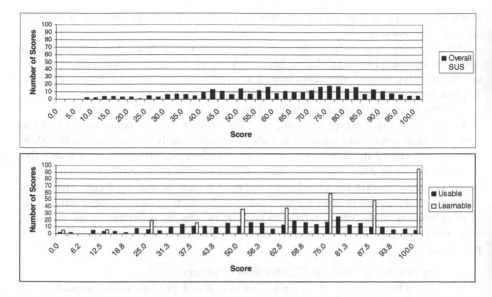

Fig. 2. Distributions of the Overall SUS, Usable, and Learnable scores from the current data set

5 Discussion

5.1 Benefit of an Improved Understanding of the Factor Structure of the SUS – A Cleaner and Possibly Quicker Usability Scale

In the 23 years since the introduction of the SUS, it has certainly stood the test of time. The results of the current research show that it would be possible to use the new Usable scale in place of the Overall SUS. The scales had an extremely high correlation (.985), and the reduction in reliability in moving from the 10-item Overall SUS to the 8-item Usable scale was negligible (.92 to .91). The time saved by dropping Items 4 and 10, however, would be of relatively little benefit compared to the advantage of getting an estimate of perceived learnability along with a cleaner estimate of perceived usability. For this reason, we encourage practitioners who use the SUS to continue doing so, but to recognize that in addition to working with the standard Overall SUS score, they can easily decompose the Overall SUS score into its Usable and Learnable components, extracting additional information from their SUS data with very little additional effort.

The difference in central tendency between the Bangor et al. [8] data and our data indicate that the two datasets may represent different types of users and products. For preliminary data on an attempt to connect SUS ratings to a 7-point adjective scale (Best Imaginable to Worst Imaginable), see Bangor et al. (pp. 586-588).

5.2 Implications for SUS Item Wording

Psychometric findings for one version of a questionnaire do not necessarily generalize to other versions. Research on the SUS and similar questionnaires has shown, however, that slight changes to item wording most often lead to no detectable differences in factor structure or reliability [10].

For example, in a study of the interpretation of the SUS by non-native English speakers, Finstad [14] found that in Item 8 ("I found the system very cumbersome to use"), all native English speakers claimed to understand the term, but half of the non-English speakers asked for clarification. When told that "cumbersome" meant "awkward", the non-English speakers indicated that this was sufficient clarification.

Bangor et al. [8] also reported some confusion (about 10% of participants) with the word "cumbersome", and replaced it with "awkward" early in their use of the SUS. They also replaced the word "system" with "product" in all items. Consequently, about 90% of their 2324 cases used the modified version of the SUS. Our 324 cases, however, used the original SUS item wording for Item 8, and used either the word "system" or the actual product name in place of "system". Despite these differences in item wording, estimates of reliability and the two-factor solutions for the two data sets were almost identical, which leads to the following two guidelines for practitioners.

- For Item 8, use "awkward" rather than "cumbersome".
- Use either "system" or "product" or the actual product name, depending on which seems more appropriate for a given test, but for consistency of presentation, use the same term in all items for any given test or across a related series of tests.

References

1. Brooke, J.: SUS: A "Quick and Dirty" Usability Scale. In: Jordan, P.W., Thomas, B., Weerdmeester, B.A., McClelland (eds.) Usability Evaluation in Industry, pp. 189–194. Taylor & Francis, London (1996)
2. Lewis, J.R.: Usability Testing. In: Salvendy, G. (ed.) Handbook of Human Factors and Ergonomics, pp. 1275–1316. John Wiley, New York (2006)
3. Sauro, J., Lewis, J.R.: Correlations among Prototypical Usability Metrics: Evidence for the Construct of Usability. In: The Proceedings of CHI 2009 (to appear, 2009)
4. Landauer, T.K.: Behavioral Research Methods in Human-Computer Interaction. In: Helander, M., Landauer, T., Prabhu, P. (eds.) Handbook of Human-Computer Interaction, pp. 203–227. Elsevier, Amsterdam (1997)
5. Nunnally, J.C.: Psychometric Theory. McGraw-Hill, New York (1978)
6. Lucey, N.M.: More than Meets the I: User-Satisfaction of Computer Systems. Unpublished thesis for Diploma in Applied Psychology, University College Cork, Cork, Ireland (1991)
7. Kirakowski, J.: The Use of Questionnaire Methods for Usability Assessment (1994), http://sumi.ucc.ie/sumipapp.html
8. Bangor, A., Kortum, P.T., Miller, J.T.: An Empirical Evaluation of the System Usability Scale. International Journal of Human-Computer Interaction 24, 574–594 (2008)
9. Tullis, T.S., Stetson, J.N.: A Comparison of Questionnaires for Assessing Website Usability. Unpublished presentation given at the UPA Annual Conference (2004), http://home.comcast.net/~tomtullis/publications/UPA2004TullisStetson.pdf

10. Lewis, J.R.: IBM Computer Usability Satisfaction Questionnaires: Psychometric Evaluation and Instructions for Use. International Journal of Human-Computer Interaction 7, 57–78 (1995)
11. Lewis, J.R.: Psychometric Evaluation of the PSSUQ Using Data from Five Years of Usability Studies. International Journal of Human-Computer Interaction 14, 463–488 (2002)
12. Cliff, N.: Analyzing Multivariate Data. Harcourt Brace Jovanovich, San Diego (1987)
13. Coovert, M.D., McNelis, K.: Determining the Number of Common Factors in Factor Analysis: A Review and Program. Educational and Psychological Measurement 48, 687–693 (1988)
14. Finstad, K.: The System Usability Scale and Non-Native English Speakers. Journal of Usability Studies 1, 185–188 (2006)

Validating a Standardized Usability/User-Experience Maturity Model: A Progress Report

Aaron Marcus[1], Richard Gunther[2], and Randy Sieffert[3]

[1] President, Aaron Marcus and Associates, Inc., 1196 Euclid Ave., Suite 1F,
Berkeley, CA, 94708 USA
Aaron.Marcus@AMandA.com
[2] Principal, Ovo Studios, LLC, 236 High St., Chagrin Falls, OH 44022
rich@ovostudios.com
[3] Principal UX/IxD Designer, Whirlpool Corp., Global Consumer Design,
1800 Paw Paw Ave., MD 6008, Benton Harbor, MI 49022
randall_h_sieffert@whirlpool.com

Abstract. The authors report on ongoing work in developing a usability/user-experience maturity model, in particular, the results of a workshop about this subject held at the Usability Professionals Association 2009 national conference.

Keywords: Business, design, experience, maturity, model, usability, user.

1 Introduction

Managers of design, analysis, and development in the human-computer interface (HCI) usability, user-experience (UX), and user-interface (UI) communities are developing or altering their programs in reaction to new business models, methods of practice, professional philosophies, and technologies. There is much turmoil, new terminology, and debate about best practices.

Recognizing the importance of understanding key terms and conceptual models of user-experience and being able to describe a model of evolving maturity in developing programs of people and resources to manage this ongoing activity, Gunther [3] organized a workshop on user-experience maturity modeling at the Usability Professionals Association (UPA) national conference in 2006 [3] and 2009 [4], at which Marcus and Sieffert took part.

Because of conflicting publishing deadlines, the exact contents cannot be provided for the HCI International (HCII) 2009 *Proceedings,* however, the results of the UPA 2009 workshop are available in summarized form from Marcus or Gunther. This paper summarizes the intentions and the expected results of the 2009 workshop and provides a high-level summary of the workshop approach. The authors will present the summary presentation of the UPA 2009 workshop and discuss its implications during this session.

2 The UPA Workshop

The UPA workshop brought together approximately 12 top-level managers and directors from large enterprise software companies who are in charge of usability/user-experience

M. Kurosu (Ed.): Human Centered Design, HCII 2009, LNCS 5619, pp. 104–109, 2009.

(U/UX) development. The authors presented these participants with a draft proposed model for assessing the maturity of a usability/user-experience organization. Based on their collective experience, they revised and refined this model. The expected outcome of the event was a revised model suitable for broader publication and review by the usability/user-experience community.

The workshop leaders themselves, individually or collectively have done research, planning, and writing pertaining to measuring the positive business effects of usability and user-centered design, including the following:

- Presented papers at UPA, CHI, and HCII conferences regarding how, when, and where to best deploy user-centered design methodologies.
- Lead/facilitated workshops and tutorials at UPA, CHI, and HCII conferences on the topics of managing user experience and the business of usability.
- Founded UPA's "Usability in the Enterprise" project, and has worked on or spearheaded all initiatives for that group thus far.
- Contributed and consulted on projects for which every penny spent on usability had to be justified and documented, have seen the ill-effects of lack of business-metric tracking in usability organizations.
- Reviewed UPA, CHI, and HCII presentation and tutorial submissions for the past 25 years.
- Members of most major U/UX organizations worldwide for up to 27 years.
- Edit/write for publications about U/UX, including *User Experience* (UPA) and Interaction (CHI).
- Authored/co-authored approximately 6 books and more than 300 publications.
- Conducted surveys about U/UX practices and trends among U/UX development leaders at major companies worldwide.
- Managed and grew UX teams through business justification even in tight economic conditions.

2.1 Previous Work on This Topic

An objective of the workshop was to add to the existing body of knowledge, both principles and techniques, regarding measuring the maturity of U/UX organizations. The report from the workshop in part collates the opinions and experiences of the participants into a collectively validated model for measuring the maturity of U/UX teams. While there have been standards published [17, 18] these have gained little traction in the UX community. There have also been a number of articles published on the topic of capability maturity models [for examples see 21, 22, 23], but there seem to have been few if any workshops or sessions on the topic at CHI, HCII, or UPA conferences.

The difference between these past presentations and the workshop was this: whereas the presentations were mostly case studies specific to an industry, methodology, or product, the workshop instead sought to develop a more general, more valid, and more reliable model, which would be backed up by the experience and knowledge of the workshop's attendees, and possibly generalizable to other types of companies. Once such a model exists, it seems possible to establish a company's level of

U/UX maturity as an independent metric, in order to better determine if there is a correlation between U/UX maturity and business performance.

2.2 Participant Selection Criteria

The authors selected participants based on their work experience, submission of a position paper, and ability/willingness to share their personal work experiences. In general, these participants were upper-level practitioners at enterprise software companies who are responsible for developing and justifying a usability team's organization, discipline-focus, process, and budget, including head count, resource needs, travel, etc. These participants did not need to be U/UX practitioners themselves, as long as they were involved in customer-centered areas of their business, the authors felt that these participants would be able to make a meaningful contribution.

In general, the authors selected participants based on the following criteria:

- Required to have at least seven years experience in the U/UX fields, or in a customer-centered area of their organization.
- Required to have held a management position regarding software development at software development companies or vertical market customer.
- Required to be able, legally, to share their work experiences, business metrics, and processes with the workshop.
- Desired mix of Manager/Directors: Manager/Directors represent a cross-section of company size, geographical location, usability team size, tenure, and budget.

The authors asked the prospective participants to submit a position paper, due one month before the workshop, and to provide details about their work experience, as well as their opinions about the following questions:

- How do you keep track of the impact of your team's work on your company's business performance?
- Has this method changed among the companies for which you have worked or the industries you have served?
- Do you think it is possible to create a standardized model for assessing the maturity of a U/UX organization?
- What do you think is the best way to go about creating such a model?

We believed that members of the U/UX community, including those who might not otherwise attend the UPA 2009 conference, would be interested in the workshop because of the resulting report. We planned to emerge from the workshop with a revision to the initially proposed draft model based on the experience and expert knowledge of the participants. Because this workshop covered a topic of great interest to the U/UX community as a whole, we also believed all who would participate in it would help publicize the results.

2.3 Pre-workshop Activities

Prior to the workshop, we provided the participants with the initial proposed model, a glossary that defines the model's terms, and a list of focus areas. We asked that the participants familiarize themselves with the model and the focus areas, and come prepared with a list of questions, comments, suggestions, or concerns. The objective of this preparation work was to have all participants share a basic familiarity of the terms and possible model taxonomy so that participants could use the workshop time most effectively.

The authors used the month leading up to the workshop to finalizing the workshop schedule, reviewing participant position papers, selecting and confirming participants, and developing handouts/presentations.

The draft U/UX Maturity Model provided to participants featured the levels of maturity (rows) against UX management practices (columns) as shown in Table 1.

Table 1. Draft U/UX Maturity Model provided to participants

Management Practice ▶ Level ▼	UX Development	Staffing Resources	Management Commitment	Organizational Alignment	Vision & Strategy
5. Optimized	Continual Process Improvement	UX Executive	Maintenance Commitment	UX part of Business Strategy Processes	Firm Level Vision and Strategy
4. Managed	Managed Process	UX Leadership	Organizational Ownership	UX Architect	Strategic Planning
3. Defined	User Data Provided to Management	Managed Engagement	Portfolio Ownership Management	Integration with Broader Business Processes	Portfolio Planning
2. Repeatable	Qualitative and Process Metrics	UX Operations	Project Manager Owns Relationship	Product Development Include UX Processes	Product Planning
1. Initial	UX Basic Practices	Staff with UX Professionals	UX Professionals Own Relationship	Localized Product Dev Team Integration	Localized Product Optimization

2.4 Workshop Sessions

In each of the work's three sessions, participants broke into groups to debate the issues that lie conceptually below each of the cells of the matrix. Each group considered one column (i.e. management practice) for a session. In the succeeding sessions, each group considered other columns in order to generate fresh ideas and to stimulate debate.

At the final wrap-up session the authors led a discussion to summarize, compare results, initially resolve conflicts, and revise terminology. The authors presented a summary of the day's results at a post-workshop presentation at the UPA 2009 conference.

Following the workshop, the authors revised the model further and posted the results on the Web as a report. This report incorporated not only the results of the workshop, but also details of the process from which those results were obtained. The report appears publicly on the Usability in the Enterprise project page of the UPA

Website (www.usabilityprofessionals.org), further plans call for announcements to be made in UPA publications to its membership as well as other venues.

3 Discussion

The authors plan to incorporate feedback from subsequent presentations, such as the session at HCII 2009, into the U/UX Maturity Model and to make a revised version public. In general, the authors hope to further articulate and generalize the maturity model so that it might be use for benchmarking U/UX across the industry.

Acknowledgements

The authors acknowledge the assistance of staff/associates from Ovo Studios, LLC, and Aaron Marcus and Associates, Inc., for providing documents and human resources to assist in the Workshop and in preparation of this paper. In particular, Marcus would like to thank Mr. Niranjan Krishnamurthi, AM+A Designer/Analyst, for his assistance in preparing this document.

References

1. European Design Centre. User-Centred Design Works (CD-ROM). Publisher: IOP Human Machine Interaction, This CD-ROM presents a case for user-centered design including case studies and information resources (2004), http://www.edc.nl
2. Cunningham, J.P., Cantor, J., Pearsall, S.H., Richardson, K.H.: Industry Briefs: AT+T. Interactions 8(2), 27–31 (2001)
3. Gunther, R.G., Butler, S.A., Dove, L., Gutierrez, P.V., Marine, L.: The Business of Usability: Developing Metrics to Justify our Existence and Budgets. In: UPA Conference, Broomfield, CO (2006)
4. Gunther, R.G., Marcus, A., Sieffert, R.: How mature is your company's UX capability?: A capability modeling workshop. In: UPA Conference, Portland, OR (2009)
5. Gunther, R., Binhack, A., Dodd, J., Mischke, A., Kerkow, D.: An Analysis of ROI Measurement and Metrics across Usability Professionals (2005), http://www.upassoc.org
6. Marcus, A.: Global/Intercultural User-Interface Design. In: Jacko, J., Spears, A. (eds.) Handbook of Human-Computer Interaction, 3rd edn., ch. 18, pp. 355–380. Lawrence Erlbaum Publishers, New York (2007)
7. Marcus, A.: Cross-Cultural User-Interface Design Patterns for Mobile Products in Japan, Taiwan, and the USA. Aaron Marcus and Associates, Inc., Internal Presentation (2007)
8. Marcus, A.: User Interface Design's Return on Investment: Examples and Statistics. In: Bias, R.G., Mayhew, D.J. (eds.) Cost-Justifying Usability, 2nd edn., ch. 2, pp. 17–39. Elsevier, San Francisco (2005)
9. Marcus, A.: What Would an Ideal CHI Education Look Like? Fast Forward Column, Interactions 12(5), 54–55 (2005)
10. Marcus, A.: Usability Grows Up: The Great Debate. Fast Forward Column, Interactions 12(4), 72–73 (2005)

11. Marcus, A.: User-Centered Design in the Enterprise. Fast-Forward Column, Interactions 13(1), 18–23 (2005)
12. Marcus, A.: Dare We Define User-Interface Design. Fast-Forward Column, Interactions 9, 31–36 (2002), http://www.acm.org
13. Marcus, A., Baumgartner, V.J.: Mapping User-Interface Design Components vs. Culture Dimensions in Corporate Websites. Visible Language Journal 38(1), 1–65 (2004)
14. Marcus, A., Gould, E.W.: Cultural Dimensions and Global Web User-Interface Design: What? So What? Now What? In: Proc., 6th Conference on Human Factors and the Web, June 19. University of Texas, Austin (2000), http://www.tri.sbc.com/hfweb
15. Marcus, A.: Graphical User-Interfaces. In: Helander, M., Landauer, T.K., Prabhu, P.V. (eds.) Handbook of Human-Computer Interaction, ch. 19, pp. 423–444. Elsevier, Amsterdam (1997)
16. Marcus, A., Baumgartner, V.J.: A Practical Set of Culture Dimension for Evaluating User-Interface Designs. In: Masoodian, M., Jones, S., Rogers, B. (eds.) APCHI 2004. LNCS, vol. 3101, pp. 252–261. Springer, Heidelberg (2004)
17. Schaeffer, E.: Institutionalization of Usability. Addison-Wesley, Reading (2004)
18. Vredenburg, K., Insensee, S., Righi, C.: User-Centered Design: An Integrated Approach. Software Quality Institute Series. Prentice Hall, Upper Saddle River (2002)
19. I.I.: 13407 Human-centred design processes for interactive systems. ISO 13407: 1999 (1999)
20. ISO: ISO 18529 Human-centred design processes for interactive systems. ISO 18529 TR: 2000 (2000)
21. Earthy, J.: Usability Maturity Model: Processes. TRUMP Technical Report (1999), http://usabilitynet.org/trump/trump/index.htm
22. Jokela, T.: Assessments of Usability Engineering Processes: Experiences from Experiments. In: IEEE: HICSS Proceedings (2003)
23. Jokela, T., Abrahamsson, P.: Modelling Usability Capability – Introducing the Dimensions. In: Bomarius, F., Oivo, M. (eds.) PROFES 2000. LNCS, vol. 1840, pp. 73–87. Springer, Heidelberg (2000)

Defining Expected Behavior for Usability Testing

Stefan Propp and Peter Forbrig

University of Rostock, Institute of Computer Science,
Albert Einstein Str. 21, 18059 Rostock, Germany
{stefan.propp,peter.forbrig}@uni-rostock.de

Abstract. Within HCI task models are widely used for development and evaluation of interactive systems. Current evaluation approaches provide support for capturing performed tasks and for analyzing them in comparison to a usability experts' captured behavior. Analyzing the amount of data works fine for the evaluation of smaller systems, but becomes cumbersome and time-consuming for larger systems. Our developed method aims at making the implicitly existing expectations of a usability expert explicit to pave the way for automatically identifying candidates for usability issues. We have enhanced a CTT-like task modeling notation with a language to express expected behavior of test users. We present tool support to graphically compose expectations and to integrate them into the usability evaluation process.

Keywords: Usability Evaluation, Task Models.

1 Introduction

Task models are widely used within the domain of Human Computer Interaction. For electing requirements task models describe the progress of task execution to accomplish a certain goal. Subsequent development stages apply task models as initial artifacts for model-based development of user interfaces [9]. Several approaches further exploit task models for usability evaluation. Examples are RemUSINE [6], ReModEl [1] and the task-based timeline visualization [4]. Those evaluation techniques capture the observed user interactions on a lower level of abstraction (e.g. mouse clicks or sensor values of user movement), which can be easily captured but the vast amount of data is difficult to interpret. Subsquently the sequence of captured events is lifted from an interaction level (e.g. button click) to a task-based abstraction (e.g. printing a document) [8], which allows interpreting the results in a more natural way. Hence a usability expert can conveniently compare the observed behavior of test users with his/her expectation of efficient task performance to reach the goal of the test case. Deviations indicate candidates for usability issues. This comparison is carried out as a manual process with tool support for visualizing a task trace, but lacking an integration of machine-readable expectations.

The approach in [6] goes a bit into this direction. It offers a comparison between two task traces: the observed trace and an "ideal path". A designer specifies this path and the degree of deviation can be visualized. However, there is no opportunity discussed to generalize the expectation to cover different expected traces. For instance

M. Kurosu (Ed.): Human Centered Design, HCII 2009, LNCS 5619, pp. 110–119, 2009.

the task "sending an email" can be accomplished in different ways. It is appropriate either to use a web interface or to start your email client. Both ways solve the task. The IBOT system [12] also provides a mechanism to capture user interactions and further automatically compares the behavior of user and designer. WebQuilt [2] visualizes the navigation path through a website. It visualizes the observed path of users in contrast to a designers' expected path, which is a comparison between two navigational pathes.

As summarized there are some approaches available for automatically comparing two captured logs with each other, which may be actually logged or even designed, but do not allow to specify some degree of freedom in a sense that a user might deviate from the expected behavior in certain aspects. Therefore existing methods work fine for the evaluation of smaller systems, but become cumbersome and time-consuming for larger systems.

We aim at overcoming this problem by defining expected user behavior in a machine-readable form with some degree of freedom for deviations. Therefore we have enhanced our CTT-like task modeling notation with a language to specify expectations. In the general case when we evaluate an artifact without an existing relationship to a task model, such a task model has to be modeled and extended to form the expectation. In certain cases when we evaluate a piece of software which was developed task-based we reuse these models and enhance them.

In section 2 we draw the bigger picture and discuss the integration of our concept into the usability evaluation process. Section 3 explains the method and provided tool support, exemplified with modeling expectations for people interacting with each other in a meeting situation. We show how to specify expectations, capture according user behavior and finally analyze the results. Section 4 gives a conclusion and future research avenues.

2   Usability Evaluation Process

Before we go into the details of specifying expectations, we give an overview of the whole usability evaluation process, where our approach fits in.

The process comprises four stages (see figure 1): modelling, test planning, test execution and analysis of test results [7].

1. The modelling stage should be carried out during product development. During requirements analysis tasks are elicted, which should be carried out by users. These tasks are put into relationship to each other. A designed task model in CTT notation can be built [5], which reflects a hierarchical decomposition of tasks into subtasks. These tasks are interconnected with temporal relations. Each task model describes how a user can achieve a certain goal. Furthermore a user may have different roles. For instance a meeting participant can switch between the roles "presenter" and "participant". Each role's available tasks are described by a respective task model. Additionally a task model for coordinating the other task models can be provided [5].

2. In the planning stage a test case is defined as it is common practice for usability evaluations. We specify for instance purpose, test objectives, a description of the environment and the evaluation measures [10]. We enhance this textual information with task models describing the possible user behavior while using the evaluated

artifact. When specifying a test case, the usability expert already has an expectation in mind, how to perform the tasks in an efficient way. This implicit knowledge should be made explicit as machine-readable expectation model.

3. In the execution stage a test case is conducted several times with different test users. Observations (like key strokes, mouse clicks, location coordinates of moving users and video streams) are captured and annotated by an expert. Our test environment provides an evaluation engine, to collect this data from the physical environment with sensors and video cameras. During evaluation the captured user behavior is compared against the expectation to discover deviations. These evaluation results are additionally captured.

4. In the analysis stage an analysis engine provides capabilities to analyze and visualize captured data. Finding usability issues within the vast amount of data is a tedious task, because often it is not intuitively clear how an issue may look like. To cope with that, we particularly emphasize on expectations, because deviations from the expected behavior can be derived automatically, through comparing expectation and accomplished task trace. The result set contains some candidates for usability issues, leading to a reduction of relevant data. Subsequently a usability expert can focus on examining data interrelated with identified situations. Interactively walking through video streams, annotations, sensor and task data, helps identifying causes of an issue and improving the underlying task models to better describe user behavior.

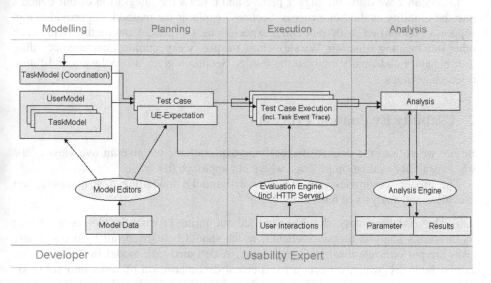

Fig. 1. Task Model-based Usability Evaluation Process

3 Specifying Expectations

3.1 Example

First we introduce a running example, which is subsequently used to describe how to specify expectations before a test starts, how to evaluate expectations during the test and how to analyze the results afterwards. Our approach aims at evaluating a wide variety of artifacts, including software and physical artifacts, where a task model can describe the interaction of a user.

According to our prototypical implementation, we consider a meeting situation within a collaborative environment. Initially no persons are present. The room is only equipped with furniture and some stationary devices, like projectors at the ceiling and movable window blinds. Before the meeting begins, three people A, B and C are populating the room, while carrying their personal devices with them. Their PDAs and laptops contain slides for the presentations and help with taking notes during listening to the other talks. All three people have to give a presentation in an arbitrary order, closing with a discussion. Finally they exit the room, carrying their devices. The room senses the loctation of the people and notices if someone takes up a device or another item in the room, like a laser pointer. The environment tries to derive which task is carried out next. For instance if a person moves to the front, connecting the laptop with projector, while the others are sitting, the environment, derives the beginning of a presentation and gives support. It shuts the window blinds at the front and moves down the appropriate projection screen. The evaluation should discover strength and weaknesses within the interactions between meeting participants and the surrounding environment. Particular usability questions to investigate within this domain are: Does the environment derive the correct user behavior from the sensor data? Are the users' performed tasks appropriately supported by the pro-active meeting assistance?

3.2 Method

To evaluate the usability of an artifact we begin with a task model which describes how users can interact with the given artifact. If we concider a software artifact, this task model may already exist from requirements elictation or task model-based development [9]. In other cases it has to be modeled first. A task model describes a set of sequences of performed tasks to reach a goal. In most cases several alternative task traces reach the goal.

A usability test typically focusses on certain functionalities of an artifact, especially when the artifact is still under development and some parts are not implemented yet. Therefore users carrying out a test case are expected to perform only tasks contained within this corresponding subset of the task model. Other tasks are possible but out of scope of the current test case, since they do not support reaching the given goal.

We distinguish between a task model describing a bunch of functionality offered by the artifact on the one hand and an expectation as subset focussing at the tested functionality. The expected task performance is further constrained to devices which have to be used, certain context conditions and maximal durations for task performance. An expectation is further described as follows:

A CTT model describes a set of possible task traces to interact with the artifact. The expectation model is build on top of this task model and comprises additional annotations to constrain these traces. To evaluate an artifact under different test conditions, for each test case a separate expectation is defined. All expectations may constrain the same model in another way depending on the designers' expected user behavior.

An expectation consists of a set of expectation statements and can be described in an EBNF-like notation:

```
expectation    = task ":" (event {"," event}) ":"
                 {statement ";"};
event          = START | END | ENABLE | DISABLE |
                 SUSPEND | RESUME | ABORT | SKIP;
statement      = classification ":" expression;
classification = PERFECT | GOOD | BAD;
```

An **expectation** is specified for a certain **task**. The example (table 1) contains several task models, one for each user role. Therefore the task has to be qualified with the respective task model as "participant.present". To evaluate expectations a task model engine was incorporated [9]. During carrying out a test case each task within the task model has a state, for instance a task begins typically as "enabled", turns into "running" and finally into "finished". State changes are triggered by events. A user can only perform leave tasks of a task model and therefore cause the task engine to fire "start" or "end" at these tasks. Each expectation is evaluated when the specified **event** is fired. Several **statements** can be associated, which are evaluated sequentially. The contained **expression** is an OCL-like expression to navigate within the task model and evaluate the tasks' attributes. Accessable attributes during runtime are for instance the states of tasks, applied devices, other involved users, the needed duration for task performance and context information. Context information depends on the available sensors. In our test environment we use mainly location sensors and RFID sensors to capture involved devices. Further context information can be annotated manually or provided via additional sensors. We use OCL [3] to specify these expressions. OCL is very expressive, while some expressions are long and difficult to read. Therefore we provide some helping functions for a more convenient navigation for the domain of task modeling, like it is discussed in [11]. The evaluation of such an expression results in a boolean value, which is interpreted in OCL as a constraint which is satified or not. We prefer a more fine grained grading. Therefore we classify a user interaction according to the degree of desirability within the current situation. We distinguish the **classification** as "perfect", "good" or "bad". For instance to perform the task "give a presentation" (a) it is goal-oriented to "load slides" (hence classified as "perfect"), (b) it is destructive to "leave room" (hence classified as "bad") and (c) optional to "open a window" to get some fresh air (hence classified as "good"). Currently we work with these three categories, but it is also possible to distinguish more or less categories. The list of **statements** is sequentially evaluated. Each OCL expression is handed over to the parser. In case of a result "true" the associated classification label is returned; in case of "false" the next statement is evaluated. Hence the first match of a statement determines the result. The statement at the end of table 1 "bad : true;" ensures the result "bad" if nothing previously matched.

Table 1. Examples for Expectation Statements

Task	Event	Classification	Expression
participant.present	start	perfect	self.device.includes (presenter_device)
		perfect	self.context.includes (presentation_zone)
		good	true
participant.present	end	perfect	self.duration() < 300
		good	self.duration() < 600
		bad	true

Table 1 exemplifies the method with two simple expectations, each comprising three statements. When a person starts to "present" the used device and location within the room is evaluated. When finishing to "present" the time of the presentation is measured. Within a test case the persons are asked to present 5 minutes (300 seconds). If the persons within the room face serious issues, preventing them from directly performing the tasks described in the test case description, often the duration needed expands. In this example a duration of more than 10 minutes is defined as threshold to mark the "present" task as potential usability issue, which has to be further investigated based on archived video and sensor material of that test session.

3.3 Graphical Tool Support

Composing statements on a textual level allows a very accurate specification of expectations. But beyond the very simple example in table 1 real world examples are much more complex and from our experience it is a very tedious task, because composing statemens manually is monotonous and error-prone due to a high degree on redundancies. Hence we provide a GUI to graphically compose expectations (figure 2) and automatically generate the according expectation statements.

On the left-hand side a tree view visualizes the task model, while the right-hand side depicts a gantt view of the timeline. The navigation within the task tree allows expanding and collapsing the task lanes at the right. Time constraints are set via drag and drop in the gantt view. The colors green and yellow mark the maximum length of a task, while red marks a task which should not be performed within the current test case. Arrows mark additional temporal dependencies. For instance if the task model allows the presentations of persons A, B and C orderindependently, an arrow from "A to B" in the expectations requires the presentation of A to be finished before B starts. If a task is not depicted as gantt lane, there is no expectation set. Normally only a few activities of interest are specified. When selecting a task details are displayed in a properties view at the bottom of the screen to adjust further parameters. Necessary devices, other involved users and certain context conditions are specified. Context parameters can be customized as necessary for the individual evaluation. Examples are the location of users within a room, touched items, light conditions, medical parameters of testers, manually annotated mental workload or categories of emotions. Arbitrary annotations are possible. Parameters can be specified for each occurrence of the task separately or globally for each repetition of a task.

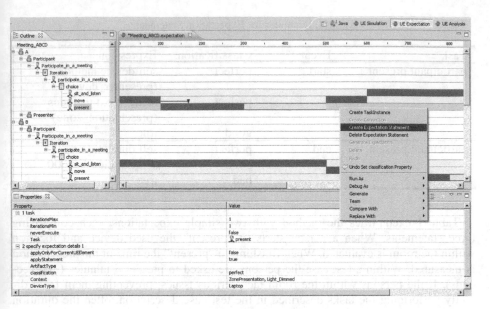

Fig. 2. Specifying Expectations

Our first approach was starting from scratch with a white gantt view, allowing a usability expert to draw task lanes and type in context parameters. To save some time we offer the possibility to load a captured test session which is close to the expected interactions. The data only needs to be adapted in certain aspects, for instance adjusting durations and deleting some unrelevant tasks while adding some missing information.

After having finished the graphical specification, expectation statements are automatically generated. For different modeled examples the expressiveness was adequate. If a usability expectation is to be defined which exceeds the opportunities of the graphical notation, the generated statements can be manually refinded to include arbitrary OCL expressions.

3.4   Test Case Execution

To test the evaluation approach we have developed an evaluation application (figure 3). Following the running example we focus on evaluating meeting situations. The lower left part depicts a bird's view of a room with some grey tables, grey chairs and the participating persons. The upper left part contains the animated task models for the three persons showing the current progress of task performance. Via drag and drop persons are moved through the room while task performance can be triggered within the animated task models. Further annotations are possible. While interactively walking through the specified environent's models the expectation are evaluated.

To replay real world data, at the upper right side a recorded test session from the physical environment can be loaded. Movements within the room and performed tasks are visualized accordingly at the left-hand side.

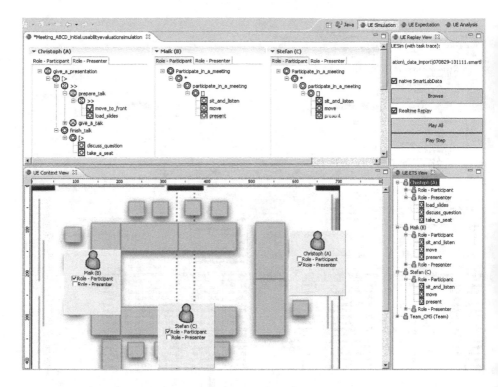

Fig. 3. Data Capturing

3.5 Analysis

After the testers have performed all tasks of the test case, the captured data has to be
analyzed. To cope with the vast amount of data, like sensor data, video streams or
annotations, we focus on results of the expectation evaluation. Figure 4 depicts the
current state of an ongoing implementation. The upper part depicts the actually fulfilled tasks of the testers as gantt timeline according to a task model at the right. The
views in the center allow interactively exploring captured data. Filtering options for
instance comprises the filtering for certain users, for tasks with very short or very long
durations and specific expectation results. Filtering for tasks which were performed
"bad" lists situations with major deviations from the expectected behavior, which
indicates candidates for usability issues. A subsequent investigation of video streams
and sensor data examines whether it is a real issue and to identify the cause. We try to
avoid that all captured data has to be examined again. Instead an expert can focus on
the automatically discovered issue candidates.

The suggested analysis has also some limitations. A prerequisite is a well defined
expectation. Otherwise real issues might be erroneously overlooked. The automatical
identification should only be a first step in analyzing evaluation results. A careful
investigation of captured data and even uncaptured details visible at the videos should
complement the presented approach.

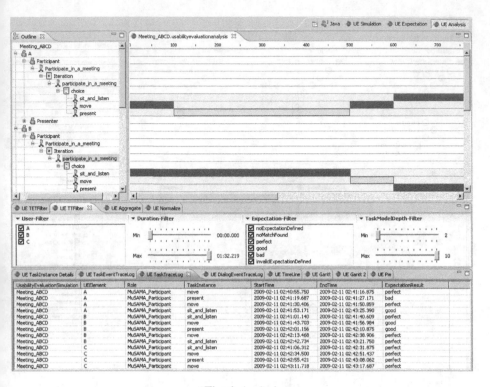

Fig. 4. Analysis

4 Conclusion

In this paper we have enhanced a task modeling notation with a language to express expectations. To ensure a better usability of the specification environment itself we have replaced the first prototypes' textual interface with a GUI to graphically compose expectations in a more convenient way. We have enhanced a task engine to evaluate these expressions during testing. The automatic identification of candidates for usability issues helps to efficiently evaluate more complex systems than supported by existing approaches. While other evaluation approaches only capture and display performed tasks, this paper presented a method to make the implicitly existing expectations explicit and exploit them for usability evaluation.

Future research avenues comprise the evaluation within a field study to discover strength and weaknesses of the approach and incorporate experiences gathered from real world data.

Acknowledgement

The work of the first author was supported by a grant of the German National Research Foundation (DFG), Graduate School 1424.

References

1. Buchholz, G., Engel, J., Märtin, C., Propp, S.: Model-Based Usability Evaluation - Evaluation of Tool Support. In: Jacko, J.A. (ed.) HCI 2007. LNCS, vol. 4550, pp. 1043–1052. Springer, Heidelberg (2007)

2. Hong, J., Landay, J.: WebQuilt: A Framework for Capturing and Visualizing the Web Experience. In: Proc. of the 10th international conference on World Wide Web, Hong Kong, China, pp. 717–724 (2001) ISBN:1-58113-348-0

3. OCL 2.0 specification of the OMG, http://www.omg.org/docs/formal/06-05-01.pdf

4. Malý, I., Slavík, P.: Towards Visual Analysis of Usability Test Logs. In: Tamodia 2006, Hasselt, Belgium, pp. 25–32 (2006)

5. Mori, G., Paternò, F., Santoro, C.: CTTE: Support for Developing and Analyzing Task Models for Interactive System Design. IEEE Trans. Softw. Eng. 28(8), 797–813 (2002)

6. Paternò, F., Russino, A., Santoro, C.: Remote evaluation of Mobile Applications. In: Winckler, M., Johnson, H., Palanque, P. (eds.) TAMODIA 2007. LNCS, vol. 4849, pp. 155–169. Springer, Heidelberg (2007)

7. Propp, S., Buchholz, G., Forbrig, P.: Task Model-based Usability Evaluation for Smart Environments. In: Forbrig, P., Paternò, F. (eds.) HCSE/TAMODIA 2008. LNCS, vol. 5247, pp. 29–40. Springer, Heidelberg (2008)

8. Hilbert, D., Redmiles, D.: Extracting Usability Information from User Interface Events. ACM Computing Surveys 32(4), 384–421 (2000)

9. Reichart, D., Forbrig, P., Dittmar, A.: Task Models as Basis for Requirements Engineering and Software Execution. In: Proc. of. Tamodia, Prague, pp. 51–58 (2004) ISBN:1-59593-000-0

10. Rubin, J.: Handbook of usability testing. In: Hudson, T. (ed.) Wiley technical communication library (1994)

11. Wurdel, M., Propp, S., Forbrig, P.: HCI-Task Models and Smart Environments. In: Proc. of HCIS 2008, Mailand, Italy (2008)

12. Zettlemoyer, L., Amant, R., Dulberg, M.: IBOTS: Agent Control Through the User Interface. In: International Conference on Intelligent User Interfaces (IUI), pp. 31–37 (1999)

Interaction Techniques for Binding Smartphones: A Desirability Evaluation

Umar Rashid and Aaron Quigley

School of Computer Science & Informatics,
University College Dublin, Ireland
{umer.rashid,aquigley}@ucd.ie

Abstract. This paper reports on the use of guided interviews to evaluate the desirability of different interaction techniques for *binding* smartphones. We demonstrate five interaction techniques using storyboard sketches and cardboard prototypes of iPhones. The participants highlight five words from a list of adjectives that best describe their experience with each technique. For comparative evaluation, we group the highlighted adjectives for all techniques into a list of nouns and let the participants rank each technique on a 5-point Lickert scale with respect to these nouns. We discuss the implications of these results for the design of interaction techniques for smartphones.

Keywords: Ubiquitous computing, spontaneous connection, smartphone, co-located collaboration, desirability evaluation.

1 Introduction

Ubiquitous computing envisions the seamless and spontaneous connection amongst computing devices prevalent in everyday life [8, 11]. To establish a connection between mobile devices that do not have *a priori* knowledge of each other's network addresses remains a challenging issue [6-12]. For the purpose of this work, we define *binding* as a way of coupling two devices by explicitly or implicitly creating a software plus network connection between them. Binding can relieve the users of the hassle of selecting the addresses/names of the devices in situations such as exchanging files or sharing photographs. The emergence of smartphones [1] has created opportunities to develop interaction techniques that bind these devices by taking advantage of their advanced features such as RFID, Bluetooth, GPS, infrared, and accelerometers. The established techniques that enable binding of devices by using physical action on part of the user include shaking [6, 9], bumping [3], touching [7, 10], pen-based stitching [4], and simultaneous button-pressing [12]. We provide an overview of these techniques in section 2 of this paper.

To date, the literature does not cite any significant activities on evaluating user satisfaction [13] with mobile binding techniques. Instead, much of the evaluation has been focused on effectiveness (can people complete the task?) and efficiency (how long do people take to complete the task?) of the said techniques. In this paper, we report on two user studies that make use of guided interviews to investigate the desirability of different interaction techniques for binding mobile phones. The first case

M. Kurosu (Ed.): Human Centered Design, HCII 2009, LNCS 5619, pp. 120–128, 2009.

study is aimed at evaluating desirability of individual techniques. It involves demonstration of techniques using storyboard sketches and paper prototypes of iPhones. For each technique, the participants highlight the top 5 adjectives in a word-list that best describe their experience with that technique. The results of this study are explained in section 3 of this paper.

As a follow-up to the first case study, we generate a list of nouns by grouping the top five adjectives selected by the participants for each technique. The participants rank each technique on a 5-point Lickert scale with respect to each noun in the list. The section 4 of the paper explains the results of this study focused on comparative desirability evaluation of these techniques. After discussing implications of our results for the design of mobile interaction techniques in section 5, we sum up our conclusions in section 6.

2 Related Work

Researchers have introduced interaction techniques to bind computing devices that do not recognize each other. Hickley et al. [3] explored the notion of *synchronous gestures* and introduced the technique of bumping two tablet computers equipped with touch sensors and two-axis linear accelerometers. Pen-based stitching [6] allows the users to couple pen-operated mobile devices with wireless networking by using pen gestures that span multiple displays. Holmquist et al. [6] implemented small embedded devices, called *Smart-It Friends*, that get connected when a user holds them together and shakes them. Expanding the idea of shaking further, Mcryhofer [9] demonstrated the coupling of two mobile phones while holding and shaking them simultaneously. Reikimoto et al. [12] introduced the "SyncTap" technique of simultaneously pressing and releasing a button on each device. Hardy et al. [5] presented an interaction technique in which a mobile phone can be touched with a large display at any position in order to establish pairing between two devices. Park et al. [10] introduced the use of intra-body communication signal [15] for touching and pairing devices.

Much of the effort in this domain has been focused on system design and implementation. On the other hand, usability evaluation of interaction techniques has received relatively little attention. Formal usability tests offer excellent tools to evaluate whether users can complete tasks (effectiveness) and how long they take to complete tasks (efficiency). However, such tests fail to measure intangible aspects of user experience (satisfaction) as often positive ratings for each question make it difficult to elicit candid or negative feedback [2, 13]. Desirability Toolkit [2] is an approach to measure satisfaction that requires the participants to sort through a series of 118 "product reaction cards" and select five cards that most closely describe their personal reaction to the system in use. The five selected cards then become the basis of a post-test guided interview. This approach has been shown to help elicit negative and critical comments from the participants and provide a better measure of desirability of the interface in test. An alternative implementation of this method is to use a simple paper checklist of adjectives [13] instead of a set of product reaction cards. This is the method we employ in our case study as it simplifies the process of sorting out the relevant words for the users.

3   Desirability Evaluation of Individual Techniques

The first part of the study deals with evaluating desirability of individual interaction techniques for binding smartphones. We included five techniques i.e. bumping, stitching, shaking, touching and simultaneous button-pressing (SBP) in this study. To give the participants an overview of the underlying mechanisms, we demonstrated these interaction techniques using cardboard prototypes of iPhones and storyboard sketches as shown in Fig. 1. We also showed them the storyboard sketches of usage scenarios that involve bindings between smartphones. With respect to each interaction technique, the participants selected a number of adjectives from a list of adjectives [13] and highlighted the top five of the selected adjectives that best described their experience. The study was conducted with 17 participants, 12 of them postgraduate students and 5 post-doctoral researchers, all in the Computer Science department. The participants comprised 14 males and 3 females in the age range of 20-35. On average, each participant spent 20-25 minutes on this part of the study. Each of them was given a small gratuity as thanks.

Fig. 1. a) Shaking b) Bumping c) Simultaneous Button Pressing d) Stitching e) Touching

3.1   Storyboard Sketching

The participants were shown storyboard sketches illustrating usage scenarios where spontaneous binding of smartphones may facilitate collocated collaboration, as follows:

Sharing. In this scenario, two users exchange digital business cards, files and photos from one mobile phone to the other, as shown in Fig. 2.

Control. There is a slave-master relationship between two mobile phones and the master mobile phone can be used to control the functions of slave mobile phone. For instance, a user can click a place on the map shown on master device and get its zoomed-in view on the slave device.

Pairing. Before coupling, each mobile phone shows a single-player map of Pacman game. After being coupled, the multi-player map of Pacman game appears on each mobile phone and the users can play game in multi-player mode on their respective phones.

Fig. 2. a) Before binding b) After binding

3.2 Guided Interviews

With respect to each interaction technique, each participant selected 5 words out of a list of 105 words. A subset of the word-list is shown in Table 1. Words selection was followed by a guided interview in which the participants explained their reasons to their selection of the words. The results of these guided interviews are explained here.

Table 1. Selected words from a list of 105 words

Accessible	Advanced	Ambiguous	Appealing	Awkward	Boring
Busy	Clean	Creative	Convenient	Efficient	Easy to Use
Frustrating	Fun	Hard to Use	Ineffective	Insecure	Misleading
New	Powerful	Professional	Reliable	Secure	Simple
Slow	Stable	Unrefined	Useful	Usable	Vague

Bumping. Bumping [3] involves striking the devices together just as clinking glasses together for a toast as shown in Fig. 1(b). With respect to bumping, the user responses are shown in Table 2.

Table 2. User responses with respect to bumping technique

Top 5 words	Selectors	Comments
Easy-to-Use	65% (11/17)	Handy approach, no complications
Fast	41% (7/17)	Ensures instant connection
Effortless	35% (6/17)	No hassle for configuring devices
Time-saving	30% (5/17)	Less time spent on connecting devices
Non-standard	24% (4/17)	Un-usual way of connecting

During the interview, many participants described bumping as a straightforward way of establishing instant connection between mobile devices. Some of them expressed their reservations about any possible physical damage to the mobile phones while using this technique. They were ambiguous about the extent of force that needs to be applied to accomplish coupling.

Stitching. Stitching [4] establishes connection between pen-operated mobile devices by using pen gestures that span multiple displays as shown in Fig. 1(d). This interaction technique recognizes devices within an arm's research and can support connection between the users sitting shoulder-to-shoulder. With respect to stitching, the user responses are shown in Table 3.

Table 3. User responses with respect to stitching technique

Top 5 words	Selectors	Comments
Difficult	35% (6/17)	Hard to use, hard to distinguish sender and receiver
Advanced	30% (5/17)	Innovative technique
Effortless	30% (5/17)	Easy to connect devices
Frustrating	30% (5/17)	Annoying, wearisome
Insecure	24% (4/17)	Prone to intrusion

During post-test interview, many participants expressed their security concerns about using this technique for spontaneous connections. They also found it cumbersome to draw pen strokes along two devices for data exchange.

Shaking. Shaking [9] is an interaction technique that enables connection between devices by holding and shaking them together as shown in Fig. 1(a). With respect to shaking, the user responses are shown in Table 4.

Table 4. User responses with respect to shaking technique

Top 5 words	Selectors	Comments
Simple	35% (6/17)	Does not require much learning
Creative	30% (5/17)	Innovative
Fun	30% (5/17)	Enjoyable, like shaking cocktail
Unattractive	24% (4/17)	Does not fascinate me
Awkward	20% (3/17)	Does not look an elegant way

Some participants find it a fun and innovative way of binding phones as it resembles to the act of shaking drinks for cocktail. However, others were apprehensive that the shaking can be unpredictable in certain situations such as two phones in a backpack may accidently get connected based on their accelerometer readings.

Touching. This technique [10] involves touching two devices and using the person's body as a medium for the signals between two devices, as shown in Fig. 1(e). It is based on Zimmerman's idea of intrabody signalling module [15] to connect devices. Table 5 shows the user responses with respect to touching technique.

Table 5. User responses with respect to touching technique

Top 5 words	Selectors	Comments
Simple	30% (5/17)	Does not require much learning
Effortless	30% (5/17)	Easy to connect devices
Time-saving	30% (5/17)	Saves time from manual configuration
Appealing	24% (4/17)	It is fascinating, I like the idea
Useful	24% (4/17)	Handy, Makes data exchange smooth and easy

Most participants find the idea of using human body as a medium for connecting two devices very appealing. They also considered it more professional-looking. However, there were some security concerns that such technique may be prone to undesirable intrusion.

Simultaneous Button Pressing (SBP). This technique involves simultaneous press and release of a button on each device [12] to establish connection as shown in Fig. 1(c). Table 6 shows the user responses with respect to SBP technique.

Table 6. User responses with respect to SBP technique

Top 5 words	Selectors	Comments
Awkward	35% (6/17)	Not an elegant way of connecting
Hard-to-Use	24% (4/17)	Difficult to synchronize
Slow	20% (3/17)	Take time to synchronize
Time-consuming	20% (3/17)	Take lot of time before making a correct choice
Dull	20% (3/17)	Not fun, no excitement

Some participants described it as awkward as both buttons need to be pressed simultaneously and it can be a big hassle to attain synchronicity between these actions. On the other hand, this technique was considered to be highly secure as the risk of intrusion seemed minimal.

4 Comparative Evaluation of Interaction Techniques

The guided interviews helped us elicit the factors that determine the desirability of interaction techniques for binding mobile devices. To conduct a comparative evaluation of interaction techniques, we grouped all the adjectives highlighted with respect to all techniques into a set of nouns. Taking assistance from the online "The Free Dictionary" (http://www.thefreedictionary.com/), we generated a list of nouns corresponding to the selected adjectives as shown in Table 7.

Table 7. Selected adjectives and corresponding nouns

Selected Adjectives	Corresponding Nouns
Easy to Use, Hard to Use, Effortless, Simple, Difficult	Ease of Use
Time-saving, Fast, Slow	Promptness
Appealing, Dull, Fun, Unattractive, Awkward, Frustrating, Useful	Appeal/Attractiveness
Advanced, Creative	Originality
Unpredictable	Reliability
Insecure	Security

We implemented a Wizard-of-Oz [14] application on the iPhone 3G, using the iPhone SDK that gives the impression of establishing a Wi-Fi connection between two iPhone devices when subjected to any of the aforementioned interaction techniques. The participants were asked to rate each of the interaction techniques with respect to the nouns (shown in Table 7) on a 5-point Lickert scale, with 5 being the highest and 1 being the lowest. Out of 17 users who participated in the first case study, 11 agreed to volunteer for the follow up study. The results of this study are shown in Table 8.

Table 8. Comparative Desirability of different interaction techniques

Features Techniques	Ease of Use	Security	Promptness	Appeal	Originality	Reliablity
Bumping	3	4	4	3	4	3
Stitching	3	3	3	2	3	3
Shaking	4	3	3	3	3	3
Touching	4	3	4	4	4	3
SBP	2	4	3	2	2	2

As shown in the table above, shaking and touching rank highest as far as ease of use is concerned. Bumping and SBP score highest in matter of security. Promptness is considered best realized in bumping and touching. As the techniques that score low in ease of use and promptness score high in security column, satisfying these apparently conflicting demands poses a key challenge for designers. Overall, touching technique receives the highest score amongst all techniques.

5 Design Implications

Based on our case study results, we believe certain safeguards should be provided while designing interaction techniques for binding smartpones. Although our participants represent a limited sample of smartphone users and, hence, we are not in a position to generalize the results and come up with an exhaustive list of implications. However, we can still deduce some useful guidelines for the design of mobile binding techniques.

5.1 Purpose of Binding

The desirability of a particular interaction technique may vary depending on the purpose of binding the mobile devices. We presented to the participants storyboard sketches showing different usage scenarios of file exchange, map zoom-in, and multi-player games. Although most participants rated each technique considering its all-purpose utility, some of them opted to select different adjectives for the same technique in different usage scenarios. For instance, they particularly preferred the touching technique for multi-player games and stitching for exchanging files. This underlines the need to assess the intended purposes of binding as well as to allow support for multiple binding techniques in a smartphone.

5.2 Social Context

Some participants mentioned social context as an important determinant for their preference for a particular technique. While exchanging business cards with a stranger, one may opt for SBP technique but while sending a photograph of wedding anniversary to one's spouse on a candle-light dinner, touching or shaking may be more attractive. This also raises the significance of considering cultural norms and social protocols (e.g. corporate events vs. social events), in addition to interpersonal relationships, while designing techniques for spontaneous device connection.

5.3 Privacy

Since mobile phones are very personal devices, containing lot of personal information and contacts, most users express their serious concerns about risks of undesired intrusions while making spontaneous connections. As shown in Table 8, interaction techniques that lead in the matters of ease of use, appeal and promptness lag behind in terms of security. To balance security with the quality of human experience is one of the key challenges in the design of these techniques.

6 Conclusions

In this paper, we described the use of guided interview method for evaluating user satisfaction with different interaction techniques for binding smartphones. We went through an exploratory phase of collecting the top 5 adjectives that best express the users experience with respect to each of the interaction techniques i.e. bumping, stitching, shaking, touching and simultaneous button pressing. After the evaluation of individual interaction techniques, we summed up the criteria for comparative evaluation by grouping all the highlighted adjectives into a list of nouns and going through guided interviews with the participants again. As shown by our study, the ease of use, security, promptness, appeal, originality and reliability are the key factors that determine the desirability of any binding technique. We also discussed the implications of purpose of binding, social context and privacy concerns on the design of interaction techniques.

In future work, we plan to build and evaluate collaborative applications on mobile phones that utilize these interaction techniques. We are also interested in exploring the social and cultural aspects that affect the desirability of interaction techniques.

Acknowledgments. This research is supported by Irish Research Council for Science, Engineering and Technology (IRCSET): funded by the National Development Plan, and co-funded by IBM.

References

1. Ballagas, R., Rohs, M., Sheridan, J., Borchers, J.: The Smart Phone: A Ubiquitous Input Device. IEEE Pervasive Computing 5(1), 70–77 (2006)
2. Benedek, J., Miner, T.: Measuring Desirability: New Methods for Evaluating Desirability in a Usability Lab Setting. In: Proc. UPA 2002 (2002)
3. Hinckley, K.: Synchronous gestures for multiple persons and computers. In: Proc. UIST 2003, pp. 149–158 (2003)

4. Hinckley, K., et al.: Stitching: Pen Gestures that Span Multiple Displays. In: Proc. AVI 2004, pp. 23–31 (2004)
5. Hardy, R., Rukizo, E.: Touch & Interact: Touch-based Interaction of Mobile Phones with Displays. In: Proc. Mobile HCI 2008, Amsertdam, Netherlands (2008)
6. Holmquist, et al.: Smart-its friends: A technique for users to easy establish connections between smart artifacts. In: Abowd, G.D., Brumitt, B., Shafer, S. (eds.) UbiComp 2001. LNCS, vol. 2201, pp. 116–122. Springer, Heidelberg (2001)
7. Iwasaki, Y., Kawaguchi, N., Inagaki, Y.: Touch-and-Connect: A connection request framework for ad-hoc networks and the pervasive computing environment. In: Proc. PerCom 2003, pp. 20–29 (2003)
8. Kindberg, T., Zhang, K.: Secure Spontaneous Device Association. In: Dey, A.K., Schmidt, A., McCarthy, J.F. (eds.) UbiComp 2003. LNCS, vol. 2864, pp. 124–131. Springer, Heidelberg (2003)
9. Mayrhofer, R., Gellersen, H.: Shake well before use: two implementations for implicit context authentication. In: Krumm, J., Abowd, G.D., Seneviratne, A., Strang, T. (eds.) UbiComp 2007. LNCS, vol. 4717, pp. 72–75. Springer, Heidelberg (2007)
10. Park, D.G., et al.: Tap: Touch and Play. In: Proc. CHI 2006, pp. 677–680 (2006)
11. Pering, T., Ballagas, R., Want, R.: Spontaneous marriages of Mobile Devices and Interactive Spaces. Communications of the ACM 48(9), 53–59 (2005)
12. Rekimoto, J., et al.: SyncTap: An Interaction Technique for Mobile Networking. In: Proc. Mobile HCI 2003, pp. 104–115 (2003)
13. Travis, D.: Measuring satisfaction: Beyond the usability questionnaire, http://www.userfocus.co.uk/articles/satisfaction.html
14. Wilson, J., Rosenberg, D.: Rapid prototyping for under interface design. In: Handbook of Human-Computer Interaction, pp. 859–875 (1988)
15. Zimmerman, T.G.: Personal area networks: near-field intrabody communication. IBM Systems Journal 35(3-4), 609–617 (1996)

A Usability Inspection of Medication Management in Three Personal Health Applications

Katie A. Siek[1], Danish Ullah Khan[1], and Stephen E. Ross[2]

[1] University of Colorado at Boulder, Department of Computer Science,
430 UCB, Boulder, Colorado, 80309-0430 USA
{ksiek,Danish.Khan}@colorado.edu
[2] University of Colorado Denver, Division of General Internal Medicine,
MS B180, Aurora, Colorado, 80045 USA
Steve.Ross@ucdenver.edu

Abstract. We present the findings of a cognitive walkthrough inspection on three Personal Health Applications (PHAs). Two of the PHAs, Google Health and Microsoft HealthVault, are general purpose PHAs that are freely available to the general public. The last PHA, Colorado Care Tablet, is a prototype PHA that was designed specifically for older adults to manage their medication information. Older adults need a way to manage medications and share this information with their caregivers and healthcare providers to avoid complications during transitions of care. PHAs provide people with the ability to collect and share health information. However, given the problems older adults have with navigating applications and web pages, we needed to inspect currently available PHAs and identify problems older adults may have when using them for medication management before conducting user studies. Based on our findings, we encourage the design community to place more of an emphasis on interface consistency and tightly coupling information with links.

Keywords: Usability Inspection Methods, Cognitive Walkthrough, Personal Health Applications, Personal Health Records.

1 Introduction

In 2006, we set out to develop a Personal Health Application (PHA) that could provide older adults an easy way to manage their medications during transitions of care. Transitions of care are broadly defined as seeing a new healthcare provider to being discharged after a long hospital stay. Older adults during transitions of care are particularly at risk for medication errors [2]. PHAs can empower patients and improve the information flow between patients, caregivers, and health professionals, however we believe that the current design of PHAs is inadequate for older adults. The design of our PHA, the Colorado Care Tablet (CO Care Tablet), was informed by a successful paper-based Personal Health Record (PHR) transitions intervention that improved the quality and safety of transitional care [2]. During the development of the CO Care Tablet, Microsoft and Google developed general purpose, web-based interoperable PHRs that provided third party developers a way to connect with their respective PHR

M. Kurosu (Ed.): Human Centered Design, HCII 2009, LNCS 5619, pp. 129–138, 2009.
© Springer-Verlag Berlin Heidelberg 2009

repositories. We did not want to reinvent the wheel by developing yet another PHA, however we wondered if Microsoft HealthVault and Google Health could meet the medication management needs of older adults given the issues older adults have with manual dexterity [3] and computer tasks [8].

The aim of this paper is to identify usability issues older adults may have using the CO Care Tablet, Google Health, and Microsoft HealthVault and share design recommendations with the usability community. To this end, our team conducted cognitive walkthroughs on each PHA with seven common medication management tasks using an older adult persona informed by our own research and the Association for the Advancement of Retired Persons (AARP) (Section 2). We found that: the most common tasks took less than five steps to complete; layouts with three columns had more usability issues; and usability issues were caused by inconsistent linking of data with the cause or action (Section 3).

2 The Study

We completed usability inspections on three PHAs: Microsoft Health Vault, Google Health, and Colorado Care Tablet. Here we discuss the rationale for PHA, inspection method, and task selection.

2.1 The PHAs

In October 2007, Microsoft launched HealthVault (http://healthvault.com) to provide users a centralized place to track health information, prepare for health professional visits, manage health-related devices (e.g., participating blood pressure monitors), and utilize third party health services to share information and receive personalized feedback (e.g., physical activity recommendations). Google Health (http://www.google. com/health) offered a PHA in May 2008 to provide Google users the ability to create online health profiles, import medical records from participating hospitals and pharmacies, find authoritative information about health issues, search for health providers, and connect to third party health services. The major differences between these systems is that Microsoft HealthVault emphasizes the ability to connect health devices to the PHA for better data tracking and personal feedback, whereas Google Health emphasizes the ability to connect to third party services (e.g., hospitals, pharmacies) and find authoritative information about issues and providers.

The CO Care Tablet is a prototype system developed as part of the Robert Wood Johnson Project HealthDesign initiative. We spent six months conducting a needs assessment with older adults and caregivers to inform the interface design. The next nineteen months were dedicated to iteratively implementing and evaluating low and high fidelity prototypes with older adults. We found older adults were more comfortable with wizard-based interfaces that lead them step-by-step with description instructions through tasks. The current prototype provides users the ability to: create medication lists; share medication lists with healthcare providers; find authoritative

Table 1. Task Complexity for common medication management tasks. The log in task was not inspected.

Task	Google Health	Microsoft HealthVault	CO Care Tablet
Log In	3	4	2
Add a Medication	7	5	7
Delete Medication	4	6	4
View Medication List	2	3	1
Find information on a Medication	4	5	4
Share Medication List	N/A	4	4
Medication Interactions	9	N/A	N/A
Two different Doses of Same Medication	24	10	14

information about medications; schedule medication reminders; send questions about their health to providers; and identify health conditions that require immediate medical attention (e.g., a fever over a specified degree). Since the goal of the CO Care Tablet was to minimize adverse medication related incidents through medication management, we spent significant amount of time decreasing the input intense nature of medication list management. We provided users with multiple input mechanisms, such as: (1) selecting medications from pharmacy dispense records; (2) scanning barcodes on medications; (3) typing in prescription numbers on medications; and (4) typing in the medication name and verifying the medication by looking at a picture of it. For this study, we chose the most input intensive method, typing in a medication name and verifying the medication by looking at a medication image, that most closely mirrored the input method used by the other two PHAs.

Although Microsoft HealthVault and Google Health are general purpose PHAs and not specific to medication management, we chose them because they are freely available to the target population and could be utilized now for medication management, whereas the CO Care Tablet is still in prototype phases and needs further development in the areas of authentication and interoperability before it is released for general use.

2.2 Inspection Method

We used the Cognitive Walkthrough (CW) usability inspection method because it thoroughly evaluates tasks based on the theory of exploratory learning [10]. CW is not without its critics, however, who have argued that CW is time consuming [7], tedious [7, 11], and inconsistent at finding usability issues [5, 7]. We continued to use CW because it is a task-driven methodology and we wanted to inspect seven specific tasks shown in Table 1. In addition, CW has been used to inspect many health systems [9].

The first two authors are similarly trained in CW and used the four metrics proposed by Wharton [13] to guide the inspection: (1) match to intent; (2) visibility; (3) labeling; and (4) indication of progress. We developed a persona based on our own needs assessment and personas developed by the AARP [1]. We assumed the older adult could use the computer and input devices (e.g., a mouse), although we understand this is a limiting factor for some older adults [6]. The older adult did not use the Internet much, but was willing to use computers to manage their medication information. We chose the tasks in Table 1 based on our needs assessment that identified the tasks older adults most wanted to do when managing their medications. The two evaluators separately conducted CWs on each PHA in one sitting and then compared results. Inconsistencies in categorization or usability issues were discussed until a consensus was met. The evaluators found discussing the thought process and reminding each other of the persona was especially helpful when discussing inconsistencies in inspections.

3 Findings

Here we present abstracted findings from the cognitive walkthrough on three PHAs. The key findings are:

- Tasks older adults want to utilize most often in managing medication lists had the least amount of steps to complete them.
- Layouts with three-columns had more usability issues than a two-column layout for PHA information.
- Visibility of medication lists and linking tasks to medications were common medication list management usability issues.
- Warning and confirmation screens were inconsistent for all of the PHAs.

3.1 Task Complexity

We show a basic measure of task complexity for each PHA in Table 1 by counting how many actions (e.g., clicks) a user would have to do to complete the eight selected medication management tasks. The happiest path for each of these tasks was developed by referencing the PHA help guides. Previous studies [9] have used actions and screen transitions as a metric for task complexity, however calculating screen transitions was difficult for these web-based PHAs because each uses overlays and pop-ups to convey information to the user that would not otherwise be defined as a screen transition. Each PHA inspected had at least one task that was not supported. Google Health did not have the functionality to share medication lists, although there was an option to save and print the lists, we were specifically looking for an electronic method to share medication lists. Users had to use a third party application on Microsoft HealthVault to check medication interactions, thus it was not part of the HealthVault PHA we were evaluating. The CO Care Tablet did not support checking for medication interactions.

Each PHA or browser had an option to save log in information, thus although log in steps are counted here for completeness, we assumed that the person who set up the PHA would save the log in so the older adult could easily access the information. Although this brings in many questions about information security and privacy, the difficulties with creating PHA accounts and logging in are beyond the scope of this paper and will not be discussed in the inspection findings.

Tasks that older adults wanted to use the most often [4], such as viewing lists, sharing lists, and getting more information about medications all took less than five actions for each PHA to complete. Unfortunately, as we will report later, these tasks had some usability issues (Section 3.3-3.5). Adding and deleting medications required more actions depending on how much automation each PHA used - discussed more in Section 3.3. Adding two different doses for the same medication had the most steps for each PHA because it required adding a medication twice, editing the dose in most cases, and verifying the doses.

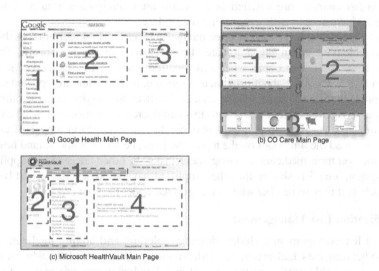

(a) Google Health Main Page

(b) CO Care Main Page

(c) Microsoft HealthVault Main Page

Fig. 1. Overview of PHA main page layouts with boxes around navigation areas

3.2 Information Layout

The two most common issues, visibility and match to intent, we identified when inspecting the PHAs were largely problematic because of the information layout and navigation areas. Labeling was the third most common issue and was usually linked to match to intent issues. The CO Care Tablet had the most indication of progress issues because, as discussed in Section 2.1, the designers used a wizard-based layout that required more navigation indicators than the other two PHAs.

Google Health and Microsoft HealthVault both used a three-column layout shown in Figure. 1 and had the most usability issues identified in the inspections (63 and 51 respectively). Google Health duplicated succinctly worded navigation links in columns 1 and 3 (Fig. 1a). Users had to look in column 3 to verify when actions were completed (e.g., adding a medication). We reasoned that older adults would have problems identifying the navigation areas, understanding the duplicated navigation links, and verifying the subtle changes based on their actions (e.g., when medications were added, they appeared with yellow highlighting in column 3) because of the succinct wording and abundance of information on one page.

Microsoft HealthVault had four navigation areas and a three-column format in the main area of the site. Navigation areas 1 and 3 duplicated information (Fig. 1c), however navigation area 1 had succinct links and navigation area 3 had more descriptive links. From our research [12], we found that older adults want more instructions and descriptive links, thus this layout would meet the needs of older adults and experienced users who want quick access to information. A possible area of confusion for older adult users would be the labeling of some of the navigation links. For example, adding a medication is categorized under the *Add, view, or edit information* link or the

Health info tab which should not be confused with the *View and update profile* link that brings the user to a general account updating page. In other areas of the application, navigation 3 becomes the main area and navigation are 4 is eliminated. All medication list changes are verified in the main area, navigation area 3, thus older adults would have an easier time verifying the updates.

The CO Care Tablet has a total of three navigation areas and 44 identified usability issues. A possible issue is that navigation area 1 only customizes the data shown in navigation area 2 (Fig. 1b). For example, when a medication is selected in the medication list in navigation area 1, the links about drug facts, schedule, and deletion are updated to name the appropriate medication in navigation area 2. Medication scheduling is the only duplicated link in navigation areas 2 and 3, thus this can decrease confusion on what each link does. However, since navigation area 2 is dependent on navigation area 1, if an older adult does not read the directions on the top of the page, he will not understand how to do many of the common medication management tasks. In other areas of the application, only navigation area 3 is shown, thus the simplified navigation scheme would be easier for older adults if they remember what is categorized under the four links.

3.3 Medication List Management

Medication list management includes viewing, deleting, and adding medications. All of the medication lists had problems with visibility. When the lists were long and went off the viewable portion of the page, it is debatable among our team if the older adult would have enough knowledge to scroll to view the rest of the list. Each PHA had a different way to order medications – Google Health ordered medications alphabetically; Microsoft HealthVault ordered medications in the order they were added with the newest on top of the list; and CO Care Tablet ordered medications in the order they were added, but with the newest on the bottom of the list. The CO Care Tablet created custom scrolling arrows to prominently display scrolling, shown in Fig. 1b, however it is unclear what part of the interface these arrows belong to.

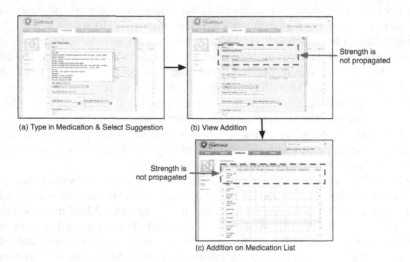

(a) Type in Medication & Select Suggestion   (b) View Addition

Strength is not propagated

Strength is not propagated

(c) Addition on Medication List

Fig. 2. Adding a medication in Microsoft HealthVault

Users must be able to view the medication they want to delete on the list in all three PHAs, thus this viewing medication list issue can pose a significant management issue. Google Health and CO Care Tablet have the delete link closely tied to the medication list. Indeed, Google Health has the delete link on the same table row as the medication listed. As discussed in Section 3.2, CO Care Tablet has the delete link connected to another navigation area, but the medication name and picture of the medication is shown near the delete link. Microsoft HealthVault had the most actions required to delete a medication because they had a list of check boxes near each medication where a user selects the medications to delete and then presses the delete button on top of the list. Although the HealthVault design provides users the ability to delete multiple medications at a time, we believe older adults would not have the knowledge necessary to connect check boxes to a delete function at the top of the page.

All three PHAs utilized different services to suggest medication spellings to decrease the complexity of medication additions. The information in these services is not always utilized completely by Microsoft HealthVault and Google Health. For example, Microsoft HealthVault provided users the ability to select the name and strength of the medication (Fig. 2a), but did not propagate the strength field with what was selected (Fig. 2b&c). This could lead to confusion of older users since they already selected the strength, but have to input it again. In addition, if a mistake is made in either selection or strength input, the older adult could become confused on what the strength is suppose to be. Google Health decreased the size of the suggested medications by only providing users with suggestions of medication spelling and *How to Take* (e.g., by mouth). This design decision increased the amount of input needed to complete a medication addition if strength and form were needed because users would have to edit each medication on the list.

In contrast, the CO Care Tablet PHA utilized a wizard configuration, shown in Fig. 3., where users selected the medication, strength, and look of the medication to add. All of the information selected was then represented in the appropriate area of the medication list. This method had more steps and problems with definite indications of progress, however it decreased the amount of user input with suggestions at each step.

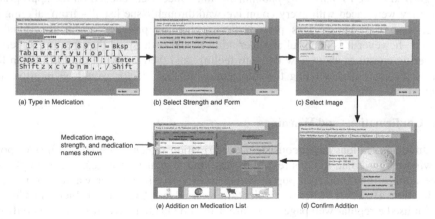

(a) Type in Medication   (b) Select Strength and Form   (c) Select Image

Medication image, strength, and medication names shown

(e) Addition on Medication List   (d) Confirm Addition

Fig. 3. Adding a medication in CO Care Tablet

3.4 Medication Information

A user could easily get information about a specific medication in the Google Health and CO Care Tablet PHAs. Similar to deleting a medication, both PHAs had medication information closely tied to the medication name. Only CO Care Tablet had an issue with labeling for getting information on a medication because of a change in terminology – *<Medication Name> Drug Facts* instead of *<Medication Name> Medication Facts*. Despite this labeling issue, our team thought that users would be able to receive medication information easily in either PHA.

There was no obvious way to get medication information using the Microsoft HealthVault PHA. A work-around we used for this task was to search for the medication name in the search box. It is unclear if the search component of the PHA was a third party application because of all of the advertisements shown when a term was searched. This method to find medication information is not intuitive because the term *search* does not answer medication related questions (e.g., What are some side effects of this medication?) that prompt older adults to look for *more information* about a medication. In addition, the search function created a new window or tab depending on the browser configuration. Thus, the user would have to understand that they are no longer in the PHA window and navigate back once the appropriate information was found.

3.5 Sharing Medication Lists

Microsoft HealthVault was the only application that prominently displayed sharing functionality. A sharing link was always shown in the tab menu (Fig. 1c, navigation area 1) and was displayed at the top of the medication list (Fig. 2c). In addition, the application described different sharing levels (view, view and modify, and custodian). The team agreed that although older adults probably would not set-up sharing, it would be easy enough to find and understand how to use for highly motivated older adults or caregivers.

CO Care Tablet provided users with the ability to share medication lists with their healthcare providers, however it was difficult to identify how to do this since it was categorized in the *Prepare for Appointment* link. Once users clicked on the appropriate link, they were faced with every cognitive walkthrough issue listed – from labeling to indications of progress. This part of the application provides useful functionality to users, however it is doubtful an older adult would understand that creating a memo, answering questions about how they are feeling, and then verifying the medication list and symptoms before sending the information to their doctor would be intuitive for older adults.

3.6 Confirmations and Warnings

All three applications had inconsistent confirmation screens – utilizing a mix of pop-up windows, overlaid screens (e.g., Fig. 2a), and regular web pages. For example, all three applications utilized a pop-up window to verify if the user wanted to delete a medication. However, when a medication was added, CO Care Tablet and Google Health used a regular web page. CO Care Tablet dedicated the entire page to confirming the medication addition, whereas Google Health only added the medication to the navigation area 3 in Fig. 1a. The Google Health method was incredibly confusing when adding two medications with different strengths because instead of listing the medication twice in navigation area 3, the interface only highlighted one instance of the medication name. Microsoft HealthVault utilized an overlaid screen when adding a medication and then showed a regular web page to verify the addition. The inconsistent confirmation screens and buttons could confuse older adults.

Google Health was the only application that had a built in medication interaction functionality. Similar to the confirmation interface, when a possible interaction (e.g., adding two blood thinner medications that could be dangerous) was on a user's medication list, a small red circle with a white exclamation point icon appeared near the *Drug Interactions* link (Fig. 1a, navigation area 1). Users could get more information about the drug interaction by clicking on the link, however no warnings were prominently shown on the medication list page or on the medication confirmation area. The subtle warning may not be visible to older adults. In addition, the information provided on the interaction warning page simply stated, *"Requires immediate attention"* – but did not give an indication of what kind of attention or action was required.

4 Summary

In this paper we have presented the results of a CW on three PHAs with an emphasis on medication management tasks. Although these results are from a usability inspection method and thorough user testing is necessary to identify more usability issues, they provide a basis for improvements to PHAs. Based on our findings, we encourage the design community to place more of an emphasis on interface consistency and tightly coupling information with links. For older adults to effectively user a consistent interface, we must provide them seamless interactions (e.g., no transitions between new windows or tabs) and uniform interface components (e.g., choose pop-ups, overlays, or web pages, but not a subset of these). In addition, we must consider how information is presented to older adults and ensure the actions (e.g., delete a medication, find information on a medication) is intuitively linked to the target item (e.g., the medication). If we can improve on these items, PHAs could be more usable for older adults and improve medication management during transitions of care.

Acknowledgments. This project was supported by Robert Wood Johnson Foundation as part of Project HealthDesign RWJ 59880 (PI Stephen E. Ross).

References

1. Chisnell, et al.: New Heuristics for Understanding Older Adults as Web Users. Technical Communication, 53, Society for Technical Communication, pp. 39–59 (2006)
2. Coleman, E.A., et al.: Preparing patients and caregivers to participate in care delivered across settings: the Care Transitions Intervention. J. Am. Ger. Soc. 52, 1817–1825 (2004)
3. Galganski, M.E., et al.: Reduced control of motor output in a human hand muscle of elderly subjects during submaximal contractions. J. Neurophysiol. 69, 2108–2115 (1993)
4. Haverhals, L., et al.: Older Adults with Multi-morbidity: Medication Information Needs and Personal Health Applications (PHA) Design Implications (in Review)

5. Hertzum, M., Jacobsen, N.: The Evaluator Effect: A Chilling Fact About Usability Evaluation Methods. International Journal of Human-Computer Interaction 13, 421–443 (2001)
6. Jacko, J., et al.: Older adults and visual impairment: what do exposure times and accuracy
   tell us about performance gains associated with multimodal feedback? In: CHI 2003: Proceedings of the SIGCHI conference on Human factors in computing systems, Ft. Lauder-
   dale, Florida, USA, pp. 33–40. ACM, New York (2003)
7. Jeffries, R., et al.: User interface evaluation in the real world: a comparison of four techniques, pp. 119–124. ACM, New York (1991)
8. Siek, K.A.: Mobile Design for Older Adults. In: Handbook of Research on User Interface
   Design and Evaluation for Mobile Technology, IGI, pp. 624–634 (2008)
9. Kaufman, D.: Usability in the real world: assessing medical information technologies in
   patients' homes. Journal of Biomedical Informatics 36, 45–60 (2003)
10. Lewis, C., et al.: Testing a walkthrough methodology for theory-based design of walk-up-
    and-use interfaces, pp. 235–242. ACM, New York (1990)
11. Spencer, R.: The streamlined cognitive walkthrough method, working around social con-
    straints encountered in a software development company, The Hague, The Netherlands,
    pp. 353–359. ACM, New York (2000)
12. Ross, S.E., et al.: Colorado Care Tablet and My Medi-Health: Personal Health Applica-
    tions for Medication Management (in Review)
13. Wharton, C., et al.: The cognitive walkthrough method: a practitioner's guide, pp. 105–
    140. John Wiley & Sons, Inc., Chichester (1994)

Designing a Lighting with Pleasure

Tyan-Yu Wu[1], Wen-chih Chang[2], and Yuan-Hao Hsu[1]

[1] Department of Industrial Design, Chang Gung University, Tao-Yuan, Taiwan
[2] Graduate School of Design, National Taiwan University of Science and Technology, Taipei, Taiwan
tnyuwu@mail.cgu.edu.tw, wchang@mail.ntust.edu.tw, eddie0614@gmail.com

Abstract. Lighting plays an important role in the enhancement of atmosphere in a house. It provides not only a luminous function, but also experiencing pleasure in the space. This study investigated the type of pleasure and its factors towards lighting. An interview was conducted to collect the responses of pleasurable feelings from 10 participants. From 250 images, 10 were extracted as stimuli for the interviews. Data analysis was used to group key sentences obtained from the responses and the results produced 7 factors which could be categorized into four types of pleasure: Appearance, Interactive, Reflective, and Novelty Pleasure. Among them, the responses related to appearance pleasure were mentioned most frequently and could elicit a consumer's pleasure, which also confirm Creusen's theory. The four pleasures can associate with Jordan and Norman's pleasure/ emotion. Particularly mentioned, novelty pleasure is distinguished from other two theories. Designer can utilize four types of pleasure in designing a lighting with pleasure as possible.

Keywords: Pleasure, Lighting design, Pleasurable product.

1 Introduction

Aristotle stated that more than anything else, men and women seek happiness. Seligman [1], the former President of the American Psychological Association (APA), has also claimed that, instead of studying only the ills of the human mind, as has been the case in past psychological research, we now should not neglect the role of happiness in human psychology and should be more aggressive in the study of positive emotions in order to help people pursue a happier life. His statement suggests the importance of happiness i.e. product pleasure in this thesis. To expend this concept to consumers, we believe that consumers should have right to seek a product, which can fulfill their desire, evoke pleasure and further bring happiness to their life. Being a designer, we should design a good product, which allows a user to have them with pleasure.

According to an article in *New York Times Good Health Magazine,* $ 15.5 million was spent on light therapy experiments in the United States by the National Institute of mental Health in 1991. The result of the study demonstrated that color and light can affect people's mood positively [2]. The similar result also showed in an American psychological report. The study illustrated that a pleasurable working environment can

M. Kurosu (Ed.): Human Centered Design, HCII 2009, LNCS 5619, pp. 139–146, 2009.
© Springer-Verlag Berlin Heidelberg 2009

improve working performance, increase problem solving skill, and harmonize relationship among people. Extensively, it is believed that a pleasurable product should also increase the life quality in terms of pleasurable using experience, since users always have an intimate relationship with light and environment. But, what are the factors, which possible affect users' pleasure and can be utilized to design. The current study attempts to identify these factors and types of pleasure.

The Oxford English Dictionary defines pleasure as 'the condition of consciousness or sensation by the enjoyment or anticipation of what is felt or viewed as good or desirable; enjoyment, delight, gratification' [3]. Jordan stated that pleasure with products can be defined as: the emotion, hedonic and practical benefits associated with products' [4]. In this paper, a pleasurable lighting is defined as one that elicits a consumer's pleasure by just looking at its appearance. The results of the study hopefully could provide designers and market people a very basic understanding of designing lighting with pleasure.

2   Why We Need a Pleasurable Product/Lighting?

Traditional lighting design mainly focuses on users' basic needs including physical function, luminance quality and its appearance. Nowadays, designing a good product, a designer has not only to create a product with good function and ergonomic, but also to understand consumers' emotion responses in order to embed these pleasurable elements to the product for pleasing users.

Desmet and Overbeeke [5] believed that product with a positive emotion/ pleasure can add extra value to the product and further benefit users. To understand product emotion, Norman [6] has commented that people are emotional and social creatures and has mentioned the role of human emotions as an influencing factor in the way people deal with and relate to objects and artifacts. To design emotional products, Norman [6] suggested to think of three levels of product emotions: visceral, behavioral, and reflective. Indeed, consumers need the product with these emotion levels as possible. A good product can help users to accomplish many tasks in everyday life. To achieve these tasks, it requires a product with a good perform. But, it is often end up with a very bore working process. It is therefore how to increase a product emotion becomes a challenge to a designer, after product function and ergonomic were accomplished well.

From marketing point of view, Jordan [4] and Marzano [7] also agreed that a product with emotional benefit is very important [8]. Noted this trendy, manufactures tend to increase product value by adding emotion into the product. As Csikszentmihalyi and Rochberg-Halton [9] comments, 'household objects' are crucial for experiencing pleasure in the home. Lighting manufacturer/ designer should also realize that a good atmosphere space rely onto have a pleasurable light.

3   Jordan's Theory

A Canadian anthropologist, Tiger [10] identified four types of pleasure: physio-pleasure, socio-pleasure, psycho-pleasure, and ideo-pleasure. Extensively, Jordan [4], based on Tiger's framework classified a products' pleasure into four categories. The first category contains products with 'physio-pleasure' which involved sensations, for example, touching or seeing something. The second contains products with 'socio-pleasure'. In this category, products can facilitate social interaction such as chatting with friends. For instance, a mobile phone can provide socio-pleasure by using it to chat with friends anywhere, anytime. The third is 'psycho-pleasure'. In terms of products, this relates to the cognitive demands and emotional reactions engendered through experiencing the product. For instance, a good icon interface on computer screen allows users to understand its meaning easily, in which evokes their psycho-pleasure when using it. The fourth is 'ideo-pleasure' which in the context of products relates to the aesthetic nature of a product and the values that it embodies. For instance, a product made with recycled material can evoke a user's ideo-pleasure, if she/he agrees with the importance of environmental issues. In general, these pleasures are mixed within the products. Users can perceive one or two, maybe more types of pleasure at once. For instance, a person's physio-pleasure may be evoked directly through perceiving a product with an elegant shape and glamorous material, as with the iMAC computer. By sharing this modern and professional-looking product with friends, the person maybe intends to show off his or her good taste and value. During this process, he or she experiences both physio- and socio-pleasure. Like many consumer products, a light product is very important to embed pleasurable elements into design in order to provide a great atmosphere for the improvement of users' living quality.

4 Norman's Theory

To design emotional products, Norman [6] stated that there are three levels of emotional response: the visceral level, the behavioral level, and the reflective level. Regarding product design, the 'visceral level' is about the initial impact of a product; its appearance, touch and feel. The 'behavioral level' is about use and experience with a product. The 'reflective level' is about the interpretation, understanding and reasoning of a product in terms of self-image, personal satisfaction and memory [6]. These three levels of product emotions basically cover whole product emotions, possibly happened to the consumers. It is therefore these three levels of emotions were applied to the data analysis as reference.

5 Research Aim

The aim of this study purposes to identify factors through designing a pleasurable light and to explore how users perceive lighting.

6 Methods

In the experiment, lighting image was used as a sample to study its pleasure. The empirical test covers an interview to collect un/pleasurable factors and questionnaire survey to confirm users' perception to the lighting. The analysis, synthesis and discussion were made as follows.

6.1 Interview

Stimuli
The stimuli chosen were limited on lighting, which include desk lighting and ceiling lighting. 250 images were collected from variety of resources such as design competition awards, magazines, catalogues, and website. They were outputted on 3x5 inch photo paper. To reduce the sample size, two designers were firstly asked to sort them and withdraw 74 images with similar or repeated style. 76 images were used later in the test. Upon commencement, the Oxford English Dictionary and Jordan's definition of 'pleasure' towards product mentioned at introduction was explained to 10 participants (i.e. four professional designers and six graduate students). To identify the representative product images, they were asked to group them into three categories: 'pleasure', 'displeasure' and 'neutral' after looking at all 76 product images. The top ten images from pleasure group were selected as stimuli for next interview and questionnaire survey (see Table 1).

Table 1. 10 pieces of pleasurable lights

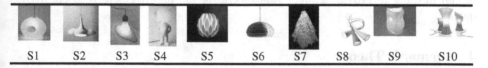

| S1 | S2 | S3 | S4 | S5 | S6 | S7 | S8 | S9 | S10 |

Procedures
An interview was conducted to collect data regarding to un/pleasure response. Ten participants, who involve the sorting work were attended the interview. In the interview, they were asked to observe ten lighting pictures one by one and then describe among the lighting samples "what are the factors which you think the easiest to evoke your pleasure?" The responses were obtained and analyzed in order to group pleasurable factors shown in Table 2.

Interview Data Analysis- Data grouping
Data collected from the interviews were transcribed into a word processing program. Data grouping is a procedure to group the key sentences having similar contents from responses. The key sentences describing the same stimulus and judged to be of the similar content were first sorted into the same group shown in column three (see Table 2). This step involved repeated and carefully reading of the transcribed data in order to have a clear understanding of the content of each response.

Secondary level labeling
Secondary level labeling is a procedure to label the group which has been identified from the last step. In order to label the group, the key word representing key sentences was first identified from each sentence and highlighted with bold text shown in column three. For instance, key words that described a product with '*round and curved*

shape', *'shape with simplicity'*, and *'repeated pattern'* were refined to shape. Hence, the secondary level label was named as 'delightful shape' to represent the key words in the same group. The labeled names were thus identified as pleasurable factors of lighting shown in column two (see Table 2).

Table 2. Pleasurable types and factors of lighting

Types of pleasure	Factors	Data grouping
Appearance Pleasure (Visceral level)	Delightful shape	The **round and curved shape** make me feel relax. A **shape with simplicity** can make me feel comfortable. The form with **repeated pattern** looks beauty. The **flow and dynamic form** make me feel soothing. **Form with light and flow style** provides a pleasant feeling. **Natural and organic form** make me pleasure. The **cute shape** demonstrates a feeling of pleasant. The **familiar form** makes me feel comfortable. **Warm and soft line** elements deliver a warm feeling which somehow evokes pleasure. **Bionic and natural pattern** looks soothing, which make me, feel great. An **interested form** catches my eye attention, which feels good. I feel much closed to the object, because of the **familiar with the shape** of light design such as liquid pattern or some geometry. The **shape of light** is familiar with me.
	Smart Material/ Texture	**Soft material** like inflatable material provides a good touch feeling. The **smooth surface** provides a tangible and comfortable feeling.
	Attractive Color	I like **bright colors**. **Light color** is great. The **color with light and flow** sense evokes pleasure. An **interested color** make light look pleasant. Somehow, **bright and primitive color** makes me feel good. I think **white color** is pure and pleasant.
Interactive Pleasure (Behavioral level)	Interaction	The **interaction with light** can evoke pleasure. A **playful** design looks fun.
	Feedback	An **interested feedback** may evoke pleasure.
Reflective Pleasure	Meaning	The light provides **a great memory** of pass. The **light with humor** make me pleasure.
Novelty Pleasure	Innovation	I am **curious** about the shape of light.

Primary level labeling

Primary level labeling is a procedure of doing cluster groups according to similarity of secondary level labels, gradually reducing the data to a higher order concept, and creating a primary level label. In this case, the primary level label was indicated a name representing the secondary level labels with similarity. The labeled names were thus identified as types of pleasurable lighting. For instance, the term, smart material/ texture, attractive color and delightful shape traits at secondary level label were associated with appearance. Thus, the 'appearance pleasure' was identified at primary level label shown in column one. The same grouping and labeled process were identified and shown in Table 2. They are four types of pleasure: appearance, interactive, reflective and novelty pleasure. Among four types of pleasure, seven factors were concluded and they are delightful shape, smart material/ texture, attractive color, interaction, feedback, meaning, and innovation.

6.2 Experiment

Participants
60 participants from Chang Gung University attended experiment. The ages range from 18-22 years.

Formulating questionnaire
Questionnaire was constructed with 19 item questionnaire which includes 17 pair adjective items and two extra questions. These two items are item 18, 'Please indicate the degree of your pleasurable response to the lightings?' and item 19, 'do you have a preference to this lighting?' 17 adjective pairs are reliable/doubtable, practical/ unpractical, humor/ serious, fun/ mature, warm/ cool, interactive/ non-interactive, high profile/ low profile, plentiful/ bore, bionic/ non-bionic, relax/ nerve, abstract/ concrete, geometry/ flow, simple/ complex, novelty/ conservative, bright/ dime, delicate/rough, slim/ bulky. Each item was indicated with 7 Liker-scale from strongly agree i.e. 7 to strongly disagree i.e. 1.

Procedures
Participants were asked to answer a 19 item questionnaire while watching each picture attached on the top of each questionnaire sheet. Participants allowed to check each item with a proper time. The data were collected and ran through SPSS statistic software.

Paired-Sample T Test
Data was analyzed by Compare Means, Paired-Samples T Test. Analyses were performed separately for each pair of items: pleasure vs. preference (Q18 vs. Q19, $p>.05$), pleasure vs. novelty (Q18 vs. Q14, $p>.05$), pleasure vs. bionics (Q18 vs. Q9, $p>.05$), and pleasure vs. warm (Q18 vs. Q5, $p>.05$). The results are shown in Table 3.

Table 3. Pairs comparison of pleasure vs. preference, pleasure vs. novelty, pleasure vs. bionics, and pleasure vs. warm. (n=60)

		Mean	SD	Significance (p)
Q18 vs.Q19	Pleasure vs.	4.36	1.44	.309
	Preference	4.31	1.51	
Q18 vs.Q14	Pleasure vs.	4.36	1.44	.409
	Novelty	4.42	1.49	
Q18 vs. Q9	Pleasure vs.	4.36	1.44	.076
	Bionic	4.50	1.63	
Q18 vs. Q5	Pleasure vs.	4.36	1.44	.942
	Warm	4.36	1.45	

P > .05

7 Results and Discussions

The discussions cover the issues of types of pleasure, pleasurable factors, consumers' preference and how product shapes affects users' pleasure feeling. It is hopefully that this research results can provide designers a reference when designing lighting for the enhancement of pleasure.

The final result from interview produced seven pleasurable factors from which four types of pleasurable were classified: Appearance Pleasure, Interactive Pleasure, Reflective Pleasure, and Novelty Pleasure (see Table 2).

Among the responses, there are more key sentences in describing pleasure evoked from appearance related factors, compared with other factors. The result implies the important of appearance of lighting design. It seems product appearance has a tendency to catch consumers' first eye attention, to invite them to explore a product function then and further to decide to purchase the product or not. This result confirms Creusen's [11] research. He commented that consumers paid more attention on product hedonic scale when evaluating the product and more focus on the product form attributes, when describing it. With this in mind, designer should note that product appearance can affect consumer's purchasing decision and need to have more effort in design.

In this study, 'appearance pleasure' pertains with pleasure derived from lighting appearance including delightful shape, smart material/ texture and attractive color factors. This pleasure is associated with physio-pleasure, which deals with sensory pleasure. 'Interactive pleasure' pertains with pleasure derived from the interaction with lighting. It can be the active process with lighting such as turn on the light, which in turn is fun to do so. 'Reflective pleasure' has to do with a product content or meaning behind the lighting. They often associate with users' pass memory or the understanding of things. These memory or special things remind them something meaningful again, which make them feel happy. At this point, to embed reflective pleasure into a lighting design, it is suggested to make a product with ability to tell a story. The story should be able to inspire users' memory and pass history. Particularly, the cultural elements are useful and meaningful to many users. However, these four types of pleasure evoked from a lighting can also associated with Norman's [6] product emotion theory. Where appearance pleasure vs. visceral level emotion, Interactive pleasure vs. behavioral level emotion, reflective pleasure vs. reflective level emotion, while novelty pleasure has a distinguished from Norman's, which led to a new interested research issue to be explore in the future.

Beside prior three pleasures (i.e. appearance, interactive and reflective) in this study, 'Novelty pleasure' deals with something new, which surprise users unexpectly. It seems to be distinguished from both Jordan's and Norman's theories. It is maybe because that novelty is very based on user's personal experience and knowledge to the product when perceiving it. Hence, the emotion/ pleasure response is very depend on the understanding of a product. For instance, if, at the first time users see a product, they may claim that the product is new in terms of novelty at the moment. But, maybe right on the next second, the same product becomes old respectively, because it is not new anymore to the users. But, for another person, the same product may still novel for them. The key point is the timing, rather then content. In this case, it is hard to define the pleasure individually.

In describing delightful shape under appearance pleasure, the key words, 'being familiar with the shape' were revealed for a few times. Participants claimed that the 'familiar shape' make them feel 'closed to the object', light in this case. For instance, the liquid pattern used at light base appears in daily life. It seems to be easier to connect with our empathic bounds to the shape of water, which evokes pleasure. To this particular comment, it seems to have a conflict to the novelty pleasure, which tends to be evoked by an unfamiliar object. However, it is an interested issue and suggested to have further study on novelty and familiar towards pleasure.

In Table 3, by using paired t Tests, it was found that the degree of a participant's pleasurable response (Q18) have no significant different from his or her preference (Q19), novelty image (Q14), bionic image (Q9) and warm image (Q5). The results indicate that participants' pleasures evoked by lighting have a positive relationship to their preference, novelty, bionic image and warm image consistently. It means that participants agree that lighting with novel, bionic, warm image has an intensity to evoke their pleasurable response better. The result also demonstrates that a user's preference to lighting has a connection to his or her pleasure response positively. It implies that participants have a preference to own a pleasurable product. This result seems also can refer to the consumer behavior in current market. Understanding the four types of pleasure and seven factors, designers should try to embed these pleasure elements into a lighting design as possible.

References

1. Seligman, E.P.M.: Time magazine, USA (February 2005)
2. Marberry, O.S., Zagon, L.: The Power of Color. John Wiley Sons, New York (1995)
3. Simpson, J.A., Weiner, E.S.C.: The Oxford English Dictionary, 2nd edn. Clarendon Press, UK (1989)
4. Jordan, P.: Designing Pleasurable Products. Tyalor and Francis, UK (2000)
5. Desmet, P., Overbeeke, K.: Designing Products with Added Emotional Value: Research through Design. The Design Journal 4(1), 32–47 (2001)
6. Norman, A.D.: Emotional design. Basic books, New York (2004)
7. Marzano, S.: Thoughts. V+K Publishing, Blaricum (1998)
8. Gotzsch, J.: Product Charisma. In: Common Ground: DRS International Conference, UK (2002)
9. Csikszentmihalyi, M., Rochberg-Halton, E.: The meaning of things. Cambridge University Press, London (1981)
10. Tiger, L.: The Pursuit of Pleasure. Transaction, London (2000)
11. Creusen, M., Snelders, D.: Product Appearance and Consumer Pleasure. Tyalor and Francis, UK (2002)

Plugging the Holes: Increasing the Impact of User Experience Evaluations

Sachin S. Yambal[1] and Sushmita Munshi[2]

[1] User Experience, Portal & ECM Lead, Accenture India,
Accenture Services Pvt. Ltd., 4/1, IBC Knowledge Park, Banerghatta Road,
Bangalore, 560079 India
[2] User Experience Thought Leadership Lead, Accenture India,
Accenture Services Pvt. Ltd. Plant 3 Godrej& Boyce Complex Off LBS Marg,
Vikhroli (West) Mumbai, 400 079 India
sachin.s.yambal@accenture.com, sushmita.munshi@accenture.com

Abstract. The principal objective of this paper is to demonstrate the APRICOT methodology that aims to streamline and increase the effectiveness of user experience initiatives within a development project and in the final solution. User Experience (UE) evaluations, both heuristic based and usability testing based are important skills and a crucial part of a practitioner's tool kit. They showcase the inadequacies in an application or system. Close inspection of projects which have used User Experience evaluations reveal that only a small percentage of User Experience recommendations actually make it into the final product. This substantially reduces the ROI for User Experience contribution. The APRICOT concept is work in progress and aims to make User Experience evaluation more effective by better integrating UE practitioners and aligning the processes and methodology with one used by development teams.

Keywords: ROI of Usability, User Experience Reviews, Institutionalizing Usability.

1 Introduction

User Experience (UE) evaluations, both heuristic based and usability testing based are important skills and a crucial part of a practitioner's tool kit. They are the most powerful way to showcase the inadequacies in an application or system. Forrester reported in 2008 that an average sized enterprise would spend about 20K USD per evaluation.

Close inspection of projects which have used User Experience evaluations reveal that only a small percentage of User Experience recommendations actually make it into the final product. This substantially reduces the ROI for User Experience contribution. Our User Experience practice has been working on developing process and guidelines to ensure User Experience reviews are more effective. The APRICOT concept is work in progress and aims to make User Experience evaluation more effective by better integrating UE practitioners and aligning the processes and methodology with one used by development teams.

M. Kurosu (Ed.): Human Centered Design, HCII 2009, LNCS 5619, pp. 147–156, 2009.
© Springer-Verlag Berlin Heidelberg 2009

This paper does not delve into methods and tools to better identify User Experience problems, report or communicate them. The authors have assumed that User Experience practitioners have reached a level of maturity and consistency in identifying and reporting these effectively.

1.1 The Challenge

Why do we need better methods to ensure User Experience recommendations get better incorporated?

A recent market trends noted that by 2009, customers will hold companies accountable for the quality of their experiences. Thus the time is ripe for User Experience/Usability to move away from a craft based individualistic approach to a more ubiquitous approach. User Experience teams are at the cusp of steep growth (Usability Services Are the Next Offshore Frontier, Partha Iyengar, Gartner Research). However it seems that teams find it tough to get their recommendations communicated and incorporated correctly to stakeholders and developers. This inability is likely to prove to be a major hurdle in making user centered design a mandatory process in the development methodology.

1.2 Additional Current State of Affairs: Unusable User Experience Recommendations

User Experience evaluations (reviews & testing) are powerful to point out usability shortcomings of a solution. Evaluations are often the entry point for engagement with new clients and the most commonly asked task from remote teams. Reams of literature have been dedicated to methodology, tools and best practices of User Experience evaluation/heuristic review/ expert evaluations. Effort has been made to demystify User Experience evaluations and codify User Experience principals to ensure that User Experience moves away from being a specialized craft and the bastion of few specialized practitioners to a more ubiquitous profession.

However, the amount of thought and research dedicated to communicate review findings and recommendations to development teams and stakeholders is limited. This has a serious impact on the perceived return of User Experience. A majority of User Experience teams are unsuccessful in ensuring their recommendations are incorporated in final products. A 2007 study by Molich, Jefferies, Dumas [1] compared User Experience insights and recommendations provided by 17 different professional User Experience teams. The study found that only 14 out of 84 recommendations (17%) were both useful and usable. The remaining findings were either not useful or presented in a manner that made them unusable.

The authors conducted an informal survey among User Experience practitioners to understand the quantum of effort and time spent by User Experience professionals in each step of a basic User Experience evaluation process. The findings were:

Knowledge transfer and understanding requirements: This step involves understanding business requirements, technology limitations, domain knowledge, etc. Most User Experience practitioners said that they spent anywhere between 1 to 5 days for a detailed knowledge transfer from the project sponsor or development team. In most cases the approach followed was that of a person to person transition including sharing of documentation and providing overview of key tasks etc. In some cases practitioners mentioned using a pre formatted questionnaire to ensure all bases are covered. A small minority of practitioners mentioned following a strict process of interviewing key stakeholders to get a 360 degree understanding of expectations.

Conducting the review: This step involves reviewing the product in question, conducting User Experience tests to gather data from end users. Most practitioners spent a bulk of their time in this step. Practitioners also mentioned that a number of reusable tools like scorecards and best practice checklists are often used. Most practitioners seemed fairly confident of their abilities of conducting reviews. On an average, the authors felt that this step is handled in a fairly mature and consistent manner.

Reporting: Most practitioners mentioned using predefined formats to report findings. However it has been noticed that report formats are often not in line with a client's expectations and are structured in a way that is not comprehensible for people with low User Experience vocabulary.

Managing and incorporating recommendation into the final solution: Very few practitioners have been hands on and involved in this step. Most practitioners expect their responsibility to be over the moment they complete their final presentation. Those who have been involved in such a process mentioned that it was primarily on their own individual interest and often their involvement evoked a fair amount of conflict between the User Experience and the development teams.

2 Reasons of Failure

Some of the key reasons of User Experience team's failure to provide deep reaching value to clients and development teams are:

2.1 Focus

Critical success factors and business objectives of a product are important to identify at the onset of a review to ensure that User Experience problems can be effectively prioritized at a later stage. McQuaid and Bishop [2] recommend categorizing User Experience insights based on similarity (using affinity diagramming), and then prioritized along the dimensions of importance (How badly a user is effected or stopped from achieving a task that has direct repercussions on the product's success or monetary gain) and how difficult it is to fix. This structured approach is important as it provides a clear ROI justification to each recommendation.

2.2 Approach and Techniques

Inappropriate use of tools and techniques is another reason for unusable User Experience recommendations. Different techniques have varied levels of depth to identify and provide pointers toward solutions. For example: Using web analytics to address reasons of drop off is speculative. This is because web analytics only provides statistics of drop offs, and does not provide any clear pointers towards the reasons for drop off. The nature of the User Experience review needs to be appropriate to address the client's doubts. Thus User Experience teams need to recommend the appropriate approach at the proposal level and educate stakeholder and development teams.

Speculative problem finding should be avoided as it reduces the credibility of User Experience. UE insights are tough to justify with stakeholders and development teams. User Experience practitioners need to back insights with either raw data or research.

2.3 Team Dynamics and Culture

User Experience teams are not appropriately integrated into development groups. Thus, they come across as external consultants. Dr. Arnie Lund [3] in his 2003 article on post-modern User Experience noted that a number of user experience teams are isolated and not well integrated with the client's business requirements and expectations. Evaluations are often very superficial; this makes clients believe that UE is primarily a skin level initiative.

In the aim to create reusable assets and codify heuristics, a number of UE teams lose the context of an evaluation and blindly apply global heuristics. Problems are identified, but there is little support or theoretical basis for specific solutions. Development teams are often left to understand the real repercussions on these insights and prioritize them.

UE teams have low understanding of the client's software development process, timelines or the approach taken to incorporate changes.

The relationship between User Experience and development teams are often very transactional. This staccato relationship does not allow UE teams to spear head decision making or be in a position to provide any strategic inputs.

In a number of companies, User Experience teams are reduced to gate keepers of User Experience or people who sign off on the User Experience of a product rather than help create it. User Experience evaluations are perceived as due diligence/quality assurance exercises.

Development teams don't learn or take away any value from these evaluations. The same mistakes seem to be made over and over again. This provides a steady stream of work for UE teams, but in the end it doesn't help the goal of enabling development teams and organizations to realize the full potential of UE.

2.4 Timing

Reviews done late in the development life cycle are validation, due diligence exercises or are commissioned to fix unforeseen fires. In both cases stakeholders and development teams want to fix problems in the least intrusive and in-expensive fashion. Late in the cycle, teams don't have the interest to make substantial changes. The aim is usually to localize and address the problem. User Experience practitioners often don't understand these limitations and fail to match the findings to the mood and expectations of the client.

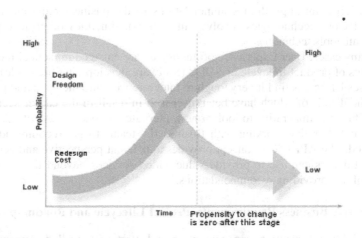

2.5 Lack of Understanding of Usability Sharks

What differentiates a *great* User Experience professional from an *effective* User Experience professional? Great UE professional design usable interfaces but Effective UE professionals respect reality and ensure the designs are implemented. Effective UE professionals usually have a better understanding of how change is perceived and addressed both at an organizational and individual level. Mobilizing a team to rework a set of features is not an easy task to push through. Having stakeholder buy in or support from key individuals is not enough. Along the way it is seen that antagonist development teams will reject UE recommendations bit by bit over a period of time. This results eventually in undoing/rejecting large chunks of recommendations made by User Experience professionals. Often User Experience professional are not around during feasibility analysis to see through their recommendations or they don't have enough clout to fight what is often called User Experience sharks. Some of the common mistakes made while proposing recommendations are:

- Problems aren't prioritized based on a deep understanding of the nature of the users and how the context of use shapes their experience. Subtle problems that don't show up as obvious errors but that impact the core value of a solution.
- Problems are presented in an incomprehensible structure.
- Terminology used to communicate insights is User Experience jargon filled.
- Insights are not correlated to business and financial implications.
- Only symptoms of problems are reported. Lack of root case analysis does not help developers understand the issues.

3   APRICOT: The APRICOT Concept Is an Acronym for Analyze, Prioritize, Customize Offerings and Team Up

It has been noted that projects in which User Experience practitioners are involved at the onset, face less challenges in getting UE recommendations implemented in the final product. However the impact of these recommendations are dependent on how well the UE team align the recommendations to the business requirements, critical success factors, technologies involved in solution, timelines of project and use of appropriate tools/techniques.

In many cases, User Experience professionals are engaged in tactical mode at various stages of product lifecycle or SDLC (Software development life cycle) phases. In such cases UE teams find it very tough to integrate well into the team and provide real value, the details of which have been discussed in detail in the earlier section of this paper. Thus it is imperative to look at new/alternate approaches, out of the box thinking and collaborative working with Business/IT teams to get recommendations implemented. The APRICOT concept provides a different perspective and activities that UE practitioners can use to provide value added services beyond the consulting approach of just providing recommendations.

3.1 Analyze Business Requirements, Project Lifecycle and Roadmap

Any usability intervention engagement should start with well documented critical success factors of assessments, clear scope definition, overall business objectives, technical constraints and organizational dynamics. This allows the UE team to focus on the burning problems that need to be addressed; other usability issues can thus be downplayed. Identifying the focus areas of assessment (strategic, navigation, interaction, information, detailed design) also helps UE teams better position their recommendations.

This written document ensures the client, and the UE and development teams have the same understanding and expectations from the engagement. The UE team should also communicate the required nature of involvement and support (should this be to) from the development team.

It is important for UE practitioners to understand the Product lifecycle (or SDLC phases for applications) used in a project at the onset of engagement. This would allow them to plan their work schedule/deliverables aligned with development team milestones. The UE teams should also look at the business/technology roadmap of the product, the release cycles and any new Business/IT initiatives that impact the product. If the application has a short term life or is due for major overhauls in the future, it is important to focus on quick wins only for reviews and help resolve issues in a timely manner.

If the applications are already in production, it will be useful to also look at the enhancements/change requests pipeline. Most of the IT projects follow a robust and streamlined process of managing bugs/enhancements through a change management process so the details are available easily.

The UE practitioner should understand the technical solution and challenges of the development teams (If the products leverages 3rd party products or packaged solutions and mostly uses out of the box features then recommending changes in those areas have a low probability of implementation).

3.2 Prioritize Recommendations

In addition to factors like ROI, technical limitations, and ease of implementation used for prioritizing UE recommendations, it is also important to consider other factors like

- **Development Methodology:** If a project uses the typical waterfall development methodology then it makes sense to do a complete review and provide recommendations aligned with milestone and release dates. The key thing is not to miss the milestone as acceptance of changes in late stages in this approach is very difficult. However for agile methodology or iterative development it is important to provide recommendations in an iterative manner and as frequently as possible (could be daily also if the iterations are weekly). In agile development, as the business/IT works with regular interactions, UE practitioners need to have closer ties with the development team.
- **Development Team's Maturity:** The more mature the team, the higher the resistance to change. Align the tone and content of your report to ensure minimal intervention and maximum benefit.
- **SDLC phase:** Assessments done late in the development cycle usually don't result in any change as teams have low propensity to amend the project at that stage. Therefore, recommendations should be prioritized based on what is realistic for implementation.

  Early Involvement
  - o Provide recommendations to improve the navigational structure.
  - o Provide reworked wireframes and before-after examples.
  - o Provide tools to help development teams reduce their UI development timelines.

  Late Involvement
  - o Provide tactical recommendations that provide maximum return for minimal change. Understand the technical limitations of the project.
  - o Do not attempt to over complicate or simplify the situation.
  - o Showcase the repercussions of User Experience issues. Create a business case to convince development teams.
  - o Be open to negotiation
- **Business/IT Alignment:** Based on the alignment of the teams in the project, the appropriate strategy should be made to push changes through IT or Business.

3.3 Customize Offerings

Earlier in the paper, the challenges of getting usability recommendations implemented have been discussed in detail. Assessments methodology and processes need to be customized to address engagement specific business and technical requirements.

- Customize heuristic checklists and best practices to the critical success factors of the engagement. Consider domain, the industry vertical the solution is made for, context of use and device of delivering the solution apart from the details of target user groups and key tasks to customize checklists.
- Conduct assessment as per focus areas identified at the onset of the engagement. Ensure these focus areas are in line with business objectives and will lead to absolute dollar returns for the client. Focused reviews also ensure teams don't have to conduct a time consuming sifting exercise. In the past, usability teams have been known to conduct exhaustive reviews, the reports of which could run into

hundreds of pages. However, as the industry has matured, clients now look for specific answers to their business problems and are rarely impressed with bulk.

- Report problem areas, identify their root cause and mention the overall repercussion on user experience.
- Provide recommendations in tandem with technical, budgetary and time constraints.
- Create reports that are easy to comprehend and use. Apply usability principals while creating reports. Ensure they are easy to read, use and learn.
- Use terms and language that your client team is comfortable with and avoid jargonized presentations.

Similarly, it is critical to evaluate all the recommendations from a project perspective and customize them as per project specifics which will help ease implementation. Some aspects to be considered are

- o   Analyze if recommendations can be inbuilt into future enhancements/bugs planned for release and align with them.
- o   Review the recommendations and group them as per business use case. Also look at feasibility of creating new a business case with proper RoI for critical recommendations.
- o   Categorize the recommendations into 2 streams: Business benefits and IT enablement. Define RoI and loss in revenue (if not implemented) for business ones. Showcase reduction in IT efforts in Productivity improvement ones.
- o   Explore alternate options of getting recommendations implemented
- o   Collaborate with business teams and explore if they can be logged as maintenance/enhancement requests.
- o   Collaborate with IT teams and explore feasibility on getting covered in any of their initiatives

3.4 Team Up and Collaborate

User Experience teams are generally not well integrated into development groups and come across as external consultants. It is imperative to team up and push for closer proximity with the development community. UE teams need to actively collaborate in each stage of the change management process to ensure recommendations are incorporated. Some of the best practices/suggestions are shared below:

- o   Evangelize User Experience. Provide User Experience training to the development community.
- o   Train User Experience team on software development methodologies
- o   Power of 'We' during presentations and discussions. This reduces the 'them and us' gap between developers and User Experience teams.
- o   Keep communication channels open. Be available for informal discussions.
- o   Share learning. Aim to empower the development community on User Experience best practices.

- o Provide tools and checklists to help development teams with recommendations
- o Empower developers with checklists to ensure upgrades and amendments are correctly handled
- o Be open to negotiation. Identify solutions that will benefit everyone.

4 Benefits

- Parts of the APRICOT concept have been implemented in a number of projects. Some of the benefits that have emerged are
- Makes the user experience review process more reliable and output predictable: The steps and guidelines of APRICOT provide a framework for feasibility analysis and decision making. This replaces adhoc discussions and general strife. Increased involvement of stakeholders and development teams during initial data gathering ensure that the recommendations are more predictable and don't shock people unnecessarily.
- Increases overall ROI of User Experience reviews: Both the perceived and the actual ROI or UE effectiveness increases. This is primarily because a larger percentage of recommendations get incorporated into the final product.
- Integrates UE as an integral part of the "Team": The APRICOT concept helps in institutionalizing usability within teams and organizations. Teams stop perceiving UE as a good to have service but more of a critical piece for a project's success. A number of developers become evangelists of UE and reduce the pressure on User Experience practitioners to push through recommendations.
- Throws up opportunities to create reusable assets and tools: Apart from providing step by step guidelines to User Experience professionals, the APRICOT process also creates opportunities of creating reusable tools and assets. This further ensures closer ties between the development and User Experience teams.
- Reduces dependency on close physical proximity of User Experience & development teams: User experience teams don't always have to be physically close to development teams to ensure that their recommendations are incorporated. The APRICOT method is an ideal tool to convince and push development teams in a dispersed set up.

5 Conclusion

If 'Need' is the mother of innovation, APRICOT is definitely 'need's' child. The concept has emerged and evolved over a period of time. However the key reason it was born was as follows – as UE practitioners, we were tired of boardroom fights to push our recommendations and see our usability insights get chopped at the proverbial editing table. The APRICOT concept is work in progress. However its early success stories and feedback from usability communities are encouraging. As next steps we are looking at rolling it out to more projects so that more concrete measures can be put in place to capture the actual dollar returns this concept brings to projects and usability practitioners. Effort is also being made to enlarge the scope of this concept by adding reusable tools and assets to this framework, thus making the process more people independent.

References

1. Molich, R., Jeffries, R., Dumas, J.S.: Making Usability Recommendations Useful and Usable (2007)
2. McQuaid, H.L., Bishop, D.: An Integrated Method for Evaluating Interfaces. In: Proceedings of UPA (2001)
3. Lund, A.M.: Post-Modern Usability (2006)
4. Nielsen, J., Mack, R.L.: Usability Inspection Methods. John Wiley & Sons, Inc., New York (1994)
5. Fadden, S., McQuaid, H.L.: Fixing what matters: Accounting for organizational priorities when communicating usability problems (2003)
6. Schaffer, E.: Institutionalization of Usability: A Step-by-Step Guide. Addison Wesley, Reading (2004)
7. Valdes, R., Gootzit, D.: Usability Drives User Experience; User Experience Delivers Business Value, Gartner Research (2007)
8. Drego. V.L.: Usability Moves Offshore Best Practices For US Firms Working With Usability Team, in India, Forrester Research (2006)

Part II

Methods and Techniques for HCD

Part II

Methods and Techniques for HCD

Elicitation of User Requirements for Mobile Interaction with Visual and RFID Tags: A Prototype-Based Exploratory Study

Margarita Anastassova[1] and Oscar Mayora-Ibarra[2]

[1] CEA, LIST, Laboratoire d'Interfaces Sensorielles, 18, route du Panorama, BP6,
FONTENAY-AUX-ROSES, F-92265 France
margarita.anastassova@cea.fr
[2] CREATE-NET International Research Center, Multimedia,
Interactions and Smart Environments Group, Via alla Cascata 56C,
Building D, 38100 Trento, Italy
oscar.mayora@create-net.org

Abstract. This paper presents a preliminary prototype-based elicitation of user requirements for mobile interaction with a public display using visual and RFID-tags. The study is based on the use of a demonstration and two applications scenarios as means for encouraging user requirements elicitation. The results show that the prototype, its demonstration and the examples of possible applications are very useful for the users: they express a large number of requirements, which, are furthermore, quite original.

Keywords: Emerging Technologies, Innovation, Mobile Interaction, Prototype Evaluation, User Requirements, Visual Tags.

1 Introduction

Relating the physical and virtual worlds using mobile phones is now a widely-accepted interaction paradigm for providing ubiquitous computing services. Several factors, related in a number of papers in the area, could explain this tendency. First, continuous innovation in mobile devices (e.g. miniaturization, diminished energy consumption) enabled the development of relatively usable technology with a lot of features and important computing power. Second, mobile devices have become extremely common and wide-spread. Third, they are usually carried by their owner and could thus be used at almost any moment of place and time. Forth, if equipped with a camera, they could read visual tags placed on physical objects and thus interact with these objects. If equipped with a Radio Frequency IDentification (RFID)-reader, mobile device could also read RFID tags. Because of all these factors, mobile phones in combination with visual tags or other identification technologies are considered as the most common alternative for accessing ubiquitous computing services and interacting with smart environments.

However, most of the research efforts in this area up to now have been concentrated either on creating new technological concepts and, eventually, services or on

M. Kurosu (Ed.): Human Centered Design, HCII 2009, LNCS 5619, pp. 159–166, 2009.
© Springer-Verlag Berlin Heidelberg 2009

evaluating the usability of existing systems. Less work has been done on the utility of these services and systems. Understandably, less work has also been done on detailed user requirements, whose elicitation and analysis is a prerequisite for providing a useful service.

The study presented in this paper is a contribution in this direction. The paper is structured as follows. We first discuss briefly the related work on mobile interaction with visual tags and on the elicitation of user requirements for emerging technologies. Then, we present the design of the study and its results. Finally, we discuss these results and open up some perspectives for our future work.

2 Related Work

As mentioned earlier, most of the work on mobile interaction with visual tags has been centered on creating new technological concepts or services. Thus, in the CoolTown project, users could read additional information about the current exhibition in a museum scanning the pieces of art using a PDA [1]. In the Aura project, a similar concept was put forward: with their PDA, users could read barcodes placed on existing products to get additional information on them [2]. Other, already commercialized, applications have been proposed, in which users could scan visual tags with their mobile phones to download mobile content [3], to retrieve bus timetables [4], and to record someone's personal details coded in a visual tag embedded in his business card [5].

Considerable work has also been done on the usability of these interaction concepts. The understandability and the intuitiveness of the concept, the speed at which the tag must be decoded by the tag reader and the speed and accuracy of user's clicks on the tag have been evaluated [6]. The authors showed that the participants understood and interacted successfully with the system after only 15 minutes of training. As for the speed, novice users could click on the visual tag quickly and accurately. Also, the usability of pointing and gestural interaction has been studied [7, 8]. It has been shown that the interaction is natural and gestures are correctly recognized and recalled if not too complex.

However, less work has been done on the utility of mobile interaction with visual tags and on the user requirements for such services. One example in this direction is an extensive field study done by [8]. It is based on 23 interviews and 370 hours of observations of field biologists' practices experimenting on tropical plants. The study gave very rich results and motivated the design of ButterflyNet, a mobile capture and access system that integrates paper notes with digital photographs captured on the field. Because of the user detailed user requirements analysis and the early user implication, the new ubiquitous computing system has been successfully adopted by field biologists.

We should note that user requirements analysis for emerging technologies such as mobile applications for ubiquitous computing differs from user requirements analysis for more traditional technologies, because future users do not know and do not imagine the technology itself and the possible services that it could provide. Consequently, they have problems when expressing requirements for such technologies and services [9]. One possible solution to these problems is the use of prototypes and scenarios which are relatively concrete representations of future emerging technologies. Their concreteness facilitates users' understanding of abstract, unfamiliar or fuzzy techno-logical concepts. In this sense, scenarios and prototypes may help people become aware of and formulate some of their unconscious requirements. Furthermore, proto-types and scenarios help building a common discussion basis for all stakeholders and thus facilitate the reaching of an agreement in the early stages of a project [10].

The above-mentioned assumptions were the starting point of the exploratory user study presented in this paper. Its major objective was to establish a number of high-level user requirements for a future technical system, based on mobile interaction with visual tags. The elicitation of user requirements was done using a low-level prototype and a number of textual scenarios.

3 Design of the Study

3.1 The Tested Prototype

The low-fidelity prototype, which has been evaluated, consisted in a public display capable of presenting both public and personalized information to people in its prox-imity (Fig. 1).

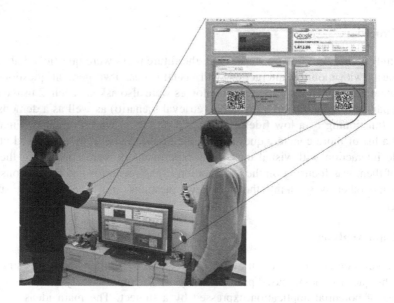

Fig. 1. Users interacting with the tested prototype

During the test, the public information presented was a weather forecast for the re-gion, in which the study was done as well as some general financial information. As for the presentation of personalized information, it was based on the user's profile [11]. The user profile includes a user-ID, some general information and layout prefer-ences (e.g. news, weather, colours, etc.), as well as some privacy and location-sensitive information hidden as a 2D visual tag on the public display.

When no identified users are nearby, the display shows general purpose location-based and weather information. When one or more users are identified in proximity of the display through their RFID contactless badge, the public display adapts its content and layout based on nearby users' number and identities. Privacy sensitive information can be viewed by users in their mobile phone by taking a snapshot of the visual tag on the screen.

3.2 Participants

Eighteen people (13M, 5F) aged between 23 and 42 years (M=34) participated in the study. They were all working in the institute developing the application. However, none of them was familiar with the concept or the tested application before the actual user tests.

Half of the participants had an engineering background and half were administrative personnel. None reported difficulties with vision or motor skills. All of the participants had personal mobile phones, using them mainly for calling and sending SMS. All but two of the participants had no prior experience with visual tags. The participants, who had seen visual tags before, had a vague idea of their application and usage.

3.3 Procedure

The study comprised 3 major steps. First, the future users were questioned about their familiarity with mobility interaction with visual codes. Two general questions about the potential applications of these technologies were also asked. Then, 2 usage scenarios (a payment scenario and a timetable retrieval scenario) as well as a demonstration of the functioning of a low-fidelity tag-based prototype were presented. Finally, we asked a list of more concrete questions of the application and the perceived utility of mobile interaction with visual tags. Two experimenters were conducting the study. One of them was focusing on the presentation of the scenarios and the demonstration, while the other was asking the interview questions and tape-recording subjects' answers.

3.4 Data Analysis

The verbal protocols obtained, transcribed verbatim, were then analysed using classical techniques for theme-based protocol analysis. The unit of analysis corresponded to an idea of potential application expressed by a subject. The main ideas evoked by future users concerned the use of visual tags for the identification of goods and people; for getting dynamically updated location-information in unknown environments; for facilitating complex data visualization using a large display; for displaying private information in shared spaces; for displaying private information in confidential contexts. Subjects' ideas were also classified according to their originality compared to the proposed scenarios, the demonstration and the participant's previous knowledge of this interaction concept.

4 Results

4.1 General Results and Types of Applications Proposed

One hundred and seventeen requirement were collected (Min=3, Max=13, M=6,5 requirements per participant). The participants with engineering background expressed almost the same number of requirements (N=66, M=7) as the participants with non-engineering background (N=51, M=6).

The big class of applications proposed most often was the one of technologies protecting privacy (in 38% of user requirements). Here are some typical examples of such applications:

In a home environment: *"I would like to use a combination of mobile phone and visual tags in a home environment. In this way, you could read some information, which should not be seen by your partner or your children...But this probably doesn't make any sense..."*

In a public space where visualisation of private information is necessary: *"One could use this technology in hospital, for visualizing personal information. I think this could be very useful. You can code the health record, which could be read only by authorized doctors and visitors. The tag could be put on the bed of the person who is ill. In this way, one can avoid that all the other people in the room could see the information. The public display could be used to present public information such as the plan of the hospital building and the location of different services."*

For personal and confidential information: *"The visual tags could be used for codifying credit cards details. I think it will be more secure than having a credit card with your details on it.'*

Another big class of applications that was proposed quite often was the one of technologies for giving additional information than the one usually presented on an object or in the physical environment (in 30% of the cases). Again, here are some typical examples of such applications:

Additional information on goods: *"Tags could be used to provide information on the content of products...We could know what we eat, could see whether we have some allergies to some substances. Another example is information on how to take a medicine. This could be very useful for some drugs which may soon be sold in supermarkets in our country."*

Additional information on services: *"In a university building or campus, the time schedule of seminars or conferences could be shown on a public display, while your private information could be encoded in visual codes."*

Additional location-based information: *"Could be interesting for...gaming. I would imagine putting the tags around the city and do some storytelling. One could use the visual tags in order to give some location-based clues which should be understood by gamers. People could go around and find the right information at a certain moment".*

Some participants proposed the use of mobile phones reading visual tags for the identification of products and people (15% of the user requirements). Examples of such applications are proposed below:

Identification of people: *"It may be used by people who have illnesses like Alzheimer and who can get lost. The code, which could be worn by the person, will encode private information like name, address and could be read only by relatives, public authorities or medical personnel".*

Identification of products: *"At home I would like to be able to install such codes on the goods that I had bought from the supermarket to know when they were bought".*

The participants also wanted to use the visual tags to encode some private information, preferences and interests in order to create personal profiles, which could be exchanged with friends or new acquaintances (11% of the requirements). In 6% of the cases, people expressed requirements for the use of the large display for displaying complex visual information. An interesting example of such an application is the following one:

"It can be somehow useful in situations in which you need to store plenty of information, rapidly and in a very "hard" environment. I'm thinking about refugees' camps or some humanitarian actions in developing countries. The advantage will be to use only one tool and spare some material resources in this way, especially for big databases where you need multiple views of the situation of each person or family."

4.2 The Role of the Scenarios and the Demonstrations

We evaluated the role of the concrete representations of an emerging technologies (i.e. application scenarios and demonstrations) based on the number of the requirements expressed by participants, the originality and the concreteness of the applications proposed and their usefulness for the redesign of the prototype. The results are presented below.

Number of Requirements Expressed. The demonstration of the prototype and the two application scenarios described by one of the experimenters seem to have a positive influence on the number of requirements expressed by the test participants. Thus, a total of only 33 requirements (28% or 2 requirements per person on the average) were expressed before the demonstration and the presentation of the scenarios (i.e. before the users had acquired a concrete idea of the application domains and the potentialities of the technology). On the contrary, 84 requirements (72% or 5 requirements per person on the average) were expressed after the scenarios presentation and the demonstration. This difference is particularly notable for the people having a non-engineering background. They expressed a total of only 8 requirements (14% of the requirements expressed by non-engineers) before having been presented with concrete examples of applications and 43 requirements (84%) after the demonstration of the prototype and the scenarios presentation. Tough less important, the contrast "before demonstration" vs. "after demonstration" is also observed for participants with an engineering background (38% of requirements before vs. 62% of requirements after demonstration).

Originality of the Proposed Applications. The originality of the applications proposed by the test participants was also evaluated by two of the three authors of the paper. In order to evaluate the originality of an idea, we compared it to the scenarios described during the test, the demonstration and the participant's previous knowledge of the interaction concept. The demonstration of the prototype and the exemplification using application scenarios do not seem to limit the participants' imagination. The ratio "original ideas vs. non original ideas" is exactly the same before and after the demonstration. Thus, before the demonstration users expressed 14 ideas of applications judged original by the authors (42%) and 20 ideas judged not very original (58%). The situation is exactly the same after the demonstration (36 ideas, 43% vs. 48 ideas, 57%).

5 Discussion and Future Work

The results show that future users express much more requirements after having getting acquainted with the application scenarios and seen the demonstration using a concrete representation of the future technology. In addition, the requirements elicitation aids do not seem to limit the participants' imagination. The ratio "original ideas vs. non original ideas" is exactly the same before and after the demonstration.

There are at least three possible interpretations of the first result showing an increase in the number of the expressed requirements after the demonstration. First, users have longer time to think about the same topic. Thus, they can have more ideas to propose. Second, these ideas clarify in the discussion with the experimenter, i.e. the requirements are constructed in the collaboration with someone who has a clearer idea about the possible applications of the technology. Third, the concreteness of the demonstration and the scenarios may have a positive influence on the understandability of the technology, and, consequently, on the number of user requirements expressed.

As for the result on the relative originality of the expressed requirements, we think that this may be due to the low-fidelity of the prototype and the openness of the proposed scenarios, which did not include any very concrete interaction sequence or task. Thus, user's imagination was supported and stimulated rather than limited to one or two possible solutions.

This preliminary study showed that scenarios and prototypes may be very useful tools for the social construction of user requirements. However, further more experimental studies are needed in order to get insight into the factors accountable for the facilitation of user requirements elicitation.

Acknowledgments

We would like to thank Iulia Ion, Boris Dragovic and Aleksandar Milosevic for designing the prototype as well as for the useful comments on this work.

References

1. Kindberg, T., Barton, J., Morgan, J., Becker, G., Caswell, D., Debaty, P., Gopal, G., Frid, M., Krishnan, V., Morris, H., Schettino, J., Serra, B., Spasojevic, M.: People, Places, Things: Web Presence for the Real World. In: 3rd IEEE Workshop on Mobile Computing Systems and Applications. IEEE Press, New York (2000)

166   M. Anastassova and O. Mayora-Ibarra

2. Smith, M., Davenport, D., Hwa, H.: Aura: a Mobile Platform for Object and Location Annotation. In: Adjunct Proceedings of International Conference for Ubiquitous Computing (2003)
3. Bango Ltd. Bango Spots Link Camera Phones Straight to Mobile Content, http://bango.net/
4. High Energy Magic Ltd. The Spotcode Framework, http://www.highenergymagic.com/
5. Masnik, M.: Mobile Barcode Scanning Catching on in Japan, http://www.thefeature.com/ article?articleid=100700&ref=5960372
6. Toye, E., Sharp, R., Madhavapeddy, A., Scott, D., Upton, E., Blackwell, A.: Interacting with Mobile Services: an Evaluation of Camera-Phones and Visual Tags. Pers. Ubi. Comp. 11, 97–106 (2007)
7. Ballagas, R., Rohs, M., Sheridan, J.: Mobile Phones as Pointing Devices. In: IEEE Pervasive 2005, Workshop on Pervasive Mobile Interaction Devices. IEEE Press, New York (2005)
8. Yeh, R.B., Liao, C., Klemmer, S.R., Guimbretière, F., Lee, B., Kakaradov, B., Stamberher, J., Paepcke, A.: ButterflyNet: a Mobile Capture and Access System for Field Biology Research. In: ACM CHI. ACM Press, New York (2006)
9. Anastassova, M., Mégard, C., Burkhardt, J.-M.: Prototype Evaluation and User-Needs Analysis in the Early Design of Emerging Technologies. In: Jacko, J.A. (ed.) HCI 2007. LNCS, vol. 4550, pp. 383–392. Springer, Heidelberg (2007)
10. Mannio, M., Nikula, U.: Requirements Elicitation Using a Combination of Prototypes and Scenarios. In: WER, Buenos Aires (2001)
11. Morales-Aranda, A.H., Mayora-Ibarra, O.: Adaptive Information Visualizations in Public Displays for Multiple User Profiles. In: Workshop on Knowledge-based User Interface (2007)

The Physiological User's Response as a Clue to Assess Visual Variables Effectiveness

Mickaël Causse[1,2,3] and Christophe Hurter[4,5]

[1] Inserm; Imagerie cérébrale et handicaps neurologiques UMR 825,
F-31059 Toulouse, France
[2] Université de Toulouse, UPS, Imagerie cérébrale et handicaps neurologiques UMR 825,
CHU Purpan, Place du Dr Baylac, F-31059 Toulouse Cedex 9, France
[3] Spatial and aeronautical center, ISAE-SUPAERO, Toulouse, F-31000 France
[4] IHCS IRIT Toulouse, F-31000 France
[5] Direction Technique de l'Iinnovation DGAC/DSNA/DTI R&D,
Toulouse, F-31000 France
mickael.causse@isae.fr,
christophe.hurter@aviation-civile.gouv.fr

Abstract. The paper deals with the introduction of Bertin's visual variables in an ATC context. The ranking of the efficiency of these variables has been experimentally verified by Cleveland, however, no studies highlight the physiological correlates of this ranking. We analyzed behavioral, physiological and subjective data recorded on 7 healthy subjects facing a visual comparison task witch involve 5 selected visual characterizations (angle, text, surface, framed rectangles and luminosity). Results showed that the observed accuracy was coherent with Mackinlay ranking of visual variables. Psychophysiological and subjective measurements are also discussed.

Keywords: Bertin's visual variables, Emotion, Mental load, Psychophysiological response.

1 Introduction

1.1 Introducing the Bertin's Visual Variable Classification in ATC

In current Air Traffic Control (ATC) environments, air traffic controllers use numerous visualization systems like radar views, timelines, electronic strips, meteorological views or supervision systems. Each of these visualizations is rich and dynamic: it displays numerous visual entities that move and evolve over time. Among its different tasks, the most important activity of the controller is the monitoring of conflicts; this refers to the process to detect a loss of safe separation between aircrafts. Conflict resolution requires quick decisions that are sometimes performed under a high uncertainty. Conflicts resolutions generate psychological stress; consequently, the ATC activity may strongly load the cognitive system. The objective of our work is a first step to evaluate the matching between the Bertin [1] classification of visual properties and the effectiveness of these visual properties in the ATC context. Bertin introduced

M. Kurosu (Ed.): Human Centered Design, HCII 2009, LNCS 5619, pp. 167–176, 2009.
© Springer-Verlag Berlin Heidelberg 2009

"la graphique" which provides rules to code information in a monosemic way (i.e. without any ambiguity in the perception of displayed information). He characterizes data to be displayed as:

- Nominal: are only equal or different to other values (aircraft's name),
- Ordered: obey an order rule (aircraft's landing number),
- Quantitative: can be manipulated by arithmetic (aircraft's altitude).

Furthermore, Bertin introduced seven visual variables (Position, Size, Shape, Orientation, Color, and Texture). Subsequently, Cleveland [2], and then Mackinlay [3] built scales of expressivity and effectiveness for Bertin's visual variables (dependent on the human perceptual capabilities). This scale depends on Bertin's data type. The quantitative data type ranking has been experimentally verified by Cleveland. This ranking was built for statistical graphs; nevertheless, this approach might be applied in many other visualization fields. Therefore in this paper we try to sort design choice from the less to the most efficient visual solutions and to validate Bertin's scale with psychophysiological data in complement to empirical experimentations.

1.2 Psychophysiological Measurements

The psychophysiological measurements provide objective measures of the state of operators, precious clues on the impact generated by a given task. In mental workload literature [4], as well as in human machine interface studies [5, 6], psychophysiological data are commonly used as an index of the level of cognitive demand generated by a task (e.g. increased temporal demand, memory loading etc.). This level is characterized by physiological changes, in particular the catabolic activity within the autonomous nervous system (ANS) and is associated with energy mobilization and the investment of mental effort to copy with a task [7]. The analysis of psychophysiological responses like the eye movements or the pupil diameter has been also successfully used in ATC, for instance to assess the workload generates by a dynamic forecast tool compared to a static one [8]. As noticed by Võ [9], whereas a substantial literature focuses on cognitive or emotional effect on ANS, little studies take into account their interactions, in focusing on workload effects, potential related psychological stress and emotion is ignored and its role in performance degradation is neglected. In consequence, the improvement of the emotional experience provided by the visual variables displayed on the screen could play a key role to facilitate the cognitive processes of the operators. Some preliminary experiments show the specifics impacts of a given cognitive activity on the ANS and the potential dissociation between emotional and cognitive effect on the ANS activity [10].

1.3 Objectives

Taking into account the research in the design field and the user's physiological considerations, we propose experiments that rely on four components:

- An ATC task based on a simplified scenario of the conflicts detection throughout judgment of vertical separation. The horizontal separation relies on the position of the aircraft on the radar; the vertical separation is classically based on a label associated with each aircraft (altitude in feet). Because the position on the screen is devoted to horizontal separation, we proposed to use different visual coding for the vertical separation.

- The behavioral measurements are the reactions times of the subjects to detect the conflicts and their accuracy to designate the concerned elements (analogies of aircraft).
- The objectives physiological measurements (ProComp Infiniti, Thought Technology) of the ANS arousal: the heart rate (HR), the galvanic skin response (GSR) and the respiratory rate (RR).
- The subjective rating of mental load, thanks to the NASA TLX, and emotional assessment: stress, anxiety and user experience quality, rated using a 9-point visual analog scale.

The overall aim of this study is to categorizing, following the Bertin's classification, the different visual variables according to their abilities to provide accurate quantitative information and highlight the workload, the psychological stress and the emotion that they generate in a context analogous to the ATC separation task. Our work is based on Cleveland previous work in wich visual variables efficiency is empirically classified. The mental load and the subjective emotional assessment should allow us to disentangle workload effect from emotional one's and their respective roles in case of performance modifications.

2 Method

2.1 Subjects

Healthy subjects (n = 7) were recruited by local advertisement. Inclusion criteria were: young (age: 28.71 ± 7.45), male, native French speakers, right-handed, under or postgraduate. Non-inclusion criteria were sensorial deficits, neurological, psychiatric or emotional disorders and/or being under the influence of any substance capable of affecting the central nervous system. All subjects received complete information on the study's goal and experimental conditions and gave their informed consent.

2.2 Visual Variables

The Semiology of Graphics describes seven visual variables that form the most basic elements of graphic composition [1]. These are position (the spatial variables), size, value (tone), texture, orientation, shape, and hue. These visual variables have perceptual lengths that can be matched to the data scales to support four styles of graphical perception. The visual variables used in our work are inspired from Bertin classification of visual properties (Fig. 1).

Mackinlay built scales of expressivity (monosemic, but dependant on a precise graphical language) and effectiveness (depend on the human perceptual capabilities) to assess alternative designs (Fig. 2). This scale depends on the data type. The visual property higher in the chart is perceived more accurately than those lower in the chart. The grey items are not relevant to that type of information. The quantitative data type ranking as been experimentally verified by Cleveland [2]. Independently of the data type, the best way to represent the data is to code it with a position on a scale. If we want to represent the speed of an aircraft (quantitative data), we can use the length of a line (speed vector). The aircraft position number in the landing sequence (Ordinal) is better coded using the color saturation than length.

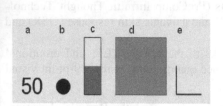

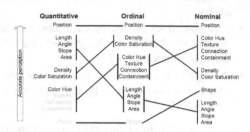

Fig. 1. The 5 selected visual characterisations. Respectively: text (a), surface (b), framed rectangles (c), luminosity (d) and angle (e)

Fig. 2. The Mackinlay ranking of perceptual task

2.3 Computerized Experimental Tasks

The task (Fig. 3) is a simplified reproduction of the air controller main activity: the monitoring of conflicts. The subject was instructed that he had to designate with the mouse, which object is in conflict with the object of reference positioned at the center of the screen. The conflicting object is the one that has the most similar visual variable setting. This task has to be performed as accurate and as quick as possible. The 5 visual variables which provide the value of the vertical information (altitude) were manipulated to assess their efficiency to draw attention to the object that entered in conflict and offer efficiency quantitative information. A total of 10 objects were simultaneously presented at the screen: nine objects to be scanned plus the reference object at the center. The visual variables coded information in a 9 step levels of graduation, for instance the angle value was manipulated from 10 degrees to 90 degrees. We did not use extreme scale values (e.g. 0 and 100 degrees, empty and full framed-rectangles etc.) to avoid as most as possible pop out effect. The target object (object that was the most similar to the center one's) was separated from only 5 graduation steps from the reference object (the other being separated from at least 10 graduations steps) and thus objects are never identical to the reference one's, this was a simple way to maintains cognitive comparisons processes and thus avoid fast visual matching of the two objects.

2.4 Experimental Design

The design was blocked, which means that a set of trial of the same visual variables was presented. Each visual variable were repeated 18 times in a block, the order in which the subject performs each block was randomized to avoid order effects. Each trial duration was of 10 seconds and the ITI (inter stimulus interval) was 10 seconds too. The stimulus display time limitation was a good way to bring temporal pressure and avoid potential longer reaction time for some subjects. The fixed 10 seconds ITI improved the quality of the phasic galvanic skin response analysis. The stimulus was masked just after the response. Each block lasted exactly 6 minutes. A training session was set up of an additional block (9 trials) and was performed before the experimental stimuli (not included in subsequent analyses) as a means of controlling the initial SCRs and HR response produced as a result of stimulus novelty. Indeed, psychophysiological measurements may be jeopardized by the stimuli novelty. The subject performance was measured by the percentage of correct answers (HITS) and the mean reaction times (RT) for correct and incorrect answers.

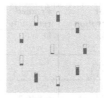

Fig. 3. A view of the task screen, here the comparison cognitive process is performed on framed rectangles. The subject had to designate the object the most similar to the center one (the reference object).

2.5 Psychophysiological Measurements

Heart rate and galvanic skin response were collected thanks to a ProComp Infiniti (©Thought Technology Ltd.). In practice, establishing mean physiological values for several groups of subjects for an entire task is meaningless because of inter-individual variability, delta values must be used (difference between working and resting states) for measuring the autonomous nervous system (ANS). Because all our subjects performed the task with the 5 five visual variables, no delta values were needed to be computed. The psychophysiological measurements were started at the launching time of the first block and continuously recorded until the end of the experiment. The block design allowed to specify relatively short ITI, indeed the visual conditions being the same in a whole block, no evoked physiological response of a given condition could overlap on a subsequent one. This is especially critical for the heart rate response that can last several minutes before coming back to its baseline level.

2.6 Subjective Assessments

The computerized version of the Nasa TLX allowed subjects to assess the mental load felted after each visual variable. The NASA-TLX provides an overall workload score based on a weighted average of ratings on six dimensions (mental demands, physical demands, temporal demands, own performance, effort and frustration). Self report of stress, anxiety (on a 1-9 scale) was collected immediately after the end of each block. User experience quality (joy to use, emotional experience, comfort etc.) was also collected.

3 Results

The main effect of the visual variables on behavioral results was tested with the Kruskal-Wallis non-parametric ANOVA. This test was also used to assess the effect of the visual variable on the psychophysiological responses and the subjective self assessments results. Kruskal-Wallis multiple comparisons were used for paired analysis. All analyses were done with Statistica 7.1 (© StatSoft).

Table 1. Mean values, standard deviation and p-values for the behavioral results across the 5 visual characterizations. Time units are in milliseconds. Kruskall-Wallis test shows significant variance: *p≤ 0.05, **p≤0.01.

Variable	Angle	Color saturation	Framed rectangle	Surface	Text	p-value
% of hits	82,53 (±10,35)	66,66 (±13,22)	88,09 (±4,99)	78,56 (±11,30)	86,49 (±2,97)	.020*
Mean RT	4428,8 (±1665,2)	3663,6 (±1414)	4202,5 (±1024,4)	3867,3 (±1274,1)	4050,4 (±1123,5)	.091
Mean Correct RT	3720, 1(±1876,7)	2441,5 (±1335,5)	3665,3 (±846,7)	3038,3 (±1393,4)	3440,7 (±1062,7)	.050*
Mean incorrect RT	826,8 (±358,8)	1222 (3±88,1)	626,75 (±317)	829 (±403,8)	609,7 (±213,5)	.010**

3.1 Behavioral Results

Kruskal Wallis ANOVA revealed several behavioral variables significantly affected by the type of visual variables (Table 1): the percentage of hits, the mean correct and incorrect response times. Wilcoxon paired analysis showed that the percentage of hits for color was significantly lower than rectangle (p=0.017), surface (p=0.046) and text (p=0.027). Concerning the mean responses time, difference were find between angle and surface (p=.023). Correct response time revealed that color generated lower RT than angle (p=.042) and rectangle (p=.017). On the contrary, incorrect response times were significantly higher for color vs. rectangle (.027) and vs. text (p=.017).

Additional analysis (Fig. 4) showed that responses times were shorter for incorrect responses vs. correct ones (p<0.01).

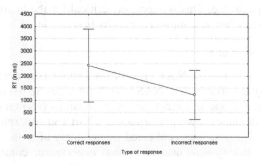

Fig. 4. Response time in milliseconds according to the type of responses, correct or incorrect

3.2 Psychophysiological Results

Thanks to the block design, tonic HR and RR could have been analyzed on a whole time course for each given visual variable. The GSR were measured in micro Siemens and analyzed off-line. The tonic GSR analysis was not well adapted to this kind of experiment because of its very long delay to come back to its baseline state, thus the measure was the magnitudes of the SCR. Responses were computer scored as a change in conductance from the pre-stimulus level to the peak of the response. Following information provided by Dawson et al. (2000) (1–4-s latency and 1–3-s rise time), the minimum level occurring within 1–3 s from word presentation was subtracted from the peak value occurring within a 3–7-s window with a minimum value of zero in the absence of a response. Kruskall-Wallis non-parametric ANOVA didn't showed overall significant difference between the five visual variables concerning the psychophysiological variables (Table 2).

Table 2. Mean values, standard deviation and p-values for the behavioral results across the 5 visual characterizations

Variable	Angle	Color saturation	Framed rectangle	Surface	Text	p-value
Mean HR (in bpm)	71,32 (±12,64)	65,99 (±3,48)	66,21 (±10,31)	67,76 (±10,74)	67,76 (±8,76)	.487
Mean RR (breath min)	15,48 (±0,72)	15,79 (±1,01)	15,41 (±0,99)	15,52 (±0,76)	15,42 (±0,7)	.296
GSR (in µS)	0,155 (±0,075)	0,192 (±0,097)	0,214 (±0,088)	0,258 (±0,144)	0,263 (±0,126)	.406

Although the ANOVA didn't show overall effect, Wilcoxon paired analysis revealed a significant difference between angle and text (Fig. 5) concerning the galvanic skin response (p=.042): the text generated significant higher galvanic skin responses.

The figure 6 is an example of a GSR recording during 3 trials (red bars). The GSR is usually characterized by the delay, the rise time and the recovery half-time.

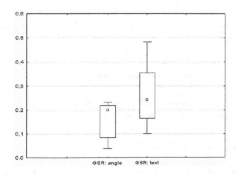

Fig. 5. Galvanic skin response (µS) during visual comparisons performed on angle and text visual variables

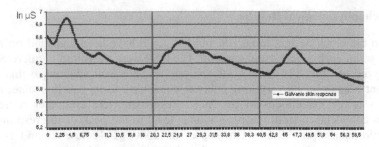

Fig. 6. Example of the galvanic skin response (µS) for one subject during the text condition. Red bars indicate the display time of the stimuli. Timeline unit is in second.

3.3 Subjective Results

Main ANOVA analysis revealed significant differences concerning the mental load rated thought the Nasa TLX and also concerning the user experience self rating (Table 3, Fig. 7).

Table 3. Median values, standard deviation and p-values for the Subjective data comparing the 5 visual characterizations. Kruskall-Wallis test shows significant variance: *$p \leq 0.05$, **$p \leq 0.01$.

Variable	Angle	Color saturation	Framed rectangle	Surface	Text	p-value
Nasa TLX mental load	51,5 (±4,47)	36,91 (±10,29)	37,33 (±9,49)	43,25 (±14,75)	31 (±11,39)	.050*
Anxiety rating	2,14 (±1,67)	2,14 (±1,21)	2,28 (±1,25)	2,42 (±1,9)	1,71 (±1,11)	.606
Stress rating	3,28 (±1,6)	3 (±1)	2,85 (±1,34)	3,28 (±1,88)	2,71 (±1,11)	.807
User XP rating	3,42 (±1,9)	3,57 (±1,51)	4,57 (±2,14)	4,42 (±1,9)	5,85 (±0,69)	.050*

Concerning the mental load (Fig. 8), the paired analysis revealed higher load for angle vs. text (p=.042) and vs. color saturation (p=.017).

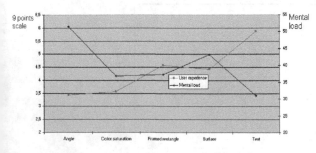
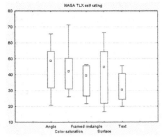

Fig. 7. Nasa TLX workload and user experience self rating according to the five visual variables

Fig. 8. Nasa TLX workload self rating according to the five visual variables

On the other hand, the user experience analysis revealed better user experience for text vs. angle (p=.027) and vs. color saturation (p=.017). The figure 9 shows user experience self rating and the GSR for the five visual variables.

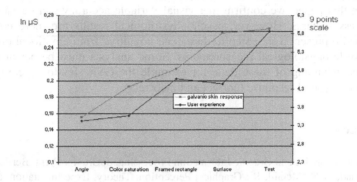

Fig. 9. Galvanic skin response (μS) and user experience self rated on a 9 points scale

4 Discussion

Our behavioral results were coherent with the Mackinlay [3] ranking of quantitative perceptual task. Indeed, the observed accuracy where ordered accordingly to its classification: length (framed rectangle) > angle > area (surface) > color Saturation. The reaction times are more complex to interpret. The correct response times may appear contradictory, indeed, the variable that provided the worst accuracy has generated the shorter reaction times. This result could be interpreted as a different strategy of the subject for this variable, unable to reach a very good accuracy; a quick visual comparison was certainly performed. This hypothesis is supported by the incorrect reaction times that were significantly lower than the correct ones concerning the 5 visual variables. Incorrect answer may be associated, at least for a part, with too low cognitive processing time.

We did not find major results concerning the psychophysiological measurements. The small number of subject had certainly contributed to limit the significance of the statistical analysis; also the ATC real activity is more stressing because of the potential impact on the security of controller's decisions. However, paired analysis showed that the text variable had a stronger effect on GSR than the angle. No significant stress or anxiety was provoked by the different visual variables, the rating remained very low. Upon this basis, the GSR observed during text may be rather interpreted as a general arousal of ANS more linked with emotional factors linked to the user experience than with the workload or a psychological stress, indeed, the text variable rated as the less loading task was the most appreciated as showed by the user experience assessment. This result validates the current design choice for air traffic controller radar screen where the altitude of each aircraft is displayed as text.

5 Conclusion

We compared the accuracy of the Bertin visual variables with the psychophysiological and subjective point of view. The results were coherent with the Mackinlay ranking. In this paper, we confirm that visual variable accuracy can be validated by empirical assessments. However, psychophysiological results are more complex to interpret and represent a first step to dissociate cognitive from emotional effects of a task on psychophysiological measurements. Some analysis must be performed for a deepest understanding of these data, for example the GSR should be compared according to the type of response (correct and incorrect).

References

1. Bertin, J.: Graphics and Graphic Information Processing. deGruyter Press, Berlin (1977)
2. Cleveland, W.S., McGill, R.: Graphical Perception: Theory, Experimentation, and Application to the Development of Graphical Methods. Journal of the American Statistical Association 79 (1984)
3. Mackinlay, J.: Applying a theory of graphical presentation to the graphic design of user interfaces. In: UIST (1988)
4. Gaillard, A.W.K.: Stress, workload, and fatigue as three biobehavioural states: a general overview. In: Hancock, P.A., Desmond, P.A. (eds.) Stress, Workload, and Fatigue, pp. 623–639. Erlbaum, Mahwah (2001)
5. Gevins, A., Smith, M.E.: Neurophysiological measures of cognitive workload during human–computer interaction. Theoretical Issues in Ergonomic Science 4(1–2), 113–121 (2003)
6. Iqbal, S., Adamczyk, P., Zheng, X., Bailey, B.: Towards an index of opportunity: understanding changes in mental workload during task execution. In: Proceedings of the SIG-CHI conference on Human factors in computing systems, Portland, Oregon, USA (2005)
7. Fairclough, S.-H., Venables, L., Tattersall, A.: The influence of task demand and learning on the psychophysiological response. International Journal of Psychophysiology 56(2), 171–184 (2005)
8. Ahlstrom, U., Friedman-Berg, F.J.: Using eye movement activity as a correlate of cognitive workload. International Journal of Industrial Ergonomics 36(7), 623–636 (2006)
9. Võ, M., Jacobs, A., Kuchinke, L., Hofmann, M., Conrad, M., Schacht, A., Hutzler, F.: The coupling of emotion and cognition in the eye: Introducing the pupil old/new effect. Psychophysiology 45, 130–140 (2008)
10. Causse, M., Pastor, J., Sénard, J.-M., Démonet, J.-F.: Unraveling emotion, difficulty and attentional demand in psychophysiological responses during executive tasks. In: First Meeting of the Federation of the European Societies of Neuropsychology, Edinburgh, Scotland (2008)

A Photo Correlation Map Using Mobile AP II for Scenario-Based Design

Yu-Li Chuang[1] and Makoto Okamoto[2]

[1] Future University-Hakodate, Graduate School of Systems Information Science,
Kamedanakano 116-2 Hakodate, 041-8655, Japan
[2] Future University-Hakodate, Kamedanakano 116-2 Hakodate,
041-8655, Japan
{g3107003,maq}@fun.ac.jp

Abstract. We aim to explore a potentially valuable tool for scenario writing, focusing on the collection of user experiences organized around photos. We have developed a tool, Mobile AP II, to produce a comprehensive context classification based on a sample scenario to demonstrate how the context of photos and their relationships can be integrated to assist with scenario-based design. An efficient scenario-building technique for concept development and design leads to a simple user-modeling framework whereby the Mobile AP II system can learn to display query results based on photo relationships provided by the designer. The resulting retrieval, browsing and visualization can adapt to the user's selection of content, context and preferences in style and interactive navigation. Our purpose is to develop rich and flexible methods and concepts that can incorporate users' descriptions and their current and potential use of a scenario into the very design reasoning about such a system.

Keywords: Activity probes, Scenario-based design, Photo category.

1 Introduction

Understanding why users do things as they do and what they are trying to achieve in the process allows us to concentrate on the human activity that is very important for interaction design. It is very difficult for a user to explain what he does or even to accurately describe how he achieves a task.

Observation is a useful data gathering method at any stage during design development. Early in design, observation helps the designer understand the users' context and goal. Observations conducted later in development may be used to investigate how well the developing prototype supports the design tasks and goals[1].

Data collected by observation is always fragmentary, inconsistent and possibly biased. The information comes from different actors and their own observations, and the designer must interpret the data to construct an evidence-based opinion about the underlying reality[2].

Narratives provide an easy and intuitive way for the designer to communicate ideas and experiences. They can be constructed from smaller anecdotes or repeated patterns that are found in the data. Narratives are used extensively in interaction design, both

M. Kurosu (Ed.): Human Centered Design, HCII 2009, LNCS 5619, pp. 177–183, 2009.
© Springer-Verlag Berlin Heidelberg 2009

to communicate the findings of investigative studies, and as the basis for further development, such as product design or system enhancements.

Narratives collected through data gathering may be used as a basis for constructing scenarios. Scenarios are hypothesized stories about people and their daily lives. They are a powerful technique for interaction design and can be used throughout the life cycle of a product or system.

Scenario-based design is a method for envisioning, clarifying and refining ideas for the design of interactions, especially between humans and machines. When using such a method, designers are required to make a wide range of observations of user activity, and to discover user states of mind and identify issues.

Using scenarios can capture more information than other methods do about the user's goals and the context the user is operating in. The context might include details about the workplace or social situation, and information about resource constraints. This provides more help in understanding why users do what they do. In much current design work the user's goals and context are often assumed implicitly, or may not be taken into account at all, when they should be used to help identify the key elements of the user experience and how they influence user activity[3].

Normally, scenarios are written as text, perhaps accompanied by some sketches, which people must then study and comprehend. It is clear that the traditional methods of investigating the user experience have serious limitations.

Using only text and sketches, it is not easy to understand a real user's experience in a scenario. Users often lack the language and introspective skill to explain their real experiences and needs, and it requires significant effort from the designer to ferret out useful design concepts from what users can communicate[4].

As technology becomes more embedded in our daily lives, the prospect is improving that a designer can achieve positive community outcomes by using scenario-based design more efficiently and creating a platform for collaboration and communication with users.

A ComScore survey in late 2007 found that 80 percent of those surveyed have taken a photo with their mobile phone [5]. Sharing camera phone photos with friends and parents has become popular, and plays an increasingly important role in everyday life in Japan. Photographs can mediate intimacy among families and friends by evoking shared memories. Where mobile phones were once mainly used as a voice communication tool, we can now have users send their observations by camera phone, providing a primary source of observational data for designers [6].

We are developing a system called Mobile AP II, which is a self-report tool to collect user experience data via hand-held devices such as camera phones to assist the process of scenario-based design.

Our intention was to create a universal platform, and encourage users to report their experiences by using a camera phone to deliver their messages to a central repository where they can support the developer in creating a scenario. This provides more accurate data than relying on the user to recall events and report on them afterwards .

2 Method

We are developing a system called Mobile AP (Activity Probes) II, which is an activity collection tool for scenario-based design. Photographs taken by a mobile camera phone can be stored to create instant special events with other users. The photos can also be edited by adding a positive or negative comment, and tagged and used as a collection of user experiences. This can enhance a designer's understanding of scenario-based design.

The purpose of this study was to determine if the use of the Mobile AP II system (Figure 1) to transmit a user's awareness photo, together with support to create a scenario, would improve the description in a scenario. The subjects were asked to take and submit awareness photos with their camera phone.

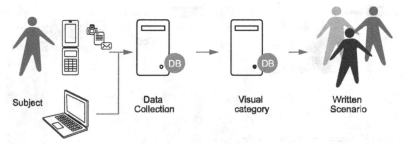

Fig. 1. Mobile AP II framework

It is possible to transmit not only a camera image, but also a user's description of the situation and his emotional reactions, to be recorded by the Mobile AP II system and made available to the designer. To facilitate research, the Mobile AP II system can collect the experiences of a user and also examine the correlations between user-sent photos of situations, so the designer can easily record, observe and categorize the experiences of the user as an aid to scenario writing. The purpose of this study is to verify whether or not classifying photographs into visual categories improves the practice of scenario-based design within an interactive design framework.

In this study, we relied heavily on the linking of the delivery method and a PC system to create a communication platform between designers and users. We gave users an exclusive camera phone mail address for each event. At the same time, we constructed the Mobile AP II system as shown in Figure 2, consisting of a receiving server and a visual categorizing tool for scenarios.

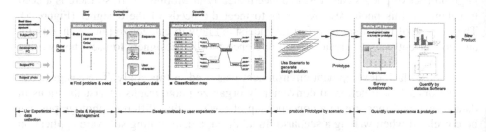

Fig. 2. System diagram of Mobile AP II system

First, we invited subjects to record everything happening in daily life or some specific activity by camera phone, and send the images to the Mobile AP system. Data were collected primarily as a means of increasing user awareness and recording special events, with photos stored for later analysis and categorization (Figure 3).

Subjects were examined individually about the events. They were asked to record and submit the objects of their attention with the camera phone, and transmit a description of the photo content plus their own positive or negative responses. This allowed the designers to communicate and share the gathered data with subjects through the Mobile AP platform. At the same time, we could examine the photos visually, as a basis for writing scenarios.

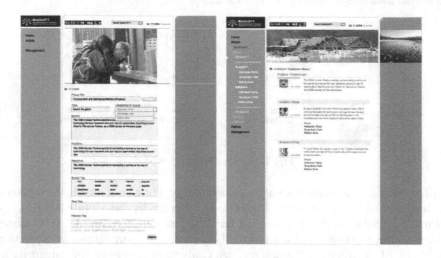

Fig. 3. Data gathering with Mobile AP II

After adding tags to the photos, the user can express either satisfaction or disappointment. Each user can thus provide a personal point of view on, for instance, cultural heritage items, with reference to information provided by experts. Having the user's point of view in the description of an item can help bring multiple users into a mutual relationship with a photo.

As scenarios can often be disorganized, a system to organize scenarios is needed. A large number of photos can be stored electronically in Mobile AP II, so there is a need to categorize this photo resource so as to meet the needs of the designer (Figure 4). The purpose of this study is to further understand how to categorize photos in a useful and convenient way, to create a good reference tool for designers to use when writing scenarios.

Adobe's Flash development environment was used in this study. The designer can use the drag-drop function to conveniently organize photographs as visualizations in data format and then create a concept map of the way that different photographs can be correlated when writing a scenario. In addition, a data mining software application was used to match tags, develop an association rule map for the photos, and to map out a scenario for new creative and collaborative tools to further aid the designer.

Users can more efficiently and effectively understand large volumes of information with the aid of visualizations (Figure 5). Our correlation map function can provide an overview of the correlation between a photograph and its tag, and can be used to satisfy various needs in the design process.

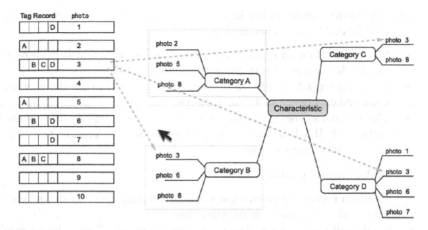

Fig. 4. A simplified diagram of the Mobile AP II system database architecture

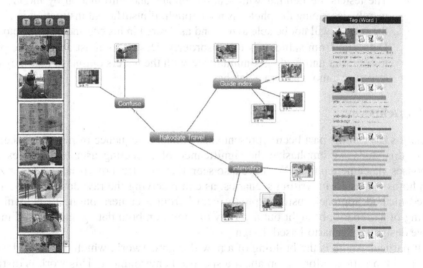

Fig. 5. Example of a visual category map of Mobile AP II

A key point to note is that Mobile AP II is a useful data gathering technique at an early stage of design; observation helps designers understand the users' consciousness and orientations. Later in development, e.g. in evaluation, Mobile AP II can be used to investigate how well the developing scenario supports these tasks and goals. map out a scenario for new creative and collaborative tools to further aid the designer.

3 Discussion

Observing users in their natural environment yields more accurate information about their behavior, and makes it possible for the designer to understand users' goals. When asked to comment on Mobile AP II in the post-study interview, the subjects were generally positive about its use in scenario development. Also, scenario-based design received favorable comments, such as:

- Mobile AP II could record observations at any time, and help show how we should use scenarios in different circumstances.
- The designer will be able to observe with ease and to get the user viewpoint and experience in the early design stage.
- The visual category tool can be more tightly integrated with photos by mobile AP II when analysis is performed on the scenario keeping the user's point of view in mind.

The study exhibits several weaknesses:

- A camera phone is a convenient way of recording data, but it can be difficult to write and observe at the same time
- People will have some preferences about the photos. These should receive more attention as a validity concern depending on the test requirements.
- The testers are familiar with sending photos and information by mobile, but merely describing the photos is not enough. If insufficient instruction is given, the subject will not be able to respond as desired in his comments on photos.
- Data sent from subjects is too disorderly. Designers must arrange the photos in order and further communicate with the testers online so that they can structure a more accurate database.

4 Conclusion

Scenarios have in the past been represented as simply a sequence of maps and scenes, but in this study we emphasize the significance of collecting user experiences and responses via camera phone. We have also seen that the effect of visually categorizing the photos can assist in writing scenarios, as can receiving the raw data quickly. This mixed-method approach, using a photo record from a camera phone to help in the writing of a scenario, brought out a variety of views on both data gathering and interactive design for scenario-based design.

Of particular note is the building of a new design network, which connects the designer with practical information about a specified environment. This work is distinct in another important way, in that our efforts were directed not to enhance the direct sharing of subjects' experiences with others, but to share a variety of media content in the form of memorable stories. The outcome of this research suggests that Mobile AP II as a framework for scenario development needs to be made more convenient for the designer. In the future, we plan to develop participatory design tools farther, and test them with higher-level design tasks.

References

1. Helen, S., Yvonne, R., Jenny, P.: Interaction Design: Beyond Human-Computer Interaction. Wiley press, West sussex (2007)
2. Jaesoon, A.: ctivity Theory for Designing Ubiquitous Learning Scenarios. Innovative Techniques in Instruction Technology. E-learning, E-assessment, and Education, 205–209 (2008)

3. Manasawee, K., Eamonn, O.: Modelling Context: An Activity Theory Approach. In: Markopoulos, P., Eggen, B., Aarts, E., Crowley, J.L. (eds.) EUSAI 2004. LNCS, vol. 3295, pp. 367–374. Springer, Heidelberg (2004)
4. Okamoto, M., Komatsu, H., Gyobu, I., Ito, K.: Participatory Design Using Scenarios in Different Cultures. In: Jacko, J.A. (ed.) HCI 2007. LNCS, vol. 4550, pp. 223–231. Springer, Heidelberg (2007)
5. ComScore, Inc.: Mobile Phone Web Users Nearly Equal PC Based Internet Users in Japan, http://www.comscore.com/press/release.asp?press=1742 (retrieved September 20, 2007)
6. Heekyoung, J., Kay, C.: Exploring Design Concepts for Sharing Experiences through Digital Photography. In: Extended Abstracts of CHI 2007. ACM Press, New York (2007)

Accelerating the Knowledge Innovation Process

Guillermo Cortes Robles, Giner Alor Hernández, Alberto Aguilar Lasserre,
and Rubén Posada Gómez

Instituto Tecnológico de Orizaba, Postgraduate Department,
Av. Oriente 9 No. 852, 94300 Orizaba, Ver. México
{gcortes,aguilar,galor,rposada}@itorizaba.edu.mx

Abstract. The generation of ideas or new concepts is the steppingstone of the innovation process. Nevertheless the transformation of those ideas in new or improved products, services or processes demands the mobilization of a huge diversity of knowledge. In this document is proposed the integration of the Theory of Inventive Problem Solving (TRIZ) and the Case-Based Reasoning (CBR) process in order to conceive a solving process capable to guide creativity while generating innovative solutions and also to store, index and reuse knowledge with the aim to accelerate the innovation process.

Keywords: Innovation, TRIZ, CBR, Knowledge, Problem Solving Process.

1 Introduction

Innovation has become the main source of value and competitiveness in nowadays market. This topic has been studied in several surveys that have explored the challenges to face when trying to innovate [1], [2]. It is then natural the effort that enterprises have dedicated to manage this complex process in order to improve its performance. This effort involves the development of organizational structures and tools for capturing and transforming ideas, concepts and knowledge in new or improved products, services or processes. It also includes the creation of technical tools to minimize time-to-market, to reduce cost of new products development and to mobilize efficiently the available knowledge within and outside the enterprise frontiers. Therefore, a structure that enables idea generation to solve the problems related to new products development but also capable to capture, store and reuse knowledge as a mean to accelerate the innovation process is highly desirable. In this document is proposed the integration of the Case-Based Reasoning process and the TRIZ theory as a tool for supporting idea generation and knowledge reutilization with the aim to support the innovation process.

Several are the reasons that impel this integration:

(1) Because knowledge must solve problems, not chance. Different from other techniques that assist idea generation, the TRIZ theory is a knowledge-based approach for problem solving that had capitalized knowledge from a vast variety of technical domains. This knowledge has been arranged in such a way that it is available when solving inventive or innovative problems. This condition produces an environment where is possible to transfer strategies, principles and problem solving heuristics that

M. Kurosu (Ed.): Human Centered Design, HCII 2009, LNCS 5619, pp. 184–192, 2009.

have proved its efficacy in other domains in order to increase the efficiency of the problem solving process [3]. (2) TRIZ mobilizes knowledge in a high level of abstraction, nevertheless an approach capable to store, index and reuse specific knowledge it is also need it. The Case-Based Reasoning approach possesses those abilities [4]. (3) TRIZ and CBR are both emulating a central human problem solving process: analogical thinking. This is according numerous authors the intrinsic human process for problem solving [5], [6].

The integration of several TRIZ concepts and the Case-Based Reasoning (CBR) is analyzed in the next three sections: first and second section briefly introduces the TRIZ theory and the Case-Based Reasoning approach, with the aim to show its complementarily. In third section a succinct description of the synergy is presented to finally describe a solved case.

2 The TRIZ Theory

The TRIZ theory (Russian acronym for "Theory of Inventive Problem Solving") has been conceived to generate solution avoiding tradeoff. This approach[1] for problem solving has its origins in the former USSR, where it was founded by G. Altshuller and other scientists [3]. One of the main advantages of TRIZ is that the solution space is not explored randomly. The TRIZ toolbox gives some directions that should be explored in order to derive a solution (the tool that should be applied depends on the nature of the problem). Consequently, TRIZ has the capacity to restrict the research space for innovative solutions and to guide thinking towards solutions or strategies that have demonstrated their efficacy in a past similar situation, besides TRIZ produces an environment where the generation of a potential solution is almost systematic [6].Of course TRIZ does not give a "ready to use" solution but it proposes some vectors that direct the search for finding innovative solutions, then it leaves place to the designer creativity. Four areas were analyzed to establish the foundations of TRIZ: (1) The global patents database (more than 3 millions of patents have been analyzed). (2) The analysis of scientific literature. (3) The analysis of psychological behavior of inventors and (4) Analysis of existing methods and tools for problem solving.

This analysis revealed the cornerstones of TRIZ that are enclosed in a set of concepts and tools that helps to solve non-routine problems or inventive problems. These concepts and tools give access to the best practices in the whole technical domain thus increasing the creative potential of designer. Among the most important conclusions and TRIZ concepts are next:

1. Problems and solutions were repeated across industries and sciences.
2. Innovations used scientific effects outside the field where they were developed.
3. A set of evolution patterns for technical systems exists. During its life cycle, a system is always evolving and this evolution is governed by objective laws. Thus, knowledge about those patterns is useful to foretell next stages of a product or technology.
4. Patterns of technical evolution were repeated across industries and sciences.

---

[1] TRIZ has been classified as a theory, methodology, tool, a set of heuristics, etc.

5. Ideality is a goal in every system. All systems evolve towards the increase of their degree of ideality. One way to measure the ideality is to use the Ideal Final Result (IFR) which is a psychological concept that allows finding the best solution for a complex problem, without taking into account cost, time, space or any problem constraints. This ideal system is often a utopian system but it guides reflection toward rarely explored directions.
6. In TRIZ, problems can be formulated in terms of contradiction. An inventive problem contains at least one contradiction, and an inventive solution overcomes totally or partially this contradiction. Several types of contradictions have been identified, but in this document only physical and technical are defined. Technical contradictions exist when any tentative to improve the performance of a useful function or characteristic in a system, produces an unacceptable deterioration in a second useful function in the system. It represents a conflict between two subsystems or characteristics. A physical contradiction occurs when a component or element in a system demands simultaneously two mutually exclusive states: a surface must be smooth and rough. It represents a conflict in the same subsystem. Contrary to classical methods for creativity stimulation (brainstorming, trial and errors, etc.), TRIZ refuses trade-off and tries to eradicate the contradiction.

Next section introduces the second component in the synergy: the Case-Based Reasoning.

3 The Case-Based Reasoning (CBR) Process

Artificial intelligence and more precisely knowledge management approaches try to use past experiences in a domain to solve new problems. The main difficulty is to find a way to store, retrieve and reuse knowledge inside an enterprise but also, to define the mechanisms to filter the knowledge in the surrounding enterprise environment. Coming from AI, CBR is a very useful approach to manage knowledge. The main idea in CBR is that similar problems have similar solutions. Basically in the CBR process users try to solve a new problem by establishing similar patterns between the initial problem and some previous experiences (solved problems). Then the CBR process uses and adapts earlier successful (or failed) solutions in order to solve the new problem. This process is at the core of everyday human problem solving.

The CBR method is a cyclic process involving at least five stages: represent, retrieve, reuse, revise and retain. The first stage or representation consists only in obtaining the relevant features that characterized a problem (i.e. components, apparatus, flow rates, pressure, temperature, etc.). With this information is possible to start the retrieving stage in which the problem to solve is compared with the cases stored in the memory with the aim to identify the most similar. Then if one or various stored cases match with the target problem, the most similar case is selected to reuse its solution. Subsequently, the derived solution must be revised, tested and repaired if necessary to increase the possibilities to obtain a satisfactory result or to avoid failure. Finally the new experiences which comprise failure or success, but also the strategies to repair and implement the final solutions (among other particular features), are retained for further utilization and the previous case memory is updated [7].

The inherent process of the CBR reduces time when solving problems because it gives an initial model for deriving solutions. Usually, it is more efficient to solve a problem from an existing starting point than to develop the whole solution from nothing. For a good performance of a CBR system, the case base must cover the whole or an important part of the problem space (all the problems that may appear in the specific domain of application). Consequently, the efficacy of the systems relies on the structure, quantity and quality of the stored cases.

Both approaches are useful to solve problems, nevertheless their objectives are different. Among the main differences between TRIZ and CBR are:

Table 1. Differences between TRIZ and CBR

CBR	TRIZ
Limited to a specific domain (specificity could be a barrier to creativity)	Transversal application (all technical fields), an environment to stimulate creativity
Routine design	Inventive design
No solution if the initial problem can not find a similar case	Gives a way of solution for each problem
Produces a solution from an initial model	Produces a solution starting from "nothing"
Posses a memory: solutions are produced rapidly. System become more efficient by learning	No memory; resolution process redeployed each time
Easy for use, thanks to its affinity with human resolution process	Difficult to use because of its particular way to tackle problems and its variety of tools

Once described the differences and complementarily between both approaches, it is possible to describe the integration of TRIZ and CBR.

4   The TRIZ-CBR Integration

Altshuller discovered that very different technical systems and processes share similarities in their evolutions. For example the same generic problem had been pointed out and solved with the same generic principle of resolution but in different technical domains and sometimes the solutions were separated by many years. Consequently, Altshuller thought that if inventors or engineers can benefit from successful solutions found in others disciplines, the innovation process will be more efficient [8]. A tool called contradiction matrix crystallize this point of view. During the patents analysis, Altshuller notices that technical contradictions can be expressed in terms of conflict between two parameters (with a limited number of parameters): one improved and the other one damaged. Only 39 parameters were extracted to describe all the contradictions encountered in patents. Representing technical contradiction as a combination of two parameters requires a broad interpretation of them, so they are generic for many engineering fields. Finally, a 39×39 matrix was built. On the line, is located the improved parameter, on the column the damaged one or the parameter that prohibit an improvement. For one contradiction, the cell at the intersection of the line and the column indicates the principle(s) to explore in order to solve it. This matrix was updated in 2003 and the number of parameters was increased to 48.

Through the contradiction matrix, TRIZ opens up the world patents bases for identifying principles that may offer possible solutions. Based on the advantages previously mentioned, the contradiction matrix was transformed in the case memory for the model schematized in next figure:

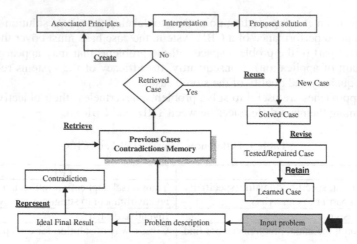

Fig. 1. The TRIZ-CBR solving problem process

The solving process starts whit the formulation and representation of the problem to solve. The process demands at least four initial features concerning the problem description: (1) the system where the problem is located; (2) the type of problem or objective: reduction/elimination of harmful function, improvement of a characteristic or new functionality; (3) the goal to reach; (4) the resources identified in the system. After this step, the ideal solution is also stated in order to offer a guide for the search direction of the future solution. Then the problem is stated as a contradiction to obtain a more robust problem description. With those data (contradiction and the other features) is possible to explore the memory and search for a similar problem. At this point of the synergy process, two different sub processes can take place:

1) The retrieval offers a similar problem or set of problems. The most similar is selected to adapt its solution (to be used as initial solution). Here the similarity between two problems is calculated with a similarity global function:

$$SIM = \frac{\sum_{i=1}^{n} w_i * sim(f_i^I, f_i^R)}{\sum_{i=1}^{n} w_i} \tag{1}$$

Where f_i^I, f_i^R represent respectively the features i for the initial problem (I) and the retrieved cases (R), *sim* the local similarity function for this feature i and w_i the weight of the feature i. [9] discuss the different manners to measure local similarity, it depends of the type of feature value: semantic, symbolic, numeric.... *SIM* represents the global similarity. If various similar cases are found, the global similarity function ranks them. Moreover, the global similarity function can be customised thanks to the weight, in order to give more importance to one feature to others, which is the case of the tool presented in next section. When calculating similarity, technical contradiction is the most important factor, then the available resources in the system followed by the type of problem and the goal to reach.

2) The memory does not have any similar solved case or sufficiently similar case (the similarity global function has a too small value). Under this condition, the system offers inventive principles associated to the contradiction, by which a satisfactory solution could be derived. The matrix finds its initial use [10].

The process exposed in figure 2 is the basis for a tool that helps to solve problems stated as a technical contradiction (this model was tested with more than 100 patents). This tool has two different processes: first one is useful to load problems in the case memory. All the cases indexed and stored in the memory are evaluated by an expert(s) to decide about its pertinence. Second process facilitates the search in the case memory. In next example is shown a case that does not have a similar case in the memory.

5   A Solved Case: Proposing a New Product

A local SME offers us the opportunity to develop a new product. The enterprise was searching ideas for developing a totally new "outside living set" line. To deal with this problem a five stages process was deployed. Stage (1) was the recognition of costumer's needs. This information was gathered –locally- from clients and other similar products. International information was also collected. Stage (2) concerns problem definition. Available information about costumer's needs was the basis for defining "the design problem" which encompasses materials, production means, ergonomic specifications and marketing among other design stakeholders. Stage (3) affects concepts development that ideally should satisfy all product dimensions. Stage (4) involves concept validation. An expert panel was responsible to evaluate prototypes and to select the most promising product. Stage (5) transforms the selected concept in product specifications and production requirements.

In this document are described only stages 2 and 3 in order to show how the problem was undertaken.

• System description: an outside living set include at minimum one table and four chairs. The main useful function is to offer a surface for resting and supporting objects. The product is available in a high diversity of materials and shapes. Next figures show typical products:

Fig. 2. Typical products

• Problem statement: Costumer's information reveals that the living set should be adaptable, lightweight -easy to move-, resistant, easy to clean, aesthetically attractive, etc. But according the enterprise perspective, the most important feature is that should be different from similar products available in the market. Then it is necessary to propose a concept enclosing all those characteristics.

- The solution must satisfy next restrictions:
  (1) To be lightweight, (2) Maximal mechanical resistance (no more than 110 kg), (3) Cost should not exceed the typical price of available products, (4) Easy to move, clean and remove when is not in use, (5) The selected prototype should be produced with the available means.
- IFR (Ideal Final Result or more desirable result): one of the costumers offer us next requirement *"the living set should be there only when I need it and disappear the rest of the time"*. Reusing and complementing this description: the living set satisfies all costumer requirements and is available only when he/she needs it.
- Resources in the system:

Substance proprieties	Metal: conductivity, rigidity, mechanical resistance, thermal expansion, among others.
	Plastic: flexibility, lightweight, weather resistance, non-expensive, among others.
	Wood: rigidity, density, humidity, among others.
Shape	Space, void, among others.
Available fields: thermal, gravitational.	

- Contradictions: numerous contradictions were identified and arranged hierarchically. The first contradiction to solve was stability versus shape (this decision was taken by the team using the analytical hierarchy process). The associated principles to this contradiction are: 1 (segmentation), 4 (Symmetry change), 35 (Parameter changes), 17 (Dimensionality change), 7 (Nested doll) and 3 (Local quality).
- Deriving solutions from inventive principles. Principle 1 suggests *"to divide an object or system into independent parts"*. The set is already fractionated. Principle 4 proposes *"Change the shape of an object or system from symmetrical to asymmetrical"*. Principle 35 recommends *"Change an object's physical state. Change the concentration or consistency. Change the degree of flexibility"* any potential solution was obtained from those principles.
- Principle 17 suggests *"Move an object or system in two- or three-dimensional space. Use a multistory arrangement of objects instead of a single-story arrangement. Tilt or reorient the object, lay it on its side, use its other side"* and principle 7 proposes *"Place one object inside another; place each object, in turn, inside the other. Make one part pass through a cavity in the other"*. Principle 17 and 7 guide the search for a conceptual solution. The resource utilized to materialize this concept was existing space.

Fig. 3. Proposed solution

This concept was the basis for several prototypes. The idea is to make the living set easy to assemble/disassemble in such a way that costumers will have the living set only when it is necessary.

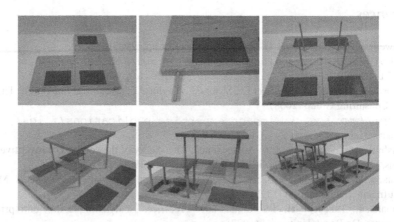

Fig. 4. Outside living set

This was the first prototype. This concept was transformed by using the evolution patterns proposing a new variety of products. The enterprise conserves those ideas.

6 Conclusions

The presented model offers a vector to transfer the solution from an identified analogous problem into a new target problem, reducing effort and time when solving inventive problems. The model also has a tool to guide creativity when there is not a case stored in the memory and the user needs to generate a completely new solution. The model combines the TRIZ ability to apply general knowledge in a very creative way and a framework that closely relates knowledge and action. But maybe the most important benefit of this model is that the model assists the learning process. The case memory transforms in a reusable way the experiences obtained while solving problems. This condition produces an environment that impulse knowledge sharing. The model exposed conserves the capacities and advantages of both components and minimize the identified disadvantages.

The main drawback of the model it is the identification of the right contradiction. The formulation of contradictions is not an exact process, is a subjective activity strongly influenced by previous experiences and knowledge. Thus, in the model it's necessary to assist users to accomplish this stage.

Acknowledgement

This work is supported by the General Council of Superior Technological Education of Mexico (DGEST). Additionally, this work is sponsored by the National Council of

Science and Technology (CONACYT) and the Public Education Secretary (SEP) through PROMEP.

References

1. Rowell A.: The Innovator's Toolbox. Aberdeen Group, (2009),
   http://www.aberdeen.com/summary/report/benchmark/
   5381-RA-empowering-difference-makers.asp
2. Andrew J., Haanaes K., Michael D., Sirkin H., Taylor A.: Measuring Innovation. The Boston Consulting Group (2008),
   http://www.bcg.com/impact_expertise/publications/files/
   Measuring_Innovation_Aug_2008.pdf
3. Altshuller, G.: Creativity as an exact science: The theory of the solution of inventive problems. Gordon and Breach Publishers (fourth printing) (1998)
4. Watson, I.: Applying Case-Based Reasoning: Techniques for enterprise systems. Morgan Kaufmann, San Francisco (1997)
5. Nonaka, I., Takeuchi, H.: La connaissance créatrice. La dynamique de l'entreprise apprenante. De Boeck Université (1997)
6. Terninko, J., Zusman, A., Zotlin, B.: Systematic Innovation: An Introduction to TRIZ. St. Lucie Press (1998)
7. Pal, S., Shiu, S.: Foundations of soft Case-Based Reasoning. John Wiley & Sons Publication, Chichester (2004)
8. Altshuller, G.: The Innovation Algorithm. Technical Innovation Center (1999)
9. Avramenko, Y., Kraslawski, A.: Similarity concept for case-based design in process engineering. Computers & Chemical Engineering 30, 548–557 (2006)
10. Cortes Robles, G., Negny, S., LeLann, J.M.: Case-based reasoning and TRIZ: A coupling for innovative conception in Chemical Engineering. In: Chemical Engineering and Processing: Process Intensification. Elsevier, Amsterdam (2008)

What Properties Make Scenarios Useful in Design for Usability?

Kentaro Go

Interdisciplinary Graduate School of Medicine and Engineering,
University of Yamanashi, 4-3-11, Takeda, Kofu 400-8511 Japan
go@yamanashi.ac.jp

Abstract. As described herein, we propose heuristics of scenario for designing usable products. From a structural viewpoint, a similarity exists between the definition of usability in ISO 9241-11 and the concept of scenario for designing a product, which suggests what elements a scenario should include and how designers should incorporate it into a human-centered design process. Particularly, this paper presents the argument that a scenario should include what a user accomplishes, sees, hears, and thinks, and how the user does them so that designers become capable of evaluating the effectiveness, efficiency, and satisfaction of goal achievement from a usability perspective.

Keywords: Guideline, heuristics, human-centered design, scenario, scenario-based design, usability.

1 Introduction

Use of scenarios for system design has been a commonly addressed issue in the field of Human–Computer Interaction (HCI) [1-7]. The HCI community has experienced scenario booms several times during the last three decades [8]. Recently, a new boom in the use of scenarios has occurred because persona-based approaches [9] have attracted the attention of the HCI community. Scenarios are likely to continue their long history of use in design.

Scenario-based design, as a technical term, encompasses a collection of methods that employ scenarios as a fundamental design representation and artifact in its design process and activities. Scenario-based design is not a single approach. Therefore, designers and practitioners in industry are confused about its use for designing a service, system, and product (artifact) with scenario-based design. They hope to have a clear usage guideline of scenarios.

This paper describes an attempt to provide design heuristics for scenarios based on our experience on work with usability and user experience divisions of major Japanese companies. It specifically examines how scenarios designed at the beginning of design process deal with a usability viewpoint. We begin our discussion about properties of scenarios from the definition of usability (ISO 9241-11 [10]). We argue the similarity between the nature of scenarios and the definition of usability.

M. Kurosu (Ed.): Human Centered Design, HCII 2009, LNCS 5619, pp. 193–201, 2009.

2 Components of the Scenario Concept and Usability Definition

To begin with, we clarify the properties of scenario concept and the definition of usability, which have similarities in their components. Analyzing those similarities and differences, we derive effective elements specified in scenarios from a usability viewpoint.

2.1 Scenario for Design

A scenario is a story. When it is used for designing an artifact such as a service, system, or product (artifact), a scenario might contain a description of its use. For example, scenarios for designing a mobile phone might include information about who is the user, what is the mobile phone, in what situation it is used, what goal or expectation the user wishes to achieve, and how it is done. To summarize, a scenario is a description that includes actors, their background information, and assumptions about the environment, actors' goals, objectives or expectations, and sequences of actions and events. Some applications might omit one element or might express it implicitly.

Scenarios can be of different forms such as narrative texts, graphics, images, videos, and prototypes [11]. For example, a typical scenario form is a narrative text in natural language like a novel. If it is prepared as a digital document, it can be shared easily with others. Another form is a storyboard used in movie production. The combination of a drawing or picture and its caption makes a story easier to understand.

Figure 1 presents a typical narrative scenario. In the scenario, the user Ichiro's sequence of actions is specified. The scenario is a command level scenario: its granularity of description is low and it specifies minute details of a user's action and response.

Ichiro double clicks on the cashbook icon on desktop to initiate its application program. A new window shows up on the screen. It contains a record entry; the last line represents the current funds remaining as 12,800 yen. Ichiro clicks on the empty entry at the bottom of the last record entry and types in 9,500 yen as the wages of his part-time job at a pizza restaurant.

Fig. 1. Typical narrative scenario in natural language

The scenario concept in general can be represented as a 4-tuple: (actors, goals and expectations, sequence of actions and events, and context of use). When the scenario concept is applied for envisioning a product, it might be extended to include the representation of a product idea. With the product, the main class of actors is that of users. They have goals to achieve. They work and act with the product or simply use it to achieve the goal under the circumstance of use. Consequently, scenario concept for designing a product can be represented as a 5-tuple:

$$(\text{product, users, goals, sequence of actions and events, context of use}). \tag{1}$$

2.2 Usability

Designers and practitioners in industry might employ the definition of usability in ISO9241-11 as a standard discipline. It defines usability as *"extent to which a product can be used by specified users to achieve specified goals with effectiveness, efficiency and satisfaction in a specified context of use."* In addition, the effectiveness, efficiency, satisfaction and context of use in the description are defined as follows.

- **Effectiveness:** Accuracy and completeness with which users achieve specified goals
- **Efficiency:** Resources expended in relation to the accuracy and completeness with which users achieve goals
- **Satisfaction:** Freedom from discomfort, and positive attitudes towards the use of the product
- **Context of use:** Users, goals, tasks, equipment (hardware, software and materials), and the physical and social environments in which a product is used

The specified elements of the usability definition can be represented as a 7-tuple:

$$(product, users, goals, effectiveness, efficiency, satisfaction, context of use). \qquad (2)$$

In a traditional development scene, a target artifact (which can be expressed as a product in formula (2)) is generally envisioned or assigned as a theme at the initial phase of development. Furthermore, users' goals might be assigned usually as target functions for the artifact. The remaining elements in the usability definition include users, effectiveness, efficiency, satisfaction, and context of use. They are categorized into two groups: subjects for research (users and context of use) and subjects for evaluation (effectiveness, efficiency, and satisfaction).

When designers discuss the usability of a product, on the one hand, the subject for research in the usability definition is the users and context of use. For example, questions such as who are the users of a target product and in which context it is used should be answered if we conduct studies. For such studies, we can use several existing techniques including focus groups, interviews, observations, surveys, and so forth.

On the other hand, the subjects for evaluation in the usability definition include the effectiveness, efficiency, and satisfaction. For example, whether users reach at a specified goal is investigated from the effectiveness viewpoint. How long it takes and what the error rate is are investigated from the efficiency viewpoint. Then how the user felt is investigated from the satisfaction viewpoint. For an evaluation, we can use several existing techniques including usability testing with a prototype.

3 Scenario Heuristics

We specifically examine the usability definition and propose a scenario description from a usability viewpoint.

3.1 Similarity of the Scenario Concept and the Usability Definition

As described earlier, a scenario for designing a product consists of a 5-tuple (product, users, goals, sequence of actions and events, context of use) and its element based on the usability definition is expressed with a 7-tuple (product, users, goals, effectiveness, efficiency, satisfaction, context of use). A similarity is readily apparent between the scenario concept and the usability definition. They share the elements: product, users, goals, and context of use (Fig. 2). In addition, the sequence of actions and events in the scenario concept corresponds to the three elements: effectiveness, efficiency, and satisfaction. The rationale behind this is that the sequence of actions and events specified in the scenario is the target for evaluation and its evaluation aspect includes effectiveness, efficiency, and satisfaction.

Scenario concept = (product, users, goals, sequence of actions and events, context of use)

Usability definition = (product, users, goals, effectiveness, efficiency, satisfaction, context of use)

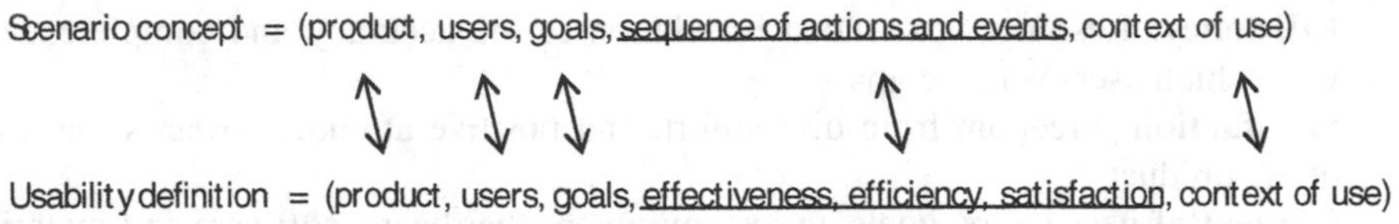

Fig. 2. Correspondence relation between the scenario concept and the usability definition

This formulation enables us to understand what and how we describe in a good scenario from a usability viewpoint. A scenario that is suitable for discussion of its usability should have its elements clarified.

Heuristic 1. Elements of scenario for designing a product should be written clearly.

To be precise, a scenario suitable for discussing usability should have its elements clearly defined, and the sequence of actions and events is specified so that it will be able to evaluate effectiveness, efficiency, and satisfaction. If designers take this heuristics and create scenarios, they could start a discussion of usability at the beginning of design process even when they have a product itself or prototypes for the target product to conduct usability tests.

3.2 Effectiveness in Scenario

When we have a clear goal description in a scenario, we can discuss its effectiveness. From an effectiveness viewpoint, there are scenarios of two types: a success scenario and a failure scenario. In a success scenario, the sequence of actions and events specified that the user reaches the goal in the scenario. A success scenario means that it is effective. It fundamentally specifies a scene that illustrates user's desirable activities with the product. Designers and developers often envision a success scenario for a product at the initial phase of design and expect that the actual user of the product will behave in such a manner in the future.

By comparison, in the failure scenario, the sequence of actions and events specified that the user does not reach the goal in the scenario. The failure scenario means that it is ineffective. It fundamentally specifies a scene that illustrates a user's undesirable activities with the product. It is not common to envision failure scenario in the design process, yet it would become an important description if designers and developers could share a boundary condition or edge case for developing the product. In addition, that description is necessary if the target product is categorized as a safe critical system.

To summarize, a heuristic for a scenario for designing a product is derived as follows.

Heuristic 2. Reachability to the goal of scenario should be clearly described to evaluate its effectiveness.

In the sample scenario portrayed in Fig. 1, the user goal is described implicitly. Consequently, the current version of the scenario is unclear in its presentation of whether the user reaches at the goal or not through the sequence of actions and events. Figure 3 portrays a modification of Fig. 1 with the added description of the goal as the first paragraph. With the description, the reader of the scenario would be able to evaluate the reachability of the goal.

Ichiro keeps his cash record with an application on his PC. He has just received the wages of his part-time job at a pizza restaurant. Therefore, he wants to record it with the application as usual.

Ichiro double clicks on the cashbook icon on desktop to initiate its application program. A new window shows up on the screen. It contains a record entry and the last line represents the current funds left as 12,800 yen. Ichiro clicks on the empty entry at the bottom of the last record entry and types in 9,500 yen as the wages of his part-time job at a pizza restaurant.

Fig. 3. An exemplary scenario specifying clear user's goal

The description of the effectiveness in a scenario enables us to analyze the other two properties of scenario: efficiency and satisfaction.

3.3 Efficiency Scenario

When we have an effective scenario, we can discuss its efficiency. Viewed from an efficiency viewpoint, two types of scenario exist: efficient and inefficient. In the efficient scenario, the sequence of actions and events specifies that the user reaches the goal in the scenario and that it takes a short time and less action. An efficient scenario means that the user reaches the goal with less effort with no mistakes. It fundamentally specifies a scene that illustrates a user's desirable activities with the product. Designers and developers often envision an efficient scenario for a product at the initial phase of design and expect that the actual user of the product behaves similarly in the future.

On the other hand, in the inefficient scenario, the sequence of actions and events specified in the scenario might take a long time and much action even though the user reaches the goal. An inefficient scenario means that it is not efficient. A typical inefficient scenario includes an error scenario in which the user makes mistakes and reattempts a task. It fundamentally specifies a scene illustrating users' undesirable activities with the product. As described in the discussion of effective scenarios, it is not common to envision inefficient scenarios in the design process.

To summarize, a heuristic for scenario for designing a product is derived as follows.

Heuristic 3. The sequence of actions and events in an effective scenario should be described sufficiently to evaluate its efficiency.

Ichiro keeps his cash record with an application on his PC. He has just received the wages of his part-time job at a pizza restaurant, so he wants to record their amount with the application as usual.

   Ichiro double clicks on the cashbook icon on the desktop to initiate its application program. A new window shows up on the screen. It contains a record entry and the last line represents the current money left as 12,800 yen. Ichiro tries to click on the empty entry at the bottom of the last record entry, but he accidentally clicks outside of the application window. The window behind the application window comes to the front. Ichiro clicks carefully on the Close button at the top right corner of the window, which brings the application window to the front. At this moment, Ichiro carefully and slowly clicks on the empty entry at the bottom of the last record entry and types in 9,500 yen as the wages of his part-time job at a pizza restaurant.

Fig. 4. Example of an inefficient scenario

The scenario presented in Fig. 3 is an example of an efficient scenario, whereas Fig. 4 portrays an inefficient scenario. In Fig. 4, the user Ichiro makes a mistake and conducts redundant actions to return to the original sequence of actions. The scenario might not derive the specification of the target artifact, but it obtains the readers' attention for usability. The readers of the scenario description can vividly understand the importance of usability for designing the target artifact.

   In addition to the discussion of the efficiency scenario, we progress to a discussion of the satisfaction scenario.

3.4   Satisfaction Scenario

Along with the efficiency scenario, we can discuss satisfaction when we have an effective scenario. From a satisfaction viewpoint, scenarios of two types exist: satisfactory and unsatisfactory. In the satisfactory scenario, the sequence of actions and events specified that the user reaches the goal in the scenario and it satisfies the user. The satisfactory scenario means that the user reaches the goal and finds satisfaction in his or her use of the product. That information is not easy to illustrate as a user's behavior; in satisfactory scenario. Therefore, a user's mental representation is more focused and described. For example, it specifies what the user sees, feels, and thinks during the course of actions and events. Designers and developers who specifically examine outside representation and functions might tend to avoid specifying a user's mental representation into scenarios. Even if they specify one, it would become a simple favorable response as a happy ending.

In contrast, in the unsatisfactory scenario, the sequence of actions and events specified in the scenario might frustrate the user even if the user reaches the goal. A typical unsatisfactory scenario includes an inefficient scenario in which the user takes a long time and/or has many steps that must be done to reach the goal. An error scenario also frustrates a user if no mechanism exists for recovering from the error with the product. Similarly to the insufficient scenario, the unsatisfactory scenario specifies a scene, which illustrates a user's undesirable activities with the product. As described in the discussion on effective scenarios, it is not common to envision unsatisfactory scenarios during the design process.

To summarize, a heuristic for a scenario for designing a product is derived as follows.

Heuristic 4. Users' mental representations should be described clearly in scenarios to evaluate users' satisfaction.

Figure 5 presents an unsatisfactory scenario, in which the user Ichiro becomes aware that he has been doing a redundant action with the application. Consequently, we might conclude that this satisfaction level is not high.

4   Scenario Properties for Usability Discussion

As described in Heuristic 1, elements of scenario for designing a product should be written clearly. In a practical development scene, the information of two elements, product and goals, might be assigned usually as a project theme and target functions for the product, respectively. The information of the two elements, users and context of use, might come from field research because they are subjects for research as discussed in 2.2. The information of sequence of actions and events might set as a hypothesis in the initial phase of design because designers and practitioners do not have the actual product. Hence, the key to make use of scenario as design for usability is what and how to specify the sequence of actions and events.

Ichiro keeps his cash record with an application on his PC. He has just received the wages of his part-time job at a pizza restaurant, so he wants to record it with the application as usual.

Ichiro double clicks on the cashbook icon on desktop to initiate its application program. A new window shows up on the screen. It contains a record entry; the last line represents the current funds left as 12,800 yen. Ichiro clicks on the empty entry at the bottom of the last record entry and types in 9,500 yen as the wages of his part-time job at a pizza restaurant. He notices that the data input is the first action every time he uses the application, so the cursor should be active in the last record entry without clicking it on.

Fig. 5. Example of unsatisfactory scenario

Table 1 summarizes scenarios for designing a product with usability. There are two views on scenario: the ideal view and the realistic view. The ideal view includes success scenario, efficient scenario, and satisfactory scenario. The realistic view contains failure scenario, inefficient scenario, and unsatisfactory scenario.

Table 1. Contrasting views on scenario for usability corresponding to scenario property: the ideal view vs. realistic view

Scenario property	Ideal view	Realistic view
Effectiveness	Success scenario	Failure scenario
Efficiency	Efficient scenario	Inefficient scenario
Satisfaction	Satisfactory scenario	Unsatisfactory scenario

What is useful for designing a product for usability? The ideal view and the realistic view are orthogonal and mutually complementary; therefore, both views in scenario contribute to improve the usability of a product. For example, the ideal view might become a useful material for usability specification [12] because it specifies the best task or activity with the product; it might work as a reference value.

From an effectiveness viewpoint, a success scenario is useful to derive specifications; in contrast, a failure scenario is useful to initiate a discussion of usability. From an efficiency viewpoint, an efficient scenario is useful to derive specifications; by comparison, an inefficient scenario is useful to initiate a discussion of usability. From a satisfaction viewpoint, a satisfactory scenario is useful to derive specifications; however, an unsatisfactory scenario is useful to initiate a discussion of usability.

The typology of scenario for usability can be used for usability education. Inexperienced designers and engineers especially who have a narrow focused representation and function of a product tend to write scenarios from the ideal view. They might not be aware of how the user and context of the product are important for usability and the user experience. In this case, to give them an assignment for envisioning scenarios from the realistic view might be significant.

5 Conclusion

As described in this paper, we proposed design heuristics for scenarios for designing a product based on our experiences obtained during collaboration with design, usability, and user experience departments in major Japanese companies. We provided several versions of scenarios.

Scenarios are flexible in form and content. The designers and practitioners in industry might have no idea what to describe as a scenario. The heuristics proposed in the paper provide a good starting point.

Acknowledgments. This project is partially supported by a Grant-in-Aid for Scientific Research (C), 19500105, 2007, from the Ministry of Education, Culture, Sports, Science and Technology, Japan.

References

1. Alexander, I.F., Maiden, N. (eds.): Scenarios, Stories, Use Cases: Through the System Development Life-Cycle. Wiley, Chichester (2004)
2. Carroll, J.M. (ed.): Scenario-Based Design: Envisioning Work and Technology in System Development. John Wiley and Sons, Chichester (1995)
3. Carroll, J.M.: Making Use: Scenario-Based Design of Human-Computer Interactions. MIT Press, Cambridge (2000)
4. Clarke, L.: The use of scenarios by user interface designers. In: Diaper, D., Hammond, N. (eds.) People and Computers VI: Proceedings of the HCI 1991 Conference, pp. 103–115 (1991)
5. Cooper, A.: The Inmates Are Running the Asylum. Sams (2004)
6. Rosson, M.B., Carroll, J.M.: Usability Engineering: Scenario-Based Development of Human-Computer Interaction. Morgan Kaufmann, San Francisco (2002)
7. Weidenhaupt, K., Pohl, K., Jarke, M., Haumer, P.: Scenario Usage in System Development: A report on Current Practice. IEEE Software 15(2), 34–45 (1998)
8. Go, K., Carroll, J.M.: The Blind Men and the Elephant: Views of Scenario-Based Design. Interactions 11(6), 44–53 (2004)
9. Pruitt, J., Tamara, A.: The Persona Lifecycle: Keeping People in Mind throughout Product Design. Morgan Kaufmann, San Francisco (2006)
10. ISO: ISO9241-11 Ergonomic requirements for office work with visual display terminals. Part 11: Guidance on Usability (1998)
11. Rolland, C., Achour, C.B., Cauvet, C., Ralyte, J., Sutcliffe, A., Maiden, N., Jarke, M., Haumer, P., Pohl, K., Dubois, E., Heymans, P.: A proposal for a scenario classification framework. Requirements Engineering 3(1), 23–47 (1998)
12. Hix, D., Hartson, H.R.: Developing user interfaces: ensuring usability through product and process. John Wiley and Sons, Chichester (1993)

A Method for Consistent Design of User Interaction with Multifunction Devices

Dong San Kim and Wan Chul Yoon

Department of Industrial & Systems Engineering,
Korea Advanced Institute of Science and Technology, 373-1,
Gu-seong, Yu-seong, Daejeon, Korea
kimdongsan@gmail.com, wcyoon@kaist.ac.kr

Abstract. Over the last decade, feature creep and the convergence of multiple devices have increased the complexity of both design and use. One way to reduce the complexity of a device without sacrificing its features is to design the UI consistently. However, designing consistent user interface of a multifunction device often becomes a formidable task, especially when the logical interaction is concerned. This paper presents a systematic method for consistent design of user interaction, called CUID (Consistent User Interaction Design), and validates its usefulness through a case study. CUID, focusing on ensuring consistency of logical interaction rather than physical or visual interfaces, employs a constraint-based interactive approach. It strives for consistency as the main goal, but also considers efficiency and safety of use. CUID will reduce the cognitive complexity of the task of interaction design to help produce devices that are easier to learn and use.

Keywords: Interaction design method, consistency, constraint-based design.

1 Introduction

Over the last decade, interactive devices such as mobile phones have become increasingly complicated despite no little efforts have been invested in usability engineering than ever before. One of the main reasons for this is *creeping featurism or feature creep*: the tendency for the number of features in a product to rise with each release of the product [1-3]. To reduce the negative effects of feature creep (e.g., complexity of both design and use), some researchers and companies emphasize the "simplicity" of a product [4-5]. In the field, however, it is often the case that removing a feature is much more difficult than adding it [3]. The primary reason for this is that consumers want products with more features even if they know that they will probably never use most of the features and products with more features will lead to decreased usability [6-8]. While most people desire more features, at the same time they also want ease of use. In general, however, the more features a product has, the higher the complexity of using the product is [7]. For this reason, many manufacturing companies are faced with a dilemma: should they make their products more capable or should they make them more usable [6-7].

M. Kurosu (Ed.): Human Centered Design, HCII 2009, LNCS 5619, pp. 202–211, 2009.

One way to resolve this dilemma – that is, to reduce the complexity of a device without sacrificing its features – is to design the device consistently. If a user interface (UI) is designed consistently, the user will benefit from what psychologists call *transfer of training*. In other words, learning to do one thing in one context will make it easier to learn how to do similar things in similar contexts [9]. However, it is often not easy to design the user interface of a multifunction device consistently for several reasons: (1) it is very cognitively demanding and error-prone, because the designer should consider hundreds of design variables and relations between them; (2) the size of the UI and the number of the given interface controls such as push buttons are getting smaller; (3) the product development environment has become fast-paced due to increasing global competition; (4) many designers collaborate on the design; and (5) product requirements typically change often during the design process.

Therefore, there is a strong need for a well-developed method that supports complex design processes. Although the importance of consistency has widely been noted and many formal models for evaluating interface consistency have been developed in the HCI literature, relatively few studies have been carried out on methods or tools for supporting the consistent design of user interaction, particularly interaction for multifunction or convergence devices.

This paper introduces a method for designing user interaction consistently, called CUID (Consistent User Interaction Design). It focuses on ensuring consistency of *logical interaction* – task procedures, operation or function availabilities in system states, and interface controls (or interaction methods) for each available operation – rather than *physical or visual interface elements* such as colors, sizes, locations, etc. In the next section, we briefly review previous research on consistency in UI design. Design problems that are dealt with in CUID and main approaches to address the problems are discussed in Section 3. In Section 4, a case study for validating CUID is presented. Finally, Section 5 concludes the paper and discusses directions for future work.

2 Related Work

Consistency is one of the most important principles in user interface design. The first rule of Ben Shneiderman's well-known eight golden rules of interface design is "Strive for consistency" [10]. Although clearly defining consistency is not simple [11], it can be loosely defined as *"do similar things in similar ways"* [14]. According to previous studies, consistency may sometimes conflict with other design principles and it is sometimes beneficial to be inconsistent [11-13]. However, it is generally agreed that consistently designed systems are easier to learn and easier to remember, reducing performance time and errors, and as a result increasing user satisfaction as compared to inconsistently designed systems [15].

Previous studies have developed a number of methods for ensuring interface consistency in three different areas: design guidelines, models or tools for evaluating interface consistency, and tools for generating consistent user interfaces. Many kinds of guidelines and standards for interface consistency have been developed thus far. Presently, most commercial applications have their own interface guidelines and most of the guidelines (e.g., [16]) include consistency as one of the major principles. Ozok and Salvendy [17] suggested twenty guidelines for the design of consistent Web-based interfaces. Examples are: *"Use same words for same items, rather than using synonyms"; "Be consistent in terms of character sizes, use of upper/lower case letters, spacing, punctuation, character colors, and wording in your text."* With these design guidelines, however, consistent design of user interaction with a complicated device is not easy. Compliance with guidelines or standards is not sufficient to ensure consistency, because the guidelines and standards leave a fair amount of leeway for the designers [15].

Therefore, there has always been much need for developing models and tools that can evaluate interface consistency. Since the 1980s, several formal models such as TAG [18], APT [19], and PROCOPE [20] have been developed for evaluating the consistency of designed user interfaces. In these models, task procedures are represented by rules, and fewer rules indicates higher consistency. Formal models, however, require much time to learn and use grammar in a fast-paced environment. In the 1990s, a few software tools for evaluating interface consistency were introduced. GLEAN [21] makes predictions about human performance on a user interface based on the GOMSL model, which can predict transfer times between tasks and identify consistency problems between similar tasks. However, it has at least two limitations. First, GOMSL regards two task goals as similar when they are different on only one of the verb and the noun in the given statements. Semantic similarity between two task goals with both different verbs and nouns is not considered (e.g. "increase the volume" and "enlarge the photo"). Second, when counting the number of similar statements between two procedures, it stops counting when it encounters two statements that are non-identical even if the remaining statements are the same. SHERLOCK [22], a family of consistency analysis tools, evaluates visual and textual properties of user interfaces. It was designed to detect inconsistencies in terminology, capitalization, typefaces, colors, button sizes, placements, etc. As it focuses only on visual and textual consistency, consistency of logical interaction is not considered.

In recent years, tools for automatically generating consistent user interfaces have also been developed. SUPPLE [23] automatically generates layouts for user interfaces using an optimization approach to choose controls and their arrangement. A branch-and-bound constrained search algorithm is used for mapping between a set of interface elements – units of information that need to be conveyed via the interface between the user and the controlled appliance or application – and available UI widgets (e.g., spinner, tab pane, slider). However, it also addresses only the design of physical or visual interfaces. UNIFORM [24] automatically generates remote control interfaces that are personally consistent, which means that the interfaces generated for each user are consistent with that particular user's past interface experiences. It can automatically identify similarities between different devices and users may specify additional similarities. The similarity information allows the interface generator to use the same type of controls for similar functions, place similar functions so that they can be found with the same navigation steps, and create interfaces that have a similar visual appearance. However, as the authors noted, its solutions are often limited because of a lack of useful semantic information about unique functions. Moreover, it takes a substantial amount of time (about 5 hours for a VCR and one week for three different VCRs for even an expert) to complete a specification for an appliance, on which similarities between different appliances are based.

3 CUID Problems and Approaches

As shown in Fig. 1, CUID deals with three kinds of problems in user interaction design, each of which can be considered as matching between two adjacent abstraction levels:

1) *Task-Procedure design*: the design of procedure(s) of each task
2) *State-Operation design*: the design of operation[1] or function availability in each system state and of the state change by each available operation
3) *Operation-Control mapping*: the mapping between available operations and interface controls (or interaction methods)

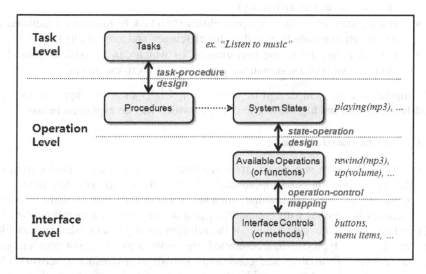

Fig. 1. Three main design problems in CUID

The three design problems pertain to the logical consistency rather than the physical or visual consistency of UIs. Among them, the first problem concerning the consistency of task procedures has been extensively studied in the HCI discipline; GOMS and TAG are representative models that address the problem. The other two design problems, however, have not received much attention in the HCI literature. Although these problems are not significant in designing a relatively simple system with a small number of features, they become important issues when designing a convergence or multifunction device.

In solving each design problem, consistency is used as the primary design principle. In this paper, the term consistency is defined more broadly than in most previous studies: *keep a relation between design elements the same at all levels*. It is beyond "do similar things in similar ways"; it deals with not only synonyms – similar things – but also antonyms – opposite pairs of things such as left and right. Therefore, for each

---

[1] In this paper, an operation represents a logical action (e.g. "delete a file") rather than a physical action (e.g. "press a push button").

design problem, at least two principles are used. For example, in operation-control mapping, the principles are "assign the same or similar interface controls to similar operations" and "assign each of a pair of controls to each of a pair of operations".

Although consistency, the primary goal of CUID, is generally desired, it is obviously not the only desirable usability characteristic. Taking consistency only as the guiding rule in design can be very risky [10-12],[14]. Therefore, CUID strives for consistency as the main goal, but it also tries to satisfy other usability characteristics such as efficiency and safety by applying the following rules:

- Frequently performed tasks should be provided with shortcut procedures (for efficiency).
- Urgent tasks, quick start of which is required, should be provided with shortcut procedures (for efficiency).
- If necessary, more than two procedures for a task or interaction methods for an operation should be possible (for efficiency and learnability).
- Critical tasks, the wrong performance of which can be critical, should include a confirmation step in the procedures (for errors/safety).

To support a consistent design of interaction, CUID uses two approaches: constraint-based design and highway-based design, which will be explained below.

3.1 Constraint-Based Design

Constraint-based design is a design approach that is widely used in engineering design [25]. In a constraint-based approach, a design problem entails determining the values of design variables while satisfying a set of constraints that are tightly coupled and sometimes in conflict with one another, and a constraint is generally defined as the bound on a single design variable or the relation among a set of design variables. Since variables are typically interconnected via constraints, a design problem constructs a network of variables and constraints known as a constraint network [26]. When a value is assigned to a variable, the design decision is propagated through the network, influencing the decisions of values of other connected variables. Constraints are generally viewed as limiting factors in design. In constraint-based design, however, constraints are not considered as negative factors but as *design drivers* that drive the exploration of the solution space [27].

To date, only a few studies have applied constraint-based approaches to UI design (e.g.,[23],[28]). A constraint-based approach is utilized for CUID because a key feature of the cognitive processes of UI designers is that they are constraint-based. In best UI design practices, designers put almost certain solutions into the partial solution set as soon as they are found during the process and use them as design constraints to narrow down solutions for the adjacent problems. Since the certain solutions are scattered in the design space, the designer will leap around problems, collecting solutions in an opportunistic manner [29].

Fig. 2 illustrates the present constraint-based design. Design variables and possible values at each abstraction level – tasks, procedures, states, operations, and interface controls – form their own network by relations among them. When the value of a design variable is determined (drawn in the solid bidirectional arrows in Fig. 2), the design decision becomes a constraint that influences the decisions of values of other variables that are connected to this variable (drawn in the dotted bidirectional arrows in Fig. 2). This constraint-based matching between two adjacent levels forms three kinds of representations: a parallelogram, triangle, and inverted triangle.

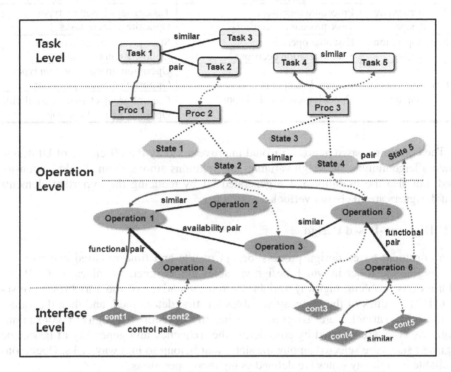

Fig. 2. An illustration of the constraint-based approach of CUID

Note that design constraints are divided into horizontal or within-level constraints – relations among design variables within the same abstraction level – and vertical or between-level constraints – relations among variables and values in different abstraction levels (decisions made by the designer). Table 1 summarizes the constraints that are used for each design problem.

Vertical constraints, i.e., variable-value relations, include both relations that are predefined and that are determined while designing. The predefined relations are external constraints for consistency with other devices and with users' mental models. These are social standards or cultural conventions. Procedure templates for each task type, essential operations that should be available in all or almost all system states, and required mappings between available operations and interface controls (e.g., "play(x)" operation and "►" button) are used as the external constraints. In this respect, CUID considers both external and internal consistency. Horizontal constraints, i.e., within-level relations, involve *groups* – that are formed by semantic or functional similarity – and *opposite pairs* that are semantically or functionally antonymous to each other. To assess the semantic similarity among variables in each design problem, taxonomies of user tasks, system states, and user operations have been developed.

Table 1. Constraints used for each design problem in CUID

Design problem	Vertical constraints (between-level)	Horizontal constraints (within-level)
Task-Procedure design	Procedure templates Task-procedure decisions	Task groups based on types Opposite pairs of tasks
State-Operation design	Essential operations State-operation decisions	State groups based on types Opposite pairs of states
Operation-Control mapping	Required mappings Operation-control decisions	Operation groups based on types Opposite pairs of operations
		Control groups based on similarity Opposite pairs of controls

The constraint-based approach would improve not only the efficiency of UI design when a large number of design variables and relations among them should be considered, but also the usability of the generated UI by reducing the important elements that designers are likely to overlook.

3.2 Highway-Based Design

In a constraint-based design process, design results and time required can vary depending on the order in which design variables are assigned to values. In CUID, to address this problem, highway variables – representatives of the variables – are selected. The values of the highway variables are first determined, and then the values of the other variables are assigned by using the decisions as highway constraints. Highway tasks are selected by considering the frequency and generality of tasks, and highway states are selected among the states that belong to highway tasks. Operations available in highway states are defined as highway operations.

4 CUID Design Process

Fig. 3 shows the CUID design process. It is comprised of three main phases – task-procedure design, state-operation design, and operation-control mapping – and each design phase is composed of three or four steps. There are feedback loops among the three phases and among the steps in a phase, and thus the overall process proceeds iteratively.

Each phase begins with an analysis of its own design variables, in which design constraints, relations among the variables are determined. Relations among tasks (i.e. task groups determined by task types and opposite pairs of tasks) are elicited by task analysis, relations among system states (state groups by state types and opposite pairs of states) by state analysis, and relations among operations (operation groups by operation types and opposite pairs of operations) and relations among interface controls (control groups and opposite pairs of controls) by operation and control analysis.

After the constraints to be used in a phase are identified, the main step of the phase – decisions of values of the design variables – follows. In these steps, procedures of the given tasks, available operations in each system state and the state change by each available operation, and mappings between operations and interface controls are determined. For each group of variables, the values of highway variables are first determined, and the values of the other variables are then assigned. The value of a variable is determined by considering the previous decisions of the values of other variables – including highway variables – that are similar or opposite to the variable. Each of the taxonomies, by which the groups of variables are determined, has two or three levels so that multi-level groups can be formed. The designer can extend or reduce the range of similar variables by changing the abstraction level.

In the last step of each phase, design decisions made in the phase are evaluated and refined. For example, the designer can assign an appropriate operation (or function) to a remaining interface control – to which no operation is assigned in a system state – by considering which operations are assigned to the control in other states. At these "refinement" steps, the designer can check if he or she missed some relations among design variables in the main steps.

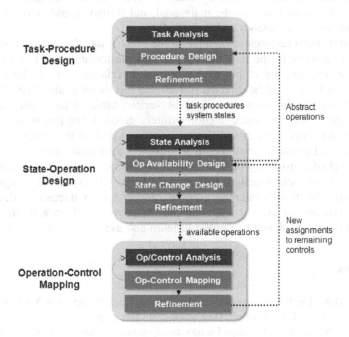

Fig. 3. The CUID design process

5 Case Study

To evaluate the usefulness of CUID, we first developed a prototype software tool that supports the CUID design process, and then the following procedure was applied: (1) the interaction part of the experimental device, a recently commercialized multifunction mp3 player, was redesigned with CUID; (2) logical inconsistency problems the device has were identified by a heuristic evaluation with six users; and (3) how many inconsistencies were resolved in the redesigned interaction was checked. Twenty-three inconsistency problems in total were collected from the six users. We found that

21 of the 23 inconsistency problems (91%) did not exist in the redesigned interaction. This indicates that CUID can be a useful tool for consistent design of user interaction.

6 Conclusions and Future Work

In this study, a method for consistent design of user interaction, called CUID, was developed. To support consistent design, CUID employs a constraint-based approach. Vertical or between-level constraints, meaning the relations between design variables and values, and horizontal or within-level constraints, meaning the relations among variables or values within an abstraction level, are both considered. CUID also uses a highway-based approach in which highway variables, representatives of variables, are selected; the values of the highway variables are first determined; and the values of the other variables are then assigned by using the decisions on the highway variables. CUID strives for consistency as the main goal, but it also considers other usability characteristics such as efficiency and errors/safety.

There is still much research to be done in the area of interface consistency, which has seen little activity in the past decade [24]. In particular, consistency of logical interaction rather than that of physical or visual interfaces needs to be addressed. Future work on CUID will be directed toward the following goals. First, CUID will be refined by redesigning user interaction for various kinds of multifunction devices such as mobile phones and multifunction printers. Second, the taxonomies of tasks, system states, and operations should also be refined, as each of the current taxonomies does not cover all possible types. Third, a consistency measure will be developed to assess the level of consistency of user interaction quantitatively. This will enable the optimization of interaction design. It is expected that CUID, for UI designers, will reduce the complexity of designing user interaction for an interactive device with many functions, and for users it will contribute to providing them with products that are consistently designed, and thus easier to learn and use.

References

1. Norman, D.A.: The Psychology of Everyday Things. Basic Books, New York (1988)
2. Lee, D.S., Woods, D.D., Kidwell, D.: Escape from Designers' Dilemma on Creeping Featurism. In: Proceedings of Human Factors and Ergonomics Society Annual Meeting, Test and Evaluation, pp. 2562–2566 (2006)
3. Surowiecki, J.: Feature Presentation. The New Yorker, May 28 (2007)
4. Philips: Our Brand Promise Sense and Simplicity,
   http://www.philips.com/about/brand/brandpromise
5. Physorg: Vodafone Simply: The More The Better? May 21 (2005),
   http://www.physorg.com/news4197.html
6. Rust, R.T., Thompson, D.V., Hamilton, R.W.: Defeating Feature Fatigue. Harvard Business Review 84(2), 98–107 (2006)
7. Norman, D.A.: Simplicity is Highly Overrated. Interactions 14(2), 40–41 (2007)
8. Norman, D.A.: Simplicity is Not the Answer. Interactions 15(5), 45–46 (2008)
9. Monk, A.: Noddy's Guide to Consistency. Interfaces 45, 4–7 (2000)

10. Shneiderman, B.: Designing the User Interface: Strategies for Effective Human-Computer Interaction, 3rd edn. Addison-Wesley, Reading (1998)
11. Grudin, J.: The Case Against User Interface Consistency. Communications of the ACM 1(1), 59–71 (1989)
12. Grudin, J.: Consistency, Standards, and Formal Approaches to Interface Development and Evaluation. ACM Transactions on Information Systems 10(1), 103–111 (1992)
13. Ketola, P., Hjelmeroos, H., Raiha, K.J.: Coping with Consistency under Multiple Design Constraints: The Case of the Nokia 9000 WWW Browser. Personal Technologies 4, 86–95 (2000)
14. Reisner, P.: What is Consistency? In: INTERACT 1990, pp. 175–181 (1990)
15. Nielsen, J.: Usability Engineering. Academic Press, Boston (1993)
16. Apple: iPhone Human Interface Guidelines. Apple Inc. (2007)
17. Ozok, A.A., Salvendy, G.: Twenty Guidelines for the Design of Web-based Interfaces with Consistent Language. Computers in Human Behavior 20, 149–161 (2004)
18. Payne, S.J., Green, T.R.G.: Task-Action Grammar: A Model of the Mental Representation of Task Languages. Human-Computer Interaction 2, 93–133 (1986)
19. Reisner, P.: APT: A Description of User Interface Inconsistency. Int. J. of Man-Machine Studies 39, 215–236 (1993)
20. Poitrenaud, S.: The PROCOPE Semantic Network: An Alternative to Action Grammars. Int. J. of Human-Computer Studies 42, 31–69 (1995)
21. Kieras, D.E., Wood, S.D., Abotel, K., Hornof, A.: GLEAN: A Computer-Based Tool for Rapid GOMS Model Usability Evaluation of User Interface Designs. In: Proceedings of UIST, pp. 91–100 (1995)
22. Mahajan, R., Shneiderman, B.: Visual Consistency Checking Tools for Graphical User Interfaces. IEEE Transactions on Software Engineering 23(11), 722–735 (1997)
23. Gajos, K., Weld, D.S.: SUPPLE: Automatically Generating User Interfaces. In: Proceedings of IUI, pp. 93–100 (2004)
24. Nichols, J., Myers, B.A., Rothrock, B.: UNIFORM: Automatically Generating Consistent Remote Control User Interface. In: Proceedings of CHI, pp. 611–620 (2006)
25. Lin, L., Chen, L.C.: Constraints Modelling in Product Design. J. of Eng. Design 13(3), 205–214 (2002)
26. Hashemian, M., Gu, P.: A Constraint-Based System for Product Design. Concurrent Engineering: Research and Applications 3(3), 177–186 (1995)
27. Kilian, A.: Design Innovation Through Constraint Modeling. Int. J. of Architectural Computing 4(1), 87–105 (2006)
28. Borning, A., Duisberg, R.: Constraint-Based Tools for Building User Interfaces. ACM Transactions on Graphics 5(4), 345–374 (1986)
29. Yoon, W.C.: Task-Interface Matching: How We May Design User Interfaces. In: Proceedings of the 15th International Ergonomics Association Triennial Congress (2003)

A Mobile Application for Survey Reports: An Evaluation

Daniel Kohlsdorf, Michael Lawo, and Michael Boronowsky

TZI Universität Bremen, Am Fallturm 1,
28359 Bremen, Germany
{dkohl,mlawo,mb}@tzi.de

Abstract. In manufacturing processes damages occur caused by humans or ma-chines. These damages have to be reported and documented, e.g. to enable a manufacturer to react in quality circles. The first part of this paper describes the process of creating survey reports. Furthermore a customized solution designed for mobile survey reports is introduced. In the second part this paper describes and discusses the advantages and disadvantages of this mobile solution in an automotive industry setting.

Keywords: Smart phone, survey reports, automobile manufacturer.

1 Introduction

Industry is interested to create fast and precise documentation techniques for survey reports as a basis for improved production processes. For this purpose most manufac-turers are using paper or notebook based solutions in combination with a digital cam-era. The disadvantage of paper-based solutions is the need for digitalization, because this information is generally input to additional processes. By using a notebook the user is forced to concentrate on his documentation task and has no possibility to pay attention to the environment around the laptop. In general paper and notebooks are difficult to handle on the shop floor for several reasons.

In cooperation with an automobile manufacturer we analyzed the survey report process and developed a concept for the survey report as an alternative to an existing notebook based solution where MS Word© form sheets are filled in complemented by photos captured by a digital camera. When observing the workers on the shop floor performing these tasks it became clear that this was either unsuitable or inappropriate. Furthermore a notebook and especially its software offer unnecessary features beyond the need of the real usage context [1]. So we decided to create a mobile solution based on a consumer smart phone adjusted to the task and already carried with for commu-nication purposes.

This paper describes and discusses the advantages and disadvantages of the mobile survey report. The smart phone application was compared to the notebook based solu-tion using the NASA TLX acceptance test.

1.1 Scenario

The process of documenting manufacturing damages during car production consists of two parts. It starts with a dialog between the persons reporting and documenting

M. Kurosu (Ed.): Human Centered Design, HCII 2009, LNCS 5619, pp. 212–220, 2009.

the damage. Most information for the documentation has to be extracted from this dialog, like who am I, whom am I talking to, where is the damage and what kind of damage is observed on the car. The second part is to examine and document the damage itself (taking pictures etc.). In the best case the whole data needed for documentation could be collected and saved on the spot, which means on the shop floor at the damaged car. From the user interviews we extracted further information about the requirements given by the environment. Due to interference issues the application environment does not allow any use of Bluetooth© or WLAN but UMTS. Thus during a session data from the phone is sent to the server via UMTS. The other more important point is the need of high-resolution pictures. The users pointed out in the interview that the pictures are the main source for documentation and high quality is thus required. However, analyzing the users work with the image data we found out that high-resolution (five or more mega pixels) was not what the users mend. The survey report is printed on a black and white printer and the resolution really used is what a printer could display in a 10 x 10 cm^2 field. With a lower resolution the latency while operating will also be lower. Furthermore, processing large images on a mid range Smartphone disturbs the workflow of the application as it is a quite lengthy task.

We decided to use a Smartphone instead of a PDA, because the people, responsible for documentation, are all equipped with a smart phone. A smart phone is a personal artifact. It refers mostly to one person and this person is responsible for the device (like keeping an eye on the battery power). By using a PDA the company has to buy new devices, so obviously many users have to share a small amount of devices. In this case it is quite difficult to specify responsibilities.

Furthermore most users already know the normal handling and behavior of smart phones. Scrolling item lists is already known because the users search in their telephone book and working on images many users know from applications for simple image manipulation tools on their phones. In contrast to the smart phone a PDA application is not as mobile and easy to learn. On a PDA text input is mostly connected with an on-screen keyboard (Windows Mobile©) or a special stylus input (Palm OS). It could also be very difficult to operate a PDA application in such an environment like a manufacturing hall because both hands are needed, one for the pen and one for the PDA.

1.2 Related Work

One important field to acquire information unobtrusively on the shop floor is wearable computing. There are many solutions in wearable computing designed to avoid paper. One of them is the "Kato Aircraft Maintenance Application" [2]. This Application supports the maintenance worker in the aircraft maintenance process where, the worker gets information like manuals or lists of technical components by a head mounted display. The system is also designed for hands free interaction [3], realized with a data glove [4].

So the user has not to use any printed documents about his task and has the possibility to focus on the maintenance work and not on interacting with a computer. This would also be a quite good solution for survey reports scenarios but first of all it is expensive due to the hardware used, like the head mounted display, and hard to get. A second obstacle for using a "true" wearable solution for survey reports is that the documentation has to be created on the spot. In case of a wearable solution the user has to wear the devices all the time during a shift.

An advantage of using a smart phone is that most users are familiar with it. Some people, like Joseph Dvorak think that the smart phone will be the heart of the next generation wearable computers [5] and for sure smart phones have many design points in common with wearable computing solutions.

2 User Study

In this study we benchmark the two solutions (smart phone and notebook) in the survey reports scenario in a laboratory setting. We created a task simulating collecting the data for a survey report. With this benchmark we point out the time effort of a mobile worker from using a customized application on a mobile device.

Furthermore the users had to use the NASA TLX form for both devices, to evaluate the mental and physical load of the two interfaces in the mobile scenario.

2.1 Task

Fig.1 shows a survey report. This is the needed output for both, Smartphone and Notebook. The needed data is described in the following. The location of the factory could be extracted from the preselected data. The report number is increased by the backend application. The collection of the other data like damage or pictures is

Survey Report

Location:Bremen	Report Nr: 0062	Damage Reporter: Peter Noname
Type of Construction:	Front	
Worker: James Nobody	Workers Tel.: 0151 12345	Date: 24.12.2008
Description: Scratch on fender		

Fig. 1. A Survey report

described in section 2.2. The users had to collect all data needed for a complete documentation [Fig. 1] as:

1. Creating a picture of the damaged area and marking the damage.
2. Creating a picture of the damage itself.
3. Completing who is creating the documentation.
4. Completing who is reporting the damage.
5. Describing the damaged area.
6. Describing the damage.

2.2 Structure

The study took place in a prepared room simulating the shop floor environment. The damaged object was placed in the middle of this room.

The study began with a dialog between the damage reporting person (local staff - reporter) and the person collecting the data and creating the documentation (user). While talking to the reporter the user had to extract the information of the damage and input it on the mobile device. For example the reporter introduced himself while the user had to input the reporter's name on his device. Somewhere in this process the user had to take two or more pictures of the damage. Once with the camera integrated in the smart phone and once with an external camera.

On the shop floor possibilities of sitting down handling different devices are quite limited, so in our study we let the user sit down on a chair but without using a table.

For both devices we took the time the user needed collecting the data with a chronograph saving the recorded time in an excel sheet.

In the following the two applications used in this study are described:

- **The first application** is the notebook based one. The user collects the data in a word form sheet and takes pictures of the damage with a digital camera. The task does not include the download of the taken pictures from the camera to the notebook, because this benchmark only refers to the time needed for collecting the data and not for creating a document. The possibilities on the shop floor to work with a notebook are very limited, so during the study the only office furniture is a chair and the users have to put the notebook on their lap.

- **The second application** is based on a consumer smart phone with the Java Micro Edition installed. The used components will be described in the following [Fig. 2]. For data collection the integrated camera of the smart phone is used. The user interacts with the normal keyboard (Text on 9 keys enabled) and the directional pad. The survey report data is transferred to a backend application on a server. This application saves the data in an archive (File system). The backend application is also able to display the survey report as an html web page or a rich text format [Fig.1] document out of the archived data. For transferring the collected data UMTS is used (see 1.1).

  The application is divided into small steps. The user is working at a time on one part only. All steps together guide the user to create the complete survey report. This is a main concept of mobile interaction [7].

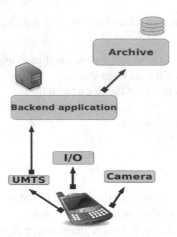

Fig. 2. Architecture of the smart phone application

In the first two steps the user can choose out of a choice of users and re-
porters. Names and contacts are taken from the smart phone address book.
In the second step the user has to take two pictures with the integrated camera, an overview picture of the damaged area and a more detailed picture of
the damage itself. Furthermore the user has to assemble the taken pictures to
one [Fig. 3]: the first step of the assembly is to move a circle, with the navigation cross, over the damage in the overview picture [Fig: 3a] and after
confirming the circle position, the detailed picture appears in the upper left
corner of the screen. The user then can place the detailed picture in one of
the corners of the screen by pressing the navigation cross to the left or the
right [Fig. 3b/3c].

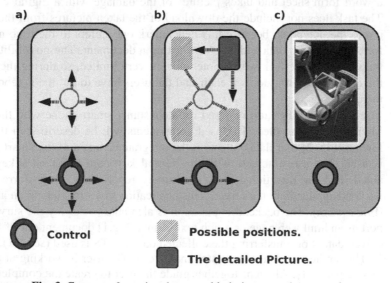

Fig. 3. Concept of creating the assembled picture on the smart phone

The damaged area has to be chosen from a graphic model of the car. By pressing the navigation cross to the left or the right, the adjacent area is highlighted, the previous area switches to normal color [Fig. 4].

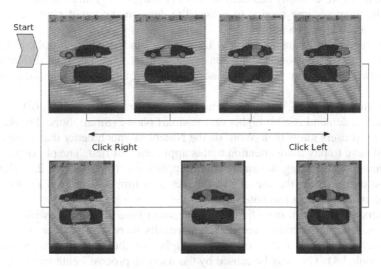

Fig. 4. Choosing the damaged Area

With the information of the area the system is able to extract the damaged components. The damage-describing phrase is built in three steps [Fig. 5, steps 2) Damage to 4) Component]. Every step is displayed as a list of elements or keywords from which the user can choose one. The first step is to choose a type of damage, the second is to choose a preposition and the third is to choose the damaged component. The list with the damaged components consists only of the extracted elements from the area information and not of all elements.

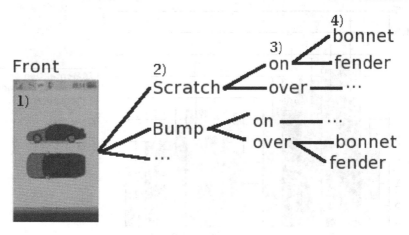

1)Area 2)Damage 3)Preposition 4)Component

Fig. 5. Menu structure for creating a damage-describing phrase

2.3 Analysis

The study was performed with 15 people for the benchmark and the NASA TLX form. The users were mostly recruited from local staff, family and friends. The difference of the average time needed to collect the data on the smart phone and on the notebook was about half a minute.

The average time on the notebook is 1.6 minutes, and on the smart phone 1.3 minutes. Using the smart phone application is 19% faster. The maximal difference between the two applications for a single user was 1min, quite a long time for handling a task.

The results of the NASA TLX forms meet our expectations (see [Fig. 6]).

The mental demand is a bit higher on the smart phone (Smartphone 2.6, Notebook 2.2). But most users know their Word on the Notebook much better than a new application and have to pay more attention to this application at first. The physical demand is quite lower while using the smart phone application (Smartphone 2.1, Notebook 3.5). Most users named the use of two devices at a time as a reason for the higher physical demand (camera and notebook).

The users also felt a bit more hurried up because of the loss of time while changing the devices. They also considered that their results were much better on the smart phone. The felt effort of the users was also higher on the smart phone (Smartphone 2.3, Notebook 3.4). This may be caused by the user- or process-centered design of the application. The frustration on the smart phone is also lower than on the notebook and the lowest point in the study (Smartphone 1.3, Notebook 2.5). In the whole the Task Load Index of the Smartphone is about 2.7 and of the Notebook solution 4.2. The Smartphone application user interface proofs to be quite smooth and easy to handle; the TLX value is 36% lower. Standard deviations are similar for the different aspects with both devices.

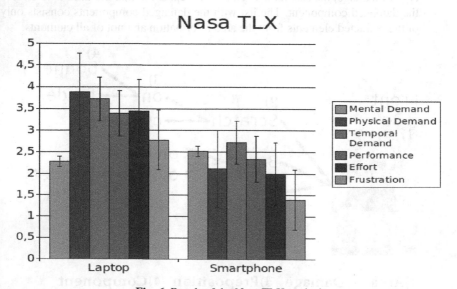

Fig. 6. Result of the Nasa TLX analysis

3 Discussion

The disadvantages of the smart phone application are small input and output devices. Some users criticized the small display and the small keyboard. While designing the application we tried to avoid typical mistakes like placing useless icons or information on the screen, confusion or loss of orientation in huge menu trees [6]. And this seemed to be a key feature of the application. Another strength is that the application is customized for the scenario. Assembling the pictures in Word or searching for the right contact data are only a few obstacles the user has by using a notebook. Also handling two devices instead of one could be quite difficult. These are actions where the users lost time on the notebook during the study.

Creating the final document is not mentioned in the study. But we built a special documentation server where the mobile application mails the collected data. The server creates a word document out of this automatically. On the notebook the user has to copy the pictures from the camera to the notebook and has to assemble the picture. So when implementing the solution the difference between notebook and smart phone would be much larger than shown in this study.

The NASA TLX results point out the advantages and disadvantages the users see in the smart phone application. Most of them had to concentrate a lot on the application during the study. Furthermore many users felt hurried up. On the other hand most of them agreed that this application helps to speed up the task and is easier to handle.

The application is an exhibit of the demo center at the mobile research center Bremen [6] where mobile and wearable solutions developed in research projects can be tested by the public [Fig. 7]. The feedback of many visitors to the smart phone application was quite positive. Most people mention other scenarios for application.

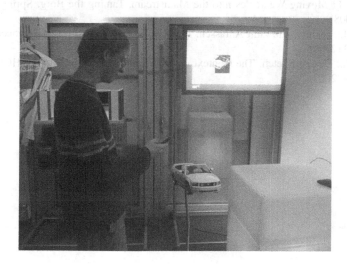

Fig. 7. Visitor of the demo center testing the application

4 Future Work

A further study is planned analyzing and evaluating the data quality of the results from both solutions, as our smart phone solution is expected to improve the quality of survey reports. For this study specialists working with the documentation are recruited. Furthermore the application will be evaluated compared to a solution based on an ultra mobile computer like the OQO.

An Ultra Mobile Personal Computer like the OQO could fix the problems with too small displays and keyboards but is linked to the problem of the camera as a second device. Maybe the most important point to do is a field test. We tried to rebuild a situation in our lab near to what is happening on the shop floor. But this is an abstraction and we have no experiences with this application in a real setting.

The application can be also transferred to other scenarios like survey reports for insurance companies in car accidents. Here adding GPS support for tagging information about where the accident happens could be advantageous. Survey reports for wind turbine maintenance are a further application we have in mind.

References

[1] Kangas, E., Kinnunen, T.: Applying User-Centered Design to Mobile Application Development. Communication of the ACM 48 (2005)
[2] Witt, H.: Human-Computer Interfaces for Wearable Computers. Vieweg + Teubner (2008)
[3] Witt, H., Nicolai, T., Kenn, H.: Designing a Wearable User Interface for Hands-free Interaction in Maintenance Applications. In: Percom proceedings, pp. 655–659 (2006)
[4] Witt, H., Leibrandt, R., Kemnade, A., Kenn, H.: Scipio, A miniaturized building block for wearable interaction devices. In: IFAWC 2006 proceedings, pp. 103–108 (2006)
[5] Dvorak, J.: Moving Wearables into the Mainstream, Taming the Borg. Springer, Heidelberg (2008)
[6] Rügge, I.: Technologie zum Anfassen. In: Mensch and Computer 2008 proceedings, pp. 219–223 (2008)
[7] Savio, N.: Design Sketch: The Context of Mobile Interaction. In: Mobile HCI 2007 proceedings (2007)

Integrating User Experience into a Software Development Company – A Case Study

Tobias Komischke

Infragistics, Inc., User Experience,
50 Millstone Road, Princeton, NJ 08520, USA
tkomischke@infragistics.com

Abstract. Establishing an user experience (UX) practice is an endeavor that more and more companies engage in to increase their market success. While there is a rich knowledge base on UX processes and methods, practical tips and tricks for starting up a corporate UX culture are harder to find. This paper summarizes best practices, recommendations and experiences from other companies and traces our own efforts and thoughts while integrating UX into our company.

Keywords: Usability, User Experience, Case Study, Organizational Aspects.

1 Introduction

Today, more and more IT companies are making user experience (UX) engineering an integral part of their development. While 10 years ago, UX engineering was perceived as a luxury, the success stories of companies that deploy user-centered design (UCD) processes have enhanced the status of this discipline substantially.

One of the drivers for UX attaining mainstream status in the corporate world is the rich literature on methods, processes and technologies that is available in books, on websites and in conference proceedings. However, from a perspective of a company wanting to start up an UX practice, it is quite a challenge to learn about experiences and best practices on implementing it within one's everyday work activities. For instance, what are the indicators that the company is on the right track and at what point can an organization claim to be user-centered? What are the steps that must be completed?

Helpful information about this topic is hard to find in traditional (paper-based) literature, but is available on blogs and forums where UX practitioners exchange their experiences.

Our company started an UX practice in 2008. There had been usability considerations before, so we did not start from zero. We established a dedicated and managed team with own goals that were tied to the company's strategy.

What could we learn from others? What kind of challenges would we face and how could we overcome them? What are all the puzzle pieces required that - as a sum - would bring us forward?

This paper summarizes knowledge and guidance that other organizations and individuals have shared about establishing a UX practice and adds our own approach and experiences.

M. Kurosu (Ed.): Human Centered Design, HCII 2009, LNCS 5619, pp. 221–229, 2009.

2 Why UX?

What motivates a company to invest in UX? While there are many answers it comes down to just one: increasing the competitive edge. This can be achieved by developing useful, usable and desirable products that yield increases in productivity, market share and sales on the one side, and save costs on the other side. Refer to [1] for more details.

Our company had been world market leader in UI controls for 20 years. The reason to form a UX team was based on two influences:

- Management was aware of the evidence that UX contributes to the success of companies. UX was a central building block to help expand our market position.
- Our customers have asked for guidance and expertise on how to design products that are geared towards their users.

The Infragistics management team saw the obvious value that would result from an investment in UX.

Our start resembled the way Adobe had established their UCD culture 10 years ago [2]. Adobe also had existed successfully in the market for 20 years. They had carried out grass-roots usability practices before a formal investment in UX. Adobe established a UX practice to achieve a competitive edge in their market, which was being crowded with competitors. The difference between Adobe and Infragistics is that at Adobe the push did not come from top management, but from two individuals within functional groups. Also, their team worked only internally to optimize their products. Our main focus was two-fold from the beginning: providing UX services to our customer base and working internally on our own products and processes.

3 Corporate Maturity

At the starting point, it was important to understand where our company was on the spectrum of companies that followed UX practices. Yet, how does one assess where a company lies on the continuum between no UCD in place at all vs. total UCD integration? Are these even meaningful semantic poles?

There are helpful references in the literature. Nielsen's corporate usability maturity model is one of them [3]. Table 1 shows 8 stages that are not totally distinct, but rather highlight characteristics of different stages of maturity. Nielsen states that companies may actually stay within the same stage for several years, so maturing from stage 1 to stage 8 can take 20 years. Obviously, it does not necessarily have to be a goal of a company to reach stage 8.

Based on Table 1 we were at stage 2, "Developer-Centered Usability" when we started establishing UX at Infragistics. Whether or not usability was considered during product development relied on team members having knowledge and experience in UCD. When this knowledge did exist, it was limited to tactical UI design best practices. There was no user research involved.

Table 1. Stages of corporate usability maturity [1]

Maturity Stage	Description
Stage 1: Hostility Toward Usability	User characteristics and needs are not considered for user interface design.
Stage 2: Developer-Centered Usability	Development teams rely on own intuition about what constitutes usability.
Stage 3: Skunkworks Usability	Usability decisions are based on user data, but are carried out ad-hoc.
Stage 4: Dedicated Usability Budget	Usability is planned for and funded, yet is limited to usability testing.
Stage 5: Managed Usability	Existence of a dedicated usability group led by a usability manager. Consistent application of usability efforts, creation of design guidelines.
Stage 6: Systematic Usability Process	Usability efforts are process-driven and applied to every project.
Stage 7: Integrated User-Centered Design	Full integration into development process, quality tracking through metrics
Stage 8: User-Driven Corporation	Company strategy is defined by user data. User experience goes beyond software.

Another way of assessing a company's UX maturity was proposed by Jared Spool's three questions for great experience design [4]. For him, vision, feedback, and culture are the three key factors determining whether or not a company is successful in UX. Each one of them can be summed up in a question:

- "Does everyone on your team know what the experience will be like interacting with your offerings five years from now?" (Vision)
- "In the last six weeks, have your team members spent at least two hours watching people experience your product or service?" (Feedback)
- "In the last six weeks, has your senior management held a celebration of a recently introduced design problem?" (Culture)

This definition is catchier, but at the same time it does not sufficiently allow for the variety of nuances between Yes and No answers.

A very similar approach that does allow for some nuance is suggested by Hurst [5] who introduced a measure called "Tesla" which stands for "time elapsed since labs attended." A Tesla is how long it has been since an individual or a company has spent time directly observing customers interacting with a product or service.

While both Spool and Hurst oversimplify the complexity of efforts and processes that are required to effectively practice UX in a company, their definitions provide focus to the fact that user research is the core of any UX effort. Also, oversimplification can sometimes help initiate discussions within companies about what efforts are required to establish UX.

After an assessment of a company's position along the UX maturity spectrum, it is important to determine if the conditions are favorable for starting up UX. When is an organization ready to adopt UX? According to [6], a company exhibits readiness when:

- Management uses UX vocabulary.
- A director or VP of UX was hired.
- Usability testing is a given.
- There is an allocated budget for hiring UX staff or consultants.
- Usability labs are in place or under discussion.
- Product management believes that UX yields strategic advantage.

In our company all but the third bullet (usability testing) and the fifth bullet (usability labs) were in place. We have already deployed and showcased usability testing. The method and the results have generated increased interest from other functional groups. We don't plan to invest in usability labs since over the past 10 years the trend has moved from formal test settings towards more informal and flexible testing environments. We believe that the value of usability testing is not the setting, but proper preparation, execution and analysis.

4 Getting Started

If the goal of a UX practice is to generate user-centered products, what factors are instrumental to achieve this?

The body of knowledge can be summarized as follows. Gartner Research [7], professionals working at Adobe [2], Apogee [6], Sage [6], WebWorld [6], Yahoo [8], Oracle [8], and Daimler [9] suggest the following:

- Find and nourish executive sponsorship.
  This is indispensible for securing funding, consideration in the company's strategic planning and gaining exposure to the right stakeholders.
- Build a talented team.
  In order to produce high-quality results, staff members must have a strong background in the methods and processes that engineer user experience. Typically, UX teams are multidisciplinary. Chapter 6 will elaborate more into this topic.
- Dedicate time and effort to communicate the value of user-centered design within the organization.
  Many people in a given organization do not understand what UX is, the return on investment from UX work and the concrete role assumed by a UX group. Communicating the value of UX serves an educational purpose and secures buy-in and funding. For this reason, at Adobe the starter group talked with both executives and financial controllers [2].
- Choose strategic projects that yield great results and demonstrate the value of UX.
  The best way to prove the return on an UX investment is to contribute to the success of key projects that (a) are of strategic importance to the company and that (b) are visible to a majority of departments and employees.
- Make your knowledge available throughout the organization.
  This helps to institutionalize UX into an organization. It can be achieved by many ways, including offering lunch and learns, providing an online repository of UX patterns and UX Do's and Don'ts on the intranet. In the end, this is about enabling non-UX people to make well-founded decisions on their own.

In our experience, working on strategic projects was a challenge at the beginning, because work was already in progress when our team was formed. Another challenge was to define the actual role of UX in a project. Since co-workers were not always aware of UCD processes, they tended to ask for our help too late for us to be able to make a significant impact in the projects. Communicating and educating UCD practices within the organization proved to mitigate this problem, because as co-workers better understood our processes and methods, they involved us earlier in their projects.

5 Organizational Models

There are three options for bringing UX expertise into a company: outsourcing, internalizing, or a mix of both.

For companies that are at the beginning of their journey towards UX maturity, it is reasonable to first hire external consultants to guide their way. This was done initially at Infragistics as consultants trained the staff on UCD and guided the management team on building internal expertise and UX strategy.

Once a company is ready to hire own staff to drive UX, there are various organizational models available [8]:

- Centralized funding model – UX staff is part of a central UX organization with central funding and own leadership.
- Client-funded model – UX staff is part of a central UX organization funded by business units and provides UX resources to their product teams.
- Distributed model – UX staff is part of individual product development teams.
- Consultancy model – UX staff is part of a central UX organization with product teams paying for UX resources on a per-project basis. The UX organization can be a cost-center (charging actual cost) or a profit-center (charging more than actual costs).
- Hybrid model—Individual business units have their own UX teams, and a centralized UX group provides infrastructure, communications, knowledge, and process.

There are advantages and disadvantages associated with each of these models. For example, a purely distributed UX organization makes it hard to harmonize and standardize UX practices as there tends to be a lack of governance on a higher level. On the other side, any central UX group – due to it catering to several business units – cannot have the same level of product or domain knowledge as UX teams that exclusively work in individual business units. There are companies (e.g. Siemens) that have both central and distributed UX organizations.

Another factor determining the organization model is the size of the organization. For small companies (100 employees or less), it is rare to even have internal UX resources. If a small company does have internal UX resources, the structure of the company is typically fairly flat and straightforward, obviating the need to find the right place for UX. As with other corporate functions like sales or marketing, when companies grow in size, they increasingly specialize and need to define effective and efficient organization models. At Infragistics, the first priority was to assist our customers with integrating UX in their projects. So, the decision was made to form the new UX team as part of our service branch. At the same time, it was clearly communicated throughout the company that the team would work internally as well to establish a user-centered culture and to optimize the UX of our products. Based on this setup our current organization can be best described as a centralized funding model for our inward-facing activities and as a consultancy model for our outward-facing activities. Like all organization models discussed in [8], the consultancy model assumes UX staff working only internally, but the model is applicable to the scenario where the service recipients are external clients. From that perspective our team is not different from a standard UX consulting company.

6 Staffing and Team Building

A team is excellent if its members are knowledgeable in the relevant topics that comprise UX (e.g. human factors, usability engineering, and qualitative research), have gathered enough experience in the industry to ramp-up quickly on any particular project, and have the enthusiasm, creativity and problem-solving skills to lead a product beyond what can be initially imagined. UX professionals are highly specialized and rare in numbers compared to other professions. Based on an informal internet research there are at least 6 million developers worldwide, while the number of UX professionals is estimated as somewhere between 200,000 and 400,000. So it is a challenge for a company without an established name in the UX community to attract talent. There are many contributing factors, including the salary that can be offered to a candidate, the geographic location of the company/workplace, and of course the overall economic climate.

If companies work with recruiters to fill the positions, the commission is typically 15% - 25% of the yearly salary of the job position. Since this payment is only due in case a person is actually hired, working with a recruiting agency is recommended. The time it takes to communicate to a recruiting company what specific skills are needed for UX should not underestimated, so one prerequisite is that it is a local agency which specializes in UX. Depending on geographic location, it can be easier or harder to find an agency knowledgeable and experienced in placing UX talent.

Another avenue for finding talent is through professional networks like special interest groups, web forums or professional associations like the UPA (Usability Professionals' Association) [10]. Prior interactions with members of these networks before their actual application for a job, provides a level of certainty that they are a good match for the position.

What should the dream team look like? UX has always been multi-disciplinary with each discipline contributing its own expertise and skills. Normally UX teams are diverse (see for example [2]) and the trend over the last 10 years has been to further broaden the range of backgrounds. A typical mix of talents that work together in an UX team are psychologists, engineers, computer scientists, and graphic designers.

At Infragistics, we built up a team of strong professionals that came from various backgrounds (development, usability engineering, human factors, visual design) and have a track record in the industry. The average job experience in UX on our team is close to 9 years, so there is a full repository of expertise. Providing a work environment that fosters excellence is a never-ending challenge, but we were lucky with at least our physical environment. Our company bought a new building that needed a completely new interior architecture. We participated in classic UCD with the architects and told them our requirements (e.g. to cater for both collaborative work spaces with an overabundance of white walls and quiet rooms to focus on detail work).

7 Community Activities

One aspect of starting a UX practice that is not often found in articles is participation in community activities. This is somewhat surprising, because UX professionals are known to be communicative and engaging. It seems that these activities, which include attending and speaking at conferences, publishing in journals and on blogs, attending and hosting local events, being active in discussion forums, etc., take place naturally and are not particularly noteworthy.

The benefits of engaging in these activities are manifold: they help the individuals to expand their professional network and increase their technical reputation. In fact, companies that employ UX experts get recognition from that fact alone. UCD must be a serious consideration if UX experts work there and help shape the culture. And last but not least, these activities help to achieve the greater goal of spreading the knowledge about UCD.

While many activities don't require explicit funding, there are costs involved in attending conferences. For instance, costs like registration and travel that quickly add up to a substantial amount of money when several conferences are attended. Consequently, to make community activities happen, there has to be a planned budget in place. Since conferences also have an educational function for the attendee, funding from a professional development budget can likewise be tapped.

Due to the recent economic downturn, our company, like most others, had to cut back on travel expenses. On the other side the need to evangelize on UX was undisputed. So we presented a list of "must go" domestic key events to our CEO who – based on our education on the relevance of UX – supported our request and made additional funding available. In terms of local events, we try to attend nearby meetings and also plan to host events in our new building.

8 Conclusions

It has been one year now since we started our UX team. Initially the team consisted of two persons. Today, we are 5 persons working together towards our goal of servicing our customer base, establishing a UX culture inside our company and actively participating in the UX community. We have successfully ramped up our customer engagements, where we provided training and consulting on UX in different market verticals. Internally, we have trained co-workers and the management team on UX and started to set up a knowledge base that is open to everyone. Due to our size we cannot participate in every ongoing project, but we selected a handful of developments where we contributed our expertise right from the beginning. We also spearheaded our own tool development thereby demonstrating the use of different UX methods. Maybe one of the most prominent indicators that we are on the right track is that our CEO continuously communicates to the entire company that UX is an integral and strategic part of who we are and what we are doing as a company.

From day one (and actually before that) we studied the lessons that other companies and individuals made while they established an UX practice. Many of our experiences with overcoming obstacles and doing the right things at the right time match those that other companies have made prior to us, which demonstrates that there is a value in sharing practical knowledge and experiences with others. Having attended UX-related conferences for more than 10 years, I believe that there is a lack of own dedicated tracks that focus on this topic. At this point I would like to acknowledge the importance of blogs. To many professionals shared online journals do not match up with technical articles in journals and conferences. While this is true for many blogs that rather reflect personal opinions on topics, there are blogs that have a peer review process and thus provide a quality assurance. Many great inputs for this paper I found on UXmatters [11], a peer-reviewed web magazine where industry thought leaders share their knowledge and experiences.

Independent from the medium, information about incorporating UX into an organization is vital for those companies that want to pursue this, it is consequently vital for the overall economy and at the end of the day, vital even for the society in general.

References

1. Bias, R.G., Mayhew, D.J. (eds.): Cost-Justifying Usability, 2nd edn. Elsevier, San Francisco (2005)
2. Ehrlich, S.: User Research at Adobe: Establishing a User-Centered Culture. In: Jacko, J., Stephanidis, C. (eds.) Human-Computer Interaction. Theory and Practice (Part I), Proceedings of HCI International, Crete, Greece, June 22-27, vol. 1, pp. 454–458. Lawrence Erlbaum Associates, London (2003)
3. Nielsen, J.: Corporate Usability Maturity,
   http://www.useit.com/alertbox/maturity.html (last retrieved 2/18/2009)
4. Spool, J.M.: The 3 Q's for Great Experience Design,
   http://www.uie.com/articles/the3qs/ (last retrieved 2/18/2009)
5. Hurst, M.: One number to grade any executive,
   http://goodexperience.com/2008/11/one-number-to-grade-a.php (last retrieved 2/18/2009)
6. Szuc, D., Sherman, P.J., Rhodes, J.S.: Selling UX (2008),
   http://uxmatters.com/MT/archives/000335.php (last retrieved 2/18/2009)
7. Valdes, R., Gootzit, D.: Usability Drives User Experience; User Experience Delivers Business Value. Gartner Research, ID Number G00152284 (2007)
8. Nieters, J., Pattison, L.: In Search of Strategic Relevance for UX Teams,
   http://uxmatters.com/MT/archives/000333.php (last retrieved 2/18/2009)

9. Metzker, E., Seffah, A.: User-centered Design in the Software Engineering Lifecycle: Organizational, Cultural and Educational Obstacles to a Successful Integration. In: Jacko, J., Stephanidis, C. (eds.) Human-Computer Interaction. Theory and Practice (Part I), Proceedings of HCI International, Crete, Greece, June 22-27, vol. 1, pp. 173–177. Lawrence Erlbaum Associates, London (2003)
10. http://www.upassoc.org/ (last retrieved 2/18/2009)
11. http://www.uxmatters.com/ (last retrieved 2/18/2009)

Full Description Persona vs. Trait List Persona in the Persona-Based sHEM Approach

Masaaki Kurosu

National Institute of Multimedia Education (NIME), Japan
masaakikurosu@spa.nifty.com

Abstract. Instead of the usual persona method (full description persona), a new type of persona (trait list persona) was proposed for the purpose of covering the wide variety of possible users. These two persona methods were then used to anticipate problems that might occur by applying sHEM (structured heuristic evaluation method) that is one of the inspection methods. In other words, it was not necessary to create the full descriptive scenario but just the possible lists of problems that are similar to the pMS (problem micro scenario) of micro scenario method were described.

Keywords: Full description persona, trait list persona, scenario, sHEM, pMS.

1 Introduction

Persona method is becoming widely used for specifying the requirements for a new design. It is usually combined with the scenario method and let designers conceive of some devices or systems that are compatible with the figure of the user and the context of use.

Usually, several personas are described to cover the diversity of users. But it is not enough to cover the whole range of diversity among users by just generating several personas. Diversity among users is far wider as can be listed in Table 1.

The purpose of this study is twofold: one is to compare the effectiveness of the full description persona (the conventional persona) and the trait list persona (the new persona method proposed in this article), and another is to confirm the effectiveness of the combination of the persona and the sHEM (structured Heuristic Evaluation Method) so that more variety of problems can be found than just describing the traditional scenario.

2 Full Description Persona vs. Trait List Persona

The full description persona is a compilation of various description of a person. For example, as is shown in Figure 1 and 2, name, gender, age, occupation, personality, birth place, current address, academic background, family, hobby, license, history of

M. Kurosu (Ed.): Human Centered Design, HCII 2009, LNCS 5619, pp. 230–238, 2009.
© Springer-Verlag Berlin Heidelberg 2009

Table 1. Diversity Among People

Traits	Context	Attitude
Biological Traits	Mental Condition	Individual Preference (Many Preferences, No Preferences, etc)
Age, Generation (Senior, Middle-aged, Young, Children, Baby)	Emotional Condition (Stable, Unstable, Urgent, etc)	Political Attitude (Left-winged, Right-winged, Neutral)
Sex, Gender (Male, Female, Gender Identity Disorder)	Level of Consciousness (Sleep, Indistinct, Aroused, Over-excited)	Religion (No Religion, Buddhism, Isram, Christianity, Newly-risen Religion, etc)
Physical Traits (Upper Limb Disorder, Lower Limb Disorder, Paralysis, Pregnancy, Hurt, Handedness, etc)	Everyday Life	Regression to Tradition (Conservative, Innovative, Radical, etc)
Cognitive Traits (Visual Disorder: Poor-sightedness, Inborn Blindness, Acquired Blindness, Color-Blindness, etc.) (Hearing Impaired) (Cognitive Impaired)	Economic Situation (Income Level, Regularity of Income, etc)	Social Attitude (Individualism, Collectivism, Antisocial, etc)
Body Dimension (Height, Weight, Hand Size, Arm Length, Leg Length, Flexibility, Hand Power, Fine Operation, etc)	Freedom (Free Situation, Staying in the Office or School, In Custody, etc)	Value Attitude (Functional Value Attitude, Usability Value Attitude, Aesthetic Value Attitude, Sensibility Value Attitude, Economic Value Attitude, Quality Value Attitude, Ethical Value Attitude)
Psychological Traits	Lifestyle (Workaholic, LOHAS, DINKS etc)	
Personality (ex. Big-Five: Openness, Conscientiousness, Extraversion, Agreeableness, Neuroticism)	Educational Background (Junior High, High School, College, Graduate School, etc)	
Mental Traits (Sensitivity, Sensibility, Psychological Disorder (Psychosis, Neurosis, Personality Disorder, Mental Retardation), etc)	Social Position (Salaried Employee, Self-Employed)	
Knowledge and Skill (Novice, Expert, Literacy)	Situational Factors	
Intelligence Type (ex. Fluid Intelligence, Crystallized Intelligence)	Urgency (Normal Situation, Urgent Situation)	
Learning Style (Systematic Learning, Ad-hoc Learning, Trial and Error, etc)	Temporary Condition (Heavy Baggage, Bulky Clothing, etc)	
	Physical Context	
	Geographical Environment (Big City, Small City, Isolated Place, etc)	
	Physical Environment (Temperature, Humidity, Illumination, Noise Level, etc)	
	Socio-Cultural Context	
	Historical Background (Ruling Class, Ruled Class, Oppressed Class, etc)	
	Cultural Background (Ethnic Culture, Nation Culture, Local Culture, Family Culture, Generation Culture, etc)	
	Language (Japanese, English, Chinese (Mandarin, Cantonese, etc), etc)	
	Civilized Level (Civilized area, developing area, etc.)	

use of something (that is the focus of the persona description), background information and picture are included. Items in this kind of persona are mutually related so that
an integrated image of the person will be described. In the case of Figure 1, for example, having the driver's license for special big cars is related to his occupation as a
farmer, living with the eldest son who has a skill of high tech devices is related to his
use of the cell phone as a first comer, etc.

The trait list persona is a list of various traits as shown in Table 2. Example list includes gender, age, occupation, personality, current resident place, educational background, family size, hobby, license, use history and use level of something (that is the
focus of the persona description), income level, etc. In the case of the trait list persona, not all the items in the list will be related to the focus. But because various traits

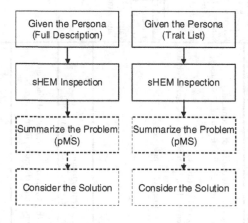

Fig. 1. Flow of experiment

are listed, the diversity among people can be covered more than in the case of full
description persona.

3   sHEM for Both Types of Persona

sHEM is the improved form of heuristic evaluation method originally proposed by
Nielsen [1,2,3]. Participants for the evaluation session are asked to detect possible
problems in each sub-session where the focus of attention changes from sub-session
to sub-session. For example, the first sub-session will be related to the ease of operation from the viewpoint of human factors engineering and the second sub-session will
be related to the ease of understanding from the viewpoint of cognitive engineering.

By using the full description scenario, participants will be asked to describe what
type of problems the persona will be confronted by changing the focus from sub-
session to sub-session. By using the trait list description, participants will be asked to
describe possible problems for the person who has the specified trait by changing the
focus from sub-session to sub-session.

List of problems obtained from this approach resembles the problem micro scenario (pMS) of micro scenario method [4,5]. Problems are described one by one as

containing one (possible) problem. By giving them the tag or the category keywords and sorting them by that tag, one can get the small set of problems having the similar problems. Considering possible solutions to the problem list each of which is the summary of the small set of problems, it is possible to consider a new concept or a new idea that can solve that problem.

4 Experiment

4.1 Method

7 male and 8 female students of Tokai University and 24 male and 1 female students of Tokyo Denki University participated in the experiment. Participants were first given Figure 2 and 3, and were asked following questions.

Name: Shoichi Nomura
Gender: Male
Age : 63
Occupation: Farmer
Personality: Serious. Accomplish anything he started. Highly cooperative and has many friends. Willingly take the role of leader for the local festival.
Birth Place Furukawa city, Miyagi Prefecture
Current Address Furukawa city, Miyagi Prefecture
Academic Background: Furukawa Agricultural High School
Family: Mother (85 yrs.), Wife (62 yrs.), Eldest son (38 yrs.), Wife of eldest son (35 yrs.), Second son (33 yrs. In Tokyo), Grandchild (7yrs. Daughter of the eldest son)
Hobby: Comic story
License: Driver's license for special big cars, Driver's license for regular cars, License for agricultural machinery, License for poisonous substance
History of Use for PC: 1 year
History of Use for Internet: 1 year
History of Use for Cell phone: 2 years. Still using the first cell phone he purchased.
Background: He started to help his family farmhouse business after graduating the Agricultural High School. Because he had a high level of knowledge and skill for the machinery and electronics, he started to introduce various machines into the farmhouse business. He obtained licenses for using his farmhouse business by just studying by himself. But he is no good for the information technology. He started to use the PC and the Internet stimulated by the eldest son, but he is not yet well-accustomed to them and has no specific purpose. Just two years ago, he started using the cell phone by the encouragement of his children. He is using only the calling function now and would like to use it more practically. His wife and eldest son also have a cell phone. Financially he is want for nothing. His mother is sickly and is lying in bed almost all the time.

Fig. 2. Full Description Scenario of Mr. Nomura

Name: Mami Sakata
Gender: Female
Age: 32
Occupation: House-wife
Personality: She is good-natured woman, but is sometimes careless for details of anything, in other words, is a-kind-of a scatterbrain. She is getting on well with her neighbors.
Birth Place: Kanazawa-city, Ishikawa Prefecture
Current Address: Koganei-city, Tokyo
Academic Background: Faculty of English Literature of Tsuda Women's College
Family: Husband (30 yrs.), Daughter (3 yrs.)
Hobby : Reading books, Listening to music, Visual arts, Cooking
License: Teacher's certificate, Driver's license for regular cars
History of Use for PC: 8 years
History of Use for Internet:8 years
History of Use for Cell phone: 10 years. Current model is her sixth one.
Background: Her parents' home is a wholesale store for clothes. She was staying in Kanazawa-city until the high school. Because she likes novels of Dickens, she came to the college in Tokyo for studying more. After the graduation, she was working for a life insurance company. She learnt how to use the PC and the internet after she started her job. Now she has her own PC at home and creates her own blog where she uploads topics on the child care and the cooking. She is carrying her cell phone almost all the time and frequently communicates with her friends by texting. She got married with her husband at the company and has spent 3 years now. At the birth of her baby, she quit the company and she has a part time job at the bread shop owned by her relative with her baby in the care of a nursery school.

Fig. 3. Full Description Scenario of Ms. Sakata

For Mr. Nomura,

Q.1 How do you think "Mr. Nomura" will use "something"?
Q.2 How do you think "Mr. Nomura" may face with the difficulty or the trouble while using "something"?

And for Ms. Sakata

Q.1 How do you think "Ms. Sakata" will use "something"?
Q.2 How do you think "Ms. Sakata" may face with the difficulty or the trouble while using "something"?

The slot "something" was iPhone in this experiment.
Participants were then given Table 2 and were asked

Q.1 How do you think a user having such a trait will use "something"?
Q.2 How do you think a user having such a trait may face with the difficulty or the trouble while using "something"?

Table 2. Trait List Persona

Category	Trait	
Gender	Male	
	Female	
Age	Younger than 10	
	20s to 30s	
	40s to 50s	
	Over 60s	
Occupation	Salaried Employee	Management Section
		Planning Section
		Research Section
		Design and Technology Section
		Manufacturing Section
		Advertisement Section
		General Affairs Section
		Sales Section
	Self Employed	Shop Owner
		Lawyer
		Medical Doctor
		Artist
		Agriculture and Fishery
		Designer
	Teacher	
	Student	
	Housewife	
	Part-timer	
	None	
Personality	Extrovert	
	Introvert	
	High Sensibility and Cooperativeness	
	Low Sensibility and Cooperativeness	
	Strong Responsibility and Faithfulness	
	Weak Responsibility and Faithfulness	
	Emotionally Stable	
	Emotionally Unstable	
	Open-minded	
	Closed-minded	
Current Resident Place	Hokkaido	
	Tohoku	
	Kanto	
	Hokuriku	
	Chubu	
	Kansai	
	Chugoku	
	Shikoku	
	Kyushu	
Educational Background	University or Graduate School	
	High School	
	Junior High School	
Family Size	More than 3 Generations	
	2 Generations (Couple and Adult Kids)	
	2 Generations (Couple and Children)	
	2 Generations (Couple and Babies)	
	1 Generation (Couple)	
	Single	

Table 2. (*continued*)

Hobby	Sports (Tennis, Swimming, etc.)
	Game (Game Machine, Chess, etc.)
	Creation (Painting, Music, etc.)
	Literary Art (Poems, Novels, etc.)
	Nature (Gardening, Flower Arrangement, etc.)
	Home (Handicrafts, Interior Design, etc.)
	Collection (Antiques, Dolls, etc.)
	Appreciation (Art, Music, etc.)
	Travel (Domestic, Foreign Countries, etc.)
	Pet (Dogs, Cats, Goldfishes, etc.)
	Expression (Dancing, Playing Music, etc.)
	Food (Cooking, Eating, etc.)
	Car (Automobiles, Motorbikes, etc.)
	Outdoor (Fishing, Camping, etc.)
	Gamble (Poker, Slot-machine, etc.)
	Information (Web Browsing, etc.)
License	Automobile
	Boating
	Special Skills
Usage History of PC	More Than Several Years
	3-5 Years
	1-2 Years
	None
Usage Level of PC	Professional (as an Occupation)
	Expert
	Middle Level
	Novice
	None
Usage History of Internet	More Than Several Years
	3-5 Years
	1-2 Years
	None
Usage Level of Internet	Professional (as an Occupation)
	Expert
	Middle Level
	Novice
	None
Usage History of Cell phone	More Than Several Years
	3-5 Years
	1-2 Years
	None
Use Level of Cell phone	Professional (as an Occupation)
	Expert
	Middle Level
	Novice
	None
Income Level	Over 100,000 USD per Year
	60,000-99,999 USD per Year
	40,000-59,999 USD per Year
	Less than 39,999 USD per Year

The procedure of the experiment is described in Figure 1. This time, the last two process of summarizing the problem and considering the solution were not included in the experiment due to the lack of time.

4.2 Result

Because of the qualitative nature of the data obtained from the experiment, statistical comparison was difficult to apply. Hence, only the qualitative description on the data will be summarized as follows.

Inspection Time
The time required to inspect the possible use and the possible problems may vary depending on the amount of description for each type of persona. But, in general, it was shorter for the full description scenario to grasp and consider the usage and the problems maybe because all the information in persona description are mutually related as representing the single personality.

Participants were required to change the mental set each time to go into the new trait in the trait list persona and this made it rather longer for participants to consider the possible usage and problems.

Contents of Inspection
Possible usage and problems were more realistic in the case of the full description persona and were fragmental in the case of the trait list persona. This result is quite natural because only the part of the information about the people is given in the trait list persona.

Coverage of Diversity
As was expected, the range of possible usage and problems was larger for the trait list persona.

5 Conclusion

The trait list persona was proposed and was compared with the traditional full description persona by asking participants to consider about the possible usage and problems. Regarding the inspection time and the contents of inspection, it was found that the full description persona brought a better result. But it was also true that the coverage of diversity was far better for the trait list persona. It is true that the information represented as a full description persona is limited to certain personalities as expressed in the limited number (usually several) of persona.

As a conclusion, it could be said that a combination of both types of persona will bring a valid set of possible usage patterns and possible list of problems.

References

1. Kurosu, M., Matsuura, S., Sugizaki, M.: Categorical Inspection Method - Structured Heuristic Evaluation (sHEM), IEEE SMC (1997)
2. Kurosu, M., Sugizaki, M., Matsuura, S.: Structured Heuristic Evaluation (sHEM). In: Usability Professional's Association 7th annual Conference Proceedings pp. 3–5 (1998)
3. Kurosu, M., Sugizaki, M., Matsuura, S.: A Comparative Study of sHEM (structured heuristic evaluation method). In: HCI International 1999, Proceedings, pp. 938–942 (1999)

4. Kurosu, M.: Micro-scenario method: a new approach to the requirement analysis. In: WWCS 2004 (2004)
5. Kurosu, M., Kakefuda, A.: Application of Micro-Scenario Method (MSM) to the Educational Problem-solving – A Case Study for the Mother Tongue Maintenance for Immigrated Children. In: CATE 2005 (2005)

Organized Reframing Process with Video Ethnography: A Case Study of Students' Design Project for New Interface Concept from Research to Visualization

Katsuhiko Kushi

Goshokaido-cho, Matsugasaki, Sakyo-ku, Kyoto,
606-8585, Japan
kushi@kit.ac.jp

Abstract. Ethnographical research is a recent trend in design profession field, but its open-ended approach does not always bring effective solutions, especially when a novice design student executes it. Inspired by Jacob Buur's video ethnography, the author revised the approach and applied it to a design project sponsored by a corporation. The project goal set by the company was vague enough for students to lose a consistent direction, however, with this method they could find underlying needs and created a set of attractive solution ideas to the client company.

This report will describe how the project went and generate results, and finally, discuss the possibility of design with video ethnography.

Keywords: Video ethnography, interface design, GUI, user need.

1 Introduction

An observation to find user needs is a typical start for a design project, especially in universities. Similar to design process in industry, it is often difficult for design students to connect research results to design solutions. If a project is focused on some specific product area, such as a mobile phone or ticket-vending machine, students execute research with confidence. In that case, they may be able to properly identify needs and make good designs based on those needs. But, if a project aims at giving a general outline of a wider product category, such as products with a small LCD display, as we were assigned to, the foci of their research and attention can easily be blurred and scattered. In such a case, the following of analysis will not give a clear definition of user needs and inevitably results in generation of poor ideas.

KOTO corporation in Kyoto city, that was planning to make wide use of a low-cost graphic chip, asked us to visualize new interface designs for daily electronic products with LCD displays since the company needed to show future design images of prospective customer manufactures' products to promote the chip. An LCD display with finger touch technology is becoming a major interface platform of daily electronic products. However, when we see beautiful and dynamic graphic design on the display, the products always have specially made graphic chips on the circuit boards of the products, and they require a large investment to make. So, although people already seem to be accustomed to seeing and using a small display interface, there are only few products that

M. Kurosu (Ed.): Human Centered Design, HCII 2009, LNCS 5619, pp. 239–246, 2009.
© Springer-Verlag Berlin Heidelberg 2009

have full color and dynamic displays, such as handheld game devices, and mobile phones including the iPhone. Other than these exceptions, common electronic products, such as fax-phones, remote controllers or microwave ovens, still have monochrome and static LCD displays. If a cheap and wide range of application graphic chip can be available for many products with small displays, how will interface design change and how will interactions between the system and people change? This question was the theme of this design project given by our client company.

2   Organized Research and Analysis

The theme above seemed to be wide and vague enough for students to fall into the typical failure. To avoid making a group of disconnected ideas, we tried to organize a consistent design process from the observation phase to prototyping. Ethnographical user research is now an emerging method in product development, owing to its potential for finding latent user needs [1]. Among those methods, the video ethnography method introduced by Jacob Buur et al. is characterized by its strength in allowing project members and observation subjects to share in the creative process of finding and interpreting problems, and defining design goals [2][3]. Although Burr's video ethnography method aims to achieve deep user participatory design environment, we focused on its collaborative aspects in regard to young team members in the design process, and applied it to this project, to establish a common interface concept for various products with small LCD displays.

The project team consisted of eight design students from Kyoto Institute of Technology (KIT) and the author, as a project leader. To cover such a wide product range in a relatively short period, we first made a list of subject products and allotted each member a portion of observations work to be conducted with video cameras. The list included 12 kinds of electronic product with small displays. To obtain a good contrast in our study, we also included 6 non-electronic products that users felt a strong attachment to. The followings are our steps in the early phase of the project.

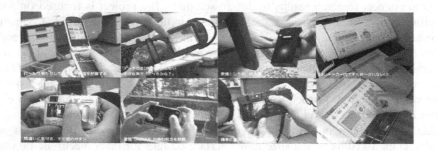

Fig. 1. Examples of subject products

Observation:

- A group of one to three design students visited the users' daily environment at, for example, their home, office, or on the street. They watched and recorded their activity using the subject products with digital video cameras.

- When possible, we asked users to try to use the latest product that was brought by us, such as a digital camera with a touch screen control panel.
- When possible, we asked users to show the product they were most attached to and show how they are using them.

Video analysis:
- A group of two to three reviewed assigned video data. A group had a combination of at least one observer who shot the video, and one non-observer of the visit.
- Watching video on a PC (Macintosh), they tried to find interesting points, including users' failures, unique usages, contextual meanings, and so on. While finding those points, they were encouraged to discuss with each other. In most cases, a discussion started from an observer's explanation of a users' situation, followed by a non-observer's questions that livened up their dialogue. Those discussions watching video data generated new findings and understandings of interactions between users and products.
- To record their findings, we made short video clips that clearly represented the points. The number of clips reached to 199 for electronic products with displays and 43 for products in the contrast group.

Collaborative editing of observation results:
- Following Buur's video ethnographical method, we played a "video card game" where we invited engineers from our client company. Necessary preparations involved making physical photo cards representing each edited clip. There were 242 cards in total.
- Before playing, we reviewed and shared all the video clips with simple explanations made by clip editors.
- Based on the Kawakita's KJ method, together with the visitors, we sorted all the cards into several groups of categories, and put names on each card group [4]. This time, we created 12 categories of findings. The followings are the names of the categories.

> **Love at first sight / Drifting / Exploring / Guessing / Stress / Feeling of wrongness / Forgetting / Compromise / Trust / Exclusive selection / Adapting / Special treatment**

- We arranged the photo card groups on a large sheet of paper. Then we could see some relationships among all groups fall along two axes. The first axis showed how much users made use of products, and the other showed how much users understood the products.
- With those categories and the 2 axes, we got an overall view of all problems found through the observation and the video analysis. Namely, we got a video card map of problematic situations. In the map, we found there was a time flow of user experience with machines.

Reframing by Needs Mining:
- From the map mentioned above, we could get an overview of a broad range of interaction problems people were facing everyday. However, it only described a visible phenomenon, and it did not indicate latent design needs for new concepts. So, we still needed more thorough interpretations of the map.

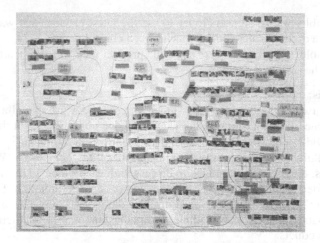

Fig. 2. The map of video card game

- In this step, each member was to collect four cards from each category to express more definite needs, rather than problems. They put four cards on a sheet of A4 paper, and wrote the title of the need at the top, and the simple explanations on the bottom. We called this sheet "a card family." Individual students made the card families and created 39 family sheets in total.
- Through the "card family" step, each team member was to find underlying realities that existed under multiple categories and product types. This activity required individual creativity and a thorough interpretation and reframing of phenomena.

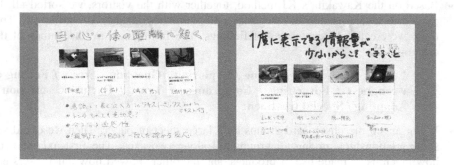

Fig. 3. 2 examples of the card family

After making card families, we put all of them on a table, and we reordered family sheets as we did in the video card game. As a result, we established 6 categories of card families. The following are the titles of the categories.

> **Small home but fun place / Minimum distance among mental model, expression, and operation / Natural expression matching with machine / Feel of the space on GUI / Message through GUI / Important GUI elements**

Design frame from understanding the structure of needs

Looking at the groups of card families, we noticed a structured relationship between them. We rearranged the group accordingly and finally made a concept diagram, which showed an interface goal with small LCD displays on various products and its constituent factors and hierarchy.

The goal we established was the 'fun-pact interface.' Fun-pact was a term coined to combine 'fun' and 'compact.' With this title, we wanted to propose fun interactions, despite the fact that the product itself was a daily appliance, and usually made at low-cost with relatively simple functions. 'Compact' does not only describe the size of the product, but also the lower complexity of the product functions and concepts.

This main keyword stems from the observations in which we saw that users had struggled to understand and manipulate small displays because of their poor communication design that uses static expressions and too many layers of command structures. So, the keyword aimed at the interface design opposite of the boring and complex current state of small interactive displays.

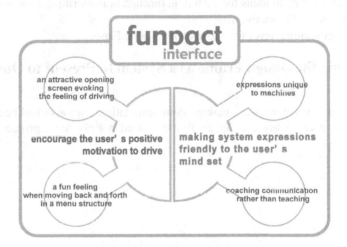

Fig. 4. The first diagram of the structure of needs

The Fun-pact interface design consists of two major factors. One factor addresses the user's experience by "encourage the user's positive motivation to drive (Driving motivation)." The second factor addresses the user's acceptance of information, by "making system expressions friendly to the user's mind set (Friendly expression)".

In the former keyword, we aimed to provide a design, which could evoke the user's positive emotional response when touching the products. This would generate the users' own customizations and rules. Users' customizations and rules were expected to make products fun to use and encouraged the users' positive attachment to them. The latter keyword - compact - was meant to achieve an interface mechanism using natural and direct graphical expressions, easily understandable to users. Reducing user's cognitive loads should be crucial to peace of mind in use.

In the concept diagram, the two major factors of the "fun-pact interface" have two additional sub-factors respectively. "Driving motivation" factor has "*an attractive opening screen evoking the feeling of driving"* and *"a fun feeling when moving back and forth in a menu structure."* "Friendly expression" factor consists of *"expressions unique to machines"* and *"coaching communication rather than teaching."*

3 Visualizing Keywords as Interface Design

Over the course of a few months, in an organized manner we gradually constructed our deeper understanding of the project theme and people's needs in terms of using daily electronic appliances with small displays.

After establishing the design frame, we moved to the phase of visualization. The following process that we executed, seemed to be based on a standard product design process. Each step is as follows.

- Sharing solution image of each design frame keyword using photo collages.
- Setting target-products to redesign.
- Setting simple user scenarios based on observations.
- Generating conceptual ideas for each item through brainstorming sessions.
- Selecting potential ideas to refine.
- Making demo-animations of the ideas with Adobe Flash.

4 Finalizing the Design Frame as a System to Present to Outside People

At the same time when we were making demo-animations, we also had been rethinking the frame itself. Since the design frame existed mainly for a project core, and

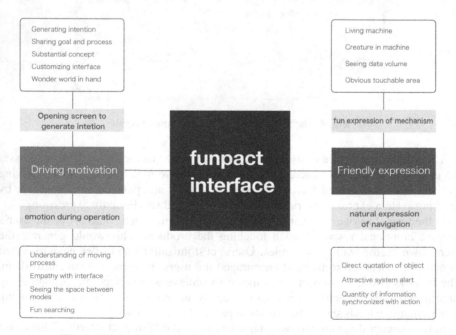

Fig. 5. The final diagram of the structure of needs with solution ideas

resources of ideation for all members inside the project, the keywords were not elaborated for outsiders to understand. We found that the demo-animation not only expressed the keywords, but also communicated each one's more concrete message. Then, we made a matrix chart and diagram to include ideas and revised keywords. In order to present ideas and their structure as a whole, we also made a viewer, where all ideas were presented using an inter-net browser.

Fig. 6. The solution ideas on the viewer

5   Effects of Video Ethnography Method in a Student Project

Rapid ethnography is regarded as a methodology used in the business field in order to explore the contextual needs of people. Usually professionals, like ethnographers, usability engineers, or interaction designers, conduct ethnographical research to see certain activities of some product domain. Those professionals are flexible in fitting methods to various circumstances. In our project, researchers were university design students who were not expected to be flexible, and the project theme was abstract. The video ethnography method introduced by Buur is definitely one of the more rapid ethnography methods. It has has a better organized approach in order to invite interdisciplinary people, including subject users, to participate in the development process. We found some merits of video ethnography in different purposes. The merits of video ethnography for a student project are:

- It can cover various people and product areas at the same time.
- Video data keeps important facts, even if an inexperienced observer overlooks them at the site.
- It is possible to explore and find needs by watching video. This feature is effective, especially when observation time is limited.

- Video analysis by means of editing short video clips by a team of observers and a non-observers gives not only a wider view of phenomena, but also a deeper understanding through their discussions.
- Video card game can give opportunities for all attendees to share all analysis results, and to collaboratively realize a broad range of problematic phenomena.
- Deeper interpretation of the phenomena, which sometimes indicates the underlying needs, can be achieved through the activity of making card family sheets.
- By sorting card families, students can find a structure of needs, and consequently, they can generate a conceptual design frame for a given project theme.

6 Conclusion

This case study has reported how video ethnography can be applied in a student project, and how can the process can provide them an environment to create GUI design solutions for small LCD displays on various daily products. From the experimental application, it can be concluded that video ethnography has a large potential to enable novice system developers, like design students, to find latent user needs and create design ideas based on these needs even with very abstract (or not well defined) theme.

In Japan, most manufacturing companies, as well as design courses in universities, do not have video ethnography specialists inside their organizations, despite the fact that the use of focus group interviews and questionnaire surveys has been well established practice. Nobody can deny the importance of triangulation, however, the fact is that only very few corporations are using ethnography for their third point of view. Since video ethnography can be started with novice observers, the author assumes that this method can be widely applicable not only to the field of design education, but also the field of business in Japan.

Acknowledgement

The author would like to thank Kozy Kubota of KOTO Corporation for giving us the chance to try this experimental approach on the project. Akitoshi Wada for providing coaching to students. I would also like to thank all of the students from Kyoto Institute of Technology who participated in this project, for their enthusiasm. This work was sponsored by the New Energy and Industrial Technology Development Organization (NEDO) in Japan.

References

1. Sherry Jr., J.F.: Ethnography, Design, and Customer Experience; An Anthropologist's Sense of it All. In: Squires, S., Byrne, B. (eds.) Creating Breakthrough Ideas, Bergin & Garvey, Foreword texts (2002)
2. Buur, J., Soendergaard, A.: Video Card Game: An Augmented environment for Use Centered Design discussions (2001)
3. Kushi, K.: Innovation notameno Design Process, ed. Research Institute of Human Engineering for Quality Life, Maruzen, pp. 104–115 (2005) (in Japanese)
4. Kawakita, J.: Hassou-Hou, Chuou-Shinsho, Chuou-Kouron-Shinsha (1967) (in Japanese)

Animated Demonstrations: Evidence of Improved Performance Efficiency and the Worked Example Effect

David Lewis and Ann Barron

University of South Florida,
13201 Bruce B. Downs Blvd MDC56, Tampa FL, 33612
dlewis3@health.usf.edu, Barron@tempest.coedu.usf.edu

Abstract. The purpose of this study was to assess the efficiency and effectiveness of animated demonstrations, to determine if those using animated demonstrations would exhibit the worked example effect [1], and a delayed performance decrement, described as Palmiter's animation deficit [2], [3]. The study measured relative condition efficiency (RCE) [4] and developed a construct called performance efficiency (PE). Results revealed the animated demonstration groups assembled the week one problem in significantly less time than the practice group, providing evidence for the worked example effect with animated demonstrations. In addition, subjects from the demonstration groups were significantly more efficient (given performance efficiency) than those from the practice group. Finally, group performance did not differ a week later, providing no evidence of Palmiter's animation deficit.

Keywords: Animation, cognitive load, performance efficiency.

1 Introduction

Sweller (1988) developed cognitive load theory, a theoretical framework for describing the actions of novices during problem solving. Sweller and his associates had found that those learners who studied worked examples outperformed those who learned by solving problems [5]. This was later described as the "worked example effect" [1]. This instructional strategy is recommended, as opposed to allowing learners to only learn through discovery problem solving [6]. For those who study human computer interaction, this may seem counter-intuitive, for we often try to "figure out" how to use software. In a computer environment, worked examples are commonly described as "demos" or animated demonstrations, and over the past few years, software providers (e.g. Microsoft), have begun using web-based animated demonstrations (demos) as a part of their products as documentation and support.

Lewis [7] proposed animated demonstrations act as animated worked-examples. This is because they allow learners to study solved problems, without actually using the software. This study considers the efficiency and effectiveness of animated demonstration as an instructional strategy, and like Touvinen and Sweller [6], the study compares two instructional strategies (animated demonstrations and discovery practice).

M. Kurosu (Ed.): Human Centered Design, HCII 2009, LNCS 5619, pp. 247–255, 2009.
© Springer-Verlag Berlin Heidelberg 2009

1.1 Animated Instruction and Procedural Learning

The animated demonstration literature extends back to the early 1990s. The most notable finding from this literature is the potential for an "animation deficit" [2], [3]. Palmiter et al. [2] found learners using animated demonstrations would acquire skills in significantly less time during an initial learning phase, but a week later, these learners had difficulty and took significantly longer to reproduce the same performance. This was later described as Palmiter's animation deficit [3]. However, other researchers were not able to replicate Palmiter's findings [3], [8]. In addition, if animated demonstrations act as worked examples as Lewis proposed [7], then Palmiter's animation deficit is in conflict with the worked example effect.

1.2 Relative Condition Efficiency (RCE) and Performance Efficiency (PE)

On some level, educational researchers use a medical model to "treat" ignorance, with instructional products that we hope are both efficient and effective. Those who study cognitive load theory analyze the differences between instructional conditions with a construct called "relative condition efficiency" [4]. To use this construct, researchers observe learners as they use one of several instructional conditions. Following the instruction, learners attempt a performance, and then rate their perceived mental effort during learning. Their performance scores and mental effort ratings are combined in a construct (See Equation 1) [4].

$$\text{Relative condition efficiency} = \frac{Z_{Performance} - Z_{MentalEffort}}{\sqrt{2}} \tag{1}$$

The resulting data is then graphed in a biplot (See Fig. 2) to allow researchers to visually compare the relative efficiency of instructional conditions [4].Since its development, relative condition efficiency has become an important basis for much of cognitive load research [9] but, this measurement relies on indirect or subjective measurements [10]. Paas and van Merriënboer [4] were aware of this, and state in their original paper that this construct should be qualified with performance data.

Efficiency and effectiveness may be described with dependent variables. Gagné described two general categories of dependent variables used during problem-solving studies [11]. He proposed most researchers consider (1) "the rate of attainment of some criterion performance" and (2) "the degree of correctness of this performance" [11], (p.295). In terms of the animated demonstration literature, these are described as performance time and accuracy [3]. So as Gagné suggests, another useful measure is performance time. While relative condition efficiency is a measure of efficiency, because it uses mental effort ratings, it does not include time in the equation, so it is difficult to analyze the efficiency of the performance given and instructional condition. Lewis [12] synthesized these ideas to develop yet another efficiency construct, called performance efficiency. Performance efficiency is similar to relative condition efficiency, but uses performance time rather than a mental effort rating (compare equations 2 and 3).

$$\text{Relative condition efficiency} = \frac{Z_{Performance} - Z_{MentalEffort}}{\sqrt{2}} \qquad (2)$$

$$\text{Performance efficiency} = \frac{Z_{Performance} - Z_{PerformanceTime}}{\sqrt{2}} \qquad (3)$$

Performance efficiency was developed to complement relative condition efficiency, and it is hoped that this metric may be used to strengthen cognitive load research. One should be aware that even though performance efficiency allows a researcher to analyze group performance, it does not include a mental effort rating, therefore it is not a measure of cognitive load. However, like relative condition efficiency, performance efficiency also may be used to analyze the relative efficiency of group performances. Performance efficiency contrasts instructional conditions in much the same manner, to combine these performance variables in a biplot with group performance times and performance scores (in this case accuracy) on the x and y axes respectively (See Fig. 1).

1.3 Hypothesis and Operational Definitions

Recall that the purpose of this study is to consider the worked example effect and Palmiter's animation deficit given animated demonstrations. This section outlines the hypotheses and operational definitions of these two effects.

The worked example effect is often defined as an improvement in learner perform ance given worked examples. Sweller and Cooper's early studies were the first to describe this effect [5]; [13]. They described this effect by saying a "decreased solution time was accompanied by a decrease in the number of mathematical errors" [5], (p.59). The dependent variables here are solution time and "a reduction of errors", or in terms of this study and the animated demonstration literature, performance time and accuracy [3]. In terms of the hypotheses of this study learner behavior may be described as:

H_a: During week one those using animated demonstrations will have a significant increase in performance over those only solving problems.

H_o: Learner performances will not differ during week one.

Lipps, et al. described Palmiter's animation deficit as "poorer retention despite faster learning following animation training" [2], (p. 1). In terms of this study and its dependent variables, an operational definition of Palmiter's animation deficit, was a significant increase in performance time and a simultaneous decrease in accuracy, one week after initial instruction given animated demonstrations as an instructional strategy. In terms of the hypotheses of this study learner behavior may be described as:

H_a: Those using animated demonstrations will have a decreased performance during week two.

H_o: Learner performances will not differ during week two.

2 Methods

The participants of this study were pre-service teachers. These undergraduates were enrolled in an introductory instructional technology course at a large university in the southeastern United States. An *a priori* power analysis for a four group MANOVA produced a sample size of $n=115$ participants. This number of participants is necessary to arrive at a power of 0.80, with a small effect size $\eta^2 = 0.125$, given an $\alpha=0.05$ ($\alpha=0.05$ is used throughout this study) [14].

2.1 Instructional Materials

Tuovinen and Sweller [6] provided subjects with an introductory overview to provide context. The overview in the current study was a short, narrated, non-animated web-based presentation (~ 2 minutes) developed with TechSmith Camtasia 4.0 [15]. All subjects viewed this overview which provided them with an introduction to graphic design and digital image editing. Thus all subjects were presented with screenshots of the required procedures. In addition subjects may have interacted with one of two animated demonstrations which were developed with Techsmith Camtasia Studio 4.0. The first animated demonstrations was identical to the practice problem used during week one (the mimic condition). A second "varied context" (collage-based) demonstration was also used. In each case, subjects were taught how to select, move, rotate, and hide layers within an Adobe Photoshop Elements [16] document.

2.2 Procedure

TechSmith Morae 1.1 [17] recorded all learner onscreen actions, as they interacted with the instruments and problem-solving scenarios. Upon completion of an initial demographics survey, all subjects were forwarded to an instructional overview. When the instructional overview concluded, a JavaScript randomly divided subjects into one of four instructional conditions. Two web-based animated demonstrations were used in these four instructional conditions:

- an animated demonstration group (demo), this group did solve the week one problem;
- an identical animated demonstration, and they practiced with the week one problem (demo+ practice);
- a different collage-based animated demonstration, with the week one problem as practice (demo2+practice);
- and a discovery-based practice condition (no animated demonstrations), with the week one problem serving as practice (practice).

Learners concluded their instructional condition with a post-treatment survey. This survey included a relative condition efficiency question, as documented by Paas and van Merriënboer [4]. Once learners finished the post-treatment survey, they were thanked for their participation, asked not to discuss their instruction with others, and not to use the software before the delayed test session.

During a second delayed session (one week later), learners returned and all groups of learners solved a different, week two problem. Learner onscreen behavior was recorded in a similar manner. Following this second set of performances, they were given a post-treatment survey which also included a relative condition efficiency question.

A researcher later reviewed the recorded video files, to score each learners attempt. Performance time and accuracy were measured using Techsmith Morae. Performance time was measured in seconds. Accuracy was scored with a rubric based upon the problem solving operators. Those who successfully completed a solution step were awarded one point, while more difficult solutions steps (flipping layers or correct layer placement) were awarded two points.

3 Results

A sample size of $n=122$ learners followed the instructions, completed all surveys, and attempted the required performances. A sample size of 115 participants was necessary for a four group MANOVA, given the power requirements. A sample size of this magnitude required data to be collected across two semesters. Therefore it was necessary to question if this pooled dataset would affect statistical tests.

The assumptions of a MANOVA were analyzed for the week two data set first, because all learners participated during this session. According to Glass and Hopkins [18] these data met the independence assumption. A SAS macro %MULTNORM [19] revealed non-normality (violating the normality assumption) since the Shapiro-Wilks' $W= 0.76$ $p<0.0001$ for accuracy, and for performance time the Shapiro-Wilks' $W=0.95$, $p=0.0015$. This violation was primarily due to a series of potential multivariate outliers. These learners were subsequently removed from the overall data set, using the %MULTNORM macro. Next the data were transformed. Box's M test was performed and X^2 (3, N=88) =4.50, $p=0.21$, $\varphi=0.23$, therefore the variance-covariance matrices were not found to be significantly different, so there was no evidence that the homoscedasticity assumption was violated. Therefore a MANOVA was used to compare the two semester subgroups (the summer and fall subgroups). The MANOVA indicated that there was not a significant difference between the two semester subsets, since Wilks' $\Lambda =0.95$, F (2, 95) = 2.47, p = 0.09, $\eta^2=0.05$.

Given it reasonable to proceed with a pool-semester solution, the week one dataset was also analyzed with a MANOVA. First the assumptions of the MANOVA were analyzed and these data met the independence assumption [18]. Multivariate non-normality was revealed when the %MULTNORM macro revealed a Shapiro-Wilks' $W= 0.62$, p<0.0001 for accuracy (AC1), and for performance time (PT1) the Shapiro-Wilks' was $W = 0.88$, p<0.0001. Mardia skewness was found to be $\beta_1 p= 66.70$, p<0.0001 and Mardia kurtosis was $\beta_2 p=3.79$, p<0.0001. Outliers were retained. Box's M test was performed, and the variance-covariance matrices were not significantly different, or homogeneous, $X^2(6, N=69) = 7.97$, p=0.24, $\varphi=0.34$.

A MANOVA was analyzed and during week one, group performances (performance time and accuracy) were found to be significantly different, since Wilks' $\Lambda =0.68$, $F (2, 68) = 6.83$, $p <0.0001$, $\eta^2=0.32$.

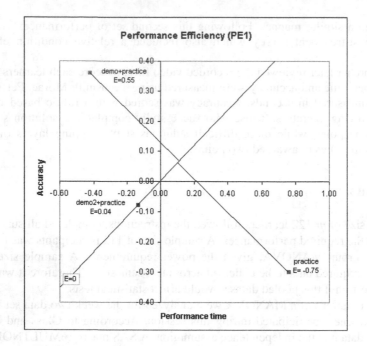

Fig. 1. Performance Efficiency

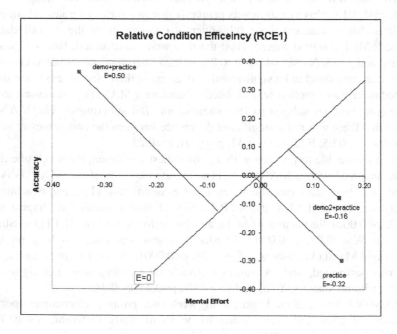

Fig. 2. Relative Condition Efficiency

Post hoc comparisons with Scheffé's test ($p<0.025$) revealed the demonstration groups (demo+practice and demo2+practice groups) assembled the problem, in significantly less time than the practice group. Also week one relative condition efficiency was considered and significant differences between conditions were revealed, since F (2, 68) = 3.69, $p=0.03$ (See Fig. 2). However, post hoc comparisons with Scheffé's test ($p<0.05$) found no differences between groups (given relative condition efficiency).

Finally, week one performance efficiency (See Equation 3) was also found to be significantly different, since F (2, 68) = 12.95, $p<0.0001$ (See Fig.1). However during week two, group differences were not found (See Table 2).

Table 1. Week one dependent variables

	demo	demo+practice	demo2+practice	practice
n	19	21	31	17
Perf. time				
M	NA	19.66	22.40	28.62
SD	NA	6.35	6.28	9.01
Accuracy				
M	NA	0.56	0.99	1.44
SD	NA	0.79	1.99	1.13

Table 2. Week two dependent variables

	demo	demo+practice	demo2+practice	practice
n	19	21	31	17
Perf time				
M	34.10	31.92	33.29	32.09
SD	3.78	4.93	4.57	3.44
Accuracy				
M	6.55	6.55	6.54	6.50
SD	0.26	0.25	0.22	0.21

4 Discussion and Conclusion

One view, the expository approach, is that learners should be guided during early instruction (instructor led). The alternative perspective is that learners should be allowed to discover problem solutions on their own (discovery learning). In short, Bruner [20] says "Practice in discovering for oneself teaches one to acquire information in a way that makes that information more readily viable in problem solving. So goes the hypothesis" [20], (p. 26). This hypothesis was tested in this study, and like many other worked example studies (e.g. [6]), it found that those learners who studied worked examples performed significantly better than their peers who learned through discovery problem solving.

Lewis [7] claimed animated demonstrations act as animated worked examples. This study justified that claim for it found that those who studied animated demonstrations assembled the week one problem in significantly less time, than those who learned through discovery problem solving; a result that is consistent with the worked example effect [1]. As expected, those who studied an identical animated demonstration (the demo or "mimic" condition) significantly out-performed their peers. However, this study also found those who studied a varied-context demonstration, significantly outperformed their problem-solving peers. Therefore this study has established animated demonstrations act as animated worked examples, and are an effective, efficient method of instruction.

Companies like Bank of America, Amazon.com, and Microsoft are all using web-based narrated animated demonstrations (demos) to train potential clients (novices) to use their products and services. The results of this study verified that this practice improves learner performance during the initial stages of learning. Therefore, it is the recommendation of this study that developers continue to use this effective and now evidence-based, e-learning strategy.

The study also found learner performance did not differ a week later, a finding that does not support Palmiter's animation deficit, further confirmation of the results by other researchers [2], [8]. However Palmiter may have been correct, that only providing learners with animation, produces mimicry; which Ausubel described as rote learning [21]. Non-narrated animation provides little guidance since learners only have self explanations in what may already be an overwhelming learning environment. The alternative is an instructor-led, narrated multimedia environment, which provides learners with a verbal narrative that simultaneously directs attention and provides an expert-level explanation. The importance of adding narration to an animated demonstration should not be underestimated, for it promotes what Mayer describes as "multimedia learning" [22].

Finally, some have proposed experience is the best teacher [20], but this position diminishes the role of an instructor. Instructors have purpose in any environment for they provide guidance and support [23]. In an e-learning environment, this role may be hindered because of an inability to communicate with an "e-learner," but the use of animated demonstrations allows an instructor to overcome the obstacles of time and space, to provide "e-learners" with guidance during "just in time" training.

References

1. Sweller, J., Chandler, P.: Evidence for Cognitive Load Theory. Cognit Instruct 8(4), 351–362 (1991)
2. Animation as documentation: A Replication with Reinterpretation, http://www.stc.org/proceedings/ConfProceed/1998/PDFs/00006.PDF
3. Palmiter, S.L., Elkerton, J., Baggett, P.: Animated demonstrations vs. written instructions for learning procedural tasks: a preliminary investigation. Int. J. Man Mach. Stud. 34, 687–701 (1991)
4. Paas, F.G.W.C., van Merrienboer, J.J.G.: The efficiency of instructional conditions: An approach to combine mental-effort and performance measures. Hum Factors 35(4), 737–743 (1993)
5. Sweller, J., Cooper, G.A.: The use of worked examples as a substitute for problem solving in learning algebra. Cognition and Instruction 2(1), 59–89 (1985)

6. Tuovinen, J.E., Sweller, J.: A comparison of cognitive load associated with discovery learning and worked examples. J. Educ. Psychol. 91(2), 334–341 (1999)
7. Lewis, R.D.: Demobank: a method of presenting just-in-time online learning. In: The Proceedings of the Association for Educational Communications and Technology (AECT) Annual International Convention, Orlando, FL, October 2005, vol. 2, pp. 371–375 (2005)
8. Waterson, P.E., O'Malley, C.E.: Using animated demonstrations in multimedia applications: Some suggestions based upon experimental evidence. In: Salvendy, G., Smith, M.J. (eds.) Human-Computer Interaction: Software and Hardware Interfaces, Proceedings of the Fifth International Conference on Human-Computer Interaction (HCI International 1993), pp. 543–548 (1993)
9. Paas, F., Tuovinen, J.E., Tabbers, H.K., Van Gerven, P.W.M.: Cognitive load measurement as a means to advance cognitive load theory. Educ. Psychol. 38(1), 63–71 (2003)
10. Brünken, R., Plass, J.L., Leutner, D.: Direct measurement of cognitive load in multimedia learning. Educ. Psychol. 38(1), 53–61 (2003)
11. Gagné, R.M.: Problem solving. In: Melton, A.W. (ed.) Categories of human learning. Academic Press, New York (1964)
12. Lewis, R.D.: The acquisition of procedural skills: An analysis of the worked-example effect using animated demonstrations. Unpublished doctoral dissertation, University of South Florida (2008)
13. Cooper, G., Sweller, J.: Effects of schema acquisition and rule automation on mathematical problem-solving transfer. J. Educ. Psychol. 79(4), 347–362 (1987)
14. Stevens, J.: Applied multivariate statistics for the social sciences. Erlbaum, Mahwah (2002)
15. Techsmith, TechSmith Camtasia Studio 4.0 [Computer program]. Okemos, MI (2006)
16. Adobe Systems: Adobe Photoshop Elements 2.0 [Computer program]. Mountain View, CA (1990-2002)
17. Techsmith: TechSmith Morae 1.0.1 [Computer program]. Okemos, MI (2004)
18. Glass, G., Hopkins, K.: Statistical methods in education and psychology, 2nd edn. Allyn and Bacon, Boston (1984)
19. Macro to test multivariate normality,
    http://support.sas.com/kb/24/983.html
20. Bruner, J.S.: The act of discovery. Harv. Educ. Rev. 31(1), 21–32 (1961)
21. Ausubel, D.P.: The psychology of meaningful verbal learning; an introduction to school learning. Grune and Stratton, New York (1963)
22. Mayer, R.: Multimedia Learning. Cambridge University Press, Cambridge (2001)
23. Kirschner, P.A., Sweller, J., Clark, R.E.: Why minimal guidance during instruction does not work: an analysis of the failure of constructivist, discovery, problem-based, experiential, and inquiry-based teaching. Educ. Psychol. 41(2), 75–86 (2006)

Personas Layering: A Cost Effective Model for Service Design in Medium-Long Term Telco Research Projects

Alessandro Marcengo[1], Elena Guercio[1], and Amon Rapp[2]

[1] Telecomitalia, Via Reiss Romoli, 274,
10148 Torino, Italy
{alessandro.marcengo,elena.guercio}@telecomitalia.it
[2] Telecomitalia, Via Reiss Romoli, 274, Università di Torino,
"Progetto Lagrange, Fondazione C.R.T.",
10148 Torino, Italy
amon.rapp@guest.telecomitalia.it

Abstract. Creating a set of Personas requires a considerable effort. Socio-psychological characteristic must be very punctually defined, needs and goals must also be well investigated and related to the service that will be designed. The accurate collection and processing of this large amount of qualitative and quantitative data represents often a huge cost. In our research area, which deals mainly with medium to long term telco projects, we developed a Personas layering model that allows us to adapt them over different contexts. The model consists in two main elements; the basic "Persona" and the "external" layer. In this paper we gave a practical application of our model within different projects developed in our research area. The benefits that arise from this model are the "durability" of Personas, their re-use in different contexts, the modularity of components and their possible recombination, thus reducing costs while maintaining excellent design insights.

Keywords: Personas, User Centred Design, Focus Groups, Ethnography.

1 Introduction

The Research and Trends area of Telecomitalia deals with scouting, evaluation and design of platforms and services that may have an interest in the TLC and Media business sector in a medium to long-term perspective of about 3-5 years. In the designing work the ambition is to set, as priority, services which combine, beside a business profitability, a predictable acceptability by the targeted users. The adoption of user centred methods is essential for design, but even more it is important to identify the approach that represents the largest cost-benefit trade-off. Our choice fell on Personas methodology [4, 6, 10, 11, 12]. A Persona is a user stereotype with specific practical and emotional goals we can use to design new services and the interaction with. This encourages the development of service concepts that address critical and real needs and that can be truly integrated into the daily lives of people, focusing on specific users and aims rather than on specific task so having the opportunity to design in a medium-long term

M. Kurosu (Ed.): Human Centered Design, HCII 2009, LNCS 5619, pp. 256–265, 2009.
© Springer-Verlag Berlin Heidelberg 2009

perspective, regardless of contingencies and technological boundaries. Indeed the main problem of working with real users in the early stages of truly innovative service lies in their inability to detach from the current technological environment and to imagine new situations that could take advantage from the service concepts offered [5]. Asking to real users a direct opinion about the design concepts can thus lead to inaccurate or distorted results. Using Personas instead manages to overcome this limitation by providing the designer a clear picture of possible desires and unmet needs that motivate different types of target audience, taking in account also their emotional and practical objectives, which, stable over time and not subject to the changing technological environment, represents a precise points of reference in the various stages of design. However, in order to make Personas a reliable starting point for the design of new services, they must be developed very carefully. They must be sufficiently different, and characterized not only by socio-demographic data, but also by personality traits and life-styles, identifying desires and values inspiring and guiding them throughout their daily life. To obtain this kind of result is necessary to combine different methods, both qualitative and quantitative, because each one provide a specific type of data useful to develop a particular aspect of the Personas individuality. The socio demographic data that characterize the individual Personas must be collected through survey or through the data analysis of national statistical offices. The socio-psychological characteristics and the identification of objectives and needs that guide their daily life must be investigated by a combination of several qualitative research techniques (i.e. interviews, ethnography, scenarios, focus groups), each one directed to take specific sides of their personalities and behaviours. This process satisfies the need to design by placing the user and its needs at the centre of the process, but in the meanwhile it also represent a high cost [i.e. 8, 10] that in the present scenario of limited resources often is an unbearable burden for some projects.

Three years ago we set as a primary goal the reduction of design costs, trying to maintain the reliability of the approach that Personas can guarantee if pursued in a complete and correct manner. So starting from our experience and best practices, we developed a model for the creation and development of a set of Personas both flexible and reusable in different application contexts. The idea behind it is that often, within the same company or the same research area, there is a single macro-user target characterized by some basic traits to whom all the research projects under development are addressed. Our research area, for example, is addressed in all its projects to a consumer target, of early adopters, with a marked propensity in the use and adoption of new technologies and services in the TLC and ICT sector. This macro target if described in its socio-demographic characteristics and its fundamental values and basic objectives that will guide its daily life, may remain valid for all the different projects led by the research area or in some cases even by the whole company. Not to forget in any case that each project is targeted to a specific context needing so to take into account, during the various stages of design, needs, desires and specific behaviours of its users highly related to the application context in which the service will enter and will impact. In our case, for example, the projects are applicable to contexts ranging from the house to the car, to the personal communication in open air environments.

The identification of these modular units, that identify various aspects of the users, the first ones commons to the different needs of the research area, and the second ones linked to the specific objectives of the singles project, is the basis of the layering model for the Personas definition described in this work. Specifically the Persona is seen as a modular stack in which the deep or internal layer, can be created once cross-project, and then reused in multiple application areas with common requirements, while the surface module, or context layer, can be developed depending on the needs of a particular project and its peculiar application context. This model allows a significant reduction in the design costs, because it promotes the reuse of "basic" Personas in various projects, lengthening their life span, and also permitting greater investments for their definition since the initial efforts aiming to accurately define the internal layers are not lost but remains a solid and reliable basis for future design.

2 Personas Layering Framework

Our model of building Personas consider the single Persona as a composite entity, consisting of several layers, partly developed once, therefore reusable, and partly developed on the basis of the application context of any project requiring a user centred perspective, therefore specific. The Persona stack is composed by:

- the "basic" Persona (reusable)
    - a core base, which identifies the central element common to the whole set of Personas. It represents the characterizing traits of the target users that a research area or a company is addressing.
    - an internal layer, fundamental and durable, that defines not only all the socio demographic characteristics, but also all the aspects that guide the Persona daily life that does not change through different contexts (its basic values, its basic needs, its "life-style", etc). It is based on quantitative data from national statistical institutes and market research and on qualitative data deriving from qualitative research methods (focus groups, interviews, scenarios, etc.).
- the external layer (specific)
    - a context "layer", modular and interchangeable. An "extension" of the internal layer based on the usage context of the services under development, which drives attitudes, needs and specific aims of that person for that context (i.e. home, car, outdoor, etc.). This "layer" is built through ad hoc qualitative research techniques (ethnography, key informants interviews, scenarios, etc.) depending on the type of information crucial for the service.

The process to develop a complete Persona is partly done once for all in order to reuse it and partly according to the research project requiring it.

2.1 The Definition Process of Reusable Personas

In this phase the Personas are built as reusable characters. The development of a set of structured modular Personas reusable trough various project proceed through this three phases:

Core Definition. The first phase includes the identification of the common characteristic of the target users of a specific area or organization. This is a basic unit that is identified according to the business needs and is not supposed to change over time. This is the *trait d'union* under which various projects of the same area can be led to a single denominator. It serves as a common baseline to all the Personas that will form the reusable set and therefore must be identified as accurately as possible, so as not to exclude sectors of the population of possible interest as targets. In our case, all the services designed in our research area are designed for a *"consumer and early adopter of ICT and telecommunications services"* user.

Set Definition. In the second stage, the general feature identified in the previous step is segmented according to the socio-demographic characteristics of the reference market. In our case it was taken into account only the Italian market. Using mainly quantitative data collected from national statistical institutions, are identified the fundamental bearers of the basic feature originating so a macro-segmentation of the target. It is important to take into account not only the socio-demographic characteristics but also those relating to life-style, use of leisure time, etc. At the end of this phase we have a set of Personas differing in it for basic types, each of whom represents a specific target of the reference population.

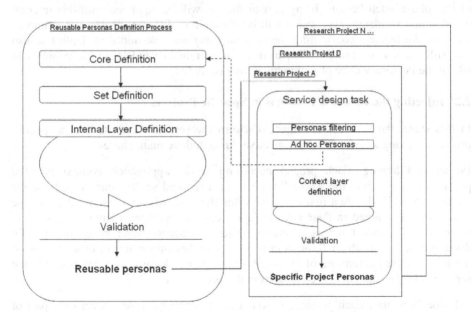

Fig. 1. Personas layering framework

Internal Layer Definition. In this step we create the internal layers for all individual Persona composing the set developed in the previous phase: this layer consists of a detailed data sheet in which there are several types of characterizations:

– basic information plotted from the quantitative data analysis performed in the previous phase. Special care must be devoted to the specification of personal data (name, surname, date of birth, civil status, etc.), socio economic data (educational qualifications, income, etc.) and data related to specific life-styles.

– practical and emotional goals guiding the daily life of the single Persona, these goals represent the reference values that guide the acting of the Persona regardless of contexts in which it is.

– desires and unmet needs that accompany Personas during their daily lives.

This internal layer do not vary over time and can be taken as stable point of reference for the design of multiple services. It is important to underline that the definition of the objectives, needs and desiderata is carried on predominantly through the use of different combined methods of qualitative inquiry (i.e. interviews, scenarios, etc.). At the end of this definition stage is particularly important the validation of the Personas set internal layer obtained through the use of methodologies to integrate and confirm the data gathered and the assumptions made by Persona designers. The use of focus groups centred on the single Persona is fundamental because all the parties share the same socio-demographic characteristics of the Persona to be validated. Focus groups allow us to confirm, correct or supplement the information that characterize each Persona bringing out, through group discussion, the values and needs that all participants share, embodying so this characterization into the specific Persona. At the end of this phase, each Persona being part of the set will be equipped with this internal detailed and validated layer which will be reusable in different projects. It is important that the features constituting the internal layer are determined in depth but also that still sufficiently broad and open in order to remain valid in all the contexts in which the Persona will be placed in his virtual daily life.

2.2 Injecting the Reusable Personas in Specific Projects

In this stage, the Personas are injected within the service design task of a specific project requiring a user centred perspective through three main phases.

Personas Filtering. Each project according to its application context has the possibility to cut from the whole Personas set a reduced set that can be suitable for the project itself. Beside a first objective filter that discards the Personas that, for the information contained in their internal layer, cannot represent a suitable target (i.e. driver license yes/not for car context), there is a second filter based on specific domain experience, which, through interviews with key informants or observations on the field, allows the removal of the profiles that are not deemed to be interested in the service solutions possibly outcoming from the project.

Ad Hoc Personas. Each project can found out that the basic set covers only part of the profiles of interest for the services to be designed. Because of this is possible through consultation with domain experts (key informants) identify new Personas that are fundamental for the specific application context of the project. In this case it is necessary to develop backward the internal layer in order to have a fully shaped Persona working like the others. The "new" Persona will enrich the reusable Personas set in future works. At the end of this phases we have a set of Personas specific for the project including only really useful profiles for the application context and the services that will be developed within.

Context Layer Definition. In the last stage, the selected subset of Personas is declined and enriched for the specific context of the given project. For each Persona a "Context Layer" has to be created containing all information concerning the behaviours, needs and objectives of that Persona in the particular context of action (i.e. car, home, office, etc.) to which the project is addressed. The research techniques used in this phase, as was in the construction phase of the Internal Layer, are many and combined together: ethnographic surveys, qualitative interviews, construction of scenarios, focus groups. Each technique responds to specific needs of data collection. If the service is developed for example in an area where is crucial the behaviour of the Persona within a given environment it will be chosen an ethnographic approach: it allows to observe the person in his natural acting, highlighting his unexpressed needs and his behavioural habits. Also the use of scenarios has his particular interest: essentially, the scenarios are stories with a plot, which concerns people and their activities [2]. Personas could be used as rounded characters allowing the designer to imagine people while acting in a specific context, highlighting what could be their unmet needs [1, 9]. Moreover scenarios can also be used as *stimuli* in focus group discussions in order to stimulate the individual participants in the comparison between what happens in the scenario, and their experiences in daily life. The context layer definition include the validation of the layer obtained through qualitative methods: in this case, for example, focus groups are fundamental in validation of the single Persona external layer. At the end of this phase, each Persona will have specific features related to a specific context of activities, revealing specific unmet needs or desires that will serve as a basis for service design activities.

The main advantage in economic terms of reusable Personas process definition is the ability to exploit, in successive service design tasks of every research project, the set of identified Personas, so dramatically cutting the definition costs dues if the process would start from scratch for every project.

The only living costs of future projects will, therefore, those relating to the context layer development or additional Personas definition, which will vary depending on the specific project and the techniques of investigation used from time to time. In this way the large investment required by the definition of the Personas set is not only justified, but also encouraged by the multiple reusing of it in future projects.

3 Reusable Personas Process in Research and Trends Area of Telecomitalia

To exemplify the personas layering framework, in this paragraph we gave a practical application of our model within different projects developed in our research area.

3.1 Building the Reusable Personas

The core definition phase was conducted through a "company business analysis" and a "company future trends analysis" identifying in "*consumer and early adopter of ICT and telecommunications services*" the common core for all the Personas that would have formed the basis of our set. Then, through analysis of quantitative data from the major italian research institutions (i.e. CENSIS [3], ISTAT [7]) we identified 11 Personas, differentiated by socio-demographic characteristics and life-styles, oriented to the adoption of new technologies in the ICT and TLC. The Internal layer definition phase was carried out by constructing, for each of the 11 Personas identified, a detailed data sheet containing all their socio-demographic data, their values and their guiding objectives, and a narrative scenario showing their daily life, in order to highlight their needs and their unexpressed desiderata. The validation of the 11 Personas was conducted with 11 focus group sessions focused on one single Persona (1 focus group of 8-10 participants for each Persona). During group discussions the participants, who shared the same social and demographic characteristics of the Persona to be validated, were able to confirm and supplement either the listed information either the developed life scenarios, through the comparison of their daily lives. The scenarios were illustrated by a series of movies in which non-professional actors were acting scenes from the daily life of the personas to be validated. The results has been values, goals and needs shared by all participants that went to confirm/disconfirm, correct and complete the assumptions made in the internal layer definition. The set of 11 Personas thus built has been reused in multiple projects over the past three years. The internal layer has maintained its validity in each new context of use requiring only to create the "external" context layer.

3.2 Three Application Examples

Telco@home. Aim of the project was to build seamless communication and interaction within a home environment laying on physical heterogeneous devices and infrastructure. In this case we selected 9 over 11 Personas among those available from the basic set considering the spending capacity and 5 over the 9 remaining trough the data obtained from key informants (structured interviews). We contextualized the 5 Personas adding the context layer according to the "home environment" usage context. For each Persona it has been developed a specific scenario reproducing his daily life, his daily aims, and his daily specific needs during his moving, playing, working, day in the home environment. Then a detailed card was built summarizing the specific characterization related to the set of services under investigation (i.e. size of house, organization of actual technology in the home environment, preferred brands, etc.). On the basis of the scenarios some short movies had been shot, showing Personas in their daily life within the home environment. Five focus groups have been carried out centering them on the 5 Personas selected, in which, with the help of the movies used as stimuli we performed the validation of the contextual layer.

Telco@car. The set of "basic" Personas was used in a second major line of activities related with in-car communication and media fruition for the driver. For this project 3 over 11 Personas has been discarded because not having the driver license and 5 more over the eight remaining has been discarded after a series of semi-structured interviews with key informants in the specific field (car dealers of different brands). In this case the key informants drove us to the discovery of an *ad hoc* Persona totally new and specific for this context application. So we worked on 4 Personas for this project. In this case, the contextual layer has been created through ethnographic sessions with

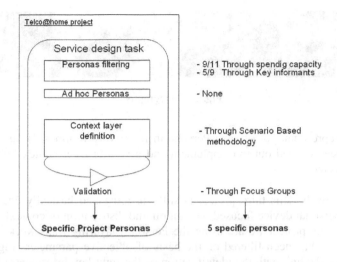

Fig. 2. Telco@home injection of reusable personas

Fig. 3. Daily life scenarios

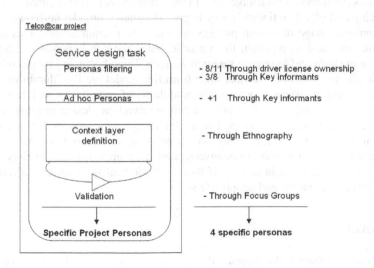

Fig. 4. Telco@car injection of reusable personas

Fig. 5. Ethnography

real people representatives of each Persona in its driving routines. Focus group sessions have been carried out to complement and give a broader context of validity to the data collected.

Personal Geoweb. This third project is still on going and basically deals with the design of a personal device focused on fruition and distribution of context aware content. Also in this project we are using the Personas layering framework. Again, the initial set of 11 has been filtered on the basis of objective parameters (age) and the project is now dealing with the identification of the right key informants. The results will drive the methodological choice of the following phases (ethnography, scenarios, etc.). An interesting issue under examination is the possible reuse of the context layer of some personas already developed in Telco@car even if not originally assumed by the framework as described in fig.1.

4 Conclusions

The accurate collection and processing of the large amount of qualitative and quantitative data needed to build a set of Personas reliable and well established for the design of new services often represents a huge cost. In the actual competitive scenario, it becomes crucial being cost effective finding a way to give Personas a broader horizon of life, that permits multiple usage in several projects whit common requirements. This approach permits on one hand to maintain the valuable but sometime expensive technique of Personas in medium and long term research projects Telco becoming on the other hand more cost effective. The benefits that arise from this model are the "durability" of Personas, their re-use in different contexts, the modularity of components and their possible recombination, thus reducing costs but maintaining excellent design insights. With this in mind we are thinking of extending our model, designing a process that allows to reuse not only the internal layer, but also some contextual layer, also if originally designed for different projects. It remains to be investigated if the interior layer developed for the reusable Personas can last in time or if there is periodical need for re-definition as a result of changes in market and people life-styles.

References

1. Borglund, E., Öberg, L.-M.: Scenario Planning and Personas as Aid to Reduce Uncertainty of Future Users. In: 30th Information Systems Research Seminar in Scandinavia (2007)
2. Carroll, J.M.: Five Reasons for Scenario-based Design. Interacting with Computers 13(1), 43–60 (2000)

3. Censis: Quinto Rapporto Censis-Ucsi sulla comunicazione in Italia 2005: 2001-2005 cinque anni di evoluzione e rivoluzione nell'uso dei media, Roma (2005)
4. Cooper, A.: The Inmates Are Running the Asylum: Why High Tech Products Drive Us Crazy and How To Restore The Sanity. Sams, Indianapolis (1999)
5. Guercio, E., Marcengo, A., Rapp, A.: How to connect user research and not so forthcoming technology scenarios – The extended home environment case study. International Journal of Social Sciences 2(4), 203–208 (2007)
6. Grudin, J., Pruitt, J.: Personas: Participatory Design and Product Development: an Infrastructure Engagement. In: Participatory Design Conference 2002, pp. 144–161. CPSR (2002)
7. Istat: La vita quotidiana nel 2005: Indagine multiscopo sulle famiglie "Aspetti della vita quotidiana" Anno 2005, Roma (2007)
8. Neiters, J.E., Subbarao, I., Iftikhar, A.: Making Personas Memorable. In: CHI 2007, pp. 1817–1823. ACM Press, New York (2007)
9. Nielsen, L.: Engaging Personas and Narrative Scenarios. Dept of Informatics vol. 17, PhD Series, Samfundslitteratur, Copenhagen (2004)
10. Pruitt, J., Grudin, J.: Personas: Practice and Theory. In: Designing for User Experiences 2003, pp. 1–15. ACM Press, New York (2003)
11. Pruitt, J., Adlin, T.: The Persona Lifecycle: Keeping People in Mind Throughout Product Design. Morgan Kaufmann, San Francisco (2006)
12. Sinha, R.: Persona Development for Information Rich Domains. In: CHI 2003, pp. 830–831. ACM Press, New York (2003)

Bridging Software Evolution's Gap: The Multilayer Concept

Bruno Merlin[1], Christophe Hurter[2], and Mathieu Raynal[1]

[1] IRIT, UPS, 118 Route de Narbonne,
31062 Toulouse CEDEX, France
{merlin,raynal}@irit.fr
[2] DSNA/DTI R&D, 7, Avenue Edouard Belin,
31055 Toulouse CEDEX, France
hurter@cena.fr

Abstract. The multilayer interface concept is used to promote the universal usability, to smooth the transition to new systems and working methods and to
help the user optimize his interface for the management of contextual situations.
In this article, we will explain how this concept can help to tackle a serious issue for R&D projects: the integration of the innovative concepts into the operational environment. To illustrate this, we will explain how we used a multilayer
interface to promote a way to integrate different concepts currently in maturation in the R&D sphere.

Keywords: Multi-layer interface, direct manipulation, working method evolution.

1 Introduction

The operational and technological concepts embedded in a system are limited by the
state of knowledge at the time of the system specification. Thus, the time span for
implementation and deployment is long, given that a system is generally a generation
late compared to the current technologies. In the field of air traffic control, where the
major systems have at least twenty years life expectancy (notably due to the intrinsic
cost of their development and their inferred costs: user and maintenance training),
important developments are conducted only with real technological leaps. In this
context, working methods and working tools have to be reconsidered. However, during the life of a system, its changes are minor and generally this system does not implement the research yield. An end-of-life system is therefore 20 to 30 years late
compared to the research field.

In order to meet the constraints of traffic incensement, the challenge of the next
system is to provide a greater responsiveness in its ability to integrate the product of
innovation. New systems should be able to quickly implement new tools in the air
traffic controller environment. However, they must also guide controllers through the
transformation of their working methods.

The goal of our approach is to show how the multi-layer interface which is HCI
design paradigm can solve a double issue: the multilayer interfaces are able to smooth

M. Kurosu (Ed.): Human Centered Design, HCII 2009, LNCS 5619, pp. 266–275, 2009.
© Springer-Verlag Berlin Heidelberg 2009

out the transition towards new working methods, and they can also be an architectural support for the management of a system's roadmap.

Through the analysis of different research projects, this article illustrates the main issue that faces research projects in order to be integrated in the controller context. Then, after introducing the concept of multilayer interface, we will show how these projects could use the multilayer concept to improve new tools and working methods integration.

2 From the Concept to the Operational Environment

The research and development yield is often awkward to implement in the operational industrial field, especially for critical activity such as nuclear power, aviation and air traffic control. Several projects, even advanced ones, struggle to reach the operational environment. To illustrate and to find out the reason of this difficulty, we will review several examples from the Air Traffic Control field.

2.1 The Appearance of the Color Screen and the Mouse

Even if the color screen technology was from ages integrated in our daily life since decades, it took years to expand them in air traffic controller working positions. Many stakeholders agreed to merge the color technology with the radar screen, but many factors had slowed down this migration. The first barriers were financial due to hardware costs, development and to technology's certification.

But, the main obstacle was the impact of the new representation design on working methods. With due reason, many controllers were worried with the new design use. The color was used to filter the radar screen by segregating information between important, secondary (shaded flights) and insignificant (hidden). This filtering was supposed to reduce the cognitive workload of controller by highlighting the main data. But the controller had difficulties to delegate this filtering to the system. They feared being induced in error and could not easily remain fully confident with the new design.

Long training periods smoothed out the brutal transition between monochromatic and colored screen. The same technological issues happened with the mouse advent.

2.2 Conflict Detection Software Issues

Medium Term Conflict Detection (MTCD) is a set of tools to assist air traffic controllers to forecast potential conflicts. Their main features are: aircraft and altitude screen filtering, supervision of relevant flight parameters and decision planning with time line tools.

Despite the significant potential brought along by such tools, their integration in the operational environment and in the current working methods raises several problems. HCI constraints, closed systems, formation costs and adaptation of controllers to new working methods make this step awkward.

2.3 ASTER Issues

The ASTER project [2, 5] focuses on the use of a vertical view in order to improve traffic management in the airport approaches (ETMA). This tool provides an electronic stripping [9, 12] using a display which gives the vertical position of the aircraft. This view facilitates the management of transverse flights in the main flow of arrivals and departures, in order to reduce the number of flight level instructions (less level instructions yield fuel savings, among other), improve the quality of traffic management in general.

Early in the design process, the issue of integration into the current control working position and in the working methods arose and became one of the main challenges. The vertical projection of traffic, proposed by ASTER, belongs to a family of electronic stripping tools. These tools aim at replacing the use of current paper strips. Air traffic controllers were involved in the user-centered design process [4]. They are offered the possibility to use these electronic strips in the same way as they did with paper strips. In addition, new design tools can offer them an opportunity to switch to new working methods.

In the light of difficulties faced by other projects, the only viable way to envision this project in terms of integration lead to the decision of developing independent tool taking the place of the existing paper strip board. This tool had to offer a consistent set of features and yet minimize interaction with the existing systems. Furthermore, it could not become an additional interaction tool redundant with existing tools or existing paradigms. The interaction with other separate tools (i.e. radar image or sequencing tools) was allowed only to provide optional features, but could not be a prerequisite for the operational viability of the system.

2.4 Results

The transition from a Research and Development (R&D) project to an operational tool is generally difficult. The projects, often forced to make sacrifices in order to fit into the existing environment, are weakened. Several factors can, however, facilitate the integration of innovative products: Structural improvements in the design and in the interactive software architecture for a greater flexibility of integration of new tools; Develop on-going training to assist users with the evolution of their working methods; Allow provision for the optimization of visual and interactive interfaces to include the handling of non nominal situations.

3 Multi-Layer Interface

The concept of multilayer interfaces (MLI), proposed by Ben Shneiderman [8, 13], aimed at promoting a more universal usage of applications. Later, the concept was extended to offer two additional services: Involve interactive systems in a continuous evolution process [10]; Adapt the software in order to be more efficient through a range of different tasks [5]. These different concepts allow us to identify where the multilayer's interfaces could also support the management of a long reaching project.

3.1 Universal Use Paradigm

The multi-layer interfaces were initially designed to promote universal use of application and allow various users (novice, amateur, expert) to use the interface efficiently, despite having heterogeneous objectives and training levels. These interfaces allow different types of uses [13] (from the most superficial to the most complex use), by activating [8] or refining the use of functions [5] and adapting visual density to the user's skills [6].

A group of functions and the corresponding visual entities define a layer. Transitions between active layers can be controlled either by the user or by the system. When layer selection is automatic, the selection is based on an analysis of the user's activity [5]. The multi-layer interface tries to help the user to gradually improve his efficiency with the software while retaining continuous control of it. To achieve this goal, we have to define guidelines for the creation of each layer.

3.2 Supporting the User along the Tool Evolutions

The concept of MLI, proposed by Schneiderman, has been extended in order to be used for the mutation of operational systems [10, 11], especially in the context of critical activity.

The new paradigm focuses on two additional issues: How to reduce the training period and to increase the application's acceptance during the transition between two systems; How to avoid brutal changes in the evolution of the applications. This property of multi-layer interfaces is achieved whenever several guide lines are followed to design the layers.

Design of Layer 1. The goal is to build the interface in a continuously evolving process and to avoid gaps between the old and the new software functions. New tools cannot suddenly challenge the working methods of the user. The interface must guide him. New tools must seduce him and progressively change his way of working.

The guide line therefore suggests establishing the old system (or at least a set of similar modality of services and interactions), as layer '1' of the new interface (see figure 1).

The direct consequence of this "conservatism" in layer 1 is that the user immediately finds a familiar environment, ideally completely similar. The training period becomes radically shortened because the user is already "layer 1" proficient. Using other layers can be done later, progressively and immediate mastery is not mandatory to use the new software.

Next layers design. The seconds guide line proposes to improve the interface in the next layers and yet to preserve an interaction redundancy with other layers. A new layer will be materialized with new functions and with new visual entities. This new layer will lead the user to consider new working methods. With these layers, the user may grasp new way to organize his workspace and new interaction paradigms that improve the accomplishment of his task.

Active layer selection. Whenever software deals with a critical activity, the user should always keep control over the system. He must be able to visualize, at any time, the active layer and be free to interact with it. This point is very important, and will help the user feel comfortable to explore new functions. At any time, the user must be left free to restore a familiar layer; bringing upfront a completely mastered interface. This provides him with the proper conditions to cope with a 'crisis' situation, at least until he feels he has got the same level of confidence with new richer layers.

Thus, the transition between two layers can be done by the user whenever he wants to, and in real-time, while using the interface. The transition must be reversible and triggered easily. The animation must be quick in case of stressful or heavy workload context. It is to be noted that this guideline clearly dismisses an automatic system which may choose to activate a layer. This strongly precludes unexpected transition in heavy workload situations.

User evolution. Suggested pattern of user progression is depicted in figure 2. Initially the user exploits lower layers (with mastered functions), and afterwards his range of used layers will slide to upper layers. The progression period of the user may vary with his assurance and his activity workload. In addition, the user's progress can accelerate with his curiosity and motivation.

While the original goal of the multi-layer interface was to encourage a heterogeneous population to use single software and to improve their skills, the goal here is to incite a homogeneous population to adopt a new working environment and new working methods at heterogeneous rhythms. The originality of the method is that it involves the user in a continuous design process of the interface. This tries to achieve the invisible integration of new tools.

3.3 Contextual Help

The first goal of the multilayer interface is to adapt the interface to a heterogeneous group of users. Its second goal is to support the user along software evolutions. A third functionality is to allow an application to optimize itself for the management of a specific task [6].

In this context, the MLI combines many different types of uses and therefore must adapt software in order to: Fulfill likeness functions with different layers; Allow users to deal more effectively with uncommon situations. The user can then optimize his interface according to his current activity.

The likeness functions context. As an example, the 'Eclipse' Integrated Development Environment (IDE) has characteristics akin to an MLI. This IDE proposes a common set of functions which help the programmer, regardless of what his programming language may be. The main menu is the same for a JAVA, C++ or UML developer. Independently of the way to use this IDE, Eclipse is identified as single software.

However, the views and some features are optimized to manage the specificity of each programming language and to back the programmer with appropriate support. The proposed functionalities (menus, buttons, tools…) are filtered to be coherent with the user activity. The layout is thus transformed to ease its exploitation.

Uncommon situation adaptation. Once again, we consider the example of software development activity. Whatever the programming language, the programmer activity is centered on two main tasks: The code edition; The execution, test and correction of the program.

To edit the code, the programmer needs to easily and quickly navigate, to be able to perform research on functions names or variables. While performing program correction and test phases, the user must be able to control the software execution.

An IDE, such as Eclipse, is able to adapt its functionality and the data it displays, and to switch between modes: navigation, edition and execution. In other words, this tool is contextualized in order to better manage momentary activity, even if the main activity is code edition.

3.4 Summary

The MLI offers a structural solution to address the issue of transition and implementation of an R&D concept. It allows: To integrate new tools with a structural theory; To offer a solution to help the user with the mutation of his working methods; To adapt a software with new tools and new visualization in the event of a momentary specific context.

Adversely, if we organized, in a particular layer, all the functionalities and all the visual entities of a given software version, the MLI can be a structural way to phase and schedule an interactive system. In other words, if each layer corresponds to a software version, the MLI becomes an architectural instrument to structure a project in the long term.

The user can quickly change his active layer. This possibility allows him to restore a better mastered layer, and then comfort his confidence with new tools in his working environment.

4 Concept Illustration

The opportunity offered by the multilayer interfaces is illustrated with the ASTER project [7] for Air Traffic Control. The MLI concept is here applied to an electronic stripping tool and serves to gradually build and design an Air Traffic Controller working position.

4.1 Electronic Stripping Features

The air traffic controller working position is intended for two collaborating persons. Each controller (radar and tactic controller) interacts with a specific part of the software application. Two radar screens, which display the current aircraft positions, are available. A strip is a standardized piece of paper used by air traffic controllers to managed information about aircraft. This abstraction of a flight is a support for coordination and communication between controllers. The strip allows specific features: Flight integration, marking, organizing and sharing aircraft instructions, work load planning, conflict avoidance, etc. (cf. videos available here http://perso.tls.cena.fr/acropole/2-1-eng.html).

The initial electronic environment is an electronic stripping (cf. figure 3) based on the DigiStrips experience [12]. This tool attempts to transpose accurately and with no modification the current working methods (for en-route and terminal sectors) into an electronic environment. All these features constitute the body of layer 1.

4.2 Working Method Transition

ASTER's initial goal was to re-organize the working method with the support of a vertical projection of the traffic. The idea was to complete the information brought by the main radar image. During the first two phases of experimentation [3, 4], despite the effectiveness of this tools (statistically assessed), the air traffic controller remained reluctant to go along with these new working methods.

The multilayer interfaces offered a solution, because they provided a smooth transition in working methods. Layers 1 to 4 have been designed to provide a smooth and continuous transition between consecutive layers. Thus, the user may work with a completely mastered environment and control his progress (cf. [10]).

4.3 New Tools and Concepts Integration

The second aim of the ASTER project is to show how the electronic stripping can be a transition step towards the future working position. It illustrates how the multilayer design paradigm could help new concepts to be integrated within an electronic environment for Air Traffic Control (ATC).

Each layer (layer 5 to 9) contains a specific tool corresponding to a new concept. Air traffic controllers may use this tool as contextual assistance. In a given context, air traffic controller is able to activate the specific layer which provides the needed functionalities.

Advanced planning tool & AMAN integration. Layer 5 is an advanced planning tool (cf. figure 5). The core of the tool is an interactive timeline (agenda), where a set of events is managed. These events can represent operational problems or advice (future conflict, catch up, flight descent etc.). The events can be created by a user or generated by different external sources such as algorithm treatments.

Quick and simple interaction enables the user to easily edit, move or delete part or all of existing events. Some urgent or critical events may trigger alarms or warnings if they have not been treated within a given time span.

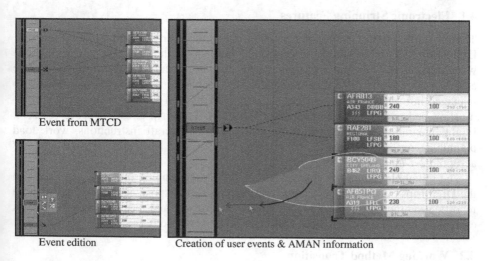

Event from MTCD

Event edition   Creation of user events & AMAN information

Fig. 1. Advanced planning tools

Layer 6 brings into the electronic strip system a complementary planning information generated by an arrival manager (green lines and dashes on figure 5). The left black columns of the timeline represent the expected sequence of arrival on two different runways (in our case, south and north runways at Roissy airport).

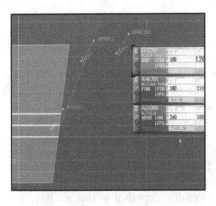

Holding management assistant. Layer 7 integrates a symbolic representation of the holding zone (the holdings are hippodrome areas where flights, separated in level, wait for descent when their destination airport is congested). by displaying the free and occupied flight levels. The representation facilitates the analysis of the planed vertical situation and enables to compare in real time the difference between the planed future situation and the actual current one. The system, informed by the controller through his management task, can identify and highlight inconsistent operations. For example: the lines highlighted in the holding area illustrate the future occupied levels. If any level is attributed to more than one flight by the controller, an alarm can be raised by the system.

Flight conformance & ASAS monitoring. Layer 8 enables controllers to delegate task to the system. At first, the controllers can program the system to monitor flight parameters (speed, level etc.) or given constraints. At the difference of the other alarms raised by the system, in this context, the controller requests explicitly to the system to the survey flight and to inform him in case of non-conform evolution of the situation. Moreover, the controllers can record an ASAS constraint (the respect of distance or a delay between flights) delegated to the pilot or to the system.

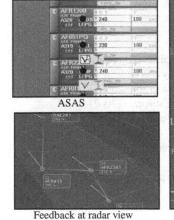

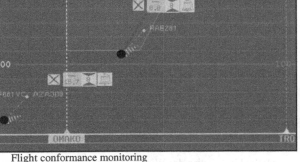

ASAS

Feedback at radar view   Flight conformance monitoring

Fig. 2. Flight conformace & ASAS monitoring

Data-link. The layer 9 integrates information and tools for data-link operation as well as a traffic monitoring system (data-link is an electronic message set of services between pilot and controllers and in the future between pilot and system or flight and system). The controllers can control the status of data link clearances and receive requests from pilots.

5 Conclusion

Our approach tried to take full advantage of the MLI concept and use it in the Air Traffic Control domain. We explained how it could ease the integration of R&D concepts and how it helped us to increase the acceptance of our product by smoothing the gap between old and new working methods. We applied and extended the multilayer interface introduced by Ben Shneiderman. Doing so, we noticed that this concept was efficient in the ATC field.

It is, however, difficult to estimate the benefits of the MLI independently from other design options applied to the tool. Lately, the tool was completely redesigned and the recent experiments are showing a better overall acceptability. But we have to acknowledge that it is difficult to measure accurately what is the relative impact of new design options, of new interaction paradigms or of the MLI on this acceptability. These different factors seem to be tightly interlinked.

In addition, the MLI philosophy implicitly guided our design choices in the sense that it forced us to: Think about the learning process in order to make it more natural and intuitive, and to render the tool itself self-explanatory; Capitalize on homogeneous and reused interaction modes [1]; And to make the tool more 'fancyful' to the user, because we had to induce him to explore the tool by himself,

We are therefore convinced that the MLI paradigm can prove a key element to simplify and accelerate the integration of innovation in an operational field. This is done by engaging the user, the working methods and the systems in a continuous mutation process.

References

1. Beaudouin-Lafon, B., Mackay, W.: Reification, polymorphism and reuse: three principles for designing visual interfaces. In: AVI 2000, Palermo, Italy (2000)
2. Benhacène, R.: A Vertical Image as a means to improve air traffic control in E-TMA. In: USA, DASC-Digital Avionics Systems Conferenc, Irvine-California (2002)
3. Benhacène, R.: As Rapid as PaperStrips? Evaluation of VertiDigi, a new control tool for Terminal Sectors. In: ATM 2005, Baltimore, USA (2005)
4. Benhacène, R., Merlin, B., Rousselle, M.P.: Tackling the problem of flight integration. In: ATM 2007, Barcelone (2007)
5. Clark, B., Matthews, J.: Deciding Layers: Adaptive Composition of Layers in a Multi-Layer User Interface. In: CHI 2005 (2005)
6. Gustavsson Christiernin, L.: Multi-Layered Design - Theoretical Framework and the Method in Practise. In: Winter Meeting 2005 Proceedings (2005)
7. http://perso.tls.cena.fr/merlin/1-4-eng.html

8. Kang, H., Plaisant, C., Shneiderman, B.: New approaches to help users get started with visual interfaces: multi-layered interfaces and integrated initial guidance. In: Proceedings of the 2003 annual national conference on Digital government research, Boston, MA (2003)
9. Mackay, W.E.: Is Paper Safer? The Role of Paper Flight Strips in Air Traffic Control. ACM Transactions on Computer-Human Interaction 6(4), 311–340 (2000)
10. Merlin, B., Hurter, C., Benhacène, R.: A solution to interface evolution issues: the multi-layer interface. In: CHI 2008, Florence, Italie (2008)
11. Merlin, B., Benhacène, R., Kapp, V.: Interface multi-layers et processus d'évolution des systèmes interactifs en activité critique. In: IHM 2007, Paris (2007)
12. Mertz, C., Chatty, S., Vinot, J.: The influence of design techniques on user interfaces: the DigiStrips experiment for air traffic control. In: HCI Aero (2000)
13. Shneiderman, B.: Promoting universal usability with multi-layer interface design. In: Proceedings of the 2003 conference on Universal usability, Vancouver, Canada (2003)

A Proposal of XB-Method, an Idea Generation System for New Services Using User Experiences

Naoka Misawa[1] and Mitsuru Fujita[2]

[1] U'eyes Design Inc., Housquare Yokohama 4F, 1-4-1,
Nakagawa Tsuzuki-ku, Yokohama 224-0001 Japan
misawa@ueyesdesign.co.jp
[2] DENSO CORPORATION, Showa-cho, Kariya-shi Aichi-ken 448-8661 Japan
mitsuru_fujita@denso.co.jp

Abstract. These days, due to diversifying standards of living, people seek such a sense of impression in products and services. We therefore developed XB-method, an idea generation system, in order to inspire the affecting experience, and that have been difficult through the conventional process in the current product planning. XB-method is a method which enables us to generate the affecting experience with multiplying the database of keywords statistically-extracted from user experiences by images of commodities in order to inspire the new product and services with the affecting experience effectively in user's perspective.

Keywords: Idea generation, Experience, Requirement definition, Emotion.

1 Introduction

These days, due to advancing technological innovation and diversifying standards of living, people demand the higher degree of satisfaction toward commodities including services. In order to fulfill such satisfaction, it is necessary to project products and services that can provide users a sense of impression [6].

We therefore develop this Cross Breeding Method (hereinafter called, XB-method) as an idea generation system. XB-method enables us to provide users products and services with the affecting experience and that have been difficult by the conventional approach in the current product planning. XB-Method is a method to generate the affecting experience with multiplying the database of keywords statistically-extracted from user experiences by images of commodities. In this study, we bring out activities of XB-method and propose it as one of idea generation systems incorporating with the fundamental principles of HCD.

2 Relevant Activity

2.1 The Current Product Planning

Table 1 shows the general process of product planning in Japan. It is necessary to continue to study the experience as users expect from the phase of research and analysis

M. Kurosu (Ed.): Human Centered Design, HCII 2009, LNCS 5619, pp. 276–283, 2009.

Table 1. Process and methods used in the current product planning

PROCESS	PURPOSE	METHODS USED
1.Research & Analysis	To find and confirm the potential need	User research: Group interview, Depth interview, Questionnaire, Contextual inquiry, Diary method, others. Positioning analysis: Factor analysis, Image mapping, others.
2.inspiration	To develop the creative concept	Divergent thinking: Brainstorming, Brain writing, Mind map, Checklist, Matrix method, Mandal-Art, Attribute listing, Forced relationship, Focused object technique, NM method, Gordon, others. Convergent thinking: KJ method, Block method, Cross method, Fishbone diagram, Story method, Morphological analysis, others.
3.optimization	To objectively-define the best concept	Idea Evaluation: Weighting evaluation method, Comparison and evaluation, others. Identification of concept requirement. Communization of specific image: persona, scenario, others.
4.installation	To make the direction for the developer and the planning	Checklist of quality, others.

through that of installation as shown below table 1 in order to project the affecting experience resonated with users.

2.2 Problems of the Current Product Planning

It however requires us various kinds of know-how to go through the process as shown in Table 1. In addition, it is not easy to select an appropriate method in every phase since each development has a different purpose depending on in which industry and business category it belongs as well as in what situation developers stand. This problem discourage us from implementing effective product planning in many cases [6]. The following 2 problems interrupt us to inspire the affecting experience resonated with users.

(1) Problem of eliminating product's utility value from one's experience

There are many cases that the developer pursues the development by screening such brand images of novelty and competitiveness in the current product development. That is, they do not identify in the requirement definition phase such as delight and impression that users can experience only by using products and services. Unless defining experience as utility value, gaps in awareness will occur among developers. As the result, it becomes impossible to implement effective brainstorming as it becomes ambiguous about the purpose of idea generation.

(2) Problem of being unable to respond to the pace of development
It is difficult to conduct surveys at every planning due to a rapid pace of development in the field of product and service planning. Actually it is still the case that developers repeatedly explore what to do by trial and error every time in many companies except the one specialized in the field.

The reduction of time and cost reduces the chance of survey, and this makes the development only with the developer's biased view. It is therefore difficult to inspire products and services resonated with users.

2.3 Positioning of Xb-Method, an Idea Generation System

We also study issues associated with the current product planning in order to make XB-method active for any developer to project products and services with the affecting experience. Table 2 shows the process and the activity of XB-method. It is characterized

Table 2. Process of product planning and the activity of XB-method

PROCESS	PURPOSE	ACTIVITIES in X-method
1-1. Research & Analysis	To find and confirm the potential need	(Collect and Analyze keywords)
1-2. Generalization of findings	To create materials for the generation	(Compile a data base of keywords)
2. Generation	To develop the creative concept	-Decide a theme and a subject to be developed. -Select a set of keywords of affecting components. -Develop images of keywords. -Create a good story by multiplying images -Exchange opinions. -Label the idea. -Organize the condition for scenario-writing. -Draw up the scenario.
3. Optimization	To objectively-define the best concept	(Evaluate the scenario in accordance with the purpose)
4. Installation	To make the direction for the developer and the plan	(Make a list of quality level in accordance with the purpose)

by the generalization of findings in order to use the result of the research and analysis efficiently and apply these effectively to the generation. This process enables us to utilize user profiles without doing a survey during the generation phase. We can define XB-method as a tool to be used in such a time-critical field of development.

3   A Proposal of an Idea Generation System

3.1   Scope of Xb-Method in the Process of Development

It is defined in HCD that in order to improve quality in use effectively, it is necessary to incorporate user profile from the upper-stream phase in the process of development. Similarly in XB-method, we consider that it is effective to use this method during exploring a direction of a product or a service in order to plan the affecting experience resonated with the user.

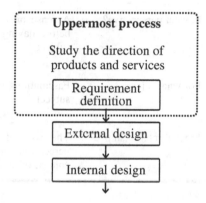

Fig. 1. The best timing of using XB-method is during the uppermost process: the earliest stage in the process of development

3.2   Scheme of Idea Generation

XB-method is a method to generate the affecting experience by multiplying keywords prepared in advance like Fig.2 by images of a product or a service to be designed. XB-method is defined as an idea forcing generation system that provides some accidental situation by multiplying keywords of images for commodity to trigger the user getting inspired a new idea.

4   Idea Generation Activity of Xb-Method

4.1   Database Used in Idea Generation

We use database prepared in advance when to generate ideas with XB-method. The database is something like Fig.2. 3 keywords are organized in a set.

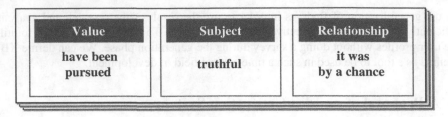

Fig. 2. The set of keyword is the result of reviewing each one of 3 components of user experience: a sense of value, subject and relationship as shown in Table 3

Table 3. Three Components of User Experience

Label of Component	Scene of appearance	Definition
Value	In what perspective?	A sense of value that people cherish before having a relationship with the subject
Subject	For what subject?	Fascinations or characteristics of the subject
Relationship	In what relationship?	The way of involvements such as perceptions and experiences, and the context of the time

The keyword is extracted from 400 affecting experiences provided through Web Questionnaire. We extract it according to the way keywords were appeared in the episode with quantification method III and then classified these into 7 patterns with cluster analysis method. The database of keywords is the data being calculated keywords with the high appearance ratio for each pattern [1].

The database is something that all the ingredients needed to generate the affecting experience are extracted so as to affect effectively each other.

4.2 Activities of Idea Development

Below is a description of the idea development activity using XB-method. This case study shows a way of developing a concept for a new car navigation system.

Activity 1. Decide a theme and a commodity to be the subject of the idea development

Decide a theme and a subject to be developed, such as a kind of interactive systems, an image of a commodity. For example, it is a case of a car navigation system.

Activity 2.   Select a set of keywords of affecting components (Use the database)
Select a set of keywords from keyword flash cards at an option. Here is an example of choosing one which indicates following 3 keywords: [have been pursued], [truthful] and [it was by a chance].

Activity 3.   Develop images of keywords
You may put the selected keywords into different words having the same meaning with visualizing some images of commodity you set. For instance, it is to develop images by putting the keyword [it was by a chance] into [it was displayed by accident] with visualizing a car navigation system.

Activity 4.   Create good stories by multiplying images
Create good short stories by multiplying images of all 3 keywords being developed at the former step. For example, in case there are keyword images of [Favorite], [Place of historical origin] and [it was displayed by accident], we can develop so as to include all the 3 components and create a story as follows: [A function to inform a historic area when riding past there if the user register a favorite in advance.]

Activity 5.   Exchange opinions each other
It is to develop images more deeply toward the good story under development by exchanging opinions with other members and to add further images in particular. Even if having no specific image of a function in the past activities, it is possible to get inspired and develop it by exchanging ideas with others.

Activity 6.   Label the idea
Select one idea out of others under development and label it in a word which expresses its content. It is the activity to organize the characteristics of the idea with reconsidering what kind of idea is created.

Activity 7.   Organize conditions for scenario-writing
It is ideal to create a scenario which represents an idea in order to store an affecting story in XB-method. In preparation for drawing up a scenario, organize and itemize the targeting user, the situation such as a time and a place and delight the idea provides during this Activity 7.

Activity 8.   Draw up a scenario
Draw up a narrative scenario to represent what kind of delight that the idea will provide. It becomes easier to create the one in user's perspective by including every component of user experience: [a sense of value], [subject] and [relationship] that have been developed until this activity.

5   Case Study

We conducted idea generation workshops with XB-method for 5 times from Sep 2007 to Oct 2008. Around 80 people were participated in the idea generation activities as shown in Table 2, and sessions required 90 min. to 120 min. Each participant conveyed the idea generation and did some brainstorming session in order to exchange opinions with others in a group of 5 people.

Participants are from various business fields such as automobile, IT service and house hold goods industries. Their professions are variable as well such as product planning, CS, usability engineer and others.

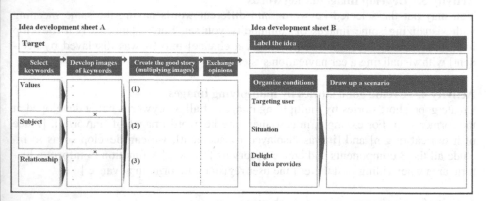

Fig. 3. Participant conveyed the idea generation using the idea developing sheet

5.1 Effect with Proven

In this study, we recognized following effects based on feedbacks from participants about the idea generation using XB-method.

First, it enables users to generate ideas effectively even for their first time. It is conducted systematically in a systemized way of multiplying keywords and it therefore is able to generate around 3 ideas per participant within 1 hour to 2 hours. We can expect an absolute performance in case of requiring a number of ideas.

Second, it is highly possible that users can meet with better-than-expected ideas. XB-method forces users to generate ideas to create good stories by multiplying keywords and offers them the condition of generation without attempts with stimuli.

Third, it enables us to inspire an idea in user's perspective. We can explore the affecting experience for users at anytime during developing 3 components of user experience: a sense of value, subject and relationship. Furthermore, it is capable to propose the relationship with users by drawing up ideas in the form of scenario.

6 Conclusion

We developed user experience-driven XB-method so that anyone can plan products and services that can provide the affecting experience.

As the result of this study involving approximately 80 people, it is proved that XB-method has advantages in product development that no other methods do. It is to generate a new product or a new service effectively based on an affecting scenario in users' perspectives. XB-method, an idea generation system in users' perspectives, can be therefore described as the one of HCD methods which is applicable to the product planning.

7 Future Activity

Regarding the database used in the idea generation, it is necessary to identify the capacity of each affecting pattern for each keyword flash card so that we can select the one effectively responding to a product or a service to be developed and users to be targeted by understanding the capacity.

References

1. Misawa, N.: The Development of AIM, an Idea Generation Support System for Products with affecting experience. KEER, Sapporo (2007)
2. Kurosu, M., Itoh, M., Tokitu, M.: Primer of User Engineering -Look at Usability- Practical Approach to ISO 13407, pp. 30–39. Kyoritsu Shuppan Co., Ltd. (1999) (in Japanese)
3. Kosaka, Y.: "KANSEI" Marketing, pp. 84–87. PHP (2006) (in Japanese)
4. Nikkei Institute of Industry and Regional Economy: Product Development of creating impression, Nikkei Inc. (2003) (in Japanese)
5. Norman, D.A.: Emotional Design, Why We Love (Or Hate) Everyday Things. Basic Books (2004)
6. Kanda, N.: 7 tools for product planning to produce hot seller at a glance. JUSE Press Ltd (2000) (in Japanese)
7. Hirano, H.: To be a mover and shaker with narrative skill. MIKASA SHOBO CO., LTD (2006) (in Japanese)
8. Peter, M., Brandon, S., David, V., Todd, W.: SUBJECT TO CHANGE: Creating Great Products and Services for an Uncertain World. Oreilly & Associates Inc. (2008)
9. Takahashi, M.: A book which encourages us to get ideas engrossingly. Chukei Publishing Company (2005) (in Japanese)

Integrating Human-Computer Interaction Artifacts into System Development

Megan Moundalexis, Janet Deery, and Kendal Roberts

Johns Hopkins University Applied Physics Laboratory,
20723 Laurel, Maryland
Megan.Moundalexis@jhuapl.edu

Abstract. This paper introduces a methodology for developing and leveraging Human Computer Interaction (HCI) artifacts into systems design within the Antisubmarine Warfare (ASW) domain. The method is structured with four integrated steps: Scenario Development, Personas, Operational Concept Documentation, and Usability. Explicit links are made between the artifacts to allow a more efficient use of design resources including legacy documentation for developers while improving the quality of design. As with current systems engineering practices this approach relies upon requirement analysis, prototyping, design iteration, and test and evaluation. Unlike current practice, however, this approach can improve the process of iteration as well as feedback on additional unanticipated requirements. Often overlooked, this process also yielded effective design team interaction. These improvements are made possible by the structured methodology that makes the HCI products attractive to systems developers: the artifacts are well organized, adaptable, and inspectable.

Keywords: Systems Design, Decision Support Systems, Human Computer Interaction.

1 Introduction

This paper introduces a methodology for systems development within the Antisubmarine Warfare (ASW) community. With reduction in defense acquisition dollars, increased technology sophistication, independent standalone legacy systems, reduced manning, and the need for improved command and control, there is a need for more effective integration and development. These opposing variables often prohibit fielding new technological capabilities to the Fleet. This is the primary challenge that our methodology aimed to overcome.

Typically, the traditional systems engineering (SE) paradigm follows a top-down approach where the project team creates designs based on the top-level requirements (TLRs). Those requirements, in essence, are understood to be the high- level goals that are completed by the user. However, this method does not lend itself to the potential design discoveries that come from interactions between real users while completing real tasks on a system. These discoveries are traditionally found by the training instructors and the fleet, but only after the system have been deployed. This is exaggerated within a

M. Kurosu (Ed.): Human Centered Design, HCII 2009, LNCS 5619, pp. 284–291, 2009.
© Springer-Verlag Berlin Heidelberg 2009

command and control network with different operator positions having varied and overlapping goals.

This paper/case proposes four design principles that the authors have used successfully in developing and evaluating the Undersea Warfare Decision Support System (USW-DSS) Build 2: ASW scenario, Click Stream Task Analysis, ASW personas, and Operational Concept Document (OCD), which are summarized here but described in greater detail below.

1. Scenario development is a fundamental element for systems development within context of user tasks. As a top-down approach, scenario development begins the requirement-gathering process.
2. Then, as a task-centered approach, a task analysis of the scenario dissects mission requirements to understand the user's mental model about the tasks that the requirements support within the context of the broader scenario.
3. Personas are HCI artifacts that contain characteristics of the user. Employing this bottom-up strategy enables the systems engineering designer/researcher to gather insight into user processes that are unique to the ASW domain.
4. Creation of an Operational Concept Document (OCD) is a documentation and evaluation process that links the cognitive and collaborative demands within the scenario to particular system capabilities that they are intended to support.

Top-down systems engineering was executed with the benefits of bottom-up analyses, which was synchronized in an end-to-end analysis (see Figure 1). In other words, a top-down project was verified using a bottom-up approach.

2 Scenario Development

Identification of a system's functional requirements often begins with scenario development, which identifies the properties and constraints of the work domain [7]. A study conducted by [4] emphasizes the need for requirements analysis in order to develop a shared vision of the system regarding the context of the design problem that we are trying to understand. Having a scenario developed also supports use case development, assessment of the design during testing and evaluation (T&E) and traceability.

As a starting point, our approach utilized an ASW scenario that must be supported by the system. This was the context of the problem for which we derived requirements to solve the problem. Functional requirements were identified that met the high-level scenario goals. Then, those requirements were decomposed into as many lower levels as were necessary, while ensuring that each level mapped back to the scenario (Figure 1). Assessing the completeness of the scenario is then a function of ensuring that allocated capabilities, functions, and requirements are captured at each subsequent level. Explicit links are made between particular system capabilities, specific collaborative demands, and external clients (e.g. interfaces) that are intended to support the scenario and operator. These linkages provide the basis for traceability and informed testing of the effectiveness of proposed designs and are captured in the operational concept document (OCD).

An issue to consider within this iterative framework is the complicated tasks that are often problematic for operators. Bottom-up task analysis provides task details. Coupled with prototyping and usability heuristic analysis, task analysis can reveal the complicated tasks that can arise during the scenario, specifically in ASW. These are areas where potential user errors might occur. Identifying these tasks early in the process affords opportunities during bottom-up design to develop sub-systems (e.g., tactical decision aids) to support the full USW-DSS system. This task-centered approach is a process of decomposing mission requirements and then designing information and control interfaces that support operator task performance in order to complete the mission. This ensures that complicated tasks are not design afterthoughts that result in interface usability problems for Fleet operators, which is traditionally a problem handed off to training instructors after the system is fielded.

While top-down and bottom-up can be used to identify system relationships as well as operational and task demands associated with them, scenario developers must consider the skills, capabilities, and training level of the operators (i.e. Personas) and any factors (e.g. chain of command, goals) that arise in the ASW dynamic high risk domain. ASW operators are capable of processing complex information. However, insufficient system design and training can lead to poor decision making. Introducing a system like USW-DSS into the decision making process will have a positive impact only if designed for the problem (e.g. scenario) and the operators are fully understood.

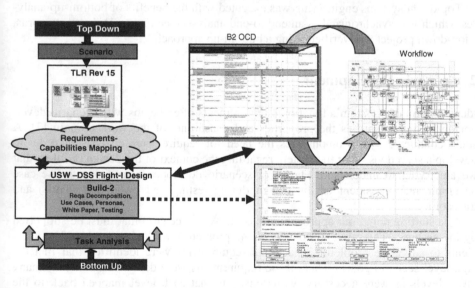

Fig. 1. System development diagram indicating scenario-to-design-to-testing

3 Personas

The authors developed personas to profile ASW users. Personas are a single page description of a specific user group [2, 3]. They do not contain information about a specific person, but is instead an amalgamation of multiple real people from the same user group. These concise user profiles are a means for the engineering team to understand the various operators who will use a system by communicating user characteristics, such as domain, responsibilities, goals, and tasks. Personas do not serve solely as deliverables. The entire process of researching, creating, validating, and sharing the user profiles engages systems developers from multiple disciplines. Reading Navy doctrine and duties to create the personas decreased the learning curve with Naval operations. Involvement with the fleet and other SMEs was essential to getting the drafts reviewed and significantly increased the engineering team's understanding of the system's purpose in the process. Engineers are then able to quickly look at the page description to get an idea of who they need to consider when developing a system.

The USW-DSS user interface (UI) working group (WG) developed 14 personas to understand various ASW roles that will interact with the system (Figure 2). Personas proved to be a valuable design research tool that created understanding between designers and Fleet stakeholders. Personas developed were for surface, subsurface, shore, and air roles including: Sea Combat Commander (SCC), Antisubmarine Warfare Evaluator (ASWE), Tactical Action Officer (TAO), Underwater Battery Fire Control System Operator (UBFCS), Computer Aided Dead Reckoning Tracker (CADRT) Plotter, Sonar Supervisor (both Surface and Subsurface), Officer of the Deck (OOD), Theater Antisubmarine Warfare Commander (TASWC), Sensor Station Operator (Sensor One), Tactical Coordinator (TACCO), Operations Specialist (OS), and Sonar Technician (ST).

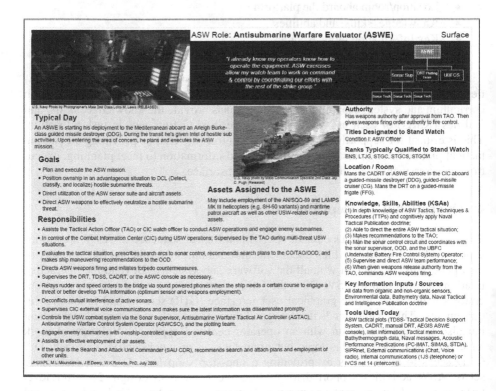

Fig. 2. Antisubmarine Warfare Evaluator (ASWE) persona

Multiple sources were referenced during the development of the persona draft versions including USW-DSS Build 1 user surveys, U.S. Navy-sponsored websites, and Naval Warfare Electronic Library (NWEL) documents. In order to ensure accuracy, the personas were then reviewed by subject matter experts (SMEs) and U.S. Navy fleet members.

Each persona includes the following information:

- Name of ASW role
- Location - e.g., surface, subsurface, shore, or air
- Operator Picture
- Quote
- Organizational Chart
- Description of operator's typical day
- Platform picture
- Goals
- Responsibilities
- Assigned assets
- Authority
- Titles designated to stand watch
- Ranks typically qualified to stand watch
- Location/room aboard the platform
- Knowledge, skills, and abilities
- Key Information inputs/sources
- Tools used today

In addition to user interface design, personas will be used to support training activities. Specifically, personas can be used to uncover potential training holes between curriculum and system capabilities. Since personas provide the knowledge, skills, and abilities (KSAs) of each user type, they can be used to find gaps in task allocation. This is critical for supporting command and control [6]. That is, providing a common tactical picture through distributing timely fused information to meet planning, execution, an assessment needs in air, surface, and subsurface.

4 Operational Concept Document

The Operational Concept Document (OCD) is a logical sequence of ASW operations supported by USW-DSS, which documents and examines the role and functionality of the operator in conjunction with all the software sub-systems. Beyond the interactive relationships internal to USW-DSS, the OCD also maps the ASW activities and operator's major tasks to the graphical user interface (GUI) screens and functional requirements. This feature lends itself to good design, not only to systems development but also test and evaluation with an operator-centered approach. Volumes of work have been published on this topic but often lack an operator-centric approach. Our OCD seeks to resolve this by providing a one-stop shop for: 1) a comprehensive set of ASW activities supported to produce the USW-DSS design, 2) lower level tasks supported by software or operator, 3) mapping of tasks to GUI and functional requirements.

As illustrated in Figure 3, the first column describes the ASW activities. The second column, entitled task name, decomposes the activity into lower level tasks mapped to the responsible actor (e.g. software configuration item, Crew, or Communications). The HCI column includes the need for and nature of the user interface (e.g. C- control, D- display, N- software only). The GUI column documents the location of the interface that supports the task. The requirements column documents the requirement fulfilled.

ASW Activity Description	Task Name	CI/Crew/Comms	HCI	GUI	Req #
Perform Initial Detect, Classify, and Localize (DCL)					
Manage Tracks					
19. **Fuse Multi-Sensor Data.** OSFCI performs background multi-sensor Integration data fusion on the received tracks when appropriate. (Note: Operator review of the fusion results is discussed in later Step X)	Fuse multi-sensor data	OSFCI	N	N	FDSS 4
20. **Update System Track Database.** TRCI updates the CTP database with system tracks. This includes the towed array, hull array, and sonobuoy tracks. This picture may also include any help-processed tracks and contact classification from known threat and signature data.	Update system track database	CMCI TRCI	N	N	FDSS 20
21. **Update Common Tactical Picture (CTP).** TDCI/CMCI continuously maintains the CTP display. This also includes maintaining ownship position, cooperating forces and all other reported tracks. The CTP supports drilldown and display of the supporting information and data.	Generate updates to CTP	TDCI/CMCI	N	N	FDSS 20
	Display updates to CTP	TDCI/CDCI	D	CTP	FDSS 8
22. **Update Non-Organic Assets via SIPRnet** All designated system tracks (what attributes) are passed to participants					
a. Receive Non-Organic Link Tracks. This is track data received via data link/SIPRnet from Cooperating Forces such as other surface ships, submarines, as well as other sources including, 8V/SIPS, National assets, etc.	Receive tracks	TRCI	N	N	FDSS 11
b. Transmit Tracks to Non-Organic Platforms The CTP will provide the capability to select and transmit fused tracks selected from the CTP to external USWDSS platforms via the TRCI	Operator select and transmit Format and forward tracks	SCC/ASWE TRCI	C, D N	CTP	FDSS 11, 12
23. **Cue Operator.** Generate operator alerts for new and previously held tracks exhibiting characteristics of interest on operator specified criteria and doctrine.	Generate alert	CMCI	N	N	FDSS 24
	Display alert	TDCI, CDCI	D	CTP	FDSS 24, 62
	Acknowledge alert	SCC/ASWE	C	CTP	FDSS 24, 62

Fig. 3. Operational Concept Document example

The OCD emphasizes the consideration of human factors at the earliest stage of the design cycle. Equally important is documentation of an artifact in a quick, efficient, and verifiable manner that lends itself to use by the software developers. This is critical because too often human factors professionals develop products that are of little use or do not impact design.

Beyond its main objective of better supporting the development of a more effective system, the OCD has additional benefits. First, the OCD contains a comprehensive list of operator tasks and will support development of training curriculum. Secondly, it will support system testing and verification. [1] proposed a test for requirements completeness based on defining required system behavior under all conditions (e.g. all scenarios). The OCD is traced to functional requirements and can easily be modified to support different scenarios to support operator, software, and usability testing.

290 M. Moundalexis, J. Deery, and K. Roberts

5 Usability Verification: Click Stream Task Analysis

In order to provide quantitative data that compared three build versions of the USW-DSS Mission Planning System (MPS), a usability testing technique called the click stream task analysis was conducted on one released version of the system (Build 2) and two proposed redesigns. The initial proposed design incorporated task flow processes used in NWP 5-01 [5]. The second proposed design incorporated feedback from a heuristic evaluation.

This task analysis method is similar to a keystroke level model (KLM) in that we compared the number of mouse clicks needed for the user to complete a specified task. However, we were not able to track the time to complete each user interaction because our evaluation was limited to screenshot display images. The efficiency of the user flow through the task was tracked (i.e. does the user flip back and forth through GUI screens). An additional benefit from analyzing display screens at the keystroke level is that it provided us with a way to fully engage with the information design and interaction design, which informed the heuristic evaluation and system redesign.

We documented an SCC's or ASWE's anticipated stream of clicks when creating a Course of Action (COA) and following it through as it becomes the mission plan ready for execution. For our analysis, we documented the task being completed (e.g. Prepare for mission, Determine mission parameters), the system object (e.g. button or pull-down menu name), the user action (e.g. click, type, select from pull-down menu), system location (e.g. Details tab, Generate Products tab, Navigation Tree), result (e.g. Populates section, Work area opens to default settings, Highlights object), and any design discussion notes. Design discussion notes included questions we had when working through the task, any unneeded functionality, and redesign ideas. Even though they were not included in the number of clicks, cognitive tasks during the COA creation were also noted to aid the development of another HCI artifact called the task diagram. If a task was not included in all three builds, it was removed from the click count total. In this way, we were able to make sure we were comparing apples to apples. This standardization was important to ensure we were comparing how many clicks it took to do the exact same set of tasks consistently across all of the builds.

Based on the three click stream task analyses, there were differences between each of the 3 builds. Build 1 required 197 clicks to create a COA while Build 2 needed 170 clicks. The JHU/APL proposed solution for Build 3 used 156 clicks to create the COA. As such, each system build decreased the number of steps needed to accomplish the same task.

6 Conclusion

ASW development challenges (e.g., reduced manning, integrating legacy tactical decision aids, command and control, etc.) led to the use of new HCI approaches to ensure value-added within the design process. Building upon earlier design work in Build 1, this approach developed HCI artifacts in a streamlined manner so that the products could support developers and thus design earlier in the cycle.

We have presented four work-centered design artifacts including the ASW scenario, persona, OCD, and usability verification click-stream task analysis. Using a consistent work-centered approach for all products ensures the ASW operator is implicitly the focus. By using a top-down approach it is possible to develop scenarios that extend the designers understanding of the domain and requirements. Equally critical is that this approach demonstrates the subsequent steps in the OCD support a complete and correct set of requirements with respect to the scenario in a language that both human factors and software developers can leverage. The process of creating the personas is a valuable experience. In addition to helping the SE acquire domain knowledge, the single-sheet personas are a quick reference for the software team to learn more and understand the customer for who they are designing the system. The click stream task analysis served as an effective quantitative analysis to compare different system builds. All together, these four HCI artifacts supported the execution of an end-to-end analysis merging the benefits of top-down systems engineering with a bottom-up approach.

Acknowledgments

Program Executive Office Integrated Warfare System 5E (Undersea Warfare Command and Control).

References

1. Carson, R.: A Set Theory Model for Anomaly Handling in System Requirements Analysis. In: Proceedings of INCOSE (1995)
2. Cooper, A.: The Inmates Are Running the Asylum. Sams, Indianapolis (1999)
3. Cooper, A.: The Origin of Personas (2003),
   http://www.cooper.com/journal/2003/08/
   the_origin_of_personas.html
4. Mar, B.W.: Back to Basics Again: A scientific Definition of System Engineering. In: Proceedings of INCOSE (1997)
5. Naval Warfare Publication: Naval Operational Planning (NWP 5-01 REV. A) (1998)
6. Roberts, W.K.: An Integrated Homeland Operating Picture: A Framework to Support Design and Analysis Among Distributed Homeland Security Teams. In: Institute of Industrial Engineering Conference (2006)
7. Vicente, K.: Cognitive Work Analysis: Toward safe, productive, and healthy computer-based work. Lawrence Erbaum Associates, Mahwah (1999)

"How Do I Evaluate THAT?" Experiences from a Systems-Level Evaluation Effort

Pardha S. Pyla, H. Rex Hartson, Manuel A. Pérez-Quiñones, James D. Arthur,
Tonya L. Smith-Jackson, and Deborah Hix

Center for Human-Computer Interaction, Virginia Tech
Blacksburg, VA 24060 USA
{ppyla,hartson,perez,arthur,smithjack,hix}@vt.edu

Abstract. In this paper we describe our experience deriving evaluation metrics
for a systems-level framework called Ripple that connects software engineering
and usability engineering life cycles. This evaluation was conducted with eight
teams of graduate students (falling under four types of development models)
competing in a joint software engineering and usability engineering course to
create a software solution for a real world client. We describe the challenges of
evaluating systems-level frameworks and the approach we used to derive metrics given our evaluation context. We conclude with the outcome of this evaluation and the effectiveness of the metrics we employed.

Keywords: Systems-level evaluation, evaluation metrics, goal-question metric.

1 Introduction and Background

1.1 Types of Evaluation

One of the fundamental activities in interaction development is evaluation. Evaluation
is multi-dimensional, and along one dimension within human-computer interaction
there are two kinds of evaluation: formative and summative. Formative evaluation,
conducted during a development effort, is used to support human-computer interaction (HCI) *practice* (identifying design flaws in the existing state of the design). On
the other hand, summative evaluations, predominantly conducted after the design is
implemented, are used to support HCI *research*.

Typically, summative studies are about comparing different systems by controlling
certain variables and observing their effects to make conclusions about which is "better" (on some measurable scale) within a narrowly defined criterion. These kinds of
studies are one of the few staple formal methods in an otherwise mostly ad-hoc
research domain of HCI. They provide formulaic answers to particular research questions where measurable independent variables can be isolated and measured. Furthermore, because these studies are relatively straightforward to conduct and work
well for such targeted contexts, they are popular in HCI literature. However, these
studies have proven difficult and impractical for answering the larger, less-
constrained research questions in systems-level research [1, 2].

M. Kurosu (Ed.): Human Centered Design, HCII 2009, LNCS 5619, pp. 292–301, 2009.
© Springer-Verlag Berlin Heidelberg 2009

1.2 Introduction to Ripple: A Systems-Level Research Effort

Interactive software systems have both functional and user interface (UI) components. UI design requires specialized usability engineering (UE) knowledge, training, and experience in topics such as psychology, cognition, perception, and task analysis. The design and development of a functional core requires specialized software engineering (SE) knowledge, training, and experience in topics such as algorithms, data structures, and database management. Given that the user interface and the functional core are two closely coupled components of a system, and that each can constrain the design of the other, there should be close connections between the two development life cycle processes. Unfortunately, the two disciplines are practiced almost independently [3-5], which results in missed opportunities to collaborate, coordinate and communicate about the overall design and often leads to project failures [6-8].

To connect the SE and UE life cycles, we designed and developed the Ripple Implementation Framework [5, 9], within which developers can define the key components, knowledge base, tool support, timelines, and development environment necessary to support an interactive-system development effort. This framework embodies a storage sub-system containing shared design representations. Here, usability and software engineers can store, access, share, and manipulate the various work products that are created during a development effort. Another important component of this framework is the constraint and dependency sub-system that has capabilities to record, propagate, and enforce the various constraints that exist between the two life cycles and their resulting work products.

1.3 Evaluating Systems-Level Frameworks

Perhaps the most scientifically rigorous way to evaluate a software development framework such as the Ripple Implementation Framework would be to use a full, formal summative experimental design in which large number of software development teams (of the same size and balanced for skills and experience) are employed to develop the same software system, using the same software development methodology, for the same client, with half the teams randomly assigned to use an instantiation of the Ripple Implementation Framework and the other half not. By controlling all other factors that could potentially impact the quality of the system or performance of the teams in the experiment, one would hope that such an experiment would make it possible to establish cause-and-effect relationships between any quality or performance indicators of the process employed by each team and the resulting product due to the use (or not) of the Ripple Implementation Framework. Also, if the number of teams is large enough, one could check for statistical significance in the measures (or indicators) of dependent variables as a causal outcome of using (or not) the Ripple Implementation Framework.

However, such a summative evaluation of a software development framework such as Ripple with a large number of real teams of professionals building the same system using different development frameworks is not practically possible. It is difficult and expensive enough for just one team to development any non-trivial software system. The most obvious reason is that such an undertaking would be ludicrously expensive given the resources necessary to conduct an experiment of such magnitude. Furthermore, given the complex nature of interactive-software development, it is impossible to control all factors (assuming it is possible to identify all factors in the first place) involved in the development process. Finally, even if somehow all these problems could be solved, any results from such an experiment will suffer from external validity problems as these results cannot necessarily be generalized to contexts where a different set of factors are at play: different sizes of development team, different skills or experience of teams, different software development methodologies in use, different type of software system being developed, different project management styles, different types of client interaction, etc.

1.4 Planning a Testbed Evaluation Environment

Given the impossibility of conducting a formal summative study to evaluate a framework like Ripple, we sought alternative approaches. After two years of planning with the Department of Computer Science at Virginia Tech, we used a joint offering of graduate-level SE and UE courses for this evaluation. Students from the UE class were trained for the usability engineer role and taught the life cycle concepts and guidelines for building user interfaces. Similarly, students from the SE class were prepared for the software engineer role and taught the life cycle concepts for functional core development.

Instantiating the Ripple Framework for a classroom setting we were able to test the framework concept before investing resources in its full software implementation. We achieved this with a kind of Wizard-of-Oz approach to simulate various aspects of the framework behind the scenes. For example, one of the authors played the role of various sub-systems in Ripple instead of actual software implementations. Email was used in place of automated Ripple messages and a password-protected, forum-based, and archived group email software was used as the shared design representation for developers to post their work products and share them with the other members in their teams.

1.5 Team Composition and Project Description

We had a total of eight teams each following one of four different models of software development (Ripple instantiation was one such model) as shown in Table 1. Semester-long projects were used to build a real-world software system for the Horticulture Club of Virginia Tech. Three members from the Horticulture club were compensated and trained to act as client representatives for all the teams. Special care (interactions with client representatives were monitored) was taken to prevent one team's interaction with the client from influencing their interaction with another team.

Table 1. Team setup and distribution

Team	Team type	SE role	UE role	Role/setup/end product
A1	Ripple	3 SE	3 UE	• Distinct roles working together • Knew counterpart team from start • Developed fully-functional system
A2		3 SE	4 UE	
B1	non-Ripple	3 SE	4 UE	• Distinct roles working independently • Counterpart team introduced post design • Developed fully-functional system
B2		3 SE	4 UE	
C	Dual experts	3 students who took both classes		• Played both roles • Developed fully-functional system
D1	UE-only team	-	5 UE	• Played usability engineer role only
D2		-	5 UE	• Had no counterpart team
D3		-	4 UE	• Developed hi-fidelity prototype only

2 Need for a Goal-Directed Approach to Evaluation

In the absence of a formal summative study, our objective became a semi-formal study, but we did not find much work in the literature about what metrics to use or how to arrive at them. Fundamentally, when setting up an evaluation, questions arise as to what to measure (i.e., metrics). In order to know this, we need to use a systematic approach to arrive at the questions that we need the data to answer. To decide that, one needs to clearly identify the goals of the evaluation.

One approach that embodies this goal-directed approach to derive metrics is the Goal-Question-Metric (GQM) paradigm, originally used in the software engineering domain [10, 11]. In this method one should first postulate the goals of evaluation and derive questions, the answers to which will indicate if the goals are met. Each of those questions is then analyzed in order to determine the measurements that are necessary to answer the question.

2.1 Applying GQM to the Ripple Framework

We started with the goal of this study, which pertained to the Ripple implementation instance:

Evaluate the effectiveness of the Ripple framework, as embodied in a specific implementation instance, to facilitate communication among developers within and between the two life cycles.

High-level questions, Q1 and Q2, can be derived from the goal:

Q1. At the highest level, what are the indicators appropriate to reflect effectiveness of framework?

Q2. To what end and in what ways is communication used in the Ripple framework?

Since an effective framework should result in better quality of process and resulting product, question Q1 can be answered as:

A1. The quality of product and process.

And, since communication in Ripple is used to facilitate various other factors, question Q2 can be answered as:

A2. Communication, the main contribution of the Ripple framework, is used to facilitate coordination, synchronization, constraint and dependency checking, and change management.

Now, one can ask, in questions Q3 and Q4, what measures are needed to provide these answers? The end result of this exercise is shown in Fig 1.

However, certain changes and adjustments had to be made to these metrics as a result of the various constraints inherent in our study. For example, it was not possible to conduct a lab-based usability evaluation of all teams' products to arrive at measures such as time on task, number of errors, and satisfaction. The reason for this is that three D teams developed high-fidelity prototypes only and did not have a functional backend. Each of these prototypes were constructed to simulate carefully scripted, but different, benchmark tasks only and therefore it was not possible to run the same set of tasks in a lab setting across all teams. Therefore we created a metric called overall value index based on product quality ratings by the client representatives (described below).

Similarly it was not possible to compare code size and complexity using an objective measure because the D teams had no functional core software. Also, in the five joint teams that created fully functional systems, they adopted different programming technologies ranging from AJAX to JSP, making it impossible to compare using code-based metrics.

2.2 Evaluation Instruments and Metrics Used

Using the approach described above, we derived the following evaluation instruments and associated metrics:

2.2.1 Value Index of a System

At the end of the semester, we conducted an eight-hour meeting in which, the experimenter and all three client representatives analyzed the prototypes for breadth of functionality covered by the UI, usability, and appropriateness for the Horticulture Club's goals. Each team's final system was analyzed in detail to arrive at an overall value index per team product. First, we compiled an exhaustive "union" list of all features across all teams (e.g. ability to search by common plant name, ability to view shopping cart at all times, feature to provide directions to plant sale, etc.). Each of these features was then ranked on a desirability scale, which indicated how important this feature was for the client's plant sale. The scale included a range of low, medium, high with fractional values in between in some cases. In other words this analysis provided a superset of all features with each feature's desirability present in all systems combined.

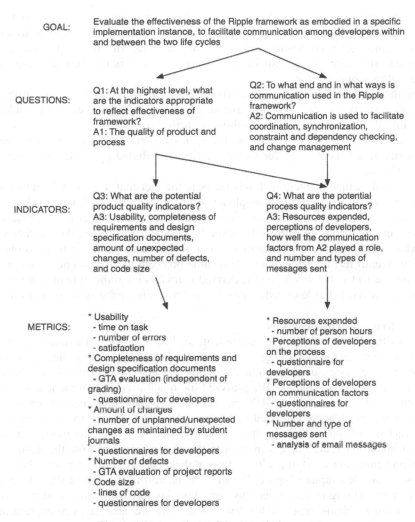

GOAL: Evaluate the effectiveness of the Ripple framework as embodied in a specific implementation instance, to facilitate communication among developers within and between the two life cycles

QUESTIONS:
Q1: At the highest level, what are the indicators appropriate to reflect effectiveness of framework?
A1: The quality of product and process

Q2: To what end and in what ways is communication used in the Ripple framework?
A2: Communication is used to facilitate coordination, synchronization, constraint and dependency checking, and change management

INDICATORS:
Q3: What are the potential product quality indicators?
A3: Usability, completeness of requirements and design specification documents, amount of unexpected changes, number of defects, and code size

Q4: What are the potential process quality indicators?
A3: Resources expended, perceptions of developers, how well the communication factors from A2 played a role, and number and types of messages sent

METRICS:
* Usability
 - time on task
 - number of errors
 - satisfaction
* Completeness of requirements and design specification documents
 - GTA evaluation (independent of grading)
 - questionnaire for developers
* Amount of changes
 - number of unplanned/unexpected changes as maintained by student journals
 - questionnaires for developers
* Number of defects
 - GTA evaluation of project reports
* Code size
 - lines of code
 - questionnaires for developers

* Resources expended
 - number of person hours
* Perceptions of developers on the process
 - questionnaire for developers
* Perceptions of developers on communication factors
 - questionnaires for developers
* Number and type of messages sent
 - analysis of email messages

Fig. 1. GQM applied to Ripple to derive metrics

After this desirability analysis, the clients also analyzed each feature of each team's product, rating it on a perceived quality scale that included values of "poor", "fair", and "good" with fractional values in between for some cases. On a matrix, each feature on the union list was marked with a one or zero to indicate its presence or absence, respectively, combined with the ratings for desirability and perceived quality. A product of these three values (feature present/absent, desirability rating, and perceived quality rating) was computed to arrive at the value index per feature. This value index per feature was aggregated to calculate the total value index per team product. We also computed an aggregate feature count per team. These two values provided an overall comparison of the different systems by each team.

One of the issues faced in comparing the value of the systems developed by the different teams in the study was their dissimilarity in terms of the "realness" of the systems. For example, the three D teams created systems without a real backend (i.e. user interface prototypes only). Therefore, the value indexes for D team systems were based on more perceived quality attributes than real ones. However, we believe this value index still serves as a basis of comparison within sets of teams that used the same development condition.

Even though the value index metric proved useful in rating the quality of different teams, one issue we noticed with this instrument was the fact that it did not provide an insight into the contribution of each sub-team in the case of A, B, and C teams. For example, a low value index indicates a failure on the part of the entire team but does not describe if the failure was due to low-quality contributed by the UE sub-team or the SE sub-team.

A potential further problem with this index is the fact that it is not tolerant to distortion due to feature creep. For example a team could accrue a high value index if its system has fewer core (essential) requirements but a large number of non-essential requirements, which would probably never be used by the client. This effect was somewhat offset by very low desirability scores for not-so-useful features, bringing down the overall feature value index. This same problem has the potential for a more pronounced effect in the second metric derived from this instrument: feature count per system. However, to our knowledge, this was not a severe problem in this analysis.

2.2.2 Activity Journal Analysis

As part of this study we required all students in both classes to maintain an online journal to record individual and group hours spent working on the project, problems encountered, strategies used, negotiations held, overall impressions of the process, and other project-related details. We analyzed the qualitative data in the journals (unexpected changes, issues with SE-UE interaction, and other experiences) in each entry of each developer's sheet and used that information as evidence to investigate and reconstruct what happened during this exploratory study. The activity journals proved to be an important source of information in the investigation of how the teams performed and aided in identifying a list of factors that seemed to play a role in the interactive-software development space. Because of the confidentiality of these journals, students were willing to share candidly their insights into what was working in the process, team problems, rationale for design decisions, etc. and provided explanations for some of the phenomenon we observed during the study.

2.2.3 Email Analysis

In order to measure the amount and nature of communication that transpired among the various members in each team, all teams were provided with custom group email IDs in which the experimenter was a silent member. The joint teams each had three email IDs: one for the SE group, one for the UE group, and one for the combined group. The students were required to use these group email IDs for all project-related communication during the semester. The entire collection of email exchanged among all groups throughout the semester was archived and analyzed. Each email was tagged with various keywords. The total number of emails per team, the frequencies of keywords, and some subjective analysis were used as metrics. This instrument provided deep insights into SE-UE sub-team dynamics. Using this instrument, we could identify the strategies different teams used, how they negotiated different problems in the team, etc. This instrument combined with the activity journal entries provided a surprising amount of nuanced information about each team's operations, their overall performance, and their working relationships.

2.2.4 Transcripts of the End-of-Semester Symposium

Given the exploratory nature of the study, it was essential for us to understand each team's final perceptions of problems inherent in the development model they were assigned, the strategies they used to overcome these problems, lessons learned as part of their experience, and (based on these experiences) any advice they have for real-world software developers. In order to facilitate this kind of debriefing and sharing of knowledge we hosted a research symposium at the end of the semester in which each team presented the key findings from their project experience. This symposium was recorded (audio and video) and the content transcribed (text files). Components of data from the transcripts were used as metrics for the study. Even though this transcription exercise took significant effort, it provided rich data about what worked for each team and what challenges they faced. This instrument provided an overview of the entire process (as opposed to the value index providing overview of the product) in the project as perceived in retrospect by the teams.

2.2.5 Group Interviews

At the end of the semester (after the symposium) each team was individually interviewed for an hour and a half. The interview was semi-structured with general questions probing the challenges faced (gleaned from the end-of-semester symposium), how prior work or academic experience in each role affected their performance in the study, how their interaction with the counterpart team affected their experience, what style of development they preferred, and other impromptu questions resulting from their answers. The interviews were recorded (audio only) and transcribed. Various parts of the data from these transcripts were used as metrics for the study. Once again, this was a tedious and time consuming exercise, but one which yielded rich qualitative data about many aspects of each team's experiences.

2.2.6 Surveys

At the end of the group interviews, each student was administered two written surveys: one with 55 questions gauging their perceptions on a wide variety of aspects of the study and another with 22 questions gauging their perceptions on various pedagogical aspects of the joint offering of SE and UE courses. The data from these surveys were aggregated to team level responses and analyzed. Also, all students were given the same two surveys regardless of the development condition they used for their group project. Therefore, for questions aimed at gauging their perceptions about other development conditions, the students had to resort to conjecture based on their experience observing the other teams in class throughout the semester and during the symposium presentation. For example, some questions in the survey asked the students about their perceptions on the importance of having periodic communication between SE and UE roles in an interactive-software development effort. However, the D teams had no first-hand knowledge with this issue as they did not have a counterpart SE role. The surveys provided some statistically significant trends about broader issues like the importance of communication in teams.

2.2.7 Subjective Feedback from Clients

After each interaction the clients had with a team (for example after their requirements gathering meetings, formative evaluation meetings, etc.) and after the end-of-semester product comparison meeting, the clients were asked if they had any general impressions on the team's performance in that meeting or their overall impression on the team's system. The feedback from clients in these situations were recorded and used as metrics. These were purely subjective assessments by the clients that provided a perspective that was not captured in any of the other instruments described above. Whereas each of the other instruments (except, perhaps, the symposium) provided unique insights into a small subset of the aspects involved in this study, this client feedback provided a holistic, albeit general, assessment of each team. Their feedback represented those aspects that are more about the overall interaction and impression they had about each team and its system.

3 Outcomes and Conclusion

In the end this evaluation proved to be a success. The data from the different metrics provided insights into different aspects of the study. In many cases insights gained using one instrument were supported by others. For example, we found strong evidence showing the effectiveness and utility of Ripple-like communication-fostering frameworks (both quantitative survey trends and qualitative discourse from journals and interviews). We also discovered that certain aspects of social dynamics of collaborative work can outweigh the effects of structured communication or its absence (from email instrument we found one SE team ignore dozens of emails from UE counterparts about UI design specifications). We found evidence for inherent conflict of interest when the same people perform both SE and UE roles and often adopt design solutions that are easy to implement rather than implementing solutions that are easy to use (gathered from client perceptions, symposium transcripts, and via team interviews). Another interesting finding was regarding UE-only D teams. Their unconstrained development context (i.e. no software developers to negotiate feasibility and development constraints) resulted in broad and rich designs that turned out to be some of the client's favorite picks (evidenced by team interviews and journal entries where team members commented about not having to face the implementation challenges other types of team were facing).

In conclusion, in this paper we described a series of measuring instruments we adopted to investigate the effectiveness of the Ripple Implementation Instance and to explore the various communication factors that could potentially impact interactive-software development. Each of these instruments provided a unique insight into the investigation of each team's performance and to the general understanding of the different factors that seem to influence the quality of interactive-software development. However, no single instrument, by itself, provided irrefutable evidence to explain beyond reasonable doubt the various factors that were at play in this exploratory study. Using an analogy of the criminal justice system, combinations of empirical "evidence", ranging from objective to subjective and from qualitative to quantitative, were pieced together to determine our understanding of what happened with each team in the study. Based on our experiences with this study we believe a broad goal-oriented and systematic approach to evaluation in general and metrics in particular provides an effective way to evaluate systems-level frameworks.

References

1. Hartson, H.R., Andre, T.S., Williges, R.C.: Criteria for evaluating usability evaluation methods. International Journal of Human-Computer Interaction 15(1), 145–181 (2003)
2. Monk, A.F.: Experiments are for small questions, not large ones like "What usability evaluation method should I use? Human-Computer Interaction 13(3), 199–201 (1998)
3. Pyla, P.S., et al.: Towards a model-based framework for integrating usability and software engineering life cycles. In: Interact 2003 Workshop on Closing the Gaps: Software Engineering and Human Computer Interaction. 2003: Université catholique de Louvain, Institut d' Administration et de Gestion (IAG) on behalf of the International Federation for Information Processing (IFIP), pp. 67–74 (2003)
4. Pyla, P.S., et al.: What we should teach, but don't: Proposal for a cross pollinated HCI-SE curriculum. In: Frontiers in Education (FIE) Conference, Savannah, Georgia, pp. S1H17–22 (2004)
5. Pyla, P.S., et al.: Ripple: An event driven design representation framework for integrating usability and software engineering life cycles. In: Seffah, A., Gulliksen, J., Desmarais, M. (eds.) Human-centered software engineering: Integrating usability in the software development lifecycle, pp. 245–265. Springer, Heidelberg (2005)
6. The Standish Group, The CHAOS Report (1994)
7. The Standish Group, Unfinished Voyages. A follow-up to The CHAOS Report (1995)
8. The Standish Group, Extreme CHAOS (2001)
9. Pyla, P.S., et al.: Evaluating Ripple: Experiences from a Cross Pollinated SE-UE Study. In: CHI 2007 Workshop on Increasing the Impact of Usability Work in Software Development, 4 pages (2007)
10. Basili, V.R., Weiss, D.: A methodology for collecting valid software engineering data. IEEE Transactions on Software Engineering SE-10(6), 728–738 (1984)
11. Fenton, N.E., Pfleeger, S.L.: Software metrics: A rigorous and practical approach, 2nd edn. International Thomson Computer Press (1997)

Changes of HCI Methods towards the Development Process of Wearable Computing Solutions

Ingrid Rügge[1], Carmen Ruthenbeck[2], and Bernd Scholz-Reiter[2]

[1] Mobile Research Center
[2] CRC 637, University of Bremen, c/o BIBA, Hochschulring 20,
28359 Bremen, Germany
{rue,rut,bsr}@biba.uni-bremen.de

Abstract. Logistics is a dynamic and heterogeneous application area for wearable computing. In this paper, wearable computing technologies are examined as basis for a support system for mobile workers at an automobile terminal under the new paradigm of autonomous controlled logistics. An appropriate wearable computing system has to fulfil different system requirements with respect to the mobile work process and the bodily conditions of the user. Therefore the requirements of wearable computing systems were defined in a participatory process with the users.

Keywords: Autonomous Control, Logistics, Mobile Usability, Mobile Work Process, Requirement-Monitoring, User-Centred Design, Wearable Computing.

1 Introduction

In the development of wearable computing solutions impressive progress has been made in the past years, especially for the use in professional business segments [1], but the breakthrough into the market has clearly fallen short of expectations. Here is an intensive study which has attempted to systematically research the innovation barriers and problems of the acceptance of mobile solutions in a broad scale technology analysis and technology assessment study, in order to trace the technical shortcomings of the components available till now, as well as the system deficits of realised applications [2]. The emphasis laid on the usability in the working world beyond the desktop. A stocktaking of technical components and realised systems was carried out, which was checked against the requirements of the potential application areas. On this basis, further configuration recommendations and methodical deductions were submitted.

This study led – amongst others – to the conclusion that wearable computing components require a standardization which would have to be similar to the manufacturing systems and especially to the serial manufacture of clothing (e. g. workwear), since the user will wear it directly on the body. Deriving from this specific bodily stipulation is the thesis that acceptance is a crucial success factor for wearable computing solutions, and the assumption that acceptance can be achieved by a stricter application of Human-Computer-Interaction (HCI) methods, as well as by adapting

M. Kurosu (Ed.): Human Centered Design, HCII 2009, LNCS 5619, pp. 302–311, 2009.

these methods to the special conditions of mobile work processes and wearable solutions [2] One new approach is "using technology as a hands-on experience", i. e. the early creation of "demonstrators" (mock-ups and prototypes) and their repeated use as a graspable basis for a participatory design dialogue [3]. This approach is being analysed using the example of developing a support system for mobile workers at an automobile terminal under the new paradigm of autonomous logistics control.

Until now, planning and control of logistic processes at automobile terminals were generally executed by centralized logistic systems, which could not cope with the high requirements needed for flexible order processing due to increasing dynamics and complexity. The main business processes at automobile terminals are planned and controlled by centralized application software systems. To deal with the increasing complexity, an innovative approach to autonomous control in automobile logistics is being studied [4]. The general idea is to develop decentralized and heterarchical planning and controlling methods – in contrast to existing central and hierarchical aligned planning and controlling approaches [5]. This research was supported by the German Research Foundation (DFG) as part of the Collaborative Research Centre 637 "Autonomous Cooperating Logistic Processes – A Paradigm Shift and its Limitations" at the University of Bremen (SFB 637) [6].

2   Characteristics of Wearable Computing Solutions

A wearable computing solution can be defined in several ways. Some of them depend on the research direction and on the application domain, most of them try to distinguish it from palm-top applications. Bradley J. Rhodes focuses on the features of the hardware: "wearable computers have many of the following characteristics: Portable while operational (…). Hands-free use (…). Sensors: In addition to user-inputs, a wearable should have sensors for the physical environment (…). "Proactive": A wearable should be able to convey information to its user even when not actively being used (…). Always on, always running (…)" [7]. The wearIT@work project focuses on the interaction between the user, the system, and the environment [1] and defines the basic idea as follows [3]: Computer systems integrated with clothing support their users in an unobtrusive way. They allow the users to perform their *primary tasks* with the assistance of a computer, but without cognitive overload. Explicit interactions with these systems have to be reduced to a minimum. Therefore, the worn system must recognise the current environment and the work situation of a user by integrated sensors. Based on the detected work context, the wearable computing solution has to push useful information to its user, e.g., how to proceed with the work by reducing probable options to a minimum. It allows the user to simultaneously interact with the system and the environment.

Components of wearable computing solutions are Ultra-Mobile PCs (UMPC) and so-called "wearable computer", head-mounted displays (HMDs), arm-worn keyboards, speech input and output, and wireless connections, as well as different types of carrying and fastening systems, amongst others vests and gloves especially developed for this purpose and augmented with technology. The stocktaking shows, that most of the documented wearable computing solutions are prototypes which serve as feasibility studies for research approaches or field studies [2]. There are hardly any products available to date. Comparing the application areas with the proposed wearable computing solutions made obvious why market penetration has not been yet achieved: The technology is neither adapted to the mobile workers, nor to their primary task beyond "computer work".

3 Characteristics of Mobile Work Processes

To achieve the required adaption, a competent conception of the term *mobile activity* or more specific *mobile work process* is essential. It is the prerequisite for understanding the particularities wearable computing solutions have to cope with when they have to be implemented, in order to open innovation potentials for working processes and to satisfy new application areas.

Assuming that activities are always performed by humans work activities are differentiated into two categories: work processes which are performed at a desk or a stationary workplace, and those which are performed in motion. The centre is the working person who has acquired knowledge and skills, who has an individual style, as well as habits and experiences, who uses tools and materials, requires information on site, who has to take measurements or record her/his work steps during her/his work process and while in motion. Mobile work processes have some common characteristics[1]:

- Performed in motion, e.g. road inspection,
- performed at different locations, e.g. industrial plant maintenance, or
- performed at one location but on varying and big or extensive objects, e.g. inventory management.

It is characteristic for mobile work processes beyond desktop work that the user's primary task and therefore her/his attention is focused on the real world. Furthermore, most mobile work processes are embedded in a more complex work process, which implies a certain level of autonomy of all entities (men, artefacts, and infrastructure) and requires a high degree of communication and cooperation between them.

However, generalizing requirements for wearable computing solutions is not possible; since the various application areas present very differing conditions of utilization which themselves require the use of different technological artefacts [2]. Therefore, the process of developing wearable computing solutions and of HCI interfaces for these technological scope is characterized by an interdependence of the components; i. e. a decision at one point has far-reaching effects, possibly on all components (see fig. 1).

At the highest level, mobile work processes require "mobile assistance systems" that can be worn on the body and that behave like a (human) assistant, e.g. a sub worker or a tutor. Therefore, wearable computing solutions are characterized as "unobtrusive", they are to be used "casually". They are to *dis*burden the user rather than to be a burden. Their use has to be simple and should not require special attention for the user interface, so ideally the user performs her/his "actual" task while at the same time the "computer work" runs automatically. The central source for the concrete requirements for an appropriate mobile assistance system is always the application area and the conditions of use (including the course of movement of the user). Their specialities determine the characteristics of the entire solution.

---

[1] For the complete scope of human activities and work processes see [2].

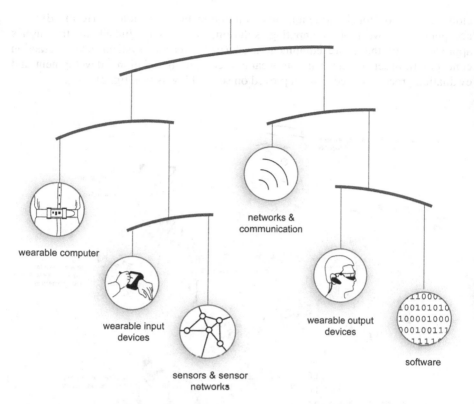

wearable computer

networks &
communication

wearable input
devices

wearable output
devices

software

sensors & sensor
networks

Fig. 1. Dependencies of the components of wearable computing solutions [2]

Most of the wearable computing solutions implemented so far have failed. The reasons are manifold, on an abstract level they conform to the points of criticism already raised by the usability research on desktop computing solutions (see i.e. [8], [9]), but there are also some other problems. There has already been research on usability [10] and wearability [11] but it also failed because of the human-computer interface. One basic mistake, which keeps reoccurring in the development of technologies, is the technology-centred perspective on the solution. Technologies are developed, which from the engineers' point of view are important and useful for the implementation of a mobile solution, without first identifying the intended users' requirements and without letting the prospective users participate in the development.

Generally, the technological approach to wearable computing solutions is not a problem. Actually, technological competence is an essential requirement for the development of complete systems. However, knowledge of the conditions within the application area and of the involved mobile workers is the crucial factor for the achievement of developing wearable solutions.

4 Design of Wearable Computing Solutions

A technology that gets this close to a person's body has to be tailor-made to optimally fit the wearer and their current task – which is always given in the case of tools or clothes, but not for information and communication technology (ICT). But the acceptance of wearable computing solutions is closely linked to the user's appreciation of the entire equipment used (incl. wearing systems and interaction concept). In order to design usable wearable computing solutions, development and evaluation processes need to be improved on several levels (see fig. 2).

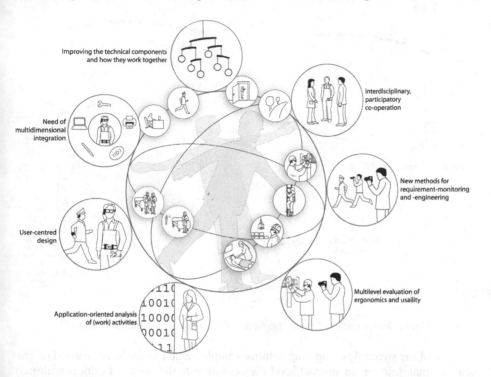

Fig. 2. Essential aspects for the development process of wearable computing solutions [2]

One very important aspect is the methodological approach at the HCI level: Until now, there are not any appropriate methods to record mobile data of mobile work processes for requirement monitoring and analysis. Additionally, most of the known usability testing methods are static, they have been developed for testing ICT used for work processes like desktop work and not for the evaluation of mobile worn technologies for mobile work processes with their dynamic changes of environmental conditions and the motion sequences of the user. To deal with mobility and wearability, additional dimensions have to be taken into consideration for requirement-monitoring and -engineering and also for usability testing. One approach to overcome these restrictions is to use recent wearable computing technology as a hands-on experience with the goal to improve the participatory design process. The basic idea is to use concrete artifacts in real work environments as "graspable arguments" for the dialogue between developers and users. The following example of a real world scenario will illustrate the method.

5 Potentials of Autonomous Control in Automobile Logistics

This paper focuses on the development process of a wearable computing solution for supporting the mobile work processes at an automobile terminal. An automobile terminal develops and provides complex services for new and used vehicles in the range of transport, handling, technical treatment and storage management (for the process see fig. 3).

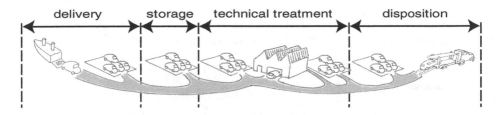

delivery   storage   technical treatment   disposition

Fig. 3. Business process at an automobile terminal

After delivery, each vehicle is identified by its vehicle identification number (VIN). The VIN allows an assignment of the vehicle to its storage and the technical treatment orders to be stored in the logistic IT-system. A handling employee moves the vehicle to the assigned storage location. After removal from stock, the vehicles possibly pass through several technical treatment stations e.g. fuel station or car wash. The sequence of the technical treatment stations is specified in the technical treatment order of the vehicle. On completion of all technical treatment tasks, the vehicle is brought to the disposition area for transportation to the automobile dealer [12].

The described storage management of vehicles at automobile terminals provides potential for improvement. The storage management is executed by a centralised planning and control system. In case of this, there is no flexible allocation of storage areas and locations considering future process steps or an immediate reaction to disturbances during order processing possible [13].

To realize the potentials for improvement a decentralized and heterarchical planning and controlling method was developed under the paradigm of autonomous control [4]. The idea is to support autonomous logistics with wearable computing technologies. This vehicle management in automobile logistic networks needs knowledge about the positions of all vehicles within the system at a given time. Only high transparency of all vehicle movements in this logistic network allows an efficient disposition of available resources at and between the automobile terminals. With wearable computing systems the needed information can be recorded directly during the working process so they can immediately be accordingly adapted. The future process steps will be planed depending on the actual storage locations and the technical treatment order of the vehicles and the stock capacity and the technical treatment stations. For these goals a wearable computing system and mobile technologies are being developed to support the terminal staff.

6 Requirement Analysis for a Wearable Computing Solution in Automobile Logistics

Objective is the development of a wearable computing system for the mobile work process at an automobile terminal under the paradigm of autonomously controlled logistics. An appropriate wearable computing system has to fulfil different system requirements. Therefore the first step was to define the requirements of wearable computing systems. The requirement analyses was divided into 6 steps (see fig.4).

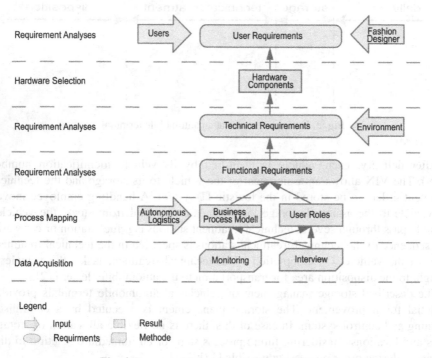

Fig. 4. Process of the analyses of the requirements

The information about the processes at an automobile terminal was gathered by interviews with the end-users and by monitoring the processes. On the basis of this information a business process model was built up, although the user roles were defined and adjusted with the business process model. Based on a decentralised controlled storage management of vehicles, functional requirements were derived from the planned processes for vehicle handling [5]. As mentioned above the vehicle identification by vehicle identification number (VIN) is necessary. The exact position of vehicles at the terminal area is needed in order to realize fast, flexible and orderly storage management. The wearable application has to support the handling employee by user guides or wizards to avoid misunderstandings. The defined functional requirements are based on the business processes and have to support the user who executes the logistic processes.

The technical requirements depend on the environment and on the functional requirements of the operational process. Technology requirements at an automobile terminal are determined by the environment and the necessarily used time. A power management is needed that reduces the energy demand for the operational time which depends on the business time in the operational process. Furthermore it is necessary to design the wearable computing solution to be robust and shock resistant to avoid damage in case of dropping. The system has to be so powerful that it is able to guarantee short response times for executing commands to reduce waiting times in the workflow. Based on these requirements and the functional requirements further technical demands on the hardware were defined. Assuming all these requirements the hardware components were selected. The technical requirements and in result the selected hardware directly affect the usability of wearable computing systems.

7 Implementation of the Wearable Computing System

The operating mode of the individual components was explained to the end-users in a collective "hands-on technology" workshop. This helped to define the users' requirements on the technical components. User requirements like the design of the graphical user interface (GUI), physical dimensions of the system, keyboard design and display quality have direct influence on the user's acceptance. They have been developed in a participatory process with the handling employees. Different wearable displays and GUIs were presented to the end-users. In this process, requirements like the GUI, the viewing distance, and an option for good readability in darkness were defined. The demand for hardware that represents no health hazard is guaranteed to users by the compliance with radiation standards [14], [15]. A fashion designer is developing with the users the design of the workwear for the wearable computing system.

The primary user in the described mobile work process is the handling employee, who moves the vehicles at the automobile terminal between storage locations, technical treatment stations and disposition areas. For this user in the project the use of the "smart jacket" has been designated. The smart jacket is a wearable computing system, which integrates technical components for identification, localisation, communication and user interaction tasks into workwear. The smart jacket includes a Ultra-High Frequency (UHF) RFID module for tag identification, a combined GPRS (General Packet Radio Service) / GPS (Global Positioning System) module for communication and localisation, a Thin-film Transistor (TFT) display with touch screen, hand free audio hardware for user-interaction and a proximity sensor for detection. All components are handled by a central software platform integrated in an embedded microcontroller. The smart jacket enables the basis for decentralised decision-making in autonomously controlled logistic systems. For the other roles on the terminal, the use of mobile technologies like a mobile data entry device (MDE) and a tablet PC is planned.

8 Usability Testing of the Smart Jacket

The first evaluation step of this wearable computing solution will be to check the operational reliability and usability of the smart jacket prototype in a laboratory environment. After implementing these results into a second prototype, the evaluation takes place in a field test in cooperation with the handling employees. In doing this, the general wearability of the system is tested by the end-users, as well as the applicability of the system in the intended operational environment. Based on mobile collected information the wearable computing system is going to be improved until it reaches the user's acceptance.

In the recurrent dialogue between users and designers, the technological artifacts serve as a communication tool as well as instances of the intended wearable computing solution. In the different evaluation phases the roles of the members of involved groups change: sometimes the user and sometimes the technology developer is the expert.

9 Summary and Next Steps

In this contribution, the specific needs of mobile work processes have been compared with the characteristics of wearable computing solutions on a general level. It has argued for changes in the approaches of requirement engineering and usability testing with regard to mobility and wearability. The proposed changes were shown by way of an example of developing a mobile supporting system for handling employees at an automobile terminal under the paradigm of autonomous control. The development of a smart jacket is still in progress and will be evaluated by means of the concept of "using technology as a hands-on experience" (see [2, p.290]).

Acknowledgements

This research was supported by the German Research Foundation (DFG) as part of the Collaborative Research Centre 637 "Autonomous Cooperating Logistic Processes – A Paradigm Shift and its Limitations" at the University of Bremen.

References

1. European Integrated Project wearIT@work – Empowering the Mobile Worker with Wearable Computing, http://www.wearitatwork.com
2. Rügge, I.: Mobile Solutions – Einsatzpotenziale, Nutzungsprobleme und Lösungsansätze. DUV/Teubner Research, Wiesbaden (2007)
3. Herzog, O., Boronowsky, M., Rügge, I., Glotzbach, U., Lawo, M.: The Future of Mobile Computing: R&D Activities in the State of Bremen. Internet Research 17(5) (2007); Special issue: TERENA conference 2007, pp. 495–504 (2007)
4. Böse, F., Windt, K.: Autonomously Controlled Storage Allocation. In: Hülsmann, M., Windt, K. (eds.) Understanding Autonomous Cooperation and Control in Logistics - The Impact on Management, Information and Communication and Material Flow, pp. 351–363. Springer, Heidelberg (2007)
5. Scholz-Reiter, B., Windt, K., Kolditz, J., Böse, F., Hildebrandt, T., Philipp, T., Höhns, H.: New Concepts of Modelling and Evaluating Autonomous Logistic Processes. In: Chryssolouris, G., Mourtzis, D. (eds.) Manufacturing, Modelling, Management and Control. IFAC Workshop Series. Elsevier Science, Amsterdam (2006)

6. Scholz-Reiter, B., Windt, K., Freitag, M.: Autonomous Logistic Processes – New De-mands and First Approaches. In: Proceeding of 37th CIRP International Seminar on Manu-facturing Systems, Budapest, pp. 357–362 (2004)
7. Rhodes, B.J.: The Wearable Remembrance Agent: A System for Augmented Memory. In: 1st International Symposium on Wearable Computers, Cambridge, pp. 123–128 (1997)
8. Beyer, H., Holtzblatt, K.: Contextual design: defining customer-centered systems. Morgan Kaufmann Publishers Inc., San Francisco (1998)
9. Nielsen, J., Mack, R.L. (eds.): Usability inspection methods. John Wiley & Sons, Inc., Chichester (1994)
10. Baber, C.: Wearable Computers: A Human Factors Review. International Journal of Hu-man-Computer Interaction 13(2), 123–145 (2001)
11. Gemperle, F., Kasabach, C., Stivoric, J., Bauer, M., Martin, R.: Design for Wearability. In: The Second International Symposium on Wearable Computers, pp. 116–122. IEEE Com-puter Society, Los Alamitos (1998)
12. Böse, F., Lampe, W., Scholz-Reiter, B.: Netzwerk für Millionen Räder. FasTEr - Eine Transponderlösung macht mobil. RFID im Blick, Sonderausgabe RFID in Bremen, pp. 20–23 (2006)
13. Böse, F., Piotrowski, J., Windt, K.: Selbststeuerung in der Automobil-Logistik. Industrie-management 20, 37–40 (2005)
14. DIN 61000-3-2, Elektromagnetische Verträglichkeit (EMV) Teil 3-2: Grenzwerte - Grenzwerte für Oberschwingungsströme (Geräte-Eingangsstrom =< 16 A je Leiter)(IEC 61000-3-2:2005). VDE Verlag, Berlin (2006)
15. DIN 61000-3-3, Elektromagnetische Verträglichkeit (EMV) Teil 3-3: Grenzwerte - Be-grenzung von Spannungsänderungen, Spannungsschwankungen und Flicker in öf-fentlichen Niederspannungs-Versorgungsnetzen für Geräte mit einem Bemessungsstrom 16 A je Leiter, die keiner Sonderanschlussbedingung unterliegen. VDE Verlag, Berlin (2006)

Combining Activity Theory and Grounded Theory for the Design of Collaborative Interfaces

Christine Rivers[1], Janko Calic[1], and Amy Tan[2]

[1] I-Lab Multimedia and DSP Research Group,
Centre of Communications Systems and Research
University of Surrey, Guildford, GU2 7XH, United Kingdom
[2] School of Architecture, Design and the Built Environment
Nottingham Trent University, NG 1 4BU, United Kingdom
{c.rivers,j.calic}@surrey.ac.uk, amy.tan@ntu.ac.uk

Abstract. In remote tabletop collaboration multiple users interact with the system and with each other. Thus, two levels of interaction human-computer interaction and human-human interaction exist in parallel. In order to improve remote tabletop systems for multiple users both levels have to be taken into account. This requires an in-depth analysis achieved by qualitative methods. This paper illustrates how a combination of Activity Theory and Grounded Theory can help researchers and designers to improve and develop better collaborative interfaces. Findings reported here are based on three video recordings that have been collected during a quasi-experiment.

Keywords: Activity theory, Grounded Theory, remote tabletop collaboration, methodological-design approach.

1 Introduction

The increasing use of collaborative technologies, in particular remote tabletop groupware, has provided a fertile ground for user interface design research. Most studies of user interface design these technologies have so far to date focused only on human-computer interaction (HCI) [1,2,3], taking only one level into account. However, in remote collaboration multiple users can interact with the system as well as with each other, understood as human-human interaction (HHI). In order to understand, improve and develop existing and new interfaces suitable for multiple users in remote collaboration, both levels should be considered.

This is a complex process and requires in-depth analysis, which can be time-consuming and expensive, as more than one researcher will be required to conduct the analysis. Mixing different research methods might be a viable solution to overcome these problems of complexity and can be beneficial rather than contradictory [4]. We believe that the combination of two qualitative approaches, activity theory [5,6,7] and Grounded Theory [8,9,10,11,12] is suitable to address these issues. An approach or the combination of approaches that enables one researcher to focus on multiple users and their collaborative behaviour is needed. Grounded Theory is commonly used to examine

M. Kurosu (Ed.): Human Centered Design, HCII 2009, LNCS 5619, pp. 312–321, 2009.

group behaviour and drawing from our previous experiences, Grounded Theory will be the primary method in our study. In our opinion, no other existing qualitative approaches have been found to be applicable to easily uncover collaborative aspects of an interface and amongst users, except Activity Theory [7]. Furthermore, the three levels that constitute Activity Theory, activities, actions and operations [7], have frequently occurred during analysis of the data collected from the study conducted in 2007 [13]. Therefore, it seems worthwhile to test the combination of Grounded Theory and Activity Theory to investigate multiple user remote tabletop collaboration.

The paper is structured as follows: The theoretical background introduces the two qualitative approaches, Activity Theory and Grounded Theory and explains the key elements of each. After having outlined the theoretical concepts of the approaches, the analytical process will be explained step-by-step to illustrate how the combination yields useful results. This is followed by a discussion of the findings reflecting the insights gained by combining the approaches. Suggestions regarding the improvement to the user interface of a collaborative system will be given. Finally, we conclude, summarize and discuss future research.

2   Theoretical Background

Mixing research methods in HCI to gain a deeper understanding regarding remote collaboration and the way technology is used, has been found to yield more in-depth and reliable results [14]. Using a single approach to study a complex phenomenon or interaction tends to only focus on a certain aspect with the danger that other elements not given the same level of focus but nonetheless equally important may easily be missed out [15]. In this section we will discuss the ideal use of activity theory and Grounded Theory and how they have been combined to study remote tabletop collaboration of mixed presence groups[1].

2.1   Activity Theory

Originally, Activity Theory is a cultural-historical theory founded by a group of revolutionary Russian psychologists Vygotsky, Rubenshtein, Leont'ev and Luria [7] in the beginning of the 20th century. It is a theoretical framework rather than an analytical technique. Its interest is the explanation of social and cultural work practices by relating them to the cultural and historic context in which the activity is being performed [16]. These two aspects are vaguely explained within the framework. The historical context could be understood as the temporal and developmental process of interaction. The collaborative context might refer to the actual interaction, taking place between the user and the system, as well as between and among users. Activity Theory gained popularity as an approach in HCI in the nineties [5, 16, 17]. Designers had difficulties identifying the nature of users problems with the system and this approach helped them to focus on the end-users' activities, actions and operations [5, 7, 16, 17]. Consequently, designers and researchers were able to study, deduce and interpret

---

[1] The term *mixed presence group* refers to mixed presence groupware, which indicates that co-located and remote users working over a shared visual workspace in real time [23].

concepts of users' needs. The user should be considered within a historical and collaborative context.

Activities, actions and operations are the three levels of activity theory [18]. These levels imply four basic principles [6]. Firstly, a hierarchical structure: activities include actions and actions require operations. Activities have motives and can be understood as series of actions. Actions are goal-oriented and part of activities. The goal is to finish the activity. Thirdly, operations are executing actions. When users start to work with a system they have to get used to the system. So, actions are carried out consciously, over time these become unconscious actions, so called operations.

The second basic principle refers to object-orientedness. Living in an object-oriented world indicates that we interact with objects and these have certain natural properties as well as cultural and social ones. Assuming that a remote tabletop also has such properties gives reason to conclude that social interaction in either co-located or remote collaboration might be influenced by these properties. The third basic principle is internalization and externalization of activities. However, it is difficult to determine and distinguish between internal and external activities, because activities constantly transform from internal to external [7]. The last basic principle [6] refers to the activity as mediated by tools and these tools are transformed during the development of the activity. Activity theory is based on these principles and understood as a conceptual framework that helps to identify actions and interaction with artifacts within a historic, cultural and social context. Bødker [19] points out that the interface can only be understood through its use in a real context and users should be analysed within the context of development. Further, the interface only becomes visible and evident when problems with the interface occur. We claim that the interface also becomes evident and visible when social interaction takes place among users interacting with the system and each other, as activities are part of social interaction.

2.2 Grounded Theory

The use of Grounded Theory in HCI is not uncommon, however, its use and validity has been criticized [20, 12]. Traditional research relies on literature, theoretical background and formulated hypothesis [21]. We believe that preconceived hypothesis and/or theories might hinder us from finding key concepts and hidden structures of interaction which may inadvertently be revealed if approached without preconceptions. Grounded Theory is an inductive approach that investigates cases as a whole and a theory emerges from the case data by analyzing it thoroughly using three coding steps as considered below. Variables, as referred to in quantitative studies are classified and labeled as codes, categories and concepts in Grounded Theory. Exploring and understanding the interrelationship between categories and concepts (theoretical sensitivity) is one of the main analytical processes of this approach [8, 10, 22].

Analysing data can be complex and time-consuming. Using Grounded Theory means that data has to be reduced and fragmented by means of three coding processes: open coding, axial coding and selective coding. The researcher starts to reduce and fragment data by identifying, naming, describing and categorizing the data. Having established codes and categories helps the researcher to relate codes, and categories. This process is called axial coding and is a combination of inductive and deductive thinking [9, 10, 21, 22]. From the reduced data a central concept will emerge. Choosing, finding or identifying this one core concept and relating all the other categories to this specific one, is understood as selective coding [22]. These insights emerge from analyzing case data which can be found in memos, or short documents by the researcher, his/her field notes or code notes (theoretical notes) made additionally to the coding processes.

In the next section we will describe how we combined and applied these two approaches and gained useful results.

3 Methodology

3.1 Data Source and Collection

The data analyzed in this study was collected from three video recordings of three mixed presence groups. In June 2008 we conducted a formative experiment of mixed presence groups performing a collaborative writing task over a remote collaboration tabletop. A mixed presence group existed of four participants, with two co-located users at each location A and B (see Figure 1).

Fig. 1. Snapshot from Video 2 (Group 2): Remote tabletop collaboration of mixed presence groups

The co-located group A had to collaborate with the other co-located group B over the shared visual workspace using an audio/video link. A digital pen and a keyboard for direct interaction have been provided at each desk location. Participants of this study were researcher from the University of Surrey (UniS) and engineers from Thales Research & Technology UK (TRT UK).

Before the two groups met virtually, each co-located group had to write a story co-located using the same tabletop groupware. The audio/video link was turned off and the same interaction tools had been provided. Both co-located groups had to write a story based on the same pictures within 30 minutes.

The task of the remote session was to merge these two stories together and produce one new story within the same time frame. The pictures provided in the co-located sessions have been displayed, as well as the two documents with the stories. In order to write the story using the keyboard and or digital pen the users had to share a word processing application. The commands cut and paste were not allowed.

All sessions have been recorded and field notes have been taken during the experiment referring to the social interaction of the users and their interaction with the system. We were interested in how the remote tabletop system hindered or facilitated collaboration of mixed presence groupware in order to deduce design criteria. Our analysis is based on the three video recordings and the field notes that formed the main case data.

3.2 Combining Activity Theory and Grounded Theory

Initial observations of the data showed that the collaborative writing task prompted users to structure the task. All three mixed presence groups proceeded in the same way; starting with planning, moving on to brainstorming and then producing. These steps have been identified as activities. Altogether, 46 video sequences have been extracted from the main data, which referred to these activities.

As stated earlier in this chapter, activity theory is based on three levels: activities, actions and operations. Approaching the data in an open manner, which is one of the fundamental ideas of Grounded Theory allowed for recognizing the emergence of activities instead of forcing activities to exist. Emergence rather than constrain is a very important feature of Grounded Theory. Having identified activities at the highest level, the next step is to investigate each activity separately and in-depth as illustrated here. During the activity *Planning* users carried out verbal and nonverbal individual actions and verbal and nonverbal collaborative actions as previously found in the dyadic remote groups analysed by means of Grounded Theory and qualitative content analysis [13]. The actions found during this activity have been categorized as seeking, viewing and preselecting. The goal of each action was to finish the activity of *Planning*. Executing these actions to achieve the goal has been defined as operations, which revealed the codes; see Table 1.

The activity *brainstorming* included two actions reviewing and selecting. In order to accomplish these, different operations have been carried out, as presented in Table 1. The same actions have been found to be applicable for the last activity *producing*. The processes of open and axial coding showed that additionally to operations, which are executing actions, cognitive processes take place. So, carrying out operations requires users' cognitive abilities. Based on these insights and memos used during the process of selective coding a central concept has been deduced: the existence of cognitive group abilities. Group cognition is understood as a process and is an outcome of group interaction. As groups are informative-processing units [26], group cognition can be described as the transmission of group-relevant knowledge and understood as a shared mental model. Unfolding group cognition can help the researcher to understand multiple users' interaction in a remote tabletop environment better, hence improving existing systems as a response to the users cognitive needs.

In this study, cognitive group abilities in remote tabletop collaboration has been found to be important in all of the video recordings analysed. Furthermore, we were able to establish the links between cognitive group abilities and pertinent actions and activities. Five cognitive group abilities have been identified as fundamental during remote tabletop collaboration: acquiring knowledge, categorizing information, associating information, creating meaning of information and drawing conclusions of information.

Table 1. Activities, Actions and Operations emerged from *Group Interaction Tables*

Activities	Actions (Categories)	Operations (Codes)
Planning	Seeking Viewing Preselecting	Opening documents from file list, using digital pen Moving documents using digital pen Arranging documents in high resolution area to read them both at the same time, using digital pen Reading documents Sharing document with word application Asking if other group has read document Asking if they had the same information to write the story Assigning roles: negotiating writer
Brainstorming	Reviewing Selecting	Suggesting ideas Scribbling notes with pen on notepad Repeating content Opening documents Pointing at information in documents with hand Pointing at information in document using the pen Using hand to express ideas
Producing	Reviewing Reselecting Editing	Pointing at information in document using whole hands Discussion about keeping content and changing content by pointing at information in documents Searching document for information using finger or pen Using index finger to point at information in document Partner uses cursor of pen to translate action Taking pen from co-located partner to edit document Giving pen to co-located partner to edit document Writer using keyboard to edit document Co-located partner takes keyboard to edit document Underlining part of information to show changes or important information Moving documents to find information Moving documents to be involved (using pen) Giving writing control function to other team

Developing a theory. Further examination of the data and in particular, focusing on the core concept revealed that cognitive group abilities were not always supported by the remote collaboration system, thus affecting the collaborative behaviour of both co-located and remote groups. Affects to co-located and remote collaborative behaviour has been found to occur either as an act of including the other remote group or excluding the other remote group. We also wanted to understand to what extent the collaborative behaviour of mixed presence groups has been influenced if the system does not support cognitive group abilities.

Four types of collaborative behaviour have been determined within our theoretical scheme: excluding co-located collaborative behaviour, excluding remote collaborative behaviour, including co-located collaborative behaviour, including remote collaborative behaviour. At this point it seemed to be interesting to know how often each type occurred during remote tabletop collaboration. For this purpose the findings qualitative results have been quantified by means of analysis of frequency, see Table 2.

Table 2. Analysis of Frequency of four types of collaborative behaviour

Collaborative Behaviour	Abbrev.	Type	Frequency
Including remote collaborative behaviour	Ic re	1	22
~~Including co-located collaborative behaviour~~	~~Ic co~~	~~2~~	~~1~~
Excluding remote collaborative behaviour	Ex re	3	28
~~Excluding co-located collaborative behaviour~~	~~Ex co~~	~~4~~	~~1~~

The two types, including and excluding co-located collaborative behaviour did not occur frequently during remote collaboration and have therefore been not included in further analysis. Interestingly, type 3 occurred more often than type 1, which indicated that the remote group has been excluded more often than included.

Focusing on the two most frequently occurring types showed that these two types took place one after another. We posit the theory that the collaborative behaviour of mixed presence groups in remote tabletop collaboration continuously transits from excluding remote collaborative behaviour to including remote collaborative behaviour and from including remote collaborative behaviour to excluding remote collaborative behaviour depending on the cognitive support a system has been designed to assist. In order to validate this theory comparative analysis has been applied for all three cases (video recordings) focusing on the structure of transition, which revealed that all video recordings reflected the same phenomenon. Moreover, post ante personal interaction showed that some of the participants constantly felt excluded and included at the same time during remote collaboration. These answers confirm the developed theory regarding the existence of excluding and including collaborative behaviour and also show that combining Grounded Theory and Activity Theory can help to understand the both levels of interaction human-human interaction and human-computer interaction. Furthermore, it helps to unfold different levels of the collaborative process: activities, actions, operations and the requirement of cognitive abilities.

4 Discussion of Findings

Our findings showed that users' cognitive abilities needed to be supported during remote tabletop collaboration. This can be achieved by incorporating an information management tool. Such an information management tool should enhance individual users' during the activities of *brainstorming* and *producing* as well as supporting the actions needed to accomplish the activities, individually and collaboratively at the same time. More importantly, an information management tool should support the cognitive group abilities of acquiring knowledge, categorizing information, associating information, creating meaning of information and drawing conclusions of information. If these cognitive group abilities are not supported by the system the co-located group A excludes the remote group B even if the intention of the co-located group A is to include the other remote group B. A theoretical framework of such an intelligent information management tool has been introduced by [13], called the *InfoManager*. Although, this idea of an intelligent information management tool has been mooted based on Activity Theory, the concept of object-orientedness has to be taken into account. Other findings of this study showed that users used different tools during certain activities. During the activity *planning* users interacted with the interface using the digital pen and the video link, whereas during *brainstorming* only the video link seemed to be important. The provided keyboard has mainly been used in conjunction with the digital pen during the activity *production*. This implies that a collaborative information management tool should allow users to use multiple and different kinds of direct and indirect interaction devices during remote tabletop collaboration. Moreover, this indicates that a remote tabletop groupware that facilitates multiple users input is highly eligible.

5 Conclusions

The presented study revealed that combining two qualitative approaches, Activity Theory and Grounded Theory, yield interesting results regarding the improvement and development of collaborative user interface design. We introduced the analytical thinking processes involved by the individual and combined approaches, presented the findings that emerged as result of the new combined method of analysis, and gave suggestions how to use findings, and lastly, how to improve the interface. The main finding was that mixed presence groups in remote tabletop collaboration transit between including and excluding remote collaborative behaviour due to the fact that their cognitive group abilities are not supported by the system. We suggest implementing an information management tool that facilitates cognitive group abilities of mixed presence groups in remote tabletop collaboration and explained how cognitive group abilities should be supported during remote tabletop collaboration.

Although, combining these two approaches has proven to be useful, further research is required to test the viability of this method specifically, using a larger sample of mixed presence groups. Additionally, implementing the proposed information management tool in an existing remote tabletop groupware would validate the outcomes and further contribute to collaborative interface design research for remote tabletop groupware.

Acknowledgement. I would like to thank Thales Research & Technology UK for providing the remote tabletops and technical support during the experiment as well as all participants of TRT UK and UniS for their time and commitment. The work presented was developed within VISNET II, a European Network of Excellence (http://www.visnet-noe.org), funded under the European Commission IST FP6 programme.

References

1. Bailey, R.W.: Human Performance Engineering: Using Human Factors/Ergonomics to achieve Computer System Usability. Prentice-Hall, Englewood Cliffs (1989)
2. Paay, J.: Form Ethnography to Interface Design. In: Lumsden, J. (ed.) User Interface Design and Evaluation for mobile Technology, vol. 1, IGI Global (2008)
3. Tarmizi, H., Payne, M., Noteboom, C., Zhang, C., Steinhauser, L., Vreede, G.-J., Zigurs, I.: Technical and Environmental Challenges of Collaboration Engineering in Distributed Environments. In: Dimitriadis, Y.A., Zigurs, I., Gómez-Sánchez, E. (eds.) CRIWG 2006. LNCS, vol. 4154, pp. 38–53. Springer, Heidelberg (2006)
4. Monk, A., Nardi, B., Gilbert, N., Mantei, M., McCarthy, J.: Mixing Oil and Water? Ethnography versus Experimental Psychology in the Study of Computer-Mediated Communication. In: Interchi 1993, pp. 3–6 (1993)
5. Nardi, B.A.: Context and Consciousness, Activity Theory and Human-Computer Interaction. MIT Press, London (1997)
6. Keptelinin, V., Nardi, B.A.: Activity Theory: Basic Concepts and Applications. In: CHI 1997 (1997),
   http://www.sigchi.org/chi97/proceedings/tutorial/bn.htm
7. Rajkumar, S.: Activity Theory (2000),
   http://mcs.open.ac.uk/yr258/act_theory
8. Alsop, G., Tompsett, C.: Grounded Theory as an Approach to Studying Student's Uses of Learning Management Systems. ALT-J 10(2), 63–76 (2002)
9. Charmaz, K.: Constructing Grounded Theory. SAGE, London (2006)
10. Glaser, B., Strauss, A.L.: The discovery of Grounded Theory. Aldine, Chicago (1967)
11. Linden, T., Cybulski, J.L.: Application of Grounded Theory to Exploring Multimedia Design Practices. In: 7th Pacific Asia Conference on Information Systems, Adelaide, South Australia, July 2003, pp. 508–522 (2003)
12. Sarker, S., Lau, F., Sahay, S.: Using an Adapted Grounded Theory Approach for Inductive Theory Build About Virtual Team Development. The DATA BASE for Advances in Information Systems 32(1), 38–56
13. Glaser, C., Tan, A., Kondoz, A.: An intelligent information management tool for complex distributed human collaboration. In: Niiranen, S., Yli-HIetanaen, L. (eds.) Open Information Managemnt: Applications of Interconnectivity and Collaboration, ch. 5, pp. 113–143 (2009)
14. Ormerod, R.J.: Mixing Methods in Practice: A Transformation-Competence Persepective. Journal of the Operational Research Society 59(1), 137–138 (2008)
15. Glaser, C., Tan, A., Kondoz, A.: Talk-in interaction of mixed-gender and same-gender Virtual Teams reflects usability of virtual collaboration systems. In: Proceeding of the third IASTED International Conference Human Computer Interaction, pp. 291–296. Innsbruck (2008)
16. Bertelsen, O.W., Bødker, S.: Information Technology in Human Activity. Scandinavian Journal of Information Systems 12, 3–14 (2000)
17. Engeström, Y.: Developmental studies of work as a testbench of activity theory. In: Chaiklin, S., Lave, J. (eds.) Understanding Practice: perspectives on activity and context, pp. 64–103. Cambridge University Press, Cambridge (1993)
18. Kuutti, K.: Activity Theory as a potential framework for human-computer interaction Research. In: Nardi, B. (ed.) Context and Consciousness, ch. 2, pp. 17–44 (1996)
19. Bødker, S.: Through the Interface: A Human Activity Approach to User Interface Design. Lawrence Erlbaum, Hillsdale (1991)

20. Qureshi, S., Liu, M., Vodel, D.: A Grounded Theory Analysis of E-Collaboration Aspects of Distributed Project Management. In: Proceedings of the 38th Hawaii International Conference on System Sciences, pp. 1–10 (2005)
21. Allan, G. (2003)
22. Strauss, A., Corbin, J.: Basics of Qualitative Research: Techniques and Procedures for Developing Grounded Theory. Sage, Thousand Oaks (1990)
23. Tang, A., Boyle, M., Greenberg, S.: Display and Presence Disparity in Mixed Presence Groups. In: Proceedings of the 5th Australasian User Interface Conference, Dunedin, NZ, vol. 28, pp. 73–82 (2004)
24. Flower, L.S., Hayes, J.R.: Problem-Solving Strategies and the Writing Process, College English. Stimulation Invention in Composition Courses 39(4), 449–461 (1977)
25. Posner, I.R., Baecker, R.M.: How People Write Together. In: Proceedings of 25th Haiwaii International Conference on Computer System Sciences vol. IV, pp. 127–138 (1992)
26. Fiore St, S.E.: Team Cognition: Understanding the Factors that Drive Processes and Performance. APA, New York (2004)

User Behavior Patterns: Gathering, Analysis, Simulation and Prediction

Lucas Stephane

International Institute of Information Technology (SUPINFO)
52 rue de Bassano, F-75008 Paris, France
lucas.stephane@supinfo.com

Abstract. This paper presents methods and tools for gathering, analyzing and predicting behavior patterns. Considered both for a single user and for groups of users, behavior patterns may impact at a local and/or global level. The first part explains how to gather behaviors in various situations and how to drill from overt behaviors into deeper cognitive processes. The rationale of cognitive modeling and guidelines to perform it are provided. The second part deals with analysis methods that enable to detect behavior patterns. Bottom-up analysis based on existing data is augmented with top-down analysis based on conceptual design choices and hypotheses. The last part emphasizes the needs of data storage and data sharing in the organization. Beyond data storage and sharing, it presents the benefits of using Adaptive Business Intelligence in order to simulate and predict possible situations as well as the appropriated behavior patterns that enable to adapt.

Keywords: Behavior patterns, modeling, Ontology systems, Multi-Agent Systems, Agent Oriented Programming, Adaptive Business Intelligence.

1 Introduction

In Human-Machine Interaction (*HMI*), the concept of *interface patterns* was massively implemented and popularized by end-user consumer devices such as Apple iPhone™ and by gaming consoles such as Sony Play Station EyeToy™ or Nintendo Wii™. The transition from classic control devices equipped with buttons, knobs and joysticks to natural-interaction based essentially on gestures is revolutionary. The evolution of technology from touch-screens to various sensor equipped interfaces make possible for the user body to become an interaction mean [30].

This induces changes in the Human Factors (*HF*) field, because new user needs have to be taken into account. The main part to consider in HF is not necessarily the novelty, but the way of dealing with criteria and functionalities as well as with contexts of use that were specific in the past. For example, not so long ago, sets of criteria and functionalities were clearly defined for mobile phones. A mobile phone was intended to communicate. Thus the screen size and the phone functionalities were homogeneous. But the introduction of new functionalities such as Internet, games and music, and also Global Positioning Systems (*GPS*) on mobile phones made mandatory to consider

M. Kurosu (Ed.): Human Centered Design, HCII 2009, LNCS 5619, pp. 322–331, 2009.

different criteria and functionalities and to combine them in new ways in order to provide the best design for new devices.

Beyond the infatuation for gesture recognition, one has to consider all the HMI interaction modalities and how to get the best benefits from studying and using them. Humans are endowed with complementary senses. In their natural environment, they have not lived based mainly on one of them. So, from a HMI perspective, taking into account these different modalities makes possible to link overt observable user behaviors to user high-level processes (what the user thinks and feels) that are covert. The complexity of users' processes depends on the complexity of the task. In the mobile phone example, the tasks that are performed may seem quite simple, even for last generation sophisticated devices, compared to the tasks users have to perform in safety critical domains. Thus, even if the HF methods and tools attempt to be generic, their use and importance differ from one domain to another. The HF and HMI challenge is to continuously adapt the methods and tools in various domains to the technological infrastructure that is transversal to these domains.

Another mandatory aspect that has to be considered is the last decades' acceleration of product life cycle. The design, implementation and commercialization of products in local and global markets arc getting shorter and shorter. This means that to succeed in new products, the whole process of product development as well as the whole Information Technology (IT) system have to be improved. In the actual global Knowledge Economy (KE), cross-domain and cross-country information flows are one of the key factors that contributes to growth and development and offers a solution for the Creative Destruction that was studied more than half of a century ago [33]. In order to ensure these information flows, it is not sufficient anymore to integrate data coming from various sources, but to perform data unification.

Business Intelligence (BI) may have lots of definitions. As it is considered in this paper, it is related to Knowledge Management (KM) and to HF. It provides a clear approach through industry proven solutions in order to perform unification of data coming from various cross-organizational data sources on one hand and on the other hand it enables to transform data in valuable cross-domain information, by providing analysis and decision making support. BI is a combination of methods and technology that provides key information to support decisions and actions in order to improve an agency's mission [3]. From a Human Factors intrinsic commitment augmented with a Human Factors vision of economics, the agency mission should focus individuals' (i.e. final users, operational people, customers) maximized utility and well-being [2], [18] rather than only on agency's maximized profits. Non-financial performance measures based on organizational strategy [17] are supported by the proven importance of innovations in the Knowledge Economy [29] as well as by the value of the Human Capital [21] that both contribute to endogenous and sustainable growth. Emerging as theory, these concepts were implemented as methods (i.e. the Baldrige National Quality Program [1], the European Foundation for Quality Management (EFQM) [13], the Knowledge Assessment Methodology [8]) that are used by worldwide organizations. Thus, the main role of BI as considered here is to improve Research and Development (R&D) local and global units [28] and the innovation processes that are essential in the organizational governance.

Simply stated, the link between user behavior patterns, BI and simulation is to create data unification, to obtain and store human behavior patterns, and finally to inject them in a Human Behavior Simulator (*HBS*). The particularity of the HBS is that it is mainly based on real patterns. The next step is to couple the HBS with technical simulators (*TS*). Technical simulators range from simulators that offer complete HMI (i.e. flight and drive simulators) to calculation simulators, based on formulae employed in mathematical and statistical models, that offer only calculation results (i.e. specific consumer behavior according to a specific economics model). The use of such HBS in research studies is wide and encompasses both studies involving few users (i.e. HF studies in aircraft cockpits with two pilots) and studies involving large samples of users (i.e. behavioral economics studies with hundreds of users, Customer Relationship Management, etc.). The final interest of using HBS based on real patterns is to participate in assessing and validating models and thus to enable predictions, optimization and co-adaptability of socio-technical systems. Prediction, optimization and adaptability are the key components of Adaptive Business Intelligence [23].

2 User Behavior Patterns

2.1 Cognitive, Emotional

The main part of the studies carried out in Human Centered Design (*HCD*) focus the cognitive aspects of users. The legitimacy of taking into account the cognitive side of users was reinforced since HF penetrated industry. In the industrial context where HF specialists had to work together with engineers, the rational aspects of users had to be dealt with in the first place. However, the cognitive side does not cover entirely humans, neither in description nor in understanding. In order to complete the missing parts of the puzzle, the emotional aspects of users started to be re- investigated, especially since neurobiology demonstrated the emotion circuits in the human brain [5], [11]. On the other hand, the market moved massively to end-user consumer products designed for entertainment. So the transition from industrial products to entertainment products was beneficial to stimulate research that focuses user emotions. Originating in Japan, Kansei Engineering is also supported in northern European countries [34].

So user behavior patterns are considered both on their cognitive (rational) as well as on their emotional sides. A simple rational behavior pattern could be: the phone rings, the user grasps the phone, presses the green button and says 'hello'. A simple emotional pattern could be: the phone rings, the user enjoys the phone design and tone (and wants to buy the same if s/he does not own it). In the first case the HMI may be addressed from a usability perspective; in the second case it may be addressed from a marketing one. It is obvious that depending on the domain, the importance of emotions and the rationale of considering them in studies may be very different (i.e. pilot fear and panic in the cockpit due to engine failure and consumer feelings related to entertainment). But in all cases, HF are involved.

2.2 Setting Up Surveys and Experiments

2.2.1 Who Are the Users?
Depending on the goals of the design-evaluation process, HF have to set-up appropriated surveys and experiments. One of the determinant factors in the study is 'who are

the users?' The answer to this question might be quite complicated for the same culture, and it gets even more complicated when users come from different cultures, as it happens currently nowadays. There are cross-cultural dissimilarities in behavior such as language, decision making styles and conflict management styles. Decision making is a transparent process, but its form is grounded in a culture's value, standards of behavior, and patterns of thinking [16], [20], [25]. So lots of intra and cross-cultural differences should be considered, especially in the actual globalization.

2.2.2 Modalities Taxonomy (Fractioned vs. Continuous)

The first category of user behaviors consists of directly observable user actions (i.e. the user presses a button) and user communications (i.e. with other users or with the machine via speech recognition). A subtler category consists of user behaviors that are not directly observable, such as eye-movements and physiologic reactivity. These last behaviors have to be measured by appropriated devices.

A main point to be emphasized is the difference between modalities (gestures, speech, eye-movements and physiologic events) in terms of distinction between fragmented and continuous series of events.

Gestures and speech are fragmented, i.e. they happen for determined periods in time, they stop and then they restart. Behavior fragmentation impacts directly the observation methods and the analysis process. Behavior events that are relevant to HMI or to interaction in general are clearly defined and separated, as well as their relevance in the context of the task to be performed or in the context of the given usability scenario.

For behaviors such as eye-movements and physiologic reactivity, behavior events occur continuously, and in parallel with other user actions. The visual modality may be considered as the supervisor of other modalities in the highly visual environments to which users are exposed. In general all the eye-movements recorded during an experiment are not directly relevant. Compared to the fragmented modalities, further work has to be accomplished in order to identify the relevant ones.

It is valuable to consider not only an observation by modality (i.e. recording the user gestures alone) but to combine the observation of several modalities in order to understand the information flows and the action flows. For example, the user gathers information from the environment. What will s/he do next? If actions are possible, then they should be observed. An event occurs, the user presses a button and then looks at the changes on the display: information gathering is a trigger for action, and the actions performed become triggers for information gathering.

The combination of several observation methods enables to determine cycles of information gathering and actions, as triggers for each other.

2.2.3 Objective and Subjective Assessment Methods

Objective Assessment methods employ various types of devices. The intrusiveness of the equipment that is used, as well as the impact it might have on users have to be considered when designing experiments. Furthermore, there are experiments that do not allow at all specific types of equipment.

A gesture means basically to move a part of the body. Observation tools such as cameras are sufficient to assess gestures. The same equipment and some more specific light conditions are necessary for the recording of facial expressions. Facial expressions could be considered as a particular category of gestures. Facial Action Coding System (*FACS*) [9] provides a complete reference to observe and encode facial expressions. Automated encoding software is also in progress [10]. Eye trackers enable to gather the gaze position on the visual scene. Eye-trackers devices are proposed in two configurations: head-mounted or remote. In order to improve the gaze accuracy, they can be completed with Magnetic Head Trackers [12].

Physiologic reactivity employs various devices of measurement such as skin-conductance patches, ECG, EEG, breath frequency, etc.. However, because of their intrusiveness, such devices are hardly usable in common experiments.

Subjective Assessment Methods employ self reporting, questionnaires and rating scales that are designed for specific fields of investigation, such as performance, workload, situation awareness, user satisfaction, human error, team assessment. Some of the methods are usable as such, other methods require a specific implementation (i.e. question design, schedule for making the questions, etc.) depending on the evaluation goals.

In order to assess user behaviors in terms of cognition and emotion, objective and subjective methods should be integrated. Cognitive and emotion assessment methods are fully described in Human Factors Methods [36] and Handbook of Emotion Elicitation and Assessment [7].

3   Quantitative Analysis, Modeling and Qualitative Interpretation

In general, data collected during experiments or surveys is analyzed using quantitative statistical methods. The results obtained are presented as histograms or graphs. The quantitative analysis focus isolated entities in terms of frequency of occurrence or in terms of variation.

A further stage that aims to extract knowledge from data and to discover relationships between entities (hidden patterns) is data mining [3], [23], [32]. Linking entities is performed via various statistic analysis and algorithms, but the linking is mainly expressed in terms of influence of an entity on another. The data mining process starts with the data sampling stage, followed by the data exploration that eventually enables to find and refine models. Such models are composed by several modules or nodes ranging from statistics to complex algorithms (i.e. decision trees, regression, neural networks) [32]. Even though it reflects a conceptual process, the model emerges bottom-up and is highly dependent on the data sample selected initially [3]. The models are statistical and algorithmic and aim to solution specific problems.

The opposite way is to build systematically a conceptual model top-down, based both on knowledge extracted from domain experts and users and on scientific information. Techniques such as knowledge elicitation are very useful when building top-down models [4], [14]. Having a conceptual model enables to reinforce the structuring in data exploration and analysis according to the entities that constitute the model. They provide also the rationale for selecting the appropriated data collection methods. In general, models employed in Human Factors are built top-down and use various conceptual inputs [15], [38]. The conceptual models aim to be generic.

Furthermore, conceptual relationships between entities that are already systematically specified in the conceptual model induce a qualitative shift in data exploration because they guide the investigations beyond isolated entities, aiming to find complex relationships at the results (instance) level. For example, an experiment that used a cognitive model guided the employment of the most appropriated analysis method [22] and enabled to find user visual patterns that link various visual items and user behaviors in complex structures. Moving form quantitative results related to isolated entities to patterns that combine several entities is valuable because the patterns obtained in this way express higher level complex cognitive processes [37]. The top-down model influences the shift from quantitative aspects to qualitative interpretation of the results.

Both bottom-up and top-down modeling have their own benefits, and both aim to understanding and prediction. Mixing both approaches improves the overall reliability of models. However, conceptual models enable to capitalize knowledge in a structured manner. Thus they should be considered and used as a top container that can hold data-driven models.

4 Storage

4.1 Bottom-Up and Top-Down Storage

Bottom-up storage is the common way of storing data. Bottom-up storage is data-centric. Results obtained via analysis of data may be stored also, but generally only the calculation functions, requests and procedures that enable to obtain results rather than the results themselves are stored. In most cases, results expressed at a self-explanatory level, such as graphs, curves, etc., are included in reports, but not stored in the data base systems. Storing only row data means that data are loosely coupled from a study to another. Data comparisons may be performed, but the understanding of differences or similarities between the data sets is limited (i.e. when comparing the levels of Situation Awareness of pilots between several experiments, the stored data show the variations but do not show directly neither why nor how such variations occurred). Moreover, collected data do not confer stability because they depend on each acquisition situation. Large variations may be observed between studies and the understanding becomes even more complicated without referring to the corresponding reports and reading lots of pages. An improvement is to store also the context (i.e. the conditions and circumstances under which data were obtained) of each study, but this is not enough.

The optimum in the storing process is to store what is generic and stable and that enables to provide a reference for understanding, traceability and reuse. Such top-down storage implies to store the model, cognitive or organizational. Top-down storage is model-driven. Such type of model-driven storage is new and rare in the software market; however an integrated implementation is proposed by Kalido. The model is expressed as meta-data and uses semantic associative relationships. One of the main advantages of this solution is that the conceptual definitions layer is separated from the physical implementations [27], [31]. But the most common means in which models are stored are ontology systems. Relational Data Base Systems (*RDBMS*) do not allow storing such models directly, but ontology systems may be plugged with RDBMS. Furthermore, there is a growing interest to store Multi-Agent

Systems using ontology systems [39]. Thus, in such an IT architecture, the top layer (ontology) contains the model and the lower layer (RDBMS) contains the contextual instances. For example, components of the cognitive model such as the visual agents and resources may be stored in the ontology system, and instances such as the time to detect visual items in a particular visual scene, the moment of occurrence of visual alerts, the scan-paths and visual patterns may be stored in the RDBMS. Using top-down storage enables to structure data in respect with the model and thus encourages not only to store the results but also to link them with the model.

4.2 Unified Storage

An important issue that is faced in organizations is that data is stored then provided by various data sources. In HF, such sources may be technical simulators and human collection methods, both subjective (i.e. forms, rating scales, surveys) and objective (i.e. physiologic devices, various analysis software). The reliable solution to this issue is the Data Warehousing (*DW*) process that uses an Extract-Transform-Load (*ETL*) stage, that enables to feed the DW with unified data, and a data preparation stage of data as Data Marts that are domain specific and user-centered (i.e. Marketing Data Marts, Financial Data Marts, etc.). DW answers cross-organizational issues by providing a single version of the 'truth' or a common reference for the whole organization [3], [24], [35].

4.3 Human Behavior Simulation

4.3.1 Rationale and Benefits

Most of the existing simulators are technical systems or artifact simulators (i.e. flight simulators, drive simulators, calculation simulators). Artificial Intelligence enables at some extent to simulate human behavior, but is mainly based on algorithms, expert systems, heuristics, fuzzy logic, neural networks, etc. that implement rationality.

Taking into account that it is possible to obtain and store real behavior patterns, it would be useful to use these patterns in a dynamic manner. This means to create a hybrid Human Behavior Simulator (*HBS*) based on one hand on the real patterns and on the other hand on Artificial Intelligence (*AI*) features.

The interest in this approach is the hybrid nature of such a HBS, because real human behaviors are combined with IA features that enable to act or play these behavior patterns dynamically. Gestural patterns were already injected into technical products (i.e. EyeToy™, Wii™ and iPhone™). From a wider research perspective that includes surveys, observations, experiments, data analysis and interpretation it would be useful to reuse static data (i.e. observed behaviors recorded on video tapes, proprietary analysis software patterns expressed as diagrams or trees, etc.) in a dynamic way. A first benefit would be to inject new patterns into new products, or to make a better use of the obtained patterns. But a step beyond would be an important contribution to prediction. Let's imagine that in the current configuration (the given technical system, the given task and procedure) specific behavior patterns were obtained. Now, in case of changes of the current configuration (that is possible to simulate via the technical simulators) what would be the behavior patterns? Without a HBS, lots of experiments or new sets of observations or surveys would be necessary. With a HBS, the possible user behavior patterns could be predicted via simulation. An HBS would improve and speed-up the HCD iterations. Furthermore, in terms of possible configurations, simulation on both technical and human sides would lead not only to prediction, but also to optimization and socio-technical co-adaptability.

4.3.2 Translating the Behavior Patterns into Agent Oriented Programming Language

The Belief Desire Intention (*BDI*) model of Multi Agent Systems (*MAS*) is inspired by a model of human behavior [6], [19]. Beyond the overall descriptive formalism itself, BDI proposes AgentSpeak(L) [26] that is a pure Agent Oriented Programming (*AOP*) language, in contrast with other efforts that implement AOP principles using Object Oriented Programming (*OOP*) languages.

Taking into account the origins of BDI, it seems natural to translate human behavior patterns into BDI AOP. AOP confers the advantage to use an understandable way to deal with large and complex systems and implement parallel processes, being suitable for process modeling. Furthermore, BDI AOP enables to preserve the links between behavior patterns and essential entities of their context of occurrence, such as active goals, triggering events, procedures and scenarios, organizational relationships. BDI AOP provides also a programming structure in terms of perception, situation assessment, belief selection, communication. The reasoning cycles that are the core interpreter of BDI AOP provide a powerful support for Decision Making modeling and implementation.

From a conceptual IT architecture perspective, it is important to emphasize that AOP enables to ensure a continuity between the model and its' implementation.

Translating behavior patterns into AOP would improve data usability and would change data nature by replacing 'static' data (diagrams, graphs) with dynamic results expressed as chunks of executable code. This would not only enrich the sources of information (i.e. interoperable format of results [3]) but would improve the final reporting by providing animated views of information instead of static bar-charts, graphs, cubes, etc.

5 Discussion

The Human Centered Design approach became central in other domains than Human Factors (*HF*) where it mainly originated and was developed. Behavioral economics, behavioral finance and more widely economics started to use human-centric methods. Beyond methods, HF may bring to these domains their expertise in socio-technical systems. In return, HF should take a better advantage of organizational means, from project sponsoring to a better integration of HF in organizations in general and in the organizational IT systems in particular. HF may create their own dedicated module in industry solutions as such modules were created for other organizational departments considered as important (i.e. Accountancy, Human Resources Management, Production, Maintenance, Marketing & Sales, etc.).

The techniques and methods described in this paper aim to an improved formalism, storage, assessment, understanding, prediction and co-adaptation of socio-technical systems, the central concept being behavior patterns. They may also contribute to improve existing solutions for the whole organization in general, or specify and implement an integrated HF technical and functional module in particular.

References

1. Baldrige National Quality Program, http://www.nist.gov
2. Bénicourt, E., Guerrien, B.: La théorie économique néoclassique. La Découverte (2008)
3. Biere, M.: Business Intelligence for the Enterprise. IBM Press (2003)
4. Boy, G.A.: Cognitive Function Analysis. Ablex (1998)
5. Bradley, M.M., Lang, P.J.: Measuring Emotion: Behavior, Feeling and Physiology. In: Cognitive Neuroscience of Emotion, pp. 242–276. Oxford University Press, Oxford (2000)
6. Bratman, M.E., Israel, D.J., Pollack, M.E.: Plans and resource-bounded practical reasoning. Computational Intelligence 4, 349–355 (1988)
7. Coan, J.A., Allen, J.B. (eds.): Handbook of Emotion Elicitation and Assessment. Oxford University Press, Oxford (2007)
8. Chen, D.H.C., Dahlman, C.J.: The Knowledge Economy, the KAM Methodology and World Bank Operations, The World Bank (2005)
9. Cohn, J.F., Ambadar, Z., Ekman, P.: Observer-Based Measurement of Facial Expression with the Facial Action Coding System. In: Coan, J.A., Allen, J.J.B. (eds.) Handbook of Emotion Elicitation and Assessment, pp. 203–221. Oxford University Press, Oxford (2007)
10. Cohn, J.F., Kanade, T.: Use of Automated Facial Image Analysis for Measurement of Emotion Expression. In: Coan, J.A., Allen, J.J.B. (eds.) Handbook of Emotion Elicitation and Assessment, pp. 222–238. Oxford University Press, Oxford (2007)
11. Damasio, A.R.: A Second Chance for Emotion. In: Cognitive Neuroscience of Emotion, pp. 12–23. Oxford University Press, Oxford (2000)
12. Duchowski, A.T.: Eye Tracking Methodology: Theory and Practice. Springer, London (2003)
13. European Fundation for Quality Management (EFQM), http://www.efqm.org
14. Fransella, F., Banister, D.: A Manual for Repertory Grid Technique, 2nd edn. John Wiley & Sons, Ltd., Chichester (2004)
15. Gray, W.D. (ed.): Integrated Models of Cognitive Systems. Oxford University Press, Oxford (2007)
16. Hofstede, G., Hofstede, G.J.: Cultures and Organizations: Software for the Mind, 2nd edn. McGraw-Hill Professional, New York (2004)
17. Hoisington, S.H., Vaneswaran, S.A.: Implementing Strategic Change : Tools for Transforming an Organization. McGraw Hill, Inc., New York (2005)
18. Kahneman, D., Krueger, A.B.: Developments in the Measurement of Subjective Well-Being. Journal of Economic Perspectives 20(1), 3–24 (Winter 2006)
19. Kinny, D., Georgeff, M.: Modeling and Design of Multi-Agent Systems. Technical Note 59. Australian Artificial Intelligence Institute (1996)
20. Li, W.C., Harris, D.: Eastern Minds in Western Cockpits: Meta-Analysis of Human Factors in Mishaps from Three Nations. Aviation Space and Environmental Medicine 78, 420–425 (2007)
21. Lucas, R.E.: On the Mechanics of Economic Development. Journal of Monetary Economics 22, 3–42 (1988)
22. Magnusson, S.M.: Discovering hidden time patterns in behavior: T-patterns and their detection. Psychonomic Society: Behavior Research Methods, Instruments & Computers 32(I), 93–110 (2000)
23. Michalewitz, Z., Schmidt, M., Michalewitz, M., Chiriac, C.: Adaptive Business Intelligence. Springer, Heidelberg (2007)
24. Moss, L.T., Atre, S.: Business Intelligence Roadmap. Addison-Wesley, Reading (2006)

25. Pendell, S., Davis, K.: Intercultural Communication Issues in Knowledge Management. In: Cunningham, P., et al. (eds.) Building the Knowledge Economy: Issues, Applications, Case Studies, Part 2, section 4, pp. 834–838. IOS Press, Amsterdam (2003)
26. Rao, A.S.: AgentSpeak(L): BDI agents speak out in a logical computable language. In: Perram, J., Van de Velde, W. (eds.) MAAMAW 1996. LNCS (LNAI), vol. 1038, pp. 42–55. Springer, Heidelberg (1996)
27. Raden, N.: Back to Business: How Business Modeling Rationalizes Data Warehousing. Hired Brains, Inc White Paper (2008)
28. Reddy, P.: The Globalization of Corporate R & D, Implications for Innovation Systems in Host Countries. Taylor & Francis, Routledge (2000)
29. Romer, P.M.: Endogenous Technological Change. The Journal of Political Economy 98(5) (Part 2: The Problem of Development: A Conference of the Institute for the Study of Free Enterprise Systems), S72–S74 (1990)
30. Saffer, D.: Designing Gestural Interfaces. O'Reilly, Sebastopol (2008)
31. Sarbanoglu, H., Ottmann, B.: Business-Model-Driven Data Warehousing. Kalido White Paper (2008)
32. SAS Institute Inc. Getting Started with SAS Enterprise Miner 5.2. SAS Institute Inc., Cary, NC (2006)
33. Schumpeter, J.: Business Cycles. McGraw-Hill Book Company, New York (1939)
34. Schütte, S., Eklund, J.: Product Design for Heart and Soul: An introduction to Kansei Engineering Methodology. Linköpings Universitet, Department for Human Systems Engineering, Uni Tryck Linköping, Sweden (2003)
35. Stackowiack, R., Rayman, J., Greenwald, R.: Oracle Data Warehousing. Wiley Publishing, Inc., Chichester (2007)
36. Stanton, N.A., Salmon, P.M., Walker, G.H., Baber, C., Jenkins, D.P.: Human Factors Methods, A Practical Guide for Engineering and Design. Ashgate (2005)
37. Stephane, L.: Visual Patterns in Civil Aircraft Cockpits. In: Proceedings of the International Conference on Human-Computer Interaction in Aeronautics, Seattle WA USA, September 20-22, pp. 208–214, Cépaduès-Editions, Toulouse (2006)
38. Stephane, L.: Cognitive and Emotional Human Models within a Multi-Agent framework. In: Harris, D. (ed.) HCII 2007 and EPCE 2007. LNCS, vol. 4562, pp. 609–618. Springer, Heidelberg (2007)
39. Tran, Q.N.N.: MOBMAS: A Methodology for Ontology-based Multi-Agent Systems Development, School of Information Systems, Technology and Management, University of New South Wales, PhD (2006)

Scenarios in the Heuristic Evaluation of Mobile Devices: Emphasizing the Context of Use

Jari Varsaluoma

Tampere University of Technology, Human-Centered Technology,
Korkeakoulunkatu 6, P.O. Box 589, FI-33101 Tampere, Finland
jari.varsaluoma@tut.fi

Abstract. Varying contexts of use make the usability studies of mobile devices difficult. The existing evaluation methods, such as Heuristic Evaluation (HE), must be redesigned in order to create more awareness of the mobile context. Through the reworking of existing heuristics and use of written use scenarios, there have already been some promising results. In this study the context of use of mobile devices was examined with written scenarios. The main target was to improve the reliability of HE by increasing the number of right predictions and reducing the number of false positives produced by the evaluators. The results seem to differ from those of a previously conducted study as the scenarios did not improve the HE regarding the numbers of false positives or accurate predictions. There is a need for more research regarding the possible benefits of different scenarios and other factors that affect the outcomes of HE.

Keywords: Heuristic evaluation, scenario, context of use, mobile device, false positive.

1 Introduction

The rapid evolution of mobile devices is constantly bringing new challenges for usability experts. Traditional evaluation methods which were mainly designed for desktop computers cannot take into account all the factors of mobility and varying contexts of use [1]. This includes Heuristic Evaluation (HE) which is a very popular evaluation method in the industry [2]. In HE a usability expert studies the product in order to find usability issues on the basis of a list of usability guidelines. Perhaps the best-known heuristics are those of Nielsen [3]. HE is considered to be quick, cheap and easy to learn [3, 4].

The use of written scenarios during the HE of a mobile device can help evaluators to become aware of the context of use [5]. However, one must note the false positives that may occur during HE. A great number of false positives reduces the reliability of the evaluation method as time and effort are wasted on correcting problems that would not actually affect the end user. The reliability of the evaluation method can be measured by verifying the predicted problems through usability testing (UT) with real users [6].

M. Kurosu (Ed.): Human Centered Design, HCII 2009, LNCS 5619, pp. 332–341, 2009.

In this study scenarios were used in order to provide contextual information for the evaluators. A better knowledge of the context of use of a mobile device is assumed to reduce the number of false positives produced by the evaluators and improve the reliability of the HE.

2 Challenges in the Heuristic Evaluation of Mobile Devices

HE has been criticized for its relatively weak ability to find usability problems and predict their actual scope and severity [5, 6, 7, 8]. Experienced evaluators place themselves in the position of an inexperienced or an expert user. In psychology it is a common opinion that introspection is not an objective method [9]. This suggests that evaluation methods based on introspection are not reliable.

An evaluator using HE, or any other evaluation method, has to bear in mind the multiple factors that can affect the outcome of the evaluation. Table 1 presents factors that have been shown to affect the results of HE.

Table 1. Empirical studies have shown that the factors presented here can affect the outcome of heuristic evaluation. Note that the target of using scenarios or a realistic environment is actually to increase the evaluator's knowledge of the context of use.

Evaluation method	Evaluator characteristics	System evaluated and its context
Heuristics used (heuristics for mobile computing) [1] Number of evaluators [10, 11] Scenarios [5] Time spent on evaluation [7]	Education or job experience in usability area [3, 11] Experience with the application domain [1, 3, 5, 11] Experience with the heuristics used [1] Knowledge of the context of use (user, environment) [5, 6]	Evaluation environment (laboratory vs. real context) [5]

Different heuristics (Nielsen vs. Gerhardt-Powals) and media used for reporting usability problems have been studied, but these did not have significant effects on the results [7]. The importance of evaluators' understanding of the heuristics used and the method itself has been discussed [6, 8]. Still, in order to maintain the fast and inexpensive implementation of the method, it is not yet clear what manner of training would be the most appropriate. The criterion for judging evaluators as experts or novices in the usability area also lacks a definition [12].

The main challenge in the usability evaluation of mobile devices is their dynamic context of use. Important context types include location, identity, time, and activity [13]. As it is not possible to cover all possible use situations during the evaluation, one must choose the ones relevant for the study.

3 Scenarios in the Heuristic Evaluation of the Mobile Device

A study by Po's et al. [5] proposed a method called Heuristic Walkthrough (HW) that combined scenarios of use with Heuristic Evaluation (HE). They created five scenarios which were located around a university campus. A handheld pocket PC was evaluated by four usability experts using the HW method and four using the HE method. Scenarios seemed to help evaluators to shift their viewpoint from a technical evaluation towards the user's point of view and also predict more critical usability problems. What was not studied was the validity of the predicted usability problems. In order to truly measure the reliability of the method, the predicted problems should be verified with real users in usability tests (UT) [6, 7].

In this study scenarios were first used in UTs with users and then by usability experts during the HE of a mobile device. The results were compared in order to verify correctly predicted problems and false positives produced by the evaluators. The detailed setting of the study is described below.

3.1 Setting for the Study

The mobile device chosen for the study was a Nokia N95 mobile phone using the Series 60 operating system [14]. Four different scenarios were designed for a university context that was familiar to the author. The scenarios were situated in different locations around the campus and the UT was carried out in these locations. The first scenario happened in a usability lab and concentrated on teaching the basic features of the N95. The second scenario was located outside in a parking lot, the third in a noisy canteen, and the fourth in a walkway along the lakeside. The last scenario involved walking while using the device. Tasks varied from simple phone calls and text messages to tasks requiring multitasking and the use of a camera, map and GPS.

Here is an example from Scenario 3: *You are Johanna, and you are 25 years old. You study at the department of computer science and information systems. It is Friday afternoon and you are in the university canteen. There is a lot of noise as people are having lunch. You are waiting for your sister to come from a lecture because you are both going to go to your friend Kati's housewarming party in Tampere. While waiting, you ponder the next week's program and remember that you have a meeting with your thesis supervisor next Monday. You decide to create an appointment in your phone's calendar.*

The study consisted of a usability test (UT) and two heuristic evaluations (H1 and H2, the latter with scenarios) as shown in Fig.1. The UT was conducted first in order to avoid the knowledge of the HE results guiding the author's observations during the UT. The UT was carried out with the users in realistic use contexts and written scenarios were used at the scene. Ten university students aged 20-28 (M=23.7; SD=2.5) participated of whom six were female. None had previous experience of using the N95, but six participants had used mobile phones with Series 40 or 60 operating systems.

The HE sessions were conducted by two groups of four usability experts. The groups were created so as to be similar on the basis of a preliminary questionnaire. The experience of HE and mobile devices and knowledge of Nielsen's heuristics were investigated. The participants were researchers from universities and usability specialists from IT companies. Everyone had experience of at least three expert evaluations and knew Nielsen's heuristics very well or quite well.

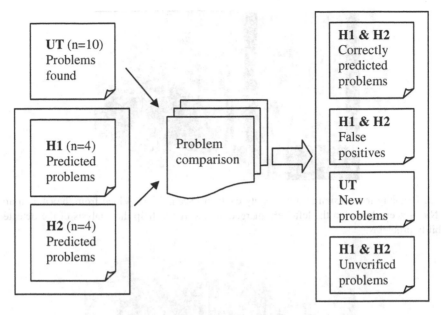

Fig. 1. The research structure. The heuristic evaluation (H1) and the heuristic evaluation with scenarios (H2) were carried out after the usability test (UT) results were analyzed. Problem comparison revealed false positives and usability problems that had been correctly predicted, had not been predicted by the evaluators (UT New problems), or were not in the scope of the tasks in the UT (H1 & H2 Unverified problems).

The first group (H1) acted as a control group and was given a simple list of features to evaluate, while the scenario group (H2) used the same written scenarios as the students in the UT. The N95 features that were evaluated were the same in both groups.

The resulting lists of predicted usability problems from the HE groups were compared to the observed problems list from the UT. Correctly predicted problems, false positives and problems found only in the UT were counted. Predictions related to N95 features that were not part of the UT could not be verified as true or false and were separated from the results as unverified problems.

3.2 Progress of the Study

Usability test with scenarios. The pretest questionnaire for choosing the participants was conducted via email. The test session started in a lab room with introductions and a questionnaire about previous experience with Series 40 or 60 phones. After a short briefing the first scenario was presented on paper. The participant was asked to think aloud during the test and was not assisted unless helpless for several minutes. An assistant videotaped the session and the author was responsible for all interaction with the user. A video feed from the N95 screen was captured by screen recording software (Fig. 2) and was sent wirelessly to a laptop that the user carried in a backpack (Fig. 3).

Fig. 2. Usability test session in a university canteen (Scenario 3). Video from the camera and the N95 screen shown on the left were merged afterwards to help the analysis of the detected usability problems.

Fig. 3. Scenario 4 included walking by the lakeside. When it was raining, the author held an umbrella for the participant. Cold weather made the usability testing challenging as users had to wear gloves while using the device.

Scenarios 2 and 4 were carried out outdoors and read aloud by the author so that the user's hands remained free to use the device (Fig. 3). At other times participants read the tasks by themselves (Fig. 2). After the last scenario the participant answered a short questionnaire about the usability of the device. The answers were talked through together and the participant was rewarded for her/his contribution.

The test videos were carefully studied by the author and all the mistakes, problems and comments were listed. Usability problems were graded as slips and problems. Slips were made by mistake (e.g. pressing the wrong button while writing) and were quick to fix (a few seconds). Other events were graded as problems. These included getting lost in menus, negative comments, and long times needed for consideration. It would have required each user to watch their own performance afterwards to judge if pressing a wrong button was a matter of the small keyboard or the result of an error of thought. Using any subjective problem ratings would also have required several researchers and discussions to have been more reliable. As this was not possible at the time, greater emphasis was placed on objectively counting the numbers of similar problems between different users.

Heuristic evaluation with scenarios. A preliminary questionnaire was administered via email to choose usability experts for both study groups. Instructions were sent at least 2 days before the HE session. The material included a list of Nielsen's ten heuristics [3] and an example of a problem reporting form. For each predicted problem the evaluator had to select the heuristics that had been violated and the severity ratings in two different scales (a 1-4 scale and a fourfold table) on the basis of those presented by Nielsen [15].

The HE session started with a questionnaire about previous experience with similar operating systems. The problem reporting form was explained in detail. Next, the scenarios or a list of features were presented to the evaluator, depending on the evaluation group. The evaluators were asked to report any usability problems they would predict that both new and experienced users would have. The author answered any questions about the N95 functions during the evaluation session. Each evaluation finished within the allocated 3 hours.

3.3 Results

The UT revealed 290 observations of problems. These were categorized as 65 different usability problems of which 10 counted as slips and 55 as problems. 36 of the total of 65 were rated as *unique* as they were experienced with only one user.

Table 2. Results from the heuristic evaluations after the predicted problems were verified by comparing them to the usability test results

	Evaluator	Correctly predicted problems	False positives	Unverified predictions	Total
H1	1	10	1	3	14
Control	2	5	2	3	10
group	3	8	1	2	11
	4	7	4	0	11
Total		30	8	8	46
Mean		7.5	2	2	
SD		2.08	1.41	1.41	
H2	5	5	5	3	13
Scenario	6	7	7	1	15
group	7	9	0	1	10
	8	4	2	2	8
Total		25	14	7	46
Mean		6.25	3.5	1.75	
SD		2.22	3.11	0.96	

The HE control group H1 reported 51 problems (M=12.75; SD=1.71) and the scenario group H2 reported 47 (M=11.75; SD=2.87). The difference was not significant (t=0.599; p>0.05). When the predicted problems were analyzed, it was noticed that a single problem reporting form could actually contain problems from several of the UT problem categories. In these cases the reported problem was divided into several predicted problems with the same severity ratings and violated heuristics. There were also 9 cases where one evaluator had reported several problems that would fit into only one of the predefined UT problem categories. For example, the problems that users had when turning on the main camera had been categorized as one problem. For these predictions the mean value was counted for the severity ratings and all the violated heuristics were included. This process was repeated for each evaluator. After the analysis there were 46 predicted problems in both groups. 15 predictions could not be verified, because they were not within the scope of the UT scenarios. These predictions included N95 features that were not used by any of the UT participants. The results after the verification of the problems are presented in Table 2.

H1 outperformed H2 but the difference between the groups was not significant in terms of correct predictions (t=0.822; p>0.05) or false positives (t=-0.878; p>0.05).

Correct predictions in the HE were studied without the problems that were rated as unique or slips in the UT. The remaining problems occurred more frequently with different users and can be considered more severe. Table 3 shows that H1 made more

Table 3. Correct predictions from the heuristic evaluation, excluding problems that were rated as unique or slips in the usability test

	Evaluator	Correctly predicted without the unique problems	Correctly predicted without the unique problems and slips
H1	1	7	7
Control	2	5	5
group	3	6	5
	4	6	3
Total		23	20
Mean		5.75	5
SD		0.96	1.63
H2	5	3	3
Scenario	6	4	4
group	7	5	4
	8	4	4
Total		16	15
Mean		4	3.75
SD		0.82	0.5

correct predictions. Without the unique problems the difference was significant (t=2.782; p=0.032). Without the unique problems and slips there was no significant difference according to the Mann-Whitney test (Z=-1.348; p>0.05).

35 problems from the UT were not predicted in the HE. Only two of these were closely related to the environment: bright sunlight made the phone screen difficult to read and the keypad was troublesome to use in cold weather with gloves on. Perhaps the scenarios did not include enough contextual cues for the evaluators to predict these problems.

Problems with the Reliability of the Heuristic Evaluation. The evaluator effect [10] could be seen in the HE results. The chosen severity ratings varied greatly between the evaluators. Only one problem was rated as critical (4) by one evaluator and another evaluator rated the same problem as small (2). None of the evaluators chose the same set of violated heuristics for the same problem. Three problems were reported without heuristics being chosen, which suggests that the heuristics used did not cover all the predicted problem types.

The control group H1 predicted 10 problems correctly (of which 3 were rated as slips) that were not reported in the scenario group H2. H2 predicted 7 problems correctly that were not reported in H1. This may be due to the evaluator effect or the scenarios might have affected the way the evaluators examined the system.

4 out of 8 false positives in H1 and 9 out of 14 in H2 were somehow related to underestimating the user's skills. It seems that the evaluators had difficulties in estimating how much the user's previous experience affects the use situation.

4 Discussion

The results of this study suggest that using written scenarios during HE does not increase the reliability of the method. The scenarios did not increase the number of accurate predictions or reduce the number of false positives. The scenarios might have had some effect on the way the evaluators examined the device, as different problems were found between the groups. It would have required more evaluators to do the evaluation to be certain that the difference was not just because of the evaluator effect [10].

During the HE some evaluators would have liked to know more about the users' previous experience with the mobile device than the scenarios described. In this study the evaluators had to consider both new and experienced users, which made the evaluation more demanding. Concentrating either on inexperienced or experienced users and providing more information about the users' skills and experience with the device being evaluated might have reduced the evaluators' mental load and led to fewer false positives.

A simple list of features to evaluate seemed to produce more reliable results than the longer descriptions of the use situations. The scenarios probably made the evaluation sessions more complex and as a result the differences between the evaluators and the number of false positives were greater in the scenario group. It is surprising that when the scenarios limited the user group to university students, the evaluators in the control group still performed better. Perhaps the chosen user group was too general and scenarios would provide more help when evaluating for more specialized user groups, such as children or aged people.

The scenarios did not improve the HE in this setting in the way they did in the earlier study [5]. This may be because of the differences in scenarios, test devices, and analytical methods. Perhaps the scenarios used in this study failed to give enough contextual cues for the evaluators.

What would be the best way to provide contextual cues for evaluators and how much information is needed? If scenarios are used, should they be in the form of text, pictures, cartoons, video, or a combination of some of these? How can the relevant information for mobile devices be chosen and how can it be made sure that this information is valid? If planning realistic scenarios requires a great deal of information to be gathered, then does this vitiate the use of HE as a rapid and inexpensive method?

Another study with a greater number of evaluators should bring more information about the merits of using scenarios with HE. Analyzing and discussing the collected UT and HE data with other usability experts would also provide more reliable results. Information is also needed on the ways that evaluators actually use the given scenarios during the evaluation session and what the best way is to present contextual information.

For HE there are still some factors whose effects on the results would be interesting to study. Does the evaluator's motivation or spryness during the evaluation affect the results? How about evaluator's personal knowledge of the user group? What about the data that are stored in the mobile device during the evaluation? Should these represent the same data as the users have when using the device? Does it matter if the system being evaluated is used for leisure or some dangerous work where lives could depend on its usability? And what if the system being evaluated is still a low-quality prototype or one that is ready for the market?

There are still plenty of questions waiting to be answered about the reliability of evaluation methods. As HE is one of the most widely used evaluation methods, improvements to its reliability are worth pursuing.

References

1. Bertini, E., Gabrielli, S., Kimani, S.: Appropriating and Assessing Heuristics for Mobile Computing. In: Proceedings of the working conference on Advanced visual interfaces, pp. 119–126. ACM Press, New York (2006)
2. Rosenbaum, S., Rohn, A.J., Humburg, J.: A Toolkit for Strategic Usability: Results from Workshops, Panels, and Surveys. In: Proceedings of SIGCHI Conference on Human Factors in Computing Systems, pp. 337–344. ACM Press, New York (2000)
3. Nielsen, J., Mack, R.L.: Usability inspection methods. John Wiley & Sons, Inc., Chichester (1994)
4. Law, E.L., Hvannberg, E.T.: Complementarity and convergence of heuristic evaluation and usability test: A case study of UNIVERSAL brokerage platform. In: Proceedings of the Second Nordic Conference on Human-Computer Interaction, pp. 71–80. ACM Press, New York (2002)
5. Po, S., Howard, S., Vetere, F., Skov, B.M.: Heuristic Evaluation and Mobile Usability: Bridging the Realism Gap. In: MobileHCI 2004, pp. 49–60. Springer, Heidelberg (2004)

6. Cockton, G., Woolrych, A.: Understanding Inspection Methods: Lessons from an Assessment of Heuristic Evaluation. In: Joint Proceedings of HCI 2001 and IHM 2001: People and Computers XV, pp. 171–192 (2001)
7. Hvannberg, E.T., Law, E.L., Lárusdóttir, M.K.: Heuristic evaluation: Comparing ways of finding and reporting usability problems. Interacting with Computers 19(2), 225–240 (2007)
8. Law, E.L., Hvannberg, E.T.: Analysis of Strategies for Improving and Estimating the Effectiveness of Heuristic Evaluation. In: Proceedings of the third Nordic conference on Human-computer interaction. ACM International Conference Proceeding Series, vol. 82, pp. 241–250. ACM Press, New York (2004)
9. Tavris, C., Wade, C.: Psychology in perspective. Prentice-Hall, Inc., Englewood Cliffs (2001)
10. Hertzum, M., Jacobsen, N.E.: The evaluator effect: A chilling fact about usability evaluation methods. International Journal of Human-Computer Interaction 13(4), 421–443 (2001)
11. Nielsen, J.: Finding usability problems through heuristic evaluation. In: Proceedings of the SIGCHI conference on Human factors in computing systems, pp. 373–380. ACM Press, New York (1992)
12. Chattratichart, J., Lindgaard, G.: A comparative evaluation of heuristic-based usability inspection methods. In: CHI 2008 extended abstracts on Human factors in computing systems, pp. 2213–2220. ACM Press, New York (2008)
13. Dey, A.K.: Providing Architectural Support for Building Context-Aware Applications. Georgia Institute of Technology (2000)
14. S60 website, http://www.s60.com/life
15. Nielsen, J.: Usability Engineering. Academic Press, London (1993)

The Proposal of Quantitative Analysis Method Based on the Method of Observation Engineering

Tomoki Wada and Toshiki Yamaoka

Design Ergonomics Lab. Wakayama Univ.
930 Sakaedani Wakayama-shi Wakayama Japan
s105064@sys.wakayama-u.ac.jp

Abstract. Observation Engineering is the logical method to obtain product's requirements. We propose the method to analyze user's behavior by Formal Concept Analysis. User's behavior was converted to category data, and peculiarity of user's behavior was obtained by FCA. Requirements were got from peculiarity. Finally, Requirements were obtained from their peculiarity, and there were verified by questionnaire.

Keywords: Observation, Formal Concept Analysis, requirement.

1   Introduction

1.1 Background

Observation Engineering is a logical method to obtain product's requirements based on viewpoints grasped relationships between human and system [1]. Researchers can observe user's behavior in true environment and obtains requirements from their ordinary or unique behavior in this method. In comparison with old usability testing done in laboratory, researchers can get real data of user's usual behavior because users are not conscious of being observed.

Moreover, Researchers can find potential requirements which can't be found by questionnaire or interview because researchers observe not only user's conscious behavior but also their unconsciously behavior in Observation Engineering. On questionnaire or interview, users answer only their clear requirements.

1.2 Objective

Objective of this paper is considering the method to analyze quantitative data obtained from observation. The quantitative analysis will enable to use statistical analysis; like Principal Component Analysis, Multiple Regression Analysis, and so on. So researchers can find potential interaction or the law among user's behavior.

M. Kurosu (Ed.): Human Centered Design, HCII 2009, LNCS 5619, pp. 342–350, 2009.
© Springer-Verlag Berlin Heidelberg 2009

1.3 Methodology

At first, observation plan was decided. Next, observation was done and user's behavior was recorded by camcorder. User's behavior was converted to category data, and it was analyzed by Correspondence Analysis and Cluster Analysis, Formal Concept Analysis (FCA). As a result, peculiarity of user's behavior was grasped and their requirements were obtained. At last, as a validity of this method, requirements were verified with questionnaire.

1.4 Results

Five-hundred users who were recorded by camcorder were divided into 72 groups by Correspondence Analysis and Cluster Analysis. Peculiarity of user's behavior (64 items) was grasped by FCA and requirements were obtained from their peculiarity. Validity of requirements was verified with questionnaire. Almost all requirements had validity.

2 Method

2.1 Decide Observation Plan

Decide purpose and subject of research. Before doing research, analyzer has to decide purpose and subject of research. The reason to decide is to limit view points and take data efficiently. If purpose and subject aren't decided enough, data may be short. Table 1 appears purpose and subject of this research.

Table 1. Purpose and subject of this research

purpose		elevator user's behavior at taking and waiting elevator
subject	product	elevator
	parts	elevator hall
	job	waiting and taking elevator
	user	general users
environment	location	elevator hall
	time	11:00~12:00, 13:00~14:00, 5days (total 10 hours)

2.2 Decide Categories

Categories are criterions to evaluate user's behavior. There are decided for purpose of research. For example, sex, age, companion, behavior (standing position to wait, operating ...), and so on.

Video data recorded at observation was converted to category data by those categories.

Table 2 appears categories of this research.

Table 2. Categories

users	sex	male, female
	age	~15, 15~30, 30~60, 60~
	pace of behavior	quickly, normal, slowly
	direction came from	left, right
conditions	having companions	yes, no
	having belongings	no, one-handed, both-handed, stroller, wheelchair, suitcase
environment	states of call button	no, only left, only right, all
	waiting time	~30sec, 30~60sec, 60sec~
operations	operating call button	no, only left, only right, left and right, right and left
	keeping door open	no, holding door, operating call button
behavior	direction of user's eyes	elevator, hand, companion, status information, floor guide, bulleting board
	direction of user's body	elevator, companion, bulleting board
	standing position (distance of elevator)	standing by elevator center, standing by elevator left, standing by elevator right, middle distance left, middle distance right, far, standing by bulleting board
	noticing elevator comes	yes, no

2.3 Observe Users and Get Video Data by Camcorder

We did observation at elevator hall. Camcorder was had installed in elevator hall and recorded user's behavior waiting and taking elevator. A sample of video data appears in figure 1.

Fig. 1. A sample of video data

2.4 Analyze

(1) Convert to category data from video data

To do FCA, data has to be converted to category data (context table).

**convert to
category data**

video data

category data
(context table)

Table 3. Example of context table

ID	sex		age				pace of behaviors			waiting time	
	male	female	~15	~30	~60	60~	quickly	normal	...	~60	60~
n1	0	1	0	0	1	0	1	0	...	0	1
n2	0	1	0	1	0	0	1	0	...	1	0
n3	1	0	0	1	0	0	1	0	...	0	1
n4	1	0	0	0	1	0	0	1	...	0	1
n5	1	0	0	1	0	0	1	0	...	1	0
n6	1	0	0	1	0	0	1	0	...	1	0
...	...	...	...	...	...	...	...	...		...	...
n49	1	0	0	0	1	0	1	0	...	0	1
n50	1	0	0	1	0	0	1	0	...	0	1

(2) Divide users into some groups

In addition, Max number of objects FCA can manage is 10~15. So they have to divided into groups (10~15 objects per a group). Correspondence of criterions and users were obtained by Correspondence Analysis and users were divided by Cluster Analysis.

divide users
Correspondence Analysis
Cluster Analysis

(3) Analyze by Formal Concept Analysis

Formal Concept Analysis (FCA) is a method to derive relationships from objects (users) and their properties (behavior). As a result of FCA, analyzer can find user's behavior tendency in the group and reason of their behavior, and finally group's peculiarity can be grasped. [2, 3]

Result of FCA is a concept lattice witch is drawn as a Hasse diagram.

Relationships between criterions and users appear on concept lattice. In addition, common behavior of all users in the group and unique behavior of a user were grasped.

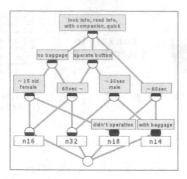

Fig. 2. Example of concept lattice

3   Results

3.1   Results of Classification

Video data we got were total 10 hours and number of users converted to category data was 500. As a result of Correspondence Analysis and Cluster Analysis, users were divided into 72 groups.

3.2   Results of FCA

We analyzed each groups by FCA. Here is an example, group-1 (6 July AM).

As example, concept lattice of Group-1 (6 July AM) appears in figure 3. Relationships between criterions and users appear on this chart.

Table 4. Table of Group-1 (6 July AM)

	sex		age				pace of behaviors			companions		belongings				waiting time			direction came from		states of call button				operating call button				standing position						direction of user's eyes					direction of user's body			noticing elevator comes		keeping door open		
	male	female	1~15	15~30	30~60	60~	quickly	normal	slowly	no	yes	no	one-handed	stroller	wheelchair	~30sec	30~60sec	60sec~	right	left	no	only left	only right	all	no	only left	only right	left and right	close center	close left	close right	middle left	middle right	far	bulleting board	EV	hand	companion	states info	EV	companion	bulleting board	yes	no	no	holding door	call button
n3	0	0	0	1	0	0	0	0	1	1	0	1	0	0	0	0	0	0	1	0	0	1	0	0	1	0	0	1	0	0	0	0	0	1	0	0	1	0	0	0	0	0	0	0	1	0	0
n6	0	0	1	0	0	0	0	0	1	0	0	1	0	0	0	0	0	0	1	0	1	0	0	0	1	0	0	0	0	0	1	0	0	0	0	0	0	0	0	0	0	0	0	1	1	0	0
n24	0	0	1	0	0	0	0	0	0	0	0	0	0	1	0	0	0	0	0	1	1	0	0	0	1	0	0	0	1	0	0	0	0	0	0	0	0	0	0	0	0	0	0	1	1	0	0
n25	0	0	0	1	0	0	0	0	1	0	1	0	1	0	0	0	0	0	1	0	0	0	1	0	0	0	0	0	1	0	0	0	0	0	1	0	0	1	0	0	0	0	0	1	1	0	0
n42	0	0	1	0	0	0	1	0	0	1	0	0	0	0	1	1	0	0	0	1	0	0	0	1	0	0	1	0	0	0	0	0	0	1	0	0	0	0	0	0	0	0	0	1	1	0	0
n43	0	0	0	1	0	0	1	0	0	1	0	1	0	0	0	0	0	0	1	0	1	0	0	0	1	0	0	1	0	0	1	0	0	0	0	0	0	0	0	1	0	0	0	1	1	0	0
n49	0	0	0	1	0	0	0	0	0	0	1	0	1	0	0	0	0	0	0	1	0	0	0	0	1	0	1	0	1	0	0	0	0	0	0	0	0	0	0	1	0	0	0	1	1	0	0

Peculiarity 1

Common behavior of all users in the group and unique behavior of a user were obtained.

"Looking at a bulletin board" is at summit. It means that all users fall under "looking at a bulletin board". In short, "all users look a bulletin board" is given as a peculiarity of this group.

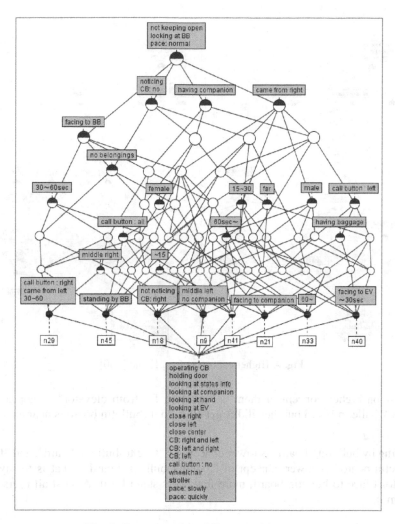

Fig. 3. Concept lattice of Group-1 (6 July AM)

Peculiarity 2

On the other hand, "standing near the elevator" is at the lowest layer. Concept at the lowest layer means "no one falls under it", so it appears "no one waits near the elevator". In addition, "no operating call button" is higher layer, so almost all users didn't operate call button.

The reason of this is that call button had been operated. Therefore, it is inferred that elevator hall is jammed, so users couldn't operate call button and come up to elevator.

Peculiarity 3

[n21] and [n40] don't fall under "facing to bulletin board", so we analyze them in detail. Figure 4 appears higher concepts of [n21] and [n40].

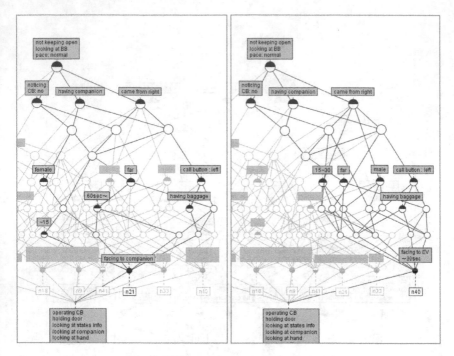

Fig. 4. Higher concepts of [n21] and [n40]

Common higher concept of them is "standing far from elevator", it appears that they look bulletin board but they didn't get close to it (bulletin board is near elevator).

Peculiarity 4
"Standing by bulleting board" is lower concept of "face to bulletin board", and "facing to bulletin board" is lower concept of "looking bulletin board". That is to say, few users don't face to bulletin board, more few users stand by it. Almost all users stand far from it.

3.3 Obtain Requirements

Requirements were obtained from peculiarity.

Requirement 1 From peculiarity 1, 3, 4
"Contents on bulletin board could be read from far".

Requirement 2 From peculiarity 2
"Users can approach elevator and call button when elevator hall is jammed".

3.4 Total Results

Finally, we got 64 requirements from 72 groups. Duplicated requirements were compiled.

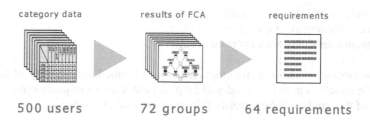

category data   results of FCA   requirements

500 users   72 groups   64 requirements

4 Verification of Results

To verify this method, we verified validity of results of this method. Survey was conducted using a questionnaire. Data were gathered from 32 collage students (male: 22, female: 10, average age: 22.3, SD=1.14).

4.1 Questionnaire

There were eleven questions (requirements) and choices were four steps, from "I think so" to "I don't think so". Figure 1 appears an example.

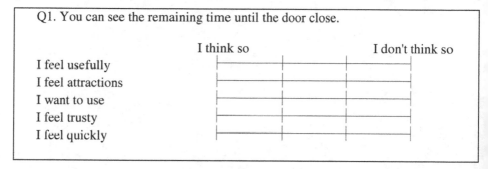

Fig. 5. An example of questionnaire

4.2 Result

Eight of the eleven requirements had significant difference between sum of "I think so" and "In think so a little" and sum of "I don't think so a little" and "I don't think so" (Direct probability of Fischer, p<0.05). As a result, almost all requirements had validity.

5 Conclusion

Objective of this paper is considering the method to analyze quantitative data obtained from observation. The method of this study is as follows.

(1) Decide observation plan and categories.
(2) Record use's behavior by camcorder.
(3) Convert to category data from video data.

(4) Divide users into some groups.
(5) Grasp peculiarity of each group by FCA.
(6) Obtain user's requirements from peculiarity.

On observation of elevator, requirements were obtained and validity of there was verified. In conclusion, this method will help to obtain user's requirements. A further direction of this study will be to verify this method in other products.

References

1. Yamaoka, T.: An Introduction to Human Design Technology, p. 87 (2003)
2. Ganter, B., Stumme, G., Wille, R.: Formal Concept Analysis: Foundations and Applications (2005)
3. The concept Explorer, http://conexp.spurceforge.net/

Translating Subjective Data to Objective Measures to Drive Product Design and Experience

Erin K. Walline and Bradley Lawrence

Dell, Inc., Experience Design Group,
One Dell Way, Round Rock, Texas, 78682
{Erin_Walline,Bradley_Lawrence}@Dell.com

Abstract. To successfully drive best-in-class human factors into product design, it is sometimes necessary to adopt more non-traditional experimental methods and reporting techniques. Within the PC industry, a traditional usability study is usually comprised of running eight to twelve participants through a set of tasks in a two-hour time period, collecting and reporting ease-of-use, success rate, time-on-task, and preference data. This traditional method is great at identifying potential usability pitfalls, but not necessarily equipped to focus on a product's visual appeal or quality perception. Two case studies are described that introduce non-traditional methods which: (1) focus on the perceived quality of specific product designs; (2) relate subjective data to concrete mechanical terms such that engineers have clear direction on how to build the products; and (3) report findings in a concise, graphical manner that is easily and quickly understood by executives and colleague functions lacking a human factors background.

Keywords: Product design, experience, usability.

1 Case Study I: Hard Drive Removal Force

1.1 Introduction

This research effort came from the side of the corporate organization where IT customers purchasing enterprise-level servers and storage devices were complaining about the feel and forces necessary to remove and replace front-access hard drives from these systems. Hard drives comprise the heart of an enterprise's storage and data and customer council feedback was indicating the perceived quality associated with these replacements was not instilling trust in the overall product line. While failure rates are very low on individual hard drives, when a data center houses over thousands, removals and replacements can occur as frequently as weekly, exacerbating the exposure to the perceived quality issue.

1.2 Method

Twenty IT professionals were recruited to participate in an investigation study where the entire study involved multiple removals and replacements of hard drives across

M. Kurosu (Ed.): Human Centered Design, HCII 2009, LNCS 5619, pp. 351–356, 2009.

several manufacturers and several product generations (depicted in Figure 1). The smoothness (defined as change in force per unit of travel) and peak force varied widely among the multiple systems tested, and there was enough resolution such that the IT professionals could detect subtle to great differences among the various designs. IT customers rated how "right" the forces and smoothness felt on a 7-point scale from "too much" to "just right" (the midpoint) to "too little." In parallel, the actual smoothness and force data were collected quantitatively.

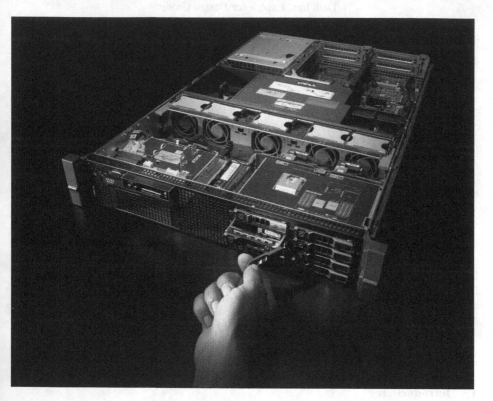

Fig. 1. Hard drive removal from a server

1.3 Results

Figure 2 shows the averages of the twenty participants' ratings of perceived smoothness and force for three representative hard drives of the multiple hard drives tested. Figure 3 shows the actual force profiles that were measured using a fixture and force gauge at every 10 mm of travel along the length of the hard drive removal for these three representative samples.

Figures 2 and 3 indicate thresholds for both smoothness and force acceptability. Figure 2 shows perceived smoothness is only acceptable for hard drive #3 (HDD 3), indicated by the bracket on Figure 3 that is just less than one unit of force across the range of travel. Figure 2 shows however that both HDD 2 and HDD 3 are in the acceptable range corresponding to two units of force (the dotted line in Figure 3).

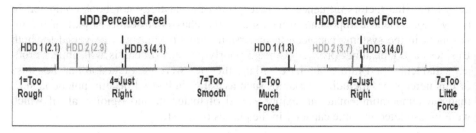

Fig. 2. Perceived hard drive removal smoothness and force

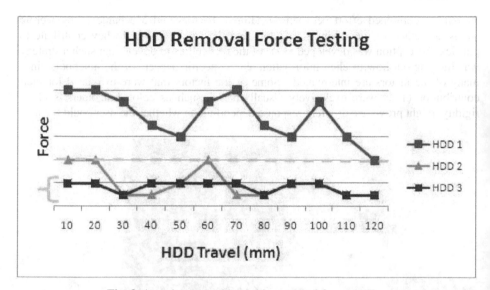

Fig. 3. Actual measured hard drive removal force profiles

When both data sources were presented together graphically in one easily digestible PowerPoint slide, the reporting technique became known across the organization as "Goldilocks" analysis and was received in a very positive light providing very clear, comprehensible direction for building the best hard drive carrier. Executives and peers across organizations have come to anticipate this type of simplified methodology and reporting in future research of this kind. Further, "Goldilocks" testing targeted nearly every subsystem and component on the server product line for hard drive carrier improvement. Perceived quality improvements were made across the board, and customer feedback from customer councils, reviews, sales calls, etc. indicate that these servers are now seen as serious enterprise hardware.

2   Case Study II: "Plasticky" Perception

2.1   Introduction

After receiving direct customer feedback, as well as media reviews, that identified products as feeling too cheap and "plasticky," an investigation was undertaken to determine what factors influence the perception of a plasticky notebook computer. In many cases, metal components were used in the chassis structure as part of the investment in the system. The need to understand this problem was universal for both Business and Consumer products, since a poorly-perceived chassis will impact product satisfaction and future purchases. Many theories were postulated about the cause of this perception, including variables such as color, industrial design, notebook size, material temperature, material texture, sound of material, and rigidity, all of which were investigated in some capacity in the phases of the study.

2.2 Method

Through a combined effort between the Human Factors and Mechanical Engineering teams, a method of identifying the factors and the extent to which they contribute to plasticky perception was developed. One of the difficulties in developing such a strategy was figuring out how to elicit information about specific aspects of the notebook, since many of the factors are interrelated. Some of the factors that were evaluated for their contribution effect were exclusively visually-based, such as color, but others, such as rigidity, might provoke a different response depending on whether the user could see it.

Fig. 4. Eliciting Non-Visual Feedback on System Factors Under Study

It was determined that the best way to get feedback on the factors was to visibly mask the systems from the user during portions of the study. This was accomplished by using a large, opaque, black cloth that was held up by stands, creating a curtain effect (Figure 4). The test coordinator placed the system being evaluated behind the cloth, allowing the user to reach under the curtain to perform the evaluation (such as touching the notebook palmrest to provide feedback on temperature) without viewing it, thus removing any visual bias. The evaluations were repeated so that the user could see the system. Although somewhat unconventional, this method led to some interesting visual vs. non-visual results, leading us to a better understanding of the impact of each factor. Users were asked to fill out rating scales after each test condition, and the order of the conditions were varied so that the users did not know if they were providing feedback on the same systems for the visual and non-visual portions.

Another difficulty in understanding the extent to which the studied factors impacted plasticky perception involved determining how to take subjective user feedback and relate it to objective measures so that engineers could build to a specification that solved the plasticky problem. Use of 10-centimeter visual analog rating scales allowed for better understanding of the perceived differences of the factors under study. Using the results, we were able to map perceived differences to physical measures that contributed to that perception. As an example, if users rated the rigidity of a specific part on a series of notebooks, we could then map those perceptions against physical measures of rigidity. This was helpful in determining if a mechanical change (and hence, investment in, for example, material thickness, cost, weight, etc.) was even perceptible by users and secondly, whether that perceived difference had an effect on acceptability or other measure of preference.

In the example in Figure 5, if it is assumed that the ratings shown are averages across a number of study participants, it could be inferred that while users could detect differences in perceived component flexibility between System A, System B, and System C, the mechanical measures of rigidity of System A and System B are not "acceptable enough". In other words, gains in perceived quality between the investments made in the component structure of System B may not be worth it, but it *may* be worth the additional component structure investment in the solution in System C in order to achieve a perception of high quality.

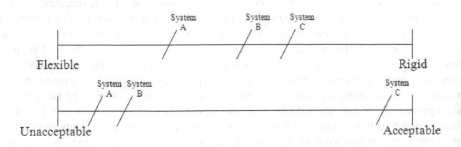

Fig. 5. Example Use of Visual Analog Scales to Determine Extent of Factor Impact

2.3 Results

Once the subjective measures had been gathered, they needed to be mapped to objective data which allows engineers to better evaluate the impact of the differences in user-generated data. Using the same example as depicted in Figure 5, each system's rigidity could be measured at a repeatable, specific point on the system in terms of some unit, such as millimeters of flexion (see Figure 6). Those values could be incorporated into a mechanical specification in such a way that the products that use the specification can be built toward a set value. An example specification statement might be: "Notebook palmrest deflection should not exceed $X - 0.6$ [mm] of deflection in the z-axis." When this type of approach of marrying subjective data to objective data is applied across multiple attributes and factors, the specifications that can be developed become more robust because they are based on user-generated data. The end goal of using the results to drive improved solutions into the products and reducing the amount of "plasticky" comments was ultimately achieved.

$$\text{System A Rigidity} = X \text{ [mm]}$$
$$\text{System B Rigidity} = X - 0.2 \text{ [mm]}$$
$$\text{System C Rigidity} = X - 0.6 \text{ [mm]}$$

Fig. 6. Example Objective Measures of Rigidity that Can Be Mapped to Subjective Measures

3 Conclusion

Oftentimes in the computer industry, human factors engineers are presented with design challenges or questions that cannot be answered with traditional usability testing but require non-traditional methods. These non-traditional methods require creative thought to pinpoint how to best solve and answer the challenge at hand. Just as important as developing the method, the human factors engineer must present the results in engineering terms by providing clear direction as to how a product should be built to meet and exceed customers' expectations rather than simply hand Likert scale ratings to an engineering organization and tell them to "make it better" because it didn't rate well. Additionally, if engineering resources are tight, schedule impact for shipping a product is in jeopardy, or the recommended solution (e.g. a change in material) increases the bill-of-material cost, an executive escalation may occur to move forward with the product design change. In this case, the results must be presentable in a highly visual and succinct way to gain favor with the time-starved executive. Both case studies presented in this paper were embraced by engineers and executives alike primarily due to the manner in which the studies were run and the data presented. While there is obviously no repeatable recipe to follow for a method that is non-traditional, creative forethought on how to design an experiment that targets the question and present the resulting data in a visually engaging and succinct fashion can be the main motivating factor for realizing the implementation of the product design improvement and the resulting positive customer experience.

Towards an Holistic Understanding of Tasks, Objects and Location in Collaborative Environments

Maik Wurdel

University of Rostock, Department of Computer Science
Albert-Einstein-Str. 21, 18059 Rostock, Germany
maik.wurdel@uni-rostock.de

Abstract. In this paper a task modeling approach is presented which tackles the integration of different kinds of models by a generic framework. The application of the framework is shown for collaborative environments, a certain sub set of ubiquitous computing environments. In these environments tasks have a close bond to the location of the executing actor as well the state of the involved objects. Therefore a location specification and a domain model are used to constrain the task execution. The language is supported by a tool, the CTML editor and simulator, which covers all steps of development from creation, editing, testing and verification. Such a model is particularly from interest for the intention recognition module of our experimental infrastructure of a collaborative environment.

Keywords: Collaborative Task Modeling, Domain Modeling, Collaborative Environments, Location Modeling.

1 Introduction

Task modeling has been accepted as mature research field within the domain of HCI. Especially in early stages of development task models specify the potential interaction to accomplish a certain goal with the software system. It can act as bridge between stakeholder and software engineer to manifest the envisioned interaction of user and software system. For desktop and mobile applications common notations such as [1, 2, 3] are well suited. However when using task models to specify the potential task performance in ubiquitous computing environments additional factors come into play. In this paper the question is tackled whether extended task models can be used to specify tasks in collaborative environments (CE), a particular sub set of ubiquitous environments.

A collaborative environment is a physical place equipped with computing devices where actors interact among each other to accomplish a certain goal. Stationary und personal devices assist the actors. Modeling tasks for such an environment is evidently much more complex as classical task modeling. A suitable modeling language has to incorporate means for diverse dependencies. We found the following dependencies particular for CE:

M. Kurosu (Ed.): Human Centered Design, HCII 2009, LNCS 5619, pp. 357–366, 2009.

1. **Cooperation.** Tasks need to synchronize across different actors. Common task model notations offer the opportunity to use temporal operators in order to specify the potential execution order but only some offer temporal constraints for multi-user task specifications.
2. **Objects.** To perform a task tools or artifacts may be needed. This has to be reflected by the modeling language to create an adequate model of the environment. Objects may need a certain state to be in and may also be manipulated (e.g. a presentation can only be given if a projector is available).
3. **Location.** In CEs the place of an actor highly influences the executability of tasks. Some task might only be enabled at a special place in the environment whereas others might be executable at a set of location (e.g. giving a talk at a conference, listening to a presentation).

Such a modeling language is particular relevant for intention recognition in collaborative systems which, based on the delivered model and sensor data provided by the environment, derive the intention of the actors. In this paper we illustrate the rationale of the Collaborative Task Modeling Language (CTML) and focus on a new component which integrates task modeling and location modeling for CEs.

The remainder of the paper is structured as follows: In Section 2 we discuss related work from different domains and highlight assets and drawbacks of the presented approaches. The following section introduces the general modeling approach of CTML. Subsequently location modeling in the context of CEs is discussed and the selected modeling technique is shown. Section 5 gives an overview of our tool environment to create, share and animate CTML specifications. Finally we conclude and give an outlook for future research avenues.

2 Background Information

In this section the reader is reminded of core concepts involved in classical task modeling and location modeling. We examine existing approaches in collaborative task model modeling and review commonly used formalisms and relevant related work.

Task analysis and task modeling is a well-established research field in HCI. Various notations for task modeling exist. Among the most popular ones are GOMS [4], HTA [5], CTT [2], and WTM [6]. Even though all notations differ in terms of presentation, level of formality and expressiveness, they assume the following common tenet: tasks are performed to achieve a certain goal. Moreover, complex tasks are decomposed into more basic tasks until an atomic level has been reached. Within the domain of HCI, CTT is the most popular notation, as it contains the richest set of temporal operators and it is supported by a tool, CTTE [7], which facilitates the creation, visualization and sharing of task models. A comprehensive overview on CTT can be found in [2].

In order to support the specification of collaborative (multi-user) interactive systems, CTT has been extended to CCTT (Collaborative ConcurTaskTrees) [7]. Similar to the corporative task modeling language presented in this paper, CCTT uses a role-based approach. A CCTT specification consists of multiple task trees. One task tree for each involved user role and one task tree that acts as a "coordinator" and specifies the collaboration and global interaction between involved user roles. The main shortcoming of CCTT is that the language does not provide means to model several actors simultaneously fulfilling the same role as well as that an actor is assumed to fulfill only one role within a CCTT specification (strict one to one mapping of actors and roles). We try to overcome this limitation by the task modeling language presented here.

Tasks are always performed within a certain context or environment and hence their interplay with the environment should also be taken into account. This issue was first tackled by Dittmar and Forbrig [8] who proposed modeling the execution environment in accordance with the task specification. The environment captures the domain entities which are manipulated, created or needed for the performance of a certain task. Unfortunately the approach by Dittmar and Forbrig is not very well integrated with standard software engineering models. We try to overcome this shortcoming by proposing UML class diagrams [9] as a notation to model the environment (domain) model while instances of the model are represented using UML object diagrams [9]. However, the environment of task execution is not limited to domain objects but several other facts can be taken into account. Present devices with their various capabilities and constraints as well as the position of the task performing actor and involved objects need to be taken into account when modeling tasks in CEs.

In [10], Molina et al. propose a generic methodology for the development of groupware user interfaces. The approach defines several interrelated models capturing roles, tasks, interactions, domain objects and presentation elements. Even though, the models cover all important aspects involved in groupware user interfaces, they are only used at the analysis stage. Subsequent development phases (e.g. requirements or design) are not covered. The methodology is not assisted by a tool which would facilitate the creation and simulation of the various models. In particular, the latter is an important shortcoming since the animation of models is an important technique to obtain stakeholder's feedback. The same disadvantages apply to the work of [11] where a modeling approach for collaborative systems is presented. The approach only covers the analysis phase. An execution semantics and tool support for the various models is not offered.

As already pointed out, the environment of task execution needs to be taken into consideration while modeling tasks. In CEs, where tasks are carried out in a physical environment, the location of the performing actors as well as the involved objects have an impact on the execution of tasks. On one hand an actor may need physical access to tools or artifacts to perform a certain task, on the other hand tasks might only be executable at a spatial position of the environment.

Different attempts have been made to model the spatial relation of objects and actors in physical environments. There are geometric models which define the spatial relation by coordinates of the objects. Applications can easily derive containment relation of objects and location. A disadvantage of those models is that properties like *is connected to* are not easy to derive.

Graph-based location models explicitly model this relation. A node specifies a location and edges represent connections between locations. Weights can be added to model distances between locations. Another type of models uses sets to specify locations and their decomposition into sub-locations. An atomic location is specified by a shape and position. Composed locations are defined by a set of existing locations. The containment relation of locations can be easily expressed using sub-set relations. Hierarchical models are also based on a set of locations which are ordered according to the containment relation. The most used types of model combine several modeling approaches to suit the special needs of the application [12].

3 The Modeling Approach: CTML Specifications

In this section we present the collaborative task modeling language (CTML). First we point out the basic structure of CTML specifications and give reason why this structure suits best. Based upon that, we go into details about the semantics of CTML specifications. As a running example a conference session will be used.

A more comprehensive description of the syntax and semantics of CTML specification can be found in [13].

The design of CTML is based on three fundamental assumptions:

1. In limited and well-defined domains the behavior of an actor can be approximated through her/his role.
2. The behavior of each role can be adequately expressed by an associated collaborative task expression.
3. The execution of tasks may depend on the current state of the environment and in turn may lead to a state modification. The state is defined by the accumulation of states given by various models (such as the domain model).

Table 1. Entities of CTML specifications

Entity	Description
Role	A role specifies a stereotypical user of the environment in the current domain (e.g. Chairman and Presenter in a conference session).
Collaborative task expression	A task expression provides a behavioral description of the assigned role. It may contain task dependencies to other task expression.
External models	External models are used to specify dependencies to relevant entities whose state can be used to define task execution constraints (such as the domain or location model).
Configuration	A configuration specifies a runtime configuration of the CTML specification. It defines the set of actors including the assigned roles as well as the domain model instance (object model). A runtime configuration is the handle to animate the CTML specification.

Based on these assumptions we define a collaborative task model as a tuple consisting of a *set of roles*, a *set of collaborative task expressions* (one for each role), a *set of external models* (such as the *domain model* or *location model*) and a *set of configuration*s (see Table 1.).

Different types of relation between these entities exist: *depends, uses*. A role may *depend* on tasks of another role which means that the task performance of the target role needs the execution of certain tasks from the source role. The *uses* relation specifies that certain modeled objects are needed to accomplish the task set of the target role. Additionally the referenced objects can be manipulated via task execution. A high level view on a CTML specification is depicted in Figure 1.

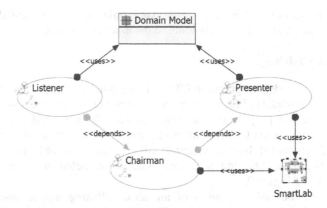

Fig. 1. CTML – Cooperation Model for "Conference Session Scenario"

As already briefly mentioned a configuration defines the runtime configuration of the model. A set of configuration may exist in a CTML specification which makes the model more flexible. A configuration contains the set of actor as well as their assignment to roles. Each actor belongs to one or more role(s). Additionally the instances of models, if necessary, are specified (the initial state of the object model corresponding to the domain model). Figure 2 shows such a configuration. The actors within the model are *Peter* and *Maik* who act as *Chairman* and/or *Presenter*.

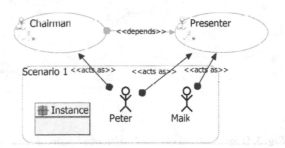

Fig. 2. Configuration *Scenario 1* of CTML Specification for "Conference Session Scenario"

The behavioral description of a role is defined by means of collaborative task expression which has the form a CTT-like task tree [2]. Besides the hierarchical structure and temporal operators a collaborative task expression consists of a set of task, each defined by an identifier, a task type (similar to CTT), a set of preconditions and effects.

Preconditions add additional execution constraints to a task as a task may only be performed if its preconditions are satisfied. Conditions can either be defined over the state of other tasks (enabled, disabled, running, suspended etc.), which potentially may be part of another task expression, or the state of external models integrated into CTML (e.g. domain and location model). An effect denotes a state change of the system or environment as a result of a task execution. Both, preconditions and effects are needed to model collaboration and synchronization across collaborative task expressions. They also denote the binding to the external models.

3.1 Domain Modeling

An instance of a model integrated in CTML is the domain model. The domain model reflects the fundamentals of the domain: the involved entity and their structure as well as their interrelation [14]. In our case we consider the involved objects needed to execute a task within the CE as domain objects. To model these objects we employ UML class diagrams [9]. Instances, representing a state of all objects, are represented by object diagrams. The initial object state is modeled and imported into a configuration (see Figure 2.).

To constrain the task execution of an actor fulfilling a role one can specify precondition using OCL syntax (Object Constraint Language of the UML [9]). The same applies for effects but a slightly adapted syntax is used since OCL offers purely querying capabilities. For more detailed information consult [15].

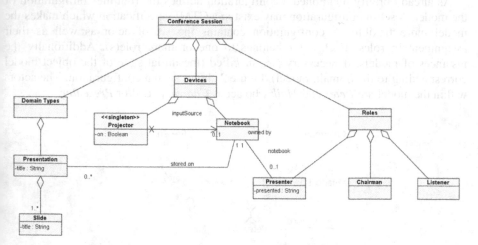

Fig. 3. Domain Model of the "Conference Session Scenario"

3.2 Execution Semantics

After defining the core concepts and involved entities we pinpoint the execution semantics of CTML specifications. A well-defined execution semantic can not only rule out ambiguities but also allow for animation. This is particular important for the creation of prototypes and derivation of lower level models used to operate the collaborative environment [16, 17].

To clarify the execution semantic the runtime model has to be investigated first. To animate a CTML specification the selected configuration has to be transformed into a CTML animation. This step is called instantiation and creates for every actor the corresponding task model animation. This animation is determined by the role assignment of the actor. For each assigned role the corresponding collaborative task expression is translated into a task animation expression. These animations are composed via the temporal operator interleaving (also known as concurrent). In this vein it is specified that roles can be fulfilled concurrently.

Whether a task is enabled during animation depends upon three different factors: (1.) the task-subtask decomposition, (2.) the defined temporal operators, and (3.) the precondition specified for the task. (1.) and (2.) are commonly known concepts from task analysis and modeling. Even though there are notations which support preconditions (3.) their usage is most often limited to informal precondition in text form. Only in [8] the authors use automatically evaluable preconditions to constraint the task execution. Our approach goes beyond that by offering different types of precondition to integrate different models. So far, there are four types of preconditions as depicted in Table 2.

Table 2. Types of Preconditions

Type	Rationale
Simple Task Precondition	This precondition enhances temporal operator by adding additional execution constraint within the collaborative task expression of the role.
Cooperative Task Precondition	This precondition addresses tasks of other roles defined in the CTML specifications. Because there may exist several actors fulfilling the same role these statements need to be quantified using the *allInstance*, *noInstance* or *oneInstance* quantifier. Those statements address multiple tasks, namely of each actor fulfilling the role.
Domain Precondition	Precondition defined in OCL are used to restrict the task execution with regard to the domain model.
Location Precondition	Precondition addressing the location of the actor performing the task (see Section 4).

In the same vein we defined different types of effects to reflect how the performance of a task changes the system state. In Figure 3 a visualization of a task modeling animation for the conference session is given.

4 Location Modeling

So far we have introduced the Collaborative Task Modeling Language and have given the reader a hint that location modeling needs to be integrated with task modeling for the application of CEs.

In the context of CTML this can be achieved by integrating location preconditions and effects into CTML specifications which was already sketched in the last section (see Table 2). In order to define proper location dependencies a conceptual model of the location needs to be introduced. On one hand the model has to be powerful enough to express complex location specifications for a CE, on the other hand specifying locations should not be a burden.

Therefore we use a local geometric model which uses a simple set of geometric figures (rectangle, ellipse, point) and their compositions. In Figure 5 a screenshot of the location editor is shown. It uses a 2-d model of our experimental infrastructure at our university as background to ease location modeling. The orange figures (located on top of the 2-d model) are simple figures (rectangle, ellipse, point) whereas blue figures (located on the right hand side) denote composed locations. Thus, complex locations can be modeled by composing simple ones. To make use of the defined locations the model has to be imported into a cooperation model and referenced via the *uses* relation (see Figure 1). This relation expresses that all defined location can be used to constrain the task execution of the role involved in the relation. Informal this means that the behavior of the role fulfilling actor depends upon the position of the actor in the CE.

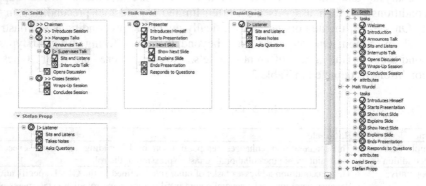

Fig. 4. Interactive Animation of the Session Scenario with four Actors

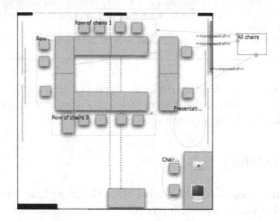

Fig. 5. Location model of Experimental CE "SmartLab" for the session scenario

Locations can either be used to define preconditions or effects. Evidently a location precondition defines that a certain task can only be performed if the actor is located within location referenced via the precondition. Performing a task can also change the position of the actor which is considered as an effect of the task execution in our model. Therefore an effect defines that an actor performing a task moves from one location to another.

5 Conclusion and Future Research

In this paper we have introduced a holistic approach to task modeling, namely the Collaborative Task Modeling Language, which offers a generic framework to integrate external models. The instruments used for integration are preconditions and effects of tasks. The integration was already accomplished for the domain and location model. Because CTML specifications can become quite complex an editor and interpreter was implemented which allows for interactive animation of CTML specification in order to ease the design and foster the understanding (see Figure 4.).

Within our research project an experimental infrastructure, namely "SmartLab", has been set up. To provide proactive assistance the intention of the users has to be determined. The intention recognition module of the system uses as input the potential execution order of tasks (an acyclic graph) plus the position data of users over time. Because CTML specifications have a well-defined semantics an acyclic graph of the potential task execution order can be derived straightforward by the use of the interpreter. Evidently task models can be created much handier than acyclic graphs [17]. We are currently setting up the interface to the system.

For future research avenues we consider the integration of a device model reflecting the potential task support by devices. At present we think that a state chart based approach is most suited [9]. A device can be defined as a class within the domain model which represents the static structure of the entity whereas the behavioral description is delivered through a state chart. States and transitions can be referenced in preconditions and effects of tasks. Due to the structure of the framework other model types can be integrated as well, such as sensor data or contextual information.

Another issue which needs further investigation is the team model. Usually individual tasks are not only influenced by the environment state and task of other actors but by a higher order model. If we consider a session at a conference, there is usually an agenda which defines when a certain talk is given. Clearly such a model is a sort of task model as well. We think that CTML specification can be used as notation for such a team model. The relation of team tasks to individual tasks will be examined in future research as well.

Acknowledgment. Supported by a grant of the German National Research Foundation (DFG), Graduate School 1424, Multimodal Smart Appliance Ensembles for Mobile Applications (MuSAMA).

References

1. van der Veer, G., Lenting, B., Bergevoet, B.: GTA: Groupware task analysis - modeling complexity. Acta Psychologica 91, 297–332 (1996)
2. Paterno, F.: Model-Based Design and Evaluation of Interactive Applications. Springer, London (1999)
3. Dittmar, A., Forbrig, P.: The Influence of Improved Task Models on Dialogues. In: CADUI, pp. 1–14 (2004)
4. John, B.E., Kieras, D.E.: The GOMS family of user interface analysis techniques: comparison and contrast. ACM Transactions on Computer-Human Interaction 3(4), 320–351 (1996)
5. Annett, J., Duncan, K.D.: Task Analysis and Training Design. Journal of Occupational Psychology 41, 211–221 (1967)

6. Bomsdorf, B.: The WebTaskModel Approach to Web Process Modelling. In: Winckler, M., Johnson, H., Palanque, P. (eds.) TAMODIA 2007. LNCS, vol. 4849, pp. 240–253. Springer, Heidelberg (2007)
7. Mori, G., Paternò, F., Santoro, C.: CTTE: Support for Developing and Analyzing Task Models for Interactive System Design. IEEE Trans. Softw. Eng. 28(8), 797–813 (2002)
8. Dittmar, A., Forbrig, P.: Higher-order Task Models. In: Jorge, J.A., Jardim Nunes, N., Falcão e Cunha, J. (eds.) DSV-IS 2003. LNCS, vol. 2844, pp. 187–202. Springer, Heidelberg (2003)
9. UML, Unified Modeling Language, http://www.uml.org/ (accessed July 10, 2008)
10. Molina, A.I., Redondo, M.A., Ortega, M., Hoppe, U.: CIAM: A Methodology for the Development of Groupware User Interfaces. Journal of Universal Computer Science 14, 1435–1446 (2008)
11. Penichet, V.M.R., Lozano, M.D., Gallud, J.A., Tesoriero, R.: Analysis models for user interface development in collaborative systems. In: CADUI 2008, Alabcete, Spain (2008)
12. Becker, C., Dürr, F.: On location models for ubiquitous computing. Personal Ubiquitous Computing 9(1), 20–31 (2005)
13. Wurdel, M., Sinnig, D., Forbrig, P.: Towards a Formal Task-based Specification Framework for Collaborative Environments. In: CADUI 2008, Albacete, Spain (2008)
14. Sommerville, I.: Software Engineering (Update), 8th edn. International Computer Science. Addison-Wesley Longman Publishing Co., Inc., Boston (2006)
15. Wurdel, M., Sinnig, D., Forbrig, P.: CTML: Domain and Task Modeling for Collaborative Environments. JUCS, 14(human-Computer Interaction) (2008)
16. Giersich, M., Forbrig, P., Fuchs, G., Kirste, T., Reichart, D., Schumann, H.: Towards an Integrated Approach for Task Modeling and Human Behavior Recognition. In: Jacko, J.A. (ed.) HCI 2007. LNCS, vol. 4550, pp. 1109–1118. Springer, Heidelberg (2007)
17. Wurdel, M., Burghardt, C., Forbrig, P.: Supporting Ambient Environments by Extended Task Models. In: Proc. AMI 2007 Workshop on MDSE for AmI Applications (2007)

Approach to Human Centered Design Innovation by Utilized Paper Prototyping

Kazuhiko Yamazaki

Faculty of Engineering, Department of Design, Chiba Institute of Technology
2-17-1 Tsudanuma Narashino-shi,
Zip 275-0016, Japan
designkaz@gmail.com

Abstract. The purpose of this study is to discover a design methodology for User Centered Design (UCD) Innovation. This paper focuses on paper prototype method for user evaluation and design. After proposing an approach to utilize paper prototype method, author proposed detail approach based on UCD process. In case study, author utilized this method for design education of design course on university. As a result, author received several innovative ideas from UCD view point.

Keywords: UCD, innovation, prototyping, design.

1 Introduction

Purpose of this study is to propose effective paper prototyping method for User Centered Design (UCD). Recently, UCD is very popular for many companies to develop a product or system form user view point. One of philosophy of UCD is to focus user at each stage of the design process and to repeat prototyping and evaluation to reach user goal.

On this paper, I focus paper prototyping method to utilize on UCD process. Snyder defined "Paper prototyping is a variation of usability testing where representative users perform realistic tasks by interacting with a paper version of the interface that is manipulated by a person "playing computer", who doesn't explain how the interface is intended to work.". And, Snyder described the benefits of paper prototyping as follows;

1. Provides substantive user feedback early in the development process-before you've invested effort in implementation.
2. Promoted rapid iterative development. You can experiment with many ideas rather the betting the farm on just one.
3. Facilitates communication within the development team and between the development team and customers.
4. Does not require any technical skills, so a multidisciplinary team can work together.
5. Encourages creativity in the product development process.

M. Kurosu (Ed.): Human Centered Design, HCII 2009, LNCS 5619, pp. 367–373, 2009.

2 Utilize Paper Prototyping Method

2.1 Utilize Paper Prototyping Method for Design and Evaluation

Snyder defined paper prototyping as a variation of usability testing. We know most designer utilize paper to study idea as a draft prototyping method. And we know one of presentation method is to act user and product to describe new product or system.

During concept stage, we utilize sketch very often. Based on my experience, following is the merit of paper prototyping to compare with sketch.

1. Real size prototyping is good tool to consider idea from human body perspective. For example, small sketch affect different feeling from real size prototyping because size is very important for human. In case of software, screen size is very important for user and paper prototyping has to be real size.
2. For future product or systems, we need to consider hardware, software and human ware as an integrated design system. The paper prototyping will be able to cover hardware, software and human ware. For example, the paper prototyping will cover new product with hardware and software.
3. In case of sketch, it is very different image by technical skill to make superior sketch. Non designer will not easy to join the idea creation by sketch. By paper prototyping, most people will be able to join the creation and promote collaboration.
4. To make innovative idea, it is important to create from user experience view point. The paper prototyping helps to feel user experience because of real size and real steps.

By considering merit of paper prototyping, I propose to utilize paper prototyping for creating idea, make presentation and evaluation as integrated tool for UCD. And I try to define paper prototyping as follow;

"Paper prototyping is an integrated tools to design interactive product and system by UCD. Paper prototyping is useful for creating idea, making presentation, design walkthrough and user testing on UCD process. Based on this definition, I categorized paper prototyping for 4 phase as follows and also I summarized this phase on Table-1.

Table 1. Utilize prototyping by each phase

Phase	Title	Purpose	Team
1	Creating idea with paper prototyping	Study idea	Designer
2	Acting out with paper prototyping	Propose idea	Actor as a user Actor as a product
3	Design walkthrough with paper prototyping	Evaluate idea in design team	Actor as a user Actor as a product Coordinator
4	Usability testing with paper prototyping	Evaluate idea by user	Actor as a user Actor as a product Coordinator Observer

1. Phase-1: Creating idea with paper prototyping
2. Phase-2: Acting out with paper prototyping
3. Phase-3: Design walkthrough with paper prototyping
4. Phase-4: Usability testing with paper prototyping

2.2 Creating Idea with Paper Prototyping

During concept design phase, paper prototyping is one of good method to create idea from user view point. We should make paper prototyping as real size based on user scenario and inspiration quickly. It is different from paper model for styling because paper prototyping has to be real size, describe controls and screen and not need fancy model. It is important to check paper prototyping by considering user experiences. (Fig.1 and Fig.2).

Fig. 1. Paper prototyping for hardware

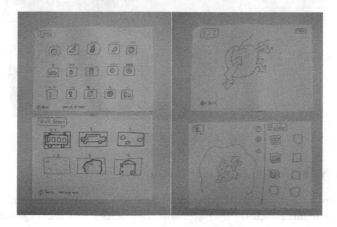

Fig. 2. Paper prototyping for software

2.3 Acting out with Paper Prototyping

During concept design phase, paper prototyping is one of good method to perform acting out for presentation. Based on user scenario, two actors will play skit with the paper prototyping. The paper prototyping has to cover all the material for the scenario such as hardware product and software interface. One actor is a user and another actor is a product with paper prototyping. By this acting out, actor and observer will find out problems or the point for modification. (Fig.3 and Fig.4).

Fig. 3. Acting out for hardware

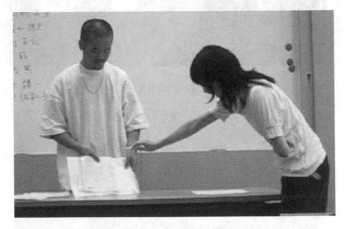

Fig. 4. Acting out for software

2.4 Design Walkthrough with Paper Prototyping

During concept evaluation phase, paper prototyping is one of good method to evaluate idea in design team. Based on user scenario, two actors and one organizer will collaborate with paper prototyping. One actor is a user, another actor is a product. One organizer directs skits and observes performance. By this design walkthrough, actors and observer will find out problems and idea from user view point.

2.5 Usability Testing with Paper Prototyping

During concept evaluation phase, paper prototyping is one of good method to evaluate idea by user. In this case, it is important to prepare several pass for user scenario for user testing. Based on user scenario, two actors, one organizer and one observer will collaborate user testing. One actor is a user, another actor is a product. One organizer directs skits and observer check the problem and findings. By this user testing, observer will find out problems and idea from user view point.

3 UCD and Paper Prototyping

3.1 Variety of Prototyping Method

On this section, I confirm variety of prototyping at each UCD process. Arnowitz described major prototyping method for software development as follows:

- Wireframes
- Storyboard
- Paper prototyping
- Digital prototyping
- Blank model prototyping
- Video prototyping
- Wizard of Oz
- Coded prototyping

In this paper, I classified prototyping for low fidelity prototyping and high fidelity prototyping to cover software development, hardware development and service development

Low fidelity prototyping is to show design concept from user view point. By low fidelity prototyping, we can see and touch the user value and user experience in draft. For example, we utilize paper prototyping as a low fidelity prototyping and it is very useful method with cheap cost. Following is the example of method for low fidelity prototyping.

- Wireframes
- Storyboard
- Paper prototyping
- Paper mockup
- Blank model prototyping
- Wizard of Oz

High fidelity prototyping is to show detail design from user view point. By high fidelity prototyping, we can see and touch the final design or detail design. For example, we utilize hardware mockup which has final color and shape, and also software prototyping by flash. High fidelity prototyping is useful method on design step or detail design step on UCD. Following is the example of method for high fidelity prototyping.

- Digital prototyping
- Video prototyping
- Coded prototyping
- Rapid prototyping
- Detailed mockup

3.2 UCD Process and Paper Prototyping

The paper prototyping is useful method to get user feedback with minimum effort. But this method has several demerit and we need to consider these demerit to utilize this method on UCD process. Following is the demerit;

1. User will be influenced by shape and color of the paper prototyping. For example, when the conceptual idea is superior for the user, user will have negative feedback because of the shape and color of the paper prototyping.
2. The paper prototyping will not cover user feedback for image and appearance. For example, rendering sketch or video image will be able to cover user feedback for image and appearance as a low fidelity prototyping.
3. For user testing by the paper prototyping, we need to consider the knowledge of prototyping. When user does not knowledge of prototyping, most users will evaluate by the image of paper prototyping. For example, most users will have negative feedback for paper prototyping. When user has experience for development, user will be able to understand the quality of paper prototyping.

By considering merit and demerit of paper prototyping, I propose to utilize paper prototyping on phase 2 (Understanding user) and 3.(Concept design) on UCD process. In case of utilizing on phase 4(Detail design), we need to consider demerit of the paper prototyping.

I summarized UCD process and paper prototyping on Table-2.

Table 2. UCD Process and paper prototyping

Phase		Prosess-1 Define Require- ment	Prosess-2 User Research	Process- 3 Concept Design	Process-4 Detail Design	Process-5 Verification	Process- 6 Life cycle
1	Creating idea with paper prototyping		Create Idea for research	Create idea	Create idea		
2	Acting out with paper prototyping		Present Idea for research	Present idea	Present idea		
3	Design walkthrough with paper prototyping		Evaluate Idea internal	Evaluate Idea internal	Evaluate Idea internal		
4	Usability testing with paper prototyping		Evaluate idea by user	Evaluate Idea by user	Utilize detailed prototyping instead of Paper prototyping		

4 Conclusion

To paper focuses on paper prototype method for user evaluation and design. After proposing an approach to utilize paper prototype method, author proposed detail approach based on UCD process. In case study, I utilize this method for design education of design course on university. As a result, I received several innovative ideas.

Reference

1. HCD-Net Website, http://www.hcdnet.org/en/index.html
2. Nielsen, J.: Usability Engineering. Academic Press, London (1993)
3. Carroll, J.M.: Scenario-based Design—envisioning work and technology in system development. Wiley, US (1996)
4. Carroll, J.M.: Making Use of Scenario-based design of human-computer interactions. MIT Press, US (2000)
5. Snyder, C.: Paper prototyping. Morgan Kaufmann, San Francisco (2003)
6. Arnowitz, J., Arent, M., Berger, N.: Effective Prototyping for Software Makers
7. Yamazaki, K., Furuta, K.: Proposal for design method considering user experience. In: 11th International Conference on Human-Computer Interaction, Las Vegas (2005)

Structured Scenario-Based Design Method

Koji Yanagida[1], Yoshihiro Ueda[2], Kentaro Go[3], Katsumi Takahashi[4],
Seiji Hayakawa[5], and Kazuhiko Yamazaki[6]

[1] Kurashiki University of Science and the Arts, Kurashiki, 712-8505, Japan
[2] Fujitsu Design, Ltd., Kawasaki, 211-8588, Japan
[3] University of Yamanashi, Kofu, 400-8511, Japan
[4] Holon Create Inc., Yokohama, 222-0033, Japan
[5] Ricoh Company, Ltd., Yokohama, 222-8530, Japan
[6] Chiba Institute of Technology, Narashino, 275-0016, Japan
yanagida@arts.kusa.ac.jp, y.ueda@jp.fujitsu.com,
go@yamanashi.ac.jp,
takahasi@hol-on.co.jp, hayakawa@rdc.ricoh.co.jp,
designkaz@gmail.com

Abstract. This paper introduces "The Structured Scenario-based Design Method", a design approach where a vision is proposed from an HCD (Human-Centered Design) perspective for use with ubiquitous computing. This method utilizes structured scenarios that are created in order to appropriately incorporate users' intrinsic needs and values into systems/products specifications at an early stage of designing. This paper discusses the method, articulating characteristics, a design process and a few case examples using a tool developed for this particular method.

Keywords: Scenario, persona.

1 Introduction

The concept of ubiquitous computing [1], introduced by Mark Weiser in 1988, is becoming a reality. Computers and sensors, embedded in the environment, are organically linked through networks, allowing users to use computers without even thinking about them. The challenge in developing systems/products for the ubiquitous age is not only solving problems with existing products and services, but also creating ideas for new products and services not yet in existence. Furthermore, we must consider users' intrinsic needs so that products and services to be developed can be truly accepted by people and society.

In 2007, our team organized a working group within the Ergonomic Design Research Group of the Japan Ergonomics Society to research a practical HCD method that will help develop future generation systems/products for the ubiquitous age. This paper outlines the accomplishments of our research; "The Structured Scenario-based Design Method." [2].

M. Kurosu (Ed.): Human Centered Design, HCII 2009, LNCS 5619, pp. 374–380, 2009.

2 The Structured Scenario-Based Design Method

2.1 Characteristics of the Structured Scenario-Based Design Method

From an HCD perspective, there are two major approaches to the development of systems/products. One is a problem-solution design approach, where usability of existing systems and products is evaluated and problems are solved. The other approach is a vision-proposal design approach where new services, not yet in existence, are envisioned based on users' intrinsic needs and values in order to create systems/products that satisfactorily meet these needs. The former is an effective approach in improving interaction problems between users and systems/products, and deals with "expected values" from a perspective of user satisfaction. The latter approach deals with "attractive values," and is intended to introduce users to new values and experiences. Either approach can be chosen according to the particular purpose of development.

The Structured Scenario-based Design Method is especially effective with a vision-proposal design approach, intended to propose visions that are securely acceptable in people's lives and society, and to further develop them. Figure 1 shows the Structured Scenario-based Design Model. The following are its characteristics.

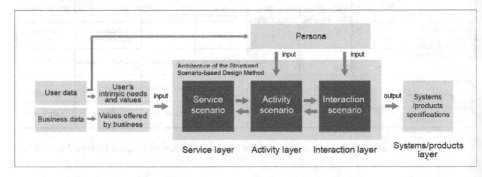

Fig. 1. Basic model for the Structured Scenario-based Design Method

1) To use a four-layered approach to exploration of systems/products development in the ubiquitous age

The Structured Scenario-based Design Method is based on architecture known as "The three layered model for ubiquitous computers society," [3] intended to facilitate understanding and discussion on products and services in the ubiquitous age. In this architecture, the upper service layer is used to discuss values that users would enjoy, the middle interaction layer is to discuss interactions between users and systems/products, and finally the lower system layer is to discuss specific technologies and specifications. In addition to the three layers, the Structured Scenario-based Design Method has added another layer, the "activity layer," between the service layer and the interaction layer in order to better identify users' activities and behaviors. The four-layered model can be used as a human-centered approach to an in-depth structural discussion on systems/products in the ubiquitous age.

2) To make use of scenarios and personas
The human-centered design approach requires that user experiences be identified to closely examine systems/products. For this reason, the Structured Scenario-based Design Method uses scenarios at every stage of products development. Also, scenarios use personas to represent users. Scenarios and personas are a useful means of sharing goals and ideas among a multidisciplinary design team in the human-centered design process.

3) To output structured scenarios and provide information stipulating systems/products specifications
In the Structured Scenario-based Design Method, users' intrinsic needs and values, values offered by business, and persona profiles are used as input information. Discussions are carried out in each functional layer, and three different structured scenarios are written and output. These three scenarios consist of a service scenario in the service layer, an activity scenario in the activity layer and an interaction scenario in the interaction layer. These scenarios are eventually used as information in order to help define systems/products specifications. Figure 2 shows characteristics of each scenario.

	Description				Evaluation criteria	Creator (team)
	User expectation Business goals	User profile User's activities	Systems/products	Technologies		
Service scenario User's intrinsic needs and values Values offered by business	Covered	Not covered	Not covered	Not covered	Business HCD(satisfaction)	Business HCD
Activity scenario Entire activity flow	Partially	Covered	Not covered	Not covered	HCD (effectiveness)	HCD
Interaction scenario interaction with systems/products	Partially	Partially	Covered	Not covered	HCD (efficiency)	HCD Technology
Systems/products specifications Realization through systems	Partially	Partially	Partially	Covered	Technology	Technology

Fig. 2. Characteristics of each scenario in the Structured Scenario-based Design Method

4) A design process that connects service and systems/products specifications through user experience
The Structured Scenario-based Design Method proposes a concept for design processes where a discussion on users' needs and values start at the service level, which is the uppermost layer of the four-layered model. Then, users' needs and values are incorporated into user activities and interactions to help realize new services. Finally, information and data are obtained to explore systems/products specifications that should meet the needs and values. This design process allows users' intrinsic needs and values to be considered throughout every stage of the systems/products development, through to the specification stage.

5) The method is based on human-centered design, and includes usability development
The Structured Scenario-based Design Method is based on a human-centered design process that is characterized by active involvement of users at an early stage, iteration of design solutions and multi-disciplinary design.

In addition, the Structured Scenario-based Design Method is intended to develop usability defined by ISO9241-11 [4] as "Extent to which a product can be used by specified users to achieve specified goals with effectiveness, efficiency and satisfaction in a specified context of use". The Structured Scenario-based Design Method uses personas to clearly define user profiles. Moreover, each scenario is provided with weighted evaluation criteria in the layers, consisting of effectiveness, efficiency and satisfaction. Service scenario focuses on "satisfaction," activity scenario on "effectiveness" and interaction scenario on "efficiency." Therefore, the Structured Scenario-based Design Method is consistent with usability as defined in ISO9241-11.

2.2 Design Process in the Structured Scenario-Based Design Method

Shown below is the design process used in the Structured Scenario-based Design Method.

1) To start the design process in the Structured Scenario-based Design Method, the following two input elements must be articulated
i) Users' intrinsic needs and values, and values offered by business
Users' intrinsic needs and values are identified by collecting user-related data such as questionnaires as a quantitative approach, or photo diaries, photo essays [5], interviews, or observations as a qualitative approach. The values business can offer are identified based on its business data, including business domain, company-owned technologies, business strategies and business environment.

ii) Persona
Personas are synthesized from data collected from users. Personas are used to develop a service scenario into an activity scenario and an interaction scenario. Personas help better articulate users' behaviors and emotions in scenarios. One or more personas can be created.

2) Create service scenario from users' intrinsic needs and values, and values offered by business
Service scenario is a scenario that demonstrates a concept of a particular service and provides a clear picture of the service. Here, specific user profiles or systems/products are not defined. Instead, a user profile is expressed as user segments or categories. Evaluation is approached from the perspectives of both business and user. From the user's perspective, a special emphasis is placed on user satisfaction in human-centered design.

3) Create activity scenario from service scenario and persona
Activity scenario is a scenario that demonstrates users' activities, articulating scenes depicted in service scenario and persona's goals to be achieved by the services, and describing user activity flow and emotions during all stages of the flow. Here specific systems/products are not described, but rather the main focus is placed on the persona's experiences. For evaluation, a special emphasis is placed on effectiveness in human-centered design.

4) Create interaction scenario from activity scenario and persona

Interaction scenario is a scenario that demonstrates users' interaction with systems/products, where each of the person's interaction developed from the activity scenarios is described in terms of time sequence as the systems/products are used. Concrete ideas for realization of systems/products are given in this scenario. For thorough and in-depth discussion, hardware, software and human ware characteristics must be addressed. For evaluation, a special emphasis is placed on efficiency in human-centered design.

5) To realize systems/products specifications, while maintaining consistency between three scenarios and systems/products specifications

Systems/products specifications are discussed through the use of three structured scenarios, which are outputs from the Structured Scenario-based Design Method. For example, use case scenarios can be used. Technical elements are described in systems/products specifications and evaluated from a technical perspective.

3 A Tool and Case Examples of the Structured Scenario-Based Design Method

In the Structured Scenario-based Design Method, templates are created as a tool to create and discuss three layered scenarios consisting of service scenario, activity scenario, and interaction scenario. Figure 3-5 show case examples created through the use of these templates.

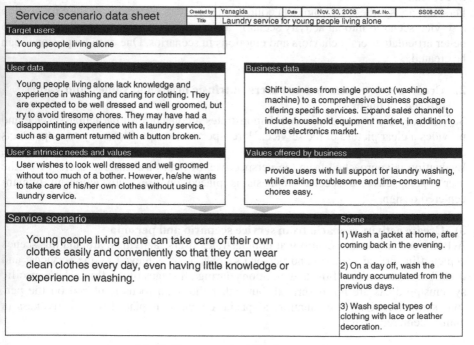

Fig. 3. Case example of service scenario

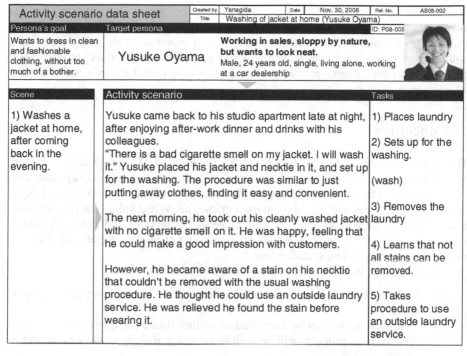

Fig. 4. Case example of activity scenario

Interaction scenario data sheet	Created by	Yanagida	Date	Nov. 30, 2008	Ref. No.	IS08-002
	Title	Wardrobe-type washing machine and collaborated laundry service				

Target persona   ID: P08-005

Yusuke Oyama

Tasks	Interaction scenario
1) Places laundry	Hardware/software/human-ware
	Yusuke opened the door of a wardrobe-type washing machine, took out
2) Sets up for the	empty clothing hangers, placed his jacket and necktie on them, and hanged
washing.	on hook. (Specific clothing data collected: material, history of washing, etc.)
(wash)	When closing the door, "Jacket: 1", "Necktie: 1" appeared on the operation
	panel of the door. He pressed the button "Wash". Then "Air wash ON", "1
3) Removes the	hour to complete" appeared. He saw that the machine sensed the
laundry	contamination level and started optimal air washing.
4) Learns that	The next morning, the operation panel of the door read "Jacket: 1:
not all stains can	Complete", "Necktie: 1: Incomplete". He touched "Necktie" and details
be removed.	appeared. The necktie had a sauce stain, although he did not notice it. He
	learned that the stain could not be removed with his air washing machine.
5) Takes	
procedure to use	The display says that if he sends the necktie to laundry service, it will be
an outside	delivered by the evening. He touched the "Call laundry service" button. He
laundry service.	followed the instruction "Place the necktie in laundry box" and put the
	necktie in the laundry delivery box at the entrance.

Fig. 5. Case example of interaction scenario

"Washing" is a theme. The current situation requires that a user either washes his laundry at home or use an outside laundry service. From users' intrinsic needs and values; "User wants to look well dressed and well groomed, without too much of a bother", a proposal is made to launch new services in the intermediate domain. Then concrete ideas for systems/products are proposed based on the user's activities and interactions.

4 Conclusion

This paper has introduced the Structured Scenario-based Design Method, a design approach where a vision is proposed from an HCD perspective in the ubiquitous age. In the Structured Scenario-based Design Method, scenarios are created for each layer of the layered architecture, based on users' intrinsic needs and values, thereby creating new visions for user experience, as well as ideas for services and systems/products which are high in user satisfaction, effectiveness and efficiency. Finally, it also provides information to define specifications.

Benefits of Structured Scenario-based Design Method include:

1) Systems/products with high usability can be developed from an HCD perspective.
2) Customer values can be identified to endure future competitiveness. Developed systems/products will be high in accuracy, making customers willing to accept and buy.
3) A focused and efficient development process is a way to reduce development lead time and achieve cost savings.
4) The direction for the future of business can be identified, contributing to effective corporate business management.

We intend to verify these benefits through application of the method proposed in this paper, and plan to improve this method to obtain a higher accuracy level.

References

1. Weiser, M.: Computer for the 21st Century. Scientific American 265(3), 94–104 (1991)
2. Yamazaki, K., Ueda, Y., Takahashi, K., Hayakawa, S., Yanagida, K., Go, K.: Universal Design Methodology for Vision Proposal. The Japanese Journal of Ergonomics 44(suppl.), 36–37 (2008)
3. Mori, H.: A Consideration Toward a Human-Centered Ubiquitous Society Based on a Three Layered Model. Journal of the Society of Instrument and Control Engineers 47, 77–81 (2008)
4. ISO 9241-11:1998 Ergonomic requirements for office work with visual display terminals (VDTs) – Part 11: Guidance on usability (1998)
5. Go, K.: Scenario-Based Design for Services and Content in the Ubiquitous Era. Journal of the Society of Instrument and Control Engineers 47, 82–87 (2008)

Facilitating Idea Generation Using Personas

Der-Jang Yu[1] and Wen-Chi Lin[2]

[1] Institute of Applied Arts, National Chiao Tung University, 1001 University Road,
300, Taiwan, R.O.C.
[2] ScenarioLab, Room 415, Building 53, ITRI, No 195, Sec. 4, Chungshin Road,
Chutung Town, Hsinchu, 31040, Taiwan, R.O.C.
djyu@scenariolab.com.tw, grace@scenariolab.com.tw

Abstract. Persona and scenario are important design tools for new concept development. Usually, scenario is used to generate ideas, and persona is for evaluation. This article proposes a new approach that embeds persona data in scenario-based design for idea generation. It includes a persona dataset, a facilitation process, and a working field. The persona dataset is a user profile collected in ethnographic research and categorized by subjects, motives, activities, goals and behaviors. The facilitation process helps designer create new ideas via re-matching the elements in the persona dataset. The working fields allow the freedom of implementing with different sizes of the designer team and persona dataset. This new approach provides a direct and effective way that materializes designers' internal experiences and persona data to create new ideas and scenarios.

Keywords: Persona, scenario-based design, ethnographic.

1 Introduction

Persona and scenario are important design tools for new concept development. Scenarios are usually synthesized from hypotheses and existing experiences, and scenario-based design is a story-telling platform for such syntheses. It is also a platform for idea leaping based on methods such as Claim Analysis. Claim Analysis, proposed by John M. Carroll, is an effective way to challenge the designers to generate new stories and drive the idea generation process forward.

Persona is majorly used as a communication tool to help design team focus on target users and prevent "self referential design" when the designers may unconsciously project their own mental models on the product design. Persona data should be extracted from field such as ethnographic researches. "It's easy to explain and justify design decisions when they're based on persona goal...", said Allen Cooper, who coined the term "persona".

The designer can create new ideas directly from the Persona data, the idea leaps will thus directly be grounded in ethnographic results. The question is how to create evolutional results using the static Persona data? This article provides the answer.

A designer can categorize the ethnographic results into S-M-O-m-o structure of the persona dataset and generate new products or service concepts via re-organizing S-M-O-m-o combination path by selecting different components from the persona dataset.

M. Kurosu (Ed.): Human Centered Design, HCII 2009, LNCS 5619, pp. 381–388, 2009.
© Springer-Verlag Berlin Heidelberg 2009

2 The Dataset

The Persona dataset includes daily activities, purposes and goals, tools and methods of the target user related to the research topic collected from observations, interviews, diary, and focus group meetings.

The persona data are categorized into five subsets. Under the subset "S"(capital S) are demographics and personality attributes such as name, age, gender, education, income, hobbies, and habits. In the activity level, the subset "O" (capital O) contains wishes, desires, or the purposes that S would like to achieve. The capital "M" subset contains S's daily or frequent activities and the services that support them. In the action level, "o" (lower case o) collects the goals that S would like to achieve, and "m" (lower case m) is the behavior framework related to "o".

Activity hierarchy	Subject	Object	Mediating Artifact
Activity level	"S": Personalities, demographics, psychographics, etc	"O": Purposes, needs	"M": Daily activities, frequent activities, services
Action level		"o": Key performance element, goals of regular actions.	"m": Framework of actions and Tools

3 Facilitation Process to Create New Ideas via Persona Dataset

The Persona Dataset reflects the user's current activities and life style. In Figure 1, for Persona S, an O-M-o-m combination is a linkage path with motives, activities, goals and actions that Persona S is used to do. If the "O-M-o-m" is an existed experience in persona's real life, there must be an existed product that supports o-m and persona applies that same product to do O-M activity in his life.

The following steps help the designer detour from the existing paths to generate different ideas.

Step 1. Select a purpose-activity pair from the "O" and "M" subsets in the activity level of the persona dataset.

Step 2. Select an effective element from "o" in the action level that supports the activity in step 1. Note that, this "o" may not be the usual o's associates with this activity, thus the path detours.

Step 3. Find a framework under "m" which may work with the element ("o") selected in step 2. By the same technique applied in step 2, this "m" may or may not be the ones that usually associated with "o".

Step 4. Create a new product or service idea based on the element-framework pair to support the activity (M) to achieve the purpose (O) in step 1.

Now the designer re-organizes the combination of path via selecting different components from the persona dataset, as the path in bold line illustrated in Figure 1.

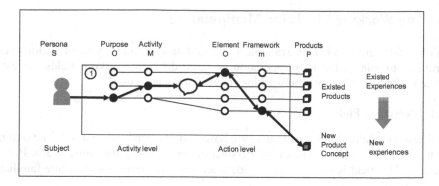

Fig. 1.

In step 4, the designer needs to figure out a new concept of product that supports the new path. The creations of new product concept and the new path are intertwined that the designer seems to play with different linkage paths and a couple of possible new ideas for an optimal result.

Changing different combinations of "O-M-o-m" paths can define an expanded coverage of possible new experiences. Since there are different choices of combinations, the designers can empathize with the persona experience during idea generation.

Here is an example to explain how to apply the above process to create a new "beauty" service for a persona of a female manager.

Step 1, she(S) usually builds up her professional image (O) by attending professional events (M).

Step 2, on the other hand, she polishes her appearance (o).

Step 3, she also refer to famous professional women (idols in the mirror , i. e. "m") to help her enhance her appearance.

Step 4, a new service concept of a hybrid service which provides business management as well as personal image consulting service may help boost her professional image (O). (Figure 2).

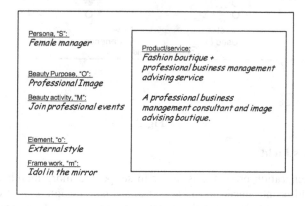

Fig. 2.

4 The Working Fields for Manipulating

With a different number of available Persona datasets, and players of various backgrounds to join in the idea generation process, different working fields are established. The followings are the working fields that we have used.

4.1 Persona's Field

In Persona's field, the S, M, O, m, o, are based on one persona's dataset. The designer can only alter the combination of Persona elements using the same user's Persona dataset. This field is useful when the ideas are required to be new but quite familiar to the persona.

4.2 Designer's Field

In the designer's field, the designer works with more than one persona datasets. He/She is given the freedom to select from all of the persona datasets the S, M, O, m and o's. Based on the integrated experience of the datasets, the designer can create new experiences and generate new ideas. The designer may eventually build a pseudo persona that synthesized from all the original personas for the following process of product development to work on.

The designer field is useful when the ethnographic study covers several research targets to meet a wide spread of demands. When a new market segment emerges, one can create a designer field from the existing persona database to work around. (Figure 3).

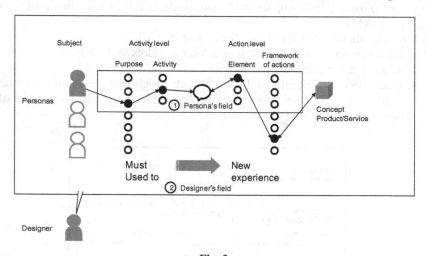

Fig. 3.

4.3 Director's Field

As the idea generation project gets larger in scope and more members are involved, Director's field is introduced to guide and coordinate design team members. It is on top of the designer's field where the director appoints roles of designers and oversees the entire idea generation process to make sure the work is in focus.

In director's field, multiple team members of different expertise share the job that one player does in the designer's field. The essence is to collect individual wisdom of the members while selecting and combining S-O-M-o-m components from persona datasets.

A sequential multi-player game, especially in the director's field, is conducted as the members select components from persona datasets in the order of S, O and M, and o and m. When a member makes a selection, the team is expected to justify the new selection with all the choices made, and thus may involve in intensive brainstorming. Players' internal wisdom will be externalized and shared with one another during the brainstorming process.

It is especially valuable while team members come from different divisions. For example, members from the marketing division can select S, as marketing personnel is sensitive in segmentation; the sales are the right persons to select O and M, as they understand the value of the activities to different users; while the designers select o and m, as the designers are familiar with user's capabilities and habits. (Figure 4).

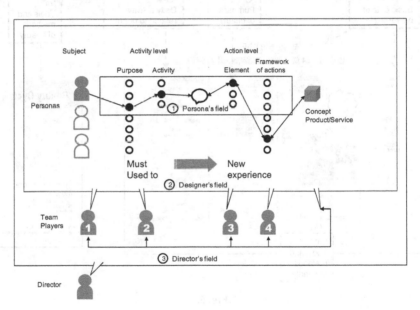

Fig. 4.

5 A Case of Facilitating Idea Generation Using Personas Dataset

This case is to propose an innovative service and/or product for the beauty industry. It was facilitated in the director's field that a 4-designer team used 24 persona datasets to create innovative products or services.

Ethnographic research of 24 users has been conducted. The collected information includes daily activities related to beauty, tools or services employed for being beautiful, the memories of beautiful moments, and their values of the beauty. The persona datasets are collected and filed in the S-O-M-o-m structure. (Figure 5).

The 24 persona datasets of action-level ethnographic data contain tons of trivial elements, which need to be grouped and consolidated into a reduced set of common key words or phrases before they are considered an effective lowercase "o" or "m". For example, skin color, skin elasticity, winkles, and age spots are grouped under "quality of skin" as one of the lowercase "o"s. "Beauty cycle", one of the lowercase "m"s, consists of actions such as applying skin care products, putting on makeup, makeup deterioration, repair makeup, and removing makeup.

After the consolidation process, a workshop in director's field was held to create product/service concepts, scenarios, and a pseudo persona.

Each 4-player team had 24 persona cards which contained S, M and O, a set of "elements of beauty" cards (lowercase "o"), and a set of "behavior framework" cards (lowercase "m"). The team brainstormed based on the cards. They were free to discuss and share viewpoints with one another.

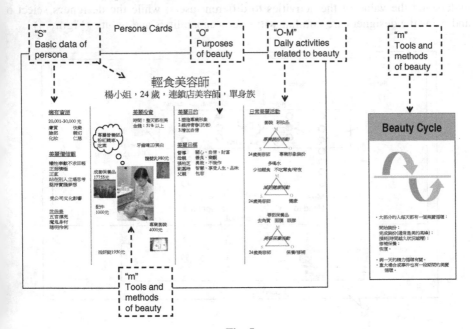

Fig. 5.

From 24 persona datasets, the first player chose a target persona. In this example, it was an office lady who worked on the computer 8 hours a day. The next player selected a motive for her, for example, "staying young". The third player spotted an activity, namely, recreation, for her. Then, the forth and last player searched among the "elements of beauty" cards (lowercase "o") and "behavior framework" cards (lowercase "m") for behaviors and elements that recreate the office lady for staying young and being attractive. For example, "Beauty cycle "("m") and "stay healthy" ("o") were chosen.

The team members were encouraged to think out loud and discuss while making selections. All team members worked together for a new product/service idea for this office lady. The process worked successfully although the team had never heard about persona before. The team generated very innovative but proper idea during 1-hour brainstorming.

Figure 6 illustrates the final proposal for this office lady to stay young at all times, a web-based alarm clock that reminds the owner time to take actions to stay beautiful. A schedule such as time to have some water, time to stretch and take a walk, time to take vitamins, and time to repair makeup, is recommended by the web-based system. This e-type alarm not only reminds things to do, it also can be backed up by a natural food company to support the consumables needed.

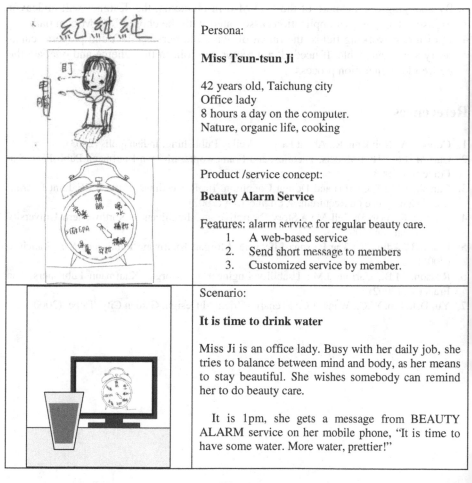

	Persona: **Miss Tsun-tsun Ji** 42 years old, Taichung city Office lady 8 hours a day on the computer. Nature, organic life, cooking

	Product /service concept:
	
	Beauty Alarm Service
	
	Features: alarm service for regular beauty care.
	    1. A web-based service
	    2. Send short message to members
	    3. Customized service by member.

	Scenario: **It is time to drink water** Miss Ji is an office lady. Busy with her daily job, she tries to balance between mind and body, as her means to stay beautiful. She wishes somebody can remind her to do beauty care. It is 1pm, she gets a message from BEAUTY ALARM service on her mobile phone, "It is time to have some water. More water, prettier!"

Fig. 6.

6   Conclusions

This article proposes a new approach which consists of a persona dataset, a facilitating process, and a working field for idea generation. The persona datasets are user profiles categorized into the structure proposed in Activity Theory. The facilitating process helps designer create new ideas; the working fields defines different sizes of designer team and persona datasets to access.

There are several benefits to apply the proposed scheme to facilitate idea generation:

1. The new service or product idea derived from the proposed scheme is rooted to the original ethnographic data; therefore is grounded with reliable facts and expected to be more acceptable in the market.
2. Any S-O-M-o-m linkage path externalizes the activity experience of a persona. It helps designers understand and create ideas easier.
3. By changing one or more of the S-O-M-o-m elements, the designer will endeavor to justify the new user applications associated with the chosen S-O-M-o-m path.
4. In different working fields, the range of available persona data and players' capability are manageable. If needed, a director can join in the scheme and oversee the entire idea generation process.

References

1. Cooper, A., Reimann, R.: About Face 2. Weiley Publishing, Indianapolis (2003)
2. Carroll, J.M.: HCI Models, Theories, and Frameworks. Morgan Kaufmann Publishers, San Francisco (2003)
3. Carroll, J.M.: Scenarios and Design Cognition. In: Proceedings of the IEEE Joint International Conference on Requirement Engineering (2002)
4. Schank, R.: (1995): Tell Me a Story: Narrative and Intelligence. Northwestern University Press, Evanston (2001)
5. Pruitt, J., Adlin, T.: The Persona Lifecycle. Morgan Kaufmann Publishers, San Francisco (2006)
6. Rosson, M.B., Carroll, J.M.: Usability Engineering. Morgan Kaufmann Publishers, San Francisco (2002)
7. Yu, D.J., Lin, W.C., Wang, J.C.: Scenario-Oriented Design, Garden City, Taipei (2000)

Part III

Understanding Diverse Human Needs and Requirements

Auditory and Visual Guidance for Reducing Cognitive Load

Hiroko Akatsu[1] and Akinori Komatsubara[2]

[1] Oki Electric Industry Co., Ltd.,
1-16-8 Chuou, Wrabi-shi, Saitama 335-8510, Japan
akatsu232@oki.com
[2] Waseda University
komatsubara.ak@waseda.jp

Abstract. Auditory and visual guidance are often used as means to make IT equipment easier to use and decrease cognitive load. However, the effective use of the guidance is not yet clarified. Accordingly, there is a case that the guidance disturbs user operation because of inappropriate use of guidance. This paper discusses the effective use of auditory and visual guidance to reduce user's cognitive loads through experiments with simulated ATM systems.

Keywords: Auditory and visual guidance, cognitive load, usability.

1 Introduction

The operation of today's IT equipment is complex due to their multifunctional nature. Therefore, many users including the elderly have difficulty using the equipment. The IT equipment must give better assistance to the elderly users. If the functions are not decreased, it is necessary to examine how cognitive characteristics (attention and memory etc.) influence operation. Then, it is important to consider about adequate attentions for decreasing the cognitive loads of users.

Voice guidance (auditory guidance) is often used as a means to improve cognition. In case of ATMs (Automatic Teller Machines), when a user does not perform the next step after a certain period of time, voice guidance that serves as prevention against operational mistakes is given from the ATM. The guidance may be repeated at a set time. However, there are times when a user did not forget the operation but are still thinking. Then the user tends to make an operational mistake when disturbed by the voice guidance [1].

Moreover, the voice guidance is sometimes not synchronized to the screen massages displayed at that time. Usually the IT equipment is made easier to use by adding visual guidance that synchronizes with the auditory guidance and induces the user to glance at the pertinent display on the screen.

This study examined the method and the effect of auditory and visual guidance using an ATM system that the elderly users find difficult to use.

M. Kurosu (Ed.): Human Centered Design, HCII 2009, LNCS 5619, pp. 391–397, 2009.

2 Guidance and Cognitive Function

It is said that the auditory guidance is effective at aiding operation. However, it has not been discussed what kind of effect auditory guidance gives to cognitive load reduction. Therefore, in this study, the effect of the auditory and visual guidance was organized based on PDS model [2]. The result is shown in the model of Fig. 1.

The user follows the P (plan), D (do), and S (see) model for each step when operating systematic equipment like the ATM, and the PDS model is repeated until all steps are completed.

It is thought that the cognitive function appeals to the PDS at perception, understanding, and judgment level, and guidance respectively promotes visual guidance, understanding and search.

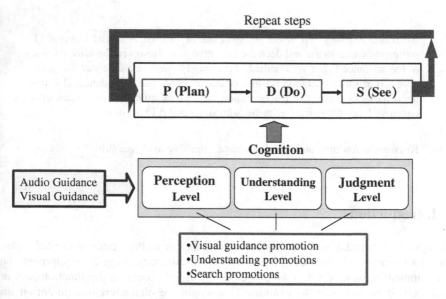

Fig. 1. Guidance and Cognitive Functions

3 Experiment

The effectiveness of guidance for the elderly and middle-aged was compared with four speed levels.

3.1 Experimental System

In this study, an ATM that presented on the screen a load of information including detailed explanation of operation and input steps was used. Not only auditory guidance but visual guidance that induces user's glance to the pertinent display on screen at the same time as the presenting auditory guidance was also shown. It was thought that the visual guidance provided better comprehension of the required information and made looking up desired items easier during searches.

When the screen is switched on, auditory guidance is presented only once and the same content as the auditory guidance is displayed on the screen. There is a case where the understanding level drops when the content of the display and voice guidance are different [3]. There is another case where the presentation timing of the voice guidance disturbs the operation [4].

In the experimental system, auditory guidance was assumed to read out all information related to the operation required on the screen as well as available choices. At that time, the pertinent word corresponding to the reading voice was enclosed with a red frame as a visual guide. The speed of the auditory and visual guidance was set at four levels (A, B, C and D) as shown in Table 1[5]. "A" represented standard speed, and "B" (1.5 times faster), "C" (2 times faster), and "D" (0.8 times slower) were prepared.

The control condition without auditory and visual guidance was defined as "no guidance". "Mora" (beat) is a unit of voice speed, and it is a phonology concept of displaying a time unit. One "Mora" equals one syllabic sound. Larger numbered "Mora" indicates faster voice speed.

3.2 Experimental Participants

The experimental participants were a group of six elderly users (three males and three females, age between 66 and 78), and a group of six middle-aged users (three males and three females, age between 50 and 53). They had negative feelings operating ATMs.

Table 1. System types with the Guidance Speed

System	Guidance Speed (Mora /second)
No guidance	—
A	6
B	9
C	12
D	5

3.3 Experimental Equipment

As an intended system, the ATM simulator was set up with a personal computer and a touch display. A video camera, tiepin-type small microphone, and recording equipments, etc., were prepared as recording tools.

Moreover, the experimental participants were asked to wear an eye camera to measure the point of their gaze.

3.4 Experimental Procedures

Each experiment was conducted by the individual participates. At first, an explanation of the experiment objectives, the use of the equipment, and preliminary questionnaires concerning the use of ATM were conducted prior to performing the tasks. A follow-up survey was given after the tasks had been completed, and additional interviews were conducted. The experimental task was "money transfer". The operational orders were counterbalanced.

4 Results and Considerations

4.1 Use of Guidance

The use of guidance while operating the ATM was counted from user's glance analysis. The elderly users used the guidance more than the middle-aged users. Each group datum was analyzed using the Chi-Square Test for the comparison of systems (Fig. 2). As a result, the elderly users were found to show no significant difference in the frequency of use, while the middle-aged users showed a significant difference ($\chi^2_{(3)}=18.211, p<.01$). It was found that the middle-aged users use the guidance of the "C" system more than the "A" system or "D" system. Therefore, the middle-aged users use the guidance of fast speed. The elderly users tend to use the less fast speed of the "B" system.

4.2 Operation Time

The bank selection step was studied to evaluate the influence of search promotion. Each group datum was analyzed by variance (Fig. 3). According to the analysis, both the elderly users and middle-aged users showed significant difference ($F_{(4,20)}=2.52, p<.10$; $F_{(4,20)}=5.74, p<.01$). Results of the multiple comparisons using LSD indicate the operation time required for "no guidance" was more than the 4-level speed guidance.

The confirmation step (confirmation of all selected items) was studied to evaluate the influence of understanding promotion. Each group datum was analyzed by variance. As a result, the middle-aged users showed no significant difference while significant difference ($F_{(4,20)}=3.79$, $p<.01$) was found in elderly users. Results of the multiple comparisons using LSD indicate the operation time required for "no guidance" was more than the 4-level speed guidance. This is the same as the bank selection step.

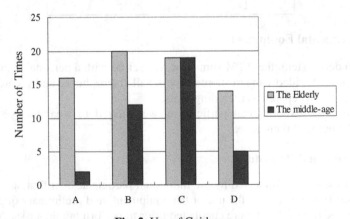

Fig. 2. Use of Guidance

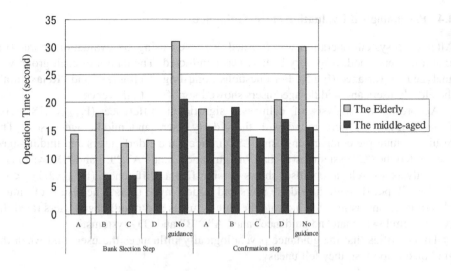

Fig. 3. Operation Time

Therefore, it can be said that the search and understanding were promoted (assisted) by the guidance.

4.3 Operational Errors

The datum of each group was analyzed using the Chi-Square Test (Fig. 4). According to the analysis, both the elderly users and middle-aged users exhibited no significant difference. Although there is no significant difference, the errors decreased with the 4-level speed guidance in the middle-aged users. The elderly users' errors were small with the above-mentioned "B" system.

Therefore, it can be said that the errors decreased with guidance.

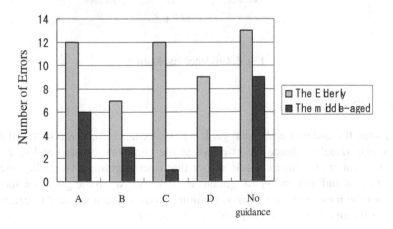

Fig. 4. Operation Errors

4.4 Psychological Evaluation of the Guidance

After each system operation was finished, a survey using six evaluation scores (for example: very good; 6 → very bad; 1) was conducted. The datum of each group was analyzed by variance (Fig.5). For questions regarding "Calmness" and "Relaxation", the elderly users and middle-aged users showed significant difference.

According to analysis of calmness, significant difference ($F_{(4,20)}$=2.78,p<.10; $F_{(4,20)}$=3.95,p<.05) were found with the elderly users and middle-aged users. The results of multiple comparisons using LSD indicate the elderly users and middle-aged users rated the "C" system lower than "no guidance" and "A", "B" and "D" systems.

Analysis of relaxation also shows a significant difference ($F_{(4,20)}$=2.35,p<.10; $F_{(4,20)}$=8.08,p<.01) with the elderly users and middle-aged users. The results of multiple comparisons using LSD indicate the elderly users and middle-aged users rated the "C" system lower than "no guidance" and "A", "B" and "D" systems.

This clarifies that the guidance psychologically influences the users and when the guidance is too fast, they felt uneasy.

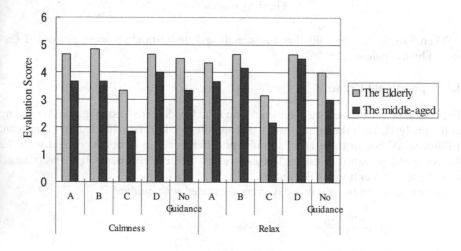

Fig. 5. Guidance Evaluation

5 Conclusion

In this paper, the auditory and visual guidance were examined. It was clarified that the auditory and visual guidance are effective at the confirmation step and search step. When user search or confirm, it was shown that operation times were shortened with the presence of auditory and visual guidance. However, when the guidance speed was too fast for the users, they felt uneasy. In future studies, it is necessary to examine the guidelines for an effective auditory and visual guidance.

References

1. Harada, T.E., Akatsu, H.: What is "Usability":A Perspective of Universal Design in An Aging Society. In: Cognitive Science of Usability. Kyoritsu Publisher (2003)
2. Komatsubara, A., Kbayashi, M.: Evaluating usability of operational sequence with "Plan Do-See" Analysis. Japan Journal of Ergonomics 31(4) (1995)
3. Tniue, N., et al.: Effect of telop in watching program process (2). In: Japanese Society for Cognitive Psychology proceeding (2003)
4. Nambu, M., et al.: Voice interface design for the elderly. In: Information Processing Society of Japan proceeding (2003)
5. JEITA TT-604, Speech Synthesizer Symbols for ITS on-Board Unit (2007)

Tailoring Interface for Spanish Language:
A Case Study with CHICA System

Vibha Anand, Paul G. Biondich, Aaron E. Carroll, and Stephen M. Downs

Children's Health Services Research, Indiana University School of Medicine Regenstrief
Institute for Health Care, Indianapolis
vanand@iupui.edu

Abstract. We developed a clinical decision support system (CDSS) – Child Health Improvement through Computer Automation (CHICA) - to deliver patient specific guidance at the point of clinical care. CHICA captures structured data from families, physicians, and nursing, staff using a scannable paper user interface - Adaptive Turnaround Documents (ATD) while remaining sensitive to the workflow constraints of a busy outpatient pediatric practice. The system was deployed in November 2004 with an English language only user interface. In July 2005, we enhanced the user interface with a Spanish version of the pre-screening questionnaire to capture information from Spanish speaking families in our clinic. Subsequently, our results show an increase in rate of family responses to the pre-screening questionnaire by 36% (51% vs. 87%) in a four month time period before and after the Spanish interface deployment and up to 32% (51% vs. 83%) since November 2004. Furthermore, our results show that Spanish speaking families, on average, respond to the questionnaire more than English speaking families (85% vs. 49%). This paper describes the design, implementation challenges and our measure of success when trying to adapt a computer scannable paper interface to another language.

1 Introduction

Computer alert and reminder systems are an effective way to improve rates of preventive services [1-7]. However, these successes have generally been limited to systems that are embedded in computerized physician order entry systems or inpatient noting systems. Unfortunately, for many outpatient preventive services, a reminder at the time of note writing or order entry is often too late, as these events frequently take place after the physician has completed the visit. "Just in time" information delivery requires that a reminder be delivered at the time the physician is making a decision, and this is often while he or she is conversing with a patient. This timeliness is even more important in pediatric practice, where preventive services often include developmental assessment, risk assessment, counseling and anticipatory guidance. Computers within exam rooms may not be a satisfactory solution, as they can be expensive and susceptible to damage by curious pediatric patients. Computers can also slow the patient encounter and negatively impact the content of physician-patient communications[8]. In fact, at our institution, which houses one of the most successful electronic

M. Kurosu (Ed.): Human Centered Design, HCII 2009, LNCS 5619, pp. 398–407, 2009.
© Springer-Verlag Berlin Heidelberg 2009

medical record systems in the world,[9] pediatricians have long been resistant to the introduction of computers in their clinics.

We developed a guideline-based decision support system that could be seamlessly integrated into the delicate workflow of a high volume pediatric clinic. We considered six essential criteria: (1) collecting data directly from patients or their parents, (2) providing reminders to nurses about age- appropriate screening data, (3) prioritizing needed preventive services (4) providing tailored prompts and reminders to physicians unobtrusively during the encounters with patients, (5) capturing data directly from physicians, and (6) requiring little or no training of staff. However, as we deployed the system in November 2004, we quickly realized that our first criterion, collecting data from patient and their families, was not adequately met because of language barriers for the high number of Spanish speaking families that our clinic serves. We added Spanish language to the scannable paper interface of our system and deployed it in July 2005. This paper describes the design, implementation and challenges of deploying a bilingual scannable paper interface and our results thus far.

2 Methods

Preliminary work by one of our investigators demonstrated the feasibility of using tailored scannable paper forms to provide patient specific reminders to physicians and capture data through optical scanning.[10, 11] We expanded this model, using advances in Optical Character Recognition (OCR) technology and international standards for knowledge representation (Arden Syntax[12]) and data communication (HL7) and developed Child Health Improvement through Computer Automation (CHICA) system [13] as an extension of the Regenstrief Medical Record System (RMRS), an inpatient and outpatient information system which contains 30 years of data and more than 300 million numeric or coded patient observations[9].

3 System Overview

The Child Health Improvement through Computer Automation (CHICA) system consists of (1) a knowledge base of guideline rules, (2) a repository of patient data, (3) a tailored document printing and scanning engine and (4) business rules to direct the communication as well as the printing and scanning of patient specific documents. At each visit CHICA generates two tailored scannable forms and additional "Just in Time" informational forms if needed for the visit. The first collects information from the patient or the parent and from the nurse before the physician encounter. The second provides reminders and collects data from the physician.

Workflow: When a patient checks into the clinic, the registration system sends an HL7 ADT message to the RMRS; the message is then routed to CHICA (Figure 1(a)). CHICA uses this "trigger" to query the RMRS for all relevant clinical data for the patient (b). Upon receipt and parsing of these data, CHICA generates a highly tailored "Pre-Screener Form" (PSF) for the parent or adolescent (c).

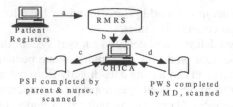

Fig. 1. Workflow

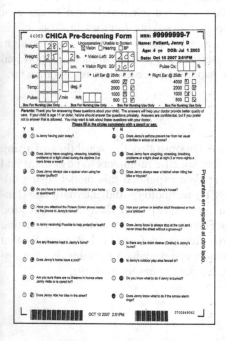

Fig. 2. Pre-screener Form (PSF)   Fig. 3. Physician Worksheet (PWS)

The PSF (Figure 2) has two sections: the top for the nursing staff and the bottom for the parent or adolescent. The parent or adolescent section consists of 20 "yes or no" questions that assess patient information, for example, common parental concerns like diet, risk behaviors like smoking, safety issues like car seats, or risk factors like lead exposure. Question selection and relevance are determined by applying the logic contained within Arden Syntax Medical Logic Modules (MLMs) to data contained within the individual's electronic medical record (EMR). Questions are based on standard national guidelines and written at a 6th grade literacy level. The parent (or the adolescent) fills out the bottom section in the waiting room. If the PSF is partially completed, the system still performs gracefully.

The nursing staff fills out the top section of the PSF. This section contains a structured template for recording measurements of height, weight, head circumference, blood pressure, and other screening tests. Based on the child's age, suggested fields are highlighted, helping the staff adhere to guidelines while avoiding unnecessary or redundant measurements. The completed PSF is scanned into the CHICA (a TIFF image of PSF is also saved) system. The software interprets the scanned data and writes all newly-recorded observations into the EMR.

Next, CHICA prints a tailored Physician Worksheet (PWS) for use by the physician (Figure 1(d)). The PWS (Figure 3) includes calculated height and weight percentiles and body mass index. It also contains a physical examination grid and areas for a hand written history, physical examination, impression and plan. The main section of the PWS form contains 6 guideline prompts for the physician. These are again selected by Arden MLMs, which query the child's EMR data, including data recently captured by the PSF. Each prompt consists of an explanation of the prompt followed by 6 checkbox responses that allow the physician to document data, procedures, or referrals. After the encounter, the PWS is scanned, and the data (with a TIFF image of the PWS) are recorded in the EMR.

Together, the PSF and the PWS implement a complete preventive services program. The PSF collects needed data from the patient/parent that informs the construction of the PWS. Because structured data are captured from both forms, the system supports multi-step guidelines and generates tailored follow-up questions and prompts at subsequent visits.

4 Challenges

When the system was initially deployed in November 2004, a large number of patient families in our clinic population did not respond to the PSF questionnaire. Our clinic serves a large Spanish speaking population. Using ethnicity (Hispanic) as a proxy for language preference we estimated 35% (Table 1) of our clinic population consists of Spanish speaking families. The questionnaire in English presented a big language barrier for these families.

Table 1. Demographic Distribution of Clinic Population

Race	% N = 10234
Black	42%
Hispanic	35%
White	16%
Asian	3%
American Indian	< 1%
Other	4%

Since PWS prompts are prioritized by guidelines (risk factors, age of child) and query the most recent data (including questionnaire responses) to assess relevance; the unanswered but scanned PSFs often produced PWS prompts that were not the most pertinent to the patient's visit. For example, an anticipatory guidance prompt may not appear to advise the parent of smoking risks to child's health because the parent did not indicate there was a smoker in the home (the questionnaire on PSF was not answered). Furthermore, the CHICA database, which is also used for retrospective data analyses and for conducting clinical studies, was not capturing family data for these families. Therefore, as the PSF plays an integral role in informing the reminders for the physician encounter (PWS), it was critical that the needs of the Spanish speaking patient population be met.

We considered asking each patient family about their language of choice at the time of registration on their first visit, but this would have required changes to our registration system and workflow processes. Additionally, it would also not have met the needs of bilingual patient families adequately and burdened the staff with tracking the patient's visit order. We also considered using demographic data such as race or family name in the patient's electronic medical record as a proxy for language. Finally, we considered failure to answer the first questionnaire in English to default their language choice to Spanish. However, none of these solutions seemed graceful enough; family names indicate ethnicity, not language preference; data derived from medical records is not most accurate surrogate for language. As we required augmenting the PSF with a Spanish version without having to alter workflow processes or requiring prior knowledge of the preferred language of the patient and their families, we looked away from human centered solutions and started to explore technically innovative solutions to the problem.

A Technical Solution. First, we changed the PSF form from a single sided form to a dual sided form to print the same set of questions on each side in English and Spanish at each visit (Figure 4a and 4b). The form remained single page but now each side had the same set of questions on the other side in the Spanish. A single page form was appropriate for various reasons – to limit the text on the questionnaire, a single page is easy to track through clinic workflow, and because the forms are answered in the waiting room by families with young children, there is a high likelihood of multiple pages being missed or not returned for scanning. This latter error would result in the OCR software not reading the form at all and a single page form can minimize the risk of unrecognized or misread forms.

Second, since the questions on the PSF are dynamically generated by a set of MLM rules, it required extending the functionality of the MLM parser - to include parsing of Spanish characters to generate Spanish language questionnaire text and to correctly substitute for phrases relating to gender such as "his/her," "him/her" or "he/she." Therefore, we modified the "action slot" and the "logic slot" of each MLM rule that generates a PSF question to include the Spanish version of the question text and to substitute any dynamic variables with correct language for gender.

Third, since there were now two copies of the questionnaire text – one for English and one for Spanish, we changed the CHICA database schema to accommodate an extra copy of the text when producing the ATD for the pre-screener form (PSF).

Finally, to print the forms dual sided, we installed duplex printers in the clinic location. With dual sided forms, the patient or the family can choose to answer either language side at a given visit. An algorithm was developed to evaluate responses from either side of PSF when it gets scanned. This algorithm records responses from the side that has the largest number of questions marked. In case of ties between the sides, it takes the responses from the English side.

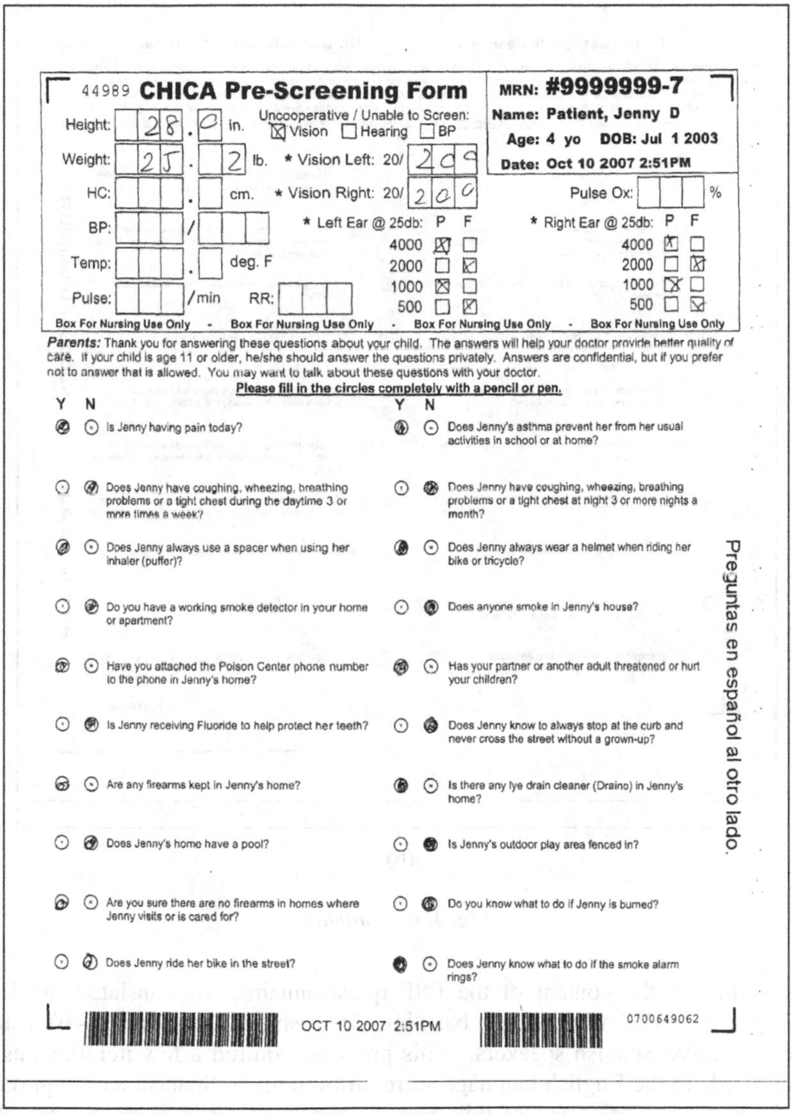
44989 CHICA Pre-Screening Form

MRN: #9999999-7
Name: Patient, Jenny D
Age: 4 yo DOB: Jul 1 2003
Date: Oct 10 2007 2:51PM

Height: 28.0 in. Uncooperative / Unable to Screen: Vision, Hearing, BP
Weight: 25.2 lb. * Vision Left: 20/ 20
HC: cm. * Vision Right: 20/ 20 Pulse Ox: %
BP: / * Left Ear @ 25db: P F * Right Ear @ 25db: P F
Temp: deg. F 4000, 2000, 1000, 500
Pulse: /min RR:

Box For Nursing Use Only - Box For Nursing Use Only - Box For Nursing Use Only - Box For Nursing Use Only

Parents: Thank you for answering these questions about your child. The answers will help your doctor provide better quality of care. If your child is age 11 or older, he/she should answer the questions privately. Answers are confidential, but if you prefer not to answer that is allowed. You may want to talk about these questions with your doctor.

Please fill in the circles completely with a pencil or pen.

Y N
Is Jenny having pain today?
Does Jenny have coughing, wheezing, breathing problems or a tight chest during the daytime 3 or more times a week?
Does Jenny always use a spacer when using her inhaler (puffer)?
Do you have a working smoke detector in your home or apartment?
Have you attached the Poison Center phone number to the phone in Jenny's home?
Is Jenny receiving Fluoride to help protect her teeth?
Are any firearms kept in Jenny's home?
Does Jenny's home have a pool?
Are you sure there are no firearms in homes where Jenny visits or is cared for?
Does Jenny ride her bike in the street?

Y N
Does Jenny's asthma prevent her from her usual activities in school or at home?
Does Jenny have coughing, wheezing, breathing problems or a tight chest at night 3 or more nights a month?
Does Jenny always wear a helmet when riding her bike or tricycle?
Does anyone smoke in Jenny's house?
Has your partner or another adult threatened or hurt your children?
Does Jenny know to always stop at the curb and never cross the street without a grown-up?
Is there any lye drain cleaner (Draino) in Jenny's home?
Is Jenny's outdoor play area fenced in?
Do you know what to do if Jenny is burned?
Does Jenny know what to do if the smoke alarm rings?

Preguntas en español al otro lado.

OCT 10 2007 2:51PM 0700649062

(a)

Fig. 4. (a) PSF in English. (b) Spanish Side of PSF.

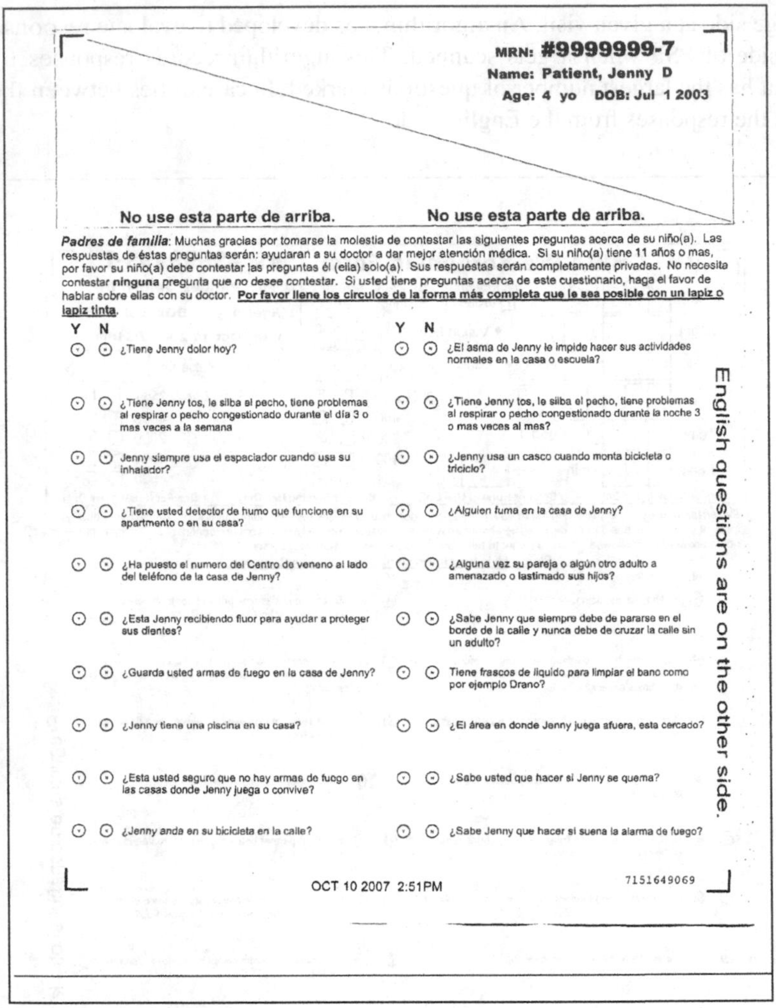

MRN: #9999999-7
Name: Patient, Jenny D
Age: 4 yo DOB: Jul 1 2003

No use esta parte de arriba. No use esta parte de arriba.

Padres de familia: Muchas gracias por tomarse la molestia de contestar las siguientes preguntas acerca de su niño(a). Las respuestas de éstas preguntas serán: ayudaran a su doctor a dar mejor atención médica. Si su niño(a) tiene 11 años o mas, por favor su niño(a) debe contestar las preguntas él (ella) solo(a). Sus respuestas serán completamente privadas. No necesita contestar **ninguna** pregunta que no desee contestar. Si usted tiene preguntas acerca de este cuestionario, haga el favor de hablar sobre ellas con su doctor. **Por favor llene los circulos de la forma más completa que le sea posible con un lapiz o lapiz tinta.**

Y	N		Y	N	
⊙	⊙	¿Tiene Jenny dolor hoy?	⊙	⊙	¿El asma de Jenny le impide hacer sus actividades normales en la casa o escuela?
⊙	⊙	¿Tiene Jenny tos, le silba el pecho, tiene problemas al respirar o pecho congestionado durante el día 3 o mas veces a la semana	⊙	⊙	¿Tiene Jenny tos, le silba el pecho, tiene problemas al respirar o pecho congestionado durante la noche 3 o mas veces al mes?
⊙	⊙	Jenny siempre usa el espaciador cuando usa su inhalador?	⊙	⊙	¿Jenny usa un casco cuando monta bicicleta o triciclo?
⊙	⊙	¿Tiene usted detector de humo que funcione en su apartmento o en su casa?	⊙	⊙	¿Alguien fuma en la casa de Jenny?
⊙	⊙	¿Ha puesto el numero del Centro de veneno al lado del teléfono de la casa de Jenny?	⊙	⊙	¿Alguna vez su pareja o algún otro adulto a amenazado o lastimado sus hijos?
⊙	⊙	¿Esta Jenny recibiendo fluor para ayudar a proteger sus dientes?	⊙	⊙	¿Sabe Jenny que siempre debe de pararse en el borde de la calle y nunca debe de cruzar la calle sin un adulto?
⊙	⊙	¿Guarda usted armas de fuego en la casa de Jenny?	⊙	⊙	Tiene frascos de liquido para limpiar el bano como por ejemplo Drano?
⊙	⊙	¿Jenny tiene una piscina en su casa?	⊙	⊙	¿El área en donde Jenny juega afuera, esta cercado?
⊙	⊙	¿Esta usted seguro que no hay armas de fuego en las casas donde Jenny juega o convive?	⊙	⊙	¿Sabe usted que hacer si Jenny se quema?
⊙	⊙	¿Jenny anda en su bicicleta en la calle?	⊙	⊙	¿Sabe Jenny que hacer si suena la alarma de fuego?

English questions are on the other side.

OCT 10 2007 2:51PM 7151649069

(b)

Fig. 4. (*continued*)

Translation. For the content of the PSF questionnaire, we translated the English questions in the CHICA knowledge base into Spanish (example Figure 4b), using the services of native Spanish speakers. This process required a few iterations as some medical words in the English language were ambiguous in Spanish and required context information. At the end of this process, we went through reverse translation, where a Spanish version of the question was translated to English, independently, by a second native speaker and a bilingual physician. This validated our translation for an ethnically diverse Spanish speaking population. A total of 176 PSF rules were translated into Spanish in our system.

We measured the PSF response rates at various points in time between Nov. 4th 2004 (system deployment date) and 4 months before and after the implementation of Spanish interface (July 28th 2005). In April 2006, we started recording actual language choice for each question on the PSF, when the PSF is scanned in the system and the question is answered.

Results. When we look at a 4 month time period before deploying the Spanish interface (July 28, 2005); of those PSF that were scanned, only 51% had a response on the questionnaire. In the same time period after deployment, the response rate increased to 87%, an improvement of 36%, when Spanish language was provided as a choice (Table 2).

Table 2. PSF Response rates 4 months before and after Spanish Language Choice deployed

Time Period	#PSF Printed	#PSF Scanned	#PSF Answered (At least 1 question answered)
Before Spanish Interface	5652	4697 (83%)	2374 (51%)
After Spanish Interface	6227	5306 (85%)	4621 (87%)

When we looked at the time period since we started measuring language choice for each question answered (April 2006) and now, the response rate has been steady at 83% (Table 3). Of those answered, 74% of the questions were answered in English and 26% were answered in Spanish.

Table 3. PSF Response rates

#PSF Printed	#PSF Scanned	#PSF Answered (At least 1 question answered)	
45,537	36,606 (80%)	30,517 (83%)	
		English	Spanish
		22,605 (74%)	7912 (26%)

Since April 2006, we also have been able to link language choice with patients' demographics. The race/ethnicity distribution in this cohort is 35% Hispanic and 65% other (Table 1). If we measure patients' language choice in this cohort (n = 10234), 79% chose to respond in English and 21% in Spanish (Table 4), suggesting that 14% of Hispanic families may be responding to our questionnaire in English. We also found that 3% of patients switched to Spanish from English, and 2% switched to English from Spanish when compared to their first visit since we deployed Spanish interface (Table 4).

Table 4. Patients' language choice

#Patients	Language Choice English	Language Choice Spanish	Language Change (1st -> Last visit) (English to Spanish)	Language Change (1st -> Last visit) (Spanish to English)
10,234	8090 (79%)	2144 (21%)	340 (3%)	245 (2%)

Finally, our results also show that Spanish speaking families respond to the PSF questionnaire 36% more (85% vs. 49%) than other families (Table 5).

Table 5. PSF Response rate by preferred language

Questions	English	Spanish
Asked	52,5164	14,9412
Answered	25,9055 (49%)	12,7018 (85%)

5 Discussion

We believe that implementing the Spanish version of the interface has essentially removed the language barrier for Spanish speaking patient families in our clinic without altering the workflow processes. We think it provides an added value for non-Spanish speaking physicians and clinic staff by providing patient answers to the question regardless of the patient's preferred language. Since deployment of the Spanish version, the response rate to the PSF has improved considerably, thus informing the physician encounters to better address the patient's needs for the visit and therefore enhancing the care process [14-16]. We also believe that higher response rates to PSF questionnaires in Spanish indicate that the needs of the Spanish speaking families to communicate with their care givers in our clinic has been adequately addressed with this solution. Finally, we believe that our methods can be generically applied to most questionnaires where language may be a barrier.

References

1. Dexter, P., et al.: A computerized reminder system to increase the use of preventive care for hospitalized patients. New England Journal of Medicine 345(13), 965–970 (2001)
2. Hunt, D., et al.: Effects of Computer-Based Clinical Decision Support Systems on Physician Performance and Patient Outcomes: A Systematic Review. JAMA 280(15), 1339–1346 (1998)
3. Johnson, M., et al.: Effects of computer-based clinical decision support systems on clinician performance and patient outcome - a critical appraisal of research. Annals of Internal Medicine 120, 135–142 (1994)
4. Kaplan, B.: Evaluating informatics applications-clinical decision support systems literature review. Int. J. Med. Inf. 64, 15–37 (2001)
5. McDonald, C.: Protocol-based computer reminders, the quality of care and the non-perfectability of man. New England Journal of Medicine 295(24), 1351–1355 (1976)
6. McDonald, C.J.: Protocol-based computer reminders, the quality of care and the non-perfectability of man. N Engl. J. Med. 295(24), 1351–1355 (1976)
7. Harris, R.P., et al.: Prompting physicians for preventive procedures: a five-year study of manual and computer reminders. Am J. Prev. Med. 6(3), 145–152 (1990)
8. Sullivan, F., Mitchell, E.: Has general practitioner computing made a difference to patient care? A systematic review of published reports. British Medical Journal 311(7009), 848–852 (1995)
9. McDonald, C., et al.: The Regenstrief Medical Record System: 20 years of experience in hospitals, clinics, and neighborhood health centers. MD Computing 9(4), 206–217 (1992)

10. Downs, S., Arbanas, J., Cohen, L.: Computer supported preventive services for children. In: Nineteenth Annual Symposium on Computer Applications in Medical Care. Hanley & Belfus, Inc., New Orleans (1995)
11. Downs, S., Wallace, M.: Mining association rules from a pediatric primary care decision support system. In: American Medical Informatics Association 2000 Annual Symposium. American Medical Informatics Association, Los Angeles (2000)
12. Hripcsak, G., et al.: Rationale for the Arden Syntax. Computers & Biomedical Research 27(4), 291–324 (1994)
13. Anand, V., et al.: Child Health Improvement through Computer Automation: the CHICA system. Medinfo. 11(Pt 1), 187–191 (2004)
14. Biondich, P.G., et al.: Automating the recognition and Prioritization of Needed Preventive Services: Early Results from the CHICA System. In: AMIA Annu. Symp. Proc., pp. 51–55 (2005)
15. Downs, S.M., et al.: Human and System Errors, Using Adaptive Turnaround Documents to Capture Data in a Busy Practice. In: Proc. AMIA Symp. (2005)
16. Downs, S.M., Uner, H.: Expected value prioritization of prompts and reminders. In: Proc. AMIA Symp., pp. 215–219 (2002)

A Personal Assistant for Autonomous Life

Alessandro Andreadis, Giuliano Benelli, and Pasquale Fedele

Department of Information Engineering
University of Siena
Via Roma, 56 – 53100 Siena, Italy
{Andreadis,Benelli,Fedele}@unisi.it

Abstract. This paper presents a design of an innovative framework to support continuous monitoring and assistance for ageing people affected by disabilities or chronic diseases during their stay in a structured environment such as home or hospital.

1 Introduction

A smart environment can be defined as "a physical world that is richly and invisibly interwoven with sensors, actuators, displays and computational elements, embedded seamlessly in the everyday objects of our lives and connected through a continuous network" (M. Weiser).

Smart environment is becoming real concept thanks to the commercial development of pervasive ICT technologies such as sensors networks and wireless sensor networks.

These technologies will improve the quality of life of all people, guaranteeing security, comfort and energy saving. However, the concept of smart environments exploits its potentialities for ageing people affected by disabilities or chronic diseases, in order to extend the time people can live independently in their environment (home) increasing their autonomy and self-confidence.

The design of smart environments can follow the paradigm of the human nervous system which implements a distributed intelligence architecture: there are sensory functions (sensors), active and reactive functions (actuators), and integrative functions to analyse the information, store it and take decisions (business intelligence).

Following the same model, the home equipment will be made of sensors and actuators, communication links, distributed intelligent systems in charge of the reflexives functions (alarms, home automation), and central functions (detection of the modes of activity, release of alarms).

A research activity, which could have significant industrial perspective and social relevance, is the utilisation of these technologies along with ambient intelligence to realise Ambient Assisting Living (AAL) aimed at augmenting autonomy of people living in their home and at monitoring care for the elderly or ill persons.

This paper presents a design of an innovative environment to support continuous monitoring and assistance of the user during his stay in a structured location such as home or hospital. Each user has a personalised set of sensors connected through a

M. Kurosu (Ed.): Human Centered Design, HCII 2009, LNCS 5619, pp. 408–415, 2009.

wireless network to a data gateway. All the measures can be released in an automatic or in a controlled way. The gateway collects data and transmits them to a control centre; users are not required to do any operation for the transmission. Some experimentations are being carried out with the Multimedia Home Platform (MHP) technology: thanks to a MHP application, the user (as well as the doctor) can view its data by using the TV remote control with four coloured buttons. An early approach to this system has been experimented in the SORRISO project founded by the Tuscany Region and will be further developed in the T-Seniority project, supported by the CIP Competitiveness and Innovation Framework Program of the European Union 2007-2013.

Aim of the framework is to permit:

- *continuous monitoring of the health status for people with special needs*, through wearable or home sensors allowing to improve independence and to control old people and persons affected by disabilities and/or chronic diseases (e.g., blood glucose for people with diabetes, blood pressure for those with hypertension, muscle tension for people with chronic pain, etc.);
- *adaptation of the home environment* according to specific needs (e.g. change of light colors and intensity for Alzheimer ills).

2 State of the Art

Today, wearable sensors and sensors with reduced dimensions able to transmit data using wireless networks (e.g., Wi-Fi, Bluetooth, Zigbee) are available off the shelf, thus permitting to monitor some "critical parameters" of the user or to measure some biological signals, such as body temperature, blood pressure, blood glucose level, movements in three dimensions to detect falls, etc.

In recent years an increasing activity was carried out in patient monitoring, in particular relating to:

- mobile telemedicine [2], [3], [4] e [5];
- home monitoring [6], [7];
- hospital-wide mobile monitoring system [8];
- Decision Support Systems (DSS) [8], [9], [10], [11] and [13].

Several works are specific for pathologies:

- wireless telemetry system for EEG epilepsy [14];
- bluetooth-based system for digitized ECGs [15];
- personal health monitors for stress monitoring [15];
- alerts [16] and [17], for elderly [18] and [19];
- real-time monitoring for disabled [20];
- autonomous intelligent agent for monitoring Alzheimer patients' health care [21];
- mobile decision making system for using symptoms of abdominal pain in children's emergency, as presented in [22].

The works on devices and sensors include:

- clothing-embedded transducers for ECG [23]
- ring-based sensor [24];
- minimally invasive wireless sensors for health-monitoring [25] and [26].

3 System Architecture

Aim of the proposed system is to foster an independent life but also an active participation to social, cultural and working activities of ageing people and/or people with
disabilities. The main modules of the system are "*Openframe*" and "*Mobile Personal
Assistant*".

Openframe

In healthcare as well as in other applications, interoperability is the ability of different
information technology systems and software applications to communicate, to exchange data accurately, effectively, and consistently, and to use the information that
has been exchanged. The OPENFRAME framework is composed of many interacting
modules and layers able to manage, in a proactive way, services and data by means of
a Service-Oriented Architecture (SOA) in which a Service Bus orchestrates many
local and remote services. It allows an easy access of new services by a Service Broker, which executes processing of service declarative descriptions in order to store
them into a Service Repository. Input services are provided in a declarative abstract
form (i.e., independently from the communication channel and/or user needs), using a
proper notation for presenting their metadata described by a suitable standard - for
example, using XML Metadata Interchange (XMI).

When such an abstract service is called by an end user through a specific delivery
cannel (i.e., ADSL, GPRS, DTT, etc.), the Service Broker retrieves the corresponding
service descriptions from the repository and sends them to a Service Adapter, which
takes charge of adaptations at two levels:

- adaptation based on characteristics of the communication and service delivery platform (the same service may have different presentations and organization when supplied through different channels);
- adaptation based on user needs, requirements and preferences.

The high level of the Openframe architecture consists of two components, the *Service Bus Layer* and the *Communication Layer*.

The **Service Bus Layer** contains the logic for processing requests and orchestrating
remote and local service. Some specialized modules manage the local services:

- *Localization Module,* permitting through GPS, wireless networks, sensor
  networks or RF-Id structure to localize the client.
- *Adaptation and Personalization Module,* permitting the configuration of the
  terminal client according to the user needs and abilities, but also to the context or the service.

- *The Authorization Module:* responsible for the efficient caching and flow of authorization information. It is also designed to make that information highly available to the service components.
- *The EHR* (Electronic Health Record) *Module*: aiming to co-ordinate the storage and retrieval of individual records both in local and distributed repositories accordingly to the main standards, as such as:
  - HL7 - a standardized messaging and text communications protocol between hospital and physical record systems, and between practice management systems
  - DICOM - an international communications protocol standard for representing and transmitting radiology (and other) image-based data, sponsored by NEMA (National Electrical Manufacturers Association)
  - ANSI X12 (EDI) - transaction protocols used for transmitting patient data. Popular in the United States for transmission of billing data.
  - CEN - CONTSYS (EN 13940), supports continuity of care record standardization.
  - CEN - IIISA (EN 12967), a services standard for inter-system communication in a clinical information environment.
  - ISO - ISO TC 215 provides international technical specifications for EHRs. ISO 18308 describes EHR architectures
  - CEN's TC/251 provides EHR standards in Europe
  - CEN - EHRcom (EN 13606), communication standards for EHR information in Europe

The **Communication layer** is responsible for communication over various Internet protocols and the marshalling and unmarshalling of messages between agents, permitting connection anywhere and any time by using the available communication wireless technology (GPRS, UMTS, Wi-Fi, etc...). It also comprises the *Synchronization Module,* allowing some applications to run offline, without an Internet connection, and the *Connector Module* responsible for communication with sensor networks or RF-Id systems.

Mobile Personal Assistant (MPA)
In the last years some experimental initiatives have been performed to monitor and assist ageing people at home to guarantee a higher independency. However, these initiatives operate only at home and they require dedicated terminals, complex installation procedures and an efficient (but costly) maintenance. MPA utilizes potentialities offered by the most recent wireless networks (Wi-Fi and Zigbee) and wireless sensor networks (MOTE) and their integration with more traditional wireless communication networks (GPRS, UMTS, WiFi, DTT). Therefore, it permits mobility, a higher independency and active participation to leisure and working initiatives at people with special needs.

MPA is a universal "always-on" device, able to operate anywhere and in different contexts, such as closed spaces (home, hospice) and open spaces (city, park).

The terminal is based on mobile phones or PDA connectivity and is specifically designed for old people, people with disabilities or users affected by chronic diseases. MPA can be used for many applications, such as:

- to monitor continuously or at certain times the health status of the user by measuring a set of parameters. Data can be stored locally or sent to a control or assistance center;
- to localize user anytime and to communicate always and anywhere through simple procedures for asking support (e.g., by simply clicking on a key or button).

In more details, the personal assistant is composed of two levels:

- the basic terminal (BT), allowing the implementation of some essential and always available features (phone, data transmission, localization, RF-Id reader). BT can be an off-the-shelf device (PC, PDA, mobile phone or a special terminal for people with special needs), equipped with Java-based software in order to allow the personalization and adaptation of interface and to support the choice of services;
- a *Wearable Sensor Terminal* (WST) designed to integrate additional sensors. In some cases, applications or services may be integrated in BT or can be a separate module. Each user may own and use different WSTs depending on her/his activities and needs. In fact, WST has a variable configuration depending on service, user and context. At the same time, WST can be integrated in different forms, such as a shirt or a bracelet. The implementation of WST is based on the MOTE system; figure 1 shows its general scheme.

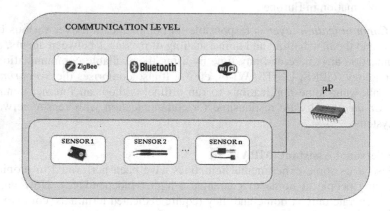

Fig. 1. General scheme of WST

However, agreements are needed with enterprises producing sensors and also communication modules, in order to integrate more sensors in WST. The goal is to achieve many different configurations of WST. As an example, a WST should integrate different sensors to detect several parameters, such as bold temperature, cardiac pulsations, pressure, pulse oximeter, EKG, EMG, accelerometer measuring movements in three dimensions, gyroscope, etc... Each WST module can be personalized according to user needs, on the basis of specific disabilities, or referring to the context or environment in which the user is at the moment.

MPA allows the user to have a continuous support and contact with trusted staff (e.g., a familiar) or with a control center, such as a voluntaries association or a sanitary structure. It is based on commercial terminal products and is characterized by the following features:

- *Always-on,* permitting connection anywhere and anytime by using the available communication wireless technology (GPRS, UMTS, Wi-Fi, DTT, ...).
- *Localization feature,* permitting through GPS, wireless networks, sensor networks or RF-Id structure to localize the user if the service requires this feature.
- *Adaptation and personalization features,* allowing the configuration of the device according to user needs and abilities, but also to the context or the specific service.

These concepts characterize the design of an innovative type of personal assistant having different configurations and characteristics. A person can use a terminal at home and the same terminal with added features during its movements outside; furthermore, sensors in a user device may be configured in different ways, in order to deal with users suffering from different diseases. To cope with usability and accessibility issues, the terminal can be based on a personal PC or a PDA, a cellular phone or a special terminal for disabled people.

4 Conclusion and Future Work

Patient monitoring will be a requirement for offering a better healthcare to patients in nursing homes and hospitals. Smart environments, in particular, will empower independent living for ageing people affected by disabilities or chronic diseases, extending the time people can live independently in their environment (home), increasing their autonomy and self-confidence.

This paper proposes the design of an innovative environment, as part of an early stage project that will be deeply defined and developed in the following months. The general architecture of the proposed system allows to orchestrate different technologies, sensors and communication modalities with the purpose to improve quality of life of ageing people and people affected by chronic diseases during their stay in a structured environment, through the easy use of existing communication devices.

References

1. Cook, D.J., Das, S.K.: Smart Environments: Technologies, Protocols, and Applications. Wiley/ IEEE (2005)
2. Anogianakis, G., Maglarera, S., Pomportsis, A.: Relief for maritime medical emergencies through telematics. IEEE Transactions on Information Technologies in Biomedicine 2 (1998)
3. Bhargava, A., Zoltowski, M.: Sensors and wireless communication for medical care. In: Proc. 14th International Workshop on Database and Expert Systems Applications (2003)

4. Pattichis, C.S., Kyriacou, E., Voskarides, S., Pattichis, M.S., Istepanian, R., Schizas, C.N.: Wireless telemedicine systems: an overview. IEEE Antenna's and Propagation Magazine 44(2) (2002)
5. Pavlopoulos, S., Kyriacou, E., Berler, A., Dembeyiotis, S., Koutsouris, D.: Novel emergency telemedicine system based on wireless communication technology—AMBULANCE. IEEE Transactions on Information Technology in Biomedicine 2(4) (1998)
6. Lee, R.G., Shen, H.S., Lin, C.C., Chang, K.C., Chen, J.H.: Home telecare system using cable television plants—an experimental field trial. IEEE Transactions on Information Technologies in Biomedicine 4(1) (2000)
7. Mendoza, G.G., Tran, B.Q.: In-home wireless monitoring of physiological data for heart failure patients. In: Proc. of the Second Joint IEEE EMBS/BMES (2002)
8. Pollard, J.K., Rohman, S., Fry, M.E.: A web-based mobile medical monitoring system. In: International Workshop on Intelligent Data Acquisition and Advanced Computing Systems: Technology and Applications (2001)
9. Shim, J.P., Warkentin, M., Courtney, J., Power, D.J., Sharda, R., Carlsson, C.: Past, present, and future of decision support technology. Decision Support Systems 33(2) (2002)
10. Bielza, C., Fernández del Pozo, J., Lucas, P.J.F.: Explaining clinical decisions by extracting regularity patterns. Decision Support Systems 44(2) (2008)
11. Brahnam, S., Chuang, C.F., Sexton, R.S., Shih, F.Y.: Machine assessment of neonatal facial expressions of acute pain. Decision Support Systems 43(4) (2007)
12. Hu, P.J.H., Wei, C.P., Cheng, T.H., Chen, J.X.: Predicting adequacy of vancomycin regimens: a learning-based classification approach to improving clinical decision making. Decision Support Systems 43(4) (2007)
13. Lin, L., Hu, P.J.-H., Sheng, O.R.L.: A decision support system for lower back pain diagnosis: uncertainty management and clinical evaluations. Decision Support Systems 42(2) (2006)
14. Modarreszadeh, S.: Wireless, 32-channel, EEG and epilepsy monitoring system. In: Proc. 19th Annual IEEE International Conference on Engineering in Medicine and Biology (1997)
15. Khoor, S., Nieberl, K., Fugedi, K., Kail, E.: Telemedicine ECG-telemetry with Bluetooth technology. In: Proc. Computers in Cardiology (2001)
16. Kafeza, E., Chiu, D.K.W., Cheung, S.C., Kafeza, M.: Alerts in mobile healthcare applications: requirements and pilot study. IEEE Transactions on Information Technologies in Biomedicine 8(2) (2004)
17. Lee, R., Chen, K., Hsiao, C., Tseng, C.: A mobile care system with alert mechanism. IEEE Transactions on Information Technology in Biomedicine 11(5) (2007)
18. Lin, C., Chiu, M., Hsiao, C., Lee, R., Tsai, Y.: Wireless health care service system for elderly with dementia. IEEE Transactions on Information Technology in Biomedicine 10(2) (2006)
19. Mikkonen, M., Vayrynen, S., Ikonen, V., Heikkila, M.O.: User and Concept Studies as Tools in Developing Mobile Communication Services for the Elderly. Springer-Verlag's Personal and Ubiquitous Computing 6 (2002)
20. Varshney, U.: Managing wireless health monitoring for patients with disabilities. IEEE IT Professional 8(6) (2006)
21. Corchadoa, J.M., Bajo, J., Paza, Y., Tapiaa, D.I.: Intelligent environment for monitoring Alzheimer patients, agent technology for health care. Decision Support Systems 44(2) (2008)

22. Michalowski, W., Rubin, S., Slowinski, R., Wilk, S.: Mobile clinical support system for pediatric emergencies. Decision Support Systems 36(2) (2003)
23. Jovanov, E., O'Donnel, A., Morgan, A., Priddy, B., Hormigo, R.: Prolonged telemetric monitoring of heart rate variability using wireless intelligent sensors and a mobile gateway. In: Proc. Second Joint IEEE EMBS/BMES Conference (2002)
24. Rhee, S., Yang, B.H., Chang, K., Asada, H.H.: The ring sensor: a new ambulatory wearable sensor for twenty-four hour patient monitoring. In: Proc. 20th Annual IEEE International Conference on Engineering in Medicine and Biology (1998)
25. Boric-Lubecke, O., Lubecke, V.M.: Wireless house calls: using communications technology for health care and monitoring. IEEE Microwave Magazine (2002)
26. Kyu, J.C., Asada, H.H.: Wireless, battery-less stethoscope for wearable health monitoring. In: Proc. the IEEE 28th Annual Northeast Bioengineering Conference (2002)

Towards a Theory of Cultural Usability: A Comparison of ADA and CM-U Theory

Torkil Clemmensen

Department of Informatics, Copenhagen Business School
Howitzvej 60, 2.10, DK, 2000 Frederiksberg C, Denmark
tc.inf@cbs.dk

Abstract. Cultural models in terms of the characteristics and content of folk theories and folk psychology have been important to social scientists for centuries. From Wilhelm Wundt's Volkerpsychologie to the distributed and situated cognition theorists in the global world of today, thinkers have seen human action as being controlled by cultural models. The study of cultural models for humans interacting with computers should thus be at the heart of the scientific study of human-computer interaction (HCI). This paper presents a theory of cultural usability that builds on the concept of Cultural Models of Use (CM-U theory). The theory is compared to existing Artifact Development Analysis (ADA) theory to identify its sensitivity to explain cultural usability phenomena. The conclusion is that a) the theory can account for empirical findings on cultural usability, and b) CM-U and ADA theories seem to fit different user populations' perception of usability.

Keywords: Cultural models, HCI, culture, usability.

1 Introduction

Cultural models in terms of the characteristics and content of folk theories and folk psychology have been important to social scientists for centuries. From Wilhelm Wundt's Volkerpsychologie to the distributed and situated cognition theorists in the global world of today, thinkers have seen human action as being controlled by cultural models. The study of cultural models for humans interacting with computers should thus be at the heart of the scientific study of human-computer interaction (HCI). In this paper, we ask the question: Which kind of theory can explain cultural usability phenomena? The answer we give is to view usability as the outcome of distributed cognitions across different kinds of culturally specific models: individual models, tool models, and situation models. The perception of cultural models as elements in distributed cognitions across individuals, tools and situations is central for much of modern cultural psychology (situated cognition, distributed cognition, cultural schema theory, activity theory, etc., see e.g. [30]). In the extension of this approach to usability, individual cultural models of use consist of the goals, actions and emotions that in traditional usability definitions constitute the effectiveness, efficiency and satisfaction of interacting with a product [1]. Tools become affordances

M. Kurosu (Ed.): Human Centered Design, HCII 2009, LNCS 5619, pp. 416–425, 2009.

[20] designed into the interactive products, and situational models of use include established usability evaluation methods [11]. This paper sees the combination of these models of use the Cultural Model theory of Usability (CM-U) theory. When compared with other theories of cultural usability, alternative understandings of quality-in-use and usability appear. This implies that what is understood as cultural usability may itself have cultural biases, and that researchers and practitioners should pay attention to which theory of cultural usability they apply.

2 Basic Assumptions about Culture and Usability

Until recently the basic assumption among HCI researchers was that cultural issues could be treated as a practical matter of occasional and peripheral interest. Depending on the actual system to be designed, designers might consider the influence on the human-computer interaction from one or more factors on a long and incomplete list of cultural variables [7, 25]. The cultural models of HCI were understood as arbitrary, i.e., they could equally well have evolved into another form [21, 24]. For example, the use of red as a warning color on a display could equally have been yellow or some other color. Most of HCI was regarded implicitly as non-cultural, and something that easily could be transferred across different cultural settings. For instance, a common assumption in HCI was that all humans could distinguish between the different colors (e.g., red, green, blue) on visual displays) [3]. Consequently, the investigation of culturally determined usability problems was inappropriate [2].

In the past few years attempts have been made to come up with new axioms for culture and usability, such as cultural dimensions [18], cultural factors [28], cultural constraints [21], and cultural usability [5, 29]. These approaches are in many ways different. What is common to them is a focus on the diversity of users and use of technology around the globe on social-cognitive approaches to usability (as opposed to psycho-physiological approaches to usability) and also on a broad understanding of the utility of human-computer interaction. This last point, namely, a broad understanding of the utility of human-computer interaction, means seriously considering the experienced utility of interactive products, and not only considering instant measures such as immediate satisfaction, efficiency and effectiveness.

A major finding from the recent literature on culture in HCI is that there are differences in usability in the East (Asia) and in the West (USA, Europe), and that these differences predict the need for localized designs [18] and for local adaptations of usability evaluation procedures [28]. Specifically, empirical studies show that Chinese users adapt a more holistic approach to using software compared to European users [27]. The definition of culture that is used in these studies is national or regional culture, see [13]. Some authors [29, 30] have suggested that in addition to studying national or regional culture, HCI research should build on the cultural-psychological assumption that historically developed ways of thinking are embedded firmly in individuals' and small groups' everyday use of interactive computer and other design products. Cultural psychologists have in many empirical studies demonstrated basic cultural-historical differences in thinking and mental-self government. For example, it has been demonstrated that Easterners (people brought up in a Confucian ethical and philosophical system) tend to be context focused in their cognitive style, while

Westerners (people brought up in an Aristotelian system) tend to be object focused. When asked to report on a scene, Easterners tend to mention the background, while Westerners tend to report the focal objects [19]. Such cross cultural differences in cognition lead us to expect cross cultural differences in HCI to be visible in usability evaluations [6].

One axiom remains unchallenged: Usability must be considered a universal phenomenon in order for HCI to move forward as a science. As HCI researchers from different countries, we cannot base our cooperation on evolutionism, i.e., the assumption that some cultures are simply more developed than others, not even in the technology-led area of usability. Relativism in its extreme form: where the concepts and theories based on research in one cultural setting which cannot at all be transposed to others settings, is also not adequate for cross cultural research on usability (although relativism could be adequate for a within-culture study of, for example, the use of symbols in Indian software). However, the sort of universalism that is needed to study cultural usability takes relativism and evolutionism into account as *empirical questions*. It thus follows the moderate universalism suggested by [23]: 1) There may or may not be cross cultural usability universals, but if not, we need empirical documentation; 2) universals in usability will be found on the level of theoretical principles rather than in the phenomena; and 3) we need to make assumptions about universals in usability to help organize data into general theories.

3 Derivation of New Theory from Basic Assumptions

This section derives eight considerations from the axioms and assumptions presented above, and then presents them as a coherent cultural model theory of usability. First, from the axioms of East-West cultural differences in human computer interaction presented above, we can see that there are several cultural backgrounds that may be relevant to a user of technology. A new theory of cultural usability must explain how users with multicultural backgrounds interact with technology. Social psychological studies of multiculturedness in a global world (see for example [4], and in particular [14]'s theory of bi-cultural frame-switching) can assist, but need to be adapted to the usability domain. The theory assume that users hold one or more cultural meaning systems, even if the systems contain conflicting cultural models of technology use. The accessibility, availability and applicability of particular cultural models of technology use will then determine the usability of a product. For example, when one writes a letter to a friend, an icon showing the Indian elephant god, Ganesh, may be available from the word processor's clipart collection; it may be accessible for those with an Indian background or knowledge of India; and it may be applicable and appropriate to use if the receivers of the letter accept a Ganesh icon in letters. In other situations it may be prudent not to use the Ganesh icon because of the belief that the readers will not appreciate that. For those with a European background, using European word processors and writing for European readers, the Ganesh icon would not be available, accessible or applicable.

Second, usability is universal in the sense that it can be seen as a folk theory, which again may be developed to different degrees in different communities or regions in the world. The usability, i.e., effectiveness, efficiency and satisfaction of an interactive product, is always an outcome of the human application of cultural models of technology use. It can be understood as a folk theory of what it means to interact with the product in one or more contexts. In one sense, a folk theory of what is an appropriate mixture of usability components for the product makes it meaningful to measure the usability of the product. In another sense, a particular folk theory may not be accessible, available or applicable to the target users and therefore leads to biased and useless usability measures. Folk theories of usability can be studied empirically.

A third consideration is that usability is universal in the sense that we want to be able to measure usability to compare across cultural settings. An accurate measure of usability should therefore build on the "culture X situation" approach [14] and consider both internal cognitions and external artifact affordances and usability evaluation situations. Internal models of use consist of the goals, actions and emotions that for an individual constitute effectiveness, efficiency and satisfaction of interacting with a product. The content and internal relations among effectiveness, efficiency and satisfaction when interacting with a product may vary across the world's population. The varying internal cognitions contribute in concert with external cultural usability models to measured usability. External cultural models of use can be distinguished into external artifacts, e.g., the affordance designed into the products themselves, and the usability evaluation situations ranging from formal usability evaluation methods to the end-users' own informal evaluation of their interactive product. The external cognitions built into the usability evaluation situation and into the computer artifacts are contributors to a measure of usability.

Fourth, the universality of usability can be explained by viewing the function of the computer artifacts as a basic characteristic of usability across cultures. Computer artifacts have, to varying degrees, built-in a model of use. An artifact can frequently be used for one specific thing in one specific way in one specific context only by a human user. This was found by the German gestalt psychologist, Karl Duncker, in 1934 and labeled 'functional fixedness,' and has been confirmed many times since then. Recently, it has been shown that universally a design's function may be a core property of an artifact concept within human memory, even in technologically sparse cultural communities [9]. This focus on artifacts' built-in models of use is also central for recent distributed and situated cognition theorists and is based on the assumption that *"the tools of thought...embody a culture's intellectual history....Tools have theories built into them, and users accept these theories—albeit unknowingly—when they use these tools"* (Resnick, 1994, pp. 476-477, in [19]). From this follows that if the ways of doing things differ in various cultures, the computer artifacts will also have to be different; conversely, the computer artifacts, to some degree, define a culture by defining ways to do things (for example the mobile phone culture).

A fifth consideration is that from the idea that usability is universal, we can deduct that usability must be built on widely accessible knowledge produced by the use of usability evaluation methods. When using an established usability evaluation method, information about usability problems are propagated across test users, evaluators, moderators, clients, notes, video screen recordings, think aloud protocols, and other units present in that concrete situation, in a way similar to the propagation of information about an airplane's speed suggested by the analysis of [15]. The usability evaluation method serves to produce and maintain culturally specific models of usability. Initially, knowledge of what problems are usability problems is embedded in meaning systems that are widely shared among the members of the cultural group doing the usability test. This 'usability problem knowledge' is frequently used in communication among members of that group and thus becomes chronically accessible within the group. In the usability test situation, where people under time pressure look for readily available and widely accepted solutions to a problem, the chronically accessible knowledge will be used, and typical cultural group conceptions of usability will emerge.

Sixth, since we know from cultural psychology that the human mind is complex and can contain conflicting cultural knowledge, usability must be seen as being primed by the computer artifact, language or other parts of the situation. It is not sufficient to have user task conditions that favor the activation of chronically accessible 'usability problem knowledge' in a usability test situation; the knowledge also must be available to the individual. Since individuals in a society increasingly are polycultural in their background and thus have more than one implicit theory of how to perceive and act in a given situation, the individuals choose or implicitly apply the theory that is available in that situation. The availability of culturally accessible knowledge is primed by culturally specific materials such a religious icons and pictures of local sights, etc. More precisely, a test of localized software applications that contains culturally specific icons and pictures may prime evaluators' and test users' culturally specific knowledge systems at a time when they complete a behavioral strategy such as a think aloud usability test.

Seventh, from cultural social psychology we derive that usability depends on what is socially appropriate. The appropriateness of applying accessible and available cultural knowledge becomes particularly questionable when evaluators and users have different socio-cultural backgrounds, for example, when they have different 'home grounds' such as China, India and Europe, but considerations of appropriateness is also relevant for other social situations (e.g., consider the nerdy technology user who tries to explain his love for his computer or his mobile phone to a technophobia-friend). Sharing knowledge of usability problems and coordinating descriptions of usability problems depend on the mutual perception of group belonginess. The participants may ask themselves implicit questions about the appropriateness of the available knowledge, such as 'if I tell them about this usability problem, will they understand that this is a problem, or will they think that I am ridiculing them?'

The eighth and final consideration is that while we must build on the idea that usability is universal, we cannot know exactly to what degree it is universal without doing empirical studies of human work and leisure in different organizational, social and cultural contexts. The usability of a computer artifact is hypothetical knowledge to be confirmed by actual use. A standard usability evaluation of a product with a particular built-in cultural model in a situation with one or more particular groups of users will result in a particular list of usability problems. To avoid these usability problems, it is not always enough to localize a product to fit cultural traits and/or demographic criteria. Because an established usability evaluation method functions as a mediator of the meanings of cultural models and the perceived reality of interactive systems, individual evaluators may find the cultural context foreign (the meaning of the cultural models), but still go on to identify well known types of usability problems (the perceived reality of the interactive system). Hence, a usability evaluation of a product for a market that is foreign to the evaluator may lead to the identification of the major usability problems that the target future users will experience, but this is not always the outcome of the usability evaluation.

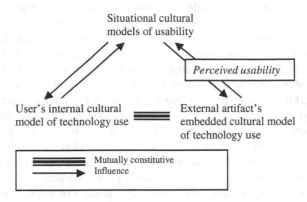

Fig. 1. Cultural Model theory of Usability (CM-U)

In the cultural model theory of usability in Figure 1, the relation between the internal cultural models of technology use (cognitive, psychological: to write a letter, do so-and-so) and the external artifact cultural models (how-to-use-this-product) is considered to be mutually constitutive (one makes no sense without the other). In contrast, the external usability evaluation situation is considered a loosely coupled mediator that creates the perception of a specific set of usability problems in much the same way as other views on technology use, such as system design methodologies or user participation approaches. The psychological sense of usability is a product of the usability evaluation-enabled communication about the references between the user's expectations of a technology and the specific artifact. A combination of specific internal cultural models and the specific artifact's cultural models may suggest a list of major usability problems, but the list may not necessarily be similar to the typical usability problems found by the established usability evaluation methods.

4 Comparison between CM-U and ADA Theory

This section discusses the CM-U theory in relation to other theories and definitions of usability to see its sensitivity to the description and explanation of the phenomenon of cultural usability. It would be beyond the scope of this paper to take on a full review of theories behind usability, but [10] can be seen as an excellent review of usability as a science. This paper focuses on comparing the cultural model theory of usability with the Artifact Development Analysis (ADA) theory of cultural usability [17], based on the idea that the evaluation criteria of a theoretical derivation is that the new theoretical derivation must be better than its best challenger [16].

ADA theory purports that conscious human behavior aims to achieve a goal, and the artifact is that which people use for achieving the goal. The artifact includes the hardware, software, and humanware. These may vary in both the time and spatial dimensions of creating a diversity of artifacts. Reasons for varying belong to three categories: goals, manufacturing, and people. The relations between people and artifacts are numerous, and one of these relations is usability. Usability is a value attitude towards the artifact. Different people may have different attitudes such as: Functional Value Attitude (put emphasis on a new function and/or the multi-functionality), Usability Value Attitude (put emphasis on the effectiveness and the efficiency), Aesthetic Value Attitude (put emphasis on the appearance and the good-looking design), Sensibility Value Attitude (put emphasis on the attachment or the emotional relationship), Economic Value Attitude (put emphasis on the cost (initial cost and maintenance cost)), Quality Value Attitude (put emphasis on the qualities such as the reliability, the safety, and the compatibility), and Ethical Value Attitude (put emphasis on the environmental aspect and the sustainability). As a value attitude, usability is composed of: 'small' usability (Jacob Nielsens and others' focus on efficiency or 'ease of use'), 'big' usability (ISO 9241-11 definition that includes utility/effectiveness/functionality), plus more subjective characteristics of users, such as emotions, motivation, values and others. All of this has to be seen also in terms of time, e.g., long term usability, and universality/diversity among users and situations. In ADA theory about usability, satisfaction is the ultimate criterion of an artifact [17]. In Table 1, CM-U theory is compared with ADA theory. The first column lists some general areas that describe cultural usability with respect to current practice.

Table 1. Evaluation of cultural usability with usage of theory

Comparison point:	ADA theory	Our proposal (CM-U theory)
Usability definition	Certain value attitude	Shared model of use
Cultural perspectives on usability	Culture as a social trait	Culture as models of use
Provide Guidelines for cultural specific design	Designers normative model includes culture as a diversity item in the design calculation	Focus on adapting usability evaluation methods to capture a diversity of models of use
Can be used to assess the extent of users experience of quality with artifact	The extent the usability of some artifact can give the core satisfaction to the user	The degree of alignment between relevant models of use
Provide a definition of the ultimate criterion of usability	Satisfaction	Alignment with models of everyday use

5 Discussion

This paper has presented the CM-U theory of cultural usability. From newer axioms for dealing with culture in HCI research and practice we have derived eight considerations. These were then presented as a coherent framework for perceived usability that can explain cultural usability phenomena. One of the potentials of CM-U theory is explaining the gap between users' and artifacts' cultural models of technology use, a gap that has been noted by other empirical studies [2]. For example, a study carried out in England of interculturally shared-systems design, asked a small group of users with diverse cultural backgrounds to participate in a think-aloud evaluation of a www system in order to identify breakdowns linked with: cultural factors in user-task interaction (language, humor, icons and jargon), user-tool interaction (understanding the tools representations), user-environment interaction (working habits, institutional practices, technological milieu) and user-user interaction (understanding the intended meaning of utterances). It was in user-tool interaction and user-task interaction that the majority of cultural breakdowns occurred. The authors proposed that cultural factors, such as religion, government, language, art, marriage, sense of humor, etc. are present in every culture, but it is the ways in which cultural factors are represented in interfaces that vary from culture to culture, and it is these that matter in HCI [2]. CM-U theory alternatively suggests that the results (the task and tool focus of usability problems) were: the effects of doing the intercultural study in one country only (all participants were primed to use their knowledge of English culture, e.g., to try to be effective) and letting all participants go through the same usability evaluation method (which biased the results in one direction, e.g., focus on foreground objects such as the interface instead of the larger work situation). Further, a nation-wide survey in a multi-cultural and multi lingual English speaking country, Botswana, [22] showed end-users having overwhelming preferences towards localized interfaces, but little need for localized icons and no agreement as to which language – not even the nationally adopted local language – was to be used for the interfaces. The little need for localized icons could be explained by the users' willing adaptation to the work environment, to the extent that they did not perceive their 'home' environment as relevant to their work environment [22]. This study supports CM-U's point of including both users' and artifacts' model of use and additionally suggests that the usability evaluation method used (at home or at work) may have had an influence on the results. Finally, a study in China [26] developed and evaluated a culturally specific metaphor (a Chinese traditional garden) to replace the western desktop metaphor for personal computing. Heuristic evaluation and user evaluation with a group of Chinese users of a metaphor based prototype suggested that background knowledge of language, logic and taboos was essential to the anticipation of user behavior in heuristic evaluation. CM-U theory points to this relation between users' model of use and the model of use built into a particular instantiation of an usability evaluation method.

The comparison of ADA and CM-U theories of cultural usability suggests differences that have implications for current usability practice. In ADA the focus is on values and social traits, and how the designer should include these as diversity items in the design process to ensure a high degree of satisfaction to the user. The CM-U focus is on shared, culturally specific models of use, and how usability evaluation methods can capture a diversity of cultural models of use to ensure a high degree of alignment between models of use involved in usability evaluation, as well as in everyday use. One issue worthy of discussion is whether the two theories of cultural usability are both relevant, albeit to a varying degree, depending on the cultural perspective from which cultural usability is seen. Empirical studies of quality-in-use and users' perception of usability show that the notion of usability is not constant across cultures. In a survey of 145 students and professionals from 30 different countries, it was found that usability professionals from various countries show different attitudes towards usability components, such as efficiency, effectiveness, and satisfaction, i.e., usability professionals from different countries have specific inclinations towards one of these components and for them any usability study primarily concerns that specific component [31]. This finding is supported by our own empirical studies. For example, Chinese users appear to be more concerned with visual appearance, satisfaction, and fun than do Danish users; Danish users prioritize effectiveness, efficiency, and lack of frustration higher than do Chinese users [8]. Danish, and to some extent, Indian end-users and system developers tend to make more use of constructs traditionally associated with usability (e.g., easy-to-use, intuitive, and liked) compared to their Chinese counterparts [12]. ADA with its focus on values and satisfaction seems to provide a good fit with Chinese users' perception of usability as visual appearance, satisfaction and fun, while CM-U with its focus on cultural models of everyday use fits with Danish users' preference for effectiveness and efficiency. Our evaluation methods may have cultural bias built-in [6], and we have to accept that there may also be a cultural bias in our theories of usability.

6 Conclusion

On basis of the comparison of CM-U and ADA, and the discussion of empirical studies of usability phenomena, we conclude that both theories are able to explain and describe cultural usability phenomena. CM-U will be best in cross cultural or cultural comparisons, if we from the beginning focus on task performance. ADA will be preferable in situations where we see emotional and aesthetic preferences as the basis for comparing usability problems. Future research should focus on collecting more evidence both on attitudes and cultural models of use, and perhaps suggest ways to combine the two theories.

Acknowledgments. This study was co-funded by the Danish Council for Independent Research (DCIR) through its support of the Cultural Usability project.

References

1. Bevan, N.: International standards for HCI and usability. International Journal of Human-Computer Studies 55(4), 533–552
2. Bourges-Waldegg, P., Scrivener, S.A.R.: Meaning, the central issue in cross-cultural HCI design. Interacting with Computers 9(3), 287
3. Carroll, J.M.: HCI Models, Theories and Frameworks - Towards a Multidisciplinary Science. Morgan Kaufmann Publishers, San Francisco (2003)
4. Chiu, C.-Y., Cheng, S.Y.Y.: Toward a Social Psychology of Culture and Globalization: Some Social Cognitive Consequences of Activating Two Cultures Simultaneously. Social and Personality Psychology Compass 1(1), 84–100
5. Clemmensen, T., Goyal, S.: Cross cultural usability testing Working paper, Department of Informatics, Copenhagen Business School, Copenhagen 20 (2005)
6. Clemmensen, T., Hertzum, M., Hornbæk, K., Shi, Q., Yammiyavar, P.: Cultural Cognition. In: The Thinking-Aloud Method For Usability Evaluation International Conference on Information Systems - ICIS 2008, Paris (2008)
7. Dix, A., Finlay, J., Abowd, G., Beale, R.: Human-Computer Interaction - third edition. Prentice-Hall, Englewood Cliffs (2004)
8. Frandsen-Thorlacius, O., Hornbæk, K., Hertzum, M., Clemmensen, T.: Non-Universal Usability? A Survey of How Usability Is Understood by Chinese and Danish Users. In: CHI 2009 (2009)
9. German, T.P., Barrett, H.C.: Functional Fixedness in a Technologically Sparse Culture. Psychological Science 16(1), 1–5
10. Gillan, D.J., Bias, R.G.: Usability science. I: Foundations. International Journal of Human-Computer Interaction 13(4), 351–372

11. Gray, W.D., Salzman, M.C.: Damaged merchandise? A review of experiments that compare usability evaluation methods. Human-Computer Interaction 13(3), 203–261
12. Hertzum, M., Clemmensen, T., Hornbæk, K., Kumar, J., Shi, Q., Yammiyavar, P.: Usability Constructs: A Cross-Cultural Study of How Users and Developers Experience Their Use of Information Systems. In: HCI International 2007, Beijing, China, July 22-27, 2007, Proceedings, Part I, pp. 317–326. Springer, Heidelberg (2007)
13. Hofstede, G.: Culture's Consequence: Comparing Values, Behaviours. Institutions and Organizations Across Nations, Sage Publications Inc. (1980)
14. Hong, Y.-Y., Mallorie, L.M.: A dynamic constructivist approach to culture: Lessons learned from personality psychology. Journal of Research in Personality 38, 59–67
15. Hutchins, E.: How a Cockpit Remembers Its Speeds. Cognitive Science 19(3), 265–288
16. Jarvinen, P.: Research Questions Guiding Selection of an Appropriate Research Method, Department of Computer Science, University of Tampere, Finland, 9 (2004)
17. Kurosu, M.: Usability and culture as two of the value criteria for evaluation the artifact. In: Clemmensen, T. (ed.) NordiCHI 2008 Workshop, Department of Informatics, Copenhagen Business School, October 19 (Working Paper; 01-2008) (2008)
18. Marcus, A., Gould, E.: Cultural Dimensions and Global User-Interface Design: What? So What? Now What? In: 6th Conference on Human Factors and the Web (2000)
19. Nisbett, R.E., Peng, K.P., Choi, I., Norenzayan, A.: Culture and systems of thought: Holistic versus analytic cognition. Psychological Review 108(2), 291–310
20. Norman, D.A.: Affordance, conventions, and design. Interactions 6(3), 38–41
21. Norman, D.A.: The Design of Everyday Things. Basic Books, New York (1988)
22. Onibere, E.A., Morgan, S., Busang, E.M., Mpoeleng, D.: Human-computer interface design issues for a multi-cultural and multi-lingual English speaking country – Botswana. Interacting with Computers 13(4), 497
23. Pepitone, A.: A social psychology perspective on the study of culture: An eye on the road to interdisciplinarianism. Cross-Cultural Research 34(3), 233–249
24. Preece, J., Rogers, Y., Sharp, H.: Interaction Design: Beyond Human-Computer Interaction. John Wiley & Sons, Chichester (2002)
25. Schneiderman, B., Plaisant, C.: Designing the User Interface. Pearson, London (2004)
26. Shen, S.-T., Woolley, M., Prior, S.: Towards culture-centred design. Interacting with Computers 18(4), 820
27. Smith, A., Dunckley, L., French, T., Minocha, S., Chang, Y.: A process model for developing usable cross-cultural websites. Interacting with Computers 16(1), 63
28. Smith, A., Yetim, F.: Global human-computer systems: Cultural determinants of usability. Editorial. Interacting with Computers 16
29. Sun, H.: Expanding the Scope of Localization: A Cultural Usability Perspective on Mobile Text Messaging Use in American and Chinese Contexts Rensselaer Polytechnic Institute Troy, New York (2004)
30. Vatrapu, R., Suthers, D.: Culture and Computers: A Review of the Concept of Culture and Implications for Intercultural Collaborative Online Learning. In: Ishida, T., R. Fussell, S., T. J. M. Vossen, P. (eds.) IWIC 2007. LNCS, vol. 4568, pp. 260–275. Springer, Heidelberg (2007)
31. Vöhringer-Kuhnt, T.: The influence of culture on Usability. Master thesis. Dept. of Educational Sciences and PsychologyFreie Universität Berlin, Berlin, Germany (2002), http://userpage.fu-berlin.de/~kuhnt/thesis/results.pdf (July 2004)

Regional Difference in the Use of Cell Phone and Other Communication Media among Senior Users

Ayako Hashizume[1], Masaaki Kurosu[2], and Toshimasa Yamanaka[1]

[1] Graduate School of Comprehensive Human Science, University of Tsukuba, Japan
[2] National Institute of Multimedia Education (NIME), Japan
hashi-aya@kansei.tsukuba.ac.jp

Abstract. In this paper, authors focused on the use of the cell phone by senior people living in the urban area and the rural area in Japan. The result of the questionnaire research showed that there are differences in the use of the cell phone and other communication media between two areas. These differences are related to the difference in the life pattern and the environmental factors in both areas.

Keywords: Communication media, usability, cell phone, senior user, regional difference.

1 Introduction

Generally, it is believed that high-tech devices are difficult to use for senior people. But it is observed some senior people are using the cell phone frequently that is one of the leading high-tech devices. It is not yet clear how much functionality they are using and how they are making the best use of the cell phone in their life. It is frequently said that there is a difference in the use of the cell phone and other communication media due to the generation where the latter includes the land line, the letter, the telegram, the cell phone mail, and the PC mail [1]. But there are few empirical evidences regarding the details of the literacy and the use of the cell phone [2]. Hence we decided to conduct the research on the selection and the use of communication media including the cell phone, especially focusing on 11 specific situations.

This paper deals with these questions by conducting the questionnaire research for senior people by comparing the result with those for young people. And the analysis was conducted from the viewpoint of Artifact Development Theory (ADT) (that was renamed now as the Artifact Development Approach (ADA)) [3, 4]. ADT is a new emerging field of user engineering that, in short, analyzes the use of artifact why "it" was selected for "that" purpose and why "others" were not selected for "that" purpose, thus clarify the reasonability of the artifact selection and the optimal design of artifact.

2 Methods

Questionnaire researches were conducted for 50 senior people living in the urban area (25 male and 25 female) and for 20 senior people living in the rural area (10 male and 10 female) all of whom possess the cell phone. The urban area is Chiba city that is

M. Kurosu (Ed.): Human Centered Design, HCII 2009, LNCS 5619, pp. 426–435, 2009.
© Springer-Verlag Berlin Heidelberg 2009

next to Tokyo and the rural area is Ishigaki city that is one of the south-most islands in Japan. Regarding the senior people, we defined them as those who are over 60, whereas WHO defines them as those over 65. It is because the senior people usually retire from their job at around 60 in Japan and their life environment changes drastically around that age. The average age of senior people living in the urban area was 68.44 with the SD of 3.67, and the average age of senior people living in the rural area was 68.60 with the SD of 4.32.

The research was conducted from October to December of 2007. The questionnaires were delivered to them and were collected after 3 weeks.

2.1 Content of the Questionnaire

The questionnaire contained following items.

2.1.1 Question 1: About the General Use of Communication Media
Informants were asked following questions on 8 types of communication media (the land line, the cell phone, the letter, the telegram, the cell phone mail, the PC mail, ask somebody to convey the message, and the face-to-face meeting).

a)   How frequently do you use it?
     (Forced choice from among: very often, sometimes, seldom, almost none)
b)   Main Purpose - For what purpose do you mainly use it? (Free answer)
c)   Major Problems - What is the major problem with it? (Free answer)
d)   If you can't use it, what would you use instead of it? (Free answer)

2.1.2 Question 2: About the Specific Use of Communication Media
Informants were asked following questions in terms of 11 situations such as "If you want to talk to somebody about everyday issue that is not urgent" and "If you want to tell family member(s) that you'll be late to return home".

(a)   What kind of media do you use? (Free choice from among: the land line, the cell phone, the letter, the telegram, the cell phone mail, the PC mail, ask somebody to convey the message, the face-to-face meeting, do nothing, other)
(b)   Why would you choose it/them? (Free answer)

3 Results

3.1 Usage Frequency of Each of 8 Types of Communication Media

Regarding the land line, the cell phone, and ask somebody to convey the message, senior people in both area showed approximately the same frequency of use. The result of Question 1 (a) is shown in Fig. 1. There was a significant difference for the use of the communication media between two areas.

Result of the chi-square test (Table 1) shows 8 types of communication media have different levels of frequency of use between two areas. The residual analysis revealed the significant differences of use at 5% level as follows. The senior people living in the urban area use the letter and the telegram infrequently whereas the senior people living in the rural area use the communication media such as the cell phone mail, the PC mail, and the face-to-face meeting almost none.

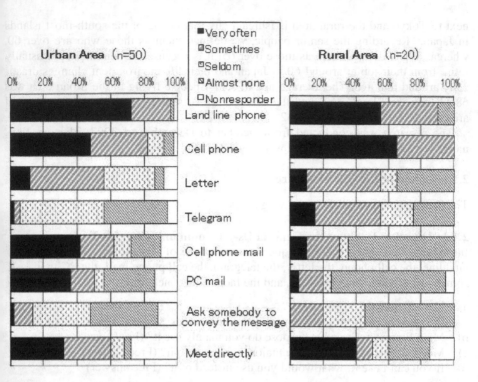

Fig. 1. Frequency of use of communication media among senior people living in the urban area (left) and the rural area (right)

Table 1. Result of the chi-square test comparing the proportion of use of each media between two areas in Figure 1

communication media	chi-square value	
Land line phone	3.785	
Cell phone	4.385	
Letter	12.416	**
Telegram	22.193	**
Cell phone mail	16.910	**
PC mail	11.623	**
Ask somebody to convey the message	4.534	
Meet directly	9.652	*

** =0.01

3.2 Selection of Communication Media in 11 Situations

Results of Question 2 (a) are shown from Fig. 2 to Fig. 12. Applying the chi-square test, there were many different usage patterns between the senior people living in the urban area and the rural area as shown in Table 2.

Table 2. Result of the chi-square test comparing the proportion of each response between two areas for each case in 11 situations

Case		Land line phone	Cell phone	Letter	Telegram	Cell phone mail	PC mail	Ask somebody to convey the message	Meet directly	Doing nothing	Other
1	Want to talk about something (not ugently)	7.343 *	0.613	0.420	0.406	2.647	5.793	0.824	0.398	0.502	0.000
2	Want to tell family menbers that you'll be late to return	1.419	4.033	0.000	0.000	5.247	0.000	0.000	0.406	0.027	0.406
3	Plan the party or something	2.240	0.824	0.726	0.000	1.756	5.920	0.027	1.744	0.955	0.000
4	Want to tell that you are urgently	10.074 **	2.835	0.000	0.000	1.815	1.697	0.027	0.824	2.536	0.000
5	Want to know the schedule of the movie, etc	9.528 **	0.575	2.536	2.536	0.345	1.779	0.463	1.477	12.468 **	0.463
6	Want to discuss about some critical issue	11.436 **	1.470	0.425	0.000	0.073	3.613	2.536	3.112	0.463	0.000
7	Want to tell something quite interesting	3.993	7.342 *	3.111	0.000	1.350	3.523	0.000	0.997	0.726	0.000
8	Want to tell thanks for the gift	7.807 *	4.126	2.337	0.000	0.406	4.131	0.406	0.451	2.536	2.536
9	Want to tell that you'll be late for the oohodulo	2.520	4.763	0.000	0.000	1.616	0.406	0.406	0.000	0.000	0.824
10	Want to borrow some money	0.220	9.890 **	0.726	0.000	0.406	0.824	0.000	4.953	3.791	0.000
11	Want to confirm if she(he) is OK	8.816 *	5.301	2.030	0.035	3.523	4.667	1.254	3.432	0.824	0.406

**P<.01, *P<.05

3.2.1 Case 1: If You Want to Talk about Everyday Issue That Is Not Urgent
For case 1, the senior people living in both areas selected the land line and the cell phone (Fig.2). It is significant at 5% level by the Chi-square test, for the media that is used very frequently, the senior people living in the urban area selected the land line.

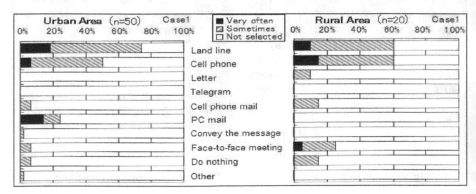

Fig. 2. Case1: media selection for the case "If you want to talk about everyday issue that is not urgent"

3.2.2 Case 2: If You Want to Tell Family Member(s) That You'll Be Late to Return Home

For case 2, the senior people living in both areas selected the cell phone (Fig.3). It seems to have a tendency that the senior people living in the urban area are using it very often and sometimes selected the land line and the cell phone mail, but it isn't significant by the Chi-square test.

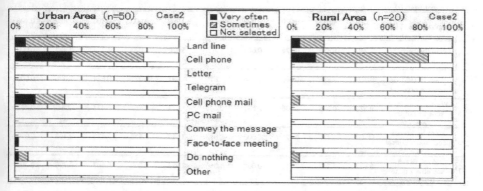

Fig. 3. Case2: media selection for the case "If you want to tell family member(s) that you'll be late to return home"

3.2.3 Case 3: If You Plan the Party or Something

For case 3, the senior people living in both areas selected the cell phone (Fig.4). There seems to be a tendency for the media that is frequently used , the senior people living in the urban area selected the land line, the cell phone mail, PC mail, and face-to-face meeting. While the media that is used for sometimes, the senior people living in the rural area selected the cell phone, but these don't have a significant result by the Chi-square test.

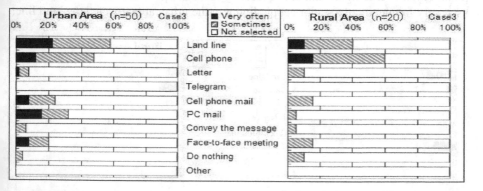

Fig. 4. Case3: media selection for the case "If you plan the party or something"

3.2.4 Case 4: If You Want to Tell That You Are in a Bad Condition

For case 4, the senior people living in the both areas selected the land line and the cell phone (Fig.5). It is significant at 1% level by the Chi-square test, for the media that is used very often and sometimes, the senior people living in the urban area selected the land line.

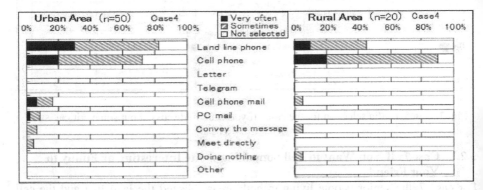

Fig. 5. Case4: media selection for the case "If you want to tell that you are in a bad condition"

3.2.5 Case 5: If You Want to Know the Schedule of the Movie, the Concert, Etc

For case 5, the senior people living in the both areas selected the land line and the cell phone (Fig.6). It is significant at 1% level by the Chi-square test, for the media that is used very often and sometimes, the senior people living in the urban area selected the land line while the people living in the rural area selected "Doing nothing". "Other" responses include the use of the "newspaper".

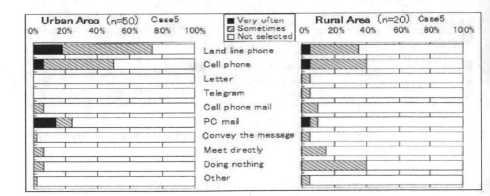

Fig. 6. Case5: media selection for the case "If you want to know the schedule of the movie, the concert, etc"

3.2.6 Case 6: If You Want to Discuss about Some Critical Issue

For case 6, the senior people living in both areas selected the land line and the cell phone (Fig.7). It is significant at 1% level by the Chi-square test, for the media that is used very often and sometimes, the senior people living in the urban area selected the land line.

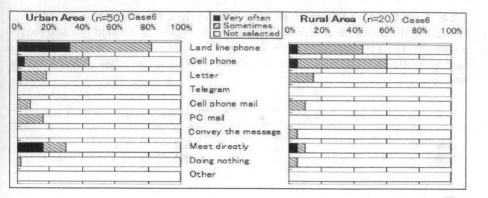

Fig. 7. Case6: media selection for the case "If you want to discuss about some critical issue"

3.2.7   Case 7: If You Want to Tell Something Quite Interesting or Funny to Your Friend

For case 7, the senior people living in both areas selected the land line and the cell phone (Fig.8). It is significant at 5% level by the Chi-square test, for the media that is used "sometimes", the senior people living in the rural area selected the cell phone.

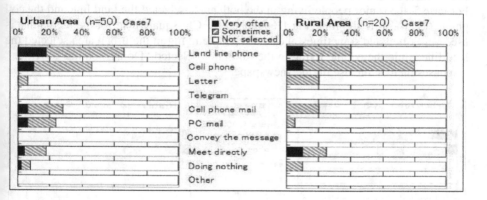

Fig. 8. Case7: media selection for the case "If you want to tell something quite interesting or funny to your friend"

3.2.8   Case 8: If You Want to Express Thanks for the Gift

For case 8, the senior people living in both areas selected the land line, the cell phone, and the letter (Fig.9). It is significant at 5% level by the Chi-square test, for the media that is used very often and sometimes, the senior people living in the urban area selected the land line. "Other" response includes sending the "reciprocal gift".

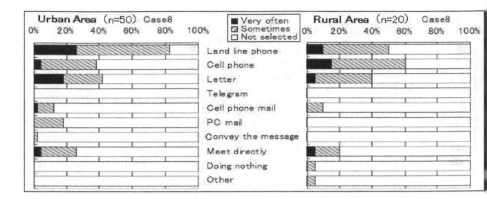

Fig. 9. Case8: media selection for the case "If you want to express thanks for the gift"

3.2.9 Case 9: If You Want Tell That You'll Be Late for the Schedule

For case 9, the senior people living in both areas selected the cell phone and the land line (Fig.10). It seems to have a tendency that the senior people living in the urban area often used the land line and the cell phone mail, but they don't have a significant result by the Chi-square test. "Other" responses include the use of the "public phone".

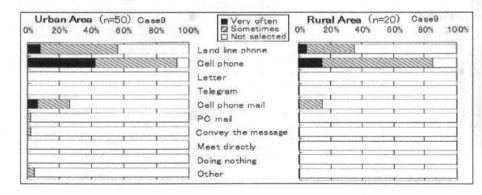

Fig. 10. Case9: media selection for the case "If you want tell that you'll be late for the schedule"

3.2.10 Case10: If You Want to Borrow Some Money from Your Friend

For case 10, the senior people living in both areas selected "Doing nothing" (Fig.11). It is significant at 1% level by the Chi-square test, for the media that is used very often and sometimes, the senior people living in the rural area selected the cell phone.

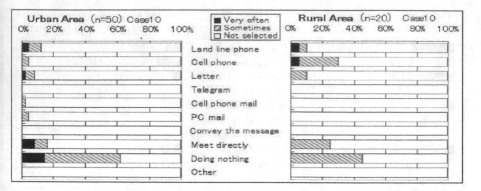

Fig. 11. Case10: media selection for the case "If you want to borrow some money from your friend"

3.2.11   Case11: If You Want to Confirm If She/He Is Ok

For case 11, the senior people living in both areas selected the land line and the cell phone (Fig.12). It is significant at 5% level by the Chi-square test, for the media that is used very often, the senior people living in the urban area selected the land line.

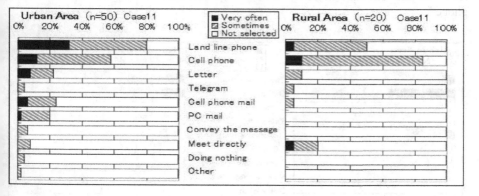

Fig. 12. Case11: media selection for the case "If you want to confirm if she/he is OK"

4   Conclusion

General tendency found in this research was that (1) the senior people use the new communication media such as the cell phone call, the cell phone mail (texting) and the PC mail less frequently compared to the young people, (2) the senior people living in the rural area use the cell phone-mail and the PC-mail less frequently compared to the senior people living in the urban area, in other words, the senior people living in rural area don't use the mail function not so much whatever the device they may use, and (3) there are some cases that the senior people living in the rural area use the cell phone more frequently than the landline.

These tendencies can be attributed to the literacy and/or the motivation. A tentative model would be that the senior people have the value attitude to accept the current life style as it is, then they may not have a motivation to learn how to use the new media, and finally they have less literacy for the new technology in general. Thus it is necessary to conduct a further research on the value attitude of senior people.

References

1. Hashizume, A., Kurosu, M., Kaneko, T.: The Choice of Communication Media and the Use of Mobile Phone among Senior Users and Young Users. In: Lee, S., Choo, H., Ha, S., Shin, I.C. (eds.) APCHI 2008. LNCS, vol. 5068, pp. 427–436. Springer, Heidelberg (2008)
2. Hashizume, A., Kurosu, M., Kaneko, T.: Regional Difference of Cell Phone Literacy Among Senior People. Bulletin of Human Centered Design Organization 4(2) (2008) (in Japanese) Human Centered Design Organization
3. Kurosu, M.: An Introduction to the Artifact Development Theory. HCD-Net 3(1) (2007) (in Japanese)
4. Kurosu, M.: The Optimality of Design from the Viewpoint of the Artifact Development Theory. Human Interface Society SIGUSE (2007) (in Japanese)

Grouping Preferences of Americans and Koreans in Interfaces for Smart Home Control

Kyeong-Ah Jeong[1], Robert W. Proctor[2], and Gavriel Salvendy[1,3]

[1] School of Industrial Engineering, Grissom Hall, Purdue University,
315 N. Grant St., West Lafayette, IN 47907, USA
[2] Department of Psychological Sciences, Purdue University,
703 Third St., West Lafayette, IN 47907, USA
[3] Department of Industrial Engineering, Tsinghua University, Beijing 100084,
People's Republic of China
{jeong7,rproctor,salvendy}@purdue.edu

Abstract. The purpose of the current study was to find the grouping principle for smart home interfaces that most closely matches the thinking styles of Americans and Koreans. The independent variables were grouping method (NO: no grouping other than alphabetical order, FS: functional and then spatial grouping, SF: spatial and then functional grouping), culture and gender. 40 American and 40 Korean students' perceptions of the interfaces and their performance times with the interfaces were measured. Both female and male Koreans preferred the SF grouping, consistent with a cognitive style favoring thematic organization and field dependence. For Americans, females preferred SF grouping but males preferred FS grouping. Thus, only American males' preferences conformed to a cognitive style favoring functional organization and field independence. Cultural differences in grouping preferences need to be taken into account in design of smart home interfaces.

Keywords: Culture, grouping, interface design, smart home.

1 Introduction

The objective of the current study was to find the grouping principles that most closely match the cognitive styles of Americans and Koreans, for the purpose of designing interfaces for smart home control. This paper describes an experiment that tests hypotheses regarding cultural differences in grouping preferences. Based on the results of the experiment, design guidelines for smart home interfaces were developed.

A smart home is defined as "a home or working environment, which includes the technology to allow for devices and systems to be controlled automatically" [1]. Briere and Hurley [2] define a smart home as a harmonious home, a collection of devices and capabilities based on home networking. The terms connected home, digital home, adaptive house, and aware home are also used to represent future homes. In a smart home environment, as the number of objects having radio-frequency identification (RFID) tags increases, it will become increasingly difficult to find a specific control

M. Kurosu (Ed.): Human Centered Design, HCII 2009, LNCS 5619, pp. 436–445, 2009.
© Springer-Verlag Berlin Heidelberg 2009

on a control-panel interface for the device or object that needs to be controlled. This problem may be resolved by grouping the objects on a control panel, remote control, or computer display in such a way that the organization matches the mental representations of smart home users. According to many studies of cultural differences, people in different cultures tend to have different thinking styles. Thus, there is a possibility that potential smart home users in different cultures, especially eastern and western cultures, may have different organizational preferences for smart home interfaces. In the current study, Americans as a representative of western cultures and Koreans as the representative of eastern cultures were compared to investigate their grouping preferences for smart home interfaces.

According to Choi et al. [3], analytic versus holistic style influences how people categorize objects. East Asians tend to perceive and reason holistically, attending to the field in which objects are embedded and attributing causality to interactions between object and the field [3, 4]. In contrast, Europeans and Americans are held to be analytic, paying attention primarily to the object, categorizing it on the basis of its attributes, and attributing causality to the object based on rules about its category memberships. Choong and Salvendy [5] showed that Chinese participants performed better with thematic organization than with functional organization, whereas American counterparts performed better with functional organization than with thematic organization, in terms of error rate. Rau et al. [6] replicated the experiment of Choong and Salvendy [5] in Taiwan and showed similar results. Since both the Koreans and Chinese can be considered as East Asians, Koreans should have a similar cognitive style and classification preference to those of Chinese people. Kim and Lee [7] provided evidence that Koreans' cognitive style can be considered the same as that of the Chinese. Hwang et al. [8], Chung and Gale [9], Yoon [10] and Kim et al. [11] also supported cultural pattern and cognitive style differences between Americans and Koreans.

In addition to culture, gender was considered in the current study because Witkin's theory [12] [13] [14] predicts that females are more likely to have a field-dependent cognitive style, whereas males more often have an analytical or field-independent cognitive style. Basically, the concept of field-dependence is similar to holistic style and the concept of field-independence seems analogous to analytic style. This relation suggests the possibility of involvement or interference of gender with culture and grouping of the objects.

To determine initial grouping facts for the current experiment that might influence the usability of smart home interfaces, studies of interface layout organizations were reviewed. Stone et al. [15] indicated that features users consider to be related should be grouped together on the user interface, or at least their association should be clearly indicated. That is, grouping of the features should reflect users' understanding of the domain and their expectations about how the user interface should be organized. Niemela and Saariluoma [16] recommended spatial grouping of items from the same semantic category. Salmeron et al. [17] studied semantic grouping, but they focused more on different user groups such as expert and novice users. Salmeron et al. found that expert users performed better than novice users in information retrieval when the items of an interface were semantically organized, but not when they were placed randomly. Mehlenbacher et al. [18] compared alphabetical ordering with functional organization across three different cues: direct match; synonym cue; iconic cue.

The alphabetic menu led to faster selection time than the functional menu under the direct match condition, whereas the opposite effect occurred under the synonym and iconic conditions. Coll et al. [19] compared three organization conditions (alphabetical menu; categorical menu; unordered menu) on performance time and the number of errors. Their results yielded a significant difference among the three organization conditions but failed to yield a significant difference between alphabetical menu and categorical menu on performance time. The average number of errors did not show statistical significance among the three organization conditions. The reviewed studies of grouping suggest that the factors of alphabetical order, functional grouping and spatial grouping are promising in terms of designing smart home interfaces. Thus, in the current experiment, these three types of layout organizations were manipulated in order to develop a more adaptable interface design for smart home context.

2 Method

2.1 Participants

40 American students (20 males and 20 females) and 40 Korean students (20 males and 20 females) at Purdue University were recruited. Korean participants were restricted to students who had spent less than 2 months in the U.S, so that they had only minimal prior exposure to the U.S. culture.

2.2 Variables

Independent variables were grouping method, culture and gender. Grouping method was a within-subject variable with three levels: functional and then spatial grouping (FS); spatial and then functional grouping (SF); no grouping other than alphabetical order (NO). Culture (American and Korean) and gender (male and female) were between-subject variables.

The dependent variables were satisfaction, ease of use, perceived performance speed, actual performance time, the perception of the number of chunked items, overall evaluation, general liking, and rank-order preference. All dependent variables except actual performance time were subjective in that they measured users' perceptions of the interfaces. All subjective responses except rank-order preference were measured using a questionnaire administered immediately after performing tasks with each grouping method. The rank order was measured after participants had completed performing tasks with all three types of grouping methods. Thus, this variable provides an overall evaluation about the three types of grouping methods. The objective variable was the actual performance time, which was defined as the time between when each participant clicked the 'start button' to start a trial and when he/she successfully found and clicked a correct control button for the targeted device (object).

2.3 Experimental Test Beds

As indicated earlier, based on the existing grouping studies, alphabetical order, functional grouping and spatial grouping seemed promising in terms of designing smart home interface. Thus, these three types of layout organizations were manipulated in order to find a more adaptable interface design for the smart home context. The first experimental test bed was NO grouping interface, for which all of the smart home devices were listed in alphabetical order. The second experimental test bed was the FS grouping, for which 'functional characteristics' was the grouping principle for the main page and 'spatial characteristics' was the grouping principle for the subordinate page. The third experimental test bed was SF grouping, for which 'spatial characteristics' was the grouping principle for the main page and 'functional characteristics' was the second grouping principle for the subordinate page.

2.4 Procedure

The experiment was conducted in the Human-Computer Interaction Lab in the School of Industrial Engineering at Purdue University. Each participant performed the experiment alone. Before starting the experiment, a brief description of what it generally involved was provided to the participant. Then, the participant was asked to fill out an informed consent form and a demographic questionnaire concerning personal characteristics.

When each participant was ready to perform the experiment, a written scenario that contained information about a situation was provided to the participant. After reading the description of the situation, the participant was told to push the 'Start' button on the computer screen to start Task 1. Once the participant pushed the 'Start' button, a short scenario of Task 1, including a device (object) to control, was presented on the computer screen. After reading the Task 1 scenario, the participant began to search for the correct control for the device (object) using one of the three grouping methods. Once the participant found the correct control for the target device, the 'ending time' was recorded by the computer. This procedure was repeated for tasks 2 and 3.

Following completion of the three tasks with the initial interface, the participant was asked to fill out a questionnaire examining satisfaction with the interface used for those tasks. After completing the questionnaire, the participant repeated the same procedure for each of the two remaining interfaces that used the other grouping methods.

Upon completion of the tasks with all three grouping methods, the participants filled out the post-experiment questionnaire. In this questionnaire, they were told to rank the interfaces from 1 to 3 (1 for most liked; 3 for east liked) based on their preferences for the three different grouping methods.

2.5 Hypotheses

The following hypotheses, based on the reviewed studies of culture and grouping, were tested in the experiment.

Hypothesis 1: Participants will prefer a smart home interface with either grouping method (either FS or SF organization) over the one with NO grouping. Overall, participants will prefer the SF organization the most, followed by the FS organization.

We examined this hypothesis to determine whether meaningful grouping that adapts to users' thinking style is better than listing items in alphabetical order on the smart home interface. We expected that SF organization would be preferred because items in a home are located spatially. Each room in a home is usually designed for a functional purpose such as sleeping and eating.

Hypothesis 2: Korean participants perceive SF organization as providing faster selection than FS organization, whereas American participants will judge FS organization as providing faster selection than SF organization.

This hypothesis was formulated to examine whether smart home interfaces should be designed differently for Americans and Koreans. To test this hypothesis, the interaction effect of grouping method and culture was examined in male and female data sets, respectively.

Hypothesis 3: A three-way interaction of grouping method, culture, and gender will exist in perception of performance speed.

This hypothesis was designed to examine whether gender differences exist within a culture. Because American culture consists of diverse sub-populations (ethnic groups), gender differences seem more likely to play a role for U.S. users than for Korean users.

3 Results and Discussion

An analysis of variance (ANOVA) was conducted on the preference and performance measures as a function of grouping method, culture, and gender. Follow-up ANOVAs were performed on the male and female data sets.

3.1 Internal Consistency

The internal consistency was measured using Cronbach's coefficient alpha. The higher the alpha is, the more consistent the measure is. Usually 0.7 and above is considered to be acceptable [21]. The internal consistency of the combined data set was 0.91. The internal consistency of American and Korean participants was 0.90 and 0.91, respectively.

3.2 Testing of Hypotheses

The means and standard deviations (SDs) of the dependent measures for each grouping method are presented in Table 1. Hypothesis 1 was statistically supported. Grouping using either FS or SF organization was preferred over NO organization on all of the satisfaction questionnaire items, the post-experiment question, and average performance time, with statistical significance ($p < .001$) and practical significance (with more than 10% difference). These results are consistent with previous findings [15, 16, 17]. Although Mehlenbacher et al. [18] and Coll et al. [19] failed to show a significant difference between alphabetical ordering and other semantic groupings such as functional or categorical, the current experiment clearly showed users' preferences toward either FS or SF organization over alphabetical ordering. Therefore, our results imply that a smart home interface should be designed using grouping that reflects users' understanding of the domain and their expectations about how the interface should be organized. In comparison to the FS grouping, the SF grouping was preferred, $F(1, 76) = 3.575$, $p = .062$, and tended to show higher agreement on perceived performance speed, $F(1, 76) = 3.059$, $p = .084$. The mean ranking for the FS grouping was worse than that for the SF grouping, and the mean rated performance speed was less for the FS grouping than for the SF grouping. These results imply that, in the context of smart home interfaces, users' general thinking style is closer to thematic organization and field dependence than to functional organization and field independence.

Table 1. Descriptive statistics for each item as a function of grouping method

Item	Grouping					
	NO		FS		SF	
	Mean	SD	Mean	SD	Mean	SD
Perceive Satisfaction	4.68	1.49	5.38	1.10	5.46	1.24
Perceived Number of Items Chunked	3.44	1.65	5.81	0.87	5.75	1.00
Perceived Performance Speed	4.68	1.52	5.28	1.10	5.58	1.18
Overall Evaluation	3.59	1.44	4.40	1.51	4.31	1.47
Overall Liking	4.16	1.50	5.23	1.26	5.35	1.39
Ease of Use	4.63	1.44	5.38	1.07	5.40	1.29
Paired Question of Perceived Satisfaction	4.35	1.52	5.38	1.12	5.35	1.30
Rank	2.68	0.63	1.79	0.71	1.54	0.64
Performance Time (seconds)	41.03	25.71	19.36	6.69	19.65	7.98

Table 2 contains the means and standard deviations for perceived performance speed as a function of grouping method, culture and gender, which are relevant to hypotheses 2 and 3. Hypothesis 2 was supported with statistical significance, $F(1, 38) = 6.001, p = .019$, for the male participants but not for the female participants. According to the results, American male participants thought that FS grouping was faster than SF grouping, whereas Korean male participants thought that SF grouping was faster than FS grouping. That is, American male users tended to consider functional characteristics of a target device or item first, whereas Korean male users tended to first consider spatial characteristics of the room in the house in which a target device or item is located. These results imply that for male users the cognitive style of Americans follows functional organization and field independence, whereas that of Koreans follows thematic organization and field dependence. The female groups did not show statistical significance in the interaction effect of grouping method and culture, but they showed statistical significance in the main effect of grouping method for perceived performance speed, $F(1, 38) = 5.282, p = .027$, and rank order, $F(1, 38) = 5.918, p = .020$. Both American and Korean females thought that SF grouping was faster than FS grouping and gave higher preference ranking to SF grouping than to FS grouping. The results from the female data set for perceived performance speed and general preference ranking imply field dependency of females. That is, female smart home interface users tend to first consider spatial/context characteristics which a target device or item belongs to, rather than the target's or item's own functional characteristics.

Table 2. Descriptive statistics of grouping, culture and gender on perceived performance speed

Culture	Gender	Grouping			
		FS		SF	
		Mean	SD	Mean	SD
American	Female	5.00	1.26	5.70	1.08
	Male	5.55	1.15	5.00	1.59
	Total	5.28	1.22	5.35	1.39
Korean	Female	5.25	0.97	5.70	1.03
	Male	5.30	1.03	5.90	0.72
	Total	5.28	0.99	5.80	0.88

Overall, in the context of interface design for smart home control, these results imply that Koreans' cognitive style tends to follow thematic organization and field dependence, regardless of gender. But, Americans' cognitive style follows functional organization and field independence only for males, possibly because females tend to be field-dependent even in the American culture.

Hypothesis 3 was also statistically supported, $F(1, 76) = 4.163$, $p = .045$, in perceived performance speed. Perceived satisfaction tended toward significance, $F(1, 76) = 3.156$, $p = .08$. All subject groups except American males showed higher agreement on SF grouping being faster than FS grouping. This outcome implies that, for Koreans, SF grouping is highly recommended for the design of smart home interfaces. For Americans, though, interface designers need to take into account the gender of the smart home user. That is, for Americans, some sort of gender adaptable interface that can change its organization of the items based on the different genders' grouping preferences is recommended because females showed preferences for SF grouping whereas males preferred FS grouping. Gender involvement can be considered as a part of a culture effect because Americans' high individualism and masculinity [22] can explain the gender differentiation of American participants.

3.3   Performance Time Measurement

As shown in Table 1, the interfaces with FS or SF grouping showed much shorter actual performance time than the interface with NO grouping. However, performance time was similar for the FS and SF groupings, which did not differ significantly. This result is consistent with the results from similar studies that measured performance time between East Asians and European Americans, even though those studies showed performance differences on recall or error rate [5] [6]. In the current experiment, recall or error rate was not measured because a popup window was continuously displayed to provide feedback to the participants when they chose a wrong category, and performance time was continuously measured until they chose the right category. We thought that the current experimental procedure might detect differences in performance in time in part due to errors increasing the time required to operate the correct control. However, the results still showed an inability to detect statistical significance on performance time. Some possible reasons for the lack of ability to detect an influence of interface organization on actual performance time include the limited number of items or objects presented on the interfaces and the limited number of participants. The inability to detect a performance time difference may be caused by the semantic groupings for the FS and SF arrangements being too similar, since these two groupings had almost the same organization except for the presentation order of the main- and sub-menus on the interface. This latter interpretation may suggest that even though the amount of time required to perform with the SF and FS interfaces was not reliably different, participants actually perceived a small difference on the interfaces and felt the impact of it.

The experiment was able to successfully capture statistical significance between alphabetical arrangement (no grouping other than alphabetical order) and semantic grouping using either FS or SF organization with more than 50% difference. That is, semantic grouping using either FS or SF grouping showed shorter performance time compared to alphabetical arrangement. This result was somewhat expected based on some of the results from Mehlenbacher et al. [18]. Since the task descriptions used in the current experiment can be considered as synonym cues, shorter performance time of semantic grouping using either FS or SF grouping seems reasonable. The statistical significance appeared between alphabetical arrangement and semantic grouping using either of FS or SF organization can extend the literature on grouping since closely related previous studies such as those of Mehlenbacher et al. [18] and Coll et al. [19] failed to find a statistical difference in performance time between alphabetical and functional or categorical arrangements. However, at the same time, there is also a possibility that this significant effect might be caused by learning because alphabetical arrangement was always showed first to the participants in the current experiment. In this regard, there is a need to conduct further research that rules out a learning effect. A traditional experimental procedure about reaction time measurement with many trial blocks might be used to further examine potential learning effect. But, any learning effect in the current experiment is likely small because alphabetical arrangement is distinct from the other two layout organizations.

4 Conclusions

None of the grouping-related studies except Chen et al.'s [23] quantitatively measured people's perceptions of the usability of the interfaces. The current experiment measured both users' perceptions and performance times with different organizations of smart home interfaces. Through the current study, more tangible ideas of how to design a smart home interface for different cultural groups (Americans and Koreans) were provided. The current study also supports the view that grouping principles should closely match users' thinking styles and mental representations. Designers of smart home control interfaces need to take cultural differences into account.

References

1. Van Berlo, A., Bob, A., Jan, E., Klaus, F., Maik, H., Charles, W.: Design Guidelines on Smart Homes. A COST 219bis Guidebook. European Commission (1999)
2. Briere, D., Hurley, P.: Smart Homes for Dummies. John Wiley & Sons, Inc., New York (2003)

3. Choi, I., Nisbett, R.E., Norenzayan, A.: Causal Attribution Across Cultures: Variation and Universality. Psychology Bulletin 125(1), 47–63 (1999)
4. Choi, I., Nisbett, R.E.: Cultural Psychology of Surprise: Holistic Theories and Recognition of Contradiction. Journal of Personality and Social Psychology 79(6), 890–905 (2000)
5. Choong, Y.Y., Salvendy, G.: Implications for Design of Computer Interfaces for Chinese Users in Mainland China. International Journal of Human-Computer Interaction 11(1), 29–46 (1999)
6. Rau, P.-L.P., Choong, Y.Y., Salvendy, G.: A cross cultural study on knowledge representation and structure in human computer interfaces. International Journal of Industrial Ergonomics 34(2), 117–129 (2004)
7. Kim, J.H., Lee, K.P.: Cultural difference and mobile phone interface design: Icon recognition according to level of abstraction. In: Proceedings of the 7th international conference on human computer interaction with mobile devices & services. ACM International Conference Proceeding Series, vol. 111, pp. 307–310 (2005)
8. Hwang, W., Jung, H.J., Salvendy, G.: Internationalization of e-commerce: a comparison of online shopping preference among Korean, Turkish, and US population. Behaviour & Information Technology 25(1), 3–18 (2006)
9. Chung, H., Gale, J.: Family Functioning and Self-Differentiation: A Cross-Cultural Examination. Contemporary Family Therapy: An International Journal 31(1), 19–33 (2009)
10. Yoon, J.W.: Searching for an image conveying connotative meanings: An exploratory cross-cultural study. Library & Information Science Research 30(4), 312–318 (2008)
11. Kim, S.W., Kim, M.J., Choo, H.J., Kim, S.H., Kang, H.J.: Cultural issues in handheld usability: Are cultural models effective for interpreting unique use patterns of Korean mobile phone users? In: Proceeding of UPA (Usability Professionals' Association) (2003), http://www.caerang.com/publications/UPA_2003.pdf (accessed on 2/15/2008)
12. Witkin, H.A.: Socialization, culture and ecology in the development of group and sex differences in cognitive style. Human Development 22, 358–372 (1979)
13. Severiens, S.E., Dam, G.T.M.: Gender differences in learning styles: A narrative review and quantitative meta-analysis. Higher Education 27(4), 487–501 (1994)
14. Fritz, R.L.: A study of gender difference in cognitive style and conative volition. In: The American Vocational Education Research Association Session at the American Vocational Association Convention, St. Louis, MO (1992)
15. Stone, D., Jarrett, C., Woodroffe, M., Minocha, S.: User Interface Design and Evaluation. Morgan Kaufmann Series in Interactive Technologies, ch. 5, p. 92 and ch. 9, p. 172 (2005)
16. Niemela, M., Saariluoma, P.: Layout attributes and recall. Behaviour & Information Technology 22(5), 353–363 (2003)
17. Salmeron, L., Canas, J.J., Fajardo, I.: Are expert users always better searchers? Interaction of expertise and semantic grouping in hypertext search tasks. Behaviour & Information Technology 24(6), 471–475 (2005)
18. Mehlenbacher, B., Duffy, T.M., Palmer, J.: Finding information on a menu: Linking menu organization to the user's goals. Human-Computer Interaction 4, 231–251 (1989)
19. Coll, J.H., Coll, R., Nandavar, R.: Attending to cognitive organization in the design of computer menus: A two-experiment study. Journal of the American Society for Information Science 44(7), 393–397 (1993)
20. Chiu, L.H.: A cross-cultural comparison of cognitive styles in Chinese and American children. International Journal of Psychology 7, 235–242 (1972)

21. Nunnally, J.C.: Psychometric Theory, 2nd edn. McGraw-Hill, New York (1978)
22. Hofstede, G.: Cultures and Organizations: Software of the Mind: Intercultural Cooperation and its Importance for Survival. McGraw Hill, New York (1991)
23. Chen, S.Y., Magoulas, G.D., Macredie, R.D.: Cognitive styles and users' responses to structured information representation. International Journal on Digital Libraries 4(2), 93–107 (2004)

User Needs of Mobile Phone Wireless Search: Focusing on Search Result Pages

Yeon Ji Kim, Sun Ju Jeon, and Min Jeong Kim

Terminal Application Development Team, Terminal Laboratory, R&D Group, KTF
P.O. Box 138-240, KTF, 7-18 Shincheon-dong, Songpa-gu, Seoul, Korea
{sweet_KIM,sjjeon,kimi}@ktf.com

Abstract. Based on understanding differences between wired and wireless search, we analyzed user needs for mobile phone wireless search. According to this research, heavy wireless search users produce more traffic searching for information than searching for downloadable contents. Through several usability tests, we can get some design guidelines for wireless search result page. Users require different results and presentation for the results of general information keyword searches to media contents keyword searches. Users preferred representative labelling of categories. In addition, it is essential to minimize navigation of the search results.

Keywords: Wireless search, wireless internet, mobile phone, usability, user satisfaction, design guidelines.

1 Introduction

The usage of wireless internet has rapidly spread in Korea where the receptive capacity of new technology is comparatively high. As of September 2007, 47.7% of mobile phone users between the ages 12 to 59 have used at least one or more wireless internet services in Korea. By service type, 46.2% have used the mobile phone wireless internet within the last 1 year while 5.2% and 2.9% have used wireless LAN and broadband respectively. Notably, almost all(92%) mobile users between ages 12 to 19 have used the mobile phone wireless internet.[1] Taking these circumstances into consideration, the potential impact of wireless applications is enormous and is rapidly growing. Just as online search engines have been a gateway to increased consumption of wired data, the wireless search using mobile devices will help meet user needs for data access at anytime and at any place. However, the usability of wireless internet is still not satisfied due to small screens, lack of flexibility and comparatively high data traffic cost, etc.

Our goal in this paper is to present

1. Understanding differences between wired and wireless search
2. Analyzing user needs for mobile phone wireless search
3. Providing design guidelines for search result pages and search routs

---

[1] NIDA(National Internet Development Agency of Korea), 2007 survey on the wireless internet use, pp 3.

M. Kurosu (Ed.): Human Centered Design, HCII 2009, LNCS 5619, pp. 446–451, 2009.

2   Usage Differences between Wireless and Wired Search

A survey was conducted in order to understand the usage differences between wired and wireless search service with 300 mobile device users who used wireless internet at least once a month.

The frequency of wireless and wired search is as follows. 40.3% of respondents accessed wireless search '1~3 times per month', followed by 'once for two days'(22.3%) and '1~3 times per week'(18.4%), while 46% accessed wired search '1~10 times per day'(46%), 'More than 10 times'(43%).

Table 1. Usage Rate of Wired-Wireless Search

Frequency	Wired Search	Wireless Search
More than 10 Times per day	123(41.0%)	12(4.0%)
From 1 to 10 Times per day	138(46.0%)	45(15.0%)
Once for 2 days	16(5.3%)	67(22.3%)
1~3 times per week	20(6.7%)	55(18.4%)
1~3 times per month	3(1.0%)	121(40.3%)

According to the results, the usage rate of wireless search is significantly lower than wired search. This result shows that the mobile phone wireless search does not give users a satisfactory experience due to complex reasons such as relatively expensive data usage charges, the small screen of mobile phones and insufficiency of search result contents, etc.

2.1   Reason for Using the Wireless Search

According to the survey, the main reasons for using mobile search was 'Information Search(21.8%)', followed by searching for 'Ringtones(15.2%)', 'terrific/GPS service(9.5%)' and 'Listening/Downloading music(8.4%)'. On the other hand, users who claimed they are using wireless search for downloading specific contents such as music, game or background image, practically accessed wireless search less than once a month. We can conclude that users produce more traffic searching for information than searching for downloadable contents.

2.2   Comparing Search Keywords between Wireless and Wired Internet

To compare the search keywords of wired search and those of wireless search, we analyzed top 100 search keywords of the KTF website and SHOW internet (KTF brand of wireless internet) from Jan. to Feb. 2008. 65% of wired search keywords were related to additional mobile services and the rest were for contents such as decorating phones. On the other hand, 75% of wireless search keywords were related to downloading contents, especially music, game and adult contents. For in-depth verification, we analyzed the KTF engine Logs for 4 days (mar.28~31.2008). The results show that wireless search is executed for an average 2.5 times per user. 88% of those users were searching for downloadable contents. Meanwhile, heavy search users who access wireless search more than 8 times tend to search for information data such as news, shopping, transaction, etc. The data shows contradicting results from the survey about the reasons for using wireless search. However, heavy user's behavior agreed with the survey results since their search habits were more information-related.

Table 2. The Search Keywords of SHOW internet (KTF brand of wireless internet)

Ranking	Search keywords	Numbers of Queries
1	Game	2273661
2	Ring Tones	860965
3	Coloring[2]	802439
4	Cyworld[3]	654659
5	Girls' Generation[4]	588540
6	Big Bang[5]	522571
7	Mini Game	479385
8	Mat-Go[6]	473945
9	Sex	469137
10	Hero SEOGGI[7]	442171

2.3 Preferred Contents on Search Result Pages

We conducted a focus group interview composing of 4 age categories(Teen Group: 6, Twenties Group: 12, Thirties and Forties group: 6, Fifties Group: 4) to compare preference of wired and wireless search. In the case of wired search results, users wanted a wide range of information from various categories, but in the case of wireless search results, they preferred location/traffic/restaurant information, specific information especially needed outdoors and everyday life. For objective analysis of preferred items in the search results, we conducted a supplementary card sorting. The results are as follows, "News" ranked first, "Regional" second, and "Traffic/Location" third. In addition, users preferred more recognizable and familiar labeling such as 'Music', 'Game', or 'Location' instead of brand names like 'Dosirak', 'G-Pang' or '**114' which are provided by KTF. We can conclude that for mobile search, information about everyday life should be provided more readily than downloadable contents.

3 Usability of Mobile Phone Wireless Search

3.1 Preferred Contents on Search Result Pages

We asked 30 users with experiences of wireless search to conduct KTF wireless search for a specifically chosen ring tone and the name of a specific service menu. While 70 percent of the users succeeded in the ring tone search, 53 percent succeeded in the

---

[2] The service title of Ring back tone.
[3] A Korean social networking website which is similar to 'Facebook'.
[4] A Korean dance group composed of nine girls.
[5] A Korean hip hop & R&B group composed of five members.
[6] The title of a famous mobile game.
[7] The title of a famous mobile game.

service menu search. Among the users who succeeded in their ring tone search, 79 percent achieved it by entering a keyword in the main domain, while only 21 percent of those users first entered search menu category to search the ring tone. In case of the service menu search, 81 percent of the users searched on the integrated search of the main domain, whereas 18 percent of the users opted for the search menu category. One reason why relatively fewer users search on the search menu category is that users are familiar with the search process using search tabs through their wired search experiences. Another reason is that users do not have a clear standard within their mental model in choosing a category containing the specific search results. It is because a user has to go through several stages before obtaining necessary information.

On the other hand, the fact that the success rate for service menu search was lower than that for the ring tone search can be explained in that it was provided as part of the contents of the category along with general information without further clues for the service menu.

3.2 Category Labeling: Individual Category Labeling vs. Representative Category Labeling

In an effort to improve the categories provided by search pages, we asked 50 people about their preference as to categories. Considering the fact that more space would be used up to distinguish each category as the number of the categories grow, we sought to merge categories for the sake of the efficient use of limited space of the search page. When merging related categories, we asked our sample group about their preference between individual labeling of categories (e.g. News/Traffic/Regional,

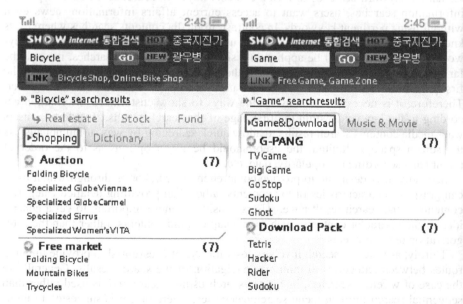

Fig. 1. 2 depth search result page (left), 3 depth result page (right).This shows the tab-style navigation. 2 depth tabs include representative categories with closely related categories 3 depth tabs include more detailed categories.

Shopping/Finance/Education, Ring tones/RingToYou/ Wallpaper) and representative labels of categories (e.g. News, Life, Music). It was found that 64 percent of the sample group preferred representative labeling and they did not depend so much on category labels at the time of wireless search. The general response was that representative labeling showing overall characteristics of the contents provided sufficient information for the purpose and more information as to the more detailed labels of the categories was needed only after 2 depth.

3.3   Navigation Method: Tab Style Navigation vs. Vertical Scrolling

Whereas in the first result page of the search the sample group users preferred the scrolling method to overview the search result as a whole, after selecting a particular category they preferred the tab-style navigation in detailed categories provided in the 2 depth. This is because, when a wrong choice of a particular category is made, this method provides economical ways to move between categories without returning to the upper-level web pages.

4   Design Guidelines for Wireless Search Result Pages

Through our research on users shown in this paper, it is concluded that the characteristics of the wireless search, such as limited screen space, pressure of the service rate and limitation in time, requires optimized and customized and keyword-specific web design so that users can find necessary information within a short period of time.

Firstly, users require different results and presentation for the results of general information keyword searches to media contents keyword searches. In case of general information searches, users want to access current affairs information/ news even without using pertinent keywords. In contrast, in media contents searches where users know the clear objective of their search, users want matching results for their keywords. Although it would be applied in a similar way in wired searches, it would be far more critical in wireless searches. If the required information is not found in the first page of the search result, most of the users tend to give up the search altogether. Therefore, it is necessary to consider the ways to show customized information according to the keyword within the first page of the search results. In addition, users want media contents search results to allow quick searching by way of thumbnails. As the screen space is limited, the users should be given options as to the order of thumbnails according to popularity, date, etc.

Secondly, it is desirable to provide comprehensive categories through which users can grasp the characteristics of the contents rather than providing detailed categories contents in the search result page. Users consider it more important to get a general idea of the particular contents, such as sitemap or paid contents, rather than the categorization of the contents.

Thirdly, as wireless search involves costs for use, it is essential to provide jumping routes between categories to minimize navigation of the search results. Generally in the case of wireless searches, integrated search using a search tab is used rather than sequential search through menu search categories. Therefore, it is suggested to minimize the categorization of contents and to provide navigation through tabs to enable movements between categories. In addition, more relative tabs should be arranged beside the chosen tab.

References

1. Buchanan, G., Farrant, S., Jones, M., Thimbleby, H., Marsden, G., Pazzani, M.: Improving Mobile Internet Usability. In: Proc of WWW, pp. 673–680 (2001)
2. Jones, M., Buchanan, G., Thimbleby, H.W.: Sorting out searching. on small screen devices. In: Proceedings of the 4th International Symposium on Mobile Human-Computer Interaction, pp. 81–94. Springer, Heidelberg (2002)
3. Buyukkokten, O., Garcia-Molina, H., Paepcke, A.: Focused web searching with PDAs. In: Proceedings of Web 9 Conference, Amsterdam (2000)
4. NIDA(National Internet Development Agency of Korea). Survey on the wireless internet use, pp. 1–18 (2007)

Why Taking Medicine Is a Chore – An Analysis of Routine and Contextual Factors in the Home

Wei Kiat Koh[1], Jamie Ng[1], Odelia Tan[1], Zelia Tay[2], Alvin Wong[2], and Martin G. Helander[3]

[1] Institute for Infocomm Research,
A*STAR (Agency for Science, Technology and Research)
1 Fusionopolis Way, #21-01 Connexis, Singapore 138632
{wkkoh,jamie,ytan}@i2r.a-star.edu.sg
[2] Singapore Institute of Manufacturing Technology
A*STAR (Agency for Science, Technology and Research)
71 Nanyang Drive, Singapore 638075
{yctay,hywong}@simtech.a-star.edu.sg
[3] School of Mechanical and Aerospace Engineering
Nanyang Technological University
50 Nanyang Avenue, North Spine, Block N3.2, Singapore 639798
martin@ntu.edu.sg

Abstract. Medication adherence is an important concern for many people. This is especially so in the older adults population where non-adherence can have serious consequences and may lead to higher healthcare cost. Non-adherence is a problem that afflicts the younger and older adults and there are many factors affecting one's medication adherence (Social/economic factors, provider-patient/health care system factors, condition-related factors, therapy-related factors, patient-related factors [1]). In this paper, we focus on patient-related factors and investigate how these factors (mainly routines in patients' daily life, their surrounding environment and their self-made systems) affect their medicine taking behaviour and their abilities to adhere to their treatment regimens. Results presented in this paper are gathered from in-depth interviews with patients during house visits and from observing how they go about handling their medication in their living space. This knowledge of how patients are currently coping with their medication will be useful for the design of an effective medication support system.

Keywords: Medication, medical adherence, reminders, older adults.

1 Introduction

Medical adherence is the degree to which patients conform to a given medication regimen. This is difficult for many people and is a problem that afflict people of all age, particularly so for elderly with ailing health. The reasons for poor medical adherence are multiple and it is implicated by declining memory and complex medication regimens among elderly.

M. Kurosu (Ed.): Human Centered Design, HCII 2009, LNCS 5619, pp. 452–461, 2009.
© Springer-Verlag Berlin Heidelberg 2009

With an ageing population, there is an increasing pressure to shift institutional healthcare to home-based healthcare. This will help reduce the public healthcare cost and cope with future increased demand. Innovation in home-based healthcare and medical compliance support has the potential to allow independent living at home while being able to manage personal medication.

Managing medication is an important activity in the lives of the elderly, besides the medical advantage will also give a sense of self reliance and independence. In our research, contextual studies were conducted to find out how the elderly manage their medication in the social context of their homes. Taking into consideration the various stakeholders in the "supply-to-intake" chain of medication, interviews with healthcare professionals were conducted to document their concerns with the use of aids for medication compliance aid.

Below we breakdown the "medication process" in several steps and present an analysis of how users can make their own personal system and routine to provide cue and medicine reminders for themselves. Various factors that contribute to poor medical adherence will also be discussed and early assessment of the effectiveness of the users "self-made system and routine" reveal possible deficiencies in their solutions. With the breakdown of the "medication process" it is hoped that designers can see opportunities to improve the "medication process" and provide a holistic solution for better medical adherence among the elderly.

2 Related Work

Many researchers have addressed the various factors promoting successful medication adherence. Klein et al. [2] proposed a generalized Adherence Loop model which explored three different nodes of adherence (Belief, Knowledge and Action). The way people loop through the model may identify areas of "Adherence Vulnerability". A number of researches have examined how technology can be developed to address the issue of medical adherence. Jay et al. [3] described a context aware system that applies patterns of past behaviour to the current context of user activity by issuing prompts at the appropriate time and place. Sachpazidis et al. [4] introduced a GSM based system to improve the medication adherence of the patients.

3 Method

We conducted an in-depth interview with elderly in their homes. The elderly were randomly selected from a pool of participants who had visited an exhibition (Silver Industry Conference and Exhibition 2008) related to ageing. During the 1.5-2hr interviews, one researcher would facilitate the interview while the other take notes. A digital voice recorder was used throughout the session. The interview recordings were transcribed and each interview was summarized in a report.

During the interview, questions related to daily activities and medications were asked. The aim was to have a better understanding of how elderly went about their daily tasks, and medication handling. Permission was obtained to take photographs and videos so that we could have visual representations of the medicine and where they were kept. Through observation of their living space, we recorded where and how they stored their medication. We uncovered self-made systems, which the elderly, their family members and caretakers employed to improve medical adherence. While the interviews and post-interview analyses are time consuming, they provide insightful details of how daily activities intertwine with medication adherence.

In addition we also interviewed healthcare professional in a private nursing home for the elderly to investigate how medication was handled differently in various settings. This allowed a better understanding of the needs and concerns of the various stakeholders in the chain of medication.

4 Medication Management

Managing medication is an important activity, which the elderly engaged in every day. To help medication schedules, they created routines and personalized systems at home to maximize their chance of taking the medication. We investigated how the elderly supported themselves by these routines, and how their social network supports them in this task.

4.1 Stakeholders

In the study at the participants' homes, we identified the various stakeholders in the "medication process" as the patient himself, the family members/caregivers, the healthcare professionals (doctors, pharmacists etc) and the pharmaceutical companies. These parties have a direct/indirect role in influencing the patient's medication adherence and they have occasionally been mentioned by the participants during our interviews. Very often, the family members appeared to play a critical role in helping the elderly gain a better understanding of the medicine they are taking. In other case, a caregiver was engaged to look after the elderly parents, specifically to look after his medication need. Given that medical adherence tends to improve as one's knowledge of his medication increases and since all the stakeholders listed above directly/indirectly influence the patients' knowledge of his medication they are therefore included in our list of stakeholders. In addition, research have also shown that medical instruction [5] given out by the doctors and pharmacists can be better remembered when they are organized in terms of the elderly's pre-existing schemes for taking medication, and pictorial aids [6] in the medical instructions can also enhance patients' understanding of how they should take their medications. For the pharmaceutical companies, packaging design improvements [7] are being examined as way to improve adherence to medication instructions.

4.2 Taking Medicine Daily

Taking medication is a routine activity that is central to many of the participants in our studies. Despite engaging this task on a daily basis and becoming habitual in the process, some of the elderly we interviewed find the process bothersome. Some of the difficulties that the participants have experienced include forgetting to take their medicine, uncertain if they have taken their medicine before, taking extra dose of medicine when they have already taken, taking the wrong types or amount of medicine, and running short of medication supply unexpectedly etc. Given that remembering to take medicine is a prospective memory task, these difficulties that they experienced are in line with past research that has indicated the elderly has a lower score on prospective memory task and this difficulty with prospective memory task increases when there are additional tasks to be performed [8]. The task of taking medicine is made more difficult by the various types of medication and the complex medication regimens they have to adhere to. In fact, in our studies, we had come across instances whereby the participants engaged caregivers to help them manage their medication as they were unable to manage it themselves. As cited by a healthcare profession in our interview at a private nursing home, one of the main reasons for the elderly to stay there is because they are unable to handle their medication independently.

Indeed, the task of taking medication is one that is tedious and fully requires one to make conscious effort to remember to take their medication. In our studies, we came across a wide range of methods which the elderly adopted to support this task. Most elderly who take medicine on a regular basis tend to associate their medication with their meals. As such it is also quite common for them to put their medication on the dinning table in full-view so that they can easily remember to take their medication on time. In addition, to solve some of the difficulties that they experienced, some of their solutions include setting up of their own systems (pillboxes with handwritten labels, transparent medicine containers etc), placing them in a specific location in the house in line with their daily activities, whereby the open display of the medicine promotes awareness for themselves and for other members in the household (e.g. children or caregivers) to help them in their medication etc.

Despite having partial knowledge of their medicine, some participants are still able to manage their medication i.e. they were able to remember when and how much medicine to take by leveraging on daily routines and the physical features of their homes. Together with their schedule at home, they have strategically placed their medicine in specific location to triggered their memory e.g. placing the medicine box near to a favourite seat in the living room for easy access. Besides writing personalized note and drawing on the pill packs, some participants also put up medical information around the house e.g. medical appointments with the doctors and when to replenish their supply of medicine on wall calendar (Fig 1.). These notes and drawing on the pill packs quickly refresh their knowledge of the medicine in their own language as all the medicine instructions on the pill packs issued by the pharmacists are in English and the small front size mean that elderly with deteriorating eyesight may have problem reading them.

Fig. 1. Medical appointments are written on wall calendar

4.3 "Self-made" Systems

While some participants left the medicine in their original packaging (with handwritten notes to reinforce their understanding of the medical instruction), others use pillboxes to manage their medication. One participant uses 3 pillboxes (for morning, afternoon and evening dose) with each pillbox compartment representing each day of the week. Every Sunday evening, he would systematically sort out the pills into the pillboxes for himself and his wife. These pillboxes are then placed in visible areas around the house and he would occasionally check on the pill compartment to see if he had taken his medicine. The pillboxes provide traceability of past actions in which the absence or presence of pills indicate past and future actions. Though he acknowledged that pill sorting is tedious, he has grown accustomed to this routine and felt that the pillbox is useful for his overall management of his medication. In our view, while the pillboxes, the handwritten label and pictorial aids on the pillboxes, their strategic placement around the house, and the medication information that he has written on papers are useful for this particular participant, the sorting of pills into the various compartments may not be error free. Given the manual nature of the sorting process, one may accidentally put in the wrong amount of pills in a pill compartment. In fact, during our interview, we observed that the participant had accidentally left an aluminum foil (Fig.2) from the blister-pack in the pill compartment and the elderly could have unknowingly taken the aluminum foil together with the pills. In addition, the same participant was able to manage his medication quite well despite not knowing what some of the medicines is treating. This reinforces the earlier point that one is able to manage his medication despite not having full knowledge of his medicine.

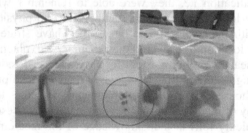

Fig. 2. Aluminum foil in one of the pill compartment

4.4 Non-routine Event

There are several occasions whereby the elderly will forget to take their medicine and anecdotal evidence indicates that on certain occasion, specifically the chances of missing a dose may be higher when unusual events happen. These occasions include time whereby they have people visiting them and they will miss their dose when the specific event falls out of sync with their usual daily routine. In one of our interview, this is exactly what happened when the participant only realized that he had forgotten to take this medicine when we touched on the topic of medication (our visit to his house was not part of his daily routine). On other occasions, they may miss a dose when they are in a hurry for an appointment, rushing to pick up an unexpected call, rushing to switch off the stove, when they are sleeping or not feeling well, or when they are engaged in other activities like watching TV. All these resonate well with research that has shown that older adults are more likely to forget to perform their originally intended action when they had to delay an intended action due to an interfering task even if the adults have rehearsed the intended action [9].

Reasons for missing their medicine when they are at home or away from home also vary. Some participants also expressed that one might tend to forget their dose more frequently when they are away from their home e.g. when they are travelling oversea. Other reasons include not having their medicine with them when they remember to take their medicine or repeatedly forgetting to take their medicine while busy at work. As such, we have come across participant who will put their medicine in pouch (Fig 3.) and bring it with her (e.g. inside her handbag) wherever she goes so that the medicine are easily accessible when she recalls to take them when she is out.

Fig. 3. User put her medicine in a pouch for use outside the house

4.5 How and Where Participants Keep Their Medicine

The exact location for the storing of medicine varies across households as it is dependent upon the particular arrangements of each domestic space, individual preferences and lifestyles etc. Common sites include the person's place at the dinning table (Fig 5.), kitchen cabinet (Fig 7.), the area he or she usually sits in the living room, outside the bedroom door etc. All these medicine storage arrangements at strategic locations scattered around the house enabled them to notice at a glance the medicine they were supposed to take and trigger off a memory recall. They were also situated at areas whereby they could be easily retrieved or accessed when needed. This ease of accessibility and high visibility of the medicine in common areas also enabled others to help remind the elderly of their medicine. Several of the participants used transparent boxes (Fig 4.) to store their medicine. This allowed them to see clearly what is inside the containers and quickly trigger their memory.

On the other hand, due to privacy concern and fear of being stereotyped, one participant hid her medicine container away from us during our visit. She would otherwise have normally left her medicine container in an open corner. There is also another participant who would place his medication on his "usual" seat at the dinning table outside of his bedroom (Fig 5.). Placing their medications in these consistent locations help them to remember to take their medication. This participant's daughter had earlier bought him a pillbox to help him manage his medication but he refused to use them and insisted on his usual way of handling his medication. This illustrates that one method that is useful for one may not be equivalently useful for another. Despite his old age and having to take a number of pills everyday, he is still able to adhere to his medication regimen. This could be due to his work nature as a tailor, before his retirement, whereby he has to remember off-hand clothes dimensions. While keeping medication in visible, consistent location helps users to remember to take their medication, the question lies in what happen if someone or the user himself unconsciously changes this physical arrangement of the medication that support the user in taking his medicine. By moving her medication out of our sight during the interview for various reasons, the participant could have altered the spatial arrangement to be out of sync with her daily activities.

Fig. 4. Transparent containers used to store medicine

Fig. 5. Medication is kept on a dinning table, outside bedroom and on the "usual" seat

Fig. 6. Medications are distributed around the house depending on the areas where the elderly is expected to be at certain time

4.6 Prescription Medicine and Over-the-Counter Medicine

In our studies, we also discovered the difference in the ways respondents handle prescribed versus over-the-counter medicines (vitamins, supplement and drugs for sudden ailment like fever, headache etc). For the over-the-counter medicine, most respondents would keep them in less visible area such as in the kitchen cabinet, away from common sight, since these medicines tend to be accessed less frequently and the timeliness of the intake may not be as critical as the prescribed medicine. In addition, these medicines are "less personalized" as they are shared among the family members and the family members do have common knowledge of where these medicine are kept. While sharing medicine in a common space seems to be a good idea, there are concerns over who is ultimately responsible to upkeep the common medicine cabinet to ensure that the medicines are not expired and its availability in time of emergency.

Though these over-the-counter medicines are less critically needed, there is a time whereby accessibility is important to prevent further deterioration of condition and the unnecessary pain that the respondents have to bear.

We found that most of the participants do not check the expiry date of their medicine. One participant had expressed that she did not check the expiry date of the prescribed medicine because she had assumed that the doctor would not issue medicine that is nearing expiry. Through one of the participant's experience, there was a time whereby he had assumed that there was enough pain-killer at home when there was none available. An IT support system to tracks the inventory and expiry date of the medicines at home would be useful in this case.

Fig. 7. OTC medication are stored separately from medication that are taken on a daily basis

4.7 Taking Medication in the Nursing Home

In our field study at a private nursing home to look into how the healthcare professionals handle medication, we found that nursing homes have standard in-house procedures which the staff strictly adhered to. As cited by one healthcare professional, one of the reasons why an elderly is staying in the nursing home is because the elderly is unable to reliably manage their medication themselves and therefore require additional assistance. The healthcare professionals at the nursing home followed a workflow process in dealing with the medication, from the doctor prescriptions, preparation of the dosages, dispensing of the medicine to the elderly, to ensuring that the elderly has taken the medication. At each stage of the process, everything ranging from timing of medication intake, person-in-charge of preparing the medicine etc will be documented.

Dealing with the medication in the nursing home is a manual and tedious process. Most of the work conducted was done mainly by human and virtually no task was automated despite some being repetitive in nature. Dosage needed by the patients was prepared at the same time everyday one day in advance. Patients at different levels of the nursing home are given different colour pillboxes. Given the manual nature of this task, it was prone to human error e.g. dispensing the wrong dosage or type of medicine. The problem is further complicated by nurses changing shift resulting in patients being attended to by different nurses who may be unfamiliar with his/her conditions and dispense the wrong type/amount of medicine as a result. It will be ideal to have the same nurse to attend to the same patient everyday though this may not be feasible due to the shifts work nature and re-allocation of task when the circumstances require. As such, automation for the dispensing of drug and the use of RFID to verify the identity of the patient and to ensure that the patient's identify and his dosage requirement tally can help to streamline this process and reduce human error.

Fig. 8. Equipment used to manage patient medication at a private nursing home

5 Conclusions

Our initial study at the elderly homes showed that people arrange and improvise things in their living environment to work around problems they faced in their daily life. People rely on the physical features of their living environments and their daily routine to help them organize and provide cues to remember to take their medication. We have shown how factors of non-compliance of medication and solutions vary across individuals and circumstances whereby pillbox may be useful for some and not others. The problem is further exacerbated as one age. The difference between individuals will widen and the needs and concerns will also differ across individual. For example, an elderly who has manual dexterity problems may require some form of automatic dispensing of the medication if he has problems sorting out his pills. While patients who have dementia may require more support with reminders. This suggests that there may not be an ideal one-size fit all solution for the problem of low medication adherence among elderly, and a system with a variety of customizable features may be useful. Knowing the interventions that the elderly adopt to improve their medication adherence, it will be useful to leverage on these existing methods to support their management of medication. An integrated IT support that provides prompts to user to take his medicine when he falls out of his temporal rhythm for his daily routine and also allow users to personalize the system features according to their preferences and lifestyles would be useful.

References

1. World Health Organization (2003)
2. Klein, D.E., Wustrack, G., Schwartz, A.: Medication Adherence: Many Conditions, A Common Problem. In: Proceedings of the Human Factors and Ergonomics Society 50th Annual Meeting 2006, Health Care, pp. 1088–1092 (2006)
3. Lundell, J., Hayes, T.L., Vurgun, S., Ozertem, U., Kimel, J., Kaye, J., Guilak, F., Pavel, M.: Continuous Activity Monitoring and Intelligent Contextual Prompting to Improve Medication Adherence. In: Proceedings of the 29th Annual International Conference of the IEEEE EMBS Cite Internationale, Lyon, France (August 2007)
4. Sachpazidis, I., Fragou, S., Sakas, G.: Medication Adherence System Using SMS Technology. In: Proceedings of the 2004 Intelligent Sensors, Sensor Networks & Information Processing Conference, pp. 571–575 (2004)
5. Morrow, D., Von, L., Andrassy, J.: Designing Medication Instructions for Older Adults. In: Human Factors and Ergonomics Society Annual Meeting Proceedings, Aging, pp. 197–201

6. Katz, M.G., Kripalani, S., Weiss, B.D.: Use of pictorial aids in medication instructions: A review of the literature. American Journal of Health-System Pharmacy 63(23), 2391–2397
7. Spillinger, A., Auerbach-Shpak, Y., Bitterman, N.: The Pill: Time for a New Look. Ergonomics in Design 16(2), 24–28 (2008)
8. Martin, M., Schumann-Hengstelar, R.: How task demands influence time-based prospective memory performance in young and older adults. International Journal of Behavioral Development 25(4), 286-391
9. McDaniel, M.A., Einsten, G.O., Stout, A.C., Morgan, Z.: Aging and maintaining intentions over delays: Do it or lose it. Psychology and Aging 18(4), 823–835

Social Robot Design

Seita Koike[1,*], Masayuki Sugawara[1], Yuki Kutsukake[1], Sayaka Yamanouchi[1], Kie Sato[1], Yoshihiro Fujita[2], and Junichi Osada[3]

[1] Tokyo City University
[2] NEC Corporation
[3] NEC Design & Promotion, Ltd.
Faculty of Environmental and Information Studies, Yokohama Campus
3-3-1 Ushikubo Nishi, Tsuzuki ward, Yokohama, Japan, 224-8551
koike@tcu.ac.jp

Abstract. The purpose of this study is to describe the network of human and non human as the personal robot in the community. We designed an original program for the personal robot and took it to the nursery school. We participated in the community of the nursery school and supported the new play group that the children's mothers set up. They used the robot for their play actively. We made new users of the robot and changed the activity of the nursery school. As a result, the design of the interface for controlling robot changed.

Keywords: Robot, society, interface, design.

1 Introduction

Bruno Latour, French sociologist of science advocated Actor Network Theory(ANT). He define the society as hybrid network of human and non human. According to ANT, artifacts as non human are not present independently. Designers or researchers usually design robots independently and improve the internal function of robots. But it is important for us to design robots as the actors of hybrid network.

1.1 Personal Robot PaPeRo

We design the performance of the robot with Partner-type Personal Robot(PaPeRo) developed by NEC Corporation. We use PaPeRo as a tool of our research. The developmental environment of the robot makes it easy to develop the robot 's action. We use the developmental environment to develop original software for the robot.

1.2 Participation in Community of the Nursery School

We designed the original contents for the robot and took it to a nursery school in Tokyo. We participate in the community of the nursery school and administered fieldwork to the community.

---

* We rent the robot from NEC Corporation and research.

M. Kurosu (Ed.): Human Centered Design, HCII 2009, LNCS 5619, pp. 462–467, 2009.
© Springer-Verlag Berlin Heidelberg 2009

Fig. 1. PaPeRo

1.3 The Specification of PaPeRo

The size of the robot is 385mm height, 262mm wide, 250mm depth and 6.0kg weight. It has microphones and speakers for speech–recognition system and speech-synthesis system and has touch sensor covered the surface of the body for interface and bumper switches of its feet for security. It can move tires at the bottom of foot.

2 Activity of the Nursery School

We participated in the community of the nursery school with the robot since 2003. At first we took the robot in classroom as a special event. We controlled the robot and the robot talked, singed, danced as a "star" in front of the children.

Fig. 2. The robot said "good morning" in the entrance

It seemed to us the robot event was successful, but at the same time we disrupted the classroom activity, because the teachers had to stop their activities because of the robot event. We changed our strategies of using the robot. We decided to make the robot act not as a star but an assistant in the classroom. The robot said "Good morning" at the entrance of the nursery school when the children came to the school.

2.1 Mother's Robot Club (Mama Robo Club)

In 2008, we understood that the teachers could not control the robot because they were busy to take care of the children. At that time, some of the mothers of the children were interested in the robot. They offered us to control the robot by themselves

Fig. 3. The member of "Mama Robo Club"

Fig. 4. The robot celebrating the children

instead of the teachers. We supported to make the play group that they control the robot for they children. We named the play group as "Mama Robo Club". six or seven mothers controlled the robot at the entrance or playroom once a week.

We supported the mothers technically to operate the robot. In nursery school there had birthday party once a month. They made the content of the robot for the birthday party. The content made the robot say congratulations to the children who reached their birthday. In the birthday party, they demonstrated the contents in front of the children.

3 Change of Network

The nursery school usually did not permit the mothers to go inside of the classroom. But the member of Mama Robo Club were permitted because they operated the robot. The robot changed the activity of mothers. We operated the robot voluntarily from 2003 to 2007. In 2008, our roll changed from the operator to supporter to the mothers. The robot had changed the network of human surrounding the robot.

4 Middle Interface for Middle Users

The mothers started to make contents of the robot by themselves. They were not either professionals like engineers nor beginners in for controlling the robot. As we made the Mama Robo Club, We made new users. We call the users as "middle users". The middle users can make the contents of the robot and play the contents by themselves. Now they are using the Software Development Kit by NEC. But it is difficult for them to use it. New users need new SDK. We designed the interface of new SDK for the mothers that we call "Middle Interface".

Fig. 5. The mothers made the contents of the robot by themselves

4.1 Required Specifications of the Middle Interface

A. The mothers made the contents of the robot in a group. When all members did not attend, they took over the task from other members step by step every week. We designed the interface that they can provide comments by using stickies.

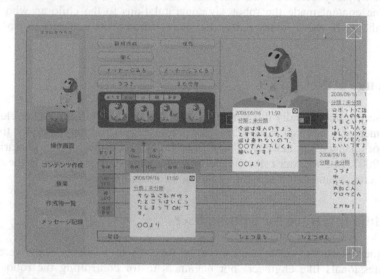

Fig. 6. The interface of taking over the task to other members

B. The mothers became anxious when the robot stopped moving. We designed the interface that they could relax by displaying of the troubleshooting and the advice for operation of the interface.

Fig. 7. The interface for resting easy

Acknowledgments. We appreciate Futako Nursery School, the mothers of Mama Robo Club and the children.

References

1. Latour, B.: Science in Action. Harvard University Press (1987)
2. Osawa, K., Kon, S., Takahashi, M.: The Robot Design as Network, gradation thesis, Musashi Institute of Technology (2006)
3. Ono, H., Sugawara, M., Yasuda, Y.: The Robot Design as Network, gradation thesis, Musashi Institute of Technology (2007)
4. Suzuki, Y., Nishikawa, T.: The Robot Design as Network gradation thesis, Musashi Institute of Technology (2008)
5. Osawa, K.: The Robot Design as Sosio-techno Network, master thesis, Musashi Institute of Technology (2008)
6. PaPeRo Information, http://www.nec.co.jp/robot/robotcenter.html

Culture and Communication Behavior:
A Research Based on the Artifact Development Analysis

Masaaki Kurosu[1] and Ayako Hashizume[2]

[1] The Open University of Japan
[2] University of Tsukuba
masaakikurosu@spa.nifty.com

Abstract. Authors focused on the use of the cell phone by senior people and young people living in the urban area and the rural area in Japan and in the US. The result of the questionnaire research showed that there are differences in the use of the cell phone and other communication media depending on the situation. These differences are related to the difference of the culture: nation culture, region culture, and generation culture.

Keywords: Communication, media, usability, cell phone, senior user, culture.

1 Introduction

Human behavior is influenced by the culture. In other words, people create the culture but at the same time the culture put constraints on various aspects of the life people. In this paper, authors focused on the relationship between the culture and the communication behavior especially on the use of the new emerging, i.e. the cell phone.

The research was conducted from the standpoint of the Artifact Development Analysis (ADA) that was proposed by one of the authors (Kurosu) in 2007. One of the characteristics of the ADA is to specify the use of specific artifact in some specific situation. There are a variety of artifacts to achieve some specific goal. The user selects one of them based on the consideration that it is the most adequate for that situation. The ADA quests the reason why the user selected that specific artifact. The culture, as the ADA presumes, is one of the strong and influential factors for that selection.

2 Culture

Usually, the term "culture" is used to mean the nation culture such as the Korean culture, Japanese Culture, French Culture, etc. Most of the researches on culture use this term as such. But if we define the term as "*the way people think, believe, behave and make and use the artifact that is shared and inherited among a specific group of people*", culture exists wherever there is a group of people whether it is geographical or virtual. Fig. 1 that is sometimes called the "onion model of culture" shows the diversity of culture. As can be seen in the figure, there are the global culture as such

M. Kurosu (Ed.): Human Centered Design, HCII 2009, LNCS 5619, pp. 468–475, 2009.

common aspects of human behavior as to celebrate the marriage, the nation culture that is specific to the nation or the country, the region culture that sometimes becomes the nation culture depending on the political situation, the family culture that is common to all of the family members, and the individual culture that is the specific behavioral pattern of the individual.

In addition to those cultures, there are the organization culture that is common to the organization such as the company or the school, the religion culture that is common to the adherent of that specific religion, the gender culture that is common to the male and the female, and the generation culture that is specific to the generation such as youth or seniors. Another important culture is the ethnic culture that is common to the specific ethnic group. This is sometimes confused with the nation culture because the nation is frequently established based on the specific ethnic group. But there are such cases as the Chinese people living in different nations.

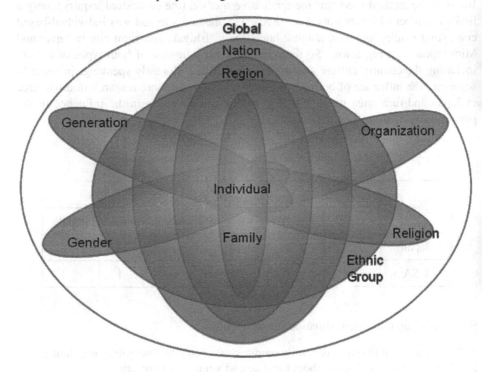

Fig. 1. Diversity of Culture

3   Research

3.1   The Method

An example of ADA approach will be shown in this paper with regard to the ICT-related behavior. It is concerned with the relationship between the goal achievement and the diversity of artifacts.

The focus of the research was the communication behavior. Historically, for the purpose of communication, various media have been developed and now we have many alternatives for achieving the goal of communicating to somebody else. The artifacts for communication that we have today include (1) letter, (2) postcard, (3) telegram, (4) PC mail (internet mail), (5) PC chat, (6) cell phone mail, (7) cell phone texting (SMS), (8) cell phone call, (9) landline, (10) fax, (11) ask somebody to convey the message, (12) leave a memo, and (13) directly speak to others. We (the author and Ayako Hashizume) conducted a research by adopting the contextual inquiry (Beyer and Holtzblatt 1998) and analyzed the data by applying the GTA (Grounded Theory Approach) (Strauss and Corbin 1998).

The research was conducted in Japan and in the US for senior people over 60 yrs old and for those who are of their 20's. Number of informants is listed in Table 1. Because the method used was the qualitative method (the contextual inquiry), only a limited number of informants were used. The method adopted was individual-based contextual inquiry and took about 2 hours long. Ishigaki is a rural city in Japan and Minneapolis is a big town. So there might be an influence of both types of culture including the country culture and the region culture. Honestly speaking, in order to segregate the influence of both types of culture, we need more research in urban area in Japan and rural area in the US. The results can also contain the influence of the gender culture and the generation culture.

Table 1. Informants (Japan: Ishigaki, US: Minneapolis)

	20's		60's		70's	
	male	female	male	female	male	female
Japan	4	4	2	2	2	2
USA	1	1	1	1	1	1

First, we set up 6 different situation setups as follows.

1. When you want to tell your family member that you'll be late going back home.
2. When you want to discuss about the date and venue of the meeting.
3. When you want to talk to someone or hear his/her voice.
4. When you want to talk to someone about the change of meeting schedule.
5. When it seems that you'll be late for the meeting.
6. When you want to express your gratitude for a gift.

During the session, we showed an informant a table of communication media (Table 2) and asked which media s/he will use in each situation and why.

And also asked the reason why the informant selected some specific artifact in any one of the situation. Alternatives for the reason are listed in Table 3.

Table 2. Means that can be used

Use the Paper-based Media	Refer to Books	Dictionary, Encyclopedia
		Technical Book (e.g. Law, Cooking, etc)
		Map, Time-table
		Product Manual, Product Guide
	Refer to the Periodical	Newspaper
		Journal
		Free Paper
	Use the Mail	Letter, Postcard
		Telegram
Use the Electronics Media	Use the Broadcasti	TV
		Radio
	Use the PC	Website Access
		e-Mail
		Chat
		Application Software
		Other Function
	Use the PDA	Website Access
		Other Function
	Use the Cellphone	Website Access
		e-Mail
		SMS
		Call
		Other Function
	Use the Landline	
	Use the Facsimile	
	Use Other means	Electronic Dictionary
		Other Device
Ask to Somebody	Ask to Your Family Member	
	Ask to Your Friend	
	Ask to Your Colleague	
	Ask to Your Neighbor	
	Ask to a Professional (e.g. Teacher, Clerk)	
	Ask to the User Support Personnel	
	Ask to Other People	
Ask to somebody else		
Use Other means		
Do nothing special		

3.2 The Result

For question 1

- In Japan, 7 out of 8 senior people (of their 60's and 70's) selected the cell phone call because of its speed, ease of operation and politeness. 2 out of 8 seniors selected the land line call (because the duplicate answer was allowed). Young people (of their 20's) are half and half for selecting the cell phone call (5 out of 8) and the cell phone mail (4 out of 8). The reason for selecting the cell phone call was its ease of operation and the reason for selecting the cell phone mail was its speed.

Table 3. Reasons why the informant selected some specific artifact

a	Ease of Use	Less workload to convey information
b	Efficiency	Less time to convey information
c	Cost	Less cost to convey information
d	Ease of Operation	Easy operation to convey information
e	Correctness	Conveyed information to be correct
f	Sureness	Conveyed information surely reach the ~~person~~
g	Politeness	No impoliteness when conveying ~~information~~
h	Other	(Please Specify)

- In the US, both generations selected the cell phone call (6 out of 6) because of its speed, ease of operation and certainty.
- Considering the uncertainty that will occur when the receiver may not notice the arrival of the mail, this answer of American informants is quire reasonable. But considering the situation where the receiver was far away from the telephone, using the cell phone mail will be more sure because the message was already sent to the receiver regardless of the situation of the receiver.

Table 4. Results for question 1

		Ishigaki Japan																Minneapolis USA					
		60's-70's								20's								60's-70's				20's	
		IS1	IS2	IS3	IS4	IS5	IS6	IS7	IS8	IT1	IT2	IT3	IT4	IT5	IT6	IT7	IT8	MS1	MS3	MS2	MS4	MT1	MT2
		64	67	75	72	60	68	79	71	28	26	21	26	27	26	26	26	63	72	69	71	27	26
		M	M	M	M	F	F	F	F	M	M	M	F	F	F	F	F	M	M	F	F	M	F
Cellphone	e-mail		f			f								abc	he	b	f						
	Call	be	f		abe	f	a	b	e	e	ae	b		be			a	bd	ahdf	ahd	ab	b	g

For question 2

Table 5. Results for question 2

		Ishigaki Japan																Minneapolis USA					
		60's-70's								20's								60's-70's				20's	
		IS1	IS2	IS3	IS4	IS5	IS6	IS7	IS8	IT1	IT2	IT3	IT4	IT5	IT6	IT7	IT8	MS1	MS3	MS2	MS4	MT1	MT2
		64	67	75	72	60	68	79	71	28	26	21	26	27	26	26	26	63	72	69	71	27	26
		M	M	M	M	F	F	F	F	M	M	M	F	F	F	F	F	M	M	F	F	M	F
PC	e-mail																	abd	af		abd	ahfb	daf
Cellphone	e-mail		a							e		b	abce		e	a	a						
	Call	a			abe	o	ao		ahe		h	b		af			a		abd		bde	
Landline		e			abe	o	ao						e,f				e	ahd				

- The result was quite contrasting. In Japan, senior people selected the cell phone call (6 out of 8) and the landline (5 out of 8) because of its certainty where young people selected the cell phone mail (6 out of 8) and the cell phone call (4 out of 8) mainly because of ease of operation.

- In the US, both informants selected the PC mail (5 out of 6) because of its ease of operation, speed and certainty. There were just a few people who selected the cell phone call and the landline but the number is 1-2 out of 6.
- Considering the possession of PC, it might be easier for American informants for using the PC than to use the cell phone. It could be attributed to the difference in the nation culture, but there is more possibility that it is related to the region culture, because Ishigaki is a rural area in Japan and more ICT devices can be possessed even by senior people in urban area such as Tokyo, Yokohama, etc.

For question 3

Table 6. Results for question 3

		Ishigaki, Japan																Minneapolis, USA					
		60's-70's								20's								60's-70's				20's	
		IS1	IS2	IS3	IS4	IS5	IS6	IS7	IS8	IT1	IT2	IT3	IT4	IT5	IT6	IT7	IT8	MS1	MS3	MS2	MS4	MT1	MT2
		64	67	75	72	60	68	79	71	28	26	21	26	27	26	26	26	63	72	69	71	27	26
		M	M	M	M	F	F	F	F	M	M	M	M	F	F	F	F	M	M	F	F	M	F
Cellphone	Call			●	●	●	●	●	●	●	●	●	●	●	●	●	●	●	●			●	●
				a,b	a	a		e,f,g		b	b			b	a	a	b	a,b,c,g			b	b	
Landline			●	●	●	●											●	●	●	●			
			a,b,e	a,b	a	a												a,b,c,d,e,f,g			b		

- The pattern of answer is quite clear that the senior people both in Japan and the US selected either of the cell phone call and the landline whereas the young people in both countries selected only the cell phone call.
- The result is reflecting the generation culture where young people are more familiar to the use of the cell phone.

For question 4

Table 7. Results for question 4

		Ishigaki, Japan																Minneapolis, USA					
		60's-70's								20's								60's-70's				20's	
		IS1	IS2	IS3	IS4	IS5	IS6	IS7	IS8	IT1	IT2	IT3	IT4	IT5	IT6	IT7	IT8	MS1	MS3	MS2	MS4	MT1	MT2
		64	67	75	72	60	68	79	71	28	26	21	26	27	26	26	26	63	72	69	71	27	26
		M	M	M	M	F	F	F	F	M	M	M	M	F	F	F	F	M	M	F	F	M	F
PC	e-mail										●							●	●			●	●
											d							f	a,b,d,e			b,d,e,f,g	
Cellphone	e-mail		●							●		●											
										e	d	a,b											
	Call	●		●	●	●	●	●	●	簡単なとき			●	●	●	●	●	●		●			●
		a		a,b,e	b	h,f	a,b,e		b,f	a		b	a,b	h,e	a	h	a						b
Landline		●	●		●	●	●							●	●			●	●			●	
		e		a,b,e	b	h,f	a,b,e							b,e	e			a,b,d,e			b,e		

- The result reflected the nation culture or the region culture in that American informants both senior and young selected the PC mail (4 out of 6) whereas Japanese informants whichever they are young or senior selected the cell phone call (7 out of 8 for senior and 7 out of 8 for young). Besides, the senior people in Japan answered to use the landline (6 out of 8).
- The answer to this question reflects the nation culture (or the region culture) and the generation culture in Japan. It is interesting that there is no generation difference in the US. And it is also interesting that the gender culture have not been seen influential to the questions including this one.

For question 5

Table 8. Results for question 5

			Ishigaki Japan																	Minneapolis USA					
			60's-70's							20's								60's-70's				20's	
			IS1	IS2	IS3	IS4	IS5	IS6	IS7	IS8	IT1	IT2	IT3	IT4	IT5	IT6	IT7	IT8	MS1	MS3	MS2	MS4	MT1	MT2
			64	67	75	72	60	68	79	71	28	26	21	26	27	26	26	26	63	72	69	71	27	26
			M	M	M	M	F	F	F	F	M	M	M	M	F	F	F	F	M	M	F	F	M	F
Cellphone	e-mail			•								ah		abf		f		•						
	Call		b		abe	b	bf	b	bef		b	ab	b	abf	a		f	b	bdg	dg		ag	bg	g

- The answer was the selection of the cell phone call and was quite the same for Japanese (14 out of 16) and American (6 out of 6) informants as well as for young (9 out of 10) and senior (13 out of 14) informants.
- The only exception was the use of cell phone mail for Japanese young informants (3 out of 8).

For question 6

Table 9. Results for question 6

			Ishigaki Japan																	Minneapolis USA					
			60's-70's							20's								60's-70's				20's	
			IS1	IS2	IS3	IS4	IS5	IS6	IS7	IS8	IT1	IT2	IT3	IT4	IT5	IT6	IT7	IT8	MS1	MS3	MS2	MS4	MT1	MT2
			64	67	75	72	60	68	79	71	28	26	21	26	27	26	26	26	63	72	69	71	27	26
			M	M	M	M	F	F	F	F	M	M	M	M	F	F	F	F	M	M	F	F	M	F
Letter, Postcard			•	f		•				•				f	f		f		g	eg		g	g	g
Cellphone	Call		•	•土圈 J.物		bf	•	f	f	f	bf	f	a	f	f	f		f	b					
					bf	f	f																
Landline			•	•		•	f	f														•		
			f		bg	bf	f	f																

- Interesting results were found to this question. The choice of the letter was frequent for Japanese senior (3 out of 8), Japanese young (3 out of 8) and American senior (3 out of 4) and American young (2 out of 2) because of the politeness. This was the only situation among 6 questions that the letter was selected as a communication media.
- And it was also interesting that both of Japanese senior (6 out of 8) and Japanese young (7 out of 8) selected the cell phone call where only 1 American senior informant selected it. Furthermore Japanese senior informants (5 out of 8) selected the landline where no Japanese young and American informants selected it.
- Answers to this question reflect the nation culture and the generation culture in Japan.

Based on above results, the artifact selected differs from situation to situation and it was revealed that there was the influence of the nation culture, the region culture, and the generation culture but the gender culture was not influential in the selection of the artifact for communication. Another finding was that the category of culture related to the selection differs depending on the type of situation.

Although this research is rather a primary one and further in-depth research is necessary, the result clearly showed that the artifact selection is closely related to the category of culture. Different patterns found in this result may be related to the future pattern of the communication behavior and will serve to the possible future development of the artifact for communication.

4 Conclusion

Depending on the situation, the nation culture and the generation culture were influential for the selection of media (artifact) for communication. The gender culture did not influence much to the situations use in this research.

What was interesting is the use of cell phone was higher in young people in both countries. This fact reflects the general tendency of high literacy among young people for new emerging technology.

References

1. Hashizume, A., Kurosu, M., Kaneko, T.: The Choice of Communication Media and the Use of Mobile Phone among Senior Users and Young Users. In: Lee, S., Choo, H., Ha, S., Shin, I.C. (eds.) APCHI 2008. LNCS, vol. 5068, pp. 427–436. Springer, Heidelberg (2008)
2. Hashizume, A., Kurosu, M., Kaneko, T.: Regional Difference of Cell Phone Literacy Among Senior People. Bulletin of Human Centered Design Organization 4(2) (2008) (in Japanese) Human Centered Design Organization
3. Kurosu, M.: An Introduction to the Artifact Development Theory. HCD-Net 3(1) (2007) (in Japanese)
4. Kurosu, M.: The Optimality of Design from the Viewpoint of the Artifact Development Theory. Human Interface Society SIGUSE (2007) (in Japanese)

Exploring the Interface Design of Mobile Phone for the Elderly

Chiuhsiang Joe Lin, Tsung-Ling Hsieh, and Wei-Jung Shiang

Department of Industrial Engineering Chung Yuan Christian University 200,
Chung Pei Rd Chung Li, Taiwan
hsiang@cycu.edu.tw, bm1129@gmail.com, wjs001@cycu.edu.tw

Abstract. This study evaluated the influences of mobile phone interface design on the operating performance of aged people. To achieve the objective, the present research adopted a 2×2 within subject experimental design to develop different experimental treatments based on two types of software interfaces and two types of hardware interfaces. A total of 20 subjects including 10 younger participants (15-30 years old) and 10 older participants (over 40 years old) were tested in this experiment. Three dependent variables were under study. One measure refers to the operating time of subjects who were requested to perform several tasks in the experiment. The second measure refers to the error frequency, defined by the number of incorrect steps that subjects make when they perform the tasks. The third variable was the subjective convenience that was measured by a seven-point Likert scale. Finally, this study discussed design directions in cell phone design for the aged people. The conclusions from this study provided a useful reference for the mobile phone designer.

Keywords: Elderly people, Mobile phone, Interface design.

1 Introduction

The distribution of cellular phones represents one of the fastest growing technological fields ever. In Taiwan, more than 69.9% of people had cellular phones [18]. In addition, following the advancement of information technology, mobile phone becomes one of the daily necessities. The trend of the present mobile phone design is to manufacture smaller and lighter ones, and probably more functions ever than in the past. Consequently, enhancing the complexity on interface of cellular phones is critical for the operation of the users. According to Huang [9], the direction key that was hardware interface of cellular phone could divide into two types such as cross and duplet roughly. Furthermore, the main menu on the screen that was software interface of cellular phone also could divide into two types such as matrix and page roughly. The main menu with page type only showed single option by symbol and words once on the screen. Users need to operate the direction key to page the other options on the screen. On the other hand, the main menu with matrix type showed nine or twelve options by symbol and words once that arrange like matrix on the screen.

The interface design is certainly one of the most important factors for users to successfully operate the mobile phone. Nevertheless, if the interface designer did not

M. Kurosu (Ed.): Human Centered Design, HCII 2009, LNCS 5619, pp. 476–481, 2009.

consider the cognition aspect of the user when they design the interface of a mobile phone, users will be more likely to encounter difficultly in using the phone [19]. Users could not only be aware all of the functions in the cellular phone quickly but also feel confused on operating easily by the software interface of matrix type [29]. Some studies of users' experiences have shown that the current design trend may be inconvenient and not as friendly for the aged users [1]; [14]; [27]; [2]. Therefore, how to find the most suitable way to show the function would be one of the meaning issues with the designers of cellular phone.

In Taiwan, the number of people who were over 45 years old jumped up from about 3.5 million in 1980 and to over 8 million in 2007 [26]. According to the above, the elderly people are increasing in the population percentage of Taiwan and the population structure of Taiwan will enter into an ageing society. Ageing means the process that is an effect of the age increasing on the physiology of human [6]. Aging affects a wide range of human behaviour. Declines in human performance started to significantly affect physiological ability of human in the 40 years old [17]. Much research indicated the intelligence, memory, and attention of elderly people would degenerate more and more while their age is increasing [12]; [8]; [4]; [24]; [11]; [7]; [25]; [28], so elderly people learn new skill difficultly [22]; [30]. However, some studies have shown that most mobile phones often tailor to what the young people want but seem to neglect the requirement of the elderly people, leaving few types and selections for the elderly [3]; [20]; [16]; [23]; [15]. Consequently, elderly people would feel frustrated when they use these cellular phones that were designed for younger. Accordingly, this study focuses on exploring the suitable hardware and software interface design of cellular phone for elderly people.

2 Method

This research adopted a 2×2 within subject experiment design to develop different experimental treatments based on different types of software interface (matrix and page) and hardware interface (cross and duplet). A user interface for cellular phone was designed, containing four user interface of varying software interface and hardware interface. This research evaluated the influences of cellular phone interface design on the operating performance of aged people.

Three independent variables were under study. One refers to user age, comparing the operating performance of younger (15-30 years) and older adults (over 40 years). The second variable was the software interface of cellular phone, defined by different types of the main menu presented on cellular phone screen. The third variable was the hardware interface of cellular phone, defined by different types of the direction key for controlling the aspect of options on software interface.

As dependent variables, the operating performances of different cellular phone interface designs were surveyed. In total, two different dependent variables were under study. One measure refers to the operating time of subjects who were requested to perform two tasks (i.e. setup alarm clock and call someone) in the experiment. Another variable was the subjective convenience that was measured by five the Likert seven-point scale items.

The experimental process was: 1) before the experiment, first explain the experiment content and related special note, and acquaint participants with this cellular phone emulation program before operating. 2) Open the experiment interface and tell user the details of experiment 3) Participants perform three tasks with four interface conditions in random order. 4) After finishing three kind of tasks, let user write "Subjective convenience" questionnaire. 5) After taking a 30-minutes break, carry on other three kinds of different software and hardware interface design tasks and repeat steps 3), 4), and 5). The experiment was finished when user finished four kinds of cellular phones. After experimental data collection, a General Linear Model (GLM) statistical analysis was used to test the effect of the subjects' age and interface designs of cellular phone on the operating time and subjective convenience, and then also to test the interactive effect of the subjects' age and interface designs.

3 Results

This study used GLM statistical analysis to explore the effect of age and cellular phone interface design on operating time and convenience of subjects. The result was shown as follow. First, there was a significantly different operating time of setup alarm clock between younger and older participants (F=10.98, P<0.05). The older participants' operating time of setup alarm clock (M=58.10) was significantly higher than younger participants' (M=25.15). Second, there was a significantly different operating time of setup alarm clock between cross and duplet hardware interface (F=6.95, P<0.05). The operating time of setup alarm clock by cross hardware interface (M=48.68) was significantly higher than duplet hardware interface (M=34.58). Third, there was a significantly interaction effect of age and hardware interface design on operating time of setup alarm clock (F=3.86, P<0.1), which results showed on the figure 1.

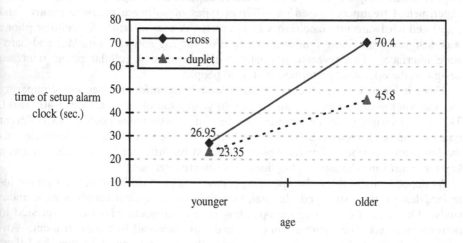

Fig. 1. The interaction effect of age and hardware interface design on operating time of setup alarm clock

The older participants' operating time of setup alarm clock by cross hardware interface (M=70.40) was significantly higher than duplet hardware interface (M=45.80). Therefore, the younger participants' operating time of setup alarm clock by cross hardware interface (M=26.95) was no significantly with duplet hardware interface (M=23.35).

Fourth, there was a significantly different operating time of call someone between younger and older participants (F=6.29, P<0.05). The older participants' operating time of call someone (M=49.90) was significantly higher than younger participants' (M=8.60). Fifth, there was a significantly different subjective convenience between younger and older participants (F=4.25, P<0.1). The older participants' subjective convenience (M=6.09) was significantly higher than younger participants' (M=5.26). Sixth, there was a significantly different subjective convenience between matrix and page software interface (F=3.20, P<0.1). The subjective convenience by page software interface (M=5.79) was significantly higher than matrix software interface (M=5.56). Finally, there was a significantly interaction effect of age and software interface design on subjective convenience (F=7.46, P<0.05), which results showed on the figure 2. The older participants' subjective convenience by page software interface (M=6.15) was higher than matrix software interface (M=6.06). Therefore, the younger participants' subjective convenience by matrix software interface (M=5.56) was higher than page software interface (M=4.96).

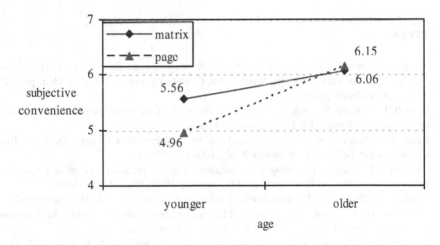

Fig. 2. The interaction effect of age and software interface design on subjective convenience

4   Discussion and Conclusion

This study evaluated the effect of age and interface design of cellular phone on operating performance. According to the results, the operating time of older participants operating time was significantly higher than that of younger participants, substantiating results of studies dealing with elderly people applying computer-based tasks [13]; [21]. Furthermore, the operating time of older participants with cross hardware interface was significantly higher than duplet hardware interface, but the operating time of younger participants with cross hardware interface was no significantly with duplet hardware interface. In addition, the subjective convenience of older participants with page software interface was higher than matrix software interface, but the subjective convenience of younger participants with matrix software interface was higher than page software interface oppositely. This study speculates that maybe because a slowing down of functions with age can be observed regarding sensory performance [10]; [31], motor performance [28], and cognitive performance [5]. Not only the cross hardware interface but also the matrix software interface would be too complex to understand and operate for older participants, so the older participants would prefer to use easily, simply, and intelligibly interface design such as duplet hardware interface and page software interface.

The conclusions from this study could act as a useful reference for the designer of cellular phone in the process of designing suitable cellular phones to elderly people. Simplifying the operating approach and decreasing the operating step would optimize the performance and subjective convenience of the elderly people.

Acknowledgments

This study is financially supported by a project from the National Science Council of Taiwan under contract No. NSC-97-2629-E-033-001.

References

1. Brodie, J., Chattratichart, J., Perry, M., Scane, R.: How age can inform the future design of the mobile phone experience. In: Stephanidis, C. (ed.) Universal Access in HCI, pp. 822–826. LEA, Mahwah (2003)
2. Carroll, J., Howard, S., Peck, J., Murphy, J.: From adoption to use the process of appropriating a mobile phone. AJIS 10(2), 38–48 (2003)
3. Child, J.: Cellular phone designers face complex requirements. Computer Design's: Electronic Systems Technology & Design 37(4), 109–113 (1998)
4. Connely, S.L., Hasher, L.: Aging and inhibition of spatial location. Journal of Experimental Psychology: Human Perception and Performance 19(6), 1238–1250 (1993)
5. Craik, F., Salthouse, T.: The Handbook of Aging and Cognition. LEA, Hillsdale (1992)
6. Hawthorn, D.: Psychophysical Aging and human computer interface design. In: Computer Human Interaction Conference 1998, Proceedings, Australasian (1998)
7. Howard, J.H., Howard, D.V.: Learning and memory. In: Fisk, A.D., Rogers, W.A. (eds.) Handbook of Human factors and the Older Adult. Academic Press, New York (1996)
8. Hoyer, W.J., Rybash, J.M.: Age and visual field differences in computing visual spatial relation. Psychology and Aging 7 (1992)
9. Huang, C.F.: Research on the Construction and Application of a Case-Based Reasoning System for Menu Design of Mobile Phones. Tatung University Department of Industrial Design Master Dissertation (2005) (in Chinese)
10. Kline, D., Scialfa, C.: Sensory and perceptual functioning: basic research and human factors implications. In: Fisk, A.D., Rogers, W.A. (eds.) Human Factors and the Older Adult, pp. 27–54 (1997)
11. Kotary, L., Hoyer, W.J.: Age and the ability to inhibit distractor information in visual selective attention. Experimental Aging Research, 21 (1995)

12. Light, L.L.: Memory and language in old age. In: Birren, J.E., Schaie, K.W. (eds.) Handbook of the Psychology of Aging, vol. 3. Academic Press, San diego (1990)
13. Lin, D.Y.: Age differences in the performance of hypertext perusal. In: Proceedings of the Human Factors And Ergonomic Society 45th Annual Meeting, Minneapolis, MN, October 8-12, pp. 211–215 (2001)
14. Maguire, M., Osman, Z.: Designing for older and inexperienced mobile phone users. In: Stephanidis, C. (ed.) Universal Access in HCI, pp. 439–443. LEA, Mahwah (2003)
15. Mallenius, S., Rossi, M., Tuunainen, V.K.: Factors affecting the adoption and use of mobile devices and services by elderly people – results from a pilot study. In: 6th Annual Global Mobility Roundtable 2007, Los Angeles, CA, May 31- June 2 (2007)
16. Mann, W.C., Helal, S., Davenport, R.D., Justiss, M.D., Tomita, M.R., Kemp, B.J.: Use of cell phones by elders with impairments: Overall appraisal, satisfaction, and suggestions. Technology and Disability 16, 49–57 (2004)
17. Merriam, S.B., Caffarella, R.S.: Learning in adult: A Comprehensive guide. Jossey-Bass, San Francisco (1991)
18. National information and communications initiative committee (NICI), http://www.nici.nat.gov.tw/index.php
19. Norman, D.A., Draper, S.W. (eds.): User Centered System Design; New Perspectives on Human-computer Interaction. Lawrence Erlbaum Associates, Inc., Mahwah (1986)
20. Oyama, H., Shiramatsu, N.: Smaller and bigger displays. Displays 23(1-2), 31–39 (2002)
21. Pak, R.: A further examination of the influence of spatial abilities on computer task performance in younger and older adults. In: Proceedings of the Human Factors and Ergonomic Society 45th Annual Meeting, Minneapolis, MN, October 8-12, pp. 1551–1555 (2001)
22. Park, D.C.: Applied cognitive aging research. In: Craik, F.I.M., Salthouse, T.A. (eds.) Handbook of Aging and Cognition. Erlbaum, Hillsdale (1992)
23. Renaud, K., van Biljon, J.: A Qualitative Study of the Applicability of Technology Acceptance Models to Senior Mobile Phone Users. In: Song, I.-Y., Piattini, M., Chen, Y.-P.P., Hartmann, S., Grandi, F., Trujillo, J., Opdahl, A.L., Ferri, F., Grifoni, P., Caschera, M.C., Rolland, C., Woo, C., Salinesi, C., Zimányi, E., Claramunt, C., Frasincar, F., Houben, G.-J., Thiran, P. (eds.) ER Workshops 2008. LNCS, vol. 5232, pp. 228–237. Springer, Heidelberg (2008)
24. Rybash, J.M., Roodin, P.A., Hoyer, W.J.: Adult Development and Aging. Brown and Benchmark, Chicago (1995)
25. Schaie, K.W.: Intellectual development in adulthood. In: Birren, J.E., Schaie, K.W. (eds.) Handbook of the Psychology of Aging, vol. 4. Academic Press, San Diego (1996)
26. Statistical yearbook of Interior in Taiwan, http://www.moi.gov.tw/stat/index.asp
27. Tuomainen, K., Haapanen, S.: Needs of the active elderly for mobile phones. In: Stephanidis, C. (ed.) Universal Access in HCI, pp. 494–498. LEA, Mahwah (2003)
28. Vercruyssen, M.: Movement control and the speed of behavior. In: Fisk, A.D., Rogers, W.A. (eds.) Handbook of Human Factors and the Older adult. Academic Press, San diego (1997)
29. Weiss, S.: Handheld Usability. John Wiley & Sons, New York (2002)
30. Welford, A.T.: Changes of performance with age. In: Charness, N. (ed.) Aging and Human Performance. Wiley, New York (1985)
31. Ziefle, M.: Aging, visual performance, and eyestrain in different screen technologies. In: Proceedings of the Human Factors And Ergonomic Society 45th Annual Meeting, Minneapolis, MN, October 8-12, pp. 262–266 (2001)

Design for China Migrant Workers:
A Case of User Research and Mobile Product Concepts Development

Xin Liu[1], Jikun Liu[1], Jun Cai[2], Ying Liu[2], and Xia Wang[2]

[1] Tsinghua University
[2] Nokia Corporation
xinl@tsinghua.edu.cn, ljk@tsinghua.edu.cn,
caijun@mail.tsinghua.edu.cn, ying.y.liu@nokia.com,
xia.s.wang@nokia.com

Abstract. The mobile user experience in China is far from optimal. The language styles, interactive modes and interfaces for most of the mobile communication products in Chinese marketplaces are just copies of those developed for Western users, which are difficult and not suitable for Asian people's thinking habits and usage customs. This especially applies to the needs of the poorly educated and users who are challenged by digital products. They are almost totally ignored by the mainstream of mobile product makers, even though a huge consumptive potential exists in those segments of consumers in the future years. This joint project of Tsinghua and Nokia, targeting the market segment of Chinese migrant workers, concentrates on user research and conceptual design for proper mobile communication products or service systems, with hopes to contribute to the corporate future design strategies and market development plans.

Keywords: Migrant workers, mobile products, user research, conceptual design.

1 Introduction

With the improvement of digital communication technologies, reduction of the communication costs and changes of people's lifestyles, the mobile communication products have become more and more popular worldwide. For Chinese consumers, mobile phones were the luxuries that only the rich could afford fifteen years ago; but now they are just communication tools that ordinary people can afford and daily objects that people increasingly depend on.

In the past, manufacturers only focused on the mainstream market demands for mobile communication products. As the market competition becomes increasingly intense, all manufacturers are seeking to add new functions and new features to their products to satisfy the specific needs of different target consumer segments.

Nokia is a world leader in the field of mobile communication and also the world's largest mobile manufacturer. In order to maintain a competitive advantage, Nokia Research Center continuously seeks innovative technology and design to provide a full range of product solutions for different groups of users. This joint research project

M. Kurosu (Ed.): Human Centered Design, HCII 2009, LNCS 5619, pp. 482–491, 2009.

of Nokia Research Center and Department of Industrial Design in Tsinghua University emphasizes user research and concept design targeting mainly the poorly educated consumer groups.

2 Research Methodology and Steps

Most methods used in this project are qualitative research approaches with quantitative analysis being only complementary. In the preliminary studies, the systematic "literature reviewing" is a guarantee of high efficient data collection and collation with proper aims; "KJ" Method has been used for a large amount of data; and "brainstorming" to stimulate the creativity of concept directions. "Questionnaires "and "interviewing" have been employed to deeply understand the real needs of the target users. For questionnaire analysis, "statistical" approaches were practiced; then, "scenario analysis" followed to sum up the data from interviewing. Finally, design evaluation was conducted with the random street interviews.

The methodology employed in this research is different from that of general product adaptive design in which it is only from the perspective of "objects" themselves. It concentrates on the perspective of "affairs" associated with the "objects". So, we comprehensively study the lifestyles of the target users, communicative modes, operation styles and the obstacles in using mobile and general electronic equipment. Through these observations, we analyze and conclude the users' physiological and psychological demands of both overt and covert, the premises of providing the target users with the best design solutions.

The project research follows the process of four steps: project definition, user research, design analysis and conceptual design, (Figure 1 below).

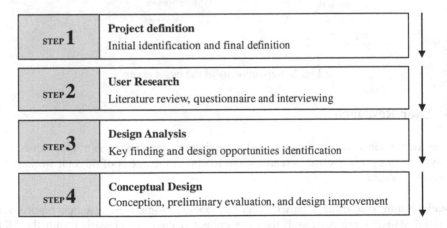

Fig. 1. The steps of the project research

3 Project Definition

Initial identification: Nokia Research Center believes that the Asian mobile user experience is far from optimal. For example, mobile communications products do not support some local languages; the user interfaces are only copies of those in the West, which are not suitable for Asian thinking habits. Especially for the poorly educated and those with obstacles in using digital products, their needs are almost always ignored by the mainstream mobile products makers.

Therefore, Nokia Research Center hopes the new products and user interactive concepts created through this research can enhance the mobile user experience for specific market segments identified by preliminary research, namely illiterate and semi-literate user groups in Chinese cities.

Final definition: Before long, the project team found during the research that the illiterate and semi-literate population in Chinese cities has been decreasing. Therefore, the target group of the research needs to be adjusted again. Upon the literature review and background investigation, the project team endows the illiterate and semi-literate with a new sense for the project, and that is the urban groups who have obstacles using the modern digital products. Migrant workers in Chinese cities are large in population, relatively poorly educated, and are typical users with obstacles for digital products, and they therefore become the target group of the project research, as shown in Figure 2.

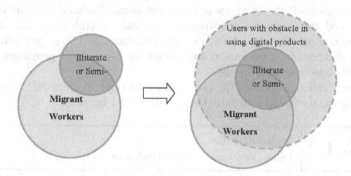

Fig. 2. Repositioning of the target group

4 User Research

Literature review, questionnaires, and interviewing are three main approaches employed in this project's user research stage to obtain the needs of migrant workers for mobile products.

Background research: Internet materials, books, magazines, research reports and official statistics are reviewed; then the collected data are classified with the "KJ" method for acquiring as much useful information as possible in a most efficient way.

Based on this information, we have a general understanding of Chinese migrant workers in their numbers, distribution, state of employment, educational level, specific work type, lifestyle, economic status, mental state and personality, etc, which supply the important support for the preparation of the next step of the questionnaire.

Questionnaire: The purpose of the questionnaire is to understand the current state of usage of migrant workers and their emerging needs for mobile products, including mobile phones, in their daily life, and filtering the samples for home interviewing.

Sixty samples of migrant workers were selected for the questionnaire from three industries of construction, manufacturing and service located around the city of Beijing. The numbers of samples from different industries vary according to the proportions of the migrant workers owning mobile phones. Therefore, the sample numbers in the service industry is slightly larger than that of the other two industries. The interviewers have taken the responsibility of filling out the questionnaires in order to keep the efficiency and avoid some migrant workers difficulties in doing it themselves.

The questionnaire totally consists of 6 parts: personal and family data, basic communication questions, basic questions for mobile phone owned, using purposes and places, problems and suggestions. The analysis of sixty questionnaires reflects the situation of migrant workers in the use of mobile products from several aspects, which provides very beneficial information for us to further understand wishes, needs and value conceptions of migrant workers.

Interviewing: The researchers identified 6 typical user samples (Figure 3 below) on the basis of the background information and analysis of the questionnaires above. They then carried out in-depth interviews at their homes, complemented by photographs and notes on the spot. A scenario framework was practiced to analyze the data gathered in the interviews.

The interviews were fully planned. The syllabus, procedures and regulations were well prepared before the interviewing, which focused not only on the sample users' activities and daily life concerning the mobile phone usage, but those concerning other relative household electronic equipment and their record of typical activities in a day as well. And the suspending problems in the questionnaires of the early period were also clarified during these in-depth interviews.

| Sample 1 | Sample 2 | Sample 3 | Sample 4 | Sample 5 | Sample 6 |

Fig. 3. Six user samples of migrant worker

The conclusion of the interviewing blends the concerns and suggestions of the interviewees and understanding of the researchers. And the direct interaction between the target sample users and research designers greatly benefits the concept generation later on.

5 Design Analysis

Design analysis is an integrated process of analyzing data and rough conclusions obtained from background research, questionnaires and in-depth interviews. Key findings are identified before design opportunities are put forward (figure 4). Techniques used here include classification, contrast, merging similarities, induction, deduction, association and brainstorming, etc.

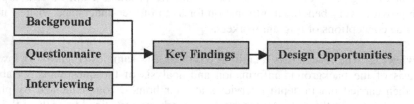

Fig. 4. Integrated design analysis process

The listing of 12 key findings is shown in table 1, and table 2 lists 5 design directions induced from the key findings and the corresponding design opportunities converted from the key findings, which will be used as starting points for the concepts generation of the next step.

Table 1. The listing of 12 key findings

Key findings

1. Demand for safety and easy to carry: The variation of living and working environments for migrant workers in different industries would point to a need for safety and easy to carry factors of the mobile phone.

2. Inconvenience to charge the phone: Mobile phones need to be charged frequently, but the living places for migrant workers lack the condition for charging the phones.

3. Problems of Identification and operation: The relatively low education level of migrant workers usually causes some difficulties in their use of a high tech device.

4. Demand for easy adjustment of volume: The shift of migrant workers among different living and working environments would lead to the high demand of easy adjustment for the ring tone and voice volume.

5. Difficulties in inputting text message: Text message are fashionable and economical for migrant workers, but most of them have difficulties inputting the text.

Table 2. (*continued*)

6. **Demand for intimate communication and privacy:** Migrant workers are away from their homes for a long time; the phone conversations with their intimate family members demand a high degree of privacy, which can't be satisfied in their poor living conditions.

7. **Need for Timer and lunar calendar:** The clock and alarm on the mobile phone are very useful for migrant workers to keep the working time in the city. The lunar calendar is in great demand for helping them get back home for help in the key farm business.

8. **Confusing for some kinds of service:** Poor education level limits the migrant workers' ability to accept new knowledge, and the capability of understanding the new "tricky and confusing" services set by the service providers.

9. **Two identities for the same phone number:** To save cost, some migrant workers save two identities in the phone for one contact number; one with the IP number for calling with a discount, and the other with a normal number for sending text messages.

10. **Demand for entertainment:** Entertainment for migrant workers is becoming an increasingly significant demand, due to their special working and living environments.

11. **Demand for learning:** In the need for a better job and further development, the migrant workers desire to learn new knowledge, new information, and new skills.

12. **"Fashionable" and "functional" are both important:** The function of mobile phones is the first concern, but the pursuit for fashion is getting stronger too. They are eager to merge into the city's living style, but not willing to be focused. The "limited fashion" is the description for most of the migrant workers' psychological state.

Table 2. Categorized key findings and design opportunities

Integration of key findings	Design opportunities
Carrying and Safety: 1- Safety and easy to carry 12- Fashionable and functional	• Shockproof, waterproof and dustproof need to be considered • Safe and convenient ways of carrying.
Identification and operation: 2- Inconvenience to charge the mobile phone 3- Identification and operation 4- Adjusting the volume 5-Difficulties in inputting text messages 6-Intimate communication and privacy	• Solar energy technique should be used in mobile phone. • Easily operated and understandable interfaces. • Using more symbols and icons, which are more easily understood. • Easy ways for volume adjustment. • Swift and convenient ways to take or reject a call. • Automatically replying text message when the user is not available at the time. • Providing graphic methods of communication, simple and easy to handle. • Providing automatic spelling correction. • More direct and graphical ways to show time. • Voice notification for time.

Table 2. (*continued*)

Entertainment & info source: 11- Demand for learning 10- Demand for entertainment	• Simple and interesting game functions. • Music and radio functions. • Camera and video functions. • Learning functions. • The information platform for migrant workers.
Economical demand: 8- Confused service 9- Two identities for one phone number	• Providing easy ways of checking money credit that is left in the phone. • A message should be automatically sent to the user notifying them of the cost and fee after a call or message. • Notifying the user to recharge the money in time. • Easy way of IP calling e.g. an IP number button.
Emotional demand: 6-Intimate communication and privacy 7- Timer and lunar calendar 12- Fashionable and functional	• Providing videoconference function to let people see their family members during the call. • Relatives' picture identification and shortcut buttons. • Relatives' birthday and important festivals notification functions. • Providing both solar and lunar calendars, as well as festival notifications. • Phone's appearance must be fashionable, but not a major focus • Outdoor device appearance is attractive for leading workers.

6 Conceptual Design

Usually, any product development must take into account user needs, market trends, technological limitations, corporate strategies and product identities, etc. But in this case, we bring forward any concepts based only on the key findings and the corresponding design opportunities from the user research in this project.

Conceptual design, preliminary evaluation and design improvement are the three sub-steps in this phase. Using the key findings with the design opportunities as the basis and starting points, the project team puts forward a number of design concepts by means of brainstorming, discussion, analysis and integration. The concepts aim at the specific concerns of migrant workers, such as maneuverability and protection, and each of them has its own emphasis.

Subsequently, preliminary assessment is conducted by on the spot interviewing. The samples are from the migrant workers randomly met on the street. The concepts are generally approved and some suggestions are proposed by the target samples.

According to the suggestions of the migrant workers, the project team conducts further improvement and optimization. Two of the final integrated concepts are depicted below.

Concept 1 - ZIPHONE: This concept, inspired by Zippo lighters popularized in cities, has a highlighted protective cover. It is easy to carry and operate, and also has an amusement function. The concept also echoes migrant workers' feelings of "low-pitched vogue", fashionable but not too focused (Figure 5).

Fig. 5. The concept 1 - ZIPHONE

The user research above finds that mobile phones of migrant workers are easily abraded on the surface when put into their pockets, or easily knocked and shocked by slipping down to the ground from their pockets during their working. And a big screen is fully necessary for their convenient operation, amusement and their feelings for urban life. Therefore, the concept product will adopt the metal slippery cover as the outer protective shell for its big screen and a metal clip in its back to make it easy and safe to carry.

The operation for the phone call is just like using a Zippo lighter, convenient and fashionable. When the phone rings, press and push the side key up. The cover will partially open and the phone number will be seen. To answer the call, just push the side key further up, the slippery cover will totally open and the phone will get through. To cut off the phone call, just close the slippery cover. Many people can't answer the phone while working. They can choose to press the shortcut key at the top and send an automatic reply message of "call you back later".

In addition, when the slippery cover opens completely, the big screen is sufficient for game playing, message sending, news reading and watching TV programs.

Concept 2 - XPOWER: The target group of this design concept is for those migrant workers with a higher income, namely the "foreman" or managers among them (Figure 6, 7). More form characteristics of outdoor objects are embedded, with a focus on urban styling and high tech feelings besides the protection and safety features.

The concept product consists of two parts: the inner and the outer. The outer part is a protective rubber shell with the function of being waterproof, dustproof, and quake-proof. The arm enclosure fixed with a rubber shell can provide convenience for carrying and use. The shell has a solar receiving plate for charging the phone temporarily, with a display of the electric value just under it.

The inner part is the body of the phone, which properly connects with the shell. The phone body with the big screen can be taken out completely from the shell to satisfy amusement and phone related operations to the greatest extent. The small and

Fig. 6. The concept 2 - XPOWER

Fig. 7. The improvement XPOWER

narrow screen on top of the body enables the user to look into the number or the name of the person calling quickly and easily. On either side of the small screen are obvious receiving or cutting off buttons which allow the user to operate conveniently and correctly with his fingers, even while wearing gloves. If the phone body is in the shell, only essential functions, like call answering, are active, saving electric power.

The simple number keyboard, which is added during the improvement stage, allows more convenient calling and faster operation in the working fields.

Concept 3 - GAME+: This concept solution shown in figure 8 is primarily suitable for the demands of the new generation of migrant workers. It has more powerful entertainment features, as well as a more fashionable style.

Fig. 8. The concept 3 - GAME+

The phone cover is made of abrasion-resistant high-strength plastic. The power switch is a metal touch key, slightly embedded in the surface. When activating phone calls, the touch-button digital keys will be faintly lit. The overall shape is round, simple and urban styled. The sunken screws increase the robustness and ruggedness of the phone, having met the needs and feelings of a new generation of migrant workers.

By flipping the phone open, the user can easily send and receive text messages, browse all kinds of information and pictures. It can also receive television programs where permitted. The most obvious feature is that the phone has very professional game handles, in which the rechargeable battery is installed, so that young migrant workers can relax and be entertained after a hard day of work.

7 Conclusion

The conceptual solutions above not only meet the basic communication needs, but also take into account both economic and fashion principles. Considering the users' special psychology, the appearance of the concepts must not be "made just for migrant workers." Many details of migrant workers' customary usage and habits are, in fact, embedded in function configurations and interactive approaches.

User research is the source of most product innovations. Only a profound insight into the users' significant potential demands can be established for the correct direction of design. This project is in accordance with the rigorous design process, through the target users positioning, background research, questionnaire surveys, home interviews and design analysis & synthesis, ultimately leading to the concept generation based on the understanding of the real needs of the migrant workers.

Acknowledgments. The project of Design for China Migrant Workers: A Case of User Research and Mobile Product Concepts Development is sponsored by Nokia Research Center, China.

Participant Graduate students: Zhenqi Tang, Miao Wang, Ying Zhao, Senli Yu, Shuo Liu. (Academy of arts and design, Tsinghua University).

We would like to thank Professor Guanzhong Liu at Tsinghua University for his insight and instructive advice.

References

1. Han, C.: The Development and Finalization of Chinese Migrant Workers 中国农民工的发展与终结. China Renmin University, Beijing (2007)
2. Holtzblatt, K., Wendell, J.B., Wood, S.: Rapid Contextual Design: A How-to Guide to Key Techniques for User-Centered Design. Morgan Kaufmann, San Francisco (2005)
3. Lam, Y. (ed.): Chopsticks Project: An Asian Life-style Study in Domestic Culinary Habits for Design. PolyU, Hong Kong (2006)
4. Liu, H.: The Problems of China Migrant Workers 中国农民工问题. People's Publishing House, Beijing (2005)
5. Zhang, X.: The Quality of Chinese Peasants Development Report: 2004 中国农民素质发展报告. 2004. Chinese Agriculture Press, Beijing (2005)

User Value Based Product Adaptation:
A Case of Mobile Products for Chinese Urban Elderly People

Jikun Liu and Xin Liu

Tsinghua University, Industrial Design Dept., Beijing, Qinghuayuan,
100084, P.R. China
ljk@tsinghua.edu.cn, xinl@tsinghua.edu.cn

Abstract. Mobile user experience in Asia is far from optimal. User interfaces and interactions are just copies of those for Western users. Product designers are not clear about the needs of many user segments in Asia. Products lack creative solutions specific for the Asian market, driving researchers to study users in Asia and create new concepts to improve the mobile user experience. Based upon the Industrial Design Value Innovation Theory developed in the Industrial Design Department of Tsinghua University, this project sponsored by the Nokia Research Center targets the market segment of urban elderly people in China, conducts the user research and concept design for proper mobile products or service systems, and hopes to contribute to the corporate future design strategies and market development plans.

Keywords: User value, product development, mobile communication, elderly people.

1 Introduction

Mobile user experience in Asia is far from optimal. User interfaces and interactions are just copies of those for Western users. Designers are not clear about the needs of many user segments in Asia, and especially lack of creative solutions specific for Asian markets.

Nokia is a worldwide leading company providing a wide range of mobile communication products, whose business covers mobile phones, multimedia, communication systems and solutions for enterprises.

In the increasingly competitive markets of mobile communications, companies are trying their best to address diverse market segments by adding new functions and features to their products. Nokia also hopes to extend their business in all areas of the segmented markets by taking advantage of its good relations with top universities around the world. This joint project between the Industrial Design Department of Tsinghua University and Nokia Research Center aims at developing concepts for Chinese urban elderly people. A research team comprised of professors, visiting scholars and students in PhD and Masters Programs has been assembled, and a user value based methodology has been proposed for research into the project.

M. Kurosu (Ed.): Human Centered Design, HCII 2009, LNCS 5619, pp. 492–500, 2009.
© Springer-Verlag Berlin Heidelberg 2009

2   Process, Methodology and Achievements

2.1   Overall Process

The user value based approach for this project is adapted from the methodology of the Industrial Design Value Innovation Theory[1] developed in the Industrial Design Department of Tsinghua University; the process is displayed in Fig. 1.

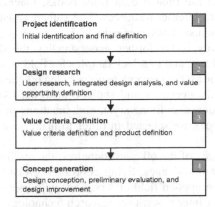

Fig. 1. Overall Research Process

2.2   Project Identification

With the help of the Nokia Research Center, the project team identified the present primary target group as the users who have difficulty using mobile products, and further focused on the segment of illiterates and semi-illiterates in Chinese cities.

After literature review, however, the researchers have found that illiterates and semi-illiterates in Chinese cities are fewer and fewer.[2] Therefore, the target of this research should extend to those groups who have physical and cognitive obstacles in using digital products; specifically the urban elderly people that finally becomes the target group to be study in this project, as seen in Fig. 2.

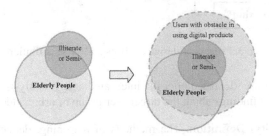

Fig. 2. Repositioning of the target group

---

[1] Jikun Liu, "From Value Analysis to Value Innovation: A Theoretical Framework of Value Innovation in Industrial Design", Ph.D. dissertation, Tsinghua University, 2007.
[2] According to the government statistics in 2001, there are about 600,000 illiterates in Beijing, 4.23% of the Beijing population, and 70% are the elderly people over 65 age of years.

2.3 Design Research

Design research is the most important stage of the project. It includes three steps: user research, integrated analysis and value opportunity identification.

User research. User research in this project consists of three perspectives: background study, questionnaire survey and filed observation. In the background study, the researchers gathered the related data from books, magazines, research reports, government statistics and Internet resources. The gathered material has been studied and classified, and a conclusion has been achieved.

A questionnaire survey is for further understanding the demand for and usage situation of mobile products, and to select samples for field observation. According to users' types and proportions, the research team selected 60 questionnaire samples, and then conducted informal interviews and statistic analysis. (Samples[3] have been selected according to gender, age, living style, and situations of using a mobile phone. Overlaps of different types are selected in order to make sure the samples are more accurately presented.)

On the basis of analysis of the 60 questionnaires, the researchers filtered 6 typical samples according to samples' ages, living conditions, vocation before retiring and income levels. They then applied field observations with user research approaches of home visits and in-depth interviews; user research technologies of still photography, videotaping, filed notes; and analysis methods of user dairies comparison and scenario analysis.

Integrated design analysis. Integrated design analysis follows the background research, questionnaire survey and field observations. In this phase, techniques of classification, comparison, similar items merging, induction, deduction and association are employed, as shown in Fig. 3.

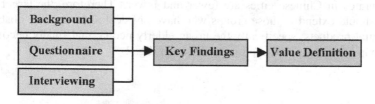

Fig. 3. Process of design analysis and value definition

The key findings are identified after the integrated design analysis sub-stage. Table 1 demonstrates 16 key findings about the target user group of urban elderly people.

Value (Opportunity) Definition. The methods of reasoning, deduction, analogy and association are used to transfer key findings into value opportunities. Induction, similar

---

[3] Credibility of research outcome is determined by sampling methods, the major principle is: the more the samples are, the less the deviation, but the higher the cost. If the amount of the samples is certain, increasing the samples' types and using variety of sampling methods are guarantee factors for high quality and low cost.

Table 1. Key Findings by Integrated Design Analysis

No.	KeyFindings
1	For the elderly with degenerating body functions, the mobile phones in the current market have too many features with complex interfaces and sophisticated operations. (For example, unclear screen displays, keys that are too small, etc. which lead to easily made mistakes.)
2	Most of the elderly fear high tech products, and are more used to a manual operation that has definite and direct feedback. And they even have regression emotion in appearance, material and texture.
3	The elderly generally pay special attention to "medical and heath care" and "medical rescue (first aid)". They are more frequently in hospital.
4	Most elderly people conduct regular exercise and entertainment activities according to their own conditions, such as walking, dancing, singing, playing ping pong, chatting, listening to storytelling, etc. The most popular is walking.
5	The elderly generally pay attention to news and current events, but often just watch several favorite TV channels. They also like to listen to radio.
6	Many elderly people are unable to use "phone book" and "shortcut keys" in mobile phones, and depend on handwriting and notebooks instead. This is principally because of the sophisticated operation, reduction of their vision, the custom of handwriting, and the complex settings of shortcut keys.
7	Elderly people often have phone calls with their family members, old friends and colleagues. The topics they are concerned about most involve health. The persons they call are relatively fixed, and the number is between 4 and 8.
8	Most of the elderly people cannot remember their own mobile phone number, but they need to exchange their phone number with new friends or doctors now and then.
9	Many elderly people cannot see the display for the amount of the power, and often forget to recharge the phone.
10	Only a few elderly people use the text message. Some are for fun, and they receive more and send less. The ability to write and edit SMS is a matter of show-off and a feeling of achievement!
11	Some of the elderly people check and reply to SMS at a fixed time. They generally check in the evening, just like others may check email, as it is not easy to hear the ring for receiving the SMS message.
12	The phone fee is a concerned issue. They are afraid of being unable to use up the deposit, while they also need to "pinch pennies."
13	Shopping outside is a regular activity for elderly people. (They shop at wet markets every day, and supermarkets once every several days.)
14	Some of the elderly people have a need for studying. (For the consideration of health care and other motivations, they learn English, new computer games, financing knowledge, health protection, etc.)
15	Some elderly people use computers, but few use the internet. They mainly play simple computer games, input text or check their email. The prospective elderly people's habits in the future will significantly change, as the people in the middle age now keep the habit of using computers for when they get older.
16	Some elderly people with healthy bodies and energy tour regularly.

items merging, clustering, and KJ Method are used to classify the value opportunities.[4] The following, Table 2, is the value opportunities, and Table 3 is the classified value opportunities.

In the process of the value opportunity classification, 32 value opportunities become 30,[5] and are classified into 4 value categories of simplifying interactions and

---

[4] Key finding listing and value definition in integrated design analysis may be conducted and listed together, to see figure 3.

[5] "Clear and direct feedback" is merged into "simplified interaction"; and two "simplified settings for shortcut keys" are combined together.

Table 2. Listing of user value opportunities

No.	Key findings	Value Opportunities
1	For the elderly with degenerating body functions, mobile phones in the current market have too many features with complex interfaces and sophisticated operations	• Simplified interaction • Automatic adjustment for voice • New communication service system for the elderly
2	Most of the elderly fear high tech products, and are more used to manual operations that have a definite and direct feedback. And they even have regressive emotions in appearance, material and texture.	• Augmented appearance appetency • Clear and direct feedback • Traditional operational ways
3	The elderly generally pay special attention to "medical and heath care" and "medical rescue (first aid)". They are more frequently in hospitals.	• Integration with first aid box or medical card • Shortcut button for first aid • New service system of medical & health care for elderly
4	Most elderly people conduct regular exercise and entertainment activities according to their own conditions.	• Functions for exercise aid: measurement for movement
5	The elderly generally pay attention to news and current events, but often just watch several favorite TV channels and like to listen to radio.	• Receiving the voice of TV programs • TV programs forecast in voice • Integration with radio, traditional operational ways
6	Many elderly people are unable to use the "phone book" and "shortcut keys" in mobile phones, and depend on handwriting and notebook.	• Simplified operation of the phone book • Simplified setting for shortcut keys
7	Elderly people often have phone calls with their family members, old friends and colleagues. The topics they are concerned about most involve health. The persons they call are relatively fixed, and the number is between 4 and 8.	• Simplified setting for shortcut keys • Fixed keys for intimates • Display for intimate photos
8	Most of the elderly people cannot remember their own mobile phone number, but they need to exchange their phone number with new friends or doctors now and then.	• Display self phone number on the wallpaper
9	Many elderly people cannot see the display for the amount of the power, and often forget to recharge the phone.	• Augmented display for power • Voice reminding for low power
10	Only a few elderly people use the text message. Some are for fun, and they receive more and send less. The ability to write and edit SMS is a matter of show-off and a feeling of achievement!	• Advanced the level of handwriting function • Augmented the contents of templates for SMS • Symbolic and iconic SMS
11	Some of the elderly people check and reply to SMS at a fixed time. They generally check in the evening, just like others may check email, as it is not easy to hear the ring for receiving the SMS message.	• Augmented reminding of SMS: voice, vibration, flash, etc.
12	The phone fee is a concerned issue. They are afraid of being unable to use up the deposit, while they also need to "pinch pennies."	• Advanced level of checking money balance • Display for balance and time left for calling
13	Shopping outside is a regular activity for elderly people. (They shop at wet markets every day, and supermarkets once every several days.)	• Integration with shopping
14	Some of the elderly people have a need for studying. (For health care and other motivation, they learn English, new computer games, financing knowledge, etc.)	• Customized learning programs
15	Some elderly people use computers, but few use the internet. This will change, as the people in the middle age now keep the habit of using computers when they get older.	• Simplified interaction between phone and computer
16	Some elderly people with healthy bodies and energy tour regularly. (The demands for tourist maps, information for traveling, hotels, dining and other activities.)	• Function for map checking and navigation • New service system of tours for the elderly

manipulations, medical rescue and health care, touring-exercise-entertainment, as well as emotional values.[6]

---

[6] Of course, value opportunities can also be classified into small groups, if necessary, to have more structures.

2.4 Value Criteria Definition

In this case, after further analysis and discussion with Nokia Research Center, the project team regards all 30 value opportunities in Table 3 as the user value criteria for the starting points of concept generation conducted later on.[7] But in this project, we

Table 3. Classification for User Value Opportunities

No.	Value Categories	Value Opportunities
1	Values for Simplifying Operation or Interaction	Simplified interaction (clear and direct feedback merged)
		Simplified operation with phone book
		Simplified setting for shortcut keys (2 combine together)
		Automatic adjustment for voice
		Augmented display for power
		Voice reminding for low power
		Augmented reminding for SMS: voice, vibration, flash
		Display for balance and time left for calling
		Advanced level of checking the balance
		Advanced the level of handwriting function
		Augmented contents of templates for SMS
		Display self phone number on the wallpaper
		Symbolic and iconic SMS content
		Simplified interaction between phone and computer
		New service system of communication for the elderly
2	Values for Medical Rescue and Health Care	Integration with first aid box or medical card
		Shortcut button for first aid
		New service system of medical & health care for elderly
3	Values for Touring, Exercise and Entertainment	Function for exercise aid: measurement for movement
		Receiving the voice of TV programs
		Forecasting TV programs in voice
		Integration with radio, traditional operational ways
		Customized learning programs
		Function for map checking and navigation
		New service system of tours for the elderly
4	Values for Emotion and Intimateness	Augmented appearance appetency
		Fixed keys for intimates
		Display for intimate photos
		Traditional operational ways
		Integration with shopping

---

[7] Generally, during the process of identifying the concept generation starting points and value criteria for evaluating concept solutions, some value opportunities would be filtered out, especially when there are too many value opportunities.

only selected several of them due to the limitations of time and energy. The selected value opportunities are marked with shades of grey color in Table 3.

2.5 Concept Development

On the basis of a deeper understanding of selected user value opportunities, the project team has developed several concept solutions described as follows, the main purpose of which is to demonstrate the relationship between the concepts and their related value opportunities.

New type of comprehensive service system for elderly people. According to the project user research, it is discovered that the elderly people feel very frustrated with sophisticated operations of mobile phones; they generally pay much attention to "medical and health care," and "medical rescue," which show an increase in demands with their increased age. According to these three key findings mentioned above, our project team proposes three value opportunities and value criteria of a new communication service system, a new touring service system, and a new medical and health care service system for the elderly people, as seen in Table 2.

During the stage of concept development, the project team found that these three service systems could be combined together to form a New Comprehensive Service System for the elderly people. Considering the technology advantage of the Nokia Corporation, the system is prepared to use the mode of mobile communication, that is, the ultimate solution will be a "New Type of Mobile Comprehensive Service System for Elderly People," the functions of which are: a) as a transfer platform providing the elderly people the service of mobile calls and short messages; b) providing the elderly people with services of first aid and health care reminding; c) providing the elderly people with household services; d) providing the elderly people with special featured touring information and traveling services. Afterwards, the project team has designed several home terminals, wearing terminals, and their wearing ways, including the function of one key medical aid switch with them. (Fig. 4).

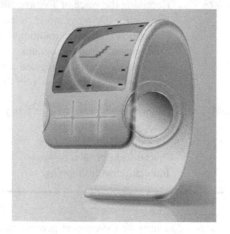

Fig. 4. 2007 M-care for Nokia

Intimate Mobile-phone. It was discovered during the research and observation that the target persons for elderly people's mobile communication is generally limited to 4 to 8, mainly family members and previous intimate friends, and therefore, intimate value is significantly important for them. The proposed Intimate Mobile-phone by the team is principally the consideration to meet this sort of emotional value and demand of the elderly people. In Table 3, emotional values consist of 5 value opportunities of appearance appetencies, fixed intimate keys, displaying intimate person's photos, relatively traditional operating modes, and an integration with purchase demands.

After serious consideration and discussion, the team determines to select the four above mentioned emotional value opportunities (except for integration with purchase demands), plus the value opportunity of simplified user interactions in the simplified operation category as the Value Criteria for concept of Intimate Mobile-phone. After the process of concept generation conducted by drawing on the wisdom of the masses, the mobile phone concept is eventually generated, which includes 12 big keys (each one can display an intimate person's head portrait), operate simply, and have a very appetent appearance. (Fig. 5).

Fig. 5. Intimate Mobile-phone in three different operative ways

3 Conclusion

This case is the demonstration for the application of the Value Innovation Theory, which has solved the problems assigned by the corporation. This is not only to make the target group positioning clear, but through background research and trends analysis, has found that sector of elderly people is an increasing proportion among the future consuming markets and is a group with a huge potential purchasing and consuming force which is well worth paying much attention to. The work for the next step is for Nokia itself to make and implement the marketing research and product (service) development plans by understanding the achievements of this project, corporate strategies and abilities within its own developing aims.

References

1. Cao, K.: Chinese Aged Audience Media Behavior Analysis. Master Thesis (2005)
2. Chu, H.X.: The Comparative Study for Leisure Life of Elderly People in Cities Between China and Japan. PhD Dissertation (2005)
3. Chinese Elderly Research Center: The Data Analysis of a One-Off Sampling Investigation for Chinese Elderly Population in Cities and Countryside. Chinese Standard Press (2003)
4. Duan, C.: Education Status Analysis for Chinese Population. Population Study. No.1 (2006)
5. Liu, J.: From Value Analysis to Value Innovation: A Theoretical Framework of Value Innovation in Industrial Design. Ph.D. dissertation, Tsinghua University, Beijing (2007)
6. Jia, Y.: The Life Stasus and Its Development Trends of Chinese Elderly Females. In: 1995-2005, Report for Gender Equity and Women Development in China. Social Science Literature Press (2006)
7. Qian, X.: Elderly People's Life and Health. Beijing Science and Technology Press (2005)
8. Ying, B.: A Research into Architecture and Demonstration for Chinese Elderly Market Segmentation Model. PhD Dissertation (2005)
9. Zhang, Y.: The Data Analysis of 5th Chinese Census, http://www.china.com.cn/zhuanti2005/txt/2001-12/25/content_5089666.htm
10. Joymain the Elderly Life Quality Index Report. Horizon Research Consultancy Group and Joymai Science & Technology Group (2005)
11. 2005 National Census by 1% Sampling of Chinese Population. Chinese National Statistic Bureau, http://www.stats.gov.cn/tjgb/rkpcgb/qgrkpcgb/t20060316_402310923.htm
12. Elderly Problems: 60 millions Chinese Aged People Suffered "High Tech Dread". Chinese Population Website, http://www.chinapop.gov.cn/rkxx/ztbd/t20060113_55164.htm
13. Two 60% Making Us Deep Reflection, the Aged in the Bead-Houses Expecting High Level Medical Care. Morning News (China), 2003.7.21, http://newspaper.jfdaily.com/xwcb/

From Novice to Expert – User's Search Approaches for Design Knowledge

Ding-Bang Luh and Chia-Ling Chang

Department of Industrial Design, National Cheng Kung University,
1, University Road, Tainan City 70101, Taiwan, R.O.C.
luhdb@mail.ncku.edu.tw, p3893105@mail.ncku.edu.tw

Abstract. As the arrival of the individual creativity era, innovation needs user creativity from the general public, nowadays, the enterprise provides the components or tools of products for users to utilize their creativity, and thus, users can be viewed as another form of designers. This research is based on the concept, "user is innovator", using LEGO bricks and its players with high design capability as the research subjects, proposes a qualitative method on users' design knowledge, the procedure includes five steps: user subject identification, status attribute classification, design knowledge categorization, search approach analysis, and knowledge model construction. This article proposed the design knowledge connotation and search approaches of four statuses of highly-involved users (junior expert, exhibition participator, business manager, award winner). When users possess the needed design knowledge and search approaches, it does not only fulfill individual creativity, also indirectly expands the creativity origin of the enterprise and increases its economic value.

Keywords: User as innovator, Design knowledge, Information search, Knowledge management.

1   Introduction

Nowadays, as we enter into Web 2.0 era, the main concept being "Let the user contribute his/her own individual value", the citizen pattern has substituted the former expert pattern [1]. Users are receptive of various information stimuli and enter into an individual creativity era, users show great interests in "design behavior," expecting to exhibit individual unique creativity through personal design. More enterprises are willing to open up more design space (for example: product components, tools or methods) for users to decide on the final appearance and function of the product [2]. Redström (2006) observed the products that have developed on the market which can realize users' own creativities, users can "utilize self-educated design knowledge" to successively design or redevelop the purchased products [3]. To be the participant in the enterprise's product development, the enterprise no longer independently produce and design a product, it creates the new product together with the users. Therefore, users can be viewed as innovators during the product development process. Sanoff

M. Kurosu (Ed.): Human Centered Design, HCII 2009, LNCS 5619, pp. 501–510, 2009.
© Springer-Verlag Berlin Heidelberg 2009

(2007) pointed out that the willingness of the enterprise to open up the design authority and the assistance of technology, allows users to participate in design and to share their design thoughts and knowledge with others at the same time, this span goes beyond the traditionally classified professional boundary [4], design knowledge does not solely belong to designers and has expand to users. In knowledge management, the user has transformed from knowledge receptor to knowledge provider. Therefore, the design knowledge connotation of the users has become the new design domain and an important topic.

Although users possess creative energy that should not be looked down upon, but realistically there is a big difference between their design capabilities, only a small portion of the users are capable of solving design difficulties. These users differ from designers with professional design training; they have acquired design capability and design knowledge on their own. Following "user-participated" idea, the design knowledge and search approaches of highly-involved users have potential to expand the creative origin of enterprises. Therefore, this research proposes a procedure to construct highly-involved users' search approaches for design knowledge, which is to discuss the needed design knowledge content and its search approaches during users' design creation process.

2   Literature Review

2.1   Design Knowledge

The two most common categorizations found in literature for knowledge are tacit knowledge and explicit knowledge [5,6]. "Explicit knowledge" means techniques and facts that can be recorded in writing, or in any tangible forms, and past down to others; and "tacit knowledge", the skills, judgment and instinct of human are hard to describe [7]. Design knowledge emphasizes on qualitative exchange and tends towards a knowledge integration of multiple disciplines. Design knowledge includes theory, judgment and design activity, as well as the various outcomes from the activity; design thinking is not independent outside of design activity that is a kind of comprehensive knowledge [8].

Recent design management researches have focused on deposit and access of design knowledge among individual designers or between designer and design company [9,10,11]. However, the more users are involved in design as the raise of user self-consciousness, the more design knowledge are needed to support it, design knowledge no longer exists between individual designers and design groups (e.g. design company), design knowledge has expanded to users. In the user knowledge management aspect, Su, Chen, & Sha (2006) proposed a knowledge management model (E-CKM model) that is applicable to innovative product development, to manage product knowledge and consumer knowledge from the perspective of the enterprise through web-based surveys and data-mining to separate consumer groups [12]. However, this consumer knowledge refers to the consumer behavior and product expectation of the consumer instead of to discuss the "design knowledge" and "knowledge search approaches" from the user's perspective. Moreover, for designers, inspiration is highly related to design presentation, Claudia & Martin (2000) believed inspiration is the most important within the design thinking process, the source of inspiration stimulates ideas, and from this the inner representation of designer's work is formed [13], thus, this research assumed that inspiration is a kind of design knowledge. According to above-mentioned reviews, this research defines design knowledge as "comprehensive knowledge implicated in design behavior," and assumes design knowledge involved in user's design process includes explicit knowledge: "product knowledge, technical knowledge, design method and skill," and tacit knowledge: experience and inspiration.

2.2 Information Searching

Wilson's (1999) information search behavior model points out that, information search behavior is caused by the user's sense of the information needs, in order to fulfill the needs, the user searches for formal and informal source of information until successfully finds the related information, or gives up [14]. Luh & Lin (2007) proposed the factors that affect designers search of design knowledge, and established designer information search behavior framework, its construction method and procedure [15]. Franke, Keinz, & Schreier (2008) discovered that in the initial self-design development stage, users can be positively encouraged in the problem-solving process when they receive design suggestions from the fellow group [16]. In the enterprise attitude aspect, Berthon, Pitt, McCarthy & Kates (2007) pointed out the importance of "customers with creative ability" to product innovation [17]; enterprise should establish proper response methods and propose a framework regarding enterprise's standpoint on consumer creativity. When "enterprise" takes the initiative and provides positive resources in supporting consumer self-design, it will be helpful to product promotion and development, such as on-line games. According to above theories, this research's information search subject developed from "taking users themselves as the core", in searching for design knowledge, will firstly learn to solve the problem by themselves, then ask for assistance from fellow groups, and find resources provided by the enterprise. Besides, design are representations of living experiences, nowadays the government or social organizations put many effort into promoting the life aesthetics activity, social resource has become one of the sources to acquire design related knowledge. Therefore, four aspects of search approaches for design knowledge in this research are self-learning, fellow group, enterprise resources, and social resources in order.

In current markets, whether it is the physical product or on-line virtual product, increasing number of enterprises provide "Design it by yourself" experiences. Luh & Chang (2006) proposed four major characteristics of this type of product and named it "User Successive Designing" product [18], LEGO is a representative product for providing ample of creation space and allowing users to utilize their individual creativity to design freely with simple components. LEGO Company has recruited works designed by users for sell. For enterprises, user creativity always brings new surprises for product and potential business opportunity at the same time.

Previous literature reviews rarely study the design knowledge and search sources belong to "users". This research aims at high-involved users, through a qualitative method, can be viewed as a preceding study in design knowledge management at user end. Through the "model of users design knowledge and search approaches," this research expects to discuss knowledge connotation and search approaches needed in course of design and creation for highly-involved users.

3   Method and Procedure

Interview is a method to acquire useful information about subject's personal experiences, opinions and thinking during a short amount of time. "Semi-structural questionnaire" is using structured questions and some open-ended questions in an attempt to induce subjects to recollect and think. With consideration of the uniqueness of this research topic, an "interview with semi-structured questionnaire" was used as the main research method. The methods and procedures are introduced as follows:

(1) User subject identification. Due to the variation in user's design ability, presently there is no clear determination standard, nowadays, to train a professional designer requires at least four years of college training, and it takes about one year after entering the work place to be able to have a stable design outcome, this article refers to "highly-involved users" as ones that have nearly normal design ability, therefore having at least five continuous years of engagement with design behavior as the basic threshold when selecting subjects. In addition, the general public thinks designers with good design works generally are invited to participate in public activities and exhibitions, received design related contest awards, and even be responsible for some management jobs or manage related businesses, because exhibitions, contests, managements and such experiences are different from simple creative work in nature, therefore, having at least one or more of the above experiences can be listed as an ideal subjects. For the above reasons, the subjects in this research, highly-involved users, and the definition must be qualified for one of the four requirements: (a) have at least five years of creation experiences, (b) have been invited to participate in public exhibitions, (c) have managed design related groups, and (d) have received related creative contest awards. Through two LEGO player groups in Taiwan, at least four persons who are qualified for each of the four requirements above were recommended for interviews. Two interviewees were selected from each requirement based on completeness of the interview content, making a total of eight subjects.

(2) Status attribute classification. Based on the subjects' design experience, they can be categorized into four types of statuses, including: Junior Expert, Exhibition Participator, Business Manager, and Award Winner. Based on their design experiences, the results of their status attributes can be generalized as below: all interviewees have at least six years of creation experience; highly-involved subjects without exhibition, organization management or competition winning experience are categorized as "Junior Experts." The ones concurrently possess two or more identities take the relatively unusual status as representation. As far as interviewees in this research are concerned, it can be found that business managers all have exhibition experiences, otherwise is untenable; award winners all possess the experiences of organization management and exhibition, otherwise is also untenable.

(3) Design knowledge categorization. According to literature reviews, the five main categories of design knowledge are further subdivided. Product knowledge is based on product component characteristic; technical knowledge is based on the component's operating characteristic and the structure principle; design method and skill are based on the design skill and basic thinking training that have been focused on during institutional design education. Moreover, design emphasizes the intuitive feelings and experiences, therefore, this research has proposed the item "object operation intuition" under the experience category, and depends on the experience relation between the different subjects, from inside out having "personal practice experience", "fellows' experience exchange", "expert instruction". In the inspiration aspect, this research has divided according to "intentionality" into "active" and "inactive" inspiration." The inactive inspiration is idea that flashes in one's brain related to the predetermined goal, which is called "intentional inspiration." The active inspiration is the idea that suddenly came across one's mind without any predetermined goals which is called "unintentional inspiration." Five categories of design knowledge, a total of 23 items are defined as Table 1. The interviewees are asked to select important knowledge items and to propose items that should be added onto the list, and to further point out the items that are comparative important (more than 50% as a rule). All interviewees stated that the list is comprehensive and all items are important. The selected results from the four types of interviewees represent the design knowledge constitution of four statuses of highly-involved users.

Table 1. Design knowledge categories and items

A. Product knowledge	B. Technical knowledge	C. Design method and skill	D. Experience	E. Inspiration
A1.Component shape	B1.Component connection method	C1.Freehand sketch representation method	D1.Object operation intuition	E1.Intentional inspiration
A2.Component color	B2.Component connection step	C2.2D representation method	D2.Personal practice experience	E2.Unintentional inspiration
A3.Component size	B3.Object construction technique	C3.3D representation method	D3.Fellows' experience exchange	
A4.Component material characteristic	B4. Auxiliary tools application method	C4.Form esthetics concept	D4.Expert instruction	
A5.Component price	B5.Structure constitution principle	C5.Color coordination skill		
		C6.Design development step		
		C7.Design thinking method		

(4) Search approaches analysis: The four aspects of searching information then subdivided accordingly: (I) self-learning is based on the normal information search approaches used by the general public; (II) fellow group is based on the information exchange and methods of interaction among LEGO-player groups; (III) enterprise resource is mainly the information search approaches provided by LEGO Company; (IV) social resource is referred to the design related information search approaches provided by the government and social organization, a total of 41 approaches are defined as Table 2. Interviewees were asked to recall the knowledge search approaches used during their design process and to check the design knowledge items that can be acquired through these search approaches.

Table 2. Design knowledge search aspects and approaches

I.Self-learning	II.Fellow group	III.Enterprise resource	IV.Social resource
I1. Instructional manual	II1. Joining fellow group	III1. Themed books	IV1.Themed books
I2. Related books	II2. Fellow exhibition	III2. Regular magazine	IV2.Contests
I3. Newspaper and magazine	II3.Themed books	III3. Membership club	IV3.Educational training
I4. Internet knowledge search	II4. Fellow group's electronic bulletin	III4. On-line simulation/test sample	IV4.Expert on-site demonstration
I5. Physical /Internet store	II5. On-line interactive forum	III5. Electronic newspaper	IV5.Expert on-line teaching
I6. Expert's blog	II6. Fellow group's blog	III6. Contest	IV6.Design exhibitions
I7. Related television program	II7. Personal communication	III7. Training course	IV7.Design related website
I8. Advertisement	II8. Group gathering	III8. Expert on-site demonstration	IV8.Themed museum
I9. Personal past experience	II9. Regular magazine published	III9. Expert on-line teaching	IV9.Creative market
I10. Leisure activity		III10. Upload system	IV10.Patent resource
		III11. Themed exhibitions	
		III12. Themed museum	
		III13. Themed park	

(5) Knowledge model construction: Collect the design knowledge items and its corresponding search approaches of four status users, and analyze design knowledge connotation based on the convergence situations to construct the model of design knowledge connotation and search approaches for high-involved users.

4   Result and Discussion

4.1   Design Knowledge Connotation of Highly-Involved User

It is difficult to discuss the standpoint of quantification statistics based on a qualitative research method, the result analysis of the significance of "the phenomenon and nature" is greater than that of "quantity and proportion." According to this framework, from the item selection situation (Table 1), within the "same status type," if the design knowledge item was selected by any subject, that means that user with this status recognizes the importance of that item, thus, this research views the knowledge items as design knowledge connotation of this particular status. The design knowledge connotation of the four types of user statuses is shown in Figure 1.

All 23 design knowledge items have been selected by subjects, that demonstrated the validity of the "design knowledge items" established by this research, and discovered that design knowledge produced four kinds of combination ways and can be the basis to determine the importance level of design knowledge. The convergence of design knowledge among the four types of statuses are called "Essential knowledge"; convergence among three types of statuses are called Main Knowledge; convergence between two types of statuses are called Secondary Knowledge; the ones without convergence are "Peculiar Knowledge", which can be viewed as the unique design knowledge exclusive for this status user. Essential knowledge coexists within four statuses of users and different statuses have different main, secondary and peculiar knowledge.

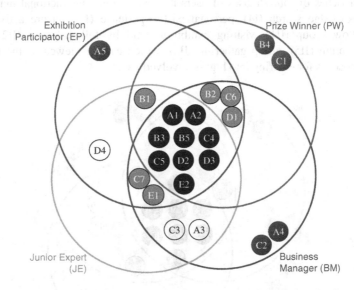

Fig. 1. Design knowledge connotation of four types of highly-involved users

"Essential knowledge" contains nine items, can be viewed as the design foundation knowledge for highly-involved users. One needs to know component shape (A1) and component color (A2); one needs to be familiar with object construction technique (B3) and structure constitution principle (B5); integrity of works mainly relies on form esthetics concept (C4) and color coordination skill (C5); personal practice experience (D2) and fellows' experience exchange (D3) are essential for designing of the highly-involved users; unintentional inspiration (E2) is the power that pushes highly-involved users to design continually. These essential knowledge dispersed throughout the five main design knowledge categories, they do not belong to any particular categories, this result demonstrates that in order to become a highly-involved user, one must grasp "interdisciplinary" design knowledge. This research found that only junior experts do not have their own exclusively peculiar knowledge, which explains that junior experts are the preliminary threshold before becoming highly-involved users.

As mentioned above, this research found "inspiration" is a design knowledge category that is valued by three statuses: junior experts, exhibition participators, and business managers. The design knowledge constitution of the four statuses of users could be acquired in Figure 1.

4.2   Design Knowledge Search Approaches of Highly - Involved User

According to Table 2, compares it with the design knowledge categorization in Figure 1, one may find different design knowledge items and search approaches of the various types of status users. Collect the four types of status users' search approaches for essential design knowledge; the extract method is also the significance of "whether there is the phenomenon" in the opinion of the subjects (Figure 2).The convergence in search approaches of four statuses of users has eight items: instructional manual (I1), internet knowledge search (I4), personal past experience (I9), leisure activity (I10), joining fellow group (II1), visiting exhibition held by fellow group (II2), on-line interactive forum (II5), group gathering (II8), which can be viewed as the necessary search approaches for entering into highly-involved users.

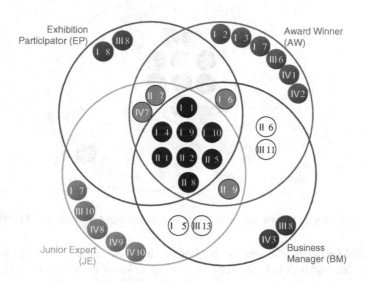

Fig. 2. Design knowledge search approaches of four types of highly-involved users

Table 3. Highly-involved user design knowledge and its search approaches

Knowledge type		Junior expert	Exhibition participator	Business manager	Award winner
Peculiar knowledge	connotation	n/a	A5	A4,C2	B4,C1
	approach	n/a	I (4,5)+II (7,8)	I (1,4,6)+ II (5,9)+IV3	I (4,6)+ II (1,2,6,9)
Secondary knowledge	connotation	A3,C3,D4	D4	A3,C3	n/a
	approach	I 4 + II (1,5,9)+ III 10+IV(1,2)	II (1,2,7)	I (1,2,4,6)+ II 5	n/a
Main knowledge	connotation	B1,C7,E1	B1,B2,C6,C7,D1 ,E1	B2,C6,C7,D1,E1	B1,B2,C6,D1
	approach	I (1,4,5,7,9)+ II (1,2,5,8)+ IV (1,2,6,7,9,10)	I (1,4,6,9,10)+ II (1,2,5,7,8)	I (1,2,4,6,9,10)+ II (1,2,5,8,9)+ III(8,11)+IV3	I (1,2,4,6,9) + II (1,5,6,9) + III 11+IV2
Essential knowledge	connotation	A1,A2,B3,B5,C4 ,C5,D2,D3,E2	A1,A2,B3,B5,C4 ,C5,D2,D3,E2	A1,A2,B3,B5,C4, C5,D2,D3,E2	A1,A2,B3,B5,C4 ,C5,D2,D3,E2
	approach	key search approaches: I (1,4,9,10) + II (1,2,5,8) other search approaches: I (2,3,5,6,7,8) + II(6,7,9)+ III(6,8,10,11,13)+IV(1,2,3,6,7,8,9,10)			

It can be summarized that the search approaches of highly-involved users centered on two aspects: "self-learning and fellow group". In other words, only if the user is willing to learn on his/her own initiative, eager to participate in activities with other players for exchange of design knowledge, then one can grasp the essential knowledge for becoming a highly-involved user. According to Figure 1, the knowledge connotation and the degree of importance for highly-involved users are shown under the knowledge type's "connotation" column in Table 3. The "approach" column's extract method is also the significance of "phenomenon" in the opinion of the subjects.

Four types of highly-involved users have nine essential knowledge items and eight key search approaches, and columns are not separated due to their commonality. For example: Junior Expert's main knowledge are B1, C7, E1, them are obtained through the fifteen search approaches: I (1,4,5,7,9)+ II (1,2,5,8)+ IV (1,2,6,7,9,10). Besides essential knowledge, four statuses of users possess some knowledge items and approaches alike; therefore, various status type users can have additional identity type as they wish in order to increase the knowledge items and know search approaches that one does not originally possesses.

5   Conclusion

Web 2.0 has facilitated an "all people expert" era, users have transformed from pure knowledge receptors to knowledge provider or knowledge developer. High-involved users are the utmost "emerging creative resource" that enterprises need to develop. In order to understand highly-involved users' design knowledge connotation and search approaches in the design process of the, this research takes LEGO bricks and its players as the subjects and proposes:

1. A procedure to construct the high-involved user design knowledge and search approaches model.
2. The questionnaire tools for qualitative research usage, including: design knowledge categorization chart, and search approaches for design knowledge.

Based on analysis of highly-involved LEGO users, the conclusions can be obtained:

1. User experience attribute categorization may be separated into four statuses: junior expert, exhibition participator, business manager, and award winner.
2. Nine essential design knowledge items and eight key search approaches of highly-involved users.
3. Junior expert does not have one's exclusive peculiar knowledge; it is the preliminary threshold to become highly-involved users.

Academically extends the "user-centered design" concept, proposes a qualitative research procedure on studying user's design knowledge. In the industry aspect, able to assist enterprise to develop new service model, become design knowledge service consultant through the understanding of "user design knowledge search behavior," wish to raise enterprises' production value, expand competitive advantage, and increase its economic value and upgrade its professional image.

Acknowledgments. This paper is partially sponsored by National Science Council, Taiwan (NSC 97-2221-E-006-162-MY2).

References

1. O'Reilly, T. (2005), http://www.oreillynet.com/pub/a/oreilly/tim/news/2005/09/30/what-is-web-20.html
2. von Hippel, E., Katz, R.: Shifting Innovation to Users via Toolkits. Manag. Sci. 48, 821–833 (2002)
3. Redström, J.: Towards User Design? On the Shift from Object to User as the Subject of Design. Des. Stud. 27, 123–139 (2006)
4. Sanoff, H.: Special Issue on Participatory Design. Des. Stud. 28, 213–215 (2007)
5. Tiwana, A.: The Knowledge Management Toolkit: Practical Techniques for Building Knowledge Management System. Prentice-Hall, Englewood Cliffs (2001)
6. Nonaka, I.: The Knowledge-Creating Company. Harv. Bus. Rev. 85, 162–171 (2007)
7. Howells, J.: Tacit Knowledge. Tech. Anal. Strat. Manag. 8, 91–106 (1996)
8. Narváez, L.M.J.: Design's Own Knowledge. Des. Issues. 16, 36–51 (2000)
9. Hsu, Y., Chang, W.C.: A Study on Knowledge Management of Design Organization. Ind. Des 28, 191–197 (2000)
10. Popovic, V.: Expertise Development in Product Design: Strategic and Domain-Specific Knowledge Connections. Des. Stud. 25, 527–545 (2004)
11. Cross, N.: Expertise in Design: an Overview. Des. Stud. 25, 427–441 (2004)
12. Su, C.T., Chen, Y.H., Sha, D.Y.: Linking Innovative Product Development with Customer Knowledge: a Data-Mining Approach. Technovation 26, 784–796 (2006)
13. Claudia, E., Martin, S.: Sources of Inspiration: a Language of Design. Des. Stud. 21, 523–538 (2000)
14. Wilson, T.D.: Models in Information Behavior Research. J. Doc. 55, 249–270 (1999)
15. Luh, D.B., Lin, T.Z.: Search Behavior of Design Information for Concept Development. In: International Association of Societies of Design Research, Hong Kong (2007) (CD ROM)
16. Franke, N., Keinz, P., Schreier, M.: Complementing Mass Customization Toolkits with User Communities: How Peer Input Improves Customer Self-Design. J. Prod Innovat Manag. 25, 546–559 (2008)
17. Berthon, P.R., Pitt, L.F., McCarthy, I., Kates, S.M.: When Customers Get Clever: Managerial Approaches to Dealing with Creative Consumers. Bus. Horiz. 50, 39–47 (2007)
18. Luh, D.B., Chang, C.L.: The Concept and Design Process of User Successive Designing. J. Des. 12, 1–13 (2007)

Leveraging User Search Behavior to Design Personalized Browsing Interfaces for Healthcare Web Sites

Malika Mahoui, Josette F. Jones, Derek Zollinger, and Kanitha Andersen

Indiana University – Purdue University Indianapolis, School of Informatics, USA
{mmahoui,jofjones,dzolling,kpa}@iupui.edu

Abstract. Understanding and leveraging user search behavior is increasingly becoming a key component towards improving web sites functionality for the health care consumer and provider. Hence, the development and improvement of any interactive browser-based information system, such as those used by digital libraries, requires consideration of the type of individuals utilizing the system, an understanding of available content and inclusion of a way to measure user interactivity. Information systems not only need to provide useful content, they must also present content in a way that results in an efficient, effective and satisfying user experience. Functional interface design is assumed to take in consideration the overall environment of the user to support users in their search tasks. Web logs – access logs and search logs - record user interactions with the interface, and as thus provide insight in user search behavior in a natural environment. The present study measures the usability of a digital library through an in depth analysis of the web logs. The study also leverages user interaction with the digital library to propose a use driven browsing interface to improve user interaction with the system.

Keywords: Log data analysis, web usage mining, search term clustering.

1 Introduction

Web-based access is becoming a standard for organization to expose valuable information to interested users. Content of data collections is made available through browsing interfaces supported by searching capabilities. Browsing interfaces also called sitemaps are supported by taxonomies that guide the user progressively into locating his/her information needs. Despite the efforts made in information architecture to facilitate user experience with web sites, it remains a fact that users often find themselves using the search capabilities as the main tool for locating relevant health information. One of the main reasons identified to explain this search behavior is that often web site design is driven by the content of the web site and the requirements set by the web site owner without too much consideration to how the user perceive his way to find the information in the web site. Formative usability evaluations enable the detection of a certain number of usability defects such as difficulty of learning and using the system, high error rate, etc. and the estimation of the degree of seriousness of the defect [1]. The usability evaluation generally is a onetime process and is only

M. Kurosu (Ed.): Human Centered Design, HCII 2009, LNCS 5619, pp. 511–520, 2009.

geared at detecting those aspects of a user interface that may cause the resulting system to have reduced usability. These techniques though, do not focus on how well the interface supports users in their search tasks nor do they take in consideration the overall environment in which the search task is performed [2, 3].

In the present study, the human computer interactions and search behaviors are studied in their natural environment. Web logs are used to detect pattern of cognitive activities as well as behaviors which occur within human-computer interactions.

Research Question
The research questions to be answered by the research team are:

• Can we observe differences in search behavior before and after a website redesign?
• Do people from the same domain name, or similar domain names search for information similarly across a website redesign?
• Do people who search for specific topics search for information similarly across a website redesign?

2 Background

The original 1979 resolution for Sigma Theta Tau International's (STTI) Virginia Henderson International Nursing Library (VHINL) called for "a national nursing library resource offering services to nurses and those interested in nursing" and soon after that an additional call for "a national clearinghouse for information regarding nurse researchers and nursing research". Ten years later, the first computer was purchased for the Library, enabling the beginnings of an electronic library. With that development, and the establishment of a database that stored findings of research studies, nursing knowledge was made available in an electronic format [5]. In 2001 a "re-visioning the library" meeting was convened and participants framed their vision in the context of the dramatic technology changes that have occurred in the 10 years since the VHINL purchased its' first computer. The call is now to extend the VHINL functions to provide access to global knowledge resources and connections made possible by the World Wide Web, as well as retaining and enhancing the rich legacy of knowledge modeling.

The development and improvement of any interactive browser-based information system, such as those used by digital libraries, requires consideration of the type of individuals utilizing the system, an understanding of available content and inclusion of a way to measure user interactivity. Information systems not only need to provide useful content, they must also present content in a way that results in an efficient, effective and satisfying user experience. These considerations guided the redesign of user-centered interfaces to extend the functionality of the Virginia Henderson International Nursing Library (VHINL) from that of a library to that of a Web-based portal for accessing nursing knowledge resources.

Formal usability evaluations of the previous version of the VHINL detected certain inadequacies that hindered navigation of the system and use of its features. A survey collected information to assess the error rate and to estimate the seriousness of identified defects. One defect contributed to what is called "low stickiness." Most user sessions (74.24 percent) lasted only a few seconds, during which page views totaled one or less. Often, as a result of ineffective strategy, search strings were poorly constructed.

In response to these findings, consideration was given to a tell-and-ask functional interface in which the user—usually a nurse—communicates with the knowledge base by making logical assertions—tell—and posing questions—ask. A prototype of this interface was also developed. Feedback was obtained from key user groups. The formal knowledge structure that currently exists within the library was examined and features needing expansion were identified along with opportunities for a controlled terminology. These analyses led to revisions of the data model and specification of an information model. The goal was that the knowledge structure and data model be complementary and reflect the search behavior of users.

A new interface and search engines have been installed and capturing of objective and subjective user data will continue. Analysis through data mining is planned to evaluate whether the revised knowledge model is consistent with the search behavior model, without unduly restricting a user's logical assertions or reasoning process in making queries.

3 Methodology

The aim of this study is to leverage web log data to improve the design of Sigma Theta Tau International's (STTI) Virgina Henderson International Nursing Library (VHINL) library. To analyze the user search behavior, two main approaches were undertaken: (1) analyze the user interaction with the web site and (2) study the search terms used to query the library.

To perform the former analysis, we leverage the fact that the VHINL web site had already undergone a previous redesign; and therefore the focus was to assess the impact of the redesign on the web site and suggest additional means to improve user interaction with the site.

To perform the latter study, the search terms used by the library visitors were the focus of a clustering analysis. The aim is to generate a browsing interface derived from the search terms deployed by the users to support the web site search capabilities and therefore better target users needs from the web site (see section 3.4).

3.1 Data Collection

Data was collected from STTI- VHINL website including both search logs and access logs. Pretest data was collected from October 1, 2002 until April 30, 2003 (7 months) while the website was operating under an Apache web server. Posttest data was collected from July 1, 2005 until September 4, 2005 (65 days) while the website was operating under an Internet Information Server (IIS) web server. In addition, the posttest data also included random popup questionnaires that complement the web log data.

3.2 Data Preparation

As in any information discovery process, a data preparation or preprocessing phase is needed in web log analysis. One aspect of data preparation is data cleaning. The main challenges faced in this study have to deal with two sets of data produced by two different applications needing to be studied using similar criteria. Furthermore, the pretest data had a high signal to noise ratio concerning the data, thus making it difficult to map web data usage as the original web design was not available to us anymore. Nevertheless it was important to see whether the redesign of the web site and the underlying web server applications had a positive impact on the site.

- Part of data cleaning involves detecting and removing web bot and web crawler' activities in order to leave only "real" user interactions with the digital library. This step resulted in discarding a large percentage of the log data. Techniques deployed to detect Bot activity involved using the browser information available in the User Agent (UA) section of the log data. The goal was to keep only log data associated with a list of compiled known browsers (white list). In the compilation of this list, efforts were made to include browsers that are longer deployed such as the old Netscape browser. Furthermore, as this filtering technique may generate some false positive we also performed a second filtering to eliminate log lines with UA information including words such as "bot" and "crawl". While the same process was attempted for the pretest data, many of the log lines did not include UA information. Therefore, another method was required to clean the data. Initially AnalyzeSpider [6] was used to help identify and remove web robots. However, further analysis showed that false positives still existed in the filtered data. As a further cleaning process, all logs that accessed the "robots.txt" file were identified. Of the identified logs, their corresponding source (IP address, domain name, etc.) were selected and used to remove all log data originating from these sources.
- Another cleaning task consisted of removing lines that are not useful for the analysis such as lines recording downloading of icons, pictures, scripts, and other files which configure the web page to be correctly presented. This cleaning was performed semi automatically using Unix grep command as well as automatically as part of the software tool WUM used for web mining the log data (see section 3).
- Format conversion is further step in data preparation to ensure that the log data can be processed by the log analysis and web mining tools we used. As most of the open source tools for server log analysis have been designed for common log file (CLF) format produced by the Apache server, IIS2Apache [7] was used to convert the posttest data from the default IIS format to CLF format.
- Furthermore to be able to analyze data using domain names, IP addresses were mapped to fully resolved domain names. We used Web utilization monitor preparation (WUMprep) [8] to perform this conversion.
- To perform the *search terms clustering analysis*, a filtering of the log data was performed to keep only records that include search queries. A list of stop keywords (e.g. Virginia Henderson, the) was also compiled and used to clean each search query. Moreover, stemming was used to be able to better identify main topics shared by the library visitors. Porter stemmer was used for this task.

3.3 Data Sampling

Data sampling was also used during the analysis to study the impact of the web site redesign based on the category of users.

- In the first group of sampling the categories selected emphasized the language used (i.e. English vs. non English), heath affiliation, government affiliation, personal utilization, and educational affiliation. Eight matching samples were therefore created from the data sets based on origin.

  - *English Speaking* sample included all visitors from Australia, Canada, New Zealand, United Kingdom, and South Africa.
  - *Non-English speaking* sample included all foreign visitors not listed in English speaking category.
  - *Government* sample included visitors from Veterans Administration and National Institute of Health.
  - *Health-care* sample included visitors from Banner Health, Carolina Health Care, Emory University Systems Health Care, Heritage Valley Health System, Methodist Health System, Partners HealthCare, Sharp HealthCare, and St. John Health Systems Detroit Hospitals.
  - *Hospitals* sample included Bassett HealthCare, Blessing Hospital, City of Hope, Cleveland Clinic, LeHigh Valley Hospital, St. Jude Children's Research Hospital, and Virginia Mason Hospital/Medical Center. ISPs included Alltcl, Ameritech, AT&T, Bellsouth, Covad, Earthlink, Mindspring, Pacbell, Qwest, and South-West Bell.
  - *Military* sample included all visitors from the .mil top-level domain (TLD).
  - *Education* sample included Harvard, Iowa, Kansas, Marquette, Maryland, Mayo Clinic, Miami, Monmouth, Purdue, University of California San Francisco, and Texas Medical Center.

The eight samples comprised about 30% of the entire population (Pretest samples 27.58%, Posttest samples 32.66%). In the remainder 70% of the population were (1) either a disproportional sampling in either the pretest or posttest data set; (2) or, the vast majority of the entries were non-existent in one or the other data set making a matched pre and post analysis not possible.

- In the second sampling method, groups were created based on the subjects searched by the users. For a given topic a list of related keywords is compiled using both scientific words and "common" words to cover for the diversity of the user population accessing the digital library. The "diabetes" topic was selected for this type of sampling. 122 keywords were used. This list included synonyms for Diabetes, such as "diabetes mellitus type 2," "non-insulin dependent diabetes mellitus," and "adult-onset diabetes." This list also included precursors and measures indicating diabetes, such as "hyperglycemia," "familial hyperproinsulinemia," "metabolic syndrome," "fasting blood glucose," and "hemoglobin A1C." Additionally, the list included common comorbidities normally associated with diabetes such as "hypertension", "nephropathy", "stroke," as well as known medications indicated for diabetes, such as "Insulin", "Humalog," and "oral antihyperglycemic." These terms were collected by a medical doctor and were as exhaustive as possible.

3.4 Analysis Method

Two main techniques were utilized to analyze the log data: traditional statistical analysis combined with web usage mining, and search term clustering.

- The aim of web usage mining is to extract knowledge from usage data representing user interaction with a web site. This knowledge can be used for several objectives including improving web site design, and supporting recommendation systems [9]. In addition of the statistical information that web usage mining provide (e.g. frequency of web page visitation), knowledge about the navigational paths that users exhibit is an integral part of the WUM process. To perform the mining process, log data is usually fragmented into users' sessions. Several criteria are proposed to define a user session including setting a time limit of a session or a time limit of non-user interaction with the system. In this study we used WUM tool [10] for web usage mining. All samples, as well as the entire population were analyzed using WUM. WUM tool creates user sessions from the raw data and measures user activities within those sessions. It also generates comprehensive reports with statistical data regarding visited pages and sessions. Report information was further post-processing to conduct statistical comparison between pretest and post data using both the whole population as well as the sample data sets. Five key markers were used to perform the comparison: Average Page Accesses Per Day, Average Visitor Sessions Per Day, Average Page Accesses Per Session, Average Unique Visitors Per Day, and Average Sessions per Unique Visitor.

- The main hypothesis behind search terms analysis is that search terms ultimately vehicle user needs from the web site; therefore they can be used to identify topics of interests shared by the users; and the topics can be combined to build a browsing interface that reflect user needs. To achieve this goal, clustering techniques were used as an unsupervised approach to identify main groups of terms shared by the visitors' web site. Experiments were performed with both k-means and hierarchical clustering using several similarity measures including cosine measure, correlation coefficient and Euclidian distance. The gCLUTO toolkit [11] was used to conduct the experiments given the useful tabular and graphical features it provides to visualize the clustered data. To build the similarity measure, one needed to represent the features to be clustered and their vector representation. For that purpose, we used WUM tool to sessionize the log data including only the search terms for both pretest and posttest data. For each session created, the list of keywords deployed during the visitors' search was identified. From there, two representations for the data was proposed: a session representation where each session has a keyword vector representation; and a keyword representation where each keyword has a session representation. With the first representation, clustering sessions is based on keyword similarity. In the second representation, clustering keywords is based on session similarity. As part of pre-filtering process, only stem keywords with a high frequency were selected for clustering. The search terms frequencies were computed, normalized and scaled for the statistical analysis.

4 Results

4.1 Web Site Usage Analysis

Tables 1 and 2 highlight some of the results obtained after comparing the pretest log data with posttest log data using the five criteria defined above:

- Visitors querying information about diabetes tend to navigate the new web page, visiting about the same amount of web pages, however there were fewer visitors interested in this web site.
- Visitor from English speaking countries tend to navigate the new web page, visiting more web pages, however there were fewer visitors interested in this web site.
- Visitors from non-English speaking countries tend to navigate the new web page, visiting more web pages, however there were fewer visitors interested in this web site.
- Visitors from government institutions tend to visit more often and more frequently, visiting more web pages when they navigated the new web site, overall there was much more interest in the web site from this group than the previous web site.
- While visitors from healthcare institutions tend to be the same, fewer institutions were aware of the newly redesigned web site, and those that did visit utilized the web site more than before.
- Visitors from hospitals tend to utilize and navigate the new web site more often, however individuals did not tend to return nearly as often.
- Visitors from Internet service providers tend to navigate the new web page, visiting more web pages, and once an individual had found the newly redesigned web page, tended to return to it more often as a resource, however there were fewer overall visitors interested in this web site.
- Visitors from military institutions tend to navigate the new web page, visiting more web pages, and once an individual had found the newly redesigned web page, tended to return to it more often as a resource, however there were fewer overall visitors interested in this web site.
- Visitors from educational institutions tend to navigate the new web page, visiting more web pages, and once an individual had found the newly redesigned web page, tended to return to it more often as a resource, however there were fewer overall visitors interested in this web site.

Overall, all visitors tend to navigate the new web page, visiting about the same amount of web pages, however there were fewer visitors interested in this web site.

When comparing each metric across the sample populations and also vis-à-vis the entire population. We can make the following observation

- The number of page access per day dropped vis-a-vis the entire population as well as for the sample populations we studied; except for the health and government related sample populations.
- Similar remark generally holds regarding the average sessions per day metric.
- While the average session length dropped for the entire population, the sample populations we studied all showed an increase in the session length. The most noticeable increase is for the hospitals population
- The number of accesses per session for each sample population follows the same trend as for the entire population. Here also the hospitals category shows a significant relative increase due to the substantially long sessions that this population presents.
- Similar remark holds for the unique visitor per day parameter with a noticeable increase also for the government sample population

- Average session per visitor for the sample populations does not follow the general trends; where we notice a noticeable drop for the English, Hospitals and Diabetes sample populations.

Table 1. Comparison of the site usage between pretest and posttest data for the sample data

Comparison metrics	Percentage change from pretest data to posttest data								
	Diabetes	English	Non English	Govern-ment	Healthcare systems	hospitals	ISPs	Military	Schools
Page Accesses Per Day	-69.08	-79.87	-2.04	359.73	50.66	154.42	-44.61	-24.73	-37.75
Average Visitor Sessions Per Day	-69.19	-84.57	-34.19	206.99	-0.48	19.09	-50.88	-50.13	-41.27
Accesses Per Session	0.37	30.39	48.84	49.76	51.38	113.64	12.76	50.93	5.98
Unique Visitors per Day	-21.50	-75.95	-32.78	192.25	-6.11	171.43	-52.59	-53.09	-44.84
Average Sessions per Visitor	-60.76	-35.83	-2.09	5.04	6.00	-56.13	3.61	6.30	6.47

Table 2. Comparison of the site usage between pretest and posttest data for the entire population

Comparison metrics	Percentage change –entire population
Page Accesses Per Day	-48.93
Average Visitor Sessions Per Day	-51.83
Accesses Per Session	6.01
Unique Visitors per Day	-51.16
Average Sessions per Visitor	-1.37

Overall, when analyzing the results of tables 1 and 2, we observe a general retraction of the web site usage in terms of the number of unique visits, how long they spend within the web site, and the number of pages they visits. However when we look at specific populations, we clearly identify that the government and health related populations have increased their interaction with the web site with more unique visitors spending longer sessions visiting more pages. In addition, we observe that the university-oriented population has decreased its activity with the web site aligning with the overall population trend. These observations suggest that the web site is becoming more professional oriented especially towards hospitals population. The results also indicate that for health and education related populations, the web site content is found useful as supported by the increase of the average user visits.

4.2 Search Terms Clustering Result

- Comparison of search terms between protest and posttest data (see Fig. 1) shows that search topics are similar with a slight higher frequency of searching of the current site (t = -4.32, p <.0001).

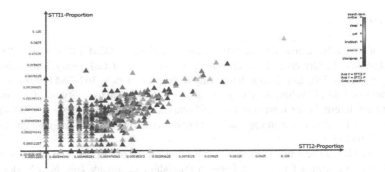

Fig. 1. Statistical comparison of search terms between protest and posttest data

- Clustering experiments using session representation (see section 3.4) allowed isolating the main topics that users are interested in (e.g. pain, nurse practitioner); which are represented in the first column describing a cluster (see fig. 2).
- Potential sub-topics are also generated and represented by the remaining columns in Fig.2. For example, we identified "manage pain" as potential sub-category for "pain" category.

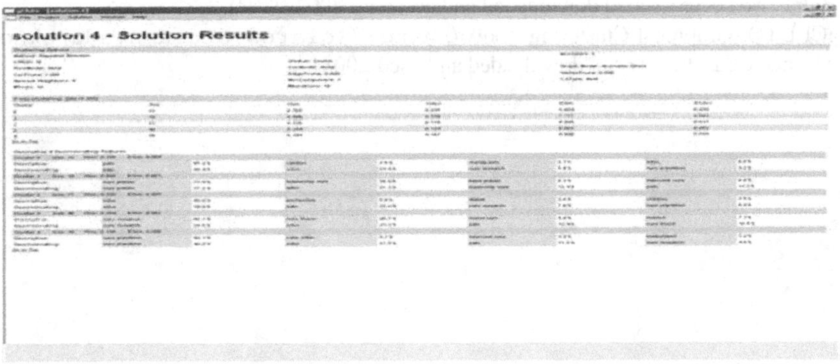

Fig. 2. Sample tabluar representation of the clustering results generated using gCluto software

5   Conclusion

In this study, we described a new approach that combines web usage mining and search terms clustering, leveraging on log data, to improve the design of STTI digital library. While the results of the web site usage mining suggest that the web site is gaining popularity within the health related institutions, it could also reveal a flaw in the current design if the audience targeted by the web site was larger than that. The results of the search terms clustering approach allowed the identification of groups of frequent topics and potentially sub-topics to be used to generate a hierarchical browsing interface for the web site. However, a more elaborated study remains to be performed to assess the utility of the user driven generated hierarchy.

References

1. Shneiderman, B.: Universal Usability. Communications of ACM 43(5), 84–91 (2000)
2. Richardson, J., Ormerod, T.C., Shepherd, A.: The role of task analysis in capturing requirements for interface design. Interacting with Computers 9, 367–384 (1998)
3. Paradowski, M., Fletcher, A.: Using task analysis to improve usability of fatigue modelling software. International Journal of Human-Computer Studies 60(1), 101–115 (2004)
4. Gruber, T.R.: A translation approach to portable ontologies. Knowledge Acquisition 5(2), 199–220 (1993)
5. Graves, J.R.: Structuring a knowledge base: the arcs© model. In: C.B. (ed.) Nursing Informatics: Education for practice. Springer Publishing Company, Inc., New York (2000)
6. Jgsoft, Analyse Spyder version 3.01. Downloaded and used 2007-2008 (2004),
   http://www.analysespider.com/analysespider.html
7. Abendschan, J.W.: "iis2apache.pl", Downloaded and used Feburary 27 (2008),
   http://www.jammed.com/~jwa/hacks/
8. WUMprep. Web Usage Mining Preparation Tool, http://hypknowsys.
   sourceforge.net/wiki/Web-Log_Preparation_with_WUMprep   (cited
   downloaded and accessed 2007-2008)
9. Mobasher, B., Cooley, R., Srivastava, J.: Automatic personalization based on web usage mining. Communications of the ACM, 142–151 (2000)
10. WUM. Web Usage Mining tool, http://hypknowsys.sourceforge.net/
    wiki/Welcome (cited downloaded and accessed 2007-2008)
11. gCLUTO. Graphical Clustering Toolkit, http://glaros.dtc.umn.edu/gkhome/
    cluto/gcluto/ (cited downloaded and used 2008)

Multimodal Corpus Analysis as a Method for Ensuring Cultural Usability of Embodied Conversational Agents

Yukiko Nakano[1] and Matthias Rehm[2]

[1] Faculty of Science and Technology, Seikei University Tokyo
[2] Faculty of Applied Informatics, University of Augsburg
y.nakano@st.seikei.ac.jp, rehm@informatik.uni-augsburg.de

Abstract. In this paper we propose the method of multimodal corpus analysis to collect enough empirical data for modeling the behavior of embodied conversational agents. This is a prerequisite to ensure the usability of such complex interactive systems. So far, the development of embodied agents suffers from a lack of explicit usability methods. In most cases, the consideration of usability aspects is constrained to preliminary user tests at the end of the development process.

Keywords: Multimodal Corpora, Embodied Conversational Agents, Cultural Usability.

1 Introduction

In this paper we are dealing with interactive systems that come in the form of virtual characters, which use verbal as well as nonverbal input and output channels, relieving the user from the burden to learn specialized control sequences and instead allowing for interacting with a complex system based on natural communicative habits. Such characters are often called Embodied Conversational Agents [3] emphasizing the available nonverbal communication channels as well as the fact that the interaction with such characters is realized as a communication between the agent and the user. At this point the question of cultural usability of such ECA systems comes into play. If people behave according to heuristics provided by their cultural groups [7], then simulating verbal and nonverbal behavior in ECAs has to adhere to such implicit cultural norms to prevent the agents from being perceived as behaving funny, weird, unnatural, annoying or even insulting.

Let's consider an example. You are staying at a hotel and discover that the WLAN is not working in your room. You go to the reception to complain about this fact and the clerk at the front desk listens carefully to you and then leans over touches your arm and assures you that he will do his utmost to fix this problem. Depending on your cultural background this might be an unwanted and unacceptable invasion of your personal space or it might just be a sign of empathy and care towards you. Such differences in spatial behavior and especially in the interpretation of spatial behavior by others have been described in [6]. The example illustrates the severe problems that can arise when quite different heuristics of how to behave "naturally" collide. Often such differences in perceiving and interpreting behavior have negative implications

M. Kurosu (Ed.): Human Centered Design, HCII 2009, LNCS 5619, pp. 521–530, 2009.

Y. Nakano and M. Rehm

leading to irritations, attribution of negative personality traits or unwanted insults. One reason might be that nonverbal behavior transports relevant non-symbolic information like feedback signals or emotions [18].

To prevent our ECA systems from failing, two challenges need to be tackled: i.) describing culturally determined differences in a principled way, and ii.) basing the design of ECA systems on reliable empirical data from different cultures. To tackle the first challenge we need a theory of culture that allows for explaining the differences that can be observed. There is one theoretical school that offers some promising ideas and defines culture as sets of norms and values which the members of a given culture have internalized (e.g. [6], [24]). Our work is based on the broadly applied dimensional model of Hofstede [7]. For the second challenge we suggest using multimodal corpus analysis (MCA) to prepare a solid empirical basis for modeling the behavior of an ECA system that adapts to the cultural background of the user.

In the remainder of this paper we describe how MCA can be employed to unravel cultural differences in behavior of two cultures that are positioned on different locations on Hofstede's dimensions (Germany and Japan) and how this information is then used to set up a model that predicts these behavior differences based on the empirical data and the dimensional theory of culture.

2   Related Work

Multimodal Corpus Analysis has been used increasingly over the last decade to decipher the specifics of nonverbal behavior in order to extract parameters for controlling the animation of virtual characters (see [19] for an overview). The general idea is to keep the intuition of the researcher at bay and at the same time to gain insights into the specifics of synchronizing different modalities like speech and gestures. Additionally, the data gathered during a corpus analysis can serve as a baseline against which the interactions between human user and embodied agent can be evaluated. [2] give an account on how the data from such a corpus can be used to directly mirror the behavior of a human speaker with an agent. This approach goes under the name of copy synthesis and is limited insofar as the agent can only directly reproduce aspects of the corpus data. A similar approach is described by [12]. Whereas [2] aim at real-time mirroring of human behavior, [12] try to extract specific behavioral data from the corpora that describe the "style" of the human speaker, which is then mimicked by the agent. A different type of approach tries to extract general behavioral information in the form of statistical data or behavioral rules that can then be employed to control an agent's behavior. [13] extract statistical rules from a corpus of natural dialogues that allow them to generate appropriate head and hand gestures for their agent that accompany the agent's utterances. An example rule would be something like "if the utterance contains a negation, shake the head". Thus, their approach exploits the relation between words and gestures. [16] concentrate on grounding phenomena in interactions with virtual characters and also extract rule-like regularities for gaze behavior from a corpus of human interactions. The same corpus is later used to judge the results of the human-agent dialogues. Instead of rules, [20] have shown how statistical information can be extracted from a multimodal corpus and used as control parameters for a virtual character. To this end they analyzed what kind of relation exists between certain types of gestures and verbal strategies of politeness.

All of the above work focuses on multimodal aspects of interaction and does not regard culture as a crucial parameter. The need to do so has been acknowledged [17] but there are few systems that actually try to tackle this challenge in a principled manner. This might easily be due to the multifaceted influences of culture that have to be regarded on different levels during the development process. Concerning the agent itself we can distinguish between cultural aspects of the agent's appearance (black, white, with French beret or an English bowler hat, etc.), cultural aspects of its verbal behavior (language, formal vs. informal, slang, etc.) as well as cultural aspects of its nonverbal behavior (use of gestures, proxemics, volume of speech, etc.), and cultural aspects of its cognitive processes (relevant features for persuasion, reaction to high status individual, etc.). Additionally, [23] gives an account on the difficulty of design-ing culturally adequate systems due to the fact that the designer's culture always inter-feres in the process by providing him with implicit assumptions about many design choices that have to be challenged actively. [9] focus on verbal and nonverbal behav-ior to manifest the cultural background of an agent, others try to simulate culture-specific behavior in order to train intercultural communication. Currently, this seems to be the main application area. [11] describe a language tutoring system that also takes cultural differences in gesture usage into account. [25] as well as [22] aim at cross-cultural training scenarios and describe ideas on how these can be realized with virtual characters. [10] present an approach to modify the behavior of characters by cultural variables relying on Hofstede's dimensions. The variables are set manually in their system to simulate the behavior of a group of characters. Most of this work is based on general claims from the literature, which brings the danger of realizing only stereotypic and cliché-like behavior in the agents. To base such systems on reliable empirical data we suggest the method of multimodal corpus analysis and in the re-mainder of this paper are going to exemplify this method for realizing culturally ade-quate behavior for German and Japanese agents.

3 Multimodal Corpus Analysis for German and Japanese Interactions

A corpus is a collection of (video) recordings of human interactive behavior that is annotated or coded with different types of information. A multimodal corpus analyses more than one modality in a single annotation, e.g. speech and gesture in order to explicate the links and cross modal relations between the different modalities. Which kind of information is coded in a given corpus is defined in an annotation scheme that specifies the coding attributes and values, for instance coding the type of a gesture along McNeill's taxonomy [15].

This short introduction already introduces a number of challenges that have to be faced if this method should be employed in a multicultural setting. A standardization of most steps in the process of recording and analyzing the data is necessary including a standardized design for the recording session, a standardized annotation scheme, and a standardized analysis of regularities in the data.

3.1 Standardized Design of Corpus Study

To ensure the replication of conditions in all cultures participating in the study, a common protocol had to be established on how to conduct the study with detailed instructions to be followed at every step. These instructions had to cover recruiting of subjects and actors, the timeline of each recording as well as "scripts" for the people conducting the experiment as well as detailed information about the necessary materials and the setup of the equipment. To produce comparable data sets it was indispensable to define technical requirements for the video recording sessions. This included the specifications for the recording equipment as well as the layout of the recording area to be able to reproduce the recording conditions.

In our study of German and Japanese behavior, dyadic interactions between human subjects were recorded in three scenarios: (i) first meeting, (ii) negotiation, (iii) interaction with status difference. One of the interaction partners in each scenario was an actor following a script for the specific situation. The rationale for using actors as interaction partners was that we would be able to elicit sufficient interactions from the subjects and to control the conditions for each participant more tightly. To control for gender effects, a male and a female actor were employed in each scenario interacting with the same number of male and female subjects. The actual number of participants differed between Germany and Japan. 21 subjects (11 male, 10 female) participated in the German data collection, 26 subjects (13 male, 13 female) in the Japanese collection. For each subject, around 25 minutes of video material was collected, 5 minutes for the first meeting, 10-15 minutes for the negotiation, and 5 minutes for the status difference. Participants were told that they take part in a study by a well-known consulting company for the automobile industry, which would take place at the same time in different countries. To attract their interest in the study, a monetary reward was granted depending on the outcome of the negotiation task. To be able to control for effects of personality on the behavior under examination, participants had to fill out a NEO-FFI personality questionnaire [12].

3.2 Standardized Annotation Schemes

The corpus study focused on nonverbal behaviors taking spatial behavior (proxemics), volume of speech, gestural expressivity, and posture into account. Initially, the analysis concentrated on expressivity and posture. Posture was annotated following the coding scheme outlined in [1], which describes posture in terms of relative positions of body parts and thus restricts interpretations to a minimum. To give an example, consider one-handed postures that require touching the other arm. These are coded in the following way: PHSr (put hand to shoulder), PHUAm (put hand to upper arm), PHEw (put hand to elbow), PHLAm (put hand to lower arm), PHWr (put hand to wrist). These hand positions are unambiguous; either the subject is touching the elbow or not, thus keeping culture-specific interpretations at a minimum. Similar codes are used for all hand, head and leg postures.

Gestural expressivity is a little more challenging. Apart from coding the type of a gesture for instance following the coding scheme described in [15], gestures provide information on a non-symbolic level by the way how they are performed. [4] has shown in a large-scale study of US-immigrants that culturally determined preferences exist on this level of granularity. To capture these differences he described the following levels of gestural activity: (i) spatio-temporal, (ii) interlocutional, and (iii) co-verbal. Co-verbal coincides with McNeill's [15] definition of co-verbal gesture usage. The interlocutional level is concerned with aspects of proxemics for instance body contact while gesturing and interacting (remember the example from the beginning). The spatio-temporal level at last describes how a gesture is performed, which we call expressivity following ideas of [5], who showed a relation between such parameters and personal style of a speaker. In [18], more details can be found on the similarities and differences of these taxonomies in describing non-symbolic gesture usage.

Gestural expressivity is analyzed with the following parameters, where each parameter was coded using a seven-point scale. On this scale, 1 denotes small values and 7 large values for the parameter: activation (number of gestures per dialog), spatial extent (space occupied for realizing the gesture), speed, power, fluidity (smooth vs. jerky). According to [4], different cultures exhibit different values of these parameters. Thus, following [7], in a culture that generally uses high spatial extent, this is also perceived as the "normal" way of doing gestures. Thus, we can expect that the baseline for attributing high or low spatial extent to a given example of gestural expressivity might depend on the coder's own culture. To prevent our coders from relying solely on their intuition in ascribing values to the expressivity parameters, a coding manual was created (accompanied by example videos) that defines high and low values based on objective criteria. The spatial extent of a gesture for instance is described by the angle between upper and lower arm.

3.3 Preliminary Analysis of Nonverbal Behavior

The results presented in this section are based on the analysis of 8 German and 8 Japanese samples to exemplify how the analysis can be done. Comprehensive results will be available soon.[1]

3.3.1 Posture Analysis

Results. The average number of head posture shifts in the German samples was 22 and in the Japanese samples it was 15.6. The average duration of each posture (how long the subjects were keeping the same posture) differed between 2.57 (German) and 2.54 (Japanese). Both differences are not statistically significant in a t-test. However, the distribution of the categories was different between the two cultures. Japanese participants generally did less head posture shifts than Germans, except for THdAP (turn head away from person). This difference was statistically significant: $\chi^2(5)=20.308$, $p<0.05$. The average number of leg posture shifts in the German samples was 9.5 and in the Japanese 16.56. A t-test revealed this to be a weak trend: $t(15)=1.764$, $p<0.1$. The average duration of each posture was 19.93 (German) vs. 24.64 (Japanese), but the difference was not statistically significant. Similar to head postures, we found that the difference in category distribution was statistically significant: $\chi^2(3)=9.205$, $p<0.05$. While LSF (lean sideways on foot' were the most frequent in both samples, Japanese people also frequently did MLP (move leg to person). The

---

[1] Please visit http://mm-werkstatt.informatik.uni-augsburg.de/projects/cube-g/ for up-to-date information.

average number of arm posture shifts in the German data was 40.38 and in the Japanese 22.8. The average duration of each posture was 7.79 (German) vs. 14.08 (Japanese). For both differences, a t-test revealed a weak trend: t(16)=1.931, p<0.1, and t(16)=2.061, p<0.1, respectively. The differences in category distributions were statistically significant in hand-to-arm (one-handed), hand-to-arm (two-handed), hand-to-head, and hand-to-cloth postures: $\chi^2(4)$=70.482, p<0.01; Fisher's Exact Test p<0.01; $\chi^2(2)$=7.208, p<0.01; $\chi^2(2)$=91.447, p<0.01, respectively. Also, a trend was found in hand-to-trunk postures: $\chi^2(2)$=5.708, p<0.1. Hand-to-head postures more frequently occurred in the Japanese data; especially PHFe (put hand to face) was the most frequent. Hand-to-arm (one-handed) postures were different depending on the culture. The most frequent category in the German samples was PHEw (put hand to elbow), and in the Japanese samples PHWr (put hand to wrist). Vice versa, German participants rarely did PHWr, and Japanese rarely did PHEw. As for hand-to-arm (two-handed) postures, the most frequent category in German data was FAs (fold arms) and that in Japanese data was JHs (join hands). Hand-to-cloth postures were rarely observed in Japanese data, but they were very frequent in German data (especially PHIPt (put hand into pocket)).

Discussion. Generally, head postures did not differ between cultures but Japanese more frequently looked away from the partner than Germans. Significant differences were found regarding arm postures. Germans more frequently changed arm postures than Japanese, and Japanese kept the same posture longer than Germans. Posture shapes also differed. Germans mainly used their arms, such as folding their arms (FAs) and putting their hands on the elbows (PHEw). In contrast to this behavior, Japanese mainly used their hands, such as joining the hands (JHs), or putting their hands on the wrists (PHWr). Moreover, Japanese frequently touched their heads, and Germans put their hands in the pockets. Although Japanese did not move their upper bodies as frequently as Germans, they used more leg postures. Additionally, the total number of posture shifts per conversation is not depending on culture. To sum up, these results suggest that the frequency of posture shifts is not different depending on culture, but the types of frequently used postures differ. Thus, the employed body parts as well as the shapes of the postures express the characteristics of each culture.

Table 1. Results of expressivity analysis

	G	JP	F
Activation	22.12	6.62	4.177[*]
Power	3.21	3.39	0.736
Speed	3.81	3.39	2.929[+]
Fluidity	4.40	2.87	68.591[**]
Repetition	1.38	2.70	50.247[**]
Spatial Extent	3.01	4.02	18.703[**]
Duration	2.39	5.13	15.461[**]

3.3.2 Expressivity Analysis
Results. In order to gain insights in the supposed differences in the use of gestures, we compared expressivity parameters of the German and the Japanese samples. Table 1 lists the results of the analysis. First of all, it has to be said that there is a significant

difference in the number of gestures that were used in the German and the Japanese samples, i.e in the overall activation. On average, German participants used three times more gestures than Japanese participants (22.12 vs. 6.62), which is shown to be statistically significant (ANOVA): $F=4.177$, $p<0.05$. For the overall comparison between the German and the Japanese sample, no significant difference can be seen for the parameter power and only a weak trend for speed ($F=2.929$, $p<0.1$). For the other parameters (fluidity, repetition, spatial extent, and duration) the difference is highly significant: $F=68.591$, $p<0.01$; $F=50.247$, $p<0.01$, $F=18.703$, $p<0.01$; $F=15.461$, $p<0.01$, respectively.

Discussion. These preliminary results show a tendency concerning the differences in how gestures are expressed in the two cultures. Germans use significantly more gestures than Japanese. On the other hand, if Japanese participants do a gesture, this takes on average twice as long compared to a German participant. But the gesture is less fluently performed that is with more interruptions. Moreover, spatial extent for the gestures is higher than in the German samples.

4 Employing the Results for Designing Enculturated Agents

In Section 2, we presented how the information derived from a MCA can be utilized to control the interactive behavior of embodied conversational agents. Here, we present a slightly different approach. Having extracted the statistical information as reported in the previous section, this is correlated with Hofstede's ideas of cultural dimensions. By setting up a Bayesian network, it becomes feasible to model the causal relations between a culture's location on Hofstede's dimensions and the observed nonverbal phenomena. Figure 1 depicts a version of such a network. The middle layer defines Hofstede's [7] five dimensions: hierarchy, identity, gender, uncertainty, and orientation. We will not go into detail here but give an example on possible correlations between dimension and behavioral heuristics. According to [8], the location on the identity dimension (individualism vs. collectivism) is for instance related to proxemics behavior. Interlocutors from individualistic cultures tend to stand further apart in face-to-face encounters than interlocutors from collectivistic cultures.

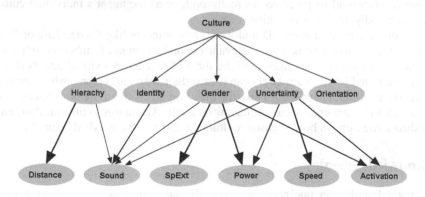

Fig. 1. Network model of causal relations between cultural dimensions and nonverbal behavior

The bottom layer consists of nodes for nonverbal behavior that can be set for a given agent. The top node which is labeled "Culture" is just for demonstration and interpretative purposes. It mainly translates the results from the dimensional representation of cultures into a probability distribution for some example cultures. The Bayesian network only presents one building block for integrating culture as a computational parameter in an agent system. Cultural influences manifest themselves on different levels of behavior generation and interpretation and thus penetrate many processing modules in a system that takes these influences into account. In [21], details of the complete system architecture are given.

5 Conclusion

In this paper, we have argued that multimodal corpus analysis can serve as a valuable method of ensuring cultural usability in systems that make use of embodied interface agents. To exemplify this method, we gave details of a study looking into culture specific nonverbal behavior in face-to-face interactions. This study was done in Germany and Japan and allows extracting statistical information about behavior routines in three prototypical scenarios. This information was then utilized to set up a Bayesian network in order to model the causal relation between a culture's position on Hofstede's dimensions and correlating nonverbal behavior.

It remains to be shown, how users from different cultures perceive the agents' behavior that is based on this network model. A first user study in Germany was very promising [21] in this respect. But research in cultural aspects of interactions with embodied agents has just recently come to the focus of attention, and thus there exist more questions than answers concerning the cultural usability of such systems. For instance, neither the importance of an agent's appearance nor of its behavior has been investigated in principle so far. There are a number of agent systems that make use of non-human characters that nevertheless exhibit human-like behavior (e.g. [16]). Others have investigated the consequences of a mismatch between the appearance of a character and its verbal and nonverbal behavior [9]. With the system described in this paper we could easily realize an agent that exhibits a severe mismatch in nonverbal behavior traits like German spatial extent, Japanese speed, and Italian proxemics. It is unclear if this would be perceived as really odd, or as the agent's individual culture, or just as a badly designed animation.

With the advent of massive 3D multiuser environments like Second Life or World of Warcraft, this question becomes relevant beyond the area of embodied interfaces. In such environments, users interact in the form of avatars, 3D virtual agents that are tightly controlled by the user. Users can create their own animations, which must not adhere to any cultural heuristics. Thus, the question arises if culture does not matter in such an environment or how culture manifests itself. An answer to this question might also shed some light on how to ensure cultural usability for embodied agents.

Acknowledgements

The work described in this paper was partially supported by the German Research Foundation (DFG) with research grant RE2619/2-1, the Japan Society for the Promotion of Science (JSPS) with a grant-in-aid for scientific research (C) (19500104), and by the European Community (EC) in the eCIRCUS project IST-4-027656-STP.

References

1. Peter, E.: Bull. Posture and Gesture. Pergamon Press, Oxford (1987)
2. Caridakis, G., Raouzaiou, A., Bevacqua, E., Mancini, M., Karpouzis, K., Malatesta, L., Pelachaud, C.: Virtual agent multimodal mimicry of humans. Language Resources and Evaluation 41, 367–388 (2007)
3. Cassell, J., Sullivan, J., Prevost, S., Churchill, E.: Embodied conversational agents. MIT Press, Cambridge (2000)
4. Efron, D.: Gesture, Race and Culture. Mouton and Co. (1972)
5. Gallaher, P.E.: Individual Differences in Nonverbal Behavior: Dimensions of Style. Journal of Personality and Social Psychology 63(1), 133–145 (1992)
6. Hall, E.T.: The Hidden Dimension. Doubleday (1966)
7. Hofstede, G.: Cultures Consequences: Comparing Values, Behaviors, Institutions, and Organizations Across Nations. Sage Publications, Thousand Oaks (2001)
8. Gert, J., Hofstede, P.B.: Pedersen, and Geert Hofstede. Exploring Culture – Exercises, Stories and Synthetic Cultures. Intercultural Press (2002)
9. Iacobelli, F., Cassell, J.: Ethnic identity and engagement in embodied conversational agents. In: Pelachaud, C., Martin, J.-C., André, E., Chollet, G., Karpouzis, K., Pelé, D. (eds.) IVA 2007. LNCS, vol. 4722, pp. 57–63. Springer, Heidelberg (2007)
10. Jan, D., Herrera, D., Martinovski, B., Novick, D., Traum, D.R.: A Computational Model of Culture-Specific Conversational Behavior. In: Pelachaud, C., Martin, J.-C., André, E., Chollet, G., Karpouzis, K., Pelé, D. (eds.) IVA 2007. LNCS, vol. 4722, pp. 45–56. Springer, Heidelberg (2007)
11. Lewis Johnson, W., Choi, S., Marsella, S., Mote, N., Narayanan, S., Vilhjálmsson, H.: Tactical Language Training System: Supporting the Rapid Acquisition of Foreign Language and Cultural Skills. In: Proc. of InSTIL/ICALL — NLP and Speech Technologies in Advanced Language Learning Systems (2004)
12. Kipp, M., Neff, M., Kipp, K.H., Albrecht, I.: Towards Natural Gesture Synthesis: Evaluating gesture units in a data-driven approach to gesture synthesis. In: Pelachaud, C., Martin, J.-C., André, E., Chollet, G., Karpouzis, K., Pelé, D., et al. (eds.) IVA 2007. LNCS, vol. 4722, pp. 15–28. Springer, Heidelberg (2007)
13. Lee, J., Marsella, S.: Nonverbal Behavior Generator for Embodied Conversational Agents. In: Gratch, J., Young, M., Aylett, R.S., Ballin, D., Olivier, P., et al. (eds.) IVA 2006. LNCS, vol. 4133, pp. 243–255. Springer, Heidelberg (2006)
14. McCrae, R.R., Allik, J. (eds.): The Five-Factor Model of Personality Across Cultures. Kluwer Academic Publishers, Dordrecht (2002)
15. McNeill, D.: Hand and Mind — What Gestures Reveal about Thought. The University of Chicago Press, Chicago (1992)
16. Nakano, Y.I., Reinstein, G., Stocky, T., Cassell, J.: Towards a Model of Face-to-face Grounding. In: Proceedings of the Association for Computational Linguistics (2003)
17. Payr, S., Trappl, R. (eds.): Agent Culture: Human-Agent Interaction in a Multicultural World. Lawrence Erlbaum Associates, London (2004)
18. Rehm, M.: Non-symbolic gestural interaction for Ambient Intelligence. In: Aghajan, H., Delgado, R.L.-C., Augusto, J.C. (eds.) Human-Centric Interfaces for Ambient Intelligence, Elsevier, Amsterdam (in press)

19. Rehm, M., André, E.: From Annotated Multimodal Corpora to Simulated Human-Like Behaviors. In: Wachsmuth, I., Knoblich, G. (eds.) ZiF Research Group International Workshop. LNCS, vol. 4930, pp. 1–17. Springer, Heidelberg (2008)
20. Rehm, M., André, E.: More Than Just a Friendly Phrase: Multimodal Aspects of Polite Behavior in Agents. In: Nishida, T. (ed.) Conversational Informatics, pp. 69–84. Wiley, Chichester (2007)
21. Rehm, M., Nakano, Y., André, E., Nishida, T.: Culturespecific first meeting encounters between virtual agents. In: Prendinger, H., Lester, J.C., Ishizuka, M. (eds.) IVA 2008. LNCS, vol. 5208, pp. 223–236. Springer, Heidelberg (2008)
22. Rehm, M., André, E., Nakano, Y., Nishida, T., Bee, N., Endrass, B., Huang, H.-H., Wissner, M.: The CUBE-G approach — Coaching culture-specific nonverbal behavior by virtual agents. In: Mayer, I., Mastik, H. (eds.) Proceedings of ISAGA 2007 (2007)
23. Ruttkay, Z.: Cultural Dialects of Real and Synthetic Facial Expressions. In: Proceedings of the IUI workshop on Enculturating Coversational Interfaces (2008)
24. Schwartz, S.H., Sagiv, L.: Identifying culture-specifics in the content and structure of values. Journal of Cross-Cultural Psychology 26(1), 92–116 (1995)
25. Warren, R., Diller, D.E., Leung, A., Ferguson, W., Sutton, J.L.: Simulating scenarios for research on culture and cognition using a commercial role-play game. In: Kuhl, M.E., Steiger, N.M., Armstrong, F.B., Joines, J.A. (eds.) Proceedings of the 2005 Winter Simulation Conference (2005)

Support Method for Improving the Ability
of People with Cerebral Palsy
to Efficiently Point a Mouse at Objects on a GUI Screen

Hiromi Nishiguchi

Tokai University,
2-3-23, Takanawa, Shinagawa, Tokyo, Japan
NAH00632@nifty.com

Abstract. Many people with cerebral palsy work in social welfare companies as data entry operators etc. Because of spastic reactions and involuntary motion, they find it difficult to use their upper limbs for movement and positioning tasks such as pointing a mouse at an object on a GUI screen. It would be of great benefit to secure the movement distance and the target size which are appropriate for people with cerebral palsy on a GUI screen, so they can perform pointing device operation effectively. However, it is not possible to increase the screen size beyond a certain limit. Therefore, ideal conditions may not be achieved. In such a situation, an effective environment for positioning tasks can be created by controlling the D/C gain, which is calculated by dividing the movement distance of the pointer by that of the input device. This study investigated the effect of changes in the D/C gain on motion time (MT) for pointing tasks and attempted to determine the D/C gain for minimizing the MT. It was found that the D/C gain for minimizing the MT could be obtained by using an appropriate combination of the target distance and target size. Further, the relation between the D/C gain and the positioning time is found to be linear or second-order curvilinear, depending on the target distance and target size.

Keywords: GUI, Mouse operation, D/C gain, User interface.

1 Introduction

People with cerebral palsy have spasticity or involuntary movements. Therefore, they often experience difficulties with regard to movement and positioning their upper limbs. When one uses a PC that runs on a CUI (Character User Interface) OS, keyboard operations are required for data entry and command inputs. When the distance between the keys is very short, movement time against unit movement distance tends to be longer. A GUI (Graphical User Interface) OS has been developed that lets the user run application software and choose a function on the menu by a positioning operation using a mouse. A positioning operation is defined as the operation in which one moves the pointer on the desktop, positions the pointer on a shortcut icon or a menu icon, and clicks the icon.

Many people with cerebral palsy work in social welfare companies performing data entry tasks and other operations. It would be of great benefit to secure a movement

M. Kurosu (Ed.): Human Centered Design, HCII 2009, LNCS 5619, pp. 531–537, 2009.
© Springer-Verlag Berlin Heidelberg 2009

distance and target size that are appropriate for them so that they can perform positioning operations effectively. This study investigated the effect of changes in the D/C gain on motion time (MT) during pointing tasks and attempted to determine the optimum D/C gain for minimizing the MT. Furthermore, the influence of the D/C gain on the MT is discussed in order to determine the best approach to minimize the MT.

2   D/C Gain

MacKenzie I.E. pointed out that an effective environment for positioning tasks can be created by controlling the D/C gain [1]. The D/C gain is calculated by dividing the movement distance of the pointer by that of the input device shown in Formula (1).

D/C gain = movement distance of the pointer/movement of the input device···

While operating a pointing device, the distance between targets or the target size can be changed by adjusting the D/C gain. Selecting a target and clicking on it is made up of the pointer operation and positioning the pointer on the target icon. Furthermore, the pointer movement time and the positioning time on the target icon will change relative to the adjustment of the D/C gain.

3   Method

3.1   Experiment Task and Equipments

In this study, the pointer operation task was created by using Microsoft Visual Basic Version 5.0. Subjects were required to operate the mouse in order to move the pointer onto the target icon and click the left button (cf. Fig. 1). The experiment task was shown on a 15-inch TFT monitor with a pixel resolution of 1024×768. An optical mouse with two buttons was used as a pointing device.

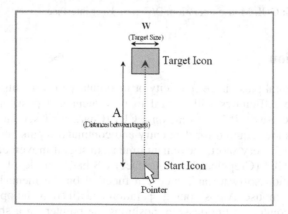

Fig. 1. The diagrammatical view of the experiment task on the desktop

3.2 Participants

Five people with cerebral palsy participated in this experiment. They were classified to the second grade in obstacle class and were employed in a social welfare company in Tokyo. All the participants operated PCs as part of their jobs and input/edited printed matter. Consequently, it can be said that the subjects were familiar with PC operation.

Before starting the experiment, the purpose of this experiment was explained by the caseworker, and only people who approved participated. Furthermore, the participants performed a number of rehearsals to familiarize themselves with the experiment task before starting the experiment.

3.3 Experiment Conditions

The movement direction of the pointer was set by only one condition of the upper direction for the following reason: the MT was not affected by the directionality of the pointer movement in the past experiment [2]. Additionally, when we use application software such as word processors, menu icons are usually displayed in the upper part of the screen.

The conditions of the target distance (A), the target size (W), and the D/C gain are described below. One pixel is equivalent to 0.3 mm on the 15-inch TFT monitor.

1) Target distance: 30 mm, 90 mm, and 150 mm (3 conditions).
2) Target size: 6 mm, 9 mm, 12 mm, and 15 mm (4 conditions).
3) D/C gain: 1.3, 2.4, 3.9, 5.1, 7.5, 9.5, and 12.1 (7 conditions).

The subjects performed every experiment task three times in each condition.

3.4 Data Analysis

The MT was measured using the "GetTickCount" function with a precision of 10 ms. The x and y coordinates of the pointer were measured by determining the pixel unit every 10 ms. The coordinate origin was set in the top left corner of the screen.

The speed wave pattern of the pointer on the GUI screen was classified into the movement aspect and the positioning aspect, and the time taken for each movement was calculated.

4 Results and Discussion

4.1 Relationship between the D/C Gain and the MT

The relationship between the D/C gain and the MT of the cerebral palsy group under the three target distance conditions (A) are shown in Figs. 2, 3, and 4. It can be seen that the MT tended to be longer when the D/C gain was low (D/C gain = 1.3) or high (D/C gain = 12.1). When the D/C gain was between the two abovementioned values, the MT was shorter.

The relationship between the D/C gain and the MT was a second-order curvilinear relationship under the following conditions (cf. Table 1):

1) W = 6 mm when A = 30 mm or A = 90 mm.
2) All W when A = 150 mm.

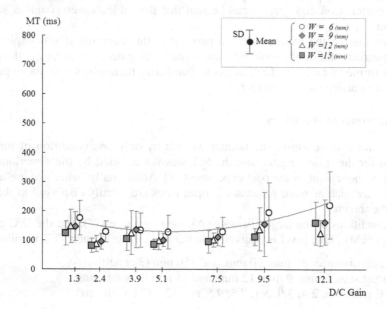

Fig. 2. Relation between D/C gain and MT (A=30mm)

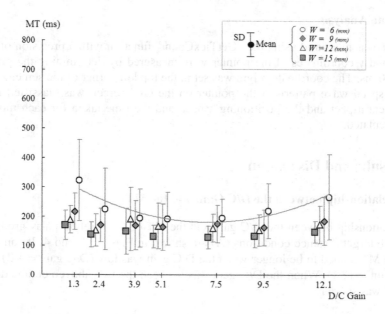

Fig. 3. Relation between D/C gain and MT (A=90mm)

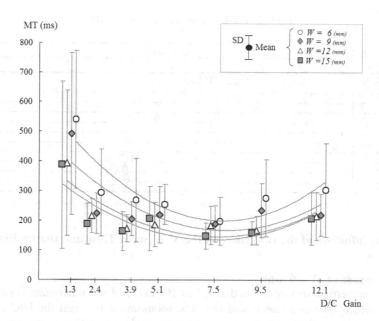

Fig. 4. Relation between D/C gain and MT (A=150mm)

Table 1. The regression line of MT of cerebral palsy group

A(mm)	W(mm)	Equation of Regression	Coefficient of Correlation	Significance
30	6	$y = 2.13\,x^2 - 22.07\,x + 186.15$	0.434	p<0.05
90	6	$y = 3.58\,x^2 - 49.31\,x + 350.19$	0.363	p<0.10
150	6	$y = 6.64\,x^2 - 101.25\,x + 584.82$	0.534	p<0.01
	9	$y = 5.17\,x^2 - 82.19\,x + 492.93$	0.511	p<0.01
	12	$y = 4.50\,x^2 - 69.94\,x + 416.77$	0.510	p<0.01
	15	$y = 4.50\,x^2 - 69.96\,x + 406.14$	0.458	p<0.05

(y : MT, x: D/C gain, *unit*: ms)

4.2   The Optimum D/C Gain for Minimizing the MT

The optimum D/C gain for minimizing the MT of the cerebral palsy group is shown in Table 2. The relationship between the D/C gain and the MT was found to be a second-order curvilinear relationship under the condition of W = 6 mm when A = 30 mm and A = 90 mm.

The optimum D/C gain for minimizing the MT was estimated to be 5.2 when A = 30 mm and 6.9 when A = 90 mm. On the other hand, second-order curvilinear relationships were observed under conditions of all W when A = 150 mm. The optimum D/C gains for minimizing the MT were estimated at 7.6 when W = 9 mm and 7.8 when W = 12 mm or 15 mm.

From these results, it was concluded that the MT could be shortened by decreasing the D/C gain when the pointer moves a short distance on the GUI screen.

Table 2. The D/C gain for minimizing MT

A (mm)	W (mm)	D/C gain 1.3	2.4	3.9	5.1	7.5	9.5	12.1
30	6				★(5.2)			
	9							
	12							
	15							
90	6					★(6.9)		
	9							
	12							
	15							
150	6					★(7.6)		
	9					★(7.9)		
	12					★(7.8)		
	15					★(7.8)		

★ The D/C gain to minimize MT.　　▨ The condition in which the D/C gain to minimize MT cannot be found.

4.3 The Influence of the D/C Gain on the Movement Time and the Positioning Time

(1) In the case of W = 6 mm

The unit movement distance was defined as 10 mm, and the movement time against unit movement distance was calculated. The relationship between the D/C gain and movement time against unit movement distance is shown in Fig. 6. As a result, the relationship between the D/C gain and the movement time was found to be linear. In the case of A = 30 mm, movement time against unit movement distance was prolonged when compared to the conditions of A = 90 mm or 150 mm.

On the other hand, the relationship between the D/C gain and the positioning time was found to be a second-order curvilinear relationship (cf. Fig. 7). As a result, the D/C gain affected the positioning time and therefore had an effect on minimizing the MT.

(2) In the case of W = 9 mm

The relationship between the D/C gain and movement time against unit movement distance is shown in Fig. 8. The relationship between the D/C gain and the movement

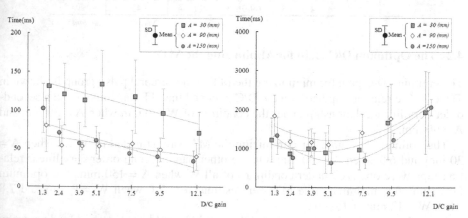

Fig. 6. Relation between D/C gain and movement time (Cerebral palsy group, W=6mm)

Fig. 7. Relation between D/C gain and positioning time (Cerebral palsy group, W=6mm)

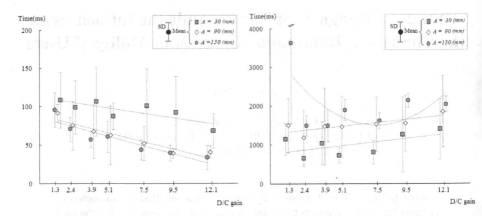

Fig. 8. Relation between D/C gain and movement time (Cerebral palsy group, W=9mm)

Fig. 9. Relation between D/C gain and positioning time (Cerebral palsy group, W=9mm)

time was found to be linear, just as in the case of W = 6 mm. In the case of A = 30 mm, movement time against unit movement distance was longer than it was in the other conditions.

Under the condition of A = 150 mm, the relationship between the D/C gain and the positioning time was a second-order curvilinear relationship, and it was observed that the positioning time was minimized by adjusting the D/C gain (cf. Fig 9). However, this relationship was linear under the conditions of A = 90 mm or A = 30 mm.

5 Conclusion

The D/C gain on which people with cerebral palsy can effectively operate a pointer on a GUI screen is usually decided by trial and error. However, the optimum D/C gain for minimizing the MT can be estimated by referring to Table 2 when the average pointer movement distance and the target size are provided. This method for minimizing the MT effectively increases the productivity of people with cerebral palsy when performing positioning operations, and it will be the effective document of the job guidance for the caseworker. Furthermore, the D/C ratio for minimizing the MT can be estimated by adding data using this experiment method when pointer operating conditions are changed by enlarging or decreasing the screen size.

References

1. MacKenzie, I.S.: Input devices and interaction techniques for advanced computing. In: Barfield, W., Furness, T.A. (eds.) Virtual Environment and Advanced Interface Design, pp. 437–470. Oxford University Press, Oxford (1995)
2. Nishiguchi, H., Saito, M.: The characteristics of mouse pointer operation in a positioning task of cerebral palsy patients on GUI screen. The Japanese Journal of Ergonomics 43(3), 124–131 (2007)

A Study of Design That Understands the Influences on the Changes of Information Processing Ability of Users

Ji Hyun Park

University of Texas at Austin,
1 University Station, D7000 Austin, TX 78712-0390
jhpark@ischool.utexas.edu

Abstract. The main goal of this study is to research new design approaches for creating interactive products by designers that take into consideration the positive long-term influence a product has on users. In recent years, users are more and more interested in products that can have a good influence on them, for example, how much of a positive change can occur to their emotional or physical health through using a product. With the further advancement of technology, a wide range of effects frequently occur between new devices and users. These effects can be considered a new experience by users. In the future, users will take into consideration the positive influences a product has on them, and the long-term experience of using a technology device. Because of these compelling reasons, research is necessary to study the factors and characteristics of influences that products have on users through objective and utility methods.

Keywords: Design factors, Users, Interaction, Information processing abilities.

1 Introduction

In recent years, many products have been developed that provide close interactions between a user and the product, resulting in both positive and negative interactions. For instance, from mobile devices to transportation-related devices, there is a flood of products that take into consideration essential design factors such as aesthetics, utility and user satisfaction. Moreover, users are more and more interested in products that can have a good influence on them, for example, how much a positive change can occur to their emotional or physical health through using a product. At the same time, this phenomenon greatly broadens the role of service-oriented fields within the design process. With the further advancement of technology, a wide range of affects frequently occur between new devices and users that can be considered by users a new experience. This trend leads to primarily focusing design on satisfying the user's needs. Many designers, however, take into consideration how to best fulfill a user's immediate needs without any additional guidelines that consider the long-term effects a product might have on a user. For that reason, designers usually borrow methods of trend research from marketing fields. That approach is effective for understanding the motivation of users for purchase products; however, it represents only a fragment of the overall situation when users select a product. In the future, users will take

M. Kurosu (Ed.): Human Centered Design, HCII 2009, LNCS 5619, pp. 538–547, 2009.

into consideration the positive influences a product has on them, such as, and the long-term experience of using a technology device. Because of these compelling reasons, research is necessary to study the factors and characteristics of positive influences products have on users through objective and utility methods.

The aim of this research is to analyze the relationship and influencing factors of a product to the user, and the research incorporates a more objective and utilitarian method than had been pursued previously. This study identified the necessity for creating a new design system that takes into consideration a product's long-term influence on a user following the use of a product.

Therefore, the research focuses on exploring new design factors and a multidisciplinary approach regarding the long-term positive effect to the users in regards to the interaction between information and humans. Through a new design approach, designers can develop interactive products that meet their intentions for providing positive long-term influences on users. Consequently, this study suggests a design approach method with a different aspect compared with previous design perspectives and a system for enhancing long-term positive effects on users.

2 Definitions and Literature Review

First, theoretical research was conducted to define the terms 'product,' 'users,' and 'interaction.' Furthermore, the relevant research was reviewed regarding the influence of interaction between humans and products.

2.1 Definitions

Users. The term users is defined as the class of people that use products in everyday life. Users experience new meanings and values by interacting with products in their daily life. In addition, users' behaviors are continually affected by using products, such as a computer and mobile phone, as required in their lives.

Products. The designing of products means that establishing a concept for all kinds of human activities including artifacts, service, information, systems and environment. That means the object of designing focuses on tangible products, for example, graphic design, product design, interior design and so on, as well as human behavior, experience, culture, and services.

For the last few decades, there are many studies regarding conceptual models that are applied to the user's recognition of product's attribute and the process of creating meaning and value by marketing experts and psychologists. According to the prior studies, consumers select their products depending on their expectations.

Interaction. Human to product communication [1]; in the context of communication between a human and an artifact, interactivity refers to the artifact's interactive behavior as experienced by the human user.

This is different from other aspects of the artifact such as its visual appearance, its internal working, and the meaning of the signs it might mediate. For example, the interactivity of an iPod is not its physical shape and color, its ability to play music, or its storage capacity—it is the behavior of its user interface as experienced by its user. This includes the way you move your finger on its input wheel, the way this allows you to select a tune in the playlist, and the way you control the volume.

An artifact's interactivity is best perceived through use. A bystander can imagine how it would be like to use an artifact by watching others use it, but it is only through actual use that its interactivity is fully experienced and "felt". This is due to the kinesthetic nature of the interactive experience. It is similar to the difference between watching someone drive a car and actually driving it.

Long-term Influences. The term long-term influences means the effects on users after the entire process of interaction with a product. This research mainly focuses on human's information processing abilities.

2.2 Literature Review

The Influences of Interaction between Users and Products. In the design discipline, it is important that designers understand the differences in the patterns of everyday life in different cultures for the success of providing people's culture-specific needs, in particular, culturally unfamiliar areas of potential business markets.

By understanding the user activity patterns in people's cultural contexts, designers are able to provide their offerings – products and services – to support specific users' behaviors in the targeted business market area. For that reason, designers' research trends shift from "products" to "activities." It is the main reason that designers currently employ anthropologic qualitative research methods [2]. As Savolainen points out, even though the ordinary processes of information utilization are ubiquitous, there is so far very little research knowledge on exactly how information is used [3].

The Relevant Studies in Various Disciplines. According to the relevant studies from other disciplines, researchers concentrate on studying factors of influences on humans while they use products. Although designers assert that they create designs based on a Human-Centered Design (HCD) philosophy, in fact, it is quite recent that designers started to focus on Human behavior more than a product itself, specifically functionality, color and shape. As much as improvement of technology is frequently shown, people also need to learn new functions and usage of products. It changes humans' behaviors in their everyday lives.

In regards to Pickering's definition of multidisciplinary research, the ideal treatment of any specific or technological development is not one that analyzes it from a single, unified perspective, but one that successfully integrates diverse component perspectives [4]. Therefore, the goal is not "unification" within a single perspective but the "integration" of several different perspectives. Multidisciplinary research probably requires more collaborative research than is the current norm in science and technology studies.

Table 1. Lists of the relevant studies, which focus on effects on humans in various disciplines

Titles	Disciplines
1 Difference of human perceptual-motor ability according to e-Sports expertise [5].	Physical science
2 Junior high school students' Multiplayer Online Role Playing Game (MMORPG) use and its effects on real interpersonal relationship cognition and communication competence [6].	Education
3 (The) Effects of ubiquitous attributes of mobile contents on perceived interactivity and behavioral outcome [7].	Business management
4 (The)Regulation process of adolescent on-line game over user's behavior [8].	Information studies
5 A study of the pattern of the value of Heart rate, Skin Conductance, Respiration depending on experience and time of On-line game [9].	Electronic engineering
6 The effects of electronic games on adolescents' self-esteem and competitive value: degree of acquired-gratification would affect adolescents' competitive value on real world competition [10].	Literature
7 The effects of computer game on Children's mathematical ability and spatial skills [11].	Domestic science

3 Factors Affecting Information Processing Abilities on Users

Two case studies were undertaken to determine the effects of human-product interactions. In terms of a researching the basic components and characteristics of humans' information processing abilities, the relevance between human interaction and the influence of information processes is provided. On the next step, through two case studies that examined effects on users after using communication devices and playing online video games, I observed situations that occurred between users and information and attempted to identify the design characteristics that needed to be considered in the future.

3.1 Two Case Studies: 91 Results of the Effects

Above all, this chapter sought a variety of influences on users. And then, through organizing the results, it found a possibility of reflecting the design factors that affect information processing abilities of users on the design. It incorporated a more objective and utilitarian method than had been pursued previously. For this research, mobile phone use and computer games were selected as products that represent the most frequently used devices that affect users.

In this process, this research sought 91 results of effects on users after computer game and mobile phone use as case studies (see, Appendix.1). Although there is a limitation of organizing results in a specific system because the results of influences vary across a wide range, the 31 influences found a common characteristic, which is a change in information processing abilities. There is more specific research regarding the changes in information processing ability indexed so as to analyze the relationship and influencing factors of a product to the user.

In order to find associations between design elements and the results of influences above, the DEMATEL method[1][12] is conducted with ten pairs of specific cases based on the thirty-one results. In the results, it is clear that user's information processing ability is the most common characteristic. Through the procedure of classifying attributes of influence factors and discussion of the range of effects on users, it becomes a background research for creating a diagram regarding user's influences by interacting with products. As the meaning of products and ranges are expanding, the characteristics of influences show that direct effects on users are increasing, in particular, user's information processing capability. Furthermore, the users follow the trends of macro changes, such as environmental and cultural changes.

Table 2. The concept applications for mobile phone use as pertain to the elements of memory capability

Elements of memory capability	Redefining with Design Perspectives	Concept applications: mobile phone use
Similarity →	When the user makes an association based on his/her experience	→ The designer naturally induces the user's perception causing a new experience that users can use to remember stored data on the mobile phone.
Paired associate learning →	When the user experiences an association of one specific meaning with another.	→ The associated image and interface are designed in a simple context to allow for easy recall of data stored on the mobile phone.
Recognition →	When the user considers cognitive and emotional familiarities that he/she has while experiencing a product.	→ Users' memories are induced through multiple-sensory interactions.
Recall →	When the user, through past experiences, can remember what they perceived and recognized during the interaction	→ The users recall information received during previous product usage.

---

[1] The DEMATEL (Decision Making Trial and Evaluation Laboratory) method is an effective method that helps in gathering group knowledge for forming a structural model, as well as in visualizing the relationship of sub-systems through a diagram (Gabus & Fontela, 1972, 1973).

3.2 Applications: Redefining the Long-Term Memory Principles as Design Factors

At first, this section applied the four characteristics of human memory capabilities [13] to concept applications for designing mobile phone interactions, including similarities, paired associate learning, recognition and recall. Secondly, the four long-term memory characteristics [13] are redefined as design factors in order to consider effects of these on humans' memory capabilities during these interactions, as seen in Table 2 and 3.

Table 3. Applications of mobile phone interface design to long-term memory

Long-term memory		Redefining with Design Perspectives		Concept applications: mobile phone interface
Episodic memory	→	Factual knowledge of a specific moment in time, place, and specific personal experiences.	→	The information contains every meaning of the context of the interaction. The design reflects the time, place, and situation at that moment
Semantic memory	→	Theoretical knowledge independent of a time and place.	→	The information has a special meaning to the user independently. The design concentrates on memorizing the message itself using images and interfaces.
Narrative memory	→	Story-telling.	→	The information has a story related with user's experience. The design induces the user's experience when he/she recalls his/her memories with story-telling in order to find information in the mobile phone.
Sequential memory	→	Sequential memory dependent of time and place, context.	→	The information is stored by the user. The design follows the sequential moment of user's interaction.

4 Three Visual Diagrams

This study detailed the creation of a design system, which takes into consideration the change of the users' information processing abilities. This study visualized diagrams, which describe the range of information processing, factors in behavioral processes, and the steps in the user's processing of information. Through the application of these three visualized diagrams, all components can be considered as synthesized principle elements for a new design system, which takes into consideration the influence of the users' information processing abilities for the purpose of achieving positive long-term effects.

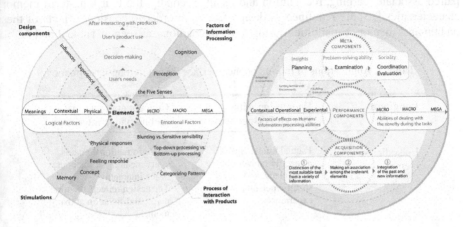

Fig. 1. The figure shows design factors that consider the changes in a user's ability to process information. Diagram A. demonstrates factors in behavioral processes (L). and Diagram B. shows the range of information processing (R).

4.1   Diagram A: Sequential Factors in User's Behavioral Processes

There are four steps in the entire process of the users' interaction with the product, which are the users' needs, decision-making (purchasing a product), interaction and influence. During each stage of the users' behavior, it is expected that the information processing elements will work distinctively.

4.2   Diagram B: The Range of User's Information Processing (Factor Analysis)

The information processing procedure that takes place while the user interacts with the product is the most crucial part. This procedure affects the users' information processing ability following product usage. This processing ability is divided into three stages based on the theory suggested by Gardner in 1993: *Meta components*, *performance components* and *acquisition components*. These three steps of human information-processing components provide criteria on how designers should look at user attributes when they are interacting with these products.

4.3   Diagram C: The Ranges of Influence Factors of User's Information
Processing Abilities

Based on the results of the factor analysis in chapter 3, Fig. 2 synthesizes the ranges of influence factors of the users' information processing abilities. At first, this research created two groups that included physical ability and mental ability of users. These groups had distinctive characteristics.

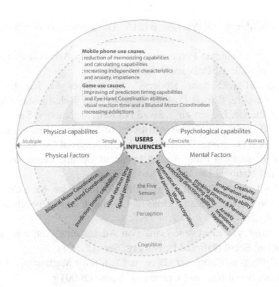

Fig. 2. Diagram C. shows the ranges of effects on users' information processing abilities

5 Conclusions

This study identified the necessity for creating a new design system that takes into consideration a product's long-term influence on a user following the use of that product. This research focuses on exploring new design factors, using a multidisciplinary approach regarding the long-term positive effects to the users regarding the interaction between information and humans. Through a new design approach, designers can develop interactive products that meet their intentions for providing positive long-term influences on users. This study suggests a design approach method with a different aspect compared with previous design perspectives, and a system for enhancing long-term positive effects on users.

The results of this study suggest that designers should take into consideration the specific positive influences that a product has on its users when they create new design concepts. With this, designers are urged to give top priority to considering the long-term physical and mental effects a product has on them. Through this study, I concluded that there needs to be additional design aspects to consider beyond usability. In the future, I believe that designers should take into consideration the human elements and the long-term effects of interaction with not only products, but also services.

References

1. Wikipedia, http://www.wikipedia.org
2. Kumar, V.: User Insights Tool: A Sharable Database for Global Research. Institute of Design, IIT, Chicago (2004)
3. Savolainen, R.: Information Behavior and Information Practice: Reviewing the "Umbrella Concepts" of Information Seeking Studies. Library Quarterly 77(2), 109–132 (2007)

4. Pickering, A.: The Mangle of Practice, p. 215. The University of Chicago Press, Chicago (1995)
5. Kim, H.: Difference of Human Perceptual-Motor Ability According to e-Sports Expertise: Dissertation: Seoul National University, Seoul (2005)
6. Oh, H.I.: Junior High School Students' Multiplayer Online Role Playing Game (MMORPG) Use and its Effects on Real Interpersonal Relationship Cognition and Communication Competence: Dissertation, Seoul National University, Seoul (2005)
7. Lee, S.H.: The Effects of Ubiquitous Attributes of Mobile Contents on Perceived Interactivity and Behavioral Outcome: Dissertation: Seoul National University, Seoul (2006)
8. Park, S.M.: The Regulation Process of Adolescent On-line Game over User's Behavior: Dissertation, Seoul National University, Seoul (2005)
9. Kim, Y.K..: A Study of the Pattern of the Value of Heart Rate, Skin Conductance, Respiration Depending on Experience and Time of On-line Game: Dissertation, Seoul National University, Seoul (2004)
10. Kim, O.T.: The Effects of Electronic Games on Adolescents' Self-esteem and Competitive Value: Dissertation: Seoul National University, Seoul (2001)
11. Lim, S.M.: The Effects of Computer Game on Children's Mathematical Ability and Spatial skills: Dissertation, Seoul National University, Seoul (2000)
12. Cho, Y.H.: A Study for Structuring Design Factors - Comparison of DEMATEL theory and KJ theory. Korea Society of Design Science 5, 49–57 (2002)
13. Kim, J.O.: The Structure of Minds (Lecture notes, unpublished). Seoul National University, Seoul (2007)

Appendix 1: The 91 Effects on Game and Mobile Phone Users

* s/o: 's' indicates a subjective variable characterized by vagueness and imprecision.
'o' represents an objective variable which can be estimated with exact numerical values.

no.	Lists of the Effects	*	no.	Lists of the Effects	*
1	Communication apprehension	s	47	Importance of user's skills and opportunities	s
2	Openness to the different cultural and social background	s	48	Self-esteem	s
3	Social experience	s	49	Eye-Hand Coordination	o
4	Leadership	s	50	Familiarity with technologies	o
5	Learning social norm, regulations	s	51	Bilateral Motor Coordination	o
6	Goal achieving management	s	52	Spatial perception	s
7	Being honest to other people	s	53	Independent thinking	s
8	Understanding	s	54	Reduction of violent actions in daily life	s
9	Sympathy	s	55	Mitigation of stress	o
10	Empathy	s	56	Motivation for learning	s
11	Limited human network	s	57	Adventure	s
12	Broad range of human network	s	58	Amusement	s
13	Preference of face to face communication	s	59	Relaxation	s
14	Indirect experience	s	60	Tension	s
15	Participation of social activities	s	61	Priority: Game is the most important thing in the world	s
16	Efficiency	s	62	Refreshing feeling, atmosphere	s

17	Adequacy	s	63	Withdrawal symptoms	s
18	Communication capability	o	64	Irregular hours of sleep	o
19	Contact capability	s	65	Possessiveness formation	s
20	Anxiety of directed communication	s	66	Formation of affection with a computer	s
21	Social experience: participation	s	67	Confusion of reality	s
22	Expression capability	s	68	Reduction of social network	s
23	Reflection	s	69	Relationship based on game	s
24	Extent of self-expression	s	70	Violent tendency	s
25	The relationship between stimulations and responses	o	71	Imitation of game actions	s
26	Fast and accurate feedback	o	72	Carpal Tunnel Syndrome	o
27	Interaction frequency	o	73	visual reaction time	o
28	Motivation of learning by the challenge, fantasy, curiosity	s	74	Aggressive attitudes	s
29	Visual distinct technique capability	o	75	Negative effect on communication ability	s
30	Development of detecting direction	o	76	Degradation of academic achievements	s
31	Spatial perception capability	o	77	Negative emotions from a game experience	o
32	Numerical representation ability	o	78	Psychological obsession	o
33	Verbal representation ability	o	79	Feeling uncomfortable, nervous and depression when I stop playing a game	s
34	Balanced association capability	o	80	Feeling comfortable and relief when I am playing a game	s
35	Rapid reflection capability	o	81	Losing interests about hobbies and activities	s
36	Sequentially thinking capability	o	82	Afraid of meeting friend in person	s
37	Focusing on the thinking and planning process	o	83	Declining hours of getting together with family	s
38	Creative thinking	o	84	Pain of fingers	s
39	Reasoning ability	o	85	Cannot remember lyrics	s
40	Problem-solving capability	o	86	Easily forget ID and password on the Internet	s
41	Visual information processing capability	o	87	Forget family's phone numbers	s
42	Logical thinking ability	o	88	Never remember friends' cell phone number	s
43	Cognitive development	o	89	Having difficulties with a basic calculation	s
44	Visual and auditory stimulation-response capability	o	90	Mis-spelled words, easily forget correct spelling	s
45	Sensory dictation capability	o	91	weak in sight	o
46	Competitive mind	s		Total: 91 cases; 31(s) and 60(o)	

Common Understanding of Graphic Image Enhance "Emotional Design"

Hisashi Shima

Lenovo Japan Ltd.,
1623-14 Shimotsuruma Yamato-city Kanagawa Japan
shima@lenovo.com

Abstract. The object of this research is to investigate the empathy of the brand design attribute to development of the product design, software screen and web site design. At early phase of the development "Empathy" is one of the important matters of emotional design. To share of the target verbal and image help to common understanding of the product characteristic, it can assume height efficiency of the development.

Keywords: Emotional design, development procedure, brand design, Tacit dimension, empathy development.

1 Introduction

It is said that a consistent brand image is important in the enterprise of recent years. And, it is also true that the user who receives it is choosing the commodity and service in the brand image. It tends to choose the commodity on an especially emotional side more than before.

However, it is not on the other hand easy to show the emotional one as well as the function and specs. It is decided by a personal idea of management, and in emotional value, this can drop emotional value and press the compromise up to designer.

User sometimes get preconception from appearance image, afterwards, they actual use impression influence to total their conclusion.

After it uses it, I tend to feel friendly for the product. It has a word of friendly appearance. It is also easy with a reformative impression for after it uses it the one of reformative externals. Not only the feature of the product but also sharing emotional value in usual development in the enterprise becomes an important factor in a development process after the fact.

Common to not only in the team but management is effective. For instance, not only understanding the user from using the persona but also the persona can share a side emotional because of the context in the team. However, there is emotional value not transmitted easily only by the persona technique. It is thought that sharing the image is effective as the technique for sharing the emotional value not transmitted easily. In this thesis, the effect of the process of sharing of the image, doing, and sharing of the people other than the profession of the design the image was questioning investigated.

M. Kurosu (Ed.): Human Centered Design, HCII 2009, LNCS 5619, pp. 548–551, 2009.

2   Methods and Procedure

2.1   Composition of Team

As for 1 that participates by the planner of the person who has the backbone of psychology, the specialist of the human engineering, the specialist of product designer, and the graphic designer product and goes through the following process), it is the one that had been decided beforehand.

2.2   Process

1. The word that is some bases is decided. This word is a word of the image to be told corporate.
2. The meaning of the word is understood about the team.
3. Two people or more took charge of each word and images were collected. A different as much as possible specialist is made to unite it.
4. The image thought to be pertinent each one of each word is pasted to the seat of one collection.
5. One seat is made in the team of two people or more. The image mutually explains be appropriate why.
6. A small each team explains all participants, and the correction is put. The seat of the image of the word decided by this first is completed. It can be assumed that this is a set of the image from which the word of the seat of the image is almost associated with the participant.
7. A big correction term eyes were not in an actual correction to management as the explanation.

The meaning of the word is confirmed again more deeply when the time made with one point person and making it from two people or more and for the selection of the image.

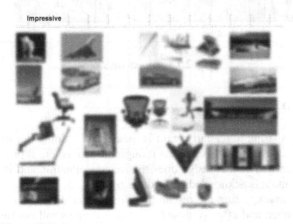

Fig. 1. Example of the images collection

3 Results and Discussion

To make the one questionnaire to the participants of this project.
About time and easiness:

1. When the first images were collected, was it easy?
2. Was the one near my image collected for the collection of images?
3. Can the meaning and the image in the word able to be included and understood the word more deeply than the place?
   Result of collection of image in team:
4. Was result of the whole team and my idea large and was the difference finally large?
5. Were a lot of compromises done to the image that I was considering and the image that made the member of the team?
6. Was deeper understanding able to be done by collecting images by seeing overall?
7. Is it useful for the product making that seems to be, for instance, Inovation in the development of the product in the future?
8. When explaining to others like management etc., is it useful?
   The questionnaire result is actually good acceptance this process and method. (fig 2)

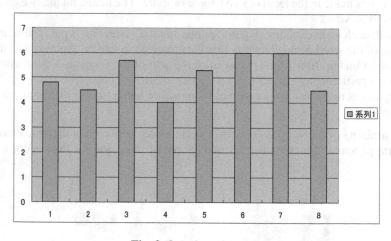

Fig. 2. Questionnaire score

4 Conclusion

It is thought that sharing in an early team is measured by doing the work done by the division of labor as a team in this process though the step where it proposes a concrete idea after the specialist of the design does the image collection and it sublimes to the proposed image and it is acknowledged is done usually.

In future, we are expecting the research of, intervener different number scale, different image number, and effectiveness of 2nd or 3rd times roll out this procedure, so on.

It turned out that this project was actually done at the same time with USA and China and a completed image was common in a lot of points though it did not become the participant of this questionnaire. As for the image, it is possible and has understood there is a possibility of the excess of the country and the culture and telling it though the seat of the image that doesn't depend on the country and the culture is in a word rough.

In future, we are expecting the research of, intervener different number scale, different image number, effectiveness of 2nd or 3rd times roll out, so on.

Acknowledgements. Our Lenovo management team helps this research and especially the Consultant; Tom Hardy, and I appreciate our team member to finish this work.

References

1. Atkinson, K., Wells, C.: Creative Therapies: A psycholodynamic approach within occupational therapy. Nelson Thornes Ltd. (2000)
2. Grice, P.: Studies in the Way of Words. President and Fellows of Harvard College (1989)
3. Merholz, P., Schauer, B., Verba, D., Wilkens, T.: Subject to change. O'Reilly Media, Inc., Sebastopol (2008)
4. Norman, D.: Emotional Design: Why We Love (or Hate) Everyday Things. Basic Books, New York (2005)
5. Norman, D.: The Design of Future Things. Basic Books, New York (2007)
6. Polanyi, M.: The Tacit Demension. Routlege & Kegan Paul Ltd., London (1966)
7. Pruitt, J.S., Adlin, T.: The Persona Lifecycle: Keeping people in Mind Throughout product Design. Elsevier Inc., Amsterdam (2006)
8. Utterback, J.M., Vedin, B.-A., Alvarez, E., Ekman, S.: Design-Inspired Innovation. World Scientific Publishing co. Pte. Ltd, Singapore (2006)
9. Verbal Visual Framework, http://www.verbal-visual.com/home.htm

Older Drivers and New In-Vehicle Technologies: Adaptation and Long-Term Effects

Anabela Simões and Marta Pereira

High Institute for Education and Sciences (ISEC), Alameda das Linhas de Torres,
179, 1750-142, Lisbon, Portugal
{anabela.simoes,msopereira}@isec.universitas.pt

Abstract. The introduction of new technologies into vehicles has been imposing new forms of interaction, being a challenge to drivers but also to HMI research. The multiplicity of on-board systems in the market has been changing the driving task, being the consequences of such interaction a concern especially to older drivers. Several studies have been conducted to report the natural functional declines of older drivers and the way they cope with additional sources of information and additional tasks in specific moments. However, the evolution of these equipments, their frequent presence in the automotive market and also the increased acceptability and familiarization of older drivers with such technologies, compel researchers to consider other aspects of these interactions: from adaptation to the long term effects of using any in-vehicle technologies.

Keywords: In-Vehicle Technologies, Older Drivers, Behavioral Adaptation, Human-Machine Cooperation.

1 Background

One of the main reasons why road safety research has devoted considerable attention to older drivers is their growing representation in the population in most industrialized countries. In fact, the proportion of older drivers is increasing in society and is expected to enlarge even more [1] [2] [3]. The elderly population represents the most growing segment of the overall population. In most of OECD countries, it is expected that by the year 2030 one person out of four will be 65 years old and over [3]. This is explained by the maturation of the "baby boom" generation, combined with a greater longevity and the decline of birth rates. Furthermore, the number of older drivers is also increasing since the twentieth century 60s and much more young men or women got their driving license.

Driving has nowadays an important role in society and is essential to determine the quality of life of older people. Driving enables easy access to activities and services, to fulfill social needs and because of that it is a good indicator of mobility, independence, good health, quality of life and well-being. The car represents to the elderly the possibility of maintaining their autonomy and self-esteem being also a symbol and a way of freedom, independence and self-reliance [4] [5]. In contrast, the fact of not being able to drive is considered a limitation in life control, mobility and independence. Stopping driving seems to have consequences not only on elderly daily life but

M. Kurosu (Ed.): Human Centered Design, HCII 2009, LNCS 5619, pp. 552–561, 2009.
© Springer-Verlag Berlin Heidelberg 2009

also a negative effect at a psychological level once it is reported to increase the feeling of isolation and loneliness [6] and depressive symptoms (as cited by Harper & Schatz in [5]). Besides the social and health disadvantages, older drivers report strong feelings about the importance of driving and extremely negative opinions about the loss of driving [7]. Ceasing driving means that their mobility is dependent of others or on public transportation. It is not surprising that many older drivers rely on their cars for most of their transportation needs, being strongly interested in keeping their own cars and licenses for as long as possible [3]. Additionally, they are sometimes reluctant to use public transportation because it doesn't fulfill all their travel needs; it is often considered as unreliable and perceived as providing inadequate personal security [5].

2 Older Drivers' Behavior and Safety Implications

Older drivers are generally considered safe and cautious drivers and it is frequently claimed that they self-regulate their driving behavior [4]. The age-related perceptual and cognitive difficulties of older people lead them to change their driving habits and often to give up driving, which affects their own mobility. Despite the age-related declines, older people can drive safely, compensating their failures by using their remaining abilities in a deeper way. In a healthy ageing process, age-related decrements don't have significant impact on task performance if previous experience can be used. Indeed, compensation based on the task knowledge and experience is the reason why some performance decrements in laboratory tests are not replicated in daily task performance. Anyway, older drivers avoid driving conditions that impose higher demands, such as driving at night, with poor weather conditions or during rush hours.

Despite an increasing individual diversity with ageing, it seems that, for the same task performance, older people develop the same compensatory behaviors for functional losses. This results in particular common patterns to compensate the age-related impaired perceptual-motor functioning. These characteristics, which act as compensation to their declines, can include the insight to one's own limitations, driving experience and some compensatory behavior [8]. Compensatory behaviors reduce the stress and anxiety felt by older drivers in some driving situations, as well as the risk of driving in these situations [9] [3]. As pre-conditions for the use of experience, a more stable and user-friendly road environment is required, which is not the case as the roadway system is more and more complex. Moreover, the driving-related interactions with new in-vehicle technologies increase the task complexity, reducing the possibility of using previous experience.

The great amount of driving experience that older drivers possess is also an important factor. The traffic experience acquired may give them the ability to anticipate some problematic situations and avoid the risk. Charlton and colleagues [4] present several examples of self-regulation behaviors: driving more slowly, travelling shorter distances, making fewer trips, avoiding driving under difficult conditions like driving at night or with a large amount of traffic, selecting for longer time gaps when turning or entering a road, and avoiding performing simultaneous activities while driving. Generally, older drivers tend to make trips shorter and closer to home (Rosenbloom as cited in [5]), avoid to drive in freeways, plan to use only routes where protected left turns can be made (for those who drive in the right hand-side of the road) and also drive with a co-pilot [10].

The major concern about older drivers' safety is not so much the potentially hazardous situations that they can cause but the risk that they are exposed to. In absolute numbers, the older drivers' crashes have been reported as lower than other age groups, especially the young drivers [11]. However, the overall number of older driver crashes may under-estimate the magnitude of the older driver problem as their total distance travelled tend to be lower when compared to younger drivers. This may have an influence on their crash rate because drivers that travel few kilometers have increased crash rates per kilometer compared to those driving more [4] [5]. In fact, when the distance travelled is taken into account, the fatality rate for the 75 years and older is more than five times higher when compared with the other drivers' group [8] [12]. This adjustment of crash rates for distance driven was reported by other studies and evidence that rates for the older drivers are not inferior to the ones observed for the younger group, giving rise to the concern of older driver population [13].

The causes of older road users' crashes are complex and there is no unique unequivocal explanation for them. Several explanations have been expressed by researchers to justify the over-representation of older drivers in fatal and serious injury crashes. Some argue that the older driver issue is mainly restricted to determined subgroups of older people, rather than related to all older drivers indiscriminately. They support that some clusters are more at risk and need to be identified in order to be done something about it [3]. One important difference between these at-risk populations is the survival rate of crashes once it is reported that seniors have lower surviving possibilities. After a motor vehicle accident, older drivers are four times as likely to be hospitalized, having also slower recoveries when compared with younger drivers (Dobbs as cited in [14]). This can also explain, at least partially, the elevated trauma of older road users. With increasing age, biological processes have the effect of reducing resilience to trauma and biomechanical tolerance to injury becomes lower. The reduction in bone and neuromuscular strength and also the diminution of the fracture tolerance make older drivers more fragile, susceptible to injury and with slower capacity to recover from trauma [4]. Seniors are more likely to have serious injuries or die from motor vehicle crashes because they are simply more medically delicate [15].

3 Older Drivers and ITS

Intelligent Transportation Systems (ITS) provide and use information about transportation conditions to improve system performance in such areas as safety, mobility, efficiency and environmental impacts [16]. ITS involve a wide variety of advanced and emerging technology applications designed with the aims of improving mobility, safety, travel efficiency, comfort and transit services, as well as reducing congestion, fuel consumption and emissions. In the context of driving, a distinction is made between two kinds of systems:

• Advanced driver assistance systems (ADAS) – devices that are designed to co-operate with the driver to achieve the trip goals, such as collision warning systems, adaptive cruise control, lane departure warning, lane change aids, parking aids, etc.,

- In-vehicle information systems (IVIS) – in-board systems that provide the driver with information and allow communications from and to the driver; some of these systems are not related to the driving task, that is, they don't provide useful information for a safe driving, being a factor of distraction; telecommunications and infotainment systems (e.g., radio, mobile phone, e-mail, Internet access) are examples of IVIS competing with the driving task; other IVIS, like navigation systems, provide useful information but can be a factor of distraction as well if they don't support user's needs and are not compatible with human capabilities and limitations.

ITS represent a potential to increase safety and possibly save lives, as older drivers are particularly vulnerable. The benefits from the use of ITS by older drivers have been identified in several studies [17] [18]. As innovative technologies, intelligent transportation systems require adequate design in order to avoid the bad outcomes of well-intentioned but poorly designed technology. Even easy-to-use and somehow intuitive, ITS might require some training, helping the driver to make a safe use of the system or understand how to cope with the limits of the technology. When using ITS, drivers must understand the appropriate level of trust to place in the system as early research has shown that sometimes people place more trust in a system than it is designed to handle [19].

Due to a generation effect, elderly people are not used to advanced technology, which lead them to avoid and even reject new technological systems. It can be argued that in the near future, the elderly population will be more familiarized with new technologies; however, there will be always a gap between their experience with new technologies and the most recent technological advances. Furthermore, there are evidences that older people present some difficulties in terms of self-learning. Actually, learning performance and learning aptitude decline with age, particularly when the information presented increases in complexity or when the speed of that presentation is beyond the control of the subject. A new technology may create an additional complexity to older people due to the lack of occasion to use previous experience, which is the main resource of the elderly. According to Stokes [20], some handicaps of elderly people are more related to the associated effects of the ageing process and the under-stimulation resulting from their retired life than to the loss of intellectual capacity.

ITS technologies offer a range of benefits to transport systems and users: from safety improvements to capacity increases and operational efficiencies, environmental preservation and the provision of information. In the context of driving, ITS include vehicle control devices (Adaptive Cruise Control, rear-view cameras, backup proximity warnings, etc.); driving assistance devices (navigation, traffic information systems); and "infotainment" and comfort devices (entertainment, Internet access and communications systems).

3.1 Behavioral Adaptation

The introduction of new artifacts means that new tasks will appear or the previous ones will change or even disappear. Therefore, design should accommodate these changes either by shaping the interface or by providing instructions and training. In more complex situations, these changes take some time to occur in consequence of the time users require to adapt to the new artifact and the modified task. The duration of this process depends on the system complexity, the frequency of use, the ease of use, etc [21]. Besides the adaptation to the new conditions, there is a previous precondition for the system use: user acceptance, which means that the systems will have to be bought, used and trusted. Different studies reported by Davidse [20] refer that in-vehicle technologies will be accepted by older drivers if the systems fit their needs in terms of safety and mobility. Furthermore, the systems design is a main request for their actual positive effects. Although ITS technologies can play a significant role in offering timely alerts, an increased sensitivity to the needs of older drivers can make these technologies even more effective.

The study of the impact of ITS on road safety is a major concern due to increasing complexity they bring about to the driver although the safety goals presiding to the design and deployment of their applications. In general terms, the safety impacts of ITS depend on the extent to which they support user's needs and are compatible with human capabilities and limitations. Actually, there are risks associated to the use of ADAS and IVIS: both can cause distraction, overload and confusion, requiring behavioral adaptation. In addition, these systems increase mobility and, consequently the drivers' exposure to hazards [22]. Some medium or long term effects like loss of skill might occur (e.g. a driver can become dependent on a navigation or collision warning system to make the appropriate decision). Therefore, the systems appropriation by the driver and behavioral adaptations regarding the use of ITS by particular groups of drivers represent some of the most relevant research needs in the field of ITS aiming at improving the systems design and establishing regulatory policy.

The term behavioral adaptation has emerged to refer to those behaviors which may occur following the introduction of changes to the road-vehicle-user system and which were not intended by the initiators of the change ([23]; p.23). This OECD report concluded that road users adapt their behavior to changes in the road transport system to increase their mobility and thereby reduce the safety impact of the change. Several studies on drivers' behavioral adaptation in response to ADAS have provided a diversity of results: (1) Behavioral changes sometimes occur and sometimes not, (2) they affect different aspects of behavior, and (3) they differ in magnitude and direction [24] [25] [26] (cited by Davidse [27]).

Apparently, behavioral changes evolve from the complex interplay of the following factors:

1) They depend on the system under investigation and so, on the type of driving function supported by the system, and hence influence driver's control processes in different ways [28]. Behavioral adaptation was also found to be mediated by changes in drivers' trust [29], situation awareness [30] [31], fatigue [32] [33], mental workload [34] , and perceived risk [35]. Adaptation occurs in a specific context, which refers to the driving situation and the related driving task demands, along with the travel conditions (purpose, duration, etc.) and refers to the social and cultural background (driver population needs, habits, attitudes, legislation, etc.).

2) Behavioral adaptation is influenced by individual driver characteristics, such as driving experience, age, gender, personality traits, attitudes and motives [36] [37].

However, many of these systems are still under development and not much research has been done on user acceptance and behavior adaptation. As a result, little can be said on whether these systems will actually be used by older drivers and will actually improve their safety. Older drivers are known to be very cautious and to drive only in controlled, safe and familiar environments. For them, a navigation system and/or a collision alert system have the potential to be useful in critical situations such as driving in non familiar environments or making a left turn at an unprotected intersection. However, there are some questions arising: 1) Following a period of adaptation, can in-vehicle systems be capable of changing their travel behavior towards driving and encourage them to drive more kilometers outside familiar areas? 2) Will the older driver rely too much on the system, becoming dependent on it and, due to the age-related declines of cognitive functioning, will lose some driving skills? If there is insufficient knowledge regarding behavioral adaptation, particularly concerning older drivers, much less exists regarding long-term effects on safety and driving skills.

3.2 From HMI to HMC

When performing a task, the human operator mobilizes different cognitive functions in order to maintain his/her cognitive stability: anticipation (preparing the reaction on the basis of the perceived information, existing knowledge and experience), interaction (reactions to an external stimulus following a previous intention or planning) and recovering (conscious actions following a diagnosis of the situation) [38]. Being the interaction centered on the subject's reaction to perceived stimuli, the interface quality remains a key point with its properties of friendliness, usability, transparency, etc. These attributes have a great importance for a perfect coupling human-machine but they are not sufficient to ensure the success of the user's actions, particularly in complex and dynamic situations. Actually, the relationship between humans and machines has evolved from a simple human-machine interaction (HMI) in which the user fully controls the machine onto more complex and dynamic interactions in which the machine processes information. In these cases, the situations are not fully controlled by the user and so, they are affected by uncertainty. Therefore, in order to manage the risks due to uncertainty, some degrees of freedom must be maintained to allow the humans and the machine to adapt to unforeseen contingencies [39] [40].

In an automated process, the machine replaces totally or partially the human to perform the task. Considering the logic of human-machine coupling, there is a cooperative relationship based on a match between the human and the machine. In this case, the technology assists the user in the task performance. In the case of driving a vehicle equipped with in-vehicle advanced technology, there is a functional cooperation between the driver and the technology to perform the driving task in a highly dynamic and uncertain environment. For a common use, the driving task is not completely automated, being the driver the element of the system who makes the final decision. Anyway, this is clearly a dynamic situation where the assistance on the driving situation diagnosis and the information or alerts provided by the system are based on a functional cooperation, allowing for an improve in safety and travel efficiency.

The assistance to the driver can be provided by different kinds of ITS: information systems (navigation, alert, traffic information, etc.), automated systems (automatic control of the vehicle) and cooperative technologies (co-pilot) [41]. A co-pilot technology is supposed to analyze the context and adapt the assistance to the driver according to the diagnosis of the situation. It can be said that driving a vehicle equipped with in-vehicle technology involves a complex human-machine relationship to perform the driving task at its different levels – strategic, tactical and operational. Being a dynamic and complex task, driving is characterized by uncertainty as it is partially controlled and is also risky as the cost of errors can be very high.

Therefore, the introduction of a new technology into the car will not necessarily represent an improvement, particularly to older drivers, as most in-vehicle systems require the driver to change the behavior patterns that have served them for decades. This change may be difficult, and the need to adopt new behaviors may deprive older drivers of one of their main advantages - the extensive driving experience they have acquired over the years. The notion that a device makes sense on the drawing board does not ensure that it will have the desired effect as it is introduced into the car. In this context, the issue is the integration of different functions in the vehicle, being designed so that it will avoid an increase of the driver's mental workload. When different technologies are installed in the car the systems should work together instead of fighting for the attention of the driver and giving him conflicting information. Therefore, the assistance provided by the systems should not have any negative safety consequences on other elements of the driving task or on the behavior of other drivers. Unless the systems interfaces and the forecasted interactions will be ergonomically designed, they will overload and confuse the driver, especially the elderly.

4 Conclusive Remarks and Further Research Needs

Nowadays older people rarely are familiarized with new technologies. This leads them to avoid, in a general way, the use of technological devices. Experience in a particular task performance is the main resource of older people to keep good scores in performing the same task. However, the use of experience requires some stability on the level of the technical and environmental conditions for the task performance. Regarding the driving task, that stability doesn't exist anymore as many changes occurred in the vehicles and road environment. The increasing complexity resulting from those changes narrow down the older driver's limits to develop compensation behaviors. Then, they start to avoid driving in more complex situations and will evolve rapidly to give up driving. ITS could help older drivers to extend their mobility and independent life, but the generation effect regarding the use of technological systems will lead them to avoid the system use. That's why an appropriate design of those systems and the provision of advice for purchasing and adequate training are so important issues.

The development of ITS is very recent and still is in progress. It is quite known that users do not necessarily accept innovation, mainly older people, who are more resisting to changes than younger. Sometimes, good intentions have bad outcomes. Actually, on the one hand, it is difficult to predict the way a device could affect driving, and, on the other hand, there are several reasons why a system that should improve safety and mobility can have smaller than expected benefits [42]:

- Users may not use the device correctly;
- The device can introduce a feeling of safety that can induce the person to take more risks;
- The device could not fit the specific driving characteristics of older drivers;
- The user may develop new behavioral patterns.

New behaviors might be induced by the use of an ITS. Older drivers are known to be very cautious and to drive only in controlled, safe and familiar environments. For them, a navigation system and/or a collision alert system will be very useful in critical situations such as driving in non familiar environments or making a left turn at an unprotected intersection. However, following a period of adaptation, can in-vehicle systems be capable of changing their behavior towards driving and encourage them to drive more kilometers outside familiar areas? Or will the older driver rely too much on the system, becoming dependent on it and, due to the age-related declines of cognitive functioning, will lose some driving skills? Actually, a cognitive understimulation in terms of collecting and memorizing spatial information could lead the older driver to a great dependence on a route guidance system, making the person unable to drive without that information. The same may occur by the use of a collision warning system, which could affect situation awareness. Therefore, the induced behaviors by the systems use should be studied in order to prevent potential decreases in fitness for driving. For each system, the induced behaviors by its use and the side effects on driving behaviors and attitudes should be identified in order to develop adequate countermeasures.

References

1. Jun, J., Ogle, H., Guensler, R., Brooks, J., Oswalt, J.: Assessing mileage exposure and speed behaviour among older drivers based in crash involvement status. In: 87th TRB Annual Meeting, Washington D.C (2007)
2. Holland, C.A.: Older Drivers: A Literature Review. Department for Transportation. Technical Report (n°. 25). London, U.K (2001)
3. OECD: Ageing and transport. Mobility needs and safety issues (2001),
   http://www.oecd.org/dataoecd/40/63/2675189.pdf
4. Charlton, J., Oxley, J., Scully, J., Koppel, S., Congiu, M., Muir, C., Fildes, B.: Self-regulatory driving practices of older drivers in the Australian Capital Territory and New South Wales. Accident Research Centre, Monash University (2006)
5. Oxley, J., Charlton, J., Fildes, B., Koppel, S., Scully, J., Congiu, M., Moore, K.: Crash risk of older female drivers. Accident research Centre, Monash University (2005)
6. Johnson, J.E.: Older rural adults and the decision to stop driving: the influence of family and friends. Journal of Community Health Nursing 15, 205–216 (1998)
7. Harris, A.: Transport and mobility in rural Victoria. Royal Automobile Club Victoria. Technical Report, PP 00/22 (2000)
8. ERSO: Older drivers. European road safety observatory (2006),
   http://www.erso.eu/knowledge/content/06_drivers/20_old/
   olderdrivers.htm
9. Kostyniuk, L.P., Trombley, D.A., Shope, J.T.: The process of reduction and cessation of driving among older drivers: a review of the literature. Michigan University, Ann Arbor, Transportation Research Institute. 50 p. Sponsor: General Motors Corporation, Warren, Mich. Report No. UMTRI-98-23. UMTRI-91238 (1998)

10. Eby, D., Molnar, L., Kartje, P.: Maintaining safe mobility in an aging society 2009. CRC Press/ Taylor & Francis, London (2009)
11. Suen, S. L., Mitchell, C.G.B.: Accessible transportation and Mobility. TRB 82nd Annual Meeting (2003) http://gulliver.trb.org/publications/millennium/00001.pdf (retrieved, 2005)
12. Baldock, M.R.J., McLean, A.J.: Older drivers: crash involvement rates and causes: The university of Adelaide (2005)
13. Brayne, C., Dufouil, C., Ahmed, A., Dening, T., Chi, L., McGee, M., Huppert, F.: Very old drivers: findings from a population cohort of people aged 84 and over. International Journal of Epidemiology 29, 704–709 (2000)
14. McGee, P., Tuokko, H.: The older & wiser driver: a self-assessment program (2003), http://www.coag.uvic.ca/documents/research_reports/Older_Wiserdriver Report.pdf (retrieved, 2007)
15. Li, G., Braver, E.R., Chen, L.H.: Fragility versus excessive crash involvement as determinants of high death rates per vehicle-mile travelled among older drivers. Accident Analysis and Prevention 35, 227–235 (2003)
16. Hu, P., Boundy, B., Truett, T., Chang, E., Gordon, S., Hu, P., Boundy, B., Truett, T., Chang, E., Gordon, S.: Cross-Cutting Studies and State-of-the-Practice Reviews: Archive and Use of ITS-Generated Data. Center for Transportation Analysis, Oak Ridge National Laboratory, Oak Ridge, Tennessee 37831–6073 (2002)
17. Suen, S.L., Mitchell, C.G.B.: Accessible transportation and Mobility. TRB 82nd Annual Meeting (2003), http://gulliver.trb.org/publications/millennium/00001.pdf (retrieved, 2005)
18. OECD. Road Safety: Impact of New Technologies. Transport, OECD, Paris (2003)
19. Stokes, G.: On Being Old: The Psychology of Later Life. The Falmer Press, Londres (1992)
20. Hollnagel, E., Woods, D.: Joint cognitive systems: Foundations of cognitive systems engineering. CRC Press, Taylor & Francis, London (2005)
21. Simões, A.: Elderly Drivers and IVIS. In: Brusque, C., Bruyas, M.-P., Carvalhais, J., Cozzolino, M., Gelau, C., Kaufmann, C., Macku, I., Pereira, M., Rehnová, V., Risser, R., Schmeidler, K., Simoes, A., Simonova, Z., Turetscheck, C., Vaa, T., Vašek, J., Zehnalová, V. (eds.): The Influence of In-Vehicle Information Systems on Driver Behaviour and Road Safety: Synthesis of Existing Knowledge – Cost 352; Les Collections de L'INRETS, Synthesis, vol. 54, pp. 37–44 (2007)
22. Okola, A.R., Walton, C.M.: Intelligent Transportation Systems to Improve Elderly Persons' Mobility and Decision Making within Departure Time Choice Framework, Centre for Transportation Research, University of Texas at Austin (2003)
23. OECD. Behavioural adaptations to changes in the road transport system. Paris:Organization for Economic Co-operation and Development (1990)
24. Dragutinovic, N., Brookhuis, H., Marchau: Behavioural effects of Advanced Cruise Control use – A meta-analytic approach. European Journal of Transport and Infrastructure Research 5(4), 267–280 (2005)
25. Eick, E.-M., Debus, G.: Adaptation effects in an automated car-following scenario. In: Underwood, G. (ed.) Traffic and transport psychology: Theory and application. Proceedings of the ICTTP 2004, pp. 243–255. Elsevier Ltd, Amsterdam (2005)
26. Saad, F., Hjälmdahl, M., Cañas, J., Alonso, M., Garayo, P., Macchi, L., et al.: Literature review of behavioural effects (Del. 1.2.1 of AIDE IST-1-507674-IP) (Retrieved from 2005), http://www.aide-eu.org/pdf/sp1_deliv/aide_d1-2-1.pdf

27. Davidse, R.: Older drivers and ADAS – Which Systems Improve Road Safety? IATSS Research 30(1) (2006)
28. Engström, J., Hollnagel, E.: A general conceptual framework for modelling behavioural effects of driver support functions. In: Cacciabue, P.C. (ed.) Modelling driver behaviour in automotive environments. Critical issues in driver interactionswith Intelligent Transport Systems, pp. 61–84. Springer, London (2007)
29. Stanton, N.A., Young, M.S.: Driver behaviour with adaptive cruise control. Ergonomics 48(10), 1294–1313 (2005)
30. Baumann, M., Krems, J.F.: Situation awareness and driving: A cognitive model. In: Cacciabue, P.C. (ed.) Modelling driver behaviour in automotive environments. Critical issues in driver interactions with Intelligent Transport Systems, pp. 253–265. Springer, London (2007)
31. Walker, G.H., Stanton, N.A., Young, M.S.: The ironies of vehicle feedback in car design. Ergonomics 49(2), 161–179 (2006)
32. Fairclough, S.H.: Mental effort regulation and the functional impairment of the driver. In: Hancock, P.A., Desmond, P.A. (eds.) Stress, workload, and fatigue, pp. 479–502. Lawrence Erlbaum Associates, Inc., Mahwah (2001)
33. Matthews, G., Desmond, P.A.: Task-induced fatigue states and simulated driving performance. The Quarterly Journal of Experimental Psychology 55A(2), 659–686 (2002)
34. Ward, N.J., Shankwitz, C., Gorgestani, A., Donath, M., De Waard, D., Boer, E.R.: An evaluation of a lane support system for bus rapid transit on narrow shoulders and the relation to bus driver mental workload. Ergonomics 49(9), 832–859 (2006)
35. Rajaonah, B., Tricot, N., Anceaux, F., Millot, P.: The role of intervening variables in driver-ACC cooperation. International Journal of Human-Computer Studies 66(3), 185–197 (2008)
36. Machin, M.A., Sankey, K.S.: Relationships between young drivers' personality characteristics, risk perceptions, and driving behaviour. Accident Analysis & Prevention 40(2), 541–547 (2008)
37. Summala, H.: Towards understanding motivational and emotional factors in driver behaviour: Comfort through satisficing. In: Cacciabue, P.C. (ed.) Modelling driver behaviour in automotive environments. Critical issues in driver interactions with Intelligent Transport Systems, pp. 189–207. Springer, London (2007)
38. Boy, G.: Documenter un artefact. In: Boy, G. (ed.) Ingénierie Cognitive. Lavoisier, Paris (2003)
39. Hoc, J.-M.: Coopération humaine et systems cooperatives. In: Boy, G. (ed.) Ingénierie Cognitive. Lavoisier, Paris (2003)
40. Hoc, J.-M.: From human - machine interaction to human–machine cooperation. Ergonomics 43(7), 833–843 (2000)
41. Bellet, T., Tattegrain-Veste, H., Chapon, A., Bruyas, M.-P., Pachiaudi, G., Deleurence, P., Guilhon, V.: Ingénierie cognitive dans le contexte de l'assistance à la conduite automobile. In: Boy, G. (ed.) Ingénierie Cognitive. Lavoisier, Paris (2003)
42. Meyer, J.: Personal Vehicle Transportation. In: Pew, R.W., Van Hemel, S.B. (eds.) Technology for Adaptive Aging. The National Academies Press, Washington (2003)

Frequency of Usage and Feelings of Connectedness in Instant Messaging by Age, Sex, and Civil Status

Michael E. Stiso

SINTEF ICT, Forskningsveien 1, 0373 Oslo, Norway
mikestiso@gmail.com

Abtract. A questionnaire was administered to determine (1) the frequency with which participants use various IM features, and (2) whether and to whom they feel more connected as a result of IM usage. Younger participants IMed more frequently than older ones, and males and younger participants were the more frequent users of the more esoteric IM activities, particularly video sharing and video chatting. Singles 25-34 were most likely to report feeling increased connections overall to people via IM. The youngest and the single ones were much more likely to feel a greater connectedness with friends than with family. In the later 20s, that focus on friends shifts to family and coworkers, resulting in relatively equivalent percentages reporting greater connectedness with each of the three categories. The suggestion is that younger, single people are using IM to fulfill a need for social interaction that would be otherwise difficult to meet.

Keywords: Age, sex, gender, civil status, instant messaging, connectedness.

1 Introduction

Instant messaging (IM) technology in the form of computer-mediated user-to-user messages has been around for at least a few decades. However, it didn't gain widespread popularity until the mid-90s, with the introduction of services such as ICQ and AOL Instant Messenger. And use is on the rise, with a Pew report finding a 29% increase in the number of U.S. IMers from 2000-2004 [1]. A third of IMers engage in the activity daily, two-thirds do so several times a week, and a quarter of the IM population now uses IM more than email [1].

Mirroring the technology's increasing popularity is an increasing interest in IM as a research topic. A relatively small but growing number of researchers find that the tool's logging capabilities, the turn-based method of communication it imposes, and the variety of interaction features associated with it make IM ideal for studying technology-mediated social interaction. [2][3] and [4], for example, looked at the use of IM in the workplace and in collaboration. [5] examined its role as a socialization tool in distance education courses. [6] investigated the social and other needs that college students use IM to fill. And [7] looked at its use among the younger generation that has grown up with the technology.

The current study builds on and combines much of the above research. It is an exploratory investigation of age, sex, and civil status differences in the use of newer IM features beyond simple text chatting — for example, webcamming, VOIPing (voice-over

M. Kurosu (Ed.): Human Centered Design, HCII 2009, LNCS 5619, pp. 562–569, 2009.
© Springer-Verlag Berlin Heidelberg 2009

IP, or internet-based phone calls), videoconferencing, online radio, and more. It will also examine whether IM provides certain people with a feeling of greater connectedness with other groups.

This research grew out of the general Uses and Gratifications approach, the basic idea of which is that people use different media to fill specific needs. The medium in this case is instant messaging, the goal being to add to the small body of literature describing motivations and habits in the use of new media tools. Doing so should facilitate the user-centered design of IM tools by (1) uncovering user motivations in relation to IM, and (2) highlighting the presence of certain kinds of IM users, allowing IM systems to anticipate user needs based on profiles and usage history.

2 Methods

The study consisted of an online questionnaire distributed to 201 compensated participants. The group was Norwegian, 108 males and 93 females, with an average age of 29 (range 15-41) and an income generally between 350,000 and 750,000 Norwegian Kroners. The age distribution was generally flat between the two extremes, and the income distribution was skewed slightly toward the higher end; the number of married participants to single ones was about even. For purposes relating to the overall project of which this study was part, all participants were required to have used an instant messenger at least two times in the past year.

The questionnaire started with a demographics section, which included sex, age, children, relationship status, education, and income. Following that section were questions asking about participants' frequency and length of IM usage, as well as the various IM features they use. (See figure 1.) Remaining questions asked participants to select whether and to whom they felt more connected via their use of IM, and to rate how often they used IM for certain reasons (5-point Likert scale with points corresponding to *never, less than monthly, monthly, weekly,* and *daily*). The majority of questions relating to IM usage were adapted from [8].

3 Results

The first part of the analysis consisted of a MANOVA[1] of the survey items dealing with IM usage, using civil status (single, married), age (three levels), and sex as independent variables. A little over half of those survey items had significantly skewed distributions, determined by comparing the skewness statistic for each to the absolute value of 2 times the standard error of skewness [9]. The problematic distributions were subjected to a log transformation, which improved all but a few of them; those few were left unchanged.

3.1 IM Usage by Sex, Age, and Civil Status

Participants were asked how often and for how long they used IM on average over the last six months, as well as the frequency with which they used various IM-based

---

[1] Note that the MANOVA is being used here for exploratory research rather than for hypothesis testing, and so generalizations and conclusions regarding the results should be treated cautiously.

tools. (See figure 1.) Results show that participants IMed on average several times a day, spending several hours a week on the activity. As might be expected, the most-used IM features were text chatting and asynchronous messaging (exchanging messages with partners not currently online), occurring about weekly across participants. Coming in second were newsreading and filesharing (documents and photos) activities, at about monthly. All other listed features occurred less than once a month.

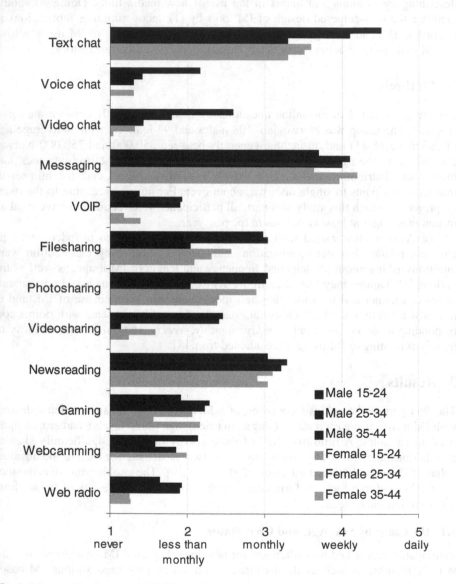

Fig. 1. Mean usage frequency of different IM-based tools across age and sex. Blacks bars are males, grey are females, and bar order represents increasing age.

Looking at age differences, the oldest age group (35-44 years) used IM less often than the younger groups, $F(2, 187) = 5.043$, p < .01, at about a few times a day compared to several times a day. They differed also in frequency of IM-based document sharing, $F(2, 187) = 3.384$, p < .05, and video sharing, $F(2, 187) = 5.536$, p < .01. The youngest and oldest groups saw a barely significant difference in video chatting $F(2, 187) = 3.020$, p = .51.

Males didn't IM with any greater frequency or amount of time than did females. They did make greater use of the more-esoteric IM features, though, including voice chatting, $F(1, 187) = 5.203$, p < .05, video chatting, $F(1, 187) = 15.374$, p < .001, video sharing, $F(1, 187) = 9.967$, p < .01, webcamming, $F(1, 187) = 4.095$, p < .05, and listening to web radio, $F(1, 187) = 4.033$, p < .05. The practical differences were minor, however, amounting to a few times a year.

Turning to civil status, the MANOVA showed that single participants did not IM significantly more or less than did married ones. It did show some interactions between status and the other two variables, but note that the sample sizes for many of the cells were quite small, particularly those for married people in the youngest group and single people in the oldest group. Hence, any such interactions found come with a great deal of uncertainty, and so they will not discussed here.

3.2 Connectedness

The survey involved in the current study asked participants whether and to whom they felt more connected as a result of IM usage. Based on [10]'s discovery of a relationship between the amount of IM use with a person and the degree of intimacy with that person, a correlational analysis in the current study compared IM usage (both frequency and usage time) with perceived connectedness. Results showed that frequency and usage time correlated highly with reported feelings of greater overall connectedness, $r(200) = .408$, p < .001 and $r(200) = .269$, p < .001, respectively. In particular, IM frequency (but not usage time) correlated with greater connectedness to family, $r(200) = .157$, p < .05, friends, $r(200) = .245$, p < .001, coworkers, $r(200) = .247$, p < .001, people who share one's hobbies, $r(200) = .179$, p < .05, and people who share one's profession, $r(200) = .236$, p < .001. Time spent IMing correlated only with hobbies, $r(200)=.281$, p < .001). Those results would seem to provide some additional support for [10]'s finding.

The connectedness data were also compared to sex, age, and civil status and subjected to chi-square tests; see figures 2 and 3. Overall, just over 40% of all participants reported feeling no change in connectedness as a result of IM; a third of participants reported greater connectedness, and very few reported feeling less connected. The people feeling more connected tended to be under 35, $c^2(2, N = 202) = 7.102$, p < .05, and single (39% vs. 22% of married), $c^2(1, N = 202) = 6.413$, p < .05); the highest percentage was participants who were single and between 26-35 age range (52% of that group vs. 38% for the next highest group).

People who felt a greater connection felt it primarily toward friends, followed by family, coworkers, people who share one's hobbies, and people who share one's profession, in that order. The likelihood of feeling more connected to friends was highest in the younger groups (80% of those under 25 vs. 60% of those over 35, $c^2(2, N = 202) = 7.246$, p < .05) and for singles (80% of singles vs. 60% of the married ones,

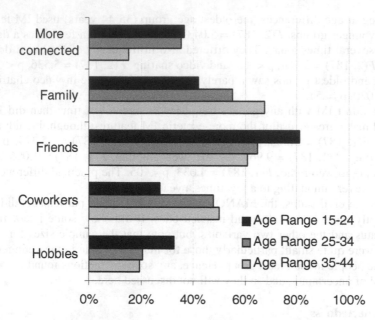

Fig. 2. Percentage of different age groups indicated whether they feel more connected overall via IM usage, and to whom they feel more connected. ("Hobbies" refers to people who share the participant's hobbies).

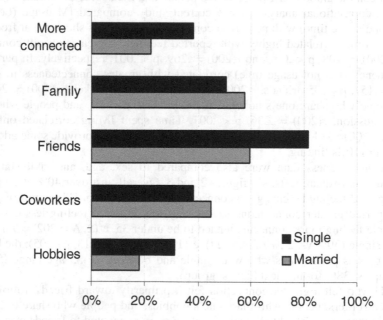

Fig. 3. Percentage of single vs. married participants indicating whether they feel more connected overall via IM usage, and to whom they feel more connected. ("Hobbies" refers to people who share the participant's hobbies).

$c^2(1, N = 202) = 8.725$, p < .01). Adding sex to the mix, however, showed that the main difference in all age groups was between married males (45% of the group) and everyone else (70-75%, depending on group). In other words, high percentages of females and single males reported feeling more connected to friends via IM; only a relatively small portion of married men felt that way.

Family was the next most popular group to which participants felt more connected. Older participants were more likely to feel that connection than younger ones, 67% (oldest group) vs. 55% (middle) vs. 42% (youngest), $c^2(2, N = 202) = 7.606$, p < .05. Similarly, females were more likely than males to feel it (61% vs. 42%), $c^2(1, N = 202) = 4.150$, p < .05.

People over 25 were more likely than younger participants to feel a greater connection to coworkers through IM usage, 47-50% of the older groups vs. 27% of those under 25, $c^2(2, N = 202) = 8.986$, p < .05. For people sharing one's hobbies, single participants were more likely to feel more connected with them than were married ones, 30% vs. 18%, $c^2(1, N = 202) = 4.196$, p < .05, as were males compared to females, 33% vs. 15%, $c^2(1, N = 202) = 9.175$, p < .01.

4 Discussion

Greater IM usage was associated with increased feelings of connectedness via IM. Just over a third of participants indicated feeling greater connections overall, with singles between 25 and 34 being the most likely to indicate such an increase. A possible explanation for that finding is that users within the 25-34 age range may have a larger network of IM contacts than older users: Younger users perhaps haven't had as much time to develop their network, and older ones are perhaps not as comfortable with the technology to have developed an extensive network. If so, the relatively large networks of people 25-34 should provide a greater number of opportunities for social interaction. In turn, their single status may provide them a greater motivation to follow up those opportunities compared to married users. That motivation could possibly lead to a greater percentage of single users in that age range reporting increased connectedness via IM usage.

The suggestion, then, is that younger, single people are using IM to fulfill a need for social interaction that would be otherwise difficult to meet. That interpretation fits with [6]'s finding that IM serves specific social needs in college students. It is still supposition in this case, of course, but future research can verify it by examining the size of one's IM network and the motivations for IMing.

Younger people engaged in IM a bit more often than older participants, and they tended to feel more connected to others as a result of their usage. The people to whom they feel connected changes with age, however. The youngest age group in this study (15-24) had a heavy focus on friends. Over 80% of them reported feeling more connected to friends, compared to just over 40% who said that they felt closer to family and 27% to coworkers. However, those differences evened out for the older groups: Connections with friends lowered to just over 60%, whereas those with family and coworkers rose to around 60% and almost 50%, respectively. In other words, as people pass their mid-20s, the focus on friends shifts to a more balanced connection with friends, family, and coworkers.

The increasing connection with coworkers over the years is likely a reflection of the presumably scant work history for participants under 25. Similarly, the increase in IM-based family connections with age may simply be a result of younger people being more likely to live at home, reducing the need for them to correspond via IM with family members. Somewhat difficult to explain, however, is why the percentage of people reporting greater connectedness with friends drops in the later 20s (81% to 68%). If friends tend to disperse geographically as they get older and leave home, one might expect IM to help them maintain contact with each other, which presumably would lead to an increase in the reports of greater connections with friends amongst older participants. As well, assuming that people gain more friends as they get older, there would seem to be a greater opportunity for IM to provide a greater sense of connection for more people. Yet, the results here suggest that relatively fewer older participants report feeling more connected with friends via IM.

One explanation is that teens and young adults experience a change in communication priorities as they get older. Another is that this is a cross-sectional study rather than a longitudinal one: Unlike the younger participants, the older ones did not grow up with IM technology; hence, older participants may be less likely to have friends with whom they can IM, which would decrease the likelihood of those participants being able to report greater connectedness with friends. As well, the finding may simply be particular to the Norwegian population that formed the sample. Future research should be able to clarify the issue by measuring attitudes over time in a longitudinal, culturally diverse (or comparative) study.

5 Conclusion

Younger participants IMed more frequently than older ones, which corresponds with previous studies. Males and younger participants were the more frequent users of the more esoteric IM activities, particularly video sharing and video chatting, though they used such features monthly at most. Regarding increased connectedness with others as a result of IM usage, singles 25-34 were most likely to report feeling it overall. Looking at specific groups of people, though, the youngest participants and the singles were much more likely to feel greater connections with friends than with family; those differences even out in the later 20s, though, suggesting that the younger, single people are using IM to fulfill a particular need for social interaction.

The results have both applied and academic implications. On the one hand, they highlight (1) IM usage frequencies as a basis for organizing the information and features in an IM tool, and (2) user profiles and usage histories as a basis for the creation of IM UIs that adapt to user needs. On the other hand, the results also support the use of IM behavior as a window into age- and sex-related differences in social interaction.

References

1. Shiu, E., Lenhart, A.: How Americans U se Instant Messaging. Pew Internet & American Life Project (2004), http://www.pewinternet.org/PPF/r/133/report_display.asp
2. Cameron, A.F., Webster, J.: Unintended Consequences of Emerging Communication Technologies: Instant Messaging in the Workplace. Computers in Human Behavior 21(1), 85–103 (2005)

3. Isaacs, E., Walendowski, A., Whittaker, S., Shiano, D.J., Kamm, C.: The Character, Functions, and Styles of Instant Messaging in the Workplace. In: Computer Supported Cooperative Work, New Orleans, Louisiana, pp. 11–20 (2002)
4. Nardi, B.A., Whittaker, S., Bradner, E.: Interaction and Outeraction: Instant Messaging in Action. In: Computer Supported Cooperative Work, Philadelphia, PA, pp. 79–88 (2000)
5. Nicholson, S.: Socialization in the "Virtual Hallway": Instant Messaging in the Asynchronous Web-Based Distance Education Classroom. The Internet and Higher Education 5(4), 363–372 (2002)
6. Flanagin, A.: IM Online: Instant Messaging Use Among College Students. Communication Research Reports 22(3), 175–187 (2005)
7. Grinter, R., Palen, L.: Instant Messaging in Teen Life. Computer Supported Cooperative Work, New Orleans, Louisiana (2002)
8. Georgia Institute of Technology, Graphics, Visualization, & Usability Center WWW User Surveying Team, GVU's 10th WWW User Survey (1998), http://www.cc.gatech.edu/gvu/user_surveys/survey-1998-10/questions/use.html
9. Tabachnick, B.G., Fidell, L.S.: Using Multivariate Statistics, 3rd edn. HarperCollins College Publishers, New York (1996)
10. Hu, Y., Wood, J.F., Smith, V., Westbrook, N.: Friendships Through IM: Examining the Relationship Between Instant Messaging and Intimacy. Journal of Computer Mediated Communication 10(1) (2004), http://jcmc.indiana.edu/vol10/issue1/hu.html

Examining Individual Differences Effects:
An Experimental Approach

Wan Adilah Wan Adnan[1], Nor Laila Md. Noor[1], and Nik Ghazali Nik Daud[2]

[1] Faculty of Information Technology & Quantitative Sciences,
Universiti Teknologi MARA Malaysia
[2] Faculty of Engineering Universiti Pertahanan Nasional Malaysia
`wan.adilah@gmail.com`, `norlaila@tmsk.uitm.edu.my`,
`nikghazali@upnm.edu.my`

Abstract. Role of individual differences has been emphasized for system success. Many researchers claim about the need for a better understanding of individual differences as empirical evidence is still very limited. The purpose of this paper is to examine the extent to which individual differences affects one's decision performance and decision processes. This study focuses on decision making style and gender as the dimensions of individual differences characteristics. Their potential effects are investigated using experimental approach. Empirical findings from this study support for the user-centered approach that emphasize on user in the design of an information system.

Keywords: Individual differences, decision making style, gender effects, human factor.

1 Introduction

The technology characteristics, task characteristics and individual differences have been empirically studied and recognized as an important element in the decision support system (DSS) and human-computer interaction (HCI) literatures. Individual characteristics have been considered as important determinants affecting one's ability to interpret information, and make effective decisions [1]. Research findings on individual difference effects suggest that different user interface and navigation supports are required to cater for different individual's characteristics and styles. Findings from a meta analysis conducted by Zhang et el. [2] indicates that less than 15% out of 350 HCI studies from year 1990 to 2002 investigate the effects of individual differences. Their findings show that only few investigate the effects of cognitive individual differences. Prior studies recommended that further examination of individual difference, including decision making styles and gender are required [3, 4].

The purpose of this paper is to examine the extent to which individual differences affects one's decision performance and decision processes. This study focuses on decision making style and gender as the dimensions of individual differences characteristics. Besides, the interaction effects of these dimensions on task characteristic and technology characteristic are also examined. Task complexity and information visualization techniques were used to represent the task and technology characteristics

M. Kurosu (Ed.): Human Centered Design, HCII 2009, LNCS 5619, pp. 570–575, 2009.
© Springer-Verlag Berlin Heidelberg 2009

respectively. Their potential effects on decision performance and process are investigated using experimental approach.

2 Theoretical Background

The theoretical background regarding the decision making style and gender is discussed in this section

2.1 Decision Making Style

HCI literature acknowledges that a clear understanding of individual characteristics including cognitive style is helpful in designing interface for a specific community of users. Te'eni [5], among other researchers, argues that for a system to be more accepted and support decision making, the design of a system need to match with the individual decision making style. Rowe and Mason [6] explain the importance of decision making style in their Decision Model. This model explains the impact of decision making style on decision behavior and decision outcome. The model indicates that decision behavior can be predicted once the decision making style is known. Meanwhile, in the field of HCI, it is known that one's cognitive rationale dictates one's physical behavior and any changes in behavior can have substantial effects on performance [7]. Though the impact of decision making style on decision performance has been acknowledged, study on decision making style effects is still lacking [8].

2.2 Gender

Research evidence suggests that gender difference individuals seek and process information using very different strategies. A prominent model on gender, the Selectivity Model by Meyers [9], supports the concept that males and females differ in the strategies they use to process information. It theorizes that females are more efficient than males on a complex task. A study by So and Smith [1] suggests that females exhibit better performance than males when working with visual representation. Besides, recent studies [10, 11, 12] also reveals that there is a significant effect of gender difference on performance in navigational task. Ford et al [13] argues that gender is a major predictor of information-seeking activities. These findings conclude that males and females require different user interface and navigation supports. In addition, Balka and Doucette [14] have argued that gender has been largely neglected, although evidence suggests that failures to consider gender have contributed to design failure.

3 Method

An Experimental study was conducted to examine individual differences effects. A decision support application regarding academic workload was used, and information visualization techniques were incorporated for examining the interaction effects. Participants were tested with three conditions of information visualization techniques: 1) zoom+pan (*Z+P*), 2) overview+detail with tree structure (*O+D Tree*), and 3) over-view+detail with Window-Explorer structure (*O+D WinExp*). These techniques differ in terms of the type of information space representation (Tree and Window Explorer structure) and the technique of organizing the information (overview+detail and Zoom+Pan).

The decision making style of participants were judged using *Decision-Style Inventory (DSI)* developed by Rowe and Boulgardes [15]. This instrument identifies four distinct categories of decision making styles: *directive, analytical, conceptual, and behavioral*. These decision making style are different in regards to their methods of perceive and evaluate information [15]. Meanwhile, decision performance was measured based on task completion time while decision process was measured using subjective workload ratings. The task completion time was derived from the system and workload rating was accessed using the NASA-Task Load Index [16] questionnaire. Participants are required to fill in the workload rating after completing both set of task types (complex and simple) regarding the Academic Workload application. This study was conducted with 16 participants who were decision-makers from the middle management level of the Universiti Teknologi MARA.

4 Analysis and Results

To facilitate the investigation of individual difference effects, a number of statistical techniques were used. A reliability test using Cronbach's alpha was conducted on the NASA-TLX scales. The Cronbach's alpha coefficient for the NASA-TLX scales was a very high with 0.828. This is expected for a well-established instrument for measuring the subjective workload assessment.

The responses from participants were recorded into two measurement variables: task completion time and subjective workload rating. Repeated measure analyses of variance (RM ANOVA) were employed to detect statistically significant differences in completion time and a non-parametric Kruskal-Wallis and Mann-Whitney U tests were employed in workload rating analysis.

The overall group means for completion time and workload rating for each of task types using three different information visualization techniques was analyzed. A lower mean for completion time indicates that shorter completion time taken to complete a task. Meanwhile, the lower mean for workload rating indicates that lesser effort or load perceived by participants in performing a task.

4.1 Decision Making Style Effects

The analysis on decision making style effects had shown that there were significant differences among decision making styles in completion time (RM ANOVA), and workload rating (Kruska-Wallis Test). As anticipated, *analytical* style which is described as task oriented and very structured in problem solving had recorded the fastest in completion time. Further investigation on the interaction effect between information visualization techniques and decision making style was conducted. The results showed that there were significant differences in completion time across information visualization techniques for different decision making style ($p<0.05$; H5). This indicates that decision making style affects the relationship between information visualization techniques and completion time. An interaction effect was also analyzed between the task complexity and decision making style on completion time. Results showed a significant interaction effects with $p<0.05$ for complex task only and not for simple task. Post hoc test for the complex task indicates that there was significant difference between *analytical* style which recorded the fastest time and *behavioral* which recorded the lowest time.

Meanwhile results for workload based on the overall mean score showed that there was significant difference among the decision making styles. *Behavioral* style exhibits the lowest workload rating and *analytical* style exhibits the highest rating. Further analysis on the interaction effects was conducted, and the result showed that there was no significant interaction effect between information visualization techniques and decision making style on the workload based on the overall mean score.

4.2 Gender Effects

Meanwhile results from gender effects have shown that there were significant differences between the males and females in both completion time (ANOVA) and workload rating (Mann-Whitney test). This finding follows similar pattern as in the decision making style effects. As anticipated females which is described as very structured in problem solving, has the fastest completion time. This result supports the Selectivity model by Meyer [9] and is consistent with So and Smith [1] findings. Further analysis was conducted for interaction effects, and results showed that there was no significant interaction effect between information visualization techniques and gender and also between task complexity and gender on completion time. The mean score showed that both gender groups perform the best using *O+D tree-based* and the worst performance was recorded using *Z+P*.

Similar to *analytical* style, females had higher workload rating which indicates that females have low tolerance to load. Males on the other hand, take longer time in completing a task and have shown higher tolerance to load or effort. Statistically, results showed that there is no interaction effect between information visualization technique and gender.

4.3 Discussions

The results indicate a significant difference across decision making style on decision performance and workload rating. The findings also illustrate that there is a tradeoff between the completion time and workload rating across decision style and gender. The style and gender group that has shorter completion time has perceived a higher workload or effort in performing a task. Results show that females who are better in performance, rates workload higher than males. Similarly, *analytical* style that has the best in performance, has perceived the highest rate in workload. These findings indicate that the decision making style and gender can be labeled as either effort reduction or performance maximization. The results indicate that the *analytical* and females emphasize on performance maximization, and the *behavioral* and males emphasize on effort reduction.

Besides, the result on the interaction effects, demonstrates that compared to gender, decision style is more significant to the variances of information visualization techniques and task complexity. These results highlight the importance of considering the effects of decision style differences in the design of decision support application.

5 Conclusions

This paper broadens up the cognitive style differences by exploring decision making style effects and strengthens the empirical evidence of prior studies that support for the gender effects. In other words, the significant differences found in decision performance and decision process between decision making style and gender groups provide additional support and evidence to the literature of HCI. This study provides empirical evidence to support that the effectiveness of a system can be improved when individual differences are considered in its design and implementation. Besides, the findings support for the user-centered approach that emphasize on user, and not only to look into the task or functional requirements but also on an individual's cognitive and perceptual needs.

References

1. So, S., Smith, M.: The impact of presentation format and individual differences on the communication of information for management decision making. Managerial Auditing Journal 18(1), 59–67 (2003)
2. Zhang, P., Li, N.: The Intellectual Development of Human-Computer Interaction Research: A Critical Assessment of the MIS Literature (1990-2002). Journal of the Association for Information System 6(11), 227–292 (2005)
3. Juvina, I., Oostendorp, H.V.: Individual differences and behavioral metrics involved in Modelling Web navigation. Journal of Association of Information System (2004), http://www.cs.Cu.nl/people/ionJuvinaandvanOoastendorpAIS2004.pdf
4. Fan, J.P., Macrediel, R.D.: Gender Differences and Hypermedia Navigation: Principles for Adaptive Hypermedia Learning Systems (2006)
5. Te'eni, D., Carey, J., Zhang, P.: Human Computer Interaction Developing Effective Organizational Information Systems. Wiley, Chichester (2007)
6. Rowe, R.O., Mason: Managing with Style: A Guide to Understanding, Assessing, and Improving Decision Making. Jossey-Bass, San Francisco (1987)
7. Gray, W.D., Fu, W.: Ignoring perfect knowledge in the worlds for imperfect knowledge in-the-head implications of rational analysis for interface design. In: Proceedings of the Computer Human Interfaces Conference (CHI 2001), pp. 112–119 (2001)
8. Richards, D.: The reuse of knowledge: a user-centered approach, international. Journal Human-Computer Studies 52, 553–579 (2000)
9. Meyers-Levy, J.: Gender Differences in Information Processing: A Selectivity Interpretation. In: Cognitive and Affective Responses to Advertising. Editors: Patricia Cafferata and Alice Tybout, Lexington, MA, pp. 219–260 (1989)
10. Campbell, K.: Gender and educational technologies: Relational frameworks for learning design. Journal of Educational Multimedia and Hypermedia 9, 131–149 (2000)
11. Large, A., Beheshti, J.: Design criteria for children's Web portals: The users speak out. Journal of the American Society for Information Science and Technology 53(2), 79–94 (2002)

12. Roy, M., Chi, M.T.C.: Gender differences in searching the Web: Gender differences in patterns of search the Web (2003)
13. Ford, N., Miller, D., Moss, N.: The role of individual differences in Internet searching: An empirical study. Journal of the American Society for Information Science and Technology 52(12), 1049–1066 (2001)
14. Balka, E., Doucette, L.: The accessibility of computers to organizations serving women in the province of Newfoundland: Preliminary study results. Electronic Journal of Virtual Culture 2(3) (1994)
15. Rowe, A.J., Boulgardes, J.D.: Managerial Decision Making. Macmillan, New York (1992)
16. NASA Task Load Index (TLX) v 1.0 Users Manual, http://iac.dtic.mil/hsiac/docs/TLX-UserManual.pdf (retrieved June 12, 2003)
17. Brusilovsky, P., Eklund, J., Schwarz, E.: Web-based education for all: A tool for developing adaptive courseware. Computer Networks and ISDN Systems 30, 291–300 (1998)

Part IV

HCD in Industry

Usability Maturity: A Case Study in Planning and Designing an Enterprise Application Suite

Jeremy Ashley and Kristin Desmond

500 Oracle Parkway, MS 2op10,
Redwood Shores, California 94065
Jeremy.ashley@oracle.com, Kristin.desmond@oracle.com

Abstract. Although user experience professionals look to the user-centered design process (UCD) as the overarching set of principles for the research, design, and testing of usable products that meet customer needs, the application of these principles varies significantly depending on the type and scale of design challenges to be solved and the level of usability maturity that a company practices. This paper describes a case study of how one organization went from a Usability Maturity Model level of *Implemented* to a level of *Integrated* while it worked through a design cycle for a large enterprise application suite. This paper also discusses lessons learned along the way.

1 The Usability Maturity Model

The Usability Maturity Model, as synthesized by Jonathan Earthy (1998), describes six levels of capability in human-centered processes at which an organization can exist. The concept of the scale is derived from earlier scales created for quality and software development. The six levels are briefly described here:

- *Unrecognized* or "ignorance": Usability is not discussed as an issue.
- *Recognized* or "uncertainty": The organization sees that it has problems with usability but does not know what to do about these problems.
- *Considered* or "awakening": The organization understands that UCD processes are necessary to improve usability and begins to implement some of these processes.
- *Implemented* or "enlightenment": The full complement of UCD processes are used and deliver good results on a per product or module basis.
- *Integrated* or "wisdom": UCD processes are included in the development lifecycle methodology, results are tracked, and performance goals are met.
- *Institutionalized* or "certainty": The organization is human centered, in both its internal function and its product design.

For this paper, we have informally assessed the levels in use in our organization; however, the complete methodology as presented by Earthy includes a more formal recording form and scale.

M. Kurosu (Ed.): Human Centered Design, HCII 2009, LNCS 5619, pp. 579–584, 2009.
© Springer-Verlag Berlin Heidelberg 2009

2 The Problem

UCD professionals, while highly trained in their individual subdisciplines, are often unprepared for the challenge of fully delivering UCD to a finished product. Occupied as they are with user research, design, prototyping, and usability testing, UCD professionals are often shocked when some or all of their work is not applied. This disconnect occurs when UCD professionals do not take into account the readiness of the organization as a whole for their work and do not undertake the non-UCD processes that ensure success.

3 A Case Study

About four years ago, Oracle started a new development cycle to design and build a next-generation integrated enterprise application suite. At the time, the company had just acquired PeopleSoft/JD Edwards and would continue to acquire significant companies during the cycle. The newly appointed applications development VP had previous positive experiences working with the UCD process and had a growing awareness that the outputs of UCD would make strategic differentiators in the suite. He decided to build a centralized team, gathering all of the enterprise applications UCD staff into one organization that reported directly to him, with dotted-line responsibility into his subteams, which were divided according to product functionality. As further evidence of his commitment to the team's success, he also granted the organization a generous number of new positions.

3.1 Initial Analysis

At the start of this process, the UCD team analyzed the historical effectiveness of its various working styles and approaches and noted areas where the organization needed to improve. While there were differences in degree, in general, the newly joined teams had all achieved a level of *Implemented* in their practices, according to the Usability Maturity Model. They had dedicated teams of UCD professionals who performed a more or less complete cycle in conjunction with individual product teams. In addition, all the teams had company-wide user interface (UI) guidelines and standards to which all products were expected to adhere. Some had managed to institute UI code reviews to ensure compliance to these standards. However, these processes were applied inconsistently across products. Some development teams were still saying that they didn't have time to incorporate all of the UCD team's recommendations. Users were still complaining that products were too technology driven and that the software was hard to learn and use. There were also numerous places in the products where the UCD teams had identified missing functionality that if implemented would greatly improve user productivity, efficiency, and satisfaction.

While the organization did not consciously use the Usability Maturity Model levels in its thinking at that time, it is clear in hindsight that the push to function fully at the next level (*Integrated*) would be a key determinant in the success of the overall effort. At the Integrated level, the following things occur:

- UCD processes are integrated with other processes in the development cycle, and UCD requirements are communicated in a way that developers understand and can implement.
- UCD processes are used to inform changes to the developing products in a timely manner, and these changes are implemented.
- Design and design solutions are treated iteratively, and goals for scope and quality are clearly stated and adhered to.

In the beginning, the technique that our organization adopted was to push harder on the UCD front. We knew that the teams needed more UCD to improve their products; however, the teams resisted the move toward this more unified approach to UCD, and we were having trouble understanding why. After all, everyone was saying that we needed to improve the products. At this discouraging moment, we decided to use some of our headcount to hire program managers and developers. We had seen that one issue that our UCD teams faced was that employees were spending too much time on process management or technical problems—not their area of expertise. We figured that by hiring product managers and developers, we could free up our UCD people to focus on their specialty. This assumption turned out to be true, but the real transformation came when these new team members changed our organization and opened up a way for us to transform the way our UCD organization functioned within the larger process.

3.2 Learning More about Our Internal Audience

We used the ethnographers in our organization who normally worked with users and customers to conduct an internal ethnographic study to help us better understand what our product management and development counterparts thought about UCD, about how our team functioned, and about how UCD could help them make better products. From this we learned three key things:

- Product managers and developers wanted a stake in the research and design process. They wanted to know that they were contributing meaningfully to the UI design. They wanted the solutions to match the requirements and to understand how the UCD team arrived at these requirements.
- Product managers and developers wanted all user experience guidelines, standards, and specifications to be in synch with delivered technology. Previously, synching user experience design standards and technology had been somewhat of a moving target, with UI designs running ahead of what was available in the technology. Developers wanted to build what they saw in prototypes and guidelines.
- Developer productivity was key. We scoured the bug database for the most annoying, time consuming, and frequently logged UI bugs. Fortunately, from one of the legacy product lines, we had data from mandatory UI reviews that we could mine. One thing we found was a problem with how key notation was generated. When developers marked fields as required, using a blue asterisk icon, they were also supposed to insert the key notation string as part of the header component. We found more than 1,000 bugs showing that developers were not

inserting the key notation because it had to be coded separately. Consider that it takes two to three hours to file and fix a bug. To optimize their process, we identified that placement of the key notation string should happen automatically when a field was marked as required, thus saving hours of extra coding or bug filing and fixing and improving overall quality and consistency. By automating the process, we enabled developers to focus on building the parts of the product that would differentiate it from our competitors.

3.3   Effect of Our Program Managers and Developers

Our program manager team members helped us understand what we needed to do to become fully integrated with the formal development process. Because the development organization was redefining this process in conjunction with a major release, the timing was ideal. In the beginning, a lot of what we had to do was to explain our process and requirements to the team responsible for defining the new development process. Once others began to understand our role and include us in the process, the tables turned, and we had to go back and look at the quality and consistency of our deliverables, ensuring that these deliverables were presented in a way that all audiences could understand, that these deliverables addressed the concerns of all stakeholders, and that all members of our organization understood how to create these deliverables consistently.

Because the development project included new UI's for every product, our team would not be able to do all the work. We needed to document the mandatory user experience deliverables in such a way that product managers could complete them as well. Had we not done the prior ethnographic work, it would have been much more difficult to do this well, and as it turned out, as teams worked through the design cycle, we did have to go through a couple of design iterations on the user experience part of the process. We were finally successful in defining a three-tier support structure to which all teams had to adhere. We then properly used the bug filing and tracking process for all user experience issues. When there were issues on which we disagreed with the design direction or approach of a particular team, the development process itself ensured that we discussed the issues and resolved them appropriately and on time. We made our process visible to other functions using the same tracking and metrics as the rest of the organization. Our program managers then helped us translate our UCD process to our executives and other team members and then translated their "language" back to us, ensuring that everyone was clear on the process.

The following illustration shows what the process evolved to over time. We were instrumental in generating this diagram, which became an effective teaching tool for all the teams involved.

Our developer and UCD architect team members participated as members of functional and technical working groups. Their work enabled us to understand how the UI would be delivered and what we needed to provide to the technical teams to ensure that we could build the UI that our users needed. Their work also gave us the opportunity to influence the technical direction, and because we were participating on the relevant working teams, our input was taken seriously, not as the agitations of clueless outsiders. We facilitated communication and cooperation among development, strategy, and product management and became in many cases the main conduit that led to the understanding of the solutions.

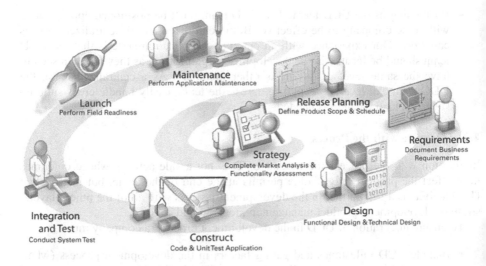

Fig. 1. The development cycle

As a result of these actions, we have successfully carried out the UCD process and avoided wasting time and effort justifying our process and resolving interorganizational problems. The design of the application suite itself, a thing of many moving parts, is now significantly more integrated in terms of the way the user experiences it. And the company now better understands the value and use of UCD. At all points in the process, UCD has a voice and a role to play.

4   Ensuring That UCD Provides Maximum Value

4.1   Get Top-Level Commitment to UCD

Top-level commitment to UCD really means that the organization recognizes that "usability" and UCD are strategic elements of the product offering. The organization may not understand all of the implications of applying these elements, but it knows enough to make an investment. Top-level commitment is demonstrated in three ways:

- Executive buy-in: Building executive buy-in takes time, unless you have a sponsor who comes in with previous successful experience with UCD. Part of your job is to increase the trust that your executives have in UCD through successful outcomes with more and more teams. The greater the number of people who have had a positive experience with UCD in an organization, the more momentum you can build around UCD.
- Adequate resources: An organization must appropriately budget for customer research, labs, test subjects, and staff for the number of necessary projects. If the major features of a product have not gone through some form of UCD, then the user experience will reflect the assumptions of the product builder, not the product users.

- Positioning of the UCD team: The UCD team must be positioned appropriately within the company to be effective. Both centralized and decentralized models can work. Our experience with an enterprise application suite is that the UCD team should be located in the development organization, as they are then seen to have the same level of buy-in as other members. As a centralized team, the UDC team can leverage expertise across the team, easily handle cross-suite issues, and load balance as needed.

4.2 Integrate with the Process

Integration with the process means that UCD is not a side activity whose outcomes may affect the product only if time permits at the end of the cycle, but instead that UCD activity is a full partner in the development process with all of the other players: strategy, development, product management, testing, and so on.

To ensure integration of UCD in the development process, a company must:

- Include UCD milestones and gating factors in the development process (whatever model your company follows), standardize deliverables, and ensure that all stakeholders understand these deliverables.
- Hire the appropriate staff to support the total effort. In order to integrate with the process, you must have people on your team who understand the perspectives and challenges of the other members of the organization Thus, you must go beyond UCD disciplines and include program management, developers, architects, and marketing specialists.

5 Conclusion

Instead of attempting to directly convert our organization to the language of UCD, we learned to communicate in the terms of product managers and developers by incorporating their disciplines on our team. We were able to meet these counterparts in the middle. Then we were able to introduce by example more detailed aspects of UCD as a normal and accepted part of the process. As a result, we are now a natural part of the process, from strategy to planning to execution. As we look forward, we see our organization beginning to function at the Institutionalized level, where the organization as a whole is human centered in its focus.

Reference

Earthy, J.: Usability Maturity Model: Human Centeredness Scale, Version 1.2 (1998),
    http://www.idemployee.id.tue.nl/g.w.m.rauterberg/
    lecturenotes/USability-Maturity-Model%5B1%5D.PDF

Developing a Scenario Database for Product Innovation

Shang Hwa Hsu and Jen Wei Chang

Department of Industrial Engineering and Management, National Chiao Tung University,
1001, Ta-Hsueh Rd, Hsinchu City 300, Taiwan ROC
shhsu@cc.nctu.edu.tw, jwchang.iem95g@nctu.edu.tw

Abstract. Introducing new product is vital for a company's survival. Scenarios
have been demonstrated as a valid tool to generate product ideas from user's
perspective. The purpose of this study is to develop a scenario data base for
product emergence and a novel method for product idea generation. The proposed scenario database is based on a product innovation database approach
that emphasizes on discovering user needs and requirements from scenarios and
incorporating them into product development. It draws on the primacy of the
idea itself as a driving force toward new product success.

Keywords: Scenario-based design, Database, Activity Theory, Product Idea
Generation, Innovation.

1 Introduction

In today's complex and dynamic competitive business environment, organizations
face serious challenges- new product life cycle shortened and only 20 percent of new
product succeeded in the market [27]. The successful products have met consumer
demands and requirements. Therefore, it is important that product innovation has to
be user-oriented rather than technical-driven.

Product innovation creates competitive advantage that allows an enterprise to differentiate itself from others [1]. In order to maintain its market position in the highly
competitive market, the enterprise requires a constant flow of product ideas [16] that
meet market demands of usability, user centered design , and time to market.

Recent research [10] indicates that the studies of product innovation can be divided
into three aspects: to emphasize the innovative people, to emphasize the process of
innovation, and to emphasize the idea generation. The studies on the innovative people focus on the issue how to obtain new innovative resources from the organization's
internal and external sources, including R & D personnel, employees, suppliers, and
schools or research organizations. The studies of innovation process emphasize on
discovering unique or special processes that allow organizations to generate new
ideas. Finally, the last one focuses on the process of product innovation that transforms creative ideas into successful new products. Among them, idea generation is
considered the most important because it is closely linked to the success of products.

Scenario-based approach has been widely adopted in product innovation because
scenarios can reveal the future viability of products [8] [13] [33], and reduce the
probability of product failure in the market. That is, scenarios provide context of use

M. Kurosu (Ed.): Human Centered Design, HCII 2009, LNCS 5619, pp. 585–593, 2009.

that describes the characteristics of the users, the activities users engage in, and the social/ cultural, technological, as well as physical environment in which they are situated. Appropriate and useful technology can thus be developed to meet user requirements [17]. In addition, users are satisfied through the interaction experience they have with the product [24]. Therefore, scenario-based design is also related to interaction design [20].

Although the scenario-based design method is beneficial to product innovation, there are several problems needed to be resolved: First of all, the cost of scenario writing and analysis is too high. The scenario analysis or inferences will consume a lot of time and manpower. The second issue is that scenario knowledge cannot be accumulated over time. A new product may have new user groups or new product application domain so that a new product development may require designers to recollect a new set of data. The third issue is that the scenario in the present form is not readily re-usable. For one thing, each scenario is considered as a unique instance of application [27]; for another, the scenario is the lack of structure [12].

To solve this problem, this study proposed an abstract structure of scenarios based on Activity Theory. The rationale underlying the abstract structure is that different activities within scenarios may share many similar elements, such as the same context and environment, the same user characteristics, or a similar goal ... and so on. By abstracting a set of scenarios, scenario genres can be generated. The abstract structure allows designers to store, to retrieve, and to reuse scenarios in product innovation.

This scenario database not only allows businesses to quickly and cost-effectively generate product ideas, but also provides enough knowledge that can tailor the past scenarios in order to meet the demand of a specific product. This will be suitable for today's highly competitive business environment in which the capability to quickly generate a large number of new product ideas determines the core competitiveness of enterprises. This database can not only assist the creative personnel in a company to innovate but also be integrated into the enterprise's own knowledge management in product innovation, helping enterprises continue to launch successful new products.

2 Related Research

2.1 Scenario Based Design

Recently scenario design method has been widely used in product development. Carroll [7] considered scenarios could be applied to three stages in product development: analysis, design, and prototype as well as evaluate. To serve its purpose, Rolland et al. [26] argued that the scenarios used in a new product development process could be divided into three types: description (descriptive), exploration (exploratory) and interpretation (explanatory). Generally, a descriptive scenario contains the description of activities [21]. More precisely, a scenario describes an event in a situation or context [34]. However, the content of a scenario is not confined to a description of the task. Verplank[32] argued that a scenario should include the people, things, activities related to all input and output of the situation, and the environment.

In the early stage of product innovation, it is desirable to develop future (exploratory) scenarios. Campbell [6] pointed out that there are four objectives of the future scenarios: to envision the use of future systems, to forecast the evolution of the function of the system, to design the product attributes or product characteristics, and to simulate the use of the product. Thus, the use of future scenarios in the development process can provide a common ground for discussion among shareholders, and make sure its future use will not deviate from the user's needs. Furthermore, Schoemaker[28] argued that in a highly uncertain environment, the use of future scenario to envision and evaluate the use of future products can assist designers in product planning. Lastly, Bardram[2,3] conceived analytical scenarios as the description of what happens, when and where the event occurs, and why and how it takes place in the scenario.

Benyon et al. [4] proposed that four kinds of scenarios should be used in various stages of the design process: the user stories (user stories) for requirement analysis, the conceptual scenarios for conceptual design, and the exact scenario (concrete scenarios) and the use cases (use cases) for detail design and evaluation. Transforming from unstructured user stories into the structured use cases in the design process requires several steps: (1)abstraction of scenarios in order to define problems identify requirements and constraints confirm design constraints and regularization of the action to define the problem, demand, resulting in the conceptual model, and finally carried out physical design. He further developed a method called PACT (people, activities, contexts, technologies) to describe people, events, scenarios and use of technology in a scenario. In addition, he used UML (Unified Modeling Language) to formalize the description. In the methods proposed by Benyon, the abstraction is an important process. By the abstraction process, sets of scenarios are classified and integrated into scenario sets (scenario corpus).

2.2 Computer Aided Idea Generation

Computer-aided idea generation is a popular field of research in the past decades [11] Many researchers use Case Based Reasoning system, TRIZ (Theory of inventive problem solving) and other computer-based product innovation support systems to generate creative ideas [19] [30]. For example, Schuring & Luijten [29] put forward a creative management system for accessing creative ideas. Creative Management System is also used to receive and capture, classification, and implementation of innovative ideas[5]. Wycoff [35] also maintained that a creative management system allows users to participate in creating, developing, restructuring, expanding, and measuring new possibilities of product ideas.

Nillson-Witell et al. [23] conceived creative management system as the system by which the user can browse creative ideas and further innovate new products. Dabhilkar & Bengtsson [8] pointed out that the structure of a creative management system should enable the majority of employees to participate in creative activities and contribute to knowledge sharing. Jamali & Boutellier [11] suggested that creative management system can collect people's ideas filter generated ideas through a collaborative process. Thus, they maintained that a creative management system is useful in enhancing the innovative capability of enterprises which is important for business growth. Riederer [25] pointed out that an innovative company, such as DaimlerChrysler, regularly use of innovative management systems to produce innovative products or services. The creative management system provides a structured approach to identifying the best potential for creativity.

Jamali & Boutellier [11] argued that a creative process management system should include four functions: (a) to produce and collect ideas; (b), the initial screening of creativity; (c), creativity to improve (Refinement); (d), screening the ultimate creative. Similarly, Turrell & Lindow [31] proposed that the creative process management system should comprise: generation, collection, development, measurement, and select the best ideas. In summary, most scholars believe that creative management system is creative and traditional methods have very different, especially the creative generation and collection process is the actual potential of new products to the creative process.

2.3 Activity Theory

Activity Theory (Activity Theory) is originated from 18,19-century German philosophy of Kant and Hegel, emphasizing that human development is be both active and constructive [18]. Activity Theory provides a framework of human activities and their context [14]. Activity Theory stressed the importance of environmental scenarios in the interaction between human activities (Interaction) and consciousness (Consciousness). Activity theory contended that the most appropriate unit of analysis is activities. Any activity components (Components) can be organized into a system of activities (Activity System) [9], as shown in Figure 1. Any activity involves the people (Subject), goals (Object), in the activities of the tools used in the (Tools), as well as the activities affect the overall output (Outcome) action (Action) and operation (Operation) [22]. According to an extended perspective proposed by Engestrom, in addition to the above-mentioned

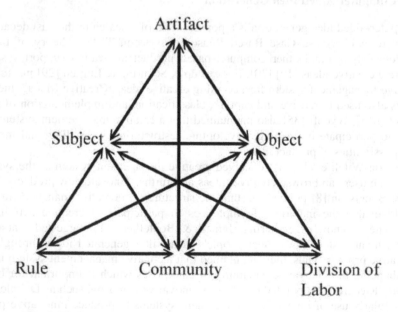

Fig. 1. The activities of the theoretical framework (Adapted from [9])

components, the norms (Rules), Community (Community) and the activities of the division (Division of Labor) should be also included. Goal is the entity that activities pursued for or a product of human mind. It can be regarded as a kind of motivation (Motivation), a problem space (Problem Space), or the output is the result of achievement of objectives. The relation between activities and objects are mediated by tools.

Kaenampornpan & O'Neill [15] suggested that the Activities Theory is helpful in identifying both key components of human activities and their relationships. It is suitable for classifying scenarios because it has several important features: (a) the framework of activities prescribed by Activity Theory can serve as a standard format to describe activities in a scenario; (b), Activity Theory relate the link between individual activities and social activities; (c), Activity Theory provides concepts of an intermediary; (d), Activity Theory models the components of human activities and the relationships between components.

3 Methods

Scenario database is a classification system of product usage scenarios. The entire development process of scenario database consists of three steps: (1) to collect scenarios, (2) to decompose scenarios into elements and then abstract to scenario genres, (3) to develop the data base, and (4)to design the functionality of the database.

3.1 Collecting Scenarios

Scenarios for an application domain (e.g., transportation) were collected. Scenario may be collected from the potential customers or end-users, designers, engineers, and marketing personnel. The methods used to collect scenarios include interviews, surveys, logs, videos, and ethnography. Typically, the content of a scenario contains a series of activities and events through which a user to achieve the final goal. The scenarios allow designers to envision user s' problems in their daily lives.

In order to ensure and validate the quality of every collected scenario, each scenario has to be examined through a scenario analysis. The main aim of scenarios analysis is to make sure the important elements of a scenario, such as the user, activities, the use of the tools, the people around, and the environment, are available in the scenario. The scenarios were analyzed in terms of 5W1H; that is, events (What is happening?), around there who (Who are the roles in scenarios?), wherever they occur (Where the scenarios take place?), when (When did the scenarios happen?), Why (Why did there scenarios happen?), and how the scenario is carried out (How did the scenarios proceed?).

3.2 The Decomposition-Abstraction Process of Scenarios

This part aims to decompose scenarios into elements and then abstract the higher-level concepts of the elements. The abstraction of higher-level concepts can yield generalized sets of scenarios (i.e., scenario genres).

- Identify the activities in Scenarios

Scenarios may cover several ongoing activities. The decomposition process should cover all the ongoing activities and their relationships. Since activity is goal-directed, abstraction of activities is to identify the activity group that shares the same goal. Accordingly, we have identify goals of activities.

- Define the Subject, Object, and Community in Scenarios

The subject attempts to achieve a goal by carrying out activities. In performing an activity, a subject (i.e., a user) has to exploit the object available to him/her. The subject in activities can be an individual or a group of people (i.e., a community or an organization). The object can not only serve as a mediator between subject and goal but also relate people within a community. A community includes the physical context (such as time, place, and weather) and social context (such as group dynamics). Therefore, in this step, we need to identify the subject, the objects, and the community in scenarios and then categorize them into subject types, object types, and community types.

- Analyze Users Motives in Scenarios

Users are driven by their motives (such as: social needs) in carrying out activities, and finally reached its desired goal (such as: friends invited to participate in the party). Analysis of the motives can allow designers to discover users' explicit and implicit needs. User needs analysis can reveal the functional and emotional requirements of the product.

- Decompose Activities in Scenarios

Activity has a hierarchical structure. It is a coordinated, integrated set of actions. Actions are targeted toward a specific goal, and are achieved through operations. Conditions will control what operations will be taken, and conditions and goals together will determine the use of the product. Based on the activity-actions-operations hierarchy and activity flow pattern, a use model is then developed to describe the mapping between product functions and activity flow. Accordingly, the decomposition of activities can lead to the mapping between functions and actions.

- Analyze the rules that govern the interaction among subjects in a community

The rules exist between the subject and the community. The rules make the subject and the community to sustain their activities. Rules may be the social rules, such as laws, customs. Analysis of the rule can led to the identification of social interactions.

- Analysis of the division of labor between Community and Object

In order for achieving a common goal, members in the community have to play certain roles and assume proper duties. Therefore, task activities have to be divided and allocated to appropriate people. This division of labor may prescribe the roles, such as the leader or be leaded, manage or be managed, and the relationship between different roles. The division of labor describes task allocation and allows designers to identify the need for team collaboration.

3.3 Develop the Database Structure

Statistical analysis is performed to establish the relationships among subject types, activity types, motives, social interaction, and team collaboration. This constitutes a database of user requirements. User requirements can then be mapped to functional requirements and

3.4 Design Scenarios Database's Functionality

This study designs the scenario database functions through case-based reasoning (CBR) mechanism. The concept of CBR is originated from psychology and became the most popular AI paradigm in idea generation domain. The theorem of CBR is to retrieve similar cases to solve the problem. CBR requires exploration of the analogical memory rather than searching it for a specific analogy. Based on CBR, the functions of the scenario database are identified as below:

- Retrieve: to search for extract the most similar scenario case, or a number of high similarity candidate in order to solve the problems.
- Re-use: to reuse the information and knowledge contained in the scenario case.
- Revise: to modify and update the information of scenario case.
- Retain: to save and add scenario case in scenario database.

By using these functions, innovation activities can be accomplished by retrieving similar scenarios cases.

4 Conclusion

Scenario Design is a widely used method in product innovation process. Scenarios describe the process and the context that people perform their activities. Through the understanding of the activities, designers can envision the demand. However, the validity of scenario-based design is questioned by opponents because only a limited number of scenarios were employed in the design. Therefore, the purpose of this study is to develop a scenario database that consists of a set of scenarios encompasses a wide range of user problems. As Nardi [21] points out that activity theory provides a framework that can tie the loosely connected elements of a scenario into a well-defined structure. Based on Activity Theory, the elements in scenarios were abstracted into generalized types and, consequently, the information in the scenario can be reused and accumulated over time.

In the database scenarios, case-based reasoning is employed as an inference mechanism for applying the database for a new product development. Case-based reasoning is a simulation of human problem-solving mechanism, with its emphasis on experience and memory through the past to solve new problems. It also has implied learning function. The construction of creative management system not only allows experts to collect, modify, and delete a scenario in the database, but helps members of the organization learn from experts how to present and solve a design problem. Through the use of this creative management system, inter-departmental co-operation (such as design and manufacturing sector) can be enhanced and so does organizational creativity.

References

1. Alves, J., Marques, M., Saur, I., Marques, P.: Building creative ideas for successful new product development. In: 9th European Conference on Creativity and Innovation (ECCI-9): 'Transformations', September 4-7 (2005)

2. Bardram, J.: Scenario-based design of cooperative systems. In: Proceedings of COOP 1998, Cannes, France (1998)
3. Bardram, J.: Scenario-based Design of Cooperative Systems: Re-designing a Hospital Information System in Denmark. In: Group Decision and Negotiation, vol. 9, pp. 237–250. Kluwer Academic Publishers, Dordrecht (2000)
4. Benyon, D., Turner, P., Turner, S.: Designing interactive systems: People, activities, contexts, technologies. Addison Wesley, Boston (2005)
5. Bessant, J., Caffyn, S., Gallagher, M.: An evolutionary model of continuous improvement behavior. Technovation, pp. 67–77 (2001)
6. Campbell, R.L.: Will the real scenario please stand up? SIGCHI, Bull. 24(2), 6–8 (1992)
7. Carroll, J.M.: Scenario-based design: Envisioning work and technology in system development. John Wiley & Sons, New York (1995)
8. Dabhilkar, D., Bengtsson, L.: Continuous improvement capability in the Swedish engineering industry. International Journal of Technology Management 37(3-4), 272–289 (2007)
9. Engestrom, Y.: Learning by expanding. Orienta-Konsultit Oy, Helsinki (1987)
10. Goldenberg, J., Mazursky, D.: Creativity in product innovation. Cambridge University Press, Cambridge (2002)
11. Jamali, N., Boutellier, R.: Idea Management System: Process. Continuous-innovation.net (2006),
    http://www.continuous-innovation.net/Members_Only/Publications/papers_7th/jamali_boutellier_cinet2006.pdf
12. Jarke, M., Bui, X.T., Carroll, J.M.: Scenario management: An interdisciplinary approach. Requirements Engineering Journal 3(3/4), 154–173 (1998)
13. Joe, P.: Scenarios as an essential tool: stories for success. Innovation Quar. J. Ind. Des. Soc. Am, 20–23 (1997)
14. Jonassen, D.H., Murphy, M.: Activity theory as a framework for designing constructivist learning environments. Annual meeting of the Association for Educational Communications and Technology 47(1), 61–79 (1998)
15. Kaenampornpan, M., O'Neill, E.: An Integrated Context Model: Bringing Activity to Context. In: Proceedings of UbiComp 2004, Tokyo, Japan (2004)
16. Kao, J.: Jamming: The Art and Discipline of Corporate Ceativity. Harper Business (Creativity), 224 (1997)
17. Kreifeldt, J.G.: Guarding snowblowers. In: Interface 1987 Proceedings, Human Implications of Product Design. Human Factors Society, Consumer Products Technical Group, Santa Monica, CA, pp. 259–304 (1987)
18. Kuutti, K.: Activity theory as a potential framework for human-computer interaction research. In: Nardi, B. (ed.) Context and consciousness: Activity theory and human computer interaction, pp. 17–44. MIT Press, Cambridge (1996)
19. Maher, M.L., Balachandran, M., Zhang, D.M.: Case-Based Reasoning in Design. Lawremce Erlbaum, Hillsdale (1995)
20. Moggridges, B.: Designing Interactions. MIT Press, Cambridge (2007)
21. Nardi, B.: The use of scenarios in design. SIGCHI Bulletin 24(4), 13–14 (1992)
22. Nardi, B.: Context and consciousness: Activity theory and human computer interaction. MIT Press, Cambridge (1997a)
23. Nilsson-Witell, L., Antoni, M., Dahlgaard, J.J.: Continuous Improvement in Product Development; Improvement Programs and Quality Principles. International Journal of Quality and Reliability Management 22(8), 753–768 (2005)
24. Norman, D.A.: The psychology of everyday things. Basic Books, New York (1998)

25. Riederer, J.P., Baier, M., Graefe, G.: Innovation Management-An Overview and some Best Practices. C-LAB Report 4(3) (2005)
26. Rolland, C., Ben, A., Cauvet, C., Ralyté, J., Sutcliffe, A., Maiden, N.A.M., Jarke, M., Haumer, P., Pohl, K., Dubois, E., Heymans, P.: A Proposal for a Scenario Classification Framework. Requirements Eng. J. 3(1), 23–47 (1998)
27. Rosson, M., Carroll, J.M.: Scenario based design. The Human-Computer Interaction Handbook: Fundamentals (2002)
28. Schoemaker, P.J.H.: Scenario planning: A tool for strategic thinking. Sloan management review 36(2), 25–40 (1995)
29. Schuring, R., Luijten, H.: Re-inventing suggestion schemes for continuous improvement, Continuous Improvement - From Idea to Reality. Twente University Press, University of Twente (1998)
30. Sun, Y., Lai, C.: A conceptual design system based on human-machine integrates intelligence. Journal of Information & Computational Science 1(2), 313–318 (2004)
31. Turrell, M., Lindow, Y.: The Innovation Pipeline. Imaginatik Research White Paper (March 2003)
32. Verplank, B., Fulton, J., Black, A., Moggridge, B.: Observatin and Invention- Use of Scenarios in Interaction Design. Tutorial notes (1993)
33. Welker, K., Sanders, E.B.-N., Couch, J.S.: Design scenarios to understand the user. Innovation Quart. J. Ind. Des. Soc. Am, 24–27 (1997)
34. Wright, P.: What's in a scenario? SIGCHI Bulletin 24(4), 11 (1992)
35. Wycoff, J.: The Big Ten Innovation Killers and How to Keep Your Innovation System Alive and Well. Innovation Network (2004)

Practice of Promoting HCD Education
by a Consumer-Electronics Manufacturer

Jun Ito[1], Akiyoshi Ikegami[2], and Tomoshi Hirayama[1]

[1] UI Technology Planning Dept., CPG UI Center, Sony Corp.
[2] SQA Dept., Software Technology Div., TV Business Group, Sony Corp.
Shinagawa INTERCITY C Tower Shinagawa Tec. 2-15-3 Konan Minato-ku,
Tokyo, Japan
{Jun.Itojun,Akiyoshi.Ikegami,Tomoshi.Hirayama}@jp.sony.com

Abstract. We conducted an internal e-learning course on the basics of HCD, starting in February 2008. E-learning is convenient in that any employee can take the course whenever they like. The e-learning course described in this report aims at preventing miscommunication regarding user interfaces. What we expect to achieve by realizing that purpose is, (1) to enable to memorization of the defined common terms, and (2) to encourage daily use of the common terms when talking about user interfaces in the workplace. Then we decided to select seven key words and three methods that make up the ten common terms.

Keywords: E-Learning, HCD, grouping, mapping, feedback, constraints, consistency, fail-safe, affordance, persona, scenarios, paper-prototyping.

1 Introduction

We conducted an internal e-learning course on the basics of HCD, starting in February 2008. E-learning is convenient in that any employee can take the course whenever they like. This report explains the e-learning course as a practice of promoting HCD education within a consumer-electronics manufacturer.

1.1 Conventional HCD-Related Education

HCD education in the company is mainly provided through on-the-job training at planning, engineering, or designing departments for those who are directly involved with user interfaces.

On the other hand, one-time and non-regular in-house seminars were held as a part of off-the-job training for those who have an interest in user interfaces.

These training curriculums focused on those who were involved directly with or who had much interest in user interfaces. Therefore, when you looked at the whole company, there was a big gap in the knowledge level of HCD. This means that those who know about HCD could learn more, while those who don't know still had no idea. However, as long as people who were engaged in the design of user interfaces

M. Kurosu (Ed.): Human Centered Design, HCII 2009, LNCS 5619, pp. 594–600, 2009.

were limited and their basic knowledge level was maintained, there was no significant problem.

2 Providing the E-Learning Course

2.1 Background of Providing the E-Learning Course

In recent years, consumer electronics have become digitized and multifunctional. As a result, the scale of user interface-related software has expanded, and more people have come to get involved in the production process accordingly. Since the number of members directly involved with user interface design has grown, people who have different knowledge levels have come to work together during the design process. When members at different knowledge levels talk about user interface design, even though they seem to be having a natural conversation, each of them actually have different image in his or her mind, and thus it turns out afterwards that they couldn't communicate with each other at all.

The e-learning course described in this report aims at preventing miscommunication regarding user interfaces. In other words, it has two main purposes; one is that this course is not only for people interested in user interface but also for all employees involved in the product design (Purpose 1), and the other is to establish and spread common terms used in talking about user interfaces in order to prevent miscommunication (Purpose 2).

2.2 Preparation of an E-Learning Course

How can we achieve these purposes? Thousands of employees are engaged in the product design process. Such a large number of people cannot be covered by non-regular seminars or training courses held at a certain place. Fortunately, as we have an e-learning system that has already been utilized for various training courses and we decided to adopt this system for the first purpose (Purpose 1).

To accomplish the second purpose (Purpose 2), establishing and spreading common terms, the following three points were taken into account: [1] the number of common terms were narrowed down to ten and were repeated again so that they were committed to memory, [2] icons corresponding to the common terms were created, so that they can be easily memorized both literally and visually, [3] and many examples were prepared to help understand their meanings.

What we expect to achieve by realizing (Purpose 2) is, (1) to enable to memorization of the defined common terms, and (2) to encourage daily use of the common terms when talking about user interfaces in the workplace. To realize this, the e-learning course is defined by the following characteristics: [1] a participant cannot finish the course unless he or she passes a brief test, and [2] the card (Fig.1) that describes the common terms with icons is given to a participant to remind him or her of the terms during daily work.

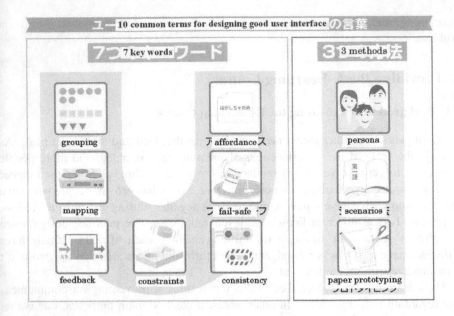

Fig. 1. The common terms reference card distributed to the learning course attendees

2.3 Contents of the E-Learning Course

We explain the common terms based on the following three aspects; (1) physically-friendly usability (ergonomics), (2) brain-friendly usability (cognitive engineering), and (3) usability to achieve a goal (human-centered design). As an example to explain the differences among these three aspects, we provide a case that is referred to through the entire course, and show the different improvements that each approach makes according to the respective concepts of ergonomics, cognitive engineering and human-centered design. Incidentally, the terms ergonomics, cognitive engineering, and human-centered design are used symbolically, and they are not necessarily used to convey academic definitions.

Now, as for the selected common terms (Fig.1), they can be mainly divided into two categories. One includes seven symbolic words; "grouping", "mapping", "feedback", "constraints", "consistency", "fail-safe", and "affordance", which are key words mainly related to cognitive engineering and also essential in talking about usability. The other includes three stereotype methods which help mainly human-centered design; "persona", "scenarios", and "paper prototyping". The seven key words and three methods make up the ten common terms.

We chose a "paper prototyping" from a number of prototyping methods because we aim to stimulate even those who have been involved with the design of user interfaces. We wanted to surprise them with an unexpected method and make them think, "I've never imagined that we could utilize such a low-tech method!"

Next, regarding the e-learning methodology, we take full advantage of the e-learning system, using a lot of videos and animations to make the program interesting and easy to understand. As shown in Fig. 2, an example of "constraints" is illustrated with the video of a situation where a 3.5-inch floppy disk can fit in the disk drive in the correct direction, but cannot fit in any other direction. The "mapping" icon represents a gas cooking stove, and the animation we used shows the scene of turning on a range after a control knob rotates (Fig. 3).

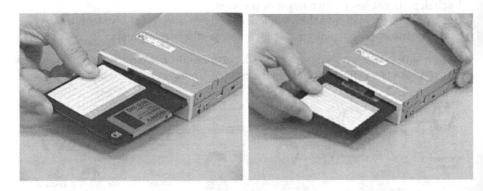

Fig. 2. Video showing the scene where a user attempts to insert a floppy disk into a disk drive: A floppy disk cannot fit in wrong direction (left). It can fit only in the correct direction(right).

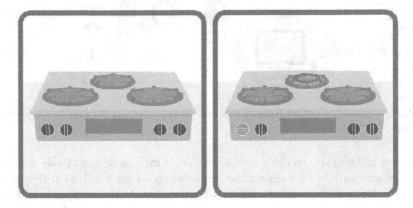

Fig. 3. An animation which shows one of the gas cooking ranges is turning on when a control knob is rotated

A paper prototyping method is explained by showing the video of a usability test using a paper prototype of an intercom and adding explanatory comments. This aims at providing a "seeing is believing" effect.

3 Persona and Scenarios Introduced in the E-Learning Course

As mentioned in the section "Background of providing the e-learning course" above, many people in various types of jobs and from various departments are involved with the design of consumer electronics products (Fig.4). Therefore, the following problems are likely to occur.

Each department is virtually independent.
Each department has its own specialized area.
Each department has its own target or position.
Each department interprets things independently to achieve its own target.
Each department is often physically away from other departments.

As a result, people working in such a department do not pay enough attention to other departments and tend to think "self-centered," not "user-centered."

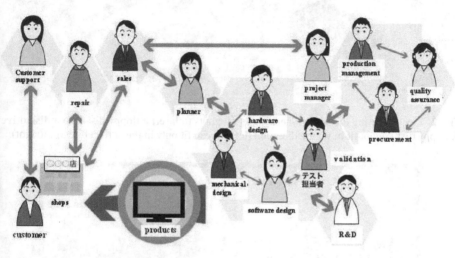

Fig. 4. Many people in various types of jobs and from various departments are involved with the design of consumer electronics products

Persona and scenario methods are introduced in the course as methods that enable people in various jobs or from various departments to get involved with and work together on design, and to share ideas of what users need. However, the methods are explained just briefly, because the primary objective of the program is to expand use of the basic terms.

Regarding persona, a basic explanation with a brief example is given; the persona is not an average person, but a specific customer image should be accurately created so that its actions are consistent with its situation.

Regarding scenario, besides an interface scenario, which can be easily imagined by engineers designed user interface. There is another method called task scenario. It is important to describe separately what a user wants to do and what he or she uses. What a user wants to do, namely a user's goal, should be described in a task scenario so that it can be shared for communication among people in various jobs or from various departments. The significance of these concepts is emphasized in the course as well.

4 Response and Effect of the E-Learning Course

4.1 Response to the E-Learning Course

Although the number of people who have taken the course reached 500, it cannot be said that it has spread widely in the company. Whereas our goal is that all employees involved with design take the course, the ratio of those who are involved directly with the design of user interfaces is still high. Part of the questionnaire results conducted at the end of the course to get feedback from those who took the course is shown in Fig.5. The course was evaluated highly both in "ease of understanding" and "satisfaction" (Fig.5). In particular, understandability was appreciated. Many of the participants answered favorably, saying that the course explains the subject and improvement approaches very well based on examples. We confirmed that we achieved one of the goals we set when preparing the course curriculums.

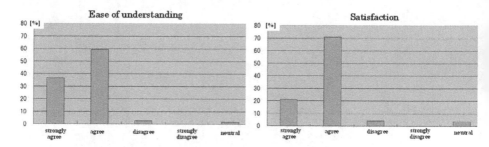

Fig. 5. Part of the questionnaire results conducted at the end of the course

On the other hand, when we look at comments from those who are already engaged in UI design, whereas many people felt that the course helped them understand logically what they had already vaguely recognized, some felt that the explanation was lengthy or repetitive. This was actually an adverse effect of the course that was designed so that entry-level personnel could memorize at least the common terms.

4.2 The Effect of the E-Learning Course

We expect that this course will encourage people to use the common terms on a daily basis to make internal discussion about user interfaces more productive, encourage them to talk about user interfaces, and let them know such terms are needed. Since not so many people have taken the course so far, it will require some time for it to have an effect.

The unexpected benefit of this course was that the contents of the learning course are very helpful for the members that prepared the course. When explaining user interface-related issues internally, they can effectively use the common terms and icons effectively. They can also quote many examples given in the course in order to supplement their conversation. The members themselves come to make full use of the course from time to time.

5 Conclusion

We would like to take measures to encourage more people to take this learning course in order to familiarize them with HCD-related common terms used internally. We are also thinking of gradually increasing the common terms. Our objective is that employees from different departments would conduct an effective and productive discussion using a rich common vocabulary for HCD or user interface design.

On a final note, we have never forgotten that it is necessary to enhance a learning course to grow those who already know much about HCD at high level.

A Survey of User-Experience Development at Enterprise Software Companies

Aaron Marcus[1], Jeremy Ashley[2], Clause Knapheide[3], Arnie Lund[4], Dan Rosenberg[5], and Karel Vredenburg[6]

[1] Aaron Marcus and Associates, Inc., 1196 Euclid Avenue,
Suite 1F, Berkeley, CA, 94708 USA
[2] Applications User Experience, Oracle, Inc., Redwood Shores, CA, USA
[3] User Interface Design Center, Siemens Corporate Research, Princeton, NJ, USA
[4] IT Relationship Experience Division, Microsoft Corporation, Redmond, WA, USA
[5] User-Experience Group, SAP, Inc., Palo Alto, CA
[6] IBM Canada, Ltd., Markham, Ontario, Canada
Aaron.Marcus@AMandA.com, Jeremy.Ashley@Oracle.com,
Claus.Knapheide@Siemens.com, Arnie.Lund@Microsoft.com,
Dan.Rosenberg@SAP.com, Karel@ca.IBM.com

Abstract. Developers worldwide wish to understand what major companies are doing in user-experience development (UXD). UXD comprises activities in user-centered design of user experience, specifically user-interface development (metaphors, mental models, navigation, interaction, and appearance) that is useful for planning, research, analysis, design, implementation, evaluation, and documentation of products/services across a wide number of platforms. This paper reports the results of a survey conducted with six enterprise software companies.

Keywords: Design, development, management, user interface, user experience.

1 Introduction

User-experience development (UXD), especially user-experience design, within major computer technology firms throughout the world is undergoing significant change in corporate product/service development. These firms themselves are attempting to determine answers to questions such as these: How does one define UXD? What is required for the UX organization? Which kind of professionals is required? What tools are useful? What methods are/should be employed?

This initial survey seeks to inquire about what is occurring among major centers of computer technology development. For this survey, Aaron Marcus contacted the other authors and a representative of Fujitsu, who agreed to participate. This survey attempts a difficult and delicate task, inquiring about UXD from busy senior management at major corporations who may not wish to divulge proprietary information or may be reluctant to divulge information that is viewed as providing their companies with strategic or tactical advantage, bordering on revealing trade secrets and other intellectual property protection.

M. Kurosu (Ed.): Human Centered Design, HCII 2009, LNCS 5619, pp. 601–610, 2009.
© Springer-Verlag Berlin Heidelberg 2009

2 Some Comments

In the survey responses, which are presented in the Appendix, comments appear throughout the survey replies to help the reader interpret those replies. Based on this survey as well as other research the first author and his firm have undertaken over the past 25 years, the following interpretations or initial conclusions may be drawn:

- Information is of interest not only to those at the contributing corporations, but among the AIGA, CHI, HCI, HFES, STC, UX, and other user-interface design communities.
- Companies vary significantly in make-up of staff and backgrounds
- Information design and information-oriented, systems-oriented visual design seem under-represented, perhaps misunderstood/undervalued in team make-up and disciplines.
- Only a few of the companies attempt to define "user experience" and other key terms. It may also be the case that only a few of the companies have a rigorous process definition and diagrammatic representation of that process.
- It seems possible that there may be "culture clashes within UX development groups due to varying backgrounds, understanding of key terms, differences of method, career goals, value placed on research and publishing *vs.* product/service deliverables, etc.
- Variations in the UX development process, methods, and objectives may be found based on local cultural preferences and habits despite an international company's preferences for uniform characteristics.

Acknowledgments. The first author acknowledges assistance of the other authors cited, in addition to Mr. Makato Morioka' Fujitsu Corporation, Technology Solution Design Dept., Fujitsu Design, Ltd., Tokyo, Japan. They provided survey replies despite busy schedules. He would also thanks Mr. Niranjan Krishnamurthi, Designer/Analyst, AM+A for contributions in editing the paper.

Appendix: Survey Questions and Replies

Note: The following questions and replies from a Survey Monkey response are edited to correct errors in spelling, grammar, ambiguity of reference, or other aspects of style.

1. What is the approximate number of full-time equivalent UX "designers" on staff (may include people whose primary job is planning, analysis, evaluation, etc.)?
Fujitsu: About 40-50. All answers are based on Fujitsu Design Limited.
IBM: Approximately 500 User Experience and Design practitioners in IBM world-wide.
Microsoft: 12.
Oracle: 150.
SAP: 180.
Siemens: 12 for the US, 30 in Germany and 10 in China in a central function to work as internal consultants. Approximately another 50 worldwide in operating companies.

2. What is the approximate number of full-time equivalent contracted UX designers?
Fujitsu: About 10-15.
IBM: Less than 10 world-wide.
Microsoft: Varies (from 1 to 6).

Oracle: 15.
SAP: 10.
Siemens: 0 in the US, 10 in Germany. The number of consultants in operating companies is not more than 12 worldwide.

3. What kinds of prototyping tools are being used?
Fujitsu: Paper prototyping, PowerPoint, Flash, HTML. Depends on the project.
IBM: PowerPoint, Visio, Visual Basic, PhotoShop, and IBM products not yet available (still in closed beta).
Microsoft: Paper prototyping, PowerPoint, HTML, Flash, WPF/Silverlight, etc.
[WPF refers to Windows Presentation Foundation. Microsoft's Website: http://www.microsoft.com/net/ wpf.aspx states: "Windows Presentation Foundation (WPF) is Microsoft's strategic presentation technology for Windows smart client user experiences. Customers can use WPF to deliver innovative user interfaces through support for multimedia and document services, hardware acceleration of vector graphics, resolution independence for different form factors, and enhanced content readability. WPF tools improve developer-designer collaboration through Microsoft Visual Studio®, Microsoft Expression® Interactive Designer, and XAML. WPF enables the developer to write user interface code once and deploy it as a stand-alone client or in a browser. Developers can incrementally take advantage of WPF through interoperability with Win32® and Windows Forms. And they can also use existing knowledge in .NET Framework, CLR languages, and Visual Studio IDE."].
Oracle: Dreamweaver, Visio, Photoshop, JDeveloper, and "homebrew."
SAP: Visio, Flex, and custom internal tools.
Siemens: Flash, WPF, Visio, Photoshop.

4. Do you plan to use some new kind of tool or technique?
Fujitsu: Not specifically.
IBM: Adoption of Agile and Lean development practices are driving forces in investigating multiple tools, techniques, methods, and assets for user research, design, and evaluation.
Microsoft: WPF/Silverlight, and we are building our own prototyping tool; plus, I like to use video simulations.
Oracle: Typically we homebrew beyond the commercial tools available.
SAP: No.
Siemens: WPF/Silverlight.

5. What is the approximate UX-development investment-ratio per one product? That is, what is the total amount of dollars spent on UX vs. total development cost dollars?
Fujitsu: Do not know.
IBM: It depends on the definition of "investment" and of "total development cost." The best figure we can provide is that the number of user-experience and design practitioners in IBM is about 2 - 3% of the total number of developers (not including technical support).
Microsoft: 5%.
Oracle: The best we can do here is that we have 150 people working on core UC supporting approximately 5000 developers. One could also include the 200 more developers who are supporting the core UX toolbox and related technologies.
SAP: Approximately 1%.
Siemens: 5%.

6. For new product development and existing product revision, what is the breakdown of development-process time for planning, research, pre-design testing, evaluation or post-design testing, and evaluation?
Fujitsu: Do not know.
IBM: There is too much variability to provide a meaningful answer. Release life cycles range from weeks to over a year, and there are wide differences across hardware, software, and service projects.
Microsoft: Varies considerably...Say 5% planning, 15% pre-design research, 10% evaluation, and the rest is the design and usability process itself from low fidelity to high.
Oracle: Depends on product. On a high value new product the design time is six months, and dev and QA 12 to 14 months.
SAP: This question does not make sense. New product development is totally different than for a revision of an existing one. I don't understand what "pre-design testing" means. You don't test something that does not exist.
Siemens: Planning 10%, research 20%, evaluation 10%; missing from your list is concept, interaction, and visual design which is 60%.

7. Do you prepare "design pattern documents"? Why?
Fujitsu: We do not prepare our original "pattern document". We will refer to the books or other available public sources.
IBM: Yes, we have been developing UI design patterns to address common design solutions across products. The value of these design patterns is to improve efficiency/reuse, promote consistency, and deliver a compelling user experience.
Microsoft: Yes, to bring consistency across systems, and to enable reusable code; we also create style guides within the specifications.
Oracle: We are very pattern oriented. So much so that we have had the design patterns integrated into the development environment so developers do not have to code each pattern from scratch each time. This increases productivity, consistency, by lowering the design time and number of bugs filed.
SAP: Yes, for both UI standards and product line specific specifications.
Siemens: Yes, to maintain consistency and foster re-use. This is part of our branding efforts.

8. Do you make use of moderated or un-moderated remote testing (via the Web)?
Fujitsu: No.
IBM: Yes.
Microsoft: We have used moderated remote testing, and are building an infrastructure for un-moderated remote testing.
Oracle: We do both. We do a lot of moderated remote testing, but we also run surveys etc.
SAP: Yes, but not that often.
Siemens: Yes, we do. Development is usually validated in all regions of the world, and our 3 local offices use the web to do that.

9. For how long have you had a UX group or department?
Fujitsu: About 15 years.
IBM: IBM has had UX practitioners for over 40 years.
Microsoft: 20 years across companies, 5 years here at Microsoft, 1 year in my current job within Microsoft.
Oracle: 13 years.
SAP: I have managed this one for 3 years, but it has existed as a corporate function for at least 13 years.
Siemens: 10 years.

10.   Do you have posted a standard definition of UX? Can you please provide it?
Fujitsu: We do not have a clear definition of UX.
IBM: Unclear whether you are referring to practitioners or to the product user experience. The product total user experience encompasses all aspects of product interactions by all stakeholders and users throughout the life cycle of the product (e.g., marketing, purchase, acquisition, installation and deployment, operation and administration, troubleshooting and technical support, etc.).
Microsoft: Not really... we argue it is not identical to UI; but it is the UI as experienced in the rich context of users attempting to achieve their goals.
Oracle: http://usableapps.oracle.com/about/index.html.
[The Website first page states the following (see next question, also): "Welcome to the Applications User Experience Web site. We created this site to share with you some of the outstanding work Oracle has been doing to define the current and future user experiences across our product lines. Through blogs and articles, we intend to enhance your understanding of our methods and processes and give you an opportunity to participate through your feedback.
SAP: Yes, refer to the SAP Design Guild site.
[The User Experience Glossary is located at the following URL: http://www.sapdesignguild.org/resources/ux_glossary.asp. Unfortunately, there is no entry for "user experience"! The following related terms are cited, however: *User:* Anyone who works directly with SAP software or applications as a regular part of his/her job – in contrast to a customer, who is typically involved in purchasing and/or supporting such software or applications – synonymous with "end user". *User-centered design (UCD):* A research and design methodology that is both a philosophy and set of methods that focuses on designing for and involving end-users in the development process. UCD is based on based on four principles: 1) focus on "real" end-users, 2) validate UI requirements and designs, 3) design and prototype iteratively, and 4) understand and design for holistic user experiences. The SAP UCD consists of 3 Phases, Understand Users, Define Interaction, and Design UI. *User interface designer (UID):* Person who consults developers with respect to the design of the user interface and interaction of an application. The UID may also design and create the user interface him- or herself. At SAP, user interface designers are organized 1) in a central group, the User Experience (Ux) group, and 2) decentralized in development groups. The Ux coordinates the information flow for the UIDs in the company and among the UIDs

and the Ux group. User profile – A part of the deliverables from Phase 1 of UCD, Understand Users, that describes users' distinguishing characteristics and aspects of their work environment that inform what the user interface should include.]

Siemens: Our mission statement says, "We enhance the value of our partner's technology by advocating human capabilities and user needs." We point to humans to capture the physical strengths and limitations of the human being and, and to users which puts our efforts into the context of Siemens products and their use at the workspace. Almost all Siemens products are capital goods, which means the customer has bought them to improve business processes. Human capabilities, even though these are often also formed by professional and cultural experience, are more general in that the phenomena are closer linked to the sensoric apparatus of the human body as well as its capabilities to move and act. Interaction design and visual design cater to both these aspects (humans, users) but in different ways. The mission also expresses that technology alone isn't enough, but that its value can be enhanced by improved User Experience. User Experience describes the functional system between the users and the system as a whole, while User Interface can be misunderstood as pointing to what 'can be seen on the screens'. We find that this would be too narrow a focus.

11. What is the mission of the UX department or group?

Fujitsu: 1. Promoting user experience of products and services to be attractive enough to contribute to the overall sales of Fujitsu. 2. As an activity of corporate social responsibility (CSR), we try to be contributory to the society. For example, we participate in the external organizations such as International Association of Universal Design (IAUD), or we have developed and provided free download web accessibility check tools for web developers. We have hosted exhibitions to public periodically.

IBM: To deliver a compelling user experience for IBM products and services.

Microsoft: Deliver successful, innovative, compelling and satisfying branded user experiences.Ground design in deep user understanding. Create a user-centered design project structure as a foundation for excellence and continuous improvement.

Oracle: Website: http://usableapps.oracle.com/about/index.html.

[The mission seems to be the following: To extend the value of your [the customer's] investment in Oracle applications. The Website first page states the following:

"Welcome to the Applications User Experience Web site. We created this site to share with you some of the outstanding work Oracle has been doing to define the current and future user experiences across our product lines. Through blogs and articles, we intend to enhance your understanding of our methods and processes and give you an opportunity to participate through your feedback.

What we do:

Enhance our customers' investment in Oracle products.

The Applications User Experience group has one goal: to extend the value of your investment in Oracle applications. We believe that the core areas of user experience that will best enhance your bottom line are increased productivity, increased insights, and increased collaboration.

Increased productivity means helping your employees become more effective at the work they already do. At the end of the day, a productive employee is one who can complete their work accurately and efficiently. We streamline task flows, increase personalization of applications, reduce unnecessary navigation, and remove unneeded clicks in order to reach this end goal.

Increased collaboration means lowering the boundaries that keep your employees from connecting. An improved user experience enables your employees to easily leverage the collective knowledge of your organization or department. We work to enhance social networking by examining integration with email and chat clients, determining how Web 2.0 features can be deployed, and considering how users make use of mobile technologies.

Increased insight means supporting rapid, effective judgments. Our design approach provides your employees with the ability to access the right information, at the right time, to make the right decisions. Throughout our product suites, we look for opportunities to make use of best practice business intelligence.

How we do it: Oracle's commitment to User Experience

Oracle has maintained a core competency in user experience for over 12 years, and most recently welcomed the PeopleSoft, JD Edwards, Siebel, and Agile user experience teams into the organization. By adding such talent to our existing expertise, we continue to increase our understanding of your business problems. We maintain usability testing laboratories around the world to monitor and respond to your changing market needs.

Oracle has taken a proactive position on ensuring a quality user experience for enterprise applications. At Oracle, "user experience" (UX) means examining how your users work today, understanding their pain points, and working closely with product management, strategy, and development to target the most effective improvements in product design. UX is an integral part of the development process for enterprise applications.

In the Applications Unlimited space, we leverage the same design techniques and usability methodologies used for all of our other applications. We also adhere to many of the same principles of improving user productivity and insight. Additionally, we place a great emphasis on ensuring each product line user experience helps reduce overall operating costs through improved usability.

Our main user experience goals for Applications Unlimited are to:

– Continue our UX investment in each product line.
– Implement best design practices by leveraging Oracle user experience guidelines, standards, and patterns.
– Evolve to the next generation of applications by using the latest innovations, such as Web 2.0, mobile experiences, and desktop integration.
– Work closely with customers through our Global Design Partners program to conduct site visits, usability tests, and other activities to ascertain and prioritize customer wants and needs.

We've spent the last three years defining world-class experiences for Fusion based on listening to you. We sent our UX professionals to your offices, branch locations, call centers, and manufacturing facilities. But we didn't stop there. As we designed our Fusion user experiences, we validated our designs with the people who would be using our products—your users. In addition, designing the Fusion user experience involved not only understanding how your end users work today, but also how they will work in the future. We continue to investigate how evolving end-user expectations will shape the future.

Fusion MiddleWare: The platform for Enterprise Applications user experiences
Oracle Fusion Middleware (FMW) is at the center of our user experience design strategy. FMW serves as the foundation for our next generation user experiences and enables them to be rendered on any supported device (laptop, desktop, mobile phone, PDA, widescreen, and so on). Oracle ADF WebCenter (a component of FMW) provides a framework for managing and maintaining context within our applications, and comprises several key underlying services, including Web 2.0 services for supporting wikis, blogs, and enhanced collaboration. Oracle ADF Design Patterns provide developers and designers with the ability to easily reuse and repurpose common design templates embedded with prebuilt services. These and other middleware innovations are at the forefront of our user experience innovations.

How to get the most from this Web site
Visit our blog regularly for a glimpse into the latest thought leadership on the user experiences of Oracle enterprise applications. We invite you to participate in those experiences by sharing your reactions and comments with us. On this Web site, we will also be revealing some of our best practice user experience methodologies, case studies, and processes in a series of articles. These articles will provide insight into how Oracle approaches user experience design in a global marketplace, on existing products, and across multiple platforms. Work with us! We work with hundreds of customers annually as part of our ongoing requirements gathering, design, and testing. Join our Global Design Partners program today to help shape the future of enterprise applications. My team and I look forward to working with you."]

SAP: To bring the product user experience up to the level of the retail (and Web) market.

Siemens: We guide our clients in identifying the user experience quality that is needed for the particular business scenario, we advise on how to reach the desired quality, and we help benchmarking against the competition. We also support the operating companies in applying the appropriate product-lifecycle management (PLM) process. And we stand for a common look and feel of Siemens products with maximum reuse and minimized re-development.

12.   What is the organizational position of your department or group (i.e., where is it located in the corporate "org chart")? For example, are you attached to engineering, or marketing, or business lines? Where exactly (for example, in relation to the designers, developers, marketers, quality assurance departments, training/educational department, etc.)?

Fujitsu: Official organization name: Product Business Support Group, Fujitsu Design Limited. Independent, but collaborates closely with engineering, marketing, business lines, quality assurance departments, and so on.

IBM: In general, IBM UX departments report to the same development executives as other developers and there are many UX practitioners who report to the same development first-line manager. The corporate UX organization is a virtual community with an indirect reporting structure (i.e., the product UX departments do not report to the corporate UX department).

Microsoft: In an IT engineering/development organization.

Oracle: Report into the SVP of Development. We are a centralized team supporting all of Applications development. My peers are the heads of each applications product area. Oracle has a number of other small UX teams and individuals who are embedded to a particular product or technology.

SAP: Reports to the Chief Technical Officer of the company.

Siemens: We are regionally organized in-house consulting groups. Our colleagues in the operating companies are always part of their PLM or R&D organization and these are also sponsors of our engagement, whether a local UX group exists or not. Sometimes our efforts get funding from marketing departments.

13.   What, specifically, is the relationship to the software development department? Is the UX group independent from software development or is it with in the software development group?

Fujitsu: Independent from the software development department.

IBM: For most IBM products, the UX department is a separate dept, but usually reports to the same executive as the development departments. However, UX departments are considered to be part of the larger software (hardware and services, too) development organization.

Microsoft: We are in the middle of it.

Oracle: Report into the SVP of Development. We are a centralized team supporting all of Applications development. My peers are the heads of each applications product area. Oracle has a number of other small UX teams and individuals who are embedded to a particular product or technology.

SAP: Independent, works like in-house design studio.

Siemens: Independent. Our internal process is very different from software engineering processes, and we generate our value from the difference between these two groups of processes. The overall process is often waterfall, internally we iterate a lot. We maintain a closer relationship to users than the SW process overall, and the UX process is typically much more diverse and multidisciplinary in our day-to-day interactions.

14.   How is the UX interpreted? That is, how does UX reflect the organization and corporate culture? For example, is "universal design" included in the mission of the UX department or group? Are usability and UX considered as essentially the same thing? Is there a separate marketing group for promoting UX?

Fujitsu: Concept of UX is following; under the concept of human centered design (HCD), we assist in improving UX through various design activities for developers and designers in the group companies. Universal design is a part of HCD, so it is included in the mission.

IBM: There is variability across products and brands. Design for globalization and accessibility are a focus of UX and development organizations, but these are important design considerations which warrant their own corporate departments in IBM. While there are individual exceptions, we believe that the vast majority of developers recognize the value of UX beyond usability evaluations. Our corporate UX leadership team promotes UX within and outside of IBM.

Microsoft: See previous entry on UX definition. Usability is just one attribute of the experience. Universal design is part of how we do our work; not a mission item. Marketing promotes the user experience; but we promote ourselves.

Oracle: Within. Superior User Experience is one of the 10 corporate values.

SAP: UX is generally interpreted as pixel level UI. UX is counter culture in every way. It is the only centralized technical organization. Everything else is line of business. Usability is more thought of as a productivity measurement. There is no internal marketing group for UX. We already have too much work to do.

Siemens: Usability is considered far too narrow to capture UX. Usability is about smart systems compensating for user limitation. UX is about the perfect fit between user profile and software profile. UX makes products attractive, desirable, competitive. Usability in our view is an important yet very limited part of UX.

15.   What is the category breakdown and number of, e.g., graphic designers, interaction designers, user-interface designers, ethnographers, analysts, etc.

Fujitsu: There is no distinction among the professions.

IBM: Most UXD practitioners in IBM fill multiple roles (e.g., user researcher, designer, evaluator), but there are some specialists. Rough breakdown: 75 visual designers, 25 user research specialists, 25 usability testing specialists, 100 interaction/UI designers, 25 UX architects, and 250 general UXD practitioners.

Microsoft: Not sure if you mean in the company or across my team. So in my team, there will be something like 1 ethnographer, 1 UX architect, 1 product planner, 3 visual designers, 3 interaction designers, and 3 user researchers.

Oracle: 60 designers, 30 usability engineers, 5 researchers, 4 graphics, 25 user assistance, 3 lab staff, 10 program managers, 10 architects, recruiters, 5 prototypers/developers.

SAP: Graphic designer 5%, interaction 60%, user-interface designers = interaction to us, ethnographers 0%, user research 20%, accessibility 10%, tools and training 5%.

Siemens: Only graphic designers need to be called out (30%); everyone else on the team has to cover ethnography/observation, analysis, interaction design and validation; based on educational background of the talent we hire.

16.   What is the role for and resulting impact of UX designers in software development?

Fujitsu: It depends on the product. Some product development has UX designers through out the development term from planning to evaluation. But there are some products where UX designers only participate in testing.

IBM: Our UX practitioners are actively involved in shaping how our software development processes are evolving from a traditional waterfall development model to an agile iterative development model. We are both learning from our early experiences and educating development organizations on the role of UX-led activities (e.g., user research, design, evaluation) in an agile development.

Microsoft: We (design and user research) own the user experience, its definition, design, and delivery (ensuring it is developed appropriately).

Oracle: We help define and are an integral part of the main applications development process, complete with UX deliverables each of the product development teams have to support.

SAP: We are trying to change the development process to allow enough time for a real design cycle upfront.

Siemens: Any profession adds to the quality of the process. When UI elements are determined and designed by SW engineering, or when SW engineers talk about users, they typically do not do so in a professional way. Expertise from UX designers means that business targets can be translated into SW, risks can be managed and results are achieved better, faster and cheaper. We often find that the SW architecture benefits from user research as the way experienced users perceive of their content is much smarter than structures that appear to be logical and complicated. Objects, their attributes and relationships are surprisingly simple and straightforward when strategies of use are considered. When UX and SW developers work together, both reap advantages, as UX design needs to consider true SW constraints from the beginning.

17. What are the definitions of these different professions within the organization: UX designers, user-interface designers, experience designers, interaction designers, product designers, human factor specialists, ethnographers, cognitive/behavioral specialists, and usability analysts?

Fujitsu: There is no clear definition in the department. Each designer has multiple roles. The person can be a graphic designer and/or an interaction designer, but can be, also, an analyst depending on the situation. Portion of the roles varies individually.

IBM: There are no official job titles for these roles in IBM, and virtually all of these titles are used by various professionals in IBM as a matter of personal preference and departmental practice. In general, we distinguish between UX practitioners who tend to focus on user research, interaction design, and usability evaluations; and visual designers, who tend to focus on visual style, branding, and the rendering of visual elements of the UI. However, many of our practitioners are skilled in both areas. We refer to the collective group of UX and Visual Design professionals as User Experience Design (UXD).

Microsoft: The usual ones.

Oracle: There is a separate document I can share with you on these definitions.

SAP: There are no specific definitions in the human-resources system. The company treats everyone as a generalist.

Siemens: We use usability analysts for testing, as these colleagues are best at devising test scenarios and at test execution and evaluation. Human factor specialists fall into the same category. Ethnographers need to be trained in non-intrusive observation and the capability to understand processes from within; so that they avoid carrying pre-conceived notions into situations. And then I would put all … designers into the same category. In this role, we need to find an appropriate response to the needs and capabilities of users. This is when innovation is happening. Ultimately, these disciplines work together, and most people on the team can assume any of these roles.

18. Do UX designers, experience designers, or interaction designers, etc., also do programming?

Fujitsu: The professionals program when making prototypes. They do not program for actual software products.

IBM: For most designers, some level of programming is required for prototyping. However, the majority of UXD practitioners do not deliver code that is shipped in the product.

Microsoft: No, although the tools we use (WPF/Silverlight) generate deployable code automatically.

Oracle: If they do it is a bonus that we encourage. We hire professional prototypers and developers in the UX team to support this skill set.

SAP: No. They only do programming to create user-interface prototypes and demos. All the code created is throw-away.

Siemens: In rare cases, but this seems to become more and more important. New technology makes this possible, and it gives UX designers more control over the results of their work.

19. Is there a role of UI programmer?

Fujitsu: There are no UI programmers in the UX department or group. There are persons who are in charge of the role depending on the project in the software development group.

IBM: While there is a *defacto* role of UI programmer in IBM, there is currently no official job classification or title.

Microsoft: Yes.
Oracle: Yes.
SAP: Only on a few newer product lines and these people (in India) only build to specifications. They are not involved in the design process.
Siemens: Yes, especially with WPF.

20.   What is the typical educational background and previous career of UX designers?
Fujitsu: Graphic design, communication design, cognitive psychology, human factors, architecture, environmental design, psychology, computer science, electrical engineering, information engineering, system engineering, etc.
IBM: Graduate degrees in HCI, human factors, computer science, psychology, and related programs. Almost all have some applied experience either through previous employment or IBM internship programs.
Microsoft: Design school, experience in product design (sometimes HCI programs and experience).
Oracle: Industrial design, interaction design, architecture, cognitive psychology, computer science with specialization in human factors and ergonomics, and media studies.
SAP: Depends on the company. Inside the US and India, most are trained designers. Everywhere else in the world most are volunteers. Some come from computer-science background, some from documentation, and some from marketing.
Siemens: Psychology, linguistics, and/or computer science. It helps if someone has a deep understanding of one or many domains. Capital goods are often difficult to understand but the knowledge of one industry makes it easier to understand the next context of use. I personally appreciate if colleagues have a background in industrial design. Many terrible SW designs would not have happened if it were understood that things on screens have a reality and are not just a bunch of pixels. And I encourage everyone in the team to look into activity theory. Every goal is part of a larger goal. The overall value system of an operation, business conditions, dependencies, etc. are reflected in UX design.

21.   Are graphical user-interface and software components developed together?
Fujitsu: Seldom.
IBM: Yes.
Microsoft: Yes.
Oracle: Yes. We design a manage them centrally to maintain consistency, and design efficiently. We have built our own repository and request tools.
SAP: Yes.
Siemens: Yes.

22.   How do you manage common components or parts (for example, icons or "widgets")?
Fujitsu: It depends on the product. It varies.
IBM: We have corporate repositories for icons, widgets, and other UI design assets such as design patterns. Development teams may create customized widgets, icons, etc., and are encouraged to contribute them back to the common repository.
Microsoft: In libraries.
Oracle: We design and manage them centrally to maintain consistency, and design efficiently. We have built our own repository and request tools.
SAP: From a central repository with source code control.
Siemens: We use pattern libraries.

23.   Are there any style guides (UI guides) or pattern documents for each product/service, or for corporate-wide development? How long are they (approximate number of pages or screens)?
Fujitsu: It depends on the product. Some products have one, but others do not. Length varies from one product to another.
IBM: There is corporate-level design guidance, as well as brand- and product-level extensions of this guidance. Each one ranges from 100 - 200 screens/pages.
Microsoft: Yes; too many to count across all the work surfaces.
Oracle: Yes, we have Applications Style and Pattern guidelines and standards. As it is a web site I will have to get back to you on its size.
SAP: Both. The corporate user-interface standard for applications is 1000 pages in length. Product guidelines based on it are usually about 100 pages.
Siemens: I do not know what you mean by style guide. I believe style guides were a trick of the 80s to keep UX designers away from SW design. Can you write a speech or a novel or an article with a dictionary? To identify the issue and find the right element to respond to user needs takes more than finding the right page in

a style guide. Design is not symmetrical. A pattern library can show exactly how something has to look like, and why it is used in a certain situation. But there is no rule by which it can be predicted which pattern is right. There is no profession that would go about their business otherwise. Whether you design a house or build a business plan, whether you write an e-mail or an article, whether you organize a party or a field trip, you can't just pull a solution from the shelf. Once the plans for the house are ready, once the numbers for the plan are lined up, once the words are written down, the data can easily be put into their target environment. But that wouldn't be called design. There were times when folks were given little books to help them write letters. This was abandoned when everyone had more than 4 years of schooling. Styles guides, in my mind, follow the wrong misunderstanding of what our profession is all about.

24. **What means of document management, content management, and user-interface component management do you employ?**
Fujitsu: Seeking the best way now....
IBM: There is a wide variability of document/content/code management systems in use in IBM today. In general, each product's UXD team will use the same system as the overall product development team since it is important to collaborate with them.
Microsoft: If you mean for our own documents, we use SharePoint.
Oracle: We have an internal UX design development website, we have the guidelines and standards website, and we use a corporate wide server based document management system (Oracle File Online).
SAP: A content management product for the user-interface specifications called 7 Steps. For user-interface resource files, the regular development build-process.
Siemens: We use the systems our sponsors have in place to manage content. Often these are the same tools that are being used by the SW development teams. Ultimately, it is irrelevant which tools you use, as long as it provides some level of structure and people can access it easily.

25. **Are document, content, and UI-component management systems developed internally or do you use commercial products? If commercial products, which do you find most useful?**
Fujitsu: Seeking the best way now....
IBM: Most of our management systems are IBM products (*e.g.*, Rational and Lotus products) specifically designed for this purpose.
Microsoft: See above.
Oracle: Developed by the company.
SAP: Both. 7 Steps publishing tool. [URL: http://www.sevensteps.nl. SevenSteps is an enterprise-oriented CMS publishing tool. The SevenSteps company Website introduces the product as follows: Dutch-based Sevensteps delivers knowledge products. These products enable companies to publish their knowledge in a variety of publication formats: from PDF to corporate intranet, from website to e-learning solutions. Seven-Steps' products prove their value across the globe at companies like: SAP, Google, Microsoft, Rabobank and the Dutch Ministry of Housing.
Siemens: Some are, some are not. This is professionalizing as well, often there are better tools on the market than what groups could build that are not specializing in document management.

26. **Are there any original or unique development methods, tools, or documents other than, for example, developing personas (user profiles), pattern documents, or user-designed components?**
Fujitsu: Not especially.
IBM: IBM has a strong culture of innovation in the area of UI design. We have 17 Master Inventors among the IBM UX Community, i.e., practitioners with a strong, sustained record of patent activity.
Microsoft: Yes.
Oracle: Yes we have these, but we cannot discuss them yet.
SAP: No.
Siemens: We consider this a trade secret and will not publish.

27. **Are there new methods you have heard about that seem worth investigating and trying out?**
Fujitsu: Not especially.
IBM: We are constantly searching for new methods to accelerate and improve the quality of our user research, design, and evaluation, especially with the move to Agile development practices.
Microsoft: I think we are trying them somewhere in the company already. We have an effort that harvests experiences and passes them around.
Oracle: Master Usability Scaling (MUS) is showing itself to be useful, as has investment of formal ethnographic studies. Other techniques we cannot discuss until we release Fusion v.1
SAP: No.
Siemens: This cannot be disclosed.

User-Experience Development

Aaron Marcus

Aaron Marcus and Associates, Inc., 1196 Euclid Avenue,
Suite 1F, Berkeley, CA, 94708 USA
Aaron.Marcus@AMandA.com

Abstract. Developers worldwide wish to understand user-experience development (UXD). UXD comprises activities in user-centered design of user experience, specifically user-interface development (metaphors, mental models, navigation, interaction, and appearance) that are useful for planning, research, analysis, design, implementation, evaluation, and documentation of products/services across a wide number of platforms. This paper summarizes some key concepts and terms.

Keywords: Design, development, management, user interface, user experience.

1 Introduction

User-experience development (UXD) is of increasing concern to developers of all computer-based products and services throughout the world. In discussing issues of UXD, it is useful to establish a basic glossary. The following terminology is not universal, but is a candidate for canonical, non-overlapping concepts that are used throughout the UXD community.

First of all, what is user-centered design (UCD)? In an earlier publication, the author wrote the following (Marcus, 2004) about UCD:

2 User-Centered Design and User-Interface Terminology

Typical steps in the UCD process are the following:

- Plan project
- Analyze needs
- Gather requirements
- Design initial solution
- Evaluate design solutions (iterate with initial and revised design steps)
- Design revised solution
- Evaluate design concepts (iterative)
- Deploy product/service
- Evaluate product/service (iterative)
- Determine future requirements/enhancements
- Maintain and improve processes
- Assess project

M. Kurosu (Ed.): Human Centered Design, HCII 2009, LNCS 5619, pp. 611–617, 2009.
© Springer-Verlag Berlin Heidelberg 2009

More specifically, user-centered development methods include the following techniques used in initial, intermediate, and final stages of development. Many of these, in turn, are the subjects of books focusing on one or more specific techniques:

- Card sorting
- Cognitive walkthroughs
- Design patterns
- Evaluation workshops
- Expert evaluation
- Field studies
- Focus groups
- Guidelines
- Heuristic evaluation
- Interviews
- Rapid and advanced prototyping
- Storyboarding
- Surveys
- Task analysis
- Use cases
- Use scenarios
- User profiles or personas
- User tests
- Workshops

Factors affecting which techniques are used may vary among the following:

- Availability of appropriate, suitable technology
- Availability of equipment, such as cameras or eye-tracking equipment
- Availability of other physical resources, such as testing rooms
- Availability of usability professionals
- Availability of users
- Budget
- Calendar schedule
- Cost of specialized skills, equipment, or tasks
- Experience of the team with different UCD methods
- Size of the project

The effects of user-centered design can be evaluated by criteria such as these:

- Customer satisfaction
- End-user engagement
- Impact on development process
- Improvements over benchmarks
- Increase of usability awareness in the development team.
- New understanding of users, their tasks, or contexts of user
- Project leader satisfaction
- Return on investment (ROI), as measured by metrics established at the beginning

Suggestions for new design

- Team satisfaction
- Usability of the developed system
- Usability problems identified

The key terms (verbs) of the user-experience development process may be considered to be as follows:

- Planning
- Research
- Analysis
- Design
- Implementation
- Evaluation
- Documentation
- Training
- Maintenance

UXD methodology involves analyzing how users of interactive systems accomplish tasks, and based on this analysis, developing designs that display and communicate information effectively. Objectives include enhancing ease-of-learning, ease-of-use, productivity, brand loyalty, and satisfaction. UIs may be said to consist of the following UI components:

- Metaphors: Easy recognition of terms, images, and concepts
- Mental Model: Appropriate organization and representation of data, functions, tasks, and roles
- Navigation: Efficient movement within the mental model through menus, dialogue boxes, and control panels
- Interaction: Effective input/output sequencing, including feedback
- Appearance: Quality perceptual characteristics (visual, verbal, acoustic, tactile/haptic, even olfactory).

3 UXD Around the World

UXD is growing rapidly as an activity throughout the world. Magazines such as *User Experience*, the member publication of the Usability Professionals Association (http://www.usabilityprofessionals.org), are devoted to the topic. Organizations have sprung up internationally, like the Human-Centered Design Network (http://www.hcdnet.org) in Japan, that foster growing awareness of "human-centered" design processes centered around the human experience of software-based technology. Major conferences in human-computer interaction (HCI), computer-human interaction (CHI), and user-interface (UI) design, and interaction design all feature sessions, if not tracks, in their programs devoted to UX and UXD. Examples include the conferences such as those of the following organization:

- ACM/SIGCHI (http://www.sigchi.org)
- American Institute of Graphic Arts (the term used is "experience design") (http://www.aiga.org)
- HCI International (http://www.hcii2009.org/)
- Society of Technical Communication (http://www.stc.org)
- Usability Professionals Association (http://www.usabilityprofessionals.org)

In addition, hundreds of specialized conferences about product/service development for many different platforms, such as mobile devices, Web 2.0, vehicle systems, etc., and specialized contents, such as animation, Web searching, virtual reality/communities, social networking, commerce, games, healthcare, travel, etc., all have sessions and tracks devoted to the topics of UXD.

In the USA, UXnet (http://www.uxnet.org) has arisen to promote communication and interaction among all organizations and professionals involved with user-experience development. Their Website, unusual among some publications, organizations, and companies, makes an effort to define "user experience:" Although this definition may be debated, it begins to provide a focus.

User Experience (abbreviated: UX) is the quality of experience a person has when interacting with a specific design. This can range from a specific artifact, such as a cup, toy or website, up to larger, integrated experiences such as a museum or an airport.

Many enterprises have made significant advances in shifting their attention to UXD. IBM has gained significant fame over the last 5-10 years for its ease of use Website. The author has also viewed the internal intranet available to IBM staff that provides an archive of documents about specifications, processes, resources, terminology, etc., that is formidable in size and scope. Several high-level IBM managers have been appointed to routinely visit centers of UXD in order to assess the quality and consistency of attention to best practices. One of them, Karel Vredenburg, has even published a book about their approach (Vredenburg et al, 2002).

In Europe, SAP has established the SAP Design Guild, which brings together a large body of information about best practices (http://www.sapdesignguild.org/). To its credit, this Website is open to all viewers/visitors, The SAP Design Guild has also published books that explain aspects of its views of best practices.

At HP, in approximately 2004, shortly before her reign as CEO of HP ended, according to anecdotes provided by HP UXD managers at the time, Ms. Carly Fiorina asked that the company Website announce a new initiative: that the company had made the "customer experience" a top priority. This signaled a significant shift in design, branding, product development, and more. During the years since then significant efforts have been made to overhaul human factors, industrial design, and interaction design teams and processes at HP. Interbrand reported in 2008 that HP (http://www.interbrand.com/portfolio_details.asp?portfolio=2564) undertook a major rebranding effort called "One Voice," which seeks "to better integrate its vast line of consumer electronics and computer hardware products. With a fresh design to the packaging, they faced a challenge to stay on brand across thousands of product lines and dozens of packaging types. This would inevitably require re-alignment of large teams of people, the establishment of templates and clear archives of canonical processes, procedures, roles, and definitions. In recent years, Mr. Sam Lucente was appointed Vice-President for Design at HP, an unusual ascension in the IT world of an executive-level person responsible for instituting superior design processes. For those interested, a YouTube 15-minute interview is available about the role of design at HP: (http://video.google.com/videoplay?docid=2924735257204340870).

Regarding AT&T, in approximately, 2003, the author learned that there had been a "user-experience engineering" group at AT&T since about 1997, or perhaps earlier. An article about that group appears in an article by Cunningham et al (2001). The abstract states: "The authors design and evaluate Internet-based applications as members of the User Experience Engineering Division of AT&T Labs. They all started in academia with doctorates in cognitive psychology, and have been at AT&T Labs for periods ranging from two to 17 years."

For about a decade, Microsoft has been buying up some of the best R+D talent it can afford to strengthen Microsoft Research. Its general user-experience development teams for existing product development seem to have lagged, and the mixed results of its product deployment have been noted in news media and the Internet. Nevertheless, Microsoft made significant advances in the early 21st century to hire anthropologists and ethnographers to join their UXD teams, so much so that a unit of Fortune magazine online named Bill Gates one of the great anthropologists of all time, somewhat tongue-in-cheek. The accolade did point to the significant ascension of anthropology and ethnography to the professionals required to achieve significant improvements in UXD.

Also of note is Microsoft's outreach to designers in its new product line and the MIX conferences, which seek to bring together software developers with designers, to raise the consciousness of both "tribes." Microsoft has appointed evangelists to communicate with the design communities. Another example of outreach and cross-disciplinary communication and interaction are the job outreach programs to design schools. A recent announcement (April 2008) from the Illinois Institute of Technology's Institute of Design, Chicago, is instructive:

"MS is footing the bill for drinks and appetizers, in exchange for Jakob Nielsen [not Jakob Nielsen of Nielsen Norman Group, but another person] doing a short pitch, as part of a national recruiting campaign for growing design, design research and UX at MS, which he heads up. ....You are invited to a special ID alumni reception, graciously sponsored by Microsoft. Free food and drink, old and new friends, plus two guests from Redmond: Jakob Nielsen, Director of User Experience, and Safiya Bhojawala (M.Des. '06), User Experience Researcher in the MS Dynamics User Experience Team..."

This kind of outreach to the design community is something that other enterprises should note and consider emulating if they wish to attract the kind of talented individuals required to generate breakthrough user experiences in the coming decades. These user experiences will require increased attention to "emotional design," "brand values," "aesthetics" and "customer culture", including issues of "identity", topics that are quite different from previous decades that focused on usability, i.e., effectiveness, efficiency, and satisfaction, according to the ISO definition.

Acknowledgments

The author acknowledges the assistance Mr. Niranjan Krishnamurthi in the preparation of this paper.

References

1. Cockton, G.: Value-Centred HCI. In: Hyrskykari, A. (ed.) Proceedings, NordiCHI 2004, Tampere Finland, October 23-27, pp. 149–160 (2004)
2. Cunningham, J.P., Cantor, J., Pearsall, S.H., Richardson, K.H.: Industry Briefs: AT+T. Interactions 8(2), 27–31 (2001)
3. European Design Centre: User-Centred Design Works (CD-ROM). IOP Human Machine Interaction (2004) This CD-ROM presents a case for user-centered design including case studies and information resources, http://www.edc.nl
4. Gaddy, C., Marcus, A.: Analyze This: A Task Analysis Primer for Web Design. User Experience 5(1), 20–23 (2006)
5. Marcus, A.: Global/Intercultural User-Interface Design. In: Jacko, J., Spears, A. (eds.) Handbook of Human-Computer Interaction, 3rd edn., ch. 18, pp. 355–380. Lawrence Erlbaum Publishers, New York (2007)
6. Marcus, A.: What Would an Ideal CHI Education Look Like? Fast Forward Column, Interactions 12(5), 54–55 (2005)
7. Marcus, A.: Usability Grows Up: The Great Debate. Fast Forward Column, Interactions 12(4), 72–73 (2005),
   http://www.uigarden.net/english/reviewsandinterviews.php/
   2005/08/20/usability_grows_up_the_great_debate
   (English),
   http://www.uigarden.net/chinese/reviewsandinterviews.php/
   2005/08/20/da_bian_lun (Chinese)
8. Marcus, A.: User-Centered Design in the Enterprise. Fast Forward Column, Interactions 13(1), 18–23 (2005)
9. Marcus, A.: User Interface Design's Return on Investment: Examples and Statistics. In: Bias, R.G., Mayhew, D.J. (eds.) Cost-Justifying Usability, 2nd edn., ch. 2, pp. 17–39. Elsevier, San Francisco (2005)
10. Schaeffer, E.: Institutionalization of Usability. Addison-Wesley, Reading (2004)
11. Vredenburg, K., Insensee, S., Righi, C.: User-Centered Design: An Integrated Approach. Software Quality Institute Series. Prentice Hall, Upper Saddle River (2002)

URLs and Email Contacts

The following URLs and email contacts, among others, are relevant to this topic:

- http://www-3.ibm.com/ibm/easy/eou_ext.nsf/publish/
- 1996
- http://www-2.cs.cmu.edu/~bam/uicourse/special/
- http://www.usabilitynet.org/tools/13407stds.htm
- http://www.usabilityprofessionals.org/upa_publications/ux_poster.html
- http://www.amanda.com/services/approach/approach_f.html
- http://www.usabilityprofessionals.org/usability_resources/guidelines_and_methods/methodologies.html

- http://www.effin.org/. EFFIN is a multi-client sponsored project to tailor user-centered methods to the development of electronic government services and to investigate these methods in practice. The project is financed through the FIFOS program of the Norwegian Research Council
- http://www.upassoc.org/upa_projects/usability_in_enterprise/index.html. The project, sponsored by the European usability community seeks to gather information on usability practice in the enterprise and evidence of return on investment (ROI) of usability.

Measurements and Concepts of Usability and User Experience: Differences between Industry and Academia

Anja B. Naumann, Ina Wechsung, and Robert Schleicher

Deutsche Telekom Laboratories, Berlin University of Technology,
Ernst-Reuter-Platz 7, 10587, Berlin
anja.naumann@telekom.de

Abstract. Usability and User experience are two central terms in the discipline of Human-Computer Interaction (HCI). The relevant literature provides a wide range of definitions and measuring methods for both concepts. This paper presents results of a survey asking usability researchers and practitioners about their views and practice on Usability and User Experience aiming to investigate the current state of the art regarding both concepts.

1 Introduction

The term "Usability" has been around for over twenty years now [1]. Although it is widely seen as an essential part when developing new applications, there is no shared agreement in the HCI-community what exactly "Usability" is beyond the ISO 9241 part 11 [2]. In the broadest sense "Usability" is used in order to indicate something is "good" from a HCI-perspective [3]. Besides the probably most common criteria *effectiveness*, *efficiency* and *satisfaction*, defined in [4], literature provides several different interpretations and criteria for Usability [5].

In contrast to "Usability", "User Experience (UX)" is a relatively new concept popularized by Don Norman in 1999 [6]. Since then, a growing and large interest in User Experience has developed [7] and the term UX has been increasingly used in the relevant literature [6,8]. Although "User Experience" is by now a common expression amongst HCI-researchers, it is used in various meanings [7] and a shared definition is still not available. According to [5], several interpretations and approaches can be found for UX.

Therefore this study was realized in order to investigate how Usability and UX are understood and measured by practitioners and researchers.

2 Method

2.1 Questionnaire

An online questionnaire (ten pages including instructions) was set up, comprising the sections (a) background, (b) measurement methods, (c) interest in and own definitions of Usability and User Experience (UX), and (d) views on UX and Usability.

M. Kurosu (Ed.): Human Centered Design, HCII 2009, LNCS 5619, pp. 618–626, 2009.
© Springer-Verlag Berlin Heidelberg 2009

With the first section, gender, age, educational background, and current field of work were assessed. After that, participants were asked if their job includes the development of new concepts, applications or prototypes and if yes, how they evaluate the Usability of their developments (via open question). Next, different measurement methods were presented. Familiar methods should be marked.

The participants were then asked if they were interested in Usability and UX. The following question, asking the primary reason for the interest in Usability and UX, was adapted from a questionnaire published in [6]. Answer options were: per se, to design better products, to sell better products, to make people happier, other. Subsequently their definitions (via open question) of Usability and UX were assessed. The respondents' understanding of Usability and UX was assessed via two open questions asking to give a definition of both terms. The answers were categorized by three raters according to criteria described in section three. Since these criteria did not cover an adequate amount of the definitions, additional criteria had to be taken into account. Disagreements were solved by using the criteria chosen by the majority of coders. The inter-rater-reliability was Cohens $\kappa=.75$ for Usability and $\kappa=.78$ for UX.

In the last section, twenty-one Usability/UX criteria resulting from a literature review were listed. The participants were given the statement: "A necessary criterion for defining the concept "Usability" is [...]" (for the criteria see Table 4). They could indicate their level of agreement on a five-point rating scale (strongly disagree – strongly agree). Furthermore, "I do not understand" was given as an additional answer option. This was repeated with replacing the word "Usability" with "User Experience" to investigate whether Usability and UX are seen as distinctive concepts and if so regarding which criteria. The last task was to state their agreement to several statements about Usability, UX and the interrelationship between them.

The questionnaire was set up in English to reach as many participants as possible. However, since the link was mainly sent out to German companies, universities and institutes, it can be assumed that most of the respondents were German native speakers.

2.2 Participants

The link and the password to the questionnaire were sent to colleagues including practitioners as well as researchers. Finally, 166 respondents accessed the questionnaire. 118 of them filled in the whole questionnaire and were included in the further analysis.

Of the valid cases (36 female, 82 male), 51 respondents stated they are working in academia, 31 were working in industry and 33 were working in both or between areas. Age ranged between 21 and 61 years (M=33.3).

3   Results

3.1   Usability Measurement Methods

88 (academia=40, both or between=25, industry=23) participants stated their job includes the development of new concepts, applications or prototypes. 70 (academia=33, both or between=19, industry=18) of them stated they conduct Usability

evaluations, most of them by carrying out user tests or user studies. Another common method likely to be used within user tests and studies are questionnaires and interviews (s. Table 1).

Table 1. Usability evaluation methods actually used and Usability evaluation methods marked as familiar

	Methods used (N)			Methods familiar (N)		
	Academia	Both or between	Industry	Academia	Both or between	Industry
Questionnaires/ interviews*	19	11	6	85	59	51
User tests/ user studies	12	11	13	**	**	**
Other	8	8	7	6	4	4
Experiment	7	1	0	46	24	20
Expert/heuristic evaluation*	5	7	7	25	10	16
Think aloud	4	0	2	32	18	21
Observation	3	0	3	29	23	19
Focus groups	2	1	2	27	18	27
Eye tracking	1	0	2	21	16	15
Log-file analysis	1	0	1	29	17	18
Task analysis	0	1	0	25	12	15
Card sorting	0	0	1	13	6	14
(Paper-) prototyping	0	0	4	24	12	18
Cognitive Walkthrough	0	0	1	29	11	17
Model-based evaluation	0	1	0	19	8	7
Review-based evaluation	0	0	0	13	5	8
Video/audio data analysis	0	0	0	26	14	20
Benchmarking	0	0	0	10	13	13
Personas	0	0	0	13	7	12
Remote usability testing	0	0	0	6	9	10
Psycho-physiological measures	0	1	0	19	8	8

* Categories were combined. Most participants did not distinguish between these two methods in the open question.
** User test/ user studies were not given as a possible answer option.

Differences between the three groups could be observed for the methods actually applied by the respondents as well as for the methods marked as known. Standard methods (e.g. questionnaires) were used and known by the majority in all groups, however, concerning less common methods (e.g. prototyping) the groups seemed to have different preferences and knowledge. "Running experiments" for example is a method used primarily in academia. Paper-prototyping on the other hand seems to be a method used in industry. Interestingly all groups use only few of the known and available methods.

Table 2. Primary interest in Usability and UX

I am interested...	in Usability						in User Experience					
	Academia		Both or between		Industry		Academia		Both or between		Industry	
	N	%	N	%	N	%	N	%	N	%	N	%
Per se	8	14.8	4	12.1	4	12.9	9	16.7	4	12.1	3	9.7
To design better prod-	30	55.6	16	48.5	21	67.7	13	24.1	10	30.3	14	45.2
To sell better products	1	1.9	4	12.1	2	6.5	2	3.7	3	9.1	5	16.1
To make people happier	5	9.3	4	12.1	4	12.9	16	29.6	9	27.3	9	29.0
Other	8	14.8	1	3.0	0	0	1	1.9	1	3.0	3	9.7

Table 3. Criteria given in initial definitions (open question) of Usability and UX and number of statements

Usability (N)	Academia	Both or between	Industry	UX	Academia	Both or between	Industry
Efficiency	26	10	9	Emotion (Fun/Joy)	7(+18)	2(+8)	2(+6)
Ease of use	18	12	7	Experience	7	8	10
Effectiveness	21	7	9	Perception/ Impression	6	4	4
Suitability for	14	6	6	Prior knowledge	8	3	4
User satisfaction	16	4	8	Ease of use	2	2	1
Other	5	4	6	Other	2	7	3
Fun/Joy	4	2	2	Context	2	2	3
Intuitiveness	2	4	1	Cognition	3	0	2
Simplicity	2	3	1	Style/Trend	2	2	1
User friendliness	4	1	0	User satisfaction	3	0	1
Learnability	4	1	0	Aesthetics/Beauty	2	0	1
Context	3	1	1	Efficiency	2	0	0
Understandibility	2	1	1	Attitude	0	0	1
User-centered design	1	1	1	Learnability	1	0	0
Usefulness	1	1	0	Simplicity	1	0	0
Accessibility	1	0	0	Impression	4	0	2
Helpfulness	0	0	1	Needs	0	1	0
Adaptibility	1	0	0	Usability	1	3	3
Self descriptiveness	1	0	0				

3.2 Interest in Usability and UX

Most of the participants in all groups were interested in Usability and UX (s. Table 2). But only in the group "industry" all participants stated an interest in both concepts. In the groups "academia" and "both or between", between 6.1% and 14.9% stated they

Table 4. Agreement ratings for Usability ("A necessary criterion for defining the concept Usability is [...]"). Lowest ranks are written in italic, highest ranks are written in bold.

Criteria	Academia			Both or between			Industry			F	df	p
	R	M	SD	R	M	SD	R	M	SD			
Accessibility	14	3.70	1.04	10	3.90	1.14	13	3.81	1.14	0.34	2,109	.713
Attitude	20	2.95	1.09	20	2.79	.93	20	2.78	.85	0.30	2,85	.742
Beauty/Aesthetics	19	3.04	1.13	18	3.16	1.19	19	3.03	1.30	0.12	2,114	.888
Consistency	5	4.37	.87	7	4.10	1.19	5	4.39	.76	1.00	2,110	.371
Context	16	3.65	1.10	15	3.56	1.15	15	3.58	.76	0.09	2,96	.916
Controllability	8	4.00	1.00	11	3.81	1.12	11	3.97	.87	0.36	2,113	.695
Effectiveness	2	4.53	.97	4	4.27	.91	3	4.55	.77	1.01	2,112	.366
Efficiency	4	4.43	.99	3	4.27	.94	2	4.58	.67	0.93	2,114	.397
Flexibility	15	3.65	1.01	17	3.44	1.01	16	3.50	.90	0.54	2,111	.585
Fun/Joy	17	3.44	1.19	16	3.52	.96	17	3.16	1.13	0.91	2,113	.406
Helpfulness	11	3.88	.95	13	3.65	1.14	14	3.77	1.06	0.51	2,110	.602
Intuitiveness	9	3.96	1.12	6	4.13	1.15	1	4.60	.62	3.75	2,107	.027
Learnability	6	4.17	.94	8	4.09	1.03	7	4.30	.88	0.38	2,112	.687
Memorability	13	3.75	.98	12	3.74	1.12	8	4.20	.76	2.37	2,106	.098
Originality	21	2.55	.83	21	2.58	1.03	21	2.41	1.12	0.26	2,108	.773
Robustness	10	3.92	1.06	14	3.58	.99	12	3.90	1.21	1.06	2,109	.349
Self-descriptiveness	7	4.08	.93	2	4.28	.96	6	4.32	.70	0.91	2,111	.404
Simplicity	12	3.79	1.21	9	3.97	1.09	10	4.14	.92	0.93	2,111	.397
Suitability for individualization	18	3.20	1.02	19	3.13	1.13	18	3.10	1.08	0.12	2,114	.891
Suitability for the task	3	4.45	.93	5	4.26	1.09	9	4.17	.95	0.91	2,113	.406
User satisfaction	1	4.56	.94	1	4.34	1.12	4	4.47	.78	0.50	2,111	.610

had no interest. As primary interest in Usability, the majority of all groups chose the answer "to design better products". The primary reason for the interest in UX differed between the groups: Again, in the groups "industry" and "both or between" the majority of participants stated their primary interest is "to design better products". The primary interest of the group "academia" was to "make people happier".

3.3 Definitions of Usability and UX

The definitions the respondents gave for Usability were for all groups strongly in accordance with the one given in the ISO standard 4. For UX *emotion* was the aspect all groups besides "industry" most frequently referred to. In contrast to Usability, pragmatic aspects like *efficiency* played a less important role.

Most participants in the group "industry" simply referred to *experience* (e.g. "The experience of the user") in their definitions. This might be an indicator for a vague understanding of the term "User Experience" as this circular definition was also given by many participants of the other groups.

Table 5. Agreement ratings for User Experience ("A necessary criteria for defining the concept User Experience is […]"). Lowest ranks are written in italic, highest ranks are written in bold.

Criteria	Academia			Both or between			Industry			F	df	p
	Rank	M	SD	Rank	M	SD	Rank	M	SD			
Accessibility	20	3.55	1.09	20	3.53	1.16	13	3.83	1.21	0.71	2,92	.494
Attitude	12	3.83	.99	14	3.75	1.11	5	4.16	.94	1.22	2, 92	.300
Beauty/Aesthetics	4	4.04	1.18	4	3.97	1.20	**3**	4.33	.99	0.00	2,107	.410
Consistency	7	3.92	1.10	8	3.94	1.01	*19*	3.73	1.20	0.34	2,108	.715
Context	17	3.72	1.05	11	3.90	1.06	12	3.83	1.09	0.26	2, 94	.770
Controllability	5	4.02	1.05	5	3.97	.93	8	3.97	.96	0.04	2,106	.962
Effectiveness	9	3.89	1.17	18	3.59	1.07	17	3.77	1.14	0.67	2,106	.524
Efficiency	11	3.88	1.15	12	3.88	.91	10	3.87	1.14	0.00	2,108	.999
Flexibility	18	3.72	1.03	17	3.63	1.07	*20*	3.70	1.18	0.07	2,105	.930
Fun/Joy	**2**	4.46	1.01	**3**	4.13	1.26	**2**	4.47	.82	1.18	2,107	.312
Helpfulness	14	3.81	1.14	**2**	4.16	1.05	14	3.83	1.18	1.03	2,106	.361
Intuitiveness	**3**	4.19	1.10	6	3.97	1.20	4	4.17	.80	0.47	2,105	.628
Learnability	8	3.90	1.12	9	3.91	1.12	6	4.10	1.06	0.36	2,107	.698
Memorability	16	3.74	1.08	10	3.90	.94	16	3.79	1.01	0.24	2,103	.788
Originality	*21*	3.49	1.23	*19*	3.58	1.12	11	3.83	1.21	0.78	2,105	.463
Robustness	13	3.81	1.12	*21*	3.42	1.06	*21*	3.57	1.22	1.20	2,106	.304
Self-descriptiveness	10	3.89	1.07	13	3.84	1.05	15	3.80	1.19	0.06	2,104	.942
Simplicity	15	3.76	1.16	15	3.75	1.16	9	3.87	1.20	0.10	2,108	.902
Suitability for individualization	*19*	3.66	1.04	16	3.75	1.19	18	3.77	1.10	0.11	2,109	.895
Suitability for the task	6	3.94	1.13	7	3.97	1.03	7	4.03	1.02	0.08	2,105	.927
User satisfaction	**1**	4.62	.87	**1**	4.34	1.12	**1**	4.83	.59	2.35	2,106	.100

The category *prior knowledge*, mentioned relatively frequently by all three groups, is interesting insofar as in German *prior knowledge* is a correct connotation of the term *experience*. However, in the context of Usability and UX, *prior knowledge* does not correspond to the meaning given in relevant HCI literature (see e.g. [5]).

3.4 Criteria of Usability

As the criteria most important for defining Usability, the groups "academia" and "both or between" rated *user satisfaction*. The group "industry" showed the highest agreement with *intuitiveness*. Furthermore, *intuitiveness* was the only criterion with ratings varying significantly between the three groups. The ranks for *effectiveness* and *efficiency*, the factors given in the ISO 9241 standard, were between two and four.

Originality was in all groups the criteria seen as least important, followed by *attitude*. *Beauty/aesthetics* was ranked as the third least important criterion by both "academia" and "industry". Detailed results are shown in Table 4.

Table 6. Agreement to different UX/Usability related statements (Max.[strongly disagree] =5/ Min.[strongly agree] =1)

Statement	Academia		Both or between		Industry		F	df	p
	M	SD	M	SD	M	SD			
Usability is not utility	4.04	0.95	3.48	1.21	3.67	1.30	2.61	2,107	.078
Usability is a necessary precondition for good user experience	4.22	0.98	3.94	1.22	4.03	1.17	0.74	2,115	.481
Usability is a necessary precondition for joy of use	3.61	1.07	3.76	1.23	3.26	1.39	1.47	2,115	.235
Usability means mainly ease of use	3.29	0.94	3.47	1.24	3.53	1.11	0.54	2,110	.586
Usability is a necessary precondition for intuitiveness	3.20	1.18	3.15	1.23	3.67	1.27	1.79	2,111	.173
User-centered design is not possible without Usability evaluation	4.25	1.04	3.68	1.08	4.16	0.93	3.20	2,112	.045
Usability is overrated	2.00	1.01	2.19	0.91	1.86	0.95	0.89	2,111	.412
Usability and user experience describe mainly the same concept	2.21	0.96	2.13	0.78	2.45	1.09	0.96	2,110	.386
Usability is irrelevant for practitioner	1.65	0.80	1.69	0.78	1.41	0.63	1.21	2,109	.303
Usability is a necessary precondition for acceptance	3.87	0.90	3.69	1.09	3.52	1.26	1.11	2,113	.334

3.5 Criteria of User Experience

All groups showed the highest agreement on *user satisfaction*. *Fun/joy* received the second highest importance ratings from the groups "industry" and "academia". In the group "both or between", *helpfulness* received the second highest ratings whereas *fun/joy* got the third highest ratings. The criteria with the third highest rating were *intuitiveness* for the group "academia" and *beauty/aesthetics* for the group "industry". As in the open definitions, pragmatic qualities were seen as less important by all groups.

The least important criteria were *robustness* for the groups "both or between" and "industry" and *originality* for the group "academia". Further details are given in Table 5.

3.6 Relationship between Usability, UX, and Related Concepts

The three groups showed major differences only for the statement regarding user centered design and Usability evaluation. Participants working "both or between" rated Usability evaluation not as necessary for user-centered design as the other groups. Another minor difference could be observed for the statement "Usability is not utility". The group „academia" showed more agreement to this statement than the other groups. Thus they might have a more complex understanding of Usability.

Over all, the results show that Usability is seen as prerequisite for UX and the re-lated concept "joy of use". Intuitiveness is not clearly seen as necessary precondition for Usability. Only for the group „industry", the mean was >3, which shows agree-ment on a scale from 1 (strong disagreement) to 5 (strong agreement). This is in line with the criteria seen as necessary for defining Usability. In the group "industry" intuitiveness got the highest ratings.

In concordance with the definition given by the respondents at the beginning of the survey, it turned out that *ease of use* is seen as a construct similar to Usability by all groups. As already indicated by the other measures (initial definitions and criteria), Usability and User Experience are seen as distinct concepts (s. Table 6).

4 Discussion

This paper reports results of a survey regarding the difference of practitioners' and researchers' views on Usability and User Experience. In the current study, practitioners and researchers showed both differences and similarities in their understanding of Usability and UX as well regarding the measurement of both concepts.

For all groups, the analysis shows that the two concepts are perceived as different, although they are overlapping. In general, the interest in Usability is focused mainly on designing better products. Concerning UX, a nearly equally often stated interest is to make people happier. This is line with the other results: UX was generally more linked to concepts with emotional content (e.g. fun and joy) and hedonic qualities.

Over all groups, the Usability definition most corresponding to both the results of the open question regarding the definition of Usability and the given criteria is the one given in the ISO standard [4] (see introduction). The majority of respondents referred to at least one criterion given in this standard. Also, the highest agreement ratings were observed for these criteria. Interestingly, the group "industry" rated intuitiveness as the most important criterion for defining Usability but did not mention intuitiveness in their initial definitions. Furthermore, the group "both or between" rated self-descriptiveness as second important but did not mention it in their initial definitions. Similar results were found for UX: The criteria *helpfulness* and *user satisfaction* (both or between) and *intuitiveness* (academia) were not mentioned in the self-written definitions but rated as very necessary for defining UX. In fact, *user satisfaction* was seen as most necessary by all groups but was rarely referred to in the initial definitions. In addition, participants agreed on all the given criteria being necessary for defining UX. Since for Usability for each of the groups five criteria were seen as not necessary (mean < 3.5) this might indicate that respondents were not sure what UX is and therefore rated everything as important. Furthermore, some participants of all groups seemed to have confused the term UX with prior knowledge, showing a connotation problem in the German language.

Overall the results show that Usability and UX are seen as distinctive concepts by all groups. However, all three groups seem to have a more exact understanding on the concept Usability than on UX. Thus, most of the differences are small. Overall, for the examined sample the results indicate that Usability/UX-research and practice are well connected. Thus the current study is in opposition to [10] where a gap between research and practice in HCI is claimed.

References

1. Bygstad, B., Ghinea, G., Brevik, E.: Software Development Methods and Usability. Perspectives from a Survey in the Software Industry in Norway. Interacting with Computers 20(3), 375–385 (2008)

2. Diaper, D., Sanger, C.: Tasks for and tasks in human-computer interaction. Interacting with Computers 18(1), 117–138 (2006)
3. van Welie, M., van der Veer, G.C., Eliens, A.: Breaking down Usability. In: Proceedings of Interact 1999 (1999)
4. ISO 9241, Ergonomics of human-system interaction. Geneva. Switzerland. International Organization for Standardization (ISO) (2006)
5. Bevan, N.: Classifying and Selecting UX and Usability Measures. In: Proc. of the 5th COST294-MAUSE Workshop on Meaningful Measures, pp. 13–18 (2008)
6. Hassenzahl, M., Lai-Chong Law, E., Hvannberg, E.T.: User Experience - Towards a unified view. In: UX WS NordiCHI 2006, Oslo, cost294.org, pp. 1–3 (2006)
7. Forlizzi, J., Battarbee, K.: Understanding experience in interactive systems. In: Proceedings of the 5th Conference on Designing interactive Systems: Processes, Practices, Methods, and Techniques. DIS 2004, pp. 261–268. ACM, New York (2004)
8. McNamara, N., Kirakowski, J.: Defining Usability: Quality of Use or Quality of Experience? In: IEEE International Professional Communication Conference Proceedings (2005)
9. COST294-MAUSE, newsletter n. 4 (2008), http://cost294.org/upload/521.pdf (retrieved: 20 Feburary 2009)
10. Parush, A.: Toward a common ground: practice and research in HCI. Interactions 13(6), 61–62 (2006)

Proactive Ergonomics in Refrigerator Concept Development

Maximiliano Romero[1], Fiammetta Costa[1], Giuseppe Andreoni[1],
Marco Mazzola[1], Juan Vargas[2], and Luigi Conenna[2]

[1] Politecnico di Milano, Dip INDACO, via Durando 38/A, 20158 Milano, Italy
[2] INDESIT Company, Fabriano Italy
{maximiliano.romero,fcosta,giuseppe.andreoni}@polimi.it,
marco.mazzola@mail.polimi.it, Luigi.Conenna@indesit.com,
Juan.Vargas@indesit.com

Abstract. Proactive Ergonomics means to pre-test the human factors features of a product in an early step of development (design or prototyping), modifying it, re-performing the test and so on. The goal is to apply an iterative process to reach a final definitive solution. This work presents a case study on refrigerator concepts development based on a Design for All approach. The first design concept was created through a participatory workshop supported by reference literature data about arthropometrical and functional parameters and by results from ethnographic observation. This also led to the design of a dedicated experimental protocol for evaluating the physical ergonomics characteristics of products and mock up through a comparative analysis. Using a bestseller's refrigerator as reference, we evaluated comparatively, our new concepts. The results were very significant and demonstrated a consistent improvement of the ergonomic quality of the concepts with respect to the standard product. The quantitative ergonomic evaluation has been validated by subjective methods.

1 Introduction

Usually to measure the effectiveness of a new concept of a product is difficult or made a posteriori through user trials with subjective questionnaires since it is very hard to involve users in projects' evaluation before the availability of a physical product. An alternative approach to integrate user requirements in the early stages of the design process is to engage them as "partners" in participatory workshops as presented in Sanders [1].

Otherwise, the ergonomic quality of the new product is measured applying anthropometrical or biomechanical data from bibliography.

Optoelectronic systems for movement analysis [2] provide to the designers a very rich source of knowledge about people strategies of movement and the biomechanical impact of design choices.

The MMGA movement analysis method, used in this study and deeply described in Andreoni et al. [3], wants to provide a quantitative measurement procedure for ergonomic assessment. This method can be applied both for the ergonomic evaluation of existing products and for the definition of design specification before its realization through mock up analysis. The qualitative analysis based on ethnographic observation

M. Kurosu (Ed.): Human Centered Design, HCII 2009, LNCS 5619, pp. 627–634, 2009.

[4] identifies the most significant motor strategy in the daily use of the home appliance and gives input for a participative design workshop, while the quantitative analysis provides a set of quantitative variables for the ergonomic classification of the observed movement strategy.

The conceptual importance of the integration of quantitative (i.e. anthropometry and movement analysis) and qualitative (i.e. questionnaires, video-ethnography and participative workshops) methods in ergonomic evaluations has been stated by several authors since only the application of an integrated set of tools can provide information on explicit, observable and tacit or latent users needs [5]. The research presented in this paper shows an applied effort in the direction of ergonomics tools integration.

Its goal is to design and evaluate prototyped proposals verifying the ergonomic improvement and, at the same time, to generate a database of natural movements useful to virtual evaluation in the preliminary steps of projects. In this way Ergonomics can really enter the early steps of the design process.

2 Methods and Process Description

The process described in the following paragraphs is based on the application of a User Centered approach as defined by Norman [6] - comprehending user involvement, interdisciplinarity of the design team and recursive evaluation - to the design of universal products.

The first preliminary phase is field analysis of user needs and actions. Following contextual design principles [7] the concept generation of a new product requires the full understanding of the constraints imposed by the specific context where the product is supposed to be used, and by the user habits and behaviours. To this purpose and to gain qualitative knowledge of different user motor strategies we observed five person with diverse ability/disability levels performing some defined tasks (storing seven products in their refrigerator, washing the drawers, regulating the temperature) in their homes.

Other data collected in this phase were anthropometrical and biomechanical parameters extracted from bibliographical sources. [8]

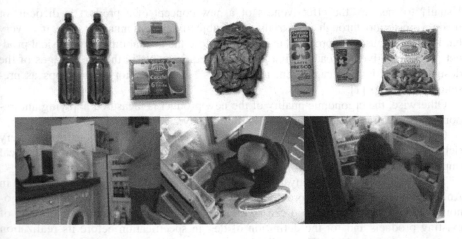

Fig. 1. Set of products provided for storing and user in action

The second phase is a one day co-design workshop with 18 participants including Indesit managers, commercial and technical staff, Politecnico researchers, designers, physiotherapists and users. We designed the workshop architecture in order to share knowledge and stimulate collective creativity [9].

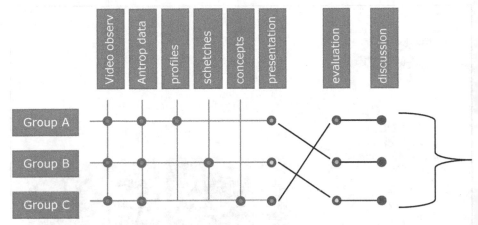

Fig. 2. Workshop architecture

The participants were divided in three groups homogeneously, i.e. all stakeholders should be present in each group.

All the groups were provided with support material as a video synthesis of the most interesting findings of the user observation and a table with a set of anthropometrical and functional parameters to be considered in the product's development. Group A worked on a specific refrigerator's critical points and received profiles of several sample users. Group B verified some proposals we sketched to this purpose through a role-play. Group C analysed some innovative concepts.

After the single work a cross-evaluation approach was adopted: group B evaluated group's A proposal, group C evaluated group's B proposal and group A evaluated group's C proposal. The workshop ended with a joint discussion on results.

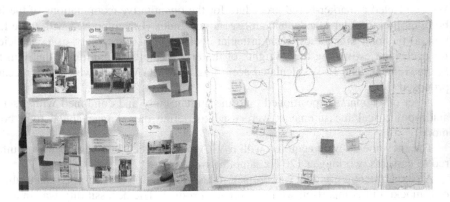

Fig. 3. Cross evaluation samples

The results were three concepts: a first radically innovative, the company considered not suitable for short term market implementation, and two others, whose integration was chosen for further development and evaluation. Prototypes were built to this aim.

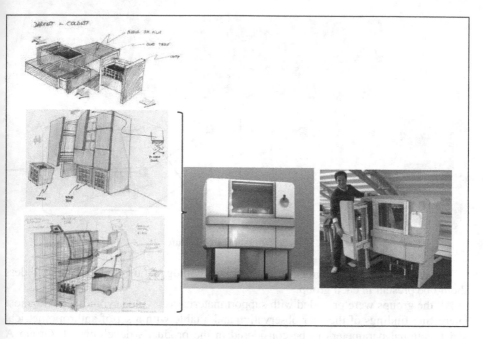

Fig. 4. Workshop results, first and second mock-up

The third phase comprehends recursive evaluation activities during which physical prototypes are submitted to the users in order to validate different aspect of the product.

2.1 Quantitative Physical Interaction Evaluation

We conducted a marker-based procedure for the quantitative examinations of the full body motion in order to calculate an Ergonomic Index (EI) composed with three factors: a) the joint kinematics, b) an articular coefficient of discomfort, c) a coefficient for each joint estimating the "weight" of the ergonomic contribution of that join for the movement. The EI is calculated according to MMGA Index which was recently proposed [3].

A set of 41 markers positioned onto the human body and combined with a set of anthropometrical measurements were used to define the biomechanical total body model.

The movements were acquired with a VICON optoelectronic system with 6 infrared M cameras, working at 120 Hz sampling frequency.

An experimental protocol was defined to test the relative accessibility and usability of a mockup in comparison with a reference product. The accessibility was intended and opening and closing doors or drawers, while usability was tested in putting/ removing objects of different sizes and weight on the internal shelves. Two test sessions were performed: one with the first mockup in the most natural situation and a second with the second more refined mock-up and a more precise protocol.

In both sessions all the subjects belonged to one of the following categories: women of the 5.th percentile, men of the 95.th percentile, subjects with functional limitations due to pathology (e.g. stroke) but living independently. All the considered subjects are skilled into the product utilization in their daily activities.

In the first session 18 subjects were asked to perform a sequentially randomized interaction activity with the standard fridge and the mock up fridge starting from different position and grasping three objects with different weight. We gave no instructions about the speed, an we let them move as naturally as they could. Three repetitions for each task were recorded.

In the second step we involved 10 subjects and the analyzed movements with the two fridges were 8 in total, 4 related to accessibility and 4 related to usability: the experimental protocol was more structured to allow for statistical comparisons. We recorded also in this case three repetitions for each task.

Fig. 5. Experimental setup for movement acquirement (first test session)

EI was calculated for each trial about with both refrigerators (reference and prototype 1 and reference and prototype 2). Finally also a global EI was calculated as the sum of the single EIs of all the tasks. High EI scores show worst situations, lower ones better situations.

	reference	mockup
EI TOT	2050	2055
ST DEV	10,4	18,4
EI MEEN	228	228

Fig. 6. Global EI for first trial

	reference	mockup
EI TOT	2588	2066
ST DEV	210	91
EI MEEN	325	256

Fig. 7. Global EI for second trial

TASK NR	TYPE	reference	mockup
1	Accessibility	224	191
2	Accessibility	240	223
3	Accessibility	208	206
4	Accessibility	450	216
5	Usability	556	239
6	Usability	229	314
7	Usability	444	385
8	Usability	237	291

Fig. 8. Detailed EI for second trial

2.2 Qualitative Subjective Evaluation

Guided interviews have been conducted in parallel with the fist test session. The questionnaire we used was built of three parts: broad information on the user, 6 questions before the trial regarding refrigerators in general and a first impact looking at the prototype and 4 questions after the trial on the interaction with the prototype.

The subjective information on the concept's perception and evaluation confirmed the results of the quantitative evaluation

The majority of the interviewed persons (16 over 18) perceived the prototype as useful in everyday life considering a more suitable accessibility due to the lower

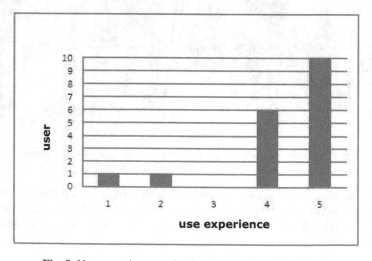

Fig. 9. User experience evaluation (1: negative, 5: positive)

height, better organizational possibility in the different spaces and a good support for disabled people because of the handle.

Also the evaluation of user experience after the trial gave satisfactory results as shown in fig. 9.

The discussion with the users gave us also the possibility to gain some not planned information like the habit to store products less frequently used in the higher part of the standard refrigerator or the idea to put a television or other devices on the top of the mock up.

3 Results

The first trail session showed small differences in the global EI evaluation (EI 2055 for the reference product and EI 2050 for our mockup), a big trade off between the results of different user categories and high standard deviation values especially for the mockup evaluation. This observation led us to the development of a more structured protocol for the second test session.

A more detailed analysis shows bad EI values for door opening in the reference product especially for people with crutches (EI over 250). This data is indirectly confirmed by the interviews: disabled people appreciated the use of the mockup's handle to remain in balance while interacting with the product.

The second trial session demonstrated a 20% improvement of the physical ergonomics index for the new concept with respect to the reference refrigerator. This result is completed by the positive subjective evaluation of user experience already registered during the first trial.

The worst scores were registered for the usability of the freezer (EI 556) and of the highest level (EI 239) of the reference product. These problems have also been highlighted by user observation and comments.

In general for each single task positive differences were shown, and the new concept also eliminated the difference in the ergonomic ranking among the different categories of users recorded in the first step. In this sense we proved that the new product is really "for All". [10]

ANOVA analysis statistically confirmed these results.

4 Conclusions

We defined an approach to Proactive Ergonomics based on data related to physical interaction between the human and new or existing products.

The method was successfully applied to refrigerators while its extension to other aspects and other product typologies is under development.

A future goal is to create a database of natural movements useful for virtual evaluation in the preliminary steps of projects. In this way Ergonomics can really be proactive entering in the early steps of the design process.

References

[1] Sanders, E.B.: Design research in 2006. Design Research Quarterly 1(1) (September 2006)

[2] Mavrikios, D., Karabatsou, V., Alexopoulos, K., Pappas, M.: An approach to human motion analysis and modeling. International Journal of Industrial Ergonomics (2006)

[3] Andreoni, G., Mazzola, M., Ciani, O., Zambetti, M., Romero, M., Costa, F., Preatoni, E.: Method for Movement and Gesture Assessment (MMGA) in Ergonomics. In: Proceedings of HCI (2009) (in press)

[4] Amit, V.: Constructing the Field: Ethnographic Fieldwork in the Contemporary World. Routledge (2000)

[5] Sanders, E.B., Dandavate, U.: Design for Experiencing: New Tools. In: Overbeeke, C.J., Hekkert, P. (eds.) Proceedings of the First International Conference on Design and Emotion, TU Delft (1999)

[6] Norman, D.A., Draper, S.W. (eds.): User centered system design: New perspectives on human-computer interaction. Lawrence Erlbaum Associates, Hillsdale (1986)

[7] Beyer, H., Holtzblatt, K.: Contextual Design: A Customer-Centred Approach to Systems Designs. Morgan Kaufmann, San Francisco (1997)

[8] Pheasant, S.: Bodyspace: Anthropometry, Ergonomics and the Design of Work, 2nd edn. Taylor and Francis, London (1996)

[9] Leonard Burton, D., Swap, W.: When sparks fly: Igniting creativity in groups. Harvard Business School Press, Boston (1999)

[10] Coleman, R.: Inclusive design - Design for all. In: Green, W.S., Jordan, P.K. (eds.) Human factors in product design. Taylor & Francis, London (1999)

Corporate User-Experience Maturity Model

Sean Van Tyne

User Experience Director, Global Architecture,
Fair Isaac, 13019 Gate Drive, San Diego, CA 92064, USA
Sean.Vantyne@Cox.net

Abstract. User experience encompasses all aspect of a persons experience with an organization's services and products [1]. Organizations may or may not be aware of their customers' experience with their services or products and give different degrees of attention to developing and managing their customers' experiences. These degrees of attentions given to their customer experience can be measured and charted by phases or stages along a continuum of dedication or maturity. Based on the Capability Maturity Model Integration [2] and the Corporate Usability Maturity [3], this paper presents a model of user experience maturity by level based on an organizations dedication of resources, budget, and process integration.

Keywords: Corporate, experience, maturity, model, user.

1 Introduction

Organizations and their products and services have a "user experience" regardless if the organization is consciously managing it. User experience encompasses all aspects of the end-user's interaction with an organization, its services, and its products. A *good* user experience delights customers – increasing adoption, retention, loyalty, and, ultimately, revenue. A poor user experience detracts customers, drives them to the competition, and, eventually, is no longer a viable source of revenue.

As organizations become more aware of their user experience and develop processes to architect, manage, and *measure* it, they gain the benefits.

User experience management varies from organizations that are just becoming aware to the concepts of user experience to organizations where user experience is one of their core distinctions if not *the* core distinction. The User Experience Maturity Model is a framework that describes an organizations maturity along this continuum. It defines where an organization is and provides the instructions to reach the next level. The model also provides a benchmark for relative comparison of organizations.

In this model, there are five levels defined along the continuum of user experience maturity starting at the initial level of no user experience management to customer focused organization. Organizations progress through a sequence of stages as their user experience management processes evolve and mature.

2 Research Approach

I have spent the tenure of my career introducing user experience processes and growing user experience departments with organizations. I have consulted and served in leadership positions for numerous companies and organizations introducing and

M. Kurosu (Ed.): Human Centered Design, HCII 2009, LNCS 5619, pp. 635–639, 2009.

integrating user experience strategies and processes. In several cases, the organizations that I have been involved had no user experience strategy, process, or resources and I grew and lead their user experience effort from no user experience process to a dedicated corporate user experience strategy.

Over the years, I have observed how different organizations progress from no user experience awareness to user experience as one of their core distinction. Much of my observation comes from first hand experience of being the change agent within the organization. I also have had numerous discussions with peers with other organizations whose user experience dedication has evolved over time. This paper is based on my research.

3 Results

The results of my research have produced this model that I presented at the Managing Innovation Conference in San Diego, May 16th, 2007 [4]. The model was inspired by the Capability Maturity Model Integration [2] and the Corporate Usability Maturity [3]. The model in Figure 1 illustrates how an organization dedication to user experience matures over time.

4 Level 0: Initial Stage

We don't know what we don't know. Initially, an organization may not be aware of the concept of user experience. Someone shares this knowledge and a grass root effort

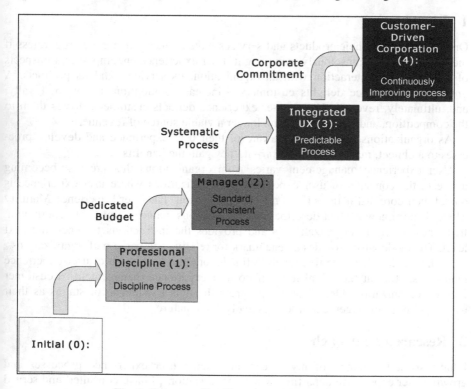

Fig. 1. Corporate User Experience Maturity Model

begins. Usually it is an ad hoc effort on a small project. If the effort is successful and the benefits are recognized then an organization may invest in user experience and advance to the next stage.

Ad hoc efforts may include a simple heuristic review to determine areas for improvement and executing the easiest ones to capture "low hanging fruit" and show immediate benefits of investing in these efforts. Sometimes, it is as easy as bringing in an expert to suggest simple changes to a process or design that can yield big returns in increase efficiency, effectiveness or satisfaction.

At this initial stage, it is typically undocumented and driven in a reactive manner by users' dissatisfaction. Not all of the stakeholders or participants may know that the effort is taking place. As a result, the new effort is likely to depend heavily on the knowledge and the efforts of relatively few people or small groups. An ad hoc effort with no approved budget may capture "low-hanging" fruit that leads to bringing in a professional in a UX discipline.

5 Level 1: Professional Discipline

Once user experience is adopted as a professional discipline then some user experience processes are repeatable with consistent results. The organization may have adopted developing wireframes as a part of their elaboration phase, found that it reduced cycle time in requirement analysis with development, and integrated this activity into their process. Or maybe they found that conducting a usability evaluation identified easy changes that increased end-user effectiveness, efficiency, and satisfaction that increased adoption and retention (and revenue). The newly introduced user experience activities to the processes may not repeat for all the projects in the organization at this stage but advocates may use some basic activities to track cost and benefits to start capturing return on investment.

At this stage, the minimum user experience process discipline is in place to repeat earlier successes on projects with similar applications and scope. The organizations project status may now include user experience deliverables to management like completion of major user experience tasks and activities at major milestones.

Consistent positive results from integrated user experience activities may promote a dedicated budget and the forming of a user experience group that develops consistent processes that lead to the next level.

6 Level 2: Managed Process

When the user experience is managed there are documented standards and process oversight. These standards and oversight are used to establish consistence performance across projects. Projects apply standards, tailored, if necessary, within similarly guidelines.

Upper management may establish and mandate these user experience standard for the organization's set of standard processes, and ensures that these objectives are appropriately addressed. The user experience roles, activities, and artifacts may be integrated into some of the organization's processes. User experience resources and tasks may be added to template project plans.

Measured results may proof a reduction in the cost of cycle time associate with definition, development, and testing along with training and support and/or an increase in customer satisfaction, retention, and adoption that captures the attention of executive management. The organization may decide that user experience must now be considered in their overall corporate strategy and integrated into their core competence which leads to the next level of user experience corporate maturity.

7 Level 3: Integrated User Experience

When an organization integrates user experience into their corporate strategy then, using metrics, they can effectively control their customers' user experience with their organization, products, and services. In particular, the organization can identify ways to adjust and adapt the process to particular projects and tailor it to fit the needs of the target market, segmentation, and customer type.

Quantitative quality user experience goals tend to be set as part of the organizations overall corporate balanced scorecard. Using quantitative, statistical techniques, process performance is measured and monitored, and becomes predictable and controllable. For example, as part of the organization's financial perspective to increase revenue, increasing customer satisfaction in the customer perspective by measuring the products usability score in the process perspective would be a part of the user experience corporate balance scorecard.

If a focus on user experience becomes a core distinction for an organization then they may enter the highest level of corporate user experience maturity.

8 Level 4: Customer-Driven Corporation

If one of the primary focuses of the organization is on continually improving the user experience process performance then the organization has become customer driven in a controlled and measured way. Quantitative user experience process improvement objectives for the organization are established. These objectives become core to the organization and are annually reviewed and revised to reflect changing market and business objectives.

The user experience process improvements are measured and evaluated against the quantitative corporate process improvement objectives including financial, customer, process, and human capital perspectives. This may include having user experience professionals involved in corporate strategies such as participating in discovering and defining new market segments or participating in third party vendor selection in terms of the overall corporate user experience integration.

9 Conclusion

Organizations products and services have a "user experience" regardless if they are aware of it. Organizations that manage and measure their user experience process gain the revenue benefits from satisfied customers. This User Experience Maturity model defines organizations' user experience maturity for organizations to understand where they are along this continuum and what they need to do to advance to the next level of maturity.

References

1. Nielsen Norman Group: User Experience - Our Definition (2007),
   http://www.nngroup.com/about/userexperience.html
2. Capability Maturity Model® Integration, http://www.sei.cmu.edu/cmmi/
3. Nielsen, J.: Corporate Usability Maturity. Alertbox, April 24 (2006),
   http://www.useit.com/alertbox/maturity.html
4. San Diego Software Managing Innovation Conference, Del Mar Marriott, San Diego, California, May 16 (2007)

Part V

HCD for Web-Based Applications and Services

Website Affective Evaluation: Analysis of Differences in Evaluations Result by Data Population

Anitawati Mohd Lokman[1], Afdallyna Fathiyah Harun[1], Nor Laila Md. Noor[1], and Mitsuo Nagamachi[2]

[1] Faculty of Information Technology and Quantitative Sciences,
University Teknologi MARA, 40450 Shah Alam, Malaysia
{anita,norlaila}@tmsk.uitm.edu.my,
afdallyna.f.harun@gmail.com
[2] International Kansei Design Institute, Japan

Abstract. Studies involving consumer studies have suggested different mechanisms of subject selections. The paper elaborates results of subject's responses by the methodology adopted from Kansei Engineering. In the research, evaluation of subject's Kansei towards website interface design was performed, targeting to measure affective quality in website design. Principal Component Analysis was performed to identify semantic structure of Kansei Words. The analyses were based on the average of evaluation results obtained from subjects. Results of PC Loadings were analyzed to see differences of determinants by size of data population. It is evident from the study that population size does not affect determinants of affective web interface design. The study makes decent contribution in determining appropriate population size in designing research instruments for future studies involving website affective evaluations.

Keywords: Consumer science, website affective evaluation, Kansei, Population size, Principal Component Analysis.

1 Introduction

Conducting a research that enables an insight into user behaviour from social standpoint can give us a lot of data to work with. However, given the nature of web development and testing, having to conduct experiments can be costly. A major part of this expense is the participant costs. Hence, it is desirable to reduce the number of participants without sacrificing the quality of the experiment. There would be a significant savings if there is a possibility of using smaller participant pool and yet get the same results as the entire pool. Therefore, for this research, we have conducted experiments with different amount of participant pool to see if the smallest pool would yield the same results as the entire population (biggest experiment population possible).

The context of web application chosen for this work is the design of online clothing websites where affective quality is assumed to be significant. Based on the result, we discuss the differences in Kansei space, concluded to determinants, as output from different size of data population.

M. Kurosu (Ed.): Human Centered Design, HCII 2009, LNCS 5619, pp. 643–652, 2009.

2 Subject Selection

Determining subjects for a study can be problematic as studies should be designed to fit the study's goals. Careful judgment need to be in place as errors in determining the number and type of participants or even the number of runs can be costly.

As this research concerns with web design, we have made reference to several usability studies theories due to its common nature of web testing. In usability studies, there is a common debate on the number of participants one must have when conducting experiments.

i. Five-User Assumption

"The best users come from testing no more than 5 users and running as many small tests as you can afford" [1]. Nielsen [1] elaborated that one user should be able to uncover a third of the findings and as more users are added, information redundancy occurs.

ii. Five Users And Beyond

According to Gilbert et al. [2], one study was done where five users were randomly chosen and only uncovered 35% of the findings, while the 13th and 18th user uncovered data that the original users missed. This result shows that if the study had been discontinued at five, those data would have been overlooked. In the same study, users 6th to 18th were able to find other new data that the original five were unable to find. This shows that, if the right users are not chosen, pertinent data can be left out [3]. Nonetheless, many has interpreted Nielsen's recommendation wrongly as Nielsen has highlighted that one should run as many tests as one can afford until the findings meet an "acceptable level".

Landesman & Perfetti [4] have conducted user testing on an e-commerce site using the recommendation "test four to five users with no more than eight". They have found this technique only yielded 35% of the problems in the system which would in return require 90 more tests runs to uncover the 600 problems in the system. From their study, it is learnt that in web testing, one need to apply a concept that fit the studies goals and needs. It must be noted that e-commerce websites have complex content which continuously and incrementally changes. Furthermore, there is a variety of e-commerce website users which implies that one sample group could not be used as a representation of the whole because each user who interacted with the system used the system differently.

Therefore, base on the reviews conducted, this study has used random sampling in selecting users and grouping them into a pool of 30, 60, 90 and 120 users. Random sampling is a chosen method for participant selection as it contains no bias and can be relatively representative of the targeted population [5]. It allows researchers to make generalizations and justification about the majority of the population by a certain level of certainty [6].

3 Affective Website Interface Design

HCI issues related to website applications were formerly focused on cognitive aspects of websites. Since the early work of Nielsen in the 1990s, the emphasis was on the qualities of usefulness and usability in producing good website design. Li & Zhang [7] cited that most studies dedicated to e-Commerce website evaluation are based on two assumptions; (i) target customers spend at least a few minutes on a website and (ii) good website features usually elicit positive cognitive evaluations and shopping experience. These assumptions have ignored the primary affective reaction or primary emotional responses towards the website. They stressed that online shopping behavior is a complex phenomena and recognized that affective reaction is one factor that promotes online shopping. This is because e-commerce websites have gone beyond the function of conveying information to the extent of providing persuasive engagement with website visitors through the lively process of perception, judgment and action. Affect has also been discussed in literatures as a factor found to influence decision-making, perception, attention, performance and cognition [8] [9].

Align with these claims, we argue that e-Commerce websites should induce desirable consumer experience and emotion that influences users' perception of the websites, to enhance visitor's stickiness that promotes consumer conversions and retentions.

Despite the gained recognition, the emotional appeal of websites is often neglected as designers tend to pay more attention to issues of usefulness and usability [10] due to the availability of established design methodology addressing usefulness and usability. Design method that incorporates emotional design requirements is lacking. In addition, numerous studies conducted on emotional design tends to look at minimizing irrelevant emotions related to usability such as confusion, anger, anxiety and frustration [9]. Therefore, it is necessary to seek for a suitable design method to handle design requirements based on emotional signatures of websites.

4 Data Population

The principles and empirical findings of behavioral science are probabilistic in nature, whereby it is possible to describe the reactions of most individuals but there is also a need to recognize that not everyone will fit the general pattern [11]. Nonetheless, the influence of a population size on results validity is critical. The common belief has always been the larger the sample, the greater the statistical power.

The solution to the dilemma of number of participants lies in ensuring that the analysis can be placed in a structural context and the research objective. The key to managing it relies on the researcher's conscious self-understanding of the research process [12]. As Ward-Schofield [13] has suggested:

"...Assumption of a qualitative research is very much influenced by the researcher's individual attributes and perspectives. The goal is not to produce a standardised set of result s[...]. Rather it is to produce a coherent and illuminating description of and perspective on a situation that is based on and consistent with detailed study of the situation."

We attempted to determine and understand the effect of number of participants to the population generalization. A design method computing analysis of 30, 60, 90 and 120 participants were put into place to view its statistic generalisation effect.

5 Kansei Engineering

Kansei Engineering (KE) is a technology that combines Kansei and the engineering realms to assimilate human Kansei into product design with the target of producing products that consumer will enjoy and be satisfied with. The focus of KE is to identify the Kansei value of products that trigger and mediate emotional response. The KE process implements different techniques to link product emotions with product properties. In the process, the chosen product domain is mapped from both a semantic and physical perspective. In terms of a design methodology, the approach of KE is to organize design requirements around the emotions that embody users' expectations and interaction [14], [15], [16]. KE has been successfully used to incorporate the emotional appeal in the product design ranging from physical consumer products to IT artifacts. Due to its success in making the connection between designers and consumers of products, KE is a well accepted industrial design method in Japan and Korea. In Europe KE is gaining acceptance but is better known as emotional design.

6 Research Method

The following fig. 1 illustrates the research method.

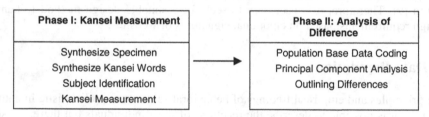

Phase I: Kansei Measurement	Phase II: Analysis of Difference
Synthesize Specimen	Population Base Data Coding
Synthesize Kansei Words	Principal Component Analysis
Subject Identification	Outlining Differences
Kansei Measurement	

Fig. 1. Research method

The research involves two phases; Phase I: Kansei Measurement, Phase II: Analysis of Difference, as shown in Figure 1. In Phase I, we adopted KE methodology to quantify website visitor's Kansei responses. Result form Phase I is then analysed statistically using Principal Component Analysis (PCA) to scrutinize the Kansei structure. The result enables the paper to conclude differences in Kansei structure by data population based on 30, 60, 90 and 120 participants. Details to the phases are described in the following sections (7 and 8).

7 Kansei Measurement

Phase I begins with selection of specific domain. It is important to control the domain and subjects as different domain will induce different Kansei. Specific target market group must be used as experiments subject, so that the intended Kansei could be measured accurately. Failing which will lead to confusion during Kansei measurement and yield invalid result. The context of web application chosen for this work is the design of e-Clothing websites where emotional appeal is assumed to be significant. Correspondingly, the selected subjects are consumer with online shopping experience. Then, the study proceeds with synthesizing specimen, synthesizing Kansei Words, Subjects Identification, and Kansei Measurement.

7.1 Synthesize Specimen

Initially, 163 online clothing websites were selected based on their visible differences in design (i.e : colors, layouts, typography). An investigation was conducted to identify detail design elements in all websites in the context of what consumer's see in the interface feature of a website. As a result, the study has identified 77 categories in design element, and 249 items as specified values in each design category identifiable from all websites.

All websites were then analyzed following a set of predefined rules in the study. From the analysis, 35 website specimens were finally used.

7.2 Synthesizing Kansei Words

Since Kansei is the state of consumer's internal sensation, the measurement process can be very challenging. In the measurement of visitor's Kansei in e-Commerce website, measurements are psychological which deals with human emotional state. Hence, the most suitable measurement method is by self-reporting system. This is done by using words that describe the emotional expression associated to e-Commerce website. In KE, this expression is called Kansei Word (KW) [15].

In the study, KWs are used to represent emotional responses and were synthesized based on web design guidebook, experts and pertinent literatures. 40 Kansei Words were then selected according to their suitability to describe website. Among the synthesized words are 'adorable', 'professional' and 'impressive'. These KWs were used to developed checklist to rate websites, organized in a 5-point Semantic Differential (SD) scale.

7.3 Subjects Identification

Deciding on the number of participants was influenced with "What information we intend to capture?" As we intend to understand the pattern of experience economy while shopping online, we have selected male and female participants ranging from the age of 20 to 25. This age group is selected as they are the second most common consumers of online shopping [17]. Furthermore, this age group is readily available in UiTM. An equal distribution of male and female participants was acquired base on Freeman et al. [18], a research concerning participant behaviours and emotions. Furthermore, Horrigan [17] has also found that there is almost an equal distribution of male and female users within the age group of 20 to 25. We have also considered the length of time required to gather and analyse information systematically due to minimal resources of labour, time and money [19].

120 undergraduate students from the Faculty of Information Technology and Quantitative Science, Faculty of Architecture, Building, Planning and Survey, Faculty of Business and Management and Faculty of Electrical Engineering from UiTM

participated in the Kansei evaluation. Exactly 30 students consisting of 15 males and 15 females were recruited from each faculty. All of them have Web experience.

7.4 Kansei Measurement

The participants were grouped according to their faculties. Four Kansei evaluation sessions were held separately for each group. During each session a briefing was given before the participants began their evaluation exercise. The 35 website specimens were shown one by one in a large white screen to all participants in a systematic and controlled manner. Participants were asked to rate their feelings into the checklist according to the given scale within 3 minutes for each specimen. They were given a break after the 15th website specimen, to refresh their minds. The order of checklist was also change to avoid bias. Each Kansei evaluation session took approximately 2 hours to complete.

8 Analysis of Difference

We analyzed the website semantic space by PCA using the averaged evaluation value for each session. In Phase II, we coded the averaged data according to population size of 30, 60, 90 and 120. This is to organize information into set of orders where Kansei semantic space can be observed. PC loadings results show the degree of Kansei affecting variables which are used to obtain KWs structure. Figure 2 shows the distribution of all data population sizes while Table 1 helps to summarize the analysis.

In Figure 2, we can observe a good distribution of variables to axis-x and axis-y, which proves that the measurement was successful. It is evident from the plot results for all population sizes that the KW that produced large negative first PC loadings (x-axis) are mostly "Beautiful", "Gorgeous", "Stylish", "Impressive" and "Appealing". The dense area of the left hand side of the chart corresponds to such KW. On the other hand, KW that produced large positive PC loadings are "Boring" and "Old-fashioned". Thus, we label this PC as the axis of "Attractiveness".

From the result, we can expect that websites with lower scores on this component are likely to have higher sense of attraction and conversely. The second PC loadings (y-axis) shows that KW with positive large loadings are "Masculine" and "Mystic"; the negative are "Cute", "Feminine", and "Chic". Thus, we label this PC as the axis of "Masculine-Feminine". We can expect that websites with high scores on this component will tend to have high characteristic of masculinity and conversely.

In Table 1, we see a contribution ratio of over 70% indicating that the first two principal components represent the total variability. Thus, most of the data structure can be captured in two underlying dimensions. This means, the KW structure are highly influenced by the first two principle components. The remaining principal components account for a very small proportion of the variability and can be ignored. Table 1 also shows that the Kansei structure on website design has two components, which are attractiveness and masculine-feminine. Blending and balancing these two components are determinants of affective website design. Furthermore, all groups of data population suggest same determinants in designing affective website. The difference in number of subjects seems to produce similar Kansei structures.

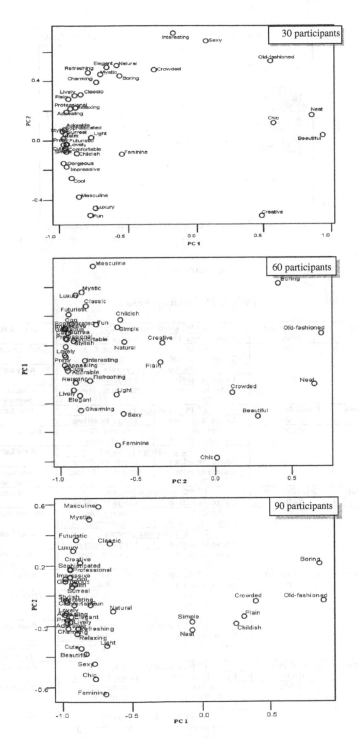

Fig. 2. PC Loadings by 30, 60, 90 and 120 data population

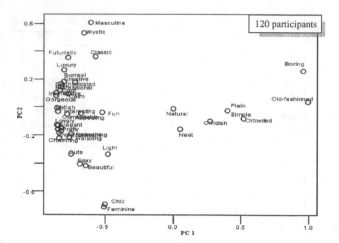

Fig. 2. (*continued*)

Table 1. PC Loadings Results

Population Size	Contribution Ratio	1st PC Loadings (x-axis)		Axis Label
		Large Negative	**Large Positive**	
30	73.0%	Beautiful, Gorgeous, Stylish	Boring, Old-fashioned	Attractiveness
		2nd PC Loadings (y-axis)		**Axis Label**
		Large Negative	**Large Positive**	
		Cute, Feminine, Chic	Masculine, Mystic	Masculine-Feminine
Population Size	**Contribution Ratio**	**1st PC Loadings (x-axis)**		**Axis Label**
		Large Negative	**Large Positive**	
60	75.0%	Beautiful, Gorgeous, Stylish	Boring, Old-fashioned	Attractiveness
		2nd PC Loadings (y-axis)		**Axis Label**
		Large Negative	**Large Positive**	
		Cute, Feminine, Chic	Masculine, Mystic	Masculine-Feminine
Population Size	**Contribution Ratio**	**1st PC Loadings (x-axis)**		**Axis Label**
		Large Negative	**Large Positive**	
60	75.0%	Beautiful, Gorgeous, Stylish	Boring, Old-fashioned	Attractiveness
		2nd PC Loadings (y-axis)		**Axis Label**
		Large Negative	**Large Positive**	
		Cute, Feminine, Chic	Masculine, Mystic	Masculine-Feminine
Population Size	**Contribution Ratio**	**1st PC Loadings (x-axis)**		**Axis Label**
		Large Negative	**Large Positive**	
90	78.7%	Impressive, Gorgeous, Appealing	Boring, Old-fashioned	Attractiveness
		2nd PC Loadings (y-axis)		**Axis Label**
		Large Negative	**Large Positive**	
		Feminine, Chic	Masculine, Mystic	Masculine-Feminine
Population Size	**Contribution Ratio**	**1st PC Loadings (x-axis)**		**Axis Label**
		Large Negative	**Large Positive**	
120	77.9%	Gorgeous, Impressive, Appealing	Boring, Old-fashioned	Attractiveness
		2nd PC Loadings (y-axis)		**Axis Label**
		Large Negative	**Large Positive**	
		Feminine, Chic	Masculine, Mystic	Masculine-Feminine

9 Conclusion

The study was performed to identify differences in Kansei structure by size of data population. The result shows that Kansei structure from different data populations are similar. With the result, we could conclude that the size of data population used in the research instruments does not affect the result of determinants in affective web design. We have shown that to first begin testing, one need to understand the type of product one intends to test, who are the users and what work with the system. Once criterions of data population are set suitable to the research methods and objectives, one can make accurate estimates about the population based on what they have learned from the sample, henceforth concluding a more accurate generalization.

Acknowledgement

The research is supported by grants from the Ministry of Science, Technology and Innovation, Malaysia, under the ScienceFund grant scheme [Project Code: 01-01-01-SF0029].

References

1. Nielsen, J.: Why You Only Need to Test With 5 Users (2000),
   http://www.useit.com/alertbox/20000319.html
   (retrieved February 2, 2007)
2. Gilbert, J.E., Williams, A., Seals, C.D.: Clustering for Usability Participant Selection. Journal of Usability Studies 3(1), 40–52 (2007)
3. Faulkner, L.: Beyond The Five-User Assumption: Benefits Of Increased Sample Sizes In Usability Testing. Behavior Research Methods, Instruments, & Computers 35(3), 379–383 (2003)
4. Landesman, L., Perfetti, C.: Eight is not Enough (2001) User Interface Engineering Website, http://www.uie.com/articles/eight_is_not_enough/
   (retrieved September 30, 2008)
5. Arteology: Sampling, http://www.uiah.fi/projects/metodi/152.htm
   (retrieved September 1, 2008)
6. Rosenstein, A.: Managing Risk with Usability Testing, Classic System Solutions (2001),
   http://www.classicsys.com/css06/cfm/article.cfm?articleid=23
   (retrieved September 1, 2008)
7. Li, N., Zhang, P.: Consumer Online Shopping Behavior, in Customer Relationship Management. In: Fjermestad, J., Romano, N. (eds.) Zwass, V (editor-in-chief). Series of Advances in Management Information Systems. M.E. Sharpe Publisher (2005)
8. Tractinsky, N., Katz, A.S., Ikar, D.: What is Beautiful is Usable. Interacting with Computers 13, 127–145 (2000)
9. Norman, D.A.: Emotional Design: Attractive Things Work Better in Interactions: New Visions of Human-Computer Interaction IX, 36–42 (2002)
10. Buchanan, R.: Good Design in the Digital Age. AIGA Journal of Design for the Network Economy 1(1), 1–5 (2000)

11. Leary, M.R.: Introduction to Behavioural Research Methods, 4th edn. Pearson, London (2008)
12. Hammersley, M., Atkinson, P.: Ethnography: Principles in practice. Tavistock Publications, London (1983)
13. Ward-Schofield, J.: Increasing the generalisability of qualitative research. In: Hammersley, M. (ed.) Social research: Philosophy, politics & practice, pp. 200–225. Open University/Sage, London (1993)
14. Anitawati, M.L., Nor Laila, M.N.: Kansei Engineering: A Study on Perception of Online Clothing Websites. In: Proceedings of the 10th International Conference on Quality Management and Operation Development 2008 (QMOD 2007). Linköping University Electronic Press, Sweden (2007)
15. Nagamachi, M.: The Story of Kansei Engineering. Japanese Standards Association 6, Tokyo (2003) (in Japanese)
16. Nagasawa, S.: Present State of Kansei Engineering in Japan. In: 2004 IEEE International Conference, vol. 1, pp. 333–338 (2004)
17. Horrigan, J.B.: Online Shopping. PEW Internet and American Life Project (2008), http://www.pewinternet.org/pdfs/PIP_Online%20Shopping.pdf (Retrieved August 20)
18. Freeman, J., Lessiter1, J., Pugh1, K., Keogh, E.: When presence and emotion are related, and when they are not. In: Proceedings of PRESENCE 2005, pp. 213–219 (2005)
19. Freedman, R., Taub, S., Silver, J.F., Pell, E., Rowe, J., Chiri, G.: Sampling: A Practical Guide for Quality Management in Home & Community-Based Waiver Programs. Thomson & Medstat (2006)

Evaluating E-Commerce User Interfaces:
Challenges and Lessons Learned

Rainer Blum and Karim Khakzar

Hochschule Fulda - University of Applied Sciences,
Marquardstr. 35, 36039 Fulda, Germany
{Rainer.Blum,Karim.Khakzar}@hs-fulda.de

Abstract. This paper presents lessons learned from a user interface evaluation - concerning the applied methodical approach. Four alternative implementations of product catalogue navigation and presentation components for online clothing shops were evaluated in a comparative study. The resulting rather complex experimentation setting revealed interesting issues for the design of similar experiments. After describing the study's methodical setup the paper analysis relevant aspects of the applied approach. Finally, lessons learned are derived that are of relevance for user interface testing methodology in related contexts.

Keywords: E-Commerce, User Interface Evaluation, Rapid Prototyping, Evaluation Methodology.

1 Introduction

Numerous scientific papers prove that not only product attributes and commercial terms and conditions are of high importance for the success of business-to-consumer e-commerce shops but also design-related characteristics of the shops themselves [1]. Concerning the latter, aspects like interactivity, usability, information presentation, media richness, accessibility, personalization, adaptation or experience-related traits (e.g. design quality, entertainment, playfulness etc.) have been identified as relevant and important. Therefore, e-commerce businesses are well advised to invest sufficient effort in an adequate consideration of these criteria when building and maintaining their online stores.

Developing e-commerce user interfaces for the B2C retail sector is a current subject in the research project SiMaKon, "Simulation of Made-to-measure and Ready-made Clothing Online for Fit Checking" at Fulda University of Applied Sciences. In doing so, the aforementioned design-related attributes are regarded as fundamental quality criteria. Concentrating on these issues, we early had to realize, that many fundamental user interface design options exist - not only for small details like user interface controls or structuring of individual pages, but even for the higher-level functional components of online shops, e.g. shopping cart or product catalogue. Due to the lack of standards or universally accepted guidelines on best practices for online shops, it was decided to conduct a controlled experiment comparing different design alternatives.

M. Kurosu (Ed.): Human Centered Design, HCII 2009, LNCS 5619, pp. 653–660, 2009.

Designing such a comparative experiment forces researchers to make challenging decisions: What shall, basically, be compared? Which attributes of the objects of comparison are suitable for a comparative consideration? How many objects can practically and reasonably be compared? Especially, when many alternatives exist - user interface design options in the case of this study - and shall be taken into account, experimentation settings may become rather complex.

These issues are, in an e-commerce setting, accompanied by other effecting factors: The audience is extensive and can become large and uncontrollable once an online shop is made public. The indispensable gathering of information about the targeted consumers is therefore not an easy task.

1.1 Objectives and Significance

Object of investigation of the research activities of the ongoing project SiMaKon is an innovative, interactive system for the support of apparel retail via Internet. The paper presents significant lessons learned in the context of a study on e-commerce user interface alternatives - concerning the chosen methodology.

As lessons learned in user interface testing are for a large part context-dependant respectively research-specific, details on the study's context are also provided. Thus, potentially, the reader is enabled to correctly interpret the findings and to transfer them to other research contexts. The results presented should help researchers working on similar problems in designing their experiments.

1.2 The SiMaKon System

Trying on and reliably appraising clothing in online shops with the help of individual virtual bodies constitutes the basic concept of the ongoing research project SiMaKon. Challenges addressed are:

- reliable fit checking and appraisal of optical characteristics in relation to one's own body respectively a virtual representation of it,
- seamless integration with product catalogue respectively product configurator,
- comprehensive interactivity, thorough usability and absorbing shopping respectively user experience,
- high customer acceptance.

Therefore, an integrated system, comprising of the functional components avatar generator, clothing pattern generator, clothing simulation, 3D rendering, e-commerce back-end and online shop front-end is developed and continuously evaluated. Some main practical implementation details are:

- virtual garment construction and sewing using real garment pattern data and realistic simulation of the garments' drape and the cloth surface,
- representation of subtle details, e.g. sewing threads and knobs, as well as high-quality lighting and shading,
- individual virtual human bodies (avatars) with configurable body dimensions, hair style, skin tone and face characteristics,

- rendered 3D content with avatar, complete clothing outfit and background scenery displayed in the Web browser, integrated in an online shop,
- customers can interactively change perspective and background scenery as well as zoom into details,
- pieces of clothing or made-to-measure garment components can be swapped directly from the product catalogue for easy comparison.

After entering the online shop customers may start by first configuring their avatar(s) or by browsing the product catalogue. There, interesting clothing can spontaneously be selected for try-on and appraisal with the currently activated avatar.

The user-related criteria is dealt with in a cyclic user-centred design process based on ISO 13407 [2], which involves rapid prototyping of design outlines. As, like already mentioned, many fundamental user interface design alternatives apply for the system's different functional components, comparative experiments with prototypic mock-ups are continuously conducted in order to come to well-founded design decisions. The functional component relevant for the study reported in this paper is the product catalogue navigation and presentation.

1.3 Modalities of Needs

One main aspect of the evaluation of the different product catalogue navigation and presentation concepts reported here was the notion of "modality of needs": In her work Michele Ambayé [3] describes three different ways e-commerce consumers usually proceed through the decision making process, depending on the consumers' current intentions. A consumer may start the shopping process with any one of the three types of needs.

Pre-knowledge driven customers know they have a need, what to look for (usually brand and product type) and where to find it on the Internet. The purchasing criteria are quite clearly defined in the consumer's mind. That means the consumers know what they are looking for. Thus, they search on limited criteria such as price and delivery. For example, a consumer has previously bought a pair of Levi's 501 jeans in a particular fabric and size online, and would like to purchase it again. When customers know the website, the price and the shop's service policy, they possess previous knowledge and act knowledge driven.

If only a vague idea of a need exists (product type), and the Internet is used for searching different e-retailers, the modality is called function driven. For example, a consumer has bought a pair of unbranded jeans before, but would like to purchase a different style now. Deciding to gather more information online, the consumer is looking for a product that fulfils a function (a pair of jeans for a particular purpose) rather than for a brand or style already well known to the customer. Thus, this mode applies if a consumer knows only what sort of product is needed, such as a suit shirt, but has not yet established specific criteria intrinsic to the product, such as for example its style or material. Therefore, the purchasing process is characterized by the need for the function provided by the product: For example, it has to be a shirt that is elegant and goes well with a suit.

Impulse driven consumers are just browsing the Internet and have no conscious knowledge of a need. Here, needs are established impulsively as a result of interacting with one or several websites. Browsing the Internet, a consumer is maybe looking for some product or not really looking for anything in particular. A website or a sales discount advertising banner may catch their eye, showing a pair of Levi's jeans at a significantly reduced price. Even though the consumer does not really need a pair of jeans, the offer is too good to miss. The purchase decision is made impulsively.

It is important to note that people may possibly switch between modalities during the purchasing process (shifts in mode). This occurs, when the consumer's current need changes. For example, consumers may start with a particular need, but while browsing on the Internet for a particular product (function-driven), an advertising banner attracts their attention for a different product and they finally buy impulsively. Therefore, they began shopping in one mode and afterwards shifted to another mode.

For e-commerce user interfaces trying to optimally support modality of needs including shifts between them may be a promising way to best serve an otherwise extensive, heterogeneous audience. Furthermore, it is likely that a particular e-commerce user interface design may serve one modality of needs better than another.

2 Method

Four different concepts, each implemented with a rapid prototype with basic functionality, were chosen for the comparative study - based on three aspects:

- support of the depicted modalities of needs
- potential for high usability and user experience for a broad spectrum of apparel consumers and,
- innovativeness.

The four prototype concepts for the product catalogue presentation used in the study are "associative structuring", "automatic adaptation", "three-dimensionality" and "classic approach".

The associative navigation structure applied for the first prototype is aimed at releasing the user from having to understand and accept the usually firmly defined navigation structure predetermined by the designer, like e.g. an ordinary hierarchical structure. Instead, the sorting of the product catalogue is permanently adapted based on similarity among different product items, with and based on every activity of the consumer. The concept is described by Peake [4] and was transferred to an online shop context for the purpose of the study reported here.

Every click on a product image sets the chosen clothing item and its four attributes brand, colour, type and pattern as the new central topics of interest - and moves this product to the middle of the screen. All other items are immediately rearranged around it: The ones more similar to the centre product and, therefore, presumably most interesting to the consumer at this moment are moved nearer, less closely related items are placed more far away from the centre product. Each of the four considered attributes is dedicated a separate part of the screen space, with all products appearing in each of these distinct areas. Items that are not visible on the screen due to their distance from the centre space can be reached via scrolling arrows provided at the four edges of the screen.

Automatic adaptation of the interface to the current modality of needs constitutes the main aspect of the second approach. Part of the navigation of the product catalogue is changed according to the customers' current mode. For this prototype a hierarchically organized product catalogue was complemented by additional navigation aids. These are

- an advanced search form with search filters for product features like colour, material or pattern for pre-knowledge driven customers,
- a navigation structure with all featured trademarks as root categories and types of garments as subcategories for function driven mode,
- a navigation structure listing and sorting the products under different thematic, lifestyle oriented aspects like winter, sports or outdoor for impulse driven consumers.

The third prototype mimics the three-dimensionality of a real store, allowing customers to explore the store's range of products similarly to real life. Separate product items are displayed in two-dimensional information spaces each located on a separate wall section. Using several virtual rooms on several floors the product range is structured spatially. Outline maps in the form of a schematic presentation of the room and level structure allow for the transfer between the separate spaces. Other navigation means, all used via mouse and keyboard, are moving forward, backward, to the right and to the left, turning around at the current position and automatically moving along a predefined path to a certain destination, while observing the crossed environment.

The fourth and last prototype concept is a classic, hierarchical presentation of the product catalogue. It was deployed for comparative purposes, to see if one of the other concepts has advantages over this traditional way to present a product catalogue. It consists of a hierarchical tree displaying the product categories on the left and a detail view in the middle. The detail view presents either the items belonging to the currently selected product category or detailed information about a chosen, single product.

Sixteen mixed-gender candidates were tested individually in laboratory conditions, one person at a time in a one hour session, supported by two test assistants. Approximately 80% of the test persons belonged to the age groups 16-24 and 25-35, with gender uniformly distributed and five persons never or infrequently (less often than every two months) buying products via the Internet.

The study employed a three (three modalities of needs) by four (four prototype concepts) factorial design. Both factors were manipulated within subjects. Modality of needs was controlled by scenario descriptions that were presented to the participants, each detailing a typical situation for the respective modality. For example, for the function driven modality it stated: "Imagine, you are currently intending to buy several pieces of clothing with quite specific characteristics, e.g. a certain type of clothing, a certain colour, a certain pattern or several attributes combined. Therefore, you are browsing the catalogue of an Internet shop for products having exactly these features". Prepared like this, the test subjects had to accomplish several given tasks with each of the prototypes. The tasks were characteristic for the current modality, e.g. "Please choose a men's shirt, basically of the colour blue with a checked pattern." After having passed through all the modality scenarios with a prototype, the next one was tested.

During task execution the test assistants collected usability-related performance measures, namely success, speed and effort of completion, occurrence and severity of difficulties. Furthermore, they noted observations and estimated user experience attributes on five-point Likert scales, with the items satisfaction, excitement, interest, joy and fun.

After completing a condition, the participants were asked to rate the tasks' complexity. Then, for each prototype, general usability-related questions were raised, involving estimation, e.g. usefulness for the given task, and open-end questions, e.g. if the subject especially likes a specific feature of this design alternative and why. This was followed by prototype-specific questions, e.g. concerning characteristic features. Finally, the participants had to appraise the same user experience attributes as mentioned above.

3 Results

Concerning this section the reader has to be aware, that it does not provide a description of the results of the described prototype study itself, but of relevant issues arisen from the methodical setup of the experiment.

The first aspect concerns the scenario descriptions presented to the participants. Each detailing a typical situation for the respective modality, they were targeted on controlling the modality of needs currently effective. Indeed, it may be questioned if "putting" a person "into" a specific modality of needs with a scenario description actually has the desired effect. No conflicting evidence occurred in the study, but, in fact, this issue was not well examined.

Then, the rapid prototypes may have been conceptually and functionally too different for a meaningful comparison. For example, the associative prototype was based on constant resorting of the product catalogue, while the 3D prototype did not offer sorting or filtering at all. Likewise, the adaptive prototype was the only alternative that "reacted" especially to the current modality.

Also, the maturity of certain aspects of the rapid prototypes may have varied to much. This concerns for example differently developed degrees of self-descriptiveness of separate interaction elements, e.g. readily identifiable scroll arrows for the associative prototype versus an abstract navigation map in case of the 3D prototype. Analysis of results also manifested that the applied principles were not clear or easily perceivable for every alternative. For example a significant number of participants did not comprehend the sorting principle of the associative prototype. Admittedly, this may put the results achieved into question. The trade-off between affordable effort and required quality proved to be an essential problem in rapid prototyping for this user interface evaluation.

Furthermore, one salient problematic aspect proved to be the prompted and observed user experience attributes. The statements and self-evaluations of the participants often differed considerably from the test assistants' notes. Clearly, definite tendencies were identified for many points and are presumably stable. Nevertheless, a significant amount of uncertainty concerning the results' validity in this domain remains, notably because neither a positive nor a negative tendency was identified by the test assistants for a great part of the participants. Insufficient qualification of test assistants or lack of quality of the observation criteria may be responsible factors. Also, it is worth discussing if these aspects may be reasonably observed at all. At least, to reduce subjective differences, the same test assistants may better have carried out all the tests. However, it is unclear if crew change really has affected the results.

Another interesting aspect is that test assistants suspected that the mood of the test candidates or their physical and mental state on the day of testing may have had a significant effect on the achieved results.

Concerning the formation of the test participants group, it should be cared for more heterogeneity: More persons of higher age groups and of different education strata should be included for the present context, online shopping of clothing. Furthermore, only those may be considered that actually buy clothes via the Internet, for the sake of significant results.

Positively, scale and one hour duration of the test seemed to be well-chosen. Conducting part of the interview after each prototype was effective as well, as the participants' impressions were still fresh and not diffused by the interaction with the alternatives.

4 Conclusions

From this work in an B2C e-commerce setting, the following lessons learned can be derived for the methodical design of user interface tests with rapid prototypes.

First of all, be sure that scenario descriptions aiming to transfer test candidates into different, mutually exclusive states respectively situations really serve their purpose. At least, try to check if the scenarios are actually "adopted" by the participants, for example via suitable verifying questions.

Ensure that in the case of comparative experiments, different test situations respectively prototypes are not too varying for a meaningful comparison. Better concentrate on one aspect of comparison. Instead, several tests on differing characteristics of a smaller extent should be conducted, though this may often be tedious. Ideally, only a small number of objects should be compared in one study in order to limit the complexity of the experimentation setting. In the present case, comparing two different versions of a 3D shop environment may have been more effective.

Then, it is important to ensure comparable maturity and functionality for all relevant aspects of the included rapid prototypes. This way, unintentional bias on results can be avoided. Furthermore, design evaluation criteria for prompted and estimated user experience attributes thoroughly. Double-check mechanisms should be applied to ensure validity of results.

Acknowledgments. This research was financially supported by the German Federal Ministry of Education and Research within the framework of the program "Forschung an Fachhochschulen mit Unternehmen (FHprofUnd)".

References

1. Chang, M., Cheung, W., Lai, V.: Literature derived reference models for the adoption of online shopping. Information & Management 42, 543–559 (2005)
2. ISO 13407, Human-centred design processes for interactive systems (1999)

3. Ambayé, M.: A Consumer Decision Process Model For The Internet. Brunel University Information Systems and Computing PhD Theses (2005)
4. Peake, R.: An Experiment in Associative Navigation (2007),
   http://www.robertpeake.com/archives/
   365-An-Experiment-In-Associative-Navigation.html

Caring and Curing by Mixing Information and Emotions in Orphan Diseases Websites: A Twofold Analysis

Maria Cristina Caratozzolo, Enrica Marchigiani, Oronzo Parlangeli,
and Marcella Zaccariello

University of Siena, Communication Sciences Department,
Via Roma, 56, 53100 Siena, Italy
cristina.caratozzolo@polimi.it,
{marchigiani,parlangeli}@unisi.it,
mzaccariello@virgilio.it

Abstract. The study reported in this article was structured as a first step in planning guidelines to design effective internet sites for Associations dealing with rare diseases. The Authors used a two-stage analysis: first, they carried out an analysis of the websites of a sample of Italian Associations on rare diseases and then they did an interview survey to identify the objectives and needs of those organizations. Results indicate that two different kinds of organizations do exist and suggest possibilities for developing guidelines aimed at improving their websites.

Keywords: Rare and orphan diseases, affective communication, patients' associations, website, usability.

1 Introduction

Orphan diseases include rare and neglected diseases. The two classes of pathologies have in common the fact that development and production of specific drugs are not profitable for the major pharmaceutical corporations (often referred to as "big Pharma"): that is why they are both named "orphan".

Rare diseases, including those of genetic origin, are life-threatening or chronically debilitating diseases which are of such low prevalence that special combined efforts are needed to address them. As a guide, low prevalence is taken as prevalence of less than 5 per 10,000 persons in the European Union and less than 200,000 individuals in the United States.

It is estimated that between 5,000 and 8,000 distinct rare diseases exist today, affecting between 6% and 8% of the population in total.

There are about 30 million people in the 25 EU Countries that have a rare disease, this means that it is not so unusual as is generally believed to be, to suffer from a rare disease [1].

The focus on rare diseases is a relatively new phenomenon in Europe. Until recently, public health authorities and policy makers largely ignored them. Clearly, it is impossible to develop a public health policy specific to each rare disease. But a global rather than a piecemeal approach could provide some solutions. A global approach to

M. Kurosu (Ed.): Human Centered Design, HCII 2009, LNCS 5619, pp. 661–670, 2009.
© Springer-Verlag Berlin Heidelberg 2009

rare diseases means that individual diseases do not fall through the net and real public health policies can be established in the areas of scientific and biomedical research, drug research and development, industry policy, information and training, social benefits, hospitalization and outpatient treatment [2].

1.1 Health Communication on Rare Diseases

People are becoming less accepting of a passive role as patients. The term, "Informed Patient", assumes that people with illnesses need information in order to be involved in their healthcare. This information is essential for seeking care, deciding on the best courses of action with their healthcare professionals, or for follow-through on an agreed course of treatment. However, as medicine becomes more complex, it is harder for an individual clinician to be able to provide all the information that a specific patient may need or want in order to make a decision.

Health is now reputedly the second most popular topic to be searched for on the internet. For traditionally underserved populations, the web can potentially unlock resources that could fundamentally improve health and wellbeing.

Each rare disease affects so few people that information about it may be difficult to find, making the situation even more traumatic and stressful. Before information was available on the web, families coping with a rare disease usually struggled alone. Support could only be found through telephone calls to other families suffering from similar diseases, and only if the names were provided by doctors.

Support groups such as the National Organization for Rare Disorders (NORD) have worked aggressively in the last 20 years to draw attention to people with rare diseases, and especially to draw attention to the lack of treatment options. New web-based support groups continue to proliferate. Not only are people receiving comfort from others with the same conditions, but they are learning from each other's experiences as well. By the late 1990s, most non-profit organizations had websites where people could ask questions and get immediate responses.

1.2 Organizations on Rare Diseases in Europe and Italy

A number of Organizations which coordinate Associations dealing with rare diseases have been set up in the past 10 years in Europe and also in Italy.

The EU, for instance, has supported several projects by EURORDIS (European Organization of Rare Diseases) which brings together more than 200 rare disease associations in 16 different Countries [3]. ORPHANET is instead a European database that deals with rare diseases and orphan drugs, with the aim of improving the diagnosis, care and treatment of patients. The database covers today about 5,000 diseases [4]. In addition, the Rare Diseases Task Force was created in January 2004 by the European Commission Public Health Directorate with the aim of advising and assisting the Directorate itself in promoting optimal prevention, diagnosis and treatment of rare diseases in Europe [5].

In Italy the National Rare Diseases Centre (CNMR, Centro Nazionale Malattie Rare) is part of the National Rare Diseases Network and carries out scientific research and public health activities, both at national and international levels.

In particular, the CNMR has created the National Rare Diseases Register which, for each pathology, identifies the number of cases and their distribution on the national territory, being the central connection of the national clinical-epidemiological network. The National Orphan Drug Network aims at activating a surveillance system for all orphan drugs reimbursed by the National Healthcare System.

The CNMR has established several profitable collaborations on various projects with the associations of individuals affected by rare diseases. In particular, it carries out studies to evaluate the accessibility to social and health services, healthcare and the quality of life of the patients affected by rare diseases and their families.

The numerous Associations of Patients living with rare diseases play a decisive role for patients and their families. The basic principle of these Associations is reciprocal assistance: this enables the development and exchange of information and education for improved access to diagnosis, therapy, rights and patient integration.

According to the guide "*Associazioni Italiane Malattie Rare*" 2008-2009 edition, there are 179 Patient Associations which deal with 435 diseases.

Some Patient Associations are very small and poorly visible, often present only in one Italian region. Other Associations are much more organized and have several affiliates in Italy.

UNIAMO ONLUS is an organization founded in 1999 with the aim of having a single organism representing all Italian Patient Associations. UNIAMO presently coordinates more than 50 Associations, dealing with more than 600 different pathologies and it is part of a European network of 16 national alliances of Patient Associations, all belonging to EURORDIS organization.

2 The Study

2.1 Objectives

The purpose of the experimental survey was to draw a representative up-to-date outline of the Italian communication activities performed through the web by Patients Associations dealing with Rare Diseases.

A twofold goal was either to define a "typical" website structure (considering both contents and appearance), and/or to find recurring website categories.

Moreover, identifying both the Associations' and the patients' (that is, the websites' final users) informative and emotional needs, was also a goal to be accomplished through an experimental investigation.

The final aim of the study was to carry out a first step in planning guidelines to design effective internet sites for Patients Associations on Rare Diseases.

2.2 Methodologies

The experimental survey was composed of two sections: firstly, a sample of 100 Italian Associations on rare diseases was selected, to give a representative snapshot of the Italian situation.

An analysis of their websites was carried out, aiming at defining differences and similarities between the websites' structures.

Secondly, an interview survey was done on a selection of 30 Associations, to identify the objectives of the organizations, their motivations and needs.

2.3 Analysis of the Associations' Websites

The analysis on the websites was completed on a sample of 100 websites of Italian Patients Associations on rare diseases. It started by the adoption of an observation grid, aimed at noticing and distinguishing the formal aspects from the content ones.
The actual structure of the grid is shown in Table 1 below.

Table 1. Sample of the websites observation grid

Website	Content		Formal		Notes
	Generic info	Medical info	Look&Feel	Usability	
...	Type of info delivered	Type of info delivered	Use and quality of graphics	Efficacy, efficiency, ease of use	Emotional aspects

It was useful to identify, collect and compare the different kinds of information present in the websites (both generic and medical) and to not their different formal aspects.

Usability issues have also been considered in the formal aspects. Given that the term "usability" refers to clarity and intelligibility with which the interaction with a computer program or a website is designed, the method used in the present study was the Usability Inspection, a rapid technique for reviewing a system based on a set of guidelines. The review was conducted by two experts who were familiar with the concepts of usability in design; a systematic inspection of the user interface of the websites was thus performed to examine and judge its compliance with recognized usability principles [6]. In particular, issues such as clarity, consistency, and error minimization were analyzed establishing their contribution to global characteristics like efficacy, efficiency, ease of use.

In addition, we have also tried to identify how much informational and formal aspects were emotionally connoted. Emotional aspects were actually present in most websites, expressed both through contents and appearance (graphics, photo, videos). Emotions appeared to affect all others components, and refer to fundamental factors connected to being a patient affected by a rare disease.

2.4 Interview Survey

The analysis carried out through the interviews was aimed at finding in depth information on the Italian Associations' websites in order to acquire knowledge on their needs, motivations and expectances on the basis of their direct experiences.

Thirty telephone-based interviews were conducted by 3 researchers who worked in parallel, interviewing Associations' representatives from different places in Italy. Nearly all the Italian regions were present in the sample.

In almost all cases, the interviewee was the President of the Association, generally a patient her/himself or a patient's relative.

The interview was divided into 3 sections. The first one aimed at gathering general information on the website, while the second one was essentially focused on finding out its main objective, the target users, the Associations' motivations and needs, and their communication strategies.

The third section was devised to gather information on how the Association support patients and their families through the website.

3 The Results

3.1 Websites Analysis

The websites analysis is synthetically shown in Table 2 below.

Contents appear to be quite homogeneous in kind and quantity all over the sites; as a matter of fact, all websites contain essentially the same typologies of general information:

- The history of the Association, information about its offices and contacts, its activities
- Demand for financial support, donations
- Legal information about rare diseases and conveniences granted to patients
- Events, initiatives
- Other: contacts, links, downloads…

A certain uniformity is also observed with regards the medical/scientific section, since a few classes of information recur in a high percentage of the monitored websites:

- A description of the pathology (causes, symptoms, consequences)
- Research: state of the art of medical research concerning the pathology
- Prevention and Therapy: information about care centres, available drugs etc.

Relatively low percentages of Associations do offer on-line support and services to users, such as:

- Medical support
- Psychological support

These are classified as "services" since they require an interaction between doctors/psychologists and patients/relatives.

A higher diffusion is found for some tools, like:

- Forums and communities, where users can register and exchange suggestions, information, feelings…
- Message boards, where users can write messages, impressions about the site, ask questions to other people, in a public and real-time space
- FAQ: Answers to Frequently Asked Questions, collected for a quick response
- Advice for dealing with frequent difficulties in everyday life.
- Newsletter and Reviews.
- Other tools, for example: chat, guestbook…

Table 2. Synthesis of results emerged in websites analysis

CATEGORY	SUB-CATEGORY	ITEM	%
Content	General information	Association	98
		Funding	98
		Legal info	89
		Events	75
	Scientific information	Disease description	85
		Research	38
		Prevention/Therapy	37
	Services	Medical support	7
		Psychological support	3
	Tools	Forum/Community	28
		Message board	5
		FAQs	8
		Advises for everyday life	25
		Newsletter/Review	17
		Other	5
Form	Usability	Easy to use	25
		Improvable	34
		Not usable	40
		Familiar language	58
		Formal language	40
		Translation	5
	Look&Feel	Sophisticated graphics	43
		"Home made" graphics	55
		Photos/Videos	48
		Images	35

The second extensive section of the analysis took into consideration the formal aspects of websites, that is: Usability and look&feel.

As for usability, a website has been judged i) usable, ii) improvable, or iii) not usable. In addition, other aspects, such as the adoption of the user's language and the possibility to obtain a translated version of the site, have been taken into account.

The observation of the look&feel focused on features like:

– Graphics, in order to define whether a sophisticated or a "home made" look had been produced
– Presence of photos and/or videos
– Use of images, drawings, animation .

Third, and last, a parameter of this part of the study, dedicated to websites, was the Emotion. In particular, the presence of a relational attitude of the Associations towards the users was detected, together with the tendency to realize an affective approach to communication.

Affective aspects are obviously crosswise, and they are detectable both in the contents and in the form. More specifically, it is possible to notice an emotional factor in:

- The way contents are expressed (language, examples, focus), the attitude towards the user (addressing him as a simple reader or as a patient/ concerned person)
- The kind of contents included (stories on patients' experiences, narratives, poems)
- The use of images, photographs, videos aimed at arousing emotion, empathy, care.

3.2 Description of the Interviews Results

The interviews analysis clearly showed that in most cases the websites of the Associations on rare diseases have been created along with the association itself, or a few years later if the Association was founded in the early 1990's. Generally, the promoter is the patient her/himself or a close relative, but in some cases even physicians and researchers cooperate to the setting up of the Associations.

Mostly the sites have been developed by the patients themselves, family members or friends with the technical advice of a web design expert.

The contents concerning general information, disease regulation and policy, activities, events, fund raising, are decided by the President of the Association or by the management team, with the support of physicians and medical specialists as far as the scientific contents are concerned.

Except for one case, all the sites of our sample are updated by volunteers on the average once a week; or in any case, whenever there is a new content to include.

In all the cases, the Associations pay special attention to patients and their families, informing and educating them on the pathology, helping them protect their social rights and providing them with an affective and psychological support, as well as, in some cases, actual practical aid.

Among others, a significant help that the site can offer is to try to create a network aimed at supporting people affected, improving health and psychological care and putting in touch persons with the same experience. People sharing the same condition, even if afar, appear willing to communicate, comparing their lives and their emotions, offering solidarity, supporting each other and exchanging information on how to solve the problems of daily life.

One interviewee, President of an Association and a patient herself, said that the site is the best way to communicate and disseminate knowledge on orphan diseases in order to avoid that other people should feel a sense of isolation, loneliness and fear of the unknown.

The Associations are often addressed to the medical community as well, especially to paediatricians, family doctors, health care facilities, caregivers, researchers and medical school final-year students. The importance is underlined of promoting scientific research, organizing and participating in conferences, seminars, workshops and fund grant research. With regard to education, the website should be a way to bring physicians and health care staff up to date and make them aware of orphan diseases in a global approach.

By reason of the lacking interoperability among different health care and health research centres, another strong point of the Associations' websites is to avoid that those centres make use of the patients on behalf of the centres themselves. So the website is also a way for the patient to become aware of the disease and to be free to choose among different possibilities of health care facilities.

In quite all the cases, fund raising is considered highly important by the Associations but it comes last on the list, after the needs of information, support, research and education. Anyway, Associations are generally very lively, organizing social events such as concerts, festivals, dinners, all aimed at collecting money for the Association itself and to promote research.

The communication strategies of the Associations' sites rely mainly on in depth information on scientific contents, their accuracy and reliability, but also on the simple way to express them, trying to make themselves understood by everyone.

An effort of the Association that is reflected in the website is to provide services such as forums, mailing lists or simply a phone contact, in order to give support and create a network of people with the same experience. Nevertheless, it would be enough to provide information on any service given in the local area or far away.

To foster a relationship among patients and medical specialists and to create a network among families and health care centres is judged to be crucial and of the greatest importance.

4 Discussion and Design Hints

The study showed that it is possible to classify the websites into two different categories, depending on the communication strategy of the correspondent Association.

Some websites are likely to be defined "institutional", for they show a complex graphic aspect (a coordinate-image, with the Association's logo), and do contain scientific information, aseptic data for generic users. They seem to address non-specific users, potential contributors or volunteers, more then patients or patient's relatives.

The other kind of websites appears more "relational". They show an inclination towards patients and families, expressed through a spontaneous graphic asset, a familiar language, patient-centred contents. The use of photos, images, drawings and narratives highlights the value assigned to emotional aspects, human relationships, empathy among sufferers and families, understanding the emotional distress and the patients' physical suffering. This finding is also noticed through the interview-survey, for all the interviewed Association representatives stressed the primary importance of giving psychological support and expressing empathy and affective involvement with sufferers and their families.

Taken together, the websites analysis and the interviews suggested some guidelines (listed in Table 3, below) to develop more efficient and effective websites for rare diseases Associations, with regard to the kind of contents and their form, the tools provided, usability and accessibility, emotional supports [7].

Internet is a very important channel to disseminate the available knowledge about orphan diseases because an early diagnosis and adequate care can reduce drastically

Table 3. Guidelines for designing websites

Contents about the pathology
Medical/scientific/ information
Communication about the research
Regulation and policy
Information about drugs
Advice for everyday life
Health education at a specific and a global level
Information about the Association
Association description, involved people and contacts
Legal information
Activities and events
Fund raising
Information about healthcare facilities
Local, National and International research and healthcare centers, hospitals
Psychological counseling
Content form
providing correct, complete, reliable and up-to-date scientific publications;
making use of a simple standard of language in expressing contents;
Tools
Forum/Community
On-line expert
FAQs
Message boards
Newsletters or Reviews
Usability and accessibility
Consistency
Visual clarity
Compatibility
Informative feedback
Explicitness
English version
Emotional support (crosswise)
Providing support through forum and on-line expert
Creating a network of patients, families, associations and health care centers in order to provide affective support and avoiding isolation
Providing information on activities and events
Provide a narrative description of patients' experiences with drawings and images

impairments and life-threatening aspects for these kinds of pathologies. As the Rare diseases Associations websites are so important for all the persons involved, we have to take into account how to express all the potentialities of this tool.

Acknowledgments. The Authors are grateful to all the Associations Representatives for their important help and contribution.

References

1. Rare diseases: Understanding this Public Health Priority. Eurordis,
   http://www.eurordis.org
2. European Commision: Public Health – Rare Diseases,
   http://ec.europa.eu/health/ph_threats/non_com/
   rare_diseases_en.htm
3. EURORDIS, http://www.eurordis.org
4. ORPHANET, http://www.orpha.net
5. European Commission: Rare Diseases Task Force,
   http://ec.europa.eu/health/ph_information/implement/wp/
   morbidity/rdtf_en.htm
6. Nielsen, J., Mack, R.L.: Usability Inspection Methods. John Wiley & Sons, New York (1994)
7. Nielsen, J.: Designing Web Usability. New Riders Publishing, Indianapolis (2000)

Eye Tracking Method to Compare the Usability of University Web Sites: A Case Study

M. Oya Çınar

Department of Biomedical Engineering, Baskent University,
06590 Ankara, Turkey
oyacinar@baskent.edu.tr

Abstract. Web sites are one of the main source which enables human computer interaction, also widely used for receiving and transmitting information. University web sites are frequently visit by their students to get some information. In today's fast life cycle these web sites has great usage, many people prefer to use them. University web sites are extremely important for the students of that institution. In the last years, usability has become a highly important research subject. Designing usable web sites is considerably important factor for the user satisfaction in our case for university students.

In this study, a new design is proposed for the engineering faculty web page and eye tracking method is used to compare the usability of it with the original design. Participants were observed while trying to finish specified tasks. In evaluation period, fixation count, fixation length and heatmaps of each website are taken into the consideration. At the end of the study showed that proposed design is more effective and efficient. Participants required fewer fixations and less time to complete the given tasks.

Keywords: Eye tracking, HCI, usability, computer interface design, design evaluation.

1 Introduction

In recent years, human-computer interaction became a really important issue because of the rapid development of computer technologies and the increase number of the web users.

With the fast developing computer technology and the foundation of the internet in the last two decades, from now on, people started to make their operations on the internet. At first, internet was started with the purpose of use of searching and gaining information. Web sites are one of the main source which enables human computer interaction, also widely used for receiving and transmitting information. University web sites are frequently visit by their students to get some information as well. In today's fast life cycle these web sites has great usage for many people.

Research in the web sites and reach correct information about their universities and becomes highly important for the people who are working or studying at the universities in this competitive and continuously growing academic world. With the help of the developments of the online technology and the world of the internet, people can

M. Kurosu (Ed.): Human Centered Design, HCII 2009, LNCS 5619, pp. 671–678, 2009.
© Springer-Verlag Berlin Heidelberg 2009

reach the resources all over the world. As a result research methods are naturally evolved from paper based catalogs to the searchable online web sites. Those university web sites become served as the main repositories of general information over the last years [1]. For this reason university web sites and their usability performance becomes very important in nowadays.

Usability is an emergent quality of an optimum design, which is reflected effective and satisfying use of information technologies. As an emergent quality, usability is implicit in the design and manifests itself through interaction with the product. Although this definition implies that usability evaluation necessarily involves a user interaction, evaluation may also be conducted on the basis of the product's features and characteristics [2]. In other words, usability, a holistic view to ergonomic and collaborative product design (in our case web page), is seen as a critical dimension of which importance is increasingly swiftly in designing stages. [3][4]. Usability is defined as effectiveness, efficiency, and satisfaction of a product for achieving specified goals for specified users in a particular environment.

Designing usable web sites is seen a company philosophy for firms in today's competitive business environment [5]. It is an important stage to observe and analyze multi dimensional web usability attributes in product design. In general, usability refers to how well users can learn and use a web sites to achieve their goals and how satisfied they are with that process. Usability, defined as that people who use the product can do so quickly and easily to accomplish their tasks. Web usability may also consider such factors as cost-effectiveness and usefulness. Usability measures the quality of a user's experience when interacting with a product or system - whether a Web site, a software application, mobile technology, or any user-operated device [3].

A key methodology for carrying out usability is called User-Centered Design. In the early 1990s Jakob Nielsen and Jeffrey Rubin pioneered the testing of web sites to determine whether they met users' needs [5][6]. They adapted usability engineering techniques developed for computer software design and applied them to Web design. Tests revealed that the way material is arranged, labeled, and presented on the Web (the site's "information architecture") has a major impact on users and their ability to operate a site effectively. Usability testing has since become the focus of considerable attention both for commercial and academic sites [7].

Eye tracking is vey successful research method that used in perception and visual research as well as the other human factors for years. Combined with conventional techniques those gather data based on users' explicitly and bluntly behavior such as speaking or mouse clicks. Eyetracking provides another layer of insight into how users process the visual information to which they respond when interacting with systems. In the literature there are many studies with related to eye tracking. [8][9][10].

In this study we evaluated existing web page design of an engineering faculty in terms of its usability, ease of search, fixation count, fixation length, and the required mouse clicks to complete the task. New web design which is proposed for the faculty was evaluated for same items. The next section of this study briefly explained the methodology which used in evaluation of the usability of the web sites. Section 3 gives the results of the experimental study and with the discussion and conclusion section ended this study.

2 Methodology

In this study, eye tracking method is used and this method is briefly and simply explained as follows. Eye tracking software follows the participants' eye movements on the web page or any other object. It's claimed that because of this its possible to work out what someone is attending to and even what they're thinking about. Eye tracking uses infra red technology that shows where a pupil is by reflecting light off the retina of the eye. It's embedded in the monitor so totally non-obtrusive.

Moreover, there are two main methods for evaluating the usability of this kind of web sites which are;

- User-based evaluation and
- Heuristic evaluation

In this study user based evaluation method is preferred. If the evaluation includes users, group tests moderated by experts have proved especially helpful. As we know the quality of a website can have different aspects such as contents, language, structure, design, navigation and accesibility [11].

Participants have selected from engineering faculty students. Ten volunteers who are using their faculty web site on average three times a week. Participants' age interval is 20-23 and five of them were male, others were female. Internet usage of the participants was changed 2-3 hours per day.

Tobii 1750 eye tracker was used in this study at METU HCI Laboratory. Device has 50 Hz. Sampling rate, 1024 x 768 pixels screen resolution. Participants were asked to use a mouse to complete their responses and given tasks.

In this experiment, the participants were asked to perform different tasks using each design to examine the websites, the original design and the proposed design. Each participant completed two blocks of tasks one block per design (5 tasks for each, total 10 tasks). These blocks of tasks presented to the participant in a random order, also websites which were evaluated in a random order to prevent the biasing and learning effects. In addition to them, there was 15 minutes unrelated mind exercise between the evaluating existing and new design websites.

Duration of this study was taken 30 minutes but all participants spent approximately 20 minutes and tried to finish given tasks. During this time period, their eye-movements and mouse movements, number of mouse clicking, consumed time on each task were tracked and recorded. All of these items allow us key points for assessment of original and proposed web sites.

Following procedure in this study is explained in below steps:

i.   Web site which is examined randomly open
ii. Random task is open, user push the 'enter' button
iii. User tried to complete given task
iv. At the end of the task F10 button is pushed
v.   When all the tasks were finished, 15 minutes unrelated exercise is started
vi. Steps 2-4 are repeated for the other web page.

3   The Results of the Experimental Study

In this part of the study, results of the evaluations are presented. Producer of this study was conducted six steps as explained at the end of the previous section. Five tasks for each web site were performed by the participants. These tasks can be explained simply like the following. In the first task participants were asked to find

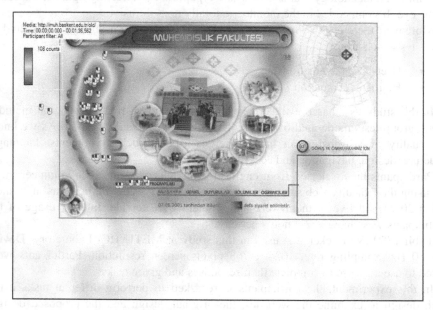

Fig. 1. Heatmaps for all tasks and all participants (original and proposed design respectively)

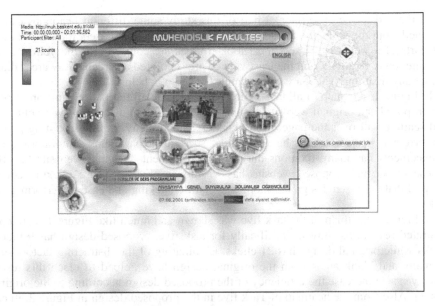

Fig. 2. Heatmaps for Task V for all participants (original and proposed design respectively)

course contents of the specifically indicated department. Then, they tried to find evaluation form about the laboratory instructor. Third task was about the computer engineering weekly schedule. In the fourth task all participants were asked to find the specific student organization in the industrial engineering department. And finally participants tried to get some information about the laboratories at the engineering faculty.

At the end of the experimental part of this study some different heatmaps can be obtained from the software. Heatmaps showed that how much users looked at different parts of a web page. Users most looked areas are colored red; the yellow areas indicate fewer fixations, and the least viewed areas colored as green. Gray areas didn't attract any fixations. There are two examples of them in below.

In Figure 1, heatmap of all task which performed by all ten participants are shown for proposed and original web site designs. In original design fixations were cluttered and scattered all over the page, even blank areas had many fixations. Listing part of the original website which has key links such as departments, laboratories, announcements has many fixations during the experiment. Original web site has 108 mouse clicks but proposed website has 47 mouse clicks. These clicking counts showed that proposed design has more concentrate clicking and ease to perform given tasks.

In Figure 2 heatmap of the task five was shown and much like Figure 1 more concentrated result was shown. Specifically for task five, proposed design has 9 mouse clicks while original design has 21 clicks. Eliminating of the distracting factors such as many and blank areas from the original design have helped to reach this result. This was also increase the efficiency of the proposed design according to the original design. Also from the heatmap of Task five in the proposed design in Figure 2, iti can be shown that little participants' tendency to read in an "F" pattern, and their focus strongly on information that is placed in 'List' part.

In this study fixation count and fixation length are also evaluated for the proposed and the original design. Fixation count and fixation lenght were observed for three different area of interests (AOI) which were heading, menu and list. These results are given in below Table 1 and 2. Fixation counts of all tasks were less in proposed design comparatevly the orginal design. Participants were completed the given tasks with less mouse clicks. Diffuculty level of Task 2 is higher than the other tasks, so in Task 2 fixation count numbers are conspicuously increase.

Table 1. Average values of the tasks for fixation count

	Proposed Design					*Original Design*			
	Heading	Menu	List	Avg.		Heading	Menu	List	Avg.
Task 1	2.500	0.333	11.167	**4.667**	Task 1	2.2	1.3	18.9	**7.467**
Task 2	4.333	1.167	32.833	**12.778**	Task 2	3.2	0.4	17.8	**7.133**
Task 3	2.222	0.222	8.889	**3.778**	Task 3	1.7	0.4	14.2	**5.433**
Task 4	1.000	0.500	11.167	**4.222**	Task 4	2.8	0.5	17	**6.767**
Task 5	0.667	0.000	5.333	**2.000**	Task 5	0.9	0.2	12.8	**4.633**
Avg.	**2.144**	**0.444**	**13.878**		**Avg.**	**2.160**	**0.560**	**16.140**	

In Table 2 which is shown below fixation length are placed. In general it can be said that participants has consumed less time while proposed design evaluating according to original design. Especially Task 2 has most time difference between the original and the proposed design even Task 2 has the most difficult one.

Table 2. Average values of the tasks for fixation length (sn.)

| | Proposed Design | | | | | Original Design | | | |
	Heading	Menu	List	Avg.		Heading	Menu	List	Avg.
Task 1	0.260	0.226	0.397	**0.294**	Task 1	0.256	0.248	0.492	**0.332**
Task 2	0.313	0.259	0.317	**0.296**	Task 2	0.284	0.308	0.548	**0.380**
Task 3	0.211	0.259	0.379	**0.283**	Task 3	0.32	0.18	0.559	**0.353**
Task 4	0.223	0.181	0.452	**0.296**	Task 4	0.265	0.279	0.479	**0.341**
Task 5	0.254	0.000	0.418	**0.294**	Task 5	0.297	0.199	0.516	**0.337**
Avg.	**0.252**	**0.185**	**0.393**		**Avg.**	**0.284**	**0.243**	**0.519**	

4 Discussion and Conclusion

In this study, an eye tracking method is used to evaluate the university web site. Original and proposed designs were evaluated by trying to complete given tasks. Five tasks were given to the all ten participants for two different web sites. All participants achieved to finish all task for each web site. All tasks include some visually search questions and needed to display correct information on the computer screen. At the end of the evaluation period results showed that proposed design was more efficient according to the eye tracking method in evaluation of usability and participants required fewer fixations to finish the task. Specifically fixation counts were differ dramatically in proposed design according to original design. Also participants consumed more time in original design according proposed design. In this study there was an exception for fixation count of Task 2. Original design has better result in that part of the study. Task 2 is most difficult and hard to display its answer on the screen. This reason is cause of this exception. In this study, an example of how eye tracking can be used to compare and improve interfaces was presented.

In future research, participant number and number of tasks in different areas can be extended, by this way more efficient, effective and accurate results can be obtained. Tasks which are used for evaluating web design can be use in cognitive mapping techniques and petri-nets methods to compare and measure cognitive complexity of web sites.

Acknowledgements

This study is performed at the Human Computer Interaction Research Laboratory at METU. The author of this article wants to thank their contributions to fulfill this research. The author would also like to sincerely thank to Dr. Ergün Eraslan for his helpfulness and guidance that making this study possible.

References

1. Asunka, S., Chae, H.S., Natriello, G.: Understanding academic information seeking habits through analysis of web server log files: The case of the teachers college library website. The Journal of Academic Librarianship 35(1), 33–45 (2009)

2. Sweeney, M., Maguire, M., Shackel, B.: Evaluating user computer interaction: a framework. J. Man-Machine Studies 38, 689–771 (1993)
3. March, A.: Usability: the new dimension of product design. Harvard Business Review 72, 144–149 (1994)
4. Han, S.H., Yun, M.H., Kim, K.J., Kwahk, J.: Evaluation of product usability: development and validation of usability dimensions and design elements based on empirical models. International Journal of Industrial Ergonomics 26, 477–488 (2000)
5. Rubin, J.: The handbook of usability Testing: How to plan, Design, and conduct effective tests. John Wiley & Sons, New York (1994)
6. Nielsen, J.: Usability Engineering. Academic Press, Boston (1993)
7. Rosenfeld, L.B., Morville, P.: Information Architecture for the World Wide Web. O'Reilly, Sebastopol (1998)
8. Cutrell, E., Guan, Z.: What are you looking for? An eye-tracking study of information usage in Web Search. In: Proceedings of CHI 2007, Human Factors in Computing Systems, pp. 407–416. ACM press, New York (2007)
9. Ewing, K.: Studying web pages using Eye Tracking, Tobii Technology, USA (2005)
10. Bojko, A.: Using Eye Tracking to Compare Web Page Designs: A Case Study. Journal of Usability Studies 3(1), 112–120 (2006)
11. Poll, R.: Evaluating the library website: Statistics and quality measures. In: World Library and Information Congress: 73rd IFLA general Conference and Council, S. Africa (2007)

User Centered Design of a Learning Object Repository

Nuria Ferran[1], Ana-Elena Guerrero-Roldán[2], Enric Mor[2], and Julià Minguillón[2]

[1] Information and Communication Science Studies
[2] Computer Science, Multimedia and Telecommunications Studies
Universitat Oberta de Catalunya, Rambla Poble Nou 156, Barcelona, Spain
{nferranf,aguerreror,emor,jminguillona}@uoc.edu

Abstract. This work outlines the design process of a user centered learning object repository. A repository should foster the development and acquisition of both generic and specific informational competencies. The results of the first stage of the user centered design process are presented which provide a clear understanding of user and task requirements and the context of use. A user study was conducted using quantitative and qualitative methodologies. A qualitative approach was performed through the content analysis of 24 in-depth interviews achieved through a random stratified sampling method. Regarding the quantitative approach, more than 5 million student navigation sessions were processed in order to know the real information behavior accomplished in the virtual campus and more specifically all the services and resources used and the search actions carried out by users. Our aim is to achieve a thorough informational behavior analysis that involves access, treatment, integration, evaluation, creation and communication of information for learning purposes which will be useful for integrating learning object repositories in virtual learning environments.

Keywords: Learning Object Repository (LOR), user centered design, log analysis, content analysis, e-learning, information-related competencies.

1 Introduction

Learning object repositories are becoming a common tool for organizing educational content, that is, all the resources used in a learning process. The main aims of a repository are, firstly, to assure access to content and its conservation and, secondly, to promote a high degree of reusability of the available resources. But learning is much more than just contents, and there is a real need to integrate the use of learning object repositories (LOR) as part of the learning process so that students take advantage of the new possibilities it offers. Using the learning object repository should be in itself a true learning experience.

Furthermore, the new European Higher Education Area (EHEA) promotes the design of learner-centered processes that focus on the acquisition and development of competences rather than on the consumption of contents. One of the goals of the EHEA framework is to 'create' professionals with appropriate skills that help them to manage in the current information society. Thus, informational-related competencies become crucial. Learning object repositories are therefore one of the most important

M. Kurosu (Ed.): Human Centered Design, HCII 2009, LNCS 5619, pp. 679–688, 2009.
© Springer-Verlag Berlin Heidelberg 2009

elements of any e-learning system and the repository interface must be designed keeping in mind that real usage must be captured in order to provide instructors with data that describes the interaction between learners and the repository, thus the abovementioned learning experience can be measured to some extent. The integration of the learning object repository as part of the virtual learning environment will not be complete unless a user-centered design approach is taken on board. This paper is organized as follows: Section 2 describes learning object repositories and their integration in e-learning systems. Section 3 describes the definition of a user-centered design project to build a learning object repository. Section 4 presents the first stage of such a project; the research of the user in the context of a virtual learning environment. Finally, the conclusions drawn from this work and future research lines are summarized in Section 5.

2 Learning Object Repositories

With the creation of the new European Higher Education Area (EHEA) [1] and for a better alignment of learning with the requirements of the knowledge society, the education sector needs a new model of learning [2]. In order to do so, teachers must change their role from dispensers of knowledge to facilitators of individual and collaborative learning and knowledge development. This means a transition from an educational model based on established information channels to a new model where there are diverse channels [3]. Textbooks, workbooks, lectures and other pre-digested information from lectures must bring about a learning process based on information resources available in the real world [4]. Students should be encouraged to select resources from the Net, use/reuse and share them with the rest of the academic community.

A Learning Object Repository (LOR) is a basic service that provides learners with the contents they need according to the learning context in any moment of the learning process [5]. Besides, it helps teachers and instructors to better manage all the available resources and to understand the real usage of these resources by learners. Furthermore, there is a huge generation of selected or created resources in each semester, therefore the LOR becomes the right tool for managing all these contents.

In addition, the academic community is acting as a curator of the quality of the chosen/created learning objects, as teachers and students are putting into practice the appropriate information competencies. In this sense, learning object repositories are perceived as an essential tool for such a collaborative learning approach. On the other hand, learners must have some basic skills for accessing the repository and the interaction with it should increase their informational skills. The basic operations related to a learning object repository (from the learner's point of view) are information searching, browsing and retrieval. Users should be empowered to have the right information, at the right time, in the right format, with the optimal quality to meet a specific information need that fulfils a learning goal.

As stated in [6], the learning repository is in itself not enough to ensure a successful learning experience. It is necessary to build a true learning community around the learning object repository, with the aim of maintaining a continuous process of creating, sharing and reusing educational resources. In a formal learning scenario, such as that defined by the new EHEA paradigm, the learning object repository cannot be just another technological service provided by the virtual learning environment, as it will be probably ignored by most of its potential users, mainly learners. Preliminary experiments [7] show that learners tend to minimize their interaction with static resources as they can be downloaded once and used locally many times. Therefore, it is important to integrate the learning object repository as an active element of the learning environment, promoting its use among learners. In order to do so, we propose to adopt a user-centered design approach, analyzing the real context where the learning object repository has to be integrated and taking into account the real user requirements from a methodological point of view.

3   User Centered Design of a Learning Object Repository

This work is part of a large project that takes place in a higher education institution, the Open University of Catalonia (UOC), with the aim of promoting the development and acquisition of competencies through the use of learning object repositories. The UOC is a purely online university that is currently evolving towards the EHEA. It has more than 40,000 students and more than 2,500 staff including instructional designers, teachers, tutors, academic and technical staff. The UOC uses a virtual campus as an integrated e-learning environment that allows students to pursue their studies purely online. We intend to design and develop a LOR that is not only useful as a mere repository but, at the same time, its use becomes an active element of the learning process, so students using the repository will achieve a set of competences.

The university established a set of organizational requirements mainly related with its technological architecture and also related with the e-learning methodological model. The second set of requirements is user defined and will be obtained through the user research described later on.

The current technological infrastructure is based on DSpace, which is already in use at the institution for publishing research results. This is the typical use of such technology, although DSpace [8] is intended to be used for storing and managing learning resources, that is, to be the core of the institutional learning object repository. One of the main drawbacks of DSpace is its user interface, which needs to be completely redesigned in order to be really usable. The default user interface reproduces the internal structure based on communities, sub-communities and collections, where each item in a collection is identified by its author, its title or its keywords. Although this may be sufficient for most typical uses, nevertheless, in the case of educational resources it is not so clear. For example, exercises do not have a clear title. On the other hand, some resources are created collaboratively during the academic semester, so the figure of the author is not clear. Therefore, we need to redefine the key elements that will be used for browsing the learning object repository, according to the desired learning goals.

As our intention is that the learning object repository is a true learning experience, its main goal cannot be just providing learners with searching and browsing capabilities. Quite the opposite, learners are expected to develop informational competences while they use the learning object repository, together with the acquisition and development of competences related to the repository thematic subject (if any, i.e. Statistics). In order to do so, we will redefine in full the user interface in order to provide a comprehensive browsing experience, which will help learners to establish a relationship between resources, topics, keywords, competences and so. In this sense, it is worth remarking the MACE project [9], which aims to provide new user interfaces for browsing. We intend to extend the capabilities of the MACE project search engine in order to accommodate the new requirements imposed by the EHEA paradigm. Furthermore, we also intend that the virtual learning environment gathers and analyzes real usage data, in order to provide learners with not only a more improved personalized system but also teachers and managers with a better understanding of the learning process built around a learning object repository.

3.1 UCD to e-Learning

Taking into account all these elements, we decided to plan and develop this project applying user-centered design (UCD) [10]. The UCD is both a design philosophy and a product development process. This discipline places the user at the center of all the process, taking into account their characteristics, needs and wishes. As a mode of philosophy, the UCD is based on the principle that a key element for the success of a product is its adaptation to the user. Adaptation is understood at different levels which include the adaptation to human characteristics and limitations, adaptation to users' needs and desires, adaptation to the context of use. A UCD product development process includes three main phases: gathering user requirements, designing the product iteratively and finally, evaluating the prototypes of each design iteration.

By applying a UCD process to the design of the LOR, we can ensure that the repository provides what it was initially conceived to provide and, at the same time, we hope to obtain important results on the Learner-Centered Design which is how the design should be done in order to guarantee a good learning experience. The project follows the principles of the ISO 13407 [11], namely the active involvement of users and a clear understanding of the user and task requirements, an appropriate allocation of functions between users and technology and the iteration of design solutions and multi-disciplinary design. This international standard describes four user-centered design activities:

1. Specification of the context of use: identify the users of the LOR and under what conditions they will use it.
2. Requirement specification: identify students' needs and goals and organizational requirements.
3. Creation and development of design solutions: these designs will take into account the information gathered in the two previous phases.
4. Design evaluation: designs are evaluated taking into account users, requirements and the context of use.

The following section presents the results of the first and second phases.

4   User Research and Specifications of the Context of Use

Keeping in mind the user context and the application needed, we have used a multiple methodology approach on the information behaviour. On the one hand, a qualitative perspective through content analysis and discourse analytical methods, and on the other, a quantitative approach through log analysis and data mining. By combining both methodologies we get, the reasons why students develop an information behaviour profile and their real navigational behaviour which provides data without the bias as their actions were transparently recorded.

The virtual campus is an integrated e-learning environment which includes all the needed services and tools. The learning object repository is one of its elements. As the main users of the learning object repository, learners are the subject of the user research.

4.1 User Qualitative Research

The virtual campus becomes a common space where students develop information behaviour; that is to say, how they execute a set of activities, such as the identification of needs, the search for, use and transfer of information [12]. And it is in this learning environment where students acquire or are able to identify, find, evaluate, organise, communicate and use the information effectively, both for solving problems and for lifelong learning [13]. Therefore, in the research carried out, we approach students in order to describe their information behaviour and their information-related competencies in the academic context in order to improve the services used to access and use the information needed for achieving the learning goals.

In this study, 24 in-depth interviews were performed on mature e-learning students from a purely virtual university (UOC). Throughout course 2006-07, there were a total of 38,842 students enrolled in undergraduate programmes, where the average age was between 26 and 35 years (58%) and more than 68% of the total students had a full time job and 55% had children.

Keeping this typology of student in mind, a stratified random selection of a sample was made with segmentation according to categories on a series of variables. We distinguished three age groups: a group of 25-35 year olds which includes the student average age; a group of 35-45 year olds and finally a group of more than 45 year olds as this age is regarded as critical for the digital divide [14]. The population was segmented in terms of gender (female/male) which can be considered proportional as there are 51% male to 49% female. Afterwards, we segmented each stratification in two groups that we called "Novice" and "Advanced" students. "Novice" referred to those students that have only 1 or 2 semesters at the university; therefore they have some knowledge and skills of the virtual campus resources. On the other hand, we used "Advanced" for those who have been enrolled for 3 or more semesters which are supposed to be more information competent. Finally, we stratified the segmentations once again in two subgroups "Experts" and "Non-experts" in terms of Information-related competencies. "Experts" was used for students that we considered to be information competent and therefore fulfilled two conditions: the first one is that they had at least once searched and retrieved an electronic article from any of the subscribed databases of the digital library and secondly, they had at least once uploaded content on the Internet, i.e. videos, photos or created a weblog.

This stratification was not proportionate, in the sense that the number of samples in each category did not necessarily correspond to their relative size in the population.

This was not regarded as a problem once the goals of the stratification are justified by the objectives of the research [15].

The initial stratified sampling was designed through telephone interviews in order to hold the face-to-face meetings. Prior to the meeting, they received a small questionnaire by email about when was the last time they searched and used information for their several needs in their academic, workplace and daily-life environment. This step helped participants to get a previous idea of what the interview was about and provide the interviewers with an incident case to be explored during the interview in order to make participants remember past real information-seeking situations [16].

Finally, the structured interviews were performed; each of them lasting from 60 to 90 minutes, between September and October 2007, in Barcelona, Spain. These interviews were recorded by audio and video, with the permission of the participants. Afterwards, the interviews were transcribed in text form and used as the raw data for the content analysis method. Afterwards, all this data was human-codified with a software tool, NVIVO 7.0 [17]. We created a codebook with all variable measures which were established following the main actions that the information behaviour manifested: access, treatment, integration, evaluation, creation and communication [18].

4.2 User Quantitative Research

The goal of the quantitative research is to collect data about how the students use the e-learning system in everyday real situations. This usage data will be processed and analyzed to obtain new evidence about system usage and about student navigation. To do that, a three-level methodology of analysis has been used [19] focusing at the first level. Working at the first level, obtaining navigation paths, requires a complex system for managing and processing log files. To obtain patterns and other results typical data mining methods are used [20].

The data set used in the analysis belongs to the spring semester of course 2006-07. Throughout the semester, the log files have been gathered, filtered and have been stored to be able to be processed in order to obtain the student navigation paths. The first step of the analysis is to obtain all the navigation sessions of the students. Afterwards, these navigation paths or sessions will be analyzed with the goal to discover new information about student behavior and system usage.

The log files do not provide information rich enough to obtain relevant information about the student behavior in the system. Therefore, to be able to obtain the navigation behaviour of the users, we decided to introduce a set of embedded marks in the system [19]. These marks leave a clear track in the log files and can be processed later. To do that, a marking strategy has been designed obtaining a map of marks that have been embedded in the e-learning system.

The spring academic semester had a duration of 136 days, the first day being the 28th February and lasted until the dates of the publication of the final marks being the 12th and 13th of July. All students share the same virtual learning environment and also the key dates of the course. The number of students registered during this semester is 29,531. During the course, the log files from the active front-end servers of the virtual campus were received and pre-processed everyday. The log files, once pre-processed thus eliminating all that redundant, incomplete or superfluous information, occupy a disk space of more than 150 GB. From this point, these files have been processed through an algorithm that allowed obtaining all the navigation sessions. These sessions have been stored in a unique file of a size of around 790 MB.

All the student navigational sessions obtained from processing the data set have been stored in one file where each line represents a single user navigation session in the virtual campus. This file contains 5,326,697 lines and, therefore, the total number of sessions of the semester. Studying carefully the data, it can be observed that there are some incomplete. We decided to eliminate them since they do not bring information about user navigation and probably were failed sessions. Once eliminated, the data set obtained contains 5,293,237 sessions which is more than 99% of the original file.

These navigation sessions are a very valuable information source that had never been obtained before. They show a lot of information about the users and how they use the virtual campus. Even so, it is interesting to continue processing and analyzing this data. As a matter of fact, these navigation paths represent the starting point of new analyses and studies about the virtual learning environment and its users.

5 Results

The analysis of the interview content and the log analysis showed two main student behavioral patterns. As a result of the qualitative analysis, one major interpretative repertoire of information behaviour appeared among mature e-learning students following the "googlizated behaviour" usually applied to teenagers [21]. Contrary to that, these students are not using the teenager's tools but paper as a preferred format, Word software as a creation tool and e-mail as their communication tool.

From the log file analysis, we can draw what we called a "blackberrized behavior" in the sense that that students access to services and tools of the campus with the aim of getting updates. The common session of a student last 7 seconds and there are several sessions each day. This finding reinforces the idea that students do not use the virtual campus for studying but for communication purposes.

Some of the recurring expressions describing both patterns found have been organized from the codification performed through the following information actions:

a) **Accessing to information:**
Google is the search tool for all the interviewers: "when I have to search anything "I do" a Google, nowadays, it is the main source of information." The common opinion is that Google is the most complete source of information and that it retrieves information in a very fast way and that it is well structured and thematically ordered.

Related to accessing the virtual learning environment, the student behavior is very constant in the sense that we can conclude that during working days, the duration of the navigation session is short and the access is frequently. 7 seconds is the time one user needs to login, load the virtual campus' home page, and make a glance to see if there are new messages. This value is important to be taken into account for interface design. If we want to capture the student's attention in the home page, the system should be capable of generating and show useful information in less than 7 seconds. The distribution of the duration of the sessions shows that there is another operation a part from see the news and to go out, and it is the one where students carry out tasks in the virtual campus and, therefore, with a longer duration.

b) Treatment of information:

Generally, people for study purposes print their selected information sources or didactical materials. Paper was considered more secure, provided a feeling of more control, something more familiar, and something more convenient. But this feature was different among the ages of interviewers and their discipline. Navigation data showed that only a small group of students use the digital library as a study habit, reinforcing the fact of not studying online and on screen.

c) Evaluation:

In the academic context, students are provided with the necessary content to carry out the evaluation tasks. If they want more information, it is something complementary, but within the virtual classroom, they got the material provided by the most reliable source which for them was the teacher. Adding to that, the analysis of the virtual campus navigation sessions showed that there is no clear pattern. Students can be grouped in several navigation profiles because they behave differently depending on the moment of the day and on the moment of the course. Therefore, educational interfaces have to take into account behavior changes throughout the semester and also the information needs of each course.

d) Creation and Integration:

The main tool for creating and integrating information is Word software. Students mostly when generating a document for evaluation purposes recognize that "I copy and paste material that I found on the Internet and then I create the essay in this process on a Word file". So this way of cutting and pasting is the most universal and familiar method of study rather than reading and digesting. Furthermore, these text files are also used for storing bibliographic sources.

e) Communication:

In order to interact with other fellows and with teachers, email is the main channel. However, they like collaboration among students from their class but they hate the "forum" or "debate" tool provided in the classroom and they also dislike teamwork as they associate it to having to meet in a face-to-face manner. As mentioned, most of the students log in on the virtual campus several times each day only to check their mail.

6   Conclusions

This present study presents the work in progress of the integration of a learning object repository in a virtual learning environment from a user- centered design approach. Learning object repositories are becoming a key element in virtual learning environments as they provide the basic infrastructure for managing all the learning resources used during the learning process. Nevertheless, it is important to ensure that learners will use the learning object repository as expected, and that such interaction will be captured in order to be further analyzed.

Once the first stage of the user-centered design process is finished and the requirements are analyzed, it is then time to proceed with generating design solutions. In the particular case of a virtual learning environment and taking into account the user research, the learning object repository must be useful not only as a simple space where learning resources are found, but as an active component providing learners with a true learning experience. Students are "googlized" and "blackberrized" and, therefore, the interface must allow them to use the repository taking into account their actual skills but should be formative enough to change their behavior and improve their skills. Obviously, this learning experience will be different depending on the nature of each subject, but some common requirements can be identified. We want the learning object repository to help learners to establish relationships between resources, in accordance with their similarities, overlapping and even user preferences. It is in this process of establishing relationships that the learner creates a mental map of the whole subject, thus improving his or her understanding of it. The browsing engine should avoid the use of Google-like search boxes and promote other interactive elements such as tag clouds and hierarchical taxonomies, among others.

Current and future research in this subject should include the development of a social layer with regard to the learning object repository, as part of its deeper integration within the learning process, in order to promote its continuous use, analyzing user behaviour in order to detect possible problems or improvements. On the other hand, as the number of learners accessing the virtual learning environment through mobile devices is increasing, hence, it is necessary to adapt some of the services provided by the learning object repository to this new learning scenario, taking into account mobility and accessibility issues. This is not only a technological issue, because the access device has different purposes depending on the context where is used. Finally, the adoption of semantic web techniques will enable better personalized services combining all the elements in the learning process, that is, users, contents and services. Personalization is one of the key aspects in providing learners with a true learning experience and which is perceived as being something real and useful.

Acknowledgments. This work has been partially supported by Spanish government projects PERSONAL(ONTO) ref. TIN2006-15107-C02 and E-MATH++ ref. EA2008-0151.

References

1. The Bologna Declaration,
   http://www.bologna-bergen2005.no/Docs/00-Main_doc/
   990719BOLOGNA_DECLARATION.PDF
2. Geser, G. (ed.): Open Educational Practices and Resources - OLCOS Roadmap 2012. Open eLearning Content Observatory Services (2007)
3. Benito Morales, F.: Nuevas necesidades, nuevos problemas. Fundamentos de la alfabetización en información. In: Gómez Hernández, J.A. (coord.) Estrategias y modelos para enseñar a usar la información. Murcia, KR (2000)
4. Breivik, P.: Student learning in the information age. ACE. Oryx Press, Arizona (1998)
5. Ferran, N., Mor, E., Minguillón, J.: Towards personalization in digital libraries through ontologies. Library Management Journal 25(4/5), 206–217 (2005)
6. McNaught, C.: Are Learning Repositories Likely To Become Mainstream In Education? In: Proceedings of the 2nd International Conference on Web Information Systems and Technologies, Setubal, Portugal, April 11-13, pp. 1S9-1S17 (2006) (Keynote address)

7. Ferran, N., Casadesús, J., Krakowska, M., Minguillón, J.: Enriching e-learning metadata through digital library usage analysis. The Electronic Library 25(2), 148–165 (2007)
8. DSpace, http://www.dspace.org/
9. Mace Project, http://portal.mace-project.eu/ProjectSearch
10. Norman, D.A., Draper, S.W.: User Centered System Design; New Perspectives on Human-Computer Interaction. Lawrence Erlbaum Associates, Inc., Mahwah (1986)
11. ISO/IEC 13407 Human centred design processes for interactive systems, http://www.usabilitynet.org/tools/13407stds.htm
12. Wilson, T.D.: Recent trends in user studies: action research and qualitative methods. Information Research 5, 3 (2000)
13. ASSL Information literacy standards for student learning, http://www.ala.org/ala/mgrps/divs/aasl/aaslproftools/informationpower/InformationLiteracyStandards_final.pdf
14. Katz, J.E., Rice, R.E.: Social consequences of Internet use. Access, involvement and interaction. The MIT Press, Cambridge (2002)
15. Neuendorf, K.A.: The content analysis guidebook. Sage Publications, Thousand Oaks (2002)
16. Talja, S., Keso, H., Pietiläinen, T.: The production of context in information seeking research: a metatheoretical view. Information Processing Management 35, 751–763 (1999)
17. Richards, T., Richards, L.: Using computers in qualitative research. In: Denzin, N.K., Lincoln, Y.S. (eds.) Collecting and Interpreting Qualitative Materials, pp. 211–245. Sage Publications, London (1998)
18. Léveillé, Y.: Les six étapes d'une démarche de recherche d'information. In: La recherche d'information à l'école secondaire, Ministère de l'Éducation du Québec, Direction des ressources didactiques (1997), http://pages.infinit.net/formanet/cs/chap2.html
19. Mor, E., Minguillón, J., Garreta-Domingo, M., Lewis, S.: A Three-Level Approach for Analyzing User Behavior in Ongoing Relationships. In: HCI Applications and Services. 12th International Conference, HCI International, Proceedings, Part IV, Beijing, China, July 22-27. LNCS. Springer-, Heidelberg (2007)
20. Duda, R.O., Hart, P.E., Stork, D.G.: Pattern classification. Wiley, New York (2001)
21. Manuel, K.: Teaching information literacy to Generation Y. Journal of Library Administration 36(1/2), 195–217 (2002)

Web Orchestration: Customization and Sharing Tool for Web Information

Lei Fu[1], Terunobu Kume[2], and Fumihito Nishino[2]

[1] Fujitsu R&D Center CO., LTD
13/F, Tower A, Ocean International Center,
No.56 Dong Si Huan Zhong Rd, Chaoyang District, Beijing, China
[2] FUJITSU LABORATORIES LTD
1-1 Kamikodanaka 4-chome, Nakahara-ku, Kawasaki 211-8588, Japan
fulei@cn.fujitsu.com, {t-kume,nishino}@jp.fujitsu.com

Abstract. In this paper, we present a tool, Web Orchestration, which allows people to customize and share the web information in a simple way. Our work is based on the web annotation and web scraping technique. It adopts B/S architecture, and has a user-friendly interface. It can be used in many aspects, such as web information monitoring, web information sharing, web information integration , recombination and so on. As an application of web 2.0 technique, it's easy to use, simple but powerful; it can enhance collaboration of each other, and make web information sharing and personalized web information customization much easier to use.

Keywords: Web annotation, web scraping, information sharing, information customization.

1 Introduction

Currently, with the spread of the World Wide Web, diverse information floods every corner of the Internet. When we want to get some information which we are interested in, we may often feel overwhelmed by the information floods. Therefore, how to manage and share the information with others on the internet gain more and more attention.

Against such a background, W3C(The World Wide Web Consortium) launches a project called "Annotea Project[1]", which aims to enhance collaboration via shared metadata based web annotations, bookmarks, and their combinations. By annotations they mean comments, notes, explanations, or other types of external remarks that can be attached to any web document or a selected part of the document without actually needing to touch the document. When the user gets the document, he or she can also load the annotations attached to it from a selected annotation server or several servers and see what his peer group thinks. One part of our work is based on this project.

The other part of our work is based on the "web scraping" technique, web scraping (sometimes called harvesting) generically describes any of various means to extract content from a website over HTTP for the purpose of transforming that content into another format suitable for use in another context. Those who scrape websites may

M. Kurosu (Ed.): Human Centered Design, HCII 2009, LNCS 5619, pp. 689–696, 2009.

wish to store the information in their own databases or manipulate the data within a spreadsheet. Others may utilize data extraction techniques as means of obtaining the most recent data possible, particularly when working with information subject to frequent changes. Investors analyzing stock prices, realtors researching home listings, meteorologists studying weather, or insurance salespeople following insurance prices are a few individuals who might fit this category of users of frequently updated data. He or she can manipulate the frequently updated data conveniently with the web scraping technique.

In this paper, we present a browser-based tool: Web Orchestration, which adopts the ideas above and provides a user-friendly view of diverse information on the internet and by which you can also comment and share the information on the internet with others conveniently. It's a light-weighted realization of web annotation and web scraping, easy to realize and easy to use. It mainly has two functions: Web Information Customization Module (WICM, for short) and Annotation Posting & Sharing Module (APSM, for short). The first one is to get and manage the information from different web sites. The other is to comment and share the information on the internet with other persons.

2   Web Information Customization Module (WICM)

This module is based on the web scraping technique, it provides a simple but powerful way for getting your desired part of information on web pages, then reorganizing them, and displaying them according to your requirement. This can be looked upon as a personalized web information customization process, the users can remix all the content they want on any web pages. It's a light-weighted realization of web scraping technique, a little similar with mash-up application.

In our method, to complete this customization procedure, firstly, we should analyze and generate the HTML DOM (Document Object Model) tree structure of the web page (Fig.1 shows an example for HTML DOM tree).

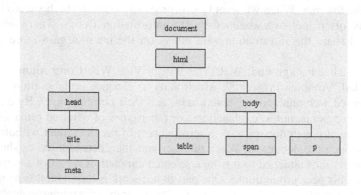

Fig. 1. HTML DOM Tree

In the DOM tree, each node corresponds to each block of the original web page seen by the terminal users. Then according to this, we can divide the DOM tree into some blocks automatically, for example, each <table>&</table>, <tr>&</tr>, <td>&</td>, <div>&</div> node or the whole <body>&</body> node in DOM tree, all these pair nodes can be a block.

Secondly, what you need to do is to choose the block you are interested in. WICM will record the path of the selected block and the URL of the web page, by the path, it's similar with XPath[2], and it's defined as the set of nodes which you have to pass if you want to reach the target node in the DOM tree. For example, if you want to reach the "table" node in Fig1. The path is "//body/table", "//" is the root of the tree. This path and the url of the page will be stored in the server.

Finally, you should choose a target page which you want to insert the selected block in. For the target page, WICM still firstly analyzes the DOM tree structure of the page, and then you can choose where to insert your interested block. The target page can be an existed page or a new page. WICM will record the insertion position, the path in the DOM tree of the target page. All the path information will be stored in the server, so when you load the page, WICM will load the insertion information of the page synchronically.

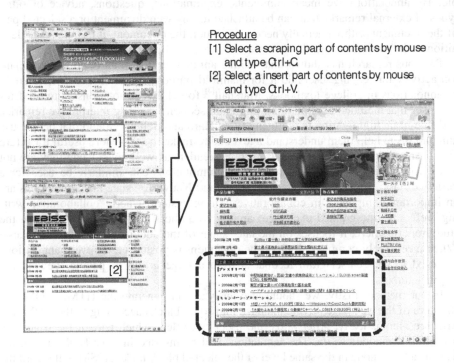

Fig. 2. WICM

Compared with the common method, WICM, as a light-weighted application of web scraping, has many advantages. Firstly, it's a dynamic process, not a simple copy & paste operation, you can choose any information you are interested in from any websites on the internet, you will not be limited to a fixed frame, you can fully express your initiative and customize the information which interests you; Secondly, all the inserted information on the target page can change with their original pages, so you can also make use of it for monitoring some pages, such as that you can follow somebody's schedule or work progress, or you can monitor the price of all the branches of a shop and so on; Thirdly, by this module, you can avoid unnecessary switches between several web sites, and read the diverse information on one page; The last one, WICM can mash up important contents on personal page. It can recombine diverse information into one existed website, with it, you can establish a purely personal page which includes all the information you are interested in. For all these advantages, the only cost is the simple selection and insertion operation on web pages, but it's once for all.

3 Annotation Posting and Sharing Module (APSM)

This module is based on the web annotation technique, it provides a simple and convenient way for posting annotations on web pages and sharing them with other people. By annotations we mean comments, explanations, questions, advice or other types of external remarks that can be attached to any web document or a selected part of the document without actually needing to touch the document, it has the same definition with "Annotea Project".

Previous research has shown that making annotation text is an important companion activity for reading, users do it for manifold purpose. In an extensive field study of annotations in college textbooks, Marshall[3] found that annotations were used for purposes that included bookmarking important sections, making interpretive remarks, and fine-grain highlighting to aid memory. O'Hara and Sellen[4] found that people use annotations to help them understand a text and to make the text more useful for future tasks. Annotations are often helpful for other readers as well, even when they are not made with others in mind. Computer-based annotations can similarly be used for a variety of task. For example, Baecker[5] et al. and Neuwirh[6] state that annotations are an important component in collaborative writing system, where "collaborative writing" refers to fine-grained exchanges among co-authors creating a document. In the study reported here, it focuses on the later stage of the document generation process. When a relatively complete draft of the document is posted on the web, annotations are used to get coarser-grain feedbacks from a large group of people (beyond the original authors).

In our method, firstly, we should also analyze and generate the HTML DOM tree structure of the web page, divide the page into some blocks according to the DOM tree structure automatically. Afterwards, the users can select which part of the page they want to put annotations on. The annotations will be inserted into the DOM tree of the page as a "div" node in the same level of the selected block node. APSM will record the path of the insertion position and the URL of the page, it's same with what WICM does. When users load and browse a web page, APSM will search for the annotations attached to it from the server, if having relative annotations, it will load and display them. All the annotations will be displayed in the form of the notepaper, different type of annotation has different background color. We use an RDF (Resource Description Framework) based annotation schema for describing annotation types, it's an extensible framework, we can define or delete types conveniently. The users can also control whether the annotations they post are public or private, once somebody loads the web page, he or she can only see all the public annotations on the page, but private ones.

Fig. 3. APSM

One of the most prominent characteristics is that it can still locate the position of the annotation when the layout of the web page changes. Let's take BBS website as an example, lots of people discuss about a same topic, when you annotate one of them, it will soon be moved to the hind page, because many people will post their opinions in a short time, the URL which contains the annotation you put will change too. In order to solve this problem, we adopt double-locating mechanism: first, besides the path of the annotation, we also consider the content which you select. Second, when you load an URL in browser, we'll generalize part of the URL to match it in annotation database of the server. By the method above, we can deal with most of the changes.

Compared with the common method for web pages sharing, such as social bookmark system. Our method, APSM, has several advantages, firstly, it enriches the content which the users share with other people, with APSM you can not only share the web pages(URLs) but also share your annotations on it. It can help you to share your unique comment or advice attached to the selected part of the page. Secondly, it provides a RDF metadata based extendible framework for rich communication about web pages while offering a simple annotation user interface. The annotation metadata can be stored locally or in one or more annotation servers and presented to the user by a client capable of understanding this metadata and capable of interacting with an annotation server with the HTTP service protocol. Last, it runs much more easily than the social bookmark service, and can reduce much more operation costs. Please refer to the Fig.4.

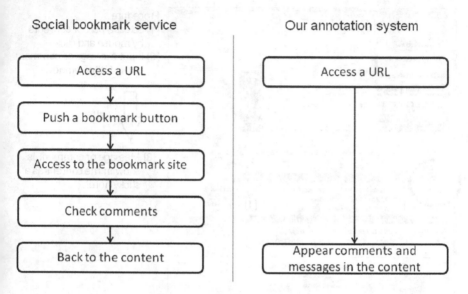

Fig. 4. Operation costs comparison

4   Features and Problems of Our Tool

Our tool has three salient features: low requirement for the system, providing an easy-to-use set of function, and has a user-friendly interface. First and foremost, the tool is based on the B/S architecture, it's just a plug-in for the explorer (IE or Firefox), so it requires less system resources, such as memory, CPU, or hard disk space and so on. It's similar with the common-used toolbar for IE, such as yahoo and Google toolbar. In the second place, although the tool is so small, it provides many practical functions. With it, you can feel some new experiences different with traditional internet surfing. You can put annotations on the selected part of the web page, and see other people's annotation on the page, if you are the administrator, you can also predefine some key words, when the users see them, it will pop up your unique annotation attached to them automatically. It also allows you to recombine many different parts of different web pages into one part of an existed page or a new page, then you can avoid frequent switches between different pages in order to browse the interested part, it will save a lot of time for you and it also makes your work much more efficiently. As a light-weighted application of web annotation and web scraping technique, it's simple but powerful. Lastly, the tool has a user-friendly interface, it's simple and easily demon-strated, it provides both push-button and right-click menu operation, similar with some common-used software, so the user can get used to it fast.

However, in spite of all the features above, it still has some problems, which are needed to be improved in the future. First of all, it can't deal with the part of the page which is generated by the script language in the running time automatically, such as the part which is generated by the javascript code, because we can't get the source code of that part when we analyze the DOM tree structure of the page. Secondly, for the WICM, when it remixes the pages which have different language encoding, it doesn't work normally. So we still have much work to do in order to improve the performance of the tool.

Furthermore, we have installed this plug-in for explorer (IE or Firefox) in our company and collect some feedbacks from the users. From the user's feedbacks, we know that they think the tool itself is useful, by it, they can gain many extra things related to their interested part of the web page, such as the other people's comment, suggestion, and recommendation, these information is very helpful for them. And they feel very convenient to recombine their interested part into one part. However, they also encountered the problems above, this influences their initiative to use it to a certain extent. They also want to annotate or remix some multimedia information, such as video, flash, which can't be dealt with by our tool now.

5 Conclusion

In this paper, we present a tool for web information customization and sharing, which is based on the web scraping and web annotation technique. It has a user-friendly interface, extendible framework and is easy to use. By it, you can customize all the information you are interested in on the web, and reorganize them according to your requirement. You can also share your unique comment or advice attached to the selected part of the page conveniently. Just as the tool's name: Web Orchestration, the tool itself is like a stage, all the web pages act as players, and the user is like a conductor, the conductor waves his baton, directs a perfect orchestration.

References

1. W3C Annotea Project, http://www.w3.org/2001/Annotea/
2. XML Path Language(XPath) Version 1.0, http://www.w3.org/TR/xpath
3. Marshall, C.: Annotation: From Paper Books to the Digital Library. In: Proceedings of the 1997 ACM International Conference on Digital Libraries (DL 1997) (1997)
4. O'Hara, K., Sellen, A.: A Comparison of Reading Paper and On Line Documents. In: Proceedings of the 1997 ACM Conference on Human Factors in Computing Systems (CHI 1997) (1997)
5. Baecker, R.M., Nastos, D., Posner, I.R., Mawby, K.L.: The user-centered iterative design of collaborative writing software. In: Proceedings of the INTERACT 1993 and CHI 1993 (1993)
6. Neuwirth, C.M., Kaufer, D.S., Chandhok, R., Morris, J.H.: Issues in the design of computer support for co-authorng and commenting ACM (1990)
7. Delicious, http://del.icio.us/

8. Luff., P., Heath, C., Greatbatch, D.: Tasksin-interaction: Paper and Screen Based Documentation in Collaborative Activity. In: Proceedings of the 1992 ACM Conference on Computer Supported Cooperative Work (CSCW 1992) (1992)

9. Cadiz, J.J., Gupta, A., Grudin, J.: Using Web Annotations for asynchronous collaboration around documents. In: Proceedings of the 2000 ACM Conference on Computer supported cooperative work (CSCW 2000) (2000)

10. Lee, K.J.: What goes around comes around: an analysis of del.icio.us as social space. In: Proceedings of the 2006 20th anniversary conference on Computer supported cooperative work (CSCW 2006) (2006)

11. Hansen, F.A.: Ubiquitous annotation system: technologies and challenges. In: Proceedings of the seventeenth conference on Hypertext and hypermedia (Hypertext 2006) (2006)

12. Kume., T., Nishino, F.: Data Matching Technique Between a Blog Content and a XPath. In: Proceedings of the Institute of electronics, information and communication engineers, Web intelligence and Interaction (WI2 2007) (2007)

Using Google Analytics to Evaluate the Usability of E-Commerce Sites

Layla Hasan, Anne Morris, and Steve Probets

Department of Information Science, Loughborough University,
Loughborough, LE11 3TU, UK
{L.Hasan2,A.Morris,S.G.Probets}@lboro.ac.uk

Abstract. The success of an e-commerce site is, in part, related to how easy it is to use. This research investigated whether advanced web metrics, calculated using Google Analytics software, could be used to evaluate the overall usability of e-commerce sites, and also to identify potential usability problem areas. Web metric data are easy to collect but analysis and interpretation are time-consuming. E-commerce site managers therefore need to be sure that employing web analytics can effectively improve the usability of their websites. The research suggested specific web metrics that are useful for quickly indicating general usability problem areas and specific pages in an e-commerce site that have usability problems. However, what they cannot do is provide in-depth detail about specific problems that might be present on a page.

Keywords: Web Analytics, Google Analytics, usability, e-commerce web sites.

1 Introduction

Ease-of-use is one of the most important characteristics of web sites, especially those provided by e-commerce organisations [1]. Norman and Nielsen, for example, stress the importance of making e-commerce sites usable. They do not regard good usability as a luxury but an essential characteristic if a site is to survive [2]. Nielsen explained the reasons behind this when he stated that the first law of e-commerce is that, if users are unable to find a product, they cannot buy it [3].

Despite the importance of good usability in e-commerce web sites, few studies were found in the literature that evaluated the usability of such sites. Those that were found, employed usability methods that involved either users or evaluators in the process of identifying usability problems. For example, Tilson et al.'s study asked sixteen users to use four e-commerce web sites and report what they liked and disliked [4]. Other studies have employed heuristic evaluation [5], user testing [6], or these two methods together in their evaluations of e-commerce sites [7]. However, little research has employed web analytic tools which automatically collect statistics regarding the detailed use of a site, in the evaluation of e-commerce web sites, although these tools have been employed to evaluate other types of web site and proved to be useful in identifying potential design and functionality problems [8,9,10].

M. Kurosu (Ed.): Human Centered Design, HCII 2009, LNCS 5619, pp. 697–706, 2009.
© Springer-Verlag Berlin Heidelberg 2009

2 Web Analytics

Web analytics is an approach that involves collecting, measuring, monitoring, analysing and reporting web usage data to understand visitors' experiences. Analytics can help to optimise web sites in order to accomplish business goals and/or to improve customer satisfaction and loyalty [11,12,13].

There are two common methods used by web analytics tools to collect web traffic data. The first involves the use of server-based log-files, and the second requires client-based page-tagging. Web analytics started with the analysis of web traffic data collected by web servers and held in log-files [14,15].

Many of the earlier studies that used web analytics to evaluate and improve different aspects of web sites used log-file based web analytics and therefore employed traditional metrics based on log-file analysis [8,9,10]. Various metrics were employed by these studies to evaluate and improve the design of the web sites with regard to four areas: content, navigation, accessibility and design. Specifically, these studies employed six metrics in evaluating and improving content (exit pages [8], search terms, referrer, search engines, top entry and exit pages and time on site [9]), three metrics to improve navigation (error pages, search terms [8], and path analysis [8,10]), four metrics to evaluate accessibility (search terms [8,9,10], search engines [9], referrer [8,10] and entry pages [8]) and two metrics to provide advice regarding the design (browser [8,10] and platform statistics [8]). These studies suggested that metrics are useful in evaluating different aspects of web sites' design [8,9,10]. However, one of these studies indicated that metrics need to be augmented by further investigation involving actual users of a web site [10]. Only one of these studies suggested a framework or matrix of metrics for evaluating web sites; Peacock's study suggested a framework of twenty log-based metrics for evaluating and improving user experience of museum web sites. This framework was an initial step towards creating an evaluation model for online museum services.

However, inaccuracies of using log-files as a data source were noticed by both web analytics vendors and customers [14]. This led to the emergence of page-tagging techniques as a new source for collecting data from web sites. The page-tagging approach involves adding a few lines of script to the pages of a web site to gather statistics from them. The data are collected when the pages load in the visitor's browser as the page tags are executed [14,15]. Page-tagging is typically much more accurate than using web server log-files and is a more informative source for web analytics applications [14,15]. There are a number of reasons for this, one is that most page tags are based on cookies to determine the uniqueness of a visitor and not on the IP address (as is the case of the web server log files), another is that non-human user agents (i.e. search engines, spiders and crawlers) are excluded from measurement and reporting because these user agents do not execute the JavaScript page tags [14,15]. An example of a Web Analytic tool that uses the page-tagging approach and which had a major effect on the web analytics' industry is Google Analytics [14]. In 2005 Google purchased a web analytics firm called Urchin software and subsequently released Google Analytics (GA) to the public in August, 2006 as a free analytics tool.

At least two studies have recognised the appearance of Google Analytics software and used this tool to evaluate and improve the design and content of web sites (a library web site [17] and an archival services web site [18]). Both used the standard reports from GA without deriving specific metrics. One of these studies used eight reports: site overlay, content by titles, funnel navigation[1], visitor segmentation, visualized summary report, information on visitors' connection speed and computer configuration [17]. The other used three reports: referrals, funnel navigation and landing pages [18]. These studies suggested that the GA tool could be a useful tool and have specific relevance to user-centred design since GA's reports enable problems to be identified quickly and help determine whether a site provides the necessary information to their visitors.

3   Aims and Objectives

The purpose of this paper is to illustrate the value and use of Google Analytics (GA) for evaluating the usability of e-commerce web sites by employing advanced web metrics. The specific objectives for this research were:

- To investigate the potential usability problem areas identified by GA software;
- To assess the main usability problem areas in three e-commerce web sites using comprehensive heuristic guidelines;
- To compare issues raised by GA software to problems identified by the web experts who evaluated the sites using heuristics approaches.

4   Methodology

This research involved three e-commerce case studies. It compared the usability findings indicated by GA software to a heuristic evaluation of the sites conducted by experts.

In order to use GA software to track usage of the e-commerce sites it was necessary to install the required script on the companies' web sites. The sites' owners identified key business processes in each site and GA was set up to assess the usability of web pages encountered by users in completing these processes. The usage of the websites was then monitored for three months.

The heuristic evaluation involved devising a set of comprehensive heuristics, specific to e-commerce websites. They were derived from a thorough review of the HCI literature and comprised six major categories: navigation, internal search, architecture, content and design, customer service, and purchasing process. A total of five web experts evaluated the sites using the heuristic guidelines.

5   Results

Thirteen key web metrics were identified that could provide an alternative to heuristic evaluation in determining usability issues. Specifically, these metrics were chosen so

---

[1] The funnel navigation report involves an analysis of the navigation paths followed by visitors to a web site while going through a number of identified steps (pages) to complete a key business process.

that, either individually or in combination, they could identify potential usability problems on e-commerce sites. These metrics are presented in Table 1 together with the results for the three sites.

Table 1. Web Metrics and Results

No	Metric		Site 1	Site 2	Site 3
1	Average page views per visit		17.00	12.56	5.62
2	Percentage of time spent visits	Percentage of low time spent visits (between 0 seconds and 3 minutes)	60.16%	76.76%	77.75%
		Percentage of medium time spent visits (between 3 and 10 minutes)	21.67%	14.48%	13.23%
		Percentage of high time spent visits (more than 10 minutes)	18.17%	7.77%	10.01%
3	Percentage of click depth visits	Percentage of low click depth visits (two pages or fewer)	31.29%	32.36%	59.20%
		Percentage of medium click depth visits (between 3 to the value of metric 1)	42.57%	40.98%	22.99%
		Percentage of high click depth visits (more than the value of metric 1)	26.14%	26.66%	17.81%
4	Bounce rate		22.77%	30.50%	47.58%
5	Order conversion rate		1.07%	0.37%	0.25%
6	Average searches per visit		0.07	0.05	NA
7	Percent of visits using search		2.14%	3.16%	NA
8	Search results to site exits ratio		0.79	0.53	NA
9	Cart start rate		5.94%	2.89%	NA
10	Cart completion rate		18.07%	12.98%	NA
11	Checkout start rate		3.63%	1.02%	1.7%
12	Checkout completion rate		29.55%	36.61%	15%
13	Information find conversion rate (ranges for the selected pages)		[0.23% to 4%]	[0% to 2.41%]	[0% to 2.71%]

Where appropriate, further explanation behind some of these metrics will be outlined in the following sections, specifically *bounce rate, information find conversion rate, order conversion rate, cart completion rate* and *checkout completion rate* will be explained in footnotes.

An analysis of the usability problems uncovered by these metrics enabled websites to be evaluated in six potential problem areas. These were: navigation, internal search, architecture, content/design, customer service and purchasing process. The following sections present the results obtained from the metrics; these are then compared to the findings obtained from the heuristic evaluators.

6 Analysis of Results

6.1 Navigation

The metrics used to investigate the general usability of a site indicated that all three sites had potential navigational problems, as shown by *bounce rate*[2] (metric 4). Site 1 had the lowest value for this metric among the three sites, whilst site 3 had the highest value. Further evidence of navigational problems on site 3 was obtained due to the low average number of page views per visit (metric 1).

However, other metrics seemed to contradict the notion of navigational problems on sites 1 and 2, for example:

- The low values for metrics 6 and 7 (*average searches per visit* and *percent of visits using search*) could suggest that these two sites either had good navigation so that a search facility was not needed or alternatively that there were problems with the search facilities (see Section 6.2).
- Metric 3 (*percentage of click depth visits*) showed that sites 1 and 2 received high percentages of medium depth visits (between 3 to 17 and 3 to 12, respectively).
- Metric 1 (*average page views per visit*) showed that site 1 and 2 had a relatively high number of pages views per visit (17 and 12.56 respectively) compared to site 3 (5.62).

The heuristic evaluators confirmed these findings; although all the sites had some navigation problems (such as misleading links) a smaller number of problems were identified on sites 1 and 2 (7 and 11 problems respectively), while a larger number of problems (42 problems) and the most serious problems were identified on site 3.

6.2 Internal Search

The metrics used to examine the usability of the internal search and the general usability of a site indicated that the internal search facilities of sites 1 and 2 had usability problems. Metric 6 (*average searches per visit*) and metric 7 (*percent of visits using search*) showed that the usage level of the internal search facilities of sites 1 and 2 was low. However, the relatively high number of pages viewed on sites 1 and 2 (metrics 1 and 3) could mean that visitors relied on navigation rather than the internal search of the sites to find what they needed. To determine if there were problems with the internal search on these sites, the value of metric 8 (*search results to site exits ratio*) for sites 1 and 2 was considered. This indicated that users were leaving the sites

---

[2] Bounce rate metric: Percentage of single page visits, i.e. visits in which a visitor left the site after visiting only the entrance page.

immediately after conducting a search and that these sites probably did have usability problems related to the inaccuracy of the search results.

The heuristic evaluators confirmed that the internal search facilities of these sites had usability problems. They identified problems with the search facilities, which were limited (site 3 did not have one), and with the results provided, which were often inaccurate.

6.3 Architecture

The metrics used to investigate the general usability of a site indicated that all the sites had potential usability problems with their information architecture. This was indicated by the large number of visitors who spent little time on the sites (i.e. their visits did not exceed 3 minutes in duration) (metric 2). Other metrics explained the significance of the architectural problems on these sites. For example, the low rate of usage of the internal search facilities of sites 1 and 2 (metrics 6 and 7), together with the high percentages of visits with medium click depth for sites 1 and 2 (metric 3) provided a potential indication that the architecture of sites 1 and 2 had fewer problems as visitors were able to navigate through these sites, implying that their search facilities may not be needed. However, the low value of the *average page views per visits* metric for site 3 (metric 1), together with the high percentage of visits with low click depth for site 3 (metric 3) provided a potential indication that site 3 had a complex architecture and that users could not navigate within it.

Although the heuristic evaluators did not report major problems with the architecture of sites 1 and 2; they did think that the order of the items on the menu of site 2 was illogical. However, as may be expected from the metrics, they found major problems with the overly complex architecture of site 3.

6.4 Content/Design

The metrics used to examine the general usability of a site indicated that the three sites had potential usability problems with some of their content. The percentages of visits in terms of the number of pages viewed (metric 3) indicated that visitors to the three sites did not appear to be interested in the content of the sites, however the degree to which content was found to be uninteresting differed among the sites. Site 3 had a high percentage of low depth visits where most visitors viewed 2 pages or fewer, indicating that most visitors were not interested in its content. Conversely, sites 1 and 2 had high percentages of medium depth visits (most visitors to sites 1 and 2 viewed between 3 and 17 pages, and between 3 and 12 pages respectively), indicating that visitors to these sites were more interested in the sites' content or products. Although more pages were viewed on sites 1 and 2, the metrics indicate that most visitors spent less than 3 minutes on all three sites (metric 2). Taken together these metrics imply that there are content problems on all three sites, but that the problems are worse on site 3.

The heuristic evaluators reinforced these findings. They identified a large number of content problems on the three sites. These included: irrelevant content, inaccurate information and missing information about products. The largest number of content problems were found on site 3 (21 problems) and the lowest on site 1 (4 problems).

The *bounce rate* metric, which is used to investigate the global design flaws in a site's page layout, also indicated that all the sites had potential usability problems in their content or design (metric 4). Bounce rate is the percentage of visits where visitors left the site after visiting only its entrance page. High bounce rate implies that either users are uninterested in the sites' content or that the design is unsuitable for the users. From the metrics it is difficult to determine if a high bounce rate is due to content or design problems. By contrast heuristic evaluation was able to identify a large number of design-specific problems with the three sites. They identified fourteen problems in sites 1 and 3 and nine in site 2. Examples of these problems include inappropriate page design and broken images. This is an area where heuristic evaluation is more precise than analytics. The analytics were able to identify potential issues, but the heuristics were able to be more specific in identifying whether problems were content or design specific.

The metrics of the top ten landing pages (*bounce rate, entrance searches* and *entrance keywords*) also identified specific pages within the sites that had possible usability problems. The top ten landing pages in each site included the home page in each site and various pages illustrating products (nine in site 1, seven in site 2 and six in site 3). The entrance keywords/searches metrics indicated that users had arrived at these pages with specific intentions, yet the high bounce rates from them suggests that the users were unimpressed with either the content or the design of the pages. The heuristic evaluators confirmed the existence of specific content and design usability problems in the product category pages and in the home pages of the three sites (i.e. irrelevant content, inappropriate page design and unaesthetic design).

6.5 Customer Service

Prior to the analysis, the customer support pages were identified by the owner of each site (12 pages for site 1, 18 for site 2 and 20 for site 3). The low *information find conversion rate*[3] metric provided evidence that visitors could not easily find and visit the customer support pages (metric 13). This suggests that either the architecture of the sites are at fault or the search facilities are poor. These findings were supported by the heuristic testing that identified navigation problems on the three sites particularly with respect to customer support links being misleading.

6.6 Purchasing Process

Metrics related to the purchasing process provided potential indications of usability problems in the overall purchasing process of the three sites. For example, the low values of the *order conversion rate*[4] metrics (metric 5) of all sites indicated that few visits resulted in an order. When viewed alongside, the relatively low values of the *percentage of high time spent visits* metrics (metric 2), this suggests that few visitors were engaged in purchasing activity on the three sites. The low *cart completion rate*[5]

---

[3] Information find conversion rate metric: Percentage of visits to a specific page that displays important customer support information.

[4] Order conversion rate metric: Percentage of visits that result in an order.

[5] Cart completion rate metric: Percentage of visits that result in an order after items have been added to a shopping cart and then the checkout process has been performed.

and *checkout completion rate*[6] metrics (metrics 10 and 12) also suggest that the three sites had usability problems in their purchasing processes.

The heuristic evaluators also experienced problems with the purchasing process of all three sites and identified usability problems regarding obstacles and difficulties that users might face while trying to make a purchase. The largest number of problems were identified on site 1 (6 problems) while four and two problems were identified on sites 2 and 3 respectively.

A similar issue was found with specific pages that make up the purchasing process. The metrics indicated that users were not only having difficulty in completing the purchasing process, but that they were also having difficulty in beginning or starting the process. Two purchasing process metrics (*cart start rate* and *checkout start rate*) and the funnel report indicated potential usability problems in this area:

- The low value of the *cart start rate* metric (which showed few users added anything to the shopping cart) (metric 9) suggests that sites 1 and 2 had usability problems on their product pages. This was confirmed by the heuristic evaluation method, which identified specific problems on these pages: navigation problems (on sites 1 and 3), design problems (on site 2), and content problems (on all three sites).
- The values of the *checkout start rate* metrics were lower than the values of the *cart start rate* metrics (metrics 11 and 9). This means that some customers, who added a product to a shopping cart, did not begin the checkout/payment process. This suggests that the pages containing the 'go to checkout' button had usability problems. This was indeed confirmed by the heuristic evaluators who experienced navigational problems on these pages in all three sites. In addition, the evaluators thought the ordering process on site 1 was too long.

More information about the purchasing process was obtained by the funnel reports, which were used to identifying possible usability problems regarding specific pages in the purchasing process of the three sites; these were confirmed by the heuristic evaluators. An example of how the funnel was used is illustrated in the following example: The statistics of the sign-in page of site 1 showed that few visitors (33%) proceeded to the next step in the purchasing process. Instead, many visitors went to the 'forgot account number' page to get their account number (18%); left the site (13%); or went to the 'login error' page by entering wrong login information (11%). Therefore, the usability problem inferred from these statistics was that it was not easy for visitors to log into the site through the sign-in page. The heuristic evaluators also had difficulties logging into site 1 because the process requires both an account number and an email address. The evaluators indicated that this is cumbersome and that users may have difficulty remembering their account details.

The funnel report provided indications of other potential usability problems on other specific pages on the three sites. These problems were also identified by the heuristic evaluators, however, the heuristic evaluators were able to provide more detail about the specific problems on these pages and how they related to navigation,

---

[6] Checkout completion rate metric: Percentage of visits that result in an order once the 'checkout' button has been selected (i.e. it does not include the process of adding items to the shopping cart).

design and content issues. Although the metrics were able to indicate possible problems, again the heuristic evaluators were able to be more specific.

It is worth mentioning that the heuristic evaluators reported other usability problems on the sites such as the lack of security and privacy, inconsistent design, and the lack of functions/capabilities/information on the sites. These problems could not be identified from the metrics.

7 Conclusion

This research identified specific web metrics that can provide, quick, easy, and cheap, indications of general potential usability problem areas on e-commerce web sites. In some instances they can also be used to identify problems on specific pages, however, not in all instances. By contrast, the results showed that the heuristic evaluators were able to identify detailed specific usability problems.

The suggested thirteen metrics can be used to provide a continual overview of a site's usability and are an important tool for indicating when potential problems may be being experienced. However, to get a more thorough appreciation of the issues other usability techniques (such as heuristic evaluation) are needed. In some aspects the web metrics have advantages - for example metrics can provide information regarding the financial performance of the site in terms of its ability to generate revenue whereas heuristic evaluators cannot provide this information.

The results offer a base for future research. The next step will be to develop a framework using GA as a first step in the process of identifying usability problems of an e-commerce web site. Other usability methods including heuristic evaluation will also be employed to identify the specific usability problems on the specific areas and pages on the web site indicated by the web metrics. The goal is to provide a framework which enables specific usability problems to be identified quickly and cheaply by fully understanding the advantages or disadvantages of the various usability methods.

References

1. Najjar, L.: Designing E-commerce User Interfaces. In: Proctor, R.W., Vu, K.-P.L. (eds.) Handbook of Human Factors in Web Design, pp. 514–527. Lawrence Erlbaum, Mahwah (2005)
2. Nielsen, J., Norman, D.: Web-Site Usability: Usability on The Web Isn't A Luxury. Information Week, http://www.informationweek.com/773/web.htm
3. Nielsen, J.: Usability 101: Introduction to Usability. Useit.com, http://www.useit.com/alertbox/20030825.html
4. Tilson, R., Dong, J., Martin, S., Kieke, E.: Factors and Principles Affecting the Usability of Four E-commerce Sites. In: 4th Conference on Human Factors and the Web (CHFW), AT&TLabs, USA (1998)
5. Chen, S.Y., Macredie, R.D.: An Assessment of Usability of Electronic Shopping: a Heuristic Evaluation. J. International Journal of Information Management 25, 516–532 (2005)
6. Freeman, M.B., Hyland, P.: Australian Online Supermarket Usability. Technical Report, Decision Systems Lab, University of Wollongong (2003)

7. Barnard, L., Wesson, J.: A Trust Model for E-commerce in South Africa. In: SAICSIT 2004, pp. 23–32 (2004)
8. Peacock, D.: Statistics, Structures & Satisfied Customers: Using Web Log Data to Improve Site Performance. In: Museums and the Web 2002, Boston (2003)
9. Xue, S.: Web Usage Statistics and Web Site Evaluation. J. Online Information Review 28(3), 180–190 (2004)
10. Yeadon, J.: Web Site Statistics. J. Vine 31(3), 55–60 (2001)
11. Malacinski, A., Dominick, S., Hartrick, T.: Measuring Web Traffic, Part1,
    http://www.ibm.com/developerworks/web/library/wa-mwt1
12. McFadden, C.: Optimizing the Online Business Channel with Web Analytics,
    http://www.webanalyticsassociation.org/en/art/?9
13. Web Analytics Association, http://www.webanalyticsassociation.org
14. Kaushik, A.: Web Analytics, an Hour a Day. Wiley Publishing, Inc., Chichester (2007)
15. Peterson, E.: Web Analytics Demystified. Celilo Group Media and CafePress (2004)
16. A Visual History of Web Measurement, Web Site Measurement,
    http://www.websitemeasurement.com
17. Fang, W.: Using Google Analytics for Improving Library Website Content and Design: A Case Study. J. Library Philosophy and Practice, 1–17 (2007)
18. Prom, C.: Understanding On-line Archival Use through Web Analytics. In: ICA-SUV Seminar, Dundee, Scotland (2007),
    http://www.library.uiuc.edu/archives/workpap/PromSUV2007.pdf

Site-it!: An Information Architecture Prototyping Tool

Atsushi Hasegawa

Concent, Inc.,
Yoyogi 1-2-9, Shibuya, 151-0053 Tokyo, Japan
hase@concentinc.jp

Abstract. *Site-it!* is a simple, powerful and cheap prototyping tool for site structure and user experience flow. This tool is a set of sticky notes printed abstract appearance of web pages. It contains 7 types of templates that express most of web pages. IAs can use it for brainstorming and discussion with clients. By using this tool, you can focus on IA discussion and cultivate understandings of user experience in the sites with team members and clients.

Keywords: Information Architecture, Workshop, Prototyping, User Experience.

1 Introduction

The web design in business is becoming increasingly important with the spread of the Internet. Along with the rise of its importance, the designing of web sites is now required to provide a solution to business needs and to assure the superiority over competition. In order to provide the solution to business need, especially to improve user satisfaction or to solve problems, human-centred design processes are utilized.

It is clearly specified that in the human-centred design processes, user needs define the requirements, and the designing is done based on those requirements. In this designing phase, preparing and comparing different several prototypes is considered to be effective for problem solution.

In the web designing and user interface designing industry, the prototyping evaluation called "paper prototyping," which uses paper-made screen images, is actively conducted in screen designing.

However, as for the design of more abstract information architecture, it is difficult to make comparison and share the process of the design. Therefore, currently such abstract information architecture is not examined very effectively.

This paper will introduce a prototyping tool that is effective in evaluating information architecture to solve the problems.

2 Web Design Process and Information Architecture Designing

2.1 Web Design Process

The general web design processes such as those for companies or services can be divided into the research phase, designing phase, and development phase, as shown in

M. Kurosu (Ed.): Human Centered Design, HCII 2009, LNCS 5619, pp. 707–711, 2009.

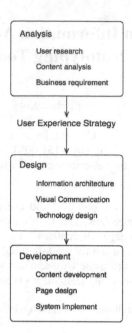

Fig. 1. One example of web design process

Fig. 1. In the designing phase, the information architecture designing occupies an important position.

The information architecture designing is essential for web sites where various users visit and that handle a large volume of information. With appropriate information architecture designing, the following solutions may be expected [1].

1. Users can find information that they want.
2. Providers can appeal information that they want to provide.
3. Manage the increase of change of the volume of information.

In general, information architecture designing consists of:

- site structure;
- inter-page navigation system;
- labeling system;
- page design.

Among them the page design is examined with methods such as paper prototyping or wireframe.

2.2 Problems in Information Architecture Designing

Meanwhile, as for the site structure designing that has an important role in the website designing and information architecture designing, there is no means for sharing the stages in the process of design, so the designing can be judged only by viewing the finished site structure documents.

The site structure is designed based on several requirements such as user experience strategy, contents classification, navigation mechanism, or page design, so the background of the site structure cannot be understood just by seeing the final map. Moreover, since the site structure document is an abstract conceptual diagram, it is difficult to imagine how the elements expressed in the structure will specifically look like.

For those reasons, problems are likely to be overlooked in the phase of the site structure designing, which may lead to a situation that a crucial requirement is found in the implementation phase and the site structure then needs to be adjusted.

In addition, at the initial stage of site structure designing, sticky notes like Post-it® is often used: those sticky notes are regarded as web pages and put on a white board to conduct the designing. This method does not cost much and allows flexible evaluation, so it is widely used. In this method, however, the roles of each sheet of sticky notes are not clarified. Therefore, participants in a workshop incorrectly recognize the role of each page, or they tend to discuss the detail of pages even though the entire picture should be the subject of discussion.

3 Development of *Site-it!*

3.1 Extraction of Page Templates

In light of the foregoing current situation, we considered that producing sticky notes on which page templates are printed in advance and using those sticky notes for designing might solve the problem.

As the template patterns to be printed, among from several types of page design pattern that have conventionally been used for designing at Concent, Inc., the most typical ones were extracted. For this extraction information architects conducted card sorting.

The thus extracted typical page patterns are shown in Fig. 2.

The page patterns are roughly classified into the "portal type" for representing a top page or category top pages; the "list type" for representing a page which contains a list of items; the "detail type" for representing the detailed information pages; and the "interactive type" for representing the dynamic pages such as a form page and a search page. We prepared variation for the list type and the detail type: users may choose to include the text only or also contain graphics. We also provided each template with space where a page title can be written down.

3.2 Workshop Using *Site-it!*

Site-it! is used in a workshop for site structure. As shown in Fig. 3, *Site-it!*s for major pages are attached on a wall or a white board. The participants can see the entire configuration of the site and compare various patterns.

Also, since *Site-it!* pieces are attached on a white board, the relationship between pages can be drawn down.

Site-it! is particularly suitable for:
- High-level site structure.
- Inter-site configuration for several websites of a single enterprise (EIA: enterprise information architecture).

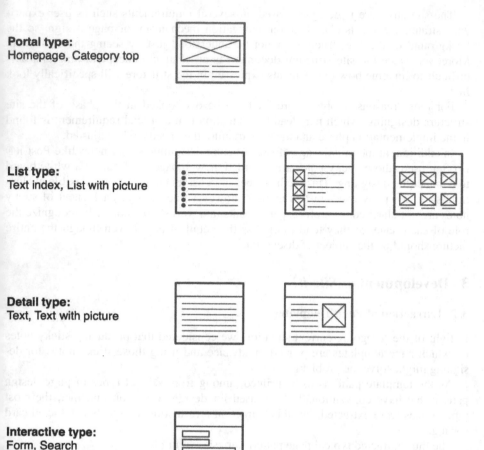

Portal type:
Homepage, Category top

List type:
Text index, List with picture

Detail type:
Text, Text with picture

Interactive type:
Form, Search

Fig. 2. *Site-it!* page templates

For a more detailed site structure, you may need applications such as Microsoft Excel or OmniGraffle.

3.3 Features of *Site-it!*

By using *Site-it!* for designing and evaluation, the process of the site structure that has seemed to be an individual task can be shared.

Moreover, the page types can be visually checked, thereby allow those who do not have much technical knowledge to catch a detailed image. At the same time, during the site structure evaluation unnecessary discussion about the detail of each page can be avoided.

In addition to those advantages, *Site-it!* uses paper pieces, on which notes can be made, and the pieces can be freely attached. Thus the site structure can be discussed anywhere at any time.

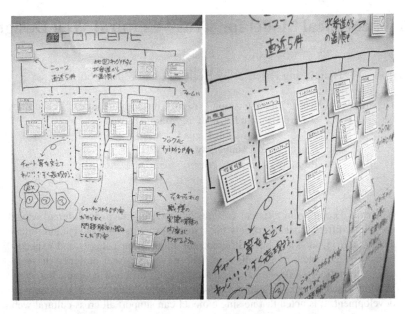

Fig. 3. View of designing workshop using *Site-it!*

Accordingly, discussion using *Site-it*! is effective at the initial stage of the site structure designing.

4 Summary

A tool using sticky notes on which typical page patterns are printed was created for the site structure phase in website designing. In order to create this tool, the templates were extracted from typical page patterns. This inexpensive, convenient tool has made it easy to hold a workshop, and to improve understanding among team member on site structure.

Reference

1. Morville, P., Rosenfeld, L.: Information Architecture for the World Wide Web. O'reilly Media, San Francisco (2006)

A Theoretical Model for Cross-Cultural Web Design

Hsiu Ching Hsieh[1], Ray Holland[2], and Mark Young[2]

[1] Department of Visual Communication Design, Chin Min Institute of Technology,
110 Hsuehfu Rd,Toufen, Miaoli, Taiwan
laurarun@gmail.com
[2] Brunel University, UB8 3PH, UK
{Ray.holland,M.young}@brunel.ac.uk

Abstract. People from different cultures use web interfaces in different ways; they hold different mental models for visual representations, navigation, interaction, and layouts, and have different communication patterns and expectations. In the context of globalisation, web developers and designers have to make adaptations to fit the needs of people from different cultures, but most previous research lacks an appropriate way to apply culture factors into the web development. It is noted that no single model can support all cross-cultural web communication but a new model is needed to bridge the gap and improve the limitations. Thus, in this paper, a thorough literature review is conducted to develop a theoretical cross-cultural model to facilitate effective communication (usability) for web design, in which the variable (cultural factors), the process of developing cross-cultural websites, and measurement criteria are identified, and two related testable hypotheses are generated.

Keywords: Web interface, globalization, cross-cultural web communication.

1 Introduction

The majority of current web-based applications assume a one-size-fits-all model (North American model), whereas people from different cultures interact and communicate according to their cultural context. North American models do not necessarily fit the needs of people from other cultures. Jagne & Smith-Atakan [13] observed the trend and stated that, "Computer software and the internet were predominately a North American skilled white male market. It has now become a worldwide commodity and the market has now grown to include all nations, creeds, gender and task use". Now, many non-English users have expanded their internet activities and have increased their utilisation of the internet.

If companies seek to expand globally, there is a growing force to provide appropriate products and services for diverse audiences (non-English users), which are increasing. Therefore, when web-based artefact developers and designers want to localise their products, they need to take the context of the target culture into account. In the past, web developments were aligned with cognition theory and computer technology. Now more companies are aware of the importance of using localisation to extend their customer base in the globalisation age. However, this research, with

M. Kurosu (Ed.): Human Centered Design, HCII 2009, LNCS 5619, pp. 712–721, 2009.

regard to applying cultural issues to web design development, needs to be considered deeper and be applied properly into the web design process.

Cultural diversity makes it impossible for designers to depend on instinctive knowledge or personal experience, therefore, many researchers have identified the need to explore cultural issues in web interface design. For example, Marcus & Gould [15] pointed out that web designers need to do much planning, research, analysis, design, evaluation, documentation, and training to deeply comprehend the requirements of the user, market, and business. Indeed, people from different cultures use web interfaces in different ways, hold different mental models for visual representations, navigation, interaction, and layouts, and have different communication patterns and expectations. In the context of globalisation, web localisation becomes a powerful strategy to acquire an audience in a global market. Therefore, web developers and designers have to make adaptations to fit the needs of people from different cultures.

Most previous research lacks an appropriate way to apply culture factors into web development. The existing culture models are not sensitive enough to the applied context of target-culture and they are too stereotypical and lack usability tests to support their claims. It is noted that no single model can support all cross-cultural web communication but a new model is needed to bridge the gap and remove the limitations. This paper presents a new model of cross-cultural web design to contribute to effective communication. According to the review of the previous research, in identifying and reducing the limitations, and bridging the gap, a new theoretical cross-cultural model for web design will be formulated.

2 Cultural Factors and Related Hypotheses

2.1 Cultural Factors

In the theoretical model proposed in this paper, Hofstede's [12] cultural model and Hall and Halls' [11] model are applied, where cultural dimensions are defined as collectivism vs. individualism, uncertain avoidance, short vs. long-term time orientation, power distance, masculinity vs. feminism, and high vs. low context culture. The web interface design characteristics which would influence web communication are defined as visual representation, multimedia, colour, navigation, layout, content and structure, links, and language.

2.2 Two Related Hypotheses

A review of the literature forms the basis of this research. Firstly, the key question "How can culture factors be incorporated into web design to facilitate communication?" is formulated, and then a theoretical cross-cultural model for web design is proposed to answer the key question, in which the variable, the process of developing cross-cultural websites, and measurement criteria are identified, and two related testable hypotheses are generated. The first hypothesis is – that there are significantly different preferences for web interface design across cultures, and the local website audit will be constructed to test this hypothesis. Furthermore, it is questioned that, if the cultural differences (significantly different preferences) do exist, can those cultural differences be applied to improve web usability? Therefore, the second related hypothesis is proposed - if the websites are embedded with culturally preferred elements and incorporated with their cultural dimension, it can be more effective in communication. To test the second hypothesis, a web experiment will be developed. Based on the two hypotheses, the proper methods, data collection instruments, and different data analysis methods are applied.

3 Cultural Models and the Existing Cultural Web Model

3.1 Hofstede's Cultural Model

According to Hofstede's [12] theory, culture can be defined as the accumulation of symbols, rituals, behaviours, customs, norms and values that distinguish a society. Symbols, heroes, rituals and values are four key terms of culture, values are the core of culture, and these terms can be applied by designers to formulate an approach to web communication. Hofstede [12] states that everyone carries their own patterns of thinking, feeling, behaviours which are accumulated from their lifetime, mostly learned from childhood. He defines the patterns of feeling, thinking, and acting as mental programs, and these vary as much as the social environments in which they were acquired. Hofstede examined IBM employees in 53 countries from 1978 to 1983. He defined patterns of differences and similarities among the replies of employees through statistical analysis of many valid data and formulated the five dimension culture theory from analysing these data. The five cultural dimensions are introduced as below.

Collectivism and Individualism Dimension. This refers to the extent to which the individuals incorporate with the group. Collectivist cultures (e.g., Taiwan and China) tend to prioritise group welfare over the individual's target. Individualistic cultures (i.e. USA, Australia) are inclined to lose ties, where everyone is expected to look after themselves.

Uncertain Avoidance Dimension. This refers to the degree to which people are comfortable with uncertain conditions. Cultures (e.g., Japan and China) with high uncertainty avoidance tend to be expressive, speaking with gestures and showing their emotions, whilst cultures with low uncertainty avoidance (countries like the USA and UK) tend to be less expressive and act without strongly showing their emotions.

Short and Long-Term Time Orientation Dimension. Long-term time orientation plays a crucial role in Asian countries (e.g., Taiwan and China) that have been influenced by Confucianism. People in these countries believe strongly that an unequal state of connection is required to keep a society stable, and virtuous behavior is identified as hard-working and perseverant. People in countries with short-term time orientation (e.g., UK and USA) tend to prefer the equal relationships.

Power Distance Dimension. Cultures with high Power Distance (e.g., Malaysia and Mexico) are characterized by hierarchies in organizations and autocratic leadership. On the contrary, cultures with low Power Distance (e.g., Austria and New Zealand) tend to have characteristics such as more equal relationships between leaders and subordinates, and elders and youngers.

Masculinity and Feminism Dimension. This refers to gender roles within a culture. Countries with Masculine cultures (for example, Japan) tend to present assertive, competitive qualities. On the contrary, countries with feminine cultures (e.g., the Scandinavian countries) tend to collapse gender distinction and present tenderness roles.

3.2 Hall and Halls' High and Low Context Culture

High context communication and Low context communication is defined by Hall [10] as, "A high context (HC) communication or message is one in which most of the information is already in the person, while very little is in the coded, explicit, transmitted part of the message. A low context (LC) communication is just the opposite". According to Hall and Hall [11], Kaplan [14], Chen and Starosta [4], and Choe [5], the attributes reflected in high and low context cultures are introduced as follows:

Communication Pattern. People from high context cultures countries are inclined to have more confidence in their non-verbal communication, and face-to-face communication is characterised by applying the non-verbal way for transferring meanings extensively. People from low context cultures are inclined to express meaning depending on content and the oral language.

Indirectness. People from high context cultures tend to use indirect and harmonious ways to communicate, whilst people from low context cultures tend to express themselves in a more direct way.

Thought Pattern. People from a high context culture tend to use an indirect strategy in their communication, usually not stating the subject directly. People from a low context culture have a strong belief that there is one objective truth which can be reached by linear exploration, so they want to meet their aims directly by applying logical thinking.

Polychronic & Monochronic Time Perception. People from high context cultures tend to be polychronic in time perception and think that everything will happen when the right time comes, whilst people from low context cultures tend to be monochromic in time perception and believe that executing a task on time..

3.3 The Existing Cultural Web Model – Theoretical Studies

Theoretical studies are constructed based on the existing cultural models. For example, Marcus & Gould [15] applied Hofstede's [12] five dimension cultural model to build up the guidelines for designing web interfaces for different countries, and outlines how these dimensions can influence components of a web interface design. Sheridan [18] also applied Hofstede's [12] cultural model to develop web interface design guidelines for localisation. Her guidelines are developed following the patterns of Marcus & Gould

[15] and predicted the tendencies of web interface design attributes in each cultural dimension. Reviewing the theoretical model built up by Marcus & Gould [15] and Sheridan [18], there is no user usability test from different countries to back up their guidelines and it is too stereotypical, with users from the same country not always conforming to Hofstede's model. Therefore, the validity of their model is questionable.

Gould, Zakaria, and Yusof [9] built up their research to compare the cultural orientations and design preferences for web interfaces between Malaysian and USA websites. Eventually they suggested design guidelines for cultural localization for Malaysia and USA based on using Hofstede's [12] cultural model and Trompenaars's [23] cross-cultural theory, but no usability test to support their claims.

3.4 The Existing Cultural Web Model – Experimental Studies

Experimental studies have been conducted using the existing cultural dimension model as a way to choose and identify samples by fitting them into a cultural category. Some empirical studies which adopt cultural dimension models are presented below.

Ford & Gelderblom [8] applied Hofstede's five dimension model to construct their empirical study. In their study, they examined whether the user's performance would be influenced by the culture variables. The results of this experiment did not provide enough evidence to support the hypothesis that cultural variables influenced the subjects' performance, but the performance levels gained revealed that the usability of the interfaces was increased for all of the subjects, as a result of incorporating the five cultural dimension attributes into the web interface design.

Smith et al. [20] constructed their study with target-culture users to determine the extent to which cultural factors influence the usability and acceptability of international websites. Based on Hofstede's [12] study of cultural dimensions, their experiment adopted the Taguchi method to investigate the differences between British and Chinese users' preferences and satisfaction within websites. Significant preferences between British users and Chinese users were found. They mentioned the preferences and perception, but did not state that satisfaction and perception were equivalent. The issue of performance was not considered in a usability test, with the focus on perception alone.

Simon [19] used Hofstede's [12] dimension as a method to examine the perception and satisfaction differences between the cultural groups and gender groups within different cultures. The analysis of this study indicates that there are differences between cultural and gender-based perception and satisfaction within different cultures. However, perception and satisfaction in this study were not defined very clearly.

3.5 The Existing Cultural Web Model – Synthesis Theoretical Works

Some researchers have seen the limitations of developing the current cross-cultural web model based on anthropologists' cultural models, so Zahedi et al. [25] and Sun [22] incorporated other theories into their cross-cultural web model. Zahedi et al. [25] combine the social construction theory with Hofstede's [12] cultural model to develop their conceptual cross-cultural web design model. The aim of their conceptual model is to analyse how the cultural and individual factors impact on the effectiveness of web designs. They claim that their conceptual framework is for web design, but actually the propositions of their study just emphasises web documents whilst other important web interface characteristics are not considered. Therefore, doubt is cast over their conceptual model, and no usability test was constructed to conclude their propositions.

Sun [22] incorporated a dynamic process and changing variables by integrating the study from previous researchers such as Hofstede [12], Hall & Hall [11], Marcus & Gould [15], and Zahedi et al. [25], but there was no usability experiment to support his claim, and it is questionable whether to validate Sun's model for cultural usability. Jagne & Smith-Atakan [13] developed a strategy for cross-cultural interface design, which combines the theory of Hofstede, the design guidelines from Marcus & Gould [15] and Barber & Badre [1], but lacks an empirical study to support their model.

4 Formulating a Theoretical Model for Cross-Cultural Web Design

4.1 The Limitations of Previous Research

Reviewing the above existing cultural web models, there are some limitations in previous research. Bourges-Waldegg and Scrivener [2] have pointed out that the existing culture models are too general and not sensitive enough to the applied context of target-culture. Most of the existing models for web interface design are too stereotypical and lack usability tests to support their claims, and one of them (the model of Zahedi et al.) just focuses on web documents, ignoring the other crucial web interface design features.

Based on reviewing previous research and criticising the drawbacks of the current model, there is a need for a new model to fill the void. Therefore, a new theoretical cross-culture web design model is proposed. It is recommended that web designers and developers should be careful that the established cultural model might be too stereotypical and does not really fit the target culture market. As Del Galdo and Nielsen [7] suggested, the web designers should get involved into the target culture directly. In Sun's study [21], it is documented how culturally preferred design elements (cultural markers) such as visuals, language, colours affect web usability by interviewing target culture users about their experiences. Also Sun [22] declared that cultures continue to develop and interact, and they are not ontologically objective. It reminds web developers and designers to maintain practical observation of the target-culture users, because culture is constantly changing, particularly in the internet era.

To avoid being too stereotypical by applying the existing cultural model and to engage the target culture directly, this new model not only adopts the established cultural model, but also applies the "Cultural Markers" [1] approach to find out the culturally preferred web interface design attributes from the target culture. A new theoretical model of cross-culture website design is formulated (see Figure. 1). The new proposed model consists of four stages and will be introduced in detail below.

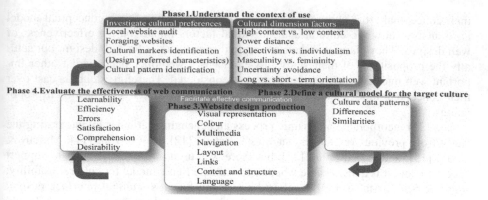

Fig. 1. Theoretical cross-cultural web design model

4.2 Understand the Context of Use

To understand the context of use, cultural models are incorporated and cultural preferences (cultural markers) are investigated. The concept of cultural markers [1] is adopted to define the web interface characteristics that reflect the signs and their meanings to match the expectations of the local culture audience. According to Smith et al. [20], to better understand how to create a website that is appropriately pitched to the target culture users, it is essential to examine the different signs or symbols (or visual representation) in a target culture, the usage of signs based on the context, and how the target culture audience interprets these signs. This can be achieved by conducting an audit of local indigenous sites. Thus, a local website will be established and comprises three steps shown below. Firstly, previous research involving cultural preferences is consulted, such as those of Barber & Badre [1], Sun [21] and Cyr & Trevor-Smith [6]. Secondly, observing the real features in websites that have been selected. Thirdly, Hofstede's [12] cultural dimension, Hall and Halls' [11] high and low context dimension, Marcus and Goulds' [15] cultural web model [15], and Würtz's [24] cross-cultural we model are incorporated, and web design characteristics (cultural markers) are identified by integration with the elements from previous research involving cultural preferences, as well as detailed inspection of the scope websites. Finally, the culturally preferred design elements are defined and comprise of eight categories: visual representations, multimedia, colour, layout, navigation, links, content & structure, and language. These elements are united to match the cultural expectations of the users from specific culture.

4.3 Defines a Cultural Model for the Target Culture

This stage defines a cultural model for the target culture and aims to identify and state a picture of differences and similarities in the observed attributes of the target-culture users' specific practice. The objective of this stage will identify the international variables needed to define a cultural model. The next step will compare and find out the similarities and significant differences in the response of the samples in order to create a pattern of the target-culture customers.

4.4 Website Design Production

Based on the results from Phase1 and 2, the website's prototype will be constructed, and the webpages will be embedded with the observed culturally preferred characteristics. The web interface's preferred design characteristics are categorised into several aspects such as visual representation, navigation, multimedia, colour, layout, language, interaction, and content and structure. This stage focuses on the production of the website's prototype.

4.5 Evaluate the Effectiveness of Web Communication

Reviewing previous research from Nielsen [16], Brink et al. [3], Preece [17], and Zahedi et al. [25], the components of web communication effectiveness can be derived from web usability. In order to measure effectiveness of each design, the evaluation criteria of web communication effectiveness are identified, which include learnability, efficiency, minimal errors, satisfaction, comprehension, and desirability. The assessment criteria are presented as follows:

- **Learnability:** Is it easy to learn? How quickly can new users learn to accurately execute the process of a task is determined by ease of learning. Usually, the fewer steps a procedure requires, the easier it is to learn.
- **Efficiency:** Is it efficient to use? Efficiency can be the assessment of the time or actions needed to carry out a task. The process of executing a task faster implies greater efficiency.
- **Errors:** Based on Nielsen [16], the evaluation criteria of errors is defined as, "users should make as few errors as possible when using computer system".
- **Satisfaction:** Is it pleasant to use? A user's perception of satisfaction can be influenced by visual graphics, layout, typography and other visual interface elements, so users' satisfaction is a combination of all of these criteria.
- **Comprehension:** Is it easy to understand? Is it readable?
- **Desirability:** Does it fit the expectation and preferences of users? The proportion of users who state that they would prefer using the web site over some specified website is used to evaluated the desirability.

Data are analyzed to modify the websites based on the results of the usability test (web experiment). A replicable process should take place subsequently by modifying the prototype website based on the results of the evaluation.

5 Conclusion

This model comprises four phases, and the process is replicable. In cross-cultural web design development, there needs to be a strong relationship between cultural theory and practical design approaches, so that an improved web product can be obtained by a replicable process of design, evaluation and reflection on theory.

This theoretical model is the initial phase of a multi-phase empirical research program, and the next step is to conduct the local websites audit to test the first related hypothesis, while a web experiment will be conducted to collect data, as well as data analysis to validate this proposed theoretical model. A team should be organized for the localization within the target culture to meet the needs of the communication pattern from the target culture.

This model has the potential to contribute to the need of localization and help web developers and designers develop their web products as culturally appropriate.

References

1. Barber, W., Bardre, A.: Culturability: The Merging of Culture and Usability. In: Proceedings of the 4th Conference on Human Factors and Usability (1998),
   http://zing.ncsl.nist.gov/hfweb/att4/proceedings/barber
2. Bourges-Waldegg, P., Scrivener, S.A.R.: Meaning, the central issue in cross-cultural HCI design. Interacting with Computers 9(1998), 287–309 (1998)
3. Brinck, T., Gergle, D., Wood, S.D.: Designing Web sites that work: usability for the Web. Morgan Kaufmann, San Francisco (2002)
4. Chen, G., Starosta, W.: Foundations of Intercultural Communication. Allyn and Bacon, Boston (1998)
5. Choe, Y.: Intercultural conflict patterns and intercultural training implications for Koreans. In: The16th Biennal World Communication Association Conference, Cantabria, Spain (2001)
6. Cyr, D., Trevor-Smiths, H.: Localization of Web design: An empirical comparison of German, Japanese, and United States Web site characteristics. Journal of the American Society for Information Science and Technology 55(13), 1199–1208 (2004)
7. Del Galdo, E.: Culture and Design. In: Del Galdo, E., Nielsen, J. (eds.) International User Interfaces, pp. 74–87. John Wiley and Sons, Inc., Chichester (1996)
8. Ford, G., Gelderblom, H.: The effects of culture on performance achieved through the use of human computer interaction. In: Proceedings of the 2003 annual research conference of the South African institute of computer scientists and information technologists on Enablement through technology. ACM International Conference Proceeding Series, vol. 47, pp. 218–230 (2003)
9. Gould, E.W., Zakaria, N., Yusof, S.A.M.: Applying Culture to Website Design: A comparison of Malaysian and US Websites. In: ACM Special Interest Group for Design of Communications archive Proc. IEEE professional communication society international professional communication conf. and Proc. 18th annual ACM international conf. Computer documentation: technology & teamwork, pp. 161–171 (2000)
10. Hall, E.T.: Beyond Culture. Doubleday, New York (1976)
11. Hall, E., Hall, M.: Understanding cultural differences. Intercultural Press, London (1990)
12. Hofstede, G.: Cultures and Organizations: Software of the Mind. McGraw-Hill, London (2005)
13. Jagne, J., Smith-Atakan, A.S.G.: Cross-cultural interface design strategy. Universal Access in the Information Society 5(3), 299–305 (2006)
14. Kaplan, R.: Cultural thought patterns in intercultural education. Languge Learning 16, 1–20 (1966)
15. Marcus, A., Gould, E.W.: Cultural Dimensions and Global Web User-Interface Design. Interactions, 33–46 (July/August 2000)
16. Nielsen, J.: Usability Engineering. Academic Press, San Francisco (1993)
17. Preece, J.: A guide to usability: human factors in computing. Addison-Wesley, New York (1993)

18. Sheridan, E.F.: Cross-cultural Web Site Design. Considerations for developing and strategies for validating locale appropriate on-line content, MultiLingual Computing & Technology #43,12(7) (2003) http://www.multilingual.com
19. Simon, S.J.: The Impact of Culture and Gender on Web Sites: An Empirical Study. The Database for Advances in Information Systems 32(1), 18–37 (2001)
20. Smith, A., Dunckley, L., French, T., Minocha, S., Chang, Y.: A process model for developing usable cross-cultural websites. Interacting with Computers 16(1), 63–91 (2004)
21. Sun, H.: Building a culturally-competent corporate web site: an exploratory study of cultural markers in multilingual web design. In: Proceedings of the 19th annual international conference on Computer documentation, pp. 95–102 (2001)
22. Sun, H.: Exploring Cultural Usability. In: Proceedings of IEEE International Professional Communication Conference, Portland OR, September 2002, pp. 319–330 (2002)
23. Trompenaars, F., Hampden-Turner, C.: Riding the Waves of Culture: Understanding Cultural Diversity in Business. Nicholas Brealey, London (1997)
24. Würtz, E.: A cross-cultural analysis of websites from high-context cultures and low-context cultures. Journal of Computer-Mediated Communication 11(1), article 13 (2005), http://jcmc.indiana.edu/vol11/issue1/wuertz.html
25. Zahedi, F., Van Pelt, W., Song, J.: A conceptual framework for international web design. IEEE Transactions on Professional Communication 44(2), 83–103 (2001)

An Investigation of User's Mental Models on Website

Hui-Jiun Hu[1,2] and Jen Yen[1]

[1] Graduate School of Design, National Yunlin University of Science and Technology,
Yunlin, Taiwan
[2] Department of Information Management, Transworld Institute of Technology,
Yunlin, Taiwan
momo@tit.edu.tw

Abstract. Since mid 1990s Internet has been developing rapidly to become the most booming and emerging media in recent history and played an important role in human livelihood. People's demands on website interface interaction have thus been increasing. How to make a website interface easy to learn and easy to use? It has thus become an important issue pertaining to Human Computer Interaction (HCI). In this paper, we use the Interactive Qualitative Analysis (IQA) approach to conduct a qualitative data-gathering, analysis and examination made by 9 expert participants. The result show the affinity of 11 website user is thus produce. And, we can make some suggestions such as website user is speed & efficiency-oriented and negative images of web advertiser. In addition, the affinity of User Requirement is an important affinity to keep in website user.

Keywords: Mental model, website user, Interactive Qualitative Analysis (IQA).

1 Introduction

Internet is a media for info exchange and experience transmission. The purpose of a website is to conduct the most efficient communication with the biggest group. For most of people nowadays, Internet is a livelihood necessity instead of a professional noun. Rosenfeld & Morville [19] stated that a website can be assumed as successful or a building can be regarded as good is dependent on enabling users getting satisfaction instead of outer shape or internal technological products. Garrett [5] also pointed out that User Experiences have surpassed other factors to be the most important key for users' decision making. Krug [11] even emphasized that the first principle for usage among various Internet design factors is "Don't make me think". Therefore, Internet users' behavior can't afford being ignored.

In the notions of cognitive psychology, it is assumed that human mental model is the important factor influencing behavior. Fodor [4] acclaimed that mental express is a language for thinking. If we know users think what affinities are necessary for a good website, we should be able to develop systems (website) that would help users form correct mental models and consequently, improve usability.

As this study adopts Interactive Qualitative Analysis (IQA) derived from qualitative research method for searching users' mental models. Focus Group is used for data collection, category coding. The explanations are then made for the relation

M. Kurosu (Ed.): Human Centered Design, HCII 2009, LNCS 5619, pp. 722–728, 2009.
© Springer-Verlag Berlin Heidelberg 2009

between causes and effects concerning various factors and also mindmap developed from website user. Positive suggestions and directions for futuristic website design are then provided.

2 Mental Models

Mental models are representations of reality that people use to understand specific phenomena. They were first postulated by the Scottish psychologist Kenneth Craik [3], who assumes that the human will establish Small-scale Models for external physical according to past experiences and also use such a model to predict futuristic situations as to make the most appropriate and safest responses regarding the situations faced by them [3]. Johnson-Laird [9] submitted that mental model a process pertaining to human's problems solving and logic reasoning and is also the concepts and operation performance upon the interaction between the human and complicated system.

Norman [15] also believed that mental model is personalized intrinsic experience model formed by interaction phenomena between the human and nature and those phenomena provide us the basis for predicting and explaining interactive behavior. That is the human transform external world into intrinsic symbols to create mental model. Languages, worlds or other symbols are used for communications among social groups by model operation. The mental model interactions among the human thus develop the models commonly accepted by social groups. The notion of mental model has also applied and also research by such fields as psychology, computer science, artificial intelligence and education [18, 20, 22].

Mental model can be separate into Design Mode, User's Model and System Image. When having a design model in mind, the designer will start to a system which can be used, learn and function according to design model. The user will use this system according to previous experiences and one's mental model. Ideally, the user's model and the design model are equivalent [16]. Having made design in one's point of views directly, instead of communicating with a user directly, the designer thus neglects that there is differences between the designer's and the user's cognition on design object. Nielsen [13] acclaimed in the research that users often experience some problems of which designer never have thought. Lazar [12] also pointed out in research that websites are not user-friendly currently, as Internet providers adopt designer center, different goals of both sides cause serious gap between usage experience for both sides.

3 Research Method

Interactive Qualitative Analysis (IQA) is an analysis characterized by classified structure and system and has been applied in various fields [1, 2, 7, 10](Lin, Fang, & Tu, 2007). The IQA methodology intends to make thinking integration by extracting tree most important concepts, correspondence, coherence and constructiveness constitute a group of participants' collective opinions as to find the causes-and-effects relationships.

IQA also a process characterized by qualitative data collection and analysis completed by a group of participants' focus group and brainstorming. The participants' focus group can provide diversifying experience and ideas pertaining to our research problem and can establish their interpretation on such affinity by using interactive discussion. IQA can find the causes-and-effect relationship and draw a picture of the System Influence Diagram (SID). The SID also called a mindmap, is a visual representation of an entire system of influences and outcomes and is created by representing the information present in the Interrelation Diagram (IRD) as a system of affinities and relationships among them.

The whole IQA process takes about 3~4 hours. IQA starts with proceeding focus group. The host explains the research purpose and actualization methods and each participant fill in the cards by silent brainstorming and each participant fill in about 25 cards in about 1 hour. The cards are then classified and named and inductive coding is conducted on all cards. After inductive coding category naming is finished, axial coding is conducted. Theoretical coding is conducted then. Each participant fills the mutual affected arrow head direction among each affinity into the Affinity Relationship Table (ART). Pareto Principle is used for deciding which relationships shall be included into Interrelation Diagram (IRD). Finally System Influence Diagram (SID) is drawn according to the driver or outcome derived from between IRD and each affinity.

4 Data Collection and Analysis

The participants' age range is set be between 16~25 years old and the users shall be expert users who are familiar with using Internet and have used internet for at least 8

Table 1. Focus group-Users' Affinities & Definition

Axial Code(Affinity)	Definition
User Interface (UI)	Being equipped with web page's sense of beauty, design, operation and features to enable users to understand every design feature easily
Web Function WF)	Enabling users to understand this website's functions with super link and website menu
Back-End Management (BEM)	Relevant with back-end setting, security setting and maintenance
User Requirement (UR)	Being designed or improved according to the functions demanded by users
User Help (UH)	Helping users solve any problem or difficulty when users have problems
Classified Information (CI)	Clear website menu and also banners to make users seeing clearly
Useful Tools (UT)	Useful tools, such as translation software....searching engines...etc
Content Enrichment (CE)	Frequently updating content accuracy and enrich contents
Web Advertiser (WA)	Pop-up ad and ad-blocks in web page
Multimedia (Mu)	Animation AV functions
Speed & Efficiency (SE)	Mainframe in website's service can execute web page download fast

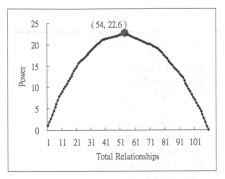

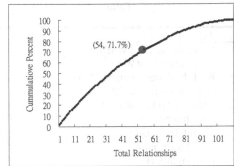

Fig. 1. Power Analysis-Total Relationships diagram of the website user

Fig. 2. Cumulative Frequency Percent-Total Relationships diagram of the website user

years. There are 9 users (5 males and 4 females). And, the problem submitted is "According to your Internet usage experience, what are the necessary factors (attributes) for a good website that you think?" The affinity of 11 users (see Table 1) is thus produced after data coding.

We find that the first 54(49.1% of total) relationships (Fig. 1) account for 71.7% of the total variation (Fig. 2). The power curve peaks at a value of 22.6 (Fig. 1), which is associated with 71.7% of the total variance.

We decide which relationships could be retained by MinMax criterion. The retained relationships are then used to organize Interrelationship Diagram (IRD) and arranged in a descending sequence according to delta value to get a design team's IRD (see Table. 2). The tentative System Influence Diagram (SID) assignment of each affinity is then defined (see Table. 3). Tentative SID assignments can be used to decide the relationship between affinities and SID. Finally, we draw out SID of the web design team (Figure. 3). The SID is drawn with all Drivers on the left and all

Table 2. Tabular IRD of website user

	UI	WF	BEM	UR	UH	CI	UT	CE	WA	Mu	SE	OUT	IN	Δ
BEM	↑	↑		↑	↑	←	↑	↑		↑	←	7	2	5
CE	↑	↑	←	↑		↑				↑		5	1	4
SE			↑	↑						↑		3	0	3
CI			↑	↑	↑			←				3	1	2
WA				↑						↑		2	0	2
UI		↑	←		↑			←			←	2	3	-1
UT		←	←	↑								1	2	-1
WF	←		←	↑			↑	←			←	2	4	-2
UH	←		←	↑		←						1	3	-2
Mu	↑	↑	←	←				←	←		←	2	5	-3
UR		←	←		←	←	←	←	←	↑	←	1	8	-7

*Count the number of up arrows (↑) or Outs.
*Count the number of up arrows (←) or Ins.
*Subtract the number of Ins from the Outs to determine the (Δ) deltas.
*Δ=Out-In.

Table 3. Tentative SID Assignments of website user

Affinity Name	Determinant
Speed & Efficiency(SE)	Primary Driver
Web Advertiser(WA)	Primary Driver
Back-End Management(BEM)	Secondary Driver
Content Enrichment(CE)	Secondary Driver
Classified Information(CI)	Secondary Driver
User Interface(UI)	Secondary Outcome
Useful Tools(UT)	Secondary Outcome
Web Function(WF)	Secondary Outcome
User Help(UH)	Secondary Outcome
Multimedia(Mu)	Secondary Outcome
User Requirement(UR)	Secondary Outcome

Outcomes on the right. The result shows that Speed & Efficiency (SE), Web Advertiser (WA), Back-End Management (BEM), Content Enrichment (CE) and Classified Information (CI) are drivers in the SID. In addition, User Interface (UI), Useful Tools (UT), Web Function (WF), User Help (UH), Multimedia (Mu) and User Requirement (UR) are outcomes in the SID. And, this study finds 3 loops. First, BEM, CE and CI constituting a mutually influencing loop. UR, Mu, WF and UT constituting second mutually influencing loop. Finally, UR, Mu, UI and UH also constituting a mutually influencing loop. The affinity of UR has two loops of website users' focus group. Therefore, User Requirement (UR) is a very important influencing factor for users.

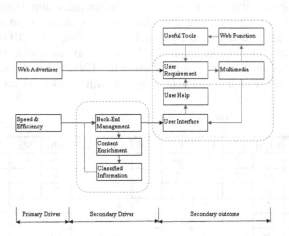

Fig. 3. SID of website user

5 Conclusion

The research presented herein demonstrates a methodology for extracting and analyzing mental models from IQA approach. We suggest that the methodology used in this study can find the causes-and-effect relationship and draw a picture of the mind-map (SID). In this study, we found the affinity of 11 users is thus produce, Speed &

Efficiency (SE), Web Advertiser (WA), Back-End Management (BEM), Content Enrichment (CE) and Classified Information (CI) are drivers in the SID. In addition, User Interface (UI), Useful Tools (UT), Web Function (WF), User Help (UH), Multimedia (Mu) and User Requirement (UR) are outcomes in the SID.

And, the website user' mindmap is derived from speed & efficiency-orientated. Relevant researches indicated that Internet download speed is a very important factor for successful website design [6, 13]. System lag will cause negative feelings. Such feelings will also influence the attitudes toward e-commerce websites [8]. Internet users can tolerate with a lag of only "8 seconds and 3 clicks" [21]. Therefore, Speed & Efficiency are very important for website users.

An interesting phenomenon is that users in this study bear negative images toward every aspect of web advertiser mainly because they don't want to see too many ads or don't want to be bothered by ads. One of the top 10 Internet design error submitted by Nielsen [14] is "Anything that looks like an advertisement". The web pages are thus filled with various ads, users could be bothered easily during using browsers or thus become unhappy after being seduced to click.

Finally, we also found the affinity of Users Requirement (UR) has two loops. User Requirement is an important affinity to keep in website user. Lazar [12] mentioned in the research that a website would be not user-friendly as website provider adopts designer center or could be dominated by corporate organization. Pearrow [17] also deduced that design team ignores users' demands in favor of emphasis on functions and features. This shows the importance of User-Centered Design (UCD) on website construction.

References

1. Bann, E.E.: Effects of media representations of a cultural ideal of feminine beauty on self body image in college-aged women: an interactive qualitative analysis. Unpublished doctoral dissertation, University of Texas at Austin (2001)
2. Burrow, B.A.: The vertical extension of Florida's community college system: A case study of politics and entrepreneurial leadership. Unpublished doctoral dissertation, University of Texas at Austin (2002)
3. Craik, K.: The Nature of Explanation. Cambridge University Press, Cambridge (1943)
4. Fodor, J.A.: The Language of Thought. Harvard University Press, Thomas Crowell (1975)
5. Garrett, J.J.: The Elements of User Experience: User-centered Design for the Web. New Riders, Indianapolis (2003)
6. Gehrke, D., Turban, E.: Determinants of Successful Website Design:Relative Importance and Recommendation for Effectiveness. Paper presented at the Proceedings of the 32nd Hawaii International Conference on Systems Sciences (1999)
7. Gray, S.H.: The role of community college president in vision building for rural community development. Unpublished dissertation, University of Texas at Austin (2003)
8. Guynes, J.L.: Impact of System Response Time on State Anxiety. Communications of the ACM, 342–347 (1988)
9. Johnson-Laird, P.N.: Mental Models - Towards a Cognitive Science of Language, Inference and Consciousness. Harvard University Press, Cambridge (1983)

10. Knezek, E.J.: Supervision as a selectional leadership behavior of elementary principals and student achievement in reading. Unpublished dissertation, University of Texas at Austin (2001)
11. Krug, S.: Don't Make Me Think!: A Common Sense Approach to Web Usability. New Riders (2005)
12. Lazar, J.: User-centered Web Development. Jones & Bartlett Publishers (2001)
13. Nielsen, J.: Usability Engineering. Morgan Kaufmann Publishers Inc., San Francisco (1993)
14. Nielsen, J.: Top Ten Mistakes in Web Design (2007),
    http://www.useit.com/alertbox/9605.html
15. Norman, D.A.: Some Observation on Mental Model. Lawrence Erlbaum, NJ (1983)
16. Norman, D.A.: User centered system design: New perspectives on human-computer interaction. In: Norman, D.A., Draper, S.W. (eds.) Lawrence Erlbaum Associates, Hillsdale (1986)
17. Pearrow, M.: Web Site Usability Handbook. Charles River Media Inc., Rochkand (2000)
18. Preece, J.: A Guide to Usability - Human Factors in Computing. Addison-Wesley, England (1993)
19. Rosenfeld, L., Morville, P.: Information Architecture for the World Wide Web: Designing Large-Scale Web Sites, 2nd edn. O'Reilly, CA (2002)
20. Senge, P.: The fifth discipline: The art and practice of the learning organization. Doubleday/ Currency, New York (1990)
21. Translated by Wu, K.-L. 8 seconds and 3 clicks Internet psychology (April 2008),
    http://www.cheers.com.tw/doc/
    page.jspx?id=402881e8134e403a01134e4d390d0aa0. Cheers.
22. Velodf, J., Beavers, K.: Going mental Tackling mental models for the online library tutorial. Research stategies 18, 3–20 (2001)
23. Yang, X., Ahmed, Z.U., Ghingold, M., Boon, G.S.: Consumer preferences for commercial Web site design: an Asia-Pacific perspective. The Journal of Consumer Marketing 20(1), 10–27 (2003)

Using Measurements from Usability Testing, Search Log Analysis and Web Traffic Analysis to Inform Development of a Complex Web Site Used for Complex Tasks

Caroline Jarrett[1], Whitney Quesenbery[2], Ian Roddis[3], Sarah Allen[3], and Viki Stirling[3]

[1] Effortmark Ltd, 16 Heath Road, Leighton Buzzard,
LU7 3AB, United Kingdom
[2] Whitney Interactive Design, 78 Washington Avenue,
High Bridge, New Jersey 08829, USA
[3] The Open University, Walton Hall, Milton Keynes, United Kingdom
caroline.jarrett@effortmark.co.uk, whitneyq@wqusability.com,
{i.roddis,s.j.allen,v.l.stirling}@open.ac.uk

Abstract. In this case study, we describe how we use measurements taken from web analytics and search log analysis with findings from usability testing to inform the development of web site. We describe an example of triangulating data taken from all three sources to help make design decisions; an example of drawing on web analytics and search log analysis to inform our choices of tasks during a measurement usability evaluation; and an example of using search log data to decide whether a new feature was worth investigating further. The context is enquirers making decisions about whether to pursue a course of study at a distance learning university: a long-term, complex problem.

Keywords: Measurement, web analytics, usability testing, search analysis, triangulation, multi-measurement, online prospectus, university, enquirers.

1 Introduction

The Open University is the UK's largest university and the only one dedicated solely to distance learning. Its 220,000 students include more than 40,000 who are studying from outside the UK. Its online prospectus "Study at the OU" is a key tool in attracting and retaining students, and is also important as the sole route to online course registrations: more than £100 million (equivalent to US$150 million) of online registrations are taken each year.

The usability of the University's online prospectus is clearly important to the University, and has been the subject of user research and usability studies for several years now, for example [1].

The overall responsibility for development of the Open University's web presence is led by Ian Roddis, Head of Online Services in the Communications team. He

M. Kurosu (Ed.): Human Centered Design, HCII 2009, LNCS 5619, pp. 729–738, 2009.
© Springer-Verlag Berlin Heidelberg 2009

co-ordinates the efforts of stakeholder groups, including developers, usability consultants, the academics, and many others.

The team is committed to user-centred design, both by involving users directly in usability tests, participatory design sessions and other research, and indirectly through a variety of different data sets, including search logs and web tracking.

In this paper, we describe three examples of the way we use measurement to inform development:

- Triangulating between web analytics, search logs and usability testing
- Drawing on data from web analytics and search logs to inform our choices of what to measure in summative testing
- Using search logs to establish whether a new feature is important for usability.

2 Triangulating between Web Analytics, Search Logs, and Usability Testing

There long been discussion in the usability community about the 'right' number of participants for a usability test; to give just three of the contributions to the arguments, there is the claim that five users is enough [3], the rejoinders that five users are nowhere near enough [4], and discussion of the mathematics that can help you to discern how many users you need [5].

We prefer to think in terms of iteration between usability testing, typically with five to 12 users, and the use of other data sources – a view supported by Lewis [6]. Each of these methods informs the others, providing direction about possible usability problems and design solutions. In addition, when we see consistent insights from both analytics and qualitative evaluations we have greater confidence in the results of both methods.

For example, in one round of usability testing on the prospectus we noticed a problem. Users found it difficult to work with the list of subjects:

- The list was quite long (50 subjects)
- When viewed on a typical screen at that time, some of the list was 'below the fold' and not visible to the user
- The list was presented in alphabetical order, which meant that some related subjects (e.g. Computing and Information Technology) were separated from each other.

We could have done more testing with more participants to measure exactly how much of a problem this was, but instead we opted instead to note that it *was* a problem and to look for ways to understand the behavior in more depth. We decided to use web analytics to look at the relative numbers of visits to the different subjects, and the likelihood that a visitor would combine exploration of two different subjects in a single visit.

For example, in Figure 3 we see that 37% of visits that involved Information Technology also involved Computing, but that only 27% of visits that involved Computing also involved Information Technology. In addition, we found that Computing was

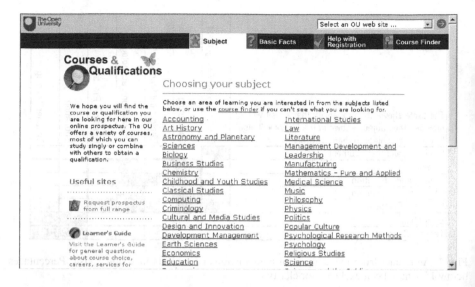

Fig. 1. The original list of subjects on the prospectus, as seen on a typical screen

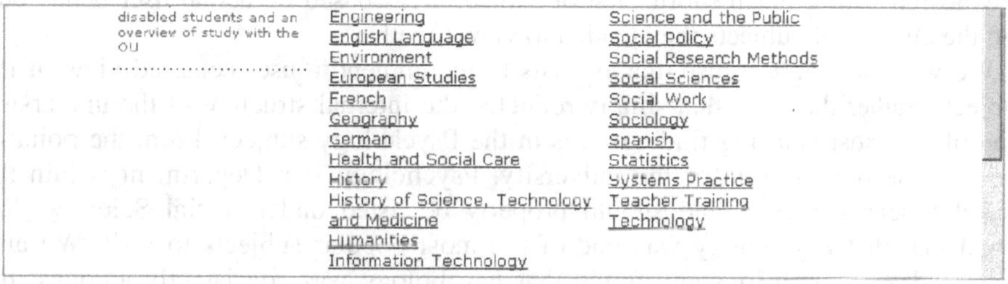

Fig. 2. Scrolling down revealed 'missing' subjects, such as Information Technology, more sciences, Social Work and Teacher Training

receiving 33% more visitors than Information Technology. This was likely caused by the user interface: we had also seen that our usability test participants were more likely to click on Computing (above the fold) than on Information Technology (below the fold).

We looked at the content of these two subjects and discovered that visitors should really think about both of them before choosing either. The interface, however, did not present them in a way that encouraged this comparison. Indeed, some visitors might not understand the difference between these two subjects, as taught at the Open University.

We also see from Figure 3 that there are clusters of related subjects. For example, visits to Education or Teacher Training were unlikely to also include visits to Computing or Information Technology.

The table shows the probability that a visit which browses the subject shown in the ROW will also browse the subject in the COLUMN.	Business and Management	Computing	Design and Innovation	Information Technology	Systems Practice	Education	Teacher Training
Business and Management	100%						
Computing		100%		27%			
Design and Innovation	19%	23%	100%	23%			
Information Technology	16%	37%		100%			
Systems Practice	30%	37%	18%	41%	100%		15%
Education						100%	26%
Teacher Training						27%	100%

Fig. 3. An extract from the analysis of combinations of subjects in a single visit. Percentages below 15% have been excluded for clarity.

From this type of analysis, across the entire list of subjects, we recommended a new design with a much shorter list of subject areas based on actual user behaviour, and the clusters of subjects they tended to view together.

We were also able to use this analysis to recommend a user-centered view of the subjects, rather than one that simply reflected the internal structure of the university. One of the most startling findings was in the Psychology subject. From the point of view of the organisation of the university, Psychology is a Department within the Social Sciences faculty and should properly be listed under Social Sciences. We found that that Psychology was one of the most popular subjects to visit. We also examined the search logs and found that Psychology was consistently amongst the popular search terms. We recommended that Psychology should be listed on its own as a subject. The University chose to back the user-centred approach and continues to list Psychology separately from Social Sciences.

Once a new list of subjects was designed, we ran usability tests to determine whether they improved the ability of visitors to find the subject that best matched their goals for study at the OU. These usability tests allowed us to continue to test and refine the list of subjects, adding a deeper understanding of how visitors were interpreting the terminology we chose. For example, our initial analysis showed that visitors tended to group 'Criminology' with 'Law', so we grouped them together. Usability testing showed that participants interpreted this as implying that the two subjects were closely related, whereas in fact Criminology as taught at the Open University is about the sociology of crime.

3 Drawing on Data from Web Analytics and Search Logs to Inform Our Choices of What to Measure in Summative Testing

We wanted to establish a baseline measurement of the usability of the prospectus before a major new release. The choice of technique seemed obvious: conduct summative usability testing, asking representative users to attempt an appropriate range of tasks.

3.1 What Tasks Should Be Measured?

Broadly, the online prospectus has to support the user task "Find out if study at the Open University will allow me to meet my educational goals, and if so sign up". This task is a complex one.

- **A mixture of sub-tasks:** Users have a variety of levels of understanding of their own needs, ranging from a vague concept like "I want to work with children and I think some studying at a university will help me to get there" through to highly specific tasks such as "I want to register on M248 Analysing Data". Table 1 illustrates the range of goals expressed by participants in one of our usability tests.
- **No clear time pattern:** It is unusual for a user to sign up for a long programme of study based on a single visit to the web site. Enquirers may take years to make up their minds to sign up.
- **A mixture of online and offline activity:** The University regards it as a success for the web site if the user opts to order a paper prospectus, telephones to discuss options in more detail, or elects to attend a face-to-face course choice event. We find this mirrors the needs expressed by participants in usability studies.
- **A variety of end points:** Clearly, a desire to register for a specific course has an end point of achieving registration, but the less clearly articulated tasks may have many different end points – or none.
- **A mixture of entry points:** Users may arrive from the Open University home page, from search, or from many other areas of the University's web presence (it has over 2000 web sites) such as the BBC/Open University web site associated with its popular television programs.
- **A wide range of options:** Most universities in the UK offer named degrees with a relatively fixed programme of study: A student might sign up for French, say, and then study only French for three years. The Open University model is much more like a typical USA programme: over 600 courses in different subjects that can be put together in various ways to make up over 200 qualifications, each with its own rules.
- **A wide range of levels:** As its name implies, the "Open" University offers many starting points that have no entry requirements; other courses are restricted to those with degrees; some are aimed at people with specific prior experience or working in particular types of employments; some are advanced courses aimed at graduates who will go on to PhD studies.

Clearly this is a complex task, so we looked to the literature for the related domain of complexity in software. One of Mirel's suggestions on this point is to "Describe the task landscapes that users construct for their patterns of inquiry and subgoals" [7]. In

Table 1. A selection from the educational goals expressed by usability test participants planning to pursue university-level education within the next 18 months

Add to current nursing course, possibly get a degree
BSc Psychology degree- always interested it and wants a more academic degree
Childhood and Youth degree
Compete in marketplace, get ahead in journalism
Get ahead, maybe in counselling
Marketable office skills
Masters in Education; wants to teach in primary schools
Postgraduate certificate in Health Studies
Pull existing study credits into a degree
Pursue degree in an area of interest; "learn with an adult frame of mind"
Pursue dream of teaching
Second degree in Business Studies with Economics; Sponsored by employer
Secondary school teacher of Italian or maybe French
Some sort of marketing course
Something flexible when the kids are at school - maybe accountancy
Use interest in Lit/Arts for possible degree
Wants to improve English
Work with kids

terms of measurement, we interpret this as breaking the complex task into smaller tasks that relate to the whole. Two obvious ones were:

- **Order a print prospectus:** a defined success point
- **Extract simple information from a course description:** a basic sub-task that contributes to users' overall decisions.

These two hardly seemed enough to capture the richness of the full task. But many of the other tasks were highly specific, relying on an interest in a particular subject area. The challenge we faced was how to select tasks that were both appropriate for measurement, but also created an overall picture that could stand as a proxy for the richness of the full task.

3.2 Tasks Extracted from Search Log Analysis

When we examine our search logs, we find that subjects, courses and specific jobs dominate, as can be seen from the list of top search terms in Table 2 below. As is usual with search analysis, we find that the top few search terms are strongly indicative of the searches in general [8].

We also found that external search (search terms entered in Google) is dominated by terms such as 'open university', which we interpret as markers of the visitor's intention of getting to the Open University web site specifically [8]. Stripping out those markers, we found that search will bring visitors to many different entry points in the web presence, as shown in table 3.

Table 2. Most popular terms used on the Open University's internal search

Term	Rank
psychology	1
courses	2
short courses	3
credit transfer	4
jobs	5
photography	6
law	7
creative writing	8
mba	9
social work	10

Table 3. Most popular terms used in external searches that bring visitors to the Open University web site

Notes: Markers such as 'Open University' have been ignored
                We have combined similar terms such as 'distance learning' and 'home learning'.

Term	Rank	OU site that is the target of the first link
courses	1	Study at the OU, the online prospectus
students	2	StudentHome: the extranet for students
openlearn	3	Openlearn: publishes selected course material for free
distance learning	4	New to the OU: explains the Open University to new visitors
ireland	5	The Open University in Ireland
london	6	The Open University in London
business school	7	The Open University Business School
mba	8	The Open University Business School's description of an MBA
jobs	9	Jobs at the Open University
cheri	10	Centre for Higher Education Research and Information

3.3 Using Web Traffic Analysis a Source of Tasks

Our next source of data was the traffic analysis. We wanted to find out whether the users tended to stay on their arrival point within the University's web presence, or whether they tended to move across to the part of the web site that we particularly wanted to measure: the prospectus, Study at the OU.

Figure 4 illustrates traffic flows into the "Study at the OU" prospectus sub-site from external sites (shaded area) and other parts of the OU's web presence. We found that the two biggest flows of traffic into "Study at the OU" are from Google and from the Open University's home page, as we expected. But we also found important flows from many other web sites within the overall OU web presence, and we found that a wide range of web sites had some flow. Broadly, any visitor to the OU's web presence was likely to end up on the prospectus at some point during their visit.

This added a third element to our mixture of tasks:

- Visitors to many of the OU's web sites end up on the prospectus.

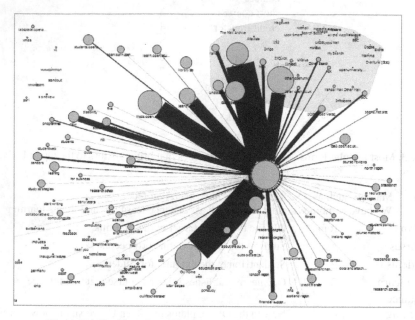

Fig. 4. Traffic flows into "Study at the OU" (the online prospectus)

The size of the circle is in proportion to the number of visitors to that site; the size of the line is in proportion to the flow of visitors from the site to the prospectus. Sites in the shaded are external to the Open University.

Table 4. A selection from the tasks used for a recent baseline summative test of the Open University's web site

Task	Entry point
Reading a course page (Creative Writing) to find information about the timing and requirements for the course	Search for "creative writing" in Google
Find a course on the psychology of children.	Faculty site
Find a first course in psychology if you haven't studied recently	Home page
Find a section of the OU web site that offers advice on using your education for your career.	New to the OU

3.4   The Set of Tasks Used for Our Measurement

The final set of tasks reflected all of these considerations:

- Different user goals
- The relative popularity of different subjects
- A range of different entry points

By using the site and search analytics to construct the tasks we could be confident that the summative test would reflect typical behavior. It also meant that the tasks were relevant to many of the participants, making their behavior more realistic.

4 Using Search Logs to Establish Whether a New Feature Is Important for Usability

A third way we use measurement is to establish whether something is important enough to require further attention.

For example, Google is a crucial source of traffic for this site, as for so many others. For some time, Google provided a set of site-specific links as part of the results for selected sites. In 2008, Google introduced a new feature into its results for selected large domains: a site-constrained search box (figure 5, below).

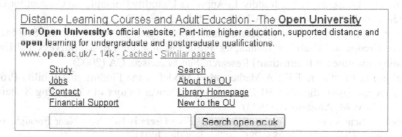

Fig. 5. The site-specific links and site-constrained search box within Google results

We wanted to know whether this box had affected user behaviour. Were visitors using it? Did we need to think about exploring it in our next round of usability testing?

4.1 Use of the 'Site' Box Is Negligible

The site-constrained box has the same effect as using the Google 'site:' **feature** in their advanced search: it performs a Google search, restricted to a particular domain. In March, these searches were referred with 'site:www.open.ac.uk' appended to the search term. By July, Google was referring them with 'site:open.ac.uk' appended.

We therefore analysed search logs from three different weeks: before Google introduced the search box, immediately after, and some months later. We looked for search terms that included the indicator terms "site:www.open.ac.uk" or "site:open.ac.uk".

Table 5. Percentages of visits that included the indicator search terms, before and after Google introduced the site-specific search box

% of visits in a week that included...	Before the change	After the change	Three months later
site:www.open.ac.uk	0.001%	0.004%	0.007%
site:open.ac.uk	0.015%	0.018%	0.038%

The analysis showed that use of the site-constrained search box was a negligible proportion of total visits, well under a tenth of a percent (fewer than one in a thousand), and that we could safely ignore the feature for the moment.

5 Conclusion

Unsurprisingly, our conclusion is that we get the best insights when we combine data from whatever sources we can lay our hands on, and we continue to iterate between different approaches according to what we find and the questions that we want to answer. And it can be just as valuable to find out what we can safely ignore.

References

1. Jarrett, C., Quesenbery, W., Roddis, I.: Applying Usability Principles to Content for Diverse Audiences. In: Proceedings of HCI 2006, vol. 2, pp. 98–102 (2006)
2. Jarrett, C., Roddis, I.: How to Obtain Maximum Insight by Cross-Referring Site Statistics, Focus Groups and Usability Techniques. Presentation at Web Based Surveys and Usability Testing, Institute for International Research, San Francisco, CA (2002)
3. Nielsen, J., Landauer, T.K.: A Mathematical Model of the Finding of Usability Problems. In: Proceedings of the SIGCHI Conference on Human factors in Computing Systems, pp. 206–213. ACM, Amsterdam (1993)
4. Spool, J., Schroeder, W.: Testing Web Sites: Five Users is Nowhere Near Enough. In: Proceedings of CHI 2001, pp. 285–286. ACM, Seattle (2001)
5. Lewis, J.R.: Sample Sizes for Usability Tests: Mostly Math, not Magic. Interactions 13(6), 29–33 (2006)
6. Lewis, J.R.: Usability testing. In: Salvendy, G. (ed.) Handbook of Human Factors and Ergonomics, pp. 1275–1316. John Wiley and Sons, Hoboken (2006)
7. Mirel, B.: Interaction Design for Complex Problem Solving: Developing Useful and Usable Software. Morgan Kaufmann, San Francisco (2003)
8. Quesenbery, W., Jarrett, C., Roddis, I., Allen, S., Stirling, V.: Search Is Now Normal Behavior. What Do We Do about That? In: Proceedings of the Usability Professionals' Association Conference (electronic proceedings) (2008)

User-Centered Design Meets Feature-Driven Development: An Integrating Approach for Developing the Web Application myPIM

Torsten Krohn[1], Martin Christof Kindsmüller[2], and Michael Herczeg[2]

[1] Itemis AG, Schauenburger Str. 116, 24118 Kiel, Germany
[2] Institute for Multimedia and Interactive Systems, University of Lübeck,
Ratzeburger Allee 160, 23538 Lübeck, Germany
torsten.krohn@itemis.de,
{mck,herczeg}@imis.uni-luebeck.de

Abstract. In this paper we show how a user-centered design (UCD) method can be successfully combined with an agile software development approach, namely feature-driven development (FDD), to develop the web-based information management system myPIM. This system supports users' workflow requirements in research and teaching/learning contexts. It provides bookmark, file, and reference archives, as well as possibilities for exchanging information with colleagues and students. By describing the system and its development process we show how this combination of methodologies supported our development process to create a service that truly assists the target audience and is easy to use.

Keywords: World Wide Web, online community, feature-driven development, folksonomy, information management, information sharing, internet-based collaboration, social bookmarking, social software, tagging, user-centered design.

1 Introduction

Users in research and teaching/learning contexts are highly dependent on resources that can be found in the World Wide Web. The massive growth of online information continually increases in complexity and needs both to be managed efficiently and resourcefully. Particularly, the paradigm shift from mere consumption to a new digital lifestyle of vigorous contribution of content through weblogs, wikis or the like, adds to the rapid increase of information in the WWW.

Numerous projects in the past have addressed managing large amounts of information, but with limited success. The GAB (Group Asynchronous Browsing) approach by Wittenburg et al. [1] suffers from the small overlap of the users' interests. Vistabar [2], a tool to support the users' handling with web resources, failed mainly because of the long-winded installation routine and the limited categorization options. A similar system, WebTagger [3], failed because of insufficient compatibility with many web pages. Kanawati and Maleks [4] tried to classify bookmarks automatically with multiple agents. Most people rejected the system, since this automatism did not work

M. Kurosu (Ed.): Human Centered Design, HCII 2009, LNCS 5619, pp. 739–748, 2009.

properly for them. Recent approaches (e.g. del.icio.us) are apparently better accepted by the users but still have usability problems and are lacking some functionality [5].

The most important lesson learned from the failures of these tools is that the users' needs and preferences should not be neglected in the initial design of the tool. For this reason, we employed a user-centered design (UCD) [6] process which focuses on the users' needs as well as their tasks and goals.

2 User-Centered Design and Feature-Driven Development

To achieve a high usability, we started by interviewing researchers, educators, and students. The results were correlated to generate eleven prototypical use cases. Each use case describes the interaction between a user and the system in respect to achieving a certain goal.

2.1 Interviews

To get an impression about dealing with online resources, the following topics were discussed: current use of bookmarks, research for preparing lectures, lecturing and research workflows, social bookmarking with integrated contact management, teamwork, support for course studies, rating of potential features, and experience with existing solutions. Especially the first topic gives us hints to conclude the relevance and importance of the interviewees' statements.

The procedure and resulting answers of a verbal interview are not completely predictable [7]. An interviewee could, without being aware, talk a lot about a certain topic and this could take much more time as intended. Furthermore, unexpected but interesting topics could arise. Hence, we conducted semi-structured interviews (guided but not completely determined) to create an environment to facilitate fruitful interview sessions.

2.1.1 Searching, Storing and Recovering

One central question the interviewees were asked is how useful information can be found quickly and efficiently. How can we recover this information at a later point in time? How can we find similar information about this topic? The workflows of students, lecturers, and researchers were studied by focusing on their handling of web resources in order to conclude the best way to support them. This support allows them to quickly and efficiently give or attend lectures, create or complete assignments or lab work, do research, and communicate with colleagues. It was concluded that tagging bookmarks is much more efficient than filing them into a single category: If an individual wants to remember an item (article, image, book reference), they have to go through a multi-stage process [8]: After the individual makes a decision to remember something (stage 0), multiple concepts are usually activated simultaneously (stage 1), and then, one of these multiple concepts has to be chosen (stage 2): The information is then stored in the appropriate category. It might be difficult to access this information in the future, because the user has to remember what concept was finally chosen maybe months or even years ago. The underlying cognitive process for tagging differs from the process for categorizing, making the retrieval of information relatively simple. The user can recover the bookmark by choosing any of the concepts that was activated in the tagging process.

2.1.2 Finding Something Similar

Frequently, the user likes to access related sources. It seems reasonable to provide an opportunity for users to benefit from the other users' knowledge. In a real world group of people, e. g. a business seminar, the participants of the seminar can share and communicate face-to-face. They know each other and their respective field of work.

Nevertheless, a conversation face-to-face or via e-mail is not always possible. Some interviewees tried to browse strangers' bookmarks, but they could not extract the information they needed. Because every user has a unique knowledge structure (cf. [9]) and every information structure created by this user depends on this unique knowledge structure, it is difficult to interpret other users' information structures without "knowing" this person. A better support of what Surowiecki [10] called the "wisdom of the crowd" in online communities seems to be obvious: "under the right circumstances, groups are remarkably intelligent and are often smarter than the smartest people in them". But how can this be exploited?

2.1.3 Information Gathering Workflows

Most people said that they often neither know the right space, nor the right time to save interesting information. It is also important to remember, that we often encounter information we have not actively searched for at that particular moment in time. Sometimes we come across interesting hints during work, which could be useful for some private interests. Then again, we find helpful articles for our job during surfing the web at home. Thus, the search results are still useful, but in a different context. Cutrell et al. [11] call this encountered information. This unexpected, but possibly useful information often interrupts the users' current workflows.

Moreover, it is possible to find multiple contexts by using certain tools. For example, somebody wants to call on a medical practitioner that a friend recommended a week ago. He does not remember the name or the telephone number of the practitioner, but he knows that he has saved it somewhere. Was it an email or an instant message? So, it must be in the inbox, or in his instant messaging logfile. Or, maybe it is already in his calendar's to-do-list. A useful PIM tool would deal with these multiple contexts e.g. by searching them automatically.

2.2 Workshop

Summing up, by studying all of the above-mentioned information, it is possible to determine meaningful use cases. For instance, "research new resources for lectures" or "find similar resources" can be considered as typical use cases. Based on these use cases, we derive 30 features for our tool, e. g. getting reminders of work to do or store bibliographical references in a repository. During a workshop with all interviewees, the features are ranked by importance. Afterwards, all features are grouped into feature sets. Each set consists of related features, i. e. all features of a set have to serve a joint purpose. These purposes are subject to dependencies, and therefore affect the order in which they can be put into practice.

2.3 Integrating Development Processes

By applying UCD, the potential users and their tasks are right in the center of the design process of the future system. Combining scientific information, users' opinions from interviews, information derived from the prioritization of the features and natural dependencies, a sorted list of feature sets emerge. To work through the list, we need a software development process which supports such an approach. The iterative and incremental process of feature-driven development (FDD) [12, 13] complements our UCD process in the development process.

Hence, we can start by implementing the most important features and quickly come to a running prototype, which then can be used and evaluated by the target users. The evaluation results are prioritized according to level of importance and when changes should be done. Less important features need not be dropped, but are left to be implemented later on in the development process.

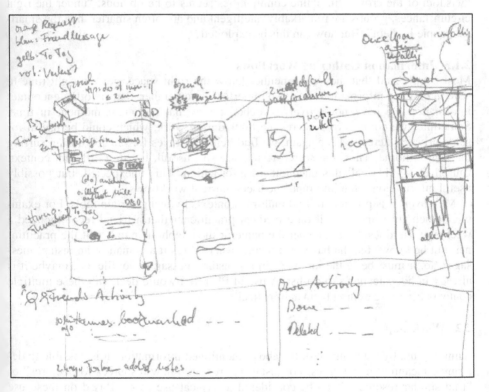

Fig. 1. Paper sketch of the dashboard

3 Design

Our myPIM tool consists of a core system and several modules. The core system implements the main functionality like entering, displaying, retrieving and distributing information. All further feature sets are implemented as modules, which can dock to this core system. By using an FDD process as our software engineering paradigm, we were able to deploy the basic features of the core system quickly, without negative effect on the expandability of the system later on.

3.1 From Paper-Mockup to Final Version

According to the FDD process "plan by feature", we gear towards the feature sets and iterate through their features. Thus, for example, in the feature list "user interface and navigation" we start with a paper sketch of the interface (Fig.1) to get a general idea of the look and feel.

Adding more and more ideas leaves this sketch cluttered in the end. But, nevertheless, it is a useful instrument to organize the individual's thoughts. Screen mockups specify these ideas and lay the groundwork for the final implementation of the user interface (Fig. 2).

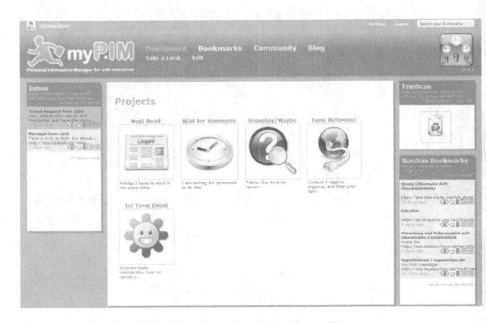

Fig. 2. Screen mockup of the signup dialog

3.2 Methodology's Benefit

By combining UCD and FDD, we can not only present first impressions of the look-and-feel, but also collect hands-on-experience of prospective community members at an early stage. This is especially important for systems that foster online communities because these communities only emerge under two conditions [14]. Either the system has to provide a so-called killer feature, which comes with immediate benefit for the users as soon as they use it – even if no one else uses it. Or, if no such feature can be implemented, the system has to provide awareness and so-called community memory [15, 16] for the users: (1) they need to be able to learn to get to know each other, (2) they need to be able to see the others' histories, (3) they need to feel confident to meet each other again in the future within the community and (4) the code of conduct needs to be made obvious.

No matter if one aims to design for a killer feature or if one has to compensate for the lack of it, our methodology of combining UCD and FDD helps to iteratively create a product, which addresses the users' needs and their task requirements, and any defect can be counteracted in good time.

4 My Personal Information Manager

After presenting the integrated development process we introduce the core system basically with screen shots of the implemented main features.

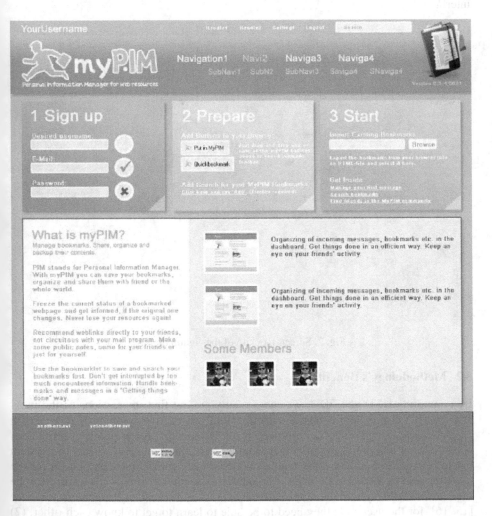

Fig. 3. Dialog for signing up in three steps

4.1 User Interface and Navigation

A very important design goal was to allow the user to setup and work with myPIM instantly and easily. Therefore we designed the registering and setup procedure as a three-step process displayed on a single page (Fig. 3). Unlike plain HTML pages, many actions can be performed without (re-)loading a whole webpage. This asynchronous data transfer between browser and server is done with the help of the Java-Script frameworks Ajax, Prototype and script.aculo.us. For instance, the cross and the checkmark indicating the concept of failure and success are changed according to a continuous evaluation while the user is typing (cf. Fig. 3).

Moreover, there are three main areas: dashboard, bookmarks and community. On the dashboard, the user finds logs about the latest activities of his community's friends as well as of himself. All projects and the newest tasks that are to be completed can be seen on the dashboard. As summarized by Sinha [8], tags are good as fast accessible pointers to personal knowledge. On the other hand plain tags are not optimum for organizing, restructuring and working with knowledge. To counteract these limitations, we provided some basic project management functions for entities like bookmarks, notes and tasks.

The bookmark area is used to search both individual and other members' bookmarks, as well as to insert, edit and delete bookmarks. The user can customize the user avatar, title, tags, URL and bookmark listing system. To enhance community awareness, the latest comment regarding a certain bookmark is displayed in the large view.

The community area provides the standard functionality for maintaining relationships and tasks, e.g. friend requests, "breaking up friendships" as well as the list of conducted conversations.

4.2 Bookmarking and Workflow

To search for a bookmark, the user enters the search terms into the input field (Fig. 4), checks the desired search parameters (tags, title, notes, URL), specifies the scope (own, friends', and others' bookmarks) and hits the return key or clicks the button. The results are displayed in sorted order, which is specified in the upper right corner. For a fast search without visiting the myPIM web service first, users can add a searchlet to their browsers. This is usually done during the registration process. No matter which way myPIM is accessed, the search function always returns the same kind of results (Fig. 4).

4.3 Searching and Community

The principle of managing all relevant information in one context was combined with Allen's [17] time- and self-managing method called "Getting Things Done". With regard to the tasks that our system is designed to support, these activities deal with URLs, notes and messages from other users. myPIM provides context-related to-do lists so that users do not have to worry about forgetting something. The remaining

---

[1] A searchlet is a plugin that extends the browser based search functionality by adding a new search service to the browser search field.

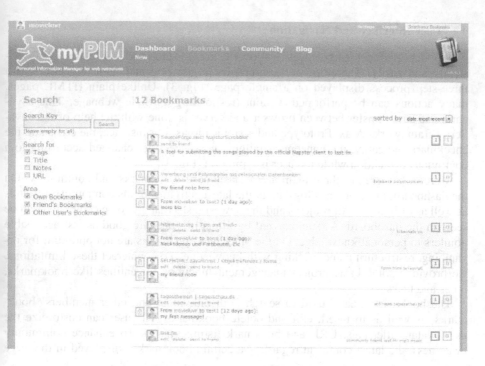

Fig. 4. Bookmark search and results

feature set "searching and community" complements our solution for the development of the tool myPIM. As stated above, we designed additional feature sets which are currently under development.

4.4 Future Work

So far we presented a description of the concept and the core features of our system myPIM. Now, we will briefly introduce further features that are part of the advanced system.

Because the supported resources of myPIM are manifold, they are made distinguishable by their types (e.g. PDF, web page, image) and can be sorted and grouped accordingly. The document types can be indicated by icons or thumbnail representations of the web resources. Furthermore, to save time, the resources to be displayed after a search takes place can be performed asynchronously on-the-fly.

The migration from previous (browser-based or online) bookmark archives to myPIM can be performed with an import function. All up-to-date browsers and online bookmarking services can export their bookmarks to a text file according to the Netscape standard. This file can be imported and integrated into the existing myPIM resource structure. If desired, the user can tag each bookmark within this process or create a reminder to do this at a later point in time.

To combine bookmarks and a file archive, we need a repository. During the bookmarking process, the bookmarked resource is saved in the repository and therefore

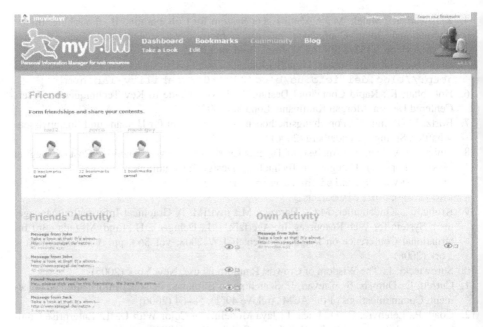

Fig. 5. Community dialog displaying friends and their activities (details negligible)

always available (mirroring). This makes local backups of web resources redundant. Additionally, the user could snapshoot the web page to freeze a certain state (versioning) [18].

5 Conclusion

The first evaluations show that not only the intended target audience of our system (computer scientists within research and teaching/learning context) can benefit from myPIM, but it can be easily extended to other target audiences. Our integrated UCD and FDD processes helped create a system than can be regarded as a general tool for users in all contexts.

References

1. Wittenburg, K., Das, D., Hill, W.C., Stead, L.: Group Asynchronous Browsing on the World Wide Web. In: 4th International World Wide Web Conference, pp. 51–62. O'Reilly, Boston (1995)
2. Marais, H., Bharat, K.: Supporting Cooperative and Personal Surfing with a Desktop Assistant. In: 10th Annual ACM Symposium on User Interface Software and Technology, pp. 129–138. ACM Press, New York (1997)
3. Keller, R.M., Wolfe, S.R., Chen, J.R., Rabinowitz, J.L., Mathe, N.: A Bookmarking Service for Organizing and Sharing URLs. Computer Networks and ISDN Systems, 1103–1114 (1997)

4. Kanawati, R., Malek, M.: A Multi-Agent System for Collaborative Bookmarking. In: 1st International Joint Conference on Autonomous Agents and Multiagent Systems (AAMAS), pp. 1137–1138. ACM Press, Bologna (2002)
5. Hood, S.: del.icio.us Blog on Usability Lab,
   http://blog.del.icio.us/blog/2007/07/usability-lab.html
6. Holtzblatt, K.: Rapid Contextual Design: A How-to Guide to Key Techniques for User-Centered Design. Morgan Kaufmann, London (2005)
7. Bortz, J., Döring, N.: Forschungsmethoden und Evaluation für Human- und Sozialwissenschaftler. Springer, Heidelberg (2006)
8. Sinha, R.: A Cognitive Analysis of Tagging (or How the Lower Cognitive Cost of Tagging Makes It Popular). Thoughts on Technology, Design & Cognition,
   http://www.rashmisinha.com/archives/05_09/
   tagging-cognitive.html
9. Krohn, T., Kindsmüller, M.C., Herczeg, M.: myPIM: A Graphical Information Management System for Web Resources. In: Ågerfalk, P.J., Delugach, H., Lind, M. (eds.) 3rd International Conference on Pragmatic Web. ICPW 2008, vol. 363, pp. 3–12. ACM, New York (2008)
10. Surowiecki, J.: The Wisdom of Crowds. Random House, New York (2005)
11. Cutrell, E., Dumais, S., Teevan, J.: Searching to Eliminate Personal Information Management. Communications of the ACM Archive 49(1), 58–64 (2006)
12. Coad, P., Lefebvre, E., De Luca, J.: Java Modeling in Color With UML: Enterprise Components and Process. Prentice Hall, Upper Saddle River (1999)
13. Palmer, S.R., Felsing, J.M.: A Practical Guide to Feature-Driven Development. Prentice-Hall, Upper Saddle River (2002)
14. Kindsmüller, M.C., Melzer, A., Mentler, T.: Online Community Building. In: Khosrow-Pour, M. (ed.) Encyclopedia of Information Science and Technology, pp. 2899–2905. Information Science Publishing, Hershey (2009)
15. Kollock, P.: The Economies of Online Cooperation. Gifts and Public Goods in Cyberspace. In: Smith, M.A., Kollock, P. (eds.) Communities in Cyberspace. Routledge, London (1999)
16. Kollock, P., Smith, M.: Managing the Virtual Commons: Cooperation and Conflict in Computer Communities. In: Herring, S. (ed.) Computer-Mediated Communication: Linguistic, Social, and Cross-Cultural Perspectives, pp. 109–128. John Benjamins, Amsterdam (1996)
17. Allen, D.: Getting Things Done. The Art of Stress-Free Productivity. Penguin, London (2003)
18. Reinke, C.: Optimierung von Reload-Strategien im Web durch Komponenten-Tracking. Institute for Informationsystems, University of Lübeck (2007)

The Effects of Information Architecture and Atmosphere Style on the Usability of an Ecology Education Website

Chao-jen Ku[1,2], Ji-Liang Doong[2], and Li-Chieh Chen[3]

[1] Department of Visual Design, Hsuan Chuang University,
No. 48, Hsuan Chuang Rd. Hsinchu, Taiwan, 300
[2] Graduate Institute in Design Science, Tatung University,
No. 40, Chung Shan North Rd., Sec. 3, Taipei, Taiwan, 104
[3] Department of Media Design, Tatung University,
No. 40, Chung Shan North Rd., Sec. 3, Taipei, Taiwan
chaojenku@gmail.com, jldo@mail2000.com.tw, lcchen@ttu.edu.tw

Abstract. Ecology education is an important issue nowadays. But not everyone has equal opportunities to learn relevant topics through direct access to the great nature. In such a case, the platform of Web becomes the potential channel for people to learn ecological topics. Therefore, the key to success is to enhance the platform so that the performance is close to that of experiencing the great nature in person. In this study, expert interview was first conducted. The participants pointed out the differences in information architectures, atmosphere styles, the differences in learners' backgrounds were important factors needed to be considered while designing such a system. Therefore, several experiment websites were constructed based on different atmosphere styles. The results revealed that there were significant differences in browsing behaviors and the usability of websites between people from rural and urban. Websites with breadth architecture and natural atmosphere could reduce user's pressure and perceived workload.

Keywords: Ecology Education, World Wide Web, Information Architecture, Layout, Usability.

1 Background and Motivation

Ecological education is an important issue nowadays. But not everyone has equal opportunities to learn relevant topics through direct access to the great nature. In such a case, e-learning platform becomes the potential channel for those who are eager to learn ecological topics in classroom. How to enhance the platform so that the performance of e-learning in classroom is close to that of experiencing the great nature in person is the key to success. In addition, the differences in learners' growing-up environments are also important factors needed to be considered while constructing such a system. Therefore, the objective of this study is to examine whether different personal backgrounds or different webpage design styles and information architecture have significant influences on the performances of learning.

M. Kurosu (Ed.): Human Centered Design, HCII 2009, LNCS 5619, pp. 749–757, 2009.
© Springer-Verlag Berlin Heidelberg 2009

2 Literature Review

2.1 Information Architecture of Websites

In the design factors of web site, the information architecture of web site directly affects the overall practicability of web site. Consequently, how to choose the appropriate link architecture of web pages is the challenge of the designers.

The link architecture of web pages is also called as topology. Some scholars have categorized the topology into six categories (Brinck et al., 2002): hierarchical or tree, linear or sequence, matrix or grid, full mesh, arbitrary network, and hybrid. The hierarchical is the most common architecture. Its advantage is more speedy browsing and easy to expand new information. The linear or sequence is appropriate for procedure, story, or relate to sequent information guiding browse. The matrix is good to reveal two-dimension space. The full mesh can link to all of other web pages from every page, therefore it can provide most speedy guiding browse. The designer can use arbitrary network if the information architecture is less organized. The hybrid architecture integrates more than two categories, and is the more common category in reality. When one website is designed for education, some researchers had the following findings. For the low prior knowledge learners, the results showed that the hierarchical structure supported better free recall performance and reduced feelings of disorientation. In contrast, the high prior knowledge learners performed better and followed more coherent reading sequences in the network structure (Amadieu et al., 2009).

2.2 The Aesthetic and Emotional Aspects of Websites

The aesthetic and emotional aspects of websites or e-learning systems have obtained much attention recently (Hartmann et al., 2007; Stenalt & Godsk, 2006; Light, 2004). It was recognized that when the look and feel of an interface is pleasing, users are likely to be more tolerant, e.g. they may be more prepared to wait a few more seconds for a website to download (Sharp et al., 2007).

In addition, during the design process, thought should be given to maintaining a consistent expressive style from the top to the bottom of the page, showing its distinguishing features and giving users a unique impression (Nielsen, 2000). Such design styles and forms, including background, images, colors, interactive guide, data...etc., are able to combine the webpage's theme with the user's cognitive experiences, and aren't just a pursuit for new and popular elements, such as brilliant colors and dynamic changes.

The design that combines concepts of the natural ecology website with users' backgrounds should also consider the design style of simulated scenery, which is the presentation of true images and sounds from nature, allowing users to achieve better learning results.

3 Study of Available Websites

The method used for this study is based on considerations of the research purpose; literature is reviewed and the current webpage is analyzed. Further discussions on the webpage's architecture and design style, teaching method of ecology education, and learning achievements of media are carried out via interviews with three specialists and used as reference for modifying the interactive website's design. Afterwards, an experiment is designed using the new and old version web pages, in which the subjects are observed for reactions and habits when using the webpage, the subjects then fill in a questionnaire that is used for finding which version and users from which background achieved better learning results.

3.1 Website Description

Design style analysis of the original Taiwan natural ecology website: In the Government Information Office's natural ecology website (www.gio.gov.tw/info/ecology/Chinese/), the homepage uses Flash and provides two options, one enters an audiovisual introduction based on text and the other enters the main menu. The main menu shows Taiwan's terrain and altitudes (Fig. 1).

Fig. 1. The loading page and main page of the original ecology education website

When the mouse is over text on the main menu, users can understand Taiwan's terrain by watching the picture change to show which parts belong to the altitude the mouse is located over. Clicking on text or other items on the main menu will link to text descriptions of the terrain and ecology of the altitude, text links below the descriptions link to animals and plants within the altitude (Fig. 2).

Fig. 2. The main page and the pages in the first two levels

After clicking on a specific species, a static image and text descriptions of the species' name, habitat, habitual behaviors...etc. are displayed. If you wish to hear the animal's sound, click on the speaker icon, or if you want to see a video of the animal, click on the camera icon and a new page will pop up to play the video (Fig. 3). The architecture of the website is illustrated in Figure 4.

Fig. 3. The detailed pages of a specific creature

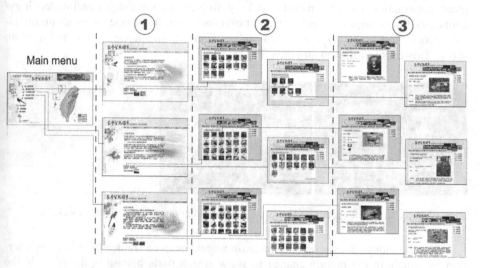

Fig. 4. The architecture of the original ecology education website

3.2 Interviews with Specialists

The opinions of three specialists, which are from different fields, on the current web-page's learning achievements are recorded on tape:

1. Web designer Chen Hsiu-chen: The architecture of websites is generally divided into Hierarchical and Network...etc.; hierarchical architecture has the advantage of clear categorization and easy to find data, but its disadvantages include causing the user to lose patience and learning capacity; network architecture has the advantage of rapidly finding targets, but will easily cause users to become lost. The interactive design of most learning websites are based on functional requirements; not many learning websites include aesthetic elements, simulate scenarios, combine text with images for easy search, and are integrated with databases. Such elements are more common in company websites, where the image and theme are emphasized on and visual and audio elements are well thought out.
2. High school biology teacher Chu Fang-lin: Students easily become tired of textbooks and explanations in class and hope to learn outside of school, such as zoos and botanical gardens, learning by seeing and hearing the real thing is very

effective. In the computer classroom, websites are a very important tool for aiding teaching because they allow students to see recent news, such as fire ants, and learn related knowledge. Biology classes should be able to achieve even better learning results if a professional room and audiovisual equipment is provided to simulate scenarios for students to learn in.

3. Ecology photographer Pan Chien-hung: Tolerance and respect for the environment and nature can only result from the transfer of knowledge, as well as understanding and cognition. Ecology education is best presented in the form of videos, and will achieve the best results if shown on TV or in class using a DVD produced from a professional director, screenwriter, filming and editing (such as National Geographic and Discovery Animal Planet series).

The three specialists mentioned above were asked to use several existing natural ecology websites, they were observed and recorded by combining the Think Aloud method with screen capturing and sound recording, and their opinions are organized as follows: Taiwan's natural ecology websites have too much data, clicking on text to link to species' subkingdom, class, order, family, genus... requires too much professional knowledge, and the design emphasizes more on functionality and less on audiovisual and scenario creation, which makes searches difficult and prevents better learning achievements; dividing Taiwan into different altitudes is a better way for searching animals and plants.

4 Experiment Design

4.1 The Atmosphere

4.1.1 The Design of a New Website with Natural Atmosphere

Based on the results of literature review and the opinions of specialists, a 16:9 wide screen ratio is used to simulate a broader view, images and text information of the old version website are used a basic elements, and images and sounds of nature are enhanced to let users feel the atmosphere, becoming the website's unique style. The homepage uses a beautiful natural atmosphere to attract the user's attention, plays music and sounds of nature at the same time, and shows a menu that includes animals, plants, theme videos, and links to other websites.

After a species is selected, the image of a forest first appears, to the right is a diagram with text that indicates the altitude, semi-transparent animal images appear for each altitude, the normal colors of the animal appear when the mouse is moved over the image, and the animals name is shown to the bottom of the image.

The background turns dark, but the altitude diagram and animal images can still be used, text contents include the animal's name, scientific name, English name, characteristics, habitual behaviors and habitat.

The user can click on the speaker icon to listen to the animal's sound. When the camera icon is clicked on, the static animal image plays a dynamic video. The trangle arrow on the upper left corner allows the user to move back to the previous page until the main menu.

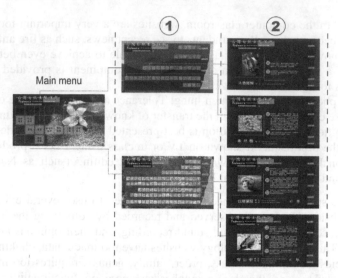

Fig. 5. The new website with breadth architecture and natural atmosphere

4.1.2 Comparative Evaluation

In the experiment, each participant was required to carry out typical tasks that interact with the interface prototypes. After that, the participant was asked to complete tests about the learning contents and system usability scale (SUS) and NASA-TLX questionnaire. The dependent variables in this research were the learners' performance, including the time consumed and the correctness of answering questions, and system usability evaluation.

The tasks are to search for pictures, text, sound effects and videos, and then draw them or compare them. Task one is to search for the shapes of leaves from two plants that grow on different altitudes: Ebony tree and Yellow water lily, and then draw them on the test sheet. Task two is to search for the Highland red-bellied swallowtail butterfly and write down its habitat and habitual behaviors. Task three is to compare the sounds of Formosan rock-monkey and Formosan reeve's muntjac. The screen capture software Cantasia is used to record how users operate the website, allowing the way users complete a task to be observed while comparing the accuracy of their answers. Subjects fill in a questionnaire with SUS and NASA-TLX on a 9-level Likert scale immediately after completing the tasks. Cross examination of positive and negative questions are used to find the subjects true subjective experience. The SUS part has 8 questions in total on easiness to learn, pictures and text, architecture and style. The NASA-TLX part has 6 questions in total on mental load, physical load, time load, energy consumption, performance and frustration.

Two groups of junior high school students participated in the experiments. Their growing-up environments were classified into "rural" and "urban." The urban group had 16 first-year students from Nei-Hu Junior High School in Taipei. They used World Wide Web frequently. The rural group had 16 first-year students from Bei-Nan Junior High School in the rural areas of Taitung. Internet access for them was not as convenient as that of the urban group. Therefore, they seldom used World Wide Web.

Each group was further divided into two subgroups, one for experiencing the breadth -hierarchy sites, the other for the depth-hierarchy sites, respectively. The learning processes of the participants were observed and recorded through video and further converted into qualitative and quantitative data, such as browsing behaviors and time spent for learning.

4.1.3 Results

Test data were analyzed using two-way ANOVA. The results revealed that there were significant differences in browsing behaviors and the usability of websites between people from rural and urban. The students in the urban group tended to able to find the required information in the depth-hierarchy websites in shorter time. Students in the rural group tended to rely on the interface with natural atmosphere, and easily got frustrated in depth-hierarchy websites. Evidences showed that participants perceived less workload when using the website with natural atmosphere and situational music, no matter they were from rural or urban. In addition, evidences showed students using 16:9 wide-ratio webpage got better satisfaction then using the layout with traditional 4:3 ratio. Compared to traditional layout of websites, wide-screen layout provided better landscape view to simulate human's perception of the real world.

Table 1. The mean and standard deviation of task completion time

Atmosphere	User Background	
	City	Rural
Natural Atmosphere	287.50(16.96)	451.88(21.26)
Traditional Atmosphere	384.50(19.61)	424.25(20.60)

Table 2. The mean and standard deviation of user satisfaction

Atmosphere	User Background	
	City	Rural
Natural Atmosphere	8.09(1.01)	7.30(0.89)
Traditional Atmosphere	6.72(1.21)	6.83(1.61)

Table 3. The mean and standard deviation of rating in reducing workload

Atmosphere	User Background	
	City	Rural
Natural Atmosphere	8.33(1.18)	7.60(1.13)
Traditional Atmosphere	6.94(1.08)	6.46(1.04)

4.2 The Information Architecture

4.2.1 The Modification of the Traditional Website
In the second stage, two websites with same atmosphere but different information architectures were compared. The traditional website was modified by changing the

screen ratio into 16:9 and by adding situational sounds from the great nature as those in the new website discussed in the previous section. The website with the depth architecture separates the menu of altitudes and the links to animals and plants navigation pages into two different pages. The website the breadth architecture combines the menu of altitudes and the navigation menus for animals and plants in the same page.

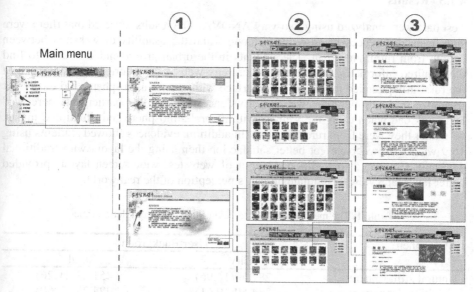

Fig. 6. The modified website with depth architecture and natural atmosphere

4.2.2 Comparative Evaluation
Twenty High School students participated in the experiments. They were divided into two groups. They carried out the same tasks as those in the first stage and answer the same set of questions in the post-test questionnaire.

4.2.3 Results

Test data were analyzed using t-test. The results showed that breath architecture is superior to depth architecture in reducing the steps of searching in ecology education websites.

Table 4. The mean and standard deviation of ratings

Information Architecture	Ratings	
	Rating in User Satisfaction	Rating in Reducing Workload
Depth Architecture	6.16(0.97)	6.30(1.08)
Breadth Architecture	6.43(1.29)	6.65(1.32)

5 Discussion and Conclusion

These results demonstrated that although websites with depth-hierarchy may be helpful for people who were familiar with state-of-the-art technologies while browsing, this feature did not contribute to reducing the workload of ecology learning. Websites with breadth architecture and natural atmosphere could reduce user's pressure and perceived workload.

References

1. Amadieu, F., Tricot, A., Marine, C.: Prior knowledge in learning from a non-linear electronic document: Disorientation and coherence of the reading sequences. Computers in Human Behavior 25, 381–388 (2009)
2. Brinck, T., Gergle, D., Wood, S.D.: Usability for the web. Morgan Kaufmann Publishers, San Francisco (2002)
3. Hartmann, J., Sutcliffe, A., Angeli, A.D.: Investigating attractiveness in web user interfaces. In: Proceedings of the SIGCHI conference on Human factors in computing systems, 2007, San Jose, California, USA, pp. 387–396 (2007)
4. Light, A.: Designing to persuade: the use of emotion in networked media. Interacting with Computers 16(4), 729–738 (2004)
5. Nielsen, J.: Designing Web Usability: The Practice of Simplicity. New Riders Publishing, Indiana (2000)
6. Sharp, H., Rogers, Y., Preece, J.: Interaction Design: Beyond Human-Computer Interaction, 2nd edn. Wiley, Chichester (2007)
7. Stenalt, M.H., Godsk, M.: The Pleasure of E-learning - Towards Aesthetic E-learning Platforms. In: Proceedings of the 12th International Conference of European University Information Systems, University of Tartu & EUNIS 2006, Tartu, Estonia, June 2006, pp. 210–212 (2006)

Accommodating Real User and Organisational Requirements in the Human Centered Design Process: A Case Study from the Mobile Phone Industry

Steve Love[1], Paul Hunter[2], and Michael Anaman[1]

[1] Brunel University, West London, UK
[2] Telco, UK
{steve.love,michael.anaman}@brunel.ac.uk,
paul.hunter@three.co.uk

Abstract. This paper reports on the results of a case study that investigated how different stakeholder needs within an organisation can be taken into consideration alongside the needs of real users in the human-centered design process to improve product and service design. The case study focuses on the mobile phone industry and in particular the design of a new service that was to be used in the retail stores of a major mobile phone service provider. The results indicated that by including various organizational stakeholders (such as sales and marketing teams) in an early stage evaluation of a prototype design provides valuable insight to problems (as well as suggestions to improve design) that may not otherwise come to light until a crucial time period in the project and could have a concomitant effect on sales and marketing timelines associated with the project launch.

1 Introduction

Usability is widely recognised as being a critical factor in the design of successful interactive products [1, 2, 3, 4]. In current usage, the usability of an interactive product refers to whether it is easy to learn, easy to use, and enjoyable from the user's perspective [5], and may be broken down into the following goals: effectiveness, efficiency, safety, utility, learnability, and memorability. A usable system is one that allows a user to learn how to use a system quickly, and to operate the system effectively with low rates or error, leading to improved user acceptance. A human-centered design (HCD) methodology has been advocated as an effective approach to achieve system usability [6, 7]. HCD is concerned with making user issues central in the design process, carrying out early testing and evaluation with users, and designing iteratively [6]. According to the ISO 13407 [8] standard on HCD, there are five core processes that must be undertaken in order to develop usable systems:

1. Plan the human-centred design process
2. Understand and specify the context of use
3. Specify the user and organisational requirements
4. Produce designs and prototypes
5. Carry out user-based assessment

M. Kurosu (Ed.): Human Centered Design, HCII 2009, LNCS 5619, pp. 758–764, 2009.

After the initial planning stage, the remaining four stages should be conducted iteratively until the usability objectives have been attained (see Figure 1)

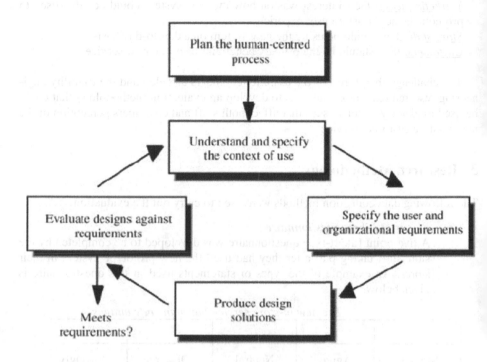

Fig. 1. Human-centred design activities (from ISO 13407)

The aim of this paper is to focus on the factors influencing the user and organisational requirements component of this process. In particular the paper will discuss the importance of considering the needs of all the stakeholders from an organisational and real user perspective. The paper will explore these issues in relation to a real-world design example based on an industrial research collaboration between Brunel University, West London UK and the mobile phone service provider known for the purposes of this paper as Telco.

2 The New Service

At the start of 2007, the organisation under study decided to investigate ways of improving its customer service. There were a number of initiatives underway in different functions, such as the contact centre and logistics areas, so it was decided that focus should be placed on the retail shops and the experience received by customers at when they entered the store.

There were a number of stakeholders identified who had an interest in the development of this new in-store PC-based customer information service that would provide details of account information and the like:

Sales team: their interest in the new system was focused on how it could be used to increase sales of products and services in-store
Marketing team: their interest was on how the new system could be advertised to promote the new in-store user experience
Store staff: they would be using the new system on a day-to-day basis
Customers: they should be the primary beneficiaries of the new service

The challenge therefore, for the usability engineers at Telco and the usability engineering research team at Brunel was to develop an evaluation methodology that could be used to identify factors/issues that affect both staff and customers perception of the new in-store customer service.

3   Research Methodology

The following data collection methods were used to carry out the evaluation:

- **Post-trial Likert questionnaire**
  A five-point Likert-type questionnaire was developed to be completed by the store staff taking part after they had used the new prototype system in their stores. An example of the types of statements used in this questionnaire is given below:

  The new in-store service has many advantages

Strongly agree	Agree	Neutral	Disagree	Strongly disagree

  Store staff were also asked to explain the answers that they provided for each Likert statement in order to gain more detailed information on their attitude towards the current in-store service

- **Online attitude questionnaire to be completed by customers in store**
  The attitude of customers towards the new service was also evaluated by a five-point Likert-type attitude questionnaire. These data were collected in-store via an online questionnaire that customers accessed via a PC. An example of the types of statements used in this questionnaire is given below:

  I would be happy to use the new in-store service again

Strongly agree	Agree	Neutral	Disagree	Strongly disagree

- **Observations of customers and staff in the trial stores**
  A researcher also visited the stores on a weekly basis to observe customers and staff using the new service. The customers were always asked in advance if they could be observed using the new in-store service with the store staff and assured that all data that were collected would be anonymised.

- *Post-trial interviews with store staff*
  A series of in-depth interviews were conducted with a cross-section of the store staff who took part in the trial. This was to explore any issues highlighted in the observations made during the trial and salient factors identified from the analysis of the Likert attitude data.
- *Post-store visit telephone interviews with customers*
  Customers who completed the online questionnaire were asked if they could be interviewed by the researchers at a later date. The idea behind this form of data collection was to explore issues arising from an analysis of the Likert attitude data. In addition, it was felt that customers may be more relaxed discussing their attitude towards the new prototype service away from the actual store environment.

3.1 Procedure

The trial lasted for four weeks. The sales and marketing teams identified five stores to take part in the evaluation of the new in-store customer service prototype. Before the trial began, store staff were given a training session on how to use the new prototype service. Once the trial began, store staff were asked to get customers to complete the online attitude questionnaire after they had used the new in-store customer service. In addition, the researcher visited the stores on a weekly basis to try to observe customers and staff using the new service. Store staff were also asked to collect details on customers who had agreed to be contacted and interviewed by the researcher at a later date. After the trial had finished, store staff were interviewed again by the researcher to find out what they thought of the new in-store customer service in relation to the old in-store customer service provision.

4 Results

4.1 Store Staff Attitude towards the New In-Store Service

A total of ten staff from the five trial stores that took part in the study agreed to be interviewed. There were five females and five males and they were all in the 18-24 age group.

Table 1. Store staff attitude towards the new in-store service

Attitude statement	Attitude score
The new in store customer service has many advantages	4.1
The new in store service needs a lot of improvement	3
I can easily explain to the customer how to use the new in store service	4.4.
Customers are dissatisfied with the new in store customer service	3.8
The new in store service helps me do my job more effectively	3.9

As can be seen from Table 1 above, the store staff had a positive attitude towards the new in-store customer service system. These scores were backed up in the post trial interviews and observations made by the researcher during the trial as can be seen from the following examples:

The new service looks more professional in the eyes of the customer

It would be great to have more functionality, such as being able to pay bills

Yes, it's easy to use and navigate around

Customers are pleased that issues can be explained in store, like bill queries

4.2 Customer Attitude towards the New In-Store Service

A total of 35 customers completed the online attitude questionnaire. However, not all the customers provided details on their age group or gender. From the data collected, there was an almost 50-50 split between male and female customers. In terms of age, there was a good spread between age groups (18-24, 25-34, 35-44, 45-55) apart from the last age group (55-64). There were no recorded entries for this group.

Customer attitude towards the new in-store prototype service are given below in Table 2.

Table 2. Customer attitude towards the new in-store service

Attitude statement	Attitude score
Happy to use new service again	4.1
Sales person knowledgeable about service	4.2
Feel confident using the new service	3.7
New service was friendly	4.0
New service was helpful	3.9

As can be seen from Table 2 above, customers generally had a positive attitude towards the new in-store prototype service.

4.3 Customer Telephone Interviews

There were only a few customers (five in total) who agreed to be contacted by the researcher to discuss their in-store experience. Of those five, only three were available for a telephone interview. For two of these customers, one customer had gone to the store to enquire about the possibility of changing their make of phone as they had been experiencing problems with it. The second customer appeared to have problems with their sim card. As a result of this, the customer had a large monthly phone bill. Both these customers felt that their in-store experience had been a good one and that the staff, using the new in-store where appropriate, had helped to resolve these issues.

The third customer, who was interviewed, stated that they found store staff to be very helpful and preferred going into the store to get an issue resolved rather than phoning the call centre. The reason for this is that they had, on previous occasions, spent a long time on the phone to the call centre without getting the issue resolved to their satisfaction.

5 Discussion

From a customer perspective the new in-store service, in general, offered them the opportunity to have an enhanced in store customer experience. For example, issues that may have required them to contact the call centre in the past could now be resolved in-store. This can be seen the high average scores obtained for statements such as *"I thought the sales person was knowledgeable about the new service"* and *"Overall, I enjoyed my experience in the store today"*

In terms of store staff perception, many felt that the introduction of the new service increased their ability to provide an improved in store experience for customers. This can be seen in comments such as *"Customers are pleased that issues can be explained in store, like bill queries"* and *the new service looks more professional in the eyes of the customer. Don't always need to get them to phone through to customer services to get their problems sorted"*. Another observation that was made during this trial was that both customers and the store staff felt as if they were being involved in the design of a new service by taking part in this trial. They believed that it was the right for the company to take into consideration the views, attitudes and observations of the people who would be using the new service on a regular basis.

In terms of the human-centred design process, this trial highlighted an important factor for the company to consider: the development and design of evaluation metrics for products and services at an early stage of the development and design process. This early stage evaluation should provide valuable insight to problems (as well as suggestions to improve design) that may not otherwise come to light until a crucial time period in the project and could have a concomitant effect on sales and marketing timelines associated with the project launch.

Overall, in terms of the core processes involved in the human-centered design cycle, this study found that specifying organizational requirements in relation to product or service design requires time in order to accommodate the needs of the different groups, as far as possible, in the proposed design solution and prototype development. In addition, the HCD approach also provides the different organizational groups with the opportunity to assess if evaluation results meets their requirements to launch the product or service into the business and if not, it also provides the framework to go through as further iteration of the product development and evaluation.

References

1. Shackel, B.: The concept of usability. In: Proceedings of IBM Software and Information Usability Symposium, pp. 1–30 (1981)
2. Eason, K.D.: Towards the experimental study of usability. Behaviour and Information Technology 3, 133–143 (1984)
3. Whiteside, J., Bennett, J., Holtzblatt, K.: Usability engineering: our experience and evolution. In: Helander, M. (ed.) Handbook of Human-Computer Interaction, pp. 791–817. Elsevier, Amsterdam (1988)

4. Nielsen, J.: Usability laboratories. Behaviour and Information Technology 13, 1–2 (1994)
5. Preece, J.J., Rogers, Y.R., Sharp, H.: Interaction Design: Beyond Human-Computer Interaction. John Wiley and Sons, Chichester (2002)
6. Preece, J.J., Rogers, Y.R., Sharp, H., Benyon, D.R., Holland, S., Carey, T.: Human-Computer Interaction. Addison-Wesley, Reading (1994)
7. Maguire, M.C.: Methods to support human-centred design. International Journal of Human-Computer Studies 55(4), 587–634 (2001)
8. ISO 13407:1999: Human-centered design processes for interactive systems (1999)

Affectively Intelligent User Interfaces for Enhanced E-Learning Applications

Fatma Nasoz and Mehmet Bayburt

School of Informatics, University of Nevada,
Las Vegas, NV 89154
fatma.nasoz@unlv.edu, bayburtm@unlv.nevada.edu

Abstract. In this article we describe a new approach for electronic learning applications to interact with their users. First we discuss our motivation to build affectively intelligent user interfaces that can recognize learning related emotions and adapt to these through user modeling. In the remainder of the paper we describe the experiment we designed to elicit learning related emotions from students in order collect their physiological signals while they are experiencing those emotions and to classify those physiological signals into emotional states with pattern recognition algorithms.

1 Introduction

In the recent years there has been an increasing attempt to develop computer systems and interfaces that recognize the affective states of their users, learn their preferences and personality, and adapt to these accordingly (Bianchi and Lisetti, 2002; Conati and Maclaren, 2009; D'Mello et al., 2008; Hudlicka and McNeese, 2002; Nasoz and Lisetti, 2007; Scheirer et al., 2002). Studies conducted in this direction started with the birth of a new field in Computer Science: Affective Computing (Picard, 1997).

The main motivation behind all this research is that the current theories of cognition suggest there is a strong correlation between affect and cognition (Bower, 1981; Colquitt et al,. 2000; Damasio, 1994; Derryberry and Tucker 1992; Ledoux, 1992) and people emote while they are interacting with computers (Reeves and Nass, 1996); therefore machine perception needs to be able to capture such phenomenon in order to enhance our interaction with computers.

Main focus of our research is to recognize the emotions that users experience while they are interacting with computers and develop Affectively Intelligent User Interfaces that can adapt to these emotions for optimal Human-Computer Interaction. Objectives of our research in general include

- Designing experiments to elicit emotions from participants; collecting and measuring their physiological signals; and recording their facial expressions;
- Analyzing measured physiological signals with pattern recognition and machine learning algorithms to find unique patterns of physiological signals for each emotion;
- Using these patterns to recognize users' emotions in real-time;

M. Kurosu (Ed.): Human Centered Design, HCII 2009, LNCS 5619, pp. 765–774, 2009.
© Springer-Verlag Berlin Heidelberg 2009

- Creating affective models of users that take their emotional states, personality, and preferences into account;
- Develop affectively intelligent user interfaces that combine i) emotion recognition through physiology and ii) affective user modeling to give appropriate feedback to the user about their emotional state and interact with them accordingly.

2 Motivation

There are several possible applications where emotions play an important role and where it is desirable and necessary to develop affective interfaces with an intelligent agent that can recognize and adapt to users' emotional state in the current context.

One of these possible applications is *learning* since it is one of the cognitive processes that is affected by people's emotional states (Goleman, 1995). *Frustration*, for example, leads to a reduction in the ability to learn (Lewis and Williams, 1989). It can also lead to negative attitudes towards the training environment and reduce a person's belief in his or her ability to do well in the learning or the training task. As a result, frustration can hamper learning.

Rozell and Gardner's study (2000) pointed out that when people have negative attitudes towards computers, their self-efficacy toward using them decreases, which then reduces their chances of performing computer-related tasks (when compared to those with positive attitudes towards computers). This research also emphasized that individuals with more positive affect exert more effort on computer-related tasks.

Another emotion that influences learning is *anxiety*. In training situations, anxiety is presumed to interfere with the ability to focus cognitive attention on the task at hand because that attention is preoccupied with thoughts of past negative experiences with similar tasks, in similar situations (Martocchio, 1994; Warr and Bunce, 1995). It follows that learning may be impaired when trainees are experiencing high levels of anxiety during training. Indeed, with a sample of university employees in a microcomputer class, Martocchio (1994) found that anxiety was negatively related to scores on a multiple choice knowledge test at the end of training. In addition, individuals who had more positive expectations prior to training had significantly less anxiety than individuals who had negative expectations of training.

Anxiety also appears to influence reactions to training. For example, with a sample of British junior managers enrolled in a self-paced management course, Warr and Bunce (1995) found that task anxiety was positively related to difficulty reactions in training. Individuals who experienced high task anxiety perceived training to be more difficult than individuals who experienced low task anxiety. In this study, interpersonal and task anxiety were assessed prior to training. Task anxiety was significantly higher than interpersonal anxiety and only task anxiety was associated with difficulty reactions. Similarly, in their meta-analytic path analysis, Colquitt et al. (2000) reported that anxiety was negatively related to motivation to learn, pre-training self-efficacy, post-training self-efficacy, learning, and training performance.

In summary, the most consistent findings are that frustration and anxiety are negatively related to self-efficacy, motivation, learning, and training performance. In addition, social anxiety may influence training outcomes when trainees are taught new tasks as a team. Furthermore, facilitating a mastery orientation towards the task may help to reduce the anxiety (e.g., attitude change) experienced during training and allow trainees to focus their cognitions on the task at hand, resulting in better learning (Martocchio, 1994).

In traditional classroom environments students are educated by live human teachers. This setting allows maximum level of natural interaction between the students and the teachers. Students have the opportunity to ask questions in real-time when they need more clarification or more examples. Teachers, on the other hand, can actually tell when their students are anxious, confused, frustrated, or bored and adapt their interaction with the students or adjust the pace of their teaching accordingly to accommodate different students.

In conventional electronic learning (e-learning henceforth) applications however there is no mechanism to assess students' various emotional states that may negatively affect their learning experience. There are several e-learning applications that employ user modeling in order to cater to the specific needs of different students with varying degrees of existing knowledge, skill levels, goals, and motivation levels (Barket et al., 2002; Corbett et al, 2000; Millan and Perez-de-la Cruz, 2002; Selker, 1994). However learning is a cognitive process that can be affected by variety of factors and a very important factor that these e-learning applications do not include in their user models is the students' affective state.

The strong correlation between students' learning performance and some of the affective states they experience (such as frustration and anxiety) makes it necessary to develop a mechanism for e-learning applications to recognize their users' affective states and interact accordingly. The main objective of the research discussed in this paper is to create affectively intelligent computer systems that can recognize learning related emotions of students through physiology; learn their preferences and personality through user modeling; and interact with the students by adapting to those emotions and student-dependent factors.

Affectively intelligent user interfaces we aim to develop for e-learning applications will enhance presence and co-presence for students and trainees in the learning environment. For example when the system recognizes that the learner is anxious, in response, it might provide encouragement in order to reduce anxiety and allow the individual to focus more attention on the task. Similarly, when the system recognizes the learner as being frustrated or bored it might adjust the pace of the training accordingly so that the optimal level of arousal for learning is achieved. Finally, when the system recognizes that a person is confused it might clarify the information just presented. All these adaptation techniques will improve the learner's sense of being in a real classroom environment where a live instructor would typically recognize these same emotions and respond accordingly.

3 Research Methodology

Our research methodology is two-fold:

- Recognizing students' emotional states by measuring their physiological signals and analyzing them with pattern recognition algorithms; and
- Through user modeling adapting the e-learning interface to the negative emotion of the student by also considering other student-dependent factors such as motivation, personality, preferences, knowledge and skill levels.

Our current method of emotion recognition is through analyzing Autonomic Nervous System (ANS) arousal. Students' physiological signals such as heart rate/blood volume pressure, galvanic skin response, temperature, and respiration are measured with wearable computers and they are classified into their corresponding emotion classes via pattern recognition algorithms. This is our choice of emotion recognition method because although people might be able to control their facial expressions, their vocal intonation, or natural language; they have minimal control over their ANS arousal, which makes it a trustable mode of input.

3.1 Experiment Design

The first step in developing the affectively intelligent user interfaces for e-learning applications is building the mechanism and the algorithms that will recognize students' emotional states. For this purpose we designed an experiment that aims to find a mapping between students' physiological signals and the learning-related emotions that they experience. Non-invasive wearable computers are used to measure participants' physiological signals such as heart rate/blood volume pressure, galvanic skin response, temperature, and respiration; and pattern recognition algorithms are employed to classify these physiological signals into learning related emotions.

Procedure. Participants of this experiment are University of Nevada, Las Vegas (UNLV) students who have taken a specific course within the last year. For this experiment a user interface is developed to administer an electronic test, which is consisted of multiple choice problems from the topics of that course they have taken within the last year. The students are given this test and for incentive, they are told that they will be given compensation in the form of a check for every problem they solve correctly and prove their answer on paper.

With this study we aim to elicit frustration from the participants so that we can measure their physiological signals while they are experiencing this emotion. Students are told that they will be compensated for each problem only if they choose the correct answer for that problem and prove their answer on paper. In order to frustrate them, the list of possible answers for some questions won't include the correct answer; therefore even if the students solve the problem correctly they won't be able to find the answer in the choices listed, which will lead to frustration. In addition, the use interface that is administering the multiple choice questions is running a faulty algorithm and is programmed to be occasionally and randomly non-responsive, in order to frustrate students even further.

Before the study, each participant is given an informed consent form to read. After they complete reading the informed consent form, they are presented with the experiment set-up, shown the non-invasive wearable computer that will be used to collect physiological data and explained how it works, and are informed about the compensation process. Then, they are given the opportunity to ask any questions they might have regarding the study and the procedures.

If they agree to participate in this experiment, then:

- They are asked to sign the informed consent form;
- They are asked to fill out the pre-experiment questionnaire;
- They are asked to put on the wearable computer that will collect their physiological signals during the experiment;

- They are presented with the test that is consisted of several problems from the topics of the specific course they have taken within the last year. They are given a specific time limit to solve each problem. Students are told that those problems are multiple choice questions with one correct answer listed among the choices. They are also told that they are required to solve the problem and prove their answer on paper to be compensated for that problem;
- Students' facial expression are recorded during the experiment with a standard web camera for annotation purposes only;
- Once they complete the test, they are asked to fill out the post-experiment questionnaire;

After completing the experiment students are fully debriefed and thanked for their time and participation.

3.2 Placement of Electrodes for All Measures

Participants' physiological signals including heart rate/blood volume pressure, galvanic skin response, temperature, and respiration are measured with one of the two non-invasive wearable computers: BodyMedia SenseWear armband or ProComp Infiniti 8.

BodyMedia SenseWear Armband: BodyMedia SenseWear armband is a completely wireless armband that is placed around the upper arm and can measure galvanic skin response and skin temperature. It also works in compliance with Polar WearLink coded chest transmitter, which collects heart rate data and communicates with the armband wirelessly. The chest strap is placed around the chest. The participants are shown how to wear the chest strap and they are given their privacy so that they can put it on themselves.

ProComp Infiniti 8: ProComp Infiniti 8 is a biofeedback and neurofeedback system that has 8 protected pin sensor inputs with two channels sampled at 2048 samples per second and six channels sampled at 256 samples per second. The sensors are connected to device on one end and the participant on the other.

Heart Rate/Blood Volume Pressure Sensor:
It is placed against the palmar surface of a fingertip with an elastic strap (not tight so as to cut off blood blow) or a small length of adhesive tape. It is very movement sensitive; therefore it can also be placed to an ear lobe with double sided adhesive tape.

Skin Conductance (Galvanic Skin Response Sensor):
The skin conductance sensor has two short leads that extend from the circuit box. At the end of each lead is a sensor snap similar to those on the extender cables. Each sensor strap is fastened around a fingertip tightly enough so the sensor surface is in contact with the finger pad but not so tightly that it limits blood circulation. The electrodes face against the palmar surfaces of the fingertips.

Temperature Sensor:
The temperature sensor is a 0.125 inch bead thermistor that can detect the temperature of the tissue (skin) on which it is applied. It is placed on a finger pad and held lightly in place by a hook and loop fastener ring.

Respiration Sensor:
The respiration sensor is sensitive to stretch. It is strapped around the chest or abdomen and it converts the expansion and contraction of the rib cage or abdominal area, to a rise and fall of the signal. It can be placed on top of clothing as long as clothing is not too bulky.

3.3 Software Used in the Experiment

In this experiment each participant is given an electronic test, which is consisted of several multiple choice problems from a UNLV course that they have taken within the last year. Students are given a specific time limit for each problem and they are told that they will be compensated for every problem they answer correctly and prove their answers on paper. Students are presented with the problems and choose and submit their answers through a graphical user interface (Figure 1). They will also be asked to show all their work and prove their answers on paper and told that compensation cannot be awarded otherwise.

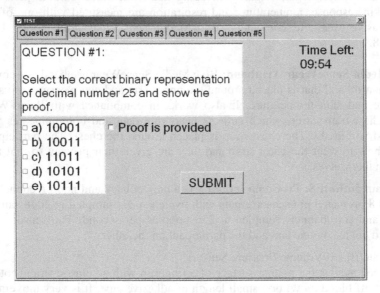

Fig. 1. Graphical User Interface Used to Administer the Electronic Test

In order to frustrate the students some of the problems do not have the correct answer among the list of provided answers. Also students are given limited amount of time for each problem. Once the time is up for a problem the system automatically displays the next problem and does not allow the student to go back to the previous one.

Remaining time for each question blinks on the interface to pressure the students and to frustrate them as the time passes. The interface will also has a faulty working algorithm, where sometimes the student needs to click several times (random number of times up to six) to select the desired choice. Students are subjected to similar difficulty while using the tabs to select a different problem and while clicking on the submit button to submit their answers, which are all designed to elicit further frustration from students.

3.4 Deception/Debriefing

In this study, misleading information is presented to the participants with the informed consent form in order to ensure the elicitation of frustration from them. This misleading information includes:

- Every multiple choice problem that will be presented to the participants has the correct answer listed among the list of possible answers;
- The software that is used to administer the test runs properly;
- Participants will be given compensation only for problems they solve correctly and prove their solution on paper.

Where in fact,

- For some problems in the test, the correct answer will not be listed among the list possible answers in order to deliberately frustrate the participants;
- The software that is used to administer the test has a faulty algorithm, which occasionally and randomly causes the interface be non-responsive;
- Participants will be given compensation for i) every question they submit an answer for and show/prove corresponding solution on paper and ii) all deceptive problems they attempt to solve.

For this experiment, misleading the participants is a necessary action in order to properly elicit the targeted learning related emotions. If the participants knew that the correct answers were not included in the list of choices for some of the problems or that the user interface is deliberately programmed to be non-responsive, not being able to find the correct answer or not being able use the interface efficiently would not frustrate them. Similarly, if they knew that they would be compensated for every problem they attempt to solve and submit an answer for, they would not try to solve the problems correctly, in turn would not get frustrated when they couldn't solve them.

Although misleading the participants makes this a deceptive research study, this deception does not expose the participants to more than minimal risk. The biggest expected effect of this deception is eliciting frustration from the participants, which is the desired outcome of this study.

In order to compensate for the deceptive nature of this study, at the end of the procedure the participants are fully debriefed with the correct information and they are given the opportunity to ask any questions they might have regarding this misleading information that was given to them. After their questions are answered they are also given the option to withdraw their participation and data from the study without affecting their compensation for the study. If they choose to do so, all data collected from that participant is destroyed immediately.

3.5 Emotion Recognition with Pattern Recognition Algorithms

We implemented pattern recognition and neural network algorithms, including k-nearest neighbor algorithm (Mitchell, 1997), discriminant function analysis (Nicol and Pexman,

1999), Levenberg-Marquardt backpropagation algorithm (Hagan and Menhaj, 1994), and resilient backpropagation algorithm (Riedmiller and Braun, 1993), to analyze the physiological data that is collected with the wearable computers while the participants are interacting with the system. The participants are observed while they are solving the questions in order to time stamp the points where they become frustrated because they cannot solve a particular question. The points where the faulty algorithm of the interface randomly becomes unresponsive are also time stamped.

After determining the time slots corresponding to the events in the experiment where the intended emotion was most likely to be experienced, physiological data corresponding to those time slots are determined. Features including minimum, maximum, average, median, and standard deviation are extracted from that data. Values for the extracted features of the physiological signals are then normalized and stores in a in a three-dimensional array. The three dimensions are the participant; the time stamped events during the experiment and extracted features. Extracted features are normalized using the following sample formula, which shows how average value of galvanic skin response is normalized:

$$normalized\_avg\_GSR = \frac{emo\_avg\_GSR - relax\_avg\_GSR}{relax\_avg\_GSR} \quad (1)$$

Normalized extracted features are then inputted to the pattern recognition algorithms as a 20-tuple (i.e. 4*5, where 4 is the number of types physiological signals, which are heart rate/blood volume pressure, galvanic skin response, skin temperature, and respiration; and 5 is the number extracted features, which are minimum, maximum, average, median, and standard deviation). We initially look at the collection of data samples in order to classify them into emotional states as opposed to analyzing each data signal individually. Further analysis will be performed to look at the effect of different emotional states on each physiological signal.

In this study, we try to differentiate both between emotions and between relaxed, non-aroused, non-emotional states versus tense, aroused, emotional states. Percentage of accuracy is expected to get lower when distinguishing between finer grained emotions.

3.6 User Modeling

Once the system recognizes the emotional states of students, it adapts to those states through students' user models that combine student dependent specifics such as knowledge, goals, and personality. User models in our affectively intelligent e-learning application are developed using Bayesian Belief Networks (BBN) (Pearl 1988). BBNs are our choice of user modeling approach due to the fact that unlike the traditional rule-based expert systems, BBNs are able to represent and reason with uncertain knowledge and they can update a belief in a particular case when new evidence is provided.

4 Conclusion

With the affectively intelligent learning interfaces we are developing, we aim to enhance presence and co-presence in learning environments by designing the system to recognize the students' affective states and adapt its interaction in order to aid their learning. For example once the system learns a student's skills, goals, personality, and preferences and recognizes her/his anxiousness during the interaction with the e-learning application, in response, it can provide their preferred style of encouragement, thus potentially reducing anxiety and allowing the them to focus more attention to the task at hand. Similarly, when the system recognizes that the learner is becoming frustrated or bored, it could adjust the pace of the training accordingly so that the optimal level of arousal for that student's learning is achieved. In this manner, the system will provide assistance to students in order to foster positive attitudes and emotions toward the e-learning application therefore will enhance their learning experience. All these adaptation techniques will improve the student's sense of being in a real classroom environment where a live instructor would typically recognize these same emotions and respond accordingly, thus enhancing presence.

References

Barker, T., Jones, S., Britton, J., Messer, D.: The Use of a Cooperative Student Model of Learner Characteristics to Configure a Multimedia Application. User Modeling and User-Adapted Interaction 12, 207–224 (2002)

Bianchi, N., Lisetti, C.L.: Modeling Multimodal Expression of User's Affective Subjective Experience. User Modeling and User-Adapted Interaction 12(1), 49–84 (2002)

Bower, G.: Mood and Memory. American Psychologist 36(2), 129–148 (1981)

Colquitt, J.A., LePine, J.A., Noe, R.A.: Toward an integrative theory of training motivation: A meta-analytic path analysis of 20 years of research. Journal of Applied Psychology 85, 678–707 (2000)

Conati, C., Maclaren, H.: Empirically Building and Evaluating a Probabilistic Model of User Affect. User Modeling and User-Adapted Interaction (to appear, 2009)

Corbett, A., McLaughlin, M., Scarpinatto, S.C.: Modeling Student Knowledge: Cognitive Tutors in High School and College. User Modeling and User Adapted Interaction 10, 81–108 (2000)

Damasio, A.: Descartes' Error. Avon Books, New York (1994)

Derryberry, D., Tucker, D.: Neural Mechanisms of Emotion. Journal of Consulting and Clinical Psychology 60(3), 329–337 (1992)

D'Mello, S., Jackson, T., Craig, S., Morgan, B., Chipman, P., White, H., Person, N., Kort, B., el Kaliouby, R., Picard, R.W., Graesser, A.: AutoTutor Detects and Responds to Learners Affective and Cognitive States. In: Workshop on Emotional and Cognitive Issues at the International Conference of Intelligent Tutoring Systems, Montreal, Canada, June 23-27 (2008)

Goleman, D.: Emotional Intelligence. Bantam Books, New York (1995)

Hagan, M.T., Menhaj, M.B.: Training Feedforward Networks with the Marquardt Algorithm. IEEE Transactions on Neural Networks 5(6), 989–993 (1994)

Hudlicka, E., McNeese, M.D.: Assessment of User Affective and Belief States for Interface Adaptation: Application to an Air Force Pilot Task. User Modeling and User-Adapted Interaction 12, 1–47 (2002)

Ledoux, J.: Brain Mechanisms of Emotion and Emotional Learning. Current Opinion in Neurobiology (2), 191–197 (1992)

Lewis, V.E., Williams, R.N.: Mood-congruent vs. mood-state-dependent learning: Implications for a view of emotion. Special issue of the Journal of Social Behavior and Personality 4, 157–171 (1989)

Martocchio, J.J.: Effects of conceptions of ability on anxiety, self-efficacy, and learning in training. Journal of Applied Psychology 79, 819–825 (1994)

Millan, E., Perez-de-la Cruz, J.L.: A Bayesian Diagnostic Algorithm for Student Modeling and its Evaluation. User Modeling and User-Adapted Interaction 12, 281–330 (2002)

Mitchell, T.M.: Machine Learning. McGraw-Hill Companies Inc., New York (1997)

Nasoz, F., Lisetti, C.L.: Affective User Modeling for Adaptive Intelligent User Interfaces. In: Proceedings of 12th International Conference on Human-Computer Interaction, Beijing, China (July 2007)

Nicol, A., Pexman, P.M.: Presenting Your Findings: A Practical Guide for Creating Tables. American Physiological Association, Washington (1999)

Pearl, J.: Probabilistic Reasoning in Expert Systems: Networks of Plausible Inference. Morgan Kaufmann Publishers, Inc., San Francisco (1988)

Reeves, B., Nass, C.I.: The Media Equation: How People Treat Computers, Television, and New Media Like Real People and Places. Cambridge University Press, Cambridge (1996)

Picard, R.W.: Affective Computing. MIT Press, Cambridge (1997)

Riedmiller, M., Braun, H.: A direct adaptive method for faster backpropagation learning: The RPROP algorithm. In: Proceedings of the IEEE International Conference on Neural Networks, San Francisco, CA (1993)

Rozell, E.J., Gardner, W.L.: Cognitive, motivation, and affective processes associated with computer-related performance: A path analysis. Computers in Human Behavior 17, 199–222 (2000)

Scheirer, J., Fernandez, R., Klein, J., Picard, R.W.: Frustrating the User on Purpose: A Step Toward Building an Affective Computer. Interacting with Computers 14(2), 93–118 (2002)

Selker, T.: A Teaching Agent that Learns. Communications of the ACM 37(7), 92–99 (1994)

Warr, P., Bunce, D.: Trainee characteristics and the outcomes of open learning. Personnel Psychology 48, 347–475 (1995)

Design of a Web-Based Symptom Management Intervention for Cancer Patients

Christine M. Newlon[1], Chin-Chun A. Hu[1], Renee M. Stratton[2],
and Anna M. McDaniel[1,2]

[1] Indiana University School of Informatics, 535 W. Michigan, Indianapolis, IN 46202
[2] Indiana University School of Nursing, 1111 Middle Drive NU 340, Indianapolis, IN 46202
{cnewlon,chinghu,rstratto,amcdanie}@iupui.edu

Abstract. The discipline of Human-Computer Interaction design has potential for significant benefit to the field of health informatics. This paper describes the design approach used to develop a web-based interface to help cancer patients manage their chemotherapy side effects. Previous versions of this intervention utilizing telephone technology had been efficacious, but limited. The paper discusses the design decisions made in order to leverage the potential benefits of the Internet in supporting patients while avoiding the potential pitfalls that the patients may encounter with a web-based approach.

Keywords: Human-computer interaction, iterative design, cancer, web-based intervention, symptom management, evidence-based practice, reading level, continuous evaluation.

1 Introduction

Human-computer interaction (HCI), as a design philosophy, is an approach likely to yield significant benefits to the growing field of health informatics. This interactive design technique has increased our ability to develop new interventions and translate existing interventions, which can be either delivered within the health care system or extended through the Internet into people's homes. During an HCI design process, designers and users collaborate to identify the requirements of an application. The interface is then designed and implemented through an iterative process with feedback from the users at each stage [1]. Current conditions favor such application development in the field of healthcare. Behavioral scientists have already developed and tested efficacious interventions to support patients and families before, during, and long after their diagnosis [2, 3, 4, 5]. The next essential step is to translate these interventions for patients and families struggling with disease, and ultimately for the general population, through the use of interaction design and web-based technology. This paper describes the human computer interaction design approach used to develop one such application, a web-based interface to help cancer patients manage their chemotherapy side effects.

M. Kurosu (Ed.): Human Centered Design, HCII 2009, LNCS 5619, pp. 775–784, 2009.
© Springer-Verlag Berlin Heidelberg 2009

1.1   A Challenge in Symptom Management

Cancer is one of the most prevalent chronic illnesses in this country. Management of patient symptoms during chemotherapy is a challenge in clinical oncology, requiring a significant amount of time and effort to communicate with patients and families. Self-management of symptoms during cancer treatment can be a considerable burden that can negatively impact quality of life for cancer patients and their families. Therefore, an efficacious intervention to support symptom management has the potential to increase the efficiency and effectiveness of clinical care [6].

1.2   Symptom Management Interventions

The symptom management intervention described in this paper has grown through several iterations, each carefully studied following the principles of evidence-based practice. Initially this symptom management concept was implemented both through a series of phone calls by a nurse, and through an Automated Telephone Symptom Management (ATSM) system [6]. In this clinical trial, 435 patients were randomized to receive either automated calls or a similar intervention delivered over the telephone by trained cancer nurses. After reporting their symptoms over the phone, patients were referred to a paper-based "toolkit" of self care strategies for managing those symptoms that were reported to be above a pre-established threshold. Both arms produced significant reduction in symptoms over baseline while the ATSM intervention was more cost effective than the telephone-based nurse counseling [6]. However, the automated telephone system posed limitations, both from a technical stand point and from a usability perspective (Table 1.) Advances in web technology could offer potential solutions for these problems. The current project begins a new iteration, designed to translate the self-management intervention to a web-based platform.

Table 1. Limitations posed by ATSM system [6]

Perspective	Limitations
Technical	• An increased number of phone lines were needed as patients were added.
Effectiveness	• Patients were required to follow the pace of the system. • There was no way to change incorrect responses.
Efficiency	• Option choices were repeated, which increased call time. • Resumption of symptom reporting after interruptions was complicated. • Patients were required to remember/write down a list of significant symptoms and refer to a secondary source (toolkit) for self-care tips.
Accessibility	• Patients were required to be available at a predetermined time and number. • The contact method was limited to one patient-provided number. • It was hard to carry around the paper-based toolkit, limiting access to the symptom management tips. • Symptom history/progress could be monitored only by the health care providers.

2 Background

2.1 Potential Uses of the Internet for Cancer Care

The Internet offers a new paradigm for cancer care delivery. Cancer is one of the most prevalent chronic illnesses in this country. Scientific and medical advances in the detection, prevention, and treatment of cancers have resulted in a decreased incidence of new cancers and an overall decline in cancer death rates, thus increasing the number of people living with cancer. Survivors and their families have significant information needs concerning such things as diagnosis and treatment, management of symptoms, coordination of care, and prevention of further problems [7].

Web technology has the potential to dramatically improve the delivery of cancer care. The Internet can provide extensive information regarding treatment and diagnosis. Patients actively seek such information soon after the diagnosis of cancer and before starting treatment [8]. They typically want to know as much information as possible, and may not be satisfied with the amount of information from health care professionals [8]. Therefore, cancer patients and their relatives turn to the Internet as an alternative information source [9]. Their reasons for using the Internet include finding information quickly and easily, sharing their experiences with others, and obtaining health-related information confidentially [9, 10].

Research has demonstrated that increased knowledge of cancer enhances coping in cancer patients [11]. Newly diagnosed cancer patients who access the Internet for health information report feeling empowered to make decisions about their health and the management of their disease [12]. Patients report higher levels of satisfaction with health information obtained on the Internet compared to that from other media such as television, newspaper, magazine, and radio [8].

Therefore, it follows that one potential application of web technology in cancer care is through its use to monitor symptoms and deliver interventions that can improve the quality of life for cancer patients and their families. Cancer patients experience a number of life altering symptoms during treatment. Management of chemotherapy-related symptoms is a significant burden to cancer patients and their families. Information about strategies to reduce symptoms can be complex and difficult to manage. The Internet can be an important source of information and support for symptom management [9, 10, 13]. Additionally, there is a strong positive relationship between Internet use and cancer patients' self-efficacy for participation in care [11].

2.2 Issues of Concern in Internet-Based Intervention

Although web-based sources of cancer-related information and support are increasingly common and have been shown to have a positive impact on patient outcomes [8, 14], there are important issues to consider in designing a cancer support application. First, even though most patients prefer to have as much information as possible to inform their health care decision making, some cancer patients have reported being overwhelmed with too much information [8, 15]. In addition, much of the available information on the Internet is not evidence-based [6] and may be inaccurate or misleading [16]. Unfortunately, little of web-based cancer information is peer reviewed, and few sites differentiate information according to source, quality, or accuracy.

A second issue is the readability of cancer information. Currently, most online information is at a high school (Grade 11-12) or collegiate level (Grade 13+), which is much higher than the average reading level (Grade 8) [16]. An inappropriate reading level could be a barrier to accessing the health information on the Internet or in print. Web-based multimedia technology allows information to be presented in a variety of formats (e.g., audio, and video), independent of reading ability. Furthermore, the Internet has the potential to facilitate the delivery of customized information, which has been shown to increase responsiveness to health messages and may lead to more positive behavior changes [17].

3   A Web-Based Symptom Management Toolkit

It is against this background of both potential and concern that we have undertaken the next iteration of symptom management intervention. Our proposed solution to address the limitations of the ATSM intervention is a secure Symptom Management Toolkit deployed on a web-based platform, which includes a secure server to host the application, a database for content storage, Flash programming to present the interface, algorithms to control sequencing of events, and an authentication system for user identification. Using the Internet for cancer care delivery not only makes possible powerful functions that were previously unavailable with other technology, but also allows patients to access information and perform their symptom management activities anytime and anywhere convenient for them. We describe below how we addressed the potential issues of concern through our design.

3.1   Cancer Patients as the Center of Our Design

It was our intention to develop an easy-to-follow system with helpful functionalities – one that could fully utilize technology to support cancer patients' symptom management effort and increase their awareness of their health condition. To this end, user characteristic variances such as reading level, computer literacy, and cognitive ability, were considered in the design. We also identified and addressed three central user needs: help with symptom management, improved patient-provider communication, and increased coordination of care. The system was intended to be used by patients outside the clinic setting. Therefore, the system's operation is intuitive, requiring a minimum amount of patient learning (e.g., clicking the next button will advance system to the next screen.) Since the system will be used by patients in a real world context, interruptions while performing tasks are possible. The design allows patients to easily resume system activity without having to start from the beginning.

3.2   System Functionality

The system was iteratively designed, beginning with a series of meetings with cancer survivors and health care providers to assess user needs and identify requirements, followed by usability testing of the system prototype to evaluate and improve the design. Functional requirements elicited through this process were translated into interface components. In addition, we designed a feedback component that integrates continuous evaluation into the normal operation of the system.

Symptom Assessment. The system begins each weekly session by sending patients an automated symptom assessment request with an embedded link to the login screen. (If a patient fails to log into the system within 24 hours, the system sends a reminder message.) After logging in, patients are required to assess 9 symptoms commonly experienced by people undergoing chemotherapy (Fig. 1). Through the instructing and conversing interaction modes, patients proceed at their own pace to report symptoms they have experienced, rating the severity of each symptom, and the extent to which it interfered with their enjoyment of life, relationship with others, daily activity, and mood. The assessment is controlled in a linear fashion to ensure data integrity. Then, patients have the opportunity to report more symptoms by selecting from a secondary list of symptoms. Input is progressively saved into the database. Patients can stop at anytime and easily resume from where it was left off the next time they login the system. After assessing their symptoms, patients are presented a Symptom Summary (a bar graph summary of the data entered) from which they can review and edit information they have entered. [NOTE: Patients are instructed to call their oncologist if they are experiencing severe symptoms or have a medical concern.]

Custom Toolkit. Upon assessment completion, the system presents patients with a Custom Toolkit containing self-care strategies for symptoms assessed at a level of 4 or higher out of 10 (Fig. 2). This threshold was derived from multiple studies using the Symptom Experience Inventory items in clinical practice [18]. Content in the toolkit is quality-filtered, evidence-based cancer information. It is written at a 5th grade reading level to support readability. For patients wanting to increase the font size, a text-resizing function is offered. In addition, a print function is offered allowing a printable, text-only version of the symptom management information. [NOTE: Patients may also access information on any of the symptoms from the Full Toolkit, even if they are not experiencing those symptoms at the time.]

Symptom History. Web technology has made it possible to retrieve and display symptom assessment data in a visualized format, allowing both patients and health care providers to easily monitor changes in symptoms over time. Graphical representations of the symptom history are presented in order of the highest to lowest severity rating for the latest assessment to help with decisions on which symptoms to attend to first (Fig. 3).

Provider Alerts. Another function of the system is to immediately notify providers by e-mail when severe symptoms have been reported (rated 7 or higher out of 10). (Since e-mail transmissions are not considered HIPAA compliant, no patient names are included in the e-mail notification.) After receiving an email alert and logging in, providers can review the symptom history of the patient with severe symptoms to determine follow-up actions. They also have the option of printing the report for further follow-up as would be part of their clinical practice patterns when addressing a patient concern.

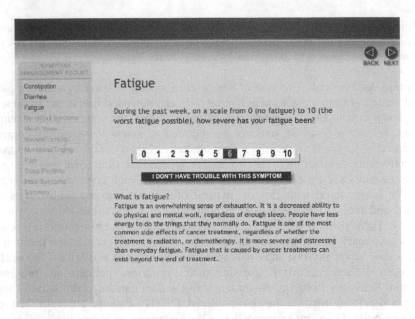

Fig. 1. Symptom Assessment

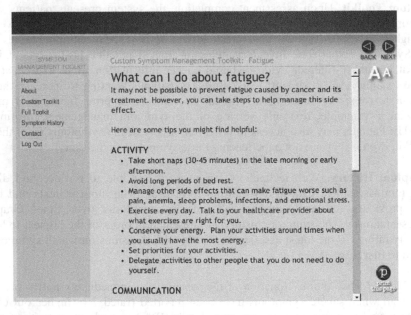

Fig. 2. Custom Toolkit

Continuous Evaluation. Two days after the assessment request, another automated email message is sent asking patients to update the current status of their significant symptoms (those rated at 4 or higher, on a scale of 0-10 in the latest assessment.) This evaluation function allows patients to report symptom status in a three option response format (i.e., worse, about the same, or better) using a radio button interface. This design is intended to emulate a follow-up encounter with a health care provider who is monitoring symptom severity. In addition to the symptom status update, the evaluation function uses this opportunity to gather information from the patients about the utility of the toolkit, and its ease of use. This stream of information will provide the system's designers with the continuous feedback they need to monitor the effectiveness of the toolkit in meeting patient needs.

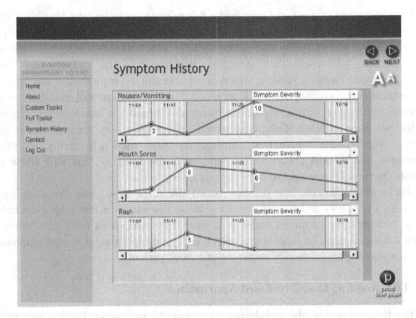

Fig. 3. Symptom History

4 Conclusions

Since the web system essentially serves as a platform for the "translation" of the previously studied ATSM intervention, and that intervention has been shown to be efficacious in a randomized clinical trial [6], the results of that study support our use of technology-driven cancer control interventions. In addition, a recent study by Basch and colleagues showed that online reporting of toxicity during chemotherapy is feasible and is acceptable to cancer patients [19], lending further support for our approach. However, the success of the system will still depend on how successful we have been at avoiding the potential pitfalls of web design. Even with the tremendous potential of the Internet, the process of designing the human-computer interaction still depends on iteration and testing.

4.1 Insights from the Design Process

While our usability test results were good overall, with a slight upward trend as the site was improved, it was instructive to see the results on specific features. For

782   C.M. Newlon et al.

instance, we discovered we needed a button saying, "I do not have this symptom." Otherwise, several of our testers felt obligated to enter a non-zero number for every possible symptom. For them, not having a symptom might be rated as a "2" on our "0-10" scale. This sort of insight is impossible to achieve without having actual users test the application.

4.2 Next Steps

As this paper went to press, we launched a full pilot study to test the feasibility of this web-based symptom monitoring and management system. In this pilot study, the system is being put in use by real patients and in real context. During the eight week study, patients will assess their symptoms and evaluate the system once a week. Throughout the duration, patients will receive customized information that may help relieve their chemotherapy-related symptoms. Their providers will monitor symptom severities, and may further customize treatment based on assessments. From the results, we hope to learn if the customized information was presented in a usable format, whether the customized information was used effectively to manage symptoms, and overall patient satisfaction with the system. This will allow us to validate the new system against our previous interventions.

System enhancements are planned based on responses from the usability test follow-up interviews and the pilot study results. For instance, the responses indicate that a customizable reminder sheet is desired that users can print and bring to a doctor's appointment. In future iterations, a caregiver application is also planned that will provide support for family members of cancer patients.

4.3 Transcending Usability-Based Approaches

For the first version of the web-based Symptom Management Toolkit, our primary focus was usability, and only the most critical functions, supported by minimum visual elements, were incorporated into the system's design. Noting the increased importance of human factors in product creation, we plan to go beyond traditional usability-based approaches in the next version and explore affective approaches to understand cancer patients holistically. In addition to cognitive and physical abilities, we will consider all of the possible hedonic, emotional, and practical benefits offered by this system by further investigating the role it plays in patients' everyday and working lives [20]. Ultimately, our goal is to design a system that is not only usable but also provides comfort to cancer patients.

Acknowledgements. The authors would like to thank Dr. Anthony Faiola at the Indiana University School of Informatics, the development team, Robert S. Comer and Shielly Hartanto, and content experts, Drs. Charles W. and Barbara Given. The project is the result of collaboration between the IU Melvin and Bren Simon Cancer Center, Walther Cancer Foundation, and Indiana University Schools of Nursing, Informatics, and Medicine.

References

1. Helander, M. (ed.): Handbook of human-computer interaction. Elsevier, New York (1997)
2. Kurtz, M.E., Kurtz, J.C., Given, C.W., Given, B.: Symptom cluster among cancer patients and effects of a cognitive behavioral intervention. Cancer Therapy 5, 105–112 (2007)
3. Sherwood, P., Given, B.A., Given, C.W., Champion, V.L., Doorenbos, A.Z., Azzouz, F., Kozachik, S., Wagler-Ziner, K., Monahan, P.O.: A cognitive behavioral intervention for symptom management in patients with advanced cancer. Oncol. Nurs. Forum. 32(6), 1190–1198 (2005) [Online]
4. Rawl, S., Given, B., Given, C.W., Champion, V., Kozachik, S., Barton, D.: Interventions to improve psychological functioning for newly diagnosed patients with cancer. Oncol. Nurs. Forum. 29(6), 967–975 (2002)
5. Given, C.W., Given, B., Rahbar, M., Jeon, S., McCorkle, R., Cimprich, B., et al.: Effect of a cognitive behavioral intervention on reducing symptom severity during chemotherapy. J. Clin. Oncol. 22(3), 507–516 (2004)
6. Sikorskii, A., Given, C.W., Given, B., Jeon, S., Decker, V., Decker, D., Champion, V., McCorkle, R.: Symptom management for cancer patients: a trial comparing two multimodal interventions. J. Pain Symptom Manage. 34(3), 253–264 (2007)
7. Hewitt, M., Grennfield, S., Stoval, E. (eds.): From cancer patient to cancer survivor: Lost in Transition. National Academies Press, Washington (2006)
8. Eysenbach, G.: The impact of the Internet on cancer outcomes. CA Cancer J. Clin. 53(6), 356–371 (2003)
9. National Cancer Institute. About Cancer Survivorship Research: Survivorship Definitions, http://dccps.nci.nih.gov/ocs/definitions.html
10. Satterlund, M.J., McCaul, K.D., Sandgren, A.K.: Information gathering over time by breast cancer patients. J. Med. Internet Res. 5(3), e15 (2003)
11. Kyngas, H., Mikkonen, R., Nousiainen, E.M., Rytilahti, M., Seppänen, P., Vaattovaara, R., Jämsä, T.: Coping with the onset of cancer: coping strategies and resources of young people with cancer. Eur. J. Cancer Care. 10(1), 6–11 (2001)
12. Bass, S.B., Ruzek, S.B., Gordon, T.F., Fleisher, L., McKeown-Conn, N., Moore, D.: Relationship of Internet health information use with patient behavior and self-efficacy: experiences of newly diagnosed cancer patients who contact the National Cancer Institute's Cancer Information Service. J. Health Commun. 11(2), 219–236 (2006)
13. Raupach, J.C., Hiller, J.E.: Information and support for women following the primary treatment of breast cancer. Health Expect. 5(4), 289–301 (2002)
14. Mills, M.E., Sullivan, K.: The importance of information giving for patients newly diagnosed with cancer: a review of the literature. J. Clin. Nurs. 8(6), 631–642 (1999)
15. Leydon, G.M., Boulton, M., Moynihan, C., Jones, A., Mossman, J., Boudioni, M., McPherson, K.: Cancer patients' information needs and information seeking behaviour: in depth interview study. BMJ 320(7239), 909–913 (2000)
16. Friedman, D.B., Hoffman-Goetz, L., Arocha, J.F.: Health literacy and the world wide web: Comparing the readability of leading incident cancers on the Internet. Med. Inform. Internet in Med. 31(1), 67–87 (2006)
17. Kreuter, M.W., Strecher, V.J., Glassman, B.: One size does not fit all: the case for tailoring print materials. Ann. Behav. Med. 21(4), 276–283 (1999)

18. Kurtz, M.E., Kurtz, J.C., Given, C.W., Given, B.: Effects of a symptom control intervention on utilization of health care services among cancer patients. Med. Sci. Monit. 12(7), CR319–24 (2006)
19. Basch, E., Artz, D., Dulko, D., Scher, K., Sabbatini, P., Hensley, M., Mitra, N., Speakman, J., McCabe, M., Schrag, D.: Patient online self-reporting of toxicity symptoms during chemotherapy. J. Clin. Oncol. 23(15), 3552–3561 (2005)
20. Jordan, P.W.: Designing pleasurable products. Taylor & Francis, London (2000)

A Preliminary Usability Evaluation of Hemo@Care: A Web-Based Application for Managing Clinical Information in Hemophilia Care

Vasco Saavedra[1], Leonor Teixeira[1,2], Carlos Ferreira[1,3], and Beatriz Sousa Santos[4,5]

[1] Department of Economics, Management and Industrial Engineering,
University of Aveiro Portugal
[2] Governance, Competitiveness and Public Politics (GOVCOPP),
University of Aveiro Portugal
[3] Operational Research Centre (CIO),
University of Lisbon Portugal
[4] Department of Electronics, Telecommunications and Informatics,
University of Aveiro Portugal
[5] Institute of Electronics and Telematics Engineering of Aveiro (IEETA)
Portugal
{vsaavedra,lteixeira,carlosf,bss@ua.pt}@ua.pt

Abstract. In this work, an overall description of the methods used and the results obtained in the on-going evaluation of *hemo@care* is presented. To help understanding the methods and results, we first give an overview of the main functionalities of *hemo@care*, which is a web application to manage the clinical information in hemophilia care, developed to be used by hematologists, nursing staff and patients suffering from hemophilia. Following we described the methods used in this particular evaluation, and finally we present the main results and general conclusions of these preliminary usability evaluation.

Keywords: Health information system, *Hemo@care*, Usability evaluation.

1 Introduction

The main objective of usability testing is to identify usability deficiencies and, at the same time, to create functional products that are easy to use [1]. We considered these ideas when evaluating our application, adapting usability testing techniques to obtain feedback from users while they performed a set of tasks in a health information system (HIS), called *hemo@care*, currently under development. *Hemo@care* is a web application to manage the clinical information in hemophilia care, developed to be used by hematologists, nursing staff and patients suffering from hemophilia. It incorporates an extensive dataset including medical information: medical history, physical examination results, laboratory data, detailed information on the primary diagnosis, symptoms and manifestations, treatments, potential complications, etc., and non-medical information: demographic information, socio-psychological background, etc. In a nutshell, it provides healthcare professionals (HCP) the tools to manage all the essential information regarding patients' data and treatments. Patients are deeply involved in this process too,

M. Kurosu (Ed.): Human Centered Design, HCII 2009, LNCS 5619, pp. 785–794, 2009.
© Springer-Verlag Berlin Heidelberg 2009

as the system also provides the tools for managing their individual treatments, delegating to them the responsibility for data accuracy.

During the development process of this complex web application we used a human-centered design approach within an iterative process, which started by recognizing the potential users, their contexts of use and the tasks they need to perform [2, 3]. After this, the design continued by using formative evaluation along different phases of the development cycle of *hemo@care* with strong end-user participation [4]. Moreover, several well-known usability methods, as heuristic evaluation, observation, questionnaires and interviews, were used in the latter phases of the process: basically we used heuristic evaluation followed by testing with users in a formal evaluation and finishing with informal evaluation, conducting to design revision.

In this paper we present the results of this preliminary usability evaluation and describe their impact in a revision version of *hemo@care*. To help understanding the methods and results presented, we first give an overview of the *hemo@care*, as well as the main modules and functionalities; then we describe the evaluation methods and methodology applied and the main results obtained, and finally we present conclusions and future work.

2 Overview of Hemo@care

The *hemo@care* web application was designed to provide a tool to respond a specific problem regarding management information process in a Hemophilia Treatment Center (HTC). The existence of parallel data creation processes (as home treatment records by patients) followed by the inexistence of secure storing and data forwarding mechanisms to the HTCs is one of these problems. Part of the hemophilia treatments are made in the patient's homes and these treatments data have to be correctly stored and forward to the HTC for patient's data control. Traditionally, patients annotate this data in specific paper datasheets and send them to HTC by traditional mail, fax or deliver in person. On one hand, due to the lack of security of this process, there is a strong possibility of resulting in redundant or incorrect data. On the other hand, the lack of knowledge of this data by the clinical staff may adversely affect clinical decisions concerning the patient's treatment.

Moreover, the lack of automatic stock management of treatment products (Coagulation Factor Concentrate (CFC)) was also another motivation to develop the present tool. The CFC is extremely expensive and to minimize potential waste, patients are advised to control the expiration date of their personal stock, returning to their HTC all products which expiration date is approaching. In fact this process is very hard accomplished without automatic mechanisms.

Aiming to respond to these specific needs, we developed a web application solution (*hemo@care*) that aggregates three different modules: Patient Clinical Data Management (PCDM), Treatment Data Management (TDM) and CFC Stock Data Management (CFCSDM).

- *Patient Clinical Data Management (PCDM) Module* - This module manages the patient's clinical data. Although parts of this data are available from other Information Systems (IS), it is very difficult to obtain aggregated reports to support clinical treatment analysis. Moreover, there is a large number of data related to the specifications of the hemophilia pathology that are not supported by other ISs, and are stored in paper files or in isolated ISs. This module (PCDM) aims to provide physicians with the tools to manage the patient's clinical data, also providing the means to transform the flat information records stored in other ISs in relevant information which can be aggregated in several different views.

- *Treatment Data Management (TDM) Module* - This module is responsible for the management of the patient treatment information lifecycle, which data is generated in the scope of CFC treatments. There are three actors that interact with this module: the nurse, the patient and the physician, Fig.1.

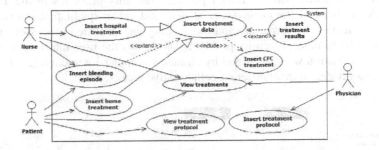

Fig. 1. Generic use case diagram of the TDM module

The nurse and the patient are responsible by inserting treatments result data, respectively, in a hospital and homemade regime, including the bleeding episode associated and the CFC administrated in the scope of that treatment. The physician controls the treatment evaluation process, consulting the related information and is responsible for the insertion of the specific treatment protocol.

- *CFC Stocks Data Management (CFCSDM) Module* - In a nutshell, this module provides the tools for managing the stocks of the products used in the hemophilia treatment, specifically the CFC. This module is integrated with the TDM, providing an automatic management of CFC products. Two actors interact with this module: the nurse and the patient, Fig.2.

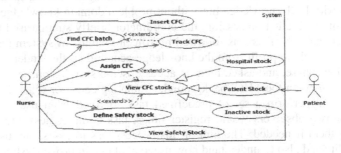

Fig. 2. Generic use case diagram of the CFCSDM module

These three different modules (meant to be used by physician, nurses and patients) were evaluated in the latter phases of the development process using different approaches. While, the physician and nurse components were evaluated in the field with a small group of actual end-users (hematologists and nurses) through observation and interviews, the user interface corresponding to the patient and nurse component was evaluated in the laboratory using various methods.

3   Study Protocol and Evaluation Methods

The aim of this study was to identify potential usability problems in the TDM and CFCSDM modules of the *hemo@care* system and gather qualitative information to propose guidelines for correcting and redesigning the identified processes and interfaces. In order to detect usability problems and assess the user interface learnability [5-8], this evaluation was performed by Human-Computer Interaction (HCI) students in a process divided in three different collaborative phases, Fig.3.

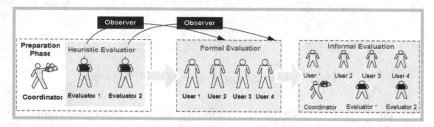

Fig. 3. Evaluation process and participants involved

- The preparation phase involved a coordinator (research collaborator) and two students. Firstly the students acquired a general understanding of the application and next they inspected the system evaluating it based on heuristic evaluation [9]. Each identified usability problem was recorded in a written report, including the problem description, a clear explanation of why it doesn't conform to the usability principals, and a severity rate.
- The second phase of the evaluation (formal evaluation) involved a small number of students with HCI knowledge but without problem domain knowledge. At the beginning of the evaluation session, the coordinator briefly summarised the application functionalities. This was followed by fifteen minutes of system testing by the users. We decide to profit from the knowledge acquired by the students involved in the previous phase, and asked them to act as observers in this phase. Four students were chosen given their HCI background, simulating end-user roles. The latter students had to perform a set of predefined tasks derived from use case diagram, while the two observers would register times, task completeness, etc., as well as assisting them if needed. The aim of this process was to assess the user interface learnability and also to understand how usage could be improved. After all students completed the tasks, a specific questionnaire was presented. This questionnaire was prepared to collect basic demographic data, and a group of questions to assess user reactions and general opinion about the system.

- The last phase of the evaluation involved all six participants (4 users and 2 observers) as well as the coordinator. The main purpose of this phase was to have an informal conversation where students would give their opinions and suggestions on how to improve system usability.

We chose students of HCI course due to their knowledge about usability evaluation that made them reasonably suitable as subjects for the evaluation the user interfaces.

3.1 Tasks in the Formal Evaluation

In the formal evaluation, the tasks to be carried by the users were chosen to represent the most common activities in *hemo@care* system performed by nurse and patient actors. We defined 15 tasks for the users to perform during the observation session. These tasks refer to the components of the treatment data management (TDM) and CFC stock data management (CFCSDM) modules, which will have to be performed by patients and nurses and are described in use case diagrams presented in section 2.

Table 1. Use case and tasks mapping

	Actor	Use Case	Task #	Task Specification
CFCSDM	Nurse	Insert CFC	1	Insert an order of CFC products in the system
	Nurse	Define Safety Stock	2	Define the safety stock of a specific product
	Nurse	View CFC Stock	3	View the details of a specific CFC product
	Nurse	View Safety Stock	4	View all the products which stock level is bellow the safety stock
	Nurse	Track CFC	5	View the tracking details of a specific CFC batch
	Nurse	Track CFC	6	Analyze the tracking details to know in which patients the product was used
	Nurse	Track CFC	7	Analyze the tracking details to know in which treatments the product was used
	Nurse	View Patient Stock	10	View all the patients with CFC products in home
	Nurse	Assign CFC	11	Assign two CFC products to a specific patient
TDM	Patient Nurse	View Treatments	8	View the details of a specific treatment
	Patient	Insert home treatment	9	Insert a post operation treatment with specific orthopedic equipment
	Patient Nurse	View Treatments	12	List all treatments that meet a specific criteria
	Nurse	View Treatments	13	Verify the batch number used in a specific patient in the scope of a specific treatment.
	Nurse	Insert hospital treatment	14	Insert the details of a hospital treatment given to a specific patient.
	Nurse	Insert CFC Treatment	15	Insert the CFC products used in the scope of the task number 14.

Table 1 presents a brief summary of the task performed by users that participated in this evaluation, the corresponding use case and involved actors.

Each task had to be completed in a given time window, defined regarding the user's profile and the perceived reasonable time for the execution of the task. These tasks were monitored by an observer and, throughout the evaluation session. For each task the observer had to register the some data concerning user performance: the time spent performing the task; if the user finished the task; if the user asked for help to perform the task; if the user had made a mistake; the difficulty level perceived by the observer in the user while executing the task and any additional observation considered relevant. Regarding the difficulty level, users' and the observers' answers were measured using a four level scale, level one being the least favorable (very difficult) and level four the most favorable answer (very easy).

3.2 Questionnaire

After performing all tasks, users were invited to answer a brief questionnaire, giving their global opinion about the *hemo@care* system. Beyond demographic data, the questionnaire included a large number of questions regarding the system being evaluated. These questions were aimed to know the users opinion regarding the system use (easiness of use, content consistency, adequate and perceivable functionalities, etc.) and specific aspects of the system (layout, easiness of navigation, clear and appropriated error and information messages, etc.). The questionnaire was elaborated having as base the QUIS (*Questionnaire for User Interaction Satisfaction*) developed in the University of Maryland [10].

4   Results and Discussion

We had the collaboration of six (masculine gender) students. The median age of this group was 22 years. Two of these students were trained for the observer role and to perform a heuristic evaluation, and four of them performed the predefined tasks. The results of users' reactions to *hemo@care* are presented in Table 2.

Table 2. Users opinion on general and specific aspects of *hemo@care* in a scale with five levels (1 – complete disagreement; 5 – complete agreement)

Feature	U1	U2	U3	U4
1 – It is easy to find the information	1	2	3	2
2 – The terminology used is consistent	4	5	5	4
3 – Help is needed to perform some tasks	2	3	3	2
4 –The characters (text) are easy to read	4	4	4	5
5 –The information layout is adequate	3	3	5	4
6 –The icons used are intuitive	4	3	4	4
7 –The interface is visually appealing	4	4	4	4
8 –The system is pleasant to use	5	5	5	4

These results convey a positive reaction to the application; except for feature 1 (It is easy to find information) and maybe the feature 3 (Help is needed to perform some tasks).

As a result of the heuristic evaluation, several minor problems were detected. Table 3 presents a brief summary of these problems, the corresponding heuristic and a suggestion to correct the problem and/or an observation.

Table 3. General problems, heuristic and action considered

Problem	Heuristic	Action / Observation
Due to the extremely technical nature of the medical terms, several words and concepts were unfamiliar to the user	**Match between system and the real world**	All the words/descriptions used in the system were stored in specific properties file that was developed with the help of the domain experts (i.e., healthcare professionals).
Several errors were reported when submitting the data	**Error prevention**	All the errors reported were due to a prototype limitation. The system will have two types of error prevention mechanisms: client and server side. The evaluated prototype only had server side validation implemented.
When an error occurred, technical data (error code and description) were displayed.	**Help users recognize, diagnose, and recover from errors**	This was due to a prototype limitation. On an error occurrence, the system will display non technical and informative information, automatically logging the technical information for further analysis.
Although the system displayed an help icon, it doesn't have any functionality	**Help and documentation**	Not implemented in the prototype.

The data collected from tasks observation, revealed minor difficulties in performing some tasks, namely tasks 9, 11 and 15 (Table 1) that presented, in general, a high time of execution. These tasks are concerned with 'insert treatments (9)', 'assign CFC product (11)' and 'insert CFC treatment (15)'. Analyzing the tasks and considering users' comments in the final discussion (informal session), we concluded that for some tasks, the specific operation flow was not the most correct. This aspect was also confirmed by the answers to the questionnaire concerning feature 1 (It is easy to find information) and feature 3 (Help is needed to perform some tasks).

Based on the qualitative analysis of the data obtained through different evaluation components of this study, the *operations' flow complexity* was maybe the most relevant usability problem identified during this preliminary evaluation. To improve it, those flows were reformulated, leading to simpler and easy to use interfaces. In order to demonstrate how this redesign was performed, we present the modifications performed on task 11 – "assign CFC". Fig.4 presents the corresponding operation flow for task 11, before and after reformulation.

As a result of this reformulation, task forms were redesigned too. Fig.5 and Fig.6 present the corresponding forms for task 11, before and after reformulation.

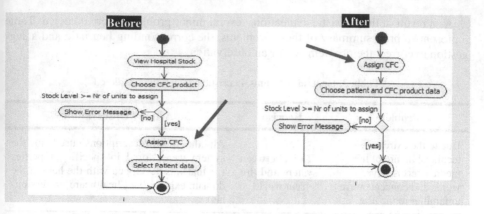

Fig. 4. Example of *operations' flow* for task 11 before and after reformulation

Operation 1 – The user selects "view hospital stock" in the left menu. The system presents the next form where the user performs operation 2;

Operation 2 – The user views all the available CFC and, based on this information, selects the product and clicks in "assign to patient". The system presents the next form where the user performs operation 3;

Operation 3 – The user introduces the patient name (patient number is automatically filled) as well as the amount of CFC delivered to the patient.

Fig. 5. Form sequence to perform task "assign CFC" before reformulation

We concluded that we could achieve the same result by aggregating all the previous information in a single interface, in accordance with of *operations' flow* represented in the Fig.4.

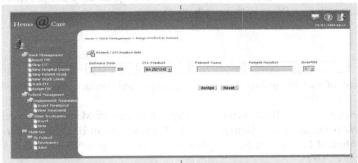

The user can selects directly the task "assign CFC". The three operations presented before are now aggregated in one single operation, Fig. 6.

Fig. 6. Single form to perform the task "Assign CFC" after reformulation

5 Conclusion

In this paper we describe the on-going evaluation of a health information system (HIS), called *hemo@care,* currently under development. *Hemo@care* is a web application to manage the clinical information in hemophilia care, developed to be used by hematologists, nursing staff and patients suffering from hemophilia. During the development process of this HIS we used different techniques and methods illustrating the results and evaluating the design along different phases of the iterative development cycle. The requirements were constantly refined and new requirements were identified. Usually in this type of approach, usability studies start early in the design process, with techniques of formative evaluation, and continue along the development cycle until the end of product development through summative evaluation. In the latter phases of the process we applied several usability evaluation methods (heuristic evaluation, task observation and questionnaire) in order to find usability problems and elicit new ideas that will, eventually, allow us to improve the usability of the system. This latter was the main subject of this work, where we presented the design and results of the preliminary user evaluation and also briefly described the implication of these results in a new version of the *hemo@care* system.

The evaluation results of the data collected from tasks observation revealed minor difficulties in performing some tasks having this aspect also been confirmed by the answers to the questionnaire. Based on the qualitative analysis of the data obtained through different evaluation components including the final informal conversation, the operations' flow complexity was, in fact, the most relevant usability problem identified during this preliminary evaluation.

The questionnaire gave us a general positive idea about the usability of the *hemo@care* system, except in the 'ease of use' dimension.

These results contributed to increment some improvement in new version of *hemo@care* system, namely in the operation flow and tasks. In general form, we are pleased with results obtained, still, we intend to evaluate the new version of *hemo@care* system with a large group of users.

Acknowledgements

The authors wish to thank all students of the Human-Computer Interaction course, of the University of Aveiro, that have graciously collaborated, emphasizing the precious help of the *Hugo Félix* and *José Pinto* in this work.

References

1. Rubin, J.: Handbook of usability testing: how to plan, design, and conduct effective tests. John Wiley, New York (1994)
2. Teixeira, L., Ferreira, C., Santos, B.S.: Web-enabled System Design for Managing Clinical Information. In: Wickramasinghe, N., Geisler, E. (eds.) Encyclopedia of Healthcare Information Systems. Medical Information Science Reference, vol. III, pp. 1398–1406. IDEA Group Inc., Hershey (2008)
3. Teixeira, L., Ferreira, C., Santos, B.S., Martins, N.: Modeling a Web-based Information System for Managing Clinical Information in Hemophilia Care. In: Proceedings of 28th Annual International Conference of the IEEE Engineering in Medicine and Biology Society (EMBS), pp. 2610–2613. IEEE CNF, NY (2006)
4. Teixeira, L., Ferreira, C., Santos, B.S.: Using Task Analysis to Improve the Requirements Elicitation in Health Information System. In: Proceedings of 29th Annual International Conference of the IEEE Engineering in Medicine and Biology Society (EMBS), pp. 3669–3672. IEEE CNF, Lyon (2007)
5. Dix, A., Finley, J., Abowd, G., Russell, B.: Human Computer Interaction. Pearson Education, Harlow (2004)
6. Mitchell, P.: A Step-by-step Guide to Usability Testing. iUniverse (2007)
7. Peute, L.W.P., Jaspers, M.W.M.: The significance of a usability evaluation of an emerging laboratory order entry system. International Journal of Medical Informatics 76, 157–168 (2007)
8. Preece, J., Rogers, Y., Sharp, H., Benyon, D., Holland, S., Carey, T.: Human Computer Interaction. Addison-Wesley, UK (1994)
9. Nielsen, J.: Ten Usability Heuristics (2005), http://www.useit.com/papers/heuristic/heuristic_list.html (visited on 21/02/2009)
10. Shneiderman, B.: Designing the User Interface - Strategies for effective Human-Computer Interaction. Addison-Wesley, Reading (1998)

Fundamental Studies on Effective e-Learning Using Physiology Indices

Miki Shibukawa[1], Mariko Funada[1], Yoshihide Igarashi[2], and Satoki P. Ninomija[3]

[1] Hakuoh University, 1117 Daigyouji, Oyama, Japan
[2] Professor Emeritus, Gunma University, 1-5-1, Tenjin-cho, Kiryu, Japan
[3] Professor Emeritus, Aoyama Gakuin University, 5-10-1 Fuchinobe, Sagamihara, Japan

Abstract. In order to apply individual learning methods to an e-learning system, we need some appropriate measures to know the quantitative evaluation for the learning progress of each individual. The ratio of the number of correct answers to the number of questions is a simple measure of the achievement of the learner. However, such a simple measure may not accurately reflect the real progress of the learner. Event Related Potentials (ERPs for short) are measured from electroencephalograms (EEGs for short). We consider that ERPs may contain meaningful information about the level of the learner's achievement. We had experiments measuring ERPs of subjects learning chemical formulae on an e-learning system. We try to characterize the relation among the learner's achievement, hardness of learning, and the waveforms of his ERPs. This kind of characterizations may be useful for evaluating the learner's achievement.

Keywords: EEG, event related potential, achievement, learning, chemical formulae.

1 Introduction

In the modern educational world, various learning systems have been introduced. The effectiveness of these learning systems has been intensively studied. We consider that there might be suitable learning methods for individual cases. In order to apply such individual learning methods, we need some appropriate measures to know the quantitative evaluation for the learning progress of each individual. The ratio of the number of correct answers to the number of questions is a simple measure of the achievement of the learner. However, this simple measure may not accurately reflect the real progress of the learner. In general, we cannot decide by such a simple measure whether the learner chose a correct answer with confidence or without confidence.

ERPs are measured from electroencephalograms (EEGs for short) [2]. We consider that ERPs contain meaningful information about the level of the learner's achievement [1][3][4]. We had experiments of measuring ERPs of some subjects learning chemical formulae on an e-learning system. By analyzing the data of ERPs obtained in the experiments, we discuss some fundamental features that may be useful to find suitable and effective methods for individual learners.

M. Kurosu (Ed.): Human Centered Design, HCII 2009, LNCS 5619, pp. 795–804, 2009.
© Springer-Verlag Berlin Heidelberg 2009

2 Experiments

2.1 Experimental Methods

The subjects of the experiments learnt the atomic weight of each of commonly known atoms in advance. In each experiment, given chemical formulae in a monitor display, the subjects calculate the molecular weight of each chemical formula. The details of the experiments are as follows:

(1) Subjects: Three adults (male, 20-21 years old) are involved in the experiments. The discussions in this paper are mainly based on the experimental data of *subject a*, one of the three subjects.
(2) Laboratory: We prepared a laboratory shielded from external stimuli so that the subjects could concentrate on the given tasks during the experiments.
(3) A task: Given chemical formulae (Table 1) in the display, the subjects calculate the molecular weight of each chemical formula. For example, C (image 1), O_2 (image 2) and CO_2 (image 3) are displayed sequentially in this order (see Fig. 1 and Table 1). Then the subjects calculate the molecular weight of CO_2 as a sequence of calculation, the molecular weight of C, the molecular weight of O_2, and the molecular weight of CO_2. The last one is calculated by adding the molecular weights of the first two.

C	O_2	CO_2
image 1	image 2	image 3

Fig. 1. Three images of a chemical formula sequentially displayed

Table 1. Chemical formulae used in tasks

No	image1	image 2	image 3
1	C	O_2	CO_2
2	N	H_3	NH_3
3	H_2	S	H_2S
:	:	:	:
15	Na	Cl	NaCl

(4) Task display: Around the center of a 19-inch monitor, a set of the images are displayed for one second each at a random interval, ranging from 750-1250 msec. A subject is asked to sit at the position 60-80 cm away from the monitor. The subject is able to see the display without eye movements. He gives each molecular weight using a 10-key pad. The time duration from the start of a task shown in the display to the entry of his answer is recorded. This time duration is called the response time of the subject.
(5) Repetition: In each set of experiments, 15 chemical formulae are displayed in random order. The display of each chemical formula is given in the order of image 1, image 2 and image 3. There is one minute time interval between two consecutive

sets of experiments. During the time interval the subject may take a rest. An experimental set consisting of 15 chemical formulae is carried out 5 times a day.

(6) Experimental duration: The time duration for displaying an image is 3 seconds. Each chemical formula is shown in the display as a sequence of 3 images. A set of experiments includes the task for a subject to calculate the molecular weights of 15 chemical formulae. Therefore, each set of experiments needs approximately 135 seconds.

(7) Electroencephalography: According to the International 10-20 system, A_1 and A_2 are used as reference electrodes. Fp_1, Fp_2, C_3, and C_4 are unipolar leads for EEG measurement. Neurofax EEG8310 (Nihon Kohden) is used to measure EEGs. Its cut-off frequency and time constant are 60 Hz and 0.3 seconds, respectively. The experimental data are processed in real time by a Gateway G7-600 computer annexing an A/D converter board.

(8) A/D converter: The sampling frequency for EEG measurement is 1 kHz. The sampled data is digitalized within a second by the A/D converter. Then the digital data is loaded to the computer.

2.2 Methodology of Data Analysis

Initially, we eliminate high-frequency noise as well as low-frequency noise by an adaptive filter. The cut-off frequency of the adaptive filter changes time to time. Each EEG is normalized. Then we obtain the ERPs of a set of experiments by taking the average of 30 normalized waveforms. Furthermore, we average the normalized ERPs of 10 sets. The value obtained in this way is determined to be the representative ERP for the day of the experiments. For each set of experiments we calculate the ratio of correct answers, and plot it in a graph.

3 Experimental Results

For 4-days experiments of *subject a*, ERPs measured at the stage of image 1 are shown in Fig. 2. The horizontal axis of the graph is the time duration (msec) from the start of the task display, while the vertical axis is the amplitude (μV) of the measured potential. Amplitude plotted vertically direction is the average potential of 30 normalized ERPs. We can observe waveforms with P_{100}, N_{200}, P_{300} and N_{400}, where P_{100} and P_{300} are positive peaks and N_{200} and N_{400} are negative peaks. The time duration from the start of the task to a peak in the waveform is called the latency for the peak. We notice that the latency for P_{100}, N_{200} and P_{300} are shortened by repeating experiments. This means that the results can be improved by repeated learning. The average ERPs for image 1, image 2, and image 3 are shown in Fig 3. In Fig. 3, N_{400} and P_{500} do not appear clearly on the plotted curve for image 1. We judge image 3 to be harder than image 1 and image 2, since it requests more additions than others. From the same reason image 2 is harder than image 1. That is, a sequence of images appears in the display in the order of easy one, harder one and hardest one. In particular, the hardness of image 3 reflects the plotted curve from P_{300} to N_{600}.

The change of the ratio of the number of correct answers to the number of questions is shown in Fig. 4. Each vertical value in Fig. 4 shows the correctness ratio for an experimental set. We can notice the tendency of the improvement of the correctness ratio by repeated learning. The correctness ratio remarkably improves from the 4th day. This improvement is due to the fact that the subject usually notices his calculation errors around the 4th experimental day. The correctness ratio is already about 80% at the initial part of the experiments. This means that the subject has already memorized the molecular weights of some well known chemical formulae.

The response time of the subject also changes by repeated learning. This tendency is shown in Fig. 5. It remarkably improves after the 19th day of the experiments, since the subject finds how to simplify a way of giving his answers to the computer around the 11th day. For lack of concentration of the subjects, the subjects vary in their response time to the tasks. The average of all ERPs is shown in Fig. 6. We can observe the peaks, P_{100}, N_{200}, P_{300}, N_{400}, P_{500} and N_{600}, in the waveforms there.

The average of normalized ERPs for each image during the first 3 days and the last 3 days of the experiments is shown in Fig. 7 and Fig. 8, respectively. Since the subject easily recognizes such a simple chemical formula given at the stage of image 1, the amplitude of P_{100} is larger than other peaks. The waveforms for image 2 and image 3 in Fig. 7 and Fig. 8 resemble each other. Since the calculation for image 2 and image 3 requires some numerical additions, N_{400} and P_{500} appear in the waveforms for these images.

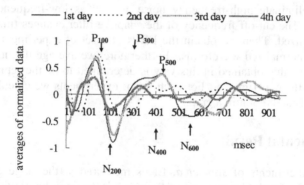

Fig. 2. ERPs of *subject a* measured at C_3

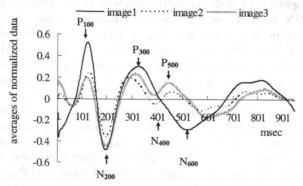

Fig. 3. The average of ERPs of *subject a* for image 1, image 2 and image 3 measured at site C_3

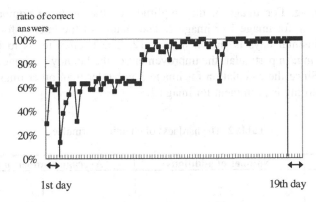

Fig. 4. Correct answer ratio

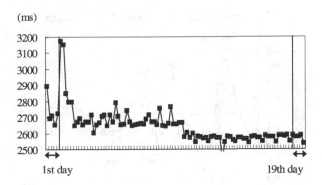

Fig. 5. Answering time

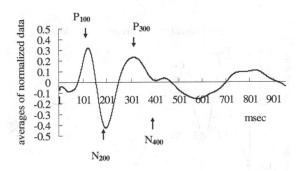

Fig. 6. The average of all ERPs

The averages of normalized ERPs for image 1 during the first 3 days, the intermediate 3 days, and the last 3 days are shown in Fig. 9. The corresponding averaged data for image 2 and image 3 are shown in Fig. 10 and Fig. 11, respectively. For every image, we can notice the tendency that the latency is getting shorter by

800   M. Shibukawa et al.

repeated learning. For image 3, the amplitude of P_{300} is large compared with the amplitude of P_{300} for image 1 or image 2. The change of the latency for N_{400} through all the experimental days is shown in Fig. 12. The latency tends to be shorter by repeated learning. In particular, the improvement of the latency for image 2 and image 3 is notable. Since the calculation for image 3 is harder than other images, the effect of iterative learning is prominent for image 3.

Table 2. The hardness of chemical formulae

images	Number of additions	Average number of digits
image 1	2	2.1
image 2	7	1.9
image 3	15	2.3

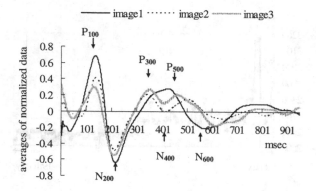

Fig. 7. The average of ERPs for each image during the first 3 days

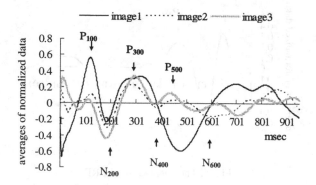

Fig. 8. The average of ERPs for each image during the last 3 days

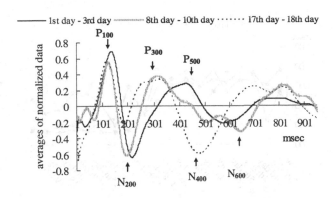

Fig. 9. The average of ERPs for image 1 during each of the first 3 days, the intermediate 3 days, and the last 3 days

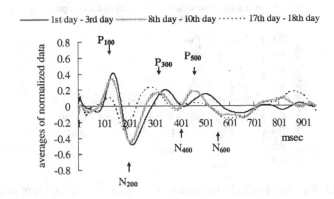

Fig. 10. The average of ERPs for image 2 during each of the first 3, days the intermediate 3 days, and the last 3 days

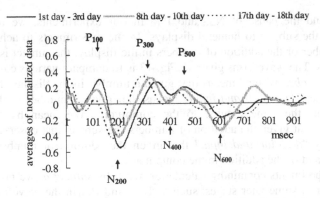

Fig. 11. The average of ERPs for image 3 during each of the first 3 days, the intermediate 3 days, and the last 3 days

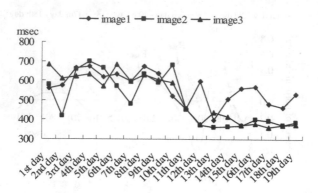

Fig. 12. Change of the latency for N_{400}

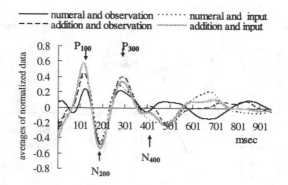

Fig. 13. Waveforms of ERPs for numeral displays and some actions of subject a

4 Discussions

To understand the cause of N_{400} and P_{500} in our experiments, we investigate the response of the subject to numeral displays. In the experiments to achieve this aim, given a number or the addition of numbers in the display, the subject is asked to take some actions. The waveforms given in Fig. 13 is to compare various cases. In Fig. 13, "*numeral and observation*" means that given a number in the display the subject just observes it, and "*numeral and input*" means that given a number in the display the subject inputs the same number to the computer. We mean by "*addition and observation*" that given the addition of numbers the subject just observe the addition. We mean by "*addition and input*" that given the addition of numbers the subject inputs the result of the addition to the computer.

In the experiments containing calculation work by *subject a*, we notice that some peaks appear at some later stages, such as N_{400} and P_{500} in the waveforms of ERPs. The waveforms of ERPs for image 1 of a chemical formula and for numeral input case resemble each other. However, there is the small difference between these waveforms. For image 1 of a chemical formula, a small swelling appears just after

P_{300} in the waveforms, but such a swelling does not appear in the numeral input case. We notice that this small swelling is caused by the calculation work for converting the chemical formula to its molecule weight. The calculation work for image 1 is almost trivial, and N_{400} does not appear in the waveforms for image 1. From this fact, we can consider that for simple tasks requiring just almost trivial calculation, N_{400} does not clearly appear in the waveforms. For image 2 and image 3 of chemical formulae, the subject converts each formula to its molecule weight. This is the main reason why N_{400}, P_{500} and N_{600} appear in the waveforms of ERPs for image 2 and image 3. In the case of *addition and input*, the subject need not the converting work. The latency for this case is shorter than the latency for image 2 and image 3 of chemical formulae. Since tasks for image 3 are harder than other tasks, the amplitude of ERPs for image 3 is larger than others.

We believe that N_{400} and P_{500} in Fig. 10 and Fig. 11 are caused by the addition work in the molecule weight calculation. This belief is based on the comparison among the waveforms shown in Fig. 13. These peaks appear more clearly while the subject engages in calculating molecule weights for image 2 and image 3 than in calculating numeral additions from the numeral values shown in the display.

5 Concluding Remarks

The experimental results discussed in this paper are summarized as follows:

(1) In the experiments, the calculation of the molecule weights consists of three stages (image 1, image 2, and image 3). The ERPs of the subjects are measured at each stage. We can notice some difference of the amplitude and the latency of the waveforms of ERPs among these stages.

(2) For the stages of image 2 and image 3, the subjects are involved in calculation work for addition. During each of these stages, multiple peak potentials appear in the time range between 300 ms to 600 ms from the start of the task.

(3) The order of hardness among the tasks at these stages is image 1, image 2 and image 3 in increasing order.

(4) We can observe eminent potentials, P_{100} and N_{200} in the waveforms of ERPs for image 1. These seem to be visual evoked potentials.

(5) The latency of P_{300} for image 1 is clearly improved by repeated learning. On the other hand, the effectiveness by repeated learning does not clearly appear at the improvement of the latency of P_{300} for image 2 and image 3. For these images, the subjects are involved in some work for numeral addition.

(6) From the discussions about the analysis of ERPs, we can predict that the level of achievement of the learner can be evaluated by his ERPs.

As described above, from ERP measurements in our experiments, we can observe some difference among achievements of learning chemical formulae by the subjects. Our experimental results suggest a possibility that information obtained from ERPs can be directly applied to the problem of how to know the brain status of the learner. This kind of information may be useful to develop the effective e-learning methods. However, our experiments shown in this paper have some serious problems. For example, we did not unify the hardness of the chemical formulae used in the tasks.

We did not also clarify the relationship between the hardness of a task and ERPs measured from subjects. It might be worthy to resolve these problems by further investigation.

References

1. Funada, M., Shibukawa, M., et al.: Comparison between Event Related Potentials Obtained by Syllable Recall Tasks and by Associative Recall Tasks. In: Stephanidis, C. (ed.) UAHCI 2007 (Part II). LNCS, vol. 4555, pp. 838–847. Springer, Heidelberg (2007)
2. Piction, T.W., Bentin, P., Donchin, E., Hillyard, S.E., Johnson Jr., R., Miller, G.A.W., Ruchikin, D.S., Rugg, M.D., Taylor, M.J.: Guidelines for Using Human Event-Related Potentials to Study Cognition: Recording Standards and Publication Criteria. Psychophysiology 37, 128–152 (2000)
3. Shibukawa, M., Funada, M., Ninomija, S.P.: An Analysis of the Relationship between Event Related Potentials and Response Time of Iterative Learning of Chinese Characters. Japanese Journal of Physiological Anthropology 11(2), 1–13 (2006)
4. Shibukawa, M., Funada, M., Ninomija, S.P.: A Comparison between a Computer Aided Learning and a Conventional Method for Leaning Chinese Characters by Event Related Potentials. Japanese Journal of Physiological Anthropology 12(1), 25–36 (2007)

Culture Design of Information Architecture for B2C E-Commerce Websites

Wan Abdul Rahim Wan Mohd. Isa, Nor Laila Md. Noor, and Shafie Mehad

Department of System Sciences, Faculty of Information Technology and
Quantitative Sciences, Universiti Teknologi MARA (UiTM),
40450 Shah Alam, Selangor, Malaysia
wrahim2@gmail.com, {norlaila,shafie}@tmsk.uitm.edu.my

Abstract. Culture is widely treated as an essential factor for the success of
e-commerce, yet the concept itself is still clouded in bewilderment. Furthermore, there has been little research on usage behavior in the context of developing countries; e.g., Islamic countries. By using Islamic culture as the case study,
this study highlights the website information architecture practical design indication and reports the partial analysis of the investigation on how culture design
of information architecture (IA) for B2C e-commerce website will has a positive affect to the user performance tasks (browsing, searching and purchasing
books activities). Analyses of one-way between-groups multivariate analysis of
variance (one-way MANOVA) and paired-samples t-test were performed. The
result showed that the task time performance of the Middle East and the Malaysian users are different and faster when using the culture centred e-commerce
website. Thus, provides empirical evidence on the positive influence of culturally design website to performance.

Keywords: Website Information Architecture, Culture Centred Website, Muslim Online User, Islamic Culture.

1 Introduction

The relevance of cultural issues to the design of global human-computer systems [22]
necessitates website designers to have a thorough understandings of the cultural constraints in usage behavior. However, there has been little research on usage behavior
in the context of developing countries such as from Islamic countries. This lack of
understanding of the cultural issues that are unique to that country also contributes to
the large scale diffusion for e-commerce in developing countries [12] [29]. This may
also partially explain the lacking of research or priorities given for Islamic websites
[24] from Arabic or Islamic countries on user interface and website IA in the online
environment.

The number of Muslim users are steadily growing and this growth has provided the
opportunities for rapid diffusion of e-commerce due to the nature of the environment
and culture. In 2002, United Nations Conference on Trade Development reported that
the Arab e-commerce market value was expected to grow to five billion dollars by
2005, with Saudi Arabia and Egypt as the major players [11]. As Middle East region

M. Kurosu (Ed.): Human Centered Design, HCII 2009, LNCS 5619, pp. 805–814, 2009.
© Springer-Verlag Berlin Heidelberg 2009

predominantly associated to Islam, this expected growth has provides the motivation for this study to investigate the usage behavior of Muslims in culture-centred e-commerce website towards creating a better understanding on Muslims behavior. However, culture remains difficult to study even to HCI practices and there are still many unsolved problems concerning the extent to which culture may affect the usability [22].

This study will also highlights practical guideline and empirical evidence on the positive influence of culturally adapted design to user performance. The understanding and results put forward by this study may be useful for those who want to reach out to Muslim users. The paper is organized as follow. The next section provides the literature review of the current studies related to e-commerce and culture. This section will also highlight an analytical framework on the adaptation of the cultural values on the website IA design by using Islamic culture as the case study. Then, the method used in this study is described and the results are reported, which is followed by conclusion, limitations and future implication.

2 Literature Review

2.1 E-Commerce and Culture

The degree of cultural adaptation exhibited on the country-specific websites is still considered at an early stage [21]. However, in recent years, there were emerging studies oriented towards enhancing the understanding for the role of culture in e-commerce. For instance, studies related to the e-commerce in Thai Culture [4], e-commerce for Chinese users [6] [14], e-commerce use for the Egyptian consumers [9], e-commerce differences between US and China [2] [18] [21], e-commerce acceptance in Malaysia and Algeria [30] and cultural considerations in India and Australia for Internet shopping [1] that mold the understanding to the intersection between culture and e-commerce. Generally, it is widely recommended that researcher identify cultural traits corresponding to the use of e-commerce. The growing usage of systems in diverse cultural contexts put forward the extent to which the use of such system is actually a matter of culture [22].

Various outcomes can be seen if cultural issues to be put in parallel with the development of design for e-commerce. For example, culturally-adapted web content enhances usability, accessibility and web interactivity with relevant cultural groups [20]. There are also indicators that culture may influence attitudes about e-commerce [19], affects the behavior [3] [16] [20], preferences of people [5] [16], satisfaction [5], loyalty [5] [20] and success of e-commerce [17]. It is important that the existence of the relevant cultural features alone does not fully guarantee the success for a particular website [20]. For example, trust also has been identified as one critical issue [5] [6] [8] [20] and essential element for success of e-commerce [23]. Individual culture may also play the role in forming the foundation for the development of trust [23].

According to Smith and Yetim [22], effective strategies that are critical to the success of a particular information system should address cultural issues in both, (1) the product and; (2) the process of information systems development. Pertaining to the product of development, cultural variation in actions, signs, conventions, values, norms or meanings raise novel research issues ranging from methodological, ethical and usability issues of culture [22]. Relative to the process of development, culture variation affects the manner in which users participate in design and act as subjects in evaluation studies [22]. Thus, this study attempts to reduce the gap and issues related to the product and process of e-commerce development by highlighting practical design guideline and providing modest empirical contribution on how culture design of information architecture (IA) for B2C e-commerce websites will has a positive affect to the user performance in performing tasks *(browsing, searching and purchasing activities)*. The next section highlights the analytical cultural framework and the associated IA website designs derived by using Islamic culture as the case study.

2.2 Cultural Framework and Associated IA Design Prescriptions

Cultural dimensions are often used for creating the understanding for the intersection among e-commerce, culture and online consumer behavior [20]. Hence, related design prescriptions to Islamic culture dimensions as theorized by Wan Abdul Rahim et al. (2008) are as seen in Table 1, to reflect practical design indications of the associated cultural values [27]. The application of cultural dimensions can be used in practical terms in designing more effective websites for Muslim users. The justification for the usage of the designs were also based on our prior theoretical works on website IA [25][26].

Table 1. Adapted IA design prescriptions to Islamic dimensions [26] [27]

Islamic Dimension	IA Dimension	Design Prescriptions
High Uncertainty Avoidance	Navigation	- Navigation schemes to prevent users from lost - Simple clear metaphor, limited choice & restricted data - Local & contextual navigational system - Include customer service, navigation local stores, local terms, free trial
	Content	- Mental model and help systems on reducing "user error" - Redundant cues (color, typography and sound, etc) to reduce ambiguity and chunk info by topic / modular. - Include tradition themes, local stores & local terms, customer service and navigation, free trials & download
Universalism	Navigation	- Global & local navigational system
	Content	- Chunk information by task or topic
Polychronics	Navigation	- Local & contextual navigational system
	Content	- Chunk information by topic or modular
Masculinity (MAS) & Femininity	Navigation	- (MAS) Navigation oriented to exploration & control
	Content	- Content focused on truth & certainty of beliefs. - Rules as a source of information & credibility.
Individualism (IND) &	Navigation	- (IND) Global & customizable navigational system - (COL) Contextual navigational system

Table 1. (*continued*)

Collectivism (COL)	Content	- (IND) Chunk information by task
		- (COL) Chunk information by modular
		- (COL) Include family theme, clubs or chatrooms, loyalty programs, community relations, symbols of group identity, newsletter & links to local websites
High Power Distance	Content	- Include hierarchy info & pictures of important people with title.
		- Include quality assurance, awards, vision statements & appeal in pride of ownership.
		- Tall hierarchy in mental models
		- Highly structured access to information
	Context	- Significant, frequent emphasis on the social & moral order (e.g. portrayal of nationalism/religion) and symbols
High Context	Navigation	- Local and contextual navigational system
	Content	- Chunk information by topic or modular
		- Use politeness, soft sell approach in message delivery
	Context	- Strong preference for visual
		- Use implicit cultural marker like visual & color
		- Emphasize on aesthetics value

3 Research Method

The main research questions (RQ) formulated for this study are:

RQ1: Do Muslim (Middle East and Malaysian) users performance differs when performing tasks (*browsing, searching and purchasing*) in targeted B2C e-commerce websites for their culture?

RQ2: Do Muslim (Middle East and Malaysian) users perform better when performing tasks (*browsing, searching and purchasing*) in targeted B2C e-commerce websites for their culture?

An experimental session with 44 Middle East and 44 Malaysian postgraduate students was conducted. The full discussion on the experimental method and the initial result was published in our prior work which focuses on the task of browsing and searching for specific books [28]. However, here, we provide the subsequent analysis on the partial findings that involve the tasks of user browsing, searching and purchasing the specified books. The dimensions of 'power distance' and 'uncertainty avoidance' were used to incorporate the IA cultural designs in the two business-to-consumer (B2C) e-commerce prototypes. In general, 'power distance' refers to the condition where weaker member accepts inequality in power distribution whereas 'uncertainty avoidance' dimension relates to the condition where a society feels vulnerable of taking risks in unpredictable situation [13].

The first website ('Iqra Book Store 1') reflects the high direction for the dimensions of 'power distance' and 'uncertainty avoidance'. The designs of the Iqra Book

Store 1' reflects the cultural dimensions of Islamic culture as theorized by Wan Abdul Rahim et al. [27]. The second website ('Iqra Book Store 2') however reflects the opposite direction the dimensions of 'power distance' and 'uncertainty avoidance'. This was used for comparative purpose in the experimental session. The summary of design on website prototypes are as shown in Table 2, which is being derived from Table 1.

Table 2. Summary of design on website prototypes

Cultural Dimension	Power Distance	Uncertainty Avoidance
Iqra Book Store 1 (High)	- Religion symbol / color - Highly structured access to information *(with sub- categories)*	- Link open within window - Symmetrical layout
Iqra Book Store 2 (Low)	- Basic logo / non-religion related color - Low structure access to information *(without sub-categories)*	- Link open new window - Non-symmetrical layout

Two website prototypes were used; in two separate treatments, for the experimental sessions. A three minutes basic training was given. Then, in both separate treatments, subjects were instructed to perform the following specified tasks:

(1) Registering with the website
(2) Seek the price information of two given books:
   − "Belief of the Beliefs" and "Hadith: Summarized Sahih-Ul-Bukhari" *('Iqra Book Store 1') (1st treatment)*
   − "Provisions for the Seekers: A Manual of Prophetic Hadiths" and "The Book of Beliefs" *('Iqra Book Store 2') (2nd treatment)*
   Then, find out the total purchase amount.
(3) By using a shopping cart tool, add the two books. Then, 'check out' and make purchase of the books by using the provided 'dummy' credit card number, expired date and 'Card Verification Value' (CVV) number.
(4) Finally, click 'log out'.

The results and discussion in next section was done based from the complete time performance recorded for respondents doing all specified tasks; (1), (2), (3) and (4).

4   Results and Discussion

Data transformation has been made to the 'timesite1' variable *(performance time recorded while browsing and purchasing book using 'Iqra Book Store 1')* and 'timesite2' variable *(performance time recorded while browsing and purchasing book using 'Iqra Book Store 2')*. This was done to meet the assumption of normal distribution, as suggested by Field (2005) [7]. A hypothesis was developed to provide the answer for the RQ1. The hypothesis is as shown as follows:

H_1: Muslim *(Middle East and Malaysian)* users time performance differs in doing tasks *(browsing, searching and purchasing books)* when using websites of their own cultures.

An analysis of one-way between-groups multivariate analysis of variance (one-way MANOVA) was performed to investigate the Muslims (Middle East and Malaysian) users, performing tasks (browsing, searching and purchasing books) in two different websites. This separate analysis was done to investigate whether sub-cultural groupings of Muslims (Middle and Malaysian) users are different. The analysis is important to show that the characteristics of sub-cultural grouping of Muslims are heterogeneous for between-groups but homogenous in within-groups. The differences may be due to other cultural factors such as nationality, gender and others. However, this assumption need to be tested in a different study. Two dependent variables were used: 'timesite1' and 'time-site2'. The independent variable was 'region group' Muslims (by Middle East or Malaysian) users. Preliminary testing was conducted for normality, linearity, univariate and multivariate outliers, homogeneity of variance-covariance matrices and multicollinearity, with no serious violations. There was a statistically significant difference between the Muslims (Middle East and Malaysian) users on the combined dependent variables: $F(2, 85)=38.15$, $p=.000$; Wilks' Lambda=.53; partial Eta squared = .47 as shown in Table 3.

Table 3. Summary of multivariate tests

Effect	Value	F	Hypo df	Error df	Sig.	Partial Eta Square
Intercept						
Pillai's Trace	.981	2215.32	2.00	85.00	.00	.981
Wilks' Lambda	.019	2215.32	2.00	85.00	.00	.981
Hotelling's Trace	52.13	2215.32	2.00	85.00	.00	.981
Roy's Largest Root	52.13	2215.32	2.00	85.00	.00	.981
Region						
Pillai's Trace	.473	38.15	2.00	85.00	.00	.473
Wilks' Lambda	.527	38.15	2.00	85.00	.00	.473
Hotelling's Trace	.898	38.15	2.00	85.00	.00	.473
Roy's Largest Root	.898	38.15	2.00	85.00	.00	.473

Table 4. Summary of multivariate tests

Source	Dependent Variable	Type III Sum of Square	df	Mean Square	F	Sig.	Partial Eta Square
Corrected	Time_Site1	2.968	1	2.968	49.56	.00	.37
Model	Time_Site2	3.209	1	3.209	40.92	.00	.32
Intercept	Time_Site1	133.18	1	133.18	2224.18	.00	.96
	Time_Site2	237.03	1	237.03	3022.84	.00	.97
Region	Time_Site1	2.968	1	2.968	49.56	.00	.37
	Time_Site2	3.209	1	3.209	40.92	.00	.32
Error	Time_Site1	5.150	86	.060			
	Time_Site2	6.744	86	.078			
Total	Time_Site1	141.30	88				
	Time_Site2	246.99	88				
Corrected	Time_Site1	8.117	87				
Total	Time_Site2	9.952	87				

When the results for the dependent variables were considered separately, both of the dependent variables reached the statistical significance using a Bonferroni adjusted alpha level of .025. We will only consider our results significant if the probability value (Sig.) is less than 0.25. The first dependent variable was 'timesite1': $F(1,86)= 49.56$, $p=.00$, partial Eta squared=.37 as shown in Table 4. The second dependent variable was 'timesite2': $F(1,86)= 40.92$, $p=.00$, partial Eta squared=.32.

The results supported H_1 that the time performance recorded for Muslim (Middle East and the Malaysian) users differs when using the websites targeted for their own culture ('Iqra Book Store 1'). Although, there was significant support from previous hypothesis that the Muslim users performed better in culture-centred website, however, the performance time between the Middle East and Malaysian users was different. This was suggested based on the findings of the inspection of the mean scores which showed that the Muslims (Middle East users reported higher levels of time performance for the 'Iqra Book Store 1' ($M = 1.41$, $SD = .26$) than the Malaysian users ($M = 1.05$, $SD = .23$) as shown in Table 5. This was also suggested based on the findings of the inspection of the mean scores which showed that the Middle East users reported higher levels of time performance for the 'Iqra Book Store 2' ($M= 1.83$, $SD = .29$) than the Malaysian users ($M = 1.45$, $SD = .27$).

Table 5. Mean score for time-on-task performance

	Group	Mean	Std. Deviation	N
Timesite1	Middle East	1.414	.260	44
	Malaysian	1.047	.229	44
Timesite2	Middle East	1.832	.294	44
	Malaysian	1.450	.266	44

A hypothesis was developed to provide the answer for the RQ2. The hypothesis is as shown as follows:

H_2: Muslims *(Middle East and Malaysian)* users perform faster for doing tasks *(browsing, searching and purchasing books)* in website targeted for their own culture.

A paired-samples t-test was done separately to the Muslims (Middle East and Malaysian) users. First, a paired-samples t-test was conducted to evaluate the cultural dimensions impact on the time performance recorded for the 'Iqra Book Store 1' and the 'Iqra Book Store 2' for the Muslim (Middle East) group users. There was a statistically significant increase in the time performance score recorded for the 'Iqra Book Store 1'; 'timesite1' ($M = 1.41$, $SD = .26$) to the time performance score recorded for the 'Iqra Book Store 2'; 'timesite2' ($M = 1.83$, $SD = .29$), $t(43) = -7.924$, $p<0.0005$ in doing tasks *(browsing, searching and purchasing books)*. The Eta squared statistics (0.59) showed a large effect size.

Second, a paired-samples t-test was conducted to evaluate the cultural dimensions impact on the time performance recorded for the 'Iqra Book Store 1' and the 'Iqra Book Store 2' for the Muslim (Malaysian) group users. There was a statistically significant increase in the time performance score recorded for the 'Iqra Book Store 1'; 'timesite1' ($M = 1.05$, $SD = .23$) to the time performance score recorded for the 'Iqra

Book Store 2'; 'timesite2' (M = 1.45, SD = .27), t(43) = -8.208, p<0.0005 in doing tasks *(browsing, searching and purchasing books)*. The Eta squared statistics (.61) showed a large effect size.

The results support H_2. Both empirical results showed that the Muslim (Middle East and Malaysian) groups performance were much better for the 'Iqra Book Store 1' compared to the 'Iqra Book Store 2', in doing tasks *(browsing, searching and purchasing books)*. This assumption was based on the time performance score which recorded an increase for the 'Iqra Book Store 2' for the Muslim (Middle East) (M = 1.83, SD = .29), t(43) = -7.924, p<0.0005 and also for Muslim (Malaysian) users (M = 1.45, SD = .27), t(43) = -8.208, p<0.0005 in doing tasks *(browsing, searching and purchasing books)*.

5 Conclusion

This study highlights practical design guideline and empirical evidence on the positive influence of culturally adapted design to user performance for performing tasks *(browsing, searching and purchasing books)* in e-commerce website. The understanding and results put forward by this study is useful for those who wants to reach out to Muslim users. However, this study has several limitations. Culture may not be the only reason that affects the usage behavior of users. This research focused only on the website category of online bookstore of website prototypes. The role of cultural dimensions and prescriptions imposed in this study might potentially posed different results for other categories of products and services. However, this assumption needs to be tested in a different study. Further possible extensions of the study by having more updated measure for cultural diversity across nations as well as their purchasing behavior may serve as a better indicator of how truly people behave in e-commerce environment. On the practical side, this study provides valuable information to companies that are considering on adapting the e-commerce website to cultural design. This study can also serve as point for future study in the emerging area of cultural usability.

References

1. Adapa, S.: Adoption of Internet Shopping: Cultural Considerations in India and Australia. Journal of Internet Banking & Commerce 13(2), 1–17 (2008)
2. Bin, Q., Chen, S.-J., Sun, S.Q.: Cultural Differences in E-Commerce: A Comparison Between U.S. and China. In: Hunter, G., Tan, F. (eds.) Advanced Topics in Global Information Management, vol. 3, pp. 19–26. Idea Group Publishing, Hershey (2004)
3. Chau, P.Y.K., Cole, M., Massey, A.P., Montoya-Weiss, M., O'Keefe, R.M.: Cultural Differences in the Online Behavior of Consumers. Commun. ACM 45(10), 138–143 (2002)
4. Chieochan, O., Lindley, D., Dunn, T.: The Adoption of Information Technol-ogy: A Foundation of E-Commerce Development in Thai Culture. In: Thanasankit, T. (ed.) E-Commerce and Cultural Values, ch. II, pp. 17–50. Idea Group Inc. (2003)
5. Cyr, D., Bonanni, C., Ilsever, J.: Design and E-loyalty Across Cultures in Electronic Commerce. In: 6th International Conference on Electronic Commerce, pp. 351–359. ACM Press, New York (2004)

6. Efendioglue, A.M.: E-Commerce Use by Chinese Consumers. In: Khosrowpour, M. (ed.) Encyclopedia of E-Commerce, E-Government and Mobile Commerce, vol. I, A-J. IGI Publishing (2006)
7. Field, A.: Discovering Statistics Using SPSS. SAGE Publication Ltd., Thousand Oaks (2005)
8. Gefen, D., Heart, T.: On the Need to include National Culture as a Central Issue in E-commerce Trust and Beliefs. In: Hunter, M., Tan, F. (eds.) Handbook of Research on Information Management and the Global Landscape, Information Science Reference, pp. 185–208 (2009)
9. Ghada Refaat, E.S., Galal, H.G.-E.: The role of culture in e-commerce use for the Egyptian consumers. Business Process Management Journal 15(1), 34–47 (2009)
10. Gong, W.: National culture and global diffusion of business-to-consumer e-commerce. Cross Cultural Management: An International Journal 16(1), 83–101 (2009)
11. Ha, L., Ganahl, R.J.: Webcasting Worldwide. Routledge (2007)
12. Hafez, M.M.: Role of Culture in Electronic Business Diffusion in Developing Countries. In: Kamel, S. (ed.) Electronic Business in Developing Countries: Opportunities and Challenges, pp. 34–44. Idea Group Inc. (2006)
13. Hofstede, G., Hofstede, G.J.: Cultures and Organization. McGraw-Hill, New York (2005)
14. Hsu, J.: Targeting E-Commerce to Chinese Audiences and Markets: Managing Cultural and Regional Challenges. In: Lytras, M.D., De Pablos, P.O. (eds.) Emerging Topics and Technologies in Information Systems, Idea Group Inc. (2009)
15. Huff, T.E.: Globalization and the Internet: The Malaysian Experience. In: Schabler, B., Stenberg, L. (eds.) Globalization and the Muslim World, pp. 138–152. Syracuse University Press (2004)
16. Kim, J., Lee, I., Choi, B., Hong, S.J., Tam, K.Y., Naruse, K.: Towards Reliable Metrics for Cultural Aspects of Human-Computer Interaction: Focusing on the Mobile Internet in Three Asian Countries. In: Galletta, D., Zhang, P. (eds.) Human - Computer Interaction and Man-agement Information Systems: Applications, M. E. Sharpe Inc. (2006)
17. Le, T.T., Rao, S.S., Truong, D.: A Managerial Perspective on E-Commerce: Adoption, Diffusion and Culture Issues. In: Gunasekaran, A., Khalil, O., Rahman, S.M. (eds.) Knowledge and Information Technology Management: Human and Social Perspectives, ch. 14. IGI Publishing (2003)
18. Liao, H., Proctor, R.W., Salvendy, G.: Chinese and US Online Consumers' Preferences for Content of E-Commerce Websites: A Survey. Theoretical Issues in Ergonomics Science 10(1), 19–42 (2009)
19. Sagi, J., Carayannis, E., Dasgupta, S., Thomas, G.: Globalization and E - Commerce – A Cross Cultural Investigation of User Attitudes. In: Hunter, M.G., Tan, F.B. (eds.) Advanced Topics in Global Information management, vol. 5 (2006)
20. Singh, N., Pereira, A.: Culturally Customized Web Site: Customizing Web Sites for the Global Marketplace. Butterworth-Heinemann (2005)
21. Singh, N., Zhao, H., Hu, X.: Cultural Adaptation on the Web - A Study of American Companies' Domestic and Chinese Websites. In: Hunter, M.G., Tan, F.B. (eds.) Advanced Topics in Global Information Management, vol. 4. IGI Publishing (2005)
22. Smith, A., Yetim, F.: Global human-computer systems: cultural determinants of usability. Interacting with Computers 16(1), 1–5 (2004)
23. Tan, F.B., Sutherland, P.: Online Consumer Trust: A Multi - Dimensional Model. In: Khosrowpour, M. (ed.) Advanced Topics in Electronic Commerce, vol. 1. Idea Group Publishing (2005)

24. Wan Abdul Rahim, W.M.I., Nor Laila, M.N., Shafie, M.: Towards Conceptualization of Islamic User Interface for Islamic Website: An Initial Investigation. In: Proceedings of Interna-tional Conference on Information & Communication Technology for the Muslim World (ICT4M), Malaysia (2006a), http://www.tmsk.uitm.edu.my/wrahim2/Wan_ICT4M.pdf
25. Wan Abdul Rahim, W.M.I., Nor Laila, M.N., Shafie, M.: Towards a Theoretical Frame-work for Understanding Website Information Architecture. In: Proceedings of the 8th International Arab Conference on Information Technology (ACIT 2006), Jordan (2006), http://www.tmsk.uitm.edu.my/~wrahim2/Wan_ACIT06.pdf
26. Wan Abdul Rahim, W.M.I., Nor Laila, M.N., Shafie, M.: Incorporating the Cultural Dimensions into the Theoretical Framework of Website Information Architecture. In: Aykin, N. (ed.) HCII 2007. LNCS, vol. 4559, pp. 212–221. Springer, Heidelberg (2007)
27. Wan Abdul Rahim, W.M.I., Nor Laila, M.N., Shafie, M.: Inducting the Dimensions of Islamic Culture: A Theoretical Building Approach and Website IA Design Application. In: Khong, C.W., Wong, C.Y., Niman, B.V. (eds.) 21st International Symposium Human Factors in Telecommunication: User Experience of ICTs, pp. 89–96. Prentice Hall, Englewood Cliffs (2008)
28. Wan Abdul Rahim, W.M.I., Nor Laila, M.N., Shafie, M.: The Information Architecture of E-Commerce: An Experimental Study on User Performance and Preference. In: Papado-poulos, G.A., Wojtkowski, W., Wojtkowski, W.G., Wrycza, S., Zupancic, J. (eds.) Information Systems Development: Towards a Service Provision Society, vol. 2. Springer, New York (2009)
29. Yap, A., Das, J., Burbridge, J., Cort, K.: A Composite-Model for E-Commerce Diffusion: Integrating Cultural and Socio-Economic Dimensions to the Dynamics of diffusion. In: Tan, F. (ed.) Global Information Technologies: Concepts, Methodologies, Tools and Applications, pp. 2905–2928. Information Science Reference (2008)
30. Zakariya, B., Syed, A.W.: Cultural Interpretation of E - Commerce Acceptance in Develop-ing Countries: Empirical Evidence from Malaysia and Algeria. In: Rouibah, K., Khalil, O., Ella Hassanien, A. (eds.) Emerging Markets and E-Commerce in Developing Economies, pp. 193–209. Information Science Reference (2008)

Influence and Impact Relationship between GIS Users and GIS Interfaces

Hongmei Wang

Northern Kentucky University, Department of Computer Science, Nunn Drive,
Highland Heights, KY 41099, USA
wangh1@nku.edu

Abstract. A Geographic Information System (GIS) is a computer-based system
for managing and processing geospatial data. GIS has been an important tool in
science, government agencies, private agencies and the public since the 1960s.
To make GIS more usable and useful to GIS users, the GIS community has paid
increasing attention to GIS interfaces. This paper provides an individual level
analysis of influence and impact relationship between GIS users and GIS interfaces based on the level analysis framework developed by Korpela et al. The
analysis results show that, on one hand, different physical and mental characteristics of the end GIS users have influenced design of GIS interfaces; on
the other hand, technologies developed in GIS interfaces have impacted how
the GIS users understand and use the GIS in different ways.

Keywords: GIS, Influence, Impact, Users, Interfaces, Individual level of analysis, Level analysis framework.

1 Introduction

A Geographic Information System (GIS) is a computer-based system for managing
and processing geospatial data [1]. It has been an important tool in science, government agencies, private agencies and the public since the 1960s. The user interacts
with the system through the user interface, which directly influences the usability and
usefulness of GIS. As a result, the GIS community has paid increasing attention to
GIS interfaces. The interfaces of commercial GIS software have evolved from early
command-line interfaces to current Graphical User Interfaces (GUI). More recently,
to further improve the efficiency of GIS and enable broader use of GIS, the GIS
community has proposed natural interfaces (e. g. speech and gesture enabled interfaces) to GIS [2, 3] and have developed some experimental natural interfaces for
geographic information use [4-6].

The GIS interface evolution process usually involves an influence and impact relationship between GIS users and GIS interfaces. On one hand, design of GIS interfaces
usually presents a user-centered paradigm, and designers of GIS interfaces usually
consider the end GIS users' needs, skills, habits, and working environments [7]. On
the other hand, advances in GIS interfaces have impacted the GIS users in different
ways. However, up to date, no systematic studies on the influence/impact relationship
between the GIS users and technologies of GIS interfaces have been conducted.

M. Kurosu (Ed.): Human Centered Design, HCII 2009, LNCS 5619, pp. 815–824, 2009.
© Springer-Verlag Berlin Heidelberg 2009

The goal of this paper is to analyze the influence and impact relationship between GIS users and GIS interfaces. This paper presents an individual level of analysis on the influence/impact relationship by following the level analysis framework [8, 9]. The influence/impact relationship is analyzed and discussed based on existing literature in the GIS field. The human context in which the GIS interface technologies are discussed in this paper is individual GIS users.

2 Research Approach

One of the major concerns in the Information Systems (IS) research is interaction relation, e. g. influence/impact relationship, between Information Technology (IT) and the human context [10, 11]. The interaction relationship between IT and the human context has been analyzed in different levels [8, 9] depending on the human context.

The framework of 2 X 4 integrative levels of analysis [8] is a commonly applied framework in the IS research field. In this framework (Table 1), levels of analysis in the IS research are divided into four major categories, including individuals, group, organization, and society. Each level is further divided into two viewpoints: intra (single case) and inter (multiple cases, relational, comparative).

Table 1. 2X4 integrative levels of analysis [8]

Level of analysis	Intra-viewpoint	Inter-view point (an example)	Theories, frameworks, names (examples)
Societal	Country/culture entity	Trans-national service chain	Sociology, political economy, cross-culture-cultural studies
Organizational	Organization	Business between organizations	Organizational theories, economics, resource-based theory
Group	Activity	Service chain between activities	Work research, activity theory, actor network theory
Individual	Person	Men/women, doctor/nurses, etc.	Social psychology, gender studies, HCI

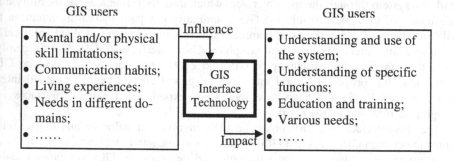

Fig. 1. Influence/impact model of GIS users and GIS interfaces

It is impossible to cover all levels of analysis on the influence/impact relationship between the GIS interface technology and the human context in a single study. This paper focuses on an intra-individual level of analysis on the influence/impact relationship. By following the principles in the framework of 2 X 4 integrative levels of analysis, in particular, those for the intra-individual level of analysis and being based on existing GIS literature, this paper analyzes the influence/impact relationship between GIS users and GIS interface technology (Fig. 1.).

3 Influences of GIS Users on GIS Interfaces

As a critical component of GIS, the user interface directly influences the usability and usefulness of the system. Human factors have been important to consider since people began to use tools [12]. There is no exception in design of GIS interfaces. Most current commercial GIS software has GUI based interfaces, such as ArcGIS Desktop 9 and MapInfo 9. Design of such interfaces usually adopts general user interface design principles (e. g. the widely accepted *Human Interface Guidelines* by *Apple Computer* [13]), which usually involve consideration of the user characteristics. Such general interface design principles show that the GIS interface design must consider physical and mental characteristics of the users, such as the end users' needs, skills, and habits [7], as well as other factors related to the user tasks, system functionalities, and the physical environment. Therefore, individual GIS users have directly influenced design and development of GIS interfaces.

3.1 Mental and/or Physical Skill Limitations

The individual user's mental and/or physical skill limitations drive development of GIS interfaces requiring less mental/physical loads. These limitations lead to design of GIS interfaces from command-line interfaces, to GUI-based interfaces, and then to natural interfaces.

Before the early 1980s, GIS interfaces had been dominated by the command-line interfaces [14]. To execute a command in this kind of GIS, the user needs to remember that command and all its required parameters. Only GIS expert users, who have received a lot of training, can operate this kind of GIS due to large memory load requirements.

After the late 1980s, GUI interfaces have been prevalent in commercial GIS [14]. The user interacts with GIS through visual interfaces, such as the interfaces characterized by Windows-Icon-Menu-Pointer (WIMP). Such visual interfaces allow the user directly to manipulate the GIS commands and get visual aids directly from the interfaces, and the user does not need to memorize so many commands and parameters as he/she does through the command-line interfaces.

Operating WIMP-based GIS still requires the user remember how to manipulate the GIS interface while executing a domain task, while natural interface based GIS does not. Further reducing the user's memory and mental loads required while using GIS is one of major reasons to develop natural interface based GIS [15]. By talking to or gesturing on the GIS, the user can focus on the domain task, instead of thinking how to manipulate the system.

3.2 Communication Habits

To improve GIS usability, the GIS designers have been considering human communication habits in design of GIS software.

Considering the users' traditional habit of using maps to deal with spatial information, most GIS functions of commercial GIS software respond to the users' spatial information request by computer screen maps and other graphics, and the users can also directly request spatial information based on existing maps/graphics on the computer screen [14, 16]. For example, the GIS can return a map of Florida counties to a user's spatial information request about "Show me all counties in Florida." Based on the GIS map response, the user can require other information. For instance, he/she can ask for "what are the populations of these counties" by circling or clicking the intended counties on the screen map.

Usually, humans communicate with each other through natural communication modalities, such as speech, gesture, gaze, facial expression, etc. Considering the user's natural communication habits, the GIS community proposed development of natural interfaces for geographic information use [2, 3] so that non-expert users can directly use this kind of GIS by talking to or gesturing on the system, with no training required. Natural interfaces have been developed for decades for geographic information use along with advances in speech processing and capturing human gestures. The most recent natural interface based GIS enables use of speech and natural free-hand gestures for geographic information access, e. g. *Dave_G*, which uses both speech and free-hand gestures [6].

In task-oriented human-human communication, it is common for communicators to use dialogues and feedback to build a shared context and further reduce uncertainties [15]. Such collaborative behavior between human speakers have been proposed to be applied for GIS interface design [15, 17]. Simulating users' collaborative behavior in task-oriented dialogues leads to development of new interaction styles for GIS, delegation [14]. For example, *Dave_G* can assist the user to reach the user's spatial information needs as the common goal in human-GIS collaboration [18], and enable collaborative dialogues with the users to handle some uncertain communication problems, e. g. the vagueness problem [19].

3.3 Living Experiences

The users' experiences have influenced design of metaphors in GIS interfaces. A good metaphor is familiar to the users, is internally coherent, and supports intuitive reasoning [16]. Metaphors developed in GIS interfaces are based on the GIS users' living experiences in the real world, from the GIS users' languages, their work regulations, documentation of existing technologies, and other representations of how the users think and act when they do their work [16].

The design of the metaphors in GIS interfaces usually map familiar concepts in the users' experiences to abstract geographical concepts in GIS. For example, the icon showing a palm indicates the function of "pan the map," which maps the GIS users' operation of moving a paper-based map into this GIS function. The general design principle of human-computer interaction through GIS interfaces is based on the users' experience of interviewing another person [14]. The interfaces to GIS query functions is designed based on the user's experience of asking the other person by providing some desired data characteristics [14]. The visual and transparent interfaces for overlay functions in GIS are designed based on the metaphor that people analyze data by stacking transparent sheets (containing datasets) on top of each other on a light table [20]. In virtual reality-based interfaces to GIS, design of digital equivalents for human ways of interacting with the real world is based on the users' living experience in the real world, such as turning one's head, walking, driving, flying, gesturing, and manipulating objects [14].

3.4 User Needs

GIS users have different needs from different perspectives, such as needs in different application domains, needs for different interaction modalities, and for different functionalities. Such user needs drive design and development of different GIS interfaces.

The GIS users have different information needs in different application domains [21], which lead to development of different user interfaces for GIS. One example is visual interfaces developed for terrain data display [22]. To facilitate the user's interaction with terrain data and support the user's needs of visualizing terrain data, Kavouras [22] proposed and developed the visual interface with advanced terrain visualization functions. Another example is multimedia interfaces developed for GIS [23-25]. Animation and video in the multimedia GIS are used to meet the user's needs in city planning to visualize the spatial-temporal city development process, or fly over the studied area [24], as well as the user's needs for video data [23].

Users' needs for sharing GIS data and functions over the Internet lead to development of web-based GIS applications [26]. Broad GIS applications lead to wide spread of GIS datasets on the Internet. However, the user needs GIS software to manipulate such GIS datasets. Some GIS projects involve collaboration of participants in different locations, who need to be able to manipulate not only their own GIS datasets, but also other partners' GIS datasets. The recent development of web-based GIS applications, e. g. such as ESRI's ArcIMS, AutoDesk's MapGuide, and Intergraph's Geomedia WebMap, provides a potential solution to the challenge of sharing GIS data and functionalities over the Internet [26].

4 Impact of GIS Interfaces on GIS Users

This section details the impact of the GIS interface technologies on the GIS users from three different perspectives, including the GIS users' understanding and use of the system, understanding of specific GIS functions, education and training requirements, and various needs.

4.1 Understanding and Use of the System

Different metaphors used in GIS interfaces and different types of GIS interfaces can lead to GIS users' different understandings of the overall system [27].

The screen map is the major method for spatial information representation in most GIS. Interaction with GIS is strongly map-oriented through use of maps [28]. This type of information representation method has impacted the GIS users' understanding and use of GIS [22]. For example, some users consider GIS as a mapping system and use GIS as substitutes of paper-based maps. The storage of spatial data has been regarded as storage of maps in digital form, and GIS users usually understand GIS data as map-like data [28].

Metaphors in GIS transcend specific tasks and shape the user's understanding of the overall system, which have been called "use paradigms" [27]. These metaphors allow the user to understand GIS in general. For instance, some users understand using GIS as interviewing people; some regard using GIS as analyzing tools. Some paradigms allow the user to understand GIS as "using GIS is programming," and some make the user think "using GIS is interacting with spatial phenomena" [29].

Using GIS with WIMP-based interfaces, some users consider GIS as repositories of geographic data that they can retrieve information from [27]. Users using a natural interface-based GIS may feel they were interacting with an assistant agent [18, 30]. Using GIS with multimedia interfaces, some users take GIS as a model of reality in which they want to find what they are interested in by experiencing in GIS. Virtual reality-based interfaces for GIS enable the GIS users to feel that they are living in the virtual world, not just looking at it [14].

4.2 Understanding of GIS Functions

Many metaphors are implemented in GIS interfaces to facilitate the users' understanding and use of some functions. When the users use these functions through these metaphors, the metaphors can impact the users' understandings of corresponding GIS functions.

Query functions in most GIS are based on the metaphor that the user accesses data by specifying certain properties of desired results [14]. When the GIS user operates the queries in GIS, the system uses the metaphor to activate the user's understanding of the queries from general database systems so that the user can understand the GIS query function better. When the user does not know relevant properties of desired results, the current approach in GIS is to supply metadata about aggregation information of dataset properties so that the user can search interested information in the metadata. This approach enables the user to understand the interaction with the system as browsing a huge unstructured repository for spatial datasets [14].

The visual and transparent interface for the overlay functions in GIS is designed based on the metaphor that people analyze data by stacking transparent sheets on top of each other on a light table [20]. The transparent interface can lead to the user feeling that, using the overlay function, they were analyzing data by stacking transparent sheets on top of each other on a light table [20].

4.3 Education and Training Requirements

To use GIS to solve some problems, the GIS user needs to have certain general education in geographic information and training in how to use GIS [31]. The user has different education and training requirements to use GIS with different interfaces.

The GIS with command-line interfaces require the most training of all GIS interfaces, because using such GIS requires the user to remember GIS commands and their parameters. The user needs education of geographic information to operate GIS and retrieve desired spatial information.

To use most current commercial GIS with WIMP-based interfaces, the user also needs to have certain training in how to use this kind of GIS. For example, most GIS provide query functions with visual interfaces, which usually implement the Structured Query Language (SQL) [14, 32]. To use these functions through the visual interfaces, firstly, the users need to know how relevant geographic information is structured in the relational tables of GIS. Secondly, the users need to know how to express their information needs by the SQL [33]. Thirdly, not all geographical characteristics can be modeled by the relational algebra in SQL [34] and the users need to understand such limitations. Finally, the GIS users need to know how to operate these functions through the WIMP-based interfaces.

Using GIS with natural interfaces does not need so much education and training as that required to use the system with the other two types of interfaces. The major requirement for the users to use natural interface based GIS is to know how to express their spatial information needs in natural communication modalities and understand feedback from the system. The GIS users have already fulfilled such requirement from their daily communication experiences. Consequently, the users do not need much training to use such systems.

4.4 Various Needs

Different GIS interfaces have brought great convenience to GIS users and meet their needs in various ways.

The improvement of GIS interfaces from early command-line interfaces to current WIMP-based and experimental natural interfaces have resulted in GIS satisfying individual users' needs of reducing mental and/physical loads and education and training requirements. It is not difficult for us to see from the literature reviewed above that the improvement of GIS interfaces can significantly reduce individual users' memory loads and/or physical loads required when they interact with GIS (see Section 3.1), and their education and training requirements (See Section 4.3) for them to use GIS.

The improvement of GIS interfaces have also resulted in GIS satisfying more and various users' spatial information needs in broader ranges of applications [35]. For example, since its early age in the 1960s, GIS has been applied to manage natural resources (e. g. land use planning, natural hazard assessment and wildlife habitat analysis). More recently, it has been applied in more domains, such as crime data analysis, crisis management, transportation applications, business analysis, etc.

Different types of user interfaces of GIS have also satisfied GIS users' needs in various ways [21]. The literature reviewed above (see Section 3.4) show that GIS users' various needs, such as needs in different application domains, needs for different interaction modalities, and for different functionalities, have driven design and development of different GIS interfaces. In turn, these different types of GIS user interfaces, e. g. web-based interfaces [26] and multimedia-based interfaces [25], have also met the GIS users' various needs from these different perspectives.

5 Conclusion and Future Work

This paper analyzes the influence/impact relationship between individual GIS users and GIS interfaces by following the framework of 2 X 4 integrative levels of analysis. The individual level analysis is based on the review of relevant literature mainly in the GIS field.

This paper provides a systematic analysis of the influence/impact relationship between GIS users and interfaces. The analysis results can help GIS designers to have a comprehensive understanding of the relationship, which will facilitate their interface design work. The results can also be helpful for the GIS users to understand the systems better, in particular, the system design issues. By knowing how the GIS users affect design of the interfaces, the users may know better how to propose their requirements to designers of GIS interfaces.

This paper presents the relationship based on a large amount of existing literature. In the future, based on the influence/impact relationship, a study can be conducted to collect direct input from the GIS software designers and the GIS users. Such direct input can corroborate and enrich the analysis results presented in this paper.

References

1. Malczewski, J.: GIS and Multicriteria Decision Analysis. John Wiley & Sons, Inc., New York (1999)
2. Florence, J., Hornsby, K., Egenhofer, M.J.: The GIS Wallboard: interactions with spatial infor-mation on large-scale display. In: Seventh International Symposium on Spatial Data Handling (SDH 1996), pp. 8A.1-15 (1996)
3. Frank, A.U., Mark, D.M.: Language issues for GIS. In: Macguire, D., Goodchild, M.F., Rhind, D. (eds.) Geographical Information Systems: Principles and Applications, pp. 147–163. Wiley, New York (1991)
4. Neal, J.G., Thielman, C.Y., Funke, D.J., Byoun, J.S.: Multi-modal output composition for human-computer dialogues. In: Proceedings of the 1989 AI systems in Government Conference, pp. 250–257 (1989)
5. Egenhofer, M.J.: Multi-modal spatial querying. In: Kraak, J.M., Molenaar, M. (eds.) Advances in GIS Research II (Proceedings of The Seventh International Symposium on Spatial Data Handling), pp. 785–799. Taylor & Francis, London (1996)
6. MacEachren, A.M., Cai, G., Sharma, R., Rauschert, I., Brewer, I., Bolelli, L., Shaparenko, B., Fuhrmann, S., Wang, H.: Enabling Collaborative GeoInformation Access and Decision-Making Through a Natural, Multimodal Interface. International Journal of Geographical Information Science 19, 293–317 (2005)
7. Egenhofer, M.J.: User Interfaces. In: Nyerges, T.L., Mark, D.M., Laurini, R., Egenhofer, M.J. (eds.) Cognitive Aspects of Human-Computer Interaction for Geographic Information Systems, pp. 143–145. Kluwer Academic Publishers, Dordrecht (1995)
8. Korpela, M., Mursu, A., Soriyan, H.A.: Two Times Four Integrative Levels of Analysis: A Framework. In: Russo, N., Fitzgerald, B., DeGross, J.I. (eds.) Realigning Research and Practice in Information Systems Development: The Social and Organizational Perspective, pp. 367–377. Kluwer Academic Publishing, Boston (2001)
9. Walsham, G.: Globalization and IT: Agenda for Research. In: Baskerville, R., Stage, J., DeGross, J.I. (eds.) Organizational and Social Perspectives on Information Technology, pp. 195–210. Kluwer Academic Publishers, Boston (2000)
10. Orlikowski, W.J., Iacono, C.S.: Research commentary: Desperately seeking the "IT" in IT research- a call to theorizing the IT artifact. Information Systems Research 12, 121–134 (2001)
11. Trauth, E.M.: The culture of an information economy: influences and impacts in the Republic of Ireland. Kluwer Academic Publishers, Boston (2000)

12. Turk, A.: The relevance of human factors to geographical information system. In: Medyckyj-Scott, D., Hearnshaw, H.M. (eds.) Human factors in Geographical Information Systems, pp. 15–36. Belhaven Press, London (1993)
13. Apple Computer Inc.: Human Interface Guidelines: The Apple Desktop Interface. Addison-Wesley, Reading (1987)
14. Egenhofer, M.J., Kuhn, W.: Interact with GIS. In: Longley, P.A., Goodchild, M.F., Maguire, D.J., Rhind, D.W. (eds.) Geographical Information Systems, pp. 401–402. John wiley & Sonsm Inc., New York (1999)
15. Wang, H.: A collaborative dialogue approach for human-GIS communication of vague spatial concepts. Ph. D thesis, The Pennsylvania State University (2007)
16. Kuhn, W.: 7+/-2 questions and answers about metaphors for GIS user interfaces. In: Nyerges, T.L., Mark, D.M., Laurini, R., Egenhofer, M.J. (eds.) Cognitive Aspects of Human-Computer Interaction for Geographical Information Systems, pp. 113–122. Kluwer Academic Publishers, Dordrecht (1995)
17. Brodeur, J., Bédard, Y., Edwards, G., Moulin, B.: Revisiting the Concept of Geospatial Data Interoperability within the Scope of Human Communication Processes. Transactions in GIS 7, 243–265 (2003)
18. Wang, H., Cai, G., MacEachren, A.M.: GeoDialogue: A Software Agent Enabling Collaborative Dialogues between a User and a Conversational GIS. In: 20th IEEE Int'l Conference on Tools with Artificial Intelligence (ICTAI 2008), pp. 357–360. IEEE Computer Society, Los Alamitos (2008)
19. Wang, H., MacEachren, A.M., Cai, G.: Design of Human-GIS Dialogue for Communication of Vague Spatial Concepts Based on Human Communication Framework. In: Third International Conference on Geographic Information Science (GIScience 2004), pp. 220–223 (2004)
20. Steintz, C., Parker, P., Jordan, L.: Hand-drawn oeverlays: their history and prospective uses. Landscape Architecture 66, 444–455 (1976)
21. Egenhofer, M.J.: User interface. In: Nyerges, T.L., Mark, D.M., Laurini, R., Egenhofer, M.J. (eds.) Cognitive Aspects of Human-Computer Interaction for Geographical Information Systems, pp. 143–145. Kluwer Academic Publishers, Dordrecht (1995)
22. Kavouras, M.: Human-Computer interaction considerations in Terrain modelling and visualization. In: Nyerges, T.L., Mark, D.M., Laurini, R., Egenhofer, M.J. (eds.) Cognitive Aspects of Human-Computer Interaction for Geographical Information Systems, pp. 213–220. Kluwer Academic Publishers, Boston (1995)
23. Weber, C.: The representation of sptio-temporal variation in GIS and cartographic display: the case for sonification and auditory data representation. In: Egenhofer, M.J., Golledge, R. (eds.) Spatial and temporal reasoning in GIS, pp. 74–85. Oxford University Press, New York (1997)
24. Shiffer, M.J.: Geographic interaction in the city planning context: beyond the multimedia prototype. In: Nyerges, T.L., Mark, D.M., Laurini, R., Egenhofer, M.J. (eds.) Cognitive Aspects of Human-Computer Interaction for Geographical Information Systems, pp. 295–310. Kluwer Academic Publishers, Boston (1995)
25. Perez, M.R., Pons, J.M.S., Gimeno, J.B., Reig, M.N.: GIS & multimedia applications to support environmental impact assessment and local planning. In: Nyerges, T.L., Mark, D.M., Laurini, R., Egenhofer, M.J. (eds.) Cognitive Aspects of Human-Computer Interaction for Geographical Information Systems, pp. 221–238. Kluwer Academic Publishers, Boston (1995)
26. Peng, Z.-R., Tsou, M.-H.: Internet GIS: Distributed Geographic Information Services for the Internet and Wireless Networks. John Wiley and Sons, Hoboken (2003)

27. Kuhn, W.: Paradigms of GIS use. In: 5th International Symposium on Spatial Data Handling, pp. 91–103 (1992)
28. Kuhn, W.: Are displays maps or views? In: Tenth international symposium on computer-assisted cartography (Auto Carto 10), pp. 261–274 (1991)
29. Mark, D.M.: Cognitive image-schemata for geographic information: Relations to user views and GIS interfaces. In: GIS/LIS 1989, pp. 551–560 (1989)
30. Cai, G., Wang, H., MacEachren, A.M.: Communicating Vague Spatial Concepts in Human-GIS Interactions: A Collaborative Dialogue Approach. In: Kuhn, W., Worboys, M.F., Timpf, S. (eds.) COSIT 2003. LNCS, vol. 2825, pp. 287–300. Springer, Heidelberg (2003)
31. Hearnshaw, H.M.: Learning to use a GIS. In: Medyckyj-Scott, D., Hearnshaw, H.M. (eds.) Human factors in Geographical Information Systems, pp. 70–80. Belhaven Press, London (1993)
32. Melton, J.: SQL language summary. ACM Computing Surveys 28, 141–143 (1996)
33. Reisner, P.: Human factors studies of database query languages: a survey and assessment. ACM Computing Surveys 13, 13–31 (1981)
34. Egenhofer, M.J.: Spatial SQL: A Query and Presentation Language. IEEE Transactions on Knowledge and Data Engineering 6, 86–95 (1994)
35. Chang, K.-t.: Introduction to Geographic Information Systems. McGraw-Hill, Boston (2008)

Investigation of Web Usability Based on the Dialogue Principles

Masahiro Watanabe, Shunichi Yonemura, and Yoko Asano

Cyber Solutions Laboratories, Nippon Telegraph and Telephone Corporation,
1-1 Hikarinooka, Yokosuka, Kanagawa, 239-0847, Japan
{watanabe.masahiro,yonemura.syunichi,asano.yoko}@lab.ntt.co.jp

Abstract. ISO 9241-110 standard provides user-interface design rules based on 7 dialogue principles. The priority of the principles varies depending on the characteristics of the tasks, the users, and the environments. We observed the behavior of middle-aged and older novice PC users when they performed some Web navigation tasks. We also pointed out some of the problems with usability, as discerned from the observations. We found that among the dialogue principles, self-descriptiveness is the most important. The observed problems, which were associated with the dialogue principles, suggest strategies for the enhancement of Web usability.

Keywords: Web, usability, ISO 9241-110, dialogue principles, self-descriptiveness.

1 Introduction

The rapid penetration of the Internet has made Web sites one of the most important ways of passing and acquiring information today. Unfortunately, the relative newness of Web site design principles has created serious usability issues, particularly for people with disabilities and elderly users; this is especially true in Japan because it has already become an aged society. A lot of research is targeting accessibility, the ability to access information [1, 2] or is targeting usability [3, 4]. However, no Web design methods that are appropriate for elderly users have been published.

The user-centered design method stipulates the repeated cycle of user-tests followed by redesign according to the opinion of users [5, 6]. This approach is theoretically valuable but not so practical, however, since user-tests often consume too much time, effort, and cost. Web design is no exception. Web designers, who are not always ergonomics experts, decide the order of solving the problems identified by user-tests; they often modify only those problems that they can solve easily to meet the strict deadlines. Effective approaches to deciding what is important or what is not important have longed been sought.

If the level of Web usability at the start point of the cycle is high, fewer user-tests would be needed. Moreover, performing an effective evaluation in just one cycle would reduce the time, effort, and cost of user-tests. These are really useful advantages,

M. Kurosu (Ed.): Human Centered Design, HCII 2009, LNCS 5619, pp. 825–832, 2009.
© Springer-Verlag Berlin Heidelberg 2009

because only limited resources are available for developing Web sites in practice. In order to reduce the costs of user-tests, a methodology that can identify and prioritize the problems is needed.

In this study, we focus on the standard, ISO 9241-110 [7], in which dialogue is defined as "the interaction between a user and a system to achieve a particular goal". The priorities for a set of design principles depend on the characteristics of the tasks, the users, and the environments.

The final goal of our study is clarify the strategies to prioritizing these principles so as to enhance usability for elder people. Our approach secures a high level of usability at the first user-tests and avoids the haphazard repetition of redesign/retest and so makes the design process more efficient. In this paper, we discuss the strategies available for Web usability enhancement by examining practical examples from user-tests. The dialogue principles are shown below [7].

1. Suitability for the task
2. Self-descriptiveness
3. Conformity with user expectations
4. Suitability for learning
5. Controllability
6. Error tolerance
7. Suitability of individualization

We have carried out investigations into the strategies that can enhance Web usability for elder people by using the dialogue principles as shown in the below steps.

1. Collect examples for Web usability via user-tests and extract important dialogue principles.
2. Clarify the limitation of the domain in which the principles are available.
3. Clarify the strategy for enhancement of Web usability by using the principles that reduce the cost for Web site development.

2 Examination

Six PC users (4 females; 50 - 65 years old), who had some experience in PC operation, 3 - 15 years, participated in this examination. They were instructed to access a portal Web site managed by a major Japanese company for elderly users and accomplish three search tasks. We observed them and performed interviews after the tasks. We then grouped the problems into 7 categories related to the 7 dialogue principles. They were asked to perform the three tasks listed below.

1. Keyword search task: to search for articles on the Japanese hobby of "Kokedama", by using the search function of the site.
2. Browsing search task: to search for articles about "NPO: nonprofit organization" without using the search function.
3. Submission task: to apply for an event they are interested in.

3 Results

3.1 Summary of the Results

We observed a wide variety of user behavior toward the tasks and divided them into 7 categories related to the 7 dialogue principles. Table 1 lists the results in 3 contexts of use, keyword input, search results, and objective pages. The problems in contexts of use "Keyword input" and "Search result" were observed in task 1. These in context of use "objective pages" were observed in all tasks.

Table 1. Observed problems in the portal site for elder people

context of use	dialogue principles	observed problems
keyword input	self-descriptiveness	Some users could not make adequate keyword for search.
	error tolerance	Some users input sentence instead of keyword for search.
search results	self-descriptiveness	Some users did not refer to other result pages of search.
	conformity with user expectations	Some users paid too much attention to characters in red.
objective pages	self-descriptiveness	Some users did not refer to other result pages included in the main article.
	self-descriptiveness	Some users could not scroll to see all contents on main article page.

3.2 Keyword Input for Search Function

There were some problems with keyword input. The main problem was that some users could not make adequate search terms to complete the task. Some users input "How to make Kokedama." into the search box as shown in Fig. 1. Because the users input a sentence instead of a keyword, the Web system could not return useful search results. The system should provide information to users on how to select a search keyword; this depends on PC literacy level of the user.

Because the kinds of characters were restricted, some users could not input search term appropriately. In Japanese, there are several ways (characters) of indicating the same meaning. One of them could not accept Kana characters. Because she entered the term again using Kanji characters, she could get the search results. Although the Web system should accept all kinds of characters, it could not because its search system did not have the word in its dictionary. This goes against the dialogue principle "error tolerance". It is better to provide instruction on the characters available, if the Web system can not accept all kinds of characters.

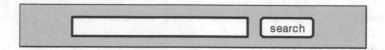

Fig. 1. Search box at the top page of the portal site

3.3 Search Results

Some users failed to examine all of the search result pages, if there were more than one search result page. A page shows page indicators at the bottom of search results page as shown in Fig.2. There was insufficient contrast between the page indicators and the background. Some users could not even identify that the numbers existed and could not refer to after the second page. This meant that they could not get an adequate result, i.e. the answer to the search task. This reflects a problem related to the dialogue principle "self-descriptiveness". The method used to refer to other pages could not be understood by the user. A better way should be provided.

Fig. 2. Page indicators for navigation (low contrast)

As shown above, some users could not identify the page indicators, see Fig.2. In the other pages, other than the search result pages, different kinds of page indicators were used as shown in Fig.3. Because these indicators had sufficient contrast, they could find the page indicators. This implied that they could not identify indicators that had insufficient contrast against the background. This is a perceptual problem for the users.

Fig. 3. Page indicators for navigation (higher contrast)

This Web site used contained 3 kinds of contents: articles, questionnaire of an event, and report of events. It was very confusing because the search results page showed 3 kinds of results simultaneously, see Fig. 4. This result shows a list of results, links to the searched pages; for example, one search result (an article) was shown and a message was also shown in grey characters, and so was not eye-catching. The others showed that there were no search results in red characters, for example there were no search results for questionnaires and reports. Almost all users were overly attracted to the words in red at the expense of the useful results which were not in red. The users could not get appropriate information and misunderstood the information presented. Although, of course, there is a problem with result presentation, this result suggests that elder users pay a lot more attention to eye-catching word styles. This is a problem for the dialogue principle "conformity with user expectations".

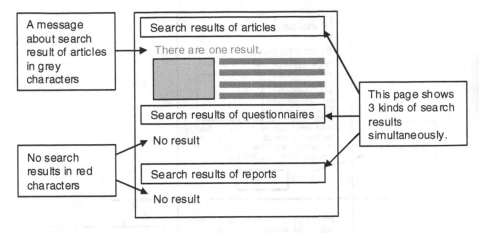

Fig. 4. Search result page shows 3 kinds of results simultaneously

3.4 Objective Page

When browsing, some users experienced the same problems as in the search result pages. Although there are page indicators at the bottom of the pages as shown in Fig.3, they could not refer to other pages in the objective contents. Although they "saw" the page numbers, they ignore them and did not use them. They answered that although they recognized the existence of the page links, they did not want to click it. They thought that the links were not to another part of the main content. They did not pay attention to the indicators. This is another problem. We call it cognitive problem with self-descriptiveness. If it is not perceived because of poor contrast, we call it perceptive problem with self-descriptiveness as shown in Fig. 2.

When the objective content consisted of 5 pages, some users read the objective pages one after the other until they reached the last page. In order to go to the first page and to refer the other contents, some of them pushed the button to move to the previous page. The previous page was shown but they could not readily detect that this page had already been examined, because no visual clues were provided. This is also a problem with self-descriptiveness because the users failed to understand the progress of the dialogue.

Some Web pages need to be scrolled in order to see all contents of the page but included a submit button above the bottom of the page as shown in Fig. 5. Some users stopped scrolling at this point and so failed to get important information below the submit button. If the submit button of the page is not placed at the bottom of the page, some users will fail to scroll through the page completely. This is also a problem in self-descriptiveness. It is a serious problem because the visualized position of the submit button is determined in part by the condition of the window, for example, window sizes and zooming level. Its position is difficult for Web designers to control.

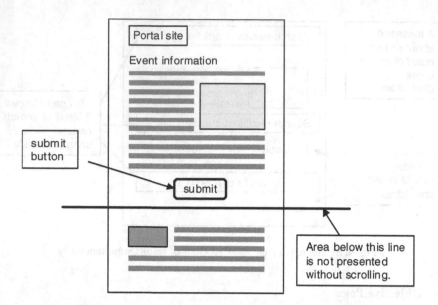

Fig. 5. Problems observed in main content. Some users stopped scrolling at this point and so failed to get important information below the submit button.

4 Discussion

The results showed that the 3 most frequently observed problems were related to the principles of "self-descriptiveness", "conformity with user expectation", and "error tolerance." The problems related to the principle "self-descriptiveness" were observed most often. The standard describes "self-descriptiveness" as "a dialogue is self-descriptive when each dialogue step is immediately comprehensible through feedback from the system or is explained to the user on request". The results suggest that self-descriptiveness is the most important principle in designing a Web site for elderly users.

Moreover, some users could not find information about self-descriptiveness, even though it was presented clearly. This suggests that there are 3 levels of problems in self-descriptiveness as shown below.

1. The lack of self-descriptiveness.
2. The perceptual problems of self-descriptiveness.
3. The cognitive problems of self-descriptiveness.

First level: the information provided did not establish self-descriptiveness. For example, some users input the wrong type of characters into the search box and the system failed to identify the error. It should clearly state the type of characters permitted. Second level: the information presented offered some form of self-descriptiveness but the amount was insufficient. For example, some users failed to notice that the search results occupied more than one page. This suggests that the contrast of the page-indicators and links to the other pages were inadequate. Third level: there are some cognitive problems although the information yielding self-descriptiveness is clearly displayed. For example, some users failed to scroll the window to read the area below the intermediate submit button. They stopped scrolling when the submit button appeared. Although the site designer thought that enough information was displayed, some problems remained. These results imply that these problems are mainly caused by the 3 common weaknesses of the elderly: poor attention span, limited short-term memory, and insufficient conceptual understanding (all were noted in the ISO standard [7]).

Our previous study showed the same results. We had performed 2 kinds of examinations [8]. One involved a search map application on a Web site served by 4 major portal sites in Japan. The other involved an application to search for information on transferring between trains on a Web page. Five PC users (4 females; 50 - 65 years old), who had some experience in PC operation, 3 - 15 years, participated in these examinations. They performed tasks similar to those described above. In the search map tasks, they input the name of a place at the beginning. When they selected a place name in the search results, a map was shown. They then had to resize and adjust the position of the map. In these contexts of use, similar results were recorded. To search for information on train transfer, they input the name of the station and the time of departure/arrival at the beginning. Then they had to confirm the name of station, the time of departure/arrival, path way and so on. After that the search results were presented to users. Although the results were divided among several pages, some users could not refer to pages other than the first page. The results confirm that the important dialogue principles strongly depend on context of use, that is, the way of Web operation. In almost all contexts of use, self-descriptiveness is important. This result matches the results reported in this paper. We think that these results are valid in the domain of Web applications and elderly users.

We think that some of the limitations of the domain are caused by the characteristics of elderly users. One problem is their inability to make useful keywords. A useful solution is for users to engage in a trial and error process with the system in order to input more appropriate keywords. For example, if there are no search results, the system should not just present the message "No search results." It should also display "Try to input the keyword in other characters". In this context of use, support for the dialogue principles of "self-descriptiveness" and "error tolerance" is needed. This may not be a problem unique to elderly users.

Another problem is the inability to identify the page indicators that have poor contrast in the search result page. It is one of the perceptual problems with self-descriptiveness. Moreover, some users could not identify the correct result because of the wrong results shown in red characters. They were attracted to the eye-catching feature of the character display. This is one of the problems with conformity with user expectations. It is thought that it is unique to elderly users because of their limited attention scope.

In the objective pages, some users could not identify the difference between the pages. This is caused by the poor short-term memory characteristics unique to elderly users. Some users misunderstood the page indicators as part of the content. Moreover, some users could not scroll beyond the submit button if it was not at the bottom of the page. These are cognitive problems with self-descriptiveness and are also unique to elderly users. It is the most serious problems among the 3 levels of problems in self-descriptiveness.

5   Conclusion

Our research shows that self-descriptiveness is the most important principle for designing Web site for elderly users. We should use the 7 dialogue principles to categorize the results of user-tests and then solve the problems associated with self-descriptiveness, especially cognitive problems, in order of their importance. This strategy makes it easy to determine what should be tackled first. We can clarify the points on which the limited design resources should be concentrated by using the standard for ergonomics for human-system interaction, dialogue principles. Problems of self-descriptiveness, especially cognitive issues, are most important. If we use the dialogue principles, we will be able to enhance the level of usability at the start of the design cycle and reduce the number of design-test cycles needed. Future works include the collection of many examples of various contexts of use and also investigate Web design methods where the focus is on the cognitive problems of elderly users.

References

1. Paciello, M.G.: Web Accessibility for People with Disabilities. CMP Books, Lawrence (2000)
2. Thatcher, J., Burks, M.R., Heilemann, C., Henry, S.L., Kirkpatrick, A., Lauke, P.H., Lawson, B., Regan, B., Rutter, R., Urban, M., Waddell, C.D.: Web Accessibility: Web Standards And Regulatory Compliance. Apress, Berkeley (2006)
3. Nielsen, J.: Designing Web Usability: The Practice of Simplicity. New Riders, Berkeley (2000)
4. Krug, S.: Don't Make Me Think! A Common Sense Approach to Web Usability, 2nd edn. New Riders, Berkeley (2000)
5. Nielsen, J.: Usability Engineering. Academic Press, San Diego (1993)
6. ISO 13407: Human-centered design processes for interactive systems (1999)
7. ISO 9241-110: Ergonomics of human-system interaction - Part 110: Dialogue principles (2006)
8. Watanabe, M., Sagata, Y., Yonemura, S., Asano, Y.: Strategies for Enhancement of Web Usability based on Dialogue Principle. Human Interface Symposium (2008) (in Japanese)

Part VI

User Involvement and Participatory Methods

Participatory Human-Centered Design: User Involvement and Design Cross-Fertilization

Guy A. Boy and Nadja Riedel

[1] Florida Institute for Human and Machine Cognition (IHMC)
40 South Alcaniz Street, FL 32502 Pensacola, USA
[2] d-Lido design + development, 5. Quergässchen 2, 86152 Augsburg, Germany
gboy@ihmc.us, nr@d-ligo.com

Abstract. Design and development of new instruments requires much attention with respect to safety, performance and comfort. Introducing new technology is a matter of taking care of past user experience on current technology and anticipating possible user experience on prototypes incrementally developed. The intricate spiral combination of prototyping and formative evaluations provides excellent support to include end-users in the design and development process. Human-centered design is also a combination of both analytical and user-centered (experimental) approaches. We cannot get rid of analyzing human-machine interaction using methods such as GOMS for example, and neither using professional design expertise. These methods provide an envelope of usability and usefulness issues; some are directly applicable, others issues require an experimental user-centered evaluation, i.e., real professional users are needed. Usability engineering is now very much used in industry and provides good results. Crucial problems are not technical any longer; they are financial, legal, social and finally relational. The various actors who will have an influence on the product being developed should participate. Participatory design enables to improve awareness of product attributes, i.e., what the product is really for, and how it should be made and used. A running example of the design of a new flight attendant panel to be included in the cabin of commercial aircraft is presented to support methodological claims and demonstrate approach soundness.

Keywords: Human factors, HCI design, user involvement.

1 Introduction

During the 1990's, practice in cognitive engineering and human-computer interaction mainly focused on usability engineering [6] where we, i.e., members of the HCI community, tried to assess user interfaces of interactive systems and develop appropriate usability testing techniques. We wanted user interfaces to be easy to learn, efficient and pleasurable to use. We also wanted that users be able to easily recover from errors, or anticipate and avoid big risks as much as possible. It seems that this is routine practice in industry. However, even if there are procedures and rules that

M. Kurosu (Ed.): Human Centered Design, HCII 2009, LNCS 5619, pp. 835–843, 2009.
© Springer-Verlag Berlin Heidelberg 2009

support these human-centered design efforts, it is mandatory to focus more on user requirements and on the integration of potential users in a real participatory way. In addition, a good design is not something that is created in a day by a single individual, but by a team of experts incrementally modifying prototypes toward a mature product. This part seems to be difficult and not fully mastered yet. Main reasons come from legal, financial, commercial, cultural, organizational and social issues. In this paper, we will not develop legal issues. Financial issues mainly deal with the cost/benefit issues of human factors integration in design activities [5]. Commercial and cultural issues are obviously related because participatory human-centered design (PHCD) cannot be carried out without cooperation between designers, manufacturers and clients. This paper will focus on the later issues where social relations are keys for the success of PHCD. PHCD still needs to incorporate these issues in a rational and systematic way because it remains a difficult and risky exercise. In this paper, we present the results of a study that was carried out over the last two years on the design of a novel interface to onboard cabin systems manipulated by flight attendants. The main goal was to harmonize the way users interact with this interface. Indeed, harmonization has become a key issue in environments that are populated by several interaction styles coming from various suppliers. The need for integration is a real central issue. Again, it is not a purely technical issue, but a social and economical one.

2 User Experience Today

If too much emphasis on current practice should not guide the design of a novel interface, it is crucial to understand the constraints and requirements that users have when they perform their work. It is important to understand why they cannot accomplish their work properly or perform it very well in a wide variety of situations. Both positive information and negative information on work practice are equally good to consider and analyze. User experience cannot be separated from the tools, methods and organizational setups that go with it. The difficult part is to access the right users, and not intermediary people who would synthesize user requirements for the design team. Obviously, all users cannot be accessed in all possible situations. However, by experience, selecting appropriate sets of users is much better than nothing! We need to remember that resulting acquired information is partial. This is why we need to have conceptual models that support interpretation and extrapolation in some cases. These conceptual models may be very loose and provided by domain experts in the form of narratives or simply active explanations of acquired information.

In our study, we have acquired such information from various flight attendants of five airlines distributed over the world. We chose these airlines because they were representative of various cultures, not only ethnical cultures but also corporate cultures. In addition, we tried to make sure that we had a variety of ages among the consulted populations. It was interesting to notice that with five airlines and roughly five flight attendants per airline, we converged toward very interesting categories of constraints and requirements. We used the Group Elicitation Method (GEM) to elicit their knowledge and knowhow. It worked perfectly once more time; GEM was used extensively and successfully in industrial settings for the last two decades [1]. The difficulty was not user-experience gathering itself. The difficulty was in the organization of the GEM sessions. Why?

First, it is mandatory to carefully prepare with the actors of the various organizations to deal with such a participatory approach, i.e., users are not necessarily prepared to participate in a design team. Once they understood our goals and the way GEM sessions work, everything went very smoothly. In addition, it worked because GEM sessions were very carefully prepared in advance. Such a session typically starts with the right question asked to the selected set of users. In fact the right question can be split into several sub-questions, and we have a set of preliminary generic questions that are used to shape and develop the final appropriate question with domain experts. Here are some of these generic questions. What is the goal of the system that we plan to analyze, design or evaluate? How is the system or its equivalent being used (current practice, observed human errors)? How would you use this system (users' requirements)? What do you expect will happen if the corresponding design is implemented (e.g., productivity, aesthetics, and safety)? How about doing the work this way (naive and/or provocative suggestions)? What constraints do you foresee (pragmatic investigation of the work environment)?

Second, the social relation between the GEM facilitator and the selected users need to be extremely professional, i.e., the facilitator needs to be an ethnographer showing the GEM participants that what they are providing is clearly understood and will be effectively used. In addition, since such information is mostly made of stories, it must be kept confidential. The difficulty is in the interpretation and simplification of these stories.

In order to cross-fertilize data gathered from expert users, it is important to develop a cognitive walkthrough (CW) that consists in exploring the user interface being designed, like a user would, to gather usability problems [11, 6]. Before running a CW session, user scenarios and associated actions have to be created. Based on the scenario the evaluators explore the interface by asking appropriate questions for each action. Spencer (2000) suggested questions such as: "Will the user know what to do at this step?" or "If the user does the right thing, will he/she know that he/she did the right thing, and is making progress towards he/she goal?" We implemented this approach.

It is interesting to notice that results were very impressive. Two main categories emerge: (1) clear user requirements and constraints; and (2) open questions that remain unexplained. Both types were checked with experts. Some open questions were answered; others needed further investigations with an extended set of domain experts. In three months, we managed to have a set of validated requirements and constraints that were never found in the related industrial community. These results were used to specify a prototype.

3 User Experience Tomorrow

It is possible to analyze some parts of possible future user experience from experts, but there are situations, configurations or initiatives that will never be possible to anticipate and therefore only a prototype-based approach will enable the elicitation of possible use patterns. First, there are user behaviors that are standard and could be anticipated because they are related to a style of interface. The more the interface conforms a standard, the more user behavior will be predictable. Standardization is therefore a great incentive for future user experience prediction. Nevertheless, when new kinds of interfaces are designed and developed, usability predictability is no longer possible without an experimental protocol that involves a set of users facing a prototype.

Prototyping is then a key requirement in participatory human-centered design. Prototypes range from low fidelity, e.g., paper and pencil mockups, to high fidelity systems, e.g., carefully implemented realistic user interfaces. Interaction analysis has to be performed at different stages of the design and development process. During the early stages, methods such as GOMS [3] can be used to rationalize various interaction envelopes. Such analyses provide a good insight of the points that need to be further investigated and the ones that are obviously right or wrong. In our harmonization study, a first prototype was developed by a third party and it became obvious very soon that some parts of the interface should be redesigned using very simple GOMS analyses. We actually used the CogTool [4]. Human-centered design is inherently iterative, but budget requirements impose that the number of steps be minimal. Without using appropriate expertise, such minimization of the number of steps is impossible for a successful result in the end. For that matter, it is often crucial to spend some more time and money on a human-centered approach during the early stages of the design and development process to gain a mature product in the end. In other words, the more we could start with appropriate and relevant high-level user requirements and constraints, the best in the end, i.e., the more mature the product.

Interaction analysis needs to be performed constantly during the design and development process. It is like a Chef cooking a memorable dish. Experience and expertise play a great role, but incremental testing is a must. We do not insist enough on this necessary capacity of domain experts to be involved and concentrated during the design process. We often talk about latent human errors [9] that are committed during the design and development process and re-appear at use time, sometimes viciously. If we take a positive approach of this problem, latent errors deal with product maturity. Product maturity is a matter of constant testing. In an ideal world, we would have to test the product and its former prototypes in all situations in order to make sure that it will be fully mature at delivery time. This is obviously impossible, but designers and engineers must remember that the more situations they will experience in the use of the product, the better. This is a practice that happens to disappear with our current industrial way of managing projects. Indeed, engineers need to fill in spreadsheets and report all the time instead of fully concentrating on their design and development tasks. It seems that reporting has become more important than actual design and development! Motivation must be kept and creativity must remain the main asset of human-centered design teams. For that matter, reporting could be used in a different way that would effectively and significantly improve design. At this stage, it is important to make a distinction between reporting for work FTE (full-time equivalent) justification and writing for improving design. Writing for design will be discussed later in this paper.

Once analytical methods such as GOMS have been used and have produced their own results, prototypes must be used to assess remaining emerging user interaction issues. We then proposed new design features based on the integration of the following aspects: usability (e.g., legibility); perception (e.g., "*gestaltgesetze*" / laws of design); and aesthetics (e.g., "joy of use"). Complex interfaces require logical and consistent definition and implementation of graphic design elements that convey design hierarchy and structure, make correlations visible, express conditions, and both clarify functions and promote workflow. Elements of design are often both structurally and functionally related. Design is in no way random; in our project, each part of the interface has been developed to fulfill a specific function. The following factors were taken into account: the use of rasters/grids, the distribution and the size of fields, the use of typography, the alignment of individual as well as multiple elements, groupings, relations, choice of colors, the size and shape of elements. Throughout this process, importance has been placed on the underlying aspect of functionality - for example, interaction elements should enable users to navigate into the system, edit, enter data and read information. The underlying structure/hierarchy should be straightforward and enable clear understanding of the correlation between elements, as well as appropriate and timely feedback.

Then, it is time to bring back the prototype to selected users in order to evaluate it. This is what we did in the cabin display harmonization study. We went back to see flight attendants of the five airlines and presented them the developed prototype. Each session included four to five flight attendants. The prototype was initially presented, i.e., a walkthrough was performed with them. They could become acquainted with the interface in a couple of hours. Three scenarios were presented. Flight attendants were required to develop these scenarios using the prototype. At this point, we would like to emphasize the importance of scenarios in participatory human-centered design. They must be simple enough to be implemented in a reasonable period of time, but they must include the necessary steps required to induce the emergence of user behaviors. Each session was recorded. The debriefing in the end was performed as a GEM session, i.e., all flight attendants performed a *brainwriting* and concepts were categorized and prioritized. Results showed that the various interaction issues discovered could not be anticipated during the analytical phase. They were used to redesign the interface.

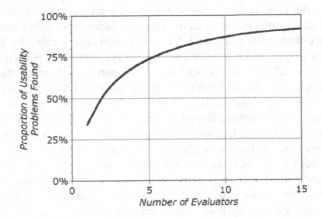

Fig. 1. Usability problems versus number of evaluators (Nielsen, 1994)

It is interesting to confirm that Nielsen and Mack's findings [7] on the number of evaluators was in line with our GEM results concerning the evaluation of the prototype (Figure 1). Across five GEM sessions of five participants each (i.e., experts users), we converged toward very consistent results with an approximate variability factor of 25%. More specifically, we elicited 24 human factors issues that were categorized into five main categories. These issues were very persistent, i.e., most human factors issue was elicited in more than 3 of 5 user groups (in our case, airlines) on 5. Some human factors issues were specific to one user group, and typically determined cultural differences.

4 Identifying Where the Real Problems Are

We already presented the -- today almost very classical -- way of how the participatory process could be carried out. However, it is as important to present where the real problems are in its implementation. First, we are talking about software engineering design. One of the most impressive attributes of software is its ease of modifiability. Since software is easy to modify, developers obviously do it with no reservation. This leads to a decrease of planning and forward thinking. The quality of high-level requirements stated in the beginning of the development process is crucial. If we do not start well, we are likely to modify the development trajectory all the time. Consequently, both product maturity and maturity of practice are directly impacted [2]. This is why user requirements and constraints are extremely useful in the definition of good high-level requirements for the product being designed. They enable to focus on practical and operational issues early enough in order to avoid later iterations on design flaws.

Industrial focus is often too much on technological solutions and not enough on what users really need to perform their work, i.e., users have to adapt to technology and not the other way around. However today, this is barely an engineering problem, it is a finance-driven management problem. Budgets are required to be reduced everywhere. But, the most important issue here is resource allocation. In fact, industry needs to hire more professionals, i.e., both human factors specialists and designers, to actually reduce their costs in the end. Such reductions are decided by finance-driven people and no longer by knowledgeable engineers. Therefore, participatory human-centered design (PHCD) budgets may look very unnecessary since they are not directly technology-centered, and immediately beneficial to the company. In addition, since PHCD has a major impact on maturity that is only observable at delivery time, finance-driven decision-makers do not see its relevance because this approach does not show immediate results, i.e., such a maturity-based approach is often perceived as a too long-term process. For that matter, this technique should be taught to top managers to get them more sensitive to what their company could gain in the end, and more importantly not loose time, money and employees' motivation by redesigning and repairing design flaws discovered at use time. Sometimes, it may happen that products are simply not acceptable to users and therefore rejected. Conversely, in our cabin display project, we clearly noticed that end users were not only pleased to be involved, but were potential champions of the PHCD approach and therefore resulting product.

Despite the expertise of designers and human factors specialists, when you design a new human-machine system, there is always a part of unknown that generate many kinds of problems. First, decision-makers who fund the innovation project do not know exactly what to expect and may only be aware of partial judgments on an ongoing design process that may be related to the social relations between actors participating in the project. When there are external actors, such as airlines in our project, problems may arise from sensitive issues that may not be related to the design itself and decisions could be made not to go in a direction despite its user-centeredness. As a generic rule, participatory human-centered design, which involve a large number of actors related by commercial and legal issues, is only possible within socially acceptable boundaries. Awareness of these boundaries is often difficult to get at all relevant levels of the design chain. Second, developing appropriate scenarios is crucial, but we never know what would be the best generic scenarios to choose without a careful and knowledgeable user-centered analysis. This is why working with expert users is crucial. However, when these expert users are difficult to access, appropriate and efficient techniques are necessary to gather their knowledge and knowhow that will lead to the development of such human-centered design scenarios. We can confirm that GEM is a great technique to this end. It has been used extensively during the project and helped us a lot in the development of scenarios by providing episodes, cases, incidents, critical issues and relations between users and technology. It was interesting to notice that what we thought important in the beginning of the project happened to be a wrong choice, i.e., developing a very general interface that enables to process very complex cases. End users agreed on simplicity and it turned out that only a very simple set of interface objects were necessary to perform the tasks that they had to perform. Again, this requirement was given by previous inputs from people who were not in the real business of operating the kind of interfaces that we were harmonizing. Therefore, it is crucial to choose the right end users.

5 Discussion

From a general standpoint, it is very important and necessary to implement several complementary techniques to cross-fertilize design concepts and converge toward a harmonized user interface. First, analytical methods are important to determine an investigation envelope, i.e., the domain of harmonization. However, we need to be aware that this acquired knowledge is full of holes. Second, it is also important and necessary to involve real end-users, flight attendants in our study, to both fill in the holes left from the previous analytical step and determine additional issues. Third, there is no participatory human-centered design without the development of a prototype. This prototype should be developed with an integrated PHCD team, and not decentralized too long, e.g., in a software development company. Prototype design mainly consists in finding out all design elements (i.e., the ontology of the user interface made of fonts, shapes, colors and so on) and various functional aspects (i.e., consistency, structure, perception and so on). As already mentioned, the prototype incrementally evolves from low fidelity productions to higher fidelity productions.

This evolution is the product of discussions with human factors specialists and inputs from user experts. It is crucial to carefully write design intentions and formative evaluation results to figure out where difficulties are: "Writing as design and design as writing" [8]. When you have difficulty to describe a concept, this is usually because this concept is not yet mature enough. Consequently, more work is required to improve the concept and return to the "drawing board" until you are able to "write" it correctly.

In the past, while developing human machine interfaces (HMI), designers, as well as human factors specialists, were not often involved in the specific teams because of technical constraints. Today, awareness and use of PHCD grows in industry. Human-computer interfaces should be seen as mediating entities that transfer information back and forth between a system and a user – it is an important part of a communication process. This is why several university programs in design focus their curriculum on "communication design" or "information design" and concentrate exactly on these topics. Students learn to include both aesthetic and functionality perspectives in their work. Therefore, design decisions are not made arbitrary but rationally. What our project brought is a combination of backgrounds including human factors, design and engineering. The job of an information-designer is editing data in a graphical way that depends on the content. We always keep in mind that the product (digital or analog interfaces) should support the user to understand meaningfully all concepts involved in the interaction. In the past, interfaces were often typography-based, i.e., to get information the user had to read the text. Today, with the specific development and definition of design-rules that support the use of color, shapes, fonts, workflows (design-language), it is possible to show existing hierarchies, properties and links.

The more complex structures, information and connections are, the more necessary the participation of a designer in the development-team is. The job of a designer, in the sense of PHCD, is not only creating the surface, it is rather to run through all parts of the developing process of a project with the inclusion of all the influencing individual aspects. For example, the definition of the problem and a detailed analysis of all considered facts are mostly the first steps in the beginning. Skills such as conceptual, systematic and analytic working methods, abstract thinking and schematic description abilities are essential to design teams to succeed with complex topics. Working in interdisciplinary teams is important for the more and more complex tasks of recent projects.

Finally, we would like to insist on two main assets that a design team should have. First, design knowledge and knowhow is mandatory to start with the right approach on consistency and rational thinking, but also to make sure that a concrete mockup and later-on a prototype could support interactions among the members of the design team. Discussions are always better supported by a common frame of reference that should be as explicit as possible. This is also true when end users are involved in the participatory design process. Second, the expertise of the cognitive engineer who will be acquiring information from end users is crucial. The cognitive engineer of course masters knowledge elicitation techniques being used. However, his/her expertise should include knowledge of the domain, e.g., aeronautics in our case, and also knowhow of how to interact and socialize with end users.

References

1. Boy, G.A.: The Group Elicitation Method for Participatory Design and Usability Testing. In: Proc. CHI 1996, the ACM Conference on Human Factors in Computing Systems, Vancouver, British Columbia, Canada, pp. 87–88 (1996)
2. Boy, G.A.: Knowledge management for product maturity. In: Proceedings of the International Conference on Knowledge Capture (K-Cap 2005), Banff, Canada. ACM Press, New York (2005)
3. Card, S.K., Moran, T.P., Newell, A.: The Psychology of Human-Computer Interaction. Lawrence Erlbaum Associates, Hillsdale (1983)
4. John, B.E.: The CogTool user guide. Carnegie Mellon University, Pittsburg, PA, USA (2008), http://www.cs.cmu.edu/~bej/cogtool/CogToolUserGuide-1_0_2.pdf
5. MoD HFI DTC. Cost arguments and evidence for human factors integration. Booklet produced by Systems Engineering & Assessment Ltd., UK (2006)
6. Nielsen, J.: Usability Engineering. Academic Press, Boston (1993)
7. Nielsen, J., Mack, R.L.: Usability Inspection Methods. John Wiley & Sons, NY (1994)
8. Norman, D.A.: Turn signals are the facial expressions of automobiles. Addisson-Wesley, Reading (1992)
9. Reason, J.: Human Error. Cambridge University Press, Cambridge (1990)
10. Spencer, R.: The Streamlined Cognitive Walkthrough Method, Working Around Social Constraints Encountered in a Software-Development Company. In: The Proceedings of CHI 2000, pp. 353–359 (2000)
11. Wharton, C., Bradford, J., Jeffries, J., Franzke, M.: Applying Cognitive Walkthroughs to more Complex User Interfaces: Experiences, Issues and Recommendations. In: The Proceedings of CHI 1992, pp. 381–388 (1992)

Playful Holistic Support to HCI Requirements Using LEGO Bricks

Lorenzo Cantoni[1,2], Luca Botturi[1], Marco Faré[2], and Davide Bolchini[3]

[1] NewMine Lab, University of Lugano, Switzerland
[2] webatelier.net, University of Lugano, Switzerland
[3] Indiana University, School of Informatics at IUPUI, USA
{lorenzo.cantoni,luca.botturi}@lu.unisi.ch,
marco.fare@lu.unisi.ch, dbolchin@iupui.edu

Abstract. This paper presents Real Time Web (RTW), a holistic method for eliciting HCI requirements and strategic design issues of web applications based on the systematic use of LEGO bricks. Capturing, understanding and expressing the requirements for the design of complex web applications can be a daunting task. This is due both to the complex nature of the tasks, and to the biased alignment of stakeholders, who often do not have an analytical understanding of their own needs and goals, and the current, mainly analytical, requirement analysis methods. The paper presents the method, its relationship with existing requirements analysis methods, and some case studies.

Keywords: Requirements analysis, web applications, LEGO bricks, informal interactions.

1 Introduction

Successful web applications fulfill or even exceed stakeholders' expectations and desires, and achieve the goals they set for them. This straightforward concept is what makes the requirements analysis phase so paramount, as the quarry from which is extracted the matter to which the creativity of designers can give shape.

Nonetheless, capturing, understanding and consistently expressing the requirements for the design of complex web applications are daunting tasks whose content often slip away like sand between the fingers of current HCI approaches. This is due both to the complex nature of the tasks itself, and to the fact that stakeholders do not always have a clear analytical understanding of their own needs and goals at an early stage of the project. The non-analytical perspective of most stakeholders is actually at odds with the analytical approach of most requirements analysis methods. It generates a tension between stakeholders' "gut feelings" or impressions, and the formats of requirements reports used by designers

This chasm between the inarticulate sense of needs and goals of the stakeholders and the structured apparatus of the conceptual tools for requirements analysis has been partially captured in the requirements engineering research community through

M. Kurosu (Ed.): Human Centered Design, HCII 2009, LNCS 5619, pp. 844–853, 2009.

the notion of "soft goals" [2]. Soft goals are goals for which there is no clear cut, necessary condition for satisfaction, but yet are essential for the success of the system under design. For example, a website goal such as "convince potential customers of the unique and rewarding experience they can get with my brand" is as relevant to the success of the communication as it is difficult to operationalize through a formal refinement of satisfying conditions. Most of requirements analysis for web communication is about these type of needs and goals, that must be elicited and addressed, and make room in the design space for creative discussion, argumentation for the design rationale.

Besides eliciting goals and needs, even more challenging is having stakeholders organization-wide develop a shared and agreed-upon expectation of how their web application should work and look like. Top-managers often have a different view from operative staff. It is a situation that can lead to disappointment, out-of-budget revisions, deadlocks and endless revisions.

The research conducted at the Università della Svizzera italiana in Lugano, Switzerland, in collaboration with Italian companies Trivioquadrivio and Kartha, took an innovative approach for tackling such issues: playfulness and a holistic approach to support the elicitation of non-analytical requirements. This paper presents Real Time Web (RTW), a holistic method for eliciting HCI requirements and strategic design issues of web applications based on the systematic use of LEGO bricks.

The next section is devoted to providing background and presenting relevant related works. Section 3 presents the method, while section 4 illustrates it with two case studies.

2 Related Work

The research tradition connected to the innovative approach presented in this paper spans over two main interrelated areas: requirements engineering and HCI design.

Current requirements engineering methods specifically developed for web applications focus mainly on the analysis and elaboration of goals and requirements for the HCI and communication design, but do not provide organization-wide elicitation strategies to engage stakeholders in out-of-the-box thinking about HCI opportunities, make tacit knowledge surface, challenge false assumptions, as well as to provide plastic and thought-provoking representations of requirements. Goal-based requirements analysis methods offer tools to identify and analyze high-level stakeholders' (including user) goals in the very early stage of the requirements management process. Some of these methods (e.g., Kaos [11] and i* [12]) emphasize the role of the responsibilities of the stakeholders involved in the process and the strategic and organizational knowledge in the early phase of requirements analysis. Recent developments in requirements analysis methods for web and hypermedia [1][2][3] focus on the structured collection and systematic analysis of the goals from different stakeholder's perspectives, but do not propose innovative strategies and techniques for eliciting tacit knowledge, unexpressed requirements or unveil false expectations. Approaches to facilitate creativity have been proposed in the requirements engineering field [6][14]. Creative workshops based on a sequence of divergent and convergent thinking sessions have been proved to be effective in designing mission-critical software systems (like Air Traffic Control Management) [6]. Brainstorming techniques and conceptual analogies to unrelated domain have been experimented and used to elicit new design ideas and unveil facts, tacit knowledge and assumptions otherwise overlooked.

In the area of HCI design, ethnographic approaches applied to interaction design and filtered through the lens of designers, have distilled some important methods such as Contextual Inquiry [13], as a response to the traditional requirements approaches. These methods address analogical thinking (e.g. the use of affinity diagrams and related techniques), as well as organization-wide brainstorming of design ideas mediated by the reflection and discussion of shared artifacts. In the same tradition, participatory design have emphasized the role of stakeholders participation in the design process as a way to elicit more salient and accurate requirements, facilitate and unleash creativity by drawing from outside the toolbox of the designers, and keep the emerging ideas more aligned with the user needs throughout the development process. It is in this tradition of approaches that the notion of make tools [7][8] recently emerged as a way to organize thoughts in participatory design sessions not only through active "doing", but through "making". Building physical artifacts or "convivial tools", as opposed to conceptual artifacts, meets the stakeholder need of creativity and introduces a playability aspect that facilitates experience sharing, engagement and openness [9][10].

A related research area which has tackled the problem of provoking, eliciting and analyzing knowledge from interested stakeholders is Collaboration Engineering. In this field, generic "facilitation" patterns (called "thinklets") [15] have been developed to support domain experts (thus non-professional facilitators) in proficiently carrying out facilitation tasks within an organization. In relationship with this research, our approach provides a domain-specific technique (the context of electronic communication artifacts) and is centered on a playful aspect of requirements elicitation, which is not central in traditional approaches.

Capitalizing on the advantages of these previous practices and research work, our approach systematically bridges the gap between current methods for web communication requirements and an emerging family of design techniques focusing on active participation of stakeholders in building and sharing meaning through playful artifacts.

3   A "Serious Play" Approach

3.1   Lego and Lego Serious Play

RTW is based on the experience of LEGO Serious Play (LSP) [4], a methodology developed more than a decade ago by LEGO and IMD, a business school based in Lausanne, Switzerland, to enter the corporate market. LSP is based on the assumption that everyone within an organization can contribute to the discussion, and help generating solutions. The main idea is developing a method that "gives your brain a hand" [4], i.e., that holistic though, supported by doing together instead of just by thinking, can enhance understanding and creativity.

For supporting creativity and expression, LSP leverages on LEGO bricks, which have the following relevant features:

- Are simple to use and do not require fine motor abilities in order to be able to build simple models
- Provide ready-made powerful symbolic pieces, such as little men and women representing many professions, skeletons, money, animals, etc.
- Are known to most people as a toy and as a joyful part of their own experience as children
- Are used in many different cultures

LSP exploits the creative power of LEGO bricks, and their intrinsic playfulness, to generate a relaxed environment where trained consultants can guide participants in team-building activities, SWOT analysis, and to the definition of simple guiding principles for advancing their projects. The key of the method is its structured sequence of timed activities, which lead participants from play and competition to modeling of complex organizational issues, to the development of what-if scenarios about alternative designs and to a systematic wrap-up of new knowledge. Each activity is based on three principles: (a) creating a model, (b) attributing a (metaphorical) meaning to it, and (c) sharing that meaning with the others as a story.

LSP currently includes activity modules for team building (Real Time Identity), project planning (Real Time Strategy) and wicked problems ("the beast"). Each module produces outcomes that can be summarized in a report, including "simple guiding principles", i.e., principles that can be applied in the team or project starting the next day. LSP heavily leverages on team interactions, so that all outcomes are negotiated and shared, preparing the ground for a smooth organizational change. LSP is being currently used in several countries by a number of LSP authorized representatives.

3.2 Real Time Web (RTW)

RTW was developed by two laboratories of the Università della Svizzera italiana in Lugano, Switzerland (webatelier.net and NewMinE: New Media in Education), in collaboration with Trivioquadrivio, an Italian consulting firm and authorized LSP representative and with the support of Kartha an Italian document management company. It is the result of a two-years effort in exploiting the strengths of LSP to support effective requirements elicitation workshops for the design of web applications.

The web application conceptual framework of the Web Communication Model (WCM) [5] was selected as backbone of the methodology, upon which a sequence of guided and timed activities was designed, following the format of other LSP modules. WCM basically understands a web application as the interaction of people (managers/administrators and users) through a web application made of content and functions, within a larger environment.

The first important step in a RTW session is the selection of participants. The RTW method, as all LSP modules, works well with up to 12 participants; when more people should be involved, it is better to split them into groups so to keep a small number. The rational here is that (a) each participant should have time to tell her/his stories and (b) the overall time of the session should not be too long. RTW builds requirements bottom-up, so that a good working group should ideally have one member for each stakeholder type (e.g., managers, IT staff, sale force, clients, etc.)

RTW, like LSP, uses a special set of Lego bricks – including connections and some particular elements and colors. RTW follows the approach of LSP, proposing a structured sequence of timed individual and collaborative activities, led by two facilitators. The activities include

1. Introduction: goals, method and warm-up activities. These activities, developed in the main LSP methodology, serve both as ice-breaker and illustration of the main steps in the methodology.
2. Individual model: your role in the project. This model represents how each participant think s/he can contribute to the project. It also allows identifying the absence of key stakeholders during the session, or even in the project at large.
3. Individual model: define users. This model represents what participants think are the main users of the web application. More than names, the models represent the main features of target users, and how/why/when they use the web application.
4. Collaborative activity: Black-box landscape. This first landscape positions all models of project stakeholders and target users around a symbolic object (the black-box) that represents the web site or application as a whole. It is a first step in collaborative thinking and team alignment.
5. Individual model: Web application content. This model represents a single content item – possibly the most important – in the web site.
6. Individual model: Web application functions. This model represents a single function – possibly the most important – in the web application.
7. Collaborative activity: Complete landscape. At this stage the black-box is removed and replaced by content and functions models. The whole landscape is then rearranged in order to fit the new situation and to make sense as a consistent narrative.
8. Connections: use and management. Connections connect stakeholders to content/functions (management) and users to content/functions (use). This allows seeing central and peripheral parts of the web application, along with unbalances (e.g., a content relevant to many users but that none cares after) or useless features (e.g., a function connected to no users).

Each activity has a precise timing (5 or 6 minutes for individual models, 10 minutes for collaborative activities), both to prompt action and to keep the session within a defined time limit. Activities are also performed on a musical background that supports lateral thinking and promotes a laid-off environment.

Of course, the RTW basic structure can be fine-tuned for specific needs. Also, it can be integrated with other LSP modules into longer sessions that include team-building, project planning and requirements analysis. However, the basic structure offers the possibility of getting non-analytical insights into stakeholder's views. Such insights have proven to be different and complementary to those more easily identified with more conventional methodologies, as it will be discussed in the case studies section.

Figure 1 presents a sample individual model, while Figure 2 presents a snapshot of a final landscape.

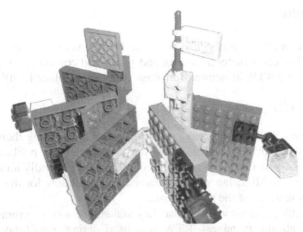

Fig. 1. RTW individual model for "many contexts" (part of content)

Fig. 2. Sample RTW final landscape

3.3 Advantages and Drawbacks

The main advantages of RTW are (a) its support to innovative and creative thinking, and (b) its ability to generate shared and agreed-upon requirements.

Its playfulness and hands-on approach generates a relaxed environment that stimulates lateral thinking and gently push people to think out-of-the-box about requirements and potential designs. This includes tearing down conventional barriers, so that often participants speak their minds more clearly and effectively than otherwise.

RTW does not only produce a requirement specification or design prototypes, but also generates a shared understanding of such requirements and the corresponding implications for the user experience. This paves the way for a more consistent development and for fewer arguments within the team – which means, in the end, a better integration of the website itself into the organization's business.

A drawback of the method is its setting. A RTW session requires a face-to-face workshop of 3 to 4 hours, where all participants interact together – and setting this up can be difficult at times. Also, while the design requirements elicited with RTW are sound, they might not always be complete. For this reason it is important to complement the outcomes with results from other (analytical) methods. Indeed, we believe RTW to be an important complement of other more formal methods.

4   Case Studies

This section presents two case studies in which RTW was used as requirements analysis technique. The cases differ as setting and goals, and consequently as implementation of the standard RTW structure. Both case studies took place in 2008.

4.1   A Large NGO

One of the largest Swiss NGOs was in the process of redesigning their website. Content was all there – institutional presentation, project reports, publications, etc., but over the years the structure had grown unordered, and it generally looked "old". Two members of the NewMinE lab were contracted as consultants for the redesign of the communication features of the new website.

After conducting interviews with two key stakeholders (the communication manager and the web site manager), RTW was used during a half-day session at the NGO's headquarter, with 10 participants from different roles. These included people who had directly to do with the website (the communication manager, the website manager, and the head of the IT staff), and with the many in-the-field projects (both from headquarter and from divisions in other states). People working in NGOs are used to non-traditional approaches, and welcome innovations – actually, most of them were interested in the method itself, as something they could adjust and apply for collaborative project planning in development projects.

This setting allowed applying the full standard RTW structure, as it was presented above.

The RTW workshop provided an opportunity to clear what appeared blurred during interviews – namely, who were the main target users for the website. Indeed, all participants were surprised to see how many and diverse they were. They included donors, sponsors, other practitioners in Switzerland and abroad, etc. Also, target user types were prioritized, and some of them were matched to other online resources from the same NGO, allowing to define a clear focus for the website.

Also, the RTW session plastically indicated the actual positioning of the "mandatory" institutional presentation of the NGO – indeed useful only for a minor, although important, part of target users. The connections exercise was particularly useful under this respect, as it clearly showed more logically central or peripheral content areas of the website.

Participants were glad to see how their ideas actually echoed, fitted or extended the others', and this generated a great commitment, resulting in (a) a stronger website team and (b) more availability of the others to give their contribution as required.

The results of the RTW session were put together into a report, that was then used to produce a detailed design and an operative plan, including content collection and revision processes, guidelines for interface and look-and-feel, and connections with other online resources. This report was presented to the website team and provided a basis for the further work.

4.2   A Cruise Company

A leading cruise company was going through the process of a general re-thinking of its online communication, with emphasis on the company's website, which had to be redesigned from scratch. Webatelier.net was involved in the definition of the new content structure.

RTW was used in a half-day session, involving 14 executives from several offices worldwide. The workshop aimed to collaboratively and creatively design the key aspects of the new website. In particular, the main goals for the session included (a) the promotion among participants of interest and commitment toward the website redesign process, emphasizing the complexity of the redesign process within a team; and (b) the elicitation of the main requirements for the new website (including, at large, web services) from the inside, i.e., from people working for the company.

In this case, the standard RTW structure was fine-tuned in order to fit the specific requirements, especially target participants and time available. Participants were asked to perform four exercises: to build three individual models, and to perform a final landscape-building collaborative activity. With respect to the basic structure, this session skipped step 4 (the black-box landscape) and 8 (connections), and merged steps 5 and 6 (content and services). The final sequence was the following:

1. Introduction: goals, method and warm-up activities
2. Individual model: your role in the project
3. Individual model: define target users
4. Individual model: Web application content/service
5. Collaborative activity: Complete landscape

With the help of the RTW workshop, participants could gain awareness and commitment to the new website project in only half-day of work. In particular, the workshop showed that some participants were deeply involved in the website project while others perceived it as being quite far from their role in the company: they were interested in getting advantages, but could not figure out how to actually contribute to it. Such outcomes provided concrete indications in managing the web site project team.

Moreover, results let emerge a sort of "hidden agenda" of all involved people. While during formal user requirements interviews, all internal stakeholders stressed the importance of travel agents, as being among the main user types, when it came to build a model of an important website user type, no one built a travel agent. This fact has been discussed at the end of the session to make sure that explicit and implicit expectations ones could be better aligned.

Results provided some of the main requirements of the website:

- Users of the website were characterized as demanding, multi generational, international, innovative, diverse, interested in planning experiences with other people, "freedom boomers"
- The website had to be multi channel, easy to be administered, constantly updated, user friendly
- The website should support easy experience planning
- The website has to cover / offer several services offered on board
- The website should clearly communicate emotions

5 Conclusions

In this paper we presented Real Time Web, a new application for Lego Serious Play. It is an innovative holistic and playful approach to requirements analysis, which allows the effective and efficient (mostly in terms of time) elicitation of requirements for complex web applications. Its side effects, which are actually central to the project, include the development of commitment and shared understanding among participants. After an analysis of related works, the method was described, including its advantages and drawbacks, and illustrated by two case studies. The case studies also illustrated the flexibility of the method, that can be tailored to different situations.

The RTW project has been an opportunity to reflect on the nature of requirements and on the relationship between the designer, or analyst, and the stakeholder. Analytic methods can lead to think that requirements "are there", in the mind of stakeholders, in organizations, or in data. RTW works the other way around, and provides a way to tear down conventional barriers and think outside the box, trying to say what one wants to say, providing a language with no words but colors and shapes. Requirements emerge from working together, and are only afterwards materialized.

The designer, or analyst, is not a detective looking for clues, but an expert that helps stakeholders look at themselves and express what they see. RTW provides a way to initiate this process and then step back, leaving room to directed thought, and for coming back later to interpret what has been said – and is now literally placed on the table – and interpret it together.

The experiences done so far, both with LSP and with RTW, indicate that this approach can be extremely powerful, especially when skillfully combined with analytical methods, in providing a more precise understanding of web application requirements. According our experience, it was like adding a third dimension to a painting: the scene gets more complex, more lively, and more engaging.

References

1. Bolchini, D., Garzotto, F., Paolini, P.: Branding and Communication Goals for Content-Intensive Interactive Applications. In: Proceedings of 15th IEEE International Conference on Requirements Engineering, New Dehli, India, pp. 173–182. IEEE Press, Los Alamitos (2007)
2. Bolchini, D., Paolini, P.: Goal-Driven Requirements Analysis for Hypermedia-intensive Web Applications. Requirements Engineering Journal 9, 85–103 (2004)
3. Bolchini, D., Garzotto, F., Paolini, P.: Value-Driven Design for 'Infosuasive' Web Applications. In: Proceedings of 17th International World Wide Web Conference, Beijing, China, pp. 745–754 (2008)
4. LEGO, LEGO Serious Play, http://www.seriousplay.com/ (retrieved October 19th, 2008)
5. Cantoni, L., Tardini, S.: Internet. Routledge, London (2006)
6. Maiden, N.A.M., Robertson, S., Gizikis, A.: Provoking Creativity: Imagine What Your Requirements Could Be Like. IEEE Software, 68–75 (September 2004)
7. Sanders, E.B.-N.: Scaffolds for building everyday creativity. In: Frascara, J. (ed.) Design for Effective Communications: Creating Contexts for Clarity and Meaning. Allworth Press, New York (2006)

8. Sanders, E.B.-N., Dandavate, U.: Design for experiencing: New tools. In: Proceedings of First International Conference on Design and Emotion, Delft, The Netherlands (1999)
9. Vaajakallio, K., Mattelmäki, T.: Collaborative Design Exploration: Envisioning Future Practices with Make Tools. In: Proceedings of DPPI 2007, Helsinki, Finland, pp. 223–238 (2007)
10. Ylirisku, S., Vaajakallio, K.: Situated Make Tools for envisioning ICTs with ageing workers. In: Proceedings of Include, Helen Hamlyn Research Center, RCA (2007), http://www.ektakta.com/include_proceedings/ (retrieved February 18th, 2009)
11. Dardenne, A., van Lamsweerde, A., Fickas, S.: Goal-Directed Requirements Acquisition. Science of Computer Programming 20(1), 3–50 (1993)
12. Yu, E.: Towards Modelling and Reasoning Support for Early-Phase Requirements Engineering. In: Proceedings of the 3rd IEEE International Symposium on Requirements Engineering, pp. 226–235. IEEE CS Press, Los Alamitos (1997)
13. Beyer, H., Holtzblatt, K.: Contextual Design. Defining Customer-Centered Systems. Morgan Kaufmann, San Francisco (1998)
14. Mich, L., Anesi, C., Berry, D.M.: Applying a pragmatics-based creativity-fostering technique to requirements elicitation. Requirements Engineering 10(4), 262–275 (2005)
15. Briggs, R.O., De Vreede, G.J., Nunamaker Jr., J.F.: Collaboration Engineering with ThinkLets to Pursue Sustained Success with Group Support Systems. Journal of Management Information Systems 19(4), 31–63 (2003)

User Research and User Centered Design; Designing, Developing, and Commercializing Widget Service on Mobile Handset

Sung Moo Hong

Termal Laboratory, KTF
7-13, Shinchon-dong, Songpa-gu, Seoul, Republic of Korea
sungmoo@ktf.com

Abstract. Mobile widget is a new paradigm for interactive idle screen service on mobile handset. Currently the standardization of widget is being discussed by W3C; Widget is globally accepted as a tiny web application. In Korea, however, mobile widget business is being expanded prior to that of web/PC widget though it has lots of issues to be solved: issues of usability, technology, and business model. In order to resolve these issues, user research and user centered design process were derived by KTF through the development of mobile widget. While using newly developed Widget, user can easily set and unset widgets by just pushing Widget key and change background images without changing one single widget. User can also personalize his/her widget adjusting its spot, size, opacity, theme, and color.

Keywords: User centered design, user research, mobile, and widget.

1 Introduction

Wallpaper download service was one of the most famous mobile services in Korea. However, thanks to phone camera, it has been replaced by the user created contents; more users are setting as their wallpaper the images taken by them. Also thanks to the bigger screen size, user can have more space on the idle screen. In 2004 observing these changes KTF, the second largest carrier in Korea, launched the first interactive idle screen service SHOW Popup[TM] which can be set and seen on the idle screen of mobile handset. A user can set an application on his/her idle screen, such as a portal, personal mobile homepage, instant messenger, stock information, and weather

| Portal | Mini Homepage | Messenger | TV Portal | Stock |

Fig. 1. Screenshots of SHOW Popup[TM], KTF idle application

M. Kurosu (Ed.): Human Centered Design, HCII 2009, LNCS 5619, pp. 854–861, 2009.
© Springer-Verlag Berlin Heidelberg 2009

information application. Using these applications, user can get information from service providers when the network is available.

2 Motivation

2.1 Problem1 Basic Scenario

After the commercial launch of SHOW Popup™, several significant usability issues were risen during the focused group discussion; it was hard for users to unset the idle application, to switch background images, and to use 4-directional arrow keys, dial key pad, and other keys while interactive idle applications were set on the idle screen.

2.1 Problem2 Basic Concept

Important insights were captured during the discussion with the content providers and developers; content providers wanted to put more than one application on user's idle screen - user only could set as the idle only one application at one time as Popup idle screen. This need seemed to meet customers' needs; they wanted not to cover entirely their own images previously set by themselves on the idle screen. Developers also wanted to access via idle applications various applications in mobile handsets, e.g., phone book, setting alarm, and media storage, etc.

3 Process of Designing Mobile Widget

3.1 Making Concept and Designing Idle Screen Application

Several design concepts which would meet the needs described on *Motivation* was derived. See Table1 for basic concepts of the idle application-to-be. Also see Figure 2 for screenshots from one of the early designs for multi application idle screen.

Table 1. Basic concept of idle application-to-be

Index	As-is	To-be
Service	One application on an idle screen	Multi-applications on an idle screen
Unset	Hard to unset; in order to unset, user should replace Popup application into other images	Easy to unset; application should have unset option in its options
Personal-ization	User's own images set on idle screen should be replaced into Popup applications	Easy to personalize user's own idle screen

Fig. 2. CUBE, one of the early designs for idle screen

3.2 Evaluating of Technology for Multiple Idle Screen Application

Three candidates of mobile platforms were evaluated whether the platform would support this concept of multiple applications on the idle screen; Platform "B" developed by Company "C", "M" by "P," and WIPI, the current platform. Finally the full scenario and design were applied on two new platforms; those platforms should have been modified. Also narrow-downed requirements were implemented into KTF's current platform. SHOW Widget™, thus, launched in June 2008.

3.3 Specifications for First Mobile Widget

Due to the lack of time and technical issue, requirements for mobile widget were not fully implemented.

Table 2. Implementation on SHOW Widget™, the first idle screen service for multi-application

Index	Feature of application-to-be	Implemented
Service	Multi-applications on an idle screen	Yes
Unset	Easy to unset; application should have unset option in its options	No
Personalization	Easy to personalize user's own idle screen	Yes

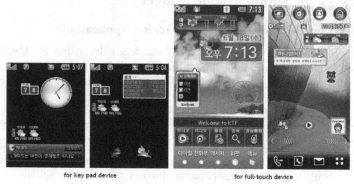

for key pad device   for full-touch device

Fig. 3. SHOW Widget™ for WIPI platform

4 User Research for Improving Mobile Widget Service

After commercial launch of SHOW Widget™, there was an opportunity to learn from the experiences. In August 2008 two kinds of surveys were executed in order to find out the customer's pattern of using the idle screen, widget, and other function. One was the on-line survey targeted 217 customers who were using full touch mobile handsets which supported SHOW Widget™. The other was in-depth interviews and task analysis with five customers using the same handsets.

4.1 On-Line Survey

The participators were asked about their satisfaction and their patterns of using idle screen and widget on PC/web, and other usability issues.

4.2 In-Depth Interviews

The participators were asked about same questionnaires with on-line survey for more detailed qualitative data. Through these researches the fact that users prefer multiple applications on the idle screen to single application was revealed.

4.3 Key Findings

4.3.1 Satisfaction

Users are satisfied more with Widget, a multiple-application for idle screen, than with Popup, a mono-application for idle screen. They evaluated the usability, function, and design of each idle screen solution.

- 29% of Widget users answered that Widget is convenient; only 6% of Popup users answered that Popup is convenient.
- 72% of Widget users answered that they find Widget has a value; only 13% of popup users answered that Popup is valuable. Also portion of users who answered that idle screen application was not valuable has been significantly reduced from 19% to 2%.
- 27% of widget users answered that they were satisfied with the design; only 19% of popup users answered so.

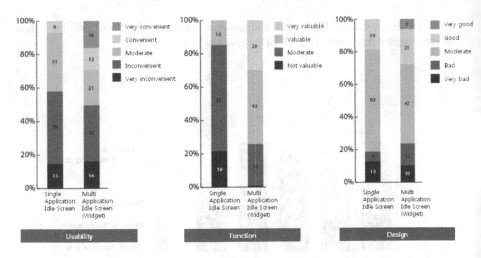

Fig. 4. Satisfaction of single application and multi application on idle screen

4.3.2 Utilizing
Widget users tend to utilize their idle screens more lively.

○ In spite of aggressive launch and promotion of Popup service, 50% of users were not live users. However, portion of passive users declined to 43%. The portion of users who utilize more lively inclined from 25% toward 30%.
○ Especially 7% of widget users answered that they change their idle screen settings every time they see the screen; none of popup users answered so.

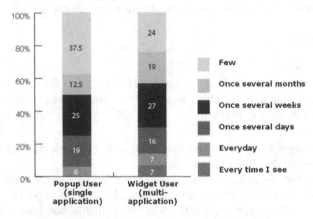

Fig. 5. Utilizing of idle application

4.3.3 PC Based Widget
PC (or web) based widget users utilize mobile widget more actively.

○ 45% of live users of PC based widget answered that they use mobile widget dozen times a day; only 23-28% of passive users answered so.

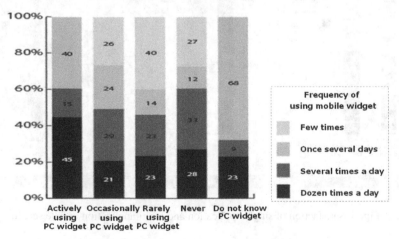

Fig. 6. Frequency of using mobile widget

4.3.4 Favorites

Users' favorite idle applications are applications of information service.

- 17.8% of popup users found that their favorite idle applications are information services. The number is the second to that of users who answered games are their favorites.
- More than 57% of users answered that their favorites are informative applications, i.e., weather, customer center, and subway information etc.

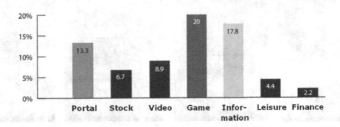

Fig. 7. Favorite Popup™ applications

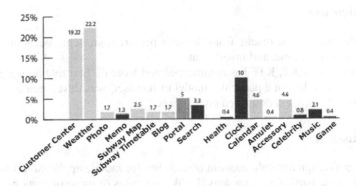

Fig. 8. Favorite SHOW Widget™ applications

Table 3. Features implemented on brand new SHOW Widget™

Index	Feature
Service	Developed seven information applications and freely embedded on the targeted device
Set-unset	Made available to set just by pushing one button and to unset via unset option which can be found on the option menus
Add-delete	Made easy to add various widgets on and delete them from idle screen
Personal-ization	Made available to personalize user's own wall paper screen without changing or unsetting user's widget

Fig. 9. Screenshots from brand new SHOW Widget™ for LG KH4500

5 Implication

After confirming of the results from the user research, strategic decision was made; putting more effort, time, and investment.

On 16 October 2008, KTF has commercialized Korea's first mobile widget service, which was embedded on a particular model of handset, was developed by carrier, and was designed by process described on this paper.

6 Discussion

User centered design and enhancement of usability for mobile application is significantly dependant on the particular technology. Unlike web sites or web applications which have global standards it should be bound tightly with particular model of mobile handset. Designing the mobile widget diverse methods of user centered design were executed such as prototyping and commercialization with iterative evaluation, focus group interview, on-line survey, task analysis, and in-depth interview, etc.

Idle screen of mobile handset is a battlefield. Several manufacturers have developed very unique features on idle screen. Some network operators also have tried creative approaches toward utilizing idle screen. Real user's need, however, should be considered more seriously. These are some points of further discussion for considering user's real need and pursuing user centered design for mobile idle screen including mobile widget.

- Shortcut vs. More information
- Personalization vs. Static
- Simplicity vs. Complexity
- Carrier vs. Manufacturer vs. 3rd party player
- Touch screen vs. Keypad

References

1. Hong, S.M., et al.: KTF Standard User Interface Requirement for handsets built on Open OS (2006)
2. Hong, S.M., et al.: KTF User Interface Requirement for Widget (2007)
3. Kim, H., Hong, S.M.: Analysis on Pattern of Using Widget (2008)

The Method of User's Requirement Analysis by Participation of the User: Constructing an Information System for Travelers

Chia-Yin Lin[1] and Makoto Okamoto[2]

[1] Future University-Hakodate, Graduate School of Systems Information Science,
116-2 Kamedanakano-cho, Hakodate, Hokkaido, 041-8655, Japan
[2] Future University-Hakodate, 116-2 Kamedanakano-cho, Hakodate, Hokkaido,
041-8655, Japan
{g3107006,maq}@fun.ac.jp

Abstract. This study attempts to capture the problems discovered on a field trip and to clarify user requirements with the participation of the user. With Mobile AP II, a platform for gathering user information and communication, user and designer can discuss issues so that user requirements can be specified from the context of the situation. In addition, we use narrative factors based on user experience and activity to make modeling scenarios easy to organize. In this study, we discovered how helpful it is to share information and to communicate via Mobile AP II, and that scenarios could be built using narrative factors to analyze the context of information systematically.

Keywords: Scenario Based Design, User Requirement, Participatory Design.

1 Introduction

With the development of a universal environment for communications, information has become easily accessible. Furthermore, with an increase in the popularity of backpacking and day trips, the need while traveling has also changed significantly. Hence, when designing an information system or device for travelers, the kind of function provided, and the method by which such information is provided, are important in meeting the requirements of a traveler. When attempting to determine the actual requirements of a traveler, designers encounter differences in the traveler's background knowledge and in the cultural aspects of communication. At times, the requirements of the traveler are implied, and at other times, even the traveler himself does not really know what he wants. Hence, a method to draw out ambiguous and implicit requirements is needed, together with an appropriate way to provide such information to travelers.

Scenario is an available method to obtain a common understanding between designer and traveler. The person who requires the information and the person who provides the needed data may reach a shared understanding by interacting through framing a scenario [1]. In addition, scenario is a simple and effective method for

M. Kurosu (Ed.): Human Centered Design, HCII 2009, LNCS 5619, pp. 862–868, 2009.
© Springer-Verlag Berlin Heidelberg 2009

describing a situation, eliciting requirements, evaluating situations, and giving opportunity to understand intentions and culture [2]. Therefore, in order for travelers and designers to better understand each other during the design process, scenarios can be used to obtain a common design language.

2 Eliciting User Requirement

Eliciting user requirement is an important starting point while developing a new design. However, the difficult problem is how to gather the important user requirements during the primary design process. When we want to discover users' requirements, the most direct method is to ask the users. Nevertheless, sometimes the user's requirements are unknown or ambiguous and not easy to express in terms of an unknown situation. In addition, not all users are able to express themselves, or are limited by their lack of experience. In such cases it is difficult to inquire of the user directly, and it is difficult to determine a user's implicit requirements and introduce new design concepts that meet the user's requirements.

User-centered design usually focuses on exploring and satisfying the user's requirements. Therefore, it is an integral part to gather the user's requirements as part of the primary design process. User's requirements would discover from various clues of user offered. These clues are the evidence of people's activities, which are an implicit communication and an important component of the design [3]. These clues are useful information in requirement elicitation, and can help the designer to analyze needs as the potential, unidentified user's requirements gradually crystallize and become definite. In the design process, when problems are put forward, user requirements are discovered, and then we can think about satisfying user's requirements and solving problems.

To facilitate understanding, the factors determining a user's emotions, behavior, environment, etc. during their activities, it is essential to develop empathy with the user. To develop empathy requires a good communication tool. The power of narrative is the ability to communicate ideas [4]. Narrating a story is a way to harness the imagination, an intuitive and easy way to begin to understand the user's activities and to develop empathy between users and designers.

3 Narrating a Scenario by Story-Telling

A story is generally considered to be a user-centered design resource in which to keep record of events and experiences that are significant to the user. Such a story provides useful information on a user's thoughts about his experiences, which is helpful in the design process for the perspective it gives the designer.

Stories from the user's life can form scenarios that focus on the things that matter to the user, including specific activities, the environment, and interpersonal interactions. Story-telling is an easy way for user and designer to communicate, so that they have the same impressions in their mind as they narrate the plot of a story, making it possible to convey meanings, particularly emotional values and experiences.

Scenarios are stories that describe people and their process of activity [5]. However, creating a scenario requires not only knowing the usual where, when, and what, but also involves probing the why and the how-to. Scenario-based design method uses story-telling to describe events involving people, objects and circumstances relevant to user activities. It provides a good context to discover problems from the user's point of view and to improve designs incorporating new concepts.

A benefit of scenario is that allows designer making stories from user reported events, obtaining experiences from user's activities, and visualizing the situations. Namely, scenario can re-create situations as the user experienced them. The method of scenario is also useful because it can facilitate forming a common consensus among users and designers, and provide the benefits of collaborative thinking while exploring appropriate user requirements.

4   The Process of Creating Scenarios

Discovering problems has an opportunity to define user requirements, and usually arise from observing user activity. For this reason, eliciting requirements could be swift when based on user self-reporting, where the scenario approach describes the user's conditions, emotions, concerns, and activities. To explore these possibilities, we investigated user's activities, experiences, and constraints through Mobile AP II.

4.1   Mobile AP II

Mobile AP II is a user self-report tool that can help a designer to understand a user's state of consciousness and activities (Figure 1). It provides support for obtaining user's life information and to documenting the particulars of the user's ongoing life, perhaps revealing whether he has improved in thinking and problem solving.

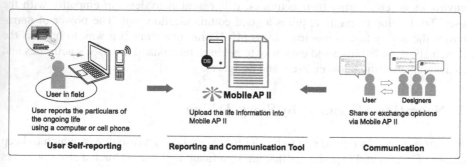

Fig. 1. Reporting and communicating via Mobile AP II

In this study, we used the Mobile AP II system to support the scenario modeling. Our aim was to use it as a tool to include user's perplexities in the design process. Using Mobile AP II is different from getting the user's information from interviews and quantitative data. The self-reporting of event by the user in the field means it will be possible to build a model of the user's own context. Self-reports are regarded as scenario-forming resources.

Mobile AP II is usable in different domains, not only gathering user reports on written scenarios, but also empowering both user and designer by putting them in communication via Mobile AP II so that they can further refine their concepts of problems.

4.2 Participation of the User

One objective of this method is to bring an extra dimension of realism to user-encountered problems and to encourage the participation of the user in analyzing problems that arise in certain situations, moreover, assist the designer to define user requirements. To achieve this, we involved Japanese students (as designers) and a traveler (as a user) from Taiwan, who used the Mobile AP II system to discuss the traveler proposed situation that the traveler was touring in Hakodate, Japan. Figure 2 illustrate the tasks of user and designers while using the Mobile AP II in design process.

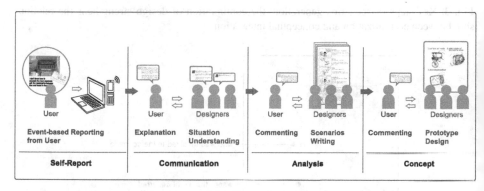

Fig. 2. The tasks of user and designers while using the Mobile AP II

The study used event-based reporting, requiring the user to provide reports of particular situation encountered. At every such event, the user could record a scene photo and a description of the situation to Mobile AP II, and discuss it with the designer using the comment function. The user and designer could jointly define the problems and requirements to determine the user's needs instead of just having the designers read a description of requirements.

4.3 The Steps of Scenarios Generation

The steps of scenarios generation are divided into the following steps (shown in Figure 3):

1. Capturing actual user experiences
2. Documenting the context of the User Story
3. Creating a Problem Scenario from User Story
4. Identifying the requirements of Requirement Scenario
5. Specifying the concepts of Solution Scenario

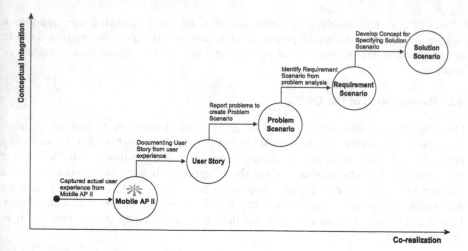

Fig. 3. Key steps of scenarios generation: the continuation of design strengthens the relationship between co-realization and conceptual integration

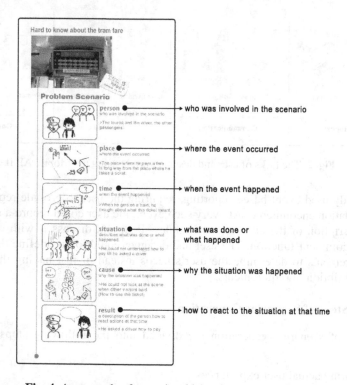

Fig. 4. An example of scenario which using narrative factors

The scenarios of this study include narrative factors: person, place, time, situation, cause and result (Figure 4). "Person" is who was involved in the scenario. "Place" is where the event occurred. "Time" is when the event happened. "Situation" describes what was done or what happened. "Cause" is why the situation was happened. "Result" is a description of the person how to react to the situation at that time. Using these factors supports the analysis of a user's activity by modeling. With these narrative factors, writing a scenario could become a simple matter of reconstructing and representing user activities and thinking.

5 Discussion

In this study, analysis of user report and discussion with user enabled designer to discover requirements that reflected user's needs gradually. Initially, user was unable to specify requirements and summarize them neatly, e.g. "I want a simpler way to find out how to pay tram fare in a foreign country." However, with discussions between user and designers, and by analyzing the context that user's offered on their experiences, the designers could gather a view of user requirements that was more specific than what the user himself might propose. For example, the tram passengers might be spared having negative emotions when they are waiting the foreign traveler to pay the fare if the designer incorporates the "quickly" requirement when designing a fare information system for travelers.

In our use of Mobile AP II to realize user's actual situation, the designers could further analyze and discuss situations with the user to determine their requirements. In addition, we have found Mobile AP II to present the following advantages:

- It is possible to conduct a survey from any distance, potentially increasing one's understanding of people of a different culture.
- It allows asynchronous communication.
- A designer can act without the need to frame a hypothesis in advance, based on problems reported from the field by the user himself.

On the other hand, a negative factor is that the users may report without adhering to a schedule, causing time-consuming. Therefore, time distribution should be considered again.

6 Conclusions and Future Work

The Mobile AP II system encourages the sharing of experiences and open communication between user and designer, leading to a consensus on key issues. In addition, it makes it easy to create scenarios based on structural narrative factors and is a simple means of communication that can represent the context of situations systematically. Furthermore, it is helpful to analyze problems and determine requirements from context with user participation.

On the other hand, while this method is suitable for exploring the problems of a single situation, combining various situations as a basis for writing scenarios with Mobile AP II is difficult. Thus, what is now desired the most is a "scenario development system". This system would be based on the user's photos of scenarios with narrative factors, and would be able to link photos of various situations to integrate them and easily create new scenarios.

References

1. Go, K.: Requirement Engineering. Kyoritsu Publisher (2002)
2. Okamoto, M., Komatsu, H., Gyobu, I., Ito, K.: Participatory Design Using Scenarios in Different Cultures. In: Jacko, J.A. (ed.) HCI 2007. LNCS, vol. 4550, pp. 223–231. Springer, Heidelberg (2007)
3. Norman, D.A.: The design of future things. A Member of the Perseus Books Group, New York (2007)
4. Cooper, A., Reimann, R., Cronin, D.: About Face 3: The Essentials of Interaction Design. Wiley Publishing Inc., Indianapolis (2007)
5. Carroll, J.M.: Making Use: Scenario-Based Design of Human-Computer Interactions. MIT Press, Cambridge (2000)

Concept Development with Real Users: Involving Customers in Creative Problem Solving

Mika P. Nieminen and Mari Tyllinen

Helsinki University of Technology, Department of Computer Science and Engineering,
P.O.Box 9210, 02015 TKK, Finland
{mika.nieminen,mari.tyllinen}@tkk.fi

Abstract. This paper describes idea generation activities in a user-centered concept development project when creating a new Enterprise Resource Planning system. With detailed statistics of the produced ideas we show that different creative problem solving methods are feasible to allow real end-users to generate ideas to improve their own ERP system. Our results show consistent success in using the various methods and a remarkably high percentage of new ideas were selected for further evaluation by the developers of the system.

Keywords: User-Centred Design, Concept Development, Creative Problem Solving, Idea Generation.

1 Introduction

This paper describes idea generation activities within a concept development project initiated to create a concept for a next generation Enterprise Resource Planning (ERP) system. The goal was to find both immediate improvements for ongoing product development and a longer-term vision to guide the future development activities in the coming five to ten years.

The used iterative concept development process includes five phases, namely 1) Project commitment, 2) User and technology research, 3) Innovation sprint, 4) Concept creation and validation and 5) Project assessment [1]. Previously this process has successfully produced concepts ranging from mixed reality demonstrators for children [2] to natural language interaction for mobile maintenance men. In these earlier concept development projects the users have not been fully involved in the actual innovation sprint as sole providers for the generated ideas.

Problem and idea finding is a critical period during the innovation sprint. It generates and rates the new ideas and features to form the basis for the new product concept. Four customer companies currently using Lean System [3] were studied during the user research phase and were asked to participate in the idea generation within the innovation sprint phase. The motivation for inviting actual users into the idea generation sessions was to explore whether groups of non-designer users, with no prior experience with formal creative problem solving, can produce feasible ideas for concept development and how the chosen idea generation methods affect the efficiency of the sessions.

M. Kurosu (Ed.): Human Centered Design, HCII 2009, LNCS 5619, pp. 869–878, 2009.

1.1 Learning the Companies Processes and Practices

In the following we recount the data collection and analysis leading to the creative problem solving tasks that identified and formulated the working hypotheses for the idea generation sessions. Half day workshops with the participating four companies were attended by employees portraying the entire production process from sales and marketing, production design and management, subcontracting and acquisitions, actual factory floor personnel, logistics and after-sales and maintenance. Using dialogical methods (for example see [4]) the companies' processes and practices were studied to find out the bottlenecks and problems in their processes and within their current ERP solution.

The processed materials included mainly audio recordings of the workshops from which interesting observations, related directly to the ERP system or the process surrounding its use, were collected. These observations were grouped and analyzed as an affinity diagram. From this analysis emerged five common themes to be used as the working hypotheses or problem statements for idea generation. For later reference these themes were:

1) Securing all *communications* into the system,
2) Availability of *documents and tools* in the system,
3) Improving visibility and traceability of *changes,*
4) *Fragmentation* of information within the system and
5) *Trust* towards the system and (real time) accuracy of its data.

1.2 Participating Companies

The study investigated four companies currently using Lean System ERP.

Company A. Company A has 2000 employees. The production process including export, acquisitions, production scheduling, manufacturing and forwarding/logistics was reviewed in the workshop which focused on the action prerequisites for delivering products in the planned schedule.

Company B. Company B has 62 employees of which 36 work in manufacturing. The production process including sales, product design, purchasing, production scheduling, manufacturing, quality control and financial administration was reviewed in the workshop which focused on successful completion of export projects.

Company C. Company C has 110 employees. The production process including sales, product development, purchasing, production scheduling, manufacturing and quality assurance was reviewed in the workshop which focused on product development projects and how gained knowledge can be utilized in serial/mass production.

Company D. Company D has 270 employees working in Finland and roughly 30 employed by affiliates in other countries. The production process including product development, custom product design, purchasing, manufacturing including subcontracting, service parts, maintenance, and financial and IT administration was reviewed in the workshop which focused on change management in the order-delivery process.

2 Creative Problem Solving

The creative problem solving (CPS) process which includes six stages, namely 1) objective finding, 2) fact finding, 3) problem finding, 4) idea finding, 5) solution finding and 6) acceptance finding [5] can be seen to have stages comparable to the iterative concept development process used. Especially the problem and idea finding stages of CPS are inherent and appropriate for the innovation sprint in concept development.

Idea generation methods can be classified into two categories, intuitive and logical, based on the quality of the process for generating ideas. A further sub-classification of intuitive methods has also been made into germinal, transformational, progressive, organizational and hybrid methods. [6].

McFadzean [7] concludes that creativity can be enhanced in five different ways: freewheeling to produce as many ideas as possible, association with already produced ideas, suspending judgment, using unrelated stimuli and applying unusual modes of expression. The latter three aspects differ by CPS method while the first two are common to all CPS methods. Based on these aspects McFadzean has developed a framework for creative techniques, called the creativity continuum, where all CPS methods can be placed. On this continuum CPS methods can be further grouped to paradigm-preserving, paradigm-stretching and paradigm-breaking.

The methods in these groups imply different considerations for the facilitator and the group using the methods in idea generation. Although paradigm-stretching and paradigm-breaking methods produce more creative and different ideas by forcing the participants to look at the problem from different perspectives and to use uncommon ways of thinking, they are more demanding for the group using them. For inexperienced groups paradigm-preserving techniques are more suitable because of their familiarity and safety [7]. Also for paradigm-stretching and paradigm-breaking methods to be effective, the group must be able to trust the other participants and the facilitator [8]. De Bono [9] according to McFadzean [10] has stated that changing paradigm requires lateral thinking, which is lateral movement to try new concepts and perceptions in relation to the problem.

The chosen methods for our workshops were classical brainstorming, a variation of the Method 635 brainwriting and six thinking hats by De Bono. Brainstorming and brainwriting are both paradigm-preserving methods [7] while six thinking hats is paradigm-stretching.

2.1 Classical Brainstorming

Classical brainstorming was introduced by Alex Osborne [11] as a creativity technique for groups. The session comprises of more than just idea generation, from statement of the problem to presentation of the result. Classical brainstorming includes four basic rules [5]: 1) Criticism is not permitted — adverse judgment of ideas must be withheld; 2) Free-wheeling is welcome — the wilder the idea the better. One should not be afraid to say anything that comes into one's mind. This complete freedom stimulates more and better ideas; 3) Quantity is required — the greater the number of ideas, the more likelihood of winners; 4) Combinations and improvements should be tried out. In addition to contributing ideas of one's own, one should suggest how ideas of others can be improved, or how two or more ideas can be joined into a still better idea.

Putman, [12] and [13] according to [14], provided the participants with additional instructions and reported a 40% increase in the number of ideas generated in brainstorming groups. The set of additional instructions was: 1) Do not tell stories or explain ideas; 2) When no one is saying ideas, restate the problem and encourage one another to generate more ideas; 3) Encourage those who are not talking to make a contribution; 4) Suggest that participants reconsider previous categories when they are not generating many more new ideas.

Including a 2-5 minute break in the middle of a 20 minute brainstorming session has been found to increase productivity after the break [15] according to [14]. Instructing the participants to think of the problem during the break and write down any additional ideas resulted in a higher increase in productivity than not giving such instructions.

2.2 Method 635

Brainwriting is a written form of brainstorming, where the same rules apply. Method 635 includes six group members writing down three ideas in a period of five minutes and then passing the papers to the adjacent person. The method was developed by Bernd Rohrbach in the 1960s [16]. Many studies have concluded that brainwriting in groups produces more ideas than brainstorming [17].

2.3 Six Thinking Hats

The method of six thinking hats by Edward de Bono [18] uses the idea of taking on different perspectives to the problem. The white hat is focused on information and data, the red hat is concerned with feelings and intuition, the black hat is about critical assessment, the yellow hat is optimistic and positive and the green hat is for creativity and growth. The blue hat is reserved for the facilitator to guide the process. These hats can be used in predefined orders to reach different goals for the session and one of those goals can be creative problem solving, for which a suitable hat sequence was used in our idea generation workshops.

3   Organized Idea Generation Workshops with Customers

Next the structure of the organized full day idea generation workshops is outlined. In all, three workshops were organized with the participating four companies. Companies A and B were both given separate workshops and companies C and D were given one common workshop. The earlier produced five design themes were assigned to the workshops according to Table 1. All workshops were attended by five employees of the company/companies from different roles in the production process. One of the authors served as the workshop leader and facilitator while several members of the project team attended the workshops as observers.

Table 1. The organized workshops with respect to the design themes and idea generation method used for each theme: B=Brainstorming, 635=Variation of method 635, 6TH =Six thinking hats

Design theme	Companies C & D	Company B	Company A
Communications	B	635	6TH
Documents and tools	635	B	
Changes	6TH		
Fragmentation		6TH	B
Trust			635

Each workshop had the following structure:

1) Introduction of themes and definition of the design problems,
2) Idea generation using each of the three methods followed by rating the ideas, and
3) Review of all top-rated ideas.

First part included three semi-structured discussions where the three design themes selected for that workshop were presented separately and a common understanding of the goal for the creative problem solving was formulated as a problem statement. These discussions lasted for a period of 50 minutes.

Each theme was then dealt by the group led by one of the authors as a facilitator with a different idea generation method each in a period of forty minutes. The used methods were classical brainstorming, a variation of method 635 and six thinking hats, in this order. All sessions started with a brief explanation of the method to be used and a short example if necessary. The problem statement formulated before was also repeated.

In brainstorming the participants were given the additional rules on paper and they were also read aloud. During the brainstorming one member of the project team wrote generated ideas on the wall for the participants to see at all times. The facilitator served to remind the participants of the rules and encourage them to generate ideas. In the middle of the brainstorming the participants were instructed to take a break and think about the problem statement and write down ideas that come to mind during the break.

In brainwriting five large sheets of paper were fastened to different walls of the room. The participants wrote on the sheet in front of them three ideas in a period of five minutes after which they moved on to the sheet of paper on their left. This was repeated so that everyone wrote on every sheet of paper once.

In six thinking hats the current mode of thinking was marked by displaying an accordingly colored paper in front of the participants. The facilitator executed the modes of thinking in the following predefined order:

1) blue hat for the facilitator to state the problem,
2) red hat to discuss the first intuitions regarding the problem,
3) green hat to generate ideas to solve the problem,
4) yellow hat to assess positive aspects of the generated ideas,
5) black hat to critically judge the generated ideas,
6) white hat for sharing and asking for more information on the problem,
7) green hat to generate more ideas to counter the earlier judgments,
8) red hat to discuss emotions evoked by the generated ideas and
9) blue hat for the facilitator to finally state the generated ideas.

Each idea generation session concluded with the participants rating the produced ideas. All participants were given ten distinct stickers to mark those ideas that he/she thought most valuable. A participant could use more than one sticker per idea to indicate the order of preference.

At the end of the workshop the top-rated ideas were further discussed with the participants in order to reach a common understanding of the ideas and their interpretations.

4 Results

The following chapters depict the analysis of the data collected from the idea generation workshops, which included video and audio recordings of group discussions, classical brainstorming and six thinking hats sessions as well as the idea sheets produced in brainwriting sessions. The results include detailed quantitative analysis based on the performed rating and idea yields with the different methods.

4.1 Analyzing the Workshops

The audio materials and the idea sheets were reviewed and distinct ideas were collected based on predefined criteria. An idea was defined as a verb-object phrase that represents a solution relevant to the problem statement. In other words, an idea expresses a thought in a meaningful, relevant and unique way. An idea was considered unique if it had not appeared earlier during the workshop or it was an elaboration of a previously stated idea.

Next the extracted idea collections were united to create a baseline for each session to compare and evaluate the effectiveness and output of all the idea generation sessions. The analysis was mainly concerned with the number of unique ideas relevant to the theme at hand and the judged quality of the ideas by the system development team.

Three members of the system development team rated the ideas together on a ten step scale which entails the idea quality together with information whether the idea was already implemented or partially implemented in the system (see Fig. 1). From

Fig. 1. System developers rating the 308 generated ideas

the system developers point of view the idea's quality was seen as the feasibility of an idea to be implemented and the level of an idea meeting the design criteria; those high on either scale were *selected*. Likewise, unrealizable or otherwise unacceptable or out-of-scope ideas were considered of poor quality and were classified as *rejected*.

The ratings using the stickers done by the participants in the idea generation sessions were not included in the analysis, because during the idea extraction the individual items were split into multiple ideas according to our definition of a unique idea and also reviewing the recordings added several dozen ideas to the lists assessed during the workshops.

4.2 Workshop Results

Table 2 presents the overall number of generated ideas and the results of the rating process. *Selected* ideas are those ideas that the members of the system development team rated as feasible and meeting the design criteria. These include already *implemented* ideas. *Rejected* ideas were those rated by the developers as unrealizable,

Table 2. The number of ideas generated in the workshops

Workshop	Quantity	Quality* Selected n	Quality* Selected %	Quality* Rejected n	Quality* Rejected %	Implemented**
A						
Group discussion	17	13	76 %	2	12 %	9
Brainstorming	39	25	64 %	3	8 %	7
Method 635	54	27	50 %	1	2 %	10
Six thinking hats	20	9	45 %	5	25 %	3
Total	130	74	57 %	11	8 %	29 (39%)
B						
Group discussion	7	3	43 %	0	0 %	1
Brainstorming	20	3	15 %	5	25 %	2
Method 635	50	31	62 %	5	10 %	9
Six thinking hats	16	11	69 %	0	0 %	7
Total	93	48	52 %	10	11 %	19 (40%)
C&D						
Group discussion	10	5	50 %	0	0 %	0
Brainstorming	21	15	71 %	1	5 %	7
Method 635	36	26	72 %	0	0 %	11
Six thinking hats	18	11	61 %	2	11 %	1
Total	85	57	67 %	3	4 %	19 (33%)
All workshops	308	179	58%	24	8%	67 (37%)

* as rated by the system developers in consensus
** in parentheses the percentage of *implemented* out of *selected*

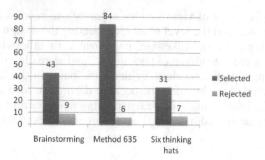

Fig. 2. Overall quantity of selected and rejected ideas produced by different idea generation methods

unacceptable or out-of-scope and thus of poor quality. *Implemented* ideas are already included in the current system or can be realized with the current system or are already approved to be included in future releases.

Several direct observations can be made from Table 2. The different methods were variably suitable for different groups. Workshop A did not excel using the six thinking hats method and produced a low number of selected ideas and a large percentage of rejected ideas. Respectively, Workshop B failed to get good results with brainstorming. Workshop C&D, combining attendants from two companies, delivered clearly the best Selected-Rejected ratio of all the groups.

Method 635 came out as a clear champion. It produced twice the amount of selected ideas compared to brainstorming, and almost three times more than the six thinking hats as shown in Fig. 2. Some of this is due to the fact that method 635 tends to produce a large almost fixed amount of finer-grained ideas, while six thinking hats seemed to assess fewer aggregate collections of ideas.

5 Conclusions

Our analysis shows that with sufficient and knowledgeable moderation non-designer groups previously unfamiliar with creative problem solving methods can be coaxed to generate coherent ideas to fuel product concept development. Even though in our case method 635 was found to be a superior idea generation method, we would recommend using several methods from both the safe and secure paradigm-preserving methods and from the more adventurous paradigm-stretching or breaking methods. This variation of methods also activated the participants kept them more engaged in consecutive idea generation sessions.

Also including members from more than one company, i.e. introducing previously unfamiliar people to a single idea generation session, improved the quality of the produced ideas. Our conclusion is that the participants had to state their ideas more clearly in order to explain their thoughts to strangers. Their more reserved approach also produced significantly less rejected ideas, while the overall number of ideas was, somewhat unexpectedly, not affected.

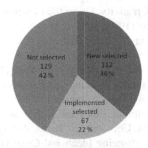

Fig. 3. Usefulness of the produced ideas based on quality and novelty

The distribution of all the produced ideas is illustrated in Fig. 3. From the total of 308 ideas 179 (58%) were selected and approved for further development. Even though a fifth of these ideas was already (partially) present in the product, a remarkably large portion of the generated ideas was still new and useful for the continuing development effort. From a research point of view it can be seen as unfortunate that the system development team's rating did not reveal the amount of completely novel ideas; however, the number of 112 (36%) ideas previously not included in the product development and as such highly potential candidates for new product features, can be considered very promising.

Acknowledgements. We would like to thank all the research and industry partners in the TuoHa II research project: Lappeenranta University of Technology (LUT), Kymenlaakso University of Applied Sciences / KymiDesign, Tieto (TietoEnator) and the participated Lean System customers. The reported research project has been funded by Finnish funding agency for technology and innovation Tekes and Tieto.

References

1. Nieminen, M.P., Mannonen, P.: User-Centered Product Concept Development. In: Karwowski, W. (ed.) International Encyclopedia of Ergonomics and Human Factors, 2nd edn., pp. 1728–1732. Taylor & Francis, New York (2006)
2. Nieminen, M.P., Viitanen, J.: Time Machine: Creating a Mixed Reality Experience for Children. In: Cunliffe, D. (ed.) Human Computer Interaction 2008, vol. 611, pp. 14–23. ACTA Press, Innsbruck (2008)
3. Tieto Lean System, http://www.leansystem.fi/ (cited on 26.2.2009)
4. Gustavsen, B.: Dialogue and Development. Theory of Communication, Action Research and the Restructuring of Working Life. Van Gorcum, Assen (1992)
5. Proctor, T.: Creative Problem Solving for Managers. Routledge, London (1999)
6. Shah, J.J., Kulkarni, S.V., Vargas-Hernandez, N.: Evaluation of Idea Generation Methods for Conceptual Design: Effectiveness Metrics and Design of Experiments. J. Mechanical Design. 122, 377–384 (2000)
7. McFadzean, E.: Critical Factors for Enhancing Creativity. J. Strategic Change. 10, 267–283 (2001)
8. McFadzean, E.: The Creativity Continuum: Towards Classification of Creative Problem Solving Techniques. J. Creativity and Innovation Management 7, 131–139 (1998)

9. De Bono, Edward.: Serious Creativity: Using the Power of Lateral Thinking to Create New Ideas. Harper Collins, London (1993)
10. McFadzean, E.: Creativity in MS/OR: Choosing the Appropriate Technique. J. Interfaces 29, 110–122 (1999)
11. Osborn, A.F.: Applied imagination. Scribner, New York (1953)
12. Putman, V.L.: Effects of Facilitator Training and Extended Rules on Group Brainstorming. Master's Thesis. University of Texas, Arlington (1998)
13. Putman, V.L.: Effects of Additional Rules and Dominance on Brainstorming and Decision Making. Doctoral Dissertation. University of Texas, Arlington (2001)
14. Paulus, P.B., Brown, V.R.: Enhancing Ideational Creativity in Groups. Lessons from Research on Brainstorming. In: Paulus, P.B. (ed.) Group Creativity: Innovation through Collaboration, pp. 110–136. Oxford University Press, Cary (2003)
15. Mitchell, K.A.C.: The Effect of Break Task on Performance During a Second Session of Brainstorming. Master's Thesis. University of Texas, Arlington (1998)
16. Rhorbach, B.: Creative nach Regeln: Methode 635, eine neue Technik zum Losen von Problemen. J. Absatzwirtschaft 12 (1969)
17. Paulus, P.B., Yang, H.-C.: Idea Generation in Groups: A Basis for Creativity in Organizations. J. Organizational Behaviour and Human Decision Processes 82, 76–87 (2000)
18. De Bono, E.: Kuusi ajatteluhattua. Mark-kustannus oy, Helsinki (1990)

Towards Fine-Grained Usability Testing: New Methodological Directions with Conversation Analysis

Marko Nieminen[1], Sari Karjalainen[2], Sirpa Riihiaho[1], and Petri Mannonen[1]

[1] Helsinki University of Technology, Department of Computer Science and Engineering,
Espoo, Finland
[2] University of Helsinki, Department of Speech Sciences, Helsinki, Finland
{Marko.Nieminen,Sirpa.Riihiaho,Petri.Mannonen}@tkk.fi,
Sari.Karjalainen@helsinki.fi

Abstract. We examine the possibilities of conversation analysis (CA) in usability testing. The goal is to examine how usability test setups serve as source for the CA analysis. We used video data from two earlier usability tests. Our results indicate that traditional test setup does not serve as a sufficient source for CA. The actions in the user interface were unclear, the user's facial reactions were not visible, and the user is occasionally having more conversation with the moderator than the system. CA approach can be taken towards two separate directions regarding usability tests: Analysis can be focused to the dialogue between the moderator and the user or on the user-system interaction. There is a need to fine-tune data gathering with detailed level recording of keypresses and system outputs. However, CA-enhanced usability testing allows in-depth analysis of usability problems as well as analysis of holistic interaction between user and system.

Keywords: Usability test, usability evaluation, conversation analysis, situated action, dialogue structures.

1 Introduction

Usability testing is perhaps the most widely applied practical usability evaluation method [1] in the development of interactive applications. Nowadays, the practical utility of the method is evident. It has long been considered as the cornerstone method for all companies that hesitate whether to apply user-centred approach in their development work. In his concept of "Discount Usability Engineering" (DUE), Nielsen [2] already promoted the method to be a good starting point: According to the idea of DUE, it is pretty easy to employ some users to test software and see how they perform with it. In our own research, we have applied usability testing both in research [3] and teaching [4] in the usability laboratory of Helsinki University of Technology. The core method has been studied and developed further in variations called Visual Walkthrough [5], Informal Walkthrough [6] and Contextual Walkthrough [3].

The main focus in traditional usability testing has been in the actions on the computer screen and the recorded speech. A key element in most usability tests is the

M. Kurosu (Ed.): Human Centered Design, HCII 2009, LNCS 5619, pp. 879–887, 2009.
© Springer-Verlag Berlin Heidelberg 2009

thinking aloud method. The core of the method has remained rather fixed with tiny variations applied to different study setup types (e.g. Apple Guidelines [7]) During the last decade, several development and consulting companies have adopted and applied the method successfully. Through the method, the researchers and practitioners have been able to discover usability problems and suggest improvements to the interface design.

The thinking aloud method reveals the user's cognitive reasoning and motivational basis behind his actions with the interface. Reflecting user's behaviour indirectly, the thinking aloud method supplements other usability evaluation methods, such as cognitive walkthrough and heuristic analysis. So far, user's other behaviour than explicitly expressed conscious thinking has not been addressed explicitly with supporting conceptual framework in the analysis of the results of the usability test. In order to find out deeper insights into user's understanding and actions, new approaches are required.

There is a need to move from straightforward summarizing type of analysis towards more interpretive and hermeneutic analysis. In searching for the new analytical insights, we are interested in dealing with the whole interaction between the user and the interface, and the variety of interactional resources used in interaction. We expect that Conversation Analysis (CA) will provide new means for analyzing the human-computer interaction in more detail and for answering practical questions, such as: Why (and how) do users end up with problems in the interaction? Why (and how) are some features fluent to use?

1.1 Conversation Analysis Method

Ethnomethodological conversation analysis (CA) is a qualitative and data driven analysis approach. Usually CA is done by analysing video data that is gathered in natural situations. During the CA, the selected data relevant for the research objectives are transcribed and analysed. The analysis proceeds without theoretical speculations of the nature of the interaction. The CA reveals how interaction is locally organized and sequentially structured of both verbal and nonverbal utterances. In spite of its name, CA is concerned with the understanding of multimodal interaction including speech and body language [8, 9].

In practice, the details of the internal structure of individual sequences are examined, for example by following Have's [10] guidelines. First, successive utterances are constructed as turns, and then the turns are organized as sequences through the recurrent structures. Characterizing the actions in the sequence include for example the participant's selection of utterance, timing and taking of turns, and how these support participant's understanding and meaning-making of the actions performed. Thus, the organized interpretations that participant himself employ, are described.

CA provides qualitative findings including detailed descriptions of the systematic structural characteristics of particular interactional phenomena. The different phenomena, then, are grouped in collections of different sequences in shorter or longer sequences of interaction. The findings concern the concepts and structures such as turn taking, repair, topic, opening and closing.

Turn-taking is a unit of conversational exchange including an initiation by A as an item which begins anew conversation and sets up an expectation of a response followed by a response by B. There may also be optional elements of exchange structure in the follow-up turns.

Repair is related to some trouble in interaction during a conversation. It includes the sequence of actions and procedures relating and following the error or misunderstanding. The repair is essential in interaction since problems are likely to arise and must be corrected if the interaction is to be successful.

Topic means the matter dealt with in the interaction. Especially the ways each participant introduces (new) topical material and provides opportunities for the other participants to introduce items are of interest. The topics flow from one to another usually through stepwise progression.

Opening and *closing* are turns where flows of interaction are initiated or brought to an end. In human-to-human conversations participants usually introduce themselves during openings and the closing statement calls for goodbye from the participant who is about to leave.

Generally the findings concern the ways in which the actions are recognized and performed by the participants in the selected sequences. Including the detailed analysis of context, CA findings offer deep micro-analytical insight of the actual phenomena of situated human action (for example [11]).

Recently, CA has been successful in exploring early human interaction with a variety of multimodal resources [12]. The analytical challenges to be dealt with reflect those arising in the HCI-data whereby the computer is examined as another participant in interaction and treated as a 'social agent'. Practically, the aim is to explore how the user makes the action through his embodied actions and how the application displays its orientation to the user's actions through certain (iconic) means that are depending on the nature of the application and its resources.

The CA approach has been contributing to both practical and theoretical issues from different aspects on the interactional settings of human-machine interface to emerging re-conceptualizations of social/material relations within HCI field (see [13]). CA has been used to study and design new technology in a variety of fields (see [14], [15], [16], [17]).

1.2 Aim of the Study

In this paper, we examine the possibilities of CA in usability testing. The main goal here is to examine, how well the traditional test setups may serve as a source of information for the CA analysis. We also try to assess the extra value that CA gives compared to traditional evaluation methods, such as heuristic evaluation, cognitive walkthrough and usability tests with thinking aloud protocol.

The video recordings of usability tests were selected as analysis material for CA since utilizing existing data would be an easy and inexpensive way to introduce CA to product development projects. The goal is not to compare the individual results, i.e. found usability problems, of different methods but to analyze whether CA could provide a useful and important additional viewpoint to usability evaluation methods in product development projects.

2 Our Experiment: Conversation Analysis with Usability Test Material

We examined the possibilities of CA by analyzing videotapes recorded in our earlier usability studies. The evaluated systems were a training application that helps in planning and monitoring goal-directed training, and a new electric payment system in a gaming slot machine. In addition to the usability tests, we had conducted a cognitive walkthrough and a heuristic evaluation to the systems.

2.1 Video Data

Since CA usually focuses on certain interaction sequences and not to whole episode, we needed to select the most prominent parts of usability test recordings for our analysis. The identified usability problems were selected for the analysis for two reasons: first, the goal of usability evaluation is to reveal usability problems and to discover the reasons, why it is a problem, and, second, the user-system interaction is "most natural" in the problem situations, since the user often stops thinking aloud. Selecting only those usability problem situations that had been also covered in the cognitive walkthroughs and heuristic evaluations further narrowed the focus of the CA.

Two examples of selected video sequences are:

- Training software usability evaluation: 3.20 min sequence of 'Analyzing the exercises'
  In the selected sequence, the user is trying to compare two training sessions. The results are illustrated graphically in a frame. The user has difficulties in finding the command for comparing. Moreover, she has difficulties in identifying which session is which in the common frame.
- Electronic payment system usability evaluation: 2.20 min sequence of 'Completing the payment'
  In the selected sequence, two test users push a button on the slot machine to finish an operation in the paying procedure. The action is quite correct, but they are not sure, where the won money went, since it was electronic money and not coins, as usual.

2.2 Observations

The original usability testing video data were not gathered Conversation Analysis in mind. Due to this, some questions concerning mainly the actions of the interface remain without answer. As one of the aims of this study is to reveal new types of findings in a usability test, the role of CA is of informative nature. Thus, instead of the detailed conversation analysis procedure, we will present here some observations regarding the ways in which such an analysis may be accomplished in terms of CA and ways in which it may contribute to an understanding of the wider issues of the CA approach.

It is observable from the video data that the users activate the test moderator to participate in the on-going sequence and conversation. In problem situations, the users often pose an explicit question expressing their thoughts and confusions. The user in the payment system poses a question that is accompanied with a gaze shift towards the moderator. The confirmative question (originally in Finnish: 'ja ne kaikki meni tilille vai?'; translated into English: 'And they all went to the account, right?') displays uncertainty about the destination of the money in the paying procedure she just accomplished.

The user's question in the training software ('mistä mä tiedän kumpi on kumpi'; 'how can I know which is which') reveals the difficulty in unclear visualization of the objects the user is interested in. The question is not accompanied with a gaze shift. The question is not necessarily directed to the moderator, but may be rather motivated by the thinking aloud procedure.

The more detailed analysis of the user's situated action in relation to the interface would give more insight into those observations. Moreover, bearing on the CA framework, the data from both sources evoke some questions that have a bit different focus than the traditional usability testing.

In the training software case, the user assumes that the two sessions are displayed within the same frame. The 'traditional' analysis of usability tests is interested in the fact that the user cannot identify which data is which in the visualization. The CA approach is also interested in how the user makes the assumption, i.e. what kind of preceding activity has resulted the user to make the assumption. Naturally the CA is also concerned with how the user finally ends up with the right operation, i.e. what kind of dialogue the user has with the system in order to accomplish the task.

In the electronic payment system case, CA opens up questions, such as: How does the application express what the next relevant action in relation to the preceding action is? Does the interface include simultaneous or competing elements in directing the user, such as 'Cash Out' and 'Bet' buttons flashing simultaneously at the current moment? Does the application provide any help or extra information to support solving potential problems in use?

Having provided some observations and simple examples to demonstrate the ways in which CA-approach may be accomplished, we now turn to discuss both our findings and more general issues concerning CA-enhanced analysis in usability testing.

3 Discussion: Towards CA-Enhanced Analysis in Usability Testing

We discuss here our findings regarding the ways in which CA may be accomplished and ways in which it may contribute to an understanding of the wider issues of the approach.

Our small experiment showed that the traditional test set up does not serve as a sufficient source for conversation analysis for two reasons. First, there is the technical problem that either the actions in the user interface were unclear or the user's facial reactions were not visible, as the user was not facing the camera. Secondly, the setting in the test is problematic, because the user is occasionally having more conversation with the moderator than the system. Therefore, we recommend some changes to the usability test settings concerning the moderator's role and data collecting.

The moderator's role in a test is to be a gracious host, be in control of how the session goes and to be a neutral observer [18]. It is important to establish and maintain a rapport with the test participants. Therefore, in our tests, the moderator practically participates in completing the tasks – in spirit, but not in practice. The moderator asks questions as a follow-up to user's expressions of problems, and gives prompts or assistance when needed. Within conversation analytic framework, the active role of the moderator contributes to more natural 'thinking aloud' by the user, being a natural conversation rather than talking aloud by oneself, which is far from being natural.

In general, conversation analysis approach can be taken towards two separate directions regarding the usability tests. Either the CA analysis can be focused to the dialogue between the moderator and the user, or the focus of the CA can be the user-system interaction.

In the first case, it would be valuable to assess the possibilities of developing the conversational interchange between the moderator and the user towards even more structured interaction. This could be realised by questions, that were prepared in advance to process some critical and potentially problematic points (interviewing-like conversation) or posing spontaneous but user-oriented clarification requests as follow-up to the user's own topical offers that are arising from his own actions with the interface.

Regarding the user-system interaction, the potential of CA is especially in situations where thinking aloud method cannot be used. For example contextual walkthrough [3] is developed for these kinds of situations and CA should be tested with it. Other way to utilize CA could be to setup laboratory tests where a user could be let to test and experiment the product without a moderator or other external persons. This would ensure that the interaction happens solely between the user and the product and make the analysis easier.

In addition, the CA could fit to earlier phases of user-centred design. CA could be used in user research to analyze the users' interaction with current devices and systems, or in design and prototyping to help designers to take into account the conversational aspects of the interaction.

In order to make use of CA as a complementary method in usability tests, there is a need to fine-tune the data gathering. The normal usability test leans heavily to users capability of explaining his or her actions by thinking aloud method. CA requires that all actions are visible to the analyser. Thus, all users key-presses and mouse gestures need to be visible in the recording. In addition CA could benefit from gaze tracking as it could bring more insight into users' activities and intentions. The same goes also for the tested system. All outputs of the system, i.e. beeps, animations and blinking of the cursor etc., need to be visible for the analyzer. All in all the CA requires a more rigorous recording of the test sessions than the normal usability test.

CA-enhanced analysis also allows for making specific collections of the video data demonstrating certain types of problems encountered in the usability testing which is often important for communication of the results.

Specifically, CA can contribute to elaborating usability test findings more fine-grained, for example as to the notion of lacking feedback that was one topic in the usability testing material we used. CA can provide us with the notion that the interaction includes also many conversational actions that are not directly related to completing the task at hand. For example repair, i.e. participants ways to correct misunderstandings and error that have happened before, is not usually used as a concept in human-computer interaction. The user interfaces include elements and functions such as undo, cancel and exit but they effectively reset the situation instead of starting a corrective dialogue with the user. CA also offers the view of the ways in which repairs are constructed (see e.g. [19]). Equally the concepts of turn-taking, topic, opening and closing allow to analyse the relationship between the system and the user, the complexity of the interaction regarding what kinds of and how many topics can be covered and how turns and roles are allocated, and the ways the interaction is initialized and ended.

CA would also contribute to supplementing the results from other methods, such as cognitive walkthrough and or heuristic analysis. The sequence examined in detail through CA-approach would also allow for more exact comparison of the empirical findings after synchronizing the current sequence with the findings from cognitive walkthrough and heuristic evaluation.

Generally, usability testing enhanced with CA would allow for in-depth analysis of found usability problems as well as the holistic interaction between the user and the system. However the benefits are not free of charge as the CA also requires stricter recording of the test sessions and takes more time than normal usability test.

4 Conclusions

In this paper, we examined the possibilities of CA in usability testing applied in the traditional usability test type of settings. CA offers an appealing and interesting framework for deeper understanding of the interaction between the user and the application.

The major question was how well the traditional test setup may serve as a source of information for the new analysis in order to find and figure out the fundamental interactional phenomena concerning the user's actions with the interface.

The current original testing data does not allow accomplishing a detailed CA procedure, but it allows for examining the possibilities and requirements for using CA and suggesting some changes to usability testing. Implementation of CA-approach into usability testing would result in different solutions in constructing usability testing set up. Most importantly, CA-approach can contribute to the nature of video data and data collecting techniques.

CA can contribute to elaborating usability test findings more fine-grained, as a single method or supplementing the results from other methods. Analysis is time consuming but it can be harnessed as a method for very narrow focusing on certain sequences extracted from the usability testing data. In the future, it should be examined how to develop CA for meeting the specific needs in the usability testing set up. CA offers also one framework for unifying concepts used in human-computer interaction.

More generally, there is no one way of 'doing' CA. The analyst may follow one's own particular interests in choosing the analytical focus. Traditionally, the usability testing has been interested in problematic phenomena. However, to contribute wider to the current needs and trends in HCI field, it may be valuable to explore also the phenomena that reveal the unproblematic interaction between the user and the application.

Promising uses for CA are situations in which thinking aloud is not possible. These test setups (Contextual walkthrough [3]) are often arranged so that the review of the walkthrough is done afterwards when the actual test event has passed. Such situations are e.g. tests that are arranged in real work settings without the possibility to interact in the situation. The organisers of the test do not interfere the progress of the workflow but analyze it later with or without users. The joint interpretation of the events with users provides an interesting way of ascertaining the interpretation made by the researcher or analyst improving the accuracy of the results.

Altogether, using CA provides interesting extensions to usability tests but not directly without modifications to the test arrangements. The recording of the test needs to be adjusted to fit the needs of both analyses. This can be achieved with reasonable amount of additional test instrumentation and work, though.

Additionally, the uses of CA are not restricted to testing and evaluation only. We expect CA to provide means for analysing and improving the existing interaction and dialogue structures towards repairing ones for improved flexibility and adaptability in less constrained interaction situations. This provides interesting directions for more fundamental consideration of human-computer dialogue.

It is our intent to further research on the inclusion of CA in our forthcoming research projects both for the improvement of usability testing and for the basic research on interaction structures.

Acknowledgments. We are grateful to Lasse Lumiaho, Ville Toivonen and Mari Tyllinen for their input to re-evaluating and analysis of the original usability data.

References

1. Bias, R., Mayhew, D.: Cost-justifying usability. Academic Press, Boston (1994)
2. Nielsen, J.: Usability Engineering. Academic Press, London (1993)
3. Riihiaho, S.: Experiences with usability evaluation menthods. Licentiate's thesis. Helsinki University of Technology, Department of computer science and engineering (2000)
4. Koivunen, M., Nieminen, M., Riihiaho, S.: Launching the Usability Approach: Experience at Helsinki University of Technology. SIGCHI Bulletin 27(2), 54–60 (1995)
5. Nieminen, M., Koivunen, M.R.: Visual Walkthrough. In: Allen, G., Wilkinson, J., Wright, P. (eds.) HCI 1995, People and Computers, Adjunct Proceedings. The School of Computing & Mathematics, The University of Huddersfield, pp. 86–89 (1995)
6. Nieminen, M.P.: Designing user interface concepts for multimedia services. Master's thesis, Helsinki University of Technology (1996)
7. Gomoll, K., Nicol, A.: Discussion of guidelines for user observation. User Observation: Guidelines for Apple Developers (1990)
8. Goodwin, C., Heritage, J.: Conversation Analysis. Annual Review of Anthropology 19, 283–307 (1990)
9. Goodwin, C.: Action and embodiment within situated human interaction. Journal of Pragmatics 32(10), 1489–1522 (2000)
10. Have, P.T.: Doing conversation analysis: a practical guide. Sage, London (1999)
11. Atkinson, M., Heritage, J.: Structures of social action: Studies in conversation analysis. Cambridge University Press, Cambridge (1984)
12. Karjalainen, S.: Multimodal Resources in Co - Constructing Topical Flow: Case of "Father's Foot". In: Esposito, A., Bratanic, M., Keller, E., Marinaro, M. (eds.) Fundamentals of verbal and nonverbal communication and the biometrical issue. IOS press, The Netherlands (2007)

13. Suchman, L.: Human-Machine Reconfigurations: Plans and Situated Actions, 2nd edn. Cambridge University press, Cambridge (2007)
14. Luff, P., Gilbert, G.N., Frohlich, D.M. (eds.): Computers and conversation. Academic Press, London (1990)
15. Button, G., Sharrock, W.: On simulacrums of conversation: toward a clarification of the relevance of conversation analysis for human-computer interaction. In: Thomas, P. (ed.) The Social and Interactional Dimensions of HumanComputer Interfaces, pp. 107–125. Cambridge University Press, Cambridge (1995)
16. Luff, P., Hindmarsh, J., Heath, C.C. (eds.): Workplace Studies: Recovering work practice and informing system design. Cambridge University Press, Cambridge (2000)
17. Woodruff, A., Aoki, P.M.: Conversation analysis and user experience. Digital Creativity 15(4), 232–238 (2004)
18. Dumas, J., Loring, B.: Moderating usability tests: Principles and practices for interacting. Morgan Kaufmann/ Elsevier (2008)
19. Raudaskoski, P.: Repair work in human computer interaction. In: Luff, P., Gilbert, G.N., Frohlich, D.M. (eds.) Computers and conversation, pp. 151–172. Academic Press, London (1990)

Possibility of Participatory Design

Makoto Okamoto

Future University Hakodate
Kamedanakano 116-2, Hakodate City, Hokkaido, 041-8655, Japan
maq@fun.ac.jp

Abstract. Participatory design has attracted attention as a design method in recent years. Scenario or Inclusive Design is one way to make users participate in
the design process. In this paper, I report on some case studies in which the
visually-impaired participated in the design process. The sighted designer
worked together with the visually-impaired wearing eye masks. The visually-impaired could tell designers some problems in their lives or their demand
which was difficult to express in words.

Keywords: Participatory Design, Inclusive Design, scenario based design, Information Design.

1 Introduction

Every artifact in life in recent years has been computerized with the development of
information technology. Coverage of designed objects has expanded from material like
products to lifestyle, the way of working and the various experiences people can get
through them. Various researchers have proposed new design methods appropriate for
such circumstances. The design method using scenario is effectual to know people's
daily lives in more depth. Inclusive Design has another effect: to make people involved
participate in the design process.

As the designed objects expanded from material to activity, it has become more
difficult for one designer to understand the region of problems by individual experience, observation and interview in a short time.

Scenario or Inclusive Design method is not to observe an object from outside, it is a
new approach in which a designer participates in the "people's world" and finds an
understanding of a problem from the same point of view as the people.

It means that a designer has a new role to develop wisdom for living with people. I
recognize that the words, participatory design, mean an approach to design with people
about a life which is being computerized. But I think that this approach will become one
of the design standards in future. In this paper, I report some cases and would like to
consider what the participatory design means.

2 Related Research of Participatory Design

Scenario-based design proposed by John M. Carroll is a very interesting method. As
scenario is to write a current activity or a designed activity in a future as an episode,

M. Kurosu (Ed.): Human Centered Design, HCII 2009, LNCS 5619, pp. 888–893, 2009.
© Springer-Verlag Berlin Heidelberg 2009

stakeholders can share a problem easily and help create an idea for the problem. In particular, one of the merits is that people can indicate that the report from either designer or observer is adequate or inadequate.

Inclusive Design is a form of participatory observation. It is the way that an observer (designer) walks into people's lives, in order to facilitate a mutual understanding of things they can't describe, or events they do not see as a problem unconsciously.

Liz Sanders created the method called Velcro Modeling to get tacit knowledge, which is difficult to understand by words alone. A subject given some task makes what he/she wants conscious by building a block which is easy to connect by Velcro. Many people can't describe clearly their request or problem but they can become aware of them in the process of concrete creative work. Additionally an observer can make a concrete question after looking at the accomplishments. Okamoto and other researchers developed a system (Mobile AP) to report people's findings using a cell-phone and its camera. The system reports requests or some kind of awareness people find in their daily life on an as needed basis. At first, the subjects were puzzled but they gave many reports after we explained that reporting a problem was not a shame, and that their request might change the products around them. We named these support environment and method Leverage Design. We report the outline and effect in a class that used Participatory Design method.

3 Create with the Handicapped

We had a participatory workshop to design an information tool for the handicapped with their help. We used Inclusive Design method.

3.1 Workshop Process

The theme was to design an information tool for the visually-impaired. But many students did not know how the visually-impaired lived or what they wanted. It was clear that design resulting from the diffidence or conjecture would frustrate them. Therefore we thought that students (designers) could understand the life of the visually-impaired and the visually-impaired could join the design process by sharing the one experience (Fig.1).

Fig. 1. Procedure of Inclusive design

At the workshop, the sighted (students) had the following experience with an eye mask (Fig.2). The visually-impaired could have an advantage with this restriction. The work proceeded keeping a partnership between the visually-impaired and designers (students).

- Eating lunch
- Walking
- Indicating where a sound comes from
- Listening to a sound (sound difference between large and small space)
- Message game

Fig. 2. Sharing of Experience

3.2 Sharing Experience and User Participation

After the experience, students were clearly different about their insight and under-standing about the issues faced by the visually-impaired compared to before they started the process. To create an information tool for the visually-impaired might not be an attractive theme for young prospective designers. For example, many students opt for the "cool" products. This was the first time for some students to talk with the visually-impaired so some of them did not know how to communicate with them. But after this experience, they amazingly made many reports on what they had found and proposed various kinds of ideas.

There was a change in their attitude for the visually-impaired. Students wore eye masks in many activities and became temporally visually-impaired. The visu-ally-impaired became good teacher who could lead students to a lightless life from the supported point of view and told students what they devised or wondered in their daily lives.

Typical reports from students are as follow.

- Planning how to lay out things to memorize easily in consideration of the next step of the work
- Sometimes communicating with an easy-to- image figurative phrase
- Do not want to use too-many tools for the visually-impaired that may inadvertently be little their ability or intelligence.
- Standard for good user interface is different between the visually-impaired and others

3.3 Proposed Design Ideas

Students made various kinds of prototypes after discussion with both the visually-impaired as well as the students.

Fig.3 is a communication device for the visually-impaired and the hearing-impaired. Communication between the visually-impaired and the hearing-impaired is very difficult. Glove-on-hand is a device to transmit the tracks drawn by fingers. Fig.4 is a device to tell "here" and "there" to the visually-impaired by the process of changing a weak ultrasonic wave to audible sound when it hits some object. Using this nature, when we aim this tool at some objects, we can hear sound saying "here".

Fig.5 is an electronic jar for the visually-impaired. To help alleviate errors when touching the buttons, each button makes a different sound. Moving Prototype they made is to transmit sound feeling.

Fig. 3. Device that transmits tracks drawn by finger

Fig. 4. Pointing Device using Ultrasonic Wave

Fig. 5. Moving Prototype – Pot with Sound Button

4 Touchable Exhibition

Students exhibited their prototype for the visually-impaired. The purpose of this presentation was so that the visually-impaired could evaluate the work. Therefore students made and exhibited the touchable prototypes (Fig.6.).

They made an oral explanation about their work touching the prototypes. The number of the visually-impaired is not so many in our city of 300,000 people, but about 50 visually-impaired and their assistants joined our workshop and presentation. As they could not look over the whole exhibition, they took a long time to touch 20 works one-by-one and gave us much feedback. Interestingly in various places many observers had a lot of things to talk about and could not stop talking.

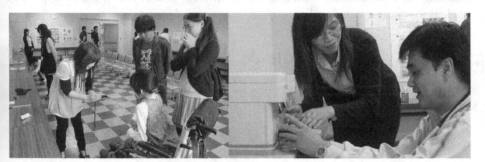

Fig. 6. Exhibition for Touching

5 Consideration

We had workshops in collaboration with the visually-impaired using Inclusive Design. It is possible to learn about understanding the disability and the type of the disability through the literature, but it is difficult to know about their individual life and feelings. We think that an approach of Participatory Design, like Inclusive Design is useful to have the same standpoint with the handicapped and understand their context.

In addition, reaction from the visually-impaired has changed the designer (student) who wore an eye mask and walked into their world.

It is not easy for the sighted to walk and eat in the blind world. The impaired became a teacher for the sighted and taught them how to understand the environment by sound. The impaired who have never thought that a new device might be made by their opinions gradually expressed what they wanted in words and started to tell it to the designers.

Naturally rapport has been established.

6 Conclusion

In order to consider the effect of participatory design, we held workshops. Students who were beginners in design could make interesting proposals easily through collaboration with the handicapped. Information systems should be designed taking the variety of people's life into consideration. Although it is difficult to understand an environment or a context of living, we consider that Participatory Design is an effective way to develop useful hypotheses.

References

1. Carroll, J.M.: Making Use. MIT Press, Cambridge (2000)
2. Okamoto, M., Komatsu, H., Gyobu, I., Ito, K.: Participatory Design Using Scenarios in Different Cultures. In: Jacko, J.A. (ed.) HCI 2007. LNCS, vol. 4550, pp. 223–231. Springer, Heidelberg (2007)
3. Go, K.: Requirement Engineering. Kyoritsu Publisher (2002)
4. Kato, S., Okamoto, M.: Tool supporting memory of visually impaired person. WIT (2006)
5. Komatsu, H., Ogawa, T., Gyobu, I., Okamoto, M.: Scenario Exchange Project: International workshop using Scenario Based Design, Human Interface 2006 (HIS 2006), Japan, pp. 503–508 (2006)

The Value of Answers without Question[s]: A Qualitative Approach to User Experience and Aging

Anna Elisabeth Pohlmeyer[1], Lucienne Blessing[2], Hartmut Wandke[3], and Julia Maue[4]

[1] Berlin Institute of Technology, Center of Human-Machine Systems,
Franklinstr. 28/29, FR 2-6, 10587 Berlin, Germany
[2] University of Luxembourg, Engineering Design and Methodology, Campus Limpertsberg,
162A, avenue de la Faïencerie, L-1511 Luxembourg
[3] Humboldt University Berlin, Engineering Psychology / Cognitive Ergonomics,
Rudower Chaussee 18, 10099 Berlin, Germany
[4] School of Art and Design Berlin-Weissensee, Bühringstr. 20, 13086 Berlin, Germany
anna.pohlmeyer@zmms.tu-berlin.de, lucienne.blessing@uni.lu,
hartmut.wandke@rz.hu-berlin.de, jmaue@aol.com

Abstract. This project investigates reasons for use and non-use of interactive products by two age groups. It was motivated by the assumption that older adults, when given the chance, report more than just usability-related aspects of interactive products. In laboratory settings, older adults are oftentimes confronted with unfamiliar technology. In this case, instrumental qualities are of primary concern. However, the picture might be different, when it is up to the participant to choose the device. Twenty younger (20-33 years) and 20 older (65-80 years) adults were provided with a disposable camera and a documentation-booklet for one week in order to photograph and describe positive as well as negative examples of interactive products in their surrounding. After this week of intensive sensitization, participants named five reasons that motivated them to use technology, and five that led to avoidant behaviour. A qualitative content analysis with an inductive development of categories was conducted.

Keywords: User Experience, Aging, Motivation, Methods, Content Analysis.

1 Introduction

In line with the current demographic shift, more and more older adults will be facing the use of technology in the upcoming years. Older adults are a target group that can clearly benefit from technology (e.g. compensating physical impairments as well as sensory and cognitive losses; facilitating communication).

However, in order to take advantage of this trend and to gain reinforcing experiences, one has to interact with the device in the first place. Many older adults underestimate their actual computer knowledge [1] and show low self-efficacy beliefs in this domain [2], which leads to a rather hesitant behaviour concerning the use of technology with increasing age. Much work has been done to address the concerns and challenges of cognitive and sensorimotor decline as people age [3].

M. Kurosu (Ed.): Human Centered Design, HCII 2009, LNCS 5619, pp. 894–903, 2009.
© Springer-Verlag Berlin Heidelberg 2009

However, hardly any efforts have been undertaken to investigate cross-generational differences concerning the appreciation and relevance of emotions in the experience of interactive products. It seems short-sighted to assume that older adults do not care about non-instrumental qualities of interactive systems such as aesthetic, symbolic, and motivational qualities, which are closely linked to emotional processing [4]. Many products designed for older users lack respectful consideration of their tastes and personal surrounding. An interactive product, even if it is of high assistive value, might be rejected if it does not fit into the person's environment or communicates "disability" to others. A design tailored to older adults' preferences could be a promising path to facilitate initial interaction with technology and therefore increase the likelihood of usage.

2 User Experience: A Paradigm Shift from Usability

According to the ISO 9241 part 11 [5], usability is defined as "the extent to which a product can be used by specified users to achieve specified goals with effectiveness, efficiency, and satisfaction in a specified context of use." This is in line with a user-centred design approach, but, in practice, heavily focused on task fulfilment and objective performance data such as time of task completion and error rates. For decades the field of ergonomics and human factors contributed with fundamental as well as applied insights to an optimization of user interface design. But granted that user' safety and functionality needs are met, the emotional involvement of the user should to be taken into account in addition. Logan [6] coined the term *emotional usability*. With an even more pronounced focus on pleasure, Hancock et al. call for a shift from *ergonomics* to *hedonomics* [7].

As noted above, user needs go beyond the efficient achievement of tasks. The wave of the information age where usability/ease of use was considered to be the differentiator is coming to an end. Nowadays, it is expected that a given device will provide the intended utility and its means to meet these goals. The still rather young, but widely accepted and evolving field of *user experience* has widened the scope of usability and its strong task-dependency in the past decade by including non-instrumental qualities of a system and its emotional responses by the user.

"Usability factors determine whether a device *can* be used, aesthetic factors determine whether a device *will* be used" [8]. This distinction is most likely an oversimplification. However, the inclusion of non-instrumental qualities of an interactive device should be acknowledged. Petersen and colleagues take a more integrative approach: the pragmatist's [9]. Accordingly, *interaction aesthetics* has an instrumental value as well by promoting the user's drive for exploration. Most likely, it is the interplay of usability and aesthetics that is of importance [10].

3 Age Differences or Age Similarities?

Usually, within the literature of human-technology interaction, the section on *aging* highlights cognitive and sensory decline in the elderly. This negative emphasis on aging is a limited view. Designing for older adults should not be restricted to sensorimotor and cognitive issues in order to complete a task. Here, a more positive focus on maintained skills (e.g. emotion-regulation) and desires has been taken. Older adults should not be underestimated in their capabilities and desires. With the increasing importance of emotion-regulation and the resulting interplay of emotion-motivation-cognition in older adults [11] the field of user experience with its emphasis on emotional involvement seems a promising path for appropriate user interface design.

3.1 Aim of the Study

The motivation for this work was the concern that in lab settings elderly people may focus on the challenges of using unfamiliar technology, but that if allowed to reflect on more familiar devices less functionally-oriented themes may arise.

In other words, the goal was to define relevant dimensions that are crucial with respect to motivational aspects of human-technology interaction. What constitutes a good interactive product, what a bad one? Davis' technology acceptance model [12,13] highlights the role of intentions, which are based on the perceived ease of use and usefulness of the product. Here, we wanted to investigate whether there were more criteria that needed to be taken into account when designing interactive products that should not only be useful and usable but actually used [14]. By this nature, we chose an open question format, not restricting the participants, in order to see what is of relevance in real-life scenarios. Instead of providing a list of questions that just needed to be checked off, we only asked two questions: (1) do you like this interactive product or not and (2) why? The answers given by the users were not pre-determined by the experimenter who might be only interested in quantitative differences. Instead, the answers revealed valuable information about qualitative differences. These answers should be the questions in studies to come.

This approach called for a qualitative content analysis. Given that inferences from a content analysis heavily depend on the participants, the task (instructions) and the chosen procedure of the analysis, these shall be outlined in the following.

4   Method

4.1   Participants

The study involved 40 participants, recruited from two age groups in order to draw cross-sectional comparisons: 20 older adults ($M = 70,8$, $SD = 5.094$), ranging from 65 to 80 years and 20 younger adults ($M = 25,2$, $SD = 4.171$), ranging from 20 to 33 years; each group consisting of ten males and ten females from all over Hamburg, Germany.

The participants were recruited from a wide range of family friends and their co-workers. However, most of the participants were unknown to the experimenters. A connection to the experimenters, despite the anonymous treatment of the collected data, encouraged to participate and helped considering the study trustworthy. Potential participants were contacted via telephone or approached personally.

One older woman and two younger men were not willing to participate after the instructional session, because the study was too time-consuming and one older woman changed her mind after the instructional session because of privacy issues (feeling uncomfortable taking pictures of personal possessions).

Participants signed an informed consent prior to the introduction session and after re-collecting the provided tools they received a voucher as a reward: Older adults received a book-voucher (10€), younger adults received a voucher for one cinema visit (~ 10€).

4.2 Apparatus and Material

Control Measures. We used a questionnaire on technology affinity [15] with the four subscales (1) enthusiasm towards electronic devices (Cronbach's $\alpha = .842$), (2) subjective competence in using electronic devices ($\alpha = .789$), and (3) perceived positive ($\alpha = .722$) and (4) negative consequences ($\alpha = .747$) connected to the use of electronic devices. This questionnaire is a 19-item self-report measurement, using 5-point Likert rating scales ranging from 1 (does not apply at all) to 5 (applies exactly) in order to express the agreement or disagreement with a given statement. Apart from the means with respect to the subscales, the authors also computed a composite score which was the sum of means, thus limited to a maximum of 20 (4 subscales * 5 maximum score each). Items with a negative connotation were reversed. As a result, the higher the score the more people feel affinity towards technology. Additionally, demographic data (family status, years of education, current occupation, health status, self-rated well-being) was collected.

Task equipment. We handed each participants a zipper-bag that included a snapshot-camera (Agfa Photo, Le Box Camera Flash single use, for in- and outdoor use, with an Agfa Vista film ISO 400 for 27 colour prints), a documentation-booklet, an instruction letter and a pen.

The documentation-booklet was designed for this purpose. It included (1) a brief description of the task demand, (2) an overview table for the number of "positive" and "negative" examples the participants photographed and the corresponding picture-number, (3) a double-page for each picture, including space for the name of the object photographed, space for marking whether it is judged to be a "positive" or "negative" example, and several lines for listing relevant reasons for this decision. An open question format was chosen to ensure that participants would not be misled by the experimenter's expectations. Apart from the name of the device and a mark indicating the judgement's valence, the list of reasons was the only question per device. This was done to avoid missing data on this crucial question. (4) At the end, two pages offered space to list five reasons that motivate the participant to use an interactive product and five reasons that keep him/her from using interactive products. (5) The last spread was meant for comments, critique, or suggestions for improvement.

4.3 Procedure of Data Collection

Data collection was divided into two phases: the instructional session with the experimenter present lasted approximately 1 to 2 hours for the young and older adults, respectively. The second phase lasted one week during which participants independently completed the task of photographing and documenting verbally good and bad examples of interactive products. The introduction sessions were executed individually at the private homes of the participants.

In the beginning of the instructional session, information on the participants' technology affinity was gathered and the demographic data obtained. We handed over the Zipper-bags containing the camera, documentation-booklet, and the instructional letter. Actively involving the participants, we thoroughly went through the instructional letter to provide a precise understanding of the task demands and to motivate the participants. Examples from a list of interactive products (from the letter) and randomly picked examples from the current environment (usually, their living room) were pointed out.

The general introduction was followed by a phase of familiarization. First, the handling of the camera was explained, including the operation of the flash and the viewfinder and information on the appropriate distance to the object when taking a picture. The participants followed the instructions while testing the provided cameras on their own: five interactive products, an Apple iPod (20 GB, 2004), a digital camera (Lumix, Panasonic, DMX-LX1, 2006), a calculator (Casio SL 300, 1995), a Nintendo Gamboy Advance (2004), and a mobile phone (Nokia 2652, 2005), were laid out in front of the participants to demonstrate the task they would later have to perform on their own. The devices differed in the status of being worn out, the designs, the brand and technical complexity. Participants were confronted with the task to evaluate the examples, selecting one device they liked and one they disliked, to take a picture of the chosen devices and to list the reasons relevant for their decisions in the documentation-booklet. This short demonstration not only exemplified a realistic training of the task and its procedure that participants would have to engage in on their own in the course of the week, but it also served as a 'framing' with respect to what we considered and outlined to them as being *interactive products*. Interactive products were described to be technical devices that had some kind of higher order structure of interactive elements. In other words, it was not sufficient to just plug the device in with no further means of interaction or selection. Although not explicitly instructed, a "product" here had the connotation of some hardware components to it (in theory, a website is a product as well). However both, analogue and digital, menu structures were considered to be valid modes.

It was stressed that the subjective opinion and experience of the participant was of interest and that there was no "right" or "wrong" answer.

4.4 Data Analysis

Subsequently, the pictures were developed as print-outs and scanned to make them digitally available. All data, including the digitalized images, were first entered and coded in Excel 2003. Final statistical procedures were conducted with SPSS 17.

The data discussed here reflect the overall reasons for use and non-use that participants named after one week of intensive sensitization to the topic by photographing and describing 24 interactive products. They were asked to name five reasons that motivate them to use an interactive product as well as five reasons that lead to avoidant behaviour regarding technology use.

Five participants (two older and three younger men) who were engaged in the entire task over the course of the week did not fill out the overall reasons. Consequently, the following analysis is based on a subsample. Since an inductive development of categories was conducted, it is especially important to know who provided the data: 18 older adults between 65-80 years (M = 70.7, SD = 4.897; 10 women) and 17 younger adults between 20-33 years (M = 25.7, SD = 4.120; 10 women). Older and younger adults did not differ with respect to years of education (M_{old} = 16.81, SE_{old} =.789, M_{young} = 16.9, SE_{young} = .666, $t(31)$ = -.089, p > .05). However, older adults showed a less pronounced technology affinity than their younger counterparts (M_{old} = 11.71, SE_{old} =.450, M_{young} = 15.16, SE_{young} = .310, $t(31)$ = -6.381, p < .001). This difference is based on less enthusiasm, less subjective competence, and more perceived negative consequences (values have been reversed) of technology in older adults. There was no age difference regarding the perception of positive consequences (see Table 1).

Table 1. Age Group Statistics of Technology Affinity (independent t-test). Item examples have been translated by the authors; participants received the scales in German.

subscale	example item	group	mean	SE	df	t	p
enthusiasm max 5	"I love to own new electronic devices"	old	2,12	,193	33	-4.784	< .001
		Young	3,32	,156			
competence max 5	"I easily learn the use of a new electronic device"	old	2,44	,196	32	-4.960	< .001
		Young	3,75	,177			
posConseq. max 5	"Electronic devices enable independence"	old	3,76	,114	32	-.875	> .05
		Young	3,89	,094			
negConseq. max 5	"Electronic devices cause stress"	old	3,15	,154	32	-5.114	< .001
		Young	4,20	,135			

In total, 375 reasons were given, of which 188 in the category of positive aspects and 187 in the category of negative characteristics. This classification was already done by the participants themselves. Due to the explorative nature of this study, an inductive development of the categories differentiating the stated reasons (e.g. ease of use vs. quality of outcome) was undertaken. The authors primarily followed the systematic, rule-guided suggestions of Mayring's content analysis [16]. The process of inductive development of the classification scheme included several iterations of

(1) Initial formulation and structuring of the reasons into distinct categories that were characterized by unique characteristics not found in other categories.
(2) This classification was revisited and modified after 20, 40, 60, 80 and 100 % of the data set. If an item could not be classified into an existing category, a new one was generated.
(3) Coding instructions were formulated, including definitions, anchor examples of positive as well as negative examples that fall into the according category, and, if appropriate specific coding rules or exceptions were pointed out. Each statement was only allowed to be assigned to one category. Thus, the coding scheme needed to be straightforward and categories easily distinguishable.

(4) A subsample of 20% (75 items, including difficult items that were challenging to classify) was presented to two colleagues, both usability professionals, together with the coding instructions. After a short introduction to the coding instructions they classified the 75 statements to the defined categories. Despite an agreement of over 90%, they were asked to critically comment their coding experience. Specifically, cases of disagreement were discussed and consensually classified. The coding scheme was refined and instructions reformulated.

(5) All 375 statements were classified accordingly.

(6) A random sample (generated by SPSS 17) of 20% (75 items) was drawn. A colleague who was 'naïve' to this study and with a background in computer science was provided with the coding instructions and a short introduction to the study. In the following step, she coded the 75 items on her own to the provided 33 categories (see Results). This classification served to determine the inter-coder reliability using Krippendorff's α statistic [17]. The agreement among coders (inter-coder reliability) is one of the most important quality indicators with respect to nominal scale data developed from qualitative judgments [18].

Frequency statistics are reported in percent due to different cell sizes (there is more data available from older adults and less for motivating reasons). Consequently, relative values (percent) within groups of interest (valence x age group) are more appropriate and less misleading for the reader.

5 Results

Both, motivating and hindering, reasons were given the same chance of inclusion to a category. Put in other words, there were no categories developed that would not allow the consideration of a statement due to its valence. This decision was made to see whether reasons for use and non-use are the same, only on different ends of a scale, or whether separate dimensions are relevant.

We tried to find a justifiable balance of clustering related reasons on one hand and permitting a wider diversity of reasons than usually found in the literature on the other hand. Even in the case of small units, we decided to keep them as separate categories if they were distinguishable from the others, keeping in mind that the aimed classification scheme should serve later on as a deductive application on the reasons associated with the individual devices. As a result, we are happy to have an almost empty 'rest' category (only one item "engineers were not thinking"), but we are aware that this is partly due to the fact, that categories with small sample sizes were allowed too.

Altogether, 33 categories could be found. For the sake of simplicity, they were further allocated to 10 main categories as Table 2 illustrates.

The inter-coder reliability was found to be Krippendorff's $\alpha = .93$ (95% CI (.859, .986)) and therefore a strong support for the coding scheme [18].

Table 2. Categories and main categories (black filling). The five most frequently named reasons by valence (+ / -) x age group (O / Y) are marked accordingly.

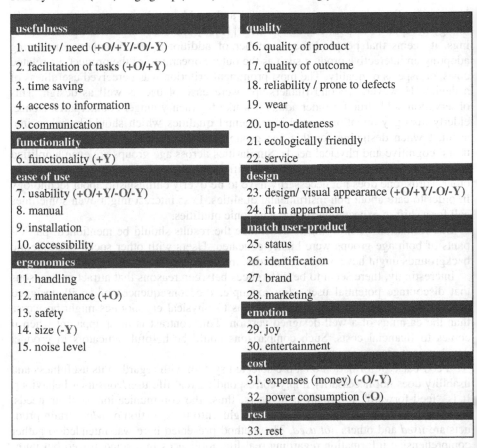

usefulness	quality
1. utility / need (+O/+Y/-O/-Y)	16. quality of product
2. facilitation of tasks (+O/+Y)	17. quality of outcome
3. time saving	18. reliability / prone to defects
4. access to information	19. wear
5. communication	20. up-to-dateness
functionality	21. ecologically friendly
6. functionality (+Y)	22. service
ease of use	**design**
7. usability (+O/+Y/-O/-Y)	23. design/ visual appearance (+O/+Y/-O/-Y)
8. manual	24. fit in appartment
9. installation	**match user-product**
10. accessibility	25. status
ergonomics	26. identification
11. handling	27. brand
12. maintenance (+O)	28. marketing
13. safety	**emotion**
14. size (-Y)	29. joy
15. noise level	30. entertainment
	cost
	31. expenses (money) (-O/-Y)
	32. power consumption (-O)
	rest
	33. rest

Table 3. Distribution of main reasons responsible for use (first two bars per category) and non-use of technology by older (continuous frame) and younger (dashed frame) adults in percent

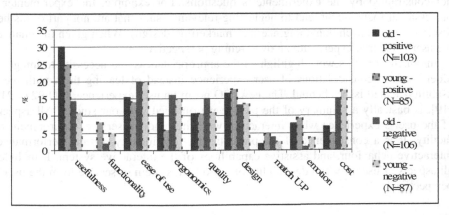

6 Discussion

Notwithstanding, the importance of useful and usable products with respect to technology acceptance as pointed out by Davis [12,13] and confirmed in the present findings, it seems that people have a number of additional criteria when considering adopting an interactive product such as the outer appearance of the product, its related costs, or aspects of quality. The most prominent criterion was perceived usefulness of a device. However, "second runners-up" were ease of use as well as design. This observation holds true for older adults just like for their younger counterparts. Hence, elderly also pay attention to non-instrumental qualities which should be taken into account when designing for this age group: technology that goes beyond meeting users' cognitive and physical needs. Similarities across age groups are even amplified when considering that the two groups significantly differed in technology affinity. In other words, one does not necessarily have to be overly enthusiastic about technology in order to care about non-instrumental qualities. Less interest might even sometimes call for a shift in expectations to more hedonic qualities.

One limitation when trying to generalize the results should be mentioned: participants of both age groups were highly educated. Users with other social or academic backgrounds might have other preferences, interests, and concerns.

Interestingly, there seem to be differences between reasons that attract and reasons that discourage potential users. For example, the consequences of an ill designed product with respect to cognitive as well as to physical ergonomics might be worse than the earnings of a well designed version. This contrast is most apparent when it comes to financial costs. Such comparisons could be helpful indicators in product development processes.

"Feed-back" information on a pre-selected system with regard to its usefulness and usability does not necessarily imply corresponding real-life user/consumer behaviour. It is "feed-forward" information of users, thus, the communication of their needs, aspirations and preferences that can shed light onto the question of *why* certain products are *used* and others *not used*. The method presented here was intended to gather comprehensive information regarding real-life appliances in a standardized but unrestricted manner. Older and younger adults were given the opportunity to express their take on things via self-documentation. In this manner, answers were given that were not constrained by the experimenter's questions. For example, the experimenter's professional focus might lead to neglecting relevant issues that are not part of his/her discipline (e.g. psychology, engineering, marketing, design). When given the chance, the user is the true expert with an overarching perspective.

In conclusion, this work highlighted the appropriateness and necessity to integrate older users in considerations of user experience instead of leaving them with mere usability-related issues behind. The new ISO norm on user experience, ISO 9241-210 [19], is basically a summary of the picture we outlined with our study: "All aspects of the user's experience when interacting with the product, service, environment or facility. It is a consequence of the presentation, functionality, system performance, interactive behaviour, and assistive capabilities of the interactive system. It includes all aspects of usability and desirability of a product, system or service from the user's perspective".

Acknowledgements. This research was supported by the German Research Foundation, DFG (nr. 1013 'Prospective Design of Human-Technology Interaction') and by the National Research Fund Luxembourg (AFR). The authors wish to thank Martin Hecht, Svenja Schiffler, and Michele Qu for their help in data screening.

References

1. Marquie, J.C., Jourdan-Boddaert, L., Huet, N.: Do older adults underestimate their actual computer knowledge? Behaviour & Information Technology 21(4), 273–280 (2002)
2. Czaja, S.J., Charness, N., Fisk, A.D., Hertzog, C., Nair, S.N., Rogers, W.A., Sharit, J.: Factors Predicting the Use of Technology: Findings From the Center for Research and Education on Aging and Technology Enhancement (CREATE). Psychology and Aging 21(2), 333–352 (2006)
3. Fisk, A.D., Rogers, W.A., Charness, N., Czaja, S.J., Sharit, J.: Designing for Older Adults: Principles and Creative Human Factors Approaches. CRC Press, Boca Raton (2004)
4. Mahlke, S.: The Diversity of Non-instrumental Qualities in Human-Technology Interaction. MMI-Interaktiv 13, 55–64 (2007)
5. ISO. ISO 9241: Ergonomic requirements for office work with visual display terminals - Part 11: Guidance on usability. ISO, Geneve (1998)
6. Logan, R.J.: Behavioral and emotional usability: ThomsonConsumerElectronics. In: Wiklund, M. (ed.) Usability in practice, pp. 59–82. Academic Press, Cambridge (1994)
7. Hancock, P.A., Pepe, A.A., Murphy, L.L.: Hedonomics: The Power of Positive and Pleasurable Economics. Ergonomics in Design 11(1), 8–14 (2005)
8. Forlizzi, J., Hirsch, T., Hyder, E., Goetz, J.: Designing Pleasurable Technology for Elders. In: Proceedings of INCLUDE 2001 (International Conference on Inclusive Design and Communications) (2001)
9. Petersen, M.G., Iversen, O.S., Krogh, P.G., Ludvigsen, M.: Aesthetic Interaction: A Pragmatist's Aesthetics of Interactive Systems. In: Proceedings of the 5th Conference on Designing Interactive Systems: Processes, Practices, Methods, and Techniques (2004)
10. Hassenzahl, M.: The Interplay of Beauty, Goodness, and Usability in Interactive Products. Human-Computer Interaction 19, 319–349 (2004)
11. Carstensen, L.L., Mikels, J.A., Mather, M.: Aging and the intersection of cognition, motivation and emotion. In: Birren, J., Schaie, K.W. (eds.) Handbook of the Psychology of Aging, 6th edn., pp. 343–362. Elsevier, Amsterdam (2006)
12. Davis, F.D.: Perceived Usefulness, Perceived Ease of Use, and User Acceptance of Information Technology. MIS Quarterly, 319–340 (September 1989)
13. Davis, F.D., Bagozzi, R.P., Warshaw, P.R.: User Acceptance of Computer Technology: A Comparison of Two Theoretical Models. Management Science 35(8), 982–1003 (1989)
14. Dix, A.: Designing for adoption and designing for appropriation. In: Talk at Berlin Institute of Technology (February 12, 2008) (unpublished)
15. Bruder, C., Clemens, C., Glaser, C., Karrer, K.: Entwicklung eines Fragebogens zur Erfassung von Technikaffinität. Diagnostica (submitted)
16. Mayring, P.: Qualitative Inhaltsanalyse. Grundlagen und Techniken, 9th edn. Beltz Verlag, Weinheim (2007)
17. Hayes, A.F., Krippendorff, K.: Answering the call for a standard reliability measure for coding data. Communication Methods and Measures 1, 77–89 (2007)
18. Krippendorff, K.: Content Analysis: An Introduction to its Methodology, 2nd edn. Sage Publications, Thousand Oaks (2004)
19. ISO CD 9241-210: Ergonomics of human-system interaction - Part 210: Human-centred design process for interactive systems. ISO (2008)

Shaping the Future with Users – Futures Research Methods as Tools for User-Centered Concept Development

Mikael Runonen and Petri Mannonen

Helsinki University of Technology, Department of Computer Science and Engineering,
P.O. Box 9210, 02015 TKK, Finland
{mikael.runonen,petri.mannonen}@soberit.hut.fi

Abstract. We have identified four problems when developing futuristic concepts. Technologies cannot be used as boundaries for concept creation, there is a lot of room for surprises, user knowledge is bound to present day, and futuristic concepts are not easily communicable. We propose three methods from Futures Studies to tackle these problems, with the emphasis in developing futuristic product and service concepts in a business-to-business context. In this paper, we introduce the methods and discuss the possible benefits gained from their use.

Keywords: Futures studies, concept development, Delphi, user-centered design, backcasting.

1 Introduction

User-centered concept development is an exploratory process that aims at creating ideas of new products. Concept development can be used to combine existing technologies and methods in a new way. Concept development can also be future-oriented, i.e. making use of ideas that are not feasible at the given moment. The aim of creating something new means that one has to take into account the situation in some future time when the new developed product or service is in use. The challenge of predicting or forecasting the future has been noticed and different strategies to tackle it have been developed (e.g. [1], [2]).

Nieminen et al. (2004) integrated analysis of emerging technologies into the user-centered concept development process [1]. However the technological focus does not take into account all changes that can occur in users' lives between the time of design and the time of product launch. In addition creating concepts for further future differs from concepts created from emerging technologies. The concepts basing on emerging technologies are incremental by nature, in other words they utilize known, existing technologies. This means that the timescale for this kind of concepts is probably somewhere between 2 to 6 years.

Salovaara and Mannonen (2005) on the other hand aimed at creating "a leap to the future" by recognizing societal and user-related trends, and stable characters in users' behaviour and context. Recognizing trends and reflecting on them makes the future

M. Kurosu (Ed.): Human Centered Design, HCII 2009, LNCS 5619, pp. 904–911, 2009.
© Springer-Verlag Berlin Heidelberg 2009

vision of the designer more realistic than basing the vision only on technological trends[2]. However, in cases where there is no suitable literature to base the trends on, this kind of approach is difficult.

2   Main Problems with Future-Oriented User-Centered Concept Development

Designers need to tackle many obstacles and problems when developing new products and services. Focusing on users is a key strategy to ensure that the designed products are usable but the user focus is not always easy to achieve. We have identified four main problems with future-oriented user-centered concept development: 1) Technologies cannot be used as boundaries for concept creation, 2) There is much room for surprises, 3) User knowledge is bound to present day, and 4) Futuristic concepts are not easily communicable.

The problems are mainly related to future-orientedness but also have significant relations to users. We do not mean to say that user-centered way of doing design is problematic; only that being user-centered can be a challenging task when doing concept development.

2.1   Technologies Cannot Be Used as Boundaries for Concept Creation

When dealing with existing technologies there are certain promised possibilities, things that can be done. These technologies can also be seen as boundaries for the concepts to be developed. This however is not the case with far-reaching concepts. Technology cannot anymore be used as a boundary as its development cannot be predicted reliably enough in order to set plausible limits for concept development.

For example predicting the impact and possibilities enabled by computer networks and World Wide Web was not possible before (if even then) Tim Berners-Lee published the idea and first versions in beginning of 1990s. In addition of difficulties in predicting what kind of technologies can and will be developed one also needs to predict which of the new technologies will succeed in markets and how the users eventually end up using them.

In spite of the evident problems on using technologies as boundaries, it is constantly done. It is very difficult to start designing products and services and not to restrict ones ideas based on the current technological possibilities and restrictions. Thus there is a need to somehow use technical development as opener of design space instead of limiter.

2.2   There Is Much Room for Surprises

Some developments can be deemed unlikely at the present but as the timescale broadens it comes more probable that at least some of those will occur. These wild cards can be for example exotic trends that have a high impact on the behaviour of users of a certain technology. In brief, there is much more room for surprises when designing products for distant future.

A classic example of an unexpected event is text messaging among young people. A more recent example is the breakthrough of social media software applications in communication, collaboration, multimedia and entertainment. An example of an unexpected, though possible, event yet to happen would be a rapid rise in electricity costs that would render web searches unprofitable, among other consequences.

2.3 User Knowledge Is Bound to Present Day

User research can be used to extract knowledge about users and dominant trends. However this knowledge is bound to present day at best and is not applicable as such for forming a base for futuristic concept development. This knowledge must be somehow projected to the future.

User-centered design studies users in their normal environments and gathers knowledge about their motivations tasks and needs. However, a lot can change in 5 to 10 years. For example during last 10 years: Digitalization has changed people's habits and viewpoints towards photographing and photographs substantially; The environmental thinking has changed people's opinions on how much energy different devices should consume; In Europe the EU has opened up the markets and changed people's possibilities of travelling and working.

Predicting these huge changes based on a couple of observations and interviews and knowledge about the tasks and needs of the users is not possible.

2.4 Futuristic Concepts Are Not Easily Communicable

Futuristic concepts utilizing non-existent technologies are not easily communicable. Points of contact with reality may be or may seem very sparse as the concepts are easily on an abstract level. Depicted changes in the ways things are done may differ so much from the accustomed that they may seem unnatural. When validating this kind of concept with a user these problems may distract the user from the key features.

In our research project we have found out that introducing social media type of services to business context introduces or promotes also substantial changes on how the companies and their employees act and thus the users can have difficulties in understanding the usage situation of a designed concept based on just description of the concept and the potential usage situation [3].

3 Futures Studies

The aim of Futures Studies is to create various views of probable, possible and preferable futures by utilizing results produced by other disciplines and by creating views of alternative futures basing on them [4]. Some of the popular methods are scenarios, the Delphi method, backcasting, and trend analysis (e.g. [4], [5]). The goal is usually to understand what is likely to change and stay the same and on what probability. Futures Studies is often applied in strategic work of companies and other organizations.

From the viewpoint of product development and user-centred design the most prominent methods of futures studies are those that either create new ideas or ground the ideas on the lives of the potential users.

We selected three futures studies methods to be tested in future-oriented concept development project. The goal of the test was to see whether futures studies methods could provide important insights of users' lives to quite concrete concept development project and if some of the main problems relating future-oriented concept development could be avoided with these methods. The selected methods were PESTEC analysis, the Delphi method, and backcasting. It is notable that none of these methods are applicable as such in product or service design and therefore they must be adjusted to meet our needs.

3.1 PESTEC

PEST (Political, Economic, Social and Technical) is an analysis framework for macro-environmental factors and is usually utilized in strategic management. PESTEC is an extension of the PEST analysis which adds Ecological and Consumer viewpoints to PEST. PESTEC is a tool to map out contextually relevant megatrends and weak signals in each corresponding category.

3.2 The Delphi Method

The Delphi method is a method to achieve consensus of opinion by a group of experts. The method utilizes questionnaires which are that are presented to an expert panel in an iterative way. For each round the facilitator produces a summary of forecast and reasons from the previous round. The answers are anonymous to other panel-lists. [6]

The Delphi method was first introduced by the RAND corporation in 1963 [7]. Delphi is a method that can be applied in a situation where statistical models are not practical or possible because of lacking data. The aim of the first implementation of the Delphi method was to estimate amounts of atomic bombs but afterwards the method has been used in many other contexts and disciplines. [6].

3.3 Backcasting

Backcasting, as in opposite of forecasting, is a method which tries to work the future backwards. Whereas forecasting is used to predict the future from past and current trends, *backcasting starts from some desirable future and identifies the steps to reach that future*. The first applications of backcasting can be found from 1976 by Amory B. Lovins, in a context of energy strategies [8].

4 Case

We create and evaluate novel social media concepts for business-to-business (B2B) context in our on-going research project. The timescale for our concepts is from 5 to 15 years. We have been using a quite general user-centered design process, i.e. iterative process with user research, concept development, and concept evaluation with users. We used a scenario-based approach with user interface sketches in developing and communicating the concepts. During the first iterations we have learned that problems arise as concepts extend further than five years.

5 Application of Futures Studies Methods

After the decision to use PESTEC we stripped the latter E, the ecological viewpoint. This viewpoint wasn't essential in our study because B2B applications are not very dependent on ecological issues. There are some ready templates but we did not use any when mapping the megatrends and weak signals (hereafter both called trends). Usually using PESTEC means gathering factors that might influence a company. In our case PESTEC was used to gather factors that have relevance in our case point of view. Using the method we were able to find 37 megatrends and 21 weak signals. In the next phase of the study we created numerical values of their probability and impact for each megatrend and weak signal.

The influence of each trend was to be evaluated and we conducted an impact analysis. The value for impact consisted of three components: the swiftness of influence, the scale of influence and the general influence in the context of B2B. Each component of the impact was evaluated on a scale from one to five. After this evaluation a cumulative impact index was created by multiplying the components. Thus the scale for the cumulative impact index is 1...125.

Each trend was given a probability. The probabilities were estimated by first dividing the trends to three categories. The categories were "improbable", "maybe" and "probable". The categories represent probability values of 0.00...0.30, 0.31...0.65 and 0.66...1.00, respectively. On the second round of estimations, each trend was given a probability estimate with the precision of 5% (0.05). A combination of the cumulative impact indices and probability estimates is shown in figure 1.

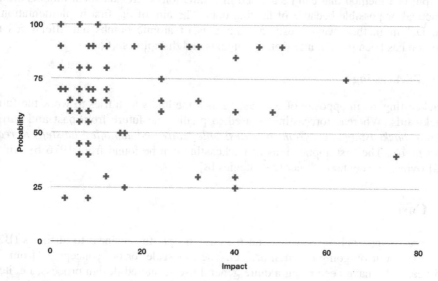

Fig. 1. Impact and probability analysis of trends gathered with PESTEC

The greatest mass of trends is concentrated on the left side of the figure 1. This is explained by the fact that there are many trends that are present today and are very likely to strengthen as time goes by. Already dominant trends are given only a small impact as they represent the world we live in. These kinds of trends are for example ubiquitous computing becoming more common, progression of mobile technologies, significance of usability, etc. On the other hand, the rightmost side is less populated. The scarcity of hits on the right is explained by our choice to use the maximum impact value of 5 (in each component) very sparingly as it would have meant the largest impact imaginable. Therefore the "sweet spot" in our analysis would be the trends having impact of 20...50 and probability more than 50. These trends represent the biggest possibilities that are not self-evident. It is though noteworthy that we did not deem the exact values for each trend very important as they will adjust in later phases in our study when applying the Delphi method.

After the trend analysis we created different combinations of trends of interest and created future scenarios basing on those. The meaning of the word 'scenario' differs somewhat between futures studies and the HCI field. Whereas HCI usually uses scenarios as stories which depict users in certain usage situations, futures studies' scenarios are depictions of the world on a larger scope. The created scenarios were of the latter kind. We also supplemented each scenario with a tentative concept idea which corresponds to the scenario. An example of a trend combination is given in figure 2.

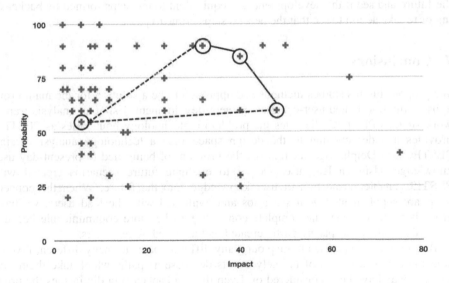

Fig. 2. An example combination of trends

The combination illustrated in figure 2 corresponds to "grown meaning of measurability", "stronger influence of the media" and "the growth in the amount of information". The dash line includes "better availability of information" as it is related with the amount of information. An example scenario that takes into account these trends describes a world where there is lots of information, companies' transparency is a

910 M. Runonen and P. Mannonen

requirement, advertisements are everywhere, there are many information suppliers but the quality of the information determines the winners and filtering is a must. A tentative concept idea for this kind of a future could be service that provides filtered and validated information about companies and separates advertisements from proper information.

6 Future Work

The next phase of the project is to evaluate the created scenarios and concept ideas with a derivative of the Delphi method. We are organizing a lead user council that consists of potential users with either personal or job related interest to new technologies. Whereas the Delphi method is usually used to achieve a consensus among experts, our version will use lead users as experts and the emphasis will be more on the feedback than on the consensus. Keeping the concept ideas and scenarios separated enables us to get feedback from both our visions of the future and the sensibility of the concept ideas.

The feedback will be used as a base to complete the concepts and to refine our future scenarios. When we will have the refined scenarios and complete concepts we will use backcasting to define the needed steps to reach the wanted future. These steps can be utilized in two separate ways. It is possible to assess situations periodically in the future and see if the developments are equivalent to the steps formed by backcasting or to take action to see that the needed steps come to pass.

7 Conclusions

We propose futures studies methods and approach to be a solution for the main problems of future-oriented user-centered concept development. Using an analysis framework such as PESTEC alleviates the problem of technology boundaries as PESTEC provides a wider opening to the design space than a technology analysis. Using PESTEC and Delphi together relieves the problem of being tied to present-day user knowledge. Using a lead user Delphi to evaluate future scenarios created with PESTEC enables projection of user knowledge into the future. When the concept ideas are supplemented with scenarios and evaluated with the lead users, we have grounds to believe that the complete concepts will be more communicable because the basis for the concepts are partly created and assessed by real users.

Backcasting can be used to map out many different paths to many different favourable futures. The effects of unlikely events decrease as paths which take them into account may have been considered of. Even though backcasting diminishes the problem, there will always be developments that cannot be anticipated.

Acknowledgements

We wish to express our gratitude to Mika Mannermaa for an introduction to Futures Studies. We also wish to thank Tekes for funding our research project.

References

1. Nieminen, M., Mannonen, P., Turkki, L.: User-centered concept development process for emerging technologies. In: 3rd Nordic Conference on Human-Computer Interaction, pp. 225–228. ACM Press, New York (2004)
2. Salovaara, A., Mannonen, P.: Use of future-oriented information in user-centered product concept ideation. In: Costabile, M.F., Paternó, F. (eds.) INTERACT 2005. LNCS, vol. 3585, pp. 727–740. Springer, Heidelberg (2005)
3. Mannonen, P., Runonen, M.: SMEs in Social Media. In: Proceedings of NordiCHI 2008 Workshop: How Can HCI Improve social media Development, Lund, pp. 86–91, October 20-22 (2008)
4. Bell, W.: Foundations of Futures Studies. Human Science for a New Era. Volume I: History, Purposes, Knowledge. Transaction Publishers, New Brunswick and London (1997)
5. Holmberg, J., Robèrt, K.-H.: Backcasting from non-overlapping sustainability principles - a framework for strategic planning. International Journal of Sustainable Development and World Ecology 74, 291–308 (2000)
6. Rowe, G., Wright, G.: The Delphi technique as a forecasting tool: issues and analysis. International Journal of Forecasting 15, 353–375 (1999)
7. Dalkey, N., Helmer, O.: An experimental application of the Delphi method to the use of experts. Management Science 9, 458–467 (1963)
8. Lovins, A.: Energy Strategy: The Road Not Taken? Foreign Affairs 55 (1976)

Empowering End Users in Design of Mobile Technology Using Role Play as a Method: Reflections on the Role-Play Conduction

Gry Seland

Department of Computer and Information Science,
Norwegian University of Science and Technology,
7491 Trondheim, Norway
gry.seland@idi.ntnu.no

Abstract. Role play as a method has several qualities that make it a profound candidate as a technique to understand user needs for mobile devices and services. However, making role-play participants act is a recognized but little discussed problem in relation to using role-play in design. This paper focuses on the how the role-play facilitator can arrange for the necessary conditions for making role-play participants act out realistic and relevant scenarios. The paper contributes with reflections on the role-play facilitator's conduction of role plays, by applying and discussing a general framework for role-play conduction on seven role play design workshops carried out in the period 2001 – 2005. The framework on role play conduction originally developed by the psychologist Yardley-Matwiejczuk [1], and has previously not been applied to role-play design workshops.

Keywords: Role play, user involvement.

1 Introduction

We are surrounded by technology in our leisure and at work, and mobile devices such as laptops and cell phones are replacing stationary technology in our society. However, many HCI methodologies for user involvement have evolved through the years to design desktop computers. The conventional techniques must be adapted and supplemented with new practice to address the specific challenged related to design of mobile devices [2-4]. For example, Kjeldskov and Stage varied the degree of physical activity in a usability test of a mobile system, and found that increased physical activity resulted in increased subjective workload [2]. Po and associates found that heuristic evaluations conducted in the field revealed some unique problems not identified in a context-free heuristic evaluation or heuristic walkthrough [4]. And Hagen and colleagues addressed methodological responses in the traditions of HCI and CSCW to the challenges of understanding the use of mobile technology [3].

Role play as a method has several qualities that make it a profound candidate as a technique to understand user needs for mobile devices and services. Dramatized

M. Kurosu (Ed.): Human Centered Design, HCII 2009, LNCS 5619, pp. 912–921, 2009.

scenarios create a situated embodied context for the technology, which enhances the participants' ability to envision needs and use of technology in specific situations. A role play can be stopped and started several times during a session, making it possible to play with different ideas of solutions, and can be conducted where it is impossible to carry out observation studies.

The objective of this paper is to provide insight to HCI practitioners on how to engage end users in design of mobile devices and services by using role play with low-fidelity prototyping as a method.

The paper contributes with reflections on the role-play facilitator's conduction of role plays, by applying and discussing a general framework for role-play conduction on seven role play design workshops carried out in the period 2001 – 2005. Our role-play workshops consisted of two parts: improvisation of scenarios and "designing on the spot" with low-fidelity prototypes. This paper focuses on understanding the first part of our workshops, the necessary conditions for making role-play participants act out realistic and relevant scenarios. The framework on role play conduction originally developed by the psychologist Yardley-Matwiejczuk [1], and has previously not been applied to role-play design workshops.

2 Background

2.1 Role Play as a Method for Involving Users

During the last decade, several researchers have explicitly stated that they have conducted role play sessions to engage users in design, e.g. [5-8]. Iacucci, Kuutti and Ranta brought in role-playing games with toys and situated and participative enactment of scenarios (SPES) as ways to make prospective users active in early concept development of mobile services and devices in the spirit of participatory design. In the role-playing games users acted themselves through a toy character with a magic mobile device [9]. Binder made simple "props" and organized user sessions with users in their own environment, with the goal of creating collaborative spaces between users and designers [6]. In later work by Iacucci and associates, the authors described how they used performances to develop scenarios [9], to communicate and test ideas, and to explore design options [10]. Brandt and Grunnet [7] used drama with props to introduces a bodily dimension into the design process, which help designers to work physically as well as intellectually. Bødker, Nielsen and Petersen worked with the development of design tools to stimulate idea generation in collaborative situations involving designers, engineers, software developers, users and usability professionals, where role-play was used as part of a larger design approach [8].

In these studies the degree of user involvement has varied from users expressing their ideas when watching designers' role-plays [7] to users improvising freely in the field [6]. There are numerous variations in how role-play has been used in design, and the usage can be described by heterogeneity at best.

And even though role play has been used regularly engage users in the last decades, it has not been used frequently. Role-play as a method has a great potential to add value to IT system and product development [11]. However, it is not a common part of system development practices yet. Why do not system developers use role-play as a method?

There may be several answers to this question, including cost concerns and lack of knowledge about the method. However, one key reason is probably tied to the role-play facilitator's jobs. Similarly to other user-centred methods which are dependent on leadership, a role-play session is not done by itself. It must be directed and facilitated, and the skills necessary to do this job must be learnt. In this paper the role-play "facilitator" is defined as the person or the people who are in charge of leading the role-play conduction in a design workshop. The facilitators have a large impact on the validity and the reliability of the scenarios and the outcomes of a role-play workshop. Thus leadership is critical.

2.2 Problems of Making Role-Play Participants Act

In an earlier presentation of lessons learnt on role-play as a system development method, we concluded that leading a role-play workshop is a relatively straightforward task [12]. Behind the conclusion was an assumption that facilitating role-play design workshops was something everybody can learn, just as paper prototyping and usability testing can be learnt. However, in that conclusion, we as researchers and role-play facilitators might have been biased by our previous experience with role-play. Both the other facilitator (DS) and myself (GS) had built up knowledge on role-play over years through experience with amateur theatre, and this understanding is not easy to grasp for someone unfamiliar with role-play. When we had conducted the first role-play design workshops and arranged the second one, we felt that it was very easy to make our role-play workshop participants enact everyday scenes. Unfortunately this might not be the rule in every project.

The problem of leading role-play workshops and making role-play participants act is recognized to a certain degree in the literature [12-14]. However, explicit discussions about the facilitator's roles are lacking. Oulasvirta and associates [14] exclaimed that "acting out was observed to be frustrating and causing costly preparations. It was speculated, however, that acting could be useful in the long run when participants can get used to the method" [14, p. 132]. Similarly, Strømberg et al. [13] described how "none of the users were enthusiastic about acting, so we ended up just talking the scenario through" [13, p. 204]. In our first workshop, *exploring the use of PDAs in hospitals,* we discovered that we as academics had problems acting health care personnel [12], and as solution to the problem, we hired an external theatre instructor to take hand of the drama in our first workshops. Similarly, Rodriquez et al. [15] stated that "practicing and managing a role-play session was seen as a critical step both for facilitators and participants who did not have previous experience with these kind of activities. Therefore, the assistance of a role-play facilitator was requested for the development of the performance session" [15]. The problems of making role-play participants act confirm the importance of the role-play facilitator's role.

3   Description of Our Role-Play Workshops

This paper is based on retrospective reflections of seven role play design workshop carried out in the period 2001 – 2005 [11, 12]. The iterative process of transforming role-play into a design method is earlier described in [12], and an evaluation of the approach as seen through the eyes of system developers is given in [11].

The role-play workshop format was developed through an iteration of seven workshops, with a total of 68 participants. The goal of the iterative process was to create an optimal structure for a workshop where role-play and low-fidelity prototyping are central for active user involvement in an early system development phases.

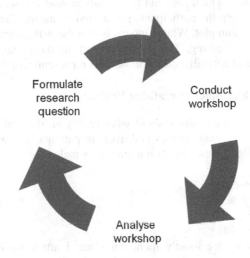

Fig. 1. Iterative development of role-play as design method

For each workshop we formulated a research question, which was implemented in the workshop and subsequently analyzed (figure 1). The analysis triggered the next research question, which again was used to guide the conduction of the next workshop.

The research questions asked in this iterative process were:

- How can role play be used to explore the potential of PDAs in hospitals?
- How can real users (nurses) be involved in a role play workshop?
- How can drama be useful in a application domain not related to health care (Petrol Service Station)
- What type of role should a system developer have in a workshop?
- What added value can field data have as input in a workshop?
- How can a role play workshop help system developers to think creatively about new concepts?
- Is it possible to teach interaction designers to conduct role-play workshops?

The role play sessions consisted of two main parts: 1) Characterization of typical work- or leisure scenarios with role play, and 2) Improvisation of ideas about new technologies and services with low fidelity prototypes as props.

The following example gives an impression on how the participants worked with developing ideas in one of our later workshops. The example is a transcription of a dialog from a workshop with nurses working on ideas about electronic patient journals.

The participants have had a brainstorming session around different possible scenarios, and have agreed to work on a scenario about a woman with breathing problems, and a possible heart problem. The example is taken from the point in time when the participants started to develop the details of their scenario. The participants have decided which nurse should have which role, but the details around the scenario are not clear. The facilitator tries to help the participants get started by asking them discuss some of the details around the main plot. What happens is that the participants started to improvise the scenario through acting. The ease by which the participants identified themselves with the roles and naturally started to act was representative for our role plays

3.1 An Illustrative Example: "Breathing Problem"

FACILITATOR: You have to think about what reality is like and what is common. In a way you are average nurses and average patients. So, what would an average patient be called? How old? Is it a female or male?

NURSE 1: Eva Antonsen
NURSE 2: About 50 years old
NURSE 3: at least 50
NURSE 2: but not too old
NURSE 1: The patients are mostly men, but since I am a woman, I need a woman's name.

FACILITATOR: And you have problems breathing?

NURSE 1 *(Breaths heavily)*: Yes, I have big problems with breathing when I came to the hospital. I cannot breathe! Ugh!
NURSE 2: Have you been short-winded for a long time?
NURSE 1: I have been like this the last few months.

NURSE 2 *(looking at the facilitator)*: We are starting to role-play already!

NURSE 1: Well, it has been like this lately, particularly the last months. During the last week it has been increasingly worse, and tonight I had to sit in my bed to be able to breathe.
NURSE 2: Have you experienced something similar during the last half a year?
NURSE 1: No, not as bad as this.
NURSE 2: Are you feeling well besides the breathing?
NURSE 1: Well, I am smoking, and I have a little asthma

    As seen in this example the role-play participant started acting as if they were the persons in the role. The acting came natural without much help from the facilitator. How did this happen?
    In our work on trying out new elements in the workshops, observing the effect and reflecting on the next steps [12], our main discoveries of how to make role-play participants act can be summarized as follows:

- The role play workshop participants must act themselves or take a role character they are very familiar with.
- The scenarios to be acted must be grounded in the participants' experiences.
- The participants must specify the details of the scenarios, such as time-of-day, place, participants and main plot.

The next section presents a general framework for role-play conduction that are useful to understand the ease by which our role-play participants acted naturally, and also explains our main lessons learnt on how to make role-play participants act.

4 The Yardley-Matwiejczuk Framework for Role-Play Conduction

To understand our experiences with the role-play part, we found the framework for role-play conduction developed by the psychologist Yardley-Matwiejczuk useful [1].

Yardley-Matwiejczuk developed a conceptual framework based on an extensive review of different uses and discussions on role-play in research, education and therapy. In her analysis, Yardley-Matwiejczuk defines role-play in a set of 8 characteristics, and provides three important principles for conduction of role-play sessions. These principles are in focus in this paper. According to Yardley-Matwiejczuk, the role-play conductor's instructions influence the participants' experience of the reality of the role-play. The role play conductor is here defined as the role-play facilitator. The principles of *particularization*, *personalization* and *presencing* are the facilitator's keys, and the attention to these principles is important for the role-play success.

4.1 Particularization, Presencing and Personalization

Particularization is defining all objects in the role-play (thus saying that a chair for example is a car), so that "all these objects are brought into awareness in order that they may be known" [1, p. 94]. This means that all objects in the role-play are made explicitly know to the participants. If an object is used as a prop in the play, all involved have to know the meaning of the object as it is used. This term is related to the attention to details in theatres: Every requisite on a stage is there for a purpose. If the prop has no purpose it should not be there. According to Yardley-Matwiejczuk, the role-play can turn to become very stereotypic if it is not particularized.

Presencing is the second key role-play induction principle proposed by Yardley-Matwiejczuk. This term is strongly related to and can be considered as an extension of the concept of particularization. In Yardley-Matwiejczuk view the particularized objects must be made present and actual in the role-plays: "so that they are perceived as 'out-there' (part of the situation or 'other person') or 'within-here' (part of the 'self')"[1, p. 95]. For the facilitator this means that he or she must use the language to emphasize that what happens in the role-play happens in present time. Instead of saying "imagine that this is the waiting room, and act as if you are waiting for the physician", the scene is made actual by saying "this *is* a waiting room, and you *are* waiting for the physician". With particularization an object is identified, and by presencing it is made familiar and actual to the participants.

Personalization is the final key role-play induction principle of importance for the perception of the role-play as realistic and real. This term is related to the degree to which the particularized objects are drawn from the subjects themselves or from the role-play facilitators. By asking the participants themselves to create the physical configuration for a role-play scene, the quality of the participants' engagement in the role-play improves.

However, it can be questioned whether these induction principles are equally important in all types of role-plays and for all types of role characters. Yardley-Matwiejczuk varied the degree of particularization, presencing and personalization in a number of experiments on role-play, and concluded that the detailing of knowledge was most important for people in roles leading the play and of less significance those who were mainly responding to the others' acting [1, p. 163].

Do they apply to our role-play design workshops, where the participants primarily act themselves to create and explore ideas about technology? To answer these questions we looked retrospectively at the seven workshop conducted in the period 2001 – 2005 to investigate the relationship between our main lessons learnt on how to lead role-play design workshops and Yardley-Matwiejczuk's three principles.

5 Application of the Principles of Particularization, Presencing and Particularization to Role-Play Design Workshops

To return to our role-plays, we found that the principles could have been very helpful in our first workshop if they had been known to us.

In our first workshop, a professional drama instructor was engaged to lead the role-play process. She was giving some freedom in how to work with the workshop theme, because we were inexperienced as role-play facilitators. The workshop theme was "to explore ideas for the use of PDAs in hospitals" and the participants were mostly academics with an interest in health informatics. Neither the participants nor the drama instructor had any experience with clinical hospital work, and the theatre instructor relied on her pervious experience on leading groups in theatre sport and improvisation when planning the workshop. In the main part of the workshop the participants were asked to improvise short scenes involving system developers and health care professionals. For example, a participant was asked to take the role of a system developer, who was to demonstrate and explain how a mobile Electronic Patient Record system works to the head nurse at a hospital ward. Nurses are usually considered to be very busy in the hospital, and to signify this, the drama instructor asked to person given the role as the nurse to fold some sheets of paper while talking to the system developer. However, the drama exercise turned out to be a superficial performance where the participants did not learn anything new about the technology.

There were several problematic aspects with this short improvisation act, which can be related to the lack of attention to Yardley-Matwiejczuk's three role-play induction principles. First of all, no details were particularized on what the mobile device could do, and in which way it could be helpful for the nurses at the hospital. Thus the person role-playing the system developer had to be creative, and become responsible for improvising the features of the mobile device. This created a pressure for performing. Similarly, the person role-playing the nurse was given the folding exercise to signify that she was busy, but it was not particularized what the folding signified. The act of busyness resulted in a situation where the person role-playing the system developer had to use all his energy on getting the attention of the person acting the nurse to convince her about the usefulness of the mobile device. The nurse-in-role was given no instructions on why she had agreed to discuss the mobile device with the system developer, and this lack of motivation resulted in a quite ignorant behavior. Because of insufficient details on both the hospital work and the purpose of the meeting, the nurse had no choice but to take a stereotypical role of busy nurse, skeptical to new technology, and not interested in listening to the system developer. The dialogue we had hoped for about possible technological solutions did not occur, and the short role-play became intimating for both the role-players and the other participants who were observing the scene.

In the second workshop the main participants had background knowledge for their role-play, but the performance nevertheless resulted in overacting and stereotypic behavior in one of the groups. The goal in this workshop was to investigate whether health care personnel could participate and develop ideas about technology in a role-play workshop. One of the two groups in this workshop consisted of three nurses, who chose to role-play a pre-round meeting. Two of the nurses took the role of physicians, and the third nurse role-played herself in her ordinary job. All three nurses were experienced and knowledgeable about their own work, the pre round meeting situation, and the physicians' work. However, due to a lack of particular instructions on who they were to act and how they should act, the two nurses who role-played physicians acted very arrogantly. They were talking to each other, came with irrelevant comments, and seemed not to be particularly interested in listening to what the nurse had to say. Instead of creating an arena for exploring information needs the role-play became a stereotypic demonstration of power relationship between nurses and physicians. This overacting could probably have been avoided by requiring the nurses to play average physicians, and by making their roles particularized and personalized.

However, when we started to state that the participants should act themselves and base their play on everyday experience, we avoided stereotypical acting. The level of detail a person has about his or her own life counter the need to take a stereotypical stance in the play. We followed up on the participants suggestions by asking questions as "what would you do in this situation?" This resulted in situations where our role-play participants had to think of details of their own behavior. This act probably counteracted the desire or need to take a stereotypical role.

The emphasize in our workshop on requesting the participants to act oneself, working with everyday scenarios and making details explicit can be related to Yardley-Matwiejczuk's concepts of particularization and personalization. We did never ask the participant to imagine that they should imagine that they were somewhere else, and our question to facilitate the role play as "How old? Is it a female or male?" can be compared to Yardley-Matwiejczuk's idea about presencing. However, is there a need for these principles?

If role play is used as a design method in a design workshop where the participants are to act themselves, Yardley-Matwiejczuk concepts can be valuable. Yardley-Matwiejczuk's concepts counter stereotypical acting because they function as tools for treating every role-play as unique. By the forcing the participants to work on the details about the setting of the role play and the characters, the role-play participants avoided simplifications and stereotypical behavior.

It is probable that some of the problems in the studies cited in section 2.2 in this paper can be traced to the fact that the participants did not have the necessary knowledge about the details of their characters to improvise naturally. In these workshops the principles of particularization, personalization and presencing would most likely have made the acting more natural.

However, it can be questioned whether these concepts are equally important for all types of role play design workshops. The key answer observed in our workshops is that the role play participants need sufficient information to play their roles. The more distant the role play theme is from the participants' experiences, the more important the principles are. A person who can act by responding to other people's moves may be able to role play a situation without a complete understanding of the setting. But for a person who has to take initiatives in the play, lack of information could result in a situation where the person feels that he/she makes a fool of oneself.

6 Conclusion

Through a retrospective view of a line of workshops conducted from 2001 – 2005 we conclude that particularization, personalization and presencing are useful terms to understand how a workshop leader must help end users with the development and rehearsal of enacted scenarios. We saw several examples of how the lack of attention to these principles in the early the workshops resulted in overacting and stereotypic behavior and the incorporation of would, if not eliminated the problem, have reduced it.

In conclusion, the three concepts from Yardley-Matwiejczuk's framework are useful for HCI practitioners who would like to use role play as method to engage users in discovering needs for new technology, and should be taken into consideration when developing and carrying out such workshops.

References

1. Yardley-Matwiejczuk, K.M.: Role play: Theory and practice. Sage, Thousand Oaks (1997)
2. Kjeldskov, J., Stage, J.: New Techniques for Usability Evaluation of Mobile Systems. Int. J. Human-Computer Studies 60, 599–620 (2004)
3. Hagen, P., et al.: Emerging Research Methods for Understanding Mobile Technology Use. In: OZCHI 2005. ACM Press, New York (2005)
4. Po, S., et al.: Heuristic Evaluation of Mobile Usability: Bridging the Realism Gap. In: MobileHCI 2004. Springer, Glasgow (2004)
5. Iacucci, G., Kuutti, K., Ranta, M.: On the Move With a Magic Thing: Role Playing in Concept Design of Mobile Services and Devices. In: DIS 2000, New York, United States (2000)
6. Binder, T.: Setting the Stage for Improvised Video Scenarios. In: CHI 1999. ACM Press, Pittsburgh (1999)
7. Brandt, E., Grunnet, C.: Evoking the Future: Drama and Props in User Centered Design. In: PDC 2000. CPSR, Palo Alto (2000)
8. Bødker, Nielsen, Petersen: Creativity, Cooperation and Interactive Design. In: DIS 2000. ACM Press, New York (2000)

9. Iacucci, G., Kuutti, K.: Everyday Life as a Stage in Creating and Performing Scenarios for Wireless Devices. Personal and Ubiquitous Computing 6, 299–306 (2002)
10. Iacucci, G., Iacucci, C., Kuutti, K.: Imagening and Experiencing in Design, the Role of Performances. In: NordiCHI. ACM, Åarhus (2002)
11. Seland, G.: System Designer Assessments of Role Play as a Design Method: A qualititive study. In: NordiCHI 2006. ACM Press, Oslo (2006)
12. Svanæs, D., Seland, G.: Putting the Users Center Stage: Role Playing and Low-Fi Prototyping Enable End Users to Design Mobile Systems. In: CHI 2003. ACM Press, Vienna (2004)
13. Strömberg, H., Pirttilä, V., Ikonen, V.: Interactive Scenarios - Building Computing Concepts in the Spirit of Participatory Design. Personal and Ubiquitous Computing 8, 200–207 (2004)
14. Oulasvirta, A., Kurvinen, E., Kankainen, T.: Understanding Context by Being There: Case Studies in Bodystorming. Personal and Ubiquitous Computing 7(2), 125–134 (2003)
15. Rodriguez, J., Diehl, J.C., Christiaans, H.: Gaining Insight Into Unfamiliar Contexts: A Design Toolbox as Input for Using Role-Play Techniques. Interacting with Computers 18(5), 956–976 (2006)

The User's Role in the Development Process of a Clinical Information System: An Example in Hemophilia Care

Leonor Teixeira[1,2], Vasco Saavedra[1], Carlos Ferreira[1,3], and Beatriz Sousa Santos[4,5]

[1] Department of Economics, Management and Industrial Engineering,
University of Aveiro Portugal
[2] Governance, Competitiveness and Public Politics (GOVCOPP),
University of Aveiro Portugal
[3] Operational Research Centre (CIO),
University of Lisbon Portugal
[4] Department of Electronics, Telecommunications and Informatics,
University of Aveiro Portugal
[5] Institute of Electronics and Telematics Engineering of Aveiro (IEETA)
Portugal
{lteixeira,vsaavedra,carlosf,bss@ua.pt}@ua.pt

Abstract. This work describes the development process of a Web-based Information System for managing clinical information in hemophilia care, emphasizing the role of the users around a human-centered development. To help understanding all this process, we first present the relevant concepts concerning human-centered design; next we describe the web application for managing the clinical information in hemophilia care, as well as, the development process followed in its development; and finally we illustrate the importance of the user's involvement in critical phases through the demonstration of some results.

Keywords: Health information system, development process, user-centered design.

1 Introduction

Healthcare Information Systems (HIS) are increasing in size, variety, complexity and sophistication. In particular Clinical Information Systems (CIS) have human safety implications and profound effects on individual patient care. Nowadays, there are a large number of HIS projects that fail, and most of these failures are not due to flawed technology, but rather due to the lack of systematic consideration of human and other non-technology issues throughout design or implementation process [1, 2]. Several studies have shown that the majority of total maintenance costs with information systems (ISs) are related to users' problems and are associated with usability problems and not with technical bugs [3]. A survey of over 8000 projects applied by the Standish Group at 350 US different companies revealed that one third of the projects

M. Kurosu (Ed.): Human Centered Design, HCII 2009, LNCS 5619, pp. 922–931, 2009.
© Springer-Verlag Berlin Heidelberg 2009

were never completed and one half succeeded with partial functionalities [4]. The same report states that the major source of such failures is related to poor requirements, namely: the lack of user involvement, incomplete requirements, changing requirements, unrealistic expectations and unclear objectives [4]. According to [3, 5], these problems are mainly due to the fact that most software engineering methodologies used in developing interactive software do not propose any mechanisms to explicitly and empirically identify and specify user needs and usability requirements; moreover, they also don't allow to test and validate requirements with end-users before implementation and during the development process. The health care domain has been particularly prone to such problems in recent years, and there are numerous examples of potentially useful systems that have failed or have been abandoned due to unanticipated human or organizational issues [6-9]. The lack of even minimum considerations of design principles centered in human factors in many HIS make them very difficult to learn and use, and this difficulty leads to strong resistance by users, namely physicians and nurses, leading in some cases to abandoning the product altogether or increasing "human error" resulting from an incorrect usage. According to [1, 2, 10, 11] the paradigm of human-centered computing offers a new look at system design including human factors and it covers more than usability engineering, human-computer interaction and human factors.

This work describes the development process of a Web-based Information System for managing clinical information in hemophilia care, emphasizing the role of the users around a human-centered development. To present this subject, this paper is structured in three parts: (i) a brief description of relevant concepts concerning human-centered design; (ii) a section describing the web application for clinical information management in hemophilia care, as well as, the development process adopted and followed in its development; and finally, (iii) brief conclusions.

2 Background: User-Centered Design and the Role of Users in the Development Process

The success of any software system depends, significantly, on how well it fits the needs of its users and its environment during the development process, normally identified in a process called requirement engineering (RE) [12]. According to [13], the RE process is considered a knowledge area of software engineering (SE) that is concerned with elicitation, analysis, specification and validation of requirements of any application. An adequate requirements specification allows developers to know what to build, and users know what to expect [14]. In order to make this connection between the software developer and users, it is possible to use different techniques and methods, among which the user-centered design (UCD) (or human-centered design (HCD)) is included. UCD allows incorporating the perspective of the users into the software development process in order to achieve a desirable and usable system [15], at the same time that it allows to find and refine the Requirements Specification Document (RSD). UCD is a complement to SE traditional methods rather than a substitute. According to [15], the key principles of UCD are basically the active involvement of users, a clear understanding of user and task requirements, and an appropriate allocation of functionalities between user and system.

ISO standard 13407 also presents a definition for HCD as a multidisciplinary activity, which incorporates human factors and ergonomics knowledge, as well as techniques to enhance effectiveness and productivity, while improving human working conditions [16].

According to [1] and [11] a HCD approach is based on four types of analyses: (i) *user analysis* - identifying the characteristics of existing and potential users; (ii) *functional analysis* - identifying a system's abstract structures of a given domain model; (iii) *representational analysis* - identifying an appropriate information display format for a given task performed by a specific type of users; and (iv) *task analysis* - identifying the procedures and actions to be carried out as well as the information to be processed to achieve task goals by using specific representations.

In general, the main goal of UCD consists in creating systems that are modeled in conformity with the characteristics and tasks of the potential users within a specific cultural, social and organizational environment. The development process of the specific clinical information system presented in this article had these aspects in consideration, making use of the different techniques and methods, based on UCD principles, to elicit user requirements and capture the cultural, social and organizational aspects, as well as to validate the result models along the iterative development process.

3   Web-Based Information System for Clinical Information Management

In this section we will describe a user-centered development process of a specific application for management of clinical information in hemophilia care; we also present some results obtained with this methodology. Firstly we present a brief contextualization of the problem and description of the solution. Then we present the main techniques and methods used in the development process, as well as the main result obtained.

3.1   Problem Contextualization and Brief Description of the Solution

The present solution was developed to support the information flow and communication among different health care professionals (HCP) that work in a specific Hemophilia Treatment Center (HTC) located in Portugal. This HTC currently uses several different Information Systems (IS) that represent generic solutions, given that they were developed to support generic requirements of Hospitals and Healthcare Centers. Moreover, the medical professionals of hemophilia care generate a lot of information when they assist their patients, and a considerable part of this information is stored in paper files. In order to help and obtain a solution that integrates all isolate ISs, as well as to incorporate other data stored in inadequate ISs (i.e. in peper files) in this specific HTC, we proposed a web-based IS solution that integrates three different modules in its back-end: Patient Clinical Data Management (PCDM), Treatment Data Management (TDM) and CFC Stock Data Management (CFCSDM) modules, Fig.1.

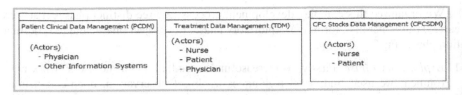

Fig. 1. Main modules that compose the back-end of the clinical information system

- *Patient Clinical Data Management (PCDM) Module* is responsible for the management of patient's clinical data and it is mostly used by physicians at the same time as using data from other ISs.
- *Treatment Data Management (TDM) Module* is responsible for the management of the patient treatment information lifecycle, which data is generated in the scope of the treatments with the Coagulation Factor Concentrate (CFC). It is used by three human actors: physicians, nurses and patients.
- *CFC Stocks Data Management (CFCSDM) Module*, which provides the tools for managing the stocks of the products used in the hemophilia treatment, specifically the CFC. It is mostly used by nurses and, indirectly, by patients, when they introduce data concerning treatments.

3.2 Development Process: Techniques and Methods

To conduct this study, we followed one approach that combined techniques derived from behavioral science disciplines, as well as, methods of Software Engineering and Human-Computer Interaction (HCI). These different approaches were used in order to understand and contextualize the problem while eliciting user requirements and capturing cultural, social and organizational aspects. Other techniques and models to represent, visualize and communicate were also used, in order to evaluate and validate de solution obtained during the development lifecycle.

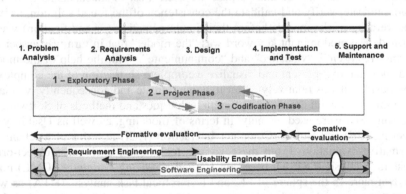

Fig. 2. Phases of development process and contribution of main disciplines [17]

We organized this complex development process in three distinct phases that were carried out in iterative form: (1) exploratory phase; (2) project phase; and (3) codification phase, Fig.2.

1. *Exploratory Phase.* Firstly, to increase understanding about the users' work environment we used traditional techniques as direct observation, open-ended interviews, documentation analysis, and focus group (expert group meetings). This phase has resulted, in practical terms, in a set of data, supporting the construction of a preliminary application model, namely the Requirements Specification Document (RSD). Besides this, a knowledge base about the clinic practice in this particular context was created that allowed to improve the communication process with the experts in this area in subsequent phases of project development, Fig.3.

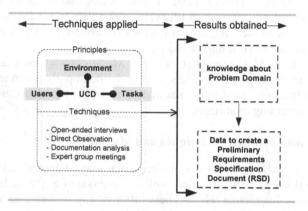

Fig. 3. Techniques applied and results obtained in the exploratory phase (adapted from [17])

2. *Project Phase.* this phase of the project was dedicated to the construction of the conceptual solution. For this purpose we used a set of mechanisms to represent, communicate, verify and validate the consistency of the found solution, as well as the veracity of the models before their implementation (or codification) was confirmed. The predominant key word was "the model" that in turn was associated to terms "represent", "visualize" and "communicate". With the help of the models it was possible to represent and visualize a complete abstraction of the complex reality, besides, it was relatively easier to communicate and consequently, validate the previous work with the users. Different techniques and methods of Software Engineering (SE) were used, namely, in terms of modeling, as well as Usability Engineering (UE) and Human-Computer Interaction (HCI) in terms of mechanisms of formative evaluation. From these various techniques, a classical object-oriented systems analysis (OOSA), based on *Unified Modeling Language* (UML) notation known more in the sphere of SE, and hierarchical task analysis (HTA) as well as Prototyping, widely used in the sphere of HCI were prominent. The order of application of the methods followed the logic of questioning "what to do", "how to do" and, finally, "how to represent"; starting with OOSA, followed by HTA and finishing with the Prototyping technique. Techniques already used in the exploratory phase were used whenever it was necessary to validate results found with each

method, namely, direct observation, open-ended interviews, and focus groups with experts in the field, as well as the frequently needed documentation analyses, Fig.4 – on the left.

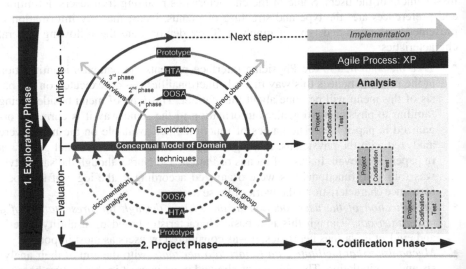

Fig. 4. Interactive and iterative process of engineering requirements in a project phase (on the left side) and implementation in a codification phase (on the right side) (adapted from [17])

    A triangulation matrix with identification of the key requirements to include in the system was achieved with results obtained with the three above mentioned methods, besides the model that defined the conceptual solution.

3. Finally, at the *Codification Phase*, we tried to follow one approach that would provide more feedback resulting from a larger involvement of the users. An approach having characteristics of evolutionary prototype proved to be adequate to projects that need to deal with rapid and unpredictable change of requirements. Agile processes [18-20] try to answer these demands, suggesting a reduced set of practices to develop projects in a simple form with mechanisms of results' evaluation in short periods through the concept of iteration and version. For this reason, an approach based on the principles of agile development was followed, using some practices of *eXtreme Programming* (XP) [21] having also characteristics of evolutionary prototype integrating the principles of User-Centered Design, Fig.4 – on the right.

    Actually, this approach led to a great deal of interaction with users, since validation and definition of the versions depended on their approval.

    In the following section some details of the developed web-based system will be presented illustrated by user interface examples.

3.3 Some Details of the Web-Based System with Illustration of User Interface

In this sub-section we will illustrate some details of the system in terms of functionalities with the help of user interface aspects.

The interface was designed according to preferences of the users involved in the development process and based on usability principles. In terms of general character- istics, namely in non-functional and usability requirements, there has been also a great involvement of the users. Some of the characteristics resulting from users definitions and preferences are: the type and structure of menus, icons, the way information is presented, definition and type of graphics. They demonstrate the following general characteristics:

- *Tree menu*, which in the Physician interface presents the patient's identification at the root, guarantees this way that all functionalities (tasks) executed on the ba- sis of this menu will be done about this patient. This type of menu besides being familiar to physicians, organizes information in the same way it is currently or- ganized in paper files. The names of functionalities available on the menu were analyzed with the physicians, making thus concepts used in traditional processes in paper file prevail. In case of menus of Patient and Nurse, though the same style was followed, functionalities were organized according to the logic of task per- formance characteristic to the work of these actors.
- *Identification of the user location in the system through the presentation of a navigation bar*. Though this is a basic characteristic based on usability criteria and evident in various heuristics, it was considered by users as most important.
- *Mechanisms that facilitate data export to other tools*, with the aim of their analy- sis and manipulation. The physicians showed great interest in this feature because it gives them freedom to create graphics and/or personalized reports. This func- tionality was introduced whenever data lists are presented making it possible to export them in three different formats (Excel, PDF and XML).
- *System flexibility* at the parameter level of some values and/or functions, as well as different forms to see specific information (textual as well as, graphic).

The characteristics corresponding to functional requirements are grouped at front-end level into sub-systems, according to the user profile that they serve. When users access the application, they face an interface to undergo the process of verification and authori- zation. Depending on user profile, they will be forwarded to a specific sub-system: sub-system of the Physician, sub-system of the Nurse or sub-system of the Patient.

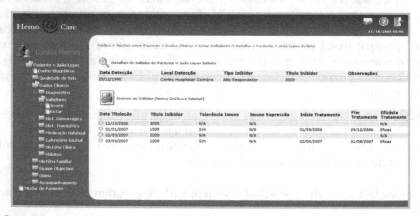

Fig. 5. Aspect of the 'Physician' sub-system user interface with textual information concerning the task 'visualize data about inhibitors'

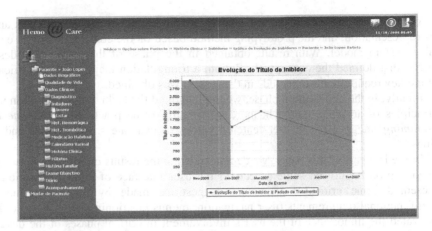

Fig. 6. Aspect of the 'Physician' sub-system user interface with graphical information concerning the task 'visualize data about inhibitors'

For example Fig.5 and Fig.6 illustrate the implementation of some non-functional requirements in a specific functionality carried out by the profile 'Physician'. The tree menu with the patient's identification at the root is located on the left side of the interface, as well as a set of functionalities that can be run. When relevant data are presented, an option to convert textual information (visible in Fig.5) in graphic format (as is depicted in Fig.6) is included.

In fact, the graphic representation was a point fairly valued by users, including physicians and nurses (perhaps by contributing to a better display of a broad range of data, thus inducing to a better understanding of information).

The same logic was applied all over the application; and the user had an important role in the definition of the functional and non-functional requirements.

4 Conclusions

This paper described the user's role within the development process of a Web-based Information System for managing clinical information in hemophilia care. Within the development process of this solution, we used different techniques and methods to elicit user requirements and capture cultural, social and organizational aspects during a development process composed by three distinct phases carried out in iterative form: exploratory, project and codification phases. The first phase (exploratory) represented the early efforts, where the problem was delimited and understood and the basic data was gathered for later development in the iterative cycles. In this phase, and in order to increase understanding about the users' work environment, we used direct observation, open-ended interviews, documentation analysis and focus group and obtained a set of data to support the construction of a preliminary application model, namely the definition of a Requirements Specification Document (RSD).

Next, in the project phase, and in order to construct the system conceptual model we used an evolutionary process of the user-centered requirements engineering based on three well-known methods: the classical OOSA method represented by use case diagrams of UML notation; the task analysis method based on hierarchical task analysis representation; and the prototyping method based on the construction of an executable system model. With results obtained in the phase of the project, besides the model that defined the conceptual solution, a triangulation matrix with identification of the key requirements to include in the system was obtained.

Finally, in the codification phase, we implemented the technological solution using principles of an agile approach, more specifically some practices of *eXtreme Programming* integrated with user-centered values to evaluate small releases and versions.

In the last part of this paper, we demonstrated some results of applying a user centered approach, showed some aspects of the user interface of the final solution and presented some critical fundamental suggestions made by users regarding the non-functional requirements (user navigation, menus positioning, etc). We also demonstrated the importance of the user's involvement in critical phases of the development process, as in the requirements engineering phase.

Based on the experience obtained through this work we conclude that the techniques and methods based on principles of user-centered design, focusing on continuous assessment (formative evaluation) in iterative processes, are most suitable for the development of complex Healthcare Information Systems.

Acknowledgements

We would like to gratefully acknowledge the contribution of the clinical professionals of *Hematology Service of Coimbra Hospital Center*, specifically to Doctor *Natália Martins* and Doctor *Ramon Salvado* for providing us access to data and information systems, as well as for all the help in the requirements analysis and discussion of the proposed model. We are also gratefully to Eng. *Igor Carreira* and *Doctor José Moreira* for fruitful discussions.

References

1. Rinkus, S., Walji, M., Johnson-Throop, K.A., Malin, J.T., Turley, J.P., Smith, J.W., Zhang, J.: Human-centered design of a distributed knowledge management system. Journal of Biomedical Informatics 38, 4–17 (2005)
2. Zhang, J.: Human-centered computing in health information systems Part 1: analysis and design. Journal of Biomedical Informatics 38(1-3) (2005)
3. Seffah, A., Desmarais, M.C., Metzker, E.: HCI, Usability and Software Engineering Integration: present and future. In: Seffah, A., Gulliksen, J., Desmarais, M.C. (eds.) Human-Centered Software Engineering - Integrating Usability in the Software Development Lifecycle, vol. 8, pp. 37–57. Springer, Netherlands (2005)
4. Standish_Group: The CHAOS report. Technical report, Standish Group (1995)
5. Seffah, A., Gulliksen, J., Desmarais, M.C.: An Introduction to Human-Centered Software Engineering: Integrating Usability in the Development Process. In: Human-Centered Software Engineering — Integrating Usability in the Software Development Lifecycle, vol. 8, pp. 3–14. Springer, Netherlands (2005)

6. Coiera, E.: Putting the technical back into socio-technical systems research. International Journal of Medical Informatics 76, S98–S103 (2007)
7. Heeks, R.: Health information systems: Failure, success and improvisation. International Journal of Medical Informatics 75, 125–137 (2006)
8. Southon, G., Sauer, C., Dampney, K.: Lessons from a failed information systems initiative: issues for complex organisations. International Journal of Medical Informatics 55, 33–46 (1999)
9. Stumpf, S.H., Zalunardo, R.R., Chen, R.J.: Barriers to telemedicine implementation: Usually it's not technology issues that undermine a project - it's everything else. Healthcare informatics 19, 45–48 (2002)
10. Zhang, J.: Human-centered computing in health information systems: Part 2: evaluation. Journal of Biomedical Informatics 38, 173–175 (2005)
11. Zhang, J., Patel, V.L., Johnson, K.A., Smith, J.W.: Designing human-centered distributed information systems. IEEE Intelligent Systems 17, 42–47 (2002)
12. Nuseibeh, B., Easterbrook, S.: Requirements engineering: a roadmap. In: Proceedings of the IEEE International Conference on Software Engineering, pp. 35–46. ICM Press, Limerick (2000)
13. Abran, A., Moore, J.W., Bourque, P., Dupuis, R.: SWEBOK - Guide to the Software Engineering Body of Knowledge. IEEE Computer Society, California (2004)
14. Saiedian, H., Dale, R.: Requirements engineering: making the connection between the software developer and customer. Information and Software Technology 42, 419–428 (2000)
15. Maguire, M.: Methods to support human-centred design. International Journal of Human-Computer Studies 55, 587–634 (2001)
16. ISO: ISO 13407 - Human-centred design processes for interactive systems. International Organization for Standardization (1999)
17. Teixeira, L.: Contribuições para o Desenvolvimento de Sistemas de Informação na Saúde: Aplicação na Área da Hemofilia (PhD Thesis - In Portuguese). Department of Economics, Management and Industrial Engineering, University of Aveiro, Aveiro (2008)
18. Blomkvist, S.: Towards a Model for bridging Agile Development and User-Centered Design. In: Seffah, A., Gulliksen, J., Desmarais, M.C. (eds.) Human-Centered Software Engineering - Integrating Usability in the Software Development Lifecycle, vol. 8, pp. 219–244. Springer, Netherlands (2005)
19. Cockburn, A., Highsmith, J.: Agile software development, the people factor. Computer 34, 131–133 (2001)
20. Larman, C., Basili, V.R.: Iterative and incremental developments. a brief history. IEEE Computer 36, 47–56 (2003)
21. Wells, D.: Extreme Programming: A gentle introduction (2006), http://extremeprogramming.org/index.html (visited on 28/07/2008)

Part VII

HCD at Work

From Tools to Teammates:
Joint Activity in Human-Agent-Robot Teams

Jeffrey M. Bradshaw, Paul Feltovich, Matthew Johnson, Maggie Breedy,
Larry Bunch, Tom Eskridge, Hyuckchul Jung, James Lott, Andrzej Uszok,
and Jurriaan van Diggelen*

Florida Institute for Human and Machine Cognition (IHMC)
40 South Alcaniz Street, Pensacola, FL 32502
{jbradshaw,pfeltovich,mjohnson,mbreedy,lbunch,teskridge,
jlott,auszok}@ihmc.us, jurriaanvandiggelen@gmail.com

Abstract. Coordination is an essential ingredient of joint activity in human-agent-robot teams. In this paper, we discuss some of the challenges and requirements for successful coordination, and briefly how we have used KAoS HART services framework to support coordination in a multi-team human-robot field exercise.

1 Introduction

Over the past several years, we have been interested in learning how to facilitate teamwork among humans, agents, and robots. To lay the groundwork for our research, we have studied how humans succeed and fail in joint activity requiring a high degree of interdependence among the participants [9; 18]. Such interdependence requires that, in addition to what team members do to accomplish the work itself, they also invest time and attention in making sure that distributed or sequenced tasks are appropriately coordinated [21].

Our research has been guided by three principles. First, we focus on situations where it is desirable for humans to remain "in-the-loop" and allow the degree and kind of control exercised by the human to vary at the initiative of the human or, optionally, with the help of adjustable autonomy mechanisms [4; 6; 17]. Second, we assure that mechanisms for appropriate robot regulation, communication, and feedback in such situations are included from the start in the foundations of system design, rather than layered on top as an afterthought [14]. Third, working in the tradition of previous agent teamwork researchers (e.g., [10; 24]), we attempt to implement a reusable model of teamwork involving a notion of shared knowledge, goals, and regulatory mechanisms that function as the glue that binds team members together. This teamwork model is to a large degree independent from and complementary to the set of domain-specific reasoners (e.g., task scheduling/optimization, spatial reasoning) that might be needed to accomplish a particular task objective.

---

* Currently at TNO, Soesterberg, The Netherlands.

M. Kurosu (Ed.): Human Centered Design, HCII 2009, LNCS 5619, pp. 935–944, 2009.

Although there are several important challenges in making automation a team player [19], in this paper we focus on only on the problem of coordination. Following a brief description of this aspect of joint activity, we describe the KAoS HART (Human-Agent-Robot Teamwork) services framework, which has been developed as a means of exploring our ideas about the role of regulatory constraints in joint activity [3; 5; 11; 14; 25; 26]. Finally, we discuss example policies from a field exercise that allowed us to implement and explore many of these capabilities. This exercise involved mixed human-robot teams whose objective was to find and apprehend an intruder hiding on a cluttered Navy pier [16].

2 Understanding Coordination

Malone and Crowston [21] defined coordination as "managing dependencies between activities." Teamwork, which by definition implies interdependence among the players, therefore requires some level of work for each party over and beyond the carrying out of task itself in order to manage its role in coordination. Part of that "extra" work involves each party doing its part to assure that relevant aspects of the agents and the situation are observable at an appropriate level of abstraction and using an effective style of interaction [1].

Although coordination is as much a requirement for agent-agent teamwork as it is for human-agent teamwork, the magnitude of the representational and reasoning gulfs separating humans from agents is much larger. Moreover, because the agent's ability to sense or infer information about the human environment and cognitive context is so limited, agent designers must find innovative ways to compensate for the fact that their agents are not situated in the human world. Brittleness of agent capabilities is difficult to avoid because only certain aspects of the human environment and cognitive context can be represented in the agent, and the representation that is made cannot be "general purpose" but must include specific representations and optimizations for the particular use scenarios the designer originally envisioned. Without sufficient basis for shared situation awareness and mutual feedback, coordination among team members simply cannot take place, and, of course, this need for shared understanding and feedback increases as the size of the team and the degree of autonomy increase.

Notwithstanding these challenges, adult humans and radically less-abled entities (e.g., small children, dogs, video game characters) are capable of working together effectively in a variety of situations where a subjective experience of collaborative teaming is often maintained despite the magnitude of their differences. Generally this is due to the ability of humans to rapidly size up and adapt to the limitations of their teammates in relatively short order, an ability we would like to exploit in the design of approaches for human-agent teamwork.

2.1 The Elements of Effective Coordination

Basic requirements. There are three basic requirements for effective coordination: interpredictability, common ground, and directability [18]:

- *Interpredictability:* In highly interdependent activities, it becomes possible to plan one's own actions (including coordination actions) only when what others will do can be accurately predicted. Skilled teams become interpredictable through shared knowledge and idiosyncratic coordination devices developed through extended experience in working together; bureaucracies with high turnover compensate for experience by substituting explicit, predesigned structured procedures and expectations.

- *Common ground:* Common ground refers to the pertinent mutual knowledge, beliefs, and assumptions that support interdependent actions in the context of a given joint activity [8]. This includes initial common ground prior to engaging in the joint activity, as well as mutual knowledge of shared history and current state that is obtained while the activity is underway. Unless I can make good assumptions about what you know and what you can do, we cannot effectively coordinate.

- *Directability:* Directability refers to the capacity for deliberately assessing and modifying the actions of the other parties in a joint activity as conditions and priorities change [7]. Effective coordination requires responsiveness of each participant to the influence of the others as the activity unfolds.

Coordination devices. People coordinate through signals and more complex messages of many sorts (e.g., face-to-face language, expressions, posture). Human signals are also mediated in many ways—for example, through third parties or through machines such as telephones or computers. Hence, direct and indirect party-to-party communication is one form of a coordination device, in this instance coordination by *agreement*. For example, a group of scientists working together on a grant proposal, may simply agree, through e-mail exchanges, to set up a subsequent conference call at a specific date and time. There are three other major types of coordination devices that people commonly employ: *convention*, *precedent*, and *situational salience* [9; 18].

Roles. Roles can be thought of as ways of packaging rights and obligations that go along with the necessary parts that participants play in joint activities. Knowing one's own role and the roles of others in a joint activity establishes expectations about how others are likely to interact with us, and how we think we should interact with them. Shoppers expect cashiers to do certain things for them (e.g., total up the items and handle payment) and to treat them in a certain way (e.g., with cheerful courtesy), and cashiers have certain expectations of shoppers. When roles are well understood and regulatory devices are performing their proper function, observers are likely to describe the activity as highly-coordinated. On the other hand, violations of the expectations associated with roles and regulatory structures can result in confusion, frustration, anger, and a breakdown in coordination.

Organizations. Collections of roles are often grouped to form organizations. In addition to regulatory considerations at the level of individual roles, organizations themselves may also add their own rules, standards, traditions, and so forth, in order to establish a common culture that will smooth interaction among parties.

Knowing how roles undergird organizations and how rights and obligations are packaged into roles helps us understand how organizations can be seen as functional or dysfunctional. Whether hierarchical or heterarchical, fluid or relatively static, organizations are functional only to the extent that their associated regulatory devices and roles generally assist them in facilitating their constituent responsibilities and their work in coordinating their actions with others when necessary.

The lesson here for mixed human-agent-robot teams is that the various roles that team members assume in their work must include more than simple names for the role and algorithmic behavior to perform their individual tasks. They must also, to be successful, include regulatory structures that define the additional work of coordination associated with that role.

3   KAoS HART Policy-Based Approach

The KAoS HART (Human-Agent-Robot Teamwork) services framework has been adapted to provide the means for *policy-based* dynamic regulation on a variety of agent, robotic, Web services, Grid services, and traditional distributed computing platforms [3; 13-15; 20; 23; 25]. In contrast to Asimov's laws of robotics—and similar in spirit to Grice's famous maxims [12]—the objective of KAoS teamwork policies is not principally to prevent harm but rather, in a positive vein, to facilitate helpful interaction among teammates. As a body, we call these policies the "Golden Rules of HART," recalling the biblical injunction: "Do unto others as you would have them do unto you" (Matthew 7:12). If robots (and people) were all sufficiently intelligent and benevolent, perhaps this abstract maxim would be the only rule needed for coordination. Since this is not the case, a number of more specific instantiations of the general principle are required as the basis of teamwork policy. Through a relatively small number of such policies, we can help assure that complex emergent activity can remain coordinated. Through policy learning mechanisms, additional constraints may emerge in an adaptive manner [22].

KAoS policies expressing these specific coordination constraints are implemented in OWL (Web Ontology Language: http://www.w3.org/ 2004/OWL), to which we have added optional extensions to increase expressiveness (e.g., role-value maps) [25]. A growing set of services for policy deconfliction and analysis are also provided [2; 25].

Policies are used to dynamically regulate the behavior of system components without changing code or requiring the cooperation of the components being governed. By changing policies, a system can be continuously adjusted to accommodate variations in externally imposed constraints and environmental conditions. There are two main types of polices; authorizations and obligations. The set of permitted actions is determined by *authorization policies* that specify which actions an actor or set of actors are permitted (*positive authorizations*) or not allowed (*negative authorizations*) to perform in a given context. *Obligation policies* specify actions that an actor or set of actors is required to perform (*positive obligations*) or for which such a requirement is waived (*negative obligations*). From these primitive policy types, we build more complex structures that form the basis for team coordination.

Rigid policy constraints and roles cannot cope with unanticipated situations in a dynamically changing environment. This is particularly important in teamwork situations where multiple agents have to cooperate to achieve a common goal. In addition to an agent's own capabilities and constraints, we also need to take into account the capabilities and constraints of team members that might help or impede a joint task. IHMC's Jung and Teng have devised a methodology to adjust a team of agents' autonomy constraints and execution plans on the fly to avoid total task failures and performance degradation in teamwork situations [4; 6; 17].

4 Coordination Policy Examples

We developed a series of human-agent-robot coordination policies to support a scenario in which an intruder must be discovered and apprehended on a cluttered Navy pier with the assistance of two humans and five robots. Issues included robot capabilities, sensor limitations, and localization, however we focused on the coordination aspects of the task. We specifically included multiple humans and robots, and more robots than a single individual could easily handle by teleoperation.[1]

The teamwork model for our coordinated operations exercise was implemented within various sets of KAoS policies. The intent of the policies is to provide information to establish and preserve common ground among both human and robotic team members, as well as helping to maintain organizational integrity. The policies are defined and enforced external to any specific robot API, so as new robots join, they automatically acquire all the teamwork intelligence possessed by the other robots.

4.1 Cohen-Levesque Notification Obligation Policy

One of the most well known heuristics in team coordination was originally formulated by Cohen and Levesque as follows: "any team member who discovers privately that a goal is impossible (has been achieved, or is irrelevant) should be left with a goal to make this fact known to the team as a whole" [19, p. 9]. We have implemented our version of this heuristic in the form of an obligation policy that can be roughly described as follows:

> A Robot is obligated to notify its Teammates when Action is Finished (whether Successfully Completed, Aborted, or Irrecoverably Failed)

For example, in our field experiments, this policy ensured that, once an intruder has been apprehended, robot and human members of all teams, are notified [13]. This obligation would be triggered as soon as one robot became aware of this fact, and each robot would begin executing the appropriate task it was designed to perform following successful completion of the team goal (e.g., return to base, resume patrolling). If, on the other hand, the team commander were to abort the task due to a higher priority objective, or if any of the robots became aware that failure was inevitable, they would let their teammates know so that the appropriate behaviors for this situation would be triggered for the other members of the team. This single policy obviated the need to write a large number of special-purpose procedures for each possible success or failure mode.

---

[1] In addition to KAoS, which is our focus in this paper, the exercise involved the Agile Computing Infrastructure (ACI), the TRIPS dialogue-based collaborative problem solving system, Kaa/Kab adjustable autonomy and backup planning components, and an advanced multimodal display capability. For an overview of the entire system and scenario, see [16]. Robin Murphy's group at the University of South Florida participated in a USV scenario that was loosely-coupled to our ground-based story.

4.2 Acknowledgements and Policy Deconfliction

We implemented a basic policy that requires robots to acknowledge requests. While this seemed a good general rule, there are important exceptions that need to be handled through KAoS policy deconfliction capabilities [7].

One reasonable exception to the acknowledgement policy is that people do not always verbally acknowledge requests, particularly when they are directly observable. Direct observability means that when a human requestor sends the communication to a robot receiver, the fact that the request was received, understood and being acted upon is observable by the requestor. For example, when a robot is told to move forward five meters, and then can be seen starting to move forward, there is normally no need for the robot to state "I have received your request to move forward and have begun." The same applies to queries. When somebody asks a robot "where are you," it is unnecessary for it to reply "I have heard your question and am about to reply", if it, alternatively, simply says "in the library." We implemented two additional policies to waive the obligation to acknowledge requests when the request is either a teleoperation command or a query.

1) *A Robot is obligated to acknowledge to the Requestor when the Robot Accepts an Action*
2) *A Robot is not obligated to acknowledge Teleoperation requests*
3) *A Robot is not obligated to acknowledge Query requests*

The two policies do indeed conflict with the original, but by assigning the more restrictive polices a higher priority (which can be done numerically or logically), it is possible to automatically deconflict these policies and achieve the desired behavior. Note an additional advantage in the use of ontologies of behaviors is the fact that we can define policies for abstract classes of actions (e.g., Action, Teleoperation requests, Query requests) that will be enforced on every specific action that falls in that class.

4.3 Role Management and Progress Appraisal

Groups often use roles to perform task division and allocation. Roles provide a membership-based construct with which to associate sets of privileges (authorizations) and expected behaviors (obligations). When an actor is assigned to a role, the regulations associated with the role automatically apply to the actor and, likewise, are no longer applicable when the actor relinquishes the role. These privileges and expectations that comprise a role may be highly domain dependent. For example the role "Team Leader" in a military domain is significantly different from "Team Leader" in sports. Roles may also specify expected behaviors. For example, if your role is a "Sentry," then you are obligated to remain at your post, and other actors will expect you to fulfill that obligation. Roles can also affect other behaviors such as expected communications. If you are assigned to be a "Sentry", you are obligated to announce any violations of your boundary and report these to your immediate superior.

Taking advantage of the extensibility and inheritance properties of OWL ontologies, we defined roles at various levels of abstraction with sub-roles refining the regulations pertinent to more generic super-roles. In this way, some high-level roles need not be domain specific or involve specific tasking, but they are still defined by their associated regulations. "Teammate" can be considered a generic role that has some of its regulations already noted. We view this level of abstraction as appropriate for expectations that facilitate coordination such as acknowledgements and progress appraisals. The obligation to acknowledge requests can be thought of as a policy associated with being a teammate. We have developed two policy sets that we feel apply generally to robots assigned to the role of "Teammate." The first is the acknowledgement policy set discussed above. The second involves progress appraisal:

1) A Robot is obligated to notify the Requestor when requested Action is Finished (includes Completed, Aborted, and Failure).

2) A Robot is not obligated to notify the Requestor when a requested Teleoperation Action is Completed.

3) A Robot is not obligated to notify the Requestor when a requested Query Action is Completed.

The first policy ensures that the requestor of a task is notified when the tasked robot encounters problems or successfully completes the task since the action status of Finished is ontologically defined as a super-class of the statuses Completed, Failed, and Aborted. The second two policies in this set are exceptions similar to those in the acknowledgement set. With knowledge that these policies are in place, human and robotic team members have the mutual expectation that these progress appraisals will be performed. This interpredictability removes the need to explicitly ask for such communication and, perhaps just as importantly, the absence of these obligatory communications becomes an indicator that additional coordination may be necessary. For example, a robot is commanded to autonomously navigate to a distant location. Since it is known that the robot would notify team members if it had arrived, or it was stuck, or had otherwise failed, the others can assume that it is still moving toward the goal. If team members were concerned with an approaching deadline or that the task was taking too long, they would query for the robot's position and create a new estimate of when it should reach the goal.

The policies outlined here are just two of several sets that we have explored, informed by previous theoretical work, simulations, and field experiments performed by ourselves and by others [1, 3, 15-17, 19, 26-28]. As we encounter new challenges in future work, we will continue to revise and expand such policy sets.

4.4 Policies Relating to Team Leaders

In contrast to our previous work on human-robot teams, where all team members were "equal," we decided to explore the role of team "leaders." Leaders not only must adhere to their own regulations, but they also impact the regulatory structure of all the other roles in the group. Peer interaction may be undirected, but Leaders tend to alter the pattern of activity, with themselves becoming the focal point. In particular we have identified several policy sets particular to leaders. The first set is about the chain of command:

1) A Robot is authorized to perform Actions requested by its Team Leader

2) A Robot is authorized to Accept Actions requested by a higher authority

3) A Robot is not authorized to perform Action requests from just any Requestor

4) A Robot is authorized to Accept Actions that are self-initiated

The first policy gives team leaders the authority to command their team. The second gives the same authority to anyone directly higher in the chain of command. The third policy explicitly restricts access to the robots from those outside of the chain of command. The fourth policy makes self initiated actions an exception to the third policy.

Another policy set was used to explore notification to help maintain common ground between the team leader and each of the team members:

1) A Robot is obligated to notify its Team Leader when an Action is requested by a higher authority

2) A Robot is obligated to notify Its Team Leader when starting a self-initiated Action

3) A Robot is obligated to notify its Team Leader when a self-initiated Action is Finished (includes statuses of Completed, Aborted, and Failure).

4.5 Team Creation and Management

The KAoS Directory Service manages organizational structure, allowing dynamic team formation and modification. Teams and subteams can be created dynamically, allowing for the creation of complex organizational structures. Agents can join and leave teams as necessary to support the desired structure. Actors can be assigned roles including Team Leader, affecting the dynamics of coordination as discussed in the previous section. Queries can be made to identify current team structure, who is on a certain team currently, or who is team leader.

5 Conclusions

In our work, the teams are not merely groupings, but provide the framework to support advanced coordination policies typical in human-human teams. When a leader is assigned, this means more then just being authorized to task other agents. For instance, it also defines the expected communication pattern among pertinent team members. As a team member, you are obligated to ensure that your leader knows you are working and to keep other members updated about pertinent information. These types of coordination, natural to humans, will enable robots to perform less like tools and more like teammates.

References

1. Bradshaw, J.M., Sierhuis, M., Acquisti, A., Feltovich, P., Hoffman, R., Jeffers, R., Prescott, D., Suri, N., Uszok, A., Van Hoof, R.: Adjustable autonomy and human-agent teamwork in practice: An interim report on space applications. In: Hexmoor, H., Falcone, R., Castelfranchi, C. (eds.) Agent Autonomy, pp. 243–280. Kluwer, Dordrecht (2003)
2. Bradshaw, J.M., Uszok, A., Jeffers, R., Suri, N., Hayes, P., Burstein, M.H., Acquisti, A., Benyo, B., Breedy, M.R., Carvalho, M., Diller, D., Johnson, M., Kulkarni, S., Lott, J., Sierhuis, M., Van Hoof, R.: Representation and reasoning for DAML-based policy and domain services in KAoS and Nomads. In: Presented at the Proceedings of the Autonomous Agents and Multi-Agent Systems Conference (AAMAS 2003), Melbourne, Australia, July 14-18, pp. 835–842 (2003)

3. Bradshaw, J.M., Beautement, P., Breedy, M.R., Bunch, L., Drakunov, S.V., Feltovich, P.J., Hoffman, R.R., Jeffers, R., Johnson, M., Kulkarni, S., Lott, J., Raj, A., Suri, N., Uszok, A.: Making agents acceptable to people. In: Zhong, N., Liu, J. (eds.) Intelligent Technologies for Information Analysis: Advances in Agents, Data Mining, and Statistical Learning, pp. 361–400. Springer, Berlin (2004)

4. Bradshaw, J.M., Feltovich, P., Jung, H., Kulkarni, S., Taysom, W., Uszok, A.: Dimensions of adjustable autonomy and mixed-initiative interaction. In: Nickles, M., Rovatsos, M., Weiss, G. (eds.) AUTONOMY 2003. LNCS, vol. 2969, pp. 17–39. Springer, Heidelberg (2004)

5. Bradshaw, J.M., Feltovich, P.J., Jung, H., Kulkarni, S., Allen, J., Bunch, L., Chambers, N., Galescu, L., Jeffers, R., Johnson, M., Sierhuis, M., Taysom, W., Uszok, A., Van Hoof, R.: Policy-based coordination in joint human-agent activity. In: Presented at the Proceedings of the IEEE International Conference on Systems, Man, and Cybernetics, The Hague, Netherlands, October 10-13 (2004)

6. Bradshaw, J.M., Jung, H., Kulkarni, S., Johnson, M., Feltovich, P., Allen, J., Bunch, L., Chambers, N., Galescu, L., Jeffers, R., Suri, N., Taysom, W., Uszok, A.: Toward trustworthy adjustable autonomy in KAoS. In: Falcone, R. (ed.) Trusting Agents for Trustworthy Electronic Societies. Springer, Berlin (2005)

7. Christofferson, K., Woods, D.D.: How to make automated systems team players. In: Salas, E. (ed.) Advances in Human Performance and Cognitive Engineering Research, vol. 2. JAI Press/ Elsevier (2002)

8. Clark, H.H., Brennan, S.E.: Grounding in communication. In: Resnick, L.B., Levine, J.M., Teasley, S.D. (eds.) Perspectives on Socially Shared Cognition. American Psychological Association, Washington (1991)

9. Clark, H.H.: Using Language. Cambridge University Press, Cambridge (1996)

10. Cohen, P.R., Levesque, H.J.: Teamwork. SRI International, Menlo Park (1991)

11. Feltovich, P., Bradshaw, J.M., Jeffers, R., Suri, N., Uszok, A.: Social order and adaptability in animal and human cultures as an analogue for agent communities: Toward a policy-based approach. In: Omicini, A., Petta, P., Pitt, J. (eds.) ESAW 2003. LNCS (LNAI), vol. 3071, pp. 21–48. Springer, Heidelberg (2004)

12. Grice, H.P.: Logic and Conversation. In: Cole, P., Morgan, J.L. (eds.) Syntax and Semantics 3: Speech Acts. Academic Press, New York (1975)

13. Johnson, M., Chang, P., Jeffers, R., Bradshaw, J.M., Soo, V.-W., Breedy, M.R., Bunch, L., Kulkarni, S., Lott, J., Suri, N., Uszok, A.: KAoS semantic policy and domain services: An application of DAML to Web services-based grid architectures. In: Presented at the Proceedings of the AAMAS 2003 Workshop on Web Services and Agent-Based Engineering, Melbourne, Australia (July 2003)

14. Johnson, M., Bradshaw, J.M., Feltovich, P., Jeffers, R., Uszok, A.: A semantically-rich policy-based approach to robot control. In: Proceedings of the International Conference on Informatics in Control, Automation, and Robotics, Lisbon, Portugal (2006)

15. Johnson, M., Feltovich, P.J., Bradshaw, J.M., Bunch, L.: Human-robot coordination through dynamic regulation. In: Proceedings of the International Conference on Robotics and Automation (ICRA), Pasadena, CA (2008) (in press)

16. Johnson, M., Intlekofer Jr., K., Jung, H., Bradshaw, J.M., Allen, J., Suri, N., Carvalho, M.: Coordinated operations in mixed teams of humans and robots. In: Marik, V., Bradshaw, J.M., Meyer, J. (eds.) Proceedings of the First IEEE Conference on Distributed Human-Machine Systems (DHMS 2008), Athens, Greece (2008) (in press)

17. Jung, H., Bradshaw, J.M., Kulkarni, S., Breedy, M.R., Bunch, L., Feltovich, P., Jeffers, R., Johnson, M., Lott, J., Suri, N., Taysom, W., Tonti, G., Uszok, A.: An ontology-based representation for policy-governed adjustable autonomy. In: Proceedings of the AAAI Spring Symposium. AAAI Press, Stanford (2004)
18. Klein, G., Feltovich, P.J., Bradshaw, J.M., Woods, D.D.: Common ground and coordination in joint activity. In: Rouse, W.B., Boff, K.R. (eds.) Organizational Simulation, pp. 139–184. John Wiley, New York City (2004)
19. Klein, G., Woods, D.D., Bradshaw, J.M., Hoffman, R., Feltovich, P.: Ten challenges for making automation a "team player" in joint human-agent activity. IEEE Intelligent Systems 19(6), 91–95 (2004)
20. Lott, J., Bradshaw, J.M., Uszok, A., Jeffers, R.: Using KAoS policy and domain services within Cougaar. In: Presented at the Proceedings of the Open Cougaar Conference 2004, New York City, July 20, pp. 89–95 (2004)
21. Malone, T.W., Crowston, K.: What is coordination theory and how can it help design cooperative work systems? In: Presented at the Conference on Computer-Supported Cooperative Work (CSCW 1990), Los Angeles, CA, October 7-10, pp. 357–370 (1990)
22. Rehak, M., Pechoucek, M., Bradshaw, J.M.: Representing context for multiagent trust modeling. In: Proceedings of the 2006 IEEE/WIC/ACM International Conference on Intelligent Agent Technology, Hong Kong (2006)
23. Sierhuis, M., Bradshaw, J.M., Acquisti, A., Van Hoof, R., Jeffers, R., Uszok, A.: Human-agent teamwork and adjustable autonomy in practice. In: Presented at the Proceedings of the Seventh International Symposium on Artificial Intelligence, Robotics and Automation in Space (i-SAIRAS), Nara, Japan, May 19-23 (2003)
24. Tambe, M., Shen, W., Mataric, M., Pynadath, D.V., Goldberg, D., Modi, P.J., Qiu, Z., Salemi, B.: Teamwork in cyberspace: Using TEAMCORE to make agents team-ready. In: Presented at the Proceedings of the AAAI Spring Symposium on Agents in Cyberspace, Menlo Park, CA (1999)
25. Uszok, A., Bradshaw, J.M., Johnson, M., Jeffers, R., Tate, A., Dalton, J., Aitken, S.: KAoS policy management for semantic web services. IEEE Intelligent Systems 19(4), 32–41 (2004)
26. Uszok, A., Bradshaw, J.M., Breedy, M.R., Bunch, L., Feltovich, P., Johnson, M., Jung, H.: New developments in ontology-based policy management: Increasing the practicality and comprehensiveness of KAoS. In: Proceedings of the 2008 IEEE Conference on Policy, Palisades, NY (2008)

Capturing and Restoring the Context of Everyday Work: A Case Study at a Law Office

Gaston R. Cangiano and James D. Hollan

Distributed Cognition and Human-Computer Interaction Laboratory
Cognitive Science, University of California at San Diego
{gaston,hollan}@cogsci.ucsd.edu

Abstract. Real-world activity is complex and increasingly involves use of multiple computer applications and communication devices over extended periods of time. To understand activity at the level of detail required to provide natural and comprehensive support for it necessitates appreciating both its richness and dynamically changing context. In this article, we (1) summarize field work in which we recorded the desktop activities of workers in a law office, (2) analyze interview data in detail to show the effects of context reinstatement when viewing video summaries of past desktop activity. We conclude by discussing the implications of our results for the design of software tools to assist work in office settings.

Keywords: User behavior, empirical study, screen recording, summarization.

1 Introduction

Observational studies of office workers reveal that real-life work is highly fragmented [1,2,3,4]. For instance, Mark et al. [1] found that during the course of a typical day information workers spend an average of only 12 minutes on any given task and most continuous (uninterrupted) "events" average about 3 minutes in duration. Furthermore, the nature of information-intensive work requires people to manage a complex mix of multiple tasks and activities, each one frequently requiring different collections of applications, devices, and other resources. Typically each individual application provides only partial support for the real task it is being used to help accomplish. For example, a task might require access to a conversation, consisting of a history of emails and instant messages between groups of individuals, interleaved with examining spreadsheets, notes, and bookmarked websites relevant to this task. But each application only provides support for a single aspect of the task. Users must decompose tasks into components appropriate for various applications, assemble the resulting outcomes, and maintain the overall context needed to complete the real task.

An interesting challenge for researchers today is posed by a seeming paradox in personal computing: while most applications are designed to support a single, often simplified task, most real tasks and activities necessitate coordinated use of multiple applications. As a consequence, the information needed to conduct daily life is increasingly spread across a variety of digital resources, resulting in fragmentation of our activities and increased complexity.

M. Kurosu (Ed.): Human Centered Design, HCII 2009, LNCS 5619, pp. 945–954, 2009.
© Springer-Verlag Berlin Heidelberg 2009

While individual applications, such as word processing, continue to accrue myriad new features, there is extremely limited support for aspects of tasks that span applications. For example, while change-tracking is a useful facility for highlighting the history of editing changes in a document, it captures none of the activity accomplished outside the application. That activity is necessary for a full picture of the history of the document's evolution. Even with application suites like Microsoft Office or Open Office, integration is typically document-centric. If we are to move towards understanding and supporting real activities, what is needed is access to *episodic* views of activity within and across applications and resources.

There are, of course, advantages and reasonable software engineering motivations for developing applications that focus on specific components of tasks. The longevity and continuing evolution of applications to support tasks like document authoring, email, browsing, and search are testimony to their usefulness and to this specific decomposition of activities. On the other hand, this focus may have created the paradox to which we refer to above and may be a factor motivating increasing interest in task-support tools [5,6,7].

The goal common to task-support tools is assisting task switching and restoring associated contexts. The underlying idea dates to the seminal work on the Rooms workspace manager [8], but there is renewed interest in this area due to new empirical evidence on the nature of modern office work [1,3,4] and widespread concern about the cognitive costs of multitasking [22]. But most approaches are still based on the assumption that task context can be represented as a static state without access to the history of interactions that led to that state. In this paper we question this assumption and argue for the importance of providing episodic access to the temporal dynamics of past interaction as well as associated activities that might otherwise be deemed interruptions. The motivation for questioning this assumption derives from data of desktop activity captured in a law office and interviews with the participants while they viewed their past activity.

Below we discuss related work, describe the details of our study and the methods we used to collect data, summarize analysis of the desktop activity data and interviews with participants, and conclude with a discussion on the direct implications of our analysis for developing tools to support reestablishing context and easing continuation of past activities.

2 Related Work

Logging Activity. There is a long history and recent resurgence of interest in recording personal activity. The history dates at least to Vannevar Bush's Memex [20] and recently on a series of ACM CARPE workshops [10] on capturing, archiving, and retrieval of personal experiences. There has been little empirical evaluation of any of these systems. One exception is recent work by Sellen and her colleagues [9]. Using a personal recording device (SenseCam http://research.microsoft.com/sensecam/) they were able to quantitatively test recall performance using self-recorded images. The images employed were taken from a first- person point-of-view, recorded either at random intervals or initiated by the participants. They tested recall of discrete items over a period raging from 3 days to 4 months. Overall, they found positive recall with the use of images, but the significant differences in memory performance were only revealed after making a distinction between *remembering* and *knowing*. The former process is that by which people re-experience the past. The latter has to do with inferring that something happened, based on cues available or habitual patterns. They found significant differences between short and long-term knowing when using images to recall the past.

Task Support Systems. Research on task support has been gaining momentum in the last five years in HCI. The main notion is that a task can be defined as a discrete group of resources, such as documents in the file system, URLs, emails and so on. There is an implicit assumption in this area of research that these static groups are a sufficient representation of the *context* for a given task. The main approach therefore has been to support grouping windows, since window can represent any number of types of documents [5,6,7]. For instance the SWISH project (part of the Microsoft Vibe Technologies group http://research.microsoft.com/vibe/) applies machine learning technology to cluster windows based on their title and temporal proximity. Even though this kind of approach greatly facilitates switching and restoring sets of windows, it only represents a *partial* and *static* view of the context. More may be needed to reconstruct the mental representation that was active at the time the activity took place. Part of the reason why this view of context is incomplete is that it does not contain the *history* of interaction nor is flexible enough to include sporadic items that can act as memory triggers. We elaborate this further below as we describe our data in detail.

Other task support approaches provide users with ways to organize unstructured data, such as conversations and interactions, without restrictions on the type of resource involved (e.g., email, documents, IM, web pages, etc). For instance the Haystack project [13] provides a sophisticated user interface that can access almost any element from users' activities. Haystack provides the facilities to view emails, chats, web pages and documents under user-defined categories. In principle, this type of approach could solve the support challenges we described so far, especially if histories of interactions are available for navigation. But there is an important caveat: users are required to explicitly define the parameters of the associations between resources (i.e., the categories). Having to manually define every *context* that a user is engaged in a work environment (e.g., by tagging or annotating), significantly increases the cost-benefit ratio of the approach. In addition, people don't have conscious access to the complete context of their activity nor know the memory triggers and cues needed to reinstate it, further constraining the impact of this approach.

3 Overview of Study

Over the course of the last year, we have conducted an ethnographic study of a law office. The main goal was to understand the culture, work practices, and demands of this information-intensive work setting. Our longer term goal is to design and prototype applications to assist office workers. To ground design in the real activity of the office, we built a data-collecting tool called *ActivityTrails* that enables the capture of all desktop activity. One motivation for this tool was to be able to replay specific episodes of desktop activity to encourage participants to talk in detail about their work and the way they think about it. In addition, we conjectured that summaries and visual representations of the activity have promise in helping users to reestablish the context of previous work and to assist in resuming it.

We conducted a series of detailed interviews to learn about the tools and practices of this office setting and to explore the use of video summaries as an aid to recall of past activity.

The Office Setting. The field site for this study is an Immigration Law firm located in the San Francisco Bay Area. The firm has a staff of eight and has been in operation for over 3 years. Its members have worked together as a team for almost a decade, starting as employees of a large immigration law firm before leaving to start their current office. Shaped by their history of working together they have well established work practices for assisting clients with immigration issues. The work flow of the office is smooth and efficient.

The division labor at this office involves two main echelons: *paralegals and attorneys*. The role of attorneys is to address legal questions from the paralegals and clients, and to keep up with legal research, the industry, and government legislation. Paralegals' main job is to compile cases for filing. Compiling cases means to gather all the required documentation and information about foreign nationals in order to fill out the required forms for each visa category. It also means following up on filing deadlines, including sending notifications to clients. In addition, paralegals are responsible for maintaining updated client profiles, entering all communication, forms, and deadlines into an online database.

Because this is an Immigration Law firm, billing is primarily done on a filing basis. Unlike other types of law firms that have a need to keep track of billable hours, billing in this office is done on a per visa or work permit category basis. The focus is on keeping track of deadlines, client communication, and documentation. It is important for paralegals to keep case logs, not with the time spend on each case, but rather with a timeline of chronological interactions with clients, filing deadlines and follow up alerts. Because of the emphasis on "time-lining" rather than "time-tracking", both paralegals and attorney often multitask more frequently than they might if specific time was being charged to each client. They can address issues of many different clients in an interleaved fashion, not having to worry how much time they spend on each email or phone call. We mention this because it affects their work practice. A common pattern that emerges is that everyone usually works on multiple cases at the same time; the only exception is when a case is being compiled prior to a filing deadline or when a new case is opened and substantial data entry is needed.

4 Data Collection

Researchers from many disciplines are taking advantage of increasingly inexpensive digital video and storage facilities to assemble extensive data collections of human activity captured in real-world settings. Hollan and Hutchins [21] have argued that the ability to record and share such data has created a critical moment in the practice and scope of behavioral research. The use of video recording for the study of workplace activity is increasingly central to advancing HCI research [15]. Lately, workplace ethnography has incorporated the use of computer screen recording [16]. One of the advantages of using screen capture is that it provides researchers with detailed records of activity. Our study makes use of screen recordings as a means to elicit data about users' perception of workflow. We employed a technique known as "auto-confrontation", proposed as an effective tool for participatory design [17].

4.1 Interviews

Our research consisted of two interview phases with four participants: one attorney and three paralegals. During the first phase the primary goal was to learn about the work practices in the law office. We recorded over fifty hours of workstation activity spanning a period of several days, and also videotaped hour long interview sessions. During the interviews, we showed participants visual summaries of their desktop activity taken from recordings (recorded not more than three days prior and edited manually to create video summaries). The interviews in the first phase were bound with a set of open-ended questions to elicit descriptions of a participant's work activities. During the second phase of interviews, our goal was to obtain qualitative data about how they described their activities and associated context while viewing visual summaries. We subsequently recorded the same participants for several days and conducted another round of interviews a week after these recordings were made.

Video summaries were generated from snapshots of the desktop taken every five seconds. These were played at one frame per second. Therefore time is compressed by a factor of five. The viewing experience is that of a slightly speeded-up movie but still very comprehensible. The figure below shows the current interface used for displaying video summaries using ActivityTrails. The bottom portion of the interface shows the playback area (a full capture of the desktop). The top area of the interface shows a timeline composed of thumbnails indexing specific episodes. Users can scrub through the video or access specific points in it by using the control at the bottom of the window, or jump in time by double-clicking on a thumbnail at the top.

The following interview excerpts are representative exemplars from our first phase of recorded interviews. The interviewer portions of the transcript and some other irrelevant data have been removed to keep the excerpt short. Line numbers are kept for referencing purposes and ellipses inserted to indicate where text was left out.

The first except involves a senior attorney at the firm. She uses two computer screens on her desk. On one screen is her email inbox; on the other one are several instant messaging (IM) windows. She is writing several instant messages while simultaneously going back and forth her inbox.

```
1 this is one of my ahh legal assistant paralegals asking me a legal
  question
3 that's related to an email I had received that I am supposed to give re-
  sponse to this is another paralegal asking me … and I just responded she
  usually just writes hi to see whether I am there and then waits to my re-
  sponse
5 … here she is telling me to politely to review a file that she can file
  and so then I am going into the email to check for
7 for that the email that she wants me to respond to
9 here she's letting me know about the client I am starting to go through
  her emails so ah this is another legal assistant so I am am dual tasking?
  between IM and email
```

We see from the transcription that she is engaged in talking to three different people, while also going through her email. According to her report, this is something that she typically does each morning. We refer to this activity as "getting situated". She spends a significant portion of her mornings getting situated by instant messaging everyone else in the office and looking over her inbox to see "what is going on". Instant messaging facilitates the ability to have multiple conversations at the same time and to quickly get a glimpse of everyone's concern.

It is interesting she uses the term "dual tasking" on line 9 above, adopting a common conception of multitasking. An alternative is to characterize it not as multitasking but as an instance of her typical morning routine of getting situated. This makes a difference in what might be relevant aspects of the context. The former case fits with an application-centric view of activity, while the later highlights her routine as central.

As a thought experiment, imagine that a month later she needs to recall the reasons she gave a particular piece of legal advice that morning. What aspects of the context of her activity from earlier would be most useful in helping her to remember or reconstruct those reasons? Should the "interruptions" and other aspects of multitasking be removed or might they actually be integral elements to assist her memory? Perhaps her legal advice came after reading an email and discussing another related issue with a paralegal. If she only has separate application-based access to individual emails or IM conversations, her ability to effectively recall her reasoning might be diminished. It takes considerable extra effort to bring the separate pieces of information together, and even then she would not have access to perceptual cues associated with the unfolding of the original episode in time to help trigger her memory.

A second excerpt involves a paralegal. In this segment, she is using her Outlook calendar, drafting an email, and switching to her browser repeatedly to navigate the case database. It turns out that she uses her calendar initially as a visual reference to know what to do next. She realizes that she needs to work on a case, but before she starts working on it (by drafting an email) she also realizes she needs some information from another case that shares a similar legal resolution to use as a reference before she can draft the email.

```
1 I am going through my calendar cuz i probably had to schedule a call
  that's why i went the calendar
2 ahhhh hmmmm then I went into another case which was the one that i was
  looking previously for a resolution
3 and i guess i FOUND a resolution for what i needed and that's why i went
  to the actual case and looked at the database and started working on the
  case
```

This is an example of a typical use of multiple tools to accomplish a single task, in this case drafting a client email to schedule a conference call. We can see from the narrative above how she gradually recalls the entire episode while narrating the sequence of her interactions with the different tools: her calendar, her email folders and a web-based database. The history of interaction between these different tools seems fundamental to reestablishing context.

From the point of view of design, these and other episodes in our data highlight the importance of cross application and temporal cues in participants' descriptions of meaningful episodes of their activity. For example in this case, the fact that her task was initiated by looking at her calendar, and that she then realized she needed to look at a different case for reference purposes, is information not available from access to individual applications separately.

4.2 Visual Summaries

The following interviews excerpts were obtained during the second phase of our study. We recorded continuous desktop activity for periods of a week to a week and a half. We interviewed participants using the same auto-confrontation method as in the earlier phase, one week after the recordings were made. We employed the same type of visual summaries but the guiding questions for the interviews were different. The goal for these second interviews focused on participants' reaction to and recall from the visual summaries.

The first excerpt corresponds to a paralegal describing how she resumes prior client communication. The images show her drafting an email with several interruptions from an IM conversation window. She recalls during her narrative how she was prompted by a client fax to resume prior communication.

```
1 sometimes I have to figure where i left off, and I am like should I do
  something new?
2 but then should I wait for her and then you kinda go back
3 NOW I AM emailing on so i got a fax i emailed XXX (attorney's name) or
  IM'ed XXX about the fax
4 and now i am emailing them saying (pause) call to discuss the fax
```

The capital letters in the excerpt indicate emphasis in speech. The switch to the present tense ("I AM") and use of the word "NOW" in line 3 may signal the cognitive reinstatement of an episode and its associated context. We noticed that as soon as she *shifts* tense and begins to speak of herself as if *being there*, she is able to recall the causes and details of why and how she wrote that email.

The next excerpt is from a paralegal recalling a case filing. The screen shows her having a lengthy IM conversation while having a Yahoo Portal window open in the background. She then switches to the web browser and navigates to the finance section and starts reading information about a specific company.

```
1 ok, so this guy calls or I called him so then the form or the forms came
  in that Fedex that we were waiting so then I called him to get a paystub
  so we can file the case today
2 and then he is saying about the XXX (company name) bought out his company
  XXX (other company name) or whatever
3 Interviewer: sooo what made you remember that?
4 that's just my memory I have a very odd memory cuz I just saw the word
  "paystub" and then i remember that this guy works for XXX (company name)
5 (silence pause) CAUSE I AM ON THE PHONE WITH HIM i'm on the phone with him
  and IM'ing XXX (attorney's name) and then because he wanted me
```

In line 4 above she says that seeing word "paystub" on the screen reminded her about the situation surrounding her contact with the client. This visual cue together with the name of the company showing on the screen (in the opened background window) facilitated her recall of the episode. We see in line 5 that after a pause, she raises the tone of her voice when she realizes that she was actually on the phone with the client *while* having an IM conversation and researching the company online. The full context of this episode, in this case extending to the surrounding physical context, is gradually available to her through the combination of visual and temporal cues. These types of cues appear to be a powerful and effective mechanism to reinstate a past episode, as we can see from her even recalling her physical context. Our data contains several instances similar to this, where a participant recalls the physical surroundings of an episode from viewing only images of the desktop. For example, one of our participants recalled someone walking into her office; another participant recalled a courier service employee walking up to her desk and dropping a package.

The following excerpt involves an attorney while she researches legal websites and drafts a related email.

```
1 the news come daily so ahm I look through the emails regularly i don't
  scan their websites regularly but since we are in a marketing mode right
  now and I am thinking of redoing my website some this was a good occasion
  to do that
3 this is a website i really like (silence pause) oh and then I sent an
  email because I really liked it I sent an email to somebody who wants to
  do the marketing for me more regularly to let her know that this is one
4 (pause) [in lower voice] I don't remember getting a response to that I
  guess come to think of it
```

In the excerpt above we see the participant browsing through a website, following links, reading and then drafting an email. As the narrative develops, her recall indicates a rich recollection of details. For example, she remembers that she had completely forgotten to follow up on the email she drafted.

In information-intensive work environments like this one, not everything we are exposed to gets annotated, saved, bookmarked or filed. It may therefore be exceedingly valuable to have access to what we see and do on a daily basis without the need for explicit tagging or saving. Of course, much work remains on how to summarize and provide access to such rich activity records.

5 Conclusions and Future Work

Technological advances have dramatically changed work practices. Everyday work increasingly involves switching between multiple applications and responding to myriad "interruptions." To understand these practices and thus be able to provide comprehensive support for them requires detailed characterizations. One contribution of the work reported here is a case-study description of everyday work practice in a legal office. We also briefly describe ActivityTrails, a tool developed to transparently record desktop activity and enable navigation and replaying of episodes.

Examining the fabric of everyday work activity reveals a dialectic between what people want to accomplish and how it can be accomplished within the constraints of the facilities and applications available. Although work activity from an application perspective can appear fragmented and disjoint, it is important to note that our participants still perceived themselves as working on a *single* activity in a number of those same episodes. Similarly, the many instances of self-initiated "interruptions" we saw in the data may constitute the very fabric of work patterns and function as key components of the activity context.

The increasing complexity of work and frequent use of multiple independently designed applications limit the ability of developers of any single application to provide comprehensive support for work practices. This results in increased cognitive load and fragmented contexts. We think there is promise in providing users with facilities to record activity and mechanisms to exploit those records to help restore previous context. While we have only begun to develop such facilities we are encouraged by the data we report here that demonstrates how visual summaries of past activity influence recall not only of the activity but its wider context.

Human memory is finely tuned to reconstruct the past. We argue that it is important to understand the power of visual and temporal cues to evoke the meaningful structure and context of past activity. Activity summaries, to be effective, must cut across artificial application boundaries and the resulting fragmented workflow to engage episodic memory to help reinstate the full rich context of previous activity. Such summaries will, of course, need to do more than provide access to raw activity recordings. Understanding how to create effective summaries is a key direction for future research.

We are quite a way from being able to provide the facilities needed for this scenario. Our current plan is to first continue with data collection in order to develop a quantitative measure of the effects of visual summaries for context reinstatement. Surprising little work has been done in this area and we think it is crucial to understanding how to design effective visual summaries. We are also starting to explore algorithmic mechanisms to automatically generate activity summaries at different timescales. Identifying natural activity breakpoints and meaningful scales are important steps towards summarization. In parallel with developing automatic algorithms, we will investigate the effectiveness of human identified activity breakpoints. We are especially interested in comparing how breakpoints differ when viewing one's own activity and when others are viewing it. We expect, as we saw in our study, that viewing one's own activity provides particularly rich access to the context and purpose of the activity.

We have argued in this paper for the importance of capturing and examining activity beyond the boundaries established by individual applications. We believe better understandings of the rich dynamics of human behavior in complex settings and of the dialectic nature of interaction between new digital tools and the co-evolution of new work practices is fundamental to HCI advancement. We are encouraged by the data from our initial field study that shows that reviewing activity episodes seems to trigger vivid recollections of the past. While there are many challenges involved in understanding how to effectively represent such activity histories, we see it as an exceedingly promising direction for our future research.

References

1. Mark, G., Gonzalez, V.M., Harris, J.: No task left behind?: examining the nature of fragmented work. In: Proc. CHI 2005, pp. 321–330 (2005)
2. Brush, A.B., Meyers, B.R., Tan, D.S., Czerwinski, M.: Understanding memory triggers for task tracking. In: Proc. CHI 2007, pp. 947–950 (2007)
3. Czerwinski, M., Horvitz, E., Wilhite, S.: A diary study of task switching and interruptions. In: CHI 2004, pp. 175–182 (2004)
4. Iqbal, S.T., Horvitz, E.: Disruption and recovery of computing tasks: field study, analysis, and directions. In: Proc. CHI 2007, pp. 677–686 (2007)
5. Oliver, N., Smith, G., Thakkar, C., Surendran, A.C.: SWISH: semantic analysis of window titles and switching history. In: Proc. IUI 2006, pp. 194–201 (2006)
6. Dragunov, A.N., Dietterich, T.G., Johnsrude, K., McLaughlin, M., Li, L., Herlocker, J.L.: TaskTracer: a desktop environment to support multi-tasking knowledge workers. In: Proc. IUI 2005, pp. 75–82 (2005)

7. Rattenbury, T., Canny, J.: CAAD: an automatic task support system. In: Proc. CHI 2007, pp. 687–696 (2007)
8. Card, S.K., Henderson, A.: Rooms: A multiple, virtual-workspace interface to support user task switching. In: Proc. CHI 1987, pp. 53–59 (1987)
9. Sellen, A.J., Fogg, A., Aitken, M., Hodges, S., Rother, C., Wood, K.: Do life-logging technologies support memory for the past?: an experimental study using sensecam. In: Proc. CHI 2007, pp. 81–90 (2007)
10. Gemmell, J., Williams, L., Wood, K., Lueder, R., Bell, G.: Passive capture and ensuing issues for a personal lifetime store. In: Proc. CARPE 2004, pp. 48–55 (2004)
11. Czerwinski, M., Horvitz, E.: An Investigation of Memory for Daily Computing Events. In: Proc. British HCI Group Annual Conference (2002)
12. Lamming, M., Brown, P., Carter, K., Eldridge, M., Flynn, M., Louie, G., Robinson, P., Sellen, A.: The Design of a Human Memory Prosthesis. The Computer Journal 37(3), 153–163 (1994)
13. Karger, D., Bakshi, K., Huynh, D., Quan, D., Sinha, V.: Haystack: a customizable general-purpose information management tool for end users of semistructured data. In: CIDR 2003 (2003)
14. Ruhleder, K., Jordan, B.: Capturing Complex, Distributed Activities: Video-Based Interaction Analysis as a Component of Workplace Ethnography. In: Lee, A.S., Liebenau, J., DeGross, J.I. (eds.) Information Systems and Qualitative Research, pp. 246–275. Chapman and Hall, London (1997)
15. Tang, J.C., Liu, S.B., Muller, M., Lin, J., Drews, C.: Unobtrusive but invasive: using screen recording to collect field data on computer-mediated interaction. In: Proc. CSCW 2006, pp. 479–482 (2006)
16. Kuorinka, I.: Tools and means of implementing participatory ergonomics. In: Papers from the Nordic Ergonomics Society Conference, April 1997, vol. 19(4), pp. 267–270 (1997)
17. Redish, J., Wixon, D.: Task Analysis. In: Jacko, J.A., Sears, A. (eds.) The Human Computer Interaction Handbook, pp. 922–940. Lawrence Erlbaum Associates, Mahwah (2003)
18. Rekimoto, J.: Time-machine computing. In: 12th annual symposium on user interface software and technology, Asheville (1999)
19. Chau, D.H., Myers, B., Faulring, A.: What to Do When Search Fails: Finding Information by Association. In: Proc. CHI 2008 (2008)
20. Bush, V.: As We may Think. The Atlantic Monthly, Boston (1973)
21. Hollan, J.D., Hutchins, E.L.: Opportunities and Challenges for Augmented Environments: A Distributed Cognition Perspective. In: Lahlou, S. (ed.) Designing User Friendly Augmented Work Environments: From Meeting Rooms to Digital Collaborative Spaces. Springer, Heidelberg (in press, 2009)
22. Bailey, B.P., Iqbal, S.T.: Understanding changes in mental workload during execution of goal-directed tasks and its application for interruption management. ACM Trans. Comput.-Hum. Interact. 14(4), 1–28 (2008)

Development of CSCW Interfaces from a User-Centered Viewpoint: Extending the TOUCHE Process Model through Defeasible Argumentation

María Paula González[1,2,4], Victor M.R. Penichet[3], Guillermo R. Simari[2], and Ricardo Tesoriero[3]

[1] National Council of Scientific and Technical Research (CONICET), Argentina
[2] Department of Computer Science and Engineering, Universidad Nacional del Sur
Av Alem 1253, 8000 Bahía Blanca, Argentina
[3] Computer Systems Department, Universidad Castilla-La Mancha
02071 Albacete, Spain
[4] GRIHO Research Group, Universitat de Lleida
25001 Lleida, Spain
{mpg,grs}@cs.uns.edu.ar, victor.penichet@uclm.es,
ricardo@dsi.uclm.es

Abstract. The *Task-Oriented and User*-Centered *Process Model for Developing Interfaces for Human-Computer-Human Environments* (TOUCHE) is aimed to build up user interfaces for groupware applications under a Human-Computer Interaction perspective. It includes a large set of well known formal models like Class Diagrams, Organizational Structure Diagrams, Task Diagrams, Collaboration Diagrams and Abstract Interaction Objects among others. Most of such models, however, suffer from a number of limitations when formalizing users' commonsense. Over the last few years, Argumentation Systems have been gaining importance in several areas of Artificial Intelligence, mainly as a vehicle for facilitating rationally justifiable decision making when handling incomplete and potentially inconsistent information. This paper sketches a Proof of Concept to show how defeasible argumentation techniques can be embedded within the TOUCHE. The final goal is to enhance the capability of development process models for CSCW systems by including a rule-based approach for efficient reasoning with incomplete and inconsistent information.

1 Introduction and Motivations

Nowadays Computer Supported Collaborative Work (CSCW) is a challenging research field focused on developing groupware applications. In particular, the Task-Oriented and User-Centered Process Model for Developing Interfaces for Human-Computer-Human Environments (TOUCHE) [1,2] was formally defined, aimed to clearly describe design decisions when deploying CSCW interfaces under a Human-Computer Interaction (HCI) perspective. However, most of TOUCHE models suffer from a number of limitations when formalizing User-Centered commonsense and qualitative reasoning. Indeed, TOUCHE models only accounts for describing design

M. Kurosu (Ed.): Human Centered Design, HCII 2009, LNCS 5619, pp. 955–964, 2009.
© Springer-Verlag Berlin Heidelberg 2009

issues omitting the record of the associated decision making process. As a consequence, only results of the discussion between members of the development team are traceable. But being able to support and document this decision making process can enhance TOUCHE capabilities, especially those related to maintenance and scalability. In this settings, TOUCHE can rely on argumentation techniques to solve the above problem by means of a rational justified procedure.

This paper describes how the TOUCHE model can be empowered through argumentation, in which knowledge representation and inference are captured in terms of Defeasible Logic Programming, a general-purpose defeasible argumentation formalism. We show how argument-based reasoning can be integrated in a TOUCHE System Requirement Document in order to provide a qualitative perspective capable to support and document part of the associated decision making process automatically.

First, some related works are described. Then, the TOUCHE process model is briefly described. Section 4 introduces some key concepts of defeasible argumentation. Next, Section 5 sketches a Proof of Concept to show how defeasible argumentation can be used in TOUCHE to model incomplete and probably inconsistent information, focusing the Case Study in the two first steps of the model. Finally, Section 6 concludes.

2 Related Work

To our knowledge there is not similar proposal for integrating defeasible argumentation and TOUCHE as described in this paper. An alternative approach is proposed by Design Rationale. Even though some models for Design Rationale are based on the use of arguments [3] none of them include an embedded engine capable of carrying out the automatic computation of defeasible arguments as in DeLP. The use of defeasible argumentation during requirement elicitation is shown in [4, 5]. These papers present the argument-based Goal Argumentation Method for justifying modelling decisions in goal-oriented requirements engineering, while our approach is based on a task-oriented methodology. The integration of argumentation techniques using DeLP within CSCW systems was proposed in [6]. Another example that shows the possibility of using argumentation to enhance CSCW capabilities is presented in [7], where a tool for drawing arguments was introduced.

Finally, it must be remarked that recent research has led to some interesting results to model dialectical discussions and negotiation in CSCW scenarios. For example, [8] have proposed a mechanism to manage dialectical discussions when a group of people want to collaboratively define requirements in natural language. However, their proposal is based on constraint satisfaction techniques, and does not consider the use of argumentation

3 The TOUCHE Process Model

TOUCHE is a process model and a methodology for the development of user interfaces for groupware applications from the requirements gathering up to the implementation stage [1,2]. It includes four development stages, namely: Requirements Gathering, Analysis, Design, and Implementation. The first stage (based on [9]) gathers the requirements of the system to be developed. Novel requirement gathering templates which include some metadata that are important for the specification of groupware applications have been developed, as well as the System Requirement Document DRS [1]. In this first stage those templates provides the information to describe the organizational structure of the users in the system, the different participant (groups, users, individuals, and agents), the system objectives, the functional and non-functional requirements, and information requirements. All this information is provided according to CSCW criteria where the user as a member of a group of users is the main aspect to take into account.

The Analysis stage [2] studies the problem domain. Roles and tasks are identified and described from a structural perspective using Class Diagrams and the Organizational Structure Diagram; and from a behavioural perspective by means of the Task Diagram or TD (using CTT notation [10]). Besides, the Co-interaction Diagrams or CDs are included to identify relationships among the actors of the system. The Design stage provides a way to present the information (visualization, entries, controls, etc.) to the final user. All the information gathered up to now is processed and translated to a software representation. Users' awareness should be considered in order to get a good CSCW design. Abstract Interaction Objects (AIOs) are used to design Abstract User Interfaces (AUIs). We use the UsiXML conceptual scheme [11]. The model is enriched with a new AIO and several facets which provide more expressiveness to represent CSCW systems. Finally, the Implementation stage deals with the generation of the UI from the previous AIOs. It is a reification process from every component to more concrete elements according to implementation and platform details. The Cameleon process is follow. Several specific CIOs for groupware applications are proposed. The traceability between the defined models and between the different stages is also considered. [2].

As it can be seen, several formal models have been included within the TOUCHE methodology. However, those models expressed only final decision of the development team omitting the possibility of dealing with incomplete and possible inconsistent information, especially those associated with the decision making process. These limitations turn out to be critical especially in the first and second stage of the TOUCHE model, as decisions made during these two stages strongly condition the final product characteristics. As we will see next, defeasible argumentation -a sound setting formalization to model incomplete and possible inconsistent information- can cope with the problem described below.

4 Defeasible Argumentation

Argumentation Systems (AS) are increasingly being considered for applications in developing software engineering tools, constituting an important component of multi-agent systems for negotiation, problem solving, and for the fusion of data and knowledge [12,13]. AS implement a dialectical reasoning process by determining when a proposition follows from certain assumptions, analyzing whether some of those assumptions can be disproved by other assumptions in our premises. AS typically refer to two kinds of knowledge: strict and defeasible knowledge. Strict knowledge (K_S) corresponds to the knowledge which is certain; typical elements in K_S are statements or undisputable facts about the world, or mathematical truths (e.g. implications of the form $(\forall x)P(x) \rightarrow Q(x)$). The strict knowledge is consistent, i.e. no contradictory conclusions can be derived from it. On the other hand, defeasible knowledge (K_D) corresponds to that knowledge which is tentative, modelled through "rules with exceptions" (defeasible rules) of the form "if P then usually Q" (e.g., "if something is a bird, it usually flies"). Such rules model our incomplete knowledge about the world, as they can have exceptions (e.g., a penguin, a dead bird, etc.). Syntactically, a special symbol ($\Rightarrow$) is used to distinguish "defeasible" rules from logical implications.

Example: For the sake of example, let us consider a well-known problem of non-monotonic reasoning in AI about the flying abilities of birds, recast in argumentative terms. Consider the following sentences:

> *1. Birds usually fly.*
> *2. Penguins usually do not fly.*
> *3. Penguins are birds.*

The first two sentences correspond to defeasible rules (rules which are subject to possible exceptions). The third sentence is a strict rule, where no exceptions are possible. Now, given the fact that Tweety is a penguin two different arguments can be constructed:

> *1. Argument A(based on rules 1 & 3): Tweety is a penguin. Penguins are birds. Birds usually fly. So Tweety flies.*
> *2. Argument B(based on rule 2): Tweety is a penguin. Penguins usually do not fly. So Tweety does not fly.*

AS allow the user to define a knowledge base $K = K_S \cup K_D$ involving strict and defeasible knowledge. An argument A for a claim c is basically some "tentative proof" (formally, a ground instance of a subset of K_D) for concluding c from $A \cup K_S$ [13]. Arguments must additionally satisfy the requirement of consistency (an argument cannot include contradictory propositions) and minimality (by not including repeated or unnecessary information). Conflicting arguments may emerge from K. Intuitively, an argument A attacks another argument B whenever both of them cannot be accepted at the same time, as that would lead to contradictory conclusions.

There exist two main kinds of possible "attacks" between arguments in AS: symmetric attack (arguments with opposite conclusions) and undercutting attack (an argument attacks some "subargument" in another argument). The notion of defeat comes then into play to decide which argument should be preferred. The criterion for defeat can be defined in many ways, being a partial order $\leq$ among arguments. Thus, for example, arguments can be preferred according to the source (e.g. when having arguments about weather, the argument of a meteorologist should be stronger than the argument of a layman). As a generic criterion, it is also common to prefer those arguments which are more direct or more informed. This is known as the specificity principle (see [13]). For example, in the particular situation of the previous Example two arguments arise that cannot be accepted simultaneously (as they reach contradictory conclusions). Note that argument B seems rationally preferable over argument A, as it is based on more *specific* information. This situation can easily become much more complex, as an argument may be defeated by a second argument, which in turn can be defeated by a third argument, *reinstating* the first one. Indeed, AS determine when a given argument is considered as ultimately acceptable with respect to the available knowledge by means of a dialectical analysis, which takes the form of a tree-like structure called dialectical tree. The root of the tree is a given argument A supporting some claim and children nodes for the root are those defeaters $B_1, B_2, ... B_k$ for A. The process is repeated recursively on every defeater B_i, until all possible arguments have been considered. Leaves are arguments without defeaters. Some additional restrictions apply (e.g., the same argument cannot be used twice in a path, as that would be fallacious and would lead to infinite paths).

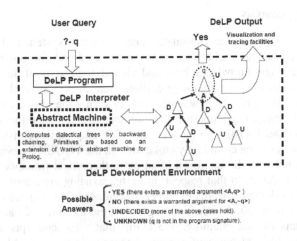

Fig. 1. A schematic view of the DeLP development environment

A marking procedure can be then performed for "marking" the nodes in the tree. Leaves will be "undefeated" nodes (or "U" nodes, for short), as they have no defeaters. Then we can propagate the marking from the leaves upward to the root as follows: an inner argument A_i in the tree will be marked as a "defeated" node ("D" node) if it has at least one "undefeated" child. Otherwise, if every child of A_i is a "D" node, then A_i will be marked as "U" node. If the root of a dialectical tree (the argument Arg) turns out to be marked as "U" node, then it is ultimately undefeated (given the knowledge available), so that the argument Arg (and its conclusion) is said to be warranted (i.e. ultimately accepted). Given a knowledge base K, AS automatically compute the dialectical tree associated with any particular claim (provided by the user as an input). In this context, Defeasible Logic Programming (DeLP)[1] is a general-purpose AS which has been particularly successful in real-world applications [14, 15, 16], providing an integrated environment for defining a knowledge base and solving user queries (claims) interactively. For any claim the DeLP engine automatically computes and visualizes the emerging dialectical tree, which acts as an explanation facility for the user, helping him to understand why the given claim is warranted or not (Figure 1). As we have seen in this Section, warranted arguments support beliefs that are accepted beyond dispute on the basis of the available knowledge. This notion can be applied in different contexts as, in particular, in multiple-party reasoning,

---

[1] See http://lidia.cs.uns.edu.ar/delp_client

where a group of several people participate on the basis of some common knowledge (e.g. members of a software development team). In the next section we will analyze how this idea can be integrated in the TOUCHE process model to characterize and document decision making by using of a general-purpose argumentation system like DeLP as a support tool.

5 Proof of Concept

5.1 Feasibility Analysis

First, our proposal requires an automated argumentation system to be integrated in TOUCHE, which should include an appropriate front-end for posing queries, and facilities for defining a knowledge base and visualizing results of the computation of the underlying argumentation engine (e.g. dialectical trees). Several of such kinds of platforms are freely available nowadays [16], covering the above expectations with reasonable costs (including economical resources, time consumption, etc.). Thus, AS (and particularly DeLP [16]) can be seen as a first step on the construction of argument-based modules to be completely embedded in the TOUCHE environment.

Costs associated with the inclusion of an extra theoretical framework in TOUCHE must be also considered. In that respect, note that existing argument-based platforms are not specially oriented towards experts on argumentation, but rather towards general users with a conceptual understanding about the meaning of facts, rules and inference by means of rule chaining. We claim that these concepts are suitable for TOUCHE users. The existence of graphical front-ends in some AS platforms [17] minimizes the complexity of text input for rules and facts as well as the interpretation of obtained results. Finally, from the TOUCHE viewpoint, it must be remarked that the integration of Artificial Intelligence and argument techniques within CSCW systems has proven to be fruitful, resulting in novel proposals [6,7,8] and systems such as I-MINDS [15] or SCALE [16] for example.

5.2 Case Study

The next example sketches how defeasible argumentation can be used in the first and second stages of the TOUCHE model to deal with inconsistent or incomplete information. Let us suppose we are developing the interface of a groupware application which allows several authors to create the same document through the Internet. When the authors of the document have written a draft, one of them is responsible for sending, through the same application, the document which is candidate to be published to some reviewers. Then, the reviewers discuss about the document and give their own opinion on whether it should be published or not. A published document can be read by all the users of the system, even if they are not authors or reviewers.

In the above settings, Table 1 shows part of the specification of a functional requirement called *document edition* (DC) by means of the metadata of the general template and the extensions introduced to consider the specific features concerning CSCW systems.

Table 1. Description of part of the functional requirement called *Document edition* with the proposed template (only relevant rows for the current case study are included)

RF-8	Document Edition
... ...	
Awareness issues	The following actors should be aware of this requirement: • #G-1 (AUTHORS): - *What*: an actor is modifying part of the current document - *How*: current modification is showed graphically - *When*: in real-time - *Where*: in the same workspace, in the same window - *Why*: to know who is modifying what and not to interfere - *What*: an actor modified a document - *How*: a past modification is showed by e-mail - *When*: after saving the current version, asynchronously - *Where*: in the actor's intranet and by e-mail - *Why*: to know who modified the document and what part
... ...	
CSCW description	Because of the collaborative nature of the current requirement: • Notifications are necessary for user awareness • Insertion, modification, and modification in a document are issues to be careful. Awareness in real time is important. Some actions such as deleting an image could be too fast for the rest of authors to be aware. They should be aware in some way. • Real-time feeling in the document elaboration is important but not vital.
... ...	

Note that some points of the Awareness Issue part of DC are in conflict with other consideration of the CSCW description. For example, the statements $s_1 =$ *"Awareness in real time is important"*, $s_2 =$ *"Some actions such as deleting an image could be too fast for the rest of authors to be aware"* are actually in conflict, while statements like $s_3 =$ *"Real-time feeling in the document elaboration is important but not vital"* are somehow vague or incomplete, and consequently their interpretation will be probably biased by the development team later on. At this point defeasible argumentation notation can be embedded into the template on Table 1 to enhance the comprehension of DC. For example, the following arguments can be associated with s_1, s_2 and s_3:

% Defeasible rules (Commonsense knowledge)
%W stands for an arbitrary writer, ¬ stands for "not" and T stands for "text"

- author (W) $\Rightarrow$ show_real_time (awareness, W)
- author (W), Deleting (W, T) $\Rightarrow$ show_real_time (awareness, W)
- author (W), Deleting (W,T), Image (T) $\Rightarrow$¬ show_real_time (awareness, W)
- coordinator (W) $\Rightarrow$¬ show_real_time (awareness, W)
- coordinator (W), author (W) $\Rightarrow$ show_real_time (awareness, W) %the coordinator is one of the authors

Later on, during the 2nd stage in TOUCHE, roles and tasks will be instantiated and modelled by means of some diagrams and descriptions (see Section 3). For example, DC will be designed by means of the framework shown in Figure 1. This time, arguments describing defeasible rules like those showed above can be used to decide how to link elements in the diagrams of the second stage, especially in those cases where more than one solution must be taken into account. Here is when the potential of having an automated engine like DeLP, capable to compute arguments automatically comes into play. By means of the analysis of real Use Case Diagrams (see Figure 2 for the Use Case Diagram of the DC requirement) that can be easily collected and instantiated, members of the development teams can rely on AS like DeLP to compute alternative sets of facts automatically. This way, DeLP answers can be used to analyze alternative design responses for requirement descriptions minimizing the subjectiveness and the cultural bias present in the decision making process.

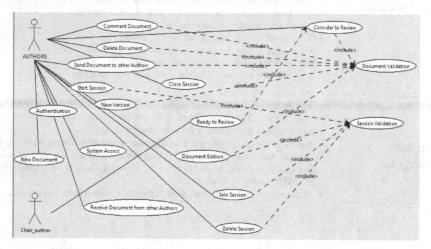

Fig. 2. Use Case Diagram for the requirement expressed in Table 1

6 Conclusions and Future Work

This paper proposed the integration of Argumentation Systems (AS) in the Task-Oriented and User-Centered Process Model for Developing Interfaces for Human-Computer-Human Environments (TOUCHE), aimed to build up CSCW interfaces from the Human-Computer Interaction viewpoint. The final goal is to enhance the capability of development process models for CSCW systems by including a rule-based approach for efficient reasoning with incomplete and inconsistent information. In our approach defeasible argumentation was captured in terms of Defeasible Logic Programming (DeLP), a general-purpose AS which has been particularly successful in real-world applications, providing an integrated environment for defining a knowledge base and solving user queries (claims) interactively. For any claim the DeLP engine automatically computes and visualizes the emerging dialectical tree, which acts as an explanation facility for the user, helping him to understand why the given claim is warranted or not.

As a first step in a novel research line, this work sketched a Proof of Concept to show how DeLP can enhance the TOUCHE first and second stages (namely Requirements Gathering and Analysis) respecting their capability to deal with incomplete, uncertain and possible contradictory information. Part of our future work includes the development of a graphical module based on DeLP to be incorporated into the TOUCHE Case Tool.[2] Based on this incorporation, future work will focus on performing a set of experiments beyond this Proof of Concept. Indeed, it will be necessary to carry out a variety of complete CSCW interface developments under TOUCHE –comparing the results of including and omitting the use of DeLP– in order to validate the real scope of the current proposal. Work in this direction is being pursued.

Acknowledgments. We would like to acknowledge to CONICET (Argentina) and the projects PGI 24/N020, PGI 24/N023 (UNS, Argentina), PIP-CONICET 112-200801-02798 (CONICET, Argentina), CICYT TIN2008-06596-C02-01 (Spain) and the Junta de Comunidades de Castilla-La Mancha PAI06-0093-8836 for funding this work.

References

1. Penichet, V., Lozano, M., Gallud, J., Tesoriero, R.: Requirement Gathering Templates for Groupware Applications. In: Macías, J.A., Granollers, T., Latorre, P.M. (eds.) New Trends on HCI. Research, Development, New Tools and Methods. Springer, Heidelberg (2009)
2. Penichet, V., Lozano, M., Gallud, J., Tesoriero, R.: User Interface Analysis for Groupware Applications in the TOUCHE Process Model. International Journal Advances in Engineering Software (ADES) (in press, 2009) ISSN: 0965-9978, doi:10.1016/j.advengsoft.2009.01.026
3. Burge, J., Brown, D.C.: Reasoning with design rationale. In: AI in Design 2000, pp. 611–629. Kluwer Academic Publishers, Dordrecht (2000)
4. Jureta, J., Faulkner, S., Schobbens, P.: Clear justification of modelling decisions for goal-oriented requirements engineering. Requirements Eng. 13, 87–115 (2008)
5. Jureta, I.J., Faulkner, S.: Tracing the Rationale Behind UML Model Change Through Argumentation. In: Parent, C., Schewe, K.-D., Storey, V.C., Thalheim, B. (eds.) ER 2007. LNCS, vol. 4801, pp. 454–469. Springer, Heidelberg (2007)
6. González, M.P., Chesñevar, C., Collazos, C., Simari, R.: Modelling Shared Knowledge and Shared Knowledge Awareness in CSCL Scenarios through Automated Argumentation Systems. In: Haake, J.M., Ochoa, S.F., Cechich, A. (eds.) CRIWG 2007. LNCS, vol. 4715, pp. 207–222. Springer, Heidelberg (2007)
7. Kirschner, P., Buckingham, S., Carr, C. (eds.): Visualizing Argumentation: Software Tools for Collaborative and Educational Sense-Making. Springer, London (2003)
8. Gervasi, V., Zowghi, D.: Reasoning about inconsistencies in natural language requirements. ACM Trans. Softw. Eng. Methodol. 14(3), 277–330 (2005)
9. Durán, A.: Applying Requirements Engineering. Catedral Publicaciones, Spain (2003) ISBN: 84-96086-06-2
10. Paternò, F.: Model-based Design and Evaluation of Interactive Applications. Springer, Heidelberg (1999)
11. Limbourg, Q., et al.: USIXML: A Language Supporting Multi-path Development of User Interfaces. In: Bastide, R., Palanque, P., Roth, J. (eds.) DSV-IS 2004 and EHCI 2004. LNCS, vol. 3425, pp. 200–220. Springer, Heidelberg (2005)
12. Rahwan, I., Parsons, S., Reed, C. (eds.): Argumentation in Multi-Agent Systems. LNCS (LNAI), vol. 4946. Springer, Heidelberg (2008)
13. Chesñevar, C.I., Maguitman, A., Loui, R.: Logical Models of Argument. ACM Computing Surveys 32(4), 337–383 (2000)

---

[2] To access to the TOUCHE Case Tool consult www.penichet.net

14. Chesñevar, C., Maguitman, A., Simari, G.: Argument-Based Critics and Recommenders: A Qualitative Perspective on User Support Systems. Data & Knowledge Engineering 59(2), 293–319 (2006)
15. Brena, R., Aguirre, J., Chesñevar, C., Ramirez, E., Garrido, L.: Knowledge and Information Distribution Leveraged by Intelligent Agents. In: Knowledge and Information Systems (KAIS), vol. 12(2), pp. 203–227. Springer, Heidelberg (2007)
16. García, A., Simari, G.: Defeasible Logic Programming: An Argumentative Approach. Theory and Practice of Logic Programming 4(1), 95–138 (2004)
17. Reed, C., Rowe, G.: Araucaria: Software for Argument Analysis, Diagramming and Representation. Int. J. on Artificial Intelligence Tools 13(4), 961–979 (2004)
18. Liu, X., Zhang, X., Soh, L.-K., Al-Jaroodi, J., Jiang, H.: A Distributed, Multiagent Infrastructure for Real-Time, Virtual Classrooms. In: Proc. ICCE 2003, pp. 640–647 (2003)
19. Soller, A., Guizzardi, R., Molani, A., Perini, A.: SCALE: supporting community awareness, learning, and evolvement in an organizational learning environment. In: Proc. of the 6th international conference on Learning sciences, pp. 489–496 (2004)

Ergonomic Approach for the Conception of a Theatre Medical Regulation System

William Guessard[1], Alain Puidupin[1], Richard Besses[1], Paul-Olivier Miloche[1], and Aurélie Sylvain[2]

[1] Service de santé des armées,
section technique de l'armée de terre,
CS 90701, 78013 Versailles cedex, France
[2] Délégation générale pour l'armement
william.guessard@stat.terre.defense.gouv.fr,
alain.puidupin@sante.defense.gouv.fr, besses.richard@wanadoo.fr,
dr.pom@free.fr, aurelie.sylvain@dga.defense.gouv.fr

Abstract. This paper is a reflection for the conception of an overseas operations' computerised medical regulation system. After a short description of problem-solving and human error cognitive mechanisms, these concepts are used for the conception of a human centred theatre's medical regulation system.

Keywords: Medical regulation, problem-solving, human error, human reliability.

1 Introduction

Medical regulation consists in matching the needs of the patient or casualty with the most appropriate medical capabilities, whether in terms of equipment, personnel or technical support centre.

The service de santé des armées (defence medical services) is currently developing a medical regulation concept for the management of troops injured in overseas operations. Such a concept cannot reuse existing solutions designed to handle a limited number of casualties (as "SAMU" type regulation software) or to a medical organisation that strongly differs from that deployed in overseas operations ("plan rouge" management software for example). Indeed, on an operational theatre, institutional armed forces, likely to conduct coercive actions, are deployed against institutional or non-institutional, controllable or non-controllable forces. Besides, this deployment occurs in a complex international diplomatic framework and under a media pressure that takes place in real time, since populations are reactive to it. Available means are at once limited and can only be adjusted via urgent political decisions under media pressure. Without these extraordinary decisions, available means are systematically degraded in an uncontrolled way by enemy forces. Stress and sleepiness rapidly add their effect to jeopardize a medical regulation that has to simultaneously consider a disaster situation and an emergency situation. Although it does have similarities with emergency medical regulations (SAMU) and "plan rouge" management software, medical regulation on an operational theatre is very unique and has to be considered as such as a complement to the previous two.

M. Kurosu (Ed.): Human Centered Design, HCII 2009, LNCS 5619, pp. 965–971, 2009.
© Springer-Verlag Berlin Heidelberg 2009

2 Ergonomic Issue

The core of medical regulation lies at the level of information gathering, processing and integration required to its implementation. Quantitative and qualitative factors combined with the time factor determine the critical character of situations (need to make choices), decisions having to be made by vulnerable, stressable and fatigable men. Human decision in medicine cannot be replaced but the level of human performance is fluctuating, both naturally circadian and likely to be rapidly degraded under constraint. Man has thus to be employed for what he is irreplaceable, his abilities have to be conserved and he must be released from what only concerns automatism. In order to avoid the pitfall of a software program imposing an action not desired by the operator, we have thus wanted to put him in the centre of the system. To that end, we have searched for the information relevant to such a regulation and the links between them in order to propose consistent information to the physician in charge of regulation.

2.1 Cognitive Bases of Decision-Making

Medical regulation is part of problem-solving processes. The physician in charge of regulation has to make a decision in view of the coordination of the patient management means, based on data available to him, which is necessarily fragmentary and changes as the patient status evolves.

Rasmussen [3] described three levels of operator action: the skill-based level, the rule-based level and the knowledge-based level. The level concerned here is the rule-based level (assuming an experienced operator and a "simple" case: rescue organisation complies with specific rules based on the patient's status), but most often, actions are based on the implementation of declarative knowledge: the physician in charge of regulation makes his decision based on what he imagines of the patient's situation and of the capacities of the rescue teams.

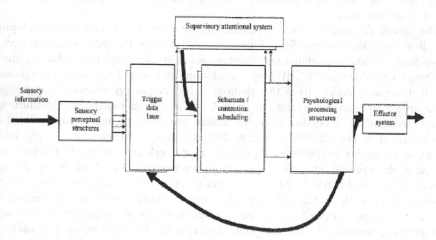

Fig. 1. Norman and Shallice's information processing model. Sensory information triggers an action after processing by psychological processing structures. These structures are activated by the schemata and contention scheduling system and the data base trigger. The supervisory attentional system handles the activation of these processing structures.

The information processing model proposed by Norman and Shallice [5] in 1980 reminds us that the implementation of that knowledge is done after sensory information have been processed by data base trigger and psychological processing structures, via a schemata and contention scheduling system, itself under control of the supervisory attentional system.

These various information processing and decision-making mechanisms generate risks of errors by the physician in charge of regulation. These errors can be fatal to the patient but also, on a theatre of operations, to the rescue teams.

2.2 Error-Production Mechanisms

Reason [4] showed how a system could allow or limit the production of errors by operators. Thus, in the specific case of problem-solving, errors can be due to the application of erroneous rules (the rule used by the operator is incomplete or incorrect: get a psychological casualty to a surgical structure for example) or to the erroneous application of right rules (the operator chooses to implement a solution that, although it works in other situations, is not adapted to the particular case of the current problem: it may be disastrous for a casualty suffering lung damage to be evacuated by air because although aeromedical evacuation is much faster, it may worsen the problem).

According to Amalberti [1], these errors are linked to bounded rationality phenomena: an individual cannot apprehend all the environment information, and all the consequences of current actions. The operator has to make choices based on his fragmentary perception of information and consequences of his actions. To that must be added the time constraint, as in many other situations, that forces the physician in charge of regulation to act "under emergency" and not wait for the situation to stabilize.

However, in spite of his weaknesses, the human operator makes the decision-making process more flexible. Indeed, he is able to make choices, bets, in spite of fragmentary and even erroneous information, while a computer system, more rigid, will request complementary information before proposing a solution. The information automatic processing by systems can only be based on a structured (form) and well formed (e. g.: XML format) piece of information corresponding to a pre-established meaning (codes). The use of metadata does enable automatic cross-check and interpretation actions that offset the lack of consistence or the gaps and fuzzy logic techniques make the algorithmic process more flexible but the man maintains an edge on computer systems. He is able to exploit the implicit information contained in any communication and to use his metaknowledge (the knowledge of his own abilities). This exploitation is immediate, even if the interpretation entails risks. It has the advantage of saving time even if the postponement of a rational decision would not necessarily be better since only built on explicit information that is incomplete and sometimes wrong.

For us, rule making does not consist in forcing the physician in charge of regulation to follow a pre-established algorithm but in assisting him to balance entities he will have to use to make his decision. Balance the relevance of upstream information, balance the useful availability of deployed means, balance the risk-taking calculated based on known elements.

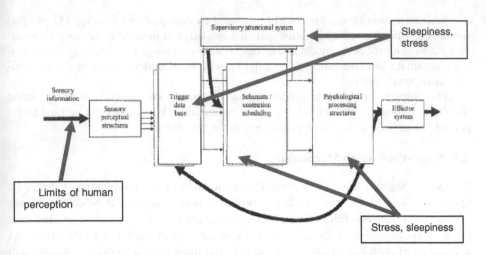

Fig. 2. Error-production mechanisms based on Norman and Shallice's information processing model

Thus, designing a theatre medical regulation system as we imagine it does not aim at developing a software program imposing ready-made solutions to an operator who will just be a mere executant. It aims at implementing an information system that lets the man carry out the noblest tasks (genesis of information and decision-making process). Everything else is just organisation and automatism, whether the transmission of messages themselves, their addressing in structured data bases or their processing via cross-checks in view of a relevant contextual graphical presentation.

Following Hoc [2] and Amalberti [1], we will define such a concept as an ecological system. In this system, cybernetics takes on intangible automatisms, and man takes on the contact with the patient and the critical decision making.

3 The Design of a Medical Regulation System in Practice

The action of the service de santé des armées has to fall within the scope of the new Forces information and communication system that contributes to battlespace digitisation, whilst preserving the fundamental values of medical practice and in particular ethics. Human constraints such as the vulnerability of combatants, medical operators and decision-makers have thus to be taken into account to be able to contemplate the design of an information system compatible with the autonomy required for medical decision-making in degraded contexts.

There is great temptation to believe that decision-making can be optimised if human errors are removed and man enslaved to the system since compelled to apply procedures. The human factor has long been considered the primary cause of any error. In the interhuman "adversity" situation characterising the operational theatre situation, human will and the meaning of committed, or not committed acts, have as much weight as the reality of their consequences. The cognitive workload weighing on decision-makers has thus to be alleviated in order to preserve their ability to take on their technical and ethical responsibility.

3.1 The Context: Medical Management on a Military Operational Theatre

Medial regulation in overseas operations is characterized by the extremely variable number of potential casualties likely to be managed by medical teams on an action zone that may be vast (several units can simultaneously suffer casualties) and in a non stabilised space-time context (combat operations may continue over time and space, which limits the action capabilities of medical teams or compels them to a quick management of casualty/ies).

The physician in charge of regulation, the system's cornerstone, has thus to be able to remotely build a reliable representation of the casualties' status, the context in which medical teams will have to act, the dispatching in the field of the latter, etc. Besides, he shares information with the people in charge of logistics, conduct of operations and health resources on the military theatre to develop a strategy compatible with conditions of medical management. It is important to note that the context in which the regulation activity takes place will strongly impact the regulation process. As a consequence, analogies made with civil situations encountered by the services d'aide médicale d'urgence (SAMU) and those acting in a disaster situation will have to be cautiously analysed.

3.2 Technological Environment to Support Action

The material support of regulation activities is all the information and communication technologies that are more and more an everyday reality for the troop and the civilian population as well. An optimized medical management will require resorting to "mobile" computer technologies (personnel digital assistant, storage and digitisation of information on a physical medium) and to the organisation of information exchange through centred networks. In absolute terms, these concepts allow the combatant, as a basic tactical unit, to work in network with the higher hierarchical echelons. Medical management on an operational theatre involves the same principles. The current stake is that health actors in the field can communicate in a unique "information infrastructure" to develop a shared representation of the situation.

For a few years, tests have been carried out to approach the issue of medical information digitisation in the field, in France and overseas. Those various contexts of use allow us to identify the opportunities opened by digital technologies to meet our requirement for military regulation.

3.3 Variety of Actors and Levels

The variety of actors (from the paramedic in the field to the physician commanding medical service elements on the theatre of operation) and the distance between these various actors (an isolated paramedic may have to manage a casualty few hundreds kilometres away from the physician) are what makes our system unique.

The system under development will ensure medical regulation at the local level (the chief medical officer of the medical team on site) as well as at higher levels. Available information is not the same and only some pieces of information, that will have to be clearly identified, will be useful to the medical regulation. Its role will consist here in sending a team of varying medicalization level according to the patient's state, or only in guiding the paramedic on site in the actions to be taken.

For example, an analysis of the use of existing regulation software (SAMU regulation software, "plan rouge" monitoring software, etc.) allows us to identify the relevance of the information gathered and their exploitation during an intervention. This work is supplemented by interviews with the various actors of medical regulation in overseas operations (physicians commanding medical service elements, resuscitators in surgical team, surgeons, etc.). The library of characteristic activities thus built allows us to extrapolate a need for relevant information, whether medical (concerning casualty/ies) or operational (concerning logistic and more strategic data).

3.4 Purposes: Assistance and Advice on the Design of an Information System

The analysis of the results gained at the end of the usage studies on existing systems and of our interviews with operators in order to formalize the need shall allow us:

- to determine the information used during medical regulation by medical actors
- to identify the information exchanged that is necessary to the development of a shared representation of a given situation
- to understand a part of the problem-solving and decision-making processes by the operator in charge of regulation.

All these elements are essential to determine the information architecture required to manage a patient or a war casualty, or, in more practical terms, to decide on the organisation of data on the screens of the medical regulation system. The aim is to design an information system that allows the physician in charge of regulation to get a very good picture of the situation in order to make the decision that seems the most adapted to him based on his knowledge of tactical data, teams present in the field and their skills.

4 Conclusion

The purpose of this paper was to share our thoughts on the design of an operator-centred software program, which aims at enabling him to make a decision he will be able to follow and modify over time, based on the representation of the situation he has built, but which also takes account of his own abilities. This design compels us to reuse decision-making and error-production theoretical bases, in order to accurately identify the pitfalls we may encounter that will be challenges to be addressed by the end user. Thus, the software program that will be designed will have to include an ecological interface making the representation of the situation easier, proposing a decision support for the physician in charge of regulation but not imposing a rigid solution that could be a source of difficulty for the operator and entailing a risk of causing him to lose control of the situation.

This analysis will have to be faced with the field reality in order to evaluate a prototype that will be validated by future users during different exercises simulating various and realistic employment conditions.

Acknowledgements

The authors wish to thank Anne-Lise BEUGNET for her translation.

References

1. Amalberti, R.: La conduite des systèmes à risques. Presses universitaires de France, Paris (2002)
2. Hoc, J.M.: Conditions méthodologiques d'une recherche fondamentale en psychologie ergonomique et validité des résultats. Le travail humain, tome 56(2-3), 171–184 (1993)
3. Rasmussen, J., Jensen, A.: Mental procedures in real-life tasks: A case study of electronic troubleshooting. Ergonomics 17, 293–307 (1974)
4. Reason, J.: L'erreur humaine. Presses universitaires de France, Paris (1993)
5. Shallice, T.: Symptômes et modèles en neuropsychologie. Des schémas aux réseaux. Presses universitaires de France, Paris (1995)

Use of Nursing Management Minimum Data Set (NMMDS) for a Focused Information Retrieval

Josette Jones[1], Eric T. Newsom[1], and Connie Delaney[2]

[1] Indiana University School of Informatics- School of Nursing
535 West Michigan Street, Indianapolis, IN 46202, USA
[2] University of Minnesota, School of Nursing
308 Harvard Street SE, Minneapolis, MN 55455, USA
jofjones@iupui.edu, enewsom@iupui.edu, delaney@umn.edu

Abstract. Evidence-based nursing (EBN) is central to the knowledge base for nursing practice, and evidence based interventions are considered as one of the best avenues to achieve maximum outcomes [1]. These interventions typically are implemented through the portals of nursing management forming the context for delivery of nursing practice [2]. Hence, understanding how to provide high quality nursing care efficiently and making management decisions based in evidence is of increasing importance. EBN management requires a research-driven approach for identifying patient, professional, and setting characteristics that affect the processes of care at micro and meso levels. Yet nurse managers cite lack of time and skills; limited access to search engines or poor understanding of research language and most importantly a paucity of management research articles as a barrier to EBN. This study explored the possibility of using the NMMDS to retrieve research related to nursing management.

Keywords: Information Retrieval, Search Queries, Terminology, Evidence-Based Practice.

1 Introduction

The impetus for this study came from a previous study showing a significantly lower frequency of nursing management research abstracts compared to nursing clinical research [3]. However, the researchers observed that mapping keywords of research abstracts to the NMMDS was possible, though only for few abstracts. The lower frequency of nursing management research provides a suitable atmosphere to test two important measurements for information retrieval, recall, and precision. This relative paucity of management research and less effective search skills and strategies require more time and effort impeding the implementation of best practices [4]. Providing search and filtering tools that increase precision and recall and that are mindful of key aspects of the work process is suggested to facilitate the integration of research into practice [5]. The NMMDS, currently the only nursing management terminology, is assumed to capture nursing management concepts and as thus it can be hypothesized also being feasible to be mapped to semantic propositions describing management

M. Kurosu (Ed.): Human Centered Design, HCII 2009, LNCS 5619, pp. 972–978, 2009.

practices. If successful, the approach developed can be applied to and automated to accurately filter the unstructured research databases for specific content.

In order to electronically extract semantic propositions, a terminology needs a concept-oriented design to process complex concepts expressed through sometimes ambiguous vocabulary[6]. C. K. Ogden's [6] semiotic triangle pbell, et al., 1998) provides the theoretical foundation for a concept-oriented terminology capable of processing contextual semantic content. A central premise in Ogden's theory to the human experience of understanding meaning within a concept-oriented world lies with the human capability to expand on a symbol's meaning by understanding the context and environment in which the symbol is used. This capability contrasts with a machine's capability, and the challenge to nursing terminology developers is to achieve that human capability in system terminology processing that lacks a non-verbal and an environmental support to arrive at meaning. To accomplish this, terminologies become semantic by identifying relationships between terms and variables and by adhering to specific relationship rules [7]. One fundamental rule is the one term to one concept rule (1:1) whereby a specific definition establishes meaning for a concept represented by a term.

While the last published research on the NMMDS alluded to the complications of conceptual definition agreement to achieve a 1:1 rule, it reported the NMMDS to be polished and ready for further testing (Huber et al., 1997). Based on that recommendation, this study tested the meaning of term labels of the NMMDS to serve as conceptual propositions to guide extraction of key phrases from abstracts to correlate and index matching concepts between the NMMDS (table 1) and an online database of research abstracts.

The article describes the feasibility testing of using the NMMDS as a concept-oriented terminology to semantically index articles, with the intent to lessen the gap between research evidence and what happens in practice.

Table 1. NMMDS Categories

Code	Description	Environment	Nursing Care Resources	Financial resources
01	Unit/Service Unique Identifier	x		
02	Type of Nursing Delivery Unit/Service	x		
03	Patient/Client Population	x		
04	Volume of Nursing Delivery Unit/Service	x		
05	Care Delivery Structure and Outcomes	x		
06	Patient/Client Accessibility	x		
07	Clinical Decision Making Complexity	x		
08	Environmental Complexity	x		
09	Autonomy	x		
10	Nursing Delivery Unit/Service Accreditation	x		
11	Management Demographic Profile		x	
12	Staff Demographic Profile		x	
13	Staffing		x	
14	Satisfaction		x	
15	Payer Type			x
16	Reimbursement			x
17	Nursing Delivery Unit/Service Budget			x
18	Expenses			x

2 Study Design

This study used investigator and method triangulation. Several methods of semantic proposition extraction and mapping have been investigated to eliminate confounding factors impacting the feasibility of using NMDS as an information retrieval strategy. Three raters with different educational backgrounds and nursing management expertise are employed to measure consensus of meaning of propositions and labels of the NMMDS. Consensus was measured by inter-rater reliability score at each phase of the study, which would signify a key element or step in creating a potential algorithm for successful indexing and information retrieval. At each phase, each rater used the same natural language processing technique, and developmental research analysis [8] served as guiding framework for method triangulation based on the mapping results and direct subsequent phases of the study The NMMDS Survey (Huber & Delaney, 2005) served as the primary source of concepts, randomly selected abstracts from the Virginia Henderson International Nursing Library – Sigma Theta Tau International (STTI) served as the *extraction* data source for semantic propositions.

3 Methodological Approach

In the first step, two student-raters, one undergraduate in nursing and one graduate informatics student reviewed 25 randomly selected nursing research abstracts. Each rater summarized in one sentence/proposition the main conceptual meaning of the abstract as either clinical or nursing management using a natural language processing technique (Mani, 2001). In addition Think-Aloud was applied to capture the rater's tacit processes in arriving at his/her meaning for purpose of future algorithm development. The conceptual meanings of the abstracts were compared for agreement and inter-rater reliability was calculated. The following formula was used to determine the percent agreement:

$$P=(N_A- N_D)/ \text{ total}$$

where N_A = number of agreements, N_D= number of disagreement, and total =number of abstracts. Since reliability exceeding .70 to .80 is considered indicative of high reliability [9], the acceptance rate was set at .70 or more. For this phase, the inter-rater reliability was calculated at .64, below the set acceptance rate. Hence, the decision was made to re-use this set for training purposes.

In the second phase, a new randomly selected set of 25abstracts, were reviewed and discussed by the research team on content and meaning. Experience gained from previous set provided a framework for extracting conceptual meaning from the these abstracts [10]. Using this approach, as expected the inter-rater reliability increased (.92), emphasizing the importance of appropriate training, familiarity, and knowledge with the process of extracting semantic propositions (but not necessarily knowledge of nursing management) while preserving independent thinking.

The third phase of the study, only those abstracts from the initial set for which agreement was obtained were re-used; the conceptual meaning as defined in phase 1 were reviewed and discussed by the team. Similarly each category and term in the NMMDS terminology was 'translated' in a conceptual description. Next, the NMMDS labels were mapped to the semantic propositions describing the meaning within each abstracts. Each rater independently mapped the concepts, describing their rationale for the mapping. The Think-Aloud process of this phase is regarding to the coding is reported in figure 1.

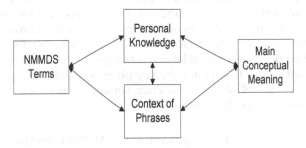

Fig. 1. Mapping NMMDS -Semantic Concepts

Personal knowledge and experience, which is recognized in investigator triangulation and the incorporation of contextual meaning in semantic propositions, simulate systems process of mapping. Under supervision of the investigator, the coding was compared on agreement. Despite the agreement on semantic meaning and on NMMDS coding, it soon was recognized that NMMDS codes failed to capture all nursing management concepts and processes within the abstract. It was concluded that even though a conceptual meaning and related NMMDS codes could possibly identify a nursing management focused abstract, it failed at providing a full and in-depth representation of the knowledge described in the research abstract.

In the fourth phase, a graduate nursing student with nursing management expertise joined the research team as a third rater. Also, in an attempt to facilitate identification of nursing management reseach, the 18 categories of the NMMDS were converted into a set of Yes/No questions to be answered when reviewing an abstract. Each rater answered the question on the list and supported their decision with key phrases from the abstract. The Think-Aloud process for this phase is reported in Figure2.

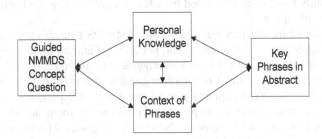

Fig. 2. NMMDS Questions – Key Phrases

Agreement between the raters on the answers to the probing questions and the selected key phrases was inconsistent, yet the use of directed questions identified more NMMDS categories and codes than previous approaches. Also, while not significant, it was noted that the two raters with nursing background more frequently agreed confirming the domain specificity of NMMDS terminology.

Since the question and answer approach appeared to identify more nursing management phrases, the fifth phase attempted to evaluate the effectiveness of the question and answer technique to identify nursing management research from a set research abstracts . Raters independently evaluated seven nursing abstracts (3 management, 1 behavioral, and 3 clinical) selected by the investigator from the integral STTI database. Identifying information and keywords were removed before reviewing; the research focus of each selected abstract was only known by the investigator. The raters reviewed the abstracts and marked key phrases that supported and mapped to the probing questions. Based upon the key phrases found in the articles, the three raters unanimously identified one of the three nursing management focused abstracts (Table 5) but failed to identify the other two.

Table 2. Semantic Mapping of Key Phrase Codified to NMMDS Using Question Answer Tool

Concept Question	Answer to Question (Supporting Phrase)	NMMDS Concept (Code)
Does the abstract discuss extent of professional outcome(s) that relate to compensation, career growth, achievements, or on the job injury for those who provide delivery of care?	'the success of new nurse managers'	Outcomes— Professional, Career mobility and Expansion (05.22)

4 Results

In reiterating the premise of this study to explore the feasibility of the NMMDS to facilitate search and retrieval of nursing management interventions research the results of this study were inconclusive. While failure to meet the set expectations at all phases, the study results –as illustrated by a few examples- revealed some trends that need further exploration.

If the content of the abstract was mapped to NMMDS categories 1 (unit/service), 2 (type of nursing delivery), 3 (patient, client, population) and /or to category 05.3 (clinical outcomes) the abstract did not have a management focus and had most likely a clinical focus. Conceptual categories 1 to 3 capture general classification and environment of nursing services, while 05.3 captures clinical outcomes that pertain to mortality rate, length of stay, adverse reactions, complications, pain management, and skin integrity.

Hence if the content can be mapped to NMMDS categories 4 (volume of nursing delivery) and 6 (accessibility) or 18 (expenses), and 05.1 (structure of care delivery) and 05.2 (outcomes-professional), most often the abstract had a nursing management focus. Likewise, it could be suggested that content mapped to: 9 (autonomy), 12 (staff demographic profile), and 15 (payer type) pointed to nursing management research. However, when reviewing the abstract in its context, the abstract definitely described not nursing management or administration but a clinical research intervention. The issue presented by this abstract showed that reviewers had only captured independent concepts but not the relationships and/or constraints within the abstract to capture the content.

While the NMMDS terminology's intent is to document and codify nursing management practices, the reviewers systematically (1) failed the application of one term/symbol to one concept rule in a concept-oriented design and (2) were not able to correctly represent the semantics of the NMMDS terms without extensive management experience. The reviewers also struggled over clarity and granularity of terms representing certain NMMDS concepts and their relationships. For example, NMMDS code 03.3X specifies one of four interaction focus types of individual, family, group, or community/population of the 'Patient/Client Population' (code 03) concept. This concept implies that an abstract should have only one type of focus because at the NMMDS level of unit/service, a focus type should be on an individual, a family, a group, or a community/population. Mapping and codifying key phrase(s) frequently, according to each reviewer's personal meaning, potentially represented more than one type (see Table 6a). When an abstract contains such phrases as 'African American,' 'males,' and 'females,' reviewers raised questions as to how to codify terms to interaction focus.

5 Discussion

Our limited study showed that NMMDS has a specialty/focused scope. As such it is most understood and used by nurse managers and administrators. Within our study NMMDS variables became labels, not concepts with hierarchical relationships among these concepts to express a formal representation of a specific domain's language that is sharable, reusable, adaptable to knowledge changes, and formidable to development of transferable algorithms capable of other domain operations. The latter suggests the need for a formal knowledge model and a related ontology to represent knowledge contained in nursing management practice and research. The current NMMDS with proven use in documenting nursing unit management practices could constitute the skeleton for such a model. NMMDS study is in the process of mapping to the Logical Observation Identifier Names and Codes (LOINC), a biomedical database. (LOINC) aims to facilitate the nursing outcomes by providing a set of universal codes and names from the perspective of laboratory and clinical observations. LOINC has proven method for data collection that is similar and easily adaptable to NMMDS without information loss. The care level of nursing is measured in individual or group or family or population, with quantitative result" in similar manner like LOINC.

6 Conclusion

Evidence based nursing management practice involves complex and conscientious decision-making which is based not only on the available evidence in the literature but also on unit characteristics, situations, and organizational cultures. The available evidence for nursing management is limited compared to clinical nursing and educational nursing research. Our study proposed to test the feasibility of using a nursing management terminology to deploy search and filtering tools to facilitate the retrieval of management focused research. The results of the study were inconclusive. Extracting semantic meaning of the abstracts and mapping those concepts to NMMDS terms provided some NMMDS labels but lacked to capture relationships and structure. In addition no conclusion could be made on the focus of the content based on the frequencies of mapping, type of categories/terms mapped or combinations of the type and frequencies.

References

1. Ball, M.: Enabling technologies: Transforming healthcare current and future impact on: Patient safety culture and process. Indianapolis, IN: Indiana Chapter HIMSS (2008)
2. Huber, D., Schumacher, L., Delaney, C.: Nursing management minimum data set (NMMDS). Journal of Nursing Administration 27(4), 42–48 (1997)
3. Greene, M., Marcellus, J., Deprez-Kreifels, M., Thompson-Bagley, C., Coenen, A., Jones, J.: A collaborative effort to map variables from the Virginia Henderson International Library to SNOMED-CT and ICNP. MNRS, Omaha, Nebraska (2007)
4. Weaver, C.A., Warren, J.J., Delaney, C.: Bedside, classroom and bench: Collaborative strategies to generate evidence-based knowledge for nursing practice. International Journal of Medical Informatics 74(11-12), 989–999 (2005)
5. Lang, N.M., Hook, M.L., Akre, M.E., Kim, T.Y., Berg, K.S., Lundeen, S.P.: Translating knowledge-based nursing into referential and executable applications. In: Weaver, C.A., Delaney, C.W., Weber, P., Carr, R.L. (eds.) Nursing and informatics for the 21st century: An international look at practice, trends, and the future, pp. 291–304. HIMSS Publishing, Chicago (2006)
6. Campbell, K.E., Oliver, D.E., Spackman, K.A., Shortliffe, E.H.: Representing Thoughts, Words, and Things in the UMLS. J. Am. Med. Inform. Assoc. 5(5), 421–431 (1998)
7. Harris, M.R., Graves, J.R., Solbrig, H.R., Elkin, P.L., Chute, C.G.: Embedded structures and representation of nursing knowledge. J. Am. Med. Inform. Assoc. 7(6), 539–549 (2000)
8. Lijnse, P.L.: "Developmental research" as a way to an empirically based didactical structure of science. Science Education 79(2), 189–199 (1995)
9. Neuendorf, K.A.: The Content Analysis Guidebook. Sage Publications Inc., Thousand Oaks (2002)
10. Zhou, L., Parsons, S., Hripcsak, G.: The Evaluation of a Temporal Reasoning System in Processing Clinical Discharge Summaries. J. Am. Med. Inform. Assoc. 15(1), 99–106 (2008)

HCD Case Study for the Information Security Training System

Akira Kondo[1] and Makoto Yoshii[2]

[1] Hitachi Intermedix Co., Ltd. 2-1-5 Kandanishikicho, Chiyodaku, Tokyo, 101-0054 Japan
[2] IST Co.,Ltd. Nagase-building 201, 2-15 MinamiChuocho, Kitaku, Okayamashi,
Okayama, 700-0837, Japan
kondo@hipri.com, makoto.yoshii@ist-japan.co.jp

Abstract. We proposed organization persona as persona scenario method for business to business content creation process.This paper introduces three projects cases which were based on HCD process. We improved design process practically and enhanced persona for organization as company.

1 Introduction

In recent years, the case of using HCD (Human Centered Design) process is increasing for digital-content creation and system development such as website construction. Especially, development process of the persona scenario method has been attracted attention. On the other hand, when developing new service with contents and systems, user's goal and purpose may not be clear. Especially for development process of contents for BtoBtoC, there are not many opportunities to hear directly from each other between information provider and receiver, and misleading can be occurred between required knowledge and delivered information. First of all, when users do not have any interests of the content which provider wants to deliver them, planning to rouse the interest and let them use is the most important concern.

In this paper, we describe an application of the HCD process to develop a learning system for a general user who is not specialist in the vulnerability of information systems, with the past projects of creating enlightenment contents.

2 Challenge of Persona Scenario Method in Learning System Development

In developing a common learning system, a learner is set up as a target user and the learning system is designed based on the interest and the knowledge level of the persona. However, this learning system is aimed at raising the knowledge level in the vulnerability of information systems for small sized organizations. We need to create each persona based on the nature of the stakeholders with a comprehensive scenario, not for each user's learning and acquiring knowledge about the vulnerability of information systems but for an organization's learning and preparing for threats.

M. Kurosu (Ed.): Human Centered Design, HCII 2009, LNCS 5619, pp. 979–985, 2009.
© Springer-Verlag Berlin Heidelberg 2009

3   Target of This Project

The content of the learning system is the vulnerability of information systems, which general users are not aware. Moreover, the content provider has structured technical information but they do not recognize clear goal for the information security, even though it is becoming a social issue.

Furthermore, we did not have clear measures in industry type and scale to specify small companies which is the target of the project, and their information systems were different in each company. Because of that, it was necessary to clarify a concrete target image. In order to extract the model of small company, we reviewed the White Paper on Small and Medium Enterprises in Japan and the report of IPA (Information-technology Promotion Agency, Japan), decided to assume the following small companies as typical companies in Japan, and set them as target in the project.

Type of industry: manufacturing industry, construction industry, circulation and retail trade

Scale: a few to around hundred employees

Resource of Information System: no CIO and no high IT skilled worker.

4   The Development Process of the Past Project

(1) The Development Process of "Do You Know? The Information System Vulnerability"

This project intends to explain ten typical vulnerabilities on website (weak points in the security in software) based on "the Method of Making Safe Websites" provided by IPA. The target users are broad range of people related to website administration. We made contents by animation of dialogue using characters to make it easy to understand not only for engineers.

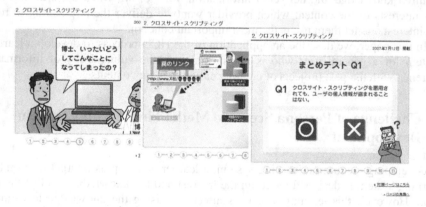

Fig. 1. Screen shot of "Do You Know? The Information System Vulnerability"

The problem of the HCD process in the project was that because the target users were set up very broadly, the project members could not share the clear user's goal, so each member had different direction about necessary technical level of information and appropriate content expression. Moreover, evaluation result we could get was limited only from users who were interested in information security, instead of various target users. Although the evaluation result was positive, verification was not proper enough in whether general users, who are target users at the beginning, could recognize the vulnerability as a problem. As a result, the specification process of "grasp of use situation" and "clear requirement of user and organization" were inadequate, and the result of "evaluation of design for requirement" was ambiguous.

(2) The Development Process of "Introduction to Safety Web Sites"

We developed simulation software, not only to explain an outline of vulnerability but also to be able to learn how to operate actual business with realistic cases.

In the project, developers understood technical information but did not understand user's situation. It was difficult mostly for a user to imagine the total image of service, even if user investigation was conducted to specify "grasp of a use situation" and "clear requirement of user and organization." Because of that, the process of extracting requirements was used in many cases by a hypothesis from current situation survey, prototype creation, and user's evaluation. In this case, however, there were too may types of users and they could not clarify requirements for organizations, so we decided to establish the adviser committee which consists of specialists and to carry out HCD process.

Members of the adviser committee were selected on the following conditions.

Specialists who have sufficient knowledge about the level of the information which should be provided to users.

Specialists who can provide appropriate information to users to utilize system.

Users who can express opinions about the contents of the learning system.

Project result shows that the adviser committee contributes for creating ideas and developing specification in each stage of development processes from early concept building to concretization of content planning from the viewpoint of user as recipient and specialist as provider.

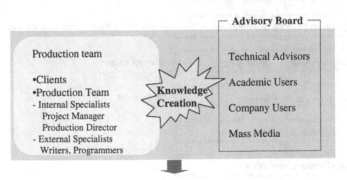

Service with Experiential Value for Users through Knowledge Creation Process

Fig. 2. Activity of the production team and the adviser committee

As a future agenda, we should build general effective method in the HCD process to share and communicate opinions between the adviser committee and users. After this environment is established, we have to build the comprehensive process for a service design with users.

5 Outline of the Project in the 2008 Fiscal Year; "The Study Tool for Small and Medium-Sized Enterprises in Vulnerability"

This latest project was planned for more on the actual condition. New learning tool corresponded to various needs for the security countermeasures of small companies. We decided to develop a customizable simulation learning tool which could optimize based on various types of industries and target users (employees in various positions, ranks, occupations, and ages.)

The project should be concerned about design of the contents and the system, information security technology and management in small companies. Therefore, we established the adviser committee to discuss from professional viewpoints like the last project, and developed the system based on the process of human centered design.

6 Process of Persona Development

Idea of Organization Persona

Ordinary Persona Method extracts individual, but in this project, people's relation in an organization broadly influences a scenario. Because of that, we clarified the role of every worker and their relations in the company before creating individual persona, so that we could evaluate not only each individual but also a company persona in more organizations.

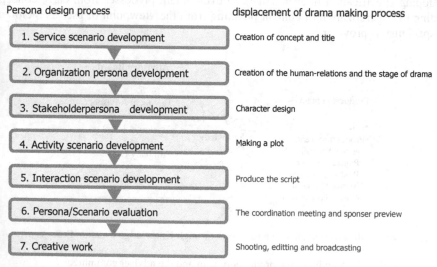

Persona design process	displacement of drama making process
1. Service scenario development	Creation of concept and title
2. Organization persona development	Creation of the human-relations and the stage of drama
3. Stakeholderpersona development	Character design
4. Activity scenario development	Making a plot
5. Interaction scenario development	Produce the script
6. Persona/Scenario evaluation	The coordination meeting and sponser preview
7. Creative work	Shooting, editting and broadcasting

Fig. 3. Persona design process

Actual Development Process

Usually, in a process of persona development, two or more targets are selected out from segmented assumed targets. Common actions are extracted in depth interview of the selected target users and user image is established. Only facts should be described in story style so that all people can imagine user's action patterns.

In this project, we decided to develop not only an individual persona but also a company persona which can imagine how it acts as an organization. We explain each document by transposing it to the meaning of drama making.

(1) Service Scenario

Service Scenario corresponds to title or concept in drama making; making a basic plan and collecting necessary information for scenario. It is subsequent work for smooth development by describing request of users and the conditions of joy.

(2) Organization Persona

Organization Persona corresponds to the human-relation correlation diagram of the stage and characters in drama making. In the organization persona, the human relations in a company and company situation are described based on the above-mentioned service scenario. Especially in the project, human relations in the company influence the correspondence of security incident, so we also summarized emotional relations and business roles.

(3) Stakeholder Persona

Stakeholder Persona corresponds to characters. When creating subsequent scenarios, an indispensable user image is packed along with the service scenario and the organization persona sheet.

(4) Activity Scenario

Activity Scenario corresponds to outline in drama. It is summarized in an itemized statement that how service influences the persona (stakeholders) and what the final goal is.

(5) Interaction Scenario

Interaction Scenario corresponds to script in drama. Based on service scenario, persona sheets, and activity scenario, more detailed story is created. The interaction scenario should be created so that the project participants can understand the big picture of the project; what the service do and what it can provide users.

(6) Evaluation and Development Process

In drama making process, actual creation starts when members gain a consensus. In this stage, main project members discuss the scenario and persona which have been completed so far and check correspondence to the purpose and feasibility.

7 Evaluation

The service scenario sheet, the organization persona sheet, stakeholder persona sheet, and interaction scenario sheet were evaluated by project members in the following aspects;

Effectiveness of the Development Process

Even in the stage where service was not concrete yet, we could get practical opinions and utilize it for the compatibility evaluation.

Effectiveness of Creating User Characteristic

Without the persona method, the type of user may turn into ambiguous because experience and viewpoint of participants vary, which make the discussion failure. With defining an organization persona and a stakeholder persona, a person who should participate could speak in the discussion and people could help other participants who did not have enough knowledge to understand the discussion. Especially in the kickoff meeting, the persona method was very effective to activate discussion even without sufficient opinion exchange beforehand. It was also helpful to prevent from getting off the subject.

Effectiveness of the Communication Tool

It was very effective to extract opinions when interviewing not only the project members but also the users who should do a monitor test for accuracy of interaction scenario. Especially, because interviewed users did not have technical knowledge of the information security, it might be difficult to hear their real opinions by using technical terms. In this project, we explained the interaction scenario when they had enough sympathy with the organization persona and the stakeholder persona. Therefore, they could understand the scenario and could talk about the issue in the same level with interviewers.

Consensus with Development Team

It was necessary to adjust slightly when developing the tool. However, because the big picture had been defined, what we should correct was clear, so we could avoid wasting time. Moreover, we could pass on user's true voice to the development team which usually did not have enough opportunity to hear it, so we got very good result.

Contribution for the Final Tool

Because HCD methods, such as persona and scenario, were used for making the tool with assuming the actual use scene, satisfied result is expectable after tool completion. Especially, user expressed the expectation of the tool to improve situation in the pre-research interview. We believe it means that the tool is not for just filling functions but for meeting the usage demand.

8 Future Development

Our project team has following plans in content creation and system development.

Application to Content Scenario Development

The project team plans to develop a content scenario which meets the actual user's usage and situation with useful results of persona/scenario method. It is applicable to the content creation more efficiently in the actual condition.

Evaluation of the Established Tool

We will check the correspondence of the scenario to practical scene and if it does not match, we will look for the cause of its divergence. Because future action may vary based on the cause of divergence, we will need to clarify where the divergence exists in future system evaluation.

Other Results

Two or more companies, which we interviewed for create scenario this time, carried out full security check. When users understand the persona and empathize with the scenario, they can regard it as own problem. Moreover, because their interest of security became very high, we believe the organization persona and the stakeholder persona scenario is effective when they implement the created learning tool.

Acknowledgment

I would like to express my sincere gratitude to Mr. Kobayashi and member of the Security Center in Information-technology Promotion Agency, Japan for giving me a research opportunity and useful comments in developing the method.

I want to thank the project member.
I would also like to extend my indebtedness to Naoko Kondo for her help.

References

1. Nielsen, J.: Usability Engineering. Academic Press, US (1993)
2. Carroll, J.M.: Scenario-based Design envisioning work and technology in system development. Wiley, US (1996)
3. Carroll, J.M.: Making Use of Scenario-based design of human-computer interactions. MIT Press, US (2000)
4. Yamazaki, K., Furuta, K.: Proposal for design method considering user experience. In: 11th International Conference on Human-Computer Interaction, Las Vegas (2005)
5. Kameoka, A.: Service Science. NTS, Inc., Japan (2007) (Japanese)
6. HCD-Net Website, http://www.hcdnet.org/en/index.html
7. Kurosu, M., Horibe, Y., Hirasawa, N., Miki, H.: ISO13407. Ohmsha, Ltd., Japan (2001) (Japanese)
8. Hiroyuki, T.: Persona tsukutee sorekara dousuruno. Softbank Creative (2008) (Japanese)
9. Pruitt, J.S.: Persona senryaku, Diamond (2007) (Japanese)
10. Yamazaki, K., Takahashi, K., Ueda, Y., Go, K., Hayakawa, S., Yanagida, K.: Universal Design Methodology for Vision Proposal. The Japanese Journal of Ergonomics 44(suppl.), 36–45 (2008)
11. Yoshii, M., Yamazaki, K., Yanagida, K.: Design Approach of Presence for Vision - Service Scenarios for Produce Web Site. In: Proceedings of the 38th Annual Meeting of Kanto-Branch, pp. 29–30. Japan Ergonomics Society (2008)

Driving and Situation Awareness:
A Cognitive Model of Memory-Update Processes

Josef F. Krems[1] and Martin R.K. Baumann[2]

[1] Chemnitz University of Technology, Institute of Psychology,
Wilhelm-Raabe-Str. 43, 09120 Chemnitz, Germany
[2] German Aerospace Center DLR, Institute for Transportation Systems,
Lilienthalplatz 7, 38108 Braunschweig, Germany
krems@phil.tu-chemnitz.de, martin.baumann@dlr.de

Abstract. Safe driving requires a mental representation of objects and situational features relevant to the driver's behavior. This includes the generation of predictions of how the situation will develop in the near future. These processes are summarized under the term "situation awareness", previously proposed in the aviation domain. By now the cognitive mechanisms underlying situation awareness are far from being understood properly. In this paper we propose a theory that is based on results from studies in language understanding [1] and attention [2] and that is applied to the driving context. Mechanisms for the construction of a situation model and for the selection of actions are outlined. Finally, predictions of the model concerning the effect of experience, relevance, and criticality on the drivers' mental representation are investigated. In a second study the effects of cognitive tasks on predicting events in traffic are focused.

Keywords: Situation awareness, driver modeling, cognitive processes.

1 Introduction

Drivers have to correctly perceive, identify, and interpret the relevant objects and elements of the current traffic situation and they have to consider these elements in planning and controlling their behavior. Such elements may be other traffic participants, traffic signs or weather conditions. Drivers do not just have to perceive these elements but they have to understand them according to their relevance with regard to safe and efficient driving. An additional requirement is that drivers must also make predictions about the future actions or states of these entities. A concept - *situation awareness* - that was developed in the aviation domain has recently become rather popular also in traffic psychology. It aims to describe and integrate these different cognitive processes [3], [4], [5]. In this paper we will outline a cognitive model of memory-update processes connected to the comprehension of driving situations that is described in more detail in [6]. Finally we will summarize results from experimental studies.

In Endsley's theory [5] of situation awareness three levels are distinguished. The first level involves the perception of the status, the attributes, and the dynamics of the

M. Kurosu (Ed.): Human Centered Design, HCII 2009, LNCS 5619, pp. 986–994, 2009.
© Springer-Verlag Berlin Heidelberg 2009

relevant situation elements. The second level, the comprehension level, involves integrating the different situation elements into a holistic picture of the situation, resulting in the comprehension of the meaning of the different elements. The third level involves the generation of assumptions about the future behavior of the elements on the basis of the comprehension of the situation.

However, Endsley's [5] model does not specify the cognitive mechanisms underlying cognitive performance. It is unclear, how one achieves the comprehension of a situation from the perception of the elements of this situation, and how assumptions about their future behavior are generated. In [6] and [7] a model of how situation awareness is "constructed" while driving is proposed. This model assumes that the comprehension of the meaning of perceived situation elements is achieved through comprehension processes that are similar to those involved in language comprehension, in particular in understanding text, e.g., [1]. By these comprehension processes a mental representation of the current situation is constructed and constantly updated with new information. This situation model is used as context for the interpretation of new information.

2 Basic Mechanisms

The level of concern for models of situation awareness is the *algorithmic* level, in Marr's [8] or Newell's [9] terminology, in contrast to the functional or implementation level. At the algorithmic level of analysis an information processing system is described in terms of information processing steps and a functional architecture. The goal at this level is to identify the connection between the functional description of the information processor and its implementation. This is achieved when the computational steps an information processing system takes can be described in terms of primitive functions that are implemented in the system's hardware. In this sense the three levels of situation awareness in Endsley's [5] model can be viewed as different functions of situation awareness. The goal of our research is to move the algorithmic analysis and description of situation awareness a step further and to come closer to a more complete algorithmic analysis of situation awareness.

Our model views the construction and maintenance of situation awareness as a *comprehension* process. Essentially it assumes that the comprehension process for understanding the meaning of different elements in a situation is comparable to the comprehension process for understanding language or discourse. In both cases an integrated mental representation of the perceived and processed pieces of information is constructed. This representation reflects the understanding of these elements. Fundamental to our approach is Kintsch's [1] Construction-Integration theory of text comprehension. As was explained above, situation awareness is a dynamic representation that is influenced both by the situation and the actions of the person. Therefore, for a model of situation awareness to be complete it has to specify how situation awareness is translated into action. A theory, that can be used here is Norman and Shallice's [2] theory of action selection. With the combination of these two theories it is possible to describe in more detail how situation awareness is constructed, maintained, and used as a basis for action selection.

2.1 Comprehending the Situation

According to [1], comprehending new information involves two phases. In the construction phase, the perception of new pieces of information activates knowledge structures in long-term memory, such as propositions and schemata that are associated with these pieces of information. This activation process is undirected and follows already established associations among these different pieces of knowledge. The result of this automatic activation process is a rather unstructured activation of knowledge in long-term memory. This activated knowledge then becomes integrated during the integration phase. Integration is accomplished through inhibitory and excitatory processes: knowledge that was activated in the first phase and is compatible to already activated knowledge (i.e., to the existing situation model), will remain activated and at the same time will activate other knowledge compatible with it. Simultaneously, incompatible pieces of knowledge inhibit each other.

The result of these two phases is an episodic memory representation. [1] distinguishes between two components of the episodic memory representation: the textbase and the situation model. The textbase consists of information that is directly derived from the text. The situation model results from the connection between the text representation and the comprehender's world knowledge. Translated into the domain of driving and situation awareness we propose that the episodic memory representation similarly consists of two components: a situation specific representation and the situation model. The situation specific representation consists of information that is directly perceived from the environment. This mainly includes the situation elements the driver perceives and information about the status of these elements that can be directly perceived, such as one's own speed. It also might include inferences that can be directly made from the perception of the situation elements and their status, such as that one's own speed is greater than the speed of the lead vehicle. This representation might reflect what [10] assigns to the first level of situation awareness. The situation model probably reflecting Endsley's second and third level of situation awareness then results from the connection between this situation specific representation and the driver's world knowledge. Through this connection the perceived configuration of situation elements becomes connected to the mental representation of previously experienced situations with similar configurations of elements resulting in an interpretation of the current situation. In this sense, the interpretation of a current traffic situation greatly depends on the driver's available knowledge about previously encountered traffic situations. This might at least in part explain the advantage of experienced drivers compared to novice drivers in hazard perception [11], [12], [13]. Experienced drivers have a much greater chance of having stored in long-term memory the relevant information that identifies the current situation as dangerous because their database of experienced traffic situations is much greater than that of novices. But the situation model does not only provide interpretations of the current situation, such as whether it is dangerous or not. It also contains expectations about the future development of the current situation and the future behavior of the relevant elements. These expectations are also derived from prior experiences that get connected to the situation specific representation. For the construction process of a coherent situation model it is necessary that when new information is perceived those parts of the situation model relevant for the interpretation of this new piece of information have to be available in working memory [14], [15]. Because of the amount and the complexity of the information that has to be represented in the situation model, such as the position and speed of other vehicles near to the own vehicle, traffic rules defined by signs, the status of the road surface and so on, it is assumed that the situation model contains much more information than can be kept active in working memory. Therefore a mechanism is necessary that allows to keep this information stored in long-term memory but makes this information reliably available for processing when it becomes relevant.

The theory of long-term working memory (LT-WM) describes a mechanism [16] that could provide such a function. According to LT-WM theory "cognitive processes are viewed as a sequence of stable states representing end products of processing [...] acquired memory skills allow these end products to be stored in long-term memory and kept directly accessible by means of retrieval cues in short-term memory" ([16] p.211). Such set of retrieval cues are arranged into stable retrieval structures that can have many different forms depending on the domain. In the case of driving such retrieval structures might be based on the huge amount of driving situations an experienced driver encountered. The experienced driver possibly developed many highly differentiated schemata of driving situations that allow him to easily identify many different types of driving situations. Such a mechanism might, for example, at least in part explain why experienced drivers are much better and much faster in identifying dangerous traffic situations than less experienced drivers [17], [18], [19].

In general, we assume that the different levels of situation awareness are highly interconnected as they are part of an unitary episodic memory representation. By using Kintsch's [1] Construction-Integration theory of text comprehension the model specifies a mechanism of how this representation is constructed starting from the perception of elements of the current situation. It is accomplished by a two-phase process. The first phase consists of an unstructured activation process, the second phase is a constraint-satisfaction process. This constraint-satisfaction process keeps the representation of the current situation coherent through the activation of compatible knowledge and inhibition of incompatible knowledge.

2.2 Action Selection and the Control of Behavior

While building up the representation of the situation, not only knowledge relevant to the interpretation of the situation and its future development gets activated, but also other actions that are appropriate in the current situation. According to the framework of [2], actions are represented as schemata. These schemata are procedures for the control of routine tasks. Schemata are organized in hierarchies. There are also connections to functionally related schemata or to schemata that are incompatible. The connections are either excitatory, like the connections from a superordinate schema to its subordinates, or inhibitory if the schemata are incompatible with each other, such as accelerating or decelerating when facing a yellow traffic light. Each schema is connected with trigger conditions that activate the schema if present in the current situation representation. So, the activation of situation appropriate actions becomes part of the construction of the situation representation. Experience is the major factor that connects these trigger conditions to action schemata. Successfully performing an action in a certain situation will establish excitatory links between relevant elements of the situation and this action schema. Afterwards, these elements will then become the trigger conditions for this schema. A failure in performing this action in this situation may reduce the strength of already existing excitatory connections or might even establish inhibitory links between relevant situation elements and the action schema. If more than one schema is activated a selection has to be made. Norman and Shallice [2] propose a mechanism called contention scheduling which operates on two processes: competition, the different action schemata compete with one another for their activation value, and selection, which takes place on the basis of the activation alone. Contention scheduling is sufficient to control well learned, simple action sequences. It resolves competition for selection, prevents competitive use of common resources, and negotiates cooperative use of common resources when possible.

If a schema is not available, or when the task is novel or complex, another control structure is necessary. [2] call this structure the Supervisory Attentional System (SAS). It allows for voluntary, attentional control of performance. The SAS is an attentional system that influences the selection of schemata in the contention scheduling process by providing additional activation and/or inhibition to schemata in order to bias their selection. The selection process itself is always based on the highest activation values of the competing schemata. This system allows for top-down influences in the action selection process.

To summarize, two well established theories, Kintsch's [1] Construction-Integration theory of text comprehension and Norman and Shallice's [2] theory of action selection are applied in the context of driving. We combine these two theories into a comprehension based model of situation awareness. By this it is possible to make much more detailed assumptions about the processes that are involved in the construction and maintenance of situation awareness and its use as a basis for action selection in driving.

3   Predictions of the Model and Empirical Results

A series of experiments was conducted to test the basic mechanisms of the model. In this paper two lines of research will be summarized.

3.1   The Effect of Experience, Relevance, and Criticality on Drivers' Mental Representation of a Traffic Situation

As stated above for safe driving the driver must have available a great amount of information that has to be integrated into the driver's situation model. As this amount of information clearly exceeds the capacity limitations of working memory a mechanism must be available to the driver that allows to keep this amount of information available despite the working memory capacity limits. We suggested as mechanisms the theory of LT-WM [16]. When applying this theory to situation awareness it is possible to make specific predictions about the effect of different factors on the availability of the driver's situation model. First, driver's experience should be important. More experienced drivers should recall more from a traffic situation than inexperienced drivers. The greater and more differentiated knowledge of experienced drivers about traffic situations should allow them first to identify the relevant information of a traffic situation more easily and second to encode this information faster and more reliable than less experienced drivers.

Second, as the use of LT-WM means to store results of cognitive processes in long-term memory these results should be available for retrieving even after longer delays when the appropriate retrieval cues are presented. As the applicability of the LT-WM mechanism depends on the person's experience in the relevant domain experienced drivers should be able to use this mechanism much more effective than less experienced drivers. Therefore, providing retrieval cues should facilitate recall of traffic situation elements much more for experienced than for less experienced drivers. Third, providing retrieval cues that are related to relevant situation elements of a previous traffic situation should help experienced drivers more than less experienced drivers. Forth, given that experienced drivers are able to use the LT-WM mechanism more effectively than less experienced drivers experienced drivers should show less influence of the delay between the actual traffic situation and a recall test for this situation.

Baumann, Franke and Krems [20] tested these predictions in a driving simulator experiment where participants had to drive on a three-lane highway. The simulation was interrupted repeatedly and participants were asked about the number of cars on different locations around the participant's car, for example on the left lane behind the participant. Two groups of drivers were tested: experienced and less experienced drivers. The length of the delay between interruption and recall, the criticality of the traffic situation at the time of interruption, the type of retrieval cue presented at recall, and the queried location's relevance to the participant's driving task were manipulated in the experiment.

The results of this experiment, described in more detail in [20], give some indication that LT-WM plays an important role in situation awareness while driving. Experienced drivers recall more information especially when provided with retrieval cues than less experienced drivers. Especially they recall more relevant information. Therefore, it seems that experienced drivers possess retrieval structures that allow them to efficiently and effortlessly encode traffic relevant information while driving to keep this information available for later use. It seems also that the retrieval structures that experienced drivers develop to efficiently encode and process information about the traffic situation are adapted to the high dynamics of the driving task. That means, that they are not appropriate for holding information for longer time periods what is different to other domains where the LT-WM theory was successfully applied [16]. This at least holds true for traffic situation information related to planning and executing driving manoeuvres. There might by other kinds of information that need to be stored for longer time periods, such as the travel destination, road characteristics, and where the retrieval structures are therefore also allow the long-term storage of information. This has to be considered in the development of measures for situation awareness while driving.

3.2   The Effects of Cognitive Tasks on Predicting Events in Traffic

According to [21] working memory is not a single structure but consists of different parts: a phonological buffer with an articulatory loop, a visuo-spatial buffer, an episodic buffer and the central executive. Recent studies indicate that the central executive serves different functions that seem to be highly relevant for the comprehension and projection processes of situation awareness as described above. One of these central executive functions is the retrieval of information from memory, a process essential in the comprehension process of situation awareness. Another is the maintenance of information in a non-modality specific buffer, the episodic buffer, and the control of working memory contents. We assume that the effect of cognitive loading tasks on driving is at least in part due to the interference of these tasks with these central executive processes. Baumann et al. [22] tested this prediction. The specific aim of their experiment was to test whether specific kinds of cognitive load that are designed to interfere with specific central executive processes interfere especially with the prediction function of situation awareness. In a simulator study the participants drove through a scenario that contained both predictable and non-predictable events. These events were designed to be exactly equivalent besides that in the predictable version a warning sign warned the driver of the upcoming event. The reaction to the event when the participant was warned was compared to the reaction when the driver was not warned.

The results confirm our hypothesis that the central executive function of controlling working memory content is highly involved in the construction of situation awareness. Interfering with this function by a secondary task leads to a detrimental effect in the driver's ability to predict the future development of a traffic situation. This results help to clarify the cognitive mechanisms that underlie situation awareness.

4 Summary

The goal of our approach is to establish the cognitive basis of situation awareness in order to be able to apply it to the driving task. We assume that the construction of situation awareness is basically a comprehension process that yields a mental representation of the meaning of the different elements of a traffic situation and the situation as a whole, that is, the situation representation. This situation representation is the basis for planning future behaviour that in turn alters the situation representation again. According to this perspective, driver knowledge plays a key role in determining the significance of events and elements in a traffic situation. Viewing situation awareness as a comprehension process also highlights the role of working memory in the process of constructing and maintaining situation awareness. The processes necessary to interpret new pieces of information, to determine their consequences for the current situation model, to integrate new pieces of information into the situation model, and to remove irrelevant information from the situation model must take place in working memory. If the resources of working memory are occupied by other tasks these processes will be impaired leading to a degraded situation awareness. This is important to consider for the design of tasks to be performed by the driver while driving, such as interacting with an in-vehicle information system (IVIS). Based on a detailed theory of situation awareness one should be able to develop methods that allow a detailed assessment of the effects of such systems on situation awareness. Such methods are highly needed for the development of systems suitable to be used while driving. Currently the process model of situation awareness described in this chapter is used to develop a procedure that should allow the evaluation of the effects of IVIS tasks on situation awareness based on an assessment of the visual and working memory demands of these IVIS tasks. The idea is that knowing the visual and working memory demands of IVIS tasks allows to make predictions about the effects of these IVIS tasks on situation awareness on the background of a detailed theory of the construction, maintenance, and updating of situation awareness.

References

1. Kintsch, W.: Comprehension: A paradigm for cognition. Cambridge University Press, New York (1998)
2. Norman, D.A., Shallice, T.: Attention to action: Willed and automatic control of behaviour. Reprinted in revised form. In: Davidson, R.J., Schwartz, G.E., Shapiro, D. (eds.) Consciousness and self-regulation, vol. 4, pp. 1–18. Plenum Press, New York (1986)
3. Adams, M.J., Tenney, Y.J., Pew, R.W.: Situation awareness and the cognitive management of complex systems. Human Factors 37, 85–104 (1995)
4. Bailly, B., Bellet, T., Goupil, C.: Driver's mental representations: Experimental study and training perspectives. In: Dorn, L. (ed.) Driver behavior and training, Aldershot, UK, Ashgate (2003)
5. Endsley, M.R.: Towards a theory of situation awareness in dynamic systems. Human Factors 37, 32–64 (1995a)
6. Baumann, M., Krems, J.: Situation Awareness and Driving: A Cognitive Model. In: Cacciabue, P.C. (ed.) Modelling Driver Behavior in Automotive Environments, pp. 253–265. Springer, London (2007)
7. Baumann, M.R.K., Rösler, D., Krems, J.F.: Situation awareness and secondary task performance while driving. In: Harris, D. (ed.) HCII 2007 and EPCE 2007. LNCS, vol. 4562, pp. 256–263. Springer, Heidelberg (2007)
8. Marr, D.: Vision: A computational investigation into the human representation and processing of visual information. W.H. Freeman, San Francisco (1982)
9. Newell, A.: Unified Theories of Cognition. Harvard University Press, Cambridge (1990)
10. Endsley, M.R.: Measurement of situation awareness in dynamic systems. Human Factors 37, 65–84 (1995b)
11. Chapman, P.R., Underwood, G.: Visual search of driving situations: Danger and experience. Perception 27, 951–964 (1998)
12. Groeger, J.A.: Understanding Driving: Applying Cognitive Psychology to a complex everyday task. Psychology Press, Hove (2000)
13. McKenna, F.P., Crick, J.: Hazard perception in drivers: a methodology for testing and training. Contractor Report 313. Transport Research Laboratory, Crowthorne (1994)
14. Fischer, B., Glanzer, M.: Short-term storage and the processing of cohesion during reading. The Quarterly Journal of Experimental Psychology 38A, 431–460 (1986)
15. Glanzer, M., Nolan, S.D.: Memory mechanisms in text comprehension. In: Bower, G.H. (ed.) The psychology of learning and motivation, pp. 275–317. Academic Press, New York (1986)
16. Ericsson, K.A., Kintsch, W.: Long-term working memory. Psychological Review 102, 211–245 (1995)
17. Crundall, D., Chapman, P.R., Phelps, N.R., Underwood, G.: Eye movements and hazard perception in police pursuit and rapid response driving. Journal of Experimental Psychology: Applied 9, 163–174 (2003)
18. Crundall, D., Underwood, G.: Effects of experience and processing demands on visual information acquisition in drivers. Ergonomics 41, 448–458 (1998)

19. Underwood, G., Chapman, P., Brocklehurst, N., Underwood, J., Crundall, D.: Visual attention while driving: Sequences of eye fixations made by experienced and novice drivers. Ergonomics 46, 629–646 (2003)
20. Baumann, M., Franke, T., Krems, J.F.: The effect of experience, relevance, and interruption duration on drivers' mental representation of a traffic situation. In: de Waard, D., Flemisch, F., Lorenz, B., Oberheid, H., Brookuis, K. (eds.) Human Factors for assistance and automation, pp. 141–152. Shaker Publishing, Maastricht (2008)
21. Baddeley, A.: The episodic buffer: A new component of working memory? Trends in Cognitive Sciences 4, 417–423 (2000)
22. Baumann, M., Petzoldt, T., Groenewoud, C., Hogema, K.J.F.: The effect of cognitive tasks on predicting events in traffic. In: Brusque, C. (ed.) Proceedings of the European Conference on Human Interface Design for Intelligent Transport Systems. Humanist Publications, Lyon (2008)

Redefining Architectural Elements by Digital Media

Kai-hsiang Liang

Graduate Institute of Architecture, NCTU
1001 Ta Hsueh Road, Hsinchu City 300
Rexkai@arch.nctu.edu.tw

Abstract. An architectural element is a unit of a construction. Architects and researchers can understand how to design by defining architectural elements. Design media act as helping roles between the designers' abstract concept and the concrete composition. Today, architectures designed by different media have different outcomes. This study discusses the relationship between architectural elements and design media, and treats whether digital media impact the existing architectural elements. In conclusion, digital media can not only use architectural elements from non-digital media, but also create three digital architectural elements, which are "curvy-surface", "multiple-functions" and "slit-opening."

Keywords: Digital architecture, Design media, Architectural element.

1 Introduction

Architectural elements have variety of different definitions by different times or field. The reviews of elements and their element-defining principle include the follows, the functions and usability of objects [1], solid and void and surface [2], the geometry [3] and [4], and the systematic classification by spatial dimensions [5]. The reviews in recent years include the follows, the architect's viewpoint [6], relative positioning and the dimensioning [7] and [8]. This study analyzes the architectural elements of the research, and concludes three basic factors of architectural elements, "basic geometry," "interface" and "function."

Design media act as helping roles between the designers' abstract concept and the concrete composition [9]. The earliest design media in architecture is simply as a tool to record [9]. Then architects started to make physical models to aid design thinking of 3D space [10]. In the twentieth century, in order to create new forms, some architects, such as Antoni Gaudi, used home-made media. In the twenty-first century, digital media are used by architects to not only record but also aid design of 2D and 3D, and are much precise than media in the past [11]. Today, architects just give computer enough design rules, and computer will feed back design possibilities.

From previous studies, the architectural elements are never given new explanations or definitions because of the different media. This study tries to discuss the latest media of architecture, digital media, to understand whether the digital media will affect the existing architecture elements or not? Compared with the physical media, the previous studies about discussing the relation between architecture elements and design media are rare. This study tries to find out that whether architects' usage of different media, digital media and non-digital media, would create the physical architectural

M. Kurosu (Ed.): Human Centered Design, HCII 2009, LNCS 5619, pp. 995–1002, 2009.

elements that couldn't defined by the existing ones, and compares the differences between these elements, the existing ones and the defined ones. And then induce and define appropriate architecture element to the era of digital architecture.

1.1 Methodology and Steps

This study chooses to analyse the cases to get the architectural elements, and aims at the architectural elements which created by digital and non-digital media. This study takes the cases in magazine a+u (Architecture and Urbanism) from 1977 to 2007, and analyses the cases of the architects who are top five published-times in a+u.

1.2 Analytic Factors

This study concludes three factors from the previous studies, which are "basic geometry", "interface", and "function."

(1) Basic geometry
Basic geometric forms are circle, triangle and rectangle [3]. The factor basic geometry, can be modify to generate sub-factors, which are "addition", "penetration", "bending" and "intercept" [5]. There are three steps of analysing the factor. First, trace every line of architectural drawings. Second, extend the traced-line to be reference lines, and make each intersected. Third, analyse the forms shaped by the reference lines' intersection, and compare the forms with the sub-factors of basic geometric.

(2) Interface
The interface of architectural elements is indispensable [12]. The sub-factors of interface include "corner", "surface", "opening" and "site planning". The main analysis of interface is based on photos. There are four steps of analyzing factor interface. First, analyse corners of the architecture, and measure the angle of corner. Second, analyse the architectural surface. This study divides the sub-factor, surface, into three levels: plane, fold and curved surface. Third, analyse the architectural openings. This study bases on the position of openings and divides the sub-factor, opening, into three levels: entrance, window and skylight. Fourth, analyse site planning focused on the relationship between the site and the architecture.

(3) Function
Definitions of architectural elements are generally classified by their functions [1]. Every architectural element has its unique and one function. The sub-factors of function include "door", "window", "column", "beam", "slab", and "wall." The main analysis of function is based on architectural drawings, plan and section. There are two steps of analysing the factor interface. First, mark every category with colours, "door" as blue; "window" as green; "column" as orange; "beam" as purple; "slab" as yellow; "wall" as red. Second, compare functional elements with the whole architecture.

2 Non-digital Architecture

2.1 Gemini G.E.L

The name of this project is "Gemini G.E.L", and the architect is Frank O. Gehry. It is finished in 1979. This case is a combination of the basic geometric forms. The layout of first-floor-plan drawing is based on rectangles, and it is applied the sub-factor, intercept, to change rectangular layout into the L-shaped. The layout of third-floor-plan drawing is applied the sub-factor, penetration, to place the skylight. The included angle of the skylight's and the site's reference lines is because the skylight- placed is applied sub-factor, bending. Other sub-factors are not applied, and this case can be fully deconstructed by basic geometry. According to the photos, the sub-factor, corner, of this case is fully orthogonal. The sub-factor, surface, is fully plane. The sub-factor, opening, has three, entrance, window and skylight. The sub-factor, site planning, is vertical line and the included angle of ground line and the walls is right angle. This case can be fully deconstructed by interface. Mark every category with colors. This case can be fully deconstructed by function.

2.2 House VI

The name of this project is "House VI", and the architect is Peter Eisenman. It is finished in 1975. This case is a combination of the basic geometric forms. The layout of plan drawing is based on rectangles, and it is applied the sub-factor, addition, to form the layout of plan, section and elevation drawings. Other sub-factors are not applied, and this case can be fully deconstructed by basic geometry. According to the photos, the sub-factor, corner, of this case is fully orthogonal. The sub-factor, surface, is fully plane. The sub-factor, opening, has two, entrance and window. The sub-factor, site planning, is vertical line and the included angle of ground line and the walls is right angle. This case can be fully deconstructed by interface.

Mark every category with colours. The peculiarity of this case is the sub-factor, column because the columns of this case are not independent. The columns look like parts of the walls. This case can be fully deconstructed by function.

2.3 Two Patio Villas

The name of this project is "Two Patio Villas", and the architect is Rem Koolhaas. It is finished in 1988. This case is a combination of the basic geometric forms. The layout of plan drawing is based on rectangles, and it is applied the sub-factors, addition and penetration, to form the layout of plan, section and elevation drawings. Other sub-factors are not applied, and this case can be fully deconstructed by basic geometry. According to the photos, the sub-factor, corner, of this case is fully orthogonal. The sub-factor, surface, is fully plane. The sub-factor, opening, has three, entrance, window and skylight, and there are two entrance of different heights because of the landforms. The sub-factor, site planning, is vertical line and the included angle of ground line and the walls is right angle. This case can be fully deconstructed by interface. Mark every category with colours. This case can be fully deconstructed by function.

2.4 House in Magomezawa

The name of this project is "House in Magomezawa", and the architect is Toyo Ito. It is finished in 1986. This case is a combination of the basic geometric forms, circle and rectangle. The layout of plan drawing is based on rectangles, and it is applied the sub-factors, intercept, to form the layout of plan. It is applied the sub-factors, addition, to form the layout of section and elevation drawings. Other sub-factors are not applied, and this case can be fully deconstructed by basic geometry. According to the photos, the sub-factor, corner, is fully orthogonal. The sub-factor, surface, is fully plane besides the roof. The sub-factor, opening, has two, entrance and window. The sub-factor, site planning, is vertical line. This case can be fully deconstructed by interface. Mark every category with colours. This case can be fully deconstructed by function.

2.5 Vitra Fire Station

The name of this project is "Vitra Fire Station", and the architect is Zaha Hadid. It is finished in 1990. This case is a combination of the basic geometric forms. The layout of plan drawing is based on triangles and rectangles which are transformed into quadrangles, and it is applied the sub-factors, addition, to form the layout of plan, section and elevation drawings. Other sub-factors are not applied, and this case can be fully deconstructed by basic geometry. According to the photos, the sub-factor, corner, is bias. The sub-factor, surface, is fully plane. The sub-factor, opening, has two, entrance and window. The sub-factor, site planning, is bias line. This case can be fully deconstructed by interface. Mark every category with colours. This case can be fully deconstructed by function.

3 Digital Architecture

3.1 Nationale-Nederlanden Office Building

The name of this project is "Nationale-Nederlanden Office Building", and the architect is Frank O. Gehry. It is finished in 1995. This case is formed by free forms and rectangles. The layout of plan is based on free forms, especially the elevations. All sub-factors are not applied, and this case can not be deconstructed by basic geometry. According to the photos and isometric drawings, the sub-factor, corner, is curved. The sub-factor, surface, is curved, too. The sub-factor, opening, has three, entrance, window and skylight, but the columns on the ground floor are slit to form the gates which are not defined of openings. The sub-factor, site planning, is bias line. This case can be deconstructed by factor, interface. Mark every category with colours. This study finds out some elements with multiple functions, such as column and door. This case can not be deconstructed by function, and sub-factors, column and door, should be redefined (Figure 1).

Fig. 1. The elemental analysis of Nationale-Nederlanden Office Building

3.2 The Aronoff Center for Design and Art at the University of Cincinnati

The name of this project is "The Aronoff Center for Design and Art at the University of Cincinnati", and the architect is Peter Eisenman. It is finished in 1996. Because this case is to extend the old buildings, it is formed by two forms, free forms and the shape from the old building. The two forms are processed by computer-aided design to create new forms which architect can select. So, all sub-factors are not applied, and this case can be not deconstructed by basic geometry. According to the photos, the sub-factor, corner, is bias. The sub-factor, surface, is folding. The sub-factor, opening, has three, entrance, window and skylight. The sub-factor, site planning, is bias line. This case can be fully deconstructed by interface. Mark every category with colours. Because the space is formed by the computer, there are many elements which are not defined, such as column and slab. This case can not be deconstructed by function (Figure 2).

Fig. 2. The elemental analysis of The Aronoff Center for Design and Art at the University of Cincinnati

3.3 Casa da Musica

The name of this project is "Casa da Musica", and the architect is Rem Koolhaas. It is finished in 2005. This case is formed by quadrangles and rectangles, and the four angles of every quadrangle are different. It is applied to plans, sections and elevations by the same way. All sub-factors are not applied, and this case can not be deconstructed by basic geometry. According to the photos, the sub-factor, corner, is bias. The sub-factor, surface, is plane. The sub-factor, opening, has three, entrance, window and skylight. The sub-factor, site planning, is bias line. This case can be fully deconstructed by interface. Mark every category with colours. Because the continued form from slab to wall, or wall to slab, there are elements which are not defined. This case can not be deconstructed by function (Figure 3).

Fig. 3. The elemental analysis of Casa da Musica

3.4 Serpentine Gallery Pavilion 2002

The name of this project is "Serpentine Gallery Pavilion 2002", and the architect is Toyo Ito. It is finished in 2002. This case is a combination of the basic geometric forms. The layout of plan drawing is based on triangle and rectangular frames, and it is applied the sub-factors, addition, penetration and Intercept, to form the layout of plan, section and elevation drawings. Other sub-factors are not applied, and this case can be fully deconstructed by basic geometry. According to the photos, the sub-factor, corner, is fully orthogonal. The sub-factor, surface, is a double-wall which is a combination of several triangles and quadrangles. The sub-factor, opening, has one from, triangles. The sub-factor, site planning, is vertical line and the included angle of ground line and the walls is right angle. This case can be fully deconstructed by interface. Mark every category with colours. Because of the structural system, there are no column and beam. This case can be fully deconstructed by function (Figure 4).

Fig. 4. The elemental analysis of Serpentine Gallery Pavilion 2002

3.5 Ordrupgaard Museum Extension

The name of this project is "Ordrupgaard Museum Extension", and the architect is Zaha Hadid. It is finished in 2005. This case is formed by two systems, free forms and parallelograms. This study marks free forms as green, and parallelograms as red. The colors show the outer elements are free forms, and the inners are parallelogram. All sub-factors are not applied, and this case can not be deconstructed by basic geometry. According to the photos, the sub-factor, corner, is curved. The sub-factor, surface, is curved, too. The sub-factor, opening, has two, entrance and window. The sub-factor, site planning, is curved line and the case appears to be suspended. This case can be fully deconstructed by interface. Mark every category with colours. There are elements which can not be defined by all sub-factors because they have multiple functions, such as wall and door. This case can not be deconstructed by function (Figure 5).

Fig. 5. The elemental analysis of Ordrupgaard Museum Extension

4 Conclusions

Through the foregoing analysis, this study draws conclusions. In terms of Factor Basic geometry: non-digital architectural cases are primarily composed of basic geometries; digital architectural cases are difficult to disassemble using basic geometries. Non-digital architectural cases are made up mostly of rectangles, with few cases of round and triangular shapes, while digital architectural cases are mostly made up of free forms; any free lines intersect with each other at free angles. In terms of Factor Interface: the contact interfaces between surfaces in non-digital architectural cases are clear; the contact interfaces between surfaces in digital architectural cases are unclear. Surface-to-surface contacts in non-digital architectural cases are mostly at 90 degree angles, while those in digital architectural cases have varying angles. In terms of Factor

Function: the architectural element units in non-digital cases are clear, while those in digital architectural cases are not. The architecture in non-digital architectural cases can be completely disassembled based on every function; the architectural elements analysed in digital architectural cases can be disassembled according to the factor function, and it is because there are multiple functions not fit original definitions.

This study found out some components which can not be included in the existing architectural elements' definition, and redefine and called them: Digital architectural elements. Digital architectural elements include the following: curvy surface, multiple functions, and slit opening.

(1) curvy surface
This study finds some features which cannot be defined by two existing architectural element factors, "basic geometry" and "interface." According to the result of case studies, the architectural element of most digital cases is curve not straight line of basic geometry, and this study named them "curvy surface."

(2) multiple functions
This study also finds some features which cannot be defined by the existing architectural element factor, "function." According to the functional architectural elements in digital case studies are curvy, the architects have chance to merge the functions of existing elements to one element with multiple functions, such as the combination of "floors and walls", "beams and columns" and "windows and doors," and this study named them "multiple functions."

(3) slit opening
There are some features which cannot be defined by three existing architectural element factors, "basic geometry", "interface" and "function." According to the result of case studies, this study found digital architects set up the spatial function in slit of surface to create new formal entrances and windows, and named them "slit opening."

This study finds out that using digital media can not only create existing architectural elements but also the new elements that could not be defined yet. The limitation is because of the case-choosing in the international magazine a+u. If the architects who never been posted in the magazine will be the blind spot of this research. My future study will choose other domains to recertify.

References

1. Atkinson, R., Banegal, H.: Theory and Elements of Architecture. Benn, London (1926)
2. Norberg-Schulz, C.: Intentions in architecture. MIT Press, Cambridge (1963)
3. Ching, F.D.: Architecture: Form, Space and Order. Van Nostrand Reinhold, New York (1979)
4. Cha, M.Y., Gero, J.S.: Shape Pattern Recognition Using a Computable Pattern Representation. In: Gero, J.S., Sudweeks, F. (eds.) Artificial Intelligence in Design 1998, pp. 169–187 (1999)
5. Krier, R.: Elements of Architecture. AD Publications, London (1988)
6. Kalay, Y., Marx, J.: Changing the Metaphor: Cyberspace as a Place. In: CAAD Future, Taiwan, pp. 18–28 (2003)

7. De Luca, L., Florenzano, M., Veron, P.: A Generic Formalism for The Semantic Modeling and Representation of Architectural Elements. Visual Computer 23, 181–205 (2007)
8. Chevrier, C., Perrin, J.P.: Laser range data, photographs and architectural components. In: ISPRS conference, Beijing, pp. 1113–1118 (to appear) (2008)
9. Liu, Y.T.: Understanding Architecture in the Computer Era, Taiwan (1996)
10. Smith, A.C.: Architectural model as machine: A new view of models from antiquity to the present day. Architectural Press, Oxford (2004)
11. Lim, C.K.: A better digital design and construction process using CAD/CAM media, Ph. D. Thesis. Hsinchu: National Chiao Tung University (2007)
12. Unwin, S.: Analysing Architecture. Routledge, London (1997)

Cognitive Engineering for Direct Human-Robot Cooperation in Self-optimizing Assembly Cells

Marcel Ph. Mayer*, Barbara Odenthal, Marco Faber, Jan Neuhöfer, Wolfgang Kabuß, Bernhard Kausch, and Christopher M. Schlick

Institute of Industrial Engineering and Ergonomics at RWTH Aachen University,
Bergdriesch 27, 52062 Aachen, Germany
{m.mayer,b.odenthal,m.faber,j.neuhoefer,w.kabuss,
b.kausch,c.schlick}@iaw.rwth-aachen.de

Abstract. In a work system with direct human robot cooperation the conformity of the operator's expectation with the behavior of the robotic device is of great importance. In this contribution a novel approach for the numerical control of such a system based on human cognition and a cognitive engineered approach for the encoding of the system's a priori knowledge is introduced. The implementation using an established method in the field of design of cognitive systems is compared to a schema describing human decision making. Finally, simulation results of the implementation are compared to empirical tests with individuals.

Keywords: Cognition, HRC, Automation.

1 Introduction

Future manufacturing systems will include human operators and robotic devices jointly working together in a cooperative way. Highly automated systems are often neither efficient enough for small lot production (ideally one piece) nor flexible enough to handle products to be produced in a large number of variants. Here the robotic device could take over repetitive and dangerous tasks which are not too complex whereas the human operator could handle the variants that are hard to automate.

Concerning human-robot cooperation (HRC) in today's manufacturing systems there is a restrictive separation between automated robotic systems and the human operator (98/37/EG). This separation appears in two categories: a local separation and a temporal separation (THIEMERMANN 2005). To be precisely the term HRC in this context is misleading, for there is no cooperation in the narrow sense of the word. The robotic device as well as the human operator are performing tasks sequentially regarding one workpiece to be produced. Suspending the local separation by maintaining the temporal separation, a synchronized processing of tasks in the same working area

* The authors would like to thank the German Research Foundation DFG for the kind support of the research on human-robot cooperation within the Cluster of Excellence "Integrative Production Technology for High-Wage Countries".

M. Kurosu (Ed.): Human Centered Design, HCII 2009, LNCS 5619, pp. 1003–1012, 2009.
© Springer-Verlag Berlin Heidelberg 2009

would be possible. Suspending the temporal separation by maintaining the local separation would lead to a concurrent work on different workpieces. To realize a cooperation of human operator and robotic device both separations have to be suspended. In the following this is referred to as direct human-robot cooperation (dHRC).

Recent research aims at reducing the aforementioned separation e.g. by introducing speed reduction of the robotic device e.g. by camera surveillance of both robot and human operator (KUHN and HENRICH 2007) to reduce the risk of impacts by slowing the robotic device down or even stopping it in case of a dangerous approach. Based on an experimental direct cooperative workplace (THIEMERMANN 2005) the task allocation of human and robotic device and the optimization of such a scenario is subject to recent research (BEUMELBURG 2005).

Despite all approaches to enhance dHRC the human operator still has to follow the tact of the technical system. From an ergonomical point of view a desirable scenario would be the opposite: The technical system following the ergonomic motion cycle of the human operator. Therefore the technical system has to be flexible regarding dHRC.

2   Self-optimization Enabling Flexible Direct Human Robot Cooperation

The term flexibility regarding production systems is nothing new. Definitions of different subcategories of flexibility such as machine flexibility, process flexibility, performance flexibility just to mention a few can be found e.g. in HOFMAN (1990). Flexibility in production/manufacturing is a term mostly related to the technical equipment. Even in more up to date literature flexibility is only used in the boundaries of the technical system (e.g. MILBERG 2003, SHERIDAN 2002, CHRYSSOLOURIS 2006).

To expand flexibility to the human-machine system as a whole, the abilities of the technical system have to be enhanced towards cognitive abilities like decision-making or problem-solving which are not implemented in today's production facilities. In current production processes the human operator is the only participant who possesses cognitive abilities. On the technology side neither the interfaces nor the technical systems itself incorporate such skills. Therefore a research project has been established within the Cluster of Excellence "Integrative Production Technology for High Wage Countries" at the Faculty of Mechanical Engineering of RWTH Aachen University that aims at the development of an self-optimizing assembly cell enabling flexible automation with the possibility of dHRC. A novel design of the cell's numerical control – which will be regarded in more detail in the following chapter – forms the basis of this novel automation approach. The cell's numerical control – here referred to as cognitive control unit (CCU) – accomplishes high-level information processing based on the cognitive architecture SOAR (see LEIDEN et al. 2001). Hence to a certain extent the CCU is able to simulate rule-based behavior of the human operator. Taking into account methodological and technological limitations, knowledge-based behavior in the true sense of RASMUSSEN (1986) cannot be modeled and simulated.

2.1   Role of the Human Operator Supervising a Self-optimizing Production Cell

In classic supervisory control of automated manufacturing systems the role of the human operator includes planning of what needs to be done, teaching the system what it needs to know to manufacture, monitoring the system state during runtime to detect trends, errors or failures, intervening in the system if necessary to decide on making adjustments or changes and finally learning by evaluating the system state and behavior (SHERIDAN 2002). Regarding a cognitive automated assembly cell able to autonomously decide which part of a given task can be processed by a given system function, the classic role of the human operator has to be revised at some points. A corresponding model for task allocation between the operator and the CCU for supervisory control can be found in MAYER et al. (2008a)

When a manufacturing order is assigned to the cell a so-called master assembly schedule has to be developed by the human operator first. This master assembly schedule primarily consists of a formal description of the desired final state of the product to be assembled and a set of useful heuristics to reach the goal. Additionally constraints such as maximum energy consumption, minimum accuracy etc. can be specified. When the goal state, the heuristics and the associated constraints are submitted to the CCU, assembly procedure planning and teaching (in terms of RC programming) are carried out autonomously on the basis of stored process knowledge that is encoded in rules.

Since the CCU is able to resolve a certain class of assembly problems autonomously and therefore significantly reliefs the operator from repetitive and monotonic task processing, classic approaches for human-centered automation proclaiming a continuous involvement of the operator (BILLINGS 1991) are not very meaningful in self-optimizing assembly cells. Nevertheless, one has to carefully consider the so-called Irony of Automation (BAINBRIDGE 1987). In case of not autonomously solvable manufacturing problems or in case of human triggered intervention, the human operator needs to be in the loop and must have detailed knowledge about the state of the technical system, the already executed tasks, and the systems objectives and constraints to make good or even close to optimal decisions. To cope with this inherent complexity, an important basic requirement for the ergonomic design of the CCU is the conformity with operator's expectations, especially in low level decision-making related to the product to be assembled, where the human operator is not continuously involved.

In order to study human-machine interaction in self-optimizing assembly systems, an experimental assembly cell was designed and a manufacturing scenario was developed within the cited Cluster of Excellence at RWTH Aachen University (KEMPF et al. 2008). In the scenario one robot carries out a certain repertoire of coordinated pick and place operations with small workpieces. In this contribution only the theoretical background for the subtask of a human triggered intervention is of interest and additional interesting functions of the CCU are not considered.

2.2 Human Centered Approach for a Technical Cognitive System

As outlined in the chapter before it is crucial for the human operator to understand the plan of the CCU in the case of supervisory control. This becomes even more important in the case of direct Human-Robot Cooperation. Under the assumption that a numerical control of a technical system based on human decision making will lead to a better understanding by the human operator regarding the intention of the technical system, in a first   implementation the cognitive architecture SOAR was chosen.

SOAR belongs to the class of computational models which - compared to emergent systems like artificial neuronal networks - do not require initial training hence changes to the knowledge base can be performed quickly, which is of great interest especially for an application in an industrial environment.

To enhance the human centered approach, the question arises on how to design the symbolic representation of the knowledge base for the CCU to ensure the conformity with the operator's expectations. Proprietary programming languages that are used in conventional automation have to be learned case specific and do not necessarily match the mental model of the human operator. Focusing a human centered description to match the process knowledge to the mental model one promising approach in this special scenario is the use of motion description. These motions are familiar to human operators from manually performed tasks hence those are easier to anticipate than complex programming code. Already established methods or taxonomies e.g. from process planning can be used. Since in production systems complex handling tasks have to be broken down into fundamental elements, a promising approach here is the use of the MTM system as a library of fundamental movements, neglecting the underlying time information in a first step (MAYER et al. 2008b).

2.3 Design of the Knowledge Base

After the principal idea has been outlined, the process of designing and developing the knowledge base with its essential production rules shall be described. An established method in the field of cognitive technical systems design is used here to implement the knowledge base. The Cognitive Process Method (CP-Method; Fig. 1) introduced by PUTZER (2005) links consequently software technology and cognitive sciences by using a cognitive process as a basis. This process enables on the one hand the structured design of software code as knowledge and on the other to remain in the behavioral model. In the following sections, the a priori knowledge for the aforementioned use case which is implemented in SOAR is presented following the four steps of the static model of the CP-Method. In the actual code it is later possible, that one production rule contains elements that are part of different steps in the CP-process.

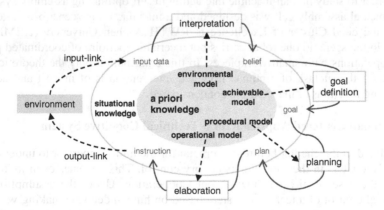

Fig. 1. The CP-Process according to PUTZER (2005)

- **Achievable Model**

  In the first step the achievable model for the cognitive system has to be defined as a goal, for all further actions are depending on this model. In this particular scenario, the achievable goal is the buildup of a structure of LEGO bricks e.g. a pyramid of identical bricks. The achievable goal contains position and rotation, color, type and the neighboring relations of each brick in the model.

- **Procedural Model**

  In the second step the knowledge for procedures to achieve the desired goal have to be implemented. Based on this knowledge, the buildup of the model is organized with respect to constraints e.g. a brick can only be positioned if either it sits on the ground or all of the parts below are already built. The elaboration rules in this particular scenario select the bricks that can be built on the workspace respective give a minor priority to the bricks that cannot be built but are part of the desired model or refuse to take bricks that are not needed. Due to the implementation with SOAR one must differentiate between elaboration rules and rules that propose an operator. The rules that propose an operator are strongly connected to rules that apply an operation but are part of the procedural model. These rules are not regarded further but there operational counterparts are.

- **Operational Model**

  The operational model is strongly linked to the procedural model, since it enables to put the generated plans of the procedural model into action. This link can be clearly seen when regarding the elements of the operational model in this scenario: The basic fundamental movements of the MTM-1 system were selected to control the robotic movement (MAYER et al. 2008b). These movements represented as rules in the operational model are REACH, GRASP, MOVE (including TURN), POSITION and RELEASE (including APPLY PRESSURE). A particular rule can only be applied if the corresponding operator was selected by the procedural model. The rules based on the fundamental MTM movements are the only elements in this scenario that can manipulate the environment.

- **Environmental Model**

  In the fourth step of the static CP-method all elements that are needed in the previous steps have to be mapped to an environmental model that can be used by the cognitive unit. In this particular scenario the gripper of the robotic device, the conveyer band with a stochastic brick lead, the workplace and the buffer are modeled. These elements are handed over to the SOAR-agent during initialization, together with the goal state.

Regarding the offset between a desired state and a given state as a problem that needs to be solved, Marshall's Schema (MARSHALL 2008) is used here to compare the implemented knowledge base to human decision making (Fig. 2). Marshall's schema describes the different kinds of human knowledge that participate in solving a certain kind of problem. Therefore the four knowledge categories are introduced here:

- **Identification Knowledge**

  This knowledge is responsible to identify "What is happening now". Compared to the introduced implementation the achievable model and the environmental model would be the parts to be found in the identification knowledge.

- **Elaboration Knowledge**
  The responsibility of this certain knowledge is to decide "What has high priority and why". Most of the rules that describe under which circumstances a certain part can be built would be part of the elaboration knowledge.
- **Planning Knowledge**
  The planning knowledge is responsible for the decision of "What needs to be done and when". The propose-rules of the MTM-operators are the only part of the planning knowledge in this particular case. These rules are strongly linked to the apply-rules. Hence the planning knowledge in this particular scenario is underrepresented.
- **Execution Knowledge**
  The execution knowledge finally decides "Who should do what". This knowledge can directly be matched to the operational model respective the apply-rules of the fundamental MTM-operators.

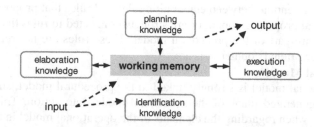

Fig. 2. Schema Model according to MARSHALL (2008)

3 Comparing the CCU to Humans

The key factor for direct HRC is the conformity with the operator's expectations regarding the performance of the robotic device. The introduced approach for a numerical control of a robotized assembly cell follows the notion that using an architecture being able to simulate rule-based human behavior on the basis of motion increases the conformity with the operator's expectations. Since at this stage of development a scenario with a meaningful dHRC is not yet possible, a comparison between the motion operations of cognitive control unit (CCU) and human behavior shall be given here instead.

Therefore two admittedly easy assembly tasks were developed, in which a model consisting 10 respective 30 identical bricks has to be built up. For the simulation only an unstructured list was given to the CCU containing position, rotation, size and color of the bricks in the model. The rules of the knowledge base are handed over during initialization. All required parts are positioned on an area within reaching distance for the gripper of the robotic device. Comparable to that only a drawing of the model to be assembled was handed over to the individuals. All parts where positioned visible within reaching distance on a defined area. With respect to the boundary conditions of the robotized assembly cell respective its simulation, the individuals had to follow the following constraints:

— No substructures are allowed to be built
— Only one brick can be picked at a time
— Only one hand can be used to built
— The model has to be assembled on a defined area

A total of 80 simulation runs of the CCU were calculated for each assembly task. To compare the results 16 individuals (13 male, 3 female) participated in the tests.

4 Results

As a first result, in all simulation runs as well as in the empirical tests the models were built up successfully. For model 1 in all simulation runs 96 steps were computed until the desired state was reached. This number is equivalent to 30 closed MTM-cycles consisting of identifying a desired block, reach, grasp, move, position and release which were counted in the empirical tests. For model 2 a total of 60 simulation steps respective 10 MTM-cycles could be counted until the desired state was reached. The build-up sequences of model 1 as well as model 2 were differing in the simulation runs although the constraints where comparable. These differences can be explained due to the implementation using SOAR and the fact that the planning knowledge regarding the Schema Model is underrepresented: In SOAR all production rules fire that match the actual state. If more than one operator is proposed, and the proposed operators are equivalent only one operator of the operational model is selected randomly. In this particular scenario all bricks that can be built are equal, hence e.g. for the first brick to be positioned all bricks in the bottom layer have the same probability. In the following only the positioning of the first brick shall be regarded exemplarily.

The theoretical probability for each position of the first brick for model 1 equals 6.25 % the probability for model 2 equals 25 %. The probability for model 1 of the position of first brick after n´=80 simulation runs can be seen in Fig. 3 on the left. The results of the individuals accordingly are displayed on the right.

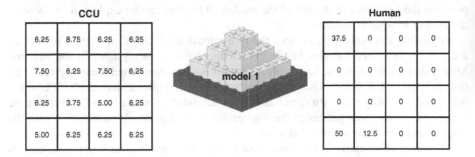

Fig. 3. Frequencies of the first brick to be set of the CCU (n´=80) (left) and the results of the empirical tests with individuals (n=16) (right)

Whereas the results of the simulation show no significant difference ($\alpha=0.05$) to the theoretical distribution, a significant difference in the way individuals start the buildup could be identified. The corresponding results of the χ^2-test can be found in Table 1.

Table 1. Results of the chi-square tests

Source	df	χ^2	p	Source	df	χ^2	p
Model 1				Model 2			
CCU	15	2.4	.98	CCU	3	4.8	.91
Individuals	15	88	.00	Individuals	3	33.5	.00
$\alpha=0.05$							

The comparison of the conditional probability of a brick to be set for model 1 between CCU and individuals (derived from results) is displayed in Fig. 4.

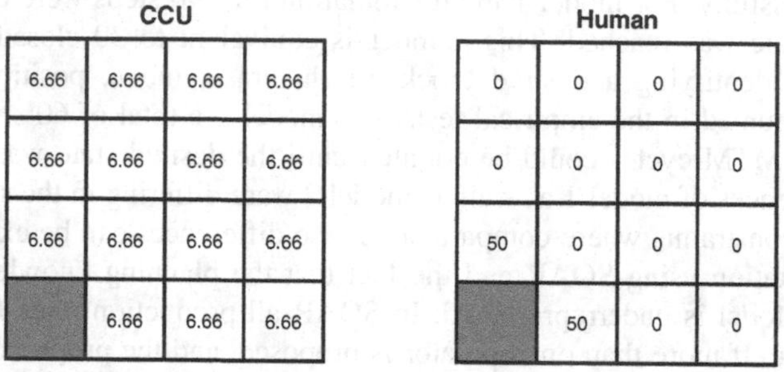

Fig. 4. Conditional probability of a brick to be set (grey fields indicate positioned bricks)

5 Conclusion

Regarding the probabilities of the position of the first brick to be set, there is a significant difference between the results of the simulation and the empirical tests. Whereas the probabilities of the simulation runs for model 1 as well as model 2 show no significant difference from the theoretical uniform distribution, the individuals tend to position the first brick at the left of the models. This fact can be explained by the way of reading in western cultures.

Regarding the buildup sequences the individuals tend to follow certain plans that are not implemented in rules in the knowledge base of the system. In the Schema Model this knowledge is referred to as planning knowledge, which cannot directly be found in the CP-Process. Using this process for the implementation of CCU for a robotized assembly cell with direct HRC, the knowledge base of the CCU has to be enhanced by rules that guarantee the conformity of the operator's expectation with the actual behavior of the technical system.

Therefore the empirical tests have to be analyzed in detail regarding the sequences the bricks are positioned. In a next step the relation of each following brick to the already positioned bricks shall be analyzed.

6 Summary and Outlook

There can be no doubt that novel methods and techniques of industrial engineering and ergonomics are necessary to improve the competitiveness of manufacturing companies in high-wage countries. A promising approach are self-optimizing manufacturing cells and systems (KLOCKE & PRITSCHOW 2004, GAUSEMEIER 2008) that rely on the unique knowledge, skills and abilities of the human operators.

In this contribution a novel approach for the numerical control of a robotic device using a cognitive architecture for the simulation of human decision making was introduced. The idea of using the fundamental movements of the MTM-system as a library for the symbolic representation of the knowledge base was outlined and two approaches for implementation were discussed. For two simple models a comparison between simulation and empirical tests with individuals was introduced and differences regarding the position of the first brick where presented. Based on these results an focusing on direct HRC it can be stated, that additional rules taking human behavior into account have to be integrated in the knowledge base of the system.

Concerning future ergonomic studies in the field of direct HRC it has to be analyzed which plans are followed by individuals and hence are expected of the technical system by the human operator when assembling a certain part. Further it has to be investigated how much information is necessary and how to present the information to get the human operator adequately informed in the case of intervention.

References

Bainbridge, L.: Ironies of Automation. In: Rasmussen, J., Duncan, K., Leplat, J. (eds.) New Technology and Human Error. Wiley, Chichester (1987)

Beumelburg, K.: Fähigkeitsorientierte Mpntageablaufplanung in der direkten Mensch-Roboter-Kooperation. In: Westkämper, E., Bullinger, H.J. (eds.) IPA – IAO Forschung und Praxis. Jost Jetter Verlag, Heimsheim (2005)

Billings, C.E.: Human-centered aircraft automation: A concept and guidelines. In: NASA Technical Memorandum 103885, NASA-Ames Research Center, Moffet Field CA (1991)

Chryssolouris, G.: Manufacturing Systems – Theory and Praxis. Springer, New York (2006)

Gausemeier, J.: From Mechatronics to Self-Optimizing Systems. In: Proceedings of the 7th International Heinz Nixdorf Symposium on Self-optimizing Mechatronic Systems: Design the Future (2008)

Hofman, P.: Fehlerbehandlung in flexible Fertigungssystemen. Oldenburg Verlag, München (1990)

Kempf, T., Herfs, W., Brecher, Ch.: Cognitive Control Technology for a Self-Optimizing Robot Based Assembly Cell. In: Proceedings of the ASME 2008 International Design Engineering Technical Conferences & Computers and Information in Engineering Conference, America Society of Mechanical Engineers, U.S (2008)

Kuhn, S., Henrich, D.: Fast Vision-Based Minimum Distance Determination Between Known and Unkown Objects. In: IEEE/RSJ 2007 International Conference on Intelligent Robots and Systems, San Diego, CA, USA, October 29-November 2 (2007)

Klocke, F., Pritschow, G. (eds.): Autonome Produktion. Springer, Berlin (2004)

Leiden, K., Laughery, K.R., Keller, J., French, J., Warwick, W., Wood, S.D.: A Review of Human Performancer Models for the Prediction of Human Error. Prepared for: National Aeronautics and Space Administration System-Wide Accident Prevention Program. Ames Research Center, Moffet Filed CA (2001)

Mayer, M., Odenthal, B., Grandt, M., Schlick, C.: Anforderungen an die benutzerzentrierte Gestaltung einer Kognitiven Steuerung für Selbstoptimierende Produktionssysteme. In: Gesellschaft für Arbeitswissenschaft e.V. (eds.) Produkt- und Produktionsergonomie – Aufgabe für Entwickler und Planer. Gfa-Press, Dortmund (2008a)

Mayer, M., Odenthal, B., Grandt, M., Schlick, C.: Task-Oriented Process Planning for Cognitive Production Systems using MTM. In: Karowski, W., Salvendy, G. (eds.) Proceedings of the 2nd International Conference on Applied Human Factors and Ergonomic (AHFE), Las Vegas, Nevada, USA, July 14-17, 2008 (2008b)

Marshall, S.P.: Cognitive Models of Tactical Decision Making. In: Karowski, W., Salvendy, G. (eds.) Proceedings of the 2nd International Conference on Applied Human Factors and Ergonomic (AHFE), Las Vegas, Nevada, USA, July 14-17 (2008)

Milberg, J.: Die agile Produktion. In: Klocke, F., Pritschow, G. (eds.) Autonome Produktion, Springer, Berlin (2003)

Putzer, H.J.: Ein uniformer Architekturansatz für kognitive Systeme und seine Umsetzung in ein operatives Framework. Köster, Berlin (2005)

Rasmussen, J.: Information Processing and Human-Machine Interaction. An Approach to Cognitive Engineering. North-Holland, New York (1986)

Sheridan, T.B.: Humans and Automation: System Design and Research Issues. John Wiley & Sons, Santa Monica (2002)

Thiemermann, S.: Direkte Mensch-Roboter-Kooperation in der Kleinteilemontage mit einem SCARA-Roboter. In: Westkämper, E., Bullinger, H.J. (eds.) IPA – IAO Forschung und Praxis. Jost Jetter Verlag, Heimsheim (2005)

Directive 98/37/EC of the European Parliament and of the Council of 22 June 1998 on the approximation of the laws of the Member States relating to machinery (June 1998)

Evaluating Design Concepts to Support Informal Communication in Hospitals through the Development of a Tool Based on an Iterative Evaluation

David A. Mejia[1], Alberto L. Morán[2], Jesus Favela[1], Sergio F. Ochoa[3], and José Pino[3]

[1] Departamento de Ciencias de la Computacion, CICESE, Ensenada, Mexico
[2] Facultad de Ciencias, UABC, Ensenada, Mexico
[3] Departamento de Ciencias de la Computación, Universidad de Chile, Santiago, Chile
{mejiam,favela}@cicese.mx, alberto_moran@uabc.mx,
{sochoa,jpino}@dcc.uchile.cl

Abstract. The evaluation of groupware systems is considered a complex activity, mainly due to the impact that this kind of tools could have in work practices, the multiples variables that influences the use and evaluation of them, as well as the expensive cost of time and resources required for an *in situ evaluation*. These reasons have complicated the generation of a generic guide for evaluating this type of tools. Some researchers in groupware evaluation have highlighted the need to evaluate groupware tools, according to the context and characteristics of those organizations in which these tools would be deployed. Thus, in this paper we present a process to evaluate a tool that supports informal collaboration in hospital. Due to nature of hospital work and the difficulty of performing an in situ evaluation, our proposal implies a multi-phase evaluation process through the development lifecycle of the tool.

Keywords: Groupware evaluation, design concepts, informal communication.

1 Introduction

Software evaluation is considered a very complex activity, especially when these systems could have direct impact on the work practices in a set of potential users. In other words, groupware is traditionally considered to be difficult to evaluate because of the effects of multiple people and the social and organizational context [6].

The understanding of these effects could be expensive in time and human efforts, since the software must be deployed in the place of the organization and used by the intended user during a period of time. In hospitals, it is not recommended to deploy a groupware application without a preliminary evaluation of the possible effects on the medical staff, since it can impact negatively in communication or work patterns and, consequently, in patient care. By improving the techniques used prior to workplace evaluations, many problems can be eliminated early on, thus improving the efficiency of evaluation as it progresses into the workplace [6].

To this date there is not a generic guide for evaluating groupware in organizations, mainly due to the broad range of organizations and users that must be dealt with [2].

M. Kurosu (Ed.): Human Centered Design, HCII 2009, LNCS 5619, pp. 1013–1022, 2009.

Thus, what is recommended is to select an evaluation approach appropriate for the actual problem or research question under consideration [1]. For this reason, in this paper we present an approach to evaluate a groupware application for supporting collaboration in hospitals. It is based on an iterative software development and evaluation cycle with multiple phases, allowing understanding the user needs from the early stages of the development process.

The remainder of the paper is organized as follows. Section 2 presents a brief background about collaboration in hospitals. Section 3 presents our proposed procedure to assess the intention of adoption of these tools at development time. In Section 4 we illustrate an instance of a groupware evaluation using the proposed approach and finally, Section 5 concludes this paper.

2 A Process for Evaluating Groupware throughout the Development Lifecycle

As a way to evaluate the possible impact of a groupware tool in an organization without having to deploy it, we are proposing an approach that involves obtaining and taking into account the needs of the users throughout the design and development process. Based on this approach, the design of a groupware tool, therefore, progresses in iterative cycles of design, evaluation and redesign. Thus, the procedure used consists in the evaluation of three main phases (see Figure 1): i) the understanding of the problem, ii) the design of the tool, and iii) the usability of the tool. A brief description of these phases follows.

2.1 Phase 1: Evaluating the Understanding of the Problem

The first phase consists in determining if the researchers' perception of the problem corresponds to the users' perception of the problem. This requires evaluating whether the users perceive as a problem the breakdowns that researchers identify in the field study. A set of interviews and a focus group session with potential users could be used to validate the field study findings.

In this phase it is recommended to show the identified breakdowns through the use of scenarios illustrating real instances of observed collaboration [in the hospital]. In this stage the discussion must be elaborated around the perceptions of the users concerning the presented scenarios. Some of the following questions could help to this purpose:

• Have you seen or experimented similar situations?
• How frequently does this occur [in hospitals]?
• Do you consider this situation as an inconvenience during the performance of your daily work tasks?
• What do you do when this situation occurs?
• Do you have some proposal for addressing this situation?

After the focus group session ends, the researchers must do a qualitative analysis to identify those misunderstandings during the analysis of the problem. This analysis is very useful, because it could help to avoid spending resources in the development of tools that will not address real breakdowns, according to the users' perception.

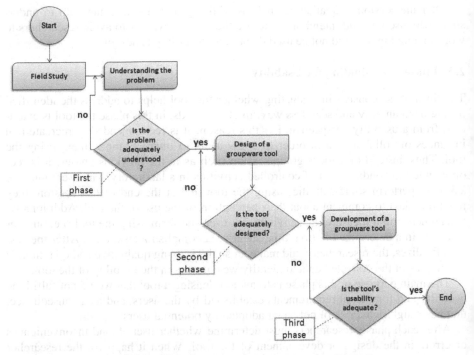

Fig. 1. Proposed iterative design and evaluation lifecycle

2.2 Phase 2: Evaluating the Design

The second phase consists in validating whether the designed functionality for the tool might help to achieve the goal of the system [augment informal communication] as identified in the field study (tool design), as well as evaluating (from the design perspective) if potential users would be interested on using the proposed tool. As in the previous phase, the use of scenarios could be very helpful to show users the envisioned support for the identified breakdown [in the hospital]. In this phase it is recommended to show two types of scenarios: the first type is aimed at showing users the problem or breakdown that is addressed. These scenarios could be very similar to those used in the previous phase. The second type instead of illustrating the problem, must envision how the proposed tool could address the specific problem.

The main focus in this phase must be to identify those weaknesses in the design of the proposed tool that could interfere with the user's intention of use. Thus, some of the following questions could help to accomplish this purpose:

- Does the tool address the problem adequately?
- Which are the advantages of this tool in comparison with the way in which users currently address this situation?
- Which are the disadvantages of this tool in comparison with the way in which users currently address this situation?
- What changes do users suggest for the tool to be useful in this situation?
- Do users intend to use this tool?

After the session, a qualitative analysis of the gathered data could help to understand why users would intend or not to use the tool, as well as to avoid researchers to work in a tool that would not be used due to errors during its design.

2.3 Phase 3: Evaluating the Usability

The third phase consists in evaluating whether the tool helps to address the identified problems in an easy and seamless way. In other words, in this phase the tool is evaluated from a usability perspective. For this reason, it is recommended to recreate real instances of collaboration in order to allow users to interact among them using the tool. Thus, instead of a focus group session such as in the previous phases, it is recommended to conduct a set of controlled activities in a laboratory. The users must be asked to perform some activities using the tool and at the end of this session, they must be asked to comment about their perception of the use of the tool. Additionally, researchers could measure the activities that users perform using the tool (i.e. number of clicks in a determined button, time spent to accomplish a task, etc.). After the session finalizes, the researches could make an analysis using qualitative and quantitative techniques of the collected data to identify weaknesses in the usability of the tool.

The main purpose of this phase is to avoid releasing a tool that would not fulfill the usability and functional requirements established by the users, and as a consequence, that has a significant risk of not being adopted by potential users.

After each phase, researchers must determine whether users found inconveniences or errors in the design or development of the tool. When it happens, the researcher must address these findings and restart the evaluation process on the current phase. Thus, at the end of this design and evaluation process, researchers will be able to provide a set of design insights and a tool informed with requirements that emerged directly from observations of the real [hospital work] situations, and validated by the potential users.

3   Illustrating the Evaluation Process through the Development   of SOLAR

To illustrate the use of the proposed evaluation process, in this section we report the main results of the evaluation of SOLAR during its development, a tool to support informal communication in hospitals that integrates an ensemble of five specialized services that includes i) location estimation of workers in the hospital displayed on the floor plan on the Smartphone, ii) awareness of relevant events and nearby colleagues and their social context, that allows them to select an appropriate moment to initiate an interaction with others, iii) transfer of files and URLs among heterogeneous devices just by clicking the right mouse button and choosing the target device, iv) screen sharing of any device in the vicinity (e.g. PDA, PC or public display) on a handheld computer, and remotely sharing the control of the device with its owner and/or other users, and v) notification of creation or modification of information, and storage of this information in a user's mobile device during a collaboration episode (capture of collaboration outcomes). [4, 5].

According to [7], the deployment of groupware tools without an adequate evaluation is not recommended, mainly because unevaluated systems tend to be unsuccessful and may fail to consider the context, stakeholders and contain errors after deployment. However, the evaluation of tools like SOLAR represents a challenge, since the evaluation process is dependent on several variables such as those imposed by the application domain (i.e. the hospital), its multiple users (i.e. physicians, medical interns and nurses), and their interaction with the provided services (i.e. how the tools affects work practices) [5]. For this reason, we decided to evaluate SOLAR through its development lifecycle as proposed previously.

3.1 Phase 1 Evaluation Results: Understanding the Problem

During the analysis of the results of the field study, we identified some breakdowns before, after and during communication that include lack of awareness of the presence of mobile colleagues, waste of opportunities for collaboration, interruption of interactions in order to gather information artifacts necessary for collaboration, and loss of information generated during collaboration.

We exemplify these breakdowns through the use of scenarios like the following:
A physician is in the Internal Medicine office when a medical specialist arrives and asks him about an X-Ray image he is consulting.

> *[Physician] What do you think about this?*
> *[Medical Specialist] (Looking the X-Ray results) Mmm, I think that he has (disease's name).*
> *[Physician] Are you sure? Because he has (medical specialist explains the symptoms of patient).*
> *[Medical Specialist] Yes, I do. I had a patient with similar symptoms.*
> *[Physician]Are you sure? I think (there) could be some differences between the symptoms of these patients.*
> *[Medical Specialist] Well, let me review the health record of my patient and I will get back to you later.*

Later, the physician and the medical specialist met opportunistically in the hallway and they restart their previous discussion.

> *[Medical Specialist] (Showing the physician a health record) Look, these are the symptoms of the patient I mentioned. His disease has the same features as those of your patient.*
> *[Physician] Ok, you are right. But, which is the treatment for this type of disease?*
> *[Medical Specialist] (Medical specialist explains the treatment for the patient). I have a book in my consulting room. Please go there later, I can lend it to you.*

In this phase we perform the evaluation with four physicians in a focus group session. We exemplified the breakdowns through three scenarios that we showed to them. It was not necessary to iterate in this first phase evaluating session, since the comments and recommendations of these physicians where only aimed at changing the medical terminology used in the scenarios.

3.2 Phase 2 Evaluation Results: Validating design

Based on the breakdowns identified in hospital collaboration, as well as on a set of design insights identified during the analysis of the data collected in the field study, we designed the functionality of SOLAR. Then, aimed to estimating the users' intention to use SOLAR through the evaluation of its functionality, we verified that SOLAR does provide support for each of these breakdowns.

The design insights that inspired SOLAR functionality include:

Use of devices that allow workers to initiate collaboration while on-the-move. Medical workers are not only constantly moving from one operation center to another, but they actually have meaningful encounters and perform work while in the hallway or bed ward. Thus, SOLAR was designed to provide support for hospital workers to allow them to collaborate when they are on the move as required by the actual situation.

Use of devices that allow workers getting into collaboration while on-the-base. Hospital workers fill administrative forms, write medical notes or analyze medical evidence in specific places that we have called "base locations" (e.g. nurse pavilion, consulting room, etc.). Furthermore, we found that hospital workers have meaningful interactions and collaborate with others colleagues when they are at those base locations. For this reason, SOLAR should be executed in devices that allow workers to collaborate while they are in their base location.

Easy access to information sources. Physicians and medical interns need to share or exchange information to accomplish the objective of their interactions. Currently, they use physical artifacts to do this, like medical records and X-Ray images, among others. For this reason, SOLAR allows them to have easy access to digital information sources anytime, anywhere.

Use of devices and services that allow interactions based on medical evidence. Physicians and medical interns often discuss clinical cases. Sometimes, they use and share information stored in physical artifacts (e.g. medical records, X-Ray images and books) and/or in electronic devices (e.g. document stored on PC, PDA or digital library) in order to explain their opinions to others. For this reason, SOLAR was designed to provide mechanisms that allow the use and manipulation of devices and information that help to enrich discussions anyplace-anytime medical workers need it.

Tracking of People. Hospital workers often interrupt their work and also walk more than 60 feet in order to locate a particular colleague. For this reason, SOLAR allows hospital workers to be aware of the presence of others, their identity and location, as well as suggesting an adequate moment for initiating collaboration with them.

Awareness of the opportunities for collaboration. In informal interactions there are two main elements that trigger people to initiate collaboration with others: the availability of a communication or interaction channel (e.g. physical proximity) and the interest or need of at least one of the participants to collaborate with the other [20]. For these reasons, SOLAR allows hospital workers to be aware of the other's presence, identity and location, as well as of an adequate moment for getting into collaboration with them.

Seamless information access and sharing. One of the purposes of informal interactions is sharing or transferring documents to others. Thus, SOLAR allows hospital workers to easily share and exchange information among them.

Capture of collaboration outcomes. Information generated during collaboration is rarely recorded in persistent artifacts. Since information could be useful in future activities or interactions, SOLAR provides mechanisms that allow users to capture the collaboration outcomes, in order to facilitate the reuse of information generated while they are collaborating.

Based on these design insights, we designed and developed five main services that conform the functionality of SOLAR: i) location estimation, ii) awareness for potential collaboration, iii) seamless information transfer, iv) remote control of heterogeneous devices and v) capture of collaboration outcomes

In order to better illustrate the features of our identified design insights, we presented the next scenario to five medical interns. This scenario projects how the situation represented in the scenario of the previous phase could be performed using a tool conceived based on the identified design insights:

A physician is in a patient's room when she realizes through a location estimation mechanism in her PDA that the medical specialist is walking down the corridor – which is close to her position and decides to ask him about an X-Ray image she was consulting in a public display.

        [Physician]   What do you think about this?

    [Medical Specialist]   (Looking at the X-Ray results) Mmm, I think that he has (disease's name).

        [Physician]   Are you sure? Because he has (medical specialist explains the symptoms of patient).

    [Medical Specialist]   Yes, I do. I had a patient with similar symptoms.

        [Physician]   Are you sure? I think (there) could be some differences between the symptoms of these patients.

    [Medical Specialist]   Let me show you the health record of my patient.

The medical specialist access the patient's health record through his PDA and transfers (seamlessly) the document to a public display.

    [Medical Specialist]   (From their PDA, using a tool for remotely controlling the public display, he explain the physician an X-Ray of the patient) Look, these are the symptoms of the patient I mentioned. His disease has the same features as those of your patient.

        [Physician]   Ok, you are right. But, which is the appropriate treatment for this type of disease?

    [Medical Specialist]   (After explaining the treatment for the patient the medical specialist accesses his office's computer through his PDA to retrieve information from a medical guide related to the medication and to the patient's condition, and transfers it to the physicians' PDA). I have this information that could be useful to you.

During the course of this interaction, the voice and textual comments are stored in order to be accessed when the hospital workers require it.

So, in this scenario we show how the services designed into SOLAR (i.e. location estimation; the use of devices that allow workers to initiate collaborations while on-the-move; the use of devices and services that allow interactions based on medical evidence; tracking people; seamless information access and sharing; and capture of collaboration outcomes) synergically work in order to augment a co-located informal interaction.

During this phase, medical workers suggested some changes in the capture of collaboration outcomes. According to our process, we made the changes to the system and conducted an additional second phase evaluation session.

3.3 Phase 3 Evaluation Results: Validating Usability

To illustrate the third phase of the evaluation, in this section we report the results of the usability evaluation of one of SOLAR services, the *Remote Control of Heterogeneous Devices*. This evaluation session was conducted in a meeting room with a public display and several personal computers. The subjects of the study were six graduate students, some of them having already used a PDA and others not. Before we started the evaluation we gave each user a PDA, simulating indeed, a real co-located meeting environment saturated with heterogeneous devices. The complete results of this evaluation are discussed elsewhere [3].

The evaluation required two subjects to collaborate in performing three different tasks, including i) checking email and filling a survey with/without the echo functionality (completed individually), ii) creating a picture slideshow accompanied with a soundtrack with the aim of making a promotional clip for a city (two users, each subject played a particular role: as a sound designer, choosing one song among ten that were available; or as a graphic designer choosing eight pictures from the thirty pictures available), and iii) following a red spot in synchronization with the tick of a metronome set at a very low speed (the priority was to follow the metronome instead of the precision, aimed at estimating the precision and the speed of the application). Thus, we randomly grouped our subjects in pairs conducting an evaluation session with each group.

We wanted to capture qualitative data and quantitative measurements of the users' inputs. Therefore, we videotaped each evaluation session capturing the users' interaction with the PDA and the verbal communication exchanges they were having while performing each task. In addition, the screen of the public display was captured using the Morae[1] software which registers all user inputs for statistics and analysis.

After the end of this session, we identified that subjects were positive about the application. The main advantages the subjects highlighted were all related to the collaborative features (pointer and control services) allowed by the application.

Additionally, we provided a list of seven improvements and we asked the subjects to choose three of them which, in their opinion, need to be attended with higher priority (see table 1).

---

[1] Available at http://www.techsmith.com/morae.asp

Table 1. Improvements for the Remote Control of Heterogeneous Devices' service of SOLAR

Improvement	*Choices*
A less jerky movement	5
Improved speed	2
Higher precision	6
Right click ability	1
Click and hold ability (to move a window for example)	3
New pointer shape	0

The results of the preliminary evaluation of the system show that the users experienced some difficulty moving the cursor in the remote screen, yet, they found the application to be useful in facilitating co-located collaborations. In addition, users seemed to require additional experience with the tool as we observed that they improved their precision and speed towards the end of the exercises. According to the proposed approach, these improvements must be introduced before repeating third phase evaluation one more time.

4 Conclusions

The evaluation of groupware tools is not a simple task, because of the multiple factors and variables that influence and define the organizations where these tools are deployed. The main mechanism to evaluate these tools would be through the deployment of the tool in the organization, followed by an *in situ* evaluation. This kind of evaluation is expensive in terms of time and human resources; perhaps its main disadvantage is that errors during the design stage would be detected very late (deployment time). Moreover, even with the deployment of the tool, there is not a specific process to evaluate a groupware tool *in situ*,

In this paper we present a process for evaluating groupware applications throughout the development lifecycle. This procedure consists of three phases that evaluate i) the understanding of the problem, ii) the design of the tool, and iii) the usability of the tool. We illustrate the use of this approach through the evaluation of SOLAR, a tool developed for supporting informal communication in hospitals. We found this procedure very useful, because it helped us to identify specific errors during the analysis and design phases. Also, based on the techniques used, this approach might help predicting how users would use and accept the proposed tool.

Acknowledgements

This work was partially supported by LACCIR under grant No. RFP0012007.

References

1. Greenberg, S., Buxton, B.: Usability Evaluation Considered Harmful (Some of the Time). In: Proceedings of the SIGCHI conference on Human Factors in computing systems, Florence, Italy, April 5-10 (2008)

2. Grudin, J.: Why CSCW applications fail: problems in the design and evaluation of organization of organizational interfaces. In: CSCW 1988, pp. 85–93. ACM Press, New York (1988)
3. Markarian, A., Favela, J., Tentori, M., Castro, L.A.: Seamless Interaction among Heterogeneous Devices in Support for Co-located Collaboration. In: Dimitriadis, Y.A., Zigurs, I., Gómez-Sánchez, E. (eds.) CRIWG 2006. LNCS, vol. 4154, pp. 389–404. Springer, Heidelberg (2006)
4. Mejia, D.A., Morán, A.L., Favela, J.: Supporting Informal Co-located Collaboration in Hospital Work. In: Haake, J.M., Ochoa, S.F., Cechich, A. (eds.) CRIWG 2007. LNCS, vol. 4715, pp. 255–270. Springer, Heidelberg (2007)
5. Mejia, D.A., Morán, A.L., Favela, J., Tentori, M., Markarian, A., Castro, L.A.: On the Move Collaborative Environments: Augmenting Face to Face Informal Collaboration in Hospitals. e-Services Journal 6(1)
6. Pinelle, D., Gutwin, C.: A review of groupware evaluations. In: WETICE 2000: 9th International Workshop on Enabling Technologies, pp. 86–91. IEEE Computer Society, Los Alamitos (2000)
7. Herskovic, V., Pino, J., Ochoa, S., Antunes, P.: Evaluation Methods for Groupware Systems. In: Haake, J.M., Ochoa, S.F., Cechich, A. (eds.) CRIWG 2007. LNCS, vol. 4715, pp. 328–336. Springer, Heidelberg (2007)

Understanding Activity Documentation Work in Remote Mobility Environments

Alberto L. Morán[1] and Raul Casillas[2]

[1] Facultad de Ciencias, UABC
[2] Ciencias de la Computacion, CICESE
alberto_moran@uabc.mx, casillas@cicese.mx

Abstract. Activity documentation is a critical part of the work of many professionals. Documents are used as a means to store personal information, remind things to do, convey and generate new meaning, and mediate contact among people. In this paper, and based on the results of an observational study, we propose a model of how activity documentation work in remote mobility environments is performed. Further, based on this model and on some identified issues that remote mobility workers face while performing activity documentation work, we propose a set of design insights that designers and developers of support systems could use to inform their designs and developments. These results allow designers and developers not only to support a single activity documentation work phase if so desired, but also to envision the creation of comprehensive services for activity documentation work throughout its complete lifecycle in a seamless, effortless and secure manner.

Keywords: Activity documentation work, remote mobility environments, activity documentation lifecycle, design implications.

1 Introduction

Documents are extensively used in the execution of professional's work and to share information with others. These professionals are specialists in the subject of work and are characterized by putting their intellectual skills, learned in systematic education and through experience, to work in organizing their work, making sense of things, and passing judgments. Hertzum [1] identifies six roles that documents play in professional work, including their use as i) personal files, ii) as reminders of things to do, iii) to share information with someone, yet withholding it from others, iv) to convey meaning, v) to generate new meaning, and vi) to mediate contacts among people. Thus, activity documentation is a critical part of many professionals' work in various application domains. Even though the application domains are different, they share basic features that could be used to characterize activity documentation.

Let us consider an approach to the classification of application domains from the perspective of "worker's mobility" and "scale of the environment" where the activity occurs. Worker's mobility refers to the property of actors to move from one location to another to get access to a resource (e.g. knowledge, artifact, or subject) to actually performing the work; while the scale of the environment refers to the property of an

M. Kurosu (Ed.): Human Centered Design, HCII 2009, LNCS 5619, pp. 1023–1032, 2009.

actual work setting that defines the spatial bounds of that environment. Both dimensions of this classification are intimately related, as actual mobility of an actor will be bounded by the limits of the environment. Particularly, we are interested in the following types of domains.

Workers with micro mobility in "at-hand" bounded environments. Examples of this domain include people working at a meeting room, classroom or office. Luff and Heath [2], refer to micro-mobility as the way in which "an artifact is mobilized and manipulated for various purposes around a relatively circumscribed, or "at-hand", domain". Here we will de-emphasize the definition of micro-mobility on the mobility of the artifact and center it into the mobility of the individuals carrying and manipulating the artifact in the circumscribed places.

Workers with local mobility within an environment with a number of workspaces. Examples of this domain include people working at an ensemble of offices and rooms, such as in hospital work. In this latter case, hospital workers move around to specialized places (e.g. bedrooms, meeting rooms and offices) that are distributed in space, and which hold specialized resources (people, artifacts and knowledge). Bardram and Bossen [3]) define local mobility as occupying "the intermediate space between working together over distance on the one hand and working face-to-face in an office or control room on the other". Further, they consider that it takes place "in cooperative settings where actors constantly are on the move to get ahead with their work". In this case, we will highlight the limited and well-defined nature of the environment, enclosing a number of defined spaces or activity centers [4], and where the local mobility of the individuals takes place.

Workers with remote mobility in wide open environments. Examples of these domains include people performing fieldwork, such as field biologists. These workers travel to particular sites (field), and once there, they move throughout the place (even walking considerable distances) in order to perform their work. This kind of mobility is referred in [2, 5] as remote mobility, although they don't emphasize the extension of the environment; the main requisite was to move around different physical locations. This type of work is characterized by periods of micro- and local- mobility work at a local setting, and periods of remote mobility work in the field.

From the previous definitions, the "mobility" and "scale" features of these workers and environments may shape the type of technology that is, or could be, available to support activity documentation in those settings.

In this work, we are interested in investigating how Information and Communication Technologies (ICT's) could bring elements of a solution to the problem of activity documentation in remote mobility environments.

The rest of this paper is organized as follows. Section 2 describes the observational study performed, while sections 3 and 4 present the results of understanding Activities in Remote Mobility Work and Activity Documentation work, respectively. Section 5 continues with some of the identified issues faced by workers at remote mobility environments while performing activity documentation. Section 6 presents some designs implications for the development of systems that aim to provide support for activity documentation work in remote mobility environments. Finally, Section 7 presents our conclusions.

2 Observational Study

Our research interest is in specific practices that compose activity documentation work in remote mobility environments. In order to provide the best register and details of the activities, remote mobility workers work individually and collaboratively to ensure that all the information regarding the performed activities is properly documented. This is also why efficient and accurate recording, organization and later access to the information about the performed activities are so important.

Methodology. We performed an observational study of a group of biologists executing an environmental impact study on the creation of a new ecological park in town, to acquire an understanding of how these remote mobility workers document their activities (e.g. how information is gathered, which media and artifacts are involved, how information is organized and how it is later accessed, among others).

Participants. Participants in our study were sixteen biologists (one teacher and fifteen students) of a master's program on the Management of Ecosystems of Arid Zones (MEZA) of the University of Baja California (UABC). There were 6 male and 9 female participants. Their ages ranged from 23 to 51 years. For clarity we will refer to the participants as biologist regardless of their being the teacher or a student.

Setting. Observations were carried out on the places where participants performed their work, both in the field (fieldwork) and at the university facilities (pre- and post-fieldwork). Fieldwork was conducted at the San Miguel Creek, the place where the construction of the ecological park is proposed. Figure 1 shows the layout of the facilities where most of the workers' local activities were performed. These include, a classroom and a meeting room, where most group work occurs before and after a fieldwork trip; the teacher's office, where the teacher worked individually and students joined her for informal discussions and consultation about the project; and the two students' cubicle areas, where most of the students' individual work occurred before and after meetings and fieldwork.

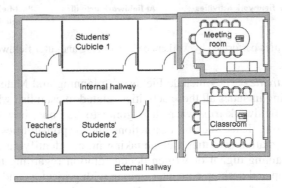

Fig. 1. Layout of the facilities where local-mobility work was performed

Procedure. To obtain a through understanding of the current activities and documentation practices of these workers, an observational study of their activities during the execution of the project was conducted. It combined direct and indirect observation of work at the university facilities during group and individual work (e.g. formal meetings and lectures, informal hallway meetings, and information gathering, processing and deliverable generation activities, among others), shadowing of three teams during fieldwork, and informal interviews. Seven observation sessions were conducted (21 hours) during pre- and post-fieldwork activities and 3 observation sessions were conducted (8 hours) during fieldwork activities. Observational data were recorded using a video camera, pen and paper. In addition to the observational videos and notes, we collected samples of documentation artifacts, such as notes, photos, and written deliverables that were used and created by workers during the study. We also asked them, whenever possible, for descriptions and explanations of their use of the artifacts, and how they used them during their work documentation activities. Particularly, we were interested in identifying the kind of artifacts they used or created during their work sessions, the way they prepared, transported and used them before, during and after fieldwork in a remote mobility environment.

3 Understanding the Activities of Remote Mobility Work

The main activities of remote mobility workers identified in the study are shown in Figure 2. They are arranged following a sequential life-line that is more or less followed (and repeated) each time fieldwork activity in remote mobility environments is conducted. These activities are also arranged to denote the activities that are performed before (pre-), during (at-) and after (post-) fieldwork activities. A brief description of them follows.

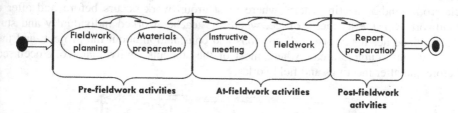

Fig. 2. Activities performed by workers before, during and after fieldwork activities

Pre-fieldwork activities. They include Fieldwork planning and Materials preparation. Fieldwork planning includes all those activities related to establish what is going to be done during the fieldwork trip, where it is going to be done, how it is going to be done, and by whom. Thus, specific instructions and responsibilities are given to the workers. Documentation activities include taking notes, identifying topics for further reading, and obtaining digital or written documentation regarding the activity to be performed. Laptop and desktop computers, along with projectors and USB devices are usually utilized to perform these activities. Fieldwork planning activity usually occurs at the classroom or meeting room during a group session.

Materials preparation includes all those activities related to gather the required material and information resources that will be used during the fieldwork trip. Concerning documentation activities, information resources are gathered in paper and in digital form from personal copies of books and journals, from the Internet, or the central library. This information is brought in paper form, either printed or hand-copied. Regarding documentation artifacts, they brought notebooks, pen and pencil for note-taking, analog and digital cameras for video and picture taking, and even tape and digital audio recorders. Materials preparation activities usually occur at the personal office/cubicle, the library, or even at the participant's home.

At-fieldwork activities. They include Instructive meeting and actual Fieldwork. Instructive meeting, as it name implies refers to an in-site meeting where up-to-the-minute information concerning the planned activity is provided and highlighted, including coordination and safety issues not previously emphasized. It is performed as a group activity. Concerning documentation activities, information resources are brought into play, distributed and stored for at-fieldwork use; note-taking is the primary method of recording the instructions and coordination issues given.

Fieldwork, as its name implies refers to actually performing the planned activity at the visited site, i.e. performing the work and documenting it. Fieldwork can be done individually or as a team. Regarding documentation activities, users take notes, annotate the materials they brought, take pictures and video of the studied subjects, and record audio of interesting events (e.g. characteristic sound of a bird).

Post-fieldwork activities. They include Report preparation and delivery. It refers to the actual organization and sense-making work on the fragmented information in order to create deliverables for the fieldwork activities. It requires individual and collaborative efforts, as each party has to contribute the part they are responsible of. Thus, it is performed both at the individual offices/cubicles and at the meeting rooms. Documentation activities include the transcription of the raw collected data, its analysis and interpretation, sharing and merging of the fragmented information (data, notes, pictures, etc.), and the writing of actual drafts and final reports.

4   Activity Documentation Work in Remote Mobility Environments

From the previous description of activities of workers in remote mobility fieldwork, we identified the flow of activities that is followed for Activity Documentation Work. This is illustrated in Figure 3.

This flow is organized following a similar pattern to that of the main activities of remote mobility work, with Pre-, At- and Post-Activity Documentation Work phases.

In the *Pre-Activity Documentation Work* phase workers basically gather initial information related to the activity that is going to be performed. Based on this initial information, they capture and search for, select and retrieve information relevant to the planned tasks of the activity, and finally, they store, organize and prepare documentation that they could take along with them to support their actual Activity Documentation Work while at the remote mobility environment.

During the *At-Activity Documentation Work* phase workers basically retrieve and consult information on the planned tasks (such as instructions and references), and capture information regarding the activities performed at the remote site (notes, pictures, audio and video, among others). The fragmented nature of the information on this phase should be highlighted.

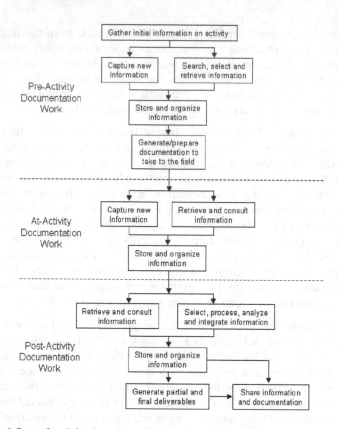

Fig. 3. General flow of activity documentation work as performed by workers in remote mobility environments

Finally, during the *Post-Activity Documentation Work* phase workers need to retrieve information from the two previous phases, transcribe the data that so requires it, select specific information to be processed, analyzed and integrated to generate additional written documentation. It should be highlighted that all data, information and documentation gathered and generated in this phase are usually shared with the other participants of the activity (and with people outside the group, e.g. stakeholders and external advisors), so that actual deliverables could be generated and submitted.

5 Issues Faced by Workers Performing Activity Documentation Work in Remote Mobility Environments

In addition to the identified general activity documentation practices, the study also allowed us to identify a set of issues that affect these documentation practices for the particular domain. Some of these issues are briefly described next.

Imbalance in the effort spent in the phases of activity documentation work. From the study we noticed that the effort required for activity documentation work is different among the three identified phases. Pre-Activity documentation required in average nearly a fifth (21%) of the total effort, while At-activity documentation required nearly one third of it (30%), and Post-activity documentation required nearly half of the total effort (49%). According to the observations a possible explanation for this has to do with the nature of the documentation work in each phase.

Need to access information and documentation products of previous phases. From the study it was clear that remote mobility workers needed to document their activities, as well as to prepare documentation to support their documentation activities across the multiple phases. These information and documents are gathered, captured and generated in one phase and are used as inputs to the processes of another phase. Thus, information and documents are accessed through the boundaries of the three identified phases (Pre-, At- and Post-Activity Documentation Work).

Information capture using non-digital media. From the study it became evident that the participants still heavily depend on non-digital media (e.g. paper and pencil, non-digital cameras) to capture and store information regarding their activities. Yeh et al. [6] argue that this is due to affordances of the non-digital media (e.g. paper notebooks that are portable, lightweight, readable outdoors, robust to harsh conditions and have infinite "battery life"). This, although convenient while performing At-Activity Documentation Work on a remote mobility environment, introduces an ensemble of limitations that become more evident at a later phase of the process, i.e. during Post-Activity Documentation Work phase. One of the main limitation of capturing information on non-digital devices is that in order for information to be easily shared with others and included into own and others reports, it has to be re-captured (e.g. transcribed). This introduces additional work overload to the documentation process. Participants reported that 49% of the effort invested into Documentation Work is spent during the Post-Activity Documentation phase. Further, as stated by one of the participants *"When capturing information using [non-digital media, such as] paper and pencil, we miss documenting fine details of the information, such as when taking notes about instructions from the teacher at the Instructing Meeting or at Fieldwork when trying to capture verbal interactions"*.

Fragmented nature of the information captured in different media and lack of effective mechanisms for linking related information. During the study, we also noticed that some information and documents that are captured or generated during Activity Documentation Work basically represent entities that are "isolated" from the other elements captured on the same media, and even more from those captured or generated on another media. Let us take for example the case of notes and [digital] photos. On one hand, notes are usually written (and stored) in a notebook or on a piece of paper (e.g. notebook). Photos, on the other hand, are taken (captured) and stored in a film-roll (or in a digital memory). Both notes and photos are sequentially organized, which allows finding series of related notes or photos. On the contrary, this sequential organization poses the problem of how to determine whether two sequential notes or pictures relate to the same or different object of study. A further problem relates to being able to link materials from different sources, such as notes and photos. Although participants might remember enough contextual information as to be able to associate specific photos with the right notes and vice versa, it is not unusual that photos and notes end up paired in an incorrect manner. A participant stated it in this way *"Sometimes, while assembling the report from the contributions of all team members, we notice that information from notes that do not correspond to pictures have been associated together"*.

Need to share the collected information among colleagues, and to collaborate in the creation of deliverables. Also, in the study we observed that information and documents are gathered or generated in different phases, most notably at the At- and Post-Activity Documentation Work phases. Even though information and documents were captured or generated by individuals or small teams, they are shared among all members of the project team, and used to document and enrich the joint production of deliverables. While in the local work area, information and documents are mainly shared by means of the facilities installed in the environment (e.g. email, and instant messaging file transfer features) or through face-to-face exchanges of USB disks and actual documents. While in the remote mobility environment, lacking a supporting infrastructure, information and documentation exchanges are limited to printed documents, handwritten notes and verbal exchanges. Regarding the joint generation of documents and deliverables, participants go beyond information exchange. In these meetings they require to establish coordination protocols, such as for the assignment of authoring responsibilities on documents and sections of documents, for the integration of partial and complete documents, or the actual production of deliverables. One example of such a protocol was the formation of three smaller teams, and the designation of a scribble within each one of these teams. The scribble entered what other participants were saying, and who was in charge of integrating the diverse fragments and sections provided by other members of the team.

6 Design Implications

In our study, we found that Activity Documentation Work in Remote Mobility Environments involves managing an ensemble of information, documents, and artifacts in several locations. Information and Communication Technologies (ICT's) could be introduced in order to supplement current practices in a way that enables or improves actual activity documentation work. Based on the proposed model and on the identified issues and features of Activity Documentation work, we identified a set of design insights for the development of applications that provide support for activity documentation work in remote mobility environments (see Table 1). A brief description of the implications of introducing supporting technology follows.

Support for fixed and mobile modes of work (insight 1). Designers and developers should consider the use of both fixed and mobile technologies, so that appropriate support is available at the right time and the right place. For instance, although PC's, laptops and projectors that are available at the students' cubicles, teacher's office, and classroom are adequate for activity documentation work during the pre- and post-activity documentation phases, these technologies won't be adequate during fieldwork at the At-Activity Documentation phase due to their specific features (e.g. weight, energy-consumption, etc.). Mobile and wearable technologies would be more appropriate for this.

Table 1. Identified insights for the provision of support for activity documentation work in remote mobility environments

Design Insights
1.- Support for fixed and mobile modes of work, according to the needs of activity documentation work phases.
2.- Seamless access to information across the phases and modes of activity documentation work.
3.- Digital capture of activity documentation information, so that even transient information (e.g. conversations between colleagues) could be persistently documented.
4.- Automatic linkage of different pieces of related information, even at capture time, to avoid additional classification work in the Post-activity documentation phase.
5.- Information sharing among colleagues, and collaborative creation of individual and group deliverables, regardless of the time and place where participants are.

Seamless access to information across the phases and modes of activity (insight 2). In a similar fashion, designers should consider the use of services that allow accessing information across the three phases of activity documentation. For instance, information gathered and generated during the pre-activity documentation work phase, is used trough out the whole process to the post-activity documentation work phase. In this case, dedicated messaging or file-transfer services among participants could suffice. However, a shared repository which provides support for intermittent or disconnected operation would be a better option as discussed later.

Digital capture of activity documentation information (insight 3). Reports and deliverables are generated in digital form, based on information gathered or generated during the three phases of activity documentation work. Designers and developers should consider the provision of services to gather, generate, organize and access information of several types, including notes, pictures, audio and video. Further, these services should manage the information in digital form from capture to integration into a final report or deliverable. The main idea behind this is trying to reduce the overload of having to re-capture (e.g. transcript or scan) the information that is not captured in digital form (e.g. handwritten notes).

Automatic linkage of different pieces of related information (insight 4). Information capture by different people, from different sources, and at different times generates information that is very fragmented. Organizing and making-sense of this kind of information usually requires an additional effort, often made during the post-activity documentation work phase. Designer and developers should consider the provision of a mechanism (e.g. GPS or semantic tagging), either manual or automatic, that allows linking related information (even) at capture time, when contextual information about the relation of the objects is certain.

Information sharing among colleagues, and collaborative support for the creation of individual and group deliverables (insight 5). Being informed and having access to all information sources across the whole lifecycle of Activity Documentation Work is vital for the creation of deliverables. Further, the success of the activity heavily depends on the participants' ability of integrating the gathered and generated information into final reports for the stakeholders of the project. For this reason, designers and developers should not only consider the inclusion of a shared repository, messaging and file transfer features to achieve this, but also, the inclusion of features of a system for computer supported collaborative writing, which assists them in the three different axes of collaborative work: communication, coordination and production of the expected deliverables.

7 Conclusions

Activity documentation is a critical part of the work of many workers in various application domains, including those that perform activity documentation in Remote Mobility Environments. In this work, aiming at understanding the practices and issues faced by activity documentation workers in these environments, we present a characterization of these activities, and present a process model for it. Also, we identified and described a set of issues faced by workers in this domain. Finally, based on these results, we identified a set of design insights and implications for the development of applications for activity documentation on remote mobility environments. The importance of these results rely in that they allow designers and developers to concentrate on essential features of an application, not only to support a single activity documentation work phase if so desired, but also to envision the creation of integrated (fixed and mobile) services for information capture, classification, retrieval and processing. Further, these services should be available and interoperable throughout the complete lifecycle of activity documentation work and this in a seamless, effortless and secure manner.

References

1. Hertzum, M.: Six Roles of Documents in Professionals' Work. In: Proc. ECSCW 1999, Copenhagen, Denmark, September 12-16. Kluwer Academic Press, Dordrecht (1999)
2. Luff, P., Heath, C.: Mobility in Collaboration. In: Proc. CSCW 1998, Seattle, Washington, USA, November 1998, pp. 305–314 (1998)
3. Bardram, J., Bossen, C.: Moving to get aHead: Local Mobility and Collaborative Work. In: Proc. of the Eighth ECSCW, Helsinki, Finland, September 2003, pp. 355–374 (2003)
4. Mejia, D.A., Morán, A.L., Favela, J., Tentori, M., Markarian, A., Castro, L.A.: On-the-move Collaborative Environments: Augmenting Face-to-Face Informal Collaboration in Hospitals. eService Journal 6(1), 98–124 (Fall 2007)
5. Bellotti, V., Bly, S.: Walking away from the Desktop Computer: Distributed Collaboration and Mobility in a Product Design Team. In: Ackerman, M.S. (ed.) Proc. CSCW 1996, Boston, MA, USA, November 16-20, pp. 209–218 (1996)
6. Yeh, R., Liao, C., Klemmer, S., Guimbretière, F., Lee, B., Kakaradov, B., Stamberger, J., Paepcke, A.: ButterflyNet: a mobile capture and access system for field biology research. In: Proc. of the SIGCHI Conf. Human Factors in Comp. Systems, Montréal, Québec, Canada, April 2006, pp. 571–580 (2006)

Human Factor's in Telemedicine: Training Surgeons by Telementoring

Dina Notte[1], Rym Mimouna[2], Guy-Bernard Cadiere[2], Jean Bruyns[2],
Michel Degueldre[2], and Pierre Mols[2]

[1] Ergodin. Sul'Wez, 1. 5524 Gerin. Belgium
[2] Chu Saint Pierre, 322, Rue Haute, 1000 Brussels. Belgium
ergodin@skynet.be, rymmimouna@gmail.com,
pierre.mols@stpierre-bru.be

Abstract. This study aim at evaluating several cases and communication options between a surgeon-mentor and a surgeon-mentoree within a telementoring context. The simulation proved that surgical telementoring was technically feasible with a satellite link simulated at 384 Kb/s, 512 Kb/s, and 768 Kb/s and with a transmission delay of 700 ms. Results show the limitations of image communication and its acceptability for telementoring and point out an important learning process for the mentoree. This seems to indicate a more efficient learning curve than with the standard training techniques (live surgical demonstration by companionship or videoconferencing), or with the use of laparoscopic simulator and surgical robots.

Keywords: Surgery, telementoring, simulation, training, usability, cognitive approach, ergonomics.

1 Introduction

Laparoscopy surgery is a new technique without contest in its first choice for some surgical indications or its extension to new indications. The view of the surgical site on a television screen, the basis of laparoscopic surgery, allows for easy transmission of image due to its intrinsically electronic nature. Laparoscopic surgery adapts particularly well to telementoring as the two surgeons (the expert and the 'counselee') have the same cognitive references (2D vision), and the live image transfer does not give unreachable problems. Therefore, the way is now open for the transfer not from teacher to student, but of the expertise itself.

While it has widely spread for ten years in industrialised countries, the concept of invasive minimal surgery has not expanded yet in developing countries. Introduction of laparoscopy in the Third World raises two problems: one is linked to acquiring communication system and imagery equipment and the other to teaching the technique due to the lack of finances. That 's why telementoring has been considered in the present study. Secondly a satellite communication system has been selected because ADSL, ISDN and optical fibre are not at disposal in most countries in Africa at present time.

M. Kurosu (Ed.): Human Centered Design, HCII 2009, LNCS 5619, pp. 1033–1041, 2009.
© Springer-Verlag Berlin Heidelberg 2009

The simulation plan aims at:

- evaluating several cases and communication options between a surgeon-mentor and a surgeon-mentoree within a telementoring context;
- testing several bandwidths for telecommunications;
- estimating the consequences of satellite delay and electronic perturbations ('bursts') on telementoring.
- appreciating the efficacy and efficiency of communications in telementoring conditions via satellite;
- highlighting the limitations of image communication and its acceptability for telementoring.

2   Methodolgy

2.1 The Simulation Device

To simulate telementoring most realistically in North/South videoconferencing, an expert surgeon, located in a production room upstairs, guides a beginning surgeon (say operating in Ngaliema) in carrying out surgical operations in operating room 5 of CHU Saint-Pierre.

2.2 Parameters to Estimate

During the simulation the two axes to test and estimate are:

1. **The technical axis** (flux variation): different flux modalities will be tested to evaluate their influence on image and sound qualities. The simulation will demonstrate the measure at which certain fluxes become insufficient to ensure good condition telementoring.
2. **The communication axis** (delay): delays linked to the used technology (codecs, satellites) will be simulated to evaluate their influence on communication between the telementor and the mentoree surgeon operating. The simulation will illustrate the measure at which the delays bias or inhibit surgeon interactions.

2.3 Tests Description

The tests happen in three phases:

Phase 0: consists in carrying out two surgical operations of the gall bladder in the operating room, following the usual training process and simulation action plan. The aim is to set up 'control' operations for classical mentoring (companionship).

Phase 1 (illustration 1): consists in simulating image transmission between videoconferencing machines to test these devices and to estimate and visualize the image and transmission qualities at varying fluxes and the transmission delays resulting from videoconferencing codecs.

Phase 2 (illustration 2): consists in simulating sound and images between videoconferencing devices to appreciate the effects of delays and error margins in data fluxes caused by satellite transmission.

The simulator contains an internal clock directly linking the two Tandberg devices with having to go through an optional switch. This stage is called satellite simulation tests.

2.4 Data Collection

The mentor his located in a room adjoining the control room. He faces a 32-inch monitor displaying events happening in the operating room, as shown on the right of illustration 3.

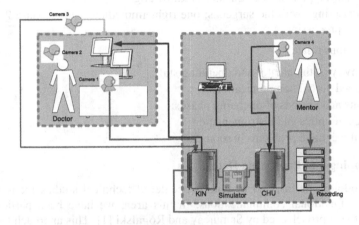

Fig. 1. Simulation Device

He has access to a remote control to switch from the endoscopic to the view lighted by the scialytic films his face. The videoconferencing microphone captures his/her voice, which operating theatre (blue line) receives. The mentor may also send an illustration or photograph from his PC to the videoconferencing device via a direct connection. A microphone and a telephone are available in the control room to communicate with operating theatre for coordination purposes.

In operating theatre, the transmission of the extracorporeal image, of the view the scialytic lamp focuses on, and of the sound captured by the surgeons' two mobile microphones (the mentoree and his/her assistant) transit to the mentor via the videoconferencing device. Monitors in operating theatre display the same images. A last Camera, which is unconnected to any circuit unlike the videoconferencing systems, gives a general view of the operating room, necessary to data-stripping.

Two VHSs located in the control room record sound and images necessary to the post-surgery evaluation of Man-Man and Man-Machine interactions. Simultaneously, the Tandberg device displays the whole of emissions necessary to data-stripping.

Four planned data collection systems are:

- Protocols followed by the technical team during the simulation;
- Surgery protocols that retrace technical problems and execution delays during the various stages of the surgical operation. These protocols are part of the general protocols, and serve as locators in order to build the stripping protocols described below;
- Stripping protocols of the two VHS and HI8 Handycam magnetic bands visualized in parallel. These protocols are genuine transcriptions of integral communications and textual or graphic information exchanged between the surgeons during the surgical operations within surgical and transmission contexts. These protocols also integrate non-verbal behaviors. These protocols allow to post-code communications and attitudes typical of the "transfer of knowledge."
- Two debriefings with the surgeons: one right immediately after phase 2, the other after data processing.

Observations relate to:

- The activity of the mentor and operating staff;
- Exchanged verbal communications;
- Moments and periods of operating phases;
- Changes of optics and instrumentation;
- Surgical technical problems in room 5;

2.5 Stripping and Processing Data

The methodology is inspired by anthropocentered technical studies focusing on the analysis of functional communications. In this area, we have been particularly inspired by the approach used by Samurçay and Rogalski [1]. This approach focuses on:

- Locating the interactive character of the activity (identification of the emitting and the receiving agents). This allows to examine the structure of communications within the surgical team and to qualify the mentor and mentoree distribution.
- Encoding communications (the nature of communications). This allows to analyze the semantic content and to prove the characteristics of the activity in the operating room.

As to the various categories of verbal interactions, we were inspired by the Savoyant[1] and Rabardel[2] approaches, which categorized communications along two concomitant axes:

1. Communications related to laparoscopic competency domains the surgeons use during interventions. These exemplify what the English-speaking world calls 'laparoscopic awareness,' which is shared consciousness that the mentor and mentoree have on the specific field.
2. Communications related to the style of exchanges in situation of distance apprenticeship.

---

[1] Distinguishes between synchronizing and planning communications.
[2] Proposes a quadripolar model of instrumented collective actions, which adds to the usual Subject-Instrument-Object relations between subjects through instruments.

3 Results

3.1 Telementoring Feasibility

The simulation proved that surgical telementoring was technically feasible with a satellite link simulated at 384 Kb/s, 512 Kb/s, and 768 Kb/s and a transmission delay of 700 ms for cholecystectomies performed through laparoscopy within simple indications (grade II gravity without anatomic complications) and respecting a codified intervention process (specific adapted strategy, routine language rules). This experiment proved itself:

- Without technical malfunctioning (transmission, link bursts).
- Without major surgical complications (hemorrhages, perforations). No major incident or surgical accident caused strategic breaks.
- Within regular skin to skin[3] intervention delays: 68' +/- 17'.
- With the permanent support of a technical team in the control room (a CHU Saint-Pierre technician in the control room for functioning and title bench use, two telecommunication specialists for the videoconferencing systems, two telecomputer specialists for the simulator).
- Following the guidance of an experienced mentor in laparoscopy, in live teleconferencing.
- With a resident candidate beginner in laparoscopy.
- With the entitled resident of the digestive surgery unit as assistant for 3 interventions and an intern beginner in laparoscopy surgery for 3 other interventions, including the last two.

3.2 Limiting Factors

Altering quality of the transmitted extracorporeal image:

- In the presence of dirt on the endoscope;
- In case of bleeding;
- During abrupt endoscopic moves;
- In case of bad endoscopic exposure of the operating field.

This leads us to focus on:

1. The necessity for the mentoree surgeon to operate with slow moves—prevents bleeding and gives the mentor time to anticipate his/her approach;
2. The importance of the assistant's role as he/she manipulates the endoscope;
3. Sound quality for the mentoree surgeon, who should normally have a headphone.
4. Need of human resources specialized in the technical organization of telementoring (control room staff, computer engineers . . .), and as a whole, of an adapted picture for the mentor and the surgical team of telementoring technical implications.
5. Preliminary necessity for the two surgeons (mentor and mentoree) to share the same cognitive references for the image and surgical process—the major part of given instructions relate to manipulations.

---

[3] Skin to skin: period separating the first abdominal incision to insert trocars and the binding of the wall after removing the gall bladder.

6. Danger of cognitive overload for the mentoree surgeon, if he/she has an inexperienced assistant that he/she would have to guide to manipulate the endoscope.
7. Possible unexpected effects linking the communicative style of the mentor surgeon
   with the mentoree surgeon, where the mentor has to adapt to a situation that he/she
   cannot totally master with the distance—anticipation, positive or negative appreciation (Man/Man interface).

Nonetheless, surprisingly, telementoring has not been influenced by:

- The frequency variation of the passing band (384, 512, 768 Kb/s or less for the
  DUO function).
- The introduction of a supplementary satellite delay (700 ms) between the mentor
  and the mentoree surgeons.
- The eruption of bursts during the operations.

3.3 Repercussion on the Training of Surgical Practice

The experiment:

- Proved that this type of distance assistance was possible at an early stage of the
  learning curve;
- Demonstrated the genuine cooperative character of training in laparoscopic surgery;
- Described an important learning process for the mentoree during the 6 operations.
  This seems to indicate a more efficient learning curve than with the standard training techniques (live surgical demonstration by companionship or videoconferencing), or with the use of laparoscopic simulator and surgical robots. A short while
  after the simulation, during his rotation, our resident candidate performed in
  laparoscopy a septicaemic appendectomy and operated a perforated ulcer alone and
  without expert assistance.

To explain this phenomenon, our hypothesis relies on the analysis of verbal and nonverbal communications exchanged during the simulation. Dominguez [2] has already
used this method, and conducted an interesting study on information elicitation[4] leading
to conversion decision during laparoscopies. The analysis of verbal protocols significantly indicated that seniors used more inferences and predictions than residents, and
expressed limitations to practise an intervention in secure conditions more frequently.
What essentially distinguished seniors related to knowledge piloting processes during an
intervention. Processes were essentially based on operative perceptual aptitudes, i.e.,
leading to action and decision. This study highlighted expertise and competence building in the form of problem solving and error reduction.

The results of our research follows the same in indicating that telementoring, despite the distance handicap, would be more efficient than standard mentoring for several reasons:

- In standard mentoring, the mentor stands next to the student and tends to resume
  the operation if the execution is hesitant, imprecise and tardy. Distance, however,

---

[4] Elicitation: encouragement, stimulation, incitation. Example: stimulation that calls up (draws
forth) a particular class of behaviors; "the elicitation of his testimony was not easy."

forbids the mentor to take the hand, which in turn compels the mentoree surgeon to issue results. This affects instilling responsibility in the mentoree, and in raising his/her attention and concentration.

- Again in standard mentoring, the mentor performs gestures, glances at the staff, particularly at the mentoree, which is impossible with a distant site. Distance compels the mentor to verbalization, which becomes the sole communication medium. It forces the mentor to take more explanatory approaches than those of live teleconferences, which pertain to demonstration—even if teleconferencing approaches include sleight-of-hand techniques, where the expert surgeon subconsciously performs in a prestidigitary manner, which the more 'limiting' conditions of telementoring forbids.

- Moreover, the expert adopts precautious and anticipation strategies, which tremendously increase from the mentoring to the telementoring modes. These gradually decrease with the increasing mastery of the mentoree. With telementoring, the expert expresses knowledge not produced usually in either standard companionship or demonstrations, which tremendously influences the mentoree's evolution of competence.

To understand the evolution mechanism of the mentoree's competencies in the simulation experiment, we refer to Rasmussen's [3] cognitive model of human reasoning. In this model, competency levels (beginner, intermediary, and expert) are identified in regard to the different stages of decision making and information processing. These three competency levels correspond to three stages of behavioral apprenticeship, which we have adapted to surgery, as follows:

- 'Skills' pertain to behaviors based on sensorimotor abilities and particularly oculomotor coordination for surgery. (Examples: endoscope positioning, knot performing, clip setting, etc.).
- 'Rules' pertain to rule-based behaviors. (Examples: standards of surgery procedures, rules for asepsis, etc.).
- 'Knowledge based behaviors' pertain to behaviors based on knowledge. (Examples: hemorrhage management or unexpected situation handling).

During training, competencies evolve. On the one hand, an expert, as opposed to a beginner, performs surgical moves 'automatically.' The expert no longer needs to go through the stages of data acquisition, then to pattern recognition [4] and identification to make a decision and execute it following rules and controlling the feedback of his/her actions. The expert realizes a cognitive economy, directly passing from stimuli to action. This is the 'magical' side of a celebrated surgeon, who can perform in a demonstration an operation renowned as complex with natural ease.

Verbalization, which is inherent to telementoring, forces the mentor to explicit his/her practising way, and therefore accelerate the mentoree's apprenticeship. By interpreting this results, we can:

- Conduct and guide analyses of human errors aiming at improving the Man-Machine System.
- Define in which critical domain to provide a technical and/or organisational assistance to the surgeon and staff. In view of recent technological advancement, rethinking the operating room of the future [5] becomes vital.

- Provide a theory model to establish training methods for certain behavioral types. Some publications mention this approach [6, 7], and recommend that these three competency types be marked by invasive minimal surgery training methods. Typically, simulators lie in the category of ability reinforcement, whereas mentoring targets more procedures and problem solving, thus decision making, which other means mentioned at the beginning of this chapter fail to address.

4 Valorization Prospects

4.1 Equipment Creation

- Codecs and laparoscopic equipment (endoscope) interface should improve.
- Codec algorithm should adapt to the needs of telemedicine.
- Telestration possibilities should be considered for videoconferencing equipment.
- Operating room infrastructure should evolve: control room, simulation, cameras, dedicated networks, microphones, equipment interconnectivity.
- Telemedicine's major limiting factor is not technology per say but the lack of interoperability, which is why all implicated actors must impose a wide interdisciplinary dialog.

4.2 Working Conditions

- Cooperative work requires strict standardization of technical norms, which we need to set up internationally.
- We can contemplate 'emergency telementoring' between a senior surgeon (mentor) and an intern surgeon (mentoree) and evaluate its results.

4.3 Surgical Apprenticeship

- We should test the effects of 'fragmented companionship' on apprenticeship in the following conditions:
  - o Live (audience-relayed telementoring);
  - o Pre-recorded (recorded telementoring).
- We should apply this technique to train trainers (mentors) to standardize telementoring for varied surgical indications.

References

1. Samurçay, R., Rogalsky, J.: Exploitation didactique des situations de simulation. Le travail humain 61(4), 333–360 (1998)
2. Dominguez, C.O.: Expertise in Laparoscopic Surgery: Anticipation and Affordances. In: Proceedings of the Fourth Conference on Naturalistic Decision Making. Warrenton, Virginia, May 29-31 (1998)
3. Rasmussen, J.: Information processing and human-machine interaction: An approach to cognitive engineering. Elsevier Science, Netherlands (1986)

4. Duda, R.O., Hart, P.E., Stork, D.G.: Pattern classification, 2nd edn. Wiley, New York (2001)
5. Haugh, R.: The future is now for surgery suites. Hospital and health network/ AHA 77(3), 50–54 (2003)
6. Dankelman, J., Wentink, M., Strassen, H.G.: Human Reliability and Training in Minimally Invasive Therapy & Allied Technologies 12(3-4), 129–135 (2003)
7. Wentink, M., Strassen, L.P.S., Alwayn, I., Hosman, R.J.A.W., Stassen, H.G.: Rasmussen's model of human behaviour in laparoscopy training. Surg Endosc 17, 1241–1246 (2003)

User Experience in Machinery Automation: From Concepts and Context to Design Implications

Jarmo Palviainen and Kaisa Väänänen-Vainio-Mattila

Human Centered Technology (IHTE),
Tampere University of Technology,
P.O. Box 589, FI-33101 Tampere, Finland
{jarmo.palviainen,kaisa.vaananen-vainio-mattila}@tut.fi

Abstract. Machinery automation (MA), e.g. different agriculture machinery, has traditionally been developed by experts in automation and in machinery engineering. As the role of interactive software is increasing, the principles and methods of human centered design (HCD) are being applied. This results in better usability of the systems particularly through efficiency of work processes and user interfaces (UIs). The user experience (UX) approach extends the HCD approach with broader motivational factors of using the systems. This paper describes the elements of UX in the MA from the interaction design perspective. After introducing the UX field, we describe the context to give an overview of the major factors affecting UX. Then we present what we consider to be the key elements of UX in MA and what implications they bring to the design of such systems. Finally we discuss the benefits and challenges of applying UX in this particular field.

Keywords: Human centered design, interaction design, MA, user centered design, user experience.

1 Introduction

Traditional usability, as defined by ISO [1] gives only little attention to what the user thinks about the system. The major concern here is how the term satisfaction is understood. The current definition does not encourage the designers to consider broadly how this satisfaction is achieved, but rather leave it in to too little attention, presuming that once the usability engineering is done right the user will be satisfied. Hassenzahl et al call this kind of tendency as usability reductionism [2]. The concept of User eXperience (UX) has emerged to bring user's motivation and thoughts more into designers' attention. UX has been researched from many perspectives, see e.g. [2, 3, 4, 5, 6, 7], but the practical applications of the theories can still be viewed as immature.

The practical applications of UX design and evaluation have been focusing on consumer products [8]. In machinery automation (MA), applications of UX principles have not been reported. By MA, we refer to the control and manipulation systems of different mobile machines, e.g. various agriculture and forestry and construction machinery.

We present the basic concepts of UX in section 2. UX is related not only to user interface (UI) design of the products but also on designing the right functionalities and the interaction mechanisms and styles to support users' tasks and values.

M. Kurosu (Ed.): Human Centered Design, HCII 2009, LNCS 5619, pp. 1042–1051, 2009.

In the 3rd section, we describe different contexts in MA and present typical product features such as the interaction styles and content in these contexts. In the 3rd and 4th sections, our findings from earlier case studies are reflected against the UX framework by Hassenzahl [3]. Our case studies and current industry practices are briefly described at the end of this introduction section.

In section 4 we present the design implications that UX perspective brings in to MA. We give examples of how the operator's motivation and development is supported in current systems and how the UX perspective can be used to create more appealing systems. These include e.g. supporting user in following and developing his/her personal growth and supporting professional identity as well as social needs.

1.1 The Case Studies Behind the Context Knowledge

During the recent five years, we have carried out more than ten case studies in the MA field. These studies vary from simple heuristic evaluations to designing a new product family. We have focused on several different user groups, e.g. maintenance, assembly and different operators. Majority of our work has involved forestry or mining industry, but we also have conducted some case studies in process automation and logistics and transportation. All of the case studies were highly product development orientated, while most of the studies contained academic research component as well. Our main focus has been institutionalizing human centered design, see e.g. [9, 10] and developing usability methods in this particular field. An integral part of the case studies has been carrying out dozens of field studies applying contextual inquiry [11]. We have extensively applied also other parts of the contextual design method [12] in our case studies.

While UX is a relatively new concept, we originally have not had it in the design or research focus in these case studies. However, with this paper we want to reflect our previous findings against UX theories and illustrate how UX perspective supports the designers in developing more engaging systems in this complex application area.

1.2 Current Design Practices in the MA Industry

In our studies during 2005-2009 we have followed and evaluated closely the design practices of five separate organizations in the MA industry. The design traditions stem strongly from sources other than software, HCD or UX design. However, most of the organizations we have covered have some usability professionals within, and their number is increasing. The design is customer oriented and the needs of internal stakeholders, e.g. maintenance and assembly are well covered. The end user is rarely represented in the design process, but the designers have relatively good understanding about the end users and their work. The concept of usability is well understood widely in the organizations, although HCD principles are not completely followed in the design process. The design processes are currently being altered to better suit to iterative design and to support software development, including more systematic requirements engineering and introduction of sets of standardized usability methods to be used during the process. For further information, see [9]. So far, the concept of UX is very new for the people in the industry, and our main concern remains in institutionalizing

HCD in the organizations. However, the usability professionals in the industry have accepted the concept of UX with enthusiasm and expect concrete design ideas from it.

2 User Experience (UX) Concepts and Frameworks

User eXperience (UX) refers to all aspects of the end-user's interaction with the company, its services, and its products [13]. From a more theoretical perspective, Hassenzahl & Tractinsky [7] have defined UX as "A consequence of a user's internal state (predispositions, expectations, needs, motivation, mood, etc.), the characteristics of the designed system (e.g. complexity, purpose, usability, functionality, etc.) and the context (or the environment) within which the interaction occurs (e.g. organizational / social setting, meaningfulness of the activity, voluntariness of use, etc.)". Furthermore, UX includes a mixture of pragmatic and hedonic experiences [7], arising from product characteristics that are functional or non-instrumental in nature. Functional characteristics support practical goal-orientation and hedonic aspects of use support users' goals such as social relatedness, development of personal skills and knowledge, and self expression i.e. users' need for joy, enjoyment, identity, and inspiration.

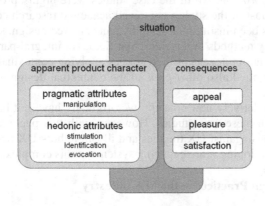

Fig. 1. The core of Hassenzahl's user experience framework [3]

Hassenzahl states that the key elements of the UX are the product features, the product character and the consequences [3], see Fig.1. The product features include content, presentation, and functionality and interaction style. The product character is a high level description - a holistic abstraction - of perceived features. This product character includes the pragmatic and hedonic attributes. The consequences refer to the judgment about the products appeal, and the emotional and behavioral consequences, e.g. pleasure and increased time spent with the product.

This paper refers to Hassenzahl's UX framework since we found it best supporting the considerations of applying UX on MA. It gives a rich picture of UX, particularly, when accompanied by the explanations about the volatility of user's expectations and about the user's goal/action mode (see [3]). In this paper, we focus on the hedonic product attributes, since the pragmatic ones are well covered by more established usability paradigms. Findings by Hassenzahl [4, 14] showed that the hedonic quality substantially contributed to the overall appeal of software prototypes for process automation control tasks.

3   The Components of the UX in MA

In Hassenzahl's framework, the UX is consisted from the product features, the product character, the consequences and the situation (context). In this section we give examples of these elements in the MA. Since UX is highly dependant on the user, we start by describing briefly the users (operators) and their values in this domain.

3.1   Operators and Their Values

The operators are in many cases highly goal oriented, particularly if their salary is based on their performance and productivity. They are aiming to achieve certain end product quality with maximum productivity. They are also concerned in keeping the systems functioning all the time and lowering the operating costs. Due to the physical risks involved, the operators are or at least they should be interested in work safety. There are also observations of less motivated operators who show no remarkable interest in using the system more than the absolute minimum. Part of the operators is very concerned about the expensive machinery and they fear that they may damage them. Sometimes working in isolated locations may add this stress, since the operators know that when facing troubles, there is no instant assistance available. Depending on the training and other experiences, the operators may or may not be concerned about the environmental issues related to their job.

The training may vary from scratch to several years of formally regulated training, even for the same particular product, when used in culturally different areas. Also the computer literacy, as well as the traditional literacy varies from illiterate to fluent.

3.2   The Context and the Situations in MA

The systems are typically equipped with a combustion motor and they are often mobile. In addition, the systems often contain hydraulic manipulators. Major part of MA is used in contexts that are extreme in many scales. Although many consumer products, like mobile phones, PDAs and sports computers are used outdoors or even underwater, there typically are not as much hazardous elements present as in the environments that are common for MA. Also, typically the context in which a consumer product is operated can be selected more freely.

Good examples of harsh and dangerous conditions can be found from forestry, mining [12] and construction. E.g. harvesters and stone crushers are operated within a temperature ranging from -50 to +60 degrees in centigrade. This is not only limiting viable technical solutions, but it also affects the human behavior and indeed is a central part of UX. Other disturbing context factor is the danger caused by the moving machinery and the processed materials, like stones or timber. In a quarry, there is a lot of dust from the stones making it difficult to see and breathe properly. The dust easily covers different surfaces and also taints the hands of the user. When using a stone crusher UI, the operator typically has to start the sequence by wiping the dust off the display. The dust covers also the physical buttons and their labels, as depicted in

(Fig. 2). The operator of a stone crusher or a harvester occasionally needs to service the equipment which also smears his/her hands, and there rarely is a place to wash them nearby. The materials and devices touched by the operator become therefore smudged, so e.g. some parts of the paper manuals may become unreadable.

Fig. 2. Detail of a dusty user interface of a stone crusher

The noise, particularly in a quarry can reach 125 dB, making it difficult to communicate with others and causing fatigue and stress to the users. In some cases the machinery vibrates and experiences high acceleration shocks, which has physically tiring effects and makes it harder to see small details from the displays or buttons attached to the machinery. Also the accuracy to use buttons and touch screens suffers due to vibration.

3.3 Tasks

Majority of our case studies have been carried out in the forestry and the stone crushing industries, which represent two extremes of MA usage. While a harvester machine operator is using the automation system UI most of the time, a mobile stone crusher operator is mainly occupied by tasks that do not involve directly monitoring or operating the system. In general, operator's task include

- Controlling and monitoring the process
- Solving problems (breakdowns, quality and efficiency problems)
- Collecting and analyzing statistics
- Using information from other systems, e.g. details of the production, customer orders, map information etc.

The operator often needs to use several different applications for different purposes, which, when used in combinations, raises the cognitive requirements of these tasks.

3.4 Product Features and Interaction Style

The systems are often a mixture of desktop applications and embedded systems. Typical UI consists of both display and numerous hardware buttons and controls for different tasks (see (Figure 3)). The content used in the system is most of the time a limited set of measurement data from the system or a selected set of parameters which the operator is using to modify the system behavior.

Fig. 3. The view from the harvester machine cabin

The navigation is based on hierarchies which are typically defined by the system structure or functionality. In many cases, there are standards and safety regulations constraining the UI design.

When considering what kind of interaction there occurs, we find a recent classification by Dubberly et al [15] useful. Interaction between the system and the user is characterized by classifying the systems in question into three different categories, i.e. linear systems, self-regulating systems, and learning systems. In our contexts, the operator may be interacting with all three types of dynamic systems. For example, when the operator controls the simplest actions of the harvester head, he is interacting with a linear system. While controlling the movements of the hydraulic actuators, the operator is interacting with a self-regulating system, i.e. a system that is using both the data from its sensors and the input from the operator to regulate its actions. When making decisions about how to cut a tree trunk, the operator is interacting with a learning system, that offers optimized suggestions based on what type of logs are needed and what the system has learned from the properties of the current location's trunks. Based on different combinations of interactions between systems, there are at least seven different interaction types [15]:

- Reacting to another system
- Regulating a simple process
- Learning how actions affect the environment
- Balancing competing systems
- Managing automatic systems
- Entertaining (maintaining the engagement of a learning system)
- Conversing

While entertaining and conversing are perhaps not in the mainstream of the current MA systems, the future systems have inevitably more of these features, since the level and sophistication of automation in these systems is increasing. With level of automation

Fig. 4. Vision of future harvester work [17] (published by permission from the author)

[16] we refer to in how extensively the system informs the operator, asks permission to actions and/or acts autonomously.

The long term trend is letting the operator focus on high level tasks and having a system to take care of simpler control tasks. An extreme example of how the interaction may change is a futuristic concept vision where the future harvester operator is controlling an army of compact forestry robots [17]. The robots autonomously take care of most of the harvesting, see Fig. 4. The operator would probably have "conversations" with the system before making important decisions, and in order to learn to use the system fully, the system has to entertain the user in some extent to motivate him/her to study and learn enough about it.

3.5   The Product Characters

When referring to the product character (see fig. 1), we omit the pragmatic characters covered by traditional usability engineering and concentrate on hedonic attributes.

According to [3], there are two major types of hedonic attributes – the ones concerning the individual's personal development or growth and the ones concerning social and societal issues. Many current systems do support both sides to some extent, but neither of them has been in the design focus so far.

Typical systems have relatively traditional online help systems explaining the functionality and the purpose of different system parameters. One recent application for harvesters assists the operator in analyzing his own performance and this way supports the operator's personal development. This can also lead to consequences such as higher motivation to use the system and to find out about other functionalities in it.

Some other features help with communicating with the customers or co-workers working e.g. in another shift. Currently, e.g. in forestry, communication with the customers is supported only through highly technical means, by delivering order and production information with a certain protocol. Social issues in general have been considered only a little in the design of current MA systems.

4 Design Implications Emerging from the UX Perspective

The harvester operators are highly concerned about their responsibility of the machinery. The harvester machines are costly and complicated systems, and the risk of doing something that might harm the equipment is bothering the operators. The operators work often in solitary locations with no one else present to help with possible technical problems. Lowering this stress by supporting them better in problematic situations and indicating the health status of the system in an understandable manner would probably improve the overall UX. Also, offering more efficient ways to communicate about the system status and about operator's personal working style with maintenance personnel or other specialists would improve the situation and help the operator to develop professionally.

The majority of the tasks done with stone crusher and harvester UI are repetitious and routine and the operator knows automatically how to do them. This leads to several implications on UX design. Firstly, the task flow has to be concentrated highly on effectiveness and efficiency, which is an area well covered by traditional usability engineering. Secondly, the novelty of new solutions for these tasks must not be exaggerated. Otherwise the users may be frustrated when they need to relearn their daily routines. Multimodal feedback and input is already in some use and it can be further exploited to offer interactions that are both efficient and offer good UX.

People need an optimal level of excitement and challenge to experience flow [18], and e.g. [3] suggests that novelty and change would assist in this. Some tasks done with the systems are monotonous, which imposes a challenge to motivate and stimulate the user without jeopardizing the efficiency and safety of the tasks.

The operators' attitude and behavior differ due to the nature of their tasks and also due to the differences in the systems. When the operator is not actively involved in the process, there is a risk of the operator loosing interest in the system and therefore being unable to notice early enough when the process is going to an unwanted state.

On the other hand, e.g. for harvester operators, the flow experience is already present in the form of competing with one self's and in the challenges of making optimal, accurate and speedy decisions about the process based on the observations from the environment and the feedback and measurement data from the system. Skilled harvester operators are like race car drivers who are concentrating on their task with maximum intensity. A chance to enhance UX in this kind of situation might be offering the user more accurate and incentive information about their performance, perhaps even developing metrics that would help them to compare their earlier and current performance under different circumstances. Later, social hedonism could be supported by enabling sharing one's results and hints to the user community.

5 Conclusions and Discussion

The UX concept is a recent addition in the usability field in its long term development from merely cognitive considerations to more social and emotional considerations (see e.g. [19, 20, 21]). There are several reasons to assume that the UX paradigm will influence the MA industry. The level of automation is increasing and the nature of work is shifting from monotonous, low level process control towards expert and team work. Also the consumer market will eventually train the operators to demand the work systems to offer better UX. Therefore we expect that the hedonic attributes will have an important role in designing the future MA products.

It will be challenging to take into account such factors as user's need to develop and express oneself and to be socially connected to others, while maintaining safety and effectiveness and compliance to different regulations typical for these contexts. Due to this fact, particularly in MA, there is a strong need for conservative solutions. Designers should avoid compromising usability e.g. in order to design more novel and exciting products. We would also like to discuss the emphasis on hedonistic attributes and avoid them being judged as if they were of "higher" value than the pragmatic ones. There are two risks. One is forgetting that for most products, there is some pragmatic reason for their existence. The other risk is, that hedonism is understood too literally as a pursuit of simple pleasure, while joy of first working hard to achieve something is an integral part of enjoying what you have achieved, reminding of the flow experience [18].

It may be a good idea to seek design ideas from the UX perspective after the fundamental usability work is in a mature state. On the other hand, there are indications that good UX may compensate for shortcomings in usability [4, 14]. UX point of view is more effective than the traditional usability perspective in helping the designers to understand and support the user's hedonic values, like the need to develop and express oneself and to be socially connected to others. Particularly, both researchers and product developers need more experience with different methods for developing, evaluating and measuring UX.

References

1. ISO/IEC. 9241-11: Ergonomic requirements for office work with visual display terminals. Part 11: Guidance on usability, International Organization for Standardization (1996)
2. Hassenzahl, M., Beu, A., Burmester, M.: Engineering joy IEEE Software 18(1), 70–76 (2001)
3. Hassenzahl, M.: The thing and I: Understanding relationship between user and product. In: Blythe, M.A., Overbeeke, K., Monk, A.F., Wright, P.C. (eds.) Funology: From Usability to Enjoyment, pp. 31–42. Kluwer Academic Publishers, Dordrecht (2003)
4. Hassenzahl, M., Platz, A., Burmester, M., Lehner, K.: Hedonic and Ergonomic Quality Aspects Determine a Software"s Appeal. In: Proc. of CHI 2000 Conference Human Factors in Computing Systems, pp. 201–208. ACM Press, New York (2000)
5. Law, E., Roto, V., Vermeeren, A.P.O.S., Kort, J., Hassenzahl, M.: Towards a shared definition of user experience. In: Proc. of ACM CHI 2008 Conference on Human Factors in Computing Systems, April 5-10, 2008, pp. 2395–2398 (2008)
6. Kankainen, A.: UCPCD: user-centered product concept design. In: Designing For User Experiences. Proc. of the 2003 conference on Designing for user experiences, pp. 1–13. ACM Press, New York (2003)
7. Hassenzahl, Tractinsky: User Experience – a Research Agenda. Behaviour and Information Technology 25(2), 91–97 (2006)

8. Väänänen-Vainio-Mattila, K., Roto, V., Hassenzahl, M.: Towards Practical User Experience Evaluation Methods. In: Proc. of the 5th COST294-MAUSE Open Workshop on Meaningful Measures: Valid Useful User Experience Measurement (VUUM 2008), June 18, University of Iceland, Reykjavik (2008)
9. Palviainen, J.: Transforming Machinery Design Traditions into User Centered Design - Principles, Challenges and Experiences. In: Proc. of Smart Systems 2008 Conference, Seinäjoki, Finland, June 4-5, pp. 13–18 (2008) ISBN 952-5598-04-7
10. Schaffer, E.: Institutionalization of Usability: A Step-by-Step Guide. Addison-Wesley, Reading (2004)
11. Palviainen, J., Leskinen, H.: User Research Challenges in Harsh Environments: A Case Study in Rock Crushing Industry. In: Proc. of The fourteenth international conference on Information Systems Development (ISD 2005), Karlstad, Sweden, August 2005, pp. 14–17 (2005)
12. Beyer, H., Holtzblatt, K.: Contextual Design: Defining Customer-Centered Systems. Morgan Kaufmann Publishers Inc., San Francisco (1998)
13. Nielsen-Norman Group, http://www.nngroup.com/about/userexperience.html (accessed on February 19, 2009)
14. Hassenzahl, M.: The Effect of Perceived Hedonic Quality on Product Appealingness. International Journal of Human-Computer Interaction 13(4), 481–499 (2001)
15. Dubberly, H., Pangaro, P., Haque, U.: ON MODELING What is interaction? Are there different types? Interactions 16(1), 69–75 (2009)
16. Sheridan, T.B.: Humans and automation: System design and research issues. Wiley Inter-Science, Chichester (2002)
17. Kokkonen, V., Kuuva, M., Leppimäki, S., Lähteinen, V., Meristö, T., Piira, S., Sääskilahti, M.: Visioiva tuotekonseptointi - työkalu tutkimus- ja kehitystoiminnan ohjaamiseen, Teknologiateollisuus ry (2005) (in Finnish)
18. Csikszentmihalyi, M.: Flow: The Psychology of Optimal Experience. Harper Collins (1990)
19. Bannon, L.J.: From Human Factors to Human Actors: The Role of Psychology and Human-Computer Interaction Studies in System Design. In: Greenbaum, J., Kyng, M. (eds.) Design at Work: Cooperative Design of Computer Systems, pp. 25–44. Lawrence Erlbaum, Hillsdale (1991)
20. Glass, B.: Swept Away in a Sea of Evolution: New Challenges and Opportunities for Usability Professionals. In: Liskowsky, R., Velickovsky, B.M., Wünschmann (eds.) Software-Ergonomie 1997. Usability Engineering: Integration von Mensh-Computer-Interaktion und Software-Entwicklung, B.G. Teubner, Stuttgart, Germany, pp. 676–680 (1997)
21. Karat, J., Karat, C.M.: The evolution of user-centered focus in the human-computer interaction field. IBM Systems Journal 42(4) (2003)

Perceived Usefulness and Perceived Ease-of-Use of Ambient Intelligence Applications in Office Environments

Carsten Röcker

RWTH Aachen University,
Human Technology Centre (HumTec), Theaterplatz 14, 52056 Aachen, Germany
roecker@humtec.rwth-aachen.de

Abstract. This paper describes a multi-national study evaluating the perceived usefulness and perceived ease of use of Ambient Intelligence (AmI) applications in office environments. In a first step, existing usage scenarios were analyzed to identify characteristic functionalities and application domains. The identified core functionalities were integrated into a representative and coherent evaluation scenario, which was presented to a target user population in a questionnaire-based study. The results of the study indicate, that the participants regard the described Ambient Intelligence functionalities as rather useful and easy to use. Nevertheless, moderate overall ratings for both factors show, that the acceptance of AmI technologies is not as high as often argued.

Keywords: Ambient Intelligence, Ubiquitous Computing, Pervasive Computing, Technology Acceptance, Study, Perceived Usefulness, Perceived Ease-of-Use.

1 Introduction

Over the last few years, companies started to show increased interest in deploying Ambient Intelligence technologies in office environments. From an economical point of view, high innovation pressure forces companies to adopt emerging technologies in an early stage in order to be competitive [18]. However, the integration of new technologies in existing business processes and work environments is always associated with high financial investments. When companies invest in new technologies and spend great amounts of resources into its integration, they usually expect a considerable increase in productivity, efficiency, and long-term benefits [3]. But in order for these benefits to occur, it is necessary, that the technology is used and also incorporated into the daily routines of the employees [29]. Empirical evidence shows, that one of the main reasons for low returns of investment is the poor usage of the installed applications (see, e.g., [6], [7] or [15]). In most cases, the potential of the implemented applications is not fully realized, due to the unwillingness of users to accept and use the systems [5]. Hence, it is important to evaluate the acceptance of future applications in an early stage of the design process in order to identify potential problems and implement appropriate countermeasures.

M. Kurosu (Ed.): Human Centered Design, HCII 2009, LNCS 5619, pp. 1052–1061, 2009.
© Springer-Verlag Berlin Heidelberg 2009

2 Technology Adoption

Predicting the adoption and use of information technology has been a key interest since the early days of information systems research [8]. The main goal of technology acceptance theory is, to explore the factors that influence the adoption and diffusion of new technologies throughout a social system [4]. Over the years, several independent theories for the acceptance as well as adoption of information technology have been developed. One of the best-established models of IT adoption and use is the Technology Acceptance Model (TAM) developed by Davis [13]. TAM is a further adaptation of the Theory of Reasoned Action (TRA) [16]. But while the Theory of Reasoned Action is a general theory of human behavior, TAM was specifically designed to model user acceptance in information systems [27]. Similar to most technology acceptance theories, it is assumed, that users could choose to employ a specific technology based on individual cost-benefit considerations (see [12]). The Technology Acceptance Model presupposes, that two particular constructs determine the user's acceptance of a technology: perceived ease-of-use (PEOU) and perceived usefulness (PU). According to the original definitions of Davis et al. [14], PEOU refers to "the degree, to which the [...] user expects the target system to be free of effort", while PU describes the individual's "subjective probability, that using a specific application system, will increase his or her job performance within an organizational context".

As shown in Figure, the Technology Acceptance Model suggests, that the user's decision to use a particular system evolves over four stages. Davis el al. [14] believe, that external variables (like individual abilities or situational constraints) indirectly influence technology usage through their impact on the perceived usefulness and perceived ease-of-use. Both factors affect a user's attitude towards the technology, which in turn influences the intention to use the technology [27]. As shown in the diagram, there is also a direct impact of perceived usefulness on the user's behavioral intention to use the technology. This is due to the fact, that even if individuals have a negative attitude towards a specific technology, this could be outweighed by a positive belief about the system's usefulness, which should finally lead to a positive usage intention.

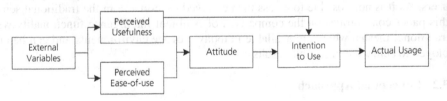

Fig. 1. Original Technology Acceptance Model [14]

The model has been tested by numerous authors, including Adams et al. [1], Chin and Todd [11], Hendrickson et al. [19], Igbaria et al. [20], Riemenschneider et al. [30], Subramanian [33], and Szajna [34]. In most of these studies, the TAM model was able to explain a reasonable amount of variance in the actual use of the technology [3]. An up-to-date review of existing TAM studies and meta analyses can be found in [25] or [26].

3   Goal and Approach

3.1   Research Goal

The majority of technology acceptance studies conducted so far analyze the adoption process of existing systems and applications, mostly with the goal of identifying the determinates that lead to the adoption. Instead of studying a real-world adoption process, the goal of this paper is to test, whether the core functionalities provided by Ambient Intelligence applications are accepted by potential users, and therefore are likely to be used in future office settings. This will be done by determining the perceived usefulness (PU) and the perceived ease-of-use (PEOU) of representative Ambient Intelligence functionalities.

Over the last decades, several studies showed, that the perceived usefulness of a system or application is a reliable predictor for its future usage. For example, Davis [13] found, that the perceived usefulness was significantly correlated with self-reported current usage (r=0,63) and self-predicted future usage (r=0,85). Similar to the perceived usefulness of a system, also the system's perceived ease-of-use proved to be a reliable indicator in numerous studies. Nevertheless, the perceived ease-of-use strongly depends on the actual implementation of the functionality and less on the functionality itself. As mentioned above, this paper aims to explore the acceptance of representative Ambient Intelligence functionalities, and therefore deliberately abstracts from concrete system implementations. In general, autonomous services are likely to receive rather high ratings regarding the perceived ease-of-use, as users are usually not required to perform any specific actions in order to benefit from a particular service. But experiences gained in previous studies (see, e.g., [31]) suggest, that users are willing to accept higher behavioral cost, in terms of additional user input, in order to gain more control over the provided services. Nonetheless, Ambient Intelligence applications are expected to be considerably easier to use than existing office applications, which greatly rely on manual user input. This means, that from the perspective of reduced behavioral costs, smart office applications will bring significant advantages over existing systems, and are therefore likely to be adopted by potential users. As it is not possible to assess the perceived ease-of-use in the traditional sense, this paper concentrates on the comparison of Ambient Intelligence functionality with traditional means, which are available in today's office environments and can be employed to achieve comparable results.

3.2   Conceptual Approach

Over the last two decades, numerous studies about technological acceptance have been conducted in different fields. The technologies and applications being tested include e-mail programs [13], internet banking [10], electronic commerce applications [28], word processors [9], electronic meeting systems [17], and tools for computer-aided software engineering [21]. Although the overall goal of this paper is quite similar to the intention of most technology adoption studies, there are two important differences that have to be taken into account. First, traditional studies investigate only the adoption of one specific technology. And second, the tested technologies exist either in form of functional prototypes or commercially available products. These two aspects do not only allow a comparably high number of different questions, but also enable participants to provide feedback on more specific aspects, as they usually gained considerable experience with the specific technology in the pre-phase of the actual evaluation. In contrast, this paper aims to explore a variety of different functionalities, which are not yet implemented in form of concrete technologies.

Therefore, it was decided to evaluate the acceptance of generic Ambient Intelligence functionalities based on a systematically constructed usage scenario. The usage of a fictive scenario provides several advantages over technical prototypes. Scenarios allow describing complicated and rich situations in meaningful and accessible terms, and thereby help to analyze and communicate the core ideas of Ambient Intelligence [22]. Especially functionalities provided by smart environments are complicated to prototype in a realistic way, and poorly implemented prototypes might significantly influence the users' perception of a functionality and its potential impact on everyday life. In addition, several technological trends may be extrapolated and combined into a single scenario to analyze the bundled effects, that these technologies could induce on users [23].

Nevertheless, it is important to note that a scenario-based evaluation approach does not allow testing the actual acceptance of functionalities, based on their concrete usage in office environments. Instead, it is only possible to explore the intended usage behavior, based on the answers gained from participants. Hence, the feedback only reflects the intention of the participants to use a specific functionality, but not the actual adoption of the functionality. But a variety of studies showed, that there is a strong correlation between the intention to use a technology and its actual usage. According to Ajzen [2] the intention of users to employ a technology defines whether they will actually use it. This assumption was also confirmed in several technology adoption studies (see, e.g., [24]). Thus, it is assumed that the stated preference of users to employ a specific functionality is a good predictor of their future adoption behavior.

4 Evaluation

4.1 Identification of Representative Functionalities

In a first step, an analysis of existing Ambient Intelligence literature was conducted to identify representative usage scenarios and application domains. The focus of this analysis was on work-related scenarios developed in Europe and the United States. In the course of the scenario analysis, 430 beneficial scenario elements were extracted from 63 scenario descriptions. In the end, 39 functional groups were identified, which described different types of Ambient Intelligence functionalities (see [32] for details).

4.2 Usage Scenario

In order to assess the usefulness and ease-of-use of the different types of functionalities, the core functionalities, identified during the scenario analysis, were integrated into a representative and coherent evaluation scenario. While it would be helpful to get feedback on all different types of functionalities, the number of scenario elements to be used in the evaluation, had to be significantly reduced in order to avoid overloading participants in the study. Therefore, it was decided to test only the functionalities, most often addressed in existing scenario descriptions. The final test scenario consisted of eight scenario elements, each illustrating an individual functionality within a smart office environment (see Table 1 and 2). The scenario incorporated the functionalities of nearly half of all the scenario elements identified during the analysis. So, even if only the functionalities of eight sub-groups were tested, these functionalities seem to be good indication about applications and services, that are likely to become part of future Ambient Intelligence environments.

4.3 Questionnaire

The scenario was presented to a target user population using a paper-based questionnaire. The participants were asked to rate each scenario element regarding the perceived usefulness and the perceived ease-of-use. Prior to the assessment of the each functionality, the corresponding scenario element was presented again in order to avoid any ambiguities. In order to capture user feedback, 10-point rating scales were used. For questions referring to the usefulness of the functionality, the endpoints were labelled 'not useful at all' and 'very useful'. Correspondingly, the endpoints of the rating scale for questions addressing the perceived ease-of-use were labelled 'more complicated' and 'easier'.

4.4 Participants

In total, 200 questionnaires were distributed to participants in Germany and the United States. For each country, 100 questionnaires were personally given out to persons with work experience in office environments. If possible, the participants were asked to hand on additional questionnaires to persons, who they regard as suitable for this study.. In total, N=161 persons.. returned their questionnaire, which resembles a return rate of 80,5%. Out of this group, N=96 came from Germany and N=65 from the United States. The overall population was nearly evenly distributed over male (49,1%) and female participants (50,9%), with slightly more males (52,1%) in Germany and slightly more female participants (55,4%) in the United States.

5 Results

5.1 Usefulness

In the first question for each scenario element the participants were asked to assess the general usefulness of the illustrated functionality. As explained above, a rating of '0' means, that a participant regards a specific functionality as not useful at all, while a '10' indicates, that this functionality is regarded to be very useful. Table 1 provides an overview over the perceived usefulness of the various scenario elements.

As shown in the table, the average rating over all scenario elements is M=6,55 on a 10-point scale. The average rating is slightly lower in the German group (M=6,44) and a little higher in the American (M=6,73). In all three groups, the average rating for each scenario element is higher than 5, which means, that all functionalities are regarded as rather useful than useless. Especially personal reminder services received relatively high ratings, which might be due to the fact, that such applications are very practical and understandable in the office context. But they are also less innovative, as similar functionalities are already implemented in existing office applications and are supported by most state-of-the-art mobile devices. Hence, the perceived usefulness could be attributed to the practicality of the application itself, or its similarity to existing office applications, which are already accepted by users as part of their daily work life. Based on the existing data, it is not possible to clearly identify the factor(s), which influence the participants' perception regarding the usefulness of personal reminders. The adaptation of the physical surrounding to enhance personal well-being is another service, receiving comparable high rating regarding its usefulness. Other prominent functionalities, which are often described in existing application scenarios, like, e.g., the adaptation of content, get rather low ratings. In the American sub-group this functionality was even rated as the least useful of all illustrated services.

Table 1. Overview over the assessment of scenario elements regarding their usefulness

Functionality	Germany		USA		Overall	
	Mean	Rank	Mean	Rank	Mean	Rank
1. Adaptation of Content	6,91	3.	5,82	8.	6,47	4.
2. Personal Well-Being	6,95	2.	7,82	1.	7,30	2.
3. Personal Encounters	6,12	6.	6,71	4.	6,36	6.
4. Speech Input	5,18	8.	6,23	7.	5,60	8.
5. Ambient Displays	6,75	4.	6,98	3.	6,84	3.
6. Personal Reminder	7,61	1.	7,57	2.	7,60	1.
7. Asynchronous Communication	5,46	7.	6,27	6.	5,78	7.
8. Public Activity Histories	6,50	5.	6,40	5.	6,46	5.
Average Usefulness	**6,44**		**6,73**		**6,55**	

5.2 Ease-of-Use

In the second question, each functionality was assessed regarding the perceived ease-of-use. Like for the previous example, a 10-point scale was used, where a rating of '0' means, that this functionality appears to be more complicated to use than existing office practices, while a '10' would represent a functionality, that is easier to use.

As shown in Table 2, the scenario elements received an average rating of M=6,18 regarding their ease-of-use. In contrast to the previous question, the average rating was higher in the German group (M=6,39) and lower in the American (M=5,87). As the participants were asked to compare the illustrated functionalities to existing office practices, the ratings indicate, that one advantage of Ambient Intelligence technologies seems to be their increased ease-of-use and user-friendliness over traditional office applications. In the overall group, all functionalities received a rating higher than 5. This means that the described functionalities are at least as easy to use as existing office technologies, even if the ratings are not as high as one might have expected. As in the previous question, the two scenario elements, describing personal reminder services and the adaptation of the physical surrounding to enhance personal well-being, received the highest ratings of all elements. While the scenario element describing speech input was rated the least useful, it received similarly low ratings regarding its ease-of-use. At least in office environments, participants from Germany as well as the United States do not seem to favor such interfaces over traditional interaction techniques. In the American sub-group, two functionalities were rated as more complicated to use than existing office technologies. The one, receiving the lowest rating regarding its ease-of-use are adaptation services, which are often seen as the key advantage of Ambient Intelligence technologies. These findings stand in a strong contrast to the development efforts that are currently put forth in the computer and telecommunication industry, to promote and establish such interaction and adaptation mechanisms.

Table 2. Overview over the assessment of scenario elements regarding their ease-of-use

Functionality	Germany		USA		Overall	
	Mean	Rank	Mean	Rank	Mean	Rank
1. Adaptation of Content	6,71	3.	4,84	8.	5,96	6.
2. Personal Well-Being	7,08	2.	7,16	1.	7,11	1.
3. Personal Encounters	6,31	6.	6,26	3.	6,29	4.
4. Speech Input	5,28	8.	5,30	6.	5,29	7.
5. Ambient Displays	6,62	4.	6,87	2.	6,72	3.
6. Personal Reminder	7,23	1.	6,21	4.	6,82	2.
7. Asynchronous Communication	5,40	7.	4,98	7.	5,23	8.
8. Public Activity Histories	6,45	5.	5,34	5.	6,00	5.
Average Ease-of-Use	**6,39**		**5,87**		**6,18**	

6 Conclusion

The results of the study indicate, that the participants regard the described Ambient Intelligence technologies as rather useful and easy to use. Nevertheless, the moderate overall ratings for both factors show, that the acceptance of Ambient Intelligence technologies is not as high as often argued. The usefulness and ease-of-use ratings of most scenario elements range between 60% and 70% of the maximal possible scores. As Table 1 and 2 show, the average overall rating regarding the usefulness is M=6,55 and the average overall rating regarding the ease-of-use is M=6,18. Those scores are not remarkable high for technologies that are often said to revolutionize the nature of office environments. Nevertheless, the results of the study also show, that there is still a considerable potential to increase the usefulness and ease-of-use of Ambient Intelligence applications. With the knowledge about general user requirements and the acceptance of specific functionalities, the next step is to identify the reasons, which caused the rather low ratings regarding the usefulness and ease-of-use of the illustrated functionalities. Only if those reasons are clearly identified, it becomes possible to revise the functionalities and application scenarios and thereby achieve higher acceptance rates.

Acknowledgments. The work presented in this paper was funded by the German Science Foundation (DFG).

References

1. Adams, D.A., Nelson, R.R., Todd, P.A.: Perceived Usefulness, Ease of Use, and Usage of Information Technology: A Replication. MIS Quarterly 16, 227–247 (1992)
2. Ajzen, I.: Residual Effects of Past on Later Behavior: Habituation and Reasoned Action Perspectives. Personality and Social Psychology Review 6(2), 107–122 (2002)
3. Alshare, K., Grandon, E., Miller, D.: Antecedents of Computer Technology Usage: Considerations of the Technology Acceptance Model in the Academic Environment. Journal of Computing Sciences in Colleges 19(4), 164–180 (2004)
4. Barnes, S.J., Huff, S.L.: Rising Sun: iMode and The Wireless Internet. Communications of the ACM 46(11), 78–84 (2003)
5. Benham, H.C., Raymond, B.C.: Information Technology Adoption: Evidence from a Voice Mail Introduction. ACM SIGCPR Computer Personnel 17(1), 3–25 (1996)
6. Brynjolfsson, E.: The Productivity Paradox of Information Technology. Communications of the ACM 36(12), 66–77 (1993)
7. Brynjolfsson, E., Hitt, L.: Paradox Lost? Firm-Level Evidence on the Returns to Information Systems Spending. Management Science 42(4), 541–558 (1996)
8. Burton-Jones, A., Hubona, G.S.: Individual Differences and Usage Behavior: Revisiting a Technology Acceptance Model Assumption. ACM SIGMIS Database 36(2), 58–77 (2005)
9. Chau, P.Y.K.: An Empirical Assessment of a Modified Technology Acceptance Model. Journal of Management Information Systems 13(2), 185–204 (1996)
10. Chau, P.Y.K., Lai, V.S.K.: An Empirical Investigation of the Determinants of User Acceptance of Internet Banking. Journal of Organizational Computing and Electronic Commerce 13(2), 123–145 (2003)
11. Chin, W.W., Todd, P.A.: On the Use, Usefulness, and Ease of Use of Structural Equation Modeling in MIS Research: A Note of Caution. MIS Quarterly 19(2), 237–246 (1995)
12. Compeau, D.R., Higgins, C.A., Huff, S.: Social Cognitive Theory and Individual Reactions to Computing Technology: A Longitudinal Study. MIS Quarterly 23(2), 145–158 (1999)
13. Davis, F.: Perceived Usefulness, Perceived Ease of Use, and End User Acceptance of Information Technology. MIS Quarterly 13(3), 318–339 (1989)
14. Davis, R.D., Bagozzi, R.R., Warshaw, P.R.: User Acceptance of Computer Technology: Comparison of Two Theoretical Models. Management Science 35(8), 982–1003 (1989)
15. Dewan, S., Min, C.: The Substitution of Information Technology for Other Factors of Production: A Firm Level Analysis. Management Science 43(12), 1660–1675 (1997)
16. Fishbein, M., Ajzen, I.: Belief, Attitude, Intention and Behavior: An Introduction to Theory and Research. Addison-Wesley, Reading (1975)
17. George, J., Nunamaker, J.F., Valacich, J.S.: Electronic Meeting Systems as Innovation: A Study of the Innovation Process. Information and Management 22(3), 187–195 (1992)

18. Görisch, F., Hennig, F.: Ambient Business - Ubiquitous Computing im wirtschaftlichen Kontext. Seminar on Novel Forms of Human Computer Interaction: Smart Artifacts and Ambient Intelligence. Department of Computer Science, Technical University of Darmstadt (2006)
19. Hendrickson, A.R., Massey, P.D., Cronan, T.P.: On the Test-Retest Reliability of Perceived Usefulness and Perceived Ease of Use Scales. MIS Quarterly 17(2), 227–230 (1993)
20. Igbaria, M., Zinatelli, N., Cragg, P., Cavaye, A.L.M.: Personal Computing Acceptance Factors in Small Firms: A Structural Equation Model. MIS Quarterly 21(3), 279–305 (1997)
21. Iivari, J.: From a Macro Innovation Theory of IS Diffusion to a Micro Innovation Theory of IS Adoption: An Application to CASE Adoption. In: Avison, D., Kendall, J.E., DeGross, J.I. (eds.) Human, Organizational, and Social Dimensions of Information Systems Development, pp. 295–315. Elsevier Science Publishers, Amsterdam (1993)
22. Kaasinen, E., Rentto, K., Ikonen, V., Välkkynen, P.: MIMOSA Initial Usage Scenarios. Deliverable D1.1 of the Project Microsystems Platform for Mobile Services and Applications (MIMOSA), IST-2002- 507045 (2004)
23. Langheinrich, M., Coroama, V., Bohn, J., Mattern, F.: Living in a Smart Environment – Implications for the Coming Ubiquitous Information Society. Telecommunications Review 15(1), 132–143 (2005)
24. Lee, Y., Lee, J., Lee, Z.: Social Influence on Technology Acceptance Behavior: Self-Identity Theory Perspective. ACM SIGMIS Database 37(2-3), 60–75 (2006)
25. Legris, P., Ingham, J., Collerette, P.: Why Do People Use Information Technology? A Critical Review of the Technology Acceptance Model. Information and Management 40, 191–204 (2003)
26. Ma, Q., Liu, L.: The Technology Acceptance Model: A Meta-Analysis of Empirical Findings. Journal of End User Computing 16, 59–72 (2004)
27. Mathieson, K., Peacock, E., Chin, W.W.: Extending the Technology Acceptance Model: The Influence of Perceived User Resources. The DATA BASE for Advances in Information Systems 32(3), 86–112 (2001)
28. McCloskey, D.: Evaluating Electronic Commerce Acceptance with the Technology Acceptance Model. Journal of Computer Information Systems 4(2), 49–57 (2003)
29. Pearson, J.M., Crosby, L., Bahmanziari, T., Conrad, E.: An Empirical Investigation into the Relationship between Organizational Culture and Computer Efficacy as Moderated by Age and Gender. Journal of Computer Information Systems 43(2), 58–70 (2002)
30. Riemenschneider, C.K., Harrison, D.A., Mykytyn, P.P.: Understanding IT Adoption Decisions in Small Business: Integrating Current Theories. Information & Management (40), 269–285 (2003)
31. Röcker, C., Janse, M., Portolan, N., Streitz, N.A.: User Requirements for Intelligent Home Environments: A Scenario-Driven Approach and Empirical Cross-Cultural Study. In: Proceedings of the Intern. Conference on Smart Objects & Ambient Intelligence (sOc-EUSAI 2005), pp. 111–116 (2005)
32. Röcker, C.: Design Requirements for Future and Emerging Business Technologies: An Empirical Cross-Cultural Study Analyzing the Requirements for Ambient Intelligence Applications in Work Environments. Verlag Dr. Driesen, Taunusstein, Germany (2009)
33. Subramanian, G.H.: A Replication of Perceived Usefulness and Perceived Ease of Use Measurement. Decision Science 25(5-6), 863–874 (1994)

34. Szajna, B.: Software Evaluation and Choice: Predictive Validation of the Technology Acceptance Instrument. MIS Quarterly 18(3), 319–324 (1994)
35. Taylor, S., Todd, P.A.: Understanding Information Technology Usage: A Test of Competing Models. Information Systems Research (6), 144–176 (1995)
36. Karahanna, E., Straub, D.W., Chervany, N.L.: Information Technology Adoption Across Time: A Cross-Sectional Comparison of Pre-Adoption and Post-Adoption Beliefs. MIS Quarterly 23(2), 183–213 (1999)

Clinical System Design Considerations for Critical Handoffs

Nancy Staggers[1], Jia-Wen Guo[1], Jacquelyn W. Blaz[1], and Bonnie M. Jennings[2]

[1] Informatics and Doctoral Students, College of Nursing,
10 S., 2000 E., University of Utah, Salt Lake City, UT 84108
[2] Healthcare Consultant, Athens, GA
nancy.staggers@hsc.utah.edu

Abstract. Change of shift report (CoSR) is a nurse-to-nurse communication event (handoff) that could potentially result in missed or incomplete information, time inefficiencies and patient errors. Although technology is touted as being amenable for this process, researchers have not yet evaluated how CoSR might be supported through computerization. This paper summarizes past research on this critical transition, describes the results of a qualitative study for shift report content on medical and surgical units in the U.S. and then outlines requirements for computerized support of the process. Three potential CoSR designs are provided and discussed: a patient summary screen, a personally-tailored design for nurses, and a problem-oriented design. Benefits and disadvantages of each are proposed.

Keywords: Handoffs, user interface design, clinical systems.

1 Introduction

Evidence from a report from the Agency for Healthcare Research and Quality (AHRQ), suggests that improving information exchange during patient care handoffs is one strategy for reducing gaps and improving patient safety.[1] Just such a gap may occur during nurses' change of shift report (CoSR). This transition at shift change is a handoff that occurs 2-3 times per day on every acute care unit in every hospital around the world. CoSR is the mechanism by which nurses learn about their patients and take responsibility for them. Nurses going off-duty synthesize and convey patient information that is received and analyzed by the oncoming nurses to set care priorities and create the immediate plan of care for the upcoming shift.

CoSR has received little attention by researchers and clinical system designers. It has not been systematically investigated on medical and surgical units in the U.S. in the last decade. This is important because findings from recent U.S. studies conducted in acute care settings reveal that handoffs in general and work complexity on medical and surgical units can interfere with patient safety.[2-5] Likewise in the past 10 years, computerized support for clinical care has greatly expanded. Authors thought technology might improve handoffs [6-7]; however, no research about technology support is available for these critical nurse-to-nurse handoffs. Consequently, the authors

M. Kurosu (Ed.): Human Centered Design, HCII 2009, LNCS 5619, pp. 1062–1069, 2009.
© Springer-Verlag Berlin Heidelberg 2009

summarize a recent study about the content of CoSR on medical and surgical units and discuss potential design considerations for computerizing change of shift report.

2 Background

2.1 Past Research about Change of Shift Report

CoSR is a current topic of great interest [6-8]. In the past, CoSR was studied in patient care areas including medical and surgical units. Primarily in the U.S. and Great Britain, investigators found the language used during report to be vague, jargon- and slang-filled, tending toward labeling patients, and at times reflecting emotions. [9-11] No mention was made, however, of the utility of this way of speaking. Medical information and technical tasks tended to dominate CoSR. [12]

The means of giving report—both verbal and non-verbal reports—have been the focus of several investigations. Sherlock [11] reported that verbal reports are often time-consuming due to the large volume of information conveyed, illogical sequencing of information, environmental noise competing for the nurses' attention, and nurses' failure to identify work priorities for the oncoming shift. Additionally, Dowding [13] questioned the effectiveness of verbal reporting based on her findings that retrospective reporting, when compared to prospective reporting, improved care planning but decreased nurses' information recall. Non-verbal reports, however, are not without problems. In 2 studies, information from documentation was richer than information exchanged verbally.[10,14] Findings were mixed from other studies assessing the effects of nonverbal report [15-16]. For example, Munkvold and colleagues [16] learned that redundancy was reintroduced, albeit at different times and places in the work process.

Wolf [17-19] completed an ethnography on a medical unit in the U.S. During report, nurses shared their clinical acumen as they passed along information about signs, symptoms, laboratory studies and diagnostic tests. The language used to share this information was filled with incorrect grammar, short phrases, slang, abbreviations and acronyms. Wolf [17] referred to this language as "professional jargon." This jargon was viewed as a way to convey the richness and complexity of nurses' work. It also fostered conformity by requiring new nurses to master the "language of nursing" [17, p. 83] which ostensibly established high standards for CoSR.

CoSR may not have complete information. Lange [20] found that information exchanged in report on 4 medical-surgical units was not sufficient for nurses to move directly from CoSR to patient care. Nurses spent just over 15 minutes afterwards searching for information that was absent from report. The top three categories of missing information related to medications (34%), physicians' orders (18%), and laboratory data (10%).

Most important, change of shift report represents complex cognitive processing. Nurses receive and manage patient information to provide continuity of care. Because human beings can process only a limited amount of information, various strategies are utilized to ease the load of information processing. One example is from Hardey, Pay and Colema [21] who conducted an ethnography on 5 acute elderly care units to discover how nurses define and communicate patient care information to each other for the delivery of care. They found that nurses use informal, unstructured records that the researchers called "scraps" to manage information. These "scraps" are commonly used in practice to address patients' needs and draw attention to relevant information to complete and also to communicate during report. Content includes basic patient information, nursing interventions and evaluations but also nurses' personal notes about patients. These informal pieces of paper, created by each individual nurse, have become a norm in the change of shift report. Interestingly, these researchers [21] called these pieces of paper "scraps" while nurses typically call their pieces of paper their "brains." Nurses' "brains" are designed differently from person to person based on personal use, so the content can best be understood by their owners. The size ranges from a 3x 5 piece of paper to a letter size with varying formats.

In summary, CoSR is a complex transition in care, cognitively intense and potentially filled with missing information, jargon, varying formats, sequences and content. Nurses rely primarily on personal, informal records to communicate during report. As yet, no research is available about existing computerized support for CoSR.

2.2 A Recent Study of CoSR on Medical and Surgical Units

Two of the authors (NS and BJ) conducted a qualitative study of CoSR on medical and surgical units in the western U.S. [22] Study data were collected from 7 units in 3 acute care facilities. A total of 53 patient reports involving 39 nurses were observed and audio-taped for face-to-face, taped and bedside types of CoSR. Purposeful sampling was used to reflect the various report types and shift times. IRB approval was obtained. The data were analyzed using conventional qualitative content analysis. [23]

CoSR Content. The average age of nurses was 34 and their modal educational level was a BSN. The content of the 53 CoSRs clustered into four themes: (a) The Dance of Report, (b) Just the Facts, (c) Professional Nursing Practice, and (d) Lightening the Load. The categories were persistent across report types. The Dance of Report refers to the choreography of reporting. It involved the discernible "movements" between report partners that were essential to the process of basic human communication. The Dance was the largest theme. [22]

Just the Facts, the second largest theme, involved exchanging non-controversial patient data such as the last time a pain medication was given or the values of a recent laboratory result. The theme Professional Nursing Practice was typified by information related to nursing actions, knowledge, reasoned judgments, and instincts combined with care decisions. Nurses' expertise involved assessments, observations and decisions. These elements were woven together in a way that represented a sophisticated integration of complex elements. Lightening the Load comprised the smallest component of coded content. This category reflected various ways in which nurses exhibited thoughtfulness toward other staff and gave helpful hints to the next shift such as a particular way that a patient liked a nitroglycerin patch applied.

CoSR formats and tools. Nurses did not use a consistent structure to deliver information across units, report types or shifts. [22] Although nurses typically began report with pertinent parameters related to patient names, ages, and bed numbers, the remainder of report followed no particular pattern. Nurses delivered content according to individual style, while perhaps considering the receivers' needs. For instance, if the receiving nurse took care of the patient previously, report was abbreviated. The patients' main problems were woven into report in a non-sequential manner. Critical information could appear at the end of report. However, pertinent information related to a specific problem, once mentioned, was typically clustered together.

The memory tools nurses used to capture pertinent information during report were exceptionally varied. These ranged from hand-writing on separate 3x5 cards for each patient, to a word-processed, 8x11 inch stylized report form with typical categories listed for note-taking during report, e.g., intake and output, vital signs, pain. Most important, no computerized tools were used during any of the observed reports and none of the nurses referred to the available clinical information systems during CoSR. [21] This occurred despite the fact that one facility had a full electronic health record (EHR) with orders, clinical documentation, and clinical decision support. The other 2 facilities had partial EHRs with available clinical results. [22]

3 Discussion of CoSR Computerized Design Considerations

Computerizing report is not a simple endeavor largely because CoSR is centered on information synthesis across at least these elements: patient or patients, unit, nurse, provider, orders, results, previous documentation, administrative concerns, and any relevant policies. Our research findings speak to the sheer mass of data and information nurses retain in their memories and to the considerable challenge nurses experience in sifting through information passed along in report as they establish shift care priorities and cull out non-critical information. This challenge is particularly great for novice nurses and those new to a particular patient care unit.

Our research findings indicate that not all content is amenable to computerization as was recommended by previous authors. [6, 22] The vast majority of information is held in nurses' memory and essentially all current synthesis still occurs in nurses' brains. The speed with which a nurse can assimilate, organize, and craft decisions about care cannot be emulated *in toto* by a clinical information system. Also, CoSR serves more than one purpose; computerization will not replace the CoSR roles for mentoring, socialization and unit cohesiveness. Perhaps the best designers can do is serve as a clever adjunct to report rather than a replacement for it. Designers must realize that a printed, thorough computerized report is not likely to replace verbal shift report either for the report giver or receiver.

Several requirements for computerization are known.

- First, the information most amenable to computerization is, logically, in the theme Just the Facts. The facts might include the latest serum glucose or the last time a pain medication was given. Other themes, such as Professional Nursing Practice and Lightening the Load, include content that will likely be more challenging to computerize.
- Second, any consistent structure would allow more complete reports and an order of information for novices and new unit staff to expect during CoSR delivery.
- Third, flagging new, critical information such as a new laboratory value or new provider order during report would be important.

- Fourth, The Joint Commission recommended the use of a physician-developed tool called SBAR (Situation, Background, Assessment, Recommendation) in 2007. [24] Intended to standardize nurse-physician communication for critical events, SBAR is being adopted for nurse-nurse communication including COSR by some institutions. The Joint Commission acknowledges that while SBAR can be used for nurse-nurse handoffs in shift transitions, it would "probably need to be adapted" [24, p. 31]
- Fifth, any design needs to be organized into support for a nurse who already "knows" the patient versus one who is new to caring for the patient. Likewise, information needs for experts and novices are distinct.
- Sixth, CoSR information support is not identical for the nurse giving report and the nurse receiving it. Different and interconnected tools will need to be developed to fit workflow.
- Last, only a very efficient design will be used by time-pressured nurses. Below, we discuss possible macro-designs for change of shift report for a nurse giving report

3.1 Patient Summary Screen Design

This approach provides a single screen that with available EHR-stored information about a patient. The information obtained during the last shift is summarized with links to more detailed data about lab results, orders, vital signs, etc. For instance, on a surgical unit, the summary screen would have the patient's temperatures for the last 24 hours, a trend of pain medications and laboratory values. The layout of the data could be somewhat customized for specific units, as well as specific nurses, providing some flexibility in the display of information. However, if the patient developed a problem, such as a pulmonary embolus, outside the usual display of issues for surgical patients nurses would not have that information on the summary screen.

The positive implication of this design is the ease of development for vendors, designers, and informaticists. Patient information currently stored in an EHR need only be grouped together in a new presentation with information pertinent to CoSR emphasized. In fact, various Health Information Technology (HIT) vendors currently support CoSR through the creation of such patient-centered summary screens. The flexibility to use these screens for different units is another advantage of this design.

The main challenge for this design is the volume of data and the lack of specificity to the particular problems for the shift. Patient information is not organized around a formal problem list or interdisciplinary treatment plan, but rather type of functional data. Thus, data pertinent to a specific problem may be displayed in several different locations. Nurses would have to know the limits of data on the summary screen and memorize what data are stored elsewhere. This will require the nurse to search several places in order to find all the salient data. Nurses will also need to be very facile with the EHR design for efficient CoSR delivery. The nurse must still synthesize the data before passing it on to the next shift. Though this type of display is slightly customizable, nurses may not have the time to fully organize the display to their liking, choosing to work with the default display that is not highly tailored to a specific patient or problems of the day.

3.2 Personally Tailored Design for Nurses

Nurses indicated that using a Kardex and other official unit documentation in hand-offs was more time consuming than using their own informal "scraps" and was, in fact, inadequate [25]. However, researchers and designers should not de-value these as "scraps." Rather they should be valued as the "brains" nurses call them, and treated as potentially valuable tools. Designers may be able to improve documentation and communication by examining these tools' construction and content. Although "brains" are not currently legal medical documents, they likely have influential meaning and information for the delivery of care.

Personally tailored designs would display patient information and the format of the information based on individual nurse's preference, perhaps consistent with the design of their current "brains." Thus, nurses' CoSR displays would take advantage of nurses' own spatial memories. The positive implication of this approach, accordingly, is that nurses can construct their "computerized brains" to be consistent with their own mental schema. This would facilitate CoSR efficiency and effectiveness. Importantly, these documents would become part of the legal medical record. A disadvantage is that nurses may not want their personal notes, currently a part of CoSR, to be included.

The development, design, implementation, and maintenance of this approach would be a challenge. First, the concept of designing a personally tailored system is quite complicated, not to mention the work and cost of maintenance. For vendors, the format and construct of "brains" is not defined or standardized and would be very time-consuming to develop. Tailoring would also be time-consuming during implementation and maintenance. Last, users may not want to use the application if the tailoring mechanism is not precise enough to fit with their mental schema.

Another disadvantage is that this design may be considered incomplete to represent patient information in handoffs. The patient problem list or interdisciplinary treatment plan may be absent and so nurses still need to synthesize and organize patients' problems and care plans in their heads.

3.3 Problem-Oriented Design

This approach would organize information around the main problems identified for the patient – with a huge caveat. The identified patient problems would only be the most pertinent ones for *that* patient, for *that* shift, for *that* unit. Thus, information would be organized as follows: if the patient has a main problem of pain post-operatively, information about pain would be pulled for display - the pain medication order(s), the last time the specific medication was given, the patient response and the condition of the operative site or any other source of pain would be clustered together. Other potential problems such as treated depression would not appear on the display because it would clutter the display with non-relevant information for that day, that shift. The treated depression problem and other potential problems could remain on a separate, comprehensive problem list. The main notion of this design is that it would connect currently disparate information across orders, documentation and the plan of care rather than treating them as separate, functional entities.

The positive implications of this design are that the CoSR support would be similar to the way humans "chunk" information about a particular topic. Rather than having to search across screens for pertinent information related to a specific problem in a clinical system, the EHR would be coded in a smart manner and organized for presentation. Other advantages include that this organization would facilitate other communication such as nurse-physician and interdisciplinary team communication. Potentially, the format would be more efficient for nurses, allowing for a very succinct transmission of the more vital, "just the facts" information and professional judgments would be woven into the discussion about pertinent problems.

The main challenges for this design are not small. For vendors, designers and informaticists, this is a complex design and highly tailored to the individual patient. Nurses may not agree on information for grouping and different nurses may wish for individualized designs rather than problem-based ones. The actual programming for this design would be extensive, the requirements determination more involved, and informaticists would need to budget time to tailor the application to units, conditions and preferences before fielding. However, once this investment was completed, the design would match the way most clinicians think about the care for patients.

4 Conclusions and Recommendations

Change of shift report is a complex transition or handoff in patient care with the majority of information synthesis still occurring in nurses' heads. Past research reveals that CoSR can be informal, unstructured and have missing information. Nurses currently document their notes on personal pieces of paper that researchers call "scraps" and nurses call their "brains." Recent research suggests that the "Just the Facts" content in CoSR is most amenable to computerization. The authors suggest three different designs, a summary screen, personally tailored to the nurse and problem-oriented as potential structures for computerization. Next steps will be to evaluate these potential structures for their viability and fit with the way nurses work and think.

References

1. Feldstein, D.H., Ray, A., Gorman, L., Schuldheis, S., Hers, W.R., Krages, K.P., Helfand, M.: The Effect of Health Care Working Conditions on Patient Safety. Evident Report/Technology Assessment Number 74. AHRQ Publication No. 03-E031. Agency for Healthcare Research and Quality, Rockville, MD (2003)
2. Ebright, P.R., Patterson, E.S., Chalko, B.A., Render, M.L.: Understanding the Complexity of Registered Nurse Work in Acute Care Settings. J. Nurs. Adm. 33, 630–638 (2003)
3. Ebright, P.R., Urden, L., Patterson, E., Chalko, B.: Themes Surrounding Novice Nurse Near-Miss and Adverse-Event Situations. J. Nurs. Adm. 34, 531–538 (2004)
4. Potter, P., Boxerman, S., Wolf, L., Marshall, J., Grayson, D., Sledge, J., et al.: Mapping the nursing process. A New Approach for Understanding the Work of Nursing. J. Nurs. Adm. 34, 101–109 (2004)
5. Potter, P., Wolf, L., Boxerman, S., Grayson, D., Sledge, J., Dunagan, C., et al.: Understanding the Cognitive Work of Nursing in the Acute Care Environment. J. Nurs. Adm. 35, 327–335 (2005)

6. Strople, B., Ottanie, P.: Can Technology Improve Intershift Report? What the research reveals. J. Prof. Nurs. 22(3), 197–204 (2006)
7. Bonomi, J.: Technology to Manage Handoffs. Hosp. Hlth. Netw. 82(8), 6 (2008)
8. Dracup, K., Morris, P.E.: Passing the Torch: The Challenge of Handoffs. Am. J. Crit. Care, 95–97 (2008)
9. Hopkinson, J.B.: The Hidden Benefit: The Supportive Function of the Nursing Handover for Qualified Nurses Caring for Dying People in Hospital. J. Clin. Nurs. 11, 168–175 (2002)
10. Sexton, A., Chan, C., Elliott, M., Stuart, J., Jayasuriya, R., Crookes, P.: Nursing Handovers: Do We Really Need Them? J. Nurs. Man. 12(1), 37–42 (2004)
11. Sherlock, C.: The Patient Handover: A Study of Its Form, Function and Efficiency. Nurs. Stan. 9(52), 33–36 (1995)
12. Ekman, I., Segesten, K.: Deputed Power of Medical Control: The Hidden Message in the Ritual of Oral Shift Reports. J. Adv. Nurs. 22, 1006–1011 (1995)
13. Dowding, D.: Examining the Effects that Manipulating Information Given in the Change of Shift Report has on Nurses' Care Planning Ability. J. Adv. Nurs. 33, 836–846 (2001)
14. Lamond, D.: The Information Content of the Nurse Change of Shift Report: A Comparative Study. J. Adv. Nurs. 31, 794–804 (2000)
15. Kennedy, J.: An Evaluation of Non-Verbal Handover. Prof. Nurs. 14, 391–394 (1999)
16. Munkvold, G., Ellingsen, G., Koksvik, H.: Formalizing Work: Reallocating Redundancy. In: Proceedings of the 2006 20th Anniversary Conference, pp. 59–68. Computer Supported Cooperative Work, Banff, Alberta, Canada (2006)
17. Wolf, Z.R.: Nursing rituals: Doing Ethnography. In: Munhall, P., Oiler-Boyd, C. (eds.) Nursing Research: A Qualitative Perspective, pp. 269–310. National League for Nursing, New York (1993)
18. Wolf, Z.R.: Learning the Professional Jargon of Nursing During Change of Shift Report. Hol. Nurs. Prac. 4(1), 78–83 (1989)
19. Wolf, Z.R.: Nurses' Work. The Sacred and the Profane, pp. 231–290. University of Pennsylvania Press, Philadelphia (1988)
20. Lange, L.L.: Information Seeking by Nurses During Beginning-of-Shift Activities. In: Frisse, M.E. (ed.) Proceedings Sixteenth Annual Symposium on Computer Applications in Medical Care, pp. 317–321. Am. Med. Inform. Assoc., Baltimore (1992)
21. Hardey, M., Payne, S., Coleman, P.: 'Scraps': hidden nursing information and its influence on the delivery of care. J. Adv. Nurs. 32(1), 208–214 (2000)
22. Staggers, N., Jennings, B.M.: The Content and Context of Change of Shift Report on Medical and Surgical Units. J. Nurs. Adm. (in press)
23. Hsieh, H.F., Shannon, S.E.: Three Approaches to Qualitative Content Analysis. Qual. Hlth. Res. 5, 1277–1288 (2005)
24. The Joint Commission. Meeting the Joint Commission's 2007 Patient Safety Goals (2007), http://books.google.com/books?id=VRr9clAyW8C&pg=
PA29&lpg=PA29&dq=Joint+Commission+sbar&source=bl&ots=
ZYtSc7yM6&sig=tY1Fge8UZ2XqWrh81Cg3y68ZSJ8&hl=en&ei=
tdOdSaDCCZC8MpjrtNML&sa=X&oi=book_result&resnum=
10&ct=result#PPA31,M1
25. Edelstein, J.: A Study of Nursing Documentation. Nurs. Manage. 21(11), 40–43, 46 (1990)

Looking for the 3D Picture: The Spatio-temporal Realm of Student Controllers

Monica Tavanti and Matthew Cooper

ITN-VITA
Linköping University, Campus Norrköping, Sweden
monica.tavanti@itn.liu.se, matt.cooper@itn.liu.se

Abstract. Employing three-dimensional displays in Air Traffic Control (ATC) has been the object of study and debates for numerous years. Although empirical studies have often led to mixed results, some preliminary evidence suggests that training could be a suitable domain of application for 3D interfaces. Little evidence, however, is available to fully support this claim. We attempted to fill this gap with a project that aims at studying and evaluating 3D displays for ATC training purposes. This paper describes the first steps of this project, by reporting and discussing the results of a study aiming at understanding whether ATC trainees form a three-dimensional image of air traffic and at comprehending what the nature of this '3D picture' is.

1 Introduction

Air Traffic Control deals with the management of air traffic by Air Traffic Controller Operators (ATCOs) working at airports and in ATC Centers. For maintaining safe separation vertically and laterally between aircraft, ATCOs use several tools, such as Flight-Progress Strips (FPS), radio communications with pilots, and information gathered through two-dimensional (2D) radar displays. In these displays, aircraft are visualized as moving 'blips', and aircraft information is presented in flight labels in the form of text and numbers. In radar displays the horizontal separation is graphically rendered by the relative position of each 'blip' on the display itself, while assessment of separation by altitude is based on the numbers shown in the flight label. Radar displays use a dynamic 2D picture enriched with symbolic information for representing constantly changing air traffic configurations along 4 dimensions (i.e. three spatial dimensions, plus time). It is not surprising, therefore, that for several years the potential of 3D for ATC has been the object of investigation; however the results have always been quite mixed. Preliminary results emerging from the literature suggest that a suitable domain of application for 3D could be ATC training, however, to our knowledge, this issue appears to have received scant attention, and little empirical evidence is available to fully support this claim. We have attempted to fill this gap with a dedicated project that aims at studying, designing, and systematically evaluating 3D displays for ATC training purposes. The present work describes the first steps taken in this project, specifically we report and discuss the results of an investigation aiming at understanding whether ATC trainees form a three-dimensional mental representation of air traffic (as claimed in some of the pertinent literature) and, if so, at comprehending the nature of this '3D picture'. The paper is organized as follows: first an

M. Kurosu (Ed.): Human Centered Design, HCII 2009, LNCS 5619, pp. 1070–1079, 2009.
© Springer-Verlag Berlin Heidelberg 2009

overview of related work entailing 3D displays and ATC is given; then the details of the study and the main results are summarized; the some implications for design and future directions are proposed.

2 Related Work

One of the first studies performed in this area was based on a survey (Burnett and Barfield, 1991) whose results revealed that ATCOs tended to prefer 3D perspective displays for "extracting immediate spatial situational and directional information"; however the results of a comparative evaluation of 2D and 3D displays across tasks entailing terrain scenarios (Wickens and May, 1994) found advantages for the 2D displays. Another comparative study involving weather formation avoidance tasks (Wickens, Campbell, Liang and Merwin, 1995) showed some speed advantages for 2D displays and differences between display types were observed in the strategies used to re-direct aircraft around weather formations. In a replication of a study carried out by Tham and Wickens (1993), Wickens and colleagues (1995) found few differences between three display formats (i.e. planar, perspective, and stereo-perspective) across a number of ATC related tasks, namely higher error rate in the perspective display for speed estimation, slower heading judgments with the stereo display, and quickest with the plan-view display, but no differences were found among displays for the conflict detection task. Brown and Slater (1997) discovered that for tasks entailing judging azimuth angle and lateral distances, 2D yielded better performance than 3D. In a series of ATC related tasks, Van Orden and Broyles (1999) discovered that performance with 2D was as good as or better than with 3D and that 3D Volumetric Display seemed particularly well suited for tasks entailing "perceiving complex, dynamic information relationships in a confined 3D space". A comparison between 2D and 3D stereoscopic display across an altitude judgment task (Tavanti, Le-Hong and Dang, 2003) showed that both controllers and ATC experts performed quicker with the 3D display, but no differences were found in accuracy. A disadvantage associated with 3D is that controllers have no familiarity with this display type, and past experience with 2D displays may play a role in the poor performance with 3D. In a few studies involving both air traffic controllers and pilots, it was observed that the costs associated with 3D were more likely to emerge with ATCOs than pilots (Wickens, 1995). ATCOs perform (and are trained to perform) their tasks with 2D planar displays, and this very experience may be a factor influencing performance whenever the nature of the artefacts in use is dramatically changed. Moreover, 3D also has typical drawbacks that can deteriorate performance: for example, ATC tasks may require precise distance judgments, potentially associated with perceptual biases arising from perspective distortions (Boyer and Wickens, 1994), a problem that challenges the fit between the nature of tasks and the chosen representation. Haskell and Wickens (1993) make a distinction between tasks that require the integration among several dimensions and the ones requiring focused attention on a single source and argue that 3D perspective displays may be viable "whenever the tasks to be performed using the display are integrated three-dimensionally". St. John, Cowen, Smallman and Oonk (2001) argue that 3D views are most useful "for tasks that require understanding the general shape of 3D objects or the layout of scenes"; whereas 2D is mostly suitable for tasks that "require judging the precise distances and angles between objects" (*ibid*). These viewpoints were supported by the results of a set of experiments (St. John et al., 2001) entailing shape understanding and relative position judgments, which suggest that 3D perspective view was superior for understanding objects shape, and 2D was advantageous for determining the relative position of objects. These results are of relevance because the idea that 3D views may support the understanding of 3D environments has some continuity with the use of 3D for ATC training. Indeed, employing 3D for ATC training is not an unprecedented concept, as preliminary evidence in support of its potential already exists. Wickens (1995) reports having observed a sort of "asymmetric transfer effect" when 3D and 2D conditions were counterbalanced: improved performances were observed with 2D when this condition followed the 3D, suggesting that 3D could enhance training, improving performance with subsequent 2D displays. Training is envisioned as a promising area of application (Monteleone, 2006; Wong et al., 2008), and during interviews carried out with ATCOs (Tavanti, 2004) it emerged that 3D could be beneficial for preparing trainees for real ATC tasks. An introductory study carried out by Akselsson and colleagues (2000) indicate that virtual reality has a great potential for teaching and explaining holding patterns operations, for enabling the understanding of the geometrical shape of an airspace sector and possibly for supporting the construction of accurate mental representations. ATC trainers, invited to examine and give feedback about an immersive 3D stereoscopic environment for ATC developed by our group (Bourgois et al., 2003; Lange et al., 2004), have commented that 3D visualizations could enhance controllers' training as these representations are similar to the constructed mental models that the trainee seeks to develop. In summary, preliminary evidence appears to point to two main claims: 1) 3D displays could assist the trainees in visualizing the actual 3D nature of the space in which air traffic operates; 2) 3D views may support the trainees in the construction of 3D mental representations required to manage air traffic. The second claim is of particular importance as, if valid, then attempting to understand the nature of this '3D picture' may give insights for the design of three-dimensional tools for training. We investigated this issue in a series of interviews involving ATC trainees, whose details and results are given in the following sections.

3 The Study

The interviews involved 9 interviewees, 8 trainees (2 females and 6 males) and one training specialist; 3 trainees were of Norwegian nationality, while 5 were Swedes; their ages ranged from 24 to 37. All of them had started their training period in August 2007. After having successfully completed the Basic Module (i.e. basic theoretical knowledge of the ATC work, with only a few practical sessions in the radar simulator) the trainees were enrolled in the rating Module 'Approach Control Surveillance with Radar Terminal Control Endorsement'. This second module (lasting about 18 weeks) enables the trainees to qualify for Approach Control Surveillance rating (with Radar and Terminal Control Endorsements). During this module, the simulator exercises are more intense and frequent, and entail the management of complex traffic scenarios (for example, handling emergencies or unusual events); they usually follow theoretical presentations and are followed by debriefing sessions with the instructors. Within this module there are several training objectives, including the handling of departing, arriving, and over-flying traffic, and cooperation between controllers. All the trainees interviewed were approaching the end of the Module (16th week out of 18). All trainees were fluent in English (English proficiency is also a prerequisite for ATC training admission), and English was the language used during the interviews. The interviews, which were semi-structured, lasted approximately 40 minutes and, upon permission from the participants they were tape-recorded. The trainees were asked to talk about the difficulties encountered during the training (up to the moment of the interview); this was done because we wanted to gain an overall understanding of ATC training from a student perspective. In addition, the students were questioned on whether they experienced forming three-dimensional mental representations of air traffic, and/or to attempt to describe these representations (if possible, even by drawing sketches on a piece of paper). The interviews were quite free, and in fact, other issues (probably characterizing the personal concerns and needs of each student) naturally emerged during the conversations. At the end of the interview, each participant was shown a short presentation composed of eight snapshots of a 3D application for ATC developed by our group at Linköping University. The snapshots illustrated the approach area around Arlanda airport (Stockholm) and displayed different visual features and textures, so as to give the flavor of the possible visualization capabilities of the application. The participants could freely inspect the snapshots and were requested to give their feedback and/or envision possible use during ATC training. Qualitative methodologies were used for the analysis of the interviews. Specifically the analysis was based on qualitative content analysis, which employs a step-by-step approach for organizing the material into content analytical units (Mayring, 2000), according to which categories are tentatively derived, further revised and reduced to main categories (*ibid*). No software package was used for the coding, which was essentially done on paper (i.e. the transcripts of the interviews); several means were used to apply the codes, ranging from annotations on the text, to color codes. Following the principles of Grounded Theory suggested by Charmaz (2006) we tried to define pertinent themes (or categories) able to summarize and explain the interviews content. In doing so, each interview was compared with the others, for example the analysis of the second interview was performed with the first interview fresh in mind; thus, as new themes were discovered within the second interview, it felt natural to go back to the first interview and check whether those themes were present, in a loop of constant comparison. Further, we attempted to link and relate the main categories in order to create a 'consistent narrative' explaining the phenomenon. In Grounded Theory, the theory informing about a certain phenomenon should naturally emerge from the available data, thus potential relationships among categories should also naturally emerge, without any forcing. This particular step was quite complex, as continuous tension was experienced between the need to explain events and facts, and the need to maintain an open attitude towards the data.

4 Results and Discussion

In order to summarize the main results of the study, we start by discussing and rectifying the meaning of the '3D picture', and more precisely to define its character. While tackling this question, we were confronted with variable patterns in the responses: whereas some students reported having experienced this '3D picture', notably 3D mental images of certain traffic arrangements in a rather clear manner (even providing drawings and sketches of these images, of which two examples are given in Fig. 1), others did not. A first hint to find a thread that could make sense of these mixed data came from the interview with the training specialist. She reported that '3D thinking' is a characteristic of every controller, and this ability is specifically tested during the initial selection of the candidates. Furthermore, according to the trainer, only if the students *"can do it* [think in 3D] *can they be controllers"*. When the trainer was asked if during the course she explicitly encouraged the students to imagine air traffic in a 3D fashion, the reply was negative. Thus, a speculation was that students are left to work out their own way to 'think in 3D'. Seen in this perspective, the data started to make sense. As a matter of fact, while examining the different definitions and examples given by the trainees to define their representations of the so-called '3D picture', it appeared that the inconsistencies present in the contents, rather than being conflicting, were probably characterizing idiosyncratic conceptualizations that each student generated in an rather individual manner. These conceptualizations may not necessarily be linked to visual imagery experiences (which clearly materialized in the descriptions reported in some interviews). For instance, expressions used by the students like "seeing the image", "a flash in your mind", or "picture in your mind" may designate the use of visual imagery; however the use of more or less conspicuous verbal expressions in discussing the issue may simply denote a personal ability in describing experienced visual imagery, rather than being an indication of its presence.

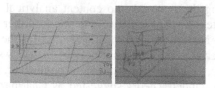

Fig. 1. Two sketches made by students

What appeared more consistently throughout all interviews were the indications hinting at the students' deep understanding of three-dimensional spatial quality of traffic scenarios, a sort of '3D awareness'. The expression '3D awareness' aims at defining the mindful understanding of the spatio-temporal relationships between aircraft, and refers to the comprehension of both current and potential (i.e. anticipated) spatial configurations. Expressions like '3D picture' and '3D thinking' simply denote the underlying awareness that air traffic has an intrinsic three-dimensional spatial quality, involving simultaneous movements of aircraft along three spatial axes. This understanding is essential, but it is not necessarily related to the (personal) propensity to create more or less vivid mental images. The following excerpts of interviews may help to illustrate the fact that the '3D picture' was evoked and described, with more or less vivid expressions and various nuances, but the understanding of the spatio-temporal qualities of air traffic seems persistent. For instance a trainee stated: *"No, I don't have it* [3D picture]...*at least not me"*; but the student's grasp on the 3D spatial character of air traffic unfolds in his words: *"I think I work more with blocks of airspace; for example, if you give an aircraft a flight level 120* [cleared to level 120] *then I just check if that is safe...which means checking that block over the airspace for conflicts, which means if is there another aircraft within...if you have a flight that is going in this direction...this block of airspace is unsafe...for another aircraft to travel in it...you can fly outside this box and everything else is safe, but as soon as you want an aircraft to fly through this box then you must be cautious, that's a conflict".* The *block* and the *box* (three-dimensional shapes) betray a reference to the three-dimensional quality of air traffic spatial relations: checking the block is foreseeing whether certain volumes of the airspace can be safely used or not. Another trainee explained: *"It is like you can see a tube... not a perfect tube, but you see OK, it* [the aircraft] *is over there, it's above, it's below...you have the 2 dimensions on the screen and then you have the information of what heights, it is almost...not a perfect...but some hints at least of some kind of 3D image...You don't have this picture all of it* [all the traffic] *like crystal clear picture, but you have the sense of it...I know when it's all right".* Another trainee declared: *"Most of the time is usually just dots on the screen and it is like you see a box like you should have a thousand feet the separation around them; but if something occurs, if they get too close, you get like a flash in your mind then you see the two airplanes...or if they are very* [much] *closer than they supposed to be, and you're keeping the minima, you just see more in 3D than you would".*

Another issue that emerged from the interviews is that while explaining their personal conceptualization of the '3D picture' the trainees made constant reference to (and provided examples about) assessments entailing current or estimated (i.e. within a certain time window) spatial proximity of aircraft. For instance, one of the students portrayed his own understanding of spatial relationships in terms of cubes and related it to proximate aircraft executing descending and climbing maneuvers, which may require the aircraft to follow crossing paths: *"it is very essential to think in three dimensions...I usually call it to make cubes, you have to make cubes, you have to think about* [that]*...I don't know how to explain but it is just that you have to think in three dimensions, you have to think that you have one* [aircraft] *that is going to go up* [climbing] *and one that is going to down* [descending] *and you have to think about how you're going to make that successful".* Another trainee explained that when aircraft tracks appear as proximate (on the radar display), then the configuration of the aircraft pair is abstracted into a mental image that encapsulates both lateral position and altitude and that describes the spatial relationships of such an arrangement. A quotation can illustrate this point: *"I had in the simulator an aircraft that was on flight level one hundred I think, and another one descending at flight level one one zero... the second one descending did not stop descending to flight level one one zero so... I saw it was on flight level one zero niner and, I was like...Maintain flight level one one zero!* [mimicking screaming on the radio] *then I got that picture, I saw the actual aircraft descending towards the other".* It is safety regulations that prescribe the minimum safety distances (vertical, lateral, and in some cases temporal) that must be kept between aircraft. Hence, the use of terms like *cube, block* or *box* employed to epitomize the '3D picture' relate to separation minima, and ultimately to safety: *"it's kind of... like a box I guess, but it is hard to explain, but it is 3D in your head. In my head it has to be 3D and so...if this is the dot* [the aircraft] *itself, I imagine it being a 3D box, so I think it is three nautical miles here, three nautical miles there and a thousand feet here, so I just keep it in my head ... I try to imagine it that always, at all time that this is a box".*

The last theme that emerged from the interviews involves a learning process that occurred during the first weeks of the rating module, the memories of which were only loosely evoked in the students' accounts relating to first experiences with radar practice became apparent. It appeared that trainees experience a sort of shift or modification in the way radar information was comprehended and interpreted. The complex spatio-temporal world, where aircraft move at different speeds within three spatial dimensions, is simplified and split into diverse, multiform representations on the radar display. For instance, the radar may employ more typically pictorial representations (for example the moving tracks and their relative positions within a 2D Cartesian space) but also symbolic (of which notable examples are aircraft level or speed, represented with numbers on the data-block). Different visual features of the radar may be variably subjected to attentive and pre-attentive processes, as some elements of the representation appear to 'stand out', dominating the scene portrayed in the radar, possibly causing confused responses. The radar representation can be perceived as inconsistent with the reality it intends to signify and describe and, for example, two tracks on the same place in the radar may be 'in reality' safely separated if the flight levels are different, using the words of a trainee: *"now it comes natural because now we have been here for such a while, but in the beginning it was new to think that you have two things [aircraft] at the same place on the radar, but in reality they are not on the same spot"*. Moreover, the numbers displayed on the aircraft data-block may assume an overshadowed role within the visual scenery; the words of a trainee made this aspect very clear: *"in the beginning for me, it was easy for me to think in two dimensions, because we see two dimensions...it's like I see this dot and where it is going, and then it was easy to forget the levels, because the level is just a number, label tagged by a number"*. Knowledge of ATC basics (covered by the Basic Module) is a pre-requisite to gain access to the rating module in which the participants were enrolled; therefore it is legitimate to state that each participant knew fairly well what the numbers in the data-block stood for. Thus, probably, the meaning of 'flight level' was correctly attributed to the numbers on the display, but the conceptual value assigned to the numbers was unlikely to be. The following quotations will help to discuss further the concept of value attribution: *"in the beginning we just had some dots with labels and the speed, and I can see physically where the dot is going, and then after a while I found that this level is quite essential... it is easy to see when they [the aircraft] are quite far apart, but when you're looking at the levels, then you have to make the levels, from a number into a level...so that came second, and then the third part was the speed"*. Thus, understanding levels is 'making levels' out of simple numbers. Progressively, the perceptual information (for example numbers) defining specific spatio-temporal properties of air traffic configurations acquire valuable and specific meaning, and are transformed into conceptual knowledge (levels) associated with a precise role. Although they may appear obvious, three issues deserve to be mentioned further. First, there is a strong temporal quality characterizing the data. The participants used a number of time-related expressions that portray their journey through the training, from the initial impact with radar representations (*in the beginning, then*), up to their current experience thereof (*then, after, now*). Second, the journey does not take place in a void: practical learning occurring during the simulator exercises is the bridge that connects the 'before' and 'nowadays' experiences of the radar representations. Third, the progressive shift towards the generation of concepts from the radar representations is a necessary step to achieve '3D awareness', which seems so crucial for ATC core tasks; but achieving this awareness by gathering air traffic information from 2D radar display is an arduous process. Only with time and practical training, does the perceptual information defining specific spatio-temporal properties of air traffic start to acquire valuable meaning and it is further transformed into conceptual knowledge. Seen in the perspective of design, the initial phases of the practical training could be a suitable niche for exploring the utility of 3D tools. A hypothesis is that 3D displays may provide the students with a more natural view of the spatio-temporal relationships inherent to air traffic, and possibly their use could foster the creation of conceptual knowledge necessary for gaining a thorough awareness of the spatio-temporal relationships of air traffic. This idea was also suggested by the students while inspecting some snapshots of a 3D-ATC application. In fact, a trainee suggested that: *"there's no way of misunderstanding this* [referring to the images of the 3D application]...*you actually see... you see everything much clearer I guess it is really helpful in the start...I think it could be very helpful in the earlier stages... because now we have seen how it works and it is not needed anymore"*.

4.1 Suggestions for Initial Designs

The results of the study along with the feedback given by the students on the potential use of 3D for training can be summarized in a set of initial suggestions for supporting and guiding the design. First, the strict relationship between 3D awareness and safety suggests that 3D representations should focus on the regulations pertaining to separation minima that prescribe procedures and rules to be employed in a number of specific air traffic cases. Suggestions given by the students indicate that using 3D for displaying holding stack management, aircraft sequences in the approach area, aircraft pathways into and/out of airports, may enhance their understanding. In addition, the students declared that sometimes it is complicated to clearly discern from the radar the actual aircraft position during climb and descent maneuvers, making it difficult to evaluate whether aircraft trajectories could cross; 3D representations could help understanding and reasoning on whether aircraft paths intercept. Also, 3D could be used to represent different aircraft behaviors while descending; to use a student's words: *"you'll never know if it* [an aircraft] *is going to* [go] *on a straight* [line], *probably it is not going to go straight up like this, so even if it is pretty fast up it might going to do like this* [mimicking with gestures a sort of stair-step climbing] *for a while...or climb at different rates of climbing, unless you tell the pilot to climb at 2000 feet per minute, then he has to go straight"*. Second, a 3D tool should be (at least at this stage) a 'trainers' device', allowing them to create, add, remove, and modify traffic scenarios on the fly. The simulator exercises usually follow theoretical presentations and explanations, and are followed by debriefing sessions with the instructors. The instructors' explanations may require illustrating and explaining specific traffic configurations on the whiteboard. Thus, the 3D tool could be used as a complement to these explanations *"to illustrate different traffic scenarios, as a demo...I mean teachers and instructors showing us different types of traffic, how you might solve different traffic situations, I mean different solutions... I think that this would be good to illustrate the demos"*. Third, 3D graphical representations should be kept as simple as possible. Students' reactions to a demo of the 3D application were positive overall, but the generous use of colors and the number of features present in the pictures were judged excessive. An idea could be to keep some continuity with the graphical representations of the radar display used by the students, in order to minimize the impact of the transfer between the two representations styles.

In summary, the students provided very helpful suggestions with respect to the potential contents and contexts of use of the 3D display to support training. However, more precise information is needed in order to define in concrete terms some design solutions for the 3D tool. In order to further address these aspects, in the near future, we will involve ATC training specialists and ATC trainees and cooperatively discuss and define in greater detail a few air traffic scenarios and some of the tool's interactive functionalities.

Acknowledgements

"This work has been co-financed by the European Organisation for the Safety of Air Navigation (EUROCONTROL) under its Research Grant scheme. The content of the work does not necessarily reflect the official position of EUROCONTROL on the matter. © 2008, EUROCONTROL and the University of Linköping. All Rights reserved." A special thanks goes to the trainer and trainees who kindly accepted to participate in this study and to Ivan Rankin for reviewing and giving precious feedback on this work.

References

Akselsson, R., Källqvist, C., Bednarek, V., Cepciansky, M., Trollås, A., Davies, R., Eriksson, J., Olsson, R., Johansson, G., Hallberg, B.-I., Håkansson, L.: Virtual Reality in Air Traffic Control. In: Proceedings of the IEA 2000/HFES 2000 Congress, San Diego, California USA (2000)

Bourgois, M., Cooper, M., Duong, V., Hjalmarsson, J., Lange, M., Ynnerman, A.: Interactive and Immersive 3D Visualization for ATC. In: Proceedings of the 6th USA-Europe ATM R&D Seminar, Baltimore, Maryland, U.S.A. (2005)

Boyer, B.S., Wickens: 3D Weather Displays for Aircraft Cockpits. Technical Report (ARL-94-11/NASA-94-4). Savoy, IL: Aviation Res (1994)

Brown, M.A., Slater, M.: Some Experiences with Three-Dimensional Display Design: An Air Traffic Control Visualisation. In: Proceedings of 6th IEEE International Workshop on Robot and Human Communication RO-MAN 1997, pp. 296–301 (1997)

Burnett, M.S., Barfield, W.: Perspective Versus Plan View Air Traffic Control (ATC) Displays: Survey and Empirical Results. In: Proceedings of the 6th International Symposium on Aviation Psychology, Columbus, Ohio State University, pp. 448–453 (1991)

Charmaz, K.: Constructing Grounded Theory: A Practical Guide Through Qualitative Analysis. Sage Publications, London (2006)

Haskell, I.D., Wickens, C.D.: Two- and Three-Dimensional Displays for Aviation: A Theoretical and Empirical Comparison. International Journal of Aviation Psychology 3(2), 87–109 (1993)

Lange, M., Cooper, M., Ynnerman, A., Duong, V.: 3D VR Air Traffic Management Project. Innovative Research Activity Report, Eurocontrol Experimental Centre, Bretigny-sur-Orge, France (2004)

Mayring, P.: Qualitative Content Analysis. Forum Qualitative Sozialforschung/Forum: Qualitative Social Research (2000),
http://www.qualitative-research.net/index.php/fqs/article/view/1089

Monteleone, A.: The 3D Technology Applied to Approach and Airport Environments in the ATC Domain. In: Proceedings of Visualization and Distributed Systems Technologies: the A4D Approach and Beyond, Innovative Research Workshop, Bretigny-sur-Orge, France (2006)

St. John, M.H., Cowen, M.B., Smallman, H.S., Oonk, H.M.: The Use of 2D and 3D Displays for Shape-Understanding versus Relative-Position Tasks. Human Factors 43(1), 79–98 (2001)

Tavanti, M., Le-Hong, H., Dang, T.: Three-dimensional Stereoscopic Visualization for Air Traffic Control Interfaces: a Preliminary Study. In: Proceedings of AIAA/IEEE 22nd Digital Avionics Systems Conference, Indianapolis Indiana, US (2003)

Tavanti, M.: On the Relative Utility of 3D Interfaces. Thesis dissertation, Uppsala University, Sweden (2004)

Tham, M., Wickens, C.D.: Evaluation of Perspective and Stereoscopic Displays as Alternative to Plan View Displays in Air Traffic Control. University of Illinois Institute of Aviation. Technical Report (ARL-93-4/FAA-93-1). Savoy, IL: Aviation Res. Lab (1993)

Van Orden, K.F., Broyles, J.W.: Visual Task Performance as a Function of Two and Three-dimensional Display Presentation Techniques. Displays 12, 17–24 (2000)

Wickens, C.D., May, P.: Terrain Representation for Air Traffic Control: A Comparison of Perspective With Plan View Displays. Technical Report (ARL- 94-10/FAA-94-2) Savoy, IL: Aviation Res. Lab (1994)

Wickens, C.D., Campbell, M., Liang, C.-C., Merwin, D.H.: Weather Displays for Air Traffic Control: the Effect of 3D Perspective. Technical Report (ARL-95-1/FAA-95-1). Savoy, IL: Aviation Res (1995)

Wickens, C.D.: Display Integration of Air Traffic Control Information: 3D Displays and Proximity Compatibility. University of Illinois Institute of Aviation. Technical Report (ARL-95-2/FAA-95-2). Savoy, IL: Aviation Res. Lab (1995)

Wong, W., Rozzi, S., Gaukrodger1, S., Boccalatte, A., Amaldi, P., Fields, B., Loomes, M., Martin, P.: Human-Centred Innovation: Developing 3D-in-2D Displays for ATC. In: Proceedings of the 3rd International Conference on Research in Air Transportation (ICRAT 2008), Fairfax (2008)

A Proposal for "Work-Effective Guidelines" for the Growth of HCD

Haruhiko Urokohara[1], Tsunehisa Yamaguchi[2], Hiroaki Nobuta[2], and Shuichi Kanda[1]

[1] U'eyes Design Inc., Housquare Yokohama 4F, 1-4-1, Nakagawa Tsuzuki-ku,
Yokohama 224-0001 Japan
urokohara@ueyesdesign.co.jp, kanda@ueyesdesign.co.jp
[2] Meidensha Corporation, 2-8-1, Osaki, Shinagawa-ku, Tokyo 141-8565 Japan
yamaguchi-ts@mb.meidensha.co.jp, nobuta-h@mb.meidensha.co.jp

Abstract. These are practical guidelines for promoting HCD in the field of design development. We review the role of guidelines that are more effective than previous methods, and which are based on the principles of HCD.

Keywords: Human centered design, Guidelines, Style Manual, Effectiveness.

1 Introduction

1.1 Effect of Human Centered Design and Expectation from Industry

It is about to be verified that HCD, Human Centered Design, is effective for development of products and services and brings favorable impacts on our business with many practical cases. At the urging of Ministry of Economy, Trade and Industry, we are coming to accept that we can not get ahead of business competition without producing products and services to be safe, secure and comfortable. In the industry, it finally started to work on developing products and services with respecting human living at around the turn of the 21st century, and introduced HCD methods into day-to-day activities.

Looking back on history of introduction of HCD into the development, it began with a visible kind of Design, such as creating Icon and Visual Interface, and secondly GUI Designing including the information design such as Operation Flow Charts and Screen Transition Diagram, and thirdly creating criteria based on Usability Evaluations and standardizing ISO 13407, and then to application of prototypes based on User Profile such as Persona and Scenario. Despite of these movements of exploring its specialty, HCD has been hard to become an interest of engineers, and therefore not adapted as a general method at work field yet.

It is because the tools for practicing HCD are not suitable for the field since these are inadequate and time-consuming in comparison to the tools for designing, manufacturing and analyzation for programming. In addition, as one of the alternatives, there was a trial implementation of Guideline in order to improve engineer's general foundation in HCD. However, the availability is still low for lack of their understanding toward HCD.

M. Kurosu (Ed.): Human Centered Design, HCII 2009, LNCS 5619, pp. 1080–1089, 2009.
© Springer-Verlag Berlin Heidelberg 2009

1.2 What Is "Work-Effective Guidelines"?

It is the practical guidelines for promoting HCD in the field of development and is established by reviewing the role of guidelines that is more effective than any other previous methods through the HCD process based on the principles of HCD.

2 What Is a Guideline?

In general, Guidelines are classified into 3 categories.

1. **Design Principles.** It is the principle and consists of 10 items at the largest. It is not for implementing HCD but for explaining the outline of HCD, in other words, what to be taken into consideration.

2. **Design Guideline.** It is the item which explains guidelines in particular for each designated business field in order to apply guidelines into development. The user is mainly the HCD learner and it requires skills of realizing contents of guidelines into concrete proposals.

3. **Style Guide.** It is the sample or the template provided for engineers and valid only for a specific project. Modification is required by each project and it costs expenses. Therefore, it is necessary to minimize these and to manage effectively in case of evolving it into several projects.

It was in the 21st century that the above kind of classification was organized. However, there was a request for obtaining some guidelines from the earliest use of GUI as symbolized by Icon Design. In fact, so called "Icon Grammar" was proposed, knowledge of Human Engineering and Cognitive Science were schematized, and many of fragmentary approaches were created. In addition, many guidelines were published in excerpted version of the each above category for general use and contributed to society with creating opportunity for beginners to learn fundamentals of development. It has been long since people considered these "Unification of Knowledge" materials (Fig. 1) as guidelines, and now many managers at sections of development still adopt these as their slogan.

Guideline turns out to be effective only when it worked as the way supposed to be most practical and the concerned goods would be in the best quality at the time of production. A good guideline enables engineers in different process to share a concept of basic design to a layout rule, and therefore the guideline can undertake a role of assuring a quality in use of concerned products, playing as a manual. On the other hand, it is neither a magic wand to make high quality of any systems, any products and services nor the one to improve skills of any engineers. We'd better attach importance to process in order to keep making products in good quality and to education and training of engineers in order to improve their skills. We can produce good results with Design Guideline or Style Manual only by identifying "Who is the user?" and "When is it in use?" within a limited purpose and limited bounds.

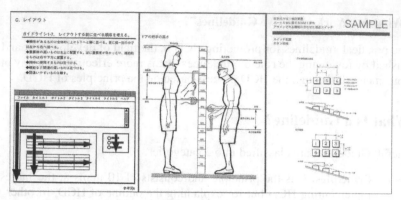

Fig. 1. Unification of Knowledge Materials

Table 1. Procedure for creating the Work-effective Guidelines

Procedure of establishing the Work-effective Guidelines
1. Establishing policy of guidelines
• Confirm the range of application   : to confirm contents of software applications developed periodically
• List up items with users' perspectives   : to select items to be examined from problems at the time of design
• List up items with administrators' perspectives   : to select items for maintaining a specific level of quality
2. Level-setting of guidelines
• General category of guidelines   : Design Manual, Design Guideline and Style Manual
• Level of effectiveness for each user of guidelines   : want to use, be able to use or may use
• Level in acceptance of items in guidelines for each process   : wholly, partially or as necessary
• Level in degree of freedom
• Level in applicable period of guidelines
• Level in achievement of brand strategies
3. Drawing up the contents of guidelines
• The general solution of subordinate items of 7 principles with 12 theories
• Cautions of the particular solution for concerned sections
4.Special items
• Prepare samples for the interaction rule
• Adjust the levels in evocativeness of cautions and alerts
5.The way to work Effective guidelines
• Select the way of providing guidelines (to htmlize etc.)
• Review the way of formatting menu
• Review information to be structured
• Review the way of link information

3 Work-Effective Guidelines

Creating guidelines tends to become a goal, although it is to provide tools for the actual operations. It is because of a bias that a guideline is a document given from the administrator to the developer one-sidedly with such an expectation that references of findings and knowledge of specialists can improve efficiency of development.

It is able to enhance effectiveness, if guidelines are proposed as needed in a proper way according to an accustomed rule, even though it is such a document so called unification of knowledge. We attempt to realize guidelines that are effective to their day-to-day work by understanding the following user information (mainly engineers) properly. Who is the user of guidelines? What kinds of constraints do they work with? How much effect do they expect to obtain from guidelines? (see Table 1).

3.1 Establishing Policy of Guidelines (Contents to Be Considered and the Whole Volume)

This is the most important phase in considering Guidelines. That is, Guidelines have to work correctly and we have to select the right items to review in consideration of HCD skill, motivation, work hours to be usable in working and business environment of actual users and constraint of development system such as the relationship among groups. Moreover, it is better to prepare 2 different types of Guidelines, one is for short term projects and another is for long term projects. It helps to sort out its priorities in case that technical or operational constraint occurred. It is better to find complaint about the present state, expectation to the effect and the workload of providing business information when carrying out the asking on real users in order to confirm their acceptance.

We start from specifying its range and man-hour requirement prior to the establishment of guidelines. It is a work for someone who is specialized in the field of User Interface Design and understands the context of concerned developmental sections. At the phase of planning, specialists divide and mark UI related problems of concerned application into around 3 categories according to the effect extent on Quality in Use based on generalities and experiences and by taking 7 principles with 12 theories. In addition, the provider of guidelines lists up candidate items of guidelines putting the mind to the operational efficiency improvement in the administration section and the whole development section which may be obtainable for designers by transfer of know-how through daily supporting service depending on its quantity, furthermore, ensuring quality in use for the end user in mind.

3.2 Level Setting of Guidelines

General Category of Guidelines. Which one of 3 levels does a guideline aim at? It is enough to make a note in order to explain it externally.

Level of Effectiveness for Each User of Guidelines. We assumed that guidelines are effective if the content is appropriate and engineers of concerned sections refer to these on time, it is able to produce high goods of quality-in-use and increases acceptance of end users.

We define the degree of effect as hereinafter called "the level in the effectiveness" and divide it into around 3 levels.

We classify the motivation of users to guidelines such as whether, 1: want to use it for all the way along, 2: be able to use it workable, or 3: may use it; in order to know especially if these are attractive enough for them to put it into practice. The user of guidelines is listed below:

1. Administrators of design (judge the effectiveness, propose and update items.)
2. GUI designers (understand guidelines and apply it into GUI practice.)

3. System engineers (understand guidelines and apply it into SE practice.)
4. Planners, Sales engineers, Attendants for customers, and company staff in general (understand guidelines and apply VOC into their practice.)

Level in Adoption of Items in Guidelines for Each Process. It is important to manage the process in concerned development sections in order to enhance the effectiveness of guidelines. HCD will never be taken into account by engineers until everyday activities start with establishing the process of development and indicating items of HCD. It is desirable to arrange the process to review HCD, however it would be different among companies, divisions and sections. In addition, it would be different among divisions and sections of the same one company since they have different ways of design among themselves.

We, therefore, list up items applicable to each process of concerned development and define the relationship among items to show how these influence each other. We then establish the acceptance level of items in guidelines for each process such as whether to accept, 1: wholly as required, 2: partially as required, or 3: as necessary along with the development.

There are many HCD factors that are applicable for development unless engineers make mistakes in following the procedure. It is devisable to frame a plan to work linking with a process flow programs and to organize HCD items in such a way as to designate "HCD items to be done today" functionally before the workday begins everyday. It is able to enhance effectiveness if guidelines are proposed in a proper way prepared as stated above.

Level in Degree of Freedom. Guideline is for users who are not familiar with HCD and is not necessary for HCD experts to comply with it. Attitude makes a difference and it is important to provide end users a comfortable operation based on the concept. There may be a requirement to express the state transition in writing, however, for example, in a case that it turns out to be more effective to express these in drawing, picture or movie, users can accept another way. What is important is to establish a new guideline for explanation.

We define the level in degree of freedom into 3 levels: high, moderate and low according to the skill of engineers.

Level in Applicable Period of Guidelines. A change in functions, devices or concerned end users requires modification of design related contents and especially lay out design of GUI. It is therefore necessary to determine the applicable period of guidelines in advance. In general, it is better to stop applying the guidelines to examine whether or not to update them in order to maintain the effectiveness at the possible time of lifecycle of the concerned product, alteration of devices or changes in ways of providing services. There are some ways of determinations regarding use-period like "for permanent", "for a certain period" and "for the concerned project only" and so on.

Level in the Achievement of Brand Strategies. There is a method called "Product Identity". Vehicles of Mercedes, for example, maintain the feature of exterior design and even general people can recognize these as Mercedes. It provides Company Identity as well.

It is able to establish such a product identity with GUI by using its components, although the design of installed OS is different respectively. In addition, it is important for the sections of managing design to aim at branding by making the best use of guidelines. The level is defined into 3 levels based on the recognition: "by end users", "by persons concerned of marketplace" and "by designers and engineers"

If the concerned product is for end users, we can re-define the "recognized by end users" into another 3 levels, for example. What is important is to establish its goal as its circumstances or conditions change.

4 Contents of Guidelines

In order to create guidelines effective, it is indispensable to organize the context of engineer's use, such as "What do engineers get flustered by?", "What kinds of advices do they need in which process of development and for which assignment of design?", "Can they understand without difficulty?" or "What is difficult to understand?" and "In which part of guidelines do they go through with interest?"

We make a prediction of these by repeating the asking to those engineers, however it is ideal to observe their moment of getting flustered by spending time with them. It trains skills of usability engineers as well for developing GUI of applications for business use through observations on real use in such cases.

5 The Way of Practical Operation

We have to review in which way to distribute guidelines to users and take the modifiability into account. It is desirable to open Web Site to provide many users with a high capability of browsing, however it is necessary to prepare paper-based media depending on the way some users may communicate with. In any cases, it is better to plan to establish guidelines with work hours of reviewing the information architecture (including the level of guidelines and the suggesting way of its operation) and the menu-driven interface consideration to the context of designing work in mind

6 Case Study: The Design Section of Meidensha Corporation

6.1 Background

It has been 15 years since they started to GUI design in tandem with the penetration of personal computers. Before that, they designed mainly hardware products styling. There is MES (Meidensha Engineering Standards) in order to standardize internal design activities, however there is a few rules on Usability such as color scheme for hardware products and layout rule for company name plates, therefore, there is no information regarding GUI development taken into account. They started to work on establishing guidelines and it became a good chance to keep up with activities in other development sections where they already worked on it.

6.2 Approach of Establishing the GUI Design Guidelines

At the time of beginning of establishing guidelines, there was no guidebook to refer to and their development work was all at sea except having an opportunity to work with specialists.

They discussed daily projects with engineers and aimed at designing a better product of the easy-to-use. As the number of GUI design increased, they became to realize the effect of what they tried. However, there are not enough designers to develop GUI and even worse they have to engage in works of lower process. This prevent them from engaging in work of upper process and therefore, they started to establish guidelines as "cribs" which enables engineers to do ,by following it, within the scope of work they are able to accomplish for themselves.

6.3 Contents Review of Guidelines

Question Asking. Guidelines must be easy to handle with for "the engineer of Meidensha Corporation" in order to exercise their worthy effects actually. Thus, it is important to decide who is the user (the section to be applied to) and to establish guidelines through the HCD process.

We gave priority to carry out Question Asking on sections that show understanding for HCD or are suitable for having Guidelines. We therefore carried out the Question Asking on System Engineers at 2 sections of advanced study at Tokyo Works / R&D Center and the other 2 sections of core business at Numazu Works.

Scenario of User Personas. Engineers of Meidensha Corporation have a lot of opportunities to work on Network Construction, TC (transmission control), SQL server such as Database Design and RDB (relational database). They have own design background of standing up for requirements of clients since they develop systems for experts, and therefore they express little interest in requirement of Usability as such outside of functionality. There are inquiries on Quality in Use from managers and someone who talk direct to customers, however the priority is decreased on Usability of their own intra-system developments. In such a business environment, they still have to provide GUI in good quality to retain the quality of software and this caused them to establish guidelines for users having less interest toward Usability.

Level-setting for Guidelines

Table 2. Level-setting for Guidelines of Meidensha Corporation

Category in general	Style Guide (including Design Guidelines)
Level of effectiveness by users	
: the design administrator	(Admitted to use)
: the design developer	(Admitted to use)
: the engineer of system development	(Admitted to use)
: the internal staff (of planning, sales, service and others)	Eliminated (use within Q & A)
: the end user	(Permitted to use) for users of unusual cases
Items for each process	
: review the Search Menu for items of guidelines	Out of touch
Level in degree of freedom	
: specify the range of merchantability	High for merchandize of other companies
	Low for merchandize of own company
Level of continuance effect	
: planning of the update time	for permanent
	for the concerned project only
Level of blanding acquisition	Middle level

7 Policy of Providing Guidelines

Although it was desirable to provide guidelines for each design process of each business unit, it was impossible to develop guidelines for a whole company. We then modified the content which is applicable, for each part of a project, for each requirement or for each assignment. (see Fig. 2)

Actual contents are made of information that engineers might request (such as the question asking for engineers and the Q & A on the actual work) and they can answer almost a half of all questions from engineers.

We focused not on the completeness of items on GUI but on encouraging actual users' lower awareness this time. As a result, there are a less and smaller items than originally scheduled. (see Fig. 3)

Fig. 2.

Fig. 3.

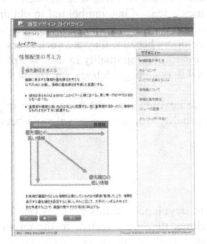

Fig. 4.

Fig. 5.

There are a lot of differences in explaining the details among the items. For example, we explained only the basic rule regarding on color by omitting details since there is a big difference in perception of understanding colors by individuals. About layout rule of screen design, we explained the details including foundations (see Fig. 4) by making the most of our know-how. We mainly listed up the items of visual effects and product identity in order to encourage engineer's understandings for design. We emphasized "things not to be done" (see Fig. 5) as a policy for all guidelines and advised engineers to ask designers for more in details.

8 Comprehensive Bounds for Engineers

For example, we have to specify colors for every case except around 12 colors that engineers can recognize the difference among these. It is necessary to look at separately with the sense of colors used in each private life such as clothes and accessories since the sense of recognizing the difference of the subtle colors would causes the model gap in such an environment that engineers can find the difference of colors they meet at work.

We added some items of justification since there were a few people who pay attention to the layout. It is mainly for designating the things to justify and not necessary to show the meaning of it. Unless an efficient environment is maintained, guidelines would not be approved since it involves immense amount of work-hours. Therefore, we utilize the way of introducing guidelines as a means to gain their attention to the items to be looked at.

9 Issues to Be Solved

In the veteran heavy electronic machinery industry, it has had a priority to functional aspects and is difficult to see potential demands for the user interface through discussions between engineers and clients. It becomes a fear of lagging behind in obtaining expectable good quality of software. It seems that they do not take guidelines as what to be done or as information for software developers because they are such users who do not pay attention to the social movement for improving quality in use. We plan to analyze the access information about the "GUI design guidelines for the software developer" published to the web.

Although the guidelines are for the software developers, we received some inquiries from the hardware developers and, we probably would provide valuable benefits on the whole.

10 Undertakings of Future

The "GUI design Guidelines for the software developer" in this paper is one of the whole idea. We are going to provide guidelines that are effective to the engineers through the work-effective process. We planned to clarify the work area and to designate an entrance to certain guidelines in order to achieve the effectiveness of guidelines, for example, this content is for the related supervisory control and that content for the related dynamic measuring. At the design section of Meidensha Corporation, they plan to establish guidelines of 3 categories below;

(1) Guidelines for product design (for each product including hardware products)
(2) Guidelines for supervisory control design (legibility focused)
(3) Guidelines for GUI design (the one in this paper)

References

1. Nonogaki, H., Kobayashi, Y., Morita, S., et al.: Fujitsu Books: The future of Human Interface – Computers that Commune with Human Being, p. 50 (1992)
2. Japan Ergonomics Society, Ergonomics Design Division & Screen Design Division: GUI Design Guidebook, pp. 215–229, KAIBUNDO (1995)
3. The Society of Instrument and Control Engineers, Human Interface Division: The collection of papers at the 12th Human Interface Symposium pp. 537–542
4. Nikkei Business Publications: Nikkei Electronics pp. 133–140 (2005)
5. Usability Handbook Editorial Committee, Kyoritsu Syuppann Co., Ltd. Usability Handbook pp. 209–215 (2007)

Working in Multi-locational Office – How Do Collaborative Working Environments Support?

Matti Vartiainen

Work Psychology and Leadership, Department of Industrial Engineering and Management,
Helsinki University of Technology, P.O.Box 5500, FIN-02015 TKK, Finland
matti.vartiainen@tkk.fi

Abstract. Multi-locational, distributed and mobile work has increased much during last years enabled by wireless connections, mobile devices and internet. This development provides possibilities to arrange work in new ways by using physical, virtual and social spaces in creative manners. There are, however, some hindrances in these very same environments that prevent achieving all of potential benefits as shown in this study. The analysis of the first phase of a developmental process shows that political decisions, organizational culture issues, costs and availability of technologies, and missing competences may slow down the implementation of the 'Multi-ocational Office Model'.

Keywords: Working in multiple places, mobility, e-governance, CWE.

1 Purpose and Research Questions

This study explores the change of a local organization into a distributed organization whose employees can work and collaborate from many locations. Operating from different locations – and while moving between them - requires using of communication infrastructures and information and communication technologies to support both solo work and collaboration with others. Therefore, it is critical to find out, what kinds of physical, virtual and social/mental spaces enable rich communication and collaboration in a network of people doing project work that often requires access to joint data stores from afar and whose contents very much require problem-defining and -solving. Supportive collaborative working environments (CWE) are defined as a combination of physical, virtual and social or organisational infrastructures supporting people in their individual and collaborative work. The research question of the study is: What kinds of requirements the change from co-located work into multi-locational work sets for collaborative working environments?

2 Working from Multiple Places

Wireless networks, mobile devices and internet provide a lot of new possibilities to organize work and collaborate from afar. The need for new working solutions is evident as the prevalence of multi-located working has increased rapidly during the last ten years and will continue to do so. In Europe, telework, including home-based telework

M. Kurosu (Ed.): Human Centered Design, HCII 2009, LNCS 5619, pp. 1090–1098, 2009.

(at least one day/week), supplementary home-based work, mobile eWork, and freelance telework from small home offices, increased from six percent in 1999 to 13 percent in 2002 [4]. The Fourth European Working Conditions Survey [13] gathered in 2005 showed that only 51 percent of the working population in the EU worked at their place of work all the time and that a total of 21 percent never worked at their workplace. This indirectly shows the increased portion of mobile working from multiple places. Furthermore, 9 percent of workers always work in locations that are outside the home and company premises. The WorldatWork 2006 Telework Trendlines report [9] shows that the sum of teleworkers (both employed and self-employed) working remotely at least one day per month in the U.S.A. had risen by 10 percent, from 26.1 million in 2005 to 28.7 million in 2006. Based on the U.S. government estimates of 149.3 million workers in the U.S. labor force, the 2006 data mean that roughly 8 percent of U.S. workers have an employer that allows them to telecommute one day per month and roughly 20 percent of the workforce engages in telework. It was estimated that 100 million U.S. workers will telework by 2010. The technological enablers are the increased use of broadband connections at home and wireless access to the internet from anywhere. Work has become multi-locational. The change into new way of working challenges not only technology that is virtual spaces but also the use of physical premises as well as social and mental spaces.

3 Types of Mobile and Multi-locational Work

Individuals working from multiple locations usually use virtual tools for collaboration with others; that is, they work in distributed virtual teams. Next, individual mobility and mobility as a feature of distributed work are discussed in more detail.

3.1 Individual Mobility

At the individual level, 'telework' and 'remote work' are terms that have been used to refer to all kinds of work and work arrangements carried out outside a main office but related to it [1,8,12,15]. The use of information and communications technologies as communication links between the teleworker and the employer was brought as a feature to the telework concept quite early, which often meant home-based telework [8]. Additionally, making full nomadicity possible by developing portable computers and communication devices was required [7]. In Continental Europe, the term 'eWork' was later used to refer to all those work practices that make use of information and communication technologies to increase efficiency, flexibility (in terms of time and place), and the sustainability of resource use. It is evident that most employees in post-industrial societies use information technologies in their work, though the degree of use varies a lot. eWork includes the following specific types of work, and one of them is mobile work [3,8]:

(1) *Home-based telework* or homeworking [5,15] is the most widely recognized and best-known type of eWork and telework. Many teleworkers divide their time between the home and the office, and they are therefore called 'alternating teleworkers'. Individuals who spend more than 90 percent of their working time at home are called 'permanent teleworkers'. 'Supplementary teleworkers' are those who spend less than one

full day per week teleworking from home. They are also called 'occasional telework-ers', to distinguish them from regular teleworkers.

(2) *Self-employed teleworkers in SOHOs* (Small Office Home Office) are private entrepreneurs, such as consultants or plumbers, working and communicating with their contractors, partners, and clients by means of new technologies. The critical difference between teleworkers in SOHOs and home-based teleworkers is their mar-ket position as self-employed.

(3) *Mobile workers* are those who "spend some *paid* working time away from their home and away from their main place of work, e.g. on business trips, in the field, travelling, or on a customer's premises" at least once per month. Lilischkis [9] calls this type of working in many places multi-locational work. Halford [5] used 'hybrid workspace' to describe the combination of organisational, i.e. 'office', and domestic, i.e. home, spaces mediated by cyberspace. Hislop and Axtell [6] added a third dimen-sion of 'locations beyond the home & office' to this concept of 'hybridity' and de-fined this type of multi-locational work as 'mobile telework'. *High-intensity mobile workers* are those who do so for 10 hours or more per week outside their primary workplace and use ICT for communication [4]. In conclusion, the terms 'mobile work' or 'multi-locational mobile work' or – why not? – 'mobile telework' could replace the traditional 'telework' in the case that work takes place with the help of ICT in and from multiple locations and while moving between them.

3.2 Mobility as a Feature of Collaborative Work

When considering mobility and the use of several places for working from the view-point of distributed group work and collaboration, mobility is just one feature, and it may concern one employee or all the team members [6,16] or the whole organization [14]. Bell and Kozlowski [2] proposed that the variety of goals and tasks, contexts, and processes needed for internal regulation "produces" different types of teams. Common goals and tasks vary according to their complexity, i.e. tasks are routine or creative, and they are interdependent to a greater or lesser extent. This results in dif-ferent communication richness needs; complex tasks require rich media. To illustrate the contextual requirements of collaborating groups, Vartiainen [17] used the follow-ing six factors, which each can be measured with several indicators, to describe and to profile the types of groups: 'location', e.g. the number and distribution of places from where team members work; 'mobility', e.g. the share of physically moving employees in a group; 'time', e.g. the degree of solo work and synchronous or asynchronous collaboration between group members; 'temporariness', e.g. the duration of coopera-tion and the number of groups each member participates in; 'diversity', e.g. differing cultural backgrounds among group members, and the 'mode of interaction', e.g. the frequency of face-to-face vs. virtual meetings for communication. Groups differ in these factors and multiple combinations are possible, producing groups and teams with different profiles and working requirements. The task content and the context characteristics of a group together create needs to communicate and organize intra-group processes in such a manner that the team can survive and prosper.

Summarizing, it can be seen that collaboration in groups and teams is complex, be-cause their purposes, tasks, working contexts, and the intra-group processes needed to adapt and work vary greatly. All these factors are inter-linked in such a way that a change in one of them influences others. Therefore, only rough categories of group types can be presented one of them being mobile and multi-locational groups [18]. Conventional groups and teams differ from distributed, virtual, and mobile teams especially in three characteristics: the geographical distance between their members, the mode of interaction, and physical mobility. Conventional groups and teams are co-located, communicate face-to-face, and work towards a joint goal here and now.

The main types of non-conventional teams are: (1) distributed; (2) virtual, and (3) mobile virtual teams. Team members working in different locations and at a geographical distance from each other make a distributed team. A team becomes virtual when group members communicate and collaborate with each other from different locations via electronic media and do not meet each other face-to-face. The physical mobility of group members adds a new feature to distributed collaboration. Mobile, virtual teams are always distributed, but not all distributed, virtual teams are mobile. Virtuality, as in the use of ICT for communication and collaboration, makes a team into a distributed virtual team or mobile virtual team. It can be said that mobile virtual teams are the most complex types of teams to lead and manage.

4 Analysis of Working in Multiple Locations

As shown above, physically mobile work is in fact fictitious, as it invariably takes place in some location, whether it is a car or a customer site. In the case study below, the requirements for the design of a new 'Multi-Locational Office' is explored by using the concept 'ba' proposed by Nonaka et al. [11] as the methodological basis for the requirement analysis. 'Ba' roughly means 'place', referring to a shared context in which knowledge is created, shared, and utilized by those who interact and communicate there. 'Ba' unifies the 'physical space', such as an office space, the 'virtual space', such as e-mail, and the 'mental' or 'social space', such as common experiences, ideas, values, and ideals shared by people with common goals as a working context. The key point is that these spaces are embedded. In this study, multiple workplaces of employees are analyzed by using these embedded space categories in the following manner [10,17]:

- A 'physical space' refers to those physical places that employees use for working while moving from one place to another. They are divided into five categories: (1) home; (2) the main workplace ('main office'); (3) moving places, such as cars, trains, planes, and ships; (4) a customer's and partner's premises or one's own company's other premises, and satellite and telework offices ('other workplaces'), and (5) hotels and cafés etc. ('third workplaces'). The use of physical places can be described by different indicators, such as their distance from each other (near – far), their number (one – many), and the frequency with which they are changed (seldom – often). The indicators can then be used to describe the degree of mobility.
- A 'virtual space' refers to an electronic working environment or virtual workspace consisting of various infrastructures, tools and media for individual employees, groups, and whole organizations. The internet and intranet provide a platform to communicate, collaborate, and find knowledge, both with different tools, such as e-mail, audioconferencing, videoconferencing, chat, group calendars, document management, and presence awareness and findability tools, and with integrated

electronic collaborative working environments, such as various groupware systems and combinations of social media such as blogs, wikis, instant messaging, chat, and other communications systems that host many-to-many interactions and support group and community interaction. The use of virtual workspaces can be analyzed and described by focusing on connections, devices, and services and on their purposes, functionality, and usability.

– A 'social space' refers to the social context and the whole social network where working takes place; that is, for example, other team members, managers, and customers. Social space creates the social capital of an organization. Network analysis is often used to explore the ties and relationships of individual members, such as "advising" and "not advising" or "helping" and "not helping".

– A 'mental space' refers to individual cognitive constructs, thoughts, beliefs, ideas, and mental states through which an employee interprets the other spaces. A mental space can be shared with others. Creating and forming joint mental spaces requires communication and collaboration, such as exchanging ideas in face-to-face or virtual dialogues. Social and mental spaces are usually studied by collecting individual perceptions, attitudes, and conceptions, and then analyzing their contents.

In conclusion, workplaces are combinations of physical, virtual, social, and mental spaces, especially in collaborative work. These spaces form a collaborative working environment, which can support or hinder working. The use of various spaces varies, depending on the type of work and the interdependence of the tasks to be done. Individual telework at home in solitude without virtual connections to others is an extreme and rather rare case. Usually, home-based teleworkers communicate sporadically with superiors and colleagues face-to-face by commuting to the main office.

When employees are working in multiple locations, the combination and emphasis of their spaces are different and variable from co-located employees, just because of the greater number of physical places they rotate through and use [6]. Still, they need not communicate virtually. The significance of virtual spaces grows when the members of a distributed team have to communicate and collaborate with each other from different locations. They are not only distributed in physical places but simultaneously use virtual places (videoconferencing and documents shared on the intranet), and are also related to other team members who must share common goals (social space) to be able to reach the aim, and possibly also share common ideas, beliefs, and values (mental space).

5 Case: Requirements for Multi-locational Office

Next, findings of a case study concerning the first phase of a change process in a government agency are shown. The agency is to move from Helsinki metropolitan area to the other part of the country based on the government's decision. The whole process of moving the office is scheduled to take place in four years 2008-2011. As there are also other agencies of state to be moved in the future, this case is used to create and test the new model of 'Multi-Locational Agency'. In order to create favorable conditions for working in new ways, a careful requirement and need analysis was first done in order to identify critical hindering and enabling factors in collaborative working environments. This knowledge can be used to design future physical premises, information and communication tools and infrastructures as well as organizational structures of the multi-locational office.

5.1 Background of the Change

The object of analysis is a governmental agency (n=~206) moving from Helsinki metropolitan area to a small city in the western part of Finland during 2008-11. The decision to move was based on the government's decision to distribute and transfer state workplaces outside the metropolitan area. The moving agency belongs to the Ministry of Agriculture and Forestry. Its task is to provide and monitor economical support to farmers all around country. Moving is planned to be finalized till 2011 when most of the employees should have their workplace in the new location. A research project was set up to explore possibilities to realize a new type of 'multi-locational office' that is working together but operating from many different locations. The study is carried out as a follow-up study and action research. In the first phase, the prerequisites of multi-locational working were studied by organizing a future workshop, by collecting existing documents, e.g. work descriptions, by interviews and a survey for all the personnel. A future workshop was meant to build a joint vision and a model of new ways of working. In the second phase 2010, the interventions and the change in all will be evaluated. The purpose of the first phase was to study hindrances and enablers to change the mode of working. This paper focuses on the physical, virtual, social and mental prerequisites of the whole work system to transform it into a new type of multi-locational office.

5.2 Hindrances and Enablers of the Multi-locational Office

Physical spaces. During the first phase of the change, the agency was already partly working in a distributed manner as its main premises are temporary and many employees work in different places. In the target city, the temporary office premises have been hired based on the needs and number of moving employees expecting the new main office building to be ready in 2010. The agency provides possibilities to some of its employees to do home-based telework. In addition, in the future there will be three areal offices in three cities for distributed work. Employees also use and will use other premises of the state employer like the Employment and Economic Development Centres (T&E Centres) around the country. Also trains and other moving vehicles are used as working places. Additionally, some employees visit farms for checking the use of monetary support. They should have access to the agency's data resources and possibilities to communicate and collaborate with their colleagues wherever they are. Moving to the new workplace will be finalized in 2010 when the new office building is ready in the new working site. Critical questions concerning the new premises are: what kinds of workplaces are needed by those who travel a lot and visit the main office only occasionally for important face-to-face meetings, as well how to create places for those who collaborate from afar? The future structure of the agency is given in Figure 1.

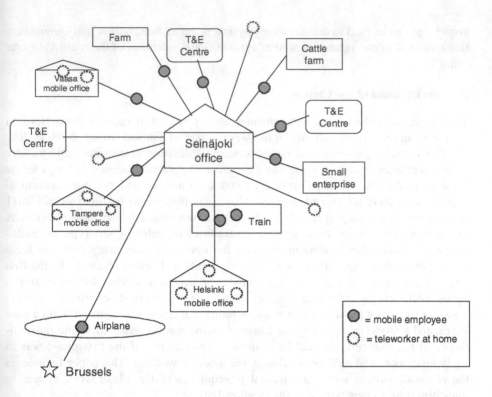

Fig. 1. The structure of the multi-locational office in the future

Virtual spaces. The agency is a workplace requiring high standards and quality of information security. IT-department consists of 25 employees responsible for electrical processing of data needed for providing economical support to farmers and monitoring its use for the state and European Union. As well the department acts as the support function for employees in their use of information and communication technologies. This role will strengthen in the future when tools are implemented. The implementation of the multi-locational office will require a new place-independent virtual private network (VPN) to guarantee secure at least partly wireless connections from remote places and for remote users. In addition to this investments are needed for such mobile devices like smart phones that guarantee access to data bases and virtual meetings from afar, for example when an employee visits a farm for inspection. Already now there are videoconferencing systems to be used for virtual meetings between sites. When new technologies are purchased its reliability is critical as well as training employees to use it.

Social spaces. The agency has a short history as it was formed just a couple of years ago by merging units from other establishments. This brought along different organizational cultures creating sub-cultures without joint identity in the existing agency: the organizational climate is seen as open and based on trust by others and as closed and bureaucratic by others. The organizational climate survey shows differences in satisfaction with leadership, support for developing, and information flow between the units of the agency. In order to guarantee a high-quality performance during the change and after it, the management of the agency has developed operating principles for telework, distributed work and work time flexibility. The telework agreement is provided for some employees who mainly work at home. The agreement defines the suitable jobs as autonomous without continuous need to be available to others and working on public documents. Because the agency handles every year about 2,2 billions EUR of monetary support, its operations are highly confidential and require high standards of information security. This limits the possibilities to telework at home in the agency, which is now two days per week. Distributed workplaces are provided in three areal offices. The employees are, however, requested to work at the main office on weekly basis. Some flexibility in working times are provided for employees during the change process: starting work between 6-10 AM, on Mondays 6-11 AM, and finishing 2-7 PM, on Fridays 1-7 PM. During the six-month period, the working time balance should be +40/-10 hours.

Mental spaces. One third of the present employees have not been willing to move, and the turnover has been 16 percent during the first 18 months of the change. As well one third is willing to telework and distributed work, others prefer an old working style. This has created the challenge of how to preserve expertise in the agency as the tasks require it. There is not only resistance to change but also missing competences of how to work in the flexible manner and how to lead and manage employees who are not under direct supervision. In all, attitudes and motivation for change are not very high.

6 Conclusion

The analysis of the first phase of the change process in this case study shows that implementing new ways of working and organizing meets many hindrances, which seem difficult to surpass. The hindrances can be found in all the four spaces that form the collaborative working environment. In practice, the new multi-locational office is a compromise of old ways or work and new possibilities enabled by ICT.

References

1. Andriessen, J.H.E., Vartiainen, M. (eds.): Mobile Virtual Work. A New Paradigm? Springer, Heidelberg (2006)
2. Bell, B.S., Kozlowski, S.W.J.: A Typology of Virtual Teams. Implications for Effective Leadership. Group & Organization Management 27, 14–49 (2002)
3. ECATT: Benchmarking Progress on New Ways of Working and New Forms of Business Across Europe (ECATT final report. IST programme, KAII: New Methods of Work and Electronic Commerce), Brussels (2000)
4. Gareis, K., Lilischkis, S., Mentrup, M.: Mapping the Mobile eWorkforce in Europe. In: Andriessen, J.H.E., Vartiainen, M. (eds.) Mobile Virtual Work. A New Paradigm?, pp. 45–69. Springer, Heidelberg (2006)

5. Halford, A.: Hybrid Workspace: Re-spatialisation of Work, Organization and Management. New Technology, Work and Employment 20, 19–33 (2005)
6. Hislop, D., Axtell, C.: The Neglect of Spatial Mobility in Contemporary Studies of Work: the Case Telework. New Technology, Work and Employment 22, 34–51 (2007)
7. Kleinrock, L.: Nomadicity: Anytime, Anywhere in a Disconnected World. Mobile Networks and Applications 1, 351–357 (1996)
8. Korte, W.B., Wynne, R.: Telework. Penetration, Potential and Practice in Europe. IOS Press, Amsterdam (1966)
9. Lilischkis, S.: More Yo-yos, Pendulums and Nomads: Trends of Mobile and Multilocation Work in the Information Society. STAR, Socio-Economic Trends Assessment for the Digital Revolution, Issue report no 36. Empirica, Germany (2003)
10. Nenonen, S.: The Nature of Workplace for Knowledge Creation. Turku Polytechnic, Research reports 19, Turku (2005)
11. Nonaka, I., Toyama, R., Konno, N.: SECI, Ba and Leadership: a Unified Model of Dynamic Knowledge Creation. Long Range Planning 33, 5–34 (2000)
12. Olson, M.H., Primps, S.B.: Working at Home with Computers: Work and Nonwork Issues. Journal of Social Issues 40, 97–112 (1984)
13. Parent-Thirion, A., Fernández Macías, E., Hurley, J., Vermeylen, G.: Fourth European Working Conditions Survey. In: European Foundation for the Improvement of Living and Working Conditions. Office for Official Publications of the European Communities, Luxembourg (2007)
14. Schaffers, H., Brodt, T., Pallot, M., Prinz, W.: The Future Workspace. Perspectives on Mobile and Collaborative Working. Enschede, Telematica Instituut (2006)
15. Sullivan, C.: What's in a Name? Definitions and Conceptualizations of Teleworking and Homeworking. New Technology, Work and Employment 18, 158–165 (2003)
16. Van der Wielen, J.M.M., Taillieu, T.C.B.: Telework: Dispersed Organizational Activity and New Forms of Spatial-temporal Coordination and Control. Paper presented at the Sixth European Congress on Work and Organizational Psychology, Alicante, Spain, April 14–17, 1993. WORC Paper 93.04.006/3 (1993)
17. Vartiainen, M.: Mobile Virtual Work – Concepts, Outcomes and Challenges. In: Andriessen, J.H.E., Vartiainen, M. (eds.) Mobile Virtual Work - A New Paradigm, pp. 13–44. Springer, Heidelberg (2006)
18. Vartiainen, M., Andriessen, J.H.E.: Virtual Team-Working and Collaboration Technologies. In: Chmiel, N. (ed.) An Introduction to Work and Organizational Psychology – a European Perspective, pp. 209–233. Blackwell, Oxford (2008)
19. WorldatWork 2006 Telework Trendlines commissioned from The Dieringer Research Group (2006), http://www.workingfromanywhere.org/ (retrieved July 19, 2008)

Human Centered Design of Mobile Machines by a Virtual Environment

Hassan Yousefi, Amir Mohssen Soleimani, and Heikki Handroos

Institute of Mechatronics and Virtual Engineering, Faculty of Tech., Lappeenranta University of Tech., Finland
Yousefi@lut.fi

Abstract. Psychomechatronics is a new holistic discipline that integrates mechatronics and cognitive science, offering innovative methods that lead to the concurrent design of human–machine systems. According to current methods, mechatronics features are designed first and human factors are considered thereafter. A problem with this approach is that it is often too late to impose significant changes in the mechatronics design at that time. Psychomechatronics, on the other hand, takes the nature of human cognition as the starting point of systems design and therefore it is a human-centered design principle. It does not adhere to the conventional sequential approach but applies simultaneously both mechatronics and cognitive science at the same conceptual stage, which optimizes the design of hybrid human–machine systems. The paper presents the key ideas of psychomechatronics design method with special reference to mobile machinery. The required virtual environment for carrying out the psychomechatronics design of mobile machines is described. The paper presents the results of applying usability test in a virtual environment for a sample mining machine.

Keywords: Human–machine systems, Human-centered design, Usability.

1 Introduction

Virtual Prototyping (VP) is becoming a commonly adopted design and validation practice in several industrial sectors. Companies are moving from expensive physical models of designs to digital (virtual) models. Compared to physical models, virtual prototypes are in general less expensive, easily configurable and support variants, and allow for several simulations to run on a single model. Moreover, tests are repeatable, and the results of validation are often immediately available for product design review. Virtual prototypes often provide insights that physical testing would not reveal. Even if VP does not completely substitute physical models, it helps optimizing and eliminating redundancy in test facilities, accelerating life testing, and reducing the overall number of physical models used in the product lifecycle. Today, VP has its main focus on the late concept and engineering analysis stages of the product development process [1].

Recent trend aims at also using VP earlier in the concept stages, when product design is not too much detailed and changes do not heavily impact on the product

M. Kurosu (Ed.): Human Centered Design, HCII 2009, LNCS 5619, pp. 1099–1108, 2009.

development process. Used in the conceptual phase, VPs offer the possibility of evaluating cognitive concepts improving the product quality, and better exploiting designers' activities. For these reasons, this practice is rapidly catching on in engineering design as well as human-centered design principle. Most advanced VP systems are based on Virtual Reality (VR) technologies. Visual techniques have rapidly evolved in the last decades, providing new devices supporting realistic rendering, stereo viewing, and immersive experiences [2]. Conversely, research and development of interactive devices have provided less innovative and effective solutions.

The three dimensional (3D) devices, like 3D mice and joysticks, support a more realistic and intuitive interaction with 3D models [3]. Since a few years, some digital design tools allow users to physically get in touch with the design while working on a computer.

Physical interaction with the new product is considered important because designers can explore the product's shape and style, and evaluate proportions. It is an intuitive modality for modeling new shapes, and for testing and evaluating product functionality and ergonomics [4].

New modeling systems are being developed which allow designers to use their usability skills while working in the virtual environment. The potential of such technologies that allow a less constrained, more natural and intuitive interaction with virtual models has increased the drive towards computer support for the whole design process, in particular for conceptual design [5, 6]. The idea of psychomechatronics design method with special reference to mobile machinery is at the basis of the research work described in this paper. This study aims at developing psychomechatronics design methods and tools for the modeling and simulation of human-operated mobile machines that perform complex tasks in unfriendly conditions. Instrumental to the success of the project is the development of a simulation platform that creates an immersive virtual environment that offers virtual user experience of the real machine.

2 Psychomechatronics

Psychomechatronics is a new holistic discipline that integrates mechatronics and cognitive science, offering innovative methods that lead to the concurrent design of human–machine systems. Psychomechatronics, on the other hand, takes the nature of human cognition as the starting point of systems design and therefore it is a human-centered design principle. It does not adhere to the conventional sequential approach but applies simultaneously both mechatronics and cognitive science at the same conceptual stage, which optimizes the design of hybrid human–machine systems. The major benefit of the psychomechatronics approach is that it enables the development of human-operated machines that enhance the performance both of operators and the machine.

2.1 Mechatronics Systems

A mechatronics system consists of mechanical and electronic components. The function of the whole system, however, cannot be explained fully by the examination of the function of each component because the final function is produced as a result of the interaction of the functions of the components [7, 8]. Fig. 1 shows the interconnections within a mechatronics human-machine system. It can clearly be concluded in the figure that a human driver forms a dynamic system with the machine through his sensory-motor skills. In semi-autonomous, it is often needed to construct artificial stimulation of human sensory functions in order to substitute the natural impacts and loadings of machine to human. Also because the modern machines are more and more fly-by-wire systems it is possible to filter out the unwanted behavior of the driver to increase safety and usability of machine.

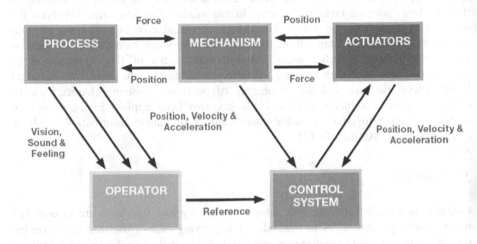

Fig. 1. Interconnections within a human operated mechatronics machine

2.2 Cognitive Science

Cognitive science may be concisely defined as the study of the nature of intelligence. It draws on multiple empirical disciplines, including psychology, philosophy, neuroscience, linguistics, anthropology, computer science, sociology and biology. The experimental research methods in cognition science range from questionnaires and statistical analyses to body tracking, pulse and blood pressure measurements [9]. The experimental research methods in cognition science range from questionnaires and statistical analyses to body tracking, pulse and blood pressure measurements [9]. According to current methods, mechatronics features are designed first and human factors are considered after that. The problem with this method is that it is often too late to impose significant changes in the mechatronics design at that time. To solve the raised problem, we suggest using the dynamics model of mechatronics system in a virtual environment and performing the usability test to find the problems of design. The modified machine is more usable for operators because of applying ergonomic and usability criteria in the design process.

Cognitive task analysis (CTA) is an emerging approach to the evaluation of mechatronics systems that represents an integration of work from the field of systems engineering and cognitive research in machine operating. It is concerned with characterizing the decision making and reasoning skills and information processing needs of subjects as they perform activities and perform tasks involving the processing of complex information [10]. There are a variety of approaches to cognitive task analysis. The approach to cognitive task analysis as described in this paper is closely related to that described by Means and Gott [11], which has been used as the basis for development of intelligent tutoring systems. The first step in such a cognitive task analysis is development of a task hierarchy describing and cataloging the individual tasks that take place during operation of a mechatronics system (with or without the aid of information technology). Once tasks have been identified, the method typically involves observation of subjects with varying levels of expertise as they perform selected tasks of interest. In our studies this has often involved the subjects carrying out the task while using a virtual environment system. Our approach, described in detail below, typically involves video recording of subjects as they work through selected tasks. An important focus of this approach is to characterize how user variation (e.g., differences in users_ educational or technical level) affects the task and the occurrence of potential problems characteristic of different types of subjects studied. CTA has also been applied in the design of systems in order to create a better understanding of human information needs in development of systems [10, 12].

3   Usability

Usability is a quality attribute that assesses how easy user interfaces are to use. The word "usability" also refers to methods for improving ease-of-use during the design process. It is important to realize that usability is not a single, one-dimensional property of a user interface. Usability has multiple components and is traditionally associated with these five usability attributes; learnability, efficiency, memorability, errors and satisfaction [13, 14]. Usability applies to all aspects of a system with which a human might interact, including installation and maintenance procedures [14]. Usability testing forms the cornerstone of most engineering systems. In the study a combination of two important protocols, think aloud and observation methods has been used.

3.1   Think-Aloud and Observation Methods

There are many possible ways of combining the various usability methods, and each new project may need a slightly different combination, depending on its exact characteristics. A combination that is often useful is that of heuristic evaluation and thinking aloud or other forms of user testing. In the study, combinations of task-based *think-aloud* and *observation* protocols were used. Thinking Aloud protocol is a popular technique used during usability testing. During the course of a test, where the participant is performing a task as part of a user scenario, the participants are asked to vocalize his or her thoughts, feelings, and opinions while interacting with the product.

Simply visiting the users to observe them work is an extremely important usability method with applications both for task analysis and for information about the true field usability of installed systems [13, 14]. Observation is really the simplest of all usability methods since it involves visiting one or more users and then doing as little as possible in order not to interfere with their work.

4   Virtual Reality in Psychomechatronics Design Process

Human-machine interaction has been under intensive research during the last decades [15]. The drawback of existing research environments is that the real products or their physical prototypes are needed in order to carry out tests with real human users. It is then expensive and slow to carry out modifications leading to better usability and user comfort and, on the other hand, better durability, lower energy consumption etc. In order to investigate the human behavior as a part of a mechatronics system such as a mobile mining machine an advanced simulation platform is required. The environment should compose at least of the following subsystems; Immersive visualization, haptic or joystick etc, Mathematical model of the machine, solver for simulating the model in virtual reality environment, sound system, and motion platform for feeling of the real machine motion.

By using the virtual environment the psychomechatronics R&D of a mobile machine can be carried out by applying the following procedure:

1. Preliminary conceptual design of the machine which is carried out by using engineering methods and psychomechatronics knowledge.
2. A real-time dynamic multi-body system model for the machine is built by appropriate formulation of equations of motion to guarantee real-time solution. This model should include description of drive, electronic control system and work process
3. Visualization and sounds from the work process and machine are created
4. A model of a preliminary interface and cabin is created in the virtual environment.
5. Test participants and the proper scenarios or tasks are selected.
6. Cognitive tests are carried out with test participants and the virtual machine.
7. The usability analyses are carried out.
8. Modifications in the interface/machine are proposed and the models are updated correspondingly.
9. Phases 6–8 are repeated until satisfactory results are obtained both from the user's point of view and the engineering point of view
10. The detailed design is carried out and using engineering methods

Fig. 2 illustrates the psychomechatronics design process of a mobile machine in a flow chart form.

It is quite obvious that during the iteration cycles in the virtual world qualitative evaluations of the machine to be developed can be given. Qualitative terms such as 'good', 'comfortable' or 'perfect' can be used in evaluating the virtual prototype. They cannot, of course, be used while using traditional engineering methods that use quantitative terms like vibration amplitude, energy consumption, stress, pressure etc. in evaluating the machine properties.

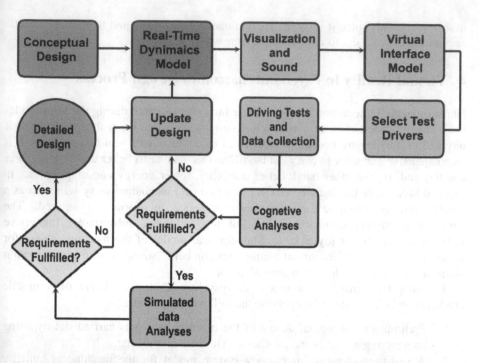

Fig. 2. Flow chart of psychomechatronics design of a mobile machine

Fig. 3. Virtual model of the loader

5 Mining Machine and Usability Test

The mining industry in the last decade has undergone significant changes with increased mechanization and development of new technologies. Despite changes, human-centered design principles are often neglected in the design and development of new equipment and technologies. This research details a usability study on the operator's of an underground loader, where human-centered design principles are proposed to be considered in the design and development of the new products. The system under study (Sandvik loader LH410) is an underground mining loader with a capacity of loading 10 ton mining stone. Figure 3 shows the virtual model of the loader in a virtual environment.

The loader was design to work in an underground mine. Two different types of buckets are used for usability tests. Figure 4 demonstrate the buckets. Both *barelip* and *shark* buckets has the same volume capacity (4.6 m^3).

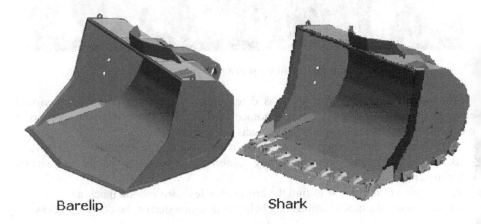

Barelip   Shark

Fig. 4. Barelip and shark buckets

The steering of loader is hydraulically operated. The power steering is controlled by double acting cylinders. The bucket and its boom are operated using hydraulics electric joystick, equipped with piston pump that delivers oil to the bucket hydraulic main valve. The oil flow from steering hydraulic pump is directed to bucket hydraulics when steering is not used.

In Figure 3 a dashed, red line outlines the operator's compartment. From this figure it can be seen that this is certainly not the norm as far as operator compartments of mining machinery go because the operator's seat is in a position perpendicular to the direction of motion and he turns his head to the left when driving forwards and to the right when driving in reverse.

Figure 5 illustrates the setup for the usability test. Three monitors were used to show the *forward*, *right side* and *rear* view of the user inside of the cabin. The joysticks and pedals could move the virtual model of the system the same as the real system in a virtual environment. The virtual simulator was design in Mevea Oy [16].

Fig. 5. Usability test in a virtual environment

The main aim of usability test was determining the success of applying virtual environment for carrying out the psychomechatronics idea in the design process of a mechatronics system. The following subjects were investigated.

A-How much virtual reality is successful in handling design process?

B-How successfully we can apply usability and human factor in virtual environment

C-Applying usability test to find the errors of a designed mobile machine

D-Suggesting the modifications that help the designer during the design process

It takes only five users to uncover 80 percent of high-level usability problems, Jakob Nielsen, [17]. The *virtual simulator* with 5 individuals pulled from the designer engineer and technician users. An entrance and exit surveys before and after each test were taken. Following five tasks were performed in a virtual environment for each individual examiner;

1. Filling loader's *barelip* bucket
2. Filling loader's *barelip* bucket (by means of a camera outside the loader)
3. Filling loader's *shark* bucket
4. Filling loader's shark bucket (using slower lifting)
5. Filling loader's bucket and moving backward to the upper level

All individuals performed each task for three times, the time and the numbers of errors were recorded and video taped. The amount of stones in the bucket was estimated after each test. Following, some of the results are provided;

Using a *shark* bucket helped most of the users to collect 10 percent more stones in their bucket compare to the *barelip* bucket. The reasons for that are the shapes of stones or the effectiveness of the *shark* buckets. Because the seats of operators are in a position perpendicular to the direction of motion and they turn their head to the left when driving forwards makes inconvenience for them, meanwhile they have not clear view of stones and bucket, so attaching a camera on the top of the cabin helped all the examiners to collect more stone in a specific time.

The speed of lifting in the designed model was so fast and the users had problem to filling their bucket. The idea of decreasing speed of lifting to 75 percent during the filling process helped all the operators to load more stone.

The idea of using virtual model of system and the environment for performing usability test is perfect. Several examiners had some comment that are very useful for future progress. In the usability the three monitors were used to show the front, right side and back views from the viewpoint of a user inside of the cabin. It is more realistic, if three screens instead of monitors are used. Meanwhile to feel the movement of the cabin a motion based platform and a real cabin are essential. During the test, the user heard the sound of engine, but it would be more realistic, if the operator can hear the sound of sands and so on.

6 Conclusions

The paper presented the systematic approach referred to as psychomechatronics for designing human operated mechatronics system. The justifications for the presented approach are clearly stated. The main ideas of psychomechatronics are presented and, finally, the virtual environment required for applying psychomechatronics in R&D of a mining machine is presented in detail. In this paper we describe human-centered methodological approach which we have applied for the evaluation of mobile machine designing in a virtual environment. The approach is strongly rooted in theories and methods from cognitive science and the emerging field of usability engineering. The focus is on assessing human machine interaction and in particular, the usability of mechatronics systems in the design process. The paper provides a review of the general area of systems evaluation with the motivation and rationale for methodological approaches underlying usability engineering and cognitive task analysis as applied to a mining machine.

References

1. Lee, K.: Principles of CAD/CAM/CAE systems. Addison-Wesley, Wokingham (1999)
2. Burdea, G.G., Coiffet, P.: Virtual reality technology. Wiley, New York (2003)
3. Bowman, D.A., Kruijff, E., LaViola, J.J., Poupyrev, I.: 3D User interfaces – theory and practice. Addison-Wesley, Wokingham (2003)
4. Yamada, Y.: Clay Modeling: techniques forgiving three dimensional form to idea. San'ei Shobo Publishing Co. (1997)
5. Bordegoni, M., Cugin, U.: Haptic modeling in the conceptual phases of product design. Virtual Reality 9, 192–202 (2006)
6. Wickens, C.D., Lee, J.D., Liu, Y., Becker, S.E.G.: An Introduction to Human Factors Engineering. Pearson Education Inc., New Jersey (2004)

7. Handroos, H., Mikkola, A., Liukkula, M.: Virtual Prototyping of Hydraulic Driven Mechatronic Machines. In: Proceedings of the 4th JHPS International Symposium on Fluid Power, Tokyo, pp. 525–530 (1999)
8. De Silva, C.: Mechatronics: An integrated approach. CRC Press, Boca Raton (2005)
9. Medin, D.L., Ross, B.H., Markham, A.B.: Cognitive Psychology, 3rd edn. Harcourt, Fort Worth (2001)
10. Vicente, K.J.: Cognitive work analysis: toward safe, productive and healthy computer-based work. Lawrence Erlbaum, New York (1999)
11. Means, B., Gott, S.: Cognitive task analysis as a basis for tutor development: articulating abstract knowledge representations. Intelligent tutoring systems: lessons learned. Lawrence Erlbaum, Hillsdale (1988)
12. Zhang, J., Patel, V.L., Johnson, K.A., Smith, J.W., Malin, J.: Designing human centered distributed information systems. IEEE Intell Systems, 42–47 (2002)
13. Rubin, J., Chisnell, D.: Handbook of Usability Testing, How to Plan, Design, and Conduct Effective Tests, 2nd edn. Wiley Publishing, Inc., Indianapolis (2008)
14. Nielsen, J.: Usability Engineering. Academic Press, London (1993)
15. Wickens, C.D., et al.: An introduction to human factors engineering. Prentice Hall, Upper Saddle River (2004)
16. Mevea Oy, http://www.mevea.com/
17. Jakob Nielsen website, http://www.useit.com/

Author Index